REVIEWS in MINERALOGY and GEOCHEMISTRY

Volume 69   2008

Minerals, Inclusions and Volcanic Processes

EDITORS

Keith D. Putirka — California State University, Fresno
Fresno, California

Frank J. Tepley III — Oregon State Univeristy
Corvallis, Oregon

ON THE COVER

Top: Plane light view of quartz with melt inclusions from a 0.5 Ma high-SiO_2 rhyolite from Kos Island in the Aegean Arc (C. Schnyder & O. Bachmann); crystals are ca. 3 mm across. *Right:* Nomarski (NDIC) image of plagioclase phenocryst and groundmass from 1985 lava flow of Arenal volcano, Costa Rica; phenocryst is 300 μm in length (M. Streck). *Bottom:* Crossed polars image of reversely zoned clinopyroxene in high-Mg andesite from Whaleback volcano, near Mt. Shasta, CA; image diameter ca. 2 mm (M. Streck). *Left:* BSE image of zoned feldspar from Mt. St. Helens, Kalama age (A. Lieuallen & F. Tepley). *Far left:* Mount St. Helens, WA (J. Rosso).

Series Editor: **Jodi J. Rosso**

**MINERALOGICAL SOCIETY OF AMERICA
GEOCHEMICAL SOCIETY**

Reviews in Mineralogy and Geochemistry, Volume 69

Minerals, Inclusions and Volcanic Processes

ISSN 1529-6466
ISBN 978-0-939950-83-6

Copyright 2008

The MINERALOGICAL SOCIETY of AMERICA
3635 Concorde Parkway, Suite 500
Chantilly, Virginia, 20151-1125, U.S.A.
WWW.MINSOCAM.ORG

The appearance of the code at the bottom of the first page of each chapter in this volume indicates the copyright owner's consent that copies of the article can be made for personal use or internal use or for the personal use or internal use of specific clients, provided the original publication is cited. The consent is given on the condition, however, that the copier pay the stated per-copy fee through the Copyright Clearance Center, Inc. for copying beyond that permitted by Sections 107 or 108 of the U.S. Copyright Law. This consent does not extend to other types of copying for general distribution, for advertising or promotional purposes, for creating new collective works, or for resale. For permission to reprint entire articles in these cases and the like, consult the Administrator of the Mineralogical Society of America as to the royalty due to the Society.

Minerals, Inclusions and Volcanic Processes

69   *Reviews in Mineralogy and Geochemistry*   **69**

FROM THE SERIES EDITOR

The chapters in this volume represent an extensive review of the material presented by the invited speakers at a two-day short course on *Minerals, Inclusions and Volcanic Processes* held prior to the annual fall American Geophysical Union meeting in San Francisco, California (December 13-14, 2008). This meeting was sponsored by the Mineralogical Society of America and the Department of Energy. The short course was also held in conjunction with an AGU-sponsored topical session at the subsequent annual fall AGU meeting.

Any supplemental material and errata (if any) can be found at the MSA website *www.minsocam.org*.

Jodi J. Rosso, Series Editor
West Richland, Washington
October 2008

SHORT COURSE SERIES DEDICATION

Dr. William C. Luth has had a long and distinguished career in research, education and in the government. He was a leader in experimental petrology and in training graduate students at Stanford University. His efforts at Sandia National Laboratory and at the Department of Energy's headquarters resulted in the initiation and long-term support of many of the cutting edge research projects whose results form the foundations of these short courses. Bill's broad interest in understanding fundamental geochemical processes and their applications to national problems is a continuous thread through both his university and government career. He retired in 1996, but his efforts to foster excellent basic research, and to promote the development of advanced analytical capabilities gave a unique focus to the basic research portfolio in Geosciences at the Department of Energy. He has been, and continues to be, a friend and mentor to many of us. It is appropriate to celebrate his career in education and government service with this series of courses.

Minerals, Inclusions and Volcanic Processes

69   *Reviews in Mineralogy and Geochemistry*   **69**

TABLE OF CONTENTS

1   Introduction to Minerals, Inclusions and Volcanic Processes
Keith D. Putirka

OVERVIEW OF THE VOLUME 1
SOME IMMEDIATE AND SOME MORE DISTANT DIRECTIONS 4
ACKNOWLEDGMENTS 5
REFERENCES CITED 5

2   Experimental Studies of the Kinetics and Energetics of Magma Crystallization
Julia E. Hammer

INTRODUCTION 9
PRE-ERUPTIVE MAGMATIC ENVIRONMENTS 10
    General approach 10
    Apparatus 10
    Evaluating experimental results 11
    Phase equilibria in volcano-petrology 12
CRYSTALLIZATION KINETICS IN SIMPLE SYSTEMS 12
    Fundamental concepts of crystal nucleation 14
    Theory development through experimentation 14
    Crystal growth 22
CRYSTALLIZATION KINETICS IN MAGMAS: METHODS 23
    Starting materials 24
    Thermal history 24
    Static vs. dynamic 25
    Constant-rate vs. variable-rate cooling 25
BASALT COOLING EXPERIMENTS 26
    Nucleation controls texture 26
    Growth rates decay with time 27
    Undercooling and the approach to equilibrium 29
    Crystallization in dynamic environments 30

FELSIC MAGMA DECOMPRESSION EXPERIMENTS	33
Effective undercooling	33
Decompression history	34
TEXTURAL QUANTIFICATION OF EXPERIMENTAL RUN PRODUCTS	37
Crystal size distribution analysis	38
Crystal aspect ratio	41
PHASE BOUNDARIES AND INTERFACES	44
Heterogeneous nucleation	44
Solid-solid interfacial energy	44
Reaction relationships	47
MELT INCLUSION ENTRAPMENT	47
Major element behavior	47
Trace element behavior	49
CONCLUSIONS	51
ACKNOWLEDGMENTS	52
REFERENCES CITED	52

3   Thermometers and Barometers for Volcanic Systems

Keith D. Putirka

INTRODUCTION	61
THEORETICAL ASPECTS	61
REGRESSION ANALYSIS & STATISTICAL CONSIDERATIONS	63
Experimental data, regression analysis & statistical considerations	63
IGNEOUS THERMOMETERS AND BAROMETERS	67
History and background	67
Calculating "liquid" and mineral components, and activities	68
"Glass" (or liquid) thermometers	71
Olivine-liquid equilibria	73
Plagioclase- and alkali feldspar-liquid thermobarometers	79
Two-feldspar thermometers	83
Orthopyroxene and orthopyroxene-liquid thermobarometers	84
Clinopyroxene and clinopyroxene-liquid thermobarometers	88
Two-pyroxene thermometers and barometers	94
The silica activity barometer	96
APPLICATIONS	98
Magma transport at Hawaii	98
The conditions of mantle partial melting	98
Temperature estimates in felsic volcanic systems	101
SOURCES OF ERROR RELATIVE TO P AND T	104
Model error	104
Error of treatment	107
CONCLUSIONS	110
ACKNOWLEDGMENTS	110
REFERENCES	111

4   Thermometers and Thermobarometers in Granitic Systems

J. Lawford Anderson, Andrew P. Barth,
Joseph L. Wooden, Frank Mazdab

INTRODUCTION 121
    Plutonic barometry of Cordilleran orogenic terranes - a historical perspective 122
SINGLE MINERAL TRACE ELEMENT THERMOMETERS
  IN IGNEOUS SYSTEMS..122
    Ti-in-quartz thermometer 123
    Ti-in-zircon and Zr-in-rutile thermometers 124
    Zr-in-sphene thermometer 128
ACCESSORY PHASE SATURATION THERMOMETRY..129
THERMOBAROMETERS APPLICABLE TO GRANITIC ROCKS................................130
    Al-in-hornblende barometry..130
    Thermobarometry for peraluminous granites...132
    Oxybarometers and other mineral reactions involving oxygen 133
    Fe-Ti oxides, pyroxenes, olivine, and quartz (QUILF) and sphene (titanite)............134
    Magnetite and ilmenite series of granitic plutons and oxygen fugacity.....................135
SUMMARY 137
ACKNOWLEDGMENT...138
REFERENCES 138

5   Fluid Inclusion Thermobarometry as a Tracer for Magmatic Processes

Thor H. Hansteen, Andreas Klügel

INTRODUCTION 143
OCCURRENCE AND PETROGRAPHY OF FLUID INCLUSIONS...............................144
INCLUSION CLASSIFICATION 144
FLUID INCLUSION FORMATION AND RE-EQUILIBRATION 147
    Trapping during crystal growth 147
    Formation of secondary inclusions..147
    Post-entrapment re-equilibration..147
MICROTHERMOMETRY 149
FLUID SYSTEMS..150
    The CO_2 system 150
    The H_2O-NaCl system 152
TRACKING VOLCANIC PLUMBING SYSTEMS USING CO_2 INCLUSIONS 156
    Rationale...158
    Determining inclusion density, pressure and depth 160
    The role of possibly missing H_2O..165
    Volumetric re-equilibration during magma ascent: a case study...............................166

PRESSURE INFORMATION FROM AQUEOUS FLUIDS IN SHALLOW MAGMA
  CHAMBERS AND PLUTONIC ROCKS 168
    Background.. 168
    Fluids in shallow magma chambers 168
    Fluid inclusions in plutonic rocks 169
SUMMARY 171
OUTLOOK 171
ACKNOWLEDGMENTS 172
REFERENCES 172

6 Petrologic Reconstruction of Magmatic System Variables and Processes

Jon Blundy, Kathy Cashman

INTRODUCTION 179
MAGMATIC VARIABLES 180
    Petrologically-determined intensive variables 181
    Linking intensive variables to the monitoring record... 181
METHODOLOGY 181
    Scanning electron microscopy (SEM)... 181
    Electron microprobe analysis (EMPA).. 184
    Secondary ion mass spectrometry (SIMS) 186
    Analysis of textures 188
GEOTHERMOMETRY 193
    Iron-titanium oxide thermometry and oxybarometry 194
    Amphibole-Plagioclase thermometry 197
    Two-pyroxene thermometry 199
GEOBAROMETRY 200
    Experimental reconstruction.. 201
    Volatile saturation pressures 203
    Haplogranite projection 207
    Crystal textures 209
COMPARISON OF PETROLOGICAL DATA WITH MONITORING SIGNALS ... 214
    Gas chemistry 214
MAGMA ASCENT AND CRYSTALLIZATION 220
    Phenocryst abundance 220
    Melt inclusion compositions 221
    Modeling volatile elements 222
    Combining degassing and crystallization 224
    Interpreting magmatic systems 225
SUMMARY 230
ACKNOWLEDGMENTS 231
REFERENCES 231

7   Magma Ascent Rates

Malcolm J. Rutherford

INTRODUCTION	241
MAGMA SOURCE AND MAGMA RESERVOIRS	242
BUBBLE FORMATION AND EVOLUTION DURING MAGMA ASCENT	243
GROUNDMASS CRYSTALLIZATION DURING MAGMA ASCENT	246
PHENOCRYST-MELT REACTIONS DURING MAGMA ASCENT	250
Reactions between volatile-bearing minerals and melt during magma ascent	250
Phenocryst-melt reactions driven by pre-eruption magma mixing	255
MAGMA ASCENT RATES FROM MASS ERUPTION RATES	257
MAGMA ASCENT FROM SEISMIC DATA	258
MAGMA ASCENT RATES BASED ON XENOLITH-MELT REACTIONS	260
XENOLITH TRANSPORT, FLOW MODELING AND MAGMA ASCENT RATES	264
SUMMARY OF MAGMA ASCENT RATE ESTIMATES	265
REFERENCES	267

8   Melt Inclusions in Basaltic and Related Volcanic Rocks

Adam J.R. Kent

INTRODUCTION	273
Previous reviews of melt inclusions in basaltic and related rocks	276
TECHNIQUES USED FOR STUDY OF MELT INCLUSIONS	276
Petrographic examination of melt inclusions	276
Preparation of melt inclusions for analysis	278
Chemical analysis of melt inclusions	281
FORMATION AND EVOLUTION OF MELT INCLUSIONS	292
Melt inclusion host minerals	292
Formation of melt inclusions	292
Evolution of melt inclusions after trapping	299
Re-equilibration with host and or external melt	302
ORIGIN OF COMPOSITIONAL VARIATIONS IN MELT INCLUSIONS	311
How representative are trapped melts?	316
MELT INCLUSIONS IN BASALTIC AND RELATED ROCKS	318
Mid-ocean ridges	318
Subduction zones	319
Oceanic islands	320
Continental flood basalts	320
CONCLUDING REMARKS	321
ACKNOWLEDGMENTS	322
REFERENCES	322

9 Interpreting H_2O and CO_2 Contents in Melt Inclusions: Constraints from Solubility Experiments and Modeling

Gordon Moore

INTRODUCTION	333
EXPERIMENTAL VOLATILE SOLUBILITY STUDIES IN NATURAL MELTS	335
The "good" solubility experiment	335
The solubility of pure H_2O and CO_2 in natural melts	337
The solubility of mixed fluids (H_2O + CO_2) in natural melts	343
VOLATILE SOLUBILITY MODELS FOR NATURAL SILICATE MELTS	345
H_2O solubility models	346
CO_2 solubility models	349
Mixed (H_2O + CO_2) solubility models	350
The compositional variation in melt inclusions and its implications for interpreting volatile contents	352
ACKNOWLEDGMENTS	355
REFERENCES	358

10 Volatile Abundances in Basaltic Magmas and Their Degassing Paths Tracked by Melt Inclusions

Nicole Métrich, Paul J. Wallace

INTRODUCTION	363
WHAT ARE MELT INCLUSIONS AND HOW DO THEY FORM?	364
POST-ENTRAPMENT MODIFICATION OF MELT INCLUSIONS	369
ANALYTICAL TECHNIQUES FOR VOLATILES	372
VOLATILE SOLUBILITIES AND VAPOR MELT PARTITIONING	373
H_2O AND CO_2 ABUNDANCES IN BASALTIC MAGMAS: CRYSTALLIZATION DEPTHS, MAGMA ASCENT AND DEGASSING PROCESSES	377
H_2O-CO_2 abundances in basaltic magmas	377
Pressures of crystallization and magma degassing	379
Sulfur and halogen degassing	384
ERUPTION STYLES AND PROCESSES – HOW DO THEY RELATE TO MELT INCLUSION DATA?	385
VOLCANIC DEGASSING: INTEGRATION OF VOLCANIC GAS COMPOSITIONS AND GAS FLUXES WITH MELT INCLUSIONS	387
Syn-eruptive magma degassing and volatile degassing budgets	387
Excessive degassing and the role of degassed but unerupted magma	388
Differential gas bubble transfer	388
GEOCHEMICAL SYSTEMATICS OF THE PRIMARY VOLATILE CONTENTS OF MAFIC MAGMAS	388
CONCLUDING REMARKS	393
ACKNOWLEGMENTS	396
REFERENCES	396

11   Inter- and Intracrystalline Isotopic Disequilibria: Techniques and Applications
Frank C. Ramos, Frank J. Tepley III

INTRODUCTION	403
MICROSPATIAL ISOTOPE ANALYSIS	404
Micromilling	406
Secondary ionization mass spectrometry (SIMS)	417
Laser ablation sampling	418
MICROANALYTE ANALYSIS	432
Single crystal analyses	432
SUMMARY	439
ACKNOWLEDGMENTS	439
REFERENCES	440

12   Oxygen Isotopes in Mantle and Crustal Magmas as Revealed by Single Crystal Analysis
Ilya Bindeman

PART I: INTRODUCTION	445
Basics of oxygen isotope variations in nature and their causes	446
Oxygen isotopes in mantle-derived rocks, normal-$\delta^{18}O$, high-$\delta^{18}O$, and low-$\delta^{18}O$ magmas	451
PART II: SINGLE PHENOCRYST ISOTOPE STUDIES	453
Modern methods of oxygen isotope analysis of crystals	453
$\delta^{18}O$ heterogeneity and $\Delta^{18}O_{crystal\text{-}melt}$ disequilibria	455
PART III: CASE STUDIES OF OXYGEN ISOTOPE DISEQUILIBRIA IN IGNEOUS ROCKS BASED ON SINGLE CRYSTAL ISOTOPE ANALYSIS	457
Basaltic igneous systems: olivine in basalts	458
Silicic igneous systems	463
Case studies of large silicic magma systems	464
Phenocryst heterogeneity in plutonic rocks	467
PART IV: ISOTOPE DISEQUILIBRIA: WHAT HAVE WE LEARNED ABOUT MAGMA GENESIS?	469
Mineral-diffusive timescales and magma residence times	469
Rapid magma genesis and high magma production rates	469
Very shallow petrogenesis and the role of hydrothermal carapaces	470
Accretion of large silicic magma bodies	471
The origin of "phenocrysts" and interpretation of their melt inclusions	472
ACKNOWLEDGMENTS	472
REFERENCES CITED	473

13   Uranium-series Crystal Ages

Kari M. Cooper, Mary R. Reid

INTRODUCTION 479
    Principles of U-series dating 479
    U-series dating of crystals 482
ANALYTICAL AND PRACTICAL CONSIDERATIONS..482
    Introduction 482
    Practical limitations: sample preparation and analysis..483
    Age calculations 490
    Effect of diffusion on crystal ages 497
WHAT DO CRYSTAL AGES TELL US?...499
    Introduction 499
    Using U-Pb and decay series ages to date eruptions 500
    Pre-eruptive crystal growth—'simple' scenarios 501
    Crystal recycling—autocrysts vs. xenocrysts. vs. antecrysts 503
    Insights from dating of subvolcanic plutons..516
RESOLVING AGES FROM COMPLEX CRYSTAL POPULATIONS:
    INCORPORATING OTHER OBSERVATIONS 519
    Introduction 519
    Combining U-Th and U-Pb SIMS dating...519
    U-Th/U-Pb spots versus mineral separates..520
    Multiple size fractions 523
    Textural perspectives 524
    Crystal-scale major element/trace element variations...527
    Crystal ages and thermometry 529
    Single-crystal or sub-crystal oxygen isotopic data 532
    Summary 534
KEY QUESTIONS AND FUTURE DIRECTIONS 535
    What can crystal ages tell us about magmatic processes?.....................................535
    What do we get from the different scales of analyses? 536
    Future directions 537
    Final thoughts 538
ACKNOWLEDGMENTS 538
REFERENCES 538

14   Time Scales of Magmatic Processes from Modeling the Zoning Patterns of Crystals

Fidel Costa, Ralf Dohmen, Sumit Chakraborty

INTRODUCTION 545
EQUATIONS TO DESCRIBE DIFFUSION PROCESSES 548
    Diffusion, flux and constitutive laws 548
    Time dependence: the continuity equation 550
MODELING NATURAL CRYSTALS..552
    Initial and boundary conditions 553

Non-isothermal diffusion..557
Variable boundary conditions – a numerical simulation using MELTS...................558
Diffusion and crystal growth – the moving boundary problem..............................563
DIFFUSION COEFFICIENTS 565
Types of diffusion coefficients..567
Multi-component diffusion in silicates...567
Parameters that determine tracer diffusion coefficients...570
DIFFUSION MODELING – POTENTIAL PITFALLS AND ERRORS..........................574
Multiple dimensions, sectioning effects and anisotropy 574
Uncertainties in the calculated times from the diffusion models 576
A CHECKLIST FOR OBTAINING TIME SCALES OF MAGMATIC PROCESSES
  BY DIFFUSION MODELING 577
TIME SCALES OF MAGMATIC PROCESSES THAT HAVE BEEN OBTAINED
  FROM MODELING THE ZONING OF CRYSTALS AND THEIR RELATION
  TO OTHER INFORMATION ON TIME SCALES OF PROCESSES 578
Time scales from diffusion modeling 579
SUMMARY AND PROSPECTS 581
ACKNOWLEDGMENTS...582
REFERENCES 582
APPENDIX I 588
APPENDIX II 591

15 Mineral Textures and Zoning as Evidence for Open System Processes

Martin J. Streck

INTRODUCTION 595
METHODS TO IMAGE TEXTURES AND QUANTIFY COMPOSITION....................595
BRIEF REVIEW OF OPEN SYSTEM PROCESSES 596
Mixing 597
Contamination (assimilation) 597
MINERAL RECORD FOR MIXING OF MAGMAS AND COUNTRY ROCK –
  PHENOCYRST, ANTECRYST, XENOCRYST 598
SUMMARY OF ZONING AND TEXTURES 599
Crystal zoning...599
Crystal textures 601
MINERAL SPECIFIC TEXTURES AND ZONING 604
Plagioclase..604
Pyroxene 607
Olivine 611
Amphibole 611
Other major minerals..613
Accessory minerals...613
MINERAL POPULATIONS AS EVIDENCE FOR OPEN SYSTEM BEHAVIOR 615
OUTLOOK AND CONCLUSIONS..618
ACKNOWLEDGMENTS...618
REFERENCES 619

16 Decryption of Igneous Rock Textures: Crystal Size Distribution Tools

Pietro Armienti

CRYSTAL SIZE DISTRIBUTIONS 623
DETERMINATION OF THE CURVES OF POPULATION DENSITY 624
    Analysis of distributions of areas 625
    Stereologic reconstruction of CSD: monodispersed distributions.........................625
    Stereologic reconstruction of CSD: poly-dispersed distributions 627
    The "best" CSD for a set of unfolded planar data 631
    Truncation effects 633
CSDs AND THE PETROGENETIC CONTEXT..635
    Continuously taped and refilled magma chamber: the Stromboli example...............636
    Cooling of a magma batch: the case of Mount Etna 639
    Estimate of mineral growth rates for Mount Etna 643
TEMPORAL EVOLUTION OF CSDs: SIZE AS FUNCTION OF TIME 644
CONCLUDING REMARKS..648
REFERENCES 648

17 Deciphering Magma Chamber Dynamics from Styles of Compositional Zoning in Large Silicic Ash Flow Sheets

Olivier Bachmann, George W. Bergantz

INTRODUCTION 651
    Types of gradients in ignimbrites 652
    Abrupt gradients 653
    Linear gradients 653
    Lack of gradients 655
MECHANISMS FOR GENERATING GRADIENTS IN SILICIC
MAGMA CHAMBERS 656
    The role of convection 656
    Origin of heterogeneity in silicic magmas: convective stirring vs.
        phase separation 661
    Mechanisms to generate abrupt gradients 662
THE CASE OF THE ABRUPTLY ZONED CRATER LAKE ERUPTION 662
    Mechanism to generate linear gradients 663
THE CASE OF THE LINEARLY ZONED BISHOP TUFF...664
    The generation of homogeneity..665
THE CASE OF THE FISH CANYON MAGMA BODY 666
THE CASE OF THE HOMOGENEOUS GRANITOIDS 667
CONCLUSIONS...667
ACKNOWLEDGMENTS...668
REFERENCES 668

Introduction to Minerals, Inclusions and Volcanic Processes

Keith D. Putirka

Department of Earth and Environmental Sciences
California State University, Fresno
2576 E. San Ramon Ave., MS/ST24
Fresno, California, 93740-8039, U.S.A.

kputirka@csufresno.edu

Minerals are intrinsically resistant to the processes that homogenize silicate liquids—their compositions thus yield an archive of volcanic and magmatic processes that are invisible at the whole rock scale. Minerals and their inclusions record diverse magma compositions, the depths and temperatures of magma storage, the nature of open system processes, and the rates at which magmas ascend. The potential for understanding volcanic systems through minerals and their inclusions has long been recognized (Sorby 1858). Sorby's (1863) study of James Hall's reversal experiments helped resolve the "basalt controversy" in favor of a volcanic origin, while Zirkel's (1863) discovery of quartz within a volcanic rock helped tip the balance in favor of a magmatic origin for granite (Young 2003). Studies of phenocrysts have also long illustrated the importance of wall rock assimilation and magma mixing (e.g., Fenner 1926; Finch and Anderson 1930; Larson et al. 1938), and the potential for geothermometry (Barth 1934). Darwin's (1844) mineralogical field-studies in the Galapagos archipelago, followed by King's (1878) studies at Hawaii, also inaugurated the establishment of fractional crystallization as an important evolutionary process (Becker 1897; Bowen 1915).

Recent advances in micro-analytical techniques open a new realm of detail, building upon a long history of mineralogical research; this volume summarizes some of this progress. Our summary focuses on volcanologic and magmatic processes, but the methods reviewed here extend well beyond terrestrial applications. Samples from the Stardust return mission, for example, show that olivine, plagioclase and pyroxene pervade the solar system (Brownlee et al. 2006)—while the topics covered here surely apply to all terrestrial-like planetary bodies, relevance may extend to a cosmic scale. Our more modest hope is that this volume will aid the study of disparate fields of terrestrial igneous systems, and perhaps provide a catalyst for new collaborations and integrated studies.

OVERVIEW OF THE VOLUME

Our review begins by tracing the origins of mineral grains, and methods to estimate pressures (P) and temperatures (T) of crystallization. Key to such attempts is an understanding of textures, and in her review, Hammer (2008) shows how "dynamic" experiments (conducted with varying P or T), yield important insights into crystal growth. Early dynamic experiments (e.g., Lofgren et al. 1974; Walker et al. 1978) have shown that porphyritic textures can result from a single episode of cooling. More recent experiments demonstrate that crystals can form during ascent due to loss of volatiles (Hammer and Rutherford 2002). Hammer (2008) describes these and other advances, and additional challenges that require new experimental research. Her review also sets the stage for the next four chapters, where the hope above all hopes is that not all is disequilibrium and discord.

Putirka (2008), Anderson et al. (2008) and Blundy and Cashman (2008) review various igneous geothermometers and geobarometers. The Anderson et al. (2008) review includes accessory phase trace-element thermometers, which can be used to estimate T for the same grains for which U-series equilibria provide age dates (Cooper and Reid 2008). Also covered among these chapters are many familiar models involving olivine, amphibole, feldspar, pyroxene, and spinel, with many new calibrations. Anderson et al. (2008) also examine the applications of thermobarometry to granitic systems—potentially the un-erupted residue of high SiO_2 volcanism (e.g., Bachman and Bergantz 2004). Hansteen and Klügel (2008) review another method for P estimation, based on densities of entrapped fluids and appropriate equations of state. Fluid inclusions appear to provide the single most precise means for P estimation in geologic systems, and so should prove useful to test other methods. As a matter of good fortune, Klügel and Klein (2005) show that P estimates obtained from fluid inclusions match those from mineral-melt equilibria in at least one instance. Both barometers thus appear to be valid—and indicate transport depths well below the base of the crust. The importance of this kind of work is well summarized by Blundy and Cashman (2008), where they note the correlation between petrologic and fumarole/seismic monitoring data at Mt. St. Helens, underscoring the applicability of such studies to forensic volcanology.

Rutherford's (2008) chapter returns us to the issue of disequilibrium, with a review of magma ascent rates. This review shows that consistent values for ascent rates are obtained from a range of textural and petrologic methods. For example, ascent rates can be ascertained from amphibole reaction rims formed in slowly ascending silicic eruptions (e.g., Browne and Gardner 2006) or H-loss profiles in melt inclusions and crystals (Peslier and Luhr 2006; Humphreys et al. 2008). A most exciting proposition is that ascent rates positively correlate with volcanic explosivity—an important forensic issue (e.g., Scandone and Malone 1985).

Our volume then moves to a review of melt inclusions, which have received much attention in the past decade. The interest lies in the fact that melt inclusions expose compositional variability far in excess of that exhibited by their host magmas (e.g., Sobolev and Shimizu 1993; Nielsen et al. 1995; Kent et al. 2002; McLennan et al. 2003). Kent (2008) shows that pre-mixed magma compositions can be preserved as inclusions, thus providing a window into pre-eruptive conditions (although such interpretations require care, e.g., Danyushevsky et al. 2002; Kent 2008). Métrich and Wallace (2008) review the volatile contents in basaltic melt inclusions and "magma degassing paths". They conclude that most phenocrysts (and hence melt inclusions) form during ascent and degassing, and so provide an invaluable record of volcanic volatile budgets, from crustal depths to the surface. Métrich et al. (2005) provide a dramatic illustration of such methods in their forensic analysis of an explosive 2003 "paroxysm" at Stromboli, as do Roggensack et al. (1997) in a forward-looking study at Cerro Negro. Such methods rely upon vapor saturation pressures, which are derived from experimentally calibrated models. Moore (2008) and Blundy and Cashman (2008) test two of the most important models, by Newman and Lowenstern (2002) (VolatileCalc) and Papale et al. (2006). Moore's review in particular shows that CO_2 and H_2O solubilities are highly sensitive to melt composition, and provides a guide to the appropriate use of these models, and their errors.

The next four chapters document insights obtained from isotopic studies and diffusion profiles. Ramos and Tepley (2008) review developments of micro-analytical isotope measurements, which now have the potential to elucidate even the most cryptic of open system behaviors. $^{87}Sr/^{87}Sr$ ratios, for example, can be matched to dissolution surfaces to identify magma recharge events (Tepley et al. 2000). And $^{87}Sr/^{86}Sr$-contrasts within and between phenocrysts allow differentiation between the roles of wall rock assimilation and enriched mantle sources to explain elevated $^{87}Sr/^{86}Sr$ (Ramos and Reid 2005). In the next chapter, Cooper and Reid

(2008) examine the time scales for such processes through U-series age dating of minerals. They show that crystals can be recycled over time scales of tens to hundreds of thousands of years before being erupted. Perhaps even more interesting are discordant ^{230}Th-^{238}U and ^{226}Ra-^{230}Th age dates at Mt. St. Helens (Cooper and Reid 2003), which imply that young magmas can mix with magmas that are tens of thousands of years older (Cooper and Reid 2008). Such mixing is not entirely unanticipated. Murata and Richter (1966) show that high-energy eruptions can dredge up and entrain masses of pre-existing crystals. And in what may be an emerging paradigm, Ryan (1988) and Kuntz (1992) emphasize that volcanic plumbing systems are best thought of as a plexus of dikes and sills, and that these in turn, as shown by Sinton and Detrick (1992) and Marsh (1995), are not filled just with liquid, but are comprised of a mush of crystals and magma (and perhaps bubbles too). Mineral age dates have spurred the introduction of a new term, "antecryst" (by Wes Hildreth; Cooper and Reid 2008), to represent minerals that are older than the magmas they inhabit (so are not "phenocrysts"), but are part of an ancestral magmatic (mush) system (so are not "xenocrysts"). Oxygen isotopes (Bindeman 2008) may be particularly useful for addressing such issues. For example, Bindeman et al. (2007) use mineral age dates and O isotopes to fingerprint the recycling of surface materials into magmas; their work reveals a volcanic cycle at the Snake River Plain that begins with partial melting of pre-existing crust, and is completed by re-melting of caldera-fill, altered by meteoric waters. Bindeman's (2008) review also reveals that at Iceland and Yellowstone, most crystals are out of isotopic equilibrium with their host magmas and are largely inherited from preexisting rocks; he suggests that true phenocrysts are rare. Costa et al. (2008) review yet another means to estimate the rates of magmatic processes, using mineral diffusion profiles. This highly versatile method has led to some very interesting insights. Age dates by Costa and Dungan (2005), for example, show that wall rock assimilation (at Tatara-San Pedro) can occur on time scales of months to years. In contrast, Klügel (2001) calculates time scales intervening between magma aggregation and eruption (at La Palma) on the order of tens to hundreds of years. Apparently, plenty of time is available for the assimilation and mixing processes that Bowen (1928) thought were unimportant.

A recurring theme in this volume is the prevalence of open system processes and magma hybridization. Though appreciated for more than a century (e.g., Bunsen 1851; Daly 1914), the open-system view fell from favor following Bowen's (1928) compelling arguments in favor of fractional crystallization, and against assimilation. Eventually, studies of mineral zoning by Macdonald and Katsura (1965), Sato (1975), and especially Eichelberger (1975), Anderson (1976) and Dungan and Rhodes (1978), precipitated the modern view that mixing and assimilation are ubiquitous. The review by Streck (2008) summarizes how zoning styles in minerals continue to be important, not just for identifying, but for quantifying open system processes. The study of Arenal in Costa Rica is illustrative, where Streck et al. (2005) show how minerals contained within "monotonous" erupted units reveal at least four magmatic episodes. In his review, Streck (2008) suggests that individual crystals can be composites of such events, so are not easily classified as phenocrysts, antecrysts or xenocrysts. Perhaps we require a return to the strictly observational origin of "phenocryst," as defined by Iddings (1889), to denote crystals that "stand out conspicuously from the surrounding crystals or glass composing the groundmass."

The volume returns to issues of kinetics with Armienti's (2008) new look at the analysis of crystal size distributions (CSD). Since Marsh (1988) introduced CSD theory to petrology, such measurements have been crucial to understanding crystal textures. Where closed systems can be identified, Armienti (2008) provides the methods by which self-consistent kinetic parameters can be derived from CSD. The CSD approach can also be extremely important for understanding open systems, as shown by Armienti et al. (2007) who use CSD, thermobarometry, and isotope-ratios in minerals at Mt. Etna, to illustrate the time scales at which magmas flush through the plumbing system. Our volume concludes with a chapter by Bachmann and Bergantz (2008) summarizing compositional zonations—not in minerals—but in erupted volcanic strata. They also review the thermal and compositional forces that drive open system behavior. A seeming departure from the first two thirds of the volume theme, this review is fundamental—our microscopic observations are of greatest value when they inform and enlarge our views of macro-level volcanic phenomena, including those manifest at the outcrop scale.

Many of the contributions in this volume emphasize the close link between technical capabilities in analytical methods and advances in understanding magmatic and volcanic systems. Descriptions of many of the most common analytical approaches are reviewed within these chapters. Such topics include: secondary ion mass spectrometry (Blundy and Cashman 2008; Kent 2008); electron microprobe (Blundy and Cashman 2008; Kent 2008; Métrich and Wallace 2008); laser ablation ICP-MS (Kent 2008); Fourier transform infrared spectroscopy (Moore 2008; Métrich and Wallace 2008); microsampling and isotope mass spectrometry (Ramos and Tepley 2008); U-series measurement techniques (Cooper and Reid 2008); Nomarski differential interference contrasts (Streck 2008); micro-Raman spectroscopy (Métrich and Wallace 2008); back-scattered electron microscopy, and cathodoluminescence (Blundy and Cashman 2008).

SOME IMMEDIATE AND SOME MORE DISTANT DIRECTIONS

The complementary nature of all the topics in this volume is not difficult to recognize. As an example, Blundy et al. (2006) track heating-degassing events during ascent at Mt. St. Helens, illustrating a marriage of thermometry to the study of melt inclusions. Works by Armienti et al. (2007) and Roggensack (2001) similarly illustrate how important insights can be obtained at the confluence of various approaches. But new allied studies are badly needed. Métrich and Wallace (2008), for example, show that vapor-saturated melt inclusions often record pressures <2 kbar and rarely above 5 kbar. In some cases, such P estimates are consistent with phase equilibria (Johnson et al. 2008), or correspond to aseismic zones (Wallace and Gerlach 1994; Spilliaert et al. 2006), suggesting a genuine link to shallow regions of magma storage. But fluid inclusions and mineral-melt equilibria often record much higher pressures, reaching 8-10 kbar (Putirka 2008; Hansteen and Klügel 2008). Hansteen and Klügel (2008) suggest that fluid inclusions can re-equilibrate rapidly, and so should have a natural bias to late-stage volcanic and magmatic processes; the same may be even more true of melt inclusions, if they are captured at (relatively) high crystallization rates, caused by ascent and degassing. In this interpretation, no methods are by necessity wrong, but different methods yield information about different parts of a volcanic plumbing system. Of course, another possibility is that some melt inclusions are not vapor saturated (see Blundy and Cashman 2008), in which case P estimates from solubility models are minimum values; however, variable CO_2 contents among many melt inclusion suites appear to require vapor saturation in certain cases (Metrich and Wallace 2008). Moore (2008) furthermore notes that VolatileCalc is not calibrated above 5 kbar.

Costa et al. (2008) and Cooper and Reid (2008) discuss an issue of time scale estimates that is parallel to that for geobarometry. They note that U-series methods provide age dates on the order of tens to hundreds of thousands of years, while diffusion profiles often yield times scales of months to years. One locality, the Bishop Tuff of CA, has been studied by both methods, and as luck would have it, U-series (Simon and Reid 2005) and diffusion profiles (Morgan and Blake 2006) yield similar time scales ($\approx$100 k.y.). But as Costa et al. (2008) note, diffusion profiles may access time scales shorter than those accessible to radioisotope methods, so as with geobarometry, different methods may inform us about different aspects of volcanic processes. Diffusion dating, for example, should inform us about the time a particular mineral has spent at high temperature, which in a thermally complex magmatic system may be substantially less than the radioisotope record of the time at which a crystal first-formed. In any case, these issues can undoubtedly be resolved by additional allied efforts.

The alliance of the sub-fields reviewed here bear upon fundamental issues of volcanology: At what depths are eruptions triggered, and over what time scales? Where and why do magmas coalesce before ascent? And if magmas stagnate for thousands of years, what forces are responsible for initiating final ascent, or the degassing processes that accelerate upward motion? To the extent that we can answer these questions, we move towards formulating tests of mechanistic models of volcanic eruptions (e.g., Wilson 1980; Slezin 2003; Scandone et al. 2007), and hypotheses of the tectonic controls on magma transport (e.g., ten Brink and Brocher 1987; Takada 1994; Putirka and Busby 2007). We undoubtedly have made some progress since Sorby (1863), Fenner (1926), Barth (1934), and many others—but as yet there is still much work to do.

ACKNOWLEDGMENTS

This volume and the associated short course were supported by a generous grant from the Department of Energy; their contribution was especially key in supporting student attendance. Debts of thanks are owed to Alex Speer for his efforts to help organize the volume and short course, and to Jodi Rosso, whose tireless efforts in assembling and editing our volume cannot be overstated—the Mineralogical Society of America is very fortunate indeed to have their services. Thanks also to my co-editor, Frank Tepley and to the co-authors of this volume for their dedication to this project.

REFERENCES CITED

Anderson AT (1976) Magma mixing: petrological process and volcanological tool. J Volcanol Geotherm Res 1:3-33
Anderson JL, Barth AP, Wooden JL, Mazdab F (2008) Thermometers and thermobarometers in granitic systems. Rev Mineral Geochem 69:121-142
Armienti P (2008) Decryption of igneous rock textures: crystal size distribution tools. Rev Mineral Geochem 69:623-649
Armienti P, Tonarini S, Innocenti F, D'Orazio M (2007) Mount Etna pyroxene as tracer of petrogenetic processes and dynamics of the feeding system. Geol Soc Am Spec Paper 418:265-276
Bachman O, Bergantz G (2004) On the origin of crystal-poor rhyolites: extracted from batholithic crystal mushes. J Petrol 45:1565-1582
Bachmann O, Bergantz GW (2008) Deciphering magma chamber dynamics from styles of compositional zoning in large silicic ash flow sheets. Rev Mineral Geochem 69:651-674
Barth TW (1934) Temperatures in lavas and magmas and a new geologic thermometer. Naturen 6:187-192
Becker GF (1897) Fractional crystallization of rocks. Am J Sci 4:257-261
Bindeman I (2008) Oxygen isotopes in mantle and crustal magmas as revealed by single crystal analysis. Rev Mineral Geochem 69:445-478
Bindeman IN, Watts KE, Schmitt AK, Morgan LA, Shanks PWC (2007) Voluminous low $\delta^{18}O$ magmas in the late Miocene Heise volcanic field, Idaho: implications for the fate of Yellowstone hotspot calderas. Geology 35:1019-1022
Blundy J, Cashman K (2008) Petrologic reconstruction of magmatic system variables and processes. Rev Mineral Geochem 69:179-239
Blundy J, Cashman K, Humphreys M (2006) Magma heating by decompression-driven crystallization beneath andesite volcanoes. Nature 443:76-80
Bowen NL (1915) The later stages of the evolution of the igneous rocks. J Geol Supp tono. 8 23:1-91
Bowen NL (1928) The Evolution of the Igneous Rocks. Dover Pub, NY
Browne BL, Gardner JE (2006) The influence of magma ascent pat on the texture, mineralogy and formation of hornblende reaction rims. Earth Planet Sci Lett 246:161-140
Brownlee D, and others (2006) Comet 81P/Wild 2 Under a microscope. Science 314:1711-1716
Bunsen R (1851) Uber die processe der vulkanischen Gesteinsbildung Islands. Annal Phys Chemie 83:197-272

Cooper KM, Reid MR (2008) Uranium-series crystal ages. Rev Mineral Geochem 69:479-554
Cooper KM, Reid MR (2003) Re-examination of crystal ages in recent Mt. St. Helens lavas: implications for magma reservoir processes. Earth Planet Sci Lett 213:149
Costa F, Dohmen R, Chakraborty S (2008) Time scales of magmatic processes from modeling the zoning patterns of crystals. Rev Mineral Geochem 69:545-594
Costa F, Dungan M (2005) Short time scales of magmatic assimilation from diffusion modeling of multiple elements in olivine. Geology 33:837-840
Daly RA (1914) Igneous Rocks and their Origin. McGraw Hill, London
Danyushevsky LV, McNeill LV, Sobolev AV (2002) Experimental and petrological studies of melt inclusions in phenocrysts from mantle-derived magmas: an overview of techniques, advantages and complications. Chem Geol 183:5-24
Darwin C (1844) Geolgoical Observations on the Volcanic Islands. London, Smith-Elder
Davis BTC, Boyd DR (1966) The join $Mg_2Si_2O_6$-$CaMgSi_2O_6$ at 30 kilobars pressure and its application to pyroxenes from kimberlites. J Geophys Res 71:3567-3576
Dungan MA, Rhodes JM (1978) Residual glasses and melt inclusions in basalts from DSDP legs 45 and 46: evidence for magma mixing. Contrib Mineral Petrol 67:417-431
Eichelberger JC (1975) Origin of andesite and dacite: evidence of missing at Glass Mountain in California and at other circum-Pacific volcanoes. Geol Soc Am Bull 86:1381-1391
Fenner RH (1926) The Katmai magmatic province. J Geol 34:673-772
Finch RH and Anderson CA (1930) The quartz basalt eruptions of Cinder Cone, Lassen Volcanic National Park, California. Cal Univ Pubs Geol Sci 19:245-273
Hammer JE (2008) Experimental studies of the kinetics and energetic of magma crystallization. Rev Mineral Geochem 69:9-59
Hammer JE, Rutherford MJ (2002) An experimental study of the kinetics of decompression-induced crystallization in silicic melt. J Geophys Res 107
Hansteen TH, Klügel A (2008) Fluid inclusion thermobarometry as a tracer for magmatic processes. Rev Mineral Geochem 69:143-177
Humphreys MCS, Menand T, Blludy JD, Klimm K (2008) Magma ascent rates in explosive eruptions: constraints form H2O diffusion in melt inclusions. Earth Planet Sci Lett 270:25-40
Iddings JP (1889) On the crystallization of igneous rocks. Bull Phil Soc Wash 11:65-113
Johnson EJ, Wallace PJ, Cashman KV, Delgado Granados, H, Kent, A (2008) Magmatic volatile contents and degassing-induced crystallization at Volcan Jorullo, Mexico: Implications for melt evolution and the plumbing systems of monogenetic volcanoes. Earth Planet Sci Lett, 269:477-486
Kent AJR (2008) Melt inclusions in basaltic and related volcanic rocks. Rev Mineral Geochem 69:273-331
Kent AJR, Baker JA, Wiedenbeck M (2002) Contamination and melt aggregation processes in continental flood basalts: constraints from melt inclusions in Oligocene basalts from Yemen. Earth Planet Sci Lett 202:577-594
King C (1878) Systematic Geology. U.S. Govt Print Office, Washington D.C.
Klügel A (2001) Prolonged reaction between harzburgite xenoliths and silica-undersaturated melt: implications for dissolution and Fe-Mg interdiffusion rates of orthopyroxene. Contrib Mineral Petrol 141:1-14
Klügel A, Klein F (2005) Complex storage and ascent at embryonic submarine volcanoes from the Madeira Archipelago. Geology 34:337-340
Kuntz MA (1992) A model-based perspective of basaltic volcanism, eastern Snake River Plain, Idaho. Geol Soc Am Mem 179:289-304
Larson ES, Irving J, Gonyer FA, Larson ES III (1938) Petrologic results of a study of the minerals from the tertiary volcanic rocks of the San Juan region, Colorado. Am Mineral 23:227-257
Lofgren GE, Donaldson CH, Williams RJ, Mullins O, Usselman TM (1974) Experimentally produced textures and mineral chemistry of Apollo 15 quartz normative basalts. Lunar Planet Sci Conf 5th, p. 549-568
Macdonald GA, Katsura T (1965) Eruption of Lassen Peak, Cascade Range, California, in 1915: example of mixed magmas. Geol Soc Am Bull 76:475-482
Maclennan J, McKenzie D, Hilton F, Gronvold K, Shimizu, N (2003) Geochemical variability in a single flow from northern Iceland. J Geophys Res 108: doi 10.1029/2000JB000142
Marsh BD (1988) Crystal size distribution (CSD) in rocks and the kinetics and dynamics of crystallization. I Theory. Contrib Mineral Petrol 99:277-291
Marsh BD (1995) Solidification fronts and magmatic evolution. Mineral Mag 60:5-40
Métrich N, Wallace PJ (2008) Volatile abundances in basaltic magmas and their degassing paths tracked by melt inclusions. Rev Mineral Geochem 69:363-402
Métrich N, Bertagnini A, Landi P, Rosi M, Belhadj O (2005) Triggering mechanism at the origin of paroxysms at Stromboli (Aeolian archipelago, Italy): the 5 April 2003 eruption. Geophys Res Lett 32:10.1029/2004GL022257

Moore G (2008) Interpreting H$_2$O and CO$_2$ contents in melt inclusions: constraints from solubility experiments and modeling. Rev Mineral Geochem 69:333-361

Morgan DJ, Blake S (2006) Magmatic residence times of zone phenocrysts: introduction and application of the binary element diffusion modeling (BEDM) technique. Contrib Mineral Petrol 151:58-70

Murata KJ, Richter DH (1966) The settling of olivine in Kilauean magma as shown by lavas of the 1959 eruption. Am J Sci 264:194-203

Newman S, Lowenstern JB (2002) VolatileCalc: a silicate melt-H$_2$O-CO$_2$ solution model written in Visual Basic for Excel. Comp Geosci 28:597-604

Nielsen RL, Crum J, Bourgeois R, Hascall K, Forsythe LM, Fisk MR, Christie DM (1995) Melt inclusions in high-An plagioclase form the Gorda Ridge: an example of the local diversity of MORB parent magmas. Contrib Mineral Petrol 122:34-50

Papale P, Moretti R, Barbato D (2006) The compositional dependence of the saturation surface of H$_2$O+CO$_2$ fluids in silicate melts. Chem Geol 229:78-95

Peslier AH, Luhr JF (2006) Hydrogen loss from olivines in mantle zenoliths from Sincoe (USA) and Mexico: mafic alkalic magma ascent rates and water budget of the sub-continental lithosphere. Earth Planet Sci Lett 242:302-319

Putirka KD (2008) Thermometers and barometers for volcanic systems. Rev Mineral Geochem 69:61-120

Putirka KD, Busby CJ (2007) The tectonic significance of high-K$_2$O volcanism in the Sierra Nevada, California. Geology 35:923-926

Ramos FC Reid MR (2005) Distinguishing melting of heterogeneous mantle sources from crustal contamination: insights from Sr isotopes at the phenocrysts scale, Pisgah Crater, California. J Petrol 46:999-1012

Ramos FC, Tepley FJ III (2008) Inter- and intracrystalline isotopic disequilibria: techniques and applications. Rev Mineral Geochem 69:403-443

Roedder E (1979) Origin and significance of magmatic inclusions. Bull Mineral 102:487-210

Roggensack K, Hervig RL, McKnight SB, Williams SN (1997) Explosive basaltic volcanism from Cerro Negro volcano: influence of volatiles on eruptive style. Science 277:1639-1642

Roggensack K (2001) Sizing up crystals and their melt inclusions: a new approach to crystallization studies. Earth Planet Sci Lett 187:221-237

Rutherford MJ (2008) Magma ascent rates. Rev Mineral Geochem 69:241-271

Ryan MP (1988) The mechanics and three-dimensional internal structure of active magmatic systems: Kilauea volcano, Hawaii. J Geophys Res 93:4213-4248

Sato H (1975) Diffusion coronas around quartz xenocrysts in andesite and basalt from Tertiary volcanic region in northeastern Shikoku, Japan. Contrib Mineral Petrol 50:49-64

Scandone R, Cashman KV, Malone SD (2007) Magma supply, magma ascent and the style of volcanic eruptions. Earth Planet Sci Lett 253:513-529

Scandone R, Malone SD (1985) Magma supply, magma discharge and readjustment of the feeding system of Mount St. Helens during 1980. J Volcanol Geotherm Res 23: 239-262

Simon JI, Reid MR (2005) The pace of rhyolite differentiation and storage in an 'archetypical' silicic magma system, Long Valley, California. Earth Planet Sci Lett 235:123-140

Sinton JM, Detrick RS (1992) Mid-ocean ridge magma chambers. J Geophys Res 97:197-216

Slezin YB (2003) The mechanism of volcanic eruptions (a steady state approach). J Volcanol Geotherm Res 122:7-50

Sobolev AV, Shimizu N (1993) Ultra-depleted primary melt included in an olivine from the Mid-Atlantic ridge. Nature 363:151-154

Sorby HC (1858) On the microscopic structures of crystals, indicating the origin of minerals and rocks. Geol Soc London Q J 14:453-500

Sorby HC (1863) On the microscopical structure of Mount Sorrel Syenite, artificially fused and slowly cooled. Proc Geol Polytech Soc W Yorkshire 4:301-304

Spilliaert N, Allard, P, Métrich N, Sobolev AV (2006) Melt inclusion record of the conditions of ascent, degassing, and extrusion of volatile-rich alkali basalt during the powerful 2002 flank eruption of Mount Etna (Italy). J Geophys Res doi:10.1029/2005JB003934

Streck MJ (2008) Mineral textures and zoning as evidence for open system processes. Rev Mineral Geochem 69:595-622

Streck MJ, Dungan MA, Bussy F, Malvassi E (2005) Mineral inventory of continuously erupting basaltic andesites at Arenal volcano, Costa Rica: implications for interpreting monotonous, crystal-rich, mafic arc stratigraphies. J Volcanol Geotherm Res 140:133-155

Takada A (1994) The influence of regional stress and magmatic input on styles of monogenetic and polygenetic volcanism. J Geophys Res 99:13563-13573

ten Brink US, Brocher TM (1987) Multichannel seismic evidence for a subcrustal intrusive complex under Oahu and a model of Hawaiian volcanism. J Geophys Res 92:13687-13707

Tepley FJ III, Davidson JP, Tilling RI, Arth JG (2000) Magma mixing, recharge and eruption histories recorded in plagioclase phenocrysts from El Chicón volcano, Mexico. J Petrol 41:1397-1411

Walker D, Powell MA, Lofgren GE, Hays JF (1978) Dynamic crystallization of a eucrite basalt. Lunar Planet Sci Conf 9: 1369-1391

Wallace PJ, Gerlach TM (1994) Magmatic vapor source for sulfur dioxide released during volcanic eruptions: Evidence from Mount Pinatubo. Science 265:497-499

Wilson L (1980) Relationships between pressure, volatile content and ejecta velocity in three types of volcanic explosion. J Volcanol Geotherm Res 8:297-313

Young DA (2003) Mind Over Magma: The Story of Igneous Petrology. Princeton Univ Press, Princeton, NJ

Zirkel F (1863) Mikroskopische Gesteinsstudien. Sitz Akad Wissen Wien, Math-Natur Klasse 47:226-270

Experimental Studies of the Kinetics and Energetics of Magma Crystallization

Julia E. Hammer
Department of Geology and Geophysics
University of Hawaii
Honolulu, Hawaii, 96822, U.S.A.

jhammer@soest.hawaii.edu

INTRODUCTION

A quantitative understanding of crystallization and solidification is central to appreciating a variety of phenomena at the interface between volcanology and petrology. Interpretation of magmatic processes from the textures of erupted lavas and pyroclasts in turn depends upon understanding how textures evolve during cooling, decompression, and devolatilization. Crystal size, shape, and compositional distribution depend upon the underlying phase equilibria, mass transport processes, and kinetics of reactions at interfaces. Applications of kinetics in physical volcanology include estimating intra-eruptive magma residence times using crystal growth rates (e.g., Mangan 1990); interpreting perturbations in the melt composition or thermal state from microtextures (e.g., Sharp et al. 1996); assessing degassing as a driver of crystal growth (e.g., Blundy and Cashman 2001) and microlite nucleation (e.g., Hammer et al. 1999) during volcanic eruptions; and interpreting magmatic conditions as they change through time using crystal population trends (e.g., Zieg and Marsh 2002).

This contribution is chiefly concerned with magmas out of chemical and textural equilibrium evolving due to changes in temperature, pressure, or composition. However, studies concerned with dynamic aspects of syn-eruptive magma evolution are predicated on an understanding of the equilibrium steady-state, with experiments typically incorporating an initial stage of equilibration, imposing magma reservoir conditions as a prelude to controlled cooling or decompression. This review likewise includes a brief overview of the phase equilibria approach for determining the "initial conditions" for volcanic eruptions. The remainder of the chapter considers kinetic aspects of crystallization, which are essential for interpreting the information encoded in the compositions and textures of volcanic materials. A section each is allotted to (1) presenting concepts and controversies in the theoretical understanding of kinetics of crystallization since the last RIMG review on this topic (Kirkpatrick 1981), (2) summarizing generic aspects of experimental methods and experiment design in geological studies, (2) reviewing established trends in basalt crystallization, highlighting recent studies that build upon these relationships, and examining pioneering research in olivine growth kinetics and melt inclusion entrapment, (3) considering constraints on kinetics of crystallization in hydrous felsic magmas due to decompression and devolatilization, (4) discussing conventional and innovative methods of characterizing crystal textures of experimental run products, (5) assessing the effects of phase interactions including epitaxial growth and heterogeneous nucleation, and (6) describing frontier research on compositional aspects of the melt inclusion entrapment process incorporating energetic and kinetic issues presented in the previous sections.

PRE-ERUPTIVE MAGMATIC ENVIRONMENTS

General approach

While magmas often contain evidence of disequilibrium (zoned phenocrysts; e.g., Bachmann and Bergantz 2008; Streck 2008), phenocryst rims and interstitial melt (glass) typically achieve equilibrium in pre-eruption magmas (e.g., Rutherford et al. 1985; Rutherford and Devine 1996). Thus, determination of pre-eruptive magma storage conditions is predicated on analyses of natural samples and reproduction of natural phase compositions in experimental charges (e.g., Rutherford et al. 1985; Webster 1987; and many others cited below). A typical approach is to constrain as many environmental variables as possible using analytical techniques applied to natural samples. Experiments are then run to establish the stability fields of the major phenocryst phases at nodes in pressure- temperature space. For example, f_{O_2} and temperature may be determined from co-existing Fe-Ti oxide minerals in natural samples (Andersen et al. 1993); H_2O and CO_2 contents may be ascertained from analyses of melt inclusions in phenocrysts (e.g., Wallace and Gerlach 1994; Wallace 2005; Kent 2008; Moore 2008; Metrich and Wallace 2008); the assumption of H_2O-saturation may follow from comparison of dissolved H_2O contents and solubility of H_2O at reasonable estimates of chamber depth. Prior knowledge of such variables drastically reduces the number of equilibrium experiments necessary to establish a compelling match between the phase assemblage in an experiment and that in the natural magma. Matrix glasses provide an important means of locating the appropriate equilibrium conditions, because melts respond to the changing proportions and abundances of crystallizing phases. Thus, a vitreous groundmass is an extremely helpful attribute in the natural materials.

Starting materials may be synthesized from reagent oxides and carbonates (less commonly gels) to match the natural matrix glass. Application of heat treatments exceeding liquidus temperatures is common practice to homogenize starting materials for phase equilibrium experiments (Costa et al. 2004). This ensures that if particles or microscopic heterogeneities remain, they are uniformly distributed. However, if natural samples are used as starting materials, care should be taken to prevent exposure of phenocryst cores during preparation, because mass sequestered inside a crystal is not necessarily in equilibrium with the matrix melt, and its exposure to melt during an experiment could measurably draw the system away from the desired pre-eruptive conditions. Selection of appropriate system boundaries to include only the "reactive magma" is discussed in depth by Pichavant et al. (2007).

Experimental investigation of processes occurring at elevated temperature and pressure usually requires assuming that the state of the system at run conditions is preserved when the material is brought quickly to low temperature, i.e., quenched. Most diffusive processes and interphase reactions of interest are sufficiently sluggish that the assumption is valid for the quench rates commonly applied (100's to 1,000's of $°C\ h^{-1}$).

Apparatus

Three apparatus have produced the vast majority of phase equilibrium studies at pressures corresponding to the Earth's surface and extending to mid-crustal depths: the 1-atmosphere gas-mixing furnace, the cold-seal (externally-heated) pressure vessel, CSPV, and the internally-heated pressure vessel, IHPV. The operating techniques are described exhaustively in several excellent texts (Ulmer 1971; Edgar 1973; Ulmer and Barnes 1987), and papers referenced in this section. Only the basic definitions, key attributes, and limitations are summarized here to provide a context for the non-experimentalist.

The gas-mixing furnace uses mixtures of H_2 and CO_2 or CO and CO_2 to reduce the oxygen content of the furnace atmosphere from Earth's surface environment (~0.2 bars O_2) to conditions prevailing in planetary interiors (10^{-7} to 10^{-17} bars O_2). For convenience, oxygen concentrations are referenced to the oxygen concentrations fixed by temperature-sensitive oxygen-buffering reactions such as $Ni + O_2 = NiO$ (NNO); fayalite + O_2 = quartz + magnetite (FMQ, alternately "QFM"); and iron + O_2 = wustite (IW), using log-unit notation. For example, QFM +1 corresponds to the oxygen fugacity one order of magnitude higher than that fixed by the QFM buffer at a specified temperature. The gas-mixing system in conjunction with a resistance furnace is a workhorse in experimental petrology for (a) its ability to generate the continuous spectrum of relevant oxygen fugacities, not just discrete buffer values, (b) its ability to generate temperatures required to melt basalts and highly refractory simple silicate compositions, and modulate temperature using programmable controllers, and (c) for rapid quenching capabilities. Of course, evaluation of the effects of volatiles in concentrations exceeding nominal values necessitates study at elevated confining pressure.

As fluid-medium pressure apparatus, the CSPV and IHPV offer pressure control advantages over solid-media apparatus such as the piston-cylinder and diamond anvil cell. The stress on a sample capsule is hydrostatic and precisely measurable (commonly less than ± 1 MPa). Attainable pressure ranges are relevant to volcanic processes in the Earth's crust (5-600 MPa for the CSPV, depending on temperature, and up to 1000 MPa for the IHPV), and may be varied during a run to simulate eruptive decompression. In the CSPV setup, a vessel of high tensile strength Mo or Ni alloy is joined to a fluid line containing either water or inert gas, and then placed inside a tube furnace. The pressure seal is made outside the furnace, giving the apparatus its name (although the junction is far from cold). Quench rates are much lower than for 1-atm gas-mixing apparatus, but various strategies (e.g., Gardner et al. 1999; Gaillard et al. 2003) have improved quench rates dramatically. The sample f_{O_2} may be controlled by the vessel itself, or in combination with metal filler rods. Alternatively, a solid-medium buffering assemblage may be included inside the sealed outer capsule along with the sample to monitor any changes in the intrinsic sample f_{O_2}. Hydrous experiments in vessels pressurized by a gas are particularly prone to oxidation.

The furnace-vessel relationship is reversed for the IHPV, with the electrical resistance furnace encircling the sample being located inside a much larger pressure vessel. The large volume in the IHPV allows multiple charges to be run simultaneously (e.g., Martel et al. 1999), or simply for larger volumes of starting material to be used. The IHPV technique is also used in seismic velocity determinations (e.g., Kern 1982), electrical resistivity measurements (Gaillard 2004), and falling-sphere viscometry (e.g., Shaw 1963; Vetere et al. 2006). IHPV technology includes apparatus capable of deforming samples to large strains in a geometry resembling simple shear (Arbaret et al. 2007; Champallier et al. 2008), with promising potential for investigating hydrodynamic effects in crystal-rich ascending magmas. Rapid quenching rates and accurate oxygen fugacity control are achieved by incorporating drop-quenching capabilities and H_2 membranes, respectively (Holloway et al. 1992; Roux and Lefevre 1992; Scaillet et al. 1992; Roux et al. 1994; Berndt et al. 2002). A sapphire-windowed IHPV recently developed by Gonde et al. (2006) offers the possibility of observing high temperature/ high-pressure processes *in situ*. Due to the technical challenges of maintaining and safely operating large-volume high-pressure gas apparatus (Lofgren 1987), the IHPV is used in fewer laboratories worldwide than CSPVs.

Evaluating experimental results

Criteria for establishing a match with natural pre-eruptive conditions include agreement between experimental run products and natural samples in terms of the phase assemblage, compositions of matrix melt (glass) and major silicate minerals, and mineral mode. Chemical equilibrium and phase stability is ascertained by evaluating crystal texture (i.e., verifying faceted surfaces), ensuring internal consistency with experiments at bracketing conditions, testing for run time-independence of outcomes, and by approaching run conditions from above and below the equilibrium crystal content. Mass balance (based on elemental compositions and abundance of all phases in the sample) may also be employed to check for sample-container interactions, and ensure the system was closed with respect to mass during the run.

Unfortunately, not all erupted magmas are suitable for a study aimed at determining pre-eruptive conditions. Holocrystalline samples pose the challenge of lacking a glass composition to compare against experimental glasses. Features such as crystal resorbtion, mingled glasses, and texturally or mineralogically distinct enclaves suggest that magmas are not in chemical equilibrium prior to eruption. The experimental technique described above is unlikely to yield meaningful results with such starting materials.

Phase equilibria in volcano-petrology

A listing of phase equilibrium studies that determine conditions prior to a specific volcanic eruption or eruptive period is presented in Table 1. The table does not include experimental studies of intrusive rocks (Clemens and Wall 1981; Clemens et al. 1986; Scaillet et al. 1995; Dall'Agnol et al. 1999; Bogaerts et al. 2006), studies that explore the effects of environmental parameters on phase equilibria (e.g., Sisson and Grove 1993; Martel et al. 1999; Brugger et al. 2003; Feig et al. 2006; Pichavant and Macdonald 2007) or other magma differentiation processes (e.g., Grove and Juster 1989; Berndt et al. 2005).

Several themes are apparent in the dataset. With the exception of the Laacher See study, all magmas are from volcanoes at convergent plate boundaries. The majority cover intermediate magma compositions (andesite to dacite bulk rock compositions) that contain highly evolved matrix melts (rhyodacite to rhyolite). The majority of studies are performed at H_2O-saturation ($P_{H_2O} = P_{total}$, i.e., the mole fraction of H_2O in the fluid phase $=1$), and at relatively oxidizing conditions ($\Delta NNO= 0$ to $+3$). Studies of crystal-poor mafic magmas, magmas rich in volatiles besides H_2O, volatile-poor magmas, and magmas equilibrated at moderately reducing conditions are scarce. Early studies of this type (Holloway and Burnham 1972; Eggler and Burnham 1973) yielded new insights of general petrological importance, but experienced oxidation-state changes and container-sample reaction. These problems remain today for experiments at oxidation states < QFM and temperatures above 1100 °C.

Table 1 allows comparison of the ranges of intrinsic conditions (temperature, pressure, oxygen fugacity, H_2O content) obtained as equilibrium conditions against the ranges that have been actually tested as experimental variables. The data sets have merit beyond fulfilling the specific volcanological objectives that motivated the research. As internally consistent probes of phase equilibria for selected starting materials, the compositions and modes are typically incorporated into the MELTS thermodynamic database (Ghiorso and Sack 1995). Apart from exceptionally cool dacite (Mt. Pinatubo, 1991) and phonolite (Laacher See volcano), the experimental results describe the expected negative correlation between bulk silica content and equilibrium temperature. Basaltic andesite (Mt. Pelée) and rhyolite (Novarupta) define the high and low P_{H_2O} endmembers, respectively, from among the H_2O-saturated magmas. However, the pre-eruption depths of equilibration are only weakly correlated with bulk composition, and entirely uncorrelated with matrix melt composition. For example, mafic andesite (SW Trident) and rhyolite (Novarupta) eruptions are staged from similarly shallow depths (50-75 MPa), while dacites erupt from both shallow (Novarupta, ≤ 50 MPa) and deep (Black Butte, 300 MPa) levels.

CRYSTALLIZATION KINETICS IN SIMPLE SYSTEMS

The vast majority of nucleation studies have been conducted in compositionally simple systems and reported in the materials science literature (reviewed recently in Fokin et al. 2006), with application to the synthesis of industrial glasses and ceramics. Materials such as $Li_2O \cdot SiO_2$ (lithium disilicate) and Na_2O-$2CaO$-$3SiO_2$ are considered analogs for geologically relevant melts (e.g., Davis and Ihinger 2002). The present treatment is limited to aspects of crystal nucleation phenomena recently evaluated with experiments (steady-state nucleation, transient-nucleation, the crystal-melt interfacial free energy) and adjustments to the classical nucleation theory (CNT). The section concludes with an overview of aspects of crystal growth theory.

Table 1. Determinations of pre-eruption magmatic conditions using the method of experimental phase equilibria.

Magma investigated						Experimental conditions				Pre-eruption equilibrium conditions				Ref.
Volcano	Eruption	Bulk Composition	Bulk SiO$_2$	Matrix Glass SiO$_2$	S.M.[1]	Pressure (MPa)	Temp. (°C)	DNNO	X$_{H_2O}$	Pressure (MPa)	Temp. (°C)	DNNO	X$_{H_2O}$	
St. Helens	1980	dacite	62.8	72.7	1A	100-320	847-1190	−1 to +4	0.5-1	220 ± 30	930 ± 10	0.4	0.5-0.7	1
St. Helens	1980	dacite	63.0-63.8	72.7	1A	220, 320	890-970	0.4	0.4-1	220	920	1.5	0.67	2
St. Helens	1480 A.D.	dacite	67.2	74.8	1A	100-350	825-875	0.67-1.17	0-1	250-350	848 ± 5	0.8	0.4-0.5	3
Mt. Pinatubo	June 15, 1991	dacite	64.76	76.44	1B	160-390	760-925	2-4	0.5-1	220 ±50	780 ±10	3	1	4
Soufriere Hills	January, 1996	andesite	59.15-59.46	n.d.	1A	50-210	800-935	1	0.5-1	115-130	820-840--> 840-880	1	1	5
W. Mexico	n.d.	andesite	~62	n.d.	1A	44.1-294.4	950-1100	1.1-2.8	1	70-150	950-975	2	1	6
W. Mexico	n.d.	basaltic andesite	~55	n.d.	1A	51.7-302.7	950-1150	1.6-5.6	1	<250	1000-1150	2	1	6
Santorini	~1645 B.C.	rhyodacite[1]	69.07	unknown	1A	50-250	800-920	0.5-1	1	210-240	825±25	0.5-1	1	7
Santorini	~1645 B.C.	rhyodacite[2]	69.07	72.1-73.9	1A	50-250	800-920	0.5-1	1	50	885	0.5-1	1	7
Santorini	~1645 B.C.	andesite	56.33	n.d.	1A	50-150	865-910	0.5-1	1	>75	885	0.5-1	1	7
Mt. Pinatubo	June 15, 1991	dacite	64.76	77.75	1C	220-390	750-900	0-2.7	0.3-1	220	760±20	1.5-1.7	>0.88	8
SW Trident	1953-1974 A.D.	andesite	55.8-58.9	≈63	1B	1000-1100	50-200	0	<1−1	90	1000±20	1	1	9
Novarupta	June 6-9, 1912	rhyolite	77.23	77.64	1A	25-200	760-900	0.5	1	40-100	800-850	0.5	1	10
Soufriere Hills	January, 1996	andesite	59.15	78.8	2	25-210	825-1100	1	1	>100	900-1000	1	1	11
Mt. Pelée	650 y.b.p.-1929	andesite	60-61	62-63	1C	200-400	850-1040	0-3	0.7-1	200 ±50	875-900	0.4-0.8	1	12
Mt. Pelée	40-19.5 ky.b.p.	basaltic andesite	53	74.0-76.5	1C	398.8-427	949-1025	0.2-3.8	0.58-1	400	950-1025	0.4	1	13
Mt. Pinatubo	June 15, 1991	dacite	64.76	78.2	1B	150-220	780	2	1	220-155	780	2	0.94-1	14
Novarupta	June 6-9, 1912	andesite	59	67.6	1B	50-225	850-1050	1	0.34-1	100-75	930-960	1	0.7-1	15
Novarupta	June 6-9, 1912	dacite	65.4	79.1	1B	15-200	820-950	0-2	0.27-1	50-25	850-880	0-2	0.9-1	15
Laacher See	12,900 y.b.p.	phonolite	58.41	57.37	1A	75-175	725-800	1	1	115-145	750-760	1	1	16
Unzen	1992	hybrid[3]	64.6	68.2	1C	100-300	750-875	0	0.4-1	300	930 hybrid[3]	0	<1	17
Stromboli	800-1600 A.D.	basalt	49.4	49.8	1C	48-417	1050-1175	0-2	0.06-0.25	100-270	1140-1160	0-2	0.15-0.65	18
Aniakchak	3430 ybp	rhyodacite	70.57	71.5	1B	50-200	820-950	0.5-1	1	125-150	870-880	0.5-1	0.8	19
Arenal	~2006	basaltic andesite	52.0-54.0	n.d.	2	500	1100-1300	−3 to −2.5	1	500	1150	−1 to 0	1	20
Black Butte	10,000 y.b.p.	dacite	64.63	67.73	1C	100-450	800-950	0-1	1	300	890-910	1	1	21

[1]Starting materials: [1A] powdered natural material, [1B] crushed natural material, [1C] fused natural material, [2] synthetic material

[2][1] Rutherford et al. 1985, [2] Rutherford and Devine 1988, [3] Gardner et al. 1995, [4] Rutherford and Devine 1996, [5] Barclay et al. 1998, [6] Moore and Carmichael 1998, [7] Cottrell et al. 1999, [8] Scaillet and Evans 1999, [9] Coombs et al. 2000, [10] Coombs and Gardner 2001, [11] Couch et al. 2001, [12] Martel et al. 1999; Pichavant et al. 2002, [13] Pichavant et al. 2002, [14] Hammer and Rutherford 2003, [15] Hammer et al. 2002, [16] Harms et al. 2004, [17] Holtz et al. 2005, [18] Di Carlo et al. 2006, [19] Larsen 2006, [20] Petermann and Lundstrom 2006, [21] McCanta et al. 2007

Fundamental concepts of crystal nucleation

Nucleation exerts the primary control on the grain size of rocks (Swanson 1977), and thus the texture and many of their mechanical properties (Kesson et al. 2001). Reviews of the first several decades of nucleation research in igneous rocks include those by Kirkpatrick (1981), Cashman (1990), and Lasaga (1998). Despite a long and distinguished history in metallurgy, ceramics, and igneous petrology, nucleation continues to lie at the frontier in materials science because it is a difficult process to study directly. Energy barriers to nucleation are strongly dependent on melt structure (Toshiya et al. 1991) and thus the thermal pre-treatment, initial volatile content, and major element composition (Davis et al. 1997) of the melt. In many cases, the preparation of the experimental "starting material" is reported here, as these steps may be as important to controlling eventual texture as subliquidus thermal history. Supercritical clusters of atoms (nuclei) are exceedingly small (10-1000 atoms, or ~1-100 nm) and ephemeral according to the classical theory of crystal nucleation (Kelton and Greer 1988). In hydrous magmatic systems relevant to crustal-level reservoirs and conduits, the process is seemingly even more inaccessible to the experimentalist because of the need to modulate the H_2O concentration in the melt. This prevents *in situ* study of nucleation at ambient pressure, as in scanning calorimetric (Davis and Ihinger 1999) and electrical resistivity approaches (Wang and Lu 2000). Nevertheless, advances in specimen analysis are allowing phenomenological studies to push theory development (Zhang et al. 2003).

Theory development through experimentation

The traditional view of nucleation is that crystal-like clusters of atoms possessing the physical and chemical properties of macroscopic crystalline solids form by random fluctuations in a melt when below its liquidus temperature (Volmer and Weber 1926; Becker and Doring 1935). The excess free energy associated with maintaining the cluster-liquid interface causes clusters below a critical size to shrink and those above it to grow into nuclei. The critical cluster size itself varies with undercooling, becoming smaller as undercooling increases. Homogeneous nucleation is rapid at the temperature where a large fraction of clusters achieve the critical cluster size. The temperature dependence of the cluster size distribution is given by a Boltzmann function: $n_r = n_o \exp(-\Delta G_r/kT)$, where n_r is the number of spherical clusters of radius r, n_o is the total number of atoms in the system, ΔG_r is the excess free energy associated with the cluster, T is temperature in Kelvin, and k is Boltzmann's constant. Porter and Easterling (1997) provide a plausible explanation for the extreme sensitivity of subliquidus nucleation behavior to duration and temperature of superliquidus heat treatments, without calling upon ephemeral heterogeneities to serve as nucleation sites: increasing temperature causes melt depolymerization, lessening the quantity of large clusters poised to become nuclei should the temperature drop below the equilibrium liquidus. If cooling after protracted superheating is rapid, the effect of unrelaxed (i.e., disequilibrium) melt structure is to inhibit nucleation at subliquidus temperatures. This deleterious effect on nucleation increases as the degree of superheating increases.

Most experimental work in nucleation employs two-stage experiments, in which crystals nucleate at low temperature, T_n, and grow to observable size in a separate "development" step at a higher (but still subliquidus) temperature, T_d. Recall that the critical size decreases with decreasing temperature. Thus, the critical size at T_n is smaller than the critical size at T_d; only nuclei that have achieved critical size at the development temperature (T_d) grow to become observable crystals. Smaller clusters have a high probability of dissolving during the development stage. Thus, not all nuclei formed in the nucleation stage survive to be counted, which could lead to underestimation of the nucleation rate at T_n. However, nucleation rates can be found through extrapolation by determining the functional dependence of the surviving crystal quantity on T_d (e.g., Fokin et al. 2006). A second important feature of nucleation phenomenon is that the steady-state (time independent) nucleation rate is not attained immediately at a given temperature. Rather, a relaxation time is required for melt to structurally adjust and to form the equilibrium distribution of clusters. An apparent lag time prior to nucleation, during which nucleation rates are much lower than at steady state, is known as the period of transient nucleation or induction time.

Steady state nucleation. According to classical nucleation theory (CNT), the steady state rate equation for the formation of spherical critical nuclei (I, m^{-3} s^{-1}) as a function of temperature has the form (James 1985):

$$I = A\exp\left[\frac{-(\Delta G^* + \Delta G_D)}{k_B T}\right]; \quad \Delta G^* = \frac{16\pi\sigma^3}{3\Delta G_V^2} S(\theta) \tag{1}$$

The pre-exponential factor A includes the frequency of attachment attempts and the specific number density of reactant atoms in the melt. $A = k_B T n_v / h$; where k_B is the Boltzmann constant, T is temperature, n_v is the volumetric concentration of reactant atoms, and h is Planck's constant. ΔG^* is the free energy required to form a spherical critical nucleus having properties of the macroscopic solid. σ is the free energy associated with the crystal-liquid interface. ΔG_V is the free energy change per volume of the transformation, defined $\Delta G_V = \Delta G/V_M$, where V_M is the molar volume of the crystallizing phase and ΔG is the bulk free energy decrease driving solidification. (Several processes can cause ΔG to decrease in natural magmas, as will be explored below. Common mechanisms include decreasing the temperature of a melt below its liquidus and changing the melt's liquidus temperature through a change in melt composition.) ΔG_D is the activation energy required for the attachment of atoms to a cluster, i.e., a kinetic barrier that includes transport of components through the fluid. θ is the wetting angle between the nucleus-wall and nucleus-liquid interfaces, and $S(\theta) = (2 + \cos\theta)(1 - \cos\theta)^2/4$.

The balance between the thermodynamic driving force (ΔG^*), which propels solidification for all temperatures below the equilibrium melting point, and the kinetic barrier (ΔG_D), which grows as temperature decreases, give rise to a nucleation function with a maximum. In the event that nucleation is homogeneous, $S(\theta) = 1$ and the thermodynamic barrier is simply a function of the volumetric free energy difference between liquid and solid (James 1985). Heterogeneous crystal nucleation is rare in metals and alloys, but common in silicate systems (James 1974; Kelton and Greer 1988).

Several simplifying assumptions are employed to compare experimental data with the CNT. The first is that activation energy of atomic jumps across the liquid-nucleus interface is the same as that of shear relaxation of the liquid, i.e., the Stokes-Einstein approximation (Ree and Eyring 1958; Dingwell and Webb 1989), allowing ΔG_D to be expressed in terms of viscosity (η). Second, the heat capacity of the liquid is assumed equal to the heat capacity of the solid, which reduces the thermodynamic driving force ΔG for crystallization to $\Delta G = \Delta H \Delta T / T_L$. Combined with the pre-exponential terms, the rate equation for homogenous crystal nucleation becomes:

$$I = \frac{A_c}{\eta} T \exp\left(\frac{-16\pi\sigma^3 T_L^2}{3\Delta H^2 \Delta T^2 k_B T}\right); \quad A_c = \frac{n_v k_B}{3\pi\lambda^3} \tag{2}$$

(although alternate formulations are proposed for the pre-exponential term, A_c; e.g., Gránásy et al. 2002).

Steady-state nucleation rate data as a function of reduced (homologous) temperature (T/T_L) are shown for several one-component systems (Fig. 1). The height and position of the maxima are parameters that determine whether given materials are likely to produce crystal-rich or glassy materials upon rapid cooling. If the kinetic barrier is decreased relative to the thermodynamic driving force, the maximum increases and shifts toward lower temperatures. The addition of H$_2$O to silicate melts produces this effect by lowering melt viscosity (Fig. 2).

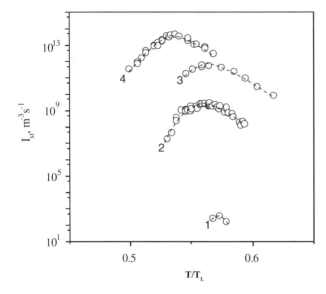

Figure 1. Steady-state nucleation rate versus reduced temperature (T_L= liquidus temperature) for crystals forming in glasses of the same composition: $3MgO \cdot Al_2O_3 \cdot 3SiO_2$ (curve 1), $Li_2O \cdot 2SiO_2$ (curve 2), $Na_2O \cdot 2CaO \cdot 3SiO_2$ (curve 3), $2Na_2O \cdot CaO \cdot 3SiO_2$ (curve 4). [Used by permission of Elsevier, from Fokin et al. (2006), J Non-Cryst Solids, Vol. 352, Fig. 11, p. 2681-2714.]

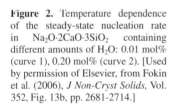

Figure 2. Temperature dependence of the steady-state nucleation rate in $Na_2O \cdot 2CaO \cdot 3SiO_2$ containing different amounts of H_2O: 0.01 mol% (curve 1), 0.20 mol% (curve 2). [Used by permission of Elsevier, from Fokin et al. (2006), J Non-Cryst Solids, Vol. 352, Fig. 13b, pp. 2681-2714.]

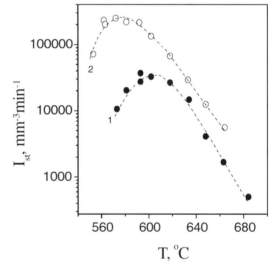

Conversely, if the free energy of forming a critical cluster is reduced by decreasing interfacial energy, e.g., Muller et al. (2000), the maximum nucleation rate is also increased, but shifted toward higher temperatures.

Transient nucleation. The transient rate of nucleation I' at time t during the period of structural relaxation and formation of the steady-state cluster distribution is given by a stochastic equation (Kashchiev 1969):

$$I' = I\left(1 + 2\sum_{n=1}^{\infty}(-1)^n \exp\left(\frac{-n^2 t}{\tau}\right)\right) \tag{3}$$

where τ is induction time and n is an integer. The nuclei density $N(t)$ is related to time according to the integrated form of Eqn. (3) obtained by Kashchiev (1969):

$$\frac{N(t)}{I\tau} = \frac{t}{\tau} - \frac{\pi^2}{6} - 2\sum_{n=1}^{\infty}\left[\frac{(-1)^n}{n^2}\right]\exp\left(-n^2\frac{t}{\tau}\right) \qquad (4)$$

Experimental data for $2Na_2O \cdot CaO \cdot 3SiO_2$ and typical $N(t)$ curves are shown in Figure 3, indicating the transient stage of nucleation, the induction time, and the steady state nucleation region.

Davis et al. (1997) report nucleation kinetics in H_2O-bearing lithium disilicate melt. Reagents and H_2O are fused at 1350 °C, quenched to glass, reheated at 435 °C for periods from 100-10,000 minutes in ambient atmosphere, and transferred to 600 °C for growth of nuclei to observable size. Small amounts (~10-1000 ppm) of H_2O in lithium disilicate glass increase nucleation rate and decrease the induction time (Fig. 4). The effects are greater than can be attributed to reduction of the kinetic barrier through decreasing viscosity alone. However, a 3% decrease in σ can account for the nucleation behavior, and is consistent with a negative correspondence between σ and dissolved H_2O content that is inferred for natural silicate melts by Hammer (2004).

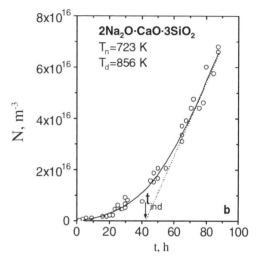

Figure 3. Typical curve of the number density of crystals forming in glass of the same composition. T_n = nucleation temperature, T_d=development (growth) temperature. [Used by permission of Elsevier, from Fokin et al. (2006), *J Non-Cryst Solids*, Vol. 352, Fig. 3, pp. 2681-2714.]

Interfacial energy. Values of the crystal-melt interfacial free energy (σ) defined in the classical nucleation theory (Eqn. 1) are exceedingly difficult to obtain independently from nucleation data. The few such values in existence were derived from measurements of macroscopic crystal-melt dihedral angles (Cooper and Kohlstedt 1982; Rose and Brenan 2001; Ikeda et al. 2002) and numerical studies of hard sphere molecular interaction potentials (Battezzati 2001; Granasy et al. 2002; Davidchack and Laird 2003). These data sets do not include many geologically-important crystal-melt combinations. One method of obtaining σ for the case of homogeneous nucleation is to assume the CNT is valid and determine its value from the slope of a plot of $(1/(T\Delta G_V^2))$ versus $\ln(I\eta/T)$ (Gonzalez-Oliver and James 1980). Interestingly, this method rarely yields constant values of σ for all experiments in a given system. As will be discussed further below, another approach is to assume that the form of the CNT is correct but that the interfacial energy is not single-valued (James 1974; Kelton and Greer 1988). It is possible to rearrange Equation (2) in order to solve for σ using nucleation rate data, thus generating a separate value for each experiment. Note that in the event of heterogeneous nucleation, it is impossible to solve for σ in this way without knowledge of θ.

Temperature and curvature dependence of σ. Recent studies critically examine the assumption that σ remains constant as clusters grow into crystal nuclei and nuclei become macroscopic crystals. One approach for determining σ that is consistent with the steady-state

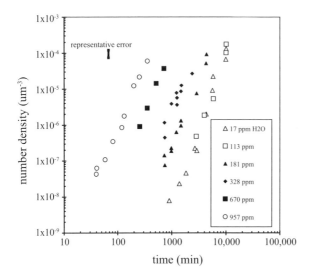

Figure 4. Number density of lithium disilicate crystals forming in variably hydrated melt of the same (anhydrous) composition at 435 °C. Symbols designate water contents of starting glasses determined by FTIR (Davis et al. 1997). [Used by permission of Elsevier, from Davies et al. (1997), *J Non-Cryst Solids*, Vol. 219, Fig. 1, pp. 62-69.]

CNT but not explicitly dependent upon it, involves fitting the transient nucleation expression (Shneidman 1988) to transient nucleation rates (Deubener and Weinberg 1998; Sycheva 1998). Three data sets obtained in this manner in the lithium disilicate system over a range of development temperatures produce positive correlation between σ and nucleation temperature (Deubener and Weinberg 1998); (Fig. 5) This result seemingly conflicts with the thermodynamic point of view in which the surface energy of a crystal having infinite radius of curvature (σ_{inf};

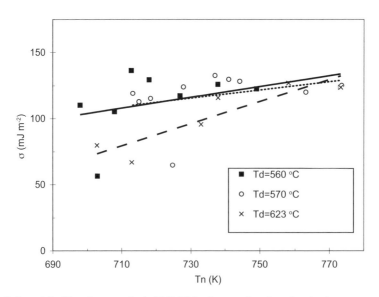

Figure 5. Crystal-liquid surface energies in $Li_2O \cdot 2SiO_2$ glass as a function of nucleation temperature (T_n) obtained at development temperatures of 560 °C (squares), 570 °C (circles), and 623 °C (x's); various sources in Deubner and Weinberg (1998). All sets include the same value at 500 °C (773 K), obtained from transient nucleation data Deubner and Weinberg (1998). Straight lines are least squares fits to the data.

i.e., a flat surface) decreases with increasing temperature (Gutzow et al. 1985; Skripov and Faizullin 2005). However, Fokin and Zonotto (2000) draw upon Tolman's equation

$$\sigma(R) = \frac{\sigma_{inf}}{\left(1 + \frac{2\delta}{R}\right)} \quad (5)$$

(Tolman 1949) to address the apparent discrepancy and separate temperature and curvature controls on σ. Tolman's parameter, δ, characterizes the width of the interface and is on the order of atomic dimensions; R is the radius of the critical cluster. Equation (5) shows how the magnitude of the curvature correction for surface energy increases as the nucleus decreases in size. Embedded in the equation is a temperature dependence of the critical radius: because the bulk free energy driving force increases with decreasing subliquidus temperature, R decreases with decreasing temperature. Near the liquidus temperature, σ closely approaches σ_{inf} because R is a maximum value. As temperature decreases, both the numerator and denominator of the ratio defining σ(R) decrease, with the value of σ dependent upon the relative magnitudes of δ and R. Incorporating Tolman's equation to the expression for the work of forming a nucleus, Fokin and Zanotto (2000) fix δ and find the range of values of σ_{inf} for which the calculated nucleation rates equal published experimental values in the Li_2O-$2SiO_2$ and Na_2O-$2CaO$-$3SiO_2$ systems. A plausible range in δ values (e.g., 2.8×10^{-10} m < δ < 3.4×10^{-10} m for Li_2O-$2SiO_2$) produces σ_{inf} values that increase with decreasing temperature (in accord with theory) and explains the correlation between σ and temperature observed by Deubener and Weinberg (1998). The analysis shows that the observed increase in σ with temperature (Fig. 5) may actually arise from the dependence of the critical radius on temperature.

Non-CNT assessment of σ. The wetting, or dihedral, angle (θ) at the contact between two crystals and melt is related to the ratio of interfacial free energies (Smith 1964):

$$\frac{\sigma_{solid\text{-}solid}}{\sigma_{solid\text{-}liquid}} = 2\cos\left(\frac{\theta}{2}\right) \quad (6)$$

However, neither $\sigma_{solid\text{-}solid}$ nor $\sigma_{solid\text{-}liquid}$ is truly single-valued for a given system at equilibrium. Euhedral crystals are nonspherical precisely because $\sigma_{solid\text{-}liquid}$ is directionally anisotropic. Sessile drop techniques, in which a drop of liquid is placed on a solid surface and allowed to mechanically equilibrate (e.g., Bagdassarov et al. 2000), confirm that different faces are usually characterized by different values of $\sigma_{solid\text{-}liquid}$, with some minerals (e.g., olivine and pyroxene) more strongly anisotropic than others (e.g., spinel structured Fe-Ti oxides; Shafer and Foley 2002). The determination of $\sigma_{solid\text{-}solid}$ interfacial free energy is complicated by the fact that solid-solid contacts include varying crystallographic misorientations among crystals, giving rise to a spectrum of equilibrium dihedral angles. Nonetheless, assessment of characteristic σ ratios through wetting angle studies is typically conducted in crystal-rich natural rocks and near-solidus melting experiments containing abundant mineral contacts (Waff and Faul 1992; Laporte and Watson 1995; Holness 2006), with obvious implications for mantle melt storage and migration. Ikeda et al. (2002) measured θ among sparse crystals in relatively melt-rich magmas in the Fe-free diopside-anorthite system to estimate the ratio of interfacial energies between diopside and melt. They observe that $\sigma_{solid\text{-}liquid}$ is ~57% of $\sigma_{solid\text{-}solid}$. Conceptually, as the crystallographic misorientation angle approaches zero and lattices of adjacent crystals come into alignment, the $\sigma_{solid\text{-}solid}$ value should decline to a minimum allowing optimal investigation of $\sigma_{solid\text{-}liquid}$ through wetting angle measurements. If variation in $\sigma_{solid\text{-}solid}$ with crystallographic misorientation is large compared with directional anisotropy of $\sigma_{solid\text{-}liquid}$, i.e., if $\sigma_{solid\text{-}liquid}$ may be assumed constant in a given crystal-melt system, then θ should increase with decreasing misorientation. Evaluating this hypothesis and determining the maximum value of $\sigma_{solid\text{-}liquid}$ as a fraction of $\sigma_{solid\text{-}solid}$ would require electron backscattered diffraction (EBSD), transmission electron microscopy (TEM) or universal stage petrography as well as accurate measurement of θ. Bearing in mind the curvature (size) effects prescribed by Equation (5), such determinations of macroscopic $\sigma_{solid-liquid}$ may yet fail to reconcile nucleation data with theory.

A novel approach to determining σ (i.e., $\sigma_{solid-liquid}$) utilizing nucleation behavior but independent from the CNT capitalizes on the observation from two-stage nucleation experiments (Fokin et al. 2000) that higher development temperatures are associated with longer induction times. The difference in induction time (Δt) scales with the difference in development temperatures because the size of the critical cluster increases, while cluster growth rate remains relatively constant. Using an analytical expression for Δt, Kalinina et al. (1976) and Fokin et al. (2000) incorporate the definition of critical size, $R = 2\sigma/\Delta G_v$, to obtain σ from measurements of Δt. The values of σ obtained are nearly an order of magnitude greater than values obtained using nucleation data in conjunction with the CNT. If the independently-derived values are used to forward calculate nucleation rate using the CNT, the prediction underestimates observed nucleation rates by many orders of magnitude. The explanation for this discrepancy put forth by Fokin et al. (2000) is that macroscopic values of thermodynamic driving force do not match the driving force for cluster formation and growth. This shift in blame from uncertainties in σ to uncertainties in ΔG is expanded in the section *Beyond the CNT*, below.

Diffuse Interface Theory. An alternative to the thermodynamic description of heterogeneous systems in which phases are separated by razor-sharp boundaries (Gibbs 1906) posits a spatial distribution of thermodynamic properties in the vicinity of clusters (Turnbull 1964). The diffuse interface theory, DIT (Spaepen 1994; Granasy et al. 1996), takes into account gradations between bulk solid and bulk liquid values of enthalpy and entropy within an interface region. This region is not a third phase in the thermodynamic sense, but rather a transitional interval that exists solely because solid and liquid surround it. The model states that relative to the bulk melt and cluster center (which possesses properties of a bulk solid), the crystal-melt interfaces are regions in which excess free energy (σ) increases to a maximum at some radial distance from the center of a subcritical cluster. In contrast to crystal-crystal boundaries, which are sharply defined regions of large enthalpy due to a high density of broken bonds, cluster-liquid interfacial regions contain a lower density of broken bonds. Bonding arrangements in this region may be somewhat similar to those in the neighboring solid (Spaepen 1994). Because undercooling a melt to produce glass is impossible without positive interfacial free energy (otherwise there would be no thermodynamic barrier to crystallization), the interfacial free energy (σ_{int}), defined as $H_{int} - TS_{int}$ must be greater than zero. Thus, the contribution of the enthalpy term (H_{int}) must increase more steeply than the entropy term (TS_{int}) with radial distance from the nucleus in order for σ to remain positive.

The central concept of the DIT is employed to interpret the observed correspondence of σ with melt H_2O content found by Hammer (2004) upon fitting the CNT with plagioclase nucleation data in hydrous silicate melt (Hammer and Rutherford 2002). The H_{int} and TS_{int} gradients are presumed sensitive to H_2O content through a control over bonding and atomic configuration in the diffuse interfacial region (Fig. 6a,c). Solubility mechanisms (Burnham 1975; Silver and Stolper 1989; McMillan 1994) and accepted views of melt structure suggest that the gradients in H_{int} and TS_{int} both shift toward the cluster as dissolved H_2O increases, while the observed σ - H_2O relationship (Fig. 6b) suggests that entropy shifts more. That is, the magnitude of the difference between H_{int} and TS_{int} decreases with the addition of H_2O to the system (Fig. 6c).

Beyond the CNT. Despite predictive strengths in some areas, the CNT fails to quantitatively describe observation for all cases in which σ is *not* used as a fit parameter (Fokin et al. 2005, 2006). A likely flaw is the assumption that critical nuclei possess the thermodynamic and surface properties of the macroscopic phase (Fokin et al. 2006, 2007). Inverting the CNT and

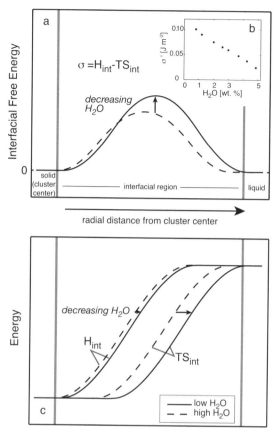

Figure 6. Application of the diffuse interface theory to experimental data (Hammer and Rutherford 2002). (a) Schematic cross section of the interfacial region between a subcritical cluster and the surrounding melt showing thermodynamic properties of a diffuse interface. Materials in "solid" and "liquid" regions possess properties of the bulk (i.e., macroscopic) phases. The region between bulk phases has interfacial free energy, $\sigma = H_{int} - TS_{int}$. Increasing σ with decreasing dissolved H_2O is inferred from (b), values obtained by back solving the CNT equation for nucleation rate using nucleation data from Hammer and Rutherford (2002). (c) Gradients in enthalpy and entropy with radial distance from a subcritical cluster for H_2O-rich (dashed curve) and H_2O-poor (unbroken curve) melts. A shift upon devolatilization of greater magnitude in the entropy gradient is conceptually consistent with the σ-H_2O relationship in (b). From Hammer (2004).

an accepted crystal growth theory (Gutzow and Schmelzer 1995), Fokin et al. (2007) solve for the diffusion coefficient for the interface region (D) two ways: (1) using experimental crystal growth data, (2) using experimentally determined nucleation induction times, literature values for σ, and an expression for the time lag for formation of critical clusters:

$$\tau = \frac{16}{3} \frac{k_B T \sigma}{\Delta G_v^2 a^2 D} \tag{7}$$

where a is a size parameter on the order of ionic radius and other variables are as defined previously. Differing values of D, where the growth-based values greatly exceed the nucleation-based values, suggest that either the definition of interfacial free energy is flawed (as explained above), or the diffusion coefficient controlling nucleation of critical clusters differs from that governing growth of macroscopic crystals. Further evidence in favor of the latter, and against the assumption of bulk properties in clusters, includes dozens of simple silicate systems studied since the 1960's in which nucleation occurs at lower thermodynamic driving force than predicted by the CNT (Granasy and James 1998). The interfacial free energy characterizing clusters may be considerably smaller than that estimated for the macrophase because clusters may more closely resemble melt in composition and/or structure (Fokin et al. 2007). Clearly, much experimental and analytical work remains in elucidating the thermodynamics and transport properties in the interfacial regions of critical clusters.

Crystal growth

The present treatment presents only the fundamentals necessary to appreciate and evaluate recent experimental literature in igneous petrology. The interested reader is pointed to excellent descriptions of crystal growth theory, Avrami phase transformation theory, numerical treatments, and related concepts by Christian (1965), Dowty (1980) and Kirkpatrick (1981). We focus on crystal growth regimes that influence crystal morphology and are defined based on the rate-limiting step of the crystallization process. If the attachment of atoms to the growing crystal face is slow compared to the migration of compatible atoms through the surrounding medium toward the surface and migration of rejected atoms away from the surface, then growth is said to be "interface-controlled." The equation describing steady-state growth (Turnbull and Cohen 1960) in either regime, with simplifying assumptions (Kirkpatrick 1981) is:

$$G = fa_o \nu \exp\left(\frac{-\Delta G'}{RT}\right) \qquad (8)$$

where G is growth rate, f is the fraction of site on the crystal surface available for attachment, a_o is the thickness of a growth layer, ν is the frequency of atom attachment attempts, R is the gas constant, T is temperature, $\Delta G'$ is the activation energy for attachment (an energy barrier), and again ΔG is the bulk chemical driving force. Growth rate is zero at the liquidus temperature because the chemical driving force is zero, increases to a maximum as undercooling increases, and then decreases as activation energy for attachment become prohibitively large. A unary (one-component) system subjected to undercooling always crystallizes in the interface-limited regime, because atomic mobility in the melt is not required for growth layers to advance. Crystallization in multicomponent systems in which the crystal and melt compositions differ (incongruent crystallization) may depart from steady state growth with increasing undercooling. If the component mobility is very slow compared to the attachment rate, then growth is "diffusion-controlled," and the growth rate evolves through time as:

$$G = k\left(\frac{D}{t}\right)^{1/2} \qquad (9)$$

where k is a constant involving correction terms, and D is the diffusion coefficient for the rate-controlling species in the melt. Each regime is associated with a relative undercooling, and gives rise to a characteristic set of growth morphologies.

Interface control. Interface-control is the prevailing case at low degrees of undercooling, where high temperatures (and/or H_2O contents) promote rapid diffusion in the melt. Crystals growing near chemical equilibrium (i.e., in response to small ΔG_v) with the melt always undergo interface-controlled growth. A simple inverse relationship between nucleation and growth rates exists if a fixed volume of solid is distributed on nuclei or pre-existing crystals per unit of time. If the area of growth substrate is small (e.g., a crystal-poor melt at low nucleation rate), distribution of a fixed volume of solid produces a greater thickness of new growth than if the same volume is distributed on a large area of substrate (Cashman 1993). Thus, for the case of crystallization at low undercooling, sluggish nucleation encourages rapid growth. The corollary is that rapid nucleation may suppress growth rates. This is essentially a mass balance argument that predicts a separation of the growth and nucleation rate maxima as functions of undercooling.

Interface-controlled growth produces faceted to euhedral crystals according to one of three mechanisms: continuous growth, advance on screw dislocations, and layer-by-layer growth. The relevant mechanism is dominantly controlled by interface roughness (Sunagawa 1981). Crystal surfaces are atomically rough if the entropy of fusion is small, and atomically smooth if the entropy of fusion is large (Jackson 1967) as is typical of silicates (with the exception of quartz). Continuous growth prevails when surfaces are atomically rough, because this signifies the attachment energy is similar for all sites on the surface to which growth units may attach. Screw dislocation and layer-by-layer growth occur on atomically smooth surfaces. Attachment energies are high and thus greatly facilitated by ledges formed where screw dislocations emerge at the crystal surface. In the absence of such defects, a new layer must be nucleated. Addition of the first growth unit to a smooth face requires substantially more energy than the addition of subsequent units, which attach at two or more sides. Thus, the advance rate of a face depends on the frequency of successful attachments of isolated growth units. Although more common at the high growth rates associated with large undercoolings, anomalously high concentrations of incompatible elements (i.e., solute trapping) may also occur in the interface-controlled regime. Elements that are incompatible in the bulk solid may be stable at the surface if the surface region contains site types not present in the lattice interior (Watson 1996). If atoms of these solute elements are slow to diffuse away from the advancing front, they may be trapped. Anisotropy of surface interface site types on different crystal faces, coupled with slow lattice diffusion is a cause of sector zoning (Watson and Liang 1995).

Through clever selection of melt compositions in the $CaO+Al_2O_3+SiO_2$ (CAS) system, Roskosz et al. (2005) investigate the importance of structural and compositional similarity between melt and crystal in assessing controls on growth rate over a range of undercoolings, all within the interface-controlled regime. They find that when the crystal and melt have the same Al/Si ratio, crystal growth rate scales inversely with melt viscosity (given by temperature as well as composition) regardless of other compositional differences between the phases. This is interpreted as a reflection that both congruent and incongruent crystallization are controlled by the rate of bond rearrangements between oxygen and network-forming cations. For crystals growing from melts of a given viscosity, growth rate varies more strongly as a function of thermodynamic undercooling than melt-crystal structural similarity.

Diffusion control. Diffusion-controlled crystal growth prevails at large undercoolings, where the chemical potential driving solidification is large yet lower temperatures and/or lower H_2O contents reduce component mobility. Morphologies of crystals grown in this regime are anhedral, progressing through a menagerie of forms including tablets, needles, hoppers, dendrites, and skeletons (Sunagawa 1981) as undercooling increases. These morphologies develop because protuberances on the corners and edges of crystals penetrate through the boundary layer that is polluted with rejected components and into an environment slightly richer in crystal-forming components.

CRYSTALLIZATION KINETICS IN MAGMAS: METHODS

"Volcano petrology" studies are broadly characterized by their practical objectives. Works that strive to elucidate the processes of crystallization, test available theories, and define relationships among variables are emphasized in this section. These investigations aim to generate reproducible relationships and internally consistent data sets. By contrast, applied studies attempt to recreate conditions preceding a specific event, and then impose changes meant to simulate volcanic processes. Selected site-specific research that tests eruption models, quantifies ascent rates, and/or reproduces natural crystallization conditions during a particular eruption is not emphasized here, although selected references to such studies are provided in the sections below.

Much of the present understanding of crystallization in silicate melts builds upon experimental work associated with lunar petrology studies, which mapped out the relationships between cooling rates, growth regimes, crystal sizes, and morphologies in lunar basalt (Lofgren et al. 1974; Walker et al. 1976; Grove and Bence 1979). Elevated pressure experiments addressed growth of plagioclase from simple hydrated plagioclase melts (Muncill and Lasaga 1988). Investigations of volatile-rich, granite-forming magmas (Fenn 1977; Swanson 1977; Swanson and Fenn 1986) concurrently filled out the silicic end of the compositional spectrum. These formative studies, reviewed thoroughly elsewhere (Dowty 1980; Kirkpatrick 1981; Cashman 1990) paved the procedural and interpretive ground guiding the recent resurgence in quantifying crystallization kinetics.

Chemical interactions in complex systems influence nucleation kinetics and growth rate in ways that are of broad volcanological interest. For example, crystal nucleation rate and density, crystal size and morphology, and the nature of multicrystalline and multiphase interactions evolve through time in response to declining undercooling. At present these issues must be considered through empirical observation by examining controlled natural samples and performing laboratory experiments.

Starting materials

A variety of starting materials are employed according to the specific study objectives. Synthetic materials emulate a natural multicomponent melt composition of interest (e.g., Martel and Schmidt 2003), or fall between end members of a compositionally simple system (e.g., Faure et al. 2003a). Lavas are commonly used as starting materials when the process of interest occurs in a magma chamber, conduit, or planetary surface (e.g., Lesher et al. 1999). Pyroclastic rocks provide the advantage that the groundmass needs only to be melted and equilibrium with crystal rims restored in order to match the chemical environment prior to eruption (e.g., Hammer and Rutherford 2002).

Thermal history

Overwhelming evidence demonstrates that nucleation at subliquidus temperatures is sensitive to the near- or superliquidus thermal history of the material (Walker et al. 1978; Lofgren 1980, 1983; Davis 2000; Davis and Ihinger 2002). The importance of thermal history is exemplified by the basalt cooling experiments of Sato (1995), in which a 20 °C difference in initial melting temperature causes a difference in plagioclase number density upon dwelling isothermally 100 °C below the liquidus of more than four orders of magnitude. Therefore, the thermal treatment selected for a particular set of experiments is governed by specific objectives. Destruction of pre-existing structure by superliquidus heating up to 100 degrees above the liquidus may be desired in order to evaluate the contribution of heterogeneities and microscopic substrates in an ensuing nucleation step (London et al. 1998). Alternatively, if the objective is to correlate crystal growth mechanisms with morphologies, starting materials may be homogenized at superliquidus temperatures and subsequently seeded with metallic impurities or mineral phases to provide substrates for heterogeneous nucleation or crystal growth (e.g., Simakin and Chevychelov 1995; Couch 2003; Faure et al. 2003a; Simakin and Salova 2004; Larsen 2005). Seeds facilitate identification of new growth if the seed composition differs appreciably from the mineral of interest. However, if the seed crystals are far from equilibrium with the surrounding melt, a reaction or dissolution may obfuscate interpretation and introduce kinetic artifacts not related to growth of the target phase. To achieve results with utmost relevance to natural volcanic processes, it may be necessary to sacrifice some of the certainty that comes from isolating the processes of crystal nucleation and growth. A preferred strategy in this case is to match pre-eruption magma conditions as closely as possible prior to imposing a decompression or cooling history, and therefore *preserve* the naturally-occurring heterogeneities and structure (e.g., degree of melt polymerization and concentration of sub-critical clusters). This is the approach taken by Lesher et al. (1999) and Hammer and Rutherford (2002), among others.

Static vs. dynamic

Experimental procedures follow one of two end members or some combination of both. "Static" experiments are composed of an initial equilibration period followed by rapid decompression or rapid cooling, executed in a single step or ramp that is short compared to the duration of the subsequent dwell period. This approach is conceptually linked to irreversible processes defined by classical thermodynamics; the response of the system against constant externally applied conditions is monitored. If these experiments are decompressions, they are called "isobaric" and if cooling "isothermal," because the response of the system to the disequilibrating event (either the pressure or temperature change) occurs at constant externally-controlled conditions. After the desired anneal period, material is quenched to preserve a snapshot of the system at elevated run temperature. A series of snapshots reveals the changes through time of the rates of crystal growth and nucleation, changes in crystal morphology, and the evolution of texture. The functional relationship between degree of undercooling and crystallization rates can only be determined from short-duration static experiments, ideally using the zero-time extrapolation of time series data. The sample undercooling thus applied is generally easily reconciled with theory, but is probably dissimilar from what occurs during most natural processes.

At the other end of the spectrum, "dynamic" experiments include continuous changes in temperature and/ or pressure, such as multi-step decompression, constant-rate cooling, and various combinations of dwell and ramp segments. Conceptually, these experiments may approach the hypothetical reversible thermodynamic process, although in practice (as described below) near-equilibrium along the path is far from assured. Continuous changes in temperature or pressure impose continually changing undercoolings, the magnitudes of which at any instant depend on the rate of process variable change and the crystallization kinetics at all previous steps as well as the kinetics at the given conditions. Thus, the undercooling through time is likely to be a sawtooth function, with unknown (and possibly varying) tooth height and spacing. The results of dynamic experiments are less easily compared with theoretical predictions than those of static experiments. However, dynamic experiments can be designed to more closely match the conditions during natural magma cooling or decompression, and are thus suited to applied research.

Constant-rate vs. variable-rate cooling

Performing constant-rate cooling experiments is a trivial procedure with modern programmable furnace power controllers. Because the cooling rate history does not vary arbitrarily, the results are not tied to a specific magma reservoir geometry or heat flow mechanism. Despite being comparatively general, the thermal profiles of constant-rate or "linear" cooling experiments have the disadvantage of not simulating natural cooling profiles. Linearly decreasing temperature may be a poor approximation of natural magma cooling, since the rate of heat transfer between the magma and surroundings strongly decreases over time and the release of latent heat is not constant through the crystallization interval. An implicit assumption of the constant-rate cooling approach is that a suite of constant-rate cooling runs spanning at least an order of magnitude in rate probably captures essential differences in the progress of crystal nucleation, crystal growth, and melt differentiation.

Bowles et al. (2007) investigated the morphological effects of variable-rate cooling by discretizing a thermal history representing cooling at arbitrary depth within a conductively cooling lava flow, incorporating release of latent heat using the solution to the Stefan problem given by Turcotte and Schubert (1982). Cooling rates thus progressively decreased from 53 °C h^{-1} to 2.2 °C h^{-1}, with a time-integrated average above 600 °C of 19 °C h^{-1}. The morphology and aspect ratio of pyroxene crystals grown in this run were compared with those in constant-rate experiments performed at 2.8, 5.7, 19, 72, and 230 °C h^{-1}. The variable-rate experiment qualitatively and quantitatively matched the constant-rate cooling experiment performed at 72 °C h^{-1}. This provisional test suggests that the early rapidly-cooled portion of the thermal profile exerts dominant control over pyroxene texture. Constant-rate cooling experiments therefore probably mimic the initial stages of magma crystallization in natural settings.

BASALT COOLING EXPERIMENTS

Nucleation controls texture

Dynamic cooling experiments on lunar and basaltic meteorite compositions in the mid to late 1970's produced a series of observations with profound implications. Since then, these results have been widely reproduced, quantified for other compositions and experimental techniques, and are now (largely) rectified with theory. One important observation was that continuous, slow cooling at a constant rate (or a declining rate), can give rise to phenocrystic textures (Lofgren et al. 1974; Walker et al. 1976; Grove and Bence 1979), thus shattering the assumption that large euhedral crystals form during slow cooling and that anhedral groundmass crystals require a second stage of crystallization at faster cooling. These observations have been replicated recently in a study of Mars analog basalt (Hammer 2006). Several factors are suggested to control whether a given melt will crystallize to produce a porphyritic texture: the degree of superheating above the liquidus (Lofgren and Lanier 1992), the quantity of phases appearing at the liquidus temperature (Lofgren et al. 1974), and the differential suppression of phase appearance temperatures among phases (Walker et al. 1978). If melts are structurally unrelaxed upon cooling, the delay in vigorous nucleation allows supersaturation to increase so that first-formed crystals grow very rapidly; conversely a granular texture results from cooling from near-liquidus temperatures because many clusters quickly achieve critical size and begin to grow. Thus, the tradeoff in nucleation and growth rate leading to qualitatively different texture is a manifestation of the aforementioned mass balance relationship, with nucleation rate determined by melt structure. A related observation is the occurrence of heterogeneous nucleation leading to fasciculate textures (intergrowths) when superliquidus heat treatments affect one phase more than another. Destruction of plagioclase clusters during superheating of Stannern meteorite basalt examined by Walker et al. (1978), for example, may have led to the development of prominent plagioclase-pyroxene intergrowths, whereas subliquidus starting temperatures produce intergranular to subophitic textures because plagioclase nucleation occurs in the absence of a pyroxene substrate. In summary, if near-critical clusters are not available at the start of cooling, then the degree of undercooling at which crystals nucleate determines growth rate and crystal morphology; if crystals nucleate from a population of near-critical clusters, then the initial number density and spatial distribution of these clusters is more important than cooling rate for determining ultimate morphology and size (Walker et al. 1978).

A related tenet of basalt crystallization developed initially for olivine (by Donaldson 1976, and replicated for other phases: Walker et al. 1978; Lofgren and Lanier 1992) is that the amount of super-liquidus heating is (a) inversely proportional to the nucleation induction time upon undercooling by a given amount, and (b) proportional to the magnitude of appearance temperature suppression at a given time following the application of supersaturation. In Donaldson's (1976) experiments, 60 °C of undercooling is required for olivine nucleation following a sudden drop in temperature from 38 °C above the olivine liquidus, while olivine doesn't appear at any temperature within 2 hours following a drop from 118 °C above the liquidus. The superheating effects diminish with longer subliquidus dwell times. Above ~4 h, the appearance temperature is suppressed similarly (<40 °C) by a wide range of superheating, 18-118 °C.

Phase appearance suppression is also observed in constant-rate cooling experiments, in which the magnitude of suppression increases with increasing cooling rate (Grove 1978; Grove and Bence 1979). The formative (1970's) experiments suggested that olivine is least-affected by superheating and rapid cooling rates. In contrast, Lesher et al. (1999) observe olivine appearance temperature in MORB suppressed to a greater extent (190 °C at 100 °C h^{-1}) than plagioclase (30 °C at 100 °C h^{-1}). Moreover, the lag time between the predicted and actual appearance time estimated for plagioclase (3600 s at 10 °C h^{-1} cooling rate) is substantially less than for olivine (12,000 s at 10 °C h^{-1}). The contrast with lunar compositions is interpreted (Lesher et al. 1999) to represent different near-liquidus melt structures and therefore different nucleation energy barriers to formation of plagioclase vs. olivine. That is, nucleation of plagioclase may be facilitated by structural similarity between plagioclase and melt and perhaps $\sigma_{plagioclase-melt}$ < $\sigma_{olivine-melt}$ in MORB. Presumably, the reverse mineral-melt structural relationship exists in lunar basalts.

Growth rates decay with time

A feature shared by all crystallization kinetics studies is a power law relationship between crystal size (assessed as the length or width of the largest crystals or the widths of dendrites) and cooling rate. Similar behavior is observed among natural cooling environments in which crystallization durations are well constrained (Cashman 1993). This trend is widely reported in the basalt dynamic cooling literature (Walker et al. 1976; Lesher et al. 1999; Kohut and Nielsen 2004; Conte et al. 2006), and continues to be interpreted as evidence that cooling rate is a paramount factor (after superliquidus heat treatments) controlling crystal growth rate. The relationship between crystal size and cooling rate may be recast as the net growth rate (defined below) versus crystallization interval duration if sufficient experimental information is reported. Data thus converted for mafic systems crystallizing dynamically and isothermally, and hydrous felsic systems crystallizing in response to multi-step and single-step decompressions, are plotted in Figures 7 and 8, respectively. Growth rates vary by at least two orders of magnitude across the data sets for any given crystallization time within the mafic and felsic groupings (probably due to differences in experimental or measurement technique as well as intrinsic differences arising from crystal and melt compositions).

It is important to note that growth rates are typically computed as a characteristic size of the observed population divided by run duration, rather than by observing progressive growth of individual crystals; one calculation of characteristic size is to find the square root of the quotient of area fraction and area number density (both obtained from analysis of backscattered electron images; e.g., Hammer and Rutherford 2002). Another method is to take the average width or length of the five largest crystals (e.g., Walker et al. 1976). Thus, an increase in the quantity of crystals in the system could produce unvarying or declining characteristic size through time. A monotonic (slope = -1) relationship on a log-log plot of growth rate and crystallization time would be generated if size were to remain constant through time. Datasets exhibiting slopes < -1 represent systems in which the characteristic size decreases with increasing crystallization time, usually due to appearance of new crystals; slopes > -1 represent systems in which the characteristic size increases with crystallization time (whether or not new crystals have been added). The majority of datasets contain deviations from the slope = 1 trend lines, revealing changes in size (and thus net growth rate) with time. Slopes describing the changing growth rate among plagioclase forming in the mafic systems range from -0.5 (Grove 1978) to -0.8 (Kohut and Nielsen 2004). In most cases, long duration experiments reflect slow cooling rates (< 10 °C h^{-1}), and these experiments produce crystals that are larger than in the quickly cooled experiments of short overall duration. For example, the large crystals produced in the Kohut and Nielsen (2004) study are the result of protracted dwell times after a cooling stage. The characteristic crystal size in these experiments is not sensitive to the rate of initial cooling. In the felsic systems, the range in slopes of individual datasets is from -0.5

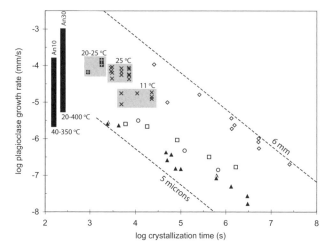

Figure 7. Crystal growth rates with time in constant-rate cooling (except as indicated) experiments. Crystallization time depends on both cooling rate and quench temperature. Diamonds: plagioclase in basalt (Kohut and Nielsen 2004). Squares: plagioclase in lunar ferrobasalt (Grove 1978). Circles: plagioclase in MORB (Grove et al. 1990). Triangles: Ca-pyroxene in lunar picritic basalt (Walker et al. 1976). Vertical black bars: plagioclase growing from H_2O-saturated plagioclase melt of same (anhydrous) composition at 200 MPa at the range of undercoolings indicated; time information not available, the x-axis location is arbitrary (Muncill and Lasaga 1988). Gray shaded boxes: isothermal experiments, crystallization time is the dwell period and ΔT as indicated. White plus symbols on black background: Ca-pyroxene crystals in hydrous hawaiite melt (Simakin et al. 2003). X's: plagioclase growing from plagioclase melt of the same composition, 1-atm (Muncill and Lasaga 1987). Dashed lines show trends produced by zero net growth rate (i.e., constant mean size through time) for sizes bounding those observed in the experimental datasets.

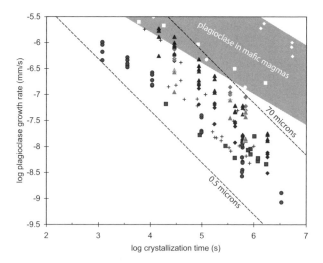

Figure 8. Plagioclase growth rates with time in single-step (filled symbols) and multi-step (open symbols) decompression experiments. Crystallization time depends on decompression rate and quench pressure. Circles: plagioclase in Pinatubo dacite (Hammer and Rutherford 2002). Diamonds: plagioclase in synthetic haplogranite (Couch 2003). Triangles: plagioclase in synthetic Soufriere Hills matrix melt (Couch et al. 2003). Squares: anorthite seeds in Aniakchak rhyodacite (Larsen 2005). Pluses: plagioclase internally nucleated in Aniakchak rhyodacite (Brugger, unpubl.). Dashed lines show trends that would be produced if size remained constant through time. Shaded region shows a portion of the plagioclase growth rate field for mafic systems.

in seeded crystal growth experiments (Larsen 2005) and slow decompressions (Brugger and Hammer, unpubl.) to < −0.9 in rapidly decompressed hydrous magma (Brugger and Hammer, unpubl.; Couch 2003), probably reflecting the varying dominance of nucleation in the overall crystallization mechanism. Of course, growth rate must eventually decline (given sufficient time and permissible kinetics) because as the system approaches equilibrium and degree of undercooling decreases, growth rates must (by definition) approach zero. Declining contrast between compositional zones in experimentally grown pyroxene crystals suggests strongly declining growth rates over time (Schwandt and McKay 2006).

Interestingly, the minimum growth rate observed in static experiments is >2 orders of magnitude faster than minimum growth rate observed in dynamic experiments, and static experiments at low to moderate undercoolings (<30 °C) exhibit growth rates that are large compared to growth rates calculated after an extended period of slow (dynamic) cooling (Fig. 7). The observation that growth rates dramatically decay with time while appearing relatively insensitive to the magnitude of initial undercooling—while theory dictates undercooling exerts the primary control over growth rate—seems to present a contradiction. The issue has implications for many volcanological applications, such as interpretation of crystal size distributions (e.g., Armienti 2008), in which growth rate is commonly assumed constant. The most likely resolution lies in the fact that growth rates have not been directly measured at very small degrees of imposed undercooling. Small growth rates evident in the long-duration static and slow rate-of-change dynamic studies (Figs. 7 and 8) suggest that natural and experimental systems achieve and maintain very small undercoolings. Future experimental work at low undercoolings will be challenging. As the magnitude of undercooling is reduced, it becomes increasingly difficult to maintain sample temperature and/or pressure within the necessary tolerances, rapid quenching becomes increasingly important, and decreases in growth layers smaller than a few microns are imperceptible by analysis methods short of high resolution TEM. Ostwald ripening experiments (e.g., Cabane et al. 2001, 2005) are a promising means of addressing growth in this context, as are extremely experiments of extremely long duration.

Undercooling and the approach to equilibrium

The compositional effects of rapid crystal growth at moderate to large undercoolings have been examined by Grove and Bence (1979) and Hammer (2006), among others. While details differ among phases, Fe-Mg partitioning between mafic phases and melt depart from equilibrium as cooling rate increases, having major implications for departure from equilibrium liquid lines of descent (Grove and Bence 1979). Partitioning of minor elements, e.g., Al and Ti in clinopyroxene, are also dependent upon crystal growth rate as modulated by cooling rate (Walker et al. 1978; Grove and Bence 1979), although the Al content of pyroxene is also affected by the Al_2O_3 content of the melt and whether plagioclase appearance has been suppressed due to rapid cooling. Further compositional effects are discussed below in the context of melt inclusion entrapment.

A kinetic effect worthy of mention is that the time scales for solidification increase with decreasing rates of cooling. Solidification of mid-ocean ridge basalt occurs ~50× faster in magma cooled rapidly (1000 °C h^{-1}) than in magma cooled slowly (10 °C h^{-1}; Lesher et al. 1999). A qualitatively similar result is obtained in an investigation of plagioclase, olivine, and pyroxene crystallization in response to fast and slow cooling of three natural alkalic lavas (Conte et al. 2006). Rapidly cooled samples more closely approach equilibrium phase proportions than the slowly cooled experiments. Within limits, higher undercoolings imposed by rapid cooling enhance the rate of crystallization. Clearly, the trend does not continue to extremely rapid cooling rates, as it would be impossible to quench a melt to glass. Finding the cooling rate at which crystallization reaches optimum efficiency may be the dynamic equivalent to mapping out individual nucleation and growth rate maxima using static experiments.

Crystallization in dynamic environments

Shaking and shearing. Novel approaches to nucleation include the Bartels and Furman (2002) examination of the effects of ultrasonic vibration frequencies on the nucleation of olivine and plagioclase in natural high-alumina basalt, and the Kouchi (1986) study of plagioclase and pyroxene nucleation in basalts undergoing shear deformation. In the Bartels and Furman (2002) study, beads of pressed powder were placed into a gas mixing furnace at sub-liquidus run temperatures then either held or vibrated at one of two frequencies (1.2 MHz, or 150 kHz) for 21-38 h. Samples were removed from the furnace and placed in a stream of air to quench. Interestingly, although the phase proportions were similar in all experiments of a given temperature, the sizes and morphologies of plagioclase differ dramatically depending on whether they were subjected to vibration; olivine texture did not differ among samples. Plagioclase crystals in the static charges were comparatively large and of high aspect ratio while the vibrated charges contained smaller, and nearly equant plagioclase crystals of high spatial number density (Fig. 9a,b). Crystal size distribution plots are linear in static experiments and concave up in the

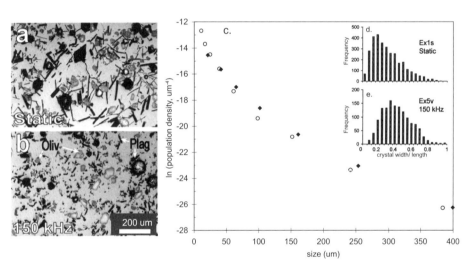

Figure 9. Qualitative and quantitative results of static and dynamic (150 kHz) basalt crystallization experiments. BSE images (a and b, same scale) reveal qualitative differences in texture; plagioclase CSDs from static (filled diamonds) experiments quantify slightly coarser grain size than in dynamic (open circles) experiments; frequency histograms of plagioclase crystal aspect ratio (d and e) quantify morphological differences. [From Bartels and Furman (2002), *Am Mineral,* Vol. 87, Figs. 4 and 5, pp. 217-226.]

vibrated experiments (Fig. 9c), reflecting the difference in number frequency at the smallest size classes. Vibration may increase the density of nuclei by assisting structural relaxation. Applicability of the results to natural volcanic systems depends on whether the observed effects are similarly important in the lower frequency range of volcano tectonic seismicity.

Kouchi (1986) applied shear forces to a disc of basaltic melt held between cylinders rotating either in the same direction at the same rate (static experiments), or in opposite directions at 9.5-70 rpm (dynamic experiments). The nucleation process and resulting textures are clearly affected by the induced flow in the dynamic experiments. At a given undercooling, the period of transient nucleation decreases as rotation rate increases. At a given observation time, the magnitude of phase appearance suppression decreases as rotation rate increases. Both findings suggest that nucleation is enhanced in the sheared melts, an observation previously noted in the industrial crystallization literature (Mullin 1980). In fact, at a given differential

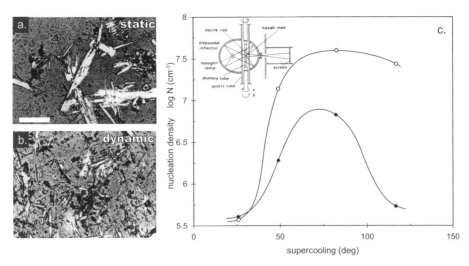

Figure 10. Qualitative and quantitative results of basalt crystallization experiments of Kouchi et al. (1986). BSE images (a and b, same scale) show textural effects of magma shearing caused by rotation (infrared furnace setup shown in c, inset) at 75 rpm. (c) Population density of plagioclase and pyroxene crystals grown in static (filled diamonds) and dynamic (open circles; 48 rpm) experiments. Curves merely guide the eye through the data points; after Kouchi et al. (1986).

rotation rate, nuclei density is greater in the dynamic experiments (Fig. 10). Kouchi (1986) suggests that nucleation enhancement in the dynamic experiments may result from a reduction in the kinetic barrier (ΔG_D of Eqn. 1) brought about by stirring and advective transport. At moderately large undercoolings (≥ 45 °C), the morphologies of plagioclase and pyroxene in the dynamic experiments are faceted and acicular, whereas crystals in the static experiments are dendritic to spherulitic. The shift in morphology suggests that the rate limiting step of crystal growth changes from diffusion-controlled to the interface-controlled. Differential motions of crystals and melt in the dynamic experiments could disrupt compositional boundary layers, promoting interface-controlled growth.

Olivine morphology and growth mechanisms. Examinations of growth and development of anhedral olivine crystals are abundant in the 1-atm basalt cooling literature (Donaldson et al. 1975; Donaldson 1976) and have been the topic of recent intensive work in the CaO+ MgO+ Al_2O_3+ SiO_2 (CMAS) system (Faure et al. 2003a,b, 2006, 2007; Faure and Schiano 2004, 2005). Faure et al. (2003a) performed dynamic cooling experiments on simple basalt analogs (51.3 wt% SiO_2) at 1-1890 °C h^{-1} on crystal-free materials that had been fused and homogenized 50 °C above the liquidus then seeded with Pt impurities and remelted at 22-124 °C above the liquidus. The quench temperatures ranged from 24-356 °C below the liquidus, a difference parameter defined as the nominal undercooling to distinguish it from the undercooling, *sensu strictu* (i.e., applied by an instantaneous drop in temperature). Olivine nucleates heterogeneously on the Pt impurities, and grows into crystals defined as polyhedral, tabular, hopper, "baby swallowtail," or swallowtail (Fig. 11) depending on nominal undercooling and cooling rate. At low cooling rates, crystal morphology is a function of both variables, while above 47 °C h^{-1} crystal morphology is sensitive only to the degree of nominal undercooling.

An important result of the crystallographic investigation is that during formation of hopper crystals, growth is slowest along [010] and fastest along [100]. Atom attachment preferentially occurs on corners and edges of (100), giving rise to melt-filled embayments that characterize the hopper morphology. These crystals are hourglass viewed along [010] and hexagonal along

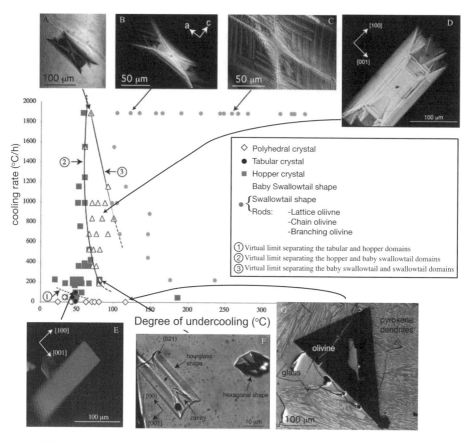

Figure 11. Morphology variations of olivine as a function of cooling rate and undercooling, emphasizing evolution in the section parallel to (010), after Faure et al. (2003a).

[100]. As nominal undercooling increases, overgrowths appear at the mouths of hopper cavities and continue to grow rapidly along [100] to create the "baby swallowtail" morphology. The overgrowths that define the faces are fine dendrites growing rapidly along [101] and [−101]. The resulting morphology is characteristically swallowtailed when viewed along [010] and rod-like when [010] is in the section plane. The elongation direction of the dendrite-bounding rhomb shifts from [100] to [010] with increasing nominal undercooling, as the morphology shifts from hourglass hopper, to hopper with dendritic overgrowths, to dendrites forming directly on Pt impurities. The authors posit that all other olivine morphologies described in the literature (e.g., Donaldson 1976; Sunagawa 1981) can be reconciled as various cuts through these four shape types.

The mechanism of olivine growth, derived from compositional and textural data, is considered by Faure and Schiano (2005) and Faure et al. (2007). Correspondence of euhedral morphology and compositional uniformity in the vicinity of crystal-melt boundaries suggest that olivine growth is controlled by interface attachment mechanisms. Furthermore, growth by screw dislocation propagation proposed to dominate at low undercoolings (Cabane et al. 2005), is consistent with the curvilinear outlines of melt inclusions observed by Faure and Schiano (2005). Compositional gradients in the melt develop at higher undercoolings when rapid crystallization is required to approach equilibrium, yet component mobility in the melt limits the growth rate of crystals. Growth is more rapid in regions of high surface area such as crystal edges and corners, leading to growth by layer spreading (layer nucleation) and development of hopper crystals. The transition to dendritic morphologies is less well understood, but may reflect a transition in mechanism to that of continuous growth. The platy spinifex olivine textures characteristic of komatiites are interpreted by Faure et al. (2006) as a special case of growth during slow cooling within a steep thermal gradient (7-35 °C cm^{-1}). Elongate dendritic morphologies suggest diffusion-control, further substantiated by the presence of compositional boundary layers between dendrites. Crystal growth occurs dominantly perpendicular to isotherms because heat is most efficiently transported along the fast-growing [100] axis.

Olivine growth processes during dynamic cooling experiments are used to interpret melt inclusion entrapment conditions in natural sparsely phyric Mid-Atlantic Ridge basalt (Faure and Schiano 2004). Cycles of cooling and mild (subliquidus) re-heating are needed to form chains of melt inclusions: growth at high undercoolings produces hopper cavities, and subsequent faceted growth at lower undercooling seals them off (Fig. 12). Of course, the applicability of forsterite growing in CMAS to natural basalts is contingent upon ideal Mg-Fe mixing in olivine and absence of interface effects arising from presence of Fe.

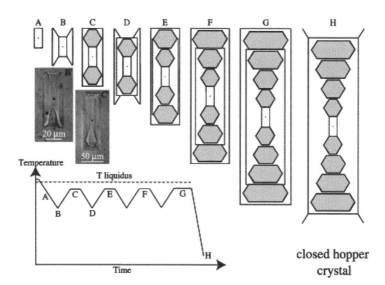

Figure 12. Formation of closed olivine hopper crystals in the section parallel to (010), with schematic illustration of the thermal fluctuations necessary to trap a chain of inclusions. Photomicrograph B′ shows a hopper crystal obtained at cooling rate of 422 °C h^{-1}. Photomicrograph C′ shows a hopper crystal formed at 437 °C h^{-1}, and inclusion entrapped after reheating for 5 minutes to subliquidus temperature. [Used by permission of Elsevier, from Faure and Schiano (2004), *Earth Planet Sci Lett*, Vol. 220, Fig. 7b, p. 331-344.]

FELSIC MAGMA DECOMPRESSION EXPERIMENTS

Effective undercooling

Formative experimental work showed that magmatic H$_2$O degassing causes crystallization of anhydrous phases (Fenn 1977; Swanson 1977; Muncill and Lasaga 1988). Because H$_2$O solubility is a strong function of pressure, magma decompression provides a driving force for solidification that is analogous to cooling of an isobaric system. An effective undercooling

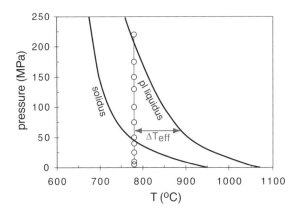

Figure 13. Plagioclase-in curve and estimated solidus curve for melt of Pinatubo matrix dacite used in decompression experiments of Hammer and Rutherford (2002). ΔT_{eff} is defined at any given pressure as the plagioclase liquidus temperature - experimental temperature (780 °C). Circles represent experimental pressures examined in single-step decompression experiments.

(ΔT_{eff}) is defined as the difference between the actual system temperature and the liquidus temperature (T_L-T) using the equilibrium P_{H_2O}-T phase diagram for the system of interest (Fig. 13). Practical applications of site-specific crystal nucleation and growth kinetics experiments (a) establish the relevance of syn-eruptive, devolatilization-driven crystallization in erupting arc magmas (Geschwind and Rutherford 1995), (b) constrain magma ascent rates during explosive and effusive eruptions (McCanta et al. 2007; Rutherford 2008), (c) determine time scales of magma mixing from strongly zoned plagioclase crystals (Larsen 2005; Martel et al. 2006), (d) consider the implications of crystal nucleation on bubble coalescence in volcanic bombs (Simakin et al. 1999), and (e) challenge the notion that magma flow rates feeding explosive eruptions exceed those of effusive eruptions (Castro and Gardner 2008). Given appropriately contextualized natural samples and targeted experiments, it is now possible to infer detailed multi-stage magma ascent histories from groundmass crystal textures (Nicholis and Rutherford 2004; Szramek et al. 2006; Suzuki et al. 2007; Blundy and Cashman 2008).

Establishing baseline crystallization kinetics of quartz, plagioclase, and alkali feldspar was the objective of a triad of process-oriented studies (Hammer and Rutherford 2002; Couch 2003; Martel and Schmidt 2003). The key experimental variables in the Hammer and Rutherford (2002) study were (a) final pressure, (b) dwell time at final pressure, and (c) decompression path. The pressure-time path consisted of either a rapid decompression in a single step, or incremental pressure drops in multiple steps. The range of ΔT_{eff} imposed by the single-step decompressions was 34-241 °C. This series is used to determine the functional dependence of feldspar nucleation and growth rates on final pressure (assessed following a fixed dwell time), and to monitor changes in crystallinity and texture through time in response to a discrete perturbation. Multi-step decompressions monitor the evolution of the system in response to stepwise decreasing pressure according to a fixed time-integrated decompression rate of 1.2 MPa h^{-1}.

Decompression history

Static experiments. Key results of the single-step experiments include: (1) Quantification of the bell-shaped crystal nucleation and growth curves as functions of P_f; shown in Figure 14. (2) Decay of growth rates (computed as the characteristic size divided by run duration) over time by 3 orders of magnitude between 0.33 and 931 h (Fig. 8). (3) Textural maturation that includes disappearance of microlites at long dwell times and initially low ΔT_{eff}. (4) Maintenance of equilibrium at low to moderate ΔT_{eff}, and departures at high ΔT_{eff}.

The general forms of the nucleation and growth rate dependence on ΔT_{eff} obtained at 168 h dwell times (Fig. 14) provide an explanation for the variable approach to equilibrium.

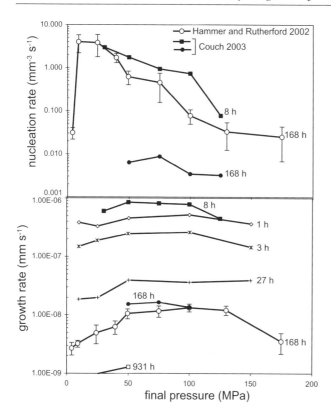

Figure 14. Plagioclase nucleation rates (a) and growth (b) rates following isothermal, single-step decompression + dwell periods of 8 and 168 h from Hammer and Rutherford (2002), open symbols, and Couch (2003), filled symbols. (a) For a given dwell time, nucleation rates vary strongly with final pressure, displaying a maximum at low pressures. Nucleation rate decays rapidly with increasing dwell time (Couch 2003). Note mismatch in nucleation rates at 168 h dwell time. (b) Growth rates display a maximum at intermediate final pressures and also decay with increasing dwell time. Growth rates at 168 h are similar in the two data sets.

Equilibrium is achieved at small to moderate ΔT_{eff} chiefly by growth of existing plagioclase and quartz crystals and sparse nuclei. Departure from equilibrium crystallinity occurs in the ΔT_{eff} interval between maximum growth rate and maximum nucleation rate. At large ΔT_{eff} (150-250 °C) nucleation rate increases and microlite formation becomes the dominant crystallization process, but it is insufficient to maintain equilibrium in the absence of rapid crystal growth.

Dynamic experiments. Striking contrasts between quasi-continuous decompression and single-step decompression run products underscore the importance of decompression history in controlling crystal texture (Fig. 15). For example, samples decompressed to final pressures >40 MPa in multiple steps were much less likely to reach equilibrium crystallinity than samples decompressed in a single step. Because crystallization during each decompression step brings the system closer to equilibrium, stepwise decompression produces smaller instantaneous ΔT_{eff} than is typical in the single-step decompressions. That the slowly-decompressed samples were *less* likely to achieve chemical equilibrium than their rapidly-decompressed counterparts is interpreted to reflect this generally lower degree of undercooling in the multistep run throughout the experiment. This observation runs contrary to conventional wisdom, which would suggest slow decompression is more likely to promote chemical equilibration than rapid decompression. However, these results and interpretations parallel those of Lesher et al. (1999) in describing the approach to equilibrium of cooling MORB. Kinetic effects, such as the functional dependence of crystal growth on ΔT_{eff} and the presence of the nucleation energy barrier, manifestly control progress toward chemical equilibrium.

Seeded experiments. Kindred static and dynamic experiments were performed in the simple system SiO_2-Al_2O_3-CaO-Na_2O-K_2O; haplogranite with 73 wt% SiO_2) by Couch

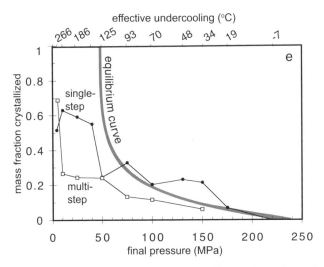

Figure 15. Comparison of equilibrium crystal content with experimental samples. Single-step decompressions followed by 168 h dwell times (closed circles) result in greater crystal contents than multi-step decompressions at similar integrated decompression rates (open squares) for all final pressured examined, although both paths fail to closely approach equilibrium near the solidus (i.e., below ~40 MPa); after Hammer and Rutherford (2002).

(2003). These experiments also examined decompression-induced crystallization by single-step and multi-step decompression paths, but differed in starting material type and preparation. After high temperature fusion and homogenization, the starting materials were held at initial conditions of H$_2$O-saturation at 160 MPa and 875 °C for 16 h. Charges were seeded with 1 vol% andesine feldspar, which were not in equilibrium with the bulk composition or any subsequent melt along the liquid line of descent. Single-step decompressions imposed ΔT_{eff} of 35-167 °C, while the incremental decompressions in the MSD set simulated a variety of average decompression rates, from 0.1-110 MPa h^{-1}.

Couch (2003) obtained a similar functional dependence of growth rate on P_f (Fig. 14b) to that of Hammer and Rutherford (2002), and nucleation and plagioclase growth rates steeply declined with increasing dwell time (Fig. 14b). However, several differences are noteworthy: (a) a 4 h nucleation lag time and cessation of nucleation after 8 h, (b) nucleation rates up to 40× lower in comparable-duration experiments, and (c) failure to reach equilibrium crystallinity at any final pressure along any decompression path. By analogy with the basalt cooling literature, the differences in nucleation rates obtained by Couch (2003) in comparison with Hammer and Rutherford (2002) likely arise from the use of synthetic starting materials and a homogenization time that may have been insufficient for melt relaxation and formation of the equilibrium cluster size distribution. The absence of equilibrium crystal substrates may have inhibited growth at low ΔT_{eff}. This inference is supported by the inference that seed crystals partially dissolved in the first 4-8 h. In combination, these factors affecting the time-dependent response of nucleation and crystal growth to application of ΔT_{eff} may explain why equilibrium crystallinities were not obtained in the Couch (2003) experiments, despite similar locations of nucleation and growth rate maxima with ΔT_{eff} for a given snapshot in time (Fig. 14).

Synthetic starting materials. Martel and Schmidt (2003) studied crystallization in H$_2$O-saturated synthetic rhyolite, focusing on both melt devolatilization and crystallization due to multi-step decompression. The starting material was patterned on melts trapped in Soufriere Hills (Montserrat) andesite crystals, normalized to the nine most abundant oxide components.

After high temperature fusion and homogenization, the starting melts were held at 860 °C and either 150 MPa or 50 MPa for 7 days, after which they were decompressed in multiple steps to final pressures of either 50 or 15 MPa (respectively). Thus, the high-pressure (150 -> 50 MPa) and low-pressure (50-15 MPa) portions of magma ascent were examined separately. The integrated decompression rates imposed by incremental stepwise decompression ranged from 3.6 - 3.5 × 10^6 MPa h^{-1}.

Consistent with the findings of Hammer and Rutherford (2002) and Couch (2003), decompression-induced crystallization at high pressure is dominated by growth of exsiting crystals, while nucleation is the dominant process at lower pressure (higher ΔT_{eff}). The compositions of plagioclase, orthopyroxene, and melt strongly depend upon both the pressure range and decompression rate. For example, plagioclase crystals forming at high pressures

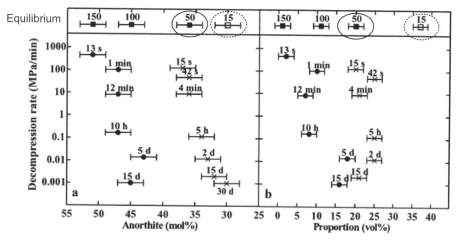

Figure 16. Plagioclase composition (a) and volume percentage (b) in equilibrium (upper panels) and multi-step decompression (lower panels) experiments. Experiment pressure (MPa) is indicated by the equilibrium symbols. Crystallization intervals (i.e., the duration of decompression) are as indicated by each symbol in the lower panels. Solid and dotted ovals enclose the equilibrium values for the high pressure (closed circles) and low-pressure (x's) experiments, respectively. After Martel and Schmidt (2003).

are more anorthitic than those growing at lower pressure, and composition was only slightly affected by varying decompression rate (Fig. 16) Pressure and decompression rate were equally important controls over the approach to the equilibrium crystal contents, with closest approach occurring in high pressure runs decompressed <360 MPa h^{-1} (0.1 MPa min^{-1}; Fig. 16). Low pressure runs and faster decompressions at high pressure trailed equilibrium due to insufficient plagioclase crystallization. An additional insight includes the observation of incomplete exsolution of H$_2$O from the melt in samples decompressed >3.6 10^5 MPa h^{-1}.

TEXTURAL QUANTIFICATION OF EXPERIMENTAL RUN PRODUCTS

Early work laid the qualitative groundwork for comparative petrography, with development of precise morphological terminology linked to a theoretical understanding of growth (Donaldson 1976; Sunagawa 1981). In the absence of temporal information for the vast majority of natural volcanic samples, the application of experimental studies is through the comparison of textures, quantified typically by crystal size (Marsh 2007) and shape.

Unfortunately, few methods and metrics have been validated experimentally to ensure that rate-dependent parameters are interpreted correctly. Investigation of several methods, including CSD analysis and 3D crystal aspect ratio, are described below.

Crystal size distribution analysis

Basalt cooling experiments. Zieg and Lofgren (2006) performed dynamic cooling experiments on synthetic chondrule compositions seeded with natural olivine, and report CSDs for six temperatures along the constant-rate (92 °C h^{-1}) cooling trajectory between 1545 and 1176 °C. Seeds were initially not in stable equilibrium with the melt at the onset of cooling; in fact most were destroyed during the 12 minute homogenization interval. However, overgrowths on the surviving seeds indicate stabilization of olivine within the first 29 minutes of cooling. Vertical zonation of crystal texture within the sample beads is extreme. Equant seed-cored crystals accumulate at the base, and skeletal overgrowths penetrate the cumulate layer toward the top of the bead. Neither of two bulk characterization parameters, mean crystal length and crystal number density, vary systematically with quench temperature. In fact, subtraction of skeletal crystals from the CSDs revealed no apparent difference in the CSDs of cumulate crystals among the runs (Fig. 17). Therefore, traditional analysis of evolving CSD slope and intercept to the experimental charges was not applied to extract characteristic growth and nucleation rates using the known run times. However, separating results from the cumulate zone and the skeletal zone, subtle differences emerge from examination of the time series of binned population density data (Fig. 18) that are modeled numerically, taking the 29-minute experiment as a baseline and using two endmember sets of assumptions. Evolution of crystal sizes within the cumulate zone is modeled assuming zero-nucleation rate and constant (i.e., size and time-independent) growth rates of 2×10^{-7}, 6×10^{-7} and 2×10^{-6} mm s^{-1}. The intermediate rate, which produces a similar frequency of small crystals, is best-fitting in view of the greater statistical certainty of population density in the smallest bins. Crystallization in the skeletal zone is modeled assuming constant growth rate (2×10^{-5} mm s^{-1}) and accelerating nucleation rate [$I(t) = 9 \times 10^{-6} \exp(4 \times 10^{-4}t)$]. This model predicts the population densities in the larger size classes, but underestimates the observed frequencies in smaller bins. The mismatch is attributed to crystal-crystal impingement and the ambiguous distinction between skeletal and non-skeletal morphologies at the top of the cumulate pile.

A recent study reporting plagioclase CSD analyses for ~40 constant-rate basalt cooling experiments in which superliquidus heat treatment temperature and duration, cooling rate

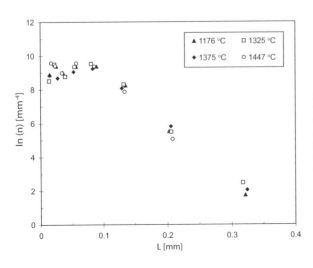

Figure 17. Crystal size distributions of non-skeletal olivine crystals in the cumulate portions of basalt cooling experimental charges cooled at a constant rate of 92 °C h^{-1} to the indicated quench temperature; after Zieg and Lofgren (2006).

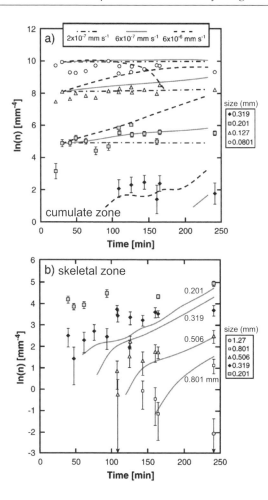

Figure 18. Comparison of modeled and observed population densities from Zieg and Lofgren (2006). (a) Calculated trajectories in the cumulate crystal zone assuming constant growth rates as indicated in top box. (b) Calculated size isopleths in the skeletal zone assuming constant growth rate (2.0 × 10^{-5} mm s^{-1}), and exponentially increasing nucleation rate. The model matches large size class data, but underestimates the frequencies in smaller classes. [Used by permission of Elsevier, from Zieg and Lofgren (2006), *J Volcanol Geotherm Res*, Vol. 154, Fig. 8, p. 74-88.]

(0.2-3.0 °C h^{-1}), and subliquidus dwell times were varied, is particularly illuminating (Pupier et al. 2008). The resulting data set (partially presented in Fig. 19) provides by far the most complete means to date of evaluating common assumptions applied in the interpretation of CSDs. Highlights include (a) nucleation rate that strongly increases with increasing cooling rate for a given experiment duration and quench temperature, in accord with expectation; (b) time-evolution of nucleation rate that cannot be modeled as exponentially increasing (as in Marsh 1988), but rather sharply decreasing; (c) fanning distributions that cannot be (in this case) the result of decreasing cooling rate through time (e.g., Zieg and Marsh 2002), because cooling rate is held constant; (d) downturn in the CSD at small sizes and flattening at large sizes along a given cooling trajectory, provisionally interpreted as synneusis (crystal-crystal attachment along (010) faces); and finally, (e) crystal growth rate that cannot be modeled as constant, either because synneusis is important, or because growth rate is crystal size-dependent (Eberl et al. 2002; Eberl and Kile 2005) or time (i.e., undercooling)-dependent (as discussed in a previous section).

Basalt heating experiments. Burkhard (2002) compares the CSDs of Fe-Ti oxides, pyroxene, and plagioclase in billets of Kilauea basalt before and after laboratory heat treatments at temperatures of 850-934 °C for 43-240 h. Representative CSDs from a fixed position within

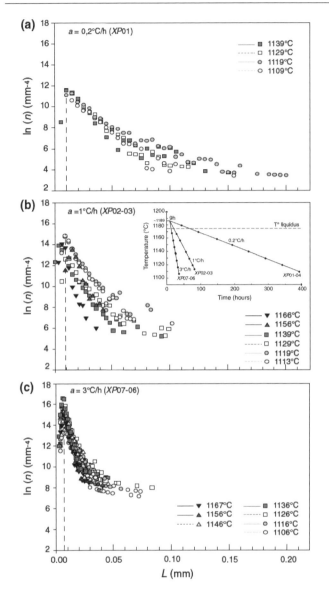

Figure 19. CSDs of crystal widths from experiments designed to evaluate cooling rate as a variable controlling plagioclase texture. Cooling rates are as indicated in each panel; experiment design shown in (b) inset. Vertical dashed lines represent the locations of maximum number densities. After Pupier et al. (2008).

the billet are shown in Figure 20. The y-intercepts of all phase CSDs increase after the heating segment suggesting an increase in the quantity of crystals. The breadth of the size distribution increases for all phases, with particularly dramatic increases for plagioclase and pyroxene, perhaps due to size-dependent (proportionate) growth (Eberl et al. 2002). While the slope of the Fe-Ti oxide CSD is unchanged by the heating step, the initially linear plagioclase and pyroxene CSDs become kinked upwards. Oddly, conventional CSD interpretive guidelines would suggest a different set of crystallization conditions and/ or magmatic processes for each phase (Higgins 1996; Marsh 1998; Higgins and Roberge 2003). For example, the transition from straight to curved CSDs evident in the plagioclase and pyroxene data imply magma mixing, while the parallel and linear Fe-Ti oxide patterns may be interpreted to reflect steady state growth and exponentially increasing nucleation. None of the minerals displays fanning

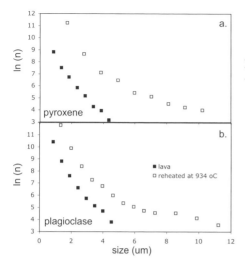
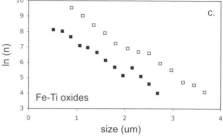

Figure 20. Crystal size distributions of pyroxene (a), plagioclase (b), and Fe-Ti oxide (c) crystals located 18mm from Kilauea lava sample edge before heating (filled symbols) and after heating (open symbols) to 934 °C for 43 h. After Burkhard (2002).

CSDs that characterize suites of natural samples that are thought to represent evolution during declining undercooling, as might be expected from the heat treatment schedule applied. The result is not surprising, perhaps, since experiments do not mimic the thermal history of natural magma cooling. Nonetheless, the ambiguous results highlight the need to further verify the efficacy of the CSD technique for interpreting crystallization processes and time scales.

Ostwald ripening experiments. Park and Hanson (1999) and Cabane et al. (2005) studied evolution of olivine crystal textures in isothermal subliquidus experiments on simple basalts in the CMAS system. Starting materials in the Park and Hanson (1999) study were prepared by fusing reagents above the liquidus, then reducing temperature and crystallizing for 48 h subliquidus (1275 °C). The ensuing ripening experiments included a series of isothermal runs at the homogenization temperature for durations of ~10 minutes to ~10 days. Identical matrix melt compositions in all experiments indicate that textural changes occur at constant crystal volume fraction. The width of the frequency distribution of crystal sizes increase by nearly an order of magnitude over the applied dwell times. This broadening distribution is interpreted to result from continuous dissolution of crystals smaller than the average size and coarsening of the largest crystals at any given instant according to an Ostwald ripening process. With increasing dwell time, the maximum frequencies of the CSDs decrease and the slopes flatten (Fig. 21), both trends indicating that large crystals grow at the expense of small crystals. On the observation that all crystals smaller than 5 microns dissolve within 5 days, the concept of an effective critical size is introduced for ripening in magmatic environments. Systems held near equilibrium for long durations will evolve so that the critical size approaches the mean size, and consequently all information pertaining to nucleation will be erased from the CSD. Cabane et al. (2005) performed similar experiments that included a series focused on plagioclase coarsening in synthetic (CMAS+Na$_2$O) andesite magma. They observe crystal size increasing with $t^{1/3}$ and positive skewness in normalized size distributions (Fig. 22), both characteristics attributable to growth according to a screw dislocation mechanism consistent with nominal undercooling.

Crystal aspect ratio

The surface area to volume ratio (3D aspect ratio) is the comparative metric selected for quantitative textural analysis of pyroxene, olivine and Fe-Ti oxides in a study of the effects of cooling rate and f_{O_2} on textures in synthetic Fe-rich Mars analog basalt (Hammer 2006). This analysis technique (Dehoff and Rhines 1968) was adopted because crystal morphologies are highly anhedral. In summary, circular test lines are digitally overlain on BSE images; locations

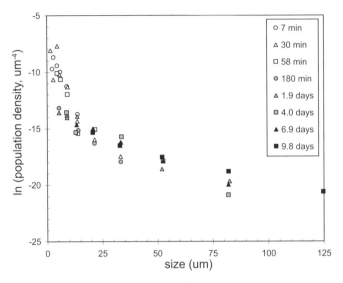

Figure 21. Crystal size distributions of forsterite in ripening experiments. Dwell times are as indicated; replotted from Park and Hanson (1999).

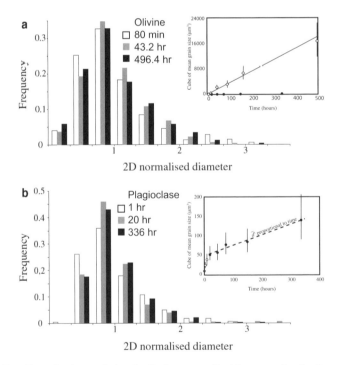

Figure 22. Two dimensional crystal size distributions normalized by mean size for three experiments focusing on olivine (a) and plagioclase (b). Insets show cube of the 2D mean grain sizes plotted against run duration. Plagioclase data (a inset) for durations ≥ 20 h (solid circles) are fitted with a line (empty circles correspond to short duration experiments ≤ 10 h). All olivine data (b inset) are fitted with a line; plagioclase mean grain sizes are shown for comparison. After Cabane et al. (2005).

where crystal-melt boundaries intersect the test line are manually identified and marked, and the quantity of marked boundary intersections per unit length of test line (N_L) is tabulated. The surface area per unit volume of a pyroxene population, S_v^P, is computed as a combination of the boundary intersection density and the volume fraction:

$$S_v^P = 2N_L/\phi \qquad (10)$$

where the area fraction of pyroxene crystals, presumed equivalent to the volume fraction ϕ, (Hilliard 1968) which is measured by standard thresholding of the characteristic range of grayscale values.

In accord with qualitative assessment of the morphlogical dependence upon cooling rate, the 3D crystal aspect ratio (hereafter "aspect ratio") is sensitive to cooling rate. The aspect ratio of pyroxene is particularly well correlated with morphology. For example, all pyroxene crystals characterized as euhedral have aspect ratios < 1000 mm^{-1}, while all anhedral morphologies have values exceeding 1000 mm^{-1}. Interestingly, the aspect ratios of separate pyroxene populations progressively diverge as cooling rate decreases (Fig. 23). This trend in textural dispersion results in the slowest-cooled materials containing populations having both the lowest and highest observed aspect ratios, while moderately-cooled experiments contain two morphologically similar populations, and fastest-cooled experiments contain only one population. In a relationship reminiscent of the power law relationship between growth rate and cooling rate, here the logarithm of the cooling rate is linearly related to the logarithm of the population aspect ratio. The coarse-grained popula-

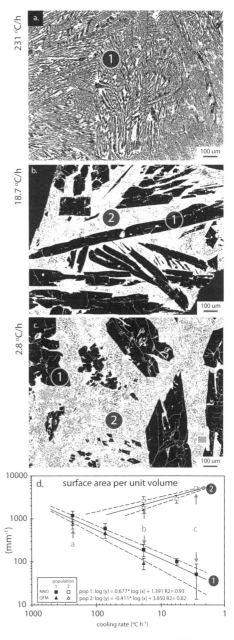

Figure 23. (a-c). Binary images of pyroxene in dynamic cooling experiments obtained from BSE images; circled numbers indicate morphologically distinct populations. (d). Quantitative relationship between cooling rate and the surface area to volume ratio of each pyroxene population. Gray arrows indicate populations represented in images (a-c). Abscissa axis values increase toward the left so that the theoretical approach to textural equilibrium increases to the right. Long dashed lines are least squares fits to 1 σ about sample means. Extension of second population trend lines to cooling rates above 18.7 °C h^{-1} is for illustrative purposes only; no experiment cooled faster than 18.7 °C h^{-1} contained more than one crystal population. Adapted from Hammer (in press).

tion decreases by two orders of magnitude and the fine-grained population doubles over the cooling rate range examined. Modest initial undercooling in slow-cooled charges probably led to efficient growth of faceted crystals by an interface-controlled mechanism, delaying buildup of sufficient driving force for a secondary nucleation event until late in the experiment. The anhedral morphologies and high Sv^P of the second generation reflect diffusion-limited growth and insufficient time for coarsening. The comparative uniformity of anhedral crystal morphologies in the most rapidly cooled charges (>100 °C h^{-1}) is probably due to rapid diffusion-controlled growth of a single generation of late (low-temperature) nuclei.

PHASE BOUNDARIES AND INTERFACES

Interface phenomena are at the frontier of both nucleation studies (described above) and multiphase dynamics. Important questions being addressed include whether interfacial free energies between crystalline or crystals and vapor bubbles play an important role in defining texture, whether crystalline phases nucleate heterogeneously on vapor surfaces, the possibility of epitaxial growth as an important process in natural compositions, and assessment of crystal clustering and synneusis in magmatic environments.

Heterogeneous nucleation

Davis and Ihinger (1998) report strong proclivity of lithium disilicate crystals to nucleate on vapor bubbles, but only following specific thermal histories (two stage fusion treatments prior to a crystal development step, with the second stage 330 °C cooler than the first) and exposure to certain vapor species (Ar and N$_2$). In keeping with observations in natural magmas, their data suggest that heterogeneous crystal nucleation does not occur on arbitrary interfaces. Rather, chemical reactions occur at specific crystal-vapor interfaces, possibly including production of a surfactant.

The opposite relationship—that of crystals providing sites for heterogeneous nucleation of vapor bubbles—has been explored by Hurwitz and Navon (1994), Gardner (2007), and Simakin et al. (2000). While both spinel-structured and rhombohedral Fe-Ti oxides apparently reduce the energy barrier to nucleation relative to the homogeneous case, major rock-forming minerals plagioclase and pyroxene apparently do not. Gardner (2007) compares the effects of melt viscosity, temperature, and water diffusivity predicted by bubble nucleation theory with the experimental data, concluding that none of these factors can account for the observed variation in nucleation rate. Rather, changes in melt structure with varying H$_2$O content are inferred to exert dominant control over the nucleation-hypersensitive $\sigma_{\text{solid-liquid}}$ and $\sigma_{\text{liquid-vapor}}$.

Solid-solid interfacial energy

The connectivity of liquid in systems near the solidus depends upon the dihedral angle θ defined in Equation (6) as a ratio between solid-solid and solid-liquid interfacial energies. Ikeda et al. (2002) measured dihedral angles among diopside crystals coexisting with melts at high melt fractions in the diopside-anorthite system, applying Equation (6) to evaluate changes in $\sigma_{\text{solid-liquid}}/\sigma_{\text{solid-solid}}$ with changing temperature and melt composition. Holocrystalline starting materials were prepared by equilibrating homogenized synthetic glass at subsolidus temperature. The crystalline materials were subsequently either (1) brought directly to a run temperature between 1280 and 1350 °C and held for 5-24 h, (2) heated at a constant rate (0.5 °C min^{-1}) from near the solidus (1280 °C) to the run temperature, or (3) cooled at constant rate (0.05 or 0.5 °C min^{-1}) from 40 °C below the liquidus (1350 °C) to the run temperature. Although sets (1) and (2) are both melting experiments, the isothermal experiments of treatment (1) are perceived to result in attainment of textural equilibrium. Set (3) represents a reversal in that equilibrium is approached from above the run temperature. Values of the dihedral angle are correlated with temperature and melt composition in all experiments (Fig. 24), although

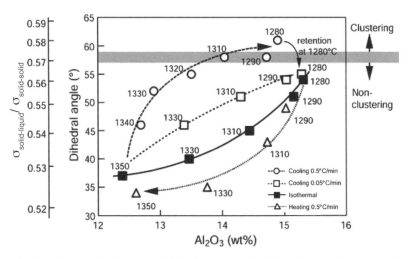

Figure 24. Variation of the dihedral angle with Al_2O_3 concentration in the melt, an indicator of crystallinity. Uncertainties in dihedral angles are ±4°. Numerical values near symbols indicate run temperatures (in isothermal experiments, solid symbols) or quench temperatures (in cooling and heating experiments, open symbols). Dihedral angles in the isothermal experiments are considered close to equilibrium values. [Used with kind permission from Springer Science+Business Media from Ikeda et al. (2002) *Contrib Mineral Petrol*, Vol. 142, Fig. 15b, p. 397-415.].

heating yields σ ratios that are lower than the isothermal values and cooling yields σ ratios that are higher. One interpretation supported by the observed trend is that the $\sigma_{solid\text{-}liquid}$ increases when the melt composition departs from the crystal composition. Formation of boundary layers enriched in rejected components around growing crystals may increase $\sigma_{solid\text{-}liquid}$, thereby providing a driving force for crystal clustering as a means of reducing the overall solid-liquid interfacial area (Fig. 25).

Crystal intergrowths are common in both experimental and natural samples. Two factors are particularly important in co-crystallization: (1) diffusion-limited crystal growth leading to stabilization of phases that incorporate rejected components, i.e., constitutional supercooling, and (2) interfacial energy reductions accommodated by coherent or semi-coherent lattice boundaries between two phases. In the latter case, nucleation may be heterogeneous, and growth epitaxial, leading to a lattice preferred orientation among phases. Interfacial energy controls are not expected to be important among minerals crystallizing from silicate melt at high undercooling. However, orientation mapping using electron backscatter diffraction (EBSD) reveals consistent textural characteristics in samples produced in 1-atm basalt cooling experiments (Hammer 2005). Clinopyroxene crystals are dendritic, with high surface area to volume ratios ($\geq 3 \times 10^3$ mm^{-1}), indicating diffusion-controlled growth. Orientation maps of contiguous regions of individual pyroxene dendrites indicate incremental rotations during crystal growth—poles to the (010) planes are strongly aligned, whereas the poles to (100) and (001) describe great circles in their respective pole figures. Titanomagnetite crystals occur preferentially on the surfaces of feathery clinopyroxene dendrites (Fig. 26a). Poles of the six symmetrically equivalent (110) planes in titanomagnetite are weakly clustered. One of these poles is always coincident with the strongly oriented (010) pole of nearby clinopyroxene crystals (Fig. 26b,c).

Dendritic pyroxene morphologies in these experiments (Fig 26a) indicate chemical potential gradients developing in a regime of constitutional supercooling. The high proportion of shared pyroxene-titanomagnetite boundaries is consistent with either constitutional super-

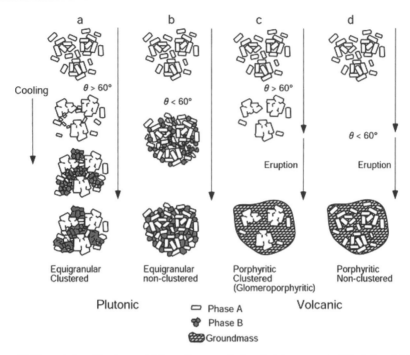

Figure 25. Textural diversification model depending upon the interfacial energy ratio (dihedral angle). The system with a large dihedral angle produces crystal clustering in both plutonic and volcanic rocks. [Used with kind permission from Springer Science+Business Media from Ikeda et al. (2002) *Contrib Mineral Petrol*, Vol. 142, Fig. 5, p. 397-415.]

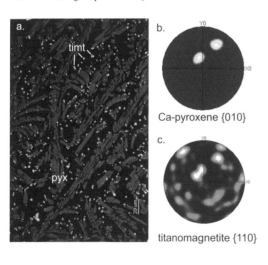

Figure 26. Titanomagnetite crystals decorate the tips of dendritic Ca-pyroxene crystals in rapidly cooled Fe-rich/ Al-poor basalt. The (010) pyroxene and (110) pole figures demonstrate coincidence of crystallographic axes suggesting epitaxial growth. (Hammer unpubl.)

cooling or an interfacial energy advantage to co-crystallization. However, the lattice preferred orientation observed among crystals of pyroxene and titanomagnetite indicates a strong control of lattice interfacial energies in the formation of titanomagnetite. The identical lattice preferred orientation has been described in oxide-pyroxene intergrowths occurring by subsolidus exsolution, presumably at low degrees of undercooling (Okamura et al. 1976) resulting from similarities in the oxygen frameworks of these minerals; i.e., the presence of a semi-coherent lattice boundary. Experiments (Hammer 2005) show that epitaxial growth and heterogeneous nucleation may also be important in highly undercooled sililcate melts in which constitutional supercooling drives multiphase intergrowths.

Reaction relationships

A final example of interphase reaction kinetics with direct application to volcanic processes is determination of magma mixing events prior to eruption and quench (e.g., Baker 1991; Coombs et al. 2003). The exposure time of dissimilar magmas thrust into contact may be estimated using experimental studies of mineral-melt reactions, given plausible estimates of temperature, melt H_2O content, and P_{H_2O}. Coombs and Gardner (2004) monitored the development of forsteritic olivine reacting with hydrous high-silica melt to form either orthopyroxene or amphibole+orthopyroxene. As may be expected, the reaction product is strongly dependent upon dissolved H_2O content. However, there is also a strong dependence of reaction *rate* on dissolved H_2O, presumably because of enhanced component mobility in the melt. Rim growth rate increases by an order of magnitude as P_{H_2O} increased from 50 to 150 MPa, despite 50 °C cooling.

MELT INCLUSION ENTRAPMENT

Crystal growth mechanisms and regime are especially important in the context of interpreting the compositions of melts trapped within growing crystals. It is extremely important to determine whether diffusion-controlled growth conditions are required to entrap melts, and whether the melts are boundary layers rather than representative of the far-field melt. As yet, very few experimental studies report information sufficient to evaluate the issue, and the extant data are intriguingly contradictory.

Major element behavior

Melt inclusion formation was studied by Kohut and Nielsen (2004) in a series of dynamic crystallization experiments using natural starting materials. MORB was placed inside capsules of An_{92}, seeded with Fo_{92} crystals, held for two hours at the computed liquidus temperature (1300 °C), cooled to 1230 or 1210 °C at rates of 1, 5, or 10 °C min^{-1}, held for dwell times ranging from 0-24 h, and then quenched. Melt inclusions were evaluated in terms of major element composition and size distribution. Evaluated in terms of Al_2O_3, CaO in olivine and Mg# and TiO_2 in plagioclase, melt inclusion compositions are not correlated with size. Importantly, melt inclusion compositions matched matrix glass composition for all runs except those quenched directly after cooling (apparently reflecting failure to equilibrate during ramped cooling). The interpreted sequence of events (for both plagioclase and olivine) is that protuberances evolve to form hopper crystals during the cooling stage, which subsequently overgrow and seal off at reduced growth rate during the dwell period. In the overwhelming majority of cases, the shift from diffusion-control to interface-controlled growth occurs before the melt inclusion is isolated from the matrix melt, and therefore any boundary layer that existed during rapid growth has been homogenized by diffusion.

Dramatically different results are obtained in a study of melt inclusion entrapment in forsterite crystallizing from synthetic basalt by Faure and Schiano (2005). In this study, reagents are fused at 1400 °C for 24 h, quenched, powdered, pressed into pellets, and reheated to superliquidus temperatures (19-52 °C above the liquidus), held for an hour, then cooled along one of four schedules. The majority of runs are either single-segment constant rate cooling ramps at 1-1890 °C h^{-1} to varying quench temperatures (representing 51-176 °C below the liquidus), or dual-segment ramped cooling where the first segment is slow (2 °C h^{-1}), the second segment is fast (1644-1730 °C h^{-1}), and quench temperatures are low (175-236 °C below the liquidus). As the charges were not seeded, all observed olivine crystals nucleated during cooling. In contrast to the Kohut and Nielsen finding that inclusion entrapment requires growth at rapid cooling followed by growth at isothermal conditions, faceted olivine crystals in the slowly cooled (1 °C h^{-1}) single-segment experiments occasionally develop embayments and trap inclusions, never having undergone rapid cooling. The near-surface melt and melt inclusion compositions formed during slow cooling are identical to far-field matrix melt in all elements. Faure and Schiano (2005) suggest that interactions among screw dislocations could promote development of embayments and ultimately seal off of melt inclusions. In contrast, skeletal and dendritic crystals produced in rapidly cooled single-segment experiments, as well as dendrite-tipped polyhedral crystals formed during the dual segment runs, are surrounded by a boundary layer of melt that is depleted in MgO and enriched in Al_2O_3 and CaO compared to the far-field melt (Fig. 27). As may be expected from such inclusions, the compositions within skeletal crystals deviate substantially from the equilibrium liquid line of descent followed by the far-field melt. Remarkably, inclusions contained within dendritic-polyhedral crystals also deviate from the liquid line of descent and far field melts, despite gross morphological similarity with slowly grown faceted crystals. The telltale sign of the latter entrapment process is elongation of inclusions parallel to (010) faces, indicating growth entrapment occurred while the crystal was skeletal (Fig. 28).

The Kohut and Nielsen (2004) and Faure and Schiano (2005) studies arrive at dramatically different conclusions regarding the melt inclusion entrapment process as a process in which far-field melts are faithfully recorded in crystals growing under a range of conditions. While it is tempting to pick apart differences in cooling rates, dwell temperatures, durations, and undercooling, the two studies differ in many respects, not the greatest of which is cooling schedule. The Kohut

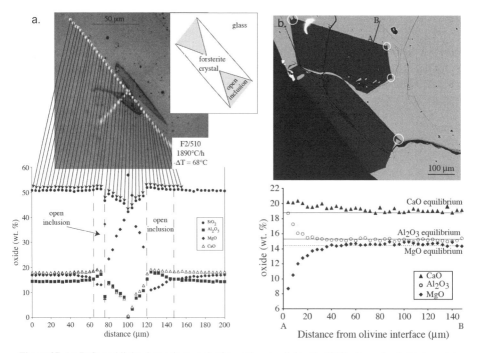

Figure 27. (a) Reflected light photomicrograph of a section parallel to the (010) plane of a skeletal crystal of olivine, showing cavities filled by glass at each end of the crystal. (b) BSE image of a dendrite-tipped polyhedral olivine crystal. Graphs show chemical gradients along the marked traverses of the olivine crystal-liquid interface. [Used by permission of Elsevier, from Faure and Schiano (2005), *Earth Planet Sci Lett*, Vol. 236, Figs. 7 and 10, pp. 882-898.]

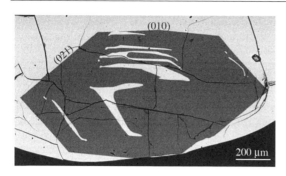

Figure 28. Back-scattered electron image of a dendritic polyhedral forsterite crystal cut nearly parallel to the (100) plane. The enclosed melt inclusions lie parallel to the external (010) or (021) faces. [Used by permission of Elsevier, from Faure and Schiano (2005), *Earth Planet Sci Lett*, Vol. 236, Fig. 9, pp. 882-898.]

and Nielsen experiments were run on natural materials with multicomponent compositions, and were seeded with natural phenocrysts. The Faure and Schiano (2005) experiments were performed unseeded in a compositionally simple system with synthetic materials. In view of nucleation and growth studies in general, it would not be surprising if starting material types and preparations were at least as important as cooling schedule in controlling melt inclusion processes. Future work that applies a variety of approaches, emphasizing characterization of the evolving melt structure and sensitivity to initial conditions, may be necessary to resolve the apparent discrepancies.

Trace element behavior

Small undercooling. Crystallization at small undercoolings results in characteristic partitioning of compatible and incompatible elements into the crystal lattice. Kennedy et al. (1993) evaluated the effect of cooling rate on partitioning of 32 elements in olivine and orthopyroxene coexisting with melt of chondrule composition, finding no correlation between the partition coefficients of incompatible elements and cooling rate, up to 100 °C h^{-1}. Similar results were obtained by Lofgren et al. (2006) in Ti-Al-clinopyroxene growing in near-isochemical melt at small undercooling. Such relationships underpin much of trace element geochemistry as a powerful interpretative tool in igneous petrology. An alternative described by Watson and Liang (1995) explains sector zoning in slowly cooled igneous minerals as the result of anisotropic enrichment at crystal surfaces coupled with sluggish lattice diffusion. That is, partitioning is an equilibrium process, but the partition coefficient for the surface of the crystals is different from (and often higher than) the bulk value.

Growing crystals can incorporate trace elements in excess of bulk partitioning equilibrium if crystal surfaces provide greater flexibility in the diversity of type of atomic sites present; trace element atoms are essentially engulfed by the advancing growth front (Watson 1996, 2004). Watson and Liang (1995) define a dimensionless growth Peclét number, Pe = $V \times l/D_i$, where V is the velocity of crystal growth, l is the half thickness of the surface layer (on the order of atomic dimensions), and D_i is the diffusivity of component i in the near surface region of the crystal. Growth entrapment is not predicted to be important in crystals growing at moderate to low undercoolings in natural magmas, as growth rates must be exceedingly rapid to overcome rapid atomic mobility in melt at magmatic temperatures (Watson 1996). However, this effect may be important at large undercoolings typical of syn-eruptive crystallization.

Large undercooling. Crystal growth at large undercoolings is more commonly associated with non-equilibrium partitioning, with at least four manifestations described in the literature. (1) Diffusion-limited crystal growth results in formation of a boundary layer enriched in rejected components, so the concentrations incorporated into the growing crystal are higher than the equilibrium values based on far-field concentrations in the melt. The observation by Grove and Raudsepp (1978) that elements with pyroxene-melt distribution coefficients greater than 1 were depleted around pyroxene phenocrysts while incompatible elements were enriched, is consistent with this process. Similarly, in their study of clinopyroxene growing in response to varying degrees of undercooling, Lofgren et al. (2006) report trace element distribution coefficients increasing to unity or above with increasing crystal growth rate. (2) Rapid, highly anisotropic growth causes entrapment of submicroscopic melt inclusions, and their boundary-layer enrichments of incompatible elements. Above 100 °C h^{-1} cooling rate, Kennedy et al. (1993) observe extreme departures from equilibrium partitioning in olivine and orthopyroxene. The element variation patterns are consistent with incorporation of tiny melt inclusions within the ion probe analyses; very small amounts of melt contamination (~5 wt%) could account for trace element patterns. The authors suggest that cooling rates 100-2000 °C h^{-1} lead to skeletal crystal morphologies for olivine and orthopyroxene, facilitating incorporation of submicroscopic melt inclusions. (3) Rapid growth causes incompatible elements to be incorporated into crystals at concentrations exceeding the bulk or surface partitioning values, in extreme cases nearing the absolute concentration in the boundary layer melt. Such partitionless growth or "solute trapping" is a disequilibrium phenomenon that affects elements selectively, based on their effective mobilities in the near-surface region. (4) Sector zoning results from differences in impurity uptake by faces advancing at different velocities as well as a strong crystal-chemical control. Shwandt and McKay (2006) describe rapidly growing pyroxene faces in the {001} sectors that incorporate fewer impurity cations than slow-growing faces (Fig. 29). The slow-growing sectors {100} and {010} cope with a boundary layer enriched in Al and impurity elements produced in the wake of the fast-growing faces by substituting octahedral cations and octahedral cations as charge-compensation couples. These slow-growing sectors are prone to incorporating differing impurity concentrations depending on whether tetrahedral and octahedral layers are exposed simultaneously as edges on {100} faces or alternating in sequence as planes parallel to the surface, on {010} faces.

Elevated P contents are observed in experimentally grown olivines, even as major elements (Fe-Mg) are distributed between melt and crystals according to equilibrium partitioning (Milman-Barris et al. 2008). The concentrations of P in the near-surface region are grossly insufficient to represent equilibrium partitioning according to process (1), above. Incorporation of tiny melt inclusions into analytical spots (2) is precluded by vastly differing Al/P ratios in olivine and boundary layer melt. Rather, crystal morphology implicates rapid

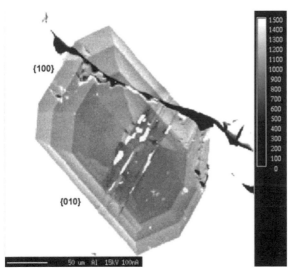

Figure 29. Aluminum map showing cooling steps and sector zoning in synthetic enstatite (shades of gray), viewed down the fast-growing [001] axis, with basaltic glass (white). The legend indicates the number of Al X-ray counts per pixel. Within one concurrent growth interval the concentration of Al is higher in the {010} sectors than the {100} sectors. The distinctiveness of all sectors decreases within each cooling step as growth slowed, prior to the start of each new cooling step, whereby the intensity of sector development is intensified.

growth in disequilibrium P trapping (3), as P-rich zones define relict hopper and skeletal morphologies within externally-faceted and euhedral crystals (Fig. 30). Given sufficient time at high temperature, high-P olivine in contact with melt dissolves and subsequently precipitates at low growth rates (controlled by slow replenishment through tortuous pathways to the crystal surface) as low-P olivine. The result is low-P interior linings of skeletal crystals. A similar dissolution/ precipitation process may produce halos of low-P olivine in the interiors of natural melt inclusions (Ikeda 1998; Danyushevsky et al. 2004). This study underscores the need to consider lattice and surface mobilities of trace element ions in the context of crystal growth mechanisms in the interpretation of natural magmatic processes, and demonstrates the need for further targeted experimental study.

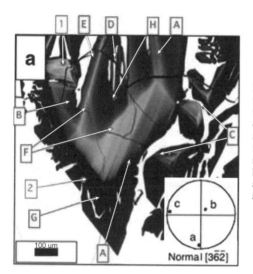

Figure 30. Phosphorus X-ray map of experimental olivine grown from basalt cooled from 1,400 to 1,190 °C at 15 °C h^{-1}. A central high-P skeleton [F] is zoned outward toward low-P rims [F] and inward toward elongate melt pools enclosed in low-P olivine [D]. Other features include fine-scale oscillatory zoning [A] and [H], knobs of high-P olivine close to interior melt pools [E] and a region of oscillatory bands [B]. [Modified after Milman-Barris et al. (2008), Fig. 7.]

CONCLUSIONS

The formative applications of experimental petrology to problems in volcanology were both simple and profound: to determine the depth, temperature, and chemical environments of restless magma prior to eruption. Many of the most important breakthroughs in volcanology, such as appreciation of magma interactions in eruption triggering and pre-eruptive gas saturation, are the fruits of interdisciplinary approaches linking field observations with experimental phase equilibria studies. Looking toward the future of volcanology, experimental studies promise to provide insights into dynamic aspects of magma transport, as the focus shifts from defining equilibrium states toward constraining the kinetics of magmatic processes. "Experimental volcanology" encompasses a variety of complementary approaches, including the use of natural silicate magma, synthetic magmas based on natural compositions, and analog materials. The strength of traditional high-pressure/temperature experiments continues to be the ability to study natural chemical processes at conditions that can be varied to emulate changing geological environments. Innovation in experimental apparatus development and improvements in the spatial and compositional resolution of analytical instruments hold the promise of increasing the applicability of experimental research in volcano hazard assessment, as well as deepening the theoretical understanding of underlying physical and chemical processes.

ACKNOWLEDGMENTS

This contribution was conceived and nurtured by discussions with M. Rutherford. C. Brugger, O.K.Neill, and L. Tatsumi are thanked for their assistance and enthusiasm. The manuscript was improved by constructive reviews from F. Faure, Y. Liang, G. Lofgren, K. Putirka, and B. Scaillet. This work was supported in part by NSF award EAR04-49888. It is SOEST publication #7501.

REFERENCES CITED

Andersen DJ, Lindsley DH, Davidson PM (1993) QUILF; a Pascal program to assess equilibria among Fe-Mg-Mn-Ti oxides, pyroxenes, olivine, and quartz. Computers Geosci 19:1333-1350

Arbaret L, Bystricky M, Champallier R (2007) Microstructures and rheology of hydrous synthetic magmatic suspensions deformed in torsion at high pressure. J Geophys Res doi:10.1029/2006JB004856

Armienti P (2008) Decryption of igneous rock textures: crystal size distribution tools. Rev Mineral Geochem 69:623-649

Bachmann O, Bergantz GW (2008) Deciphering magma chamber dynamics from styles of compositional zoning in large silicic ash flow sheets. Rev Mineral Geochem 69:651-674

Bagdassarov N, Dorfman A, Dingwell DB (2000) Effect of alkalis, phosphorus, and water on the surface tension of haplogranite melt. Am Mineral 85:33-40

Baker DR (1991) Interdiffusion of hydrous dacitic and rhyolitic melts and the efficacy of rhyolite contamination of dacitic enclaves. Contrib Mineral Petrol 106:462-473

Barclay J, Rutherford MJ, Carroll MR, Murphy MD, Devine JD, Gardner J, Sparks RSJ (1998) Experimental phase equilibria constraints on pre-eruptive storage conditions of the Soufriere Hills magma. Geophys Res Lett 25:3437-3440

Bartels KS, Furman T (2002) Effect of sonic and ultrasonic frequencies on the crystallization of basalt. Am Mineral 87:217-226

Battezzati L (2001) Thermodynamic quantities in nucleation. Mater Sci Eng A 304:103-107

Becker R, Doring W (1935) Kinetische behandlung der keim buildung in ubersattingten dampfen. Ann Physik 5:719-752

Berndt J, Koepke J, Holtz F (2005) An experimental investigation of the influence of water and oxygen fugacity on differentiation of MORB at 200 MPa. J Petrol 46:135-167

Berndt J, Liebske C, Holtz F, Freise M, Nowak M, Ziegenbein D, Hurkuck W, Koepke J (2002) A combined rapid-quench and H_2-membrane setup for internally heated pressure vessels: Description and application for water solubility in basaltic melts. Am Mineral 87:1717-1726

Blundy J, Cashman K (2001) Ascent-driven crystallisation of dacite magmas at Mount St Helens, 1980-1986. Contrib Mineral Petrol 140:631-650

Blundy J, Cashman K (2008) Petrologic reconstruction of magmatic system variables and processes. Rev Mineral Geochem 69:179-239

Bogaerts M, Scaillet B, Vander Auwera J (2006) Phase equilibria of the Lyngdal Granodiorite (Norway); implications for the origin of metaluminous ferroan granitoids. J Petrol 47:2405-2431

Bowles JA, Hammer JE, Brachfeld SA (2007) Magnetic and petrographic characterization of synthetic martian basalts. 7th International Conference on Mars, p. 3255, Pasadena, CA

Brugger C, Johnston AD, Cashman KV (2003) Phase relations in silicic systems at one-atmosphere pressure. Contrib Mineral Petrol 146:356-369

Burkhard DJM (2002) Kinetics of crystallization: example of micro-crystallization in basalt lava. Contrib Mineral Petrol 142:724-737

Burnham CW (1975) Water and magmas: a mixing model. Geochim Cosmochim Acta 39:1077-1084

Cabane H, Laporte D, Provost A (2001) Experimental investigation of the kinetics of Ostwald ripening of quartz in silicic melts. Contrib Mineral Petrol 142:361-373

Cabane H, Laporte D, Provost A (2005) An experimental study of Ostwald ripening of olivine and plagioclase in silicate melts; implications for the growth and size of crystals in magmas. Contrib Mineral Petrol 150:37-53

Cashman K (1990) Textural constraints on the kinetics of crystallization of igneous rocks. Rev Mineral Geochem 24:259-314

Cashman KV (1993) Relationship Between plagioclase crystallization and cooling rate in basaltic melts. Contrib Mineral Petrol 113:126-142

Castro JM, Gardner J (2008) Did magma ascent rate control the explosive-effusive transition at the Inyo volcanic chain, California? Geology 36:279-282

Champallier R, Bystricky M, Arbaret L (2008) Experimental investigation of magma rheology at 300 MPa: From pure hydrous melt to 76 vol.% of crystals. Earth Planet Sci Lett 267:571-583

Christian JW (1965) The Theory of Transformations in Metals and Alloys. Pergamon Press, New York

Clemens JD, Holloway JR, White AJR (1986) Origin of an A-type granite: Experimental constraints. Am Mineral 71:317-324

Clemens JD, Wall VJ (1981) Origin and crystallization of some peraluminous (S-type) granitic magmas. Can Mineral 19:111-131

Conte AM, Perinelli C, Trigila R (2006) Cooling kinetics experiments on different Stromboli lavas: Effects on crystal morphologies and phases composition. J Volc Geotherm Res 155:179-200

Coombs ML, Eichelberger JC, Rutherford MJ (2000) Magma storage and mixing conditions for the 1953-1974 eruptions of Southwest Trident Volcano, Katmai National Park, Alaska. Contrib Mineral Petrol 140:99-118

Coombs ML, Eichelberger JC, Rutherford MJ (2003) Experimental and textural constraints on mafic enclave formation in volcanic rocks. J Volcanol Geotherm Res 119:125-144

Coombs ML, Gardner JE (2001) Shallow-storage conditions for the rhyolite of the 1912 eruption at Novarupta, Alaska. Geology 29:775-778

Coombs ML, Gardner JE (2004) Reaction rim growth on olivine in silicic melts: Implications for magma mixing. Am Mineral 89:748-758

Cooper RF, Kohlstedt DL (1982) Interfacial energies in the olivine-basalt system. *In:* High-Pressure Research in Geophysics, 12. Akimoto S, Manghnani MH (eds) D. Reidel Publishing Co., Tokyo. p. 217-228

Costa F, Scaillet B, Pichavant M (2004) Petrological and experimental constraints on the pre-eruption conditions of Holocene dacite from Volcan San Pedro (36 degrees S, Chilean Andes) and the importance of sulphur in silicic subduction-related magmas. J Petrol 45:855-881

Cottrell E, Gardner JE, Rutherford MJ (1999) Petrologic and experimental evidence for the movement and heating of the pre-eruptive Minoan rhyodacite (Santorini, Greece). Contrib Mineral Petrol 135:315-331

Couch S (2003) Experimental investigation of crystallization kinetics in a haplogranite system. Am Mineral 88:1471-1485

Couch S, Sparks RSJ, Carroll MR (2001) Mineral disequilibrium in lavas explained by convective self-mixing in open magma chambers. Nature 411:1037-1039

Couch S, Sparks RSJ, Carroll MR (2003) The kinetics of degassing-induced crystallization at Soufriere Hills volcano, Montserrat. J Petrol 44:1477-1502

Dall'Agnol R, Scaillet B, Pichavant M (1999) An experimental study of a lower Proterozoic A-type granite from the eastern Amazonian Craton, Brazil. J Petrol 40:1673-1698

Danyushevsky LV, Leslie RAJ, Crawford AJ, Durance P, Niu Y, Herzberg C, Wilson M (2004) Melt inclusions in primitive olivine phenocrysts; the role of localized reaction processes in the origin of anomalous compositions. J Petrol 45:2531-2553

Davidchack RL, Laird BB (2003) Direct calculation of the crystal-melt interfacial free energies for continuous potentials:Application to the Lennard- Jones system. J Chem Phys 118:7651-7657

Davis M (2000) Effect of the growth treatment on two-stage nucleation experiments: The "flushing" effect. Glass Sci Tech 73:170-177

Davis MJ, Ihinger PD (1998) Heterogeneous crystal nucleation on bubbles in silicate melt. Am Mineral 83:1008-1015

Davis MJ, Ihinger PD (1999) Influence of hydroxyl on glass transformation kinetics in lithium disilicate melt and a re-evaluation of structural relaxation in NBS 710 and 711. J Non-Cryst Solids 244:1-15

Davis MJ, Ihinger PD (2002) Effects of thermal history on crystal nucleation in silicate melt: Numerical simulations. J Geophys Res 107 doi:10.1029/2001JB000392

Davis MJ, Ihinger PD, Lasaga AC (1997) Influence of water on nucleation kinetics in silicate melt. J Non-Cryst Solids 219:62-69

Dehoff RT, Rhines FN (1968) Quantitative Microscopy. McGraw-Hill, New York

Deubener J, Weinberg MC (1998) Crystal-liquid surface energies from transient nucleation. J Non-Cryst Solids 231:143-151

Di Carlo I, Pichavant M, Rotolo SG, Scaillet B (2006) Experimental crystallization of a high-K arc basalt: The golden pumice, Stromboli volcano (Italy). J Petrol 47:1317-1343

Dingwell DB, Webb SL (1989) Structural relaxation in silicate melts and non-Newtonian melt rheology in geologic processes. Phys Chem Mineral 16:508-516

Donaldson CH (1976) An experimental investigation of olivine morphology. Contrib Mineral Petrol 57:187-213

Donaldson CH, Usselman TM, Williams RJ, Lofgren GE (1975) Experimental modeling of the cooling history of Apollo 12 olivine basalts. Proc Lunar Sci Conf 6th, E43-87

Dowty E (1980) Crystal growth and nucleation theory. *In:* Physics of Magmatic Processes. Hargraves R (ed) Princeton University Press, Princeton, NJ. p 487-551

Eberl DD, Kile DE (2005) Crystal growth rate law identified from changes in variance of crystal size distributions. Geochim Cosmochim Acta 69:A402-A402

Eberl DD, Kile DE, Drits VA (2002) On geological interpretations of crystal size distributions: Constant vs. proportionate growth. Am Mineral 87:1235-1241

Edgar AD (1973) Experimental Petrology- Basic Principles and Techniques. Oxford University Press, New York

Eggler DH, Burnham CW (1973) Crystallization and fractionation trends in the system andesite-H_2O-CO_2-O_2 at pressures to 10 Kb. Geo Soc Am Bull 84:2517-2532

Faure F, Arndt N, Libourel G (2006) Formation of spinifex texture in komatiites: an experimental study. J Petrol 47:1591-1610

Faure F, Schiano P (2004) Crystal morphologies in pillow basalts: implications for mid-ocean ridge processes. Earth Planet Sci Lett 220:331-344

Faure F, Schiano P (2005) Experimental investigation of equilibration conditions during forsterite growth and melt inclusion formation. Earth Planet Sci Lett 236:882-898

Faure F, Schiano P, Trolliard G, Nicollet C, Soulestin B (2007) Textural evolution of polyhedral olivine experiencing rapid cooling rates. Contrib Mineral Petrol 153:405-416

Faure F, Trolliard G, Nicollet C, Montel JM (2003a) A developmental model of olivine morphology as a function of the cooling rate and the degree of undercooling. Contrib Mineral Petrol 145:251-263

Faure F, Trolliard G, Soulestin B (2003b) TEM investigation of forsterite dendrites. Am Mineral 88:1241-1250

Feig ST, Koepke J, Snow JE (2006) Effect of water on tholeiitic basalt phase equilibria: an experimental study under oxidizing conditions. Contrib Mineral Petrol 152:611-638

Fenn PM (1977) The nucleation and growth of alkali feldspar from hydrous melts. Can Mineral 15:135-161

Fokin VM, Schmelzer JWP, Nascimento MLF, Zanotto ED (2007) Diffusion coefficients for crystal nucleation and growth in deeply undercooled glass-forming liquids. J Chem Phys 126 doi:10.1063/1.2746502

Fokin VM, Zanotto ED (2000) Crystal nucleation in silicate glasses: the temperature and size dependence of crystal/liquid surface energy. J Non-Cryst Solids 265:105-112

Fokin VM, Zanotto ED, Schmelzer JWP (2000) Method to estimate crystal/liquid surface energy by dissolution of subcritical nuclei. J Non-Cryst Solids 278:24-34

Fokin VM, Zanotto ED, Schmelzer JWP, Potapov OV (2005) New insights on the thermodynamic barrier for nucleation in glasses: The case of lithium disilicate. J Non-Cryst Solids 351:1491-1499

Fokin VM, Zanotto ED, Yuritsyn NS, Schmelzer JWP (2006) Homogeneous crystal nucleation in silicate glasses: A 40 years perspective. J Non-Cryst Solids 352:2681-2714

Gaillard F (2004) Laboratory measurements of electrical conductivity of hydrous and dry silicic melts under pressure. Earth and Planet Sci Lett 218:215-228

Gaillard F, Pichavant M, Mackwell S, Champallier R, Scaillet B, McCammon C (2003) Chemical transfer during redox exchanges between H2 and Fe-bearing silicate melts. Am Mineral 88:308-315

Gardner JE (2007) Heterogeneous bubble nucleation in highly viscous silicate melts during instantaneous decompression from high pressure. Chem Geol 236:1-12

Gardner JE, Hilton M, Carroll MR (1999) Experimental constraints on degassing of magma: isothermal bubble growth during continuous decompression from high pressure. Earth Planet Sci Lett 168:201-218

Gardner JE, Rutherford M, Carey S, Sigurdsson H (1995) Experimental constraints on pre-eruptive water contents and changing magma storage prior to explosive eruptions of Mount St. Helens volcano. Bull Volcanol 57:1-17

Geschwind C, Rutherford MJ (1995) Crystallization of microlites during magma ascent: the fluid mechanics of recent eruptions at Mount St. Helens. Bull Volcanol 57:356-370

Ghiorso MS, Sack RO (1995) Chemical mass transfer in magmatic processes IV. A revised and internally consistent thermodynamic model for the interpolation and extrapolation of liquid-solid equilibria in magmatic systems at elevated temperatures and pressures. Contrib Mineral Petrol 119:197-212

Gibbs JW (1906) On the equilibrium of heterogeneous substances. *In:* The Scientific Papers of J. Willard Gibbs, 1. Ox Bow Press, Woodbridge. p 55-353

Gonde C, Massare D, Bureau H, Martel C, Pichavant M, Clocchiatti R (2006) In situ study of magmatic processes: a new experimental approach. High Pressure Res 26:243-250

Gonzalez-Oliver CJR, James PF (1980) Crystal nucleation and growth in a Na_2O-$2CaO$-$3SiO_2$ glass. J Non-Cryst Solids 38/39:699-704

Granasy L, James PF (1998) Nucleation in oxide glasses: comparison of theory and experiment. Proc R Soc London Ser A-Math Phys Eng Sci 454:1745-1766

Granasy L, Pusztai T, Hartmann E (1996) Diffuse interface model of nucleation. J Crystal Growth 167:756-765

Granasy L, Pusztai T, James PF (2002) Interfacial properties deduced from nucleation experiments: A Cahn-Hilliard analysis. J Chem Phys 117:6157-6168

Grove TL (1978) Cooling histories of Luna 24 very low Ti(VLT) ferrobasalts; an experimental study. Proc Lunar Planet Sci Conf 9:565-584

Grove TL, Bence AE (1979) Crystallization kinetics in a multiply saturated basalt magma; an experimental study of Luna 24 ferrobasalt. Lunar Planet Sci Conf X:439-478

Grove TL, Detrick RS, Honnorez J, Adamson AC, Brass GW, Gillis KM, Humphris SE, Mevel C, Meyer PS, Petersen N, Rautenschlein M, Shibata T, Staudigel H, Wooldridge AL, Yamamoto K, Bryan WB, Juteau T, Autio LK, Becker K, Bina MM, Eissen J-P, Fujii T, Grove TL, Hamano Y, Hebert R, Komor SC, Kopietz J, Krammer K, Loubet M, Moos D, Richards HG (1990) Cooling histories of lavas from Serocki Volcano. Proc Oceanic Drilling Prog 106/109:3-8

Grove TL, Juster TC (1989) Experimental investigations of low-Ca pyroxene stability and olivine pyroxene liquid equilibria at 1-atm in natural basaltic and andesitic liquids. Contrib Mineral Petrol 103:287-305

Grove TL, Raudsepp M (1978) Effects of kinetics on the crystallization of quartz normative basalt 15597; an experimental study. *In:* Petrogenic Studies; The Moon and Meteorites, vol. 1. Merrill RB (ed) Pergamon, p 585-599

Gutzow I, Kashchiev D, Avramov I (1985) Nucleation and crystallization in glass-forming melts: old problems and new questions. J Non-Cryst Solids 73:477-499

Gutzow I, Schmelzer J (1995) The Vitreous State: Thermodynamics, Structure, Rheology, and Crystallization. Springer, Berlin

Hammer JE (2004) Crystal nucleation in hydrous rhyolite: Experimental data applied to classical theory. Am Mineral 89:1673-1679

Hammer JE (2005) Strange attractors: symbiosis in magma crystallization. Eos, Trans AGU, 86, p. V12A-07

Hammer JE (2006) Influence of f_{O2} and cooling rate on the kinetics and energetics of Fe-rich basalt crystallization. Earth Planet Sci Lett 248:618-637

Hammer JE (2008) Application of a textural geospeedometer to late-stage magmatic history of MIL03346. Meteoritics Planet Sci (in press)

Hammer JE, Cashman KV, Hoblitt RP, Newman S (1999) Degassing and microlite crystallization during pre-climactic events of the 1991 eruption of Mt. Pinatubo, Philippines. Bull Volcanol 60:355-380

Hammer JE, Rutherford MJ (2002) An experimental study of the kinetics of decompression-induced crystallization in silicic melt. J Geophys Res 107 doi:10.1029/2001JB000281

Hammer JE, Rutherford MJ (2003) Petrologic indicators of preeruption magma dynamics. Geology 31:79-82

Hammer JE, Rutherford MJ, Hildreth W (2002) Magma storage prior to the 1912 eruption at Novarupta, Alaska. Contrib Mineral Petrol 144:144-162

Harms E, Gardner JE, Schmincke HU (2004) Phase equilibria of the Lower Laacher See Tephra (East Eifel, Germany): constraints on pre-eruptive storage conditions of a phonolitic magma reservoir. J Volc Geotherm Res 134:125-138

Higgins MD (1996) Magma dynamics beneath Kameni volcano, Thera, Greece, as revealed by crystal size and shape measurements. J Volc Geotherm Res 70:37-48

Higgins MD, Roberge J (2003) Crystal size distribution of plagioclase and amphibole from Soufrière Hills Volcano, Montserrat: Evidence for dynamic crystallization-textural coarsening cycle. J Petrol 44:1401-1411

Hilliard JE (1968) Measurement of volume in volume. *In:* Quantitative Microscopy. DeHoff RT, Rhines FN, (eds) McGraw-Hill, New York, p 45-76

Holloway JR, Burnham CW (1972) Melting relations of basalt with equilibrium water pressure less than total pressure. J Petrol 13:1-29

Holloway JR, Dixon JE, Pawley AR (1992) An internally heated, rapid quench, high-pressure vessel. Am Mineral 77:643-646

Holness MB (2006) Melt-Solid dihedral angles of common minerals in natural rocks. J Petrol 47:791-800

Holtz F, Sato H, Lewis J, Behrens H, Nakada S (2005) Experimental petrology of the 1991-1995 Unzen dacite, Japan. Part I: Phase relations, phase composition and pre-eruptive conditions. J Petrol 46:319-337

Hurwitz S, Navon O (1994) Bubble nucleation in rhyolitic melts: Experiments at high pressure, temperature, and water content. Earth Planet Sci Lett 122:267-280

Ikeda S, Toriumi M, Yoshida H, Shimizu I (2002) Experimental study of the textural development of igneous rocks in the late stage of crystallization: the importance of interfacial energies under non-equilibrium conditions. Contrib Mineral Petrol 142:397-415

Ikeda Y (1998) Petrology of magmatic silicate inclusions in the Allan Hills 77005 lherzolitic shergottite. Meteoritics Planet Sci 33:803-812

Jackson KA (1967) Current concepts in crystal growth from the melt. *In:* Progress in Solid State Chemistry, 4. Reiss H (ed) Pergamon, New York, p 53-80

James PF (1974) Kinetics of crystal nucleation in lithium silicate glasses. Phys Chem Glasses 15:95-105

James PF (1985) Kinetics of crystal nucleation in silicate glasses. J Non-Cryst Solids 73:517-540

Kalinina AM, Fokin VM, Filipovich VN (1976) Method for determining the parameters characterizing the formation of crystals in glasses. Fizika i Khimiya Stekla 2:298-305

Kashchiev D (1969) Solution of the non-steady state problem in nucleation kinetics. Surface Science 14:209-220

Kelton KF, Greer AL (1988) Test of classical nucleation theory in a condensed system. Phys Rev B 38:10089-10092

Kennedy AK, Lofgren GE, Wasserburg GJ (1993) An experimental study of trace element partitioning between olivine, orthopyroxene and melt in chondrules; equilibrium values and kinetic effects. Earth Planet Sci Lett 115:177-195

Kent AJR (2008) Melt inclusions in basaltic and related volcanic rocks. Rev Mineral Geochem 69:273-331

Kern H (1982) P- and S-wave velocities in crustal and mantle rocks under the simultaneous action of high confining pressure and high temperature and the effect of the rock microstructure. *In:* High-Pressure Researches in Geoscience. Schreyer W (ed) Schweizerbart, Stuttgart, p 15-45

Kesson U, Lindqvist JE, Gransson M, Stigh J (2001) Relationship between texture and mechanical properties of granites, Central Sweden, by use of image-analysing techniques. Bull Eng Geol Environ 60:277-284

Kirkpatrick RJ (1981) Kinetics of crystallization of igneous rocks. Rev Mineral Geochem 8:321-397

Kohut E, Nielsen RL (2004) Melt inclusion formation mechanisms and compositional effects in high-An feldspar and high-Fo olivine in anhydrous mafic silicate liquids. Contrib Mineral Petrol 147:684-704

Kouchi A (1986) Effect of stirring on crystallization kinetics of basalt texture and element partitioning. Contrib Mineral Petrol 93:429-438

Laporte D, Watson EB (1995) Experimental and theoretical constraints on melt distribution in crustal sources— the effect of crystalline anisotropy on melt interconnectivity. Chem Geol 124:161-184

Larsen JF (2005) Experimental study of plagioclase rim growth around anorthite seed crystals in rhyodacitic melt. Am Mineral 90:417-427

Larsen JF (2006) Rhyodacite magma storage conditions prior to the 3430 yBP caldera-forming eruption of Aniakchak volcano, Alaska. Contrib Mineral Petrol 152:523-540

Lasaga AC (1998) Kinetic Theory in the Earth Sciences. Princeton University Press, Princeton

Lesher CE, Cashman KV, Mayfield JD (1999) Kinetic controls on crystallization of Tertiary North Atlantic basalt and implications for the emplacement and cooling history of lava at Site 989, Southeast Greenland rifted margin. Proc Ocean Drilling Prog 163:135-148

Lofgren G (1983) Effect of heterogeneous nucleation on basaltic textures; a dynamic crystallization study. J Petrol 24:229-255

Lofgren G (1987) Internally heated systems. *In:* Hydrothermal Experimental Techniques. Ulmer GC, Barnes HL (eds) John Wiley & Sons, New York, p 325-332

Lofgren GE (1980) Experimental studies on the dynamic crystallization of silicate melts. *In:* Physics of Magmatic Processes. Hargraves R (ed) Princeton University Press, p 487-551

Lofgren GE, Donaldson CH, Williams RJ, Mullins O, Usselman TM (1974) Experimentally produced textures and mineral chemistry of Apollo 15 quartz normative basalts. Lunar Planet Sci Conf 5:549-568

Lofgren GE, Huss GR, Wasserburg GJ, Shearer C, Vaniman D, Labotka T (2006) An experimental study of trace-element partitioning between Ti-Al-clinopyroxene and melt; equilibrium and kinetic effects including sector zoning. Am Mineral 91:1596-1606

Lofgren GE, Lanier AB (1992) Dynamic crystallization experiments on the Angra-Dos-Reis achondritic meteorite. Earth Planet Sci Lett 111:455-466

London D, Morgan GB VI (1998) Experimental crystal growth from undercooled granitic melts; nucleation response, texture, and crystallization sequence. Eos, Trans AGU, 79:366

Mangan MT (1990) Crystal size distribution systematics and the determination of magma storage times; the 1959 eruption of Kilauea Volcano, Hawaii. J Volc Geotherm Res 44:295-302

Marsh BD (1988) Crystal size distributions (CSD) in rocks and the kinetics and dynamics of crystallization I. Theory. Contrib Mineral Petrol 99:277-291

Marsh BD (1998) On the interpretation of crystal size distributions in magmatic systems. J Petrol 39:553-599

Marsh BD (2007) Crystallization of silicate magmas deciphered using crystal size distributions. J Am Ceramic Soc 90:746-757

Martel C, Ali AR, Poussineau S, Gourgaud A, Pichavant M (2006) Basalt-inherited microlites in silicic magmas: Evidence from Mount Pelee (Martinique, French West Indies). Geology 34:905-908

Martel C, Bourdier JL, Traineau H, Pichavant M, Holtz F, Scaillet B (1999) Effects of $f(O_2)$ and H_2O on andesite phase relations between 2 and 4 kbar. J Geophys Res 104:29,453-29,470

Martel C, Schmidt BC (2003) Decompression experiments as an insight into ascent rates of silicic magmas. Contrib Mineral Petrol 144:397-415

McCanta MC, Rutherford MJ, Hammer JE (2007) Pre-eruptive and syn-eruptive conditions in the Black Butte, California dacite: Insight into crystallization kinetics in a silicic magma system. J Volc Geotherm Res 160:263-284

McMillan PF (1994) Water solubility and speciation models. Rev Mineral 30:131-156

Métrich N, Wallace PJ (2008) Volatile abundances in basaltic magmas and their degassing paths tracked by melt inclusions. Rev Mineral Geochem 69:363-402

Milman-Barris MS, Beckett JR, Baker MB, Hofmann EA, Morgan Z, Crowley MR, Vielzeuf D, Stolper E (2008) Zoning of phosphorus in igneous olivine. Contrib Mineral Petrol 155:739-765

Moore G (2008) Interpreting H_2O and CO_2 contents in melt inclusions: constraints from solubility experiments and modeling. Rev Mineral Geochem 69:333-361

Moore G, Carmichael ISE (1998) The hydrous phase equilibria (to 3 kbar) of an andesite and basaltic andesite from western Mexico: constraints on water content and conditions of phenocryst growth. Contrib Mineral Petrol 130:304-319

Muller R, Zanotto ED, Fokin VM (2000) Surface crystallization of silicate glasses: nucleation sites and kinetics. J Non-Cryst Solids 274:208-231

Mullin JW (1980) Bulk crystallization. In: Crystal Growth. Pamplin BR (ed) Pergamon, Oxford, p 521-565

Muncill GE, Lasaga AC (1987) Crystal-growth kinetics of plagioclase in igneous systems: One atmosphere experiments and application of a simplified growth model. Am Mineral 72:299-311

Muncill GE, Lasaga AC (1988) Crystal-growth kinetics of plagioclase in igneous systems: Isothermal H_2O-saturated experiments and extension of a growth model to complex silicate melts. Am Mineral 73:982-992

Nicholis MG, Rutherford MJ (2004) Experimental constraints on magma ascent rate for the Crater Flat volcanic zone hawaiite. Geology 32:489-492

Okamura FP, McCallum IS, Stroh JM, Ghose S (1976) Pyroxene-spinel intergrowths in lunar and terrestrial pyroxenes. Proc 7th Lunar Planet Sci Conf 2:1889-1899

Park Y, Hanson B (1999) Experimental investigation of Ostwald-ripening rates of forsterite in the haplobasaltic system. J Volc Geotherm Res 90:103-113

Pertermann M, Lundstrom CC (2006) Phase equilibrium experiments at 0.5 GPa and 1100-1300 degrees C on a basaltic andesite from Arenal volcano, Costa Rica. J Volc Geotherm Res 157:222-235

Pichavant M, Costa F, Burgisser A, Scaillet B, Martel C, Poussineau S (2007) Equilibration scales in silicic to intermediate magmas - Implications for experimental studies. J Petrol 48:1955-1972

Pichavant M, Macdonald R (2007) Crystallization of primitive basaltic magmas at crustal pressures and genesis of the calc-alkaline igneous suite: experimental evidence from St Vincent, Lesser Antilles arc. Contrib Mineral Petrol 154:535-558

Pichavant M, Martel C, Bourdier JL, Scaillet B (2002) Physical conditions, structure, and dynamics of a zoned magma chamber: Mount Pelee (Martinique, Lesser Antilles Arc). J Geophys Res 107 doi: 10.1029/2001JB000315

Porter D, Easterling K (1997) Phase Transformations in Metals and Alloys. Chapman & Hall, London

Pupier E, Duchene S, Toplis MJ (2008) Experimental quantification of plagioclase crystal size distribution during cooling of a basaltic liquid. Contrib Mineral Petrol 155:555-570

Ree T, Eyring H (1958) The relaxation theory of transport phenomena. In: Rheology: Theory and Applications, 2. Eirich FR (ed) Academic Press, New York

Rose LA, Brenan JM (2001) Wetting properties of Fe-Ni-Co-Cu-O-S melts against olivine: Implications for sulfide melt mobility. Econ Geol Bull Soc Econ Geol 96:145-157

Roskosz M, Toplis MJ, Richet P (2005) Experimental determination of crystal growth rates in highly supercooled aluminosilicate liquids; implications for rate-controlling processes. Am Mineral 90:1146-1156

Roux J, Holtz F, Lefevre A, Schulze F (1994) A reliable high-temperature setup for internally heated pressure vessels; applications to silicate melt studies. Am Mineral 79:1145-1149

Roux J, Lefevre A (1992) A fast-quench device for internally heated pressure vessels. Eur J Mineral 4:279-281

Rutherford MJ (2008) Magma ascent rates. Rev Mineral Geochem 69:241-271

Rutherford M, Devine J (1988) The May 18, 1980, eruption of Mount St. Helens 3. Stability and chemistry of amphibole in the magma chamber. J Geophys Res 93:11,949-11,959

Rutherford MJ, Devine JD (1996) Pre-eruption pressure-temperature conditions and volatiles in the 1991 dacitic magma of Mount Pinatubo. In: Fire and Mud: Eruptions and Lahars of Mount Pinatubo, Philippines. Newhall C, Punongbayan R (eds) University of Washington Press, Seattle, p. 751-766

Rutherford MJ, Sigurdsson H, Carey S, Davis A (1985) The May 18, 1980, eruption of Mount St. Helens 1. Melt composition and experimental phase equilibria. J Geophys Res 90:2929-2947

Sato H (1995) Textural difference between pahoehoe and A'a lavas of Izu-Oshima Volcano, Japan - an experimental study on population density of plagioclase. J Volc Geotherm Res 66:101-113

Scaillet B, Evans BW (1999) The 15 June 1991 eruption of Mount Pinatubo. I. Phase equilibria and pre-eruption P-T-fO_2-fH_2O conditions of the dacite magma. J Petrol 40:381-411

Scaillet B, Pichavant M, Roux J (1995) Experimental crystallization of leukogranite magmas. J Petrol 36:663-705

Scaillet B, Pichavant M, Roux J, Humbert G, Lefevre A (1992) Improvements of the Shaw membrane technique for measurement and control of f_{H2} at high temperatures and pressures. Am Mineral 77:647-655

Schwandt CS, McKay GA (2006) Minor- and trace-element sector zoning in synthetic enstatite. Am Mineral 91:1607-1615

Shafer FN, Foley SF (2002) The effect of crystal orientation on the wetting behavior of silicate melts on the surfaces of spinel peridotite minerals. Contrib Mineral Petrol 143:254-261

Sharp TG, Stevenson RJ, Dingwell DB (1996) Microlites and "nanolites" in rhyolitic glass; microstructural and chemical characterization. Bull Volcanol 57:631-640

Shaw HR (1963) Obsidian-H_2O viscosities at 1000 and 2000 bars in the temperature range 700-900 °C. J Geophys Res 68:6337-6343

Shneidman VA (1988) Establishment of a steady-state nucleation regime. Theory and comparison with experimental data for glasses. Soviet Phys-Tech Phys 33:1338-1342

Silver L, Stolper E (1989) Water in albitic glasses. J Petrol 30:667-709

Simakin AG, Armienti P, Epel'baum MB (1999) Coupled degassing and crystallization; experimental study at continuous pressure drop, with application to volcanic bombs. Bull Volcanol 61:275-287

Simakin AG, Armienti P, Salova TP (2000) Joint degassing and crystallization: Experimental study with a gradual pressure release. Geochem Intl 38:523-534

Simakin AG, Chevychelov VY (1995) Experimental studies of feldspar crystallization from the granitic melt with various water content. Geokhimiya 216-238

Simakin AG, Salova TP (2004) Plagioclase crystallization from a hawaiitic melt in experiments and in a volcanic conduit. Petrol 12:82-92

Simakin AG, Salova TP, Armienti P (2003) Kinetics of clinopyroxene growth from a hydrous hawaiite melt. Geochem Intl 41:1165-1175

Sisson TW, Grove TL (1993) Experimental investigations of the role of H_2O in calc-alkaline differentiation and subduction zone magmatism. Contrib Mineral Petrol 113:143-166

Skripov VP, Faizullin MZ (2005) Solid-liquid and liquid-vapor phase transitions: similarities and differences. *In:* Nucleation Theory and Applications. Schmelzer JWP (ed) Wiley. p. 4-38

Smith CS (1964) Some elementary principles of polycrystalline microstructure. Metal Rev 9:1-48

Spaepen F (1994) Homogeneous nucleation and the temperature dependence of the crystal-melt interfacial tension. Solid State Phys Adv Res Appl 47:1-32

Streck MJ (2008) Mineral textures and zoning as evidence for open system processes. Rev Mineral Geochem 69:595-622

Sunagawa I (1981) Characteristics of crystal growth in nature as seen from the morphology of mineral crystals. Bull Mineral 104:81-87

Suzuki Y, Gardner JE, Larsen JF (2007) Experimental constraints on syneruptive magma ascent related to the phreatomagmatic phase of the 2000AD eruption of Usu volcano, Japan. Bull Volcanol 69:423-444

Swanson SE (1977) Relation of nucleation and crystal-growth rate to the development of granitic textures. Am Mineral 62:966-978

Swanson SE, Fenn PM (1986) Quartz crystallization in igneous rocks. Am Mineral 71:331-342

Sycheva GA (1998) Surface energy at the crystal nucleus glass interface in alkali silicate glasses. Glass Phys Chem 24:342-347

Szramek L, Gardner JE, Larsen J (2006) Degassing and microlite crystallization of basaltic andesite magma erupting at Arenal volcano, Costa Rica. J Volc Geotherm Res 157:182-201

Tolman RC (1949) The effect of droplet size on surface tension. J Chem Phys 17:333-337

Toshiya A, Katsuo T, Ichiro S (1991) Nucleation, growth, and stability of $CaAl_2Si_2O_8$ polymorphs. Phys Chem Mineral 17:473-484

Turcotte DL, Schubert G (1982) Geodynamics: Applications of Continuum Physics to Geological Problems. John Wiley & Sons, New York

Turnbull D (1964) Physics of non-crystalline solids. *In:* Physics of Non-crystalline Solids. Proc Int Conf. Prins JA (ed) North-Holland, Deift, p 41-56

Turnbull D, Cohen M (1960) Crystallization kinetics and glass formation. *In:* Modern Aspects of the Vitreous State. MacKenzie JD (ed) Butterworth and Co., London, p 38-62

Ulmer GC (1971) Research Techniques for High Pressure and High Temperature. Springer Verlag

Ulmer GC, Barnes HL (1987) Hydrothermal Experimental Techniques. John Wiley & Sons, New York

Vetere F, Behrens H, Holtz F, Neuville DR (2006) Viscosity of andesitic melts: new experimental data and a revised calculation model. Chem Geol 228:233-245

Volmer M, Weber A (1926) Kimbildung in ubersattingten gebilden. Z Phys Chem 119:277-301

Waff HS, Faul UH (1992) Effects of crystalline anisotropy on fluid distribution in ultramafic partial melts. J Geophys Res 97:9003-9014

Walker D, Kirkpatrick RJ, Longhi J, Hays JF (1976) Crystallization history of lunar picritic basalt Sample 12002; phase-equilibria and cooling-rate studies. Geol Soc Am Bull 87:646-656

Walker D, Powell MA, Lofgren GE, Hays JF (1978) Dynamic crystallization of a eucrite basalt. Proc 9th Lunar Planet Sci Conf 1:1369-1391

Wallace PJ (2005) Volatiles in subduction zone magmas: concentrations and fluxes based on melt inclusion and volcanic gas data. J Volc Geotherm Res 140:217-240

Wallace PJ, Gerlach TM (1994) Magmatic vapor source for sulfur dioxide released during volcanic eruptions; evidence from Mount Pinatubo. Science 265:497-499

Wang YP, Lu K (2000) Crystallization kinetics of amorphous Ni80P20 alloy investigated by electrical resistance measurements. Z Metallk 91:285-290

Watson EB (1996) Surface enrichment and trace-element uptake during crystal growth. Geochim Cosmochim Acta 60:5013-5020

Watson EB (2004) A conceptual model for near-surface kinetic controls on the trace-element and stable isotope composition of abiogenic calcite crystals. Geochim Cosmochim Acta 68:1473-1488

Watson EB, Liang Y (1995) A simple model for sector zoning in slowly grown crystals: Implications for growth rate and lattice diffusion, with emphasis on accessory minerals in crustal rocks. Am Mineral 80:1179-1187

Webster JD, Holloway JR, Hervig RL (1987) Phase equilibria of a Be, U and F-enriched vitrophyre from Spor Mountain, Utah. Geochem Cosmochim Acta 51:389-402

Zhang J, Wei YH, Qiu KQ, Zhang HF, Quan MX, Hu ZQ (2003) Crystallization kinetics and pressure effect on crystallization of $Zr_{55}Al_{10}Ni_5Cu_{30}$ bulk glass. Mater Sci Eng Struct Mater Prop Microstruct Proc 357:386-392

Zieg MJ, Lofgren GE (2006) An experimental investigation of texture evolution during continuous cooling. J Volc Geotherm Res 154:74-88

Zieg MJ, Marsh BD (2002) Crystal size distributions and scaling laws in the quantification of igneous textures. J Petrol 43:85-101

Thermometers and Barometers for Volcanic Systems

Keith D. Putirka

Department of Earth and Environmental Sciences
California State University, Fresno
2576 E. San Ramon Ave., MS/ST24
Fresno, California, 93740-8039, U.S.A.

kputirka@csufresno.edu

INTRODUCTION

Knowledge of temperature and pressure, however qualitative, has been central to our views of geology since at least the early 19th century. In 1822, for example, Charles Daubeny presented what may be the very first "Geological Thermometer," comparing temperatures of various geologic processes (Torrens 2006). Daubeny (1835) may even have been the first to measure the temperature of a lava flow, by laying a thermometer on the top of a flow at Vesuvius—albeit several months following the eruption, after intervening rain (his estimate was 390 °F). In any case, pressure (P) and temperature (T) estimation lie at the heart of fundamental questions: How hot is Earth, and at what rate has the planet cooled. Are volcanoes the products of thermally driven mantle plumes? Where are magmas stored, and how are they transported to the surface—and how do storage and transport relate to plate tectonics? Well-calibrated thermometers and barometers are essential tools if we are to fully appreciate the driving forces and inner workings of volcanic systems.

This chapter presents methods to estimate the P-T conditions of volcanic and other igneous processes. The coverage includes a review of existing geothermometers and geobarometers, and a presentation of approximately 30 new models, including a new plagioclase-liquid hygrometer. Our emphasis is on experimentally calibrated "thermobarometers," based on analytic expressions using P or T as dependent variables. For numerical reasons (touched on below) such expressions will always provide the most accurate means of P-T estimation, and are also most easily employed. Analytical expressions also allow error to be ascertained; in the absence of estimates of error, P-T estimates are nearly meaningless. This chapter is intended to complement the chapters by Anderson et al. (2008), who cover granitic systems, and by Blundy and Cashman (2008) and Hansteen and Klügel (2008), who consider additional methods for P estimation.

THEORETICAL ASPECTS

At the foundation of well-founded belief lies belief that is not founded
- Ludwig Wittgenstein

In the search for useful geothermometers or geobarometers, the goal is to find some chemical equilibrium where there is a significant difference between the entropy (ΔS_r) (for a thermometer) or volume (ΔV_r) (for a barometer) of products and reactants. Consider:

$$NaO_{0.5}^{liq} + AlO_{1.5}^{liq} + 2SiO_2^{liq} = NaAlSi_2O_6^{cpx} \qquad (1)$$

where superscripts denote phase (liq = liquid; cpx = clinopyroxene); $NaAlSi_2O_6^{cpx}$ is the jadeite (Jd) content of clinopyroxene. To compare S and V, we use partial molar entropies ($\bar{S}$) and partial molar volumes ($\overline{V}$), where $\overline{S} = S/n$ and $\overline{V} = V/n$, and n = number of moles. In Equation (1), $\Delta S_r = \overline{S}_{Jd} - \overline{S}_{NaO_{0.5}^{liq}} - \overline{S}_{AlO_{1.5}^{liq}} - 2\overline{S}_{SiO_2^{liq}}$, $\Delta V_r = \overline{V}_{Jd} - \overline{V}_{NaO_{0.5}^{liq}} - \overline{V}_{AlO_{1.5}^{liq}} - 2\overline{V}_{SiO_2^{liq}}$. The equilibrium constant, K_{eq}, is:

$$K_{eq} = \frac{a_{Jd}^{cpx}}{a_{NaO_{0.5}}^{liq} \cdot a_{Al_2O_3}^{liq} \cdot \left(a_{SiO_2}^{liq}\right)^2} \quad (2)$$

where terms such as a_{Jd}^{cpx} and $a_{NaO_{0.5}}^{liq}$ represent the activities of Jd in clinopyroxene, and SiO_2 in liquid respectively. Activity is something like an 'effective concentration', and in the ideal case, is equivalent to mole fraction: $a_{Jd}^{cpx} = X_{Jd}^{cpx}$, where X_{Jd}^{cpx} is the mole fraction of Jd in a clinopyroxene solid solution. In non-ideal solutions, the two quantities can be related to one another by an activity coefficient, λ, so that $a_{Jd}^{cpx} = \lambda_{Jd}^{cpx} X_{Jd}^{cpx}$.

A "good" thermometer has a large ΔS_r and a good barometer has a large ΔV_r. But what is "good"? For metamorphic systems, Essene (1982) suggests that $\Delta S_r \geq 4.0$ J/mole·K and $\Delta V_r \geq 0.2$ J/bar; these limits perhaps also apply to igneous systems. The presence of a liquid, though, poses hurdles to the igneous petrologist. Consider the very large number of distinct equilibria used for P-T estimation in metamorphic systems (Spear 1993). Far fewer are available to igneous petrologists. Why? To begin with, igneous rocks carry fewer crystalline phases. And the numbers of crystalline phases is perhaps in turn limited by the presence of the liquid itself. A silicate liquid is a 'high variance' phase in that, compared to minerals, it is highly compressible and has no compositional stoichiometric constraints. Liquids can thus absorb significant changes in P, T and composition (X_i^j, the mole fraction of i in some phase j), by expanding, contracting or mixing, without resorting to the extreme measure of precipitating a new phase. Metamorphic rocks have no such luxury. Carrying nothing but crystalline phases, with sharp limits on their ability to expand, contract or mix, new phases readily nucleate in response to changing P-T-X_i^j. This results in sharp changes in S and V in metamorphic systems, which can be exploited to formulate geothermobarometers. By their versatility, silicate liquid-bearing systems are intrinsically less amenable to thermobarometer formulation—but all is not lost.

Perhaps the single most useful equation for petrologists is:

$$-RT \ln K_{eq} = \Delta G_r^\circ \quad (3)$$

where R is the gas constant (R = 1.9872 cal/K·mole, or 8.3144 J/dK·mole) and ΔG_r° is the Gibbs free energy change of "reaction" for a balanced equilibrium, such as Equation (1). In ΔG_r°, the superscript ° means that the Gibbs free energy is for a standard state (pure substances at 1 bar), an important distinction from ΔG_r, which is the Gibbs free energy change between products and reactants at any arbitrary P and T. At equilibrium, $\Delta G_r = 0$, by definition. But rarely is $\Delta G_r = 0$ at a standard state. So when $\Delta G_r = 0$, it is usual that $\Delta G_r^\circ \neq 0$, which means that usually $K_{eq} \neq 1$.

Despite its name, however, K_{eq} is only constant at fixed P and T. Driven by differences in ΔS_r and ΔV_r, K_{eq} can vary greatly as a function of P and T. So for example, in Equation (1), if $NaAlSi_2O_6^{cpx}$ has a smaller partial molar volume compared to the collective partial molar volumes of the reactants ($\Delta V_r > 0$), then with increased P, Equation (1) will shift to the right and K_{eq} will increase. Similarly, higher T favors whichever side of an equilibrium that has higher partial molar entropy. Equation (3) provides a quantitative footing to relate K_{eq} to P and T and is at the root of all thermometers and barometers based on 'theoretical' or thermodynamic models.

The differentiation of "empirical" and "theoretical" is not always clear. The laws of thermodynamics are rooted in fundamentally empirical expressions derived by Robert Boyle, who in 1662 determined that volume, V, is related to P by the expression: $V = C/P$, where C is a constant. In the 19th century, Jacques Charles and Joseph-Louis Gay-Lussac, further illuminated the phenomenon of thermal expansion in gases: $V = \alpha V_o T$, where α is the coefficient of thermal expansion, and V_o is some reference gas volume (see Castellan 1971). These two expressions, known today as Boyle's Law and Charles' (or Gay-Lussac's) Law are combined, with Avogadro's law (that equal volumes of gases have equal numbers of molecules), to yield the ideal gas law: $PV = nRT$ (which recognizes that the constant C in Boyle's Law is proportional to n). The relevance is that the form of Equation (3) is not accidental: it derives from the ideal gas law and an assumption that activities can be represented by fugacities or partial pressures, where we integrate $(\partial G/\partial P)_T = V$ by substituting $V = nRT/P$ (see Nordstrom and Munoz 1986). In effect, our most sophisticated and complex activity models for silicate liquids and minerals are based on the assumption that they act, in the phrase of David Walker, as a *"perverted ideal gas."* In the event, practice and experience indicate that models based on $-RT\ln K_{eq} = \Delta G_r°$ are more accurate than those based on more arbitrary relationships and so provide the greatest promise for extrapolation outside calibration boundaries.

To make use of $\ln K_{eq} = -\Delta G_r°/RT$, we use the relationship $\Delta G_r° = \Delta H_r° - T\Delta S°$ (where $\Delta H_r°$ and $\Delta S_r°$ are respectively the enthalpy and entropy differences for a chemical equilibrium at standard state), and the expressions $(\partial \Delta G_r/\partial P)_T = \Delta V_r$ and $(\partial \Delta G_r/\partial P)_P = \Delta S_r$. If P is constant, and the differences in heat capacities (C_p) between products and reactants are effectively equal ($\Delta C_p = 0$) (meaning that ΔH_r and ΔS_r are independent of T), then substitution yields the following:

$$\ln K_{eq} = -\Delta H_r°/RT + \Delta S_r°/R \quad (4)$$

If we know $\Delta H_r°$ and $\Delta S_r°$, then we can use Equations (2) and (4) to calculate T, and we have a thermometer. If $\Delta H_r°$ and $\Delta S_r°$ are unknown, we can preserve the functional relationship between K_{eq} and T in Equation (4) and use as a regression model: $\ln K_{eq} = a/T + b$, where $a = -\Delta H_r°/R$ and $b = \Delta S_r°/R$; we can then derive a and b from isobaric experimental data, where composition and T are known (and an activity model is assumed). Of course, P need not be constant, and ΔH_r and ΔS_r will vary with T if $\Delta C_p \neq 0$. Products and reactants might also exhibit significant differences in compressibility (β) or thermal expansion (α) such that, over a wide enough P interval, $\Delta \beta_r \neq 0$ and $\Delta \alpha_r \neq 0$. The following regression equation accounts for such changes (see Putirka 1998):

$$\ln K_{eq} = \frac{A}{T} + B + C\ln\left[\frac{T}{T_o}\right] + D\frac{P}{T} + E\frac{P^2}{T} + FP + GP^2 \quad (5)$$

where $A = -\Delta H_r°/R$, $B = \Delta S_r°/R$, $C = \Delta C_p/R$, $D = -\Delta V_r°/R$, $E = \Delta \beta_r \Delta V_r°/2R$, $F = \Delta \alpha_r \Delta V_r°/R$, and $G = \Delta \alpha_r \Delta \beta_r \Delta V_r°/2R$. Equation (5) contains all the terms in Equation (4), plus additional terms that correct for the cases that P is not constant, $\Delta C_p \neq 0$, etc. More terms could be added if $\Delta \alpha_r$, $\Delta \beta_r$, and ΔC_p are not constant, though for most geological purposes Equation (5) is sufficiently complex. In Equation (5), T_o is a reference temperature; in Putirka (1998) T_o was taken as the melting point of a pure mineral, T_m, in which case reference values for $\Delta H_r°$, $\Delta S_r°$, etc., are for fusions of pure substances. Equation (5) is also simplified in that the reference pressure (P_o, needed to calculate the effects of a non-zero $\Delta \beta_r$ and $\Delta \alpha_r$) is taken as 1 atm, or effectively zero on a kilobar or GPa scale. If some other reference pressure is required, or if 1 bar is large compared to the pressure of interest, then the term (P-P_o) replaces P in each instance in Equation (5).

REGRESSION ANALYSIS & STATISTICAL CONSIDERATIONS

The stars might lie but the numbers never do
- Mary Chapin Carpenter

Experimental data, regression analysis & statistical considerations

Although some geothermometers or geobarometers are developed using calorimetric and volumetric data, most derive from regression analysis of experimental data, where K_{eq}, P and T are all known. Partial melting experiments reported in the Library of Experimental Phase

Relations (LEPR) (Hirschmann et al. 2008), and elsewhere, are here used to test existing, and calibrate new thermobarometers. Regression analysis of such data yields regression coefficients that minimize error for the selected dependent variable. This approach assumes that the experiments used for regression approach equilibrium. There are several possible validity tests, but none guarantee success. One is to check whether the regression-derived coefficients compare favorably with independently determined values for quantities such as $\Delta H_r°$, $\Delta S_r°$, and $\Delta V_r°$. One can also test model results for their ability to predict P or T for natural systems where P and T might be known (e.g., a lava lake). Issues of error are discussed in detail in the penultimate section of this chapter.

Regression strategies. For the calibration of new models, no data-filters are applied at the outset, but generally, experiments that yield P or T estimates that are >3σ outside standard errors of estimate (SEE) for remaining data are considered suspect and not used for calibration or testing. And if particular studies contain multiple aberrant data, such studies are excluded. Most new calibrations utilize "global" regressions, which use all available data to determine the minimum number of parameters that are required to explain observed variability. All tests make use of "leverage plots" to ensure that a given model parameter is significant and explained by more than just one or a few potentially aberrant data (see JMP Statistics and Graphics Guide). We also employ regression models where P and T appear as dependent variables—a numerical necessity if one is to maximize precision.

Regression models. William of Ockham, a 14th century philosopher, suggested that "plurality must not be posited without necessity" (Ockham's razor), a recommendation supported by Bayesian analysis (Jefferys and Berger 1992). Jefferys and Berger (1992) reword this principle: "if a law has many adjustable parameters then it will be significantly preferred to the simpler law only if its predictions are considerably more accurate". To illustrate, they show how Galileo's quadratic expression (relating distance (x) and time (t) to the acceleration due to gravity (g), i.e., $x = x_o + vt + 1/2gt^2$) is highly superior to a 6th order polynomial fit to the same data. The problem is not whether additional coefficients in the 6th order equation have theoretical meaning: gravity changes with elevation; this fact could be used to develop additional terms beyond the familiar quadratic expression. But Galileo's original experiments have nowhere near the level of precision required to extract such information.

Similar issues apply to petrology. Using Equation (2) as an example, Figure 1a shows two regression models: 1) a two-parameter model where $\ln K_{eq} = aP/T + b$, and 2) a five-parameter model, where $\ln K = aP/T + b\ln(T) + c(P^2/T) + dP^2 + e$. All of the coefficients (a, b, c, d, e) have thermodynamic meaning (Eqn. 5), and the 5-parameter model describes a greater fraction (99.8%) of the variation of $\ln K_{eq}$ for the calibration data (ankaramites from Putirka et al. 1996). However, the 5-parameter model yields gross systematic error when predicting $\ln K_{eq}$ for "test" data (Fig. 1b; all remaining data in Putirka et al. 1996), with a slope and intercept through $\ln K_{eq}$ (measured) vs. $\ln K_{eq}$ (calculated) of 0.5 and 1.5 respectively. In contrast, our lowly 2-parameter

---

Figure 1 (*on facing page*). In (a) and (b), predicted values for $\ln K_{eq}$ based on Equation (2) are used to illustrate how model complexity, no matter how valid from theoretical considerations, can yield incorrect predictions if model parameters are not supported by data. In (a), calculated and measured values of $\ln K_{eq}$ are compared using ankaramite bulk compositions from Putirka et al. 1996. For both the 2- and 5-parameter equations, all regression coefficients have thermodynamic significance (Eqn. 5), and the 5-parameter model provides a better fit to the calibration data. However, in (b) the 5-parameter equation yields strong systematic error when used to predict $\ln K_{eq}$ for experimental data not used for regression (as illustrated by the slope and intercept of the regression line $\ln K_{eq}$(calculated) vs. $\ln K_{eq}$(measured); ideally they should be 1 and 0 respectively). In contrast, the 2-parameter model is nearly free of systematic error, and describes the 'test' data with higher precision; the 5-parameter equation is of little use beyond interpolating between calibration points, while the 2-parameter equation is a useful predictive tool. Models cannot be judged solely by their fit to calibration data, nor by their thermodynamic complexity.

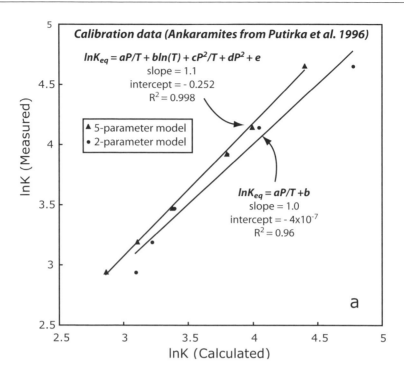

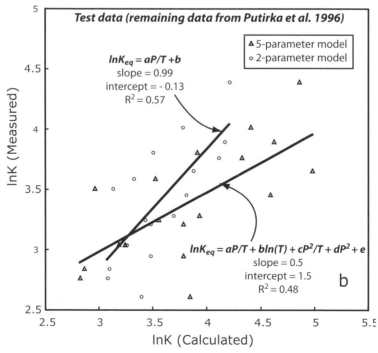

Figure 1. *caption on facing page*

model yields a higher correlation coefficient (R) for the test data, and is nearly absent systematic error, with slope and intercept values much closer to the ideal of one and zero respectively. Box and Draper (1987, p. 74) summarize the issue rather succinctly: "Remember that all models are wrong; the practical question is how wrong do they have to be to not be useful." In geothermobarometry, the answer is that only the first 4 terms of Equation (5) are usually needed.

Linear regression methods can be used to avoid the pitfalls noted in Figure 1 because unlike non-linear methods, validity tests are available for regression coefficients; parameters that are not supported by the data (no matter their thermodynamic significance), can be culled from a potential model. An occasional roadblock is that equations such as Equation (5) cannot always be rearranged so that P or T can be isolated as dependent variables. One solution is to perform a linear regression using some other quantity as the independent variable, and then rearrange the expression. But in multiple linear regression, error is minimized only for the dependent variable (regression analysis is not symmetric), so error will not be minimized for the variable of interest, and systematic error is assured. Equation (4), for example, will yield a very good model for predicting $\ln K_{eq}$, but will be less precise, and yield systematic error, if the same coefficients are rearranged to calculate T. *Regression models that use P or T as dependent variables will thus always be more precise and accurate* compared to models that minimize error on such quantities as $\ln K_{eq}$, Gibbs Free Energy, etc., an unavoidable consequence of numerical issues of regression analysis. If variations in P are not of concern, Equation (4) should be rearranged to yield:

$$\frac{1}{T} = a \ln K_{eq} + b \tag{6}$$

where $a = -R/\Delta H_r°$ and $b = \Delta S_r°/\Delta H_r°$. If variations in P are important, remaining choices are to add P as an "empirical" factor,

$$\frac{1}{T} = a \ln K_{eq} + b + cP \tag{7}$$

or apply non-linear regression methods, using:

$$T = \frac{\Delta H_r° + P\Delta V}{R \ln K_{eq} - \Delta S_r°} = \frac{a + bP}{c \ln K_{eq} + d} \tag{8}$$

Here, $\Delta C_p = 0$, and $P_o = 0$ (i.e., 1 bar); if an additional P term is required, the term $(1/2) P^2 \Delta \beta_r \Delta V_r°$ can be added to the numerator (a, b, c, and d have thermodynamic meaning as in Equation (8), but not are not identical to the coefficients in Equations (6-7)). A disadvantage to non-linear regression is that it does not yield an absolute minimum for error on the dependent variable, and there are no intrinsic validity tests for model parameters.

Activity modeling. Mixing and interaction between chemical components in an equilibrium may result in non-ideal behavior, complicating the calculation of K_{eq}. To address just such compositional dependencies, Thompson (1967) developed a polynomial model for activity coefficients, $\lambda_i = f(X_i)$. The approach uses so-called Margules parameters (W_G), which represent the energy of interaction between chemical components. Thompson's (1967) parameters can be substituted into Equation (3). For example, using the definitions for a two-parameter model (W_{G1}, and W_{G2}), describing interaction between two components, X_1 and X_2, the Margules model is: $RT\ln\lambda_1 = (2W_{G2} - W_{G1})(X_2)^2 + 2(W_{G1} - W_{G2})(X_2)^3$ and $RT\ln\lambda_1 = (2W_{G1} - W_{G2})(X_1)^2 + 2(W_{G2} - W_{G1})(X_1)^3$ (Nordstrom and Munoz 1986). Because of the extrinsic dependency of λ on T, such substitutions require non-linear regression methods for thermometers (see below). As an alternative approach, consider forsterite (Fo, Mg_2SiO_4) melting,

$$Mg_2SiO_4^{ol} = 2MgO^{liq} + SiO_4^{liq} \tag{9}$$

In the ideal case, $a_{Fo}^{ol} = X_{Fo}^{ol}$, $a_{MgO}^{liq} = X_{MgO}^{liq}$ and $a_{SiO_2}^{liq} = X_{SiO_2}^{liq}$, and so $K_{eq}^{ol-liq} = X_{Fo}^{ol}/(X_{MgO}^{liq} \cdot X_{SiO_2}^{liq})$. If the system is non-ideal and we know the values for λ_{MgO}^{liq}, λ_{Fo}^{ol}, etc., the activity coefficients can be inserted to yield $K_{eq}^{ol-liq} = \lambda_{Fo}^{ol} X_{Fo}^{ol}/(\lambda_{MgO}^{liq} X_{MgO}^{liq} \cdot \lambda_{SiO_2}^{liq} X_{SiO_2}^{liq})$. O'Nions and Powell (1977) extract such terms to yield linear coefficients in Equations such as (4) and (8). Using Equation (9) as an example,

$$\ln K_{eq} = \ln\left[\frac{\left(a_{MgO}^{liq}\right)^2 \cdot a_{SiO_2}^{liq}}{a_{Fo}^{ol}}\right] = \ln\left[\frac{\left(\lambda_{MgO}^{liq} X_{MgO}^{liq}\right)^2 \cdot \lambda_{SiO_2}^{liq} X_{SiO_2}^{liq}}{\lambda_{Fo}^{ol} X_{Fo}^{ol}}\right] \quad (10)$$

$$= \ln\left[\frac{\left(\lambda_{MgO}^{liq}\right)^2 \cdot \lambda_{SiO_2}^{liq}}{\lambda_{Fo}^{ol}}\right] + \ln\left[\frac{\left(X_{MgO}^{liq}\right)^2 \cdot X_{SiO_2}^{liq}}{X_{Fo}^{ol}}\right]$$

$$= 2\ln\left(\lambda_{MgO}^{liq}\right) + 2\ln\left(X_{MgO}^{liq}\right) + \ln\left(\lambda_{SiO_2}^{liq}\right) + \ln\left(X_{SiO_2}^{liq}\right) - \ln\left(\lambda_{Fo}^{ol}\right) - \ln\left(X_{Fo}^{ol}\right)$$

where a term such as X_i^j refers to the mole fraction of component i in phase j. A possible regression model based on Equations (6), (9) and (10) is thus:

$$\frac{1}{T} = a\ln\left[\frac{\left(X_{MgO}^{liq}\right)^2 \cdot X_{SiO_2}^{liq}}{X_{Fo}^{ol}}\right] + b + c\ln\left(\lambda_{MgO}^{liq}\right) + d\ln\left(\lambda_{SiO_2}^{liq}\right) + e\ln\left(\lambda_{Fo}^{ol}\right) \quad (11)$$

where the coefficients a and b are as in Equation (6). For ideal mixing, the λ_i^j are all 1, and c, d, and e are zero; if λ_i^j are known, then $c = d = e = 1$. For non-ideal mixing, λ_i^j can be related to mole fractions, X_i^j, using

$$\frac{1}{T} = a\ln\left[\frac{\left(X_{MgO}^{liq}\right)^2 \cdot X_{SiO_2}^{liq}}{X_{Fo}^{ol}}\right] + b + c\left(X_{MgO}^{liq}\right) + d\ln\left(X_{SiO_2}^{liq}\right) + e\ln\left(X_{Fo}^{ol}\right) \quad (12)$$

where a and b are as in Equation (6), and c, d, and e are "empirical" regression coefficients that can be related to the activity coefficients: in the case of X_{MgO}^{liq}, we presume that there is some number c such that $\lambda_{MgO}^{liq} = \exp(cX_{MgO}^{liq})$, where c is as in Equation (12). Stormer (1975) shows how coefficients to a polynomial expansion can be equated to Margules parameters. In any case, activity models are important here, only to the extent that they provide increased precision when estimating P or T.

IGNEOUS THERMOMETERS AND BAROMETERS

All models are wrong, some models are useful

- George Box

History and background

James Hall (1798) presented the first attempts to measure the melting temperatures of volcanic rocks. However, at the time there were few methods for determining temperatures above the boiling point of mercury, and his use of the Wedgwood pyrometer (based on the volume change in clays when heated) yielded widely ranging estimates (1029-3954 °C; see Hall's "Table of Fusibilities," where °W ≈ 130 °F). Later experiments had much better temperature control, with Joly (1891) using his "meldometer" and later experimentalists eventually employing the thermocouple (e.g., Doelter 1902; Day and Allen 1905; Shepherd et al. 1909).

Some of the earliest temperature-controlled experimental studies of melting temperatures for minerals and rocks (Joly 1891; Doelter 1902; Day and Allen 1905; Shepherd et al. 1909) were no doubt conducted in part to understand the thermal conditions of partial melting and crystallization. But the first study to establish, with some accuracy, the temperature interval for the crystallization of natural basaltic magmas was conducted by Sosman and Merwin (1913). They determined that the solidus and liquidus temperatures for the Palisades Sill in New York are ≈1150 °C and 1300 °C respectively. Contributions by Bowen (1913, 1915) have outshone such work. In his 1913 paper, Bowen established with accuracy the continuous solid solution of the plagioclase feldspars, and then in 1915 applied experimental results, thermodynamic theory and phase diagrams to show that a wide range of igneous rocks can be derived from basaltic parental liquids by the process of fractional crystallization (Bowen 1915). But Sosman and Merwin's estimates were nonetheless largely correct: their estimation of the 1 atm temperature interval for basalt crystallization matched well with Perret's (1913) and Jagger's (1917) direct temperature estimates of freshly formed lava at Hawaii (est. 1050-1200 °C), and were qualitatively verified by experiments in simple systems (Bowen 1915); together, these experimental and field-based studies represent the first accurate determinations of the temperatures at which basaltic rocks form. More thorough experimental studies (e.g., Tuttle and Bowen 1958; Yoder and Tilley 1962; Green and Ringwood 1967) would later establish temperatures of partial crystallization for both granitic and basaltic systems, over a range of P-T conditions.

None of these early efforts, though, established a means to *estimate T or P for any particular igneous rock*. One of the first attempts to establish a true geothermometer was by Wright and Larson (1909), using the optical and morphological properties of quartz. The first analytic geothermometer, though, was proposed by Barth (1934) and presented by Barth (1951), who developed an approach that is still in use today: the two-feldspar thermometer. Barth (1951) presented a theoretical model based on the alkali-feldspar solvus, which was retracted in proof when he discovered that his theoretical solvus did not match the experimentally determined curve of Bowen and Tuttle (1950). Barth's (1951) effort was followed by a Ti-in-magnetite thermometer by Buddington et al. (1955) and Buddington and Lindsley (1964), a two-pyroxene thermometer by Davis and Boyd (1966), a thermometer based on Ni partitioning between olivine and liquid by Hakli and Wright (1967), and Barth's own recalibrations of a two-feldspar thermometer (Barth 1962, 1968). The following decade then saw a substantial increase in the methods of igneous thermobarometry, with new calibrations of a plagioclase thermometer by Kudo and Weill (1970), a new olivine thermometer by Roeder and Emslie (1970), new calibrations of the two-feldspar thermometer (e.g., Stormer 1975) and the calibration of multiple P- and T-sensitive igneous equilibria by Nicholls et al. (1971), Nicholls and Carmichael (1972) and Bacon and Carmichael (1973), among others. Since the 1970's the number of igneous thermometers and barometers has greatly expanded.

Selection criteria. Within the space of a single chapter, it is possible to review neither all the various igneous thermometers that have been calibrated, nor to detail their thermodynamic bases. Emphasis is thus placed upon cataloging those tools that can be used widely, especially for volcanic systems. The thermometers and barometers presented here are thus based on commonly found minerals, using, for the most part, elemental components that are easily determined in most laboratories.

Calculating "liquid" and mineral components, and activities

It is absolutely crucial that all components, be they minerals or liquids, are calculated in precisely the same way as originally handled for a particular thermometer or barometer. Here, component calculations that are common to many models shown below are discussed, but care should be taken to examine specific component calculation procedures when presented below.

Several Tables are presented to illustrate nearly all the calculations presented in this chapter (Tables 1-7); readers attempting to reproduce these and other numbers should note the following: Table 1 in Putirka (1997) contains small errors relative to P and T (using the models in Putirka et al. 1996); Table A2 in Putirka et al. (2007) contains small errors relative to melt fraction F, and derivative calculations. All related equations are valid as published;

the errors result from the use of earlier versions of certain equations, later abandoned for the published (more accurate) models (the Tables were not updated). In addition, some P and T calculations presented here require numerical solution methods (when T and P depend upon one another and both are to be calculated). Here and in other works P-T estimates are based on the numerical module found within "Excel – Preferences – Calculation – Iteration". This feature may have numerical instabilities, and the "Solver" or "Goal Seek" options of Excel, or other numerical methods, should perhaps be preferred.

Liquid components. In the models that follow, various calculation schemes are used to represent the chemical components of silicate phases. For the empirical "glass" thermometers, most liquid components are entered as wt%, so no special calculations are needed, except for terms such as Mg#, which are always represented on a mole fraction basis, and calculated as Mg#(liq) = (MgO^{liq}/40.3)/[((MgO^{liq}/40.3) + (FeO^{liq}/71.85))], where terms such as MgO^{liq} represent the wt% of MgO in the liquid. For all other models, liquid components are calculated either as a mole fraction or a cation fraction (Table 1), always on an anhydrous basis (even if H$_2$O occurs in the equation or analysis, so as to avoid estimating molecular H$_2$O vs. OH$^-$). The purpose of such calculations is to be able to compare numbers of molecules, as this is how we calculate K_{eq}. Mole and cation fractions are very similar. For a mole fraction, the wt% of each oxide is divided by its molecular weight; for example, Al$_2$O$_3$ would be divided by 101.96 and Na$_2$O would be divided by 61.98; this calculation yields a "mole proportion" for the oxides, which is converted to a mole fraction by renormalizing the sum of the molecular proportions so that they sum to 1.0 (see Table 1). In a chemical equilibrium, the Al content of a liquid, for example, would be written as Al$_2$O$_3^{liq}$, and in a K_{eq}, the mole fraction would appear as $X_{Al_2O_3}^{liq}$.

Liquid components using cation fractions. For a cation fraction, each oxide is normalized by the molecular weight of the oxide having only one cation. So Al$_2$O$_3$ is divided by the molecular weight of AlO$_{1.5}$, or 50.98; similarly, the wt% of Na$_2$O is divided by the weight of NaO$_{0.5}$, or 30.99. Compared to the usual mole fraction, this approach doubles the "molecular" units for oxides that contain two cations per formula unit, such as Al$_2$O$_3$, Na$_2$O, K$_2$O, Cr$_2$O$_3$, and P$_2$O$_5$; using Al$_2$O$_3$ again as an example, we effectively divide one molecular unit of Al$_2$O$_3$ into two molecular units of AlO$_{1.5}$. For oxides such as SiO$_2$, which already have only one cation per formula unit, the wt% of SiO$_2$ is divided by 60.08, as usual. These "cation proportions" are converted to a "cation fraction" by renormalizing each oxide by the sum of the cation proportions, so that the new sum is 1.0 (Table 1). In a chemical equilibrium, the Al content of a liquid would be written as AlO$_{1.5}^{liq}$, and in a K_{eq}, would appear as $X_{AlO_{1.5}}^{liq}$.

The advantage of the cation fraction is that the relative numbers of cations can be compared directly. Ratios of Fe/Mg (or equivalently, FeO/MgO) are the same whether a "mole" or cation fraction are considered; but the cation fraction allows a more straightforward comparison of, for example, Ca/Al (CaO/AlO$_{1.5}$) or Na/Al (NaO$_{0.5}$/AlO$_{1.5}$) ratios. With a mineral composition, cation fractions can also be multiplied by the number of cations in a mineral formula so that the sum, instead of 1, is equal to the number of cations per formula unit (e.g., 2 for olivine, 4 for pyroxene, 5 for feldpsar); this procedure yields the number of cations per formula unit.

Activity models take these calculations further, recognizing that the actual chemical species in liquids and minerals probably do not occur in the form of isolated, non-interacting oxides. Two notable approaches for calculating the activities of liquid components are the two-lattice (Bottinga and Weill 1972; Nielsen and Drake 1979) and the regular solution (Ghiorso 1987) models. Beattie (1993), however, recognized that because of numerous interactions at the molecular level, silicate liquids have many more chemical species than components, which limits what activity models can reveal about liquid properties. Beattie (1993) developed an activity model that increased the precision of his olivine-liquid thermometer, and his two-lattice model appears to be more successful than any proposed thus far. Beattie's (1993) two-lattice model yields several terms (calculated from cation fractions) that appear in the models

Table 1. Sample calculations for liquid components.

	SiO$_2$	TiO$_2$	Al$_2$O$_3$	FeO	MnO	MgO	CaO	Na$_2$O	K$_2$O	total
1) Weight %	47.05	0.89	16.11	7.96	0.15	12.78	11.13	2.18	0.04	98.29
2) Molecular weights	60.08	79.9	101.96	71.85	70.94	40.3	56.08	61.98	94.2	
3) Mole proportions	0.7831	0.0111	0.1580	0.1108	0.0021	0.3171	0.1985	0.0352	0.0004	1.62
	$X^{liq}_{SiO_2}$	$X^{liq}_{TiO_2}$	$X^{liq}_{Al_2O_3}$	X^{liq}_{FeO}	X^{liq}_{MnO}	X^{liq}_{MgO}	X^{liq}_{CaO}	$X^{liq}_{Na_2O}$	$X^{liq}_{K_2O}$	
4) Mole fractions	0.4845	0.0069	0.0978	0.0685	0.0013	0.1962	0.1228	0.0218	0.0003	1.00
	SiO$_2$	TiO$_2$	AlO$_{1.5}$	FeO	MnO	MgO	CaO	NaO$_{0.5}$	KO$_{0.5}$	
5) Molecular weights	60.08	79.9	50.98	71.85	70.94	40.3	56.08	30.99	47.1	
6) Cation proportions	0.7831	0.0111	0.3160	0.1108	0.0021	0.3171	0.1985	0.0703	0.0008	1.81
	$X^{liq}_{SiO_2}$	$X^{liq}_{TiO_2}$	$X^{liq}_{AlO_{1.5}}$	X^{liq}_{FeO}	X^{liq}_{MnO}	X^{liq}_{MgO}	X^{liq}_{CaO}	$X^{liq}_{NaO_{0.5}}$	$X^{liq}_{KO_{0.5}}$	
7) Cation fractions	0.4327	0.0062	0.1746	0.0612	0.0012	0.1752	0.1096	0.0389	0.0005	1.00
	$C^{liq}_{SiO_2}$			C^{liq}_{NM}	C^{liq}_{NF}	NF				
8) Beattie (1993) components	0.4327			0.3472	0.4782	-0.7148				

The mole proportions in row 3) are equal to row 1) divided by row 2); mole fractions, row 6) are equal to row 3) divided by its total, 1.62. Cation proportions in row 6) are equal to row 1) divided by row 5); cation fractions; row 7), is row 6) divided by its total, 1.81. Beattie (1993) components are calculated from cation fractions as follows: $C^{liq}_{SiO_2} = X^{liq}_{SiO_2}$; $C^{liq}_{NM} = X^{liq}_{FeO} + X^{liq}_{MnO} + X^{liq}_{MgO} + X^{liq}_{CaO} + X^{liq}_{NiO}$; $C^{liq}_{NF} = X^{liq}_{SiO_2} + X^{liq}_{NaO_{0.5}} + X^{liq}_{KO_{0.5}} + X^{liq}_{TiO_2}$; $NF = \frac{7}{2}\ln(1 - X^{liq}_{AlO_{1.5}}) + 7\ln(1 - X^{liq}_{TiO_2})$, where X^j_i is the cation fraction of i in phase j. Here, the values of X^j_i are equal to the contents of row 7).

below (Table 1): $NF = \frac{7}{2}\ln(1-X^{liq}_{AlO_{1.5}}) + 7\ln(1-X^{liq}_{TiO_2})$, (where NF refers to "network formers," i.e., chemical constituents that comprise a silicate network in the liquid); the concentration of network modifying cations, $C^{liq}_{NM} = X^{liq}_{MgO} + X^{liq}_{MnO} + X^{liq}_{FeO} + X^{liq}_{CaO} + X^{liq}_{CoO} + X^{liq}_{NiO}$; the concentration of Si, $C^{liq}_{SiO_2} = X^{liq}_{SiO_2}$.

Mineral components. Mineral components are typically calculated using the numbers of "cations per formula unit," which can be derived from the "cation fractions" (just as for liquids), sometimes with consideration of charge balance constraints. The forsterite (Fo) content of olivine, for example, is often presented as Mg/(Mg+Fe), where Mg and Fe represent the number of cations per formula unit in olivine; because Fo is expressed as a ratio, it does not matter whether the formula is expressed as $(Mg,Fe)Si_{0.5}O_2$ or $(Mg,Fe)_2SiO_4$. This calculation is also numerically equivalent to $X_{MgO}{}^{ol}/(X_{MgO}{}^{ol}+X_{FeO}{}^{ol})$, where a term such as $X_{MgO}{}^{ol}$ represents the mole or cation fraction of MgO, just as is calculated for a liquid (Table 1). Similarly, the anorthite (An), albite (Ab) and orthoclase (Or) contents of plagioclase, can be represented as Ca/(Ca+N+K), Na/(Ca+Na+K) and K/(Ca+Na+K) respectively, and are numerically equivalent to $X_{CaO}{}^{pl}/(X_{CaO}{}^{pl}+X_{NaO0.5}{}^{pl}+X_{KO0.5}{}^{pl})$, $X_{NaO0.5}{}^{pl}/(X_{CaO}{}^{pl}+X_{NaO0.5}{}^{pl}+X_{KO0.5}{}^{pl})$ and $X_{KO0.5}{}^{pl}/(X_{CaO}{}^{pl}+X_{NaO0.5}{}^{pl}+X_{KO0.5}{}^{pl})$, where here, all X_i are cation fractions.

Pyroxene components can be calculated in a similar manner, but normative type methods are also employed, to handle charge-couple substitutions. In addition, instead of using a fixed number of cations, the cations in clinopyroxene are usually calculated on the basis of a fixed number of oxygen's (Table 2). The advantage here is that the number of cations can then be added together to test whether the analysis is valid. For example, if clinopyroxene cations are calculated on the basis of 6 oxygens per formula unit, then the sum of all cations should be 4. If the sum is very different from 4, then either the given composition is not (at least wholly) a pyroxene, or the analysis itself is in error. Because of the complexity of clinopyroxene compositions, component calculations are handled differently in different models; for this reason clinopyroxene components are discussed in conjunction with specific thermometers and barometers below.

Units and notation. It is now customary to report P in units of GPa, instead of kbar, while T units are commonly reported in either Kelvins (K) or degrees centigrade (°C). Because this review reproduces equations from earlier publications, and because unit conversions would in some cases require recalibration of regression coefficients, with some risk of introducing additional error, P-T units from original works are largely preserved; the conversions in any case are simple (10 kbar = 1 GPa; K = °C + 273.15). Additionally, since in geobarometry most pressures of interest occur within the highly convenient magnitude range of 0-10 kbar, new geobarometers are calibrated on a kbar scale, custom be damned.

For ease of calculation, all equations are presented using the same compositional notation throughout (see Tables 1-3). Terms such as $SiO_2{}^{liq}$ represent the wt% of SiO_2 (or some other oxide) in the liquid; terms such as X_i^j are either the mole or (nearly always, here) the cation fraction of i in phase j, the choice of which will be indicated as appropriate (and will be evident from oxides of Al or Na, for example, where for a mole fraction we write $X_{Al_2O_3}$ and X_{Na_2O}, and for a cation fraction we write $X_{AlO_{1.5}}$ and $X_{NaO_{1.5}}$). Terms such as NF and $C_{NM}{}^{liq}$ are calculated from cation fractions and derive from the activity model of Beattie (1993) (see Table 1). All mole and cation fractions are calculated on an anhydrous basis, from weight percent oxides that are not renormalized to 100%, even when hydrous. Compositional terms involving water are always in units of wt% H_2O.

"Glass" (or liquid) thermometers

The Helz and Thornber (1987) thermometer is simplicity itself, depending only upon the wt% of MgO in a liquid (MgO^{liq}): $T(°C) = 20.1 MgO^{liq} + 1014$ °C. Their model was updated by Montierth et al. (1995; $T(°C) = 23.0 MgO^{liq} + 1012$ °C), and Thornber et al. (2003). In spite of the simplicity and narrow calibration range, the Helz and Thornber (1987) and Montierth et

Table 2. Pyroxene calculations example part I. Calculation of pyroxenes on the basis of six oxygens, and orthopyroxene components.

	SiO$_2$	TiO$_2$	Al$_2$O$_3$	FeO	MnO	MgO	CaO	Na$_2$O	K$_2$O	Cr$_2$O$_3$
1) Opx wt%	55.7	0.03	1.88	13.82	0	27.7	1.47	0.44	0	0
2) Mol wt	60.08	79.9	101.96	71.85	70.94	40.3	56.08	61.98	94.2	152.0
3) mole prop	0.9271	0.0004	0.0184	0.1923	0.0000	0.6873	0.0262	0.0071	0	0
4) # of oxygens	2(0.9271)	2(0.0004)	3(0.0184)	0.1923	0.00	0.6873	0.0262	0.0071	0.00	3(0.00)
5) # of oxygens	1.8542	0.0008	0.0553	0.1923	0.0000	0.6873	0.0262	0.0071	0.0000	0.0000
6) sum of row 5	2.8233									
7) ORF	6/(oxy sum) = 2.1252									

Cations on the basis of 6 oxygens

	X_{Si}^{opx}	X_{Ti}^{opx}	X_{Al}^{opx}	X_{Fe}^{opx}	X_{Mn}^{opx}	X_{Mg}^{opx}	X_{Ca}^{opx}	X_{Na}^{opx}	X_{K}^{opx}	X_{Cr}^{opx}
8) cat/6 O	1.9703	0.0008	0.0784	0.4088	0	1.4607	0.0557	0.0302	0	0
9) cation sum	4.0048									

Orthopyroxene Components

	$X_{Al(IV)}^{cpx}$	$X_{Al(VI)}^{cpx}$	$X_{NaAlSi_2O_6}^{opx}$	$X_{FmTiAlSiO_6}^{opx}$	$X_{CrAl_2SiO_6}^{opx}$	$X_{FmAl_2SiO_6}^{opx}$	$X_{CaFmSi_2O_6}^{opx}$	$X_{Fm_2Si_2O_6}^{opx}$	total
10)	0.0297	0.0486	0.0302	0.0008	0.0000	0.0185	0.0557	0.8973	1.0024

To obtain row 8), the basis for P-T calculations involving pyroxenes (opx=orthopyroxene) first divide row 1) by row 2) to obtain row 3). To get row 4), multiply row 3) by the numbers of oxygens that occur in each formula unit, as indicated; results are shown in row 5). Row 6) shows the sum of row 5), and ORF, the "Oxygen Renormalization Factor", in row 7), is equal to the number of oxygens per mineral formula unit, divided by the sum shown in 6); in our case, we will calculate pyroxenes on the basis of six oxygens, yielding ORF = 2.1252. To obtain the numbers of cations on a six oxygen basis, as in row 8), multiply each entry in row 3) by the ORF and by the number of cations in each formula unit (so for SiO$_2$, X_{Si}^{opx} = (0.9271)(2.1252), and for Al$_2$O$_3$, X_{Al}^{opx} = (2)(0.0184)(2.1252)). Row 9) shows the sum of cations per six oxygens (per formula unit). For a "good" pyroxene analysis, this value should be, as in this case, close to the ideal value of 4, thus providing a test of the quality of a pyroxene analysis. Orthopyroxene components, in 10) are calculated from the cations in 8), using the following algorithm: $X_{Al(IV)}^{opx}$ = 2 − X_{Si}^{opx}; $X_{Al(VI)}^{opx}$ = X_{Al}^{opx} − $X_{Al(IV)}^{opx}$; $X_{NaAlSi_2O_6}^{opx}$ = $X_{Al(VI)}^{opx}$ or X_{Na}^{opx}, whichever is less; $X_{FmTiAlSiO_6}^{opx}$ = X_{Ti}^{opx}; $X_{CrAl_2SiO_6}^{opx}$ = X_{Cr}^{opx}; $X_{FmAl_2SiO_6}^{opx}$ = $X_{Al(VI)}^{opx}$ − $X_{NaAlSi_2O_6}^{opx}$ − $X_{CrAl_2SiO_6}^{opx}$; $X_{CaFmSi_2O_6}^{opx}$ = X_{Ca}^{opx}; $X_{Fm_2Si_2O_6}^{opx}$ = [(X_{Fe}^{opx} + X_{Mn}^{opx} + X_{Mg}^{opx}) − $X_{FmTiAlSiO_6}^{opx}$ − $X_{FmAl_2SiO_6}^{opx}$ − $X_{CaFmSi_2O_6}^{opx}$]/2. The sum of the opx components in row 10) should be close to 1.0.

al. (1995) thermometers work remarkably well. When predicting T for 1,536 olivine-saturated (± other phases) experimental data (Fig. 2), the models yield systematic error, but very high correlation coefficients (R = 0.91 for both models), capturing 84% of experimental T variation (R^2 = 0.84). A simple linear correction to the Helz and Thornber (1987) model removes the systematic error, yielding a new model:

$$T(°C) = 26.3 MgO^{liq} + 994.4 \text{ °C} \tag{13}$$

Equation (13) yields R^2 = 83 (R = 0.91) and a standard error of estimate (SEE) of ±71 °C (Fig. 2). Additional compositional terms can be added to reduce error further, yielding the following empirical expressions, one P-independent the other P-dependent,

$$T(°C) = 754 + 190.6[Mg\#] + 25.52[MgO^{liq}] + 9.585[FeO^{liq}] \\ + 14.87[(Na_2O + K_2O)^{liq}] - 9.176[H_2O^{liq}] \tag{14}$$

$$T(°C) = 815.3 + 265.5[Mg\#^{liq}] + 15.37[MgO^{liq}] + 8.61[FeO^{liq}] \\ + 6.646[(Na_2O + K_2O)^{liq}] + 39.16[P(GPa)] - 12.83[H_2O^{liq}] \tag{15}$$

where $Mg\#^{liq}$ is a molar ratio and the remaining terms are weight percent oxides in a liquid (or glass). In Equation (14), R^2 = 0.92 and SEE = 51 °C; in Equation (15) R^2 = 0.89 and SEE = 60 °C. Equations (13)-(15) are applicable to any volcanic rock saturated with olivine and any other collection of phases, over the following compositional and P-T range: P = 0.0001-14.4 GPa; T = 729-2000 °C; SiO_2 = 31.5-73.64 wt%; Na_2O+K_2O = 0-14.3 wt%; H_2O = 0-18.6 wt% (Fig. 2).

Yang et al. (1996) suggest an improvement to this type of geothermometer by requiring that additional phases be in equilibrium with the liquid, placing further constraints on compositional degrees of freedom. Their thermometer applies to liquids in equilibrium with olivine + plagioclase + clinopyroxene; a new equation similar to their equation 4 is:

$$T(°C) = -583 + 3141[X_{SiO_2}^{liq}] + 15779[X_{Al_2O_3}^{liq}] + 1338.6[X_{MgO}^{liq}] \\ - 31440[X_{SiO_2}^{liq} \cdot X_{Al_2O_3}^{liq}] + 77.67[P(GPa)] \tag{16}$$

where terms such as X_{MgO}^{liq} represent the mole fraction of MgO in the liquid. Equation (16) performs less well for liquids that are additionally saturated with other phases, such as spinel or other oxides, but for experiments where only liquid + olivine + plagioclase + clinopyroxene are reported, Equation (16) yields R^2 = 0.92 and SEE = ±19 °C (n = 73) for the calibration data and R^2 = 0.75 and SEE = 26 °C for 119 test data (Fig. 2).

Olivine-liquid equilibria

Tests for equilibrium. When using thermometers or barometers based on equilibrium constants, it is essential that equilibrium between the phases in question is approached, otherwise calculated P-T conditions have no meaning. Roeder and Emslie (1970) produced the first experiments designed to test whether olivine of a given forsterite (Fo) content might be in equilibrium with a liquid composition. Their landmark study of Fe-Mg partitioning between olivine and liquid showed that the equilibrium constant for the following reaction,

$$MgO^{ol} + FeO^{liq} = MgO^{liq} + FeO^{ol} \tag{17}$$

otherwise known as the Fe-Mg exchange coefficient, or $K_D(Fe-Mg)^{ol-liq}$ = $[(X_{Fe}^{ol} X_{Mg}^{liq})/(X_{Mg}^{ol} X_{Fe}^{liq})]$ varies little with T or composition, and so is nearly constant at 0.30±0.03. Subsequent work has shown that $K_D(Fe-Mg)^{ol-liq}$ decreases with decreasing SiO_2 or increasing alkalis (Gee and Sack 1988), and increases with increased P (Herzberg and O'Hara 1998; Putirka 2005a) (Toplis (2005) calibrates all these effects). But the Roeder and Emslie (1970)

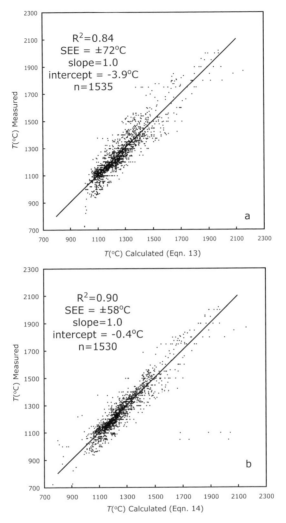

Figure 2. Calculated and measured temperatures are compared for the empirical glass geothermometers using: (a) Equation (13), (b) Equation (14), (c) Equation (15) and (d) Equation (16) from the text. In (a) – (c) all available data are used for calibration: Agee and Walker (1990), Agee and Draper (2004), Almeev et al. (2007), Arndt (1977), Baker and Eggler (1987), Baker et al. (1994), Barclay and Carmichael (2004), Bartels et al. (1991), Bender et al. (1978), Blatter and Carmichael (2001), Bulatov et al. (2002), Chen et al. (1982), Dann et al. (2001), Delano (1977), Di Carlo et al. (2006), Draper and Green (1997, 1999), Dunn and Sen (1994), Elkins and Grove (2003), Elkins et al. (2000, 2003, 2007), Falloon and Danyushevsky (2000), Falloon et al. (1997, 1999, 2001), Feig et al. (2006), Fram and Longhi (1992), Fuji and Bougault (1983), Gaetani and Grove (1998), Gee and Sack (1988), Grove and Beaty (1980), Grove and Bryan (1983), Grove et al. (1982, 1992, 1997, 2003), Grove and Juster (1989), Hesse and Grove (2003), Holbig and Grove (2008), Jurewicz et al. (1993), Juster et al. (1989), Kawamoto (1996), Kelemen et al. (1990), Kennedy et al. (1990), Kinzler and Grove (1985), Kinzler and Grove (1992a), Kinzler (1997), Kogi et al. (2005), Kogiso et al. (1998), Kogiso and Hirschmann (2001), Kushiro and Mysen (2002), Kushiro and Walter (1998), Laporte et al. (2004), Longhi (1995), Longhi (2002), Longhi and Pan (1988, 1989), Longhi et al. (1978), Maaloe (2004), Mahood and Baker (1986), McCoy and Lofgren (1999), McDade et al. (2003), Medard and Grove (2008), Medard et al. (2004), Meen (1987, 1990), Müntener et al. (2001), Mibe et al. (2005), Moore and Carmichael (1998), Morse et al. (2004), Müntener, et al. (2001), Musselwhite et al. (2006),

caption continued on facing page

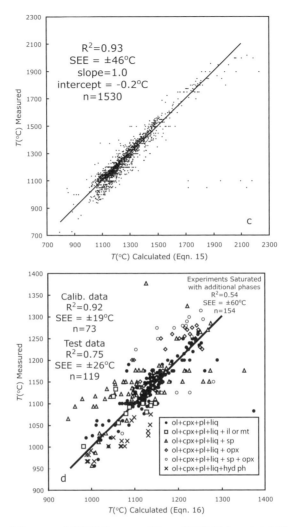

Figure 2. (continued) Parman et al. (1997), Parman and Grove (2004), Pichavant et al. (2002a,b), Pickering-Witter and Johnston (2000), Putirka (1998), Ratajeski et al. (2005), Rhodes et al. (1979b), Robinson et al. (1998), Sack et al. (1987), Salters and Longhi (1999), Scaillet and MacDonald (2003), Schwab and Johnston (2001), Scoates et al. (2006), Sisson and Grove (1993a.b), Stolper (1977), Stolper (1980), Takagi et al. (2005), Takahashi (1980), Taura et al. (1998), Thy et al. (2004), Tormey et al. (1987), Vander Auwera and Longhi (1994), Vander Auwera et al. (1998), Villiger et al. (2004), Wagner and Grove (1997, 1998), Walter (1998), Wasylenki et al. (2003), Xirouchakis et al. (2001), Yang et al. (1996). In (d) the calibration data are only the subset of experiments where liquid is in equilibrium with clinopyroxene (cpx), olivine (ol) and plagioclase (pl). In applying models with such restrictions, not only must all the relevant phases be present, but additional phases perhaps should also be absent, except for the addition of ilmenite or magnetite, Equation (16) yields poorer T estimates for liquids that in addition to cpx+liq+plag are also saturated with orthopyroxene (opx), spinel (sp) and a hydrous phase such as amphibole, or a fluid (fl). Calibration data are liquids saturated with cpx + ol + plag, and are from: Baker and Eggler (1987), Grove et al. (1992), Yang et al. (1996), Grove et al. (1997), and Feig et al. 2006). Test data are also saturated with cpx + ol + plag, and are from: Kennedy et al. (1990), Juster et al. (1989), Dunn and Sen (1994), Grove and Juster (1989), Grove and Bryan (1983), Bender et al. (1978), Baker et al. (1994), Thy et al. (2004), Medard et al. (2004), Morse et al. (2004), Pichavant et al. (2002), Mahood and Baker (1986), Meen (1990), Grove et al. (1982), Tormey et al. (1987), Almeev et al. (2007), Stolper (1980), Sisson and Grove (1993a,b), Di Carlo et al. (2006).

model, i.e., $K_D(Fe-Mg)^{ol-liq} = 0.30$, is still valid for basaltic systems generally, at $P < 2-3$ GPa (1504 experiments yield $K_D(Fe-Mg)^{ol-liq} = 0.299\pm0.053$).

J. Michael Rhodes (Dungan et al. 1978; Rhodes et al. 1979a) developed a highly elegant graphical means to apply such a test for equilibrium between olivine and a potential liquid by comparing $100 \cdot Mg\#^{liq}$ vs. $100 \cdot Mg\#^{ol}$ (where ol = olivine, and the $Mg\#^{ol}$ = Fo = $Mg^{ol}/(Mg^{ol}+Fe^{ol})$; though technically calculated on the basis of cations, it is numerically equivalent to calculate Fo from $X_{MgO}^{ol}/(X_{Mg}^{ol} + X_{FeO}^{ol})$ as in Table 1). The Rhodes's diagram (Fig. 3) simultaneously tests for equilibrium while also displaying various forms of open system behavior to explain disequilibrium.

Thermometers. Hakli and Wright (1967) were probably the first to calibrate an olivine-liquid thermometer, in their case based on Ni partitioning. Their effort was updated by Leeman and Lindstrom (1978) and subsequently expanded to other phases (e.g., Petry et al. 1997; Canil 1999; Righter 2006). Unfortunately, the numbers of experiments reporting NiO are few. Tests using 67 experimental data with NiO analyses show that the model of Arndt (1977):

$$T(°C) = \frac{10430}{\ln D_{NiO}^{ol-liq} + 4.79} - 273.15 \qquad (18)$$

yields the least systematic error, but a disappointing SEE of ±108 °C—and recalibration is of little help.

Most recent interest in olivine thermometry has revolved about the partitioning of Mg between olivine and liquid, the first such calibration of which was by Roeder and Emslie (1970). Though apparently less appreciated than their calibration of $K_D(Fe-Mg)^{ol-liq}$, their Figure 7 presents a highly elegant graphical thermometer that is still quite useful (see Putirka 2005a).

A recent review by Putirka et al. (2007) shows tests of olivine-liquid thermometers based on Mg partitioning. Of published models, that by Beattie (1993) is by far the superior. His equation, re-written here by combining terms, is:

$$T(°C) = \frac{13603 + 4.943\times10^{-7}\left(P(GPa)\times10^9 - 10^{-5}\right)}{6.26 + 2\ln D_{Mg}^{ol/liq} + 2\ln[1.5(C_{NM}^L)] + 2\ln[3(C_{SiO_2}^L)] - NF} - 273.15 \qquad (19)$$

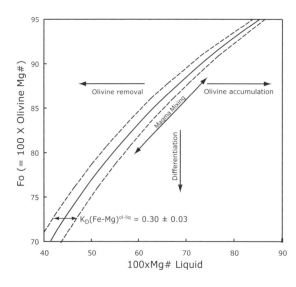

Figure 3. The Rhodes diagram yields tests of olivine-liquid equilibrium. If olivines are in equilibrium with coexisting whole rock or glass compositions, then the putative liquid-olivine pair should lie along the solid line, within some established error bound, here given as $K_D(Fe-Mg)^{ol-liq} = 0.30\pm0.03$. An additional utility of this diagram is that deviations from equilibrium can be used to explain disequilibrium, as shown by arrows. For example, if the maximum forsterite (Fo) content of a suite of olivines from a given rock were in equilibrium with the whole rock, the vertical trend would (as indicated) be consistent with (though not uniquely attributable to) closed system differentiation.

where $= X^{ol}_{MgO}/X^{liq}_{MgO}$; other components are calculated as in Table 1. Interestingly, the Beattie (1993) model works just as well when $D^{ol/liq}_{Mg}$ is calculated from the liquid composition, so like the "glass" thermometers above, Equation (19) can be used absent an olivine composition (where the thermometer retrieves the T at which a liquid would become saturated with olivine at a given P). For this calculation, Beattie's (1993) expression for $D^{ol/liq}_{Mg}$ (ignoring terms for Ni and Co content in his Table 1) must be substituted into Equation (19):

$$D^{ol/liq}_{Mg} = \frac{0.666 - \left(-0.049 X^{liq}_{MnO} + 0.027 X^{liq}_{FeO}\right)}{X^{liq}_{MgO} + 0.259 X^{liq}_{MnO} + 0.299 X^{liq}_{FeO}} \quad (20)$$

The combination of Equations (19) and (20) provide remarkably accurate predictions for olivine equilibration temperatures, and if one is interested only in anhydrous systems, at $T < 1650$ °C, no new calibrations are needed. The Beattie (1993) model yields systematic error at very high temperatures and pressures, and Herzberg and O'Hara (2002) provide a correction to eliminate such error: $T(°C) = T(°C)^{B93}_{1 bar} + 54P(GPa) - 2P(GPa)^2$, where $T(°C)^{B93}_{1 bar}$ is the temperature from Beattie (1993), as in Equation (19), but calculated for the case that $P = 1$ bar.

The Beattie (1993) thermometer also overestimates T for experiments conducted on hydrous bulk compositions. To rectify this issue, and to integrate the pressure sensitivity noted by Herzberg and O'Hara (2002) into the thermodynamic regression equations, Putirka et al. (2007) presented several new equations, the best thermometers of which are Putirka et al.'s (2007) Equations (2) and (4), as reproduced here:

$$\ln D^{ol/liq}_{Mg} = -2.158 + 55.09 \frac{P(GPa)}{T(°C)} - 6.213 \times 10^{-2} \left[H_2O^{liq}\right] + \frac{4430}{T(°C)} \quad (21)$$
$$+ 5.115 \times 10^{-2} \left[Na_2O^{liq} + K_2O^{liq}\right]$$

$$T(°C) = \{15294.6 + 1318.8 P(GPa) + 2.4834[P(GPa)]^2\}/\{8.048 \quad (22)$$
$$+ 2.8352 \ln D^{ol/liq}_{Mg} + 2.097 \ln[1.5(C^L_{NM})] + 2.575 \ln[3(C^{liq}_{SiO_2})]$$
$$- 1.41 NF + 0.222 H_2O^{liq} + 0.5 P(GPa)\}$$

In Equation (21) all compositional terms, such as Na_2O^{liq} and H_2O^{liq} are in wt%, except for $D^{ol/liq}_{Mg}$, which is a cation fraction ratio: $D^{ol/liq}_{Mg} = X^{ol}_{MgO}/X^{liq}_{MgO}$. All other components are calculated from cation fractions on an anhydrous basis, using the notation of Beattie (1993) (Table 1). Incidentally, tests conducted by Putirka et al. (2007) of the Sisson and Grove (1993b) thermometers showed that they did not perform well overall, with systematic errors for various systems, including those that are mafic or alkalic. However, their second equation: $\log_{10}[X^{ol}_{Mg}/(X^{liq}_{MgO}(X^{liq}_{SiO_2})^{0.5})] = 4129/T(K) - 2.082 + 0.0146[P(bar) - 1]/T(K)$, yields very good predictions for T for the CO_2-bearing experiments of Dasgupta et al. (2007) (regression of T^{meas} vs. T^{pred} yields $R^2 = 0.92$; SEE = 27 °C; $n = 21$), greatly outperforming models by Beattie (1993) and Putirka et al. (2007) for these special compositions. The Sisson and Grove (1993b) expression should thus be considered for peridotitic systems containing 2-25 wt% CO_2.

Generally, though, Equations (19) and (22) are by far the most precise (Fig. 4 and Putirka et al. (2007)). Equation (19) provides the best estimates for anhydrous conditions, and (22) the best estimates when water is present (H_2O ranges from 0-18.6% for the test data set). However, although these thermometers are calibrated using different experimental data, they yield similar errors using test data. This suggests that differences in their respective estimates reflect calibration error. In such a case, there is no disadvantage to using both equations and averaging the results, perhaps including the Sisson and Grove (1993b) model in the average, depending upon the extent to which the models are appropriate for the system in question.

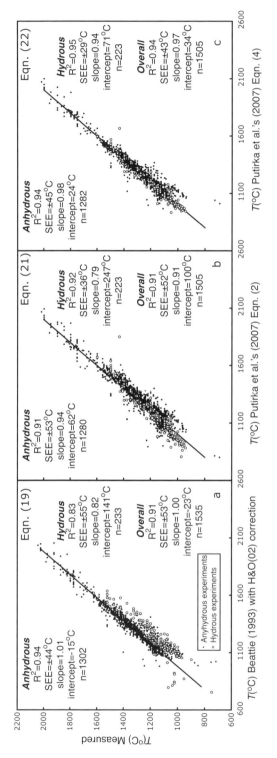

Figure 4. Tests of olivine-liquid thermometers based on the partitioning of Mg. In (a), the Beattie (1993) thermometer (Eqn. 19), using the Herzberg and O'Hara (2002) pressure correction; in (b), Putirka et al.'s (2007) Equation (2) (Eqn. 21); in (c), Putirka et al.'s (2007) Equation (4) (Eqn. 22). For anhydrous conditions, the Beattie (1993) model is best. If water is present, Putirka et al.'s (2007) Equation (4) is best. Data are as in Figures 2a-c.

Plagioclase- and alkali feldspar-liquid thermobarometers

Kudo and Weill (1970) presented the first plagioclase-liquid geothermometer. Given the common occurrence of plagioclase, it is unsurprising that plagioclase thermometry has since received much attention (e.g., Mathez 1973; Drake 1976; Loomis 1979; Glazner 1984; Ariskin and Barmina 1990; Marsh et al. 1990). Putirka's (2005b) review of plagioclase thermometers showed that the calibration of Sugawara (2001), and the MELTS/pMELTS models of Ghiorso et al. (2002) provided the most accurate predictors of T for existing models, but that these models fell short at low T (<1100 °C) and for hydrous systems. That review showed that improvements in T estimation could be obtained from a model that is simpler:

$$\frac{10^4}{T(K)} = 6.12 + 0.257 \ln\left(\frac{X^{pl}_{An}}{X^{liq}_{CaO}\left(X^{liq}_{AlO_{1.5}}\right)^2 \left(X^{liq}_{SiO_2}\right)^2}\right) - 3.166\left(X^{liq}_{CaO}\right) \quad (23)$$

$$-3.137\left(\frac{X^{liq}_{AlO_{1.5}}}{X^{liq}_{AlO_{1.5}} + X^{liq}_{SiO_2}}\right) + 1.216\left(X^{pl}_{Ab}\right)^2$$

$$-2.475\times10^{-2}\left(P(kbar)\right) + 0.2166\left(H_2O^{liq}\right)$$

Here, $X^{pl}_{An} = X^{pl}_{CaO}/\left(X^{pl}_{CaO} + X^{pl}_{NaO_{0.5}} X^{pl}_{KO_{0.5}}\right)$ and $X^{pl}_{Ab} = X^{pl}_{NaO_{0.5}}/\left(X^{pl}_{CaO} + X^{pl}_{NaO_{0.5}} X^{pl}_{KO_{0.5}}\right)$, where mineral components (all X_i^j) are calculated as cation fractions; liq = liquid and pl = plagioclase. All liquid components except H_2O^{liq} are cation fractions, calculated on an anhydrous basis without renormalization of weight percent values (Table 1); H_2O^{liq} is the wt% of H_2O in the liquid phase. Equation (23) has performed well in other tests (Blundy et al. 2006), and new tests using data from the LEPR indicate high precision (Fig. 5a), but a global regression yields a ~6 °C improvement in the SEE (Fig. 5b):

$$\frac{10^4}{T(K)} = 6.4706 + 0.3128 \ln\left(\frac{X^{pl}_{An}}{X^{liq}_{CaO}\left(X^{liq}_{AlO_{1.5}}\right)^2 \left(X^{liq}_{SiO_2}\right)^2}\right) - 8.103\left(X^{liq}_{SiO_2}\right) \quad (24a)$$

$$+4.872\left(X^{liq}_{KO_{0.5}}\right) + 1.5346\left(X^{pl}_{Ab}\right)^2 + 8.661\left(X^{liq}_{SiO_2}\right)^2$$

$$-3.341\times10^{-2}\left(P(kbar)\right) + 0.18047\left(H_2O^{liq}\right)$$

For alkali feldspars, T can be recovered from two new models based on albite-liquid equilibrium:

$$\frac{10^4}{T(K)} = 17.3 - 1.03 \ln\left(\frac{X^{afs}_{Ab}}{X^{liq}_{NaO_{0.5}} X^{liq}_{AlO_{1.5}} \left(X^{liq}_{SiO_2}\right)^3}\right) - 200\left(X^{liq}_{CaO}\right) \quad (24b)$$

$$-2.42\left(X^{liq}_{NaO_{0.5}}\right) - 29.8\left(X^{liq}_{KO_{0.5}}\right) + 13500\left(X^{liq}_{CaO} - 0.0037\right)^2$$

$$-550\left(X^{liq}_{KO_{0.5}} - 0.056\right)\left(X^{liq}_{NaO_{0.5}} - 0.089\right) - 0.078 P(kbar)$$

$$\frac{10^4}{T(K)} = 14.6 + 0.055\left(H_2O(wt\%)\right) - 0.06 P(kbar) - 99.6\left(X^{liq}_{NaO_{0.5}} X^{liq}_{AlO_{1.5}}\right) \quad (24c)$$

$$-2313\left(X^{liq}_{CaO} X^{liq}_{AlO_{1.5}}\right) + 395\left(X^{liq}_{KO_{0.5}} X^{liq}_{AlO_{1.5}}\right) - 151\left(X^{liq}_{KO_{0.5}} X^{liq}_{SiO_2}\right)$$

$$+15037\left(X^{liq}_{CaO}\right)^2$$

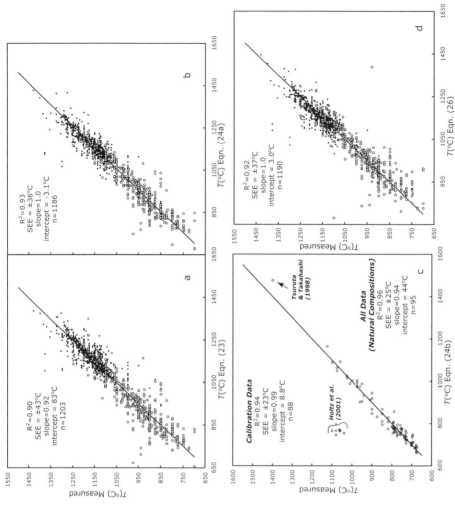

Figure 5. The plagioclase-liquid thermometer from (a) Putirka (2005b) (Eqn. 23), captures much of the variation of data from the Library of Experimental Phase Relations (LEPR); a global regression (b; Eqn. 24a) reduces error slightly; data in the subset of experiments in Figures 2a-c that are plagioclase saturated, plus data from Blundy (1997), Gerke and Kilinc (1992), Grove and Raudsepp (1978), Helz (1976), Holtz et al. (2005), Housh and Luhr (1991), Johnston (1986), Kawamoto et al. (1996), Koester et al. (2002), Martel et al. (1999), Liu et al. (1998, 2000), Minitti and Rutherford (2000), Panjasawatwong et al. (1995), Patino-Douce (1995), Patino-Douce and Beard (1995), Petermann and Hirschmann (2000), Petermann and Lundstrom (2006), Prouteau and Scaillet (2003), Rutherford et al. (1985), Scaillet and Evans (1999), Skjerlie and Patino-Douce (2002), Springer and Seck (1997), Whitaker et al. (2007). In (c), a thermometer based on alkali feldspar-liquid equilibrium (Eqn. 24b) reproduces T for experiments on natural compositions. This model was calibrated using experiments performed at $T < 1050$ °C from Gerke and Kilinc (1992), Koester et al. (2002), Patino-Douce (2005), Patino-Douce and Beard (1995), Patino-Douce and Harris (1998), Scaillet and MacDonald (2003). Test data are from these same studies, but performed at $T > 1050$ °C, with additional data from Holtz et al. (2001) and Tsuruta and Takahashi (1998). In (d), the temperature at which a liquid should become saturated with plagioclase (Eqn. 26). If this plagioclase saturation temperature is in agreement with that calculated using Equations (23) or (24a), then an approach to equilibrium is indicated,

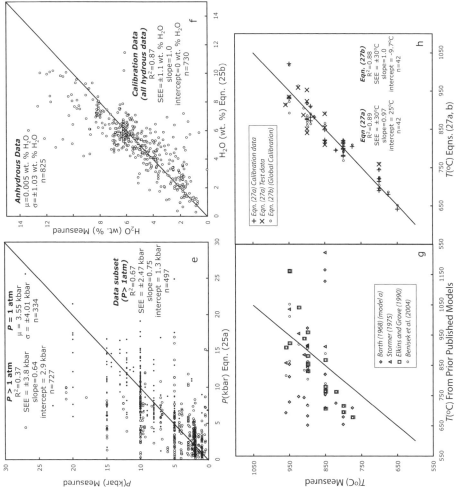

and calculated temperatures may be valid. In (e), the barometer of Putirka (2005b) yields mixed results. The model can predict P to within ±3 kbar for just under half of all data, but fails to capture P variations for the LEPR data set as a whole; the barometer appears to be work well for a subset consisting of about 2/3 of data obtained at $P > 1$ atm, from: Vander Auwera and Longhi (1994), Vander Auwera et al. (1998), Baker and Eggler (1987), Bartels et al. (1991), Bender et al. (1978), Draper and Johnston (1992), Falloon et al. (1997, 1999), Fram and Longhi (1992), Grove et al. (1992), Housh and Luhr (1991), Kinzler and Grove (1992), Meen (1987), Panjasawatwong et al. (1995), Blatter and Carmichael (2001), Feig et al. (2006), Gardner et al. (1995), Holtz et al. (2005), Martel et al. (1999), Moore and Carmichael (1998), Patino-Douce and Beard (1995), Rutherford et al. (1985), Scaillet and Evans (1999), Sisson and Grove (1993a,b), Takagi et al. (2005), Whitaker et al. (2007). In (f), a new hygrometer (based on equation (H) of Putirka (2005b)) may be useful for systems with $H_2O < 9$ wt%; 1σ model error is <±1 wt%. In (g), existing two-feldspar models are tested using experimental data not used for their calibration; newer, more complex thermodynamic models, share the same systematic error as with Barth (1962, his model a). In (h), a new two-feldspar thermometer, calibrated and tested using the simpler approach of Barth (1951, 1962) with empirical compositional corrections so as to reduce model error.

In Equation (24b), all components are calculated as for (24a); afs = alkali feldspar. The thermometer is calibrated at $T < 1050$ °C, with a calibration error of ±23 °C. The model does not perform well for experimental data for the synthetic system of Holtz et al. (2005) and efforts to include these data result in strong systematic error at $T > 1050$ °C. Although nearly all experiments are hydrous, inclusion of an H_2O^{liq} term similarly yields problems when extrapolating to $T > 1050$ °C, and incorporation of the high T data does not solve this problem. Equation (24c) yields the saturation T for alkali feldspar and is independent of an alkali feldspar composition.

As a possible test for equilibrium, the Ab-An exchange $(K_D(An-Ab)^{afs-liq} = X_{Ab}^{afs} X_{AlO_{1.5}}^{liq} X_{CaO}^{liq} / X_{An}^{afs} X_{NaO_{0.5}}^{liq} X_{SiO_2}^{liq})$ is nearly invariant with respect to T-P-H_2O, (but with a large standard deviation) yielding a mean of 0.27±0.18 for 60 experimental data. (The equilibrium constant for Or-Ab exchange is too dependent on T-P-H_2O to be useful in this regard).

Test results for the plagioclase barometer of Putirka (2005b),

$$P(\text{kbar}) = -42.2 + 4.94 \times 10^{-2} T(K) + 1.16 \times 10^{-2} T(K) \ln\left(\frac{X_{Ab}^{pl} X_{AlO_{1.5}}^{liq} X_{CaO}^{liq}}{X_{An}^{pl} X_{NaO_{0.5}}^{liq} X_{SiO_2}^{liq}}\right) \quad (25a)$$

$$-382.3\left(X_{SiO_2}^{liq}\right)^2 + 514.2\left(X_{SiO_2}^{liq}\right)^3 - 19.6\ln\left(X_{Ab}^{pl}\right) - 139.8\left(X_{CaO}^{liq}\right)$$

$$+287.2\left(X_{NaO_{0.5}}^{liq}\right) + 163.9\left(X_{KO_{0.5}}^{liq}\right)$$

are not encouraging (Fig. 5). In the original model, data from Sack et al. (1987) and Sugawara (2001) were excluded due to the potential for Na loss from liquids when using an open furnace design (see Tormey et al. 1987). However, Na loss should not be an issue for experiments conducted in sealed glass tubes, internally heated pressure vessels or in piston-cylinder apparatus. Equation (25) predicts P to within <3 kbar for new data (not used by Putirka 2005b) from Falloon et al. (1997), Scaillet and Evans (1999), Feig et al. (2006) (1-2 kbar experiments), Holtz et al. (2005), Takagi et al. (2005), Almeev et al. (2007), and Whitaker et al. (2007). However, Equation (25) yields poor results for experiments from Meen (1990), Jurewicz et al. (1993), Patino-Douce and Beard (1995), Kawamoto et al. (1996), Patino-Douce and Harris (1998), Koester et al. (2002), Pichavant et al. (2002), Skjerlie et al. (2002), Prouteau et al. (2001), Medard et al. (2004), and Villiger et al. (2004). Worse still, no global correlation appears capable of rectifying these deficiencies, nor explaining why the model can perform well for some data subsets but not for others. Indeed, with some irony, the equilibrium constant for An-Ab exchange, might be more useful as a test for equilibrium, because although it clearly varies with T, P and H_2O (e.g., Hamada and Fuji 2007), the values are normally distributed, and when divided over two temperature intervals yield $K_D(An-Ab)^{pl-liq} = X_{Ab}^{pl} X_{AlO_{1.5}}^{liq} X_{CaO}^{liq} / X_{An}^{pl} X_{NaO_{0.5}}^{liq} X_{SiO_2}^{liq} = 0.10\pm0.05$ at $T < 1050$ °C ($n = 390$; mostly hydrous), and 0.27±0.11 for $T \geq 1050$ °C ($n = 908$). New experiments, designed for the purpose of testing or developing a plagioclase-liquid barometer are needed. At best, the plagioclase barometer cannot be recommended for any other than natural compositions very similar to experimentally studied compositions where the model performs well.

While the status of plagioclase-liquid barometry is firmly in doubt, plagioclase-liquid equilibrium might be useful as a hygrometer, at least if T is well known. Equation (H) of Putirka (2005b) tends to over-predict water contents for the LEPR-derived data set, but a global calibration yields:

$$H_2O(\text{wt\%}) = 25.95 - 0.0032 T(°C) \ln\left(\frac{X_{An}^{pl}}{X_{CaO}^{liq}\left(X_{AlO_{1.5}}^{liq}\right)^2 \left(X_{SiO_2}^{liq}\right)^2}\right) \quad (25b)$$

$$-18.9\left(X_{KO_{0.5}}^{liq}\right) + 14.5\left(X_{MgO}^{liq}\right) - 40.3\left(X_{CaO}^{liq}\right) + 5.7\left(X_{An}^{pl}\right)^2 + 0.108 P(\text{kbar})$$

Equation (25b) yields water contents to within ±1.1 wt% H_2O for 730 hydrous experimental data, and recovers a mean H_2O content of 0.04±1.0 wt% for 825 anhydrous compositions not used for regression analysis (Fig. 5). Of course, Equation (25b) is something of a rearrangement of Equation (24), so as expected, estimates of water content are highly sensitive to T: just a ±38 °C error in T (~1.3σ) translates to a ±1.0 wt% error in H_2O.

As a test for plagioclase-liquid equilibrium, T from Equations (23) or (24a) can be compared to the T required for a liquid to reach plagioclase saturation. The Putirka (2005b) model predicts T to within ±48 °C for the global LEPR data set; a global regression of this data yields a 10 °C improvement in precision (Fig. 5c):

$$\frac{10^4}{T(K)} = 10.86 - 9.7654\left(X_{SiO_2}^{liq}\right) + 4.241\left(X_{CaO}^{liq}\right) - 55.56\left(X_{CaO}^{liq} X_{AlO_{1.5}}^{liq}\right) \quad (26)$$

$$+ 37.50\left(X_{KO_{0.5}}^{liq} X_{AlO_{1.5}}^{liq}\right) + 11.206\left(X_{SiO_2}^{liq}\right)^3$$

$$- 3.151 \times 10^{-2}\left(P(kbar)\right) + 0.1709\left(H_2O^{liq}\right)$$

Equation (26) yields the temperature at which plagioclase should crystallize from a silicate liquid at a given P.

Two-feldspar thermometers

Thomas W. Barth, with some fairness, could be said to be the father of modern geothermobarometry. Barth (1934) first suggested that the relative solution of albite (Ab; $NaAlSi_3O_8$) into plagioclase and alkali feldspar could be formulated as an analytic geothermometer, and he followed his own suggestion with several calibrations (Barth 1951, 1962, 1968). Stormer (1975) presented the first major revision of the two-feldspar thermometer. In Barth's (1951) original formulation, he assumed that albite solutions followed Henry's Law, i.e., that the activity of a component in a dilute solution varies linearly with composition. Stormer (1975) illustrated some "serious deficiencies" in the Barth (1962) model, at least insofar as when one desires to predict the activity of albite. Intervening years have seen the development of additional thermodynamic models, exhibiting increasing complexity (e.g., Whitney and Stormer 1977; Powell and Powell 1977; Fuhrman and Lindsley 1988; Elkins and Grove 1990; Benisek et al. 2004). These efforts, though, have not demonstrated that more complex expressions yield better thermometers. Ever since 1934, calibration efforts have been hampered by a lack of experimental data, and there are probably more published pages devoted to feldspar thermodynamic modeling than there are experimental observations; in the latest model (Benisek et al. 2004), the number of adjustable parameters (21) is more than half the total of modern experimental observations.

Here, tests and new calibrations are performed using 41 experimental data from Elkins and Grove (1990), Gerke and Kilinc (1992), Patino-Douce and Beard (1995), Patino-Douce and Harris (1998), Koester et al. (2002), Patino-Douce (2005) and Auzanneau et al. (2006). As a departure, compositions of experiments are not "adjusted" as was done by Furhman and Lindsley (1988) and Elkins and Grove (1990). In those studies, experimental input data were modified so as to minimize the residuals on parameter estimates in least squares analysis. A problem, though, is that there is no clear path by which to "adjust" the compositions of natural samples, or other experimental data that might be used for test purposes. The issue is not trivial because T estimates are highly sensitive to even small changes in X_{Ab}, X_{An} and X_{Or}. In the event, it is unclear that such adjustments are required to produce a reliable thermometer.

In spite of nearly four decades of thermodynamic modeling, our most complex efforts barely improve upon Barth's (1962) simple partitioning model (his Equation a). Indeed, Barth's (1951) original approach, though seemingly lost among more recent thermodynamic models, provides a remarkably precise (new) thermometer, when just a few empirical corrections are added:

$$\frac{10^4}{T(°C)} = 9.8 - 0.098P(\text{kbar}) - 2.46\ln\left(\frac{X_{Ab}^{afs}}{X_{Ab}^{plag}}\right) - 14.2\left(X_{Si}^{afs}\right) + 423\left(X_{Ca}^{afs}\right) \quad (27a)$$
$$-2.42\ln\left(X_{An}^{afs}\right) - 11.4\left(X_{An}^{plag}X_{Ab}^{pl}\right)$$

$$T(°C) = \frac{-442 - 3.72P(\text{kbar})}{-0.11 + 0.11\ln\left(\frac{X_{Ab}^{afs}}{X_{Ab}^{pl}}\right) - 3.27\left(X_{An}^{afs}\right) + 0.098\ln\left(X_{An}^{afs}\right) + 0.52\left(X_{An}^{plag}X_{Ab}^{pl}\right)} \quad (27b)$$

The term X_{Si}^{afs} is the cation fraction of Si in alkali feldspar, which is calculated in the same way as $X_{SiO_2}^{liq}$. Equation (27a) was calibrated from 30 of the 41 experimental data, using all data from Elkins and Grove (1990) except A″1, the sole observation from Patino-Douce (2005), all data from Patino-Douce and Harris (1998) except APD-624, samples APD570, APD571 and APD547 from Patino-Douce and Beard (1995), and sample 13-89-3 from Gerke and Kilinc (1992). Remaining data and data from Koester et al. (2002) and Auzanneau et al. (2006) were reserved for test purposes. Equation (27a) recovers T for the calibration data to ±23 °C, and ±44 °C for the test data. Equation (27b) represents a global calibration from all 41 experimental observations and recovers T to ±30 °C; these models, though much simpler than recent models, also lack the systematic error of recent efforts.

Orthopyroxene and orthopyroxene-liquid thermobarometers

Tests for equilibrium. As with olivine, the Rhodes' diagram can be used to test for equilibrium for orthopyroxene: 785 experimental data yield $K_D(\text{Fe-Mg})^{opx-liq} = 0.29\pm0.06$ (where opx = orthopyroxene). The value is independent of P or T, but decreases slightly with increased silica contents, where $K_D(\text{Fe-Mg})^{opx-liq} = 0.4805 - 0.3733X_{Si}^{liq}$ (X_{Si}^{liq} is the cation fraction of SiO_2 in the equilibrium liquid).

Calculation of orthopyroxene components. All liquid components are based on cation fractions, as illustrated in Table 1. All orthopyroxene components are based on the numbers of cations calculated on a 6-oxygen basis (Table 2). The usefulness of this procedure is that it leads easily to stoichiometric components, and the sum of such cations (ideally 4.0) yields a test for the quality of orthopyroxene analyses. Before investigating geothermobarometers, it is important to understand how certain components are partitioned into orthopyroxene. For example, how is Al charge balanced in orthopyroxene? As a Ca-Tschermak component, $CaAl_2SiO_6$? Or as $FmAl_2SiO_6$ (where Fm = Fe+Mg)? Various such components likely exist, but to write equilibria involving orthopyroxene there is some advantage to understanding which, if any, are dominant. Longhi (1976) and Thompson (1982) show how mixing trajectories can be used to make such identifications. Experimental orthopyroxenes are shown in Figure 6, with end member components calculated using Thompson's (1982) linear algebra approach and the end members: Fm_6O_6, $Ca_2Si_2O_6$, $Na_4Si_2O_6$, Ti_3O_6, Si_3O_6, Al_4O_6 and Cr_4O_6, which are projected onto the triangle Si_3O_6 - Al_4O_6 - Fm_6O_6 (Fig. 6). This projection shows that the dominant mechanism by which Al is incorporated into orthopyroxene is as $FmAl_2SiO_6$. The algorithm used to calculate orthopyroxene for thermobarometry is based on a normative scheme similar to that use for clinopyroxene (Table 2).

Thermometers and barometers. Beattie (1993) appears to present the only thermometer based on orthopyroxene-liquid equilibria. This model works well for some compositions, but over-predicts T for hydrous and low-T, nominally anhydrous compositions (Fig. 7). Two new thermometers rectify these problems:

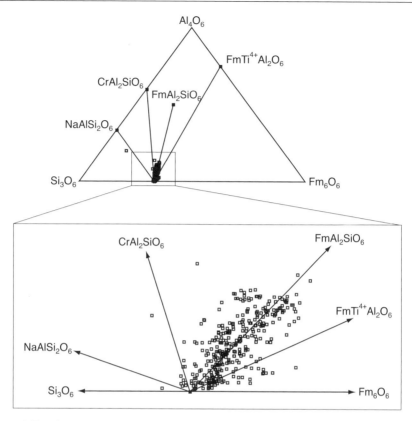

Figure 6. Experimental orthopyroxene compositions are re-cast as various oxide end members using Thompson's (1982) algebraic approach: Fm_6O_6, $Ca_2Si_2O_6$, $Na_4Si_2O_6$, Ti_3O_6, Si_3O_6, Al_4O_6 and Cr_4O_6. These end-member components are projected onto the triangle Si_3O_6 - Al_4O_6 - Fm_6O_6 (upper figure) (Fm = Fe+Mg); the data are concentrated at the base of this triangle, of which a magnified region is shown (lower figure). The projection shows that the dominant mixing vector for Al is toward the component $FmAl_2SiO_6$.

$$\frac{10^4}{T(°C)} = 4.07 - 0.329[P(GPa)] + 0.12[H_2O^{liq}]$$ (28a)

$$+ 0.567 \ln\left[\frac{X^{opx}_{Fm_2Si_2O_6}}{\left(X^{liq}_{SiO_2}\right)^2 \left(X^{liq}_{FeO} + X^{liq}_{MnO} + X^{liq}_{MgO}\right)^2}\right]$$

$$- 3.06[X^{liq}_{MgO}] - 6.17[X^{liq}_{KO_{0.5}}] + 1.89[Mg\#^{liq}] + 2.57[X^{opx}_{Fe}]$$

$$T(°C) = \frac{5573.8 + 587.9P(GPa) - 61[P(GPa)]^2}{5.3 - 0.633\ln\left(Mg\#^{liq}\right) - 3.97\left(C^L_{NM}\right) + 0.06NF + 24.7\left(X^{liq}_{CaO}\right)^2 + 0.081H_2O^{liq} + 0.156P(GPa)}$$ (28b)

For Equation (28a), all components are calculated as in Tables 1 and 2. The term $X^{opx}_{Fm_2Si_2O_6}$ is the mole fraction of $Fm_2Si_2O_6$ (or enstatite + ferrosilite, EnFs) as calculated in Table 2, where Fm = Fe + Mn + Mg; terms such as X^{opx}_{Fe} are the number of cations of the indicated element (in this case Fe) in opx, when calculated on a 6 oxygen basis (Table 2). Equation (28b) uses only liquid components to calculate the T at which a liquid should become saturated with orthopyroxene. Equation (28a) was calibrated and tested using only those experimental data

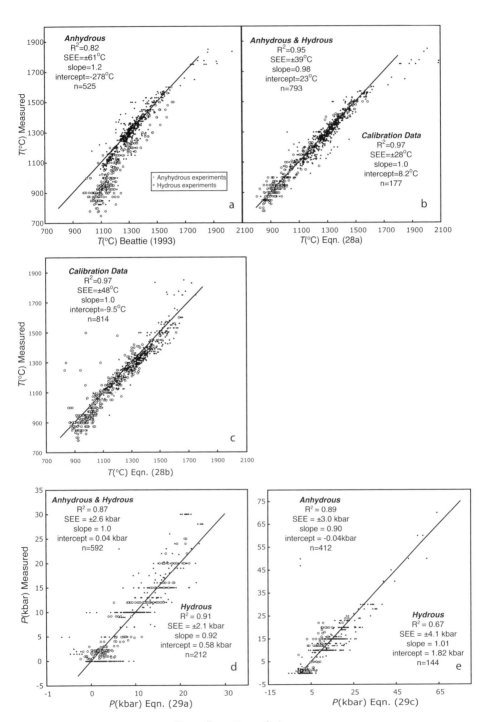

Figure 7. *caption on facing page*

whose orthopyroxenes had cations sums in the range 3.97-4.03. The expression recovers T for the calibration data to ±26 °C (slope and intercept are 0.996 and 5.17 °C; $R^2 = 0.98$; $n = 148$), and to ±41 °C for the test data (slope and intercept are 0.985 and 15.3 °C; $R^2 = 0.95$; $n = 770$). The range of applicability for (28a) is: $T = 750\text{-}1600$ °C; $P = 0.0001\text{-}11.0$ GPa; $SiO_2 = 33\text{-}77$ wt%; $H_2O = 0\text{-}14.2$ wt%. Equation (28b) has a similar range of applicability, and is the result of a global non-linear regression.

Wood (1974) was one of the first to calibrate the Al content in orthopyroxene as a barometer. His Al-in-orthopyroxene model was restricted to rocks in equilibrium with garnet. However, orthopyroxene barometry need not be restricted to such equilibria. The $FmAl_2SiO_6$ component increases slightly with increased P between 1 atm and 0.5 GPa. And as with clinopyroxene, equilibria involving a jadeite-like component $NaAlSi_2O_6$ (or Jd) shows a steep slope when plotted against P/T. Barometers calibrated using $NaAlSi_2O_6$ or $FmAl_2SiO_6$ yield errors of ca. ±2.5-3.0 GPa, and potentially useful expressions include:

$$P(\text{kbar}) = -13.97 + 0.0129 T(°C) + 0.001416 T(°C) \ln\left[\frac{X^{opx}_{NaAlSi_2O_6}}{X^{liq}_{NaO_{0.5}} X^{liq}_{AlO_{1.5}} \left(X^{liq}_{SiO_2}\right)^2}\right] \quad (29a)$$

$$-19.64(X^{liq}_{SiO_2}) + 47.49(X^{liq}_{MgO}) + 6.99(X^{opx}_{Fe})$$

$$+37.37\left(X^{opx}_{FmAl_2SiO_6}\right) + 0.748(H_2O^{liq}) + 79.67(X^{liq}_{NaO_{0.5}} + X^{liq}_{KO_{0.5}})$$

$$P(\text{kbar}) = 1.788 + 0.0375 T(°C) + 1.295\times10^{-3} T(°C) \ln\left[\frac{X^{opx}_{FmAl_2SiO_6}}{X^{liq}_{FmO} \left(X^{liq}_{AlO_{1.5}}\right)^2 X^{liq}_{SiO_2}}\right] \quad (29b)$$

$$-33.42\left(X^{liq}_{AlO_{1.5}}\right) + 9.795\left(Mg\#^{liq}\right) + 36.08\left(X^{liq}_{NaO_{0.5}} + X^{liq}_{KO_{0.5}}\right) + 0.784\left(H_2O^{liq}\right)$$

$$-26.2\left(X^{opx}_{Si}\right) + 14.21\left(X^{opx}_{Fe}\right)$$

$$P(\text{kbar}) = 2064 + 0.321 T(°C) - 343.4 \ln T(°C) + 31.52\left(X^{opx}_{Al}\right) - 12.28\left(X^{opx}_{Ca}\right) \quad (29c)$$

$$-290\left(X^{opx}_{Cr}\right) + 1.54\ln\left(X^{opx}_{Cr}\right) - 177.2\left(X^{opx}_{Al} - 0.1715\right)^2$$

$$-372\left(X^{opx}_{Al} - 0.1715\right)\left(X^{opx}_{Ca} - 0.0736\right)$$

The term X_{Al}^{opx} is the total number of Al atoms in orthopyroxene when cations are calculated on a 6 oxygen basis (equal to Al(IV) + Al(VI)). Equations (29a,b) derive from global calibrations of 592 experiments. Equation (29a) yields $R^2 = 0.83$ and SEE = ±2.6 kbar; both models underestimate P at $P > 30$ kbar. However, P for 1 atm data, even though included in the regression, yield a mean of 2.2±2.3 kbar, and their exclusion from the regression does not improve accuracy at $P > 0.001$ kbar. Remarkably, though, P is more accurately recovered for hydrous data,

---

Figure 7. (on facing page) Orthopyroxene-liquid thermometers of (a) Beattie (1993) and (b) and (c) this study are tested using orthopyroxene saturated experiments from data in Figures 2a-c, and from Barnes (1986), Keshav et al. (2004), Schmidt et al. (2004), Hammer et al. (2002), Holtz et al. (2005), Martel et al. (1999), Naney (1983), Prouteau and Scaillet (2003), Springer and Seck (1997). In (d) and (e) are tests of potential orthopyroxene barometers, based on global regressions of experimental data. For (d) the new barometer is based on Jd-liq equilibrium and was calibrated using data from Baker and Eggler (1987), Bartels et al. (1991), Elkins-Tanton et al. (2000, 2003), Falloon and Danyushevsky (2000), Falloon et al. (2001), Gaetani and Grove (1998), Grove and Juster (1989), Holbig and Grove (2007), Kinzler and Grove (1992), Kinzler (1997), Moore and Carmichael (1998), Ratajeski et al. (2005), and Wagner and Grove (1997, 1998). Equation (29c) (e) was derived from a global regression, using only orthopyroxene components as independent variables.

yielding a respectable $R^2 = 0.91$ and SEE ±2.1 kbar—perhaps reflecting more rapid orthopyroxene equilibration when H_2O is present. Because it is sometimes useful to calculate P when an equilibrium liquid is unavailable (see Nimis 1995), Equation (29c) was calibrated, using data from Longhi and Pan (1988), Grove and Juster (1989), Bartels et al. (1991), Kinzler and Grove (1992) Kinzler (1997), Wagner and Grove (1998), Walter (1998), Falloon et al. (1997, 1999), Blatter and Carmichael (2001), Gaetani and Grove (1998). The model recovers pressure for the calibration data with a $R^2 = 0.93$ and an SEE of ±2.5 kbar. Equation (29c) exhibits much less systematic error to 70 kbar for anhydrous data, but at the cost of systematic error for hydrous data; predicted pressures for a global dataset ($n = 556$) yield $R^2 = 0.84$ and an SEE of ±3.6 kbar; statistics for hydrous and anhydrous data are shown in Figure 7.

Clinopyroxene and clinopyroxene-liquid thermobarometers

Davis and Boyd (1966) introduced the first pyroxene thermometer, a formulation that requires equilibrium between clino- and orthopyroxene. The two-pyroxene approach has since received much attention, but few volcanic rocks precipitate both ortho- and clino-pyroxene phenocrysts, which limits its usefulness to volcanic systems. For this reason, thermometers and barometers were eventually developed using clinopyroxene compositions alone (Nimis 1995; Nimis and Ulmer 1998; Nimis and Taylor 2000), or clinopyroxene-liquid equilibria (Putirka et al. 1996, 2003).

Calculation of clinopyroxene components. For the models that follow, all liquid components are cation fractions, calculated on an anhydrous basis, without renormalization of wt% values (Table 1). For both the Nimis and Putirka et al. models, clinopyroxene components are based on the numbers of cations on a 6 oxygen basis. As with orthopyroxene, a "good" clinopyroxene analysis has a cation sum very close to 4. Indeed, Nimis (1995) uses cation fractions that are normalized to 4 (which can be achieved by simply multiplying cation fractions by 4). In Putirka et al. (1996, 2003, and below), cation fractions are never renormalized, and the sum is used to check for the validity of a given clinopyroxene analysis. Clinopyroxene components, such as jadeite (Jd), diopside + hedenbergite, (DiHd), and enstatite + ferrosilite (EnFs) are calculated using a normative procedure of Putirka et al. (2003); Nimis (1995) makes direct use of clinopyroxene cations calculated on a 6-oxygen basis (Table 3).

Thermometers and barometers. The highest precision and least systematic error for existing clinopyroxene barometers is Putirka et al. (1996) for anhydrous systems and Putirka et al. (2003) for those that are hydrous (Fig. 8). The Nimis (1995) model contains stronger systematic error, for both anhydrous and hydrous systems (see below). However, the Nimis (1995) model is better than Nimis (1999) (even though Nimis (1999) included corrections for thermal expansion) and Putirka et al. (1996, 2003) in recovering P for 1 atm data. For Putirka et al. (1996, 2003), except for 1 atm experiments from Grove and Juster (1989), most all other 1 atm experiments yield high P estimates, ca. 2-3 kbar (Fig. 8). And, as with other equilibria, the problem is not solved by adding 1 atm data to regression models. The problems of open furnace experiments have been discussed relative to plagioclase and are relevant here: volatile losses of Na (Tormey et al. 1987) can lead to systematically high P estimates (Putirka et al. 1996, 2003). However, although the problem is apparently obviated by the Nimis (1995) approach, which depends only upon clinopyroxene compositions, the Nimis (1995) model also yields anomalously low P estimates for high-P experiments. For example, the Putirka et al. (2003) barometer yields a mean value μ of 9.5 kbar for 184 experiments equilibrated at 10 kbar, compared to $\mu = 6.4$ kbar for Nimis (1995); for 99 experiments equilibrated at 15 kbar, Putirka et al. (2003) yields $\mu = 14.5$ kbar, compared to $\mu = 11.7$ kbar for Nimis (1995). This is not to say that the Nimis (1995) model is not useful, but rather that all silicate-based barometers have inherent error and should be used with this understanding.

To better describe hydrous samples, new barometers based on Equation (1) were calibrated using H_2O^{liq} (in wt%) as a variable:

Table 3. Pyroxene calculations example part II. Calculation of clinopyroxene components.

	SiO$_2$	TiO$_2$	Al$_2$O$_3$	FeO	MnO	MgO	CaO	Na$_2$O	K$_2$O	Cr$_2$O$_3$
1) Cpx, wt%	52.5	0.33	1.58	9.89	0.24	13.29	22.5	0.37	0	0.05
2) mol. wt	60.08	79.9	101.96	71.85	70.94	40.3	56.08	61.98	94.2	152.0
3) mol. prop.	0.8738	0.0041	0.0155	0.1376	0.0034	0.3298	0.4012	0.0060	0.0000	0.0003
4) # of O	1.7477	0.0083	0.0465	0.1376	0.0034	0.3298	0.4012	0.0060	0.0000	0.0010
5) oxy sum	2.6814									
6) ORF 6/(oxy sum) =	2.2376									

Cations on the basis of 6 oxygens

	X_{Si}^{cpx}	X_{Ti}^{cpx}	X_{Al}^{cpx}	X_{Fe}^{cpx}	X_{Mn}^{cpx}	X_{Mg}^{cpx}	X_{Ca}^{cpx}	X_{Na}^{cpx}	X_{K}^{cpx}	X_{Cr}^{cpx}
7) cat/6 Oxy	1.9553	0.0092	0.0694	0.3080	0.0076	0.7379	0.8978	0.0267	0.0000	0.0015
8) cation sum	4.0134									

Clinopyroxene Components

	$X_{Al(IV)}^{cpx}$	$X_{Al(VI)}^{cpx}$	$X_{Fe(3+)}^{cpx}$	X_{Jd}^{cpx}	X_{CaTs}^{cpx}	X_{CaTi}^{cpx}	X_{CrCaTs}^{cpx}	X_{DiHd}^{cpx}	X_{EnFs}^{cpx}	total
9)	0.0447	0.0247	0.0268	0.0247	0.0000	0.0223	0.0007	0.8747	0.0856	1.0081

To obtain row 7), the basis for calculations by Nimis (1995), first divide row 1) by row 2) to obtain row 3). To get row 4), multiply row 3) by the numbers of oxygens that occur in each formula unit. Row 5) shows the sum of row 4), and the Oxygen Renormalization Factor (ORF) in row 6), is equal to the number of oxygens per mineral formula unit, divided by the sum shown in 5); in our case, we calculate pyroxenes on the basis of six oxygens, yielding ORF = 2.2376. To obtain the numbers of cations on a six oxygen basis, as in row 7), multiply each entry in row 3) by the ORF and the number of cations in each formula unit (see Table 2). Row 8) shows the sum of cations per six oxygens (per formula unit). For a "good" pyroxene analysis, this value should be, as for orthopyroxenes, close to 4, thus providing a test of the quality of a pyroxene analysis. Clinopyroxene components in 9) are calculated from the cations in 7), using the following algorithm: $X_{Al(IV)}^{cpx} = 2 - X_{Si}^{cpx}$; $X_{Al(VI)}^{cpx} = X_{Al}^{cpx} - X_{Al(IV)}^{cpx}$; $X_{Fe(3+)}^{cpx} = X_{Na}^{cpx} + X_{Al(IV)}^{cpx} - X_{Al(VI)}^{cpx} - 2 X_{Ti}^{cpx} - X_{Cr}^{cpx}$ (Papike te al. 1974); $X_{Jd}^{cpx} = X_{Al(VI)}^{cpx}$, or X_{Na}^{cpx} whichever is less; if excess $X_{Al(VI)}^{cpx}$ remains after forming X_{Jd}^{cpx}, X_{CaTs}^{cpx} (CaAl^{IV}Al^{VI}SiO$_6$) = (CaAl^{IV}Al^{VI}SiO$_6$); if $X_{Al(IV)}^{cpx} > X_{CaTs}^{cpx} + X_{CaTi}^{cpx}$ (CaTiAl$_2$O$_6$) = [$X_{Al(IV)}^{cpx} - X_{CaTs}^{cpx}$]/2; X_{CrCaTs}^{cpx} (CaCr$_2$SiO$_6$) = X_{Cr}^{cpx}/2; X_{DiHd}^{cpx} (CaFmSi$_2$O$_6$) = $X_{Ca}^{cpx} - X_{CaTs}^{cpx} - X_{CaTi}^{cpx} - X_{CrCaTs}^{cpx}$; X_{EnFs}^{cpx} (Fm$_2Si_2O_6$) = [$X_{Fe}^{cpx} + X_{Mg}^{cpx} - X_{DiHd}^{cpx}$]/2. The sum of the cpx components in 9) should be close to 1.0.

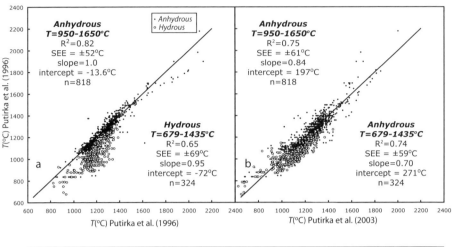
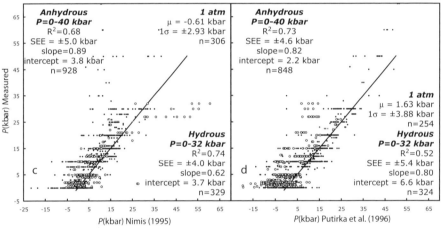
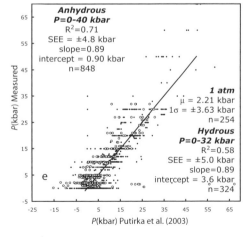

Figure 8. In (a), the clinopyroxene-liquid thermometer from Putirka et al. (1996). In (b), the clinopyroxene-liquid thermometer of Putirka et al. (2003). In (c), the clinopyroxene barometer of Nimis (1995). Also shown: clinopyroxene-liquid barometers from (d) Putirka et al. (1996) and (e) Putirka et al. (2003). Data are clinopyroxene-saturated experiments conducted at $P < 40$ kbar, from Figures 2a-c, 5a-b, and 7, plus data from Berndt et al. (2001), Hirschmann et al. (2003), Johnson (1998), Johnston (1986), Putirka et al. (1996, 2003), Takahashi et al. (1989), Tsuruta and Takahashi (1998), Ulmer and Sweeny (2002), Wood and Triglia (2001), Yasuda et al. (1994), Berndt et al. (2001), Walter and Presnall (1994).

$$P(\text{kbar}) = -48.7 + 271\frac{T(\text{K})}{10^4} + 32\frac{T(\text{K})}{10^4}\ln\left[\frac{X^{cpx}_{\text{NaAlSi}_2\text{O}_6}}{X^{liq}_{\text{NaO}_{0.5}} X^{liq}_{\text{AlO}_{1.5}} \left(X^{liq}_{\text{SiO}_2}\right)^2}\right] \quad (30)$$

$$-8.2\ln(X^{liq}_{\text{FeO}}) + 4.6\ln(X^{liq}_{\text{MgO}}) - 0.96\ln(X^{liq}_{\text{KO}_{0.5}}) - 2.2\ln\left(X^{cpx}_{\text{DiHd}}\right)$$

$$-31(Mg\#^{liq}) + 56(X^{liq}_{\text{NaO}_{0.5}} + X^{liq}_{\text{KO}_{0.5}}) + 0.76\left(\text{H}_2\text{O}^{liq}\right)$$

$$P(\text{kbar}) = -40.73 + 358\frac{T(\text{K})}{10^4} + 21.69\frac{T(\text{K})}{10^4}\ln\left[\frac{X^{cpx}_{\text{NaAlSi}_2\text{O}_6}}{X^{liq}_{\text{NaO}_{0.5}} X^{liq}_{\text{AlO}_{1.5}} \left(X^{liq}_{\text{SiO}_2}\right)^2}\right] \quad (31)$$

$$-105.7(X^{liq}_{\text{CaO}}) - 165.5(X^{liq}_{\text{NaO}_{0.5}} + X^{liq}_{\text{KO}_{0.5}})^2 - 50.15\left(X^{liq}_{\text{SiO}_2}\right)\left(X^{liq}_{\text{FeO}} + X^{liq}_{\text{MgO}}\right)$$

$$-3.178\ln\left(X^{cpx}_{\text{DiHd}}\right) - 2.205\ln(X^{cpx}_{\text{EnFs}}) + 0.864\ln(X^{cpx}_{\text{Al}}) + 0.3962\left(\text{H}_2\text{O}^{liq}\right)$$

The term X_{Al}^{cpx} is the total number of Al atoms in clinopyroxene when the formula is calculated on a 6 oxygen basis (equal to $X_{\text{Al(IV)}}^{cpx} + X_{\text{Al(VI)}}^{cpx}$). Equation (30) was calibrated using Grove and Juster (1989), Kinzler and Grove (1992), Walter and Presnall (1994), Sisson and Grove (1993a,b), Putirka et al. (1996), Scaillet and MacDonald (2003), and Patino-Douce (2005). Equation (31) was calibrated from a global regression of clinopyroxene-saturated experiments, excluding experiments performed at 1 atm (except Grove and Juster 1989) and $P > 40$ kbar (Fig. 9). A new model based on the Nimis (1995) approach was also recalibrated using experiments performed from 0.001-80 kbar:

$$P(\text{kbar}) = 3205 + 0.384T(\text{K}) - 518\ln T(\text{K}) - 5.62\left(X^{cpx}_{\text{Mg}}\right) + 83.2\left(X^{cpx}_{\text{Na}}\right) \quad (32a)$$

$$+68.2\left(X^{cpx}_{\text{DiHd}}\right) + 2.52\ln\left(X^{cpx}_{\text{Al(VI)}}\right) - 51.1\left(X^{cpx}_{\text{DiHd}}\right)^2 + 34.8\left(X^{cpx}_{\text{EnFs}}\right)^2$$

$$P(\text{kbar}) = 1458 + 0.197T(\text{K}) - 241\ln T(\text{K}) + 0.453\left(\text{H}_2\text{O}^{liq}\right) \quad (32b)$$

$$+55.5\left(X^{cpx}_{\text{Al(VI)}}\right) + 8.05\left(X^{cpx}_{\text{Fe}}\right) - 277\left(X^{cpx}_{\text{K}}\right) + 18\left(X^{cpx}_{\text{Jd}}\right) + 44.1\left(X^{cpx}_{\text{DiHd}}\right)$$

$$+2.2\ln\left(X^{cpx}_{\text{Jd}}\right) - 17.7\left(X^{cpx}_{\text{Al}}\right)^2 + 97.3\left(X^{cpx}_{\text{Fe(M2)}}\right)^2$$

$$+30.7\left(X^{cpx}_{\text{Mg(M2)}}\right)^2 - 27.6\left(X^{cpx}_{\text{DiHd}}\right)^2$$

Equation (32a) depends only upon the composition of clinopyroxene, and improves the precision of the Nimis (1995) model, but it preserves the systematic error with respect to hydrous experiments. Equation (32b) removes this systematic error, but at the cost of requiring an estimate of the H$_2$O content of the liquid in equilibrium with the clinopyroxene; it also requires use of additional calculations for the fractions of Fe and Mg occupying M1 and M2 sites, which can be found in Nimis (1995).

Equation (32c) represents a new barometer based on the partitioning of Al between clinopyroxene and liquid.

$$P(\text{kbar}) = -57.9 + 0.0475T(\text{K}) - 40.6(X^{liq}_{\text{FeO}}) - 47.7(X^{cpx}_{\text{CaTs}}) \quad (32c)$$

$$+0.676(\text{H}_2\text{O}^{liq}) - 153(X^{liq}_{\text{CaO}_{0.5}} X^{liq}_{\text{SiO}_2}) + 6.89\left(\frac{X^{cpx}_{\text{Al}}}{X^{liq}_{\text{AlO}_{1.5}}}\right)$$

Here, the last term represents the ratio of the total number of Al atoms in clinopyroxene when calculated on a 6 oxygen basis ($X_{\text{Al}}^{cpx} = X_{\text{Al(IV)}}^{cpx} + X_{\text{Al(VI)}}^{cpx}$; see Table 3), and the cation

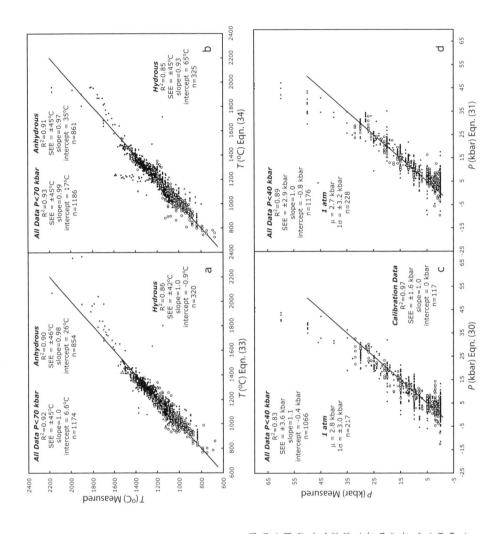

Figure 9. New clinopyroxene-liquid thermometers and barometers based on global calibrations. In (a), clinopyroxene-liquid thermometer based on Jd-DiHd exchange (Eqn. 33). In (b), clinopyroxene saturation thermometer (Eqn. 34). In (c), a barometer based on Jd-liq equilibrium, using calibration data as noted in text (Eqn. 30). In (d), a barometer as in (c) but based on a global calibration (Eqn. 31). In (e), a clinopyroxene-only thermometer of Nimis and Taylor (2000) and in (f) a new calibration for a clinopyroxene-only thermometer, using the Nimis and Taylor (2000) activity model (Eqn. 32d). In (g), a clinopyroxene-only barometer (Eqn. (32a)) and in (h) a similar calibration (32b), but incorporating an H_2O^{liq} term to rectify systematic error for hydrous data. Data are as in Figure 8.

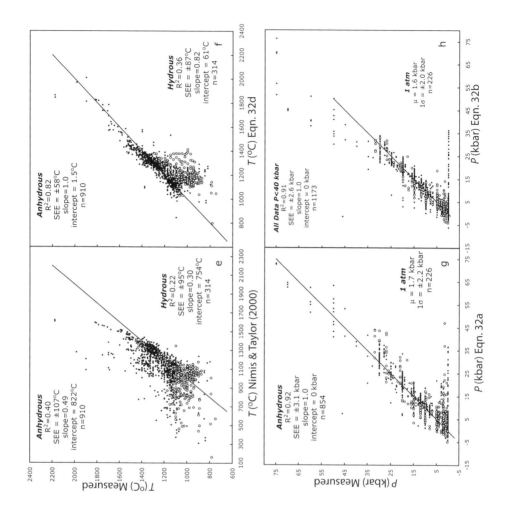

fraction of Al in liquid (see Table 1). This barometer was calibrated using data from Putirka et al. (1996), Kinzler and Grove (1992), Sisson and Grove (1993a,b) and Walter and Presnall (1994); regression statistics are: $R^2 = 0.95$, SEE = ±1.5 kbar, $n = 99$. A test using all available data ($n = 1303$) yield $R^2 = 0.82$ and SEE = ±5 kbar.

Nimis and Taylor (2000) have calibrated a clinopyroxene-only thermometer from a limited data set (Fig. 9e). Using their activity model, a new and more precise thermometer is (Fig. 9f):

$$T(K) = \frac{93100 + 544P(\text{kbar})}{61.1 + 36.6\left(X_{Ti}^{cpx}\right) + 10.9\left(X_{Fe}^{cpx}\right) - 0.95\left(X_{Al}^{cpx} + X_{Cr}^{cpx} - X_{Na}^{cpx} - X_{K}^{cpx}\right) + 0.395\left[\ln\left(a_{En}^{cpx}\right)\right]^2} \quad (32d)$$

Here, all cation fractions are on the basis of 6 oxygens (as in Eqns. 32a,b), the term X_{Al}^{cpx} refers to total Al ($X_{Al}^{cpx} = X_{Al(IV)}^{cpx} + X_{Al(VI)}^{cpx}$) and the activity of enstatite in clinopyroxene, a_{En}^{cpx}, is as in Nimis and Taylor (2000): $a_{En}^{cpx} = (1 - X_{Ca}^{cpx} - X_{Na}^{cpx} - X_{K}^{cpx}) \cdot (1 - 0.5(X_{Al}^{cpx} + X_{Cr}^{cpx} + X_{Na}^{cpx} + X_{K}^{cpx}))$.

The Jd-DiHd exchange thermometers (Putirka et al. 1996, 2003) reproduce T to ±52-60 °C. Global calibrations using experiments conducted at $P < 70$ kbar reduce these uncertainties by about 10-20 °C (Fig. 9):

$$\frac{10^4}{T(K)} = 7.53 - 0.14\ln\left(\frac{X_{Jd}^{cpx} X_{CaO}^{liq} X_{Fm}^{liq}}{X_{DiHd}^{cpx} X_{Na}^{liq} X_{Al}^{liq}}\right) + 0.07\left(H_2O^{liq}\right) - 14.9\left(X_{CaO}^{liq} X_{SiO_2}^{liq}\right) \quad (33)$$

$$-0.08\ln\left(X_{TiO_2}^{liq}\right) - 3.62\left(X_{NaO_{0.5}}^{liq} + X_{KO_{0.5}}^{liq}\right) - 1.1\left(Mg\#^{liq}\right)$$

$$-0.18\ln\left(X_{EnFs}^{cpx}\right) - 0.027P(\text{kbar})$$

$$\frac{10^4}{T(K)} = 6.39 + 0.076\left(H_2O^{liq}\right) - 5.55\left(X_{CaO}^{liq} X_{SiO_2}^{liq}\right) - 0.386\ln\left(X_{MgO}^{liq}\right) \quad (34)$$

$$-0.046P(\text{kbar}) + 2.2 \times 10^{-4}\left[P(\text{kbar})\right]^2$$

Equation (34) yields the temperature at which a liquid will become saturated with clinopyroxene at a given pressure, and so can be used as a test of temperatures derived from expressions such as Equation (33).

Tests for equilibrium. Some petrologists (e.g., Klügel and Klein 2005) test for equilibrium on the basis of Fe-Mg exchange coefficients. The danger is that Fe-Mg exchange equilibrium does not necessitate, for example, Na-Al, or Ca-Na exchange equilibrium, etc., and so Rhodes et al. (1979a) and Putirka (1999a, 2005b) suggest that equilibrium values for mineral compositions also be calculated, and compared to observed mineral compositions. Among experimental and natural data, however, deviations in observed and calculated values for $K_D(\text{Fe-Mg})^{cpx-liq}$ are highly correlated with deviations between observed and calculated values for DiHd, EnFs, etc., so tests based on Fe-Mg exchange may be sufficient. The Fe-Mg exchange coefficient derived from 1,245 experimental observations yields $K_D(\text{Fe-Mg})^{cpx-liq}$ = 0.28±0.08 (with a range from 0.04-0.68, with a roughly normal distribution). This value is not perfectly invariant, and the equation

$$\ln K_D(Fe-Mg)^{cpx-liq} = -0.107 - \frac{1719}{T(K)} \quad (35)$$

accounts for a slight temperature dependency, where $K_D(\text{Fe-Mg})^{cpx-liq}$ is recovered with a not so stunning $R^2 = 0.12$ and an SEE of 0.08.

Two-pyroxene thermometers and barometers

Sen et al. (2005) illustrate how sub-solidus equilibria can be crucial to understanding volcanic systems. Thus, since the introduction of the two-pyroxene geothermometer by Davis and Boyd (1966), much attention has been given to estimating T utilizing some form of enstatite-diopside partitioning (e.g., Wood and Banno 1973; Herzberg and Chapman 1976; Wells 1977; Lindsley 1983; Lindsley and Anderson 1983; Mercier et al. 1984; Sen 1985; Sen and Jones 1989; Brey and Köhler 1990).

As with other systems, it is important to test for equilibrium. Using Fe-Mg exchange, 311 experiments yield $K_D(\text{Fe}-\text{Mg})^{cpx-opx} = (X_{Fe}^{cpx} / X_{Mg}^{cpx}) / (X_{Fe}^{opx} / X_{Mg}^{opx}) = 1.09 \pm 0.14$. For calibration and test purposes we exclude data that lie at the outskirts of 3σ.

Of existing thermometers, the Brey and Köhler (1990) model T(BKN) is the most precise (Fig. 10a). A new global regression, based on the partitioning of enstatite + ferrosilite (= $\text{Fm}_2\text{Si}_2\text{O}_6$ = EnFs; FmO = FeO + MgO + MnO) between clinopyroxene and orthopyroxene increases precision for the available experimental data (Fig. 10b):

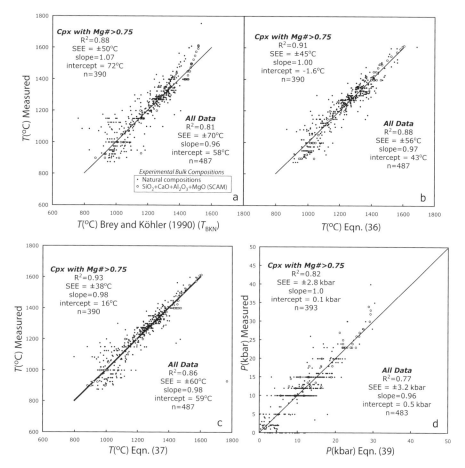

Figure 10. The Brey and Kohler (1990) two-pyroxene thermometer (a) performs reasonably well when predicting T for experimental data saturated with two pyroxenes from Figure 2a-c, plus data from Green and Ringwood (1967), Klemme and O'Neill (2000), Gudfinnsson and Presnall (1996), Milholland and Presnall (1998), and Sen (1985). A global regression (Eqn. 36) (b) provides somewhat greater precision, but precision is improved further if the calibration data base is restricted to include only Mg-rich systems, in this case defined as those cpx-opx pairs where Mg#^{cpx}>0.75. In (d), a barometer, based on the Mercier et al. (1984) approach is also calibrated using only high Mg# compositions.

$$\frac{10^4}{T(°C)} = 11.2 - 1.96 \ln\left(\frac{X_{EnFs}^{cpx}}{X_{EnFs}^{opx}}\right) - 3.3\left(X_{Ca}^{cpx}\right) - 25.8\left(X_{CrCaTs}^{cpx}\right) \tag{36}$$

$$+33.2\left(X_{Mn}^{opx}\right) - 23.6\left(X_{Na}^{opx}\right) - 2.08\left(X_{En}^{opx}\right) - 8.33\left(X_{Di}^{opx}\right) - 0.05P(kbar)$$

where the EnFs components in clino- and orthopyroxene are calculated as in Tables 2 and 3 (EnFs = Fm$_2$Si$_2$O$_6$ for orthopyroxene), and where $X_{En}^{opx} = [(X_{Mg}^{opx})/(X_{Mg}^{opx} + X_{Fe}^{opx} + X_{Mn}^{opx})](X_{Fm_2Si_2O_6}^{opx})$, and $X_{Di}^{opx} = [(X_{Mg}^{opx})/(X_{Mg}^{opx} + X_{Fe}^{opx} + X_{Mn}^{opx})](X_{CaFmSi_2O_6}^{opx})$ (with cation fractions calculated on a 6 oxygen basis). The models perform best for mafic systems where Mg#^{cpx} >0.75. A separate regression using only experiments with Mg#^{cpx}>0.75 (Fig. 10c) yields the following expression:

$$\frac{10^4}{T(°C)} = 13.4 - 3.4 \ln\left(\frac{X_{EnFs}^{cpx}}{X_{EnFs}^{opx}}\right) + 5.59 \ln\left(X_{Mg}^{cpx}\right) - 8.8\left(Mg\#^{cpx}\right) \tag{37}$$

$$+23.85\left(X_{Mn}^{opx}\right) + 6.48\left(X_{FmAl_2SiO_6}^{opx}\right) - 2.38\left(X_{Di}^{cpx}\right) - 0.044P(kbar)$$

With a similar restriction on Mg#, a new barometer (Fig. 10d) was calibrated from a global regression, utilizing the strategy of Mercier et al. (1984):

$$P(kbar) = -279.8 + 293\left(X_{Al(VI)}^{opx}\right) + 455\left(X_{Na}^{opx}\right) + 229\left(X_{Cr}^{opx}\right) \tag{38}$$

$$+519\left(X_{Fm_2Si_2O_6}^{opx}\right) - 563\left(X_{En}^{opx}\right) + 371\left(X_{Di}^{opx}\right) + 327\left(a_{En}^{opx}\right) + \frac{1.19}{K_f}$$

Equation (38) is T-independent and the term $K_f = X_{Ca}^{opx}/(1 - X_{Ca}^{cpx})$ is as in Mercier et al. (1984); $X_{En}^{opx} = (X_{Fm_2Si_2O_6}^{opx})(X_{Mg}^{opx}/[X_{Mg}^{opx} + X_{Mn}^{opx} + X_{Fe}^{opx}])$ and $X_{Di}^{opx} = (X_{CaFmSi_2O_6}^{opx})(X_{Mg}^{opx}/[X_{Mg}^{opx} + X_{Mn}^{opx} + X_{Fe}^{opx}])$; the activity of enstatite in orthopyroxene, a_{En}^{opx}, is similar to Wood and Banno (1973) (with all cations calculated on the basis of 6 oxygens):

$$a_{En}^{opx} = \left(\frac{0.5X_{Mg}^{opx}}{X_{Ca}^{opx} + 0.5X_{Mg}^{opx} + 0.5X_{Fe^{2+}}^{opx} + X_{Mn}^{opx} + X_{Na}^{opx}}\right)\left(\frac{0.5X_{Mg}^{opx}}{0.5X_{Fe^{2+}}^{opx} + X_{Fe^{3+}}^{opx} + X_{Al(VI)}^{opx} + X_{Ti}^{opx} + X_{Cr}^{opx} + 0.5X_{Mg}^{opx}}\right)$$

Here, $X_{Fe^{2+}}^{opx} = X_{Fe}^{opx} - X_{Fe^{3+}}^{opx}$, where Fe^{3+} is calculated as in Papike et al. (1974): Fe^{3+} = Al(IV) + Na − Al(VI) − Cr − 2Ti. Equation (38) recovers P to ±3.7 kbar with R^2 = 0.68 (n = 389; Fig. 11). Precision is greatly increased when T is used as input:

$$P(kbar) = -94.25 + 0.045T(°C) + 187.7\left(X_{Al(VI)}^{opx}\right) + 246.8\left(X_{Fm_2Si_2O_6}^{opx}\right) \tag{39}$$

$$-212.5\left(X_{En}^{opx}\right) + 127.5\left(a_{En}^{opx}\right) - \frac{1.66}{K_f} - 69.4\left(X_{EnFs}^{cpx}\right) - 133.9\left(a_{Di}^{cpx}\right)$$

where terms are as in Equation (38), and the activity of diopside in clinopyroxene, a_{Di}^{cpx}, is $a_{Di}^{cpx} = (X_{Ca}^{cpx})/(X_{Ca}^{cpx} + 0.5X_{Mg}^{cpx} + 0.5X_{Fe^{2+}}^{cpx} + X_{Mn}^{cpx} + X_{Na}^{cpx})$. Equation (39) recovers P to ±2.8 kbar, with R^2 = 0.82 (n = 393; Fig. 11).

The silica activity barometer

Carmichael et al. (1970) and Nicholls et al. (1971) were the first to show that because the activity of SiO$_2$ ($a_{SiO_2}^{liq}$) may be buffered at constant P and T by reactions such as

$$Mg_2SiO_4^{ol} + SiO_2^{liq} = Mg_2Si_2O_6^{opx} \tag{40}$$

that $a_{SiO_2}^{liq}$ could be used as a thermometer or barometer for igneous processes. Because most if not all basalts equilibrate with olivine and orthopyroxene in their mantle source regions, there has been much interest in using $a_{SiO_2}^{liq}$ as a barometer for mantle partial melting (though nothing precludes application to other olivine + orthopyroxene saturated liquids). For example, the

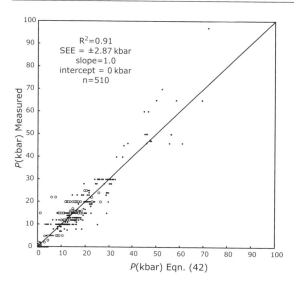

Figure 11. A silica activity barometer, using the activity for SiO$_2$ in silicate liquids devised by Beattie (1993). The data set is restricted to those liquids that are in equilibrium with both olivine and orthopyroxene (as in Nicholls et al. 1971), which include data from Figure 6, and data from Falloon et al. (1988), and Kushiro (1996).

work of Nicholls et al. (1971) is implicit in the SiO$_2$-based barometers of Klein and Langmuir (1987), Albarede (1992) and Haase (1996). Langmuir and Hanson (1980) later showed how FeO can be used as a barometer (supported by experiments by Stolper 1980), and Putirka (1999b) used Na/Ti ratios to estimate partial melting depths. But both FeO and Na/Ti are sufficiently sensitive to variations in source region composition and degree of partial melting that they do not readily lend themselves to a simple analytic expression.

The quantitative models of Albarede (1992) and Haase (1996) were calibrated using small data sets, and use wt% SiO$_2^{liq}$ as input, not $a_{SiO_2}^{liq}$; these models capture <35% of the variation in P for the larger experimental data set studied here. However, Beattie's (1993) $a_{SiO_2}^{liq}$ is highly correlated with P:

$$a_{SiO_2}^{liq} = \left(3X_{SiO_2}^{liq}\right)^{-2}\left(1-X_{AlO_{1.5}}^{liq}\right)^{7/2}\left(1-X_{TiO_2}^{liq}\right)^{7} \qquad (41)$$

When calculated in this way, $a_{SiO_2}^{liq}$ alone describes 80% of the variation in P for experimental data when Mg#^{liq} >0.75, and for 510 experiments with no restrictions on Mg#, $a_{SiO_2}^{liq}$ captures >60% of variation (over the P-T range 0.001–70 kbar and 825-2000 °C; SiO$_2$ ranges from 31.5-70%; data include hydrous compositions). Using the Beattie (1993) $a_{SiO_2}^{liq}$ a new model loosely based on Equations (5) and (41) is:

$$P(kbar) = 231.5 + 0.186T(°C) + 0.1244T(°C)\ln\left(a_{SiO_2}^{liq}\right) - 528.5\left(a_{SiO_2}^{liq}\right)^{1/2}$$
$$+103.3\left(X_{TiO_2}^{liq}\right) + 69.9\left(X_{NaO_{0.5}}^{liq} + X_{KO_{0.5}}^{liq}\right) + 77.3\left(\frac{X_{AlO_{1.5}}^{liq}}{X_{AlO_{1.5}}^{liq} + X_{SiO_2}^{liq}}\right) \qquad (42)$$

where all compositional terms are cation fractions (Table 1). Equation (42) was calibrated using a "global" database of partial melting experiments (iteratively removing data outside ±3σ) where liquids are in equilibrium with olivine and orthopyroxene (± other phases). Interestingly, calibrations on smaller data sets yield little improvement in the SEE for P, and fared poorly when predicting P for data not used in the calibration. Equation (42) recovers P to ±2.9 kbar, with R^2 = 0.91 (Fig. 11).

APPLICATIONS

Magma transport at Hawaii

Thermobarometers can be of immense use for understanding the mechanical controls of magma transport. For example, although it is reasonably well understood that density contrasts, such as at the Moho, may control magma transport (Stolper and Walker 1980), other transport models emphasize the roles of stress states within the lithosphere (ten-Brink and Brocher 1987; Parsons and Thompson 1993) or fracture toughness (Putirka 1997). These views yield contrasting predictions regarding where magmas are stored prior to eruption. Presuming that magmas undergo an episode of cooling and crystallization during storage, thermobarometers provide a test.

To illustrate, Putirka (1997) posited that magmas are stored over a wide depth range at Hawaii, and that the Moho does not provide a necessary barrier to magma transport. New data from the Hawaii Scientific Drilling project, and a re-analysis of some existing data, modify this view. Table 4 shows P-T calculations for a Pu'u O'o episode-3 flow (Garcia et al. 1992) using several new and existing models. The thermobarometers are in agreement yielding mean P-T estimates of 4.8±0.8 kbar and 1469±16 °C; standard deviations are well within model error. For Pu'u O'o episodes 1-10, depth-time estimates are shown for individual crystals whose mineral-whole rock pairs yield $K_D(\text{Fe-Mg})^{cpx-liq}$ within ±0.08 of values predicted using (35) (Fig. 12a). As in Putirka (1997), crystallization depths become shallower with time. These magmas appear to have been stored simultaneously over a range of depths, with a top-to-bottom emptying of a rather lengthy and semi-continuous and connected conduit.

However, existing data from Mauna Kea (Frey et al. 1990, 1991) and 291 new clinopyroxene-whole rock pairs from the HSDP-2 (Rhodes and Vollinger 2004; new mineral compositions from this study) alter the story line. Of these data, 120 clinopyroxenes yield $K_D(\text{Fe-Mg})^{cpx-liq}$ within ±0.08 of values determined from Equation (35). Seventy-five (73%) fall within the depth range for "high density cumulates" determined from seismic studies (Hill and Zucca 1987). This clustering, and a trend to lower T just above the Moho, was not evident in the smaller data set examined by Putirka (1997). Moreover, P-T estimates from ultramafic xenoliths (Fodor and Galar 1997; Eqns. 38, 39) (Table 5) also exhibit clustering near the Moho, and a trend towards higher (more volcanic-like) T. The volcanic phenocryst, seismic, and xenolith data thus all support the conclusions of Fodor and Galar (1997, p.1) that ultramafic xenoliths at Mauna Kea are "gravity settled and *in situ* cumulates from reservoir bottoms." However, both the volcanic phenocryst and xenolith suites also yield depth ranges as great as 35 km. A hybrid model is supported, involving a magma mush column (e.g., Ryan 1988; Marsh 1995) that extends below the Moho, possibly with the mechanical controls advocated by Putirka (1997), with transport inhibited by density contrasts near the Moho (Garcia et al. 1995; 8-14 km).

The conditions of mantle partial melting

The glass and olivine-liquid thermometers, and the silica activity barometer are calibrated to 60-70 kbar, and can be applied to understand mantle melting. For example, Morgan (1971) suggested that volcanism at Hawaii or Yellowstone is driven by plumes—thermal upwellings, which rise though the mantle. Like Hess (1962), Morgan (1971) argued that heat from the core drives mantle convection.

The mantle plume model thus predicts that "hot spots" should be hotter than ambient mantle (which feeds mid-ocean ridge basalt (MORB) volcanism). In other words, they should exhibit an "excess temperature," reflecting their derivation from a source with excess heat. Recent skepticism of the mantle plume model (e.g., Foulger and Natland 2003) has prompted efforts to estimate mantle potential temperatures (T_p) at ocean islands, to test whether hot spots are truly hot (e.g., Putirka 2005a; Putirka et al. 2007; Herzberg et al. 2007; Putirka 2008). T_p represents the T the mantle would have if it rose to the surface, adiabatically,

Table 4. Example calculations for clinopyroxene thermobarometers.

	SiO$_2$	TiO$_2$	Al$_2$O$_3$	FeO	MnO	MgO	CaO	Na$_2$O	K$_2$O	Cr$_2$O$_3$
Whl-rk (wt%)[1]	50.2	2.83	13.74	11.07	0.17	6.6	10.48	2.49	0.57	0.202
Whl-Rk (cat fr)	0.4795	0.0203	0.1547	0.0884	0.0014	0.0940	0.1072	0.0461	0.0069	0.0015
Cpx (wt%)	52.9	0.7	2.1	7.6	0	17.9	18.7	0.27	0	0.2
Cat per 6 O[2]	1.9337	0.0192	0.0905	0.2323	0.0000	0.9755	0.7323	0.0191	0.0000	0.0058
	$X_{Al^{IV}}^{cpx-6ox}$	$X_{Al^{VI}}^{cpx-6ox}$	Fe^{3+}	K$_D$(Fe-Mg)$^{cpx-liq}$	X_{Jd}^{cpx}	X_{CaTs}^{cpx}	X_{CaTi}^{cpx}	X_{CrCaTs}^{cpx}	X_{DiHd}^{cpx}	X_{EnFs}^{cpx}
	0.0663	0.0242	0.0169	0.253	0.0191	0.0051	0.0306	0.0029	0.6938	0.2570

Putirka et al. (1996)

Eqn. T2	Eqn. P1				
T(°C)	P(kbar)[3]				
1194	5.3				

Putirka et al. (2003)

T(°C)	P(kbar)				
1214	5.1				

New Models (this study)

Eqn. 30	Eqn. 31	Eqn. 32a	Eqn. 32b		
P(kbar)	P(kbar)	P(kbar)	P(kbar)		
4.9	6.3	4.0	4.1		

Eqn. 32d	Eqn. 33	Eqn. 34	Eqn. (35)		
T(°C)	T(°C)	T(°C)	K$_D$(Fe-Mg)$^{cpx-liq}$		
1199	1198	1198	0.283		

Mean estimates

	P(kbar)	T(°C)
Mean	4.95	1201
std dev	0.85	7.7

[1] "Whl-Rk" is a whole rock composition, for sample 3-88, Garcia et al. (1992).
[2] Clinopyroxene (cpx) cations (on the basis of 6 oxygens) are calculated as in Table 2.
[3] P calculated from Putirka et al. (1996) solved iteratively using Equation (T2) of Putirka et al. (1996); all other T-sensitive P estimates use T(K) from Putirka et al. (2003).

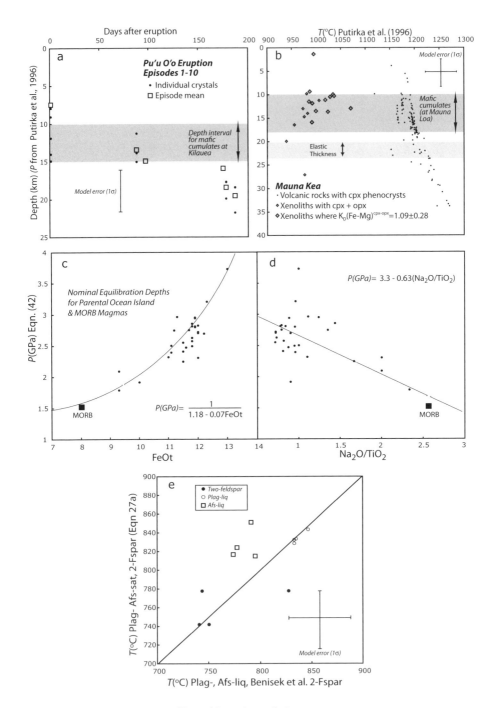

Figure 12. *caption on facing page*

without melting, and so provides a convenient reference for inter-volcano comparisons (see McKenzie and Bickle 1988). Table 6 shows calculations using parental liquid compositions for MORB and Hawaii (from Putirka 2008). For these calculations, Equations (22) and (42) are solved simultaneously to derive P and T, where T represents the temperature of olivine-liquid equilibration and P the pressure of olivine+orthopyroxene+liquid equilibration. Additional T estimates (Table 6) are derived using this P (from Eqns. 22, 42) as input. Mantle equilibration temperatures are converted to T_p by correcting for the heat of fusion, and adiabatic upwelling (Table 6). The thermometers yield highly consistent results.

The difference between T_p at Hawaii and T_p at MORB is the excess temperature, T_{ex}, which reflects some fraction of the excess heat derived from the plume source. The T_{ex} at Hawaii is ~260 °C, and the nominal depth of equilibration there is more than 50% greater than beneath mid-ocean ridges (Table 6). The calculated pressures of equilibration are called "nominal" because the reconstructed "parental liquids" used as input (Putirka 2008) are likely to represent some averaging of a range of melt compositions that exist within a melting column of considerable depth extent (Klein and Langmuir 1987). For 27 other ocean islands (Putirka 2008), average T_{ex} = 160±52 °C and average equilibration pressures and depths are 2.6±0.4 GPa and 85±11 km.

Computed inter-island variations in P may be real. Pressures calculated from $a_{SiO_2}^{liq}$ are correlated with other P-sensitive components. For example, a positive correlation between P (based on $a_{SiO_2}^{liq}$) and FeOt is expected (Fig. 12c) because increased P means that greater amounts of olivine (and FeOt) enter the liquid (Langmuir and Hanson 1980; Stolper 1980; Klein and Langmuir 1987). And a negative correlation is expected for Na_2O/TiO_2 (Fig. 12d), because the distribution coefficient for Na in clinopyroxene increases relative to that for Ti, with increased P (Putirka 1999b).

Temperature estimates in felsic volcanic systems

Unfortunately, few published studies on felsic volcanic rocks report both alkali feldspar, plagioclase feldspar and glass compositions. But such compositions from the 75 ka Toba tuffs (Beddoe-Stephens et al. 1983) indicate how some of the models presented here might be used to better understand the thermal record of such systems.

Figure 12. (on facing page) In (a) and (b) Hawaiian rock and mineral compositions are used to illustrate the use of several of the thermometers and barometers given in this study. Whole-rock-cpx pairs from the Pu'u O'o eruption of Kilauea (Garcia et al. 1989, 1992), are used in (a) to compare depth and time; the relationship indicates a top-to-bottom emptying of a continuous conduit that extends to >20 km below the Kilauea summit. In (b), depth (d)-T estimates at Mauna Kea from xenoliths (Fodor and Galar 1997) are compared to d-T estimates from volcanic rocks with clinopyroxene (cpx) phenocrysts (Frey et al. 1990, 1991; Yang et al. 1996; Rhodes and Vollinger 2004 with mineral compositions from this study). The volcanic and xenolith suites both indicate transport of magmas from as great at 35 km, but also storage and cooling of magma near the base of the volcano, in the depth range of "mafic cumulates" (Hill and Zucca 1987). In (c), depth estimates are calculated using calculated parental magma compositions for MORB and 27 ocean islands from Putirka (2008), and are compared to FeOt and Na_2O/TiO_2 contents for these magmas. There is some circularity for (c) because FeOt contents are used to estimate P. But the cross correlations are as expected if variations in SiO_2-FeOt-Na_2O/TiO_2 reflect partial melting depths, driven by differences in partial melting temperatures (see Klein and Langmuir 1987 and Putirka 1999b). In (e), temperatures are calculated for Toba Tuff ignimbrites (Beddoe-Stephens et al. 1983) using the two feldspar (2-Fspar) and plagioclase- and alkali feldspar-liquid equilibria. Glass compositions are melt inclusions in plagioclase and alkali feldspar (Beddoe-Stephens et al. 1983). The match between the two plagioclase models indicates that plagioclase most closely approaches equilibrium with included glass. Alkali feldspar T estimates are within 2σ of one another, but only one sample yields temperatures within 1σ. Though the plagioclase (plag) and alkali feldspar (afs) temperatures are within model error, plagioclase may be closer to equilibrium, and perhaps precipitated at a higher T than alkali feldspar, and temperatures derived from two-feldspar thermometry may be too low.

Table 5. Examples of two-pyroxene P-T calculations.

	SiO$_2$	TiO$_2$	Al$_2$O$_3$	FeO	MnO	MgO	CaO	Na$_2$O	K$_2$O	Cr$_2$O$_3$
Cpx (wt%)	52.3	0.7	3	5.1	0.11	16.6	21.5	0.33	0	0.58
Cat per 6 O[1]	1.9122	0.0192	0.1293	0.1559	0.0034	0.9048	0.8422	0.0234	0	0.0168
	$X_{Al^{IV}}^{cpx-6ox}$	$X_{Al^{VI}}^{cpx-6ox}$	Fe^{3+}	X_{Jd}^{cpx}	X_{CaTs}^{cpx}	X_{CaTi}^{cpx}	X_{CrCaTs}^{cpx}	X_{DiHd}^{cpx}	X_{EnFs}^{cpx}	
	0.0878	0.0414	0.0146	0.0234	0.0180	0.0349	0.0084	0.7809	0.1399	
	X_{En}^{cpx}	X_{Di}^{cpx}	a_{Di}^{cpx}							
	0.1189	0.6640	0.6050							

	SiO$_2$	TiO$_2$	Al$_2$O$_3$	FeO	MnO	MgO	CaO	Na$_2$O	K$_2$O	Cr$_2$O$_3$
Opx (wt%)	55	0.34	1.5	11.3	0.24	30.7	0.9	0.01	0	0.19
Cat per 6 O[1]	1.944	0.0090	0.0625	0.3340	0.0072	1.617	0.0341	0.0007	0	0.0053
	$X_{Al^{IV}}^{cpx-6ox}$	$X_{Al^{VI}}^{cpx-6ox}$	Fe^{3+}	X_{Jd}^{opx}	$X_{FmTiAlSiO_6}^{opx}$	X_{CrCaTs}^{opx}	$X_{FmAl_2SiO_6}^{opx}$	X_{DiHd}^{opx}	X_{EnFs}^{opx}	
	0.0563	0.00614	0.0275	0.0007	0.0090	0.0053	0.0001	0.0341	0.9576	
	X_{En}^{opx}	X_{Di}^{opx}	a_{En}^{opx}	a_{Di}^{opx}						
	0.7908	0.0281	0.6451	0.0340						

| | \multicolumn{5}{c}{New Models (this study)[2]} |
K$_D$(Fe-Mg)$^{cpx-opx}$	Eqn. 38 P(kbar)	Eqn. 39/36 P(kbar)	Eqn. 39/36 T(°C)	Eqn. 36/39 T(°C)	Eqn. 37/38 T(°C)
0.835	2.2	3.4	985	964	

[1] Clinopyroxene (cpx)-orthopyroxene (opx) pair is from sample C2 of Fodor and Galar (1997) with compositions shown in wt% and as cations on the basis of 6 oxygens.
[2] Equations (39) and (36) are solved simultaneously; Equation (38) is T-independent and Equation (36) uses 3.4 kbar as input. All clinopyroxene and orthopyroxene calculations are as in Tables 2-4 and as noted in the text.

Table 6. Examples of liquid, olivine-liquid, and mantle potential temperature calculations.

	SiO$_2$	TiO$_2$	Al$_2$O$_3$	FeO	MnO	MgO	CaO	Na$_2$O	K$_2$O	H$_2$O
MORB[1]	48.5	0.9	15.9	8	0.1	13.2	11.1	2.3	0	0.11
Hawaii[1]	47.2	1.6	9	11.6	0.2	21.4	7.3	1.4	0.2	0.65
	D_{Mg}^{ol-liq} (Eqn. 20)	Eqn. 13	Eqn. 14	Eqn. 15	Eqn. 16	Eqn. 19	Eqn. 19 w/H&O[3]	Eqn. 22	Mean T(°C)	Std. dev. T(°C)
MORB[2]	3.39	1342	1343	1359	1375	1390	1381	1374	1366	19
Hawaii[2]	2.12	1557	1575	1547	1513	1614	1579	1557	1563	31
			Eqn. 42 P(kbar)	Depth[5] (km)	T_p^{corr}	T_p	T_{ex}			
MORB[4]			15.6	52.5	33	1399	-			
Hawaii[4]			27.3	88.8	97	1660	261			

[1] Mid-ocean ridge (MORB) and Hawaiian compositions are "parental" magmas, as calculated in Putirka (2008).
[2] All temperatures are calculated in °C.
[3] "Equation (19) w/H&O" uses Equation (19) and the correction for pressure from Herzberg and O'Hara (2002); see text.
[4] T_p is mantle potential temperature (see text); T_p^{corr} is the correction required to convert an olivine equilibration T (i.e., Eqn. 22) to T_p and is given by: $T_p = T^{ol-liq} + 667F - 1.33P$(kbar), where F (melt fraction) is 0.08 for MORB and 0.20 at Hawaii (see Putirka et al. 2007) and Equation (22) is used as input.
[5] P from Equation (42) is converted to depth (d) using a regression fit based on PREM (Anderson, 1989): d(km)=14.6 + 29P(GPa) – 0.157[P(GPa) – 8.4]^2.

T estimates at Toba based on plagioclase-liquid equilibrium (Eqn. 23) and plagioclase-saturation (Eqn. 26) are nearly identical (Fig. 12e) (Table 7). Although this might reflect the higher precision of the plagioclase-based models, it might also indicate that plagioclase-hosted melt inclusions have more closely approached equilibrium with their host minerals, since all but one alkali feldspar-glass inclusion pair yield mineral-melt and mineral saturation temperatures that are >1σ apart. The one alkali feldspar-glass pair that most closely approaches equilibrium has a *T* estimate slightly less than that for plagioclase, perhaps indicating that plagioclase precipitated at a slightly higher *T* than alkali feldspar. If this is the case, temperatures from two-feldspar thermometry might not be valid, since if alkali-feldspars are not in equilibrium with their included glass, they would seem unlikely to be in equilibrium with neighboring plagioclase feldspars. In that case, *T* estimates from the two-feldspar thermometers may be too low by 40-70 °C. A greater number of samples are clearly needed to interpret these data with confidence, but the Toba examples illustrate how multiple thermometers can be used to better understand the thermal history of silicic systems.

SOURCES OF ERROR RELATIVE TO *P* AND *T*

Geothermometry and geobarometry involve two sources of error. 1) *Model error* measures the precision of a model, the magnitude of which can be identified from regression statistics. This error results from experimental error, and includes the error in reproducing *P* or *T* in the laboratory, lack of attainment of equilibrium during an experiment, and errors related to compositional measurements; the latter two are likely the most important (Putirka et al. 1996). 2) The *error of treatment* is related to the application of a model, and will trend towards greater values as natural systems deviate from experimental conditions; the most serious issue here is probably the pairing of a given mineral with its equilibrium liquid—less an issue for experiments, but problematic for natural magmas when they behave as open systems. It is also important that natural magmas and minerals have compositions and equilibration conditions similar to those studied experimentally. The best guard against error is to compare *P-T* conditions from different equilibria, which need not match precisely (olivine may precipitate before clinopyroxene) but should be correlated.

Model error

Estimates for model error can be derived from regression statistics. Usually, a "standard error of estimate" or SEE (equivalent to the root mean square error, RMSE), is reported for *T* or *P*. Like a standard deviation (1σ) in a normal distribution, the SEE (or RMSE) represents the interval ±E such that there is a 68.26% probability that the true value lies between −E and +E. If the calibration data are small in number and of high quality, the SEE will probably underestimate true model error, since by the nature of regression, coefficients are optimized for a highly specific set of circumstances, which natural samples are unlikely to mimic perfectly. If the calibration data derive from a single laboratory, it is also possible that there is systematic error related to experimental design. A truer test of model error can be derived from "test data," i.e., experimental data not used for regression.

How does one decide which experimental data should be used for test or calibration purposes? Hirschmann et al. (2008) note that many experiments are conducted at too short a time scale to allow diffusional exchange between minerals and liquid. Long-duration experiments, however, provide no assurances of equilibrium since thermal gradients in such can induce compositional gradients (Lesher and Walker 1988). In any case, errors on estimates of *P* (Eqn. 30) and *T* (Eqn. 33) are nearly independent of experimental run duration, as are values for $K_D(Fe-Mg)^{ol-liq}$, an often-used test of equilibrium (Fig. 13).

Estimates of error. Various thermometers tend to yield comparable calibration errors, on the order of ±25-30 °C, when more than a few hundred experiments are used for regression.

Table 7. Examples of feldspar calculations.

	SiO$_2$	TiO$_2$	Al$_2$O$_3$	FeO	MnO	MgO	CaO	Na$_2$O	K$_2$O	H$_2$O
Toba, R301 P [1]	73.78	0.08	12.56	1	0.04	0.02	0.89	3.01	4.47	4.0
	An	Ab	Or							
Plagioclase [1]	0.372	0.586	0.042							
Alkali Feldspar [1]	0.012	0.276	0.712							

Two Feldspar Temperatures

	Eqn 27a $T(°C)$	Eqn 27b $T(°C)$	B(2004) $T(°C)$ [2]
	743	818	778

Feldspar-liquid P-T and H$_2$O Calculations

	Eqn. 23 $T(°C)$	Eqn. 24a $T(°C)$	Eqn. 24b $T(°C)$	Eqn. 24c $T(°C)$	Eqn. 26 $T(°C)$	Eqn. 25 P(kbar)	Eqn. 25b/23 H$_2$O [3]	Eqn. 25b/24b H$_2$O [3]
	846	815	794	815	843	4.03	3.2	4.05

[1] Glass and feldspar data are from Beddoe-Stephens et al. (1983).
[2] Two feldspar thermometer of Benisek et al. (2004).
[3] H$_2$O calculations are in wt%, calculated at a pressure of 1 kbar, and at a T calculated from Equations (23) and (24b); when calculated using a pressure of 4 kbar, wt% values for H$_2$O are 0.5% less.

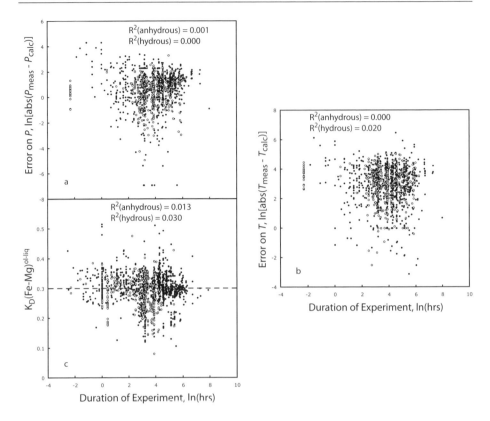

Figure 13. In (a) and (b) model errors are compared to experimental "run" duration, for P (Eqn. 30) and T (Eqn. 33) respectively. Model error is calculated as the natural log of the absolute value of the difference between predicted and experimentally reported P or T. In (c), the Fe-Mg exchange coefficient between olivine and liquid, $K_D(\text{Fe-Mg})^{ol\text{-}liq}$, is compared to experimental run duration. R^2 values are shown for hydrous (open symbols) and anhydrous (closed symbols) data. Except for some anhydrous experiments conducted at less than a few hours, increased experimental run durations do not necessarily yield improvement in model precision, or apparently even a closer approach to Fe-Mg exchange equilibrium.

Thus ±25-30 °C probably represents the smallest magnitude for model error that one can expect for any individual T estimate derived from a generalized silicate-based thermometer (regardless of whether smaller errors are reported). Similarly, the highest precision to be expected from silicate-based barometers is in the range ±1.5-2.5 kbar.

As noted in prior sections, there is hope that if experimental error is random, then more precise estimates can be derived by averaging individual P-T estimates. Because many experiments are conducted at 8, 10, 12 kbar, etc., it is easy to test this idea by calculating P for a number of experiments performed at a given pressure (say 10 kbar), then compare the mean of those estimates, and the standard error on that mean, to the measured value. Such calculations indicate that mineral-melt barometers may be as precise as ±0.5-1.0 kbar (a "standard error on the mean") when estimates are averaged for isobaric systems (e.g., Putirka et al. 1996, 2003).

For thermometers the increase in precision is not so clear. For 35 experiments on dry basalts conducted at 1200 °C, Equation (19) (Beattie 1990) yields a mean of 1225±52 °C (the ±52 °C is the standard deviation) with a standard error on the mean of ±9 °C; the mean value thus deviates from the measured value by more than twice the standard error on the mean and so the latter does not reflect greater accuracy. For the same data, Equation (22) yields the same standard error on the mean (±9 °C) with a mean value of 1213±50 °C; the mean estimate is still more than one standard error from the measured value. Although these standard errors on the means do not represent increased accuracy, the absolute differences between the mean and measured T estimates might, in which case averaging several T estimates (for isothermal systems), yield error as low as ±13-25 °C.

P-T slopes and correlated error. Mordick and Glazner (2006) note that because the clinopyroxene-liquid barometer is highly dependent on T, P-T estimates are necessarily correlated. This principle holds for any T-sensitive barometer or P-sensitive thermometer. Mordick and Glazner (2006) use a Monte Carlo technique to simulate random error and generate a P-T error array for clinopyroxene-liquid pairs to illustrate the correlation. Putirka and Condit (2003) adopt the method of Kohn and Spear (1991), using multiple analyses on a single grain to generate a similar array. However determined, a key issue is that the coefficients of any T-sensitive barometer or P-sensitive thermometer lead to a natural slope in P-T space. In the clinopyroxene-liquid example, the slope is not without meaning: the P-T arrays generated from Equations (30) or (31) yield the T at which a given liquid will become saturated with clinopyroxene at a given P—a clinopyroxene saturation surface.

Of potentially great interest is when such P-T slopes deviate from natural saturation conditions. For example, Putirka and Condit (2003) interpret near-isobaric extensions from the saturation P-T trend to represent episodes of magma stagnation, as magmas isobarically migrate down-T toward the ambient geotherm. A near-isothermal slope, in contrast, might indicate partial crystallization during a period of rapid vertical transport (Armienti et al. 2007). To make such interpretations, though, the observed P-T slope must deviate significantly from the intrinsic slope of mineral saturation. To make such a test, one could select a single equilibrium mineral-liquid pair to use as input, select an arbitrary range of temperatures, also to be used as input, and then use Equations (30) or (31) to calculate P for such conditions. The P-T array thus generated will represent the clinopyroxene saturation curve for that particular liquid. The difference between the slopes derived from different input compositions (as in Mordick and Glazner 2006) may thus also be used to derive model error on the P-T slope.

Error of treatment

Error of treatment is much more difficult to define than model error, since it depends upon how closely a natural system mimics experimental compositions and conditions. Because of the wide array of compositions that have been experimentally studied, over a very wide range of P and T, few natural conditions probably require model extrapolation. However, nearly all the models discussed in this chapter presume equilibration between a given mineral-melt or mineral-mineral pair. In natural systems, the pairing of equilibrium phases is not always clear. For example, in a lava flow, what is the liquid? Is it the whole rock? The glass? Some combination thereof? One might expect that the whole rock is the most appropriate "liquid" when considering phenocrysts. But this further presumes that the whole rock acted as a closed system.

The methods of Roeder and Emslie (1970) and Rhodes et al. (1979a) are key to testing for equilibrium. These tests are based on another assumption: two or more naturally occurring phases are equilibrated if their compositions yield exchange coefficients (e.g., $K_D(Fe-Mg)^{ol-liq}$, $K_D(Fe-Mg)^{ol-liq}$, $K_D(Fe-Mg)^{cpx-opx}$), or predicted mineral compositions (Rhodes et al. 1979a; Putirka 1999), that mimic values determined in "equilibrium" (isothermal and isobaric) experiments. Fe-Mg exchange coefficients are commonly used since they appear to be largely independent of P or T. One strategy to estimate such error, then, is to estimate the standard deviation for P-T estimates for crystals that pass some such equilibration filter. If the error is less than model error then it is plausible that the crystals either precipitated from the liquid used as input, or a liquid much like it, which in any case should provide a useful P-T estimate. The hope then is that model error approaches true error.

Compositional corrections. If mineral compositions from a single lava flow yield a wide range of *P-T* estimates, then either the conditions of crystallization were not static (and differences in *P* and *T* are real and should not be averaged), or the minerals crystallized from different melts. In the later case, mineral-melt pairs might not yield "equilibrium" values for a particular exchange coefficient, and one could make a correction for open system processing. This correction could involve using the whole rock as a starting composition, then using mass balance equations to add or subtract observed phenocryst phases until a "calculated liquid" composition is derived that yields an equilibrium value for Fe-Mg exchange (between either olivine-liquid or clinopyroxene-liquid; Putirka 1997; Putirka and Condit 2003), or a better match between observed and calculated mineral components (Putirka and Condit 2003). These corrections can result in a very large shift in estimates of *P* and *T* compared to estimates based on observed (i.e., disequilibrium) mineral-whole rock or mineral-glass pairs. The assumption is that error is decreased by such corrections, since the calculated liquid composition is, by design, consistent with tests for equilibrium. But assuredly, error is attached to the calculation. The error must be in proportion to whether the calculated liquid represents an actual liquid composition. To estimate error, one could examine the variance of *P-T* estimates using observed magma compositions that are similar to calculated compositions. And temperatures derived from different equilibria should be better correlated with one another following such corrections.

The effects of disequilibrium. Since our thermometers and barometers are based on an assumption of equilibrium, any form of disequilibrium (such as pairing a given mineral with the wrong liquid) will yield errors. But how large are such errors? And how reliable are our tests for equilibrium? Cooling rate experiments (see review by Hammer 2008), conducted at constant *P*, but with a constantly changing *T*, provide an example of forced disequilibrium. Here, Equation (32c) is tested using cooling rate experiments on a lunar ferro-basalt by Grove and Bence (1979); the experiments were conducted at 1 atm (0.001 kbar) at cooling rates of 0.5 °C/hr to >600 °C/hr.

For these tests, *P* was calculated using the initial experimental *T*, and observed mineral and melt compositions as input. For the 12 experiments conducted at cooling rates of 0.5-10 °C/hr, Equation (32c) yields a *P* range of −1.82 to 1.86 kbar and a mean of −0.5±1.4 kbar—well within model error of 0.001 kbar (Fig. 14a), and with a standard deviation similar to the model error. One lesson is that not all negative values of *P* should be ignored—they may sometimes reflect the error that results when one attempts to predict a value that is effectively zero. Another is that error is apparently random (and so the mean estimate is better than any individual estimate). Still another is that at cooling rates ≤10 °C/hr, near-equilibrium conditions are apparently achieved, since the model recovers the equilibrium *P*. Still another lesson is relative to the use of the initial *T* as input into Equation (32c). If mean *T* or final experimental *T* is used as input, the slow cooling rate experiments do not yield a mean of 1 atm. Apparently, slowly diffusing species, such as Al, record and retain initial *P-T* conditions, and will not rapidly re-equilibrate, at least at lower temperatures. This means that, in practice, clinopyroxene-based barometers might be expected to yield information about the deeper parts of a magmatic system, and will not readily yield shallow *P-T* conditions unless precipitated at such conditions.

Also interesting is that at cooling rates ≥40 °C/hr, the mean of *P* estimates using Equation (32c) increases monotonically, to a value of 4.6 kbar at cooling rates >550 °C/hr (Fig. 14a). Equation (32c) makes use of the clinopyroxene-melt partition coefficient for Al, $D_{Al}^{cpx-liq}$, and this coefficient also increases with cooling rate, which provides the source for this 4.6 kbar error (Fig. 14b). The interpretation: at low cooling rates, $D_{Al}^{cpx-liq}$ retains its equilibrium value and so provides the correct *P* (on average); at higher cooling rates, higher values of $D_{Al}^{cpx-liq}$ are disequilibrium values (see Lofgren et al. 2006). (One intriguing implication is that if *P* and *T* are known, partition coefficients might be used to estimate cooling rates.)

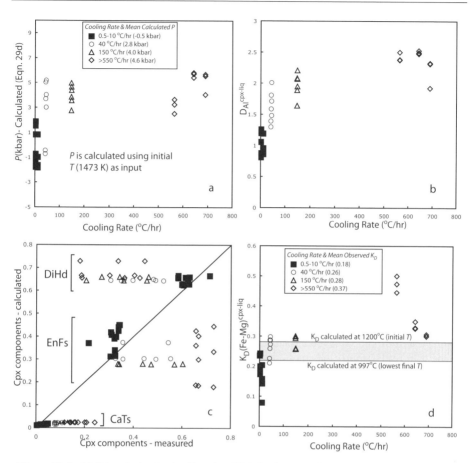

Figure 14. In (a) P is calculated using Equation (32c) for 1 atm cooling rate experiments (Grove and Bence 1979); initial experimental T (1200 °C) is used as input. Experiments conducted at cooling rates ≤10 °C/hr yield a mean value of −0.5 kbar, effectively equivalent to the actual value of 0.001 (or ca. 0) kbar. (b) As crystallization rates exceed 40 °C/hr, the error on P increases, due to the increase in the partition coefficient for Al. (c) The models of Putirka (1999) are used to predict the equilibrium values for the clinopyroxene components DiHd, EnFs and CaTs, using experimental P, T and liquid compositions as input. Only experiments conducted at cooling rates ≤10 °C/hr cluster near calculated equilibrium values. The graph is effectively identical if calculated rather than observed P (0.001 kbar for all) is used as input. (d) Observed values for $K_D(\text{Fe-Mg})^{cpx-liq}$ are compared to cooling rate. The highest cooling rate experiments most strongly deviate from expected equilibrium values, but at lower cooling rates, Fe-Mg is a less effective discriminator for equilibrium as observed values for K_D closely match expected equilibrium values. The gray box shows the range of K_D values obtained from Equation (35) when using the highest T (initial T = 1200 °C) and the lowest of the final experimental T (975 °C) as input.

Can our tests for equilibrium effectively filter these disequilibrium cases (i.e., Roeder and Emslie 1970; Putirka 1999)? Figure 14c shows results for the models of Putirka (1999) when used to predict equilibrium clinopyroxene compositions for DiHd, EnFs, and CaTs (Fig. 14c), using initial T and observed P (0.001 kbar) as input. At the lower cooling rates, the measured components match closely the predicted equilibrium values, and at higher cooling rates, measured values deviate from equilibrium. So the test appears to be effective (Figure 14c is effectively identical when calculated P, instead of the observed 0.001 kbar, is used as input).

Values for $K_D(Fe\text{-}Mg)^{cpx\text{-}liq}$ are also affected by cooling rate (see Hammer 2008), and P derived from very high cooling rates could be rejected on the basis of such values. But at cooling rates as high as 150 °C/hr, observed values are close to the equilibrium value (0.28, when using initial T as input) and, curiously, low cooling rate experiments yield values that are too low on average, even if the final experimental T is used as input into Equation (35). Thus Fe-Mg exchange appears to be somewhat less effective a discriminator than DiHd, EnFs or CaTs, which may reflect the more rapid diffusion of Fe and Mg compared to Al and Ca. Various tests for equilibrium, at least in combination, should thus be successful at filtering erroneous results (Fig. 14), but further tests of this sort should help greatly to better delimit error.

CONCLUSIONS

A great number of igneous thermometers and barometers have been calibrated since T. W. Barth's early efforts to calibrate a two-feldspar thermometer. In light of a rather expansive experimental database (e.g., Hirschmann et al. 2008) we can now confidently calculate T, and in some cases P, for nearly any volcanic or other igneous system, using relatively simple models. The geologic limits of such methods are few. Nearly all the models presented here apply to rocks equilibrated between 700-1700 °C, 1 bar- to 40 or 70 kbar, with water contents ranging from 0-10 wt%. Liquid compositions range from basalt to rhyolite, from picrite to phonolite, from tephrite to trachyte, and everything in between. Thermometers and barometers can be used to calculate T and P to a precision of ±30 °C and ±1.5 kbar, respectively, for individual mineral-melt or mineral-mineral pairs. Cooling rate (disequilibrium) experiments further indicate that existing tests for equilibrium (i.e., Roeder and Emslie 1970; Rhodes et al. 1979b; Putirka 1999) are useful as filters against non-equilibrium mineral-melt pairs.

A natural question is: Can precision be increased? More clever activity models offer a seemingly slim hope; only the activity models of Wood and Banno (1973), Beattie (1993) and Nimis and Taylor (2000) prove their worth in this regard, yielding thermometric errors of ~25-30 °C. New experiments would aid the cases of the two-feldspar and two-pyroxene models, and clarify the potential for plagioclase barometry. But absent the development of experimental techniques that circumvent or subdue the obstacle of diffusion, increased precision might instead be accomplished by the calibration of models targeted to a narrow range of rock types, or geologic problems (using a narrow set of $P\text{-}T\text{-}X_i^j$ conditions). This should obviate the need for some of the compositional corrections that are required for thermometers and barometers intended for broad application.

To increase accuracy, model tests could also be conducted using natural samples, where $P\text{-}T$ conditions are known. For example, Rosalind Helz (pers. comm.) reports that at Hawaii, sequential skylight sampling shows the growth of new silicate phases, as lavas flow and cool downstream. Geobarometers tend to yield 1 bar estimates in such cases (Putirka 2005b), even when they do not yield 1 atm estimates for all 1 atm experimental studies; additional tests of this sort can provide compelling tests of accuracy and error for experimentally derived calibrations. Undoubtedly, similar or more clever tests could be devised using well characterized systems, such as at Mt. Etna, Mt. St. Helens, or Long Valley, etc., especially where new isotopic methods (Cooper and Reid 2008) indicate the presence of long-lived systems, yielding "long duration" (near-equilibrium?) experiments, mitigated, of course, by the potential problems of open system behavior.

ACKNOWLEDGMENTS

Thanks to Alex Speer for all is his aid in organizing various aspects of the short course and volume, and to Jodi Rosso for her many editorial efforts and endless patience in answering questions. Thanks to my co-authors in this volume for their willingness to contribute. Finally, this chapter greatly benefited by thoughtful reviews and comments by Lawford Anderson, Cin-Ty Lee and Frank Tepley. This work was supported by NSF-EAR grants 0337345, 0421272 and 0313688.

REFERENCES

Agee CB, Walker D (1990) Aluminum partitioning between olivine and ultrabasic silicate liquid to 6 GPa. Contrib Mineral Petrol 105:243-254
Agee CB, Draper DS (2004) Experimental constraints on the origin of Martian meteorites and the composition of the Martian mantle. Earth Planet Sci Lett 224:415-429
Albarede F (1992) How deep to common basaltic magmas form and differentiate? J Geophys Res 97:10997-11009
Almeev RA, Holtz F, Koepke J, Parat F, Botcharnikov RE (2007) The effect of H_2O on olivine crystallization in MORB: experimental calibration at 200 MPa. Am Mineral 92:670-674
Anderson DL (1989) Theory of the Earth. Blackwell, Brookline Village, MA
Anderson JL, Barth AP, Wooden JL, Mazdab F (2008) Thermometers and thermobarometers in granitic systems. Rev Mineral Geochem 69:121-142
Ariskin AA (1999) Phase equilibria modeling in igneous petrology: use of COMAGMAT model for simulation fractionation of ferro-basaltic magmas and the genesis of high-alumina basalt. J Volcanol Geotherm Res 90:115-162
Ariskin AA, Barmina GS (1990) Equilibria thermometry between plagioclases and basalt or andesite magmas. Geochem Int 27:129-134
Armienti P, Tonarini S, Innocenti F, D'Orazio M (2007) Mount Etna pyroxene as tracer of petrogenetic processes and dynamics of the feeding system. *In:* Cenozoic Volcanism in the Mediterranean Area. Beccaluva L, Bianchini G, Wilson M (eds), Geol Soc Am Spec Paper 418:265-276
Arndt NT (1977) Partitioning of nickel between olivine and ultrabasic komatiite liquids. Carn Inst Wash Year Book 76:553-557
Auzanneau E, Vielzeuf D, Schmidt MW (2006) Experimental evidence of decompression melting during exhumation of subducted continental crust. Contrib Mineral Petrol 152:125-148
Bacon C, Carmichael ISE (1973) Stages in the P-T path of ascending basalt magma: an example from San Quintin, Baja California. Contrib Mineral Petrol 41:1-22
Baker RB, Eggler DH (1987) Compositions of anhydrous and hydrous melts coexisting with plagioclase, augite, and olivine or low-Ca pyroxene from 1 atm to 8 kbar: Application to the Aleutian volcanic center of Atka. Am Mineral 72:12-28
Baker MB, Grove TL, Price R (1994) Primitive basalts and andesites from the Mt. Shasta region, N. California: products of varying melt fraction and water content. Contrib Mineral Petrol 118:111-129
Barclay J, Carmichael ISE (2004) A hornblende basalt from western Mexico: water saturated phase relations constrain a pressure-temperature window of eruptibility. J Petrol 45:485-506
Barnes SJ (1986) The distribution of chromium among orthopyroxene, spinel and silicateliquid at atmospheric pressure. Geochim Cosmochim Acta 50:1889-1909
Bartels KS, Kinzler RJ, Grove TL (1991) High pressure phase relations of primitive high-alumina basalts from Medicine Lake volcano, northern California. Contrib Mineral Petrol 108:253-270.
Barth TW (1934) Temperatures in lavas and magmas and a new geologic thermometer. Naturen 6:187-192
Barth TW (1951) The feldspar geologic thermometers. Neues Jahrb Mineral 82:143-154
Barth TW (1962) The feldspar geologic thermometers. Norsk Geol Tidsskr 42:330-339
Barth TW (1968) Additional data for the two-feldspar geothermometer. Lithos1:21-22
Beattie P (1993) Olivine-melt and orthopyroxene-melt equilibria. Contrib Mineral Petrol 115:103-111
Becquerel AC (1826) Procédé a l'aide duquel on peut mesurer d'intensité d'un courant électrique. Ann Chim Phys 31:371
Beddoe-Stephens B, Aspden JA, Shepherd TJ (1983) Glass inclusions and melt compositions of the Toba tuffs, northern Sumatra. Contrib Mineral Petrol 83:278-287
Bender JF, Langmuir CH, Hanson GN (1984) Petrogenesis of basalt glasses from the Tamayo Region, East Pacific Rise. J Petrol 25:213-254
Benisek A, Kroll H, Cemic L (2004) New developments in two-feldspar thermometry. Am Mineral 89:1496-1504
Berndt J, Holtz F, Koepke J (2001) Experimental constraints on storage conditions in the chemically zoned phonolitic magma chamber of the Laacher See volcano. Contrib Mineral Petrol 140:469-486
Blatter DL, Carmichael ISE (2001) Hydrous phase equilibria of a Mexican high-silica andesite: A candidate for a mantle origin? Geochim Cosmochim Acta 65:4043-4065

Blundy J, Cashman K (2008) Petrologic reconstruction of magmatic system variables and processes. Rev Mineral Geochem 69:179-239

Blundy J, Cashman K, Humphreys M (2006) Magma heating by decompression-driven crystallization beneath andesite volcanoes. Nature 443:76-80

Bottinga Y, Weill DF (1972) The viscosity of magmatic silicate liquid, a model for calculation. Am J Sci 272:438-475

Bowen NL (1913) The melting phenomena of the plagioclase feldspars. Am J Sci 35:577-599

Bowen NL (1914) The ternary system: diopside-forsterite-silica. Am J Sci 38:270-264

Bowen NL (1915) The later stages of the evolution of the igneous rocks. J Geology Supp 8, 23:1-91

Bowen NL, Tuttle OF (1950) The system $NaAlSi_3O_8$-$KAlSi_3O_8$-H_2O. J Geol 489-511

Box GEP, Draper, NR (1987) Empirical Model-Building and Response Surfaces. Wiley, New York

Brey GP, Kohler T (1990) Geothermobarometry in four-phase lherzolites II. New thermobarometers, and practical assessment of existing thermobarometers. J Petrol 31:1353-1378

Buddington AF, Fahey J, Vlisdis A (1955) Thermometric and petrogenetic significance of titaniferous magnetite. Am J Sci 253:497-532

Buddington AF, Lindsley DH (1964) Iron-titanium oxide minerals and synthetic equivalents. J Petrol 5:210-357

Bulatov VK, Girnis AV, Brey GP (2002) Experimental melting of a modally heterogeneous mantle. Mineral Petrol 75:131-152

Canil D (1999) The Ni-in-garnet geothermometer: calibration at natural abundances. Contrib Mineral Petrol 136:240-246

Carmichael ISE, Nicholls J, Smith AL (1970) Silica activity in igneous rocks. Am Mineral 55:246-263

Carmichael I.S.E., Turner FJ, Verhoogen J (1974) Igneous Petrology. McGraw-Hill Book Company, New York

Castellan GW (1971) Physical Chemistry, 2nd ed. Addison-Wesley, Reading, Massachusetts

Chen HK, Delano JW, Lindsley DH (1982) Chemistry and phase relations of VLT volcanic glasses from Apollo 14 and Apollo 17. Proc Lunar Planet Sci Conf 13:A171-A181

Cooper KM, Reid MR (2008) Uranium-series crystal ages. Rev Mineral Geochem 69:479-554

Dann JC, Holzheid AH, Grove TL, McSween HY (2001) Phase equilibria of the Shergotty meteorite: Constraints on pre-eruptive water contents of Martian magmas and fractional crystallization under hydrous conditions. Meteorit Planet Sci 36:793-806

Dasgupta R, Hirschmann MM, Smith N (2007) Partial melting experiments of peridotite + CO_2 at 3 GPa and genesis of alkalic ocean island basalts. J Petrol 48:2093-2124

Daubeny C (1835) Some account of the eruption of Vesuvius, which occurred in the month of August 1834, extracted from the manuscript notes of the Cavaliere Monticelli, foreign member of the Geological Society; together with a statement of the products of the eruption and of the condition of the volcano subsequently to it. Philos Trans R Soc London 125:153-159

Davis BTC, Boyd DR (1966) The join $Mg_2Si_2O_6$-$CaMgSi_2O_6$ at 30 kilobars pressure and its application to pyroxenes from kimberlites. J Geophys Res 71:3567-3576

Day AL, Allen ET (1905) The isomorphism and thermal properties of the feldspars. Am J Sci 14:93-131

Delano JW (1977) Experimental melting relations of 63545, 76015, and 76055. Proc Lunar Planet Sci Conf 8:2097-2123

Di Carlo I, Pichavant M, Rotolo SG, Scaillet B (2006) Experimental crystallization of a high-K arc basalt: the golden pumice, Stromboli Volcano (Italy). J Petrol 47:1317-1343

Doelter C (1902) Die chemische Zusammensetzung und die Genesis der Monzonigesteine. Tscher Mineral Petrograph Mitteilung 21:191-225

Drake MJ (1976) Plagioclase-melt equilibria. Geochim Cosmochim Acta 40:457-465

Draper DS, Green TH (1997) P-T phase relations of silicic, alkaline, aluminous mantle-xenolith glasses under anhydrous and C-O-H fluid-saturated conditions. J Petrol 38:1187-1224

Draper DS, Green TH (1999) P-T phase relations of silicic, alkaline, aluminous liquids: new results and applications to mantle melting and metasomatism. Earth Planet Sci Lett 170:255-268

Draper DS, Johnston AD (1992) Anhydrous PT phase relations of an Aleutian high-MgO basalt; an investigation of the role of olivine-liquid reaction in the generation of arc high-alumina basalts. Contrib Mineral Petrol 112:501-519

Dungan MA, Long PE, Rhodes JM (1978) Magma mixing at mid-ocean ridges: evidence from legs 45 and 45-DSDP. Geophys Res Lett 5:423-425

Dunn T, Sen C (1994) Mineral/matrix partition coefficient for orthopyroxene, plagioclase, and olivine in basaltic to andesitic systems: a combined analytical and experimental study. Geochim Cosmochim Acta 58:717-733

Elkins-Tanton LT, Draper DS, Agee CB, Jewell J, Thorpe A, Hess PC (2007) The last lavas erupted during the main phase of the Siberian flood volcanic province: results from experimental petrology. Contrib Mineral Petrol 153:191-209

Elkins LT, Fernandes VA, Delano JW, Grove TL (2000) Origin of lunar ultramafic green glasses: Constraints from phase equilibrium studies. Geochim Cosmochim Acta 64:2339-2350

Elkins LT, Grove TL (1990) Ternary feldspar experiments and thermodynamic models. Am Mineral 75:544-559

Elkins-Tanton, LT, Chatterjee N, Grove TL (2003) Experimental and petrological constraints on lunar differentiation from the Apollo 15 green picritic glasses. Meteorit Planet Sci 38:515-527

Essene EJ (1982) Geologic thermometry and barometry. Rev Mineral 10:153-206

Elthon D, Scarfe CM (1984) High-pressure phase equilibria of a high-magnesia basalt and the genesis of primary oceanic basalts. Am Mineral 69:1-15

Esperança S, Holloway JR (1987) On the origin of some mica-lamprophyres; experimental evidence from a mafic minette. Contrib Mineral Petrol 95:207-216

Falloon, TJ, Danyushevsky LV (2000) Melting of refractory mantle at 1.5, 2 and 2.5 GPa under anhydrous and H_2O-undersaturated conditions: implications for the petrogenesis of high-Ca boninites and the influence of subduction components on mantle melting. J Petrol 41:257-283

Falloon TJ, Danyushevsky LV, Green DH (2001) Peridotite melting at 1 GPa; reversal experiments on partial melt compositions produced by peridotite-basalt sandwich experiments. J Petrol 42:2363-2390

Falloon TJ, Green DH (1987) Anhydrous partial melting of MORB pyrolite and other peridotite compositions at 10 kbar; implications for the origin of primitive MORB glasses. Mineral Petrol 37:181-219

Falloon TJ, Green DH, Danyushevsky LV, Faul UH (1999) Peridotite melting at 1.0 and 1.5 GPa: an experimental evaluation of techniques using diamond aggregates and mineral mixes for determination of near-solidus melts. J Petrol 40:1343-1375

Falloon TJ, Green DH, Hatton CJ, Harris KL (1988) Anhydrous partial melting of a fertile and depleted peridotite from 2 to 30 Kb and application to basalt petrogenesis. J Petrol 29:1257-1282

Falloon TJ, Green DH, O-Neill H-St, Hibberson WO (1997) Experimental tests of low degree peridotite partial melt compositions; implications for the nature of anhydrous near-solidus peridotite melts at 1 GPa. Earth Planet Sci Lett 152:149-162

Feig ST, Koepke J, Snow JE (2006) Effect of water on tholeiitic basalt phase equilibria: an experimental study under oxidizing conditions. Contrib Mineral Petrol 152:611-638

Fodor RV, Galar P (1997) A view into the subsurface of Mauna Kea Volcano, Hawaii: crystallization processes interpreted through the petrology and petrography of gabbroic and ultramafic xenoliths. J Petrol 38:581-624

Foulger GR, Natland JH (2003) Is "hotspot" volcanism a consequence of plate tectonics? Science 300:921-922

Fram MS, Longhi J (1992) Phase equilibria of dikes associated with Proterozoic anorthosite complexes. Am Mineral 77:605-616

Frey FA, Garcia MO, Wise WS, Kennedy A, Gurriet P, Albarede F (1991) The evolution of Mauna Kea volcano, Hawaii: petrogenesis of tholeiitic and alkalic basalts. J Geophys Res 96:14347-14375

Frey FA, Wise WS, Garcia MO, West H, Kwon S-T, Kennedy A (1990) Evolution of Mauna Kea volcano, Hawaii: petrologic and geochemical constraints on postshield volcanism. J Geophys Res 95:1271-1300

Fuhrman ML, Lindsley DH (1988) Ternary-feldspar modeling and thermometry. Am Mineral 73:201-215

Fujii T, Bougault H (1983) Melting relations of a magnesian abyssal tholeiite and the origin of MORBs. Earth Planet Sci Lett 62:283-295

Gaetani GA, Grove T (1998) The influence of melting of mantle peridotite. Contrib Mineral Petrol 131:323-346

Garcia MO, Ho RA, Rhodes JM, Wolfe EW (1989) Petrologic constraints on rift-zone processes: results from episode 1 of the Pu'u O'o eruption of Kilauea volcano, Hawaii. Bull Volcanol 52:81-96

Garcia MO, Hulsebosch TP, Rhodes JM (1995) Olivine-rich submarine basalts from the southwest rift zone of Mauna Loa Volcano: implications for magmatic processes and geochemical evolution. *In:* Mauna Loa Revealed: Structure, Composition, History, and Hazards. Rhodes JM, Lockwood JP (eds) AGU Geophysical Monograph 92:219-239

Garcia MO, Rhodes JM, Wolfe EW, Ulrich GE, Ho RA (1992) Petrology of lavas from episodes 2-47 of the Pu'u O'o eruption of Kilauea Volcano, Hawaii: evolution of magmatic processes. Bull Volcanol 55:1-16

Gardner JE, Rutherford M, Carey S, Sigurdsson H (1995) Experimental constraints on pre-eruptive water contents and changing magma storage prior to explosive eruptions of Mount St Helens volcano. Bull Volcanol 57:1-17

Gee LL, Sack RO (1988) Experimental petrology of melilite nephelinites. J Petrol 29:1233-1255

Gerke TL, Kilinc AI (1992) Enrichment of SiO_2 in rhyolites by fractional crystallization: an experimental study of peraluminous granitic rocks from the St. Francois Mountains, Missouri, USA. Lithos 29:273-283

Ghiorso MS (1987) Modeling magmatic systems: thermodynamic relations. Rev Mineral 17:443-465

Ghiorso MS, Hirschmann MM, Reiners PW, Kress VC III (2002) The pMELTS: a revision of MELTS for improved calculation of phase relations and major element partitioning related to partial melting of the mantle to 3 GPa. Geochem Geophys Geosyst 3:1-36, 10.1029/2001GC000217

Glazner AF (1984) Activities of olivine and plagioclase components in silicate melts and their application to geothermometry. Contrib Mineral Petrol 88:260-268

Green DH, Ringwood AE (1967) The genesis of basaltic magmas. Contrib Mineral Petrol 15:103-190

Grove TL, Beaty DW (1980) Classification, experimental petrology and possible volcanic histories of the Apollo 11 high-K basalts. Proc Lunar Planet Sci Conf 11:149-177

Grove TL, Bence AE (1979) Crystallization kinetics in a multiply saturated basalt magma: an experimental study of Luna 24 ferrobasalt. Proc Lunar Planet Sci Conf 10:439-478

Grove TL, Bryan WB (1983) Fractionation of pyroxene-phyric MORB at low pressure: an experimental study. Contrib Mineral Petrol 84:293-309

Grove TL, Donnelly-Nolan, JM Housh T (1997) Magmatic processes that generated the rhyolite of Glass Mountain, Medicine lake volcano, N. California. Contrib Mineral Petrol 127:205-223

Grove TL, Elkins-Tanton LT, Parman SW, Chatterjee N, Müntener O, Gaetani GA (2003) Fractional crystallization and mantle-melting controls on calk-alkaline differentiation trends. Contrib Mineral Petrol 145:515-533

Grove TL, Gerlach DC, Sando TW (1982) Origin of calc-alkaline series lavas at Medicine Lake Volcano by fractionation, assimilation and mixing. Contrib Mineral Petrol 80:160-182

Grove TL, Juster TC (1989) Experimental investigations of low-Ca pyroxene stability and olivine-pyroxene-liquid equilibria at 1-atm in natural basaltic and andesitic liquids. Contrib Mineral Petrol 103:287-305

Grove TL, Kinzler RJ, Bryan WB (1990) Natural and experimental phase relations of lavas from Serocki Volcano. Proc Ocean Drill Prog 106/109:9-17

Grove TL, Kinzler RJ, Bryan WB (1992) Fractionation of mid-ocean ridge basalt (MORB). *In:* Mantle Flow and Melt Generation at Mid-Ocean Ridges. Phipps Morgan J, Blackman, DK, Sinton JM (eds) AGU Geophysical Monograph Series 7:281-310

Grove TL, Raudsepp M (1978) Effects of kinetics on the crystallization of quartz normative basalt 15597: an experimental study. Proc Lunar Planet Sci Conf 9:585-599

Gudfinnsson GH, Presnall DC (1996) Melting relations of model lherzolite in the system $CaO-MgO-Al_2O_3-SiO_2$ at 2.4-3.4 GPa and the generation of komatiites. J Geophys Res 101:27701-27709

Haase KM (1996) The relationships between the age of the lithosphere and the composition of oceanic magmas: constraints on partial melting, mantle sources and the thermal structure of the plates. Earth Planet Sci Lett 144:75-92

Hakli TA, Wright TL (1967) The fractionation of nickel between olivine and augite as a geothermometer. Geochim Cosmochim Acta 31:877-884

Hamada M, Fuji T (2007) H_2O-rich island arc low-K tholeiite magma inferred from Ca-rich plagioclase-melt inclusion equilibria. Geochem J 41:437-461

Hall J (1798) Experiments on whinstone and lava. Proc R Soc Edinburgh 5:43-74

Hammer JE (2008) Experimental studies of the kinetics and energetic of magma crystallization. Rev Mineral Geochem 69:9-59

Hammer JE, Rutherford MJ, Hildreth W (2002) Magma storage prior to the 1912 eruption at Novarupta, Alaska. Contrib Mineral Petrol 144:144-162

Hansteen TH, Klügel A (2008) Fluid inclusion thermobarometry as a tracer for magmatic processes. Rev Mineral Geochem 69:143-177

Helz RT (1976) Phase relations of basalts in their melting ranges at P_{H2O} = 5kb. Part II. Melt compositions. J Petrol 17:139-193

Helz RT, Thornber CR (1987) Geothermometry of Kilauea Iki lava lake, Hawaii. Bull Volcanol 49:651-668

Herzberg C, Asimow PD, Arndt N, Niu Y, Lesher CM, Fitton JG, Cheadle MJ, Saunders AD (2007) Temperature in ambient mantle and plumes: constraints form basalts, picrites and komatiites. Geochem Geoophys Geosyst 8:10.1029/2006GC001390

Herzberg CT, Chapman NA (1976) Clinopyroxene geothermometry of spinel-lherzolites. Am Mineral 61:626-637

Herzberg C, O'Hara MJ (1998) Phase equilibrium constraints on the origin of basalts, picrites and komatiites. Earth Sci Rev 44:39-79

Herzberg C, O'Hara MJ (2002) Plume-associated ultramafic magmas off Phanerozoic age. J Petrol 43:1857-1883

Hess H (1962) History of the ocean basins. *In:* Petrologic Studies: A Volume in Honor of A.F. Buddington. Engle AE, James HL, Leonard BF (eds) GSA Special Volume,p 599-620

Hesse M, Grove TL (2003) Absarokites from the western Mexican Volcanic Belt: constraints on mantle wedge conditions. Contrib Mineral Petrol 146:10-27

Hill DP, Zucca JJ (1987) Geophysical constraints on the structure of Kilauea and Mauna Loa volcanoes and some implications for seismomagmatic processes. USGS Surv Prof Paper 1350:903-917

Hirschmann MM, Ghiorso MS, Davis FA, Gordon SM, Mukherjee S, Grove TL, Krawczynski M, Medard E, Till CB (2008) Library of experimental phase relations: a database and web portal for experimental magmatic phase equilibria data. Geochem Geophys Geosyst 9, Q03011, doi:10.1029/2007GC001894

Holbig ES, Grove TL (2008) Mantle melting beneath the Tibetan Plateau: Experimental constraints on ultrapotassic magmatism. J Geophys Res doi:10.1029/2007JB005149

Holtz F, Becker A, Freise M, Johannes W (2001) The water-undersaturated and dry Qz-Ab-Or system revisited. Experimental results at very low water activities and geological implications. Contrib Mineral Petrol 141:347-357

Holtz F, Sato H, Lewis J, Behrens H, Nakada S (2005) Experimental Petrology of the 1991-19995 Unzen Dacite, Japan. Part I: Phase relations, phase composition, and pre-eruptive conditions. J Petrol 46:319-337

Housh TB, Luhr JF (1991) Plagioclase-melt equilibria in hydrous systems. Am Mineral 76:477-492

Jaggar TA (1917) Thermal gradient of Kilauea lava lake. Wash Acad Sci 7:397-405

Jefferys WH, Berger JO (1992) Ockham's razor and Bayesian analysis. Am J Sci 80:64-72

Johnson KTM (1998) Experimental determination of partition coefficients for rare earth and high-field-strength elements between clinopyroxene, garnet, and basaltic melt at high pressures. Contrib Mineral Petrol 133:60-68

Johnston AD (1986) Anhydrous P-T phase relations of near-primary high-alumina basalt from the South Sandwich Islands. Contrib Mineral Petrol 92:368-382

JMP, Statistics and Graphics Guide (2002), Version 5, SAS Institute, Cary, NC, USA

Joly J (1891) On the determination of the melting points of minerals, Part 1. Uses of the meldometer. Proc R Irish Acad 2:38

Jurewicz AJG, Mittlefehldt DW, Jones JH (1993) Experimental partial melting of the Allende (CV) and Murchison (CM) chondrites and the origin of asteroidal basalts. Geochim Cosmochim Acta 57:2123-2139

Juster CT, Grove TL, Perfit MR (1989) Experimental constraints on the generation of FeTi basalts, andesite, and rhyodacites at the Galapagos Spreading Center, 85°W and 95°W. J Geophys Res 94:9251-9274

Kawamoto T, Hervig RL, Holloway JR (1996) Experimental evidence for a hydrous transition zone in the early Earth's mantle. Earth Planet Sci Lett 142:587-592

Kelemen PB, Joyce DB, Webster JD, Holloway JR (1990) Reaction between ultramafic rock and fractionating basaltic magma II. Experimental investigation of reaction between olivine tholeiite and harzburgite at 1150-1050 °C and 5kb. J Petrol 31:99-134

Kennedy AK, Grove TL, Johnson RW (1990) Experimental and major element constraints on the evolution of lavas from Lihir Island, Papua New Guinea. Contrib Mineral Petrol 104:722-734.

Keshav S, Gudfinnsson GH, Sen G, Fei Y (2004) High-pressure melting experiments on garnet clinopyroxenite and the alkalic to tholeiitic transition in ocean-island basalts. Earth Planet Sci Lett 223:365-379

Kinzler RJ, Grove TL (1985) Crystallization and differentiation of Archean komatiite lavas from northeast Ontario: phase equilibrium and kinetic studies. Am Mineral 70:40-51

Kinzler RJ (1997) Melting of mantle peridotite at pressures approaching the spinel to garnet transition: Application to mid-ocean ridge basalt petrogenesis. J Geophys Res 102:853-874

Kinzler RJ, Grove TL (1992) Primary magmas of mid-ocean ridge basalts 1. Experiments and methods. J Geophys Res 97:6885-6906

Kjarsgaard BA (1998) Phase relations of a carbonated high-CaO nephelinite at 0.2 and 0.5 GPa. J Petrol 39:061-2075

Klein EM, Langmuir CH (1987) Global correlations of ocean ridge basalt chemistry with axial depth and crustal thickness. J Geophys Res 92:8089-8115

Klemme S, O'Neil HSC (2000) The near-solidus transition from garnet lherzolite to spinel lherzolite. Contrib Mineral Petrol 138:237-248

Klügel A, Klein F (2005) Complex storage and ascent at embryonic submarine volcanoes from the Madeira Archipelago. Geology 34:337-340

Koester E, Pawley AR, Fernandes LA, Porcher CC, Soliani Jr E (2002) Experimental melting of cordierite gneiss and the petrogenesis of syntranscurrent peraluminous granites in southern Brazil. J Petrol 43:1595-1616

Kogi R, Müntener O, Ulmer P, Ottolini L (2005) Piston-cylinder experiments on H_2O undersaturated Fe-bearing systems: an experimental setup approaching fO_2 conditions of natural calc-alkaline magmas. Am Mineral 90:708-717

Kogiso T, Hirose K, Takahashi E (1998) Melting experiments on homogeneous mixtures of peridotite and basalt; application to the genesis of ocean island basalts. Earth Planet Sci Lett 162:45-61

Kogiso T, Hirschmann MM (2001) Experimental study of clinopyroxenite partial melting and the origin of ultra-calcic melt inclusions. Contrib Mineral Petrol 142:347-360

Kohn MT, Spear FJ (1991) Error propagation for barometers: 2. Applications to rocks. Am Mineral 76:138-147

Kudo AM, Weill DF (1970) An igneous plagioclase thermometer. Contrib Mineral Petrol 25:52-65

Kushiro I (1996) Partial melting of a fertile mantle peridotite at high pressures: an experimental study using aggregates of diamond. *In:* Earth Processes: Reading the Isotopic Code. Baso A, Hart S (eds) AGU Geophysical Monograph 95:109-122

Kushiro I, Mysen B (2002) A possible effect of melt structure on the Mg-Fe^{2+} partitioning between olivine and melt. Geochim Cosmochim Acta 66:2267-2272

Kushiro I, Walter MJ (1998) Mg-Fe partitioning between olivine and mafic-ultramafic Melts. Geophys Res Lett 25:2337-2340

Langmuir CH, Hanson GN (1980) An evaluation of major element heterogeneity in the mantle sources of basalts. Phil Trans R Soc Lond A 297:383-407

Laporte D, Toplis MJ, Seyler M, Devidal J-L (2004) A new experimental technique for extracting liquids from peridotite at very low degrees of melting: application to partial melting of depleted peridotite. Contrib Mineral Petrol 146:463-484

Leeman WP, Lindstrom DJ (1978) Partitioning of Ni^{2+} between basaltic and synthetic melts and olivines – an experimental study. Geochim Cosmochim Acta 42:801-816

Lesher CE, Walker D (1988) Cumulate maturation and melt migration in a temperature gradient. J Geophys Res 93:10295-10311

Lindsley DH (1983) Pyroxene thermometry. Am Mineral 68:477-493

Lindsley DH, Anderson DJ (1983) A two-pyroxene thermometer. J Geophys Res 88:A887-A906

Liu TC, Chen BR, Chen CH (1998) Anhydrous melting experiment of a Wannienta basalt in the Kuanyinshan area, northern Taiwan, at atmospheric pressure. Terr Atm Ocean Sci 9:165-182

Liu TC, Chen BR, Chen, CH (2000) Melting experiment of a Wannienta basalt in the Kuanyinshan area, northern Taiwan, at pressures up to 2.0 GPa. J Asian Earth Sci 18:519-531

Lofgren GE, Huss GR, Wasserburg GJ (1006) An experimental study of trace-element partitioning between Ti-Al-clinopyroxene and melt: equilibrium and kinetic effects including sector zoning. Am Mineral 91:1596-1606

Longhi J (1976) Iron, Magnesium, and Silica in Plagioclase. Ph.D. Thesis, Harvard University, 296 pp

Longhi J (1995) Liquidus equilibria of some primary lunar and terrestrial melts in the garnet stability field. Geochim Cosmochim Acta 59:2375-2386

Longhi J (2002) Some phase equilibrium systematics of lherzolite melting: I. Geochem Geophys Geosys 3:doi:10.1029/2001GC000204

Longhi J, Pan V (1988) A reconnaissance study of phase boundaries in low-alkali basaltic liquids. J Petrol 29:115-147

Longhi J, Pan V (1989) The parent magmas of the SNC meteorites. Proc Lunar Planet Sci Conf 19:451-464

Loomis TP (1979) An empirical model for plagioclase equilibrium in hydrous melts. Geochim Cosmochim Acta 43:1753-1759

Maaloe S (2004) The PT-phase relations of an MgO-rich Hawaiian tholeiite: the compositions of primary Hawaiian tholeiites. Contrib Mineral Petrol 148:236-246

Mahood GA, Baker DR (1986) Experimental constraints on depths of fractionation of mildly alkalic basalts and associated felsic rocks: Pantelleria, Strait of Sicily. Contrib Mineral Petrol 93:251-264

Marsh BD (1995) Solidification fronts and magmatic evolution. Mineral Mag 60:5-40

Marsh BD, Fournelle J, Myers JD, Chou I-M. (1990) On plagioclase thermometry in island-arc rocks: experiments and theory. *In:* Fluid-Mineral interactions: A Tribute to H.P. Eugster. Spencer RJ, Chou I-M (eds) Geochemical Society Special Publication 2:65-83

Martel C, Pichavant M, Holtz F, Scaillet B, Bourdier J, Traineau H (1999) Effects of fO_2 and H_2O on andesite phase relations between 2 and 4 kbar. J Geophys Res 104:29453 29470

Mathez EA (1973) Refinement of the Kudo-Weill plagioclase thermometer and its applications to basaltic rocks. Contrib Mineral Petrol 41:61-72

McKenzie D, Bickle MJ (1988) The volume and composition of melt generated by extension of the lithosphere. J Petrol 29:625-679

McCoy TJ, Lofgren GE (1999) Crystallization of the Zagami shergottite: an experimental study. Earth Planet Sci Lett 173:397-411

McDade P, Blundy JD, Wood BJ (2003) Trace element partitioning on the Tinaquillo lherzolite solidus at 1.5 GPa. Phys Earth Planet Interior 139:129-147

Medard E, Grove TL (2008) The effect of H_2O on the olivine liquidus of basaltic melts: experiments and thermodynamic models. Contrib Mineral Petrol 155:417-432

Medard E, Schmidt MW, Schiano P (2004) Liquidus surfaces of ultracalcic primitive melts: formation conditions and sources. Contrib Mineral Petrol 148:201-215

Meen JK (1987) Formation of shoshonites from calcalkaline basalt magmas: geochemical and experimental constraints from the type locality. Contrib Mineral Petrol 97:333-351

Meen JK (1990) Elevation of potassium content of basaltic magma by fractional crystallization: the effect of pressure. Contrib Mineral Petrol 104:309-331

Mercier J-C, Beoit V, Girardeau J (1984) Equilibrium state of diopside-bearing harzburgites from ophiolites: geobarometric and geodynamic implications. Contrib Mineral Petrol 85:391-403

Mibe K, Fuji T, Yasuda A, Ono S (2005) Mg-Fe partitioning between olivine and ultramafic melts at high pressure. Geochim Cosmochim Acta 70:757-766

Milholland CS, Presnall DC (1998) Liquidus phase relations in the CaO-MgO-Al$_2$O$_3$-SiO$_2$ system at 3.0 GPa: the aluminous pyroxene thermal divide and high pressure fractionation of picritic and komatiitic magmas. J Petrol 39:3-27

Minitti ME, Rutherford MJ (2000) Genesis of the Mars Pathfinder sulfur-free rock from SNC parental liquids. Geochim Cosmochim Acta 64:2535-2547

Montierth C, Johnston AD, Cashman KV (1995) An empirical glass-composition-based geothermometer for Mauna Loa lavas. *In:* Mauna Loa Revealed: Structure, Composition, History, and Hazards. Rhodes JM, Lockwood JP (eds) AGU Geophysical Monograph 92:207-217

Moore G, Carmichael ISE (1998) The hydrous phase equilibria (to 3 kbar) of an andesite and basaltic andesite from western Mexico: constraints on water content and conditions of phenocryst growth. Contrib Mineral Petrol 130:304-319

Mordick BE, Glazner AF (2006) Clinopyroxene thermobarometry of basalts from the Coso and Big Pine volcanic fields, California. Contrib Mineral Petrol doi 10.1007s00410-006-0097-0

Morgan WJ (1971) Convection plumes in the lower mantle. Nature 230:42-43

Morse SA, Brady JB, Sporleder BA (2004) Experimental petrology of the Kiglapait Intrusion; cotectic trace for the lower zone at 5 kbar in graphite. J Petrol 45:2225-2259

Müntener O, Kelemen PB, Grove TL (2001) The role of H$_2$O during crystallization of primitive arc magmas under uppermost mantle conditions and genesis of igneous pyroxenites: and experimental study. Contrib Mineral Petrol 141:643-658

Musselwhite DS, Dalton HA, Kiefer WS, Treiman AH (2006) Experimental petrology of the basaltic shergottite Yamato-980459: Implications for the thermal structure of the Martian mantle. Meteorit Planet Sci 41:1271-1290

Naney MT (1983) Phase equilibria of rock-forming ferromagnesian silicates in granitic Systems. Am J Sci 283:993-1033

Nicholls J, Carmichael ISE (1972) The equilibration temperature and pressure of various lava types with spine- and garnet-peridotite. Am Mineral 57:941-959

Nicholls J, Carmichael ISE, Stormer JC (1971) Silica activity and P$_{total}$ in igneous rocks. Contrib Mineral Petrol 33:1-20

Nielsen RL, Drake MJ (1979) Pyroxene-melt equilibria. Geochim Cosmochim Acta 43:1259-1272

Nimis P (1995) A clinopyroxene geobarometer for basaltic systems based on crystals-structure modeling. Contrib Mineral Petrol 121:115-125

Nimis P (1999) Clinopyroxene geobarometry of magmatic rocks. Part 2. Structural geobarometers for basic to acid, tholeiitic and mildly alkaline systems. Contrib Mineral Petrol 135:62-74

Nimis P, Taylor WR (2000) Single clinopyroxene thermobarometry for garnet peridotites. Part 1 Calibration and testing of a Cr-in-cpx barometer and an enstatite-in-cpx thermometer. Contrib Mineral Petrol 139:541-554

Nimis P, Ulmer P (1998) Clinopyroxene geobarometry of magmatic rocks. Part 1: an expanded structural geobarometer for anhydrous and hydrous basic and ultrabasic systems. Contrib Mineral Petrol 133:122-135

Nordstrom DK, Munoz JL (1986) Geochemical Thermodynamics. Blackwell, Brookline, Massachusetts

O'Nions RK, Powell R (1977) The thermodynamics of trace element distribution. *In:* Thermodynamics in Geology. Fraser DG (ed) Reidel, Dordrecht-Holland, p 349-363

Panjasawatwong Y, Danyushevsky LV, Crawford AJ, Harris KL (1995) An experimental study of the effects of melt composition on plagioclase-melt equilibria at 5 and 10 kbar: implications for the origin of magmatic high-An plagioclase. Contrib Mineral Petrol 118:420-432.

Papike JJ, Cameron KL, Baldwin K (1974) Amphiboles and pyroxenes: characterization of other than quadrilateral components and estimates of ferric iron from microprobe data. Geol Soc Am Abst Prog 6:1053-1054

Parman SW, Dann JC, Grove TL, deWit MJ (1997) Emplacement conditions of komatiite magmas from the 3.49 Ga Komati Formation, Barberton Greenstone Belt, South Africa. Earth Planet Sci Lett 150:303-323

Parman SW, Grove TL (2004) Harzburgite melting with and without H$_2$O: experimental data and predictive modeling. J Geophys Res 109:doi:10.1029/2003JB002566

Parsons T, Thompson, GA (1993) Does magmatism influence low-angle normal faulting? Geology 21:247-250

Patino-Douce AE (1995) Experimental generation of hybrid silicic melts by reaction of high-Al basalt with metamorphic rocks. J Geophys Res 100:15623-15639

Patino-Douce AE (2005) Vapor-absent melting of tonalite at 15-32 kbar. J Petrol 46:275-290.

Patino-Douce AE, Beard JS (1995) Dehydration-melting of biotite gneiss and quartz amphibolite from 3 to 15 kbar. J Petrol 36:707-738.

Patino-Douce AE, Harris N (1998) Experimental constraints on Himalayan anatexis. J Petrol 39:689-710

Perret RA (1913) Volcanic research at Kilauea in the summer of 1911. Am J Sci 4:475-483.

Petermann M, Hirschmann MM (2000) Anhydrous partial melting experiments on MORB-like eclogite: Phase relations, phase compositions and mineral-melt partitioning of major elements at 2-3 GPa. J Petrol 44:2173-2201
Petermann M, Lundstrom CC (2006) Phase equilibrium experiments at 0.5 GPa and 1100-1300 °C on a basaltic andesite from Arenal volcano, Costa Rica. J Volcanol 157:222-235
Petry C, Chakraborty S, Palme H (1997) Olivine-melt nickel-iron exchange Thermometer: cosmochemical significance and preliminary experimental results. Meteor Planet Sci 32:106-107
Pichavant M, Martel C, Bourdier J-L, Scaillet B (2002a) Physical conditions, structure, and dynamics of a zoned magma chamber: Mount Pele (Martinique, Lesser Antilles Arc). J Geophys Res 107:1-25
Pichavant M, Mysen BO, MacDonald R (2002b) Source and H_2O content of high-MgO magmas in island arc settings: an experimental study of a primitive calc-alkaline basalt from St. Vincent, Lesser Antilles arc. Geochim Cosmochim Acta 66:2193-2209
Pickering-Witter J, Johnston AD (2000) The effects of variable bulk composition on the melting systematics of fertile peridotitic assemblages. Contrib Mineral Petrol 140:190-211
Powell M, Powell R (1977) Plagioclase-alkali feldspar geothermometry revisited. Mineral Mag 41:253-256
Prouteau G, Scaillet B, Pichavant M, Maury R (2001) Evidence for mantle metasomatism by hydrous silicic melts derived from subducted oceanic crust. Nature 410:197-200
Putirka K (1997) Magma transport at Hawaii: inferences from igneous thermobarometry. Geology 25:69-72
Putirka K (1998) Garnet + liquid equilibrium. Contrib Mineral Petrol 131:273-288
Putirka K (1999a) Clinopyroxene+liquid equilibrium to 100 kbar and 2450 K. Contrib Mineral Petrol 135:151-163
Putirka K (1999b) Melting depths and mantle heterogeneity beneath Hawaii and the East Pacific Rise: Constraints from Na/Ti and REE ratios. J Geophys Res 104:2817-2829.
Putirka K (2005a) Mantle potential temperatures at Hawaii, Iceland, and the mid-ocean ridge system, as inferred from olivine phenocrysts: Evidence for thermally–driven mantle plumes. Geochem Geophys Geosys doi:10.1029/2005GC000915
Putirka K (2005b) Igneous thermometers and barometers based on plagioclase + liquid equilibria: tests of some existing models and new calibrations. Am Mineral 90:336-346
Putirka K (2008) Excess temperatures at ocean islands: implications for mantle layering and convection. Geology 36:283-286
Putirka K, Condit C (2003) A cross section of a magma conduit system at the margins of the Colorado Plateau. Geology 31:701-704
Putirka K, Johnson M, Kinzler R, Walker D (1996) Thermobarometry of mafic igneous rocks based on clinopyroxene-liquid equilibria, 0-30 kbar. Contrib Mineral Petrol 123:92-108
Putirka K, Perfit M, Ryerson FJ, Jackson MG (2007) Ambient and excess mantle temperatures, olivine thermometry, and active vs. passive upwelling. Chem Geol 241:177-206.
Putirka K, Ryerson FJ, Mikaelian H (2003) New igneous thermobarometers for mafic and evolved lava compositions, based on clinopyroxene + liquid equilibria. Am Mineral 88:1542-1554
Ratajeski K, Sisson TW, Glazner AF (2005) Experimental and geochemical evidence for derivation of the El Capitan Granite, California, by partial melting of hydrous gabbroic lower crust. Contrib Mineral Petrol 149:713-734
Rhodes JM, Dungan MA, Blanchard DP, Long PE (1979a) Magma mixing at mid-ocean ridges: evidence form basalts drilled near 22°N on the mid-Atlantic ridge. Tectonophys 55:35-61
Rhodes JM, Lofgren GE, Smith DP (1979b) One atmosphere melting experiments on ilmenite basalt 12008. Proc Lunar Planet Sci Conf 10:407-422
Rhodes JM, Vollinger MJ (2004) Composition of basaltic lavas sampled by phase-2 of the Hawaii Scientific Drilling Project: geochemical stratigraphy and magma series types. Geochem Geophys Geosys 5:doi:10.1029/2002GC00434
Righter K, Leeman WP, Hervig RL (2006) Partitioning of Ni, Co, and V between spinel-structured oxides and silicate melts; importance of spinel composition. Chem Geol 227:1-25
Robinson JAC, Wood BJ, Blundy JD (1998) The beginning of melting of fertile and depleted peridotite at 1.5 GPa. Earth Planet Sci Lett 155:97-111
Roeder PL, Emslie RF (1970) Olivine-liquid equilibrium. Contrib Mineral Petrol 29:275-289
Rutherford MJ, Sigurdsson H, Carey S, Andrew. (1985) The May 18, 1980, eruption of Mount St. Helens; 1, Melt composition and experimental phase equilibria. J Geophys Res 90:2929-2947
Ryan MP (1988) The mechanics and three-dimensional internal structure of active magmatic systems: Kilauea volcano, Hawaii. J Geophys Res 93:4213-4248
Ryan TP (1997) Modern Regression Methods. Wiley, New York
Sack RO, Walker D, Carmichael ISE (1987) Experimental petrology of alkalic lavas: constraints on cotectics of multiple saturation in natural basic liquids. Contrib Mineral Petrol 96:1-23
Salters VJM, Longhi J (1999) Trace element partitioning during the initial stages of melting beneath mid-ocean ridges. Earth Planet Sci Lett 166:15-30

Scaillet B, Evans BW (1999) The 15 June 1991 Eruption of Mount Pinatubo. I. Phase Equilibria and pre-eruption P-T-fO$_2$-fH$_2$O conditions of the dacite magma. J Petrol 40:381-411

Scaillet B, MacDonald R (2003) Experimental constraints on the relationships between peralkaline rhyolites of the Kenya rift valley. J Petrol 44:1867-1894

Schmidt MW, Green DH, Hibberson WO (2004) Ultra-calcic magmas generated from Ca-depleted mantle: an experimental study on the origin of ankaramites. J Petrol 45:531-554.

Schwab BE, Johnston AD (2001) Melting systematics of modally variable, compositionally intermediate peridotites and the effects of mineral fertility. J Petrol 42:1789-1811

Scoates JS, Lo Cascio M, Weis, D, Lindsley DH (2006) Experimental constraints on the origin and evolution of mildly alkalic basalts from the Kerguelen Archipelago, Southeast Indian Ocean. Contrib Mineral Petrol 151:582-599

Sen G (1985) Experimental determination of pyroxene compositions in the system CaO-MgO-Al$_2$O$_3$-SiO$_2$ at 900-1200 °C and 10-15 kbar using PbO and H$_2$O fluxes. Am Mineral 70:678-695

Sen G, Jones R (1989) Experimental equilibration of multicomponent pyroxenes in the spinel peridotite field: implications for practical thermometers and a possible barometer. J Geophys Res 94:17871-17880

Sen G, Keshav S, Bizimis M (2005) Hawaiian mantle xenoliths and magmas: composition and thermal character of the lithosphere. Am Mineral 90:871-887

Shepherd FS, Rankin GA, Wright FE (1909) The binary systems of alumina with silica, lime and magnesia. Am J Sci 28:293-315

Sisson TW, Grove TL (1993a) Experimental investigations of the role of H$_2$O in calc-alkaline differentiation and subduction zone magmatism. Contrib Mineral Petrol 113:143-166

Sisson TW, Grove TL (1993b) Temperatures and H$_2$O contents of low-MgO high-alumina basalts. Contrib Mineral Petrol 113:167-184

Skjerlie KP, Patino-Douce AE (2002) The fluid-absent partial melting of a zoisite-bearing quartz eclogite from 1.0 to 3.2 GPa; implications for melting in thickened continental crust and for subduction-zone processes. J Petrol 43:291-314

Sosman RB, Merwin HE (1913) Data on the intrusion temperature of the Palisade diabase. J Wash Acad 3:389-395

Spear FS (1993) Metamorphic Phase Equilibria and Pressure-Temperature-Time Paths. Mineralogical Society of America Monograph, Washington D.C.

Springer W, Seck HA (1997) Partial fusion of basic granulites at 5 to 15 kbar: implications for the origin of TTG magmas. Contrib Mineral Petrol 127:30-45

Stolper E (1977) Experimental petrology of eucritic meteorites Geochim Cosmochim Acta 41:587-611

Stolper E (1980) A Phase Diagram for Mid-Ocean Ridge Basalts: Preliminary Results and Implications for Petrogenesis. Contrib Mineral Petrol 74:13-27

Stolper E, Walker D (1980) Melt density and the average composition of basalt. Contrib Mineral Petrol 74:7-12

Stormer JC (1975) A practical two-feldspar geothermometer. Am Mineral 60:667-674

Sugawara T (2001) Ferric iron partitioning between plagioclase and silicate liquid: thermodynamics and petrological applications. Contrib Mineral Petrol 141:659-686

Takagi D, Sato H, Nakagawa N (2005) Experimental study of a low-alkali tholeiite at 1-5kbar:optimal condition for the crystallization of high-An plagioclase in hydrous arc tholeiite. Contrib Mineral Petrol 149:527-540

Takahashi E (1980) Melting relations of an alkali-olivine basalt to 30 kbar, and their bearing on the origin of alkali basalt magmas. Carneige Institute Washington Year Book 79:271-276

Takahashi E, Nakajima K, Wright TL (1998) Origin of the Columbia River basalts: melting model of a heterogeneous plume head. Earth Planet Sci Lett 162:63-80

Taura H, Yurimoto H, Kurita K, Sueno S (1998) Pressure dependence on partition coefficients for trace elements between olivine and the coexisting melts. Phys Chem Mineral 25:469-484

ten Brink US, Brocher TM (1987) Multichannel seismic evidence for a subcrustal intrusive complex under Oahu and a model for Hawaiian volcanism. J Geophys Res 92:13687-13707

Thompson JB (1967) Thermodynamic properties of simple solutions. *In*: Researches in Geochemistry, Vol 2. Abelson PH (ed) Wiley, New York, p 340-361

Thompson JB (1982) Composition space: an algebraic and geometric approach. Rev Mineral 10:1-31

Thornber CR, Heliker C, Sherrod DR, Kauahikaua JP, Miklius A, Okubo PG, Trusdell FA, Budahn JR, Ridley WI, Meeker GP (2003) Kilauea East Rift Zone magmatism: an episode 54 perspective. J Petrol 44:1525-1559

Thy P (1991) High and low pressure phase equilibria of a mildly alkalic lava from the 965 Surtsey eruption: experimental results. Lithos 26:223-243

Thy P, Lesher CE, Fram MS (2004) Low Pressure experimental constraints on the evolution of basaltic lavas from site 917, southeast Greenland continental margin. Proc Ocean Drill Prog 152:359-372

Toplis MJ (2005) The thermodynamics of iron and magnesium partitioning between olivine and liquid: criteria for assessing and predicting equilibrium in natural and experimental systems. Contrib Mineral Petrol 149:22-39

Tormey DR, Grove TL, Bryan WB (1987) Experimental petrology of normal MORB near the Kane Fracture Zone: 22°-25° N, mid-Atlantic Ridge. Contrib Mineral Petrol 96:121-139

Torrens HS (2006) The geological work of Gregory Watt, his travels with William Maclure in Italy (1801-1802) and Watt's "proto-geological" map of Italy (1804). *In:* The Origins of Geology in Italy. Vai GB, Caldwell WGE (eds) p 179-198

Tsuruta K, Takahashi E (1998) Melting study of the an alkali basalt JB-1 up to 12.5 GPa: behavior of potassium in the deep mantle. Phys Earth Planet Int 107:119-130

Tuttle OF, Bowen NL (1958) The origin of granite in the light of experimental studies in the system $NaAlSi_3O_8$-$KAlSi_3O_8$-SiO_2-H_2O. Geol Soc Am Mem 174

Vander Auwera J, Longhi J (1994) Experimental study of jotunite (hypersthene monzodiorite): constraints on the parent magma composition and crystallization conditions (P, T, fO_2) of the Bjerkreim-Sokndal layered intrusion (Norway). Contrib Mineral Petrol 118:60-78

Vander Auwera J, Longhi J, Duchesne J-C (1998) A liquid line of descent of the jotunite (hypersthene monzodiorite) suite. J Petrol 39:439-468

Villiger S, Ulmer P, Müntener O, Thompson AB (2004) The liquid line of descent of anhydrous, mantle-derived, tholeiitic liquids by fractional and equilibrium crystallization - an experimental study at 1.0 GPa. J Geol 45:2369-2388

Wagner TP, Grove TL (1997) Experimental constraints on the origin of lunar high-Ti ultramafic glasses. Geochim Cosmochim Acta 61:1315-1327

Wagner TP, Grove TL (1998) Melt/harzburgite reaction in the petrogenesis of tholeiitic magma from Kilauea volcano, Hawaii. Contrib Mineral Petrol 131:1-12

Walter MJ (1998) Melting of garnet peridotite and the origin of komatiite and depleted lithosphere of komatiite and depleted lithosphere. J Petrol 39:29-60

Walter MJ, Presnall DC (1994) Melting behavior of simplified lherzolite in the system CaO-MgO-Al_2O_3-SiO_2-Na_2O from 7 to 35 kbar. J Petrol 35:329-359

Wasylenki LE, Baker MB, Kent AJR, Stolper EM (2003) Near-solidus melting of the shallow upper mantle: partial melting experiments on depleted peridotite. J Petrol 44:1163-1191

Wells PRA (1977) Pyroxene thermometry in simple and complex systems. Contrib Mineral Petrol 62:129-139

Whitaker ML, Nekvasil, H, Lindsley DH, Difrancesco NJ (2007) The role of pressure in producing compositional diversity in intraplate basaltic magmas. J Petrol 48:365-393

Whitney JA, Stormer JC (1977) The distribution of $NaAlSi_3O_8$ between coexisting microcline and plagioclase and its effect on geothermometric calculations. Am Mineral 62:687-691

Wood BJ (1974) The solubility of alumina in orthopyroxene coexisting with garnet. Contrib Mineral Petrol 46:1-15

Wood BJ, Banno S (1973) Garnet-orthopyroxene and orthopyroxene-clinopyroxene relationships in simple and complex systems. Contrib Mineral Petrol 42:109-124

Wood BJ, Trigila R (2001) Experimental determination of aluminous clinopyroxene-melt partition coefficients for potassic liquids, with application to the evolution of the Roman province potassic magmas. Chem Geol 172:213-223

Wright FE, Larsen ES (1909) Quartz as a geologic thermometer. Am J Sci 27:421-447

Xirouchakis D, Hirschmann MM, Simpson J (2001) The effect of titanium on the silica content of mantle-derived melts. Geochem Cosmochim Acta 65:2029-2045

Yang HJ, Frey FA, Clague DA, Garcia MO (1996) Mineral chemistry of submarine lavas from Hilo Ridge, Hawaii: implications for magmatic processes within Hawaiian rift zones. Contrib Mineral Petrol 135:355-372

Yasuda A, Fujii T, Kurita K (1994) Melting phase relations of an anhydrous mid-ocean ridge basalt from 3 to 20 GPa: Implications for the behavior of subducted oceanic crust in the mantle. J Geophys Res 99:9401-9414

Yoder HS, Tilley CE (1962) Origin of basalt magmas: an experimental study of natural and synthetic rock systems. J Petrol 3:342-532

Thermometers and Thermobarometers in Granitic Systems

J. Lawford Anderson

Department of Earth Sciences, University of Southern California
Los Angeles, California, 90089-0740, U.S.A.

anderson@usc.edu

Andrew P. Barth

Department of Earth Sciences, Indiana University-Purdue University
Indianapolis, Indiana, 46202, U.S.A.

Joseph L. Wooden, Frank Mazdab

U.S. Geological Survey, 345 Middlefield Road
Menlo Park, California, 94205, U.S.A.

INTRODUCTION

Recent advances in our ability to resolve the *P-T-t* (pressure-temperature-time) paths of orogenic terranes have continued to re-emphasize the common marriage between petrology and tectonics. The importance of determining the *P-T* history of metamorphic rocks, as through chemical analysis of mineral zoning profiles and inclusions initially paved the way into this area of our science (e.g., Essene 1982, 1989; Hoisch 1991; Speer 1993). Parallel success has been demonstrated for igneous systems (see review by Anderson 1996), with the derived information being primarily related to ascent and/or cooling of the crystallizing magma. And, in multiply-intruded terranes, where the ages of intrusion span a significant portion of the orogenic deformation, dated plutons act as important crustal "nails," which offer first-order insight into the descent and/or ascent of deformed crust during the orogenic process.

Emplacement barometry can be retrieved from contact metamorphic rocks where the enclosing host assemblages contain appropriate compositions for barometric calculations. However, in pervasively intruded terranes where a range of ages of magmatic activity exists, multiple and extensive overprinting makes the use of host metamorphic mineral assemblages and mineral compositions challenging. In contrast, several igneous minerals are less prone to the loss of primary compositions, in part due to the inherent insensitivity of the rock to high-grade subsolidus reequilibration and the paucity of post-crystallization fluids. However, there are exceptions to this generalization and below we discuss an example where high-*T*, post-emplacement fluids have altered the entire margin of the Mt. Stuart batholith of Washington.

In general, many igneous rocks lack mineral assemblages suitable for thermobarometric analysis. Either the thermodynamic variance is high or, in specific cases, the phases are characterized by component exchanges easily reset during slow cooling or later thermal events. Two feldspar and iron-titanium oxide thermometry are examples of the latter case; only in shallow-emplaced plutons does feldspar or iron-titanium oxide thermometry yield consistent hypersolidus results. Nevertheless, as reviewed below, many granitic intrusions provide the means to determine temperature and pressure attending crystallization through mineral equilibria less prone to subsequent readjustment. For example, recent calibrations of Ti solubility in quartz and zircon and Zr solubility in sphene offer new means to determine crystallization temperatures in most igneous rocks. These can be also be used in concert with apatite, zircon, and allanite saturation temperatures.

For metaluminous granites, the Al-in-hornblende thermobarometer continues to be important, particularly where used in conjunction with the above thermometers or that of hornblende-plagioclase, pyroxene-hornblende, or two pyroxenes, with the caveat that the temperature-sensitive equilibria record the same conditions as that used for barometry. For peraluminous granites, the recently calibrated pressure-sensitive equilibrium of garnet-plagioclase-biotite-muscovite can be used in concert with the garnet-biotite thermometer. Other potential barometers are based on the pressure dependent Si-content of muscovite in equilibrium with biotite, alkali feldspar, quartz, and a hydrous fluid.

A basic question in igneous emplacement thermobarometry is whether or not the mineral phase compositions represent near-solidus conditions, as opposed to entrained solids grown or acquired during ascent or retrogressed during subsolidus cooling. Thorough petrography is an ever important prerequisite for the identification of solid phases in equilibrium with melt at the crustal level of final crystallization and emplacement as is the full elemental characterization of zoned minerals. Clearly, not all magmatic phases record the same history (e.g., Claiborne et al. 2006).

Plutonic barometry of Cordilleran orogenic terranes - a historical perspective

In their now classic papers, E-An Zen and Jane Hammarstrom (Zen 1985, 1989; Zen and Hammarstrom 1984; Hammarstrom and Zen 1986) demonstrated that many post-accretionary plutons within the exotic terranes of northwest North America (Idaho to British Columbia) are deep-seated with estimated emplacement depths ranging from 18 to 30 km. They founded the basis of magmatic epidote as having pressure significance, and pioneered the ever popular Al-in-hornblende barometer. For several northwest Cordilleran terranes, their work not only revealed deep levels of pluton emplacement but also demonstrated that plutonism was followed by tectonic ascent of these terranes at rates greater than 1 mm/yr. At about the same time, Anderson et al. (1988) first documented the mid-crustal origin of the lower plate of several metamorphic core complexes and, based on thermobarometry of successive plutonic events, determined tectonic ascent at rates of 2 mm/yr during core complex extension. Similarly, Barth (1990) documented the mid-crustal origin of a Mesozoic complex in the San Gabriel (Tujunga) terrane of southern California. Barth (1990) further showed that portions of the San Gabriel terrane remained at depths of approximately 22 km throughout all phases of Triassic to Late Cretaceous plutonism, followed by rapid tectonic uplift during the early Tertiary. These papers were part of a wave of renewed and continued interest in plutonic thermobarometry in response to Zen and Hammarstrom's contributions. In 1996, Anderson provided an overview of the status of thermobarometry in granitic batholiths. Now, over a decade later, this contribution serves to build on that earlier work, along with companion papers by Putirka (2008) and Blundy and Cashman (2008).

SINGLE MINERAL TRACE ELEMENT THERMOMETERS IN IGNEOUS SYSTEMS

Understanding the mechanisms that cause diversity in plutonic rocks and how these mechanisms are linked to the processes controlling explosive volcanic eruptions requires improved understanding of the thermal evolution of plutonic and volcanic rocks. However, information about this thermal evolution is most often limited to the superstructure of the youngest systems, because the evolutionary history of the older and the deeper parts of volcanic systems, and their plutonic roots is often obscured or erased by hydrothermal alteration or later metamorphism. Recent experimental calibrations, with support from empirical observations in natural systems, have shown the value of using the trace element content of zircon, quartz, and sphene for estimating temperatures and evaluating the thermal evolution of intermediate to silicic igneous rocks (Wark and Watson 2006; Watson et al. 2006; Ferry and Watson 2007; Hayden et al. 2007; Ferriss et al. 2008). These experiments are particularly important because of the near ubiquitous occurrence of quartz in intermediate and silicic plutonic rock and as phenocrysts in silicic volcanic rocks, the widespread occurrence of trace amounts of zircon and sphene in these same rocks, and the potential resistance of these minerals to later metamorphic and hydrothermal re-equilibration. New insights into deeper volcanic and plutonic processes, and variation in these processes through greater spans of geologic time, are contained in compositional variations within and among these major and trace phases.

The observation of the linear temperature dependence of trace element contents of zircon, rutile, and sphene developed in these studies is particularly powerful because these minerals are well-known hosts of the radiogenic parent isotopes of U, Th and Lu. Thus, temperature-time histories may now be constructed from combined multi-element and isotopic analyses of these phases. This is particularly relevant given recent findings confirming that many batholiths can form over several million years through a protracted plutonic history (Matzel et al. 2006; Miller et al. 2007; Walker et al. 2007). Furthermore, because zircon saturation is not always achieved during partial melting reactions and/or during the ascent of intermediate and silicic magmas, pre-magmatic zircon crystals are often incompletely dissolved, and thus may retain thermal, petrologic and geochronologic information about the magma source region. The chemistry of zircon, sphene, and quartz can, therefore, track elements from the full evolutionary path of intermediate and felsic magma formation, evolution and solidification.

Ti-in-quartz thermometer

Wark and Watson (2006) developed the following experimental calibration describing the temperature dependence of the titanium content of quartz:

$$\log \text{Ti(ppm)}_{qtz} = 5.69 - [3765/T(K)] \tag{1}$$

which expressed in T (°C) becomes

$$T(°C) = \left(-\frac{3765}{\log \text{Ti (ppm)}_{qtz} - 5.69} \right) - 273 \tag{2}$$

For rutile-undersaturated rocks (a common condition in igneous rocks), Wark and Watson (2006) offer the following:

$$T(°C) = \left(-\frac{3765}{\log\left(\text{Ti(ppm)}_{qtz}/a_{TiO_2}\right) - 5.69} \right) - 273 \tag{3}$$

Ti concentrations greater than about 15-20 ppm can be measured with a conventional electron microprobe, but quartz formed at lower temperatures requires ion microprobe or ICP-MS analyses. Although all of the experiments on which (1) is based were run at 10 kbar, comparison to limited 1 atm experiments and empirical observations (Wark and Watson 2006; Wark et al. 2007; Wiebe et al. 2007) suggests the effect of pressure on this relation is negligible for crustal rocks formed at 10 kbar or less. Analytical uncertainties are quite small in the application of this thermometer, but temperature can be seriously underestimated if $a_{TiO_2}<1$, as much as 50-75 °C over the typical temperature range of quartz growth during solidification of intermediate to silicic melts. If a_{TiO_2} is unconstrained, Equation (1) yields a minimum temperature of crystallization assuming $a_{TiO_2} = 1$. For example at $a_{TiO_2} = 1$, quartz Ti concentration of 40 ppm yields a temperature of 648 °C, and 100 ppm derives the estimate of 747 °C. At $a_{TiO_2} = 0.6$, these estimates rise to 700 and 813 °C, respectively. Reduced titanium activity can be estimated using the rutile solubility relation of Hayden and Watson (2007), or calculated independently from Equation (1) if another mineral geothermometer is available, or through sphene-based equilbria (Xirouchakis et al. 2001a, b), as described further below.

An early application of this thermometer in igneous systems is the study of quartz phenocrysts in the Bishop Tuff, east central California (Wark et al. 2007). Previous studies have estimated the temperature and compositional zonation of this unit using other equilibria (e.g. Hildreth 1981). Wark et al. (2007) showed that the quartz Ti thermometer yields temperatures in good agreement with temperatures calculated from the compositions of coexisting Fe-Ti oxides assuming an a_{TiO_2} of 0.63±0.03. Furthermore, well-preserved compositional zonation in quartz that was imaged by cathodoluminescence microscopy recorded an abrupt increase in temperature of ca. 60 °C in the later erupted (hence deeper) parts of the tuff. Wark et al. (2007) related these high-T quartz rims to renewed growth induced by an increase in T and decrease in a_{H_2O} caused by mafic magma injection immediately preceding the climactic eruption of the tuff. These results indicate that Equation (1) constitutes a robust thermometer that can recover thermal variations in evolving, compositionally-zoned silicic volcanic systems. The work also presents the issue of accurately assessing crystal growth history, given that each mineral can be stable and grow over a range of conditions and matching such to that of other minerals remains a serious challenge.

To date, this thermometer has not been tested on many granitic plutons and it may not resistant to subsolidus re-equilibration given the ease that quartz undergoes recrystallization. However, the study of Wiebe et al. (2007) on the Vinalhaven granite of Maine shows clear promise. Assuming an a_{TiO_2} of 0.6, based on the study of Wark et al. (2007) for the Bishop Tuff, their work resulted in Ti-in-quartz temperatures ranging 700-770 °C for a fine grained phase of the pluton, with quartz cores yielding a T near 810 °C. Likewise, quartz in a coarse-grained granite yielded temperatures ranging 700-840 °C, the higher range of which clearly reflect hypersolidus conditions.

Ti-in-zircon and Zr-in-rutile thermometers

Zircon in igneous systems is usually very near to the ZrSiO$_4$ end member, with 0.5 to 2.5% HfSiO$_4$ solid solution (Speer 1982). Among the many trace elements found in zircon (Hoskin and Schaltegger 2003), quadravalent Ti can substitute for Si, as indicated by the quartz experiments of Wark and Watson (2006) described above. Watson et al. (2006) developed a combined empirical and experimental calibration that describes the following temperature-dependent variation in the abundance of titanium in zircon:

$$\log Ti(ppm)_{zircon} = 6.01 - [5080/T(K)] \quad (4)$$

Rutile grown in the presence of zircon by Watson et al. (2006) preserved Zr abundances that were linearly dependent on temperature. Similarly, they offered the calibration

$$\log Zr(ppm)_{rutile} = 7.36 - [4470/T(K)] \quad (5)$$

Equation (5) is of limited utility as a thermometer in igneous rocks because of the rarity of rutile-saturated magmas (Hayden and Watson 2007), but this thermometer will find utility in metamorphic systems, particularly eclogites.

Ti concentrations can be measured in zircon formed at high temperature using a conventional electron microprobe, but ICP-MS or ion microprobe analyses are necessary for zircon crystallized at moderate to low temperatures characteristic of many intermediate and felsic igneous systems (Wooden et al. 2006).

In a study of the Spirit Mountain batholith of Nevada, Claiborne et al. (2006) adjusted Equation (4) to reflect an a_{TiO_2} of 0.7, based on conditions appropriate for sphene and titanomagnetite saturation, as the following

$$T(°C) = \frac{5080}{\left(6.01 - \log\left(0.7 \cdot \text{ppmTi}_{\text{zircon}}\right)\right)} - 273 \qquad (6)$$

Their results for zircons having Ti concentrations ranging 3.3 to 33 ppm range 680-899 °C, and are comparable to their reported zircon saturation temperatures (see below) which range from 696-878 °C.

Ferry and Watson (2007) presented an improved thermodynamic analysis of Ti substitution in zircon, demonstrating the dependence of Ti substitution on both the activities of TiO_2 and SiO_2 (using rutile and α-quartz as standard states). Their revised calibration of the thermometer emphasizes the effects of reduced activities on Ti abundance in zircon:

$$\log \text{Ti}(\text{ppm})_{\text{zircon}} = 5.711 - \left[4800 / T(K)\right] - \log\left(a_{SiO_2}\right) + \log\left(a_{TiO_2}\right) \qquad (7)$$

The differences between Equations (6) and (7) derive from additional experiments used in the later calibration, and the incorporation of coefficients to account for reduced activities of SiO_2 and/or TiO_2 in natural systems that crystallized zircon but were not quartz- and/or rutile-saturated. Ferry and Watson (2007) suggested that temperatures calculated using Equation (4) will be underestimated by ≤70 °C if $a_{TiO_2} < 1$ and overestimated by approximately the same amount if $a_{SiO_2} < 1$; these effects thus may partly compensate for each other, but care must be taken to evaluate effects of reduced activities (Heiss et al. 2008)

Equation (7) is particularly powerful for relative thermometry from mineral compositional zoning profiles and for accurate thermometry from systems where the activities are well-known. As an example, consider the complexly zoned zircons preserved in typical calc-alkalic plutonic rocks in the U.S Cordillera. Trace element analyses of zircons from a Late Cretaceous granodiorite and granite from the eastern Transverse Ranges of California illustrate the application of the zircon thermometer to gain insight into the petrogenesis of igneous rocks. Analyses of the zircons were completed by ion microprobe on the USGS SHRIMP-RG and representative analyses are given in Table 1. The ion microprobe has the advantage of lower detection limits in comparison to the electron microprobe, but it necessitates using a 20-30 μm spot size, about 10 times the diameter of a typical electron microprobe spot. The value and limitation of this technique are shown in Figure 1, where it is apparent that although the ion microprobe yields both crucial age and extensive compositional data for each spot within these complexly-zoned zircons, it clearly averages compositionally variable domains that are visually resolvable through an SEM image and would be resolvable (in part, for Hf, Y, U and perhaps some heavy rare earth elements) with an electron microprobe.

Zircon thermometry results for a sphene, hornblende, biotite granodiorite indicate crystallization of zircon over a wide temperature range, with 9000-16000 ppm Hf in solid solution (Fig. 2). Hf-rich zircon rims yield an average temperature of ~713±30 °C ($a_{SiO_2} = 1$), in good agreement with the quartz- and water-saturated granodiorite solidus, which is appropriate for this bulk composition, and in agreement with the temperature of 708±17 °C calculated from hornblende - plagioclase thermobarometry in another sample of this granodiorite (Needy et al., in press). Zircon interiors are typically Ti-enriched relative to rims and yield similar to higher temperatures at progressively lower Hf concentrations, up to an average temperature about 60 °C above the estimated temperature of zircon saturation (see also Clairborne et al. 2006, for correlation of whole-rock and zircon Hf content and fractionation). However, all of these calculated temperatures are probably a few tens of degrees too high, because phase equilibria suggest the granodiorite was not silica-saturated at these temperatures. Temperatures of zircon interiors recalculated at $a_{SiO_2} = 0.55$ are in reasonable agreement with phase equilibria and with the estimated zircon saturation temperature. Comparing core and rim temperatures and zoning profiles indicates that the zircons record solidification of a relatively cool, crystal-rich

Table 1. Representative SHRIMP-RG ion microprobe analyses of zircons from Blue granodiorite.

Spot	Sample	Hf (ppm)	Ti (ppm)	T(W)	T(FW)1	T(FW)2
JW340-5.1	rim	15610	4.1	668	698	na
JW340-6.1	core	10945	6.8	708	743	690
JW340-6.2	rim	13530	5.4	690	722	na
JW340-6.3	intermediate	9481	24.9	828	881	813

Notes:
T(W) = temperature by calibration of Watson et al (2006)
T(FW)1 = temperature by calibration of Ferry and Watson (2007), a_{SiO_2} = 1 and a_{TiO_2} = 0.7
T(FW)2 = temperature by calibration of Ferry and Watson (2007), a_{SiO_2} = 0.55 and a_{TiO_2} = 0.7

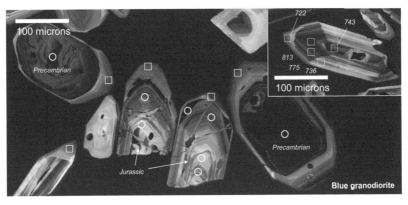

Figure 1. Scanning electron microscope (SEM) cathodoluminescence image of zircons in the Blue granodiorite. Circles are the locations of ion microprobe spot analyses of pre-magmatic cores of Precambrian and Jurassic age, and squares are locations of spot analyses of magmatic rims. Rims yield an average temperature of 713±39 °C with the Ferry and Watson (2007) calibration (a_{SiO_2}=1, a_{TiO_2}=0.7). Inset shows an oscillatory-zoned, magmatic zircon crystal preserving large temperature variations during growth, within the interval between zircon saturation and the wet solidus.

granodiorite magma that may have never been above its saturation temperature, consistent with extensive preservation of premagmatic zircon. Analyses of discrete compositional bands within the magmatic zircon indicate dynamic crystallization of the granodiorite within the temperature range bounded by the zircon saturation temperature and the water-saturated solidus.

Zircon thermometry results for a nearby pluton, the Palms biotite granite (Fig. 3), allow us to compare the crystallization histories of two calc-alkalic plutons. The Palms granite has a much more limited range in temperature between the estimated water-saturated liquidus and solidus and earlier silica saturation, illustrating the well-known near eutectic-like composition of hydrous granitic bulk compositions. Hf-rich zircon rims yield temperatures in reasonable agreement with the wet solidus (a_{SiO_2} = 1). Hf-poorer zircon interiors yield temperatures well above the nominal liquidus temperature and many yield temperatures up to 100 °C above zircon saturation. However, all of these calculated temperatures are too high because phase equilibria suggest that the granite was probably not quartz saturated over the temperature range. Recalculated temperatures of Hf-poor zircon interiors at a_{SiO_2} = 0.5 average ~800 °C, near to the estimated zircon saturation temperature; these values are more reasonable, because the granite also preserves premagmatic zircon in the cores of many grains.

Thermometers & Thermobarometers in Granitic Systems 127

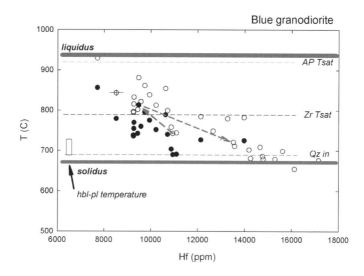

Figure 2. Hf vs. temperature plot for all analyzed magmatic zircons in the Blue granodiorite, using the thermometer of Ferry and Watson (2007). Open circles for all spot analyses ($a_{SiO_2} = 1$, $a_{TiO_2} = 0.7$), and closed circles for recalculated grain interiors ($a_{SiO_2} = 0.55$, $a_{TiO_2} = 0.7$). Representative analytical precision (box, central left) based on replicate analyses of standard zircon CZ3. Water-saturated granodiorite liquidus, Qz-in, and solidus at 4 kb from Piwinskii (1968) and Stern et al. (1975). Zircon saturation temperature from Watson and Harrison (1983) and apatite saturation temperature from Harrison and Watson (1984). Solidus temperature of 708±17 °C estimated from hornblende - plagioclase thermobarometry (Needy et al., in press). Dashed arrows connect analyses within a single zircon crystal (see Fig. 1), illustrating large temperature variations without significant zircon dissolution during solidification of the granodiorite.

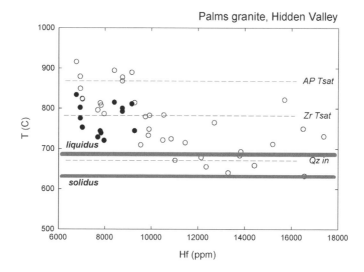

Figure 3. Hf vs. temperature plot for all analyzed magmatic zircons in Palms granite, using the thermometer of Ferry and Watson (2007). Open circles for all spot analyses ($a_{SiO_2} = 1$, $a_{TiO_2} = 0.7$), and closed circles for recalculated grain interiors ($a_{SiO_2} = 0.5$, $a_{TiO_2} = 0.7$). Water-saturated granite liquidus, Qz-in, and solidus at 4 kbar from Piwinskii (1968) and Stern et al. (1975). Zircon saturation temperature from Watson and Harrison (1983), and apatite saturation temperature from Harrison and Watson (1984).

The core to rim temperature estimates of zircon indicate that the granite and granodiorite magmas crystallized over similar temperature ranges, from near to or below zircon saturation down to the wet solidii. Most of the Palms granite zircons indicate crystallization above the water-saturated liquidus, indicating that early crystallization was probably under water-undersaturated conditions and that the magma may have been largely liquid early in its crystallization history. In contrast, even for water-saturated conditions, the granodiorite records no evidence for crystallization near its liquidus, and thus was probably a comparatively crystal-rich magma. These results, along with crystal zoning profiles, provide important limitations for models of chemical evolution and physical processes of magma transport and interaction in this calc-alkalic arc setting.

Two other recent studies (Page et al. 2007; Fu et al. 2008) have found apparent low temperatures derived from the Ti-in-zircon geothermometer in a range of settings and have suggested that parameters other than temperature and a_{TiO_2} and a_{SiO_2} may be important in controlling the Ti content of zircon. One issue may be the effect of pressure, which needs to be further evaluated. Although most of the experiments from which Equation (4) was derived were run at 10 kbar, limited experimental data combined with empirical observations from rocks crystallized at higher pressure initially suggested that the effect of pressure on Ti-in-zircon thermometry was on the order of 50 °C per 1 GPa (10 kbar) of uncertainty (Ferry and Watson 2007). In contrast, the more recent study of Ferriss et al. (2008) indicates a larger pressure dependence for the geothermometer, on the order 100 °C per 1 GPa. Clearly, this promising geothermometer needs to be used in awareness of these issues.

Zr-in-sphene thermometer

Sphene in igneous systems is usually near to the $CaTiSiO_5$ end member, exhibiting some substitution of Al and Fe^{+3} charge-balanced by coupled substitution of OH or F (Ribbe 1982). Among the many trace elements found in sphene, quadravalent Zr can substitute for Ti, as can be inferred from the rutile experiments of Watson et al. (2006) described above. Hayden et al. (2007) developed a combined empirical and experimental calibration that describes the temperature and pressure dependence of the abundance in sphene, which is as follows:

$$\log Zr(ppm)_{sphene} = 10.52 - \left[\frac{7708}{T(K)}\right] - \left[960 \times \left(\frac{P(GPa)}{T(K)}\right)\right] - \log(a_{TiO_2}) - \log(a_{SiO_2}) \quad (8)$$

The experiments upon which Equation (8) are based were run at 800-1000 °C and 10-24 kbars; addition of some natural, well-characterized sphene samples can extend the calibration to lower temperatures and pressures (Mazdab et al. 2007). Zr concentrations greater than about 150 ppm (~800 °C) can be measured with a conventional electron microprobe. Care must be taken to avoid interferences caused by overlapping peaks from Nb and Y. Although most problematic for background location at low Zr concentrations, tails from some of these interferences on the Zr peak may also be important; the significance of this is a function of the relative concentrations of Zr, Nb and Y, as well as the quality of the electron probe and analyzing crystal. Sphene formed at lower temperatures requires ion microprobe analyses (Mazdab et al. 2007). All the Zr isotopes are interfered by dimers that are largely not resolvable by any mass spectrometer, limiting the detection limit to approximately 10 ppm.

Analytical uncertainties are quite small in the application of this thermometer, but temperature can be seriously underestimated if a_{TiO_2} < 1, which would be expected to be common in igneous rocks as almost all lack rutile (see Hayden and Watson 2007) and/or if a_{SiO_2} <1. Our findings are that underestimates for the above reasons can be as much as 50-75 °C over the typical temperature range for solidification of intermediate to silicic melts. An additional important complication is that sphene typically displays sector zoning, with anomalously Zr-enriched sectors that should be recognized and avoided in using Equation (8) to calculate temperatures of crystallization (Hayden et al. 2007; Mazdab et al. 2007). The origin of Zr-enriched sectors in sphene are not known to us and offer new areas of study. However, Watson and Liang (1995) suggest that sector zoning can occur during slow growth when lattice diffusion rates for the relevant element are low, and when the diffusing element is highly charged. Their model was developed for rare earth elements, but likely may also apply to Zr in sphene.

For the typical case where rocks contain sphene and quartz, but not rutile, a_{SiO_2} can be assumed to be unity, at least in the later stages of crystallization. Finally, a_{TiO_2} can be estimated from the rutile solubility model of Hayden and Watson (2007), requiring knowledge of the composition of the melt from which the sphene crystallized. This can also be derived from electron microprobe analysis of glass coexisting with sphene or estimated from whole rock analysis.

Precise and accurate application of the Zr content of sphene as a thermometer for igneous rocks requires an independent estimate or calculation of temperature (and pressure), for example from the hornblende-plagioclase thermobarometer (Anderson and Smith 1995; see below).

ACCESSORY PHASE SATURATION THERMOMETRY

Harrison and Watson (1984) and Watson and Harrison (1983) initially calibrated compositional and temperature constraints on the saturation conditions of zircon and apatite in silicic melt systems. Subsequently, Green and Adam (2002) determined that apatite saturation is not effected by pressure up to 25 kbar. More recently, Tollari et al. (2006) have offered a revision of the apatite saturation thermometer that is particularly applicable to peraluminous granites.

Saturation thermometers, such as these presented here, offer a powerful means to assess temperature of magmatic systems. Where used in conjunction with critical petrography, they offer constraints on magmatic liquidus temperatures as well as an additional test of solidus conditions. Should one of these phases, usually zircon, be inherited, antecrystic, or of cumulate origin, derived temperatures may not have geologic significance and be in excess of the liquidus. In contrast, should zircon saturation temperatures be near-solidus, then inherited or antecrystic zircon can be expected to partially or fully dissolve. Recent applications include Hoskin et al. (2000), Miller et al. (2003), Hoskin and Schaltegger (2003), Janousek et al. (2004), Lopez-Moro and Lopez-Plaza (2004), Walker et al. (2007), and Miller et al. (2007) and all concluded their findings consistent with other forms of thermometry (see also Clairborne et al. 2006, mentioned above). However, Harrison et al. (2007) has presented evidence that the zircon saturation thermometer may at times underestimate the onset temperature of zircon crystallization, which will require further evaluation.

Figure 4 exhibits new apatite and zircon saturation temperatures for Cretaceous plutons near Yosemite National Park, California, near Cinko Lake of the Sierra Nevada batholith (Foley et al. 2007). Notably, petrography for these granitoids shows that apatite is an early phase in of the crystallization sequence relative to that for zircon. As indicated below, these plutons, based on hornblende-plagioclase thermobarometry (Anderson and Smith 1995; Holland and Blundy 1994 calibrations) were emplaced at 698-715±18 °C (error based on multiple analysis of hornblende and plagioclase rims, usually in contact with quartz). In contrast, apatite saturation temperatures for both plutons are near 900 °C, which offers an estimate of liquidus temperatures. Zircon saturation temperatures range 790±10 °C for the Harriet Lake pluton and 740±6 °C for the Fremont Lake pluton, which are above but near the solidus and are comparable to zircon saturation temperatures obtained elsewhere in the Yosemite region (Miller et al. 2007).

Klimm et al. (2008) have recently offered experimental data in support of temperature constraints on allanite saturation. As one of the epidote group minerals, the LREE enrichment of allanite expands its stability field. Unlike epidote, allanite appears as an accessory phase in many

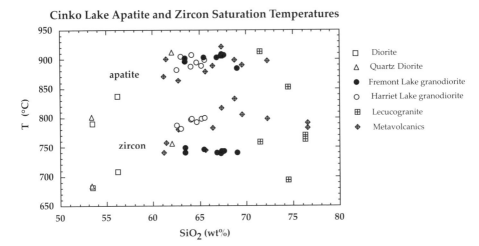

Figure 4. Apatite and zircon saturation temperatures for plutons and older metavolcanics units in the Cinko Lake region of Yosemite National Park (Foley et al. 2007).

metaluminous granites with possibly no pressure significance. In magmatic epidote-bearing plutons, allanite also can also form cores of these crystals. Thus, its onset of crystallization provides additional information on plutonic melt conditions between the liquidus and solidus. Klimm et al. (2008) have conducted melt experiments with a mid ocean ridge basalt at 750-900 °C on the Ni-NiO buffer to constrain the *T-X* (REE composition) control of allanite stability in melts having a silica range above 50% SiO_2. This accessory phase saturation thermometer has not yet been tested for plutonic rocks but offers important additional information on temperatures of magmas, both plutonic and volcanic. The pressure effect needs also to be further investigated.

THERMOBAROMETERS APPLICABLE TO GRANITIC ROCKS

Al-in-hornblende barometry

This popular barometer has yet to be fully experimentally evaluated but it has clearly served well. Anderson and Smith (1995) recalibrated the barometer to account for the effects of temperature using the experimental data of Johnson and Rutherford (1989) and Schmidt (1992). Their calibration is

$$P(\text{kbar}) = 4.76\text{Al} - 3.01 - \left[\left(\frac{T\ °C - 675}{85}\right)\right] \times \left[0.530\text{Al} + 0.005294(T\ °C - 675)\right] \quad (9)$$

where Al is the sum of $Al^{iv} + Al^{vi}$ per 13 cations. Ague (1997) has offered a thermodynamic model for the barometer and Ernst and Liu (1998) have presented new experimental data in support of a more qualitative model that emphasizes both the Al and Ti contents of amphibole. Earlier calibrations, both empirical and experimental (Hammarstrom and Zen 1986; Hollister et al. 1987; Johnson and Rutherford 1989; Schmidt 1992), which built the early formulation of this thermobarometer, lacked correction for the effects of temperature, which can be substantial (see Anderson and Smith 1995). Hence, early uses of the barometer that lacked temperature corrections should be viewed with caution and subject to revision. An appropriate thermometer is that based on plagioclase + hornblende (Holland and Blundy 1994) which offers calibrations based on two equilibria:

$$\text{edenite} + 4 \text{ quartz} = \text{tremolite} + \text{albite} \tag{10a}$$

$$\text{edenite} + \text{albite} = \text{richterite} + \text{anorthite} \tag{10b}$$

Anderson (1996) concluded that the calibration based on the second reaction (10b), which yielded lower temperatures, more precisely produced temperatures from other thermometers. It was thus the preferred calibration of the thermometer, including in conjunction with the Anderson and Smith (1995) calibration of the Al-in-hornblende barometer, which in simultaneous solution often yield P-T solutions at or above the appropriate granite to tonalite solidii. In their own investigation of the Fish Canyon tuff, Bachman and Dungan (2002) made the same conclusion. It should be noted that Holland and Blundy (1994) have an appropriate hornblende formula scheme imbedded in their two thermometers that requires its use in their application.

Interest in the Al-in-hornblende thermobarometer is quite evident in recent publications (e.g., Sial et al. 1999; Stone 2000; Barnes et al. 2001; Stein and Dietl 2001; Percival et al. 2002, Asrat et al. 2004; Janousek et al. 2004; Zhang et al. 2006; Rodriguez et al. 2007). Most of these have used the P-T information to document the depth of batholith emplacement. In volcanic systems, Manley and Bacon (2000), Lindsay et al. (2001), and Bachman et al. (2002), used the thermobarometer to determine the magma chamber depth feeding into volcanic systems. In contrast, Bachl et al. (2001), Miller and Miller, (2002), Needy et al. (2008), and Barth et al. (2008) used the thermobarometer to document tilted crustal sections enclosing plutons.

Anderson et al. (2007) applied granitic pluton thermobarometry and host metavolcanic units thermometry to present a model of upper crustal overturn attending pluton emplacement in construction of the Sierra Nevada batholith in the vicinity of Yosemite Valley. Their results depict shallow pluton emplacement at ~2.7±0.4 kbar at 698-715±18 °C (Fig. 5), leading to near thermal metamorphic equilibration of steeply-dipping and lineated host metavolcanic units at 614-712 °C, which are inferred to have descended to pluton levels at ca. 9-10 km before or during pluton emplacement. Their model of "upper crustal overturn" is based on the conclusion that surficial volcanic deposits had, subsequent to eruption, had been overturned and descended to plutonic depths of ~10 km, a topic further addressed by Paterson et al. (2008).

The Al-in-hornblende thermobarometer also had success in documenting cases of subsolidus reequilibration on a batholith scale. Figure 6 offers comparison of P-T estimates for emplacement of 14 Mesozoic plutons in the Mojave Desert region of southern California (Anderson 1996) to the Cretaceous Mt. Stuart batholith of Washington (Anderson 1997; Anderson et al. 2008). Both apply the same calibration of hornblende-plagioclase thermometer (Holland and Blundy 1994) used in concert with the Anderson and Smith (1995) calibration of the Al-in-hornblende barometer. The Mojave Desert region of California is rather unique in the western U.S. in its variable exposures of upper and middle crustal exposures of the Mesozoic arc (Anderson et al. 1992), and, it is notable in this context that all of these plutons yielded P-T solutions at or above the granite to tonalite solidii. Clearly, sizeable portions of the Mt. Stuart batholith yield erroneously low and subsolidus P-T solutions. The batholith is also isotopically zoned with respect to margin enrichment in $\delta^{18}O$ (ranging 7-8‰ upwards to over 10‰). Anderson et al. (2008) documented that where the batholith margin whole-rock oxygen isotope values of $\delta^{18}O$ are elevated, ΔPl-Qz ($\delta^{18}O$ of plagioclase vs. that of quartz) values are large (> 1.5; due largely to increased $\delta^{18}O$ of plagioclase, but in detail the $\delta^{18}O$ of both minerals have increased), which relates to a lower temperature of oxygen isotope equilibration of these minerals. They attributed this to subsolidus infiltration of high-temperature contact aureole fluids into the batholith margin, which both elevated the oxygen isotope values of all major minerals of the batholith margin and also reset hornblende and plagioclase elemental compositions. Bucher and Frost (2006) reported similar influx of contact metamorphic fluids into an anorogenic granite in Antarctica.

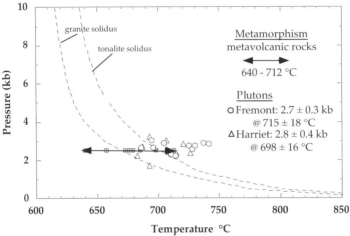

Figure 5. *P-T* estimates of pluton emplacement and contact metamorphism of host metavolcanic units, based on hornblende-plagioclase thermobarometry for Cinko Lake area of Yosemite National Park, based on the calibrations of Anderson and Smith (1995) and Holland and Blundy (1994). (Modified from Anderson et al. 2007).

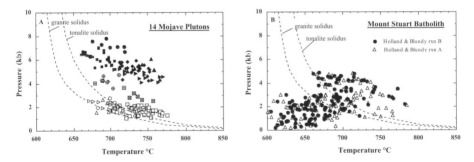

Figure 6. Results of hornblende-plagioclase thermobarometry for A) fourteen deep- to shallow-emplaced plutons exposed in the mountains of the of the Mojave Desert, California (modified from Anderson (1996; symbols represent different plutons); and B) the Mount Stuart batholith (Anderson et al. 2008). Estimates of pressure based on the Al-in-hornblende calibration of Anderson and Smith (1995) with temperature from the hornblende-plagioclase calibration (reaction B) of Holland and Blundy (1994). For the Mount Stuart, results of two calibrations of the Holland and Blundy (1994) hornblende-plagioclase thermometer are depicted.

Thermobarometry for peraluminous granites

The low molar volume of garnet introduces a significant pressure dependence to garnet-based mineral equilibria in metamorphic and igneous systems. The garnet-hornblende-plagioclase equilibrium, initially calibrated by Kohn and Spear (1990), has been revised by Dale et al. (2000). Holdaway (2001) likewise offered a revision of the GASP (garnet-Al silicate-silica-plagioclase) geobarometer in concert with an earlier revision of the garnet-biotite thermometer (Holdaway 2000). Wu et al. (2004a) presented an empirical-based barometer for garnet-biotite-plagioclase-quartz assemblages based on Holdaway's (2000, 2001) revisions. It offers a new avenue of pluton barometry in garnet, two-mica granites in comparison to prior empirical calibrations by Ghent and Stout (1981) and Hodges and Crowley (1985) for the assemblage garnet-biotite-plagioclase-muscovite.

The basic premise of these barometric reactions is that with increasing pressure, garnet becomes more enriched in the grossular component at the expense the An (anorthite) component of plagioclase. Regardless of calibration, petrographic evidence for early vs. late crystallization is of first order importance. Figure 7 depicts P-T solutions for core to rim compositions of a garnet, two mica granite using the Ghent and Stout (1981) and Ganguly and Saxena (1984) calibrations. As a reflection of mineral zoning, the data reveal lower P-T solutions from core to rim and, thus, are seen as indicative of crystallization during ascent. Some rims have also yielded subsolidus results, which is a further indication of retrogression that is reported above for this batholith.

Many peraluminous, garnet-bearing plutons also contain muscovite with a significant celadonite component $(K(Mg,Fe^{2+})(Fe^{3+},Al)[Si_4O_{10}](OH)_2$, but with important substitutions such as that for Ti, F, and Cl). The pressure-dependent solution toward the celadonite end member (Velde 1965) also provides useful pressure information. Wei and Powell (2004, 2006) have offered new thermodynamic datasets to evaluate pressure for muscovite-bearing assemblages. Their work should be evaluated for pluton barometry along with that of Berman et al (2007) on constraints on the stability of Al end members of biotite, but such is beyond the scope of this contribution.

Oxybarometers and other mineral reactions involving oxygen

In contrast to the other intensive parameters (T, P, f_{H_2O}) that affect mineral crystallization and composition in granitic magmas, oxygen fugacity stands out due to its remarkable ability to range several tens of order of magnitude even in crustal-derived magmas. Such is not only reflected in the relative abundance of magnetite (high f_{O_2}) versus ilmenite (low f_{O_2}), but also in the modal percentages of mafic minerals and their Fe/Mg composition.

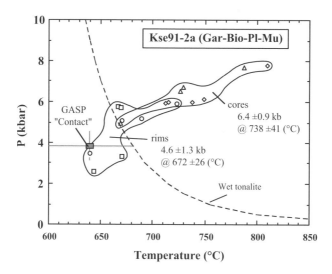

Figure 7. Estimates of emplacement conditions of a garnet, two-mica granodiorite, Mt. Stuart batholith, based on garnet-plagioclase-muscovite-biotite equilibria, using calibrations of Ghent and Stout (1981) and Ganguly and Saxena (1984). The box labeled "GASP 'Contact'" refers to estimate of the P-T of the contact aureole based on garnet-aluminosilicate- silica-plagioclase equilibria. Modified from Paterson et al. (1994).

Fe-Ti oxides, pyroxenes, olivine, and quartz (QUILF) and sphene (titanite)

The computational program and related data set termed "QUILF" for the reaction quartz + ulvospinel = ilmenite + fayalite (Andersen et al. 1993) continues to be utilized, e.g., Elliott et al. (1998), Percival and Mortensen (2002), Ren et al. (2006), and Rodriguez et al. (2007). It offers a range of T-f_{O_2} estimates for oxide-, pyroxene-, and/or olivine-bearing assemblages, even where not all included minerals are present, including olivine.

Xirouchakis et al. (2001a) expanded QUILF to include sphene and in doing so optimized standard-state properties for end member sphene (P2₁/a structure) consistent with thermodynamic data in the QUILF program. Their work confirmed the Wones' (1989) calibration of the assemblage sphene + magnetite + quartz and his prediction that the assemblage is a fundamental boundary between relatively reducing and oxidizing conditions. It is also considered to improve the precision of calculations derived through QUILF and, in addition to its use to determine T and f_{O_2}, it potentially can be used to compute the activities of silica and titania with respect to quartz and rutile in other thermometers (see above section) where these phases are lacking. Manon et al. (2008) have provided new entropy data for sphene, which should provide means to further improve this application. The current version of QUILF (QUILF95) can be downloaded from *http://www.geosciences.stonybrook.edu/people/faculty/lindsley/lindsley.html*.

Xirouchakis et al. (2001b) explored the use of sphene-based QUILF in several magmatic systems including an ilmenite-series granitic complex in Japan, the Skaergaard Intrusion of Greenland, and the Fish Canyon Tuff of Colorado. More recent applications include that of Ryabchikov and Kogarko (2006) on an alkaline complex in the Kola Peninsula of Russia.

Figure 8 shows minimum T-f_{O_2} estimates for the ilmenite-series Mt. Stuart batholith (Anderson et al. 2008) of Washington using the QUILF program indicating crystallization conditions near the quartz-fayalite-magnetite buffer. However, the more reduced determinations derive from more silica-rich portions of the batholith and are considered to reflect subsolidus

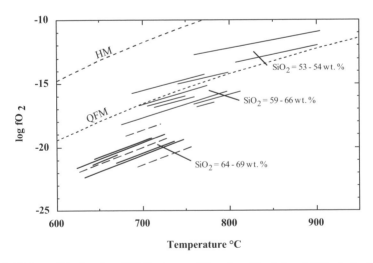

Figure 8. Minimum T-f_{O_2}, based on QUILF estimates of ilmenite solid solution stability, for the Mt. Stuart batholith at temperatures obtained from thermometry. Each line, some of which are dashed for clarity, represents individual samples. Low f_{O_2} results for more silica rich rocks of the batholith, which mostly occur near the pluton margin, are viewed as the result of reequilibration with high-T fluids derived from the contact aureole (modified from Anderson et al. 2008).

infiltration of high-T fluids from the graphite-rich units of the contact aureole, which, as indicated above, also lowered Al-in-hornblende values.

Magnetite and ilmenite series of granitic plutons and oxygen fugacity

Ishihara's (1977) recognition of magnetite- and ilmenite-series granites was another milestone that in its own right prompted renewed attention to the redox equilibria expressed in the mineralogy of granitic batholiths, particularly that of the Fe-Ti oxides. It followed the classic contributions of Buddington and Lindsley (1964) on Fe-Ti oxide oxybarometer and that of Wones and Eugster (1965) on biotite-alkali feldspar-magnetite equilibria, which was further advanced by the studies of Wones (1981, 1989), Carmichael and Ghiorso (1990), Frost (1991), Ghiorso and Sack (1991), and Andersen et al. (1993).

Frost et al. (2001) concludes the magnetite/ilmenite-series classification to be particularly applicable to continental margin-related batholiths. Ishihara's (1977) proposal was that ilmenite-series granitoids were derived and crystallized under oxygen reducing conditions in contrast to those of magnetite-series. Gastil et al. (1990) used the basis of the classification to divide the Peninsular Ranges batholith of southern and Baja California with more reduced (ilmenite-series) granitic assemblages occurring on the continental side of the arc. In contrast, Bateman et al. (1991) determined that the Sierra Nevada batholith, while mostly magnetite series, has western margin plutons that are more reduced. Such implies that batholithic arcs can have significant and variable portions that were derived from more oxidized versus more reduced source materials which provide important evidence regarding the prevailing f_{O_2} of melt origin.

The utilization of this classification remains essential as evidenced by contributions of Dall'Agnol et al. (1999), Sial et al (1999), Ishihara et al. (2000), Takagi (2004), Hart et al. (2004), Anderson and Morrison (2005), Bogaerts et al. (2006), Bucher and Frost (2006), and Dall'Agnol and Oliveria (2007).

It is the common experience that Fe-Ti oxides in plutonic rocks do not retain magmatic compositions, but exceptions occur (Anderson and Thomas 1985; Anderson 1996) and the QUILF program indicated above offers solutions to T-f_{O_2} determinations. However, the biotite, K-feldspar, and magnetite oxybarometer, although now dated (Wones 1981), is more robust given that magmatic compositions of K-feldspar and magnetite can be calculated based of compositions on plagioclase for K-feldspar and ilmenite for magnetite, with the assumption that these other phases have more likely retained magmatic compositions.

The calibration of Wones (1981) is

$$\log f_{O_2}(\text{bar}) = -2\left[\frac{4819}{T\,\text{K}}\right] + 6.69 - \log f_{H_2O} - 0.011\frac{P_{\text{bar}} - 1}{T\,\text{K}} + \log a_{\text{Annite}} - \log a_{\text{San}} - \log a_{\text{Mt}}] \quad (11)$$

It is based on the equilibrium (termed ASM):

$$\text{annite} + 1/2\ O_2 = \text{sanidine} + \text{magnetite} + H_2O \quad (12)$$

which leads to a continuous reaction where, with increasing f_{O_2}, biotite becomes more Mg-rich at the expense of components released to K-feldspar and magnetite.

Figure 9 depicts biotite compositions of Mesoproterozoic granites of the former Laurentia supercontinent (data from Anderson 1978; Anderson and Thomas 1985; Anderson and Bender 1989). This remarkable magmatic epoch, considered to be anorogenic, has been subdivided into transcontinental provinces of ilmenite-series granites (Baltic regions to Wyoming) and magnetite-series granites (mid-continent to SW USA), the latter of which also includes a peraluminous subprovince ranging from Colorado to central Arizona (Anderson and Morrison 2005). Whole-rock compositions of these A-type granites are elevated in K_2O and other

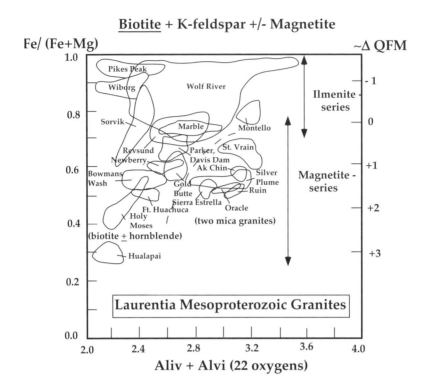

Figure 9. Biotite compositions of Mesoproterozoic ilmenite-series and magnetite-series granites (data from Anderson 1980; Anderson and Thomas 1985; Anderson and Bender 1999) and approximate f_{O_2} relative to the quartz-fayalite-magnetite (QFM) buffer (assuming $P_{H_2O} = P_{total}$) based on the calibration of Wones (1981).

incompatible elements and have high Fe/Mg ratios. As noted by Anderson and Morrison (2005), this generalization is particularly true for the Mesoproterozoic granites in the ilmenite-series province; however, these compositional parameters range lower for these granites in the magnetite-series province, particularly so for those peraluminous, presumably due to higher f_{O_2} and fH_2O of the crustal source and related increased levels of partial melting.

The biotites from these Mesoproterozoic plutons show well the control of ASM equilibria. Of these plutons, the magnetite-bearing granites have biotite with Fe/(Fe+Mg) ratios which show no relation to whole-rock composition and instead reflect intensive parameter control (T, f_{O_2}, f_{H_2O}). For example, the 1.4 Ga Holy Moses and Hualapai granites of western Arizona have typical high whole-rock FeO/(FeO+MgO) wt% ratios averaging at or above 0.85, yet their biotite Fe/(Fe+Mg) ratios average 0.45 and 0.30, respectively, and yield T-f_{O_2} estimates, based on ASM equilbria, at 2 to 3 log units above the QFM oxygen buffer. These two plutons also are characterized by striking aeromagnetic anomalies due to their high magnetite content (Anderson and Bender 1989). Hence, we suggest that oxygen fugacity is the principal factor controlling the Fe/Mg composition of biotite and other mafic phases in magnetite-series granites, as evidenced by calculated log units of f_{O_2} (calibration of Wones 1981) in Figure 9, which spans over four levels of magnitude.

For ilmenite-series granites, the ASM equilibria are incomplete (as it is a non-limiting assemblage) where magnetite is absent; hence biotite Fe/Mg compositions lack intensive parameter control and instead follow that of rock composition. Figure 10 depicts this for biotites from the ilmenite-series Wolf River batholith (Anderson 1980), which compares Fe/Mg biotite composition with that of rock composition and mineral assemblage in contrast to the few samples of the batholith that have magnetite. Where magnetite is absent in this batholith, biotite Fe/(Fe+Mg) ratios correlate well with that of whole-rock FeO/(FeO+MgO), with appropriate shifts with mineral assemblage, including presence or absence of fayalitic olivine, pyroxene, hornblende, and sphene. For the samples that contain magnetite, biotite Fe/(Fe+Mg) ratios generally depart from any such correlation, although they remain high and, thus, indicative of the oxygen-reduced nature of this ilmenite-series batholith.

SUMMARY

The ability to determine the thermal and barometric history during crystallization and emplacement of granitic plutons has been enhanced by several new calibrations applicable to granitic mineral assemblages. Other existing calibrations for granitic plutons have continued to be popular and fairly robust. Recent advances include the trace element thermometers Ti-in-quartz, Ti-in-zircon, and Zr-in-sphene (titanite), which need to be further evaluated on the roles of reduced activities due to lack of a saturating phase, the effect of pressure dependence (particularly for the Ti-in-zircon thermometer), and how resistive these thermometers are to subsolidus reequilibration. As zircon and sphene are also hosts to radiogenic isotopes, these minerals potentially also provide new insights into the temperature - time history of magmas. When used in conjunction with pressure-sensitive mineral equilibria in the same rocks, a complete assessment of the *P-T-t* (pressure-temperature-time) path is possible given that the mineralogy of plutons can reflect crystallization over a range of pressure and temperature during ascent and emplacement and that many intrusions are now seen as forming over several millions of years during the protracted history of batholith construction. Accessory mineral saturation thermometers, such as those for zircon, apatite, and allanite, provide a different

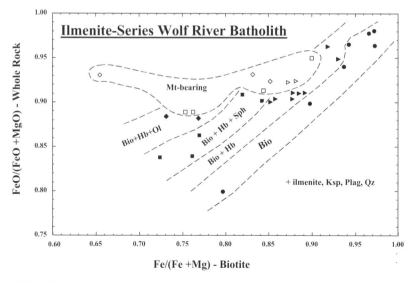

Figure 10. Biotite composition (Fe/(Fe+Mg) - atomic basis) from the ilmenite-series Wolf River batholith as a function of mineral assemblage and whole rock (FeO/(FeO+MgO), wt%) in comparison to samples that contain magnetite (modified from Anderson 1980).

and powerful perspective, specifically that of the temperature of the onset of crystallization of these minerals, which can allow an estimate of the range of temperature between the liquidus and solidus of a given pluton.

In assessment of the depth of crystallization and emplacement of granitic plutons, the Al-in-hornblende remains popular for metaluminous granites when appropriately corrected for temperature. For peraluminous granites, potential new calibrations exist for the assemblages bearing garnet, biotite, plagioclase, muscovite, and quartz.

Other thermometers, based on oxygen abundance, and including Fe-Ti oxides, pyroxene, fayalitic olivine, quartz, sphene, and/or biotite, some of which have been recently revised, can provide additional information on temperature and oxygen fugacity. Oxygen fugacity can range over several orders of magnitude in different magmatic systems and can have profound influence on the mineralogy and mineral compositions in granitic magmas. It also forms the foundation of the popular magnetite- versus ilmenite-series granite classification.

ACKNOWLEDGMENT

The authors thank the very helpful reviews by Bernard Evans, Calvin Miller, Jennifer Matzel, Keith Putirka, and Eric Essene who all made insightful comments to this contribution. The senior author also wishes to thank Scott Paterson and our students, Rita Economos, Vali Memeti, Claire Coyne, and Tao Zhang for continuing to offer advice on issues of granite thermobarometry.

REFERENCES

Ague JJ (1997) Thermodynamic calculation of emplacement pressures for batholithic rocks, California: implications for the aluminum-in-hornblende barometer. Geology 35:563 566

Andersen DJ, Lindsley DH, Davidson PM (1993) QUILF: A Pascal program to assess equilibria among Fe-Mg-Mn-Ti oxides, pyroxenes, olivine, and quartz. Computers and Geosciences 19:1333-1350

Anderson JL (1980) Mineral equilibria and crystallization conditions in the Late Precambrian Wolf River rapakivi massif, Wisconsin. Am J Sci 280:389-332

Anderson JL (1996) Status of thermobarometry in granitic batholiths. Trans R Soc Edinburgh 87:125-138

Anderson JL (1997) Regional tilt of the Mount Stuart batholith, Washington, determined using aluminum-in-hornblende barometry, implications for northward translation of Baja British Columbia: Discussion. Geol Soc Am Bull 109:1223-1225

Anderson JL, Barth AP, Young ED, Davis MJ, Farber D, Hayes EM, Johnson KA (1992). Plutonism across the Tujunga-North American terrane boundary: A middle to upper crustal view of two juxtaposed arcs. *In:* Characterization and Comparison of Ancient and Mesozoic Continental Margins - Proceedings of the 8th International Conference on Basement Tectonics. Bartholomew MJ, Hyndman DW, Mogk DW, Mason R (eds) Kluwer Academic Publishers, Dordrecht, p 205-230

Anderson JL, Barth, AP, Young ED (1988) Mid-crustal Cretaceous roots of Cordilleran core complexes. Geology 16:366-369

Anderson JL, Bender EE (1989) Nature and origin of Proterozoic A-type granite magmatism in the southwestern United States. Lithos 23:19-52

Anderson JL, Cullers RL (1978) Geochemistry and evolution of the Wolf River Batholith, a late Precambrian rapakivi massif in North Wisconsin, U.S.A. Precambrian Res 7: 287-324

Anderson JL, Foley BJ, Ball EN, Paterson SR, Memeti V, Pignotta GS (2007) Upper crustal overturn during magmatic surges – a potential Sierra-wide process. Geol Soc Am Abstr Programs 39(6):526 Paper #196-6

Anderson JL, Morrison J (2005) Ilmenite, magnetite, and peraluminous Mesoproterozoic anorogenic granites of Laurentia and Baltica. Lithos 80:45-60

Anderson JL, Morrison J, Paterson S, Francis J (2008) Post-emplacement fluids and pluton thermobarometry: Mt. Stuart batholith, Washington Cascades. J Metamorph Geol (in revision)

Anderson JL, Smith DR (1995) The effect of temperature and oxygen fugacity on Al-in-hornblende barometry. Am Mineral 80:549-559

Anderson JL, Thomas WM (1985) Proterozoic anorogenic two-mica granites: the Silver Plume and the St. Vrain Batholiths of Colorado: Geology 13:177-180

Asrat A, Barbey P, Ludden N, Reisberg L, Gleies G, Ayalew D (2004) Petrology and isotope geochemistry of the Pan-African Negash pluton, northern Ethiopia: mafic-felsic magma interactions during the construction of shallow-level calc-alkaline plutons. J Petrol 45:1147-1179

Bachl CA, Miller CF, Miller JS, Faulds JE (2001) Construction of a pluton: Evidence from an exposed cross section of the Searchlight pluton, Eldorado Mountains, Nevada. Geol Soc Am Bull 113:1213-1228

Bachmann O, Dungan MA (2002) Temperature-induced Al-zoning in hornblendes of the Fish Canyon magma, Colorado. Am Mineral 87:1062-1076

Bachmann O, Dungan MA, Lipman PW (2002) The Fish Canyon magma body, San Juan volcanic field, Colorado: Rejuvenation and eruption of an upper crustal batholith. J Petrol 43:1469-1503

Barnes CG, Burton BR, Burling TG, Wright JE, Karlsson HR (2001) Petrology and geochemistry of the late Eocene Harrison Pass pluton, Ruby Mountains core complex, northeastern Nevada. J Petrol 42:901-929

Barth AP (1990) Mid-crustal emplacement of Mesozoic plutons, San Gabriel Mountains, California, and implications for the geologic history of the San Gabriel Terrane. *In:* The Nature and Origin of Cordilleran Magmatism. Anderson JL (ed) GSA Memoir 174:33-45

Barth AP, Anderson JL, Jacobsen CE, Paterson SR, Wooden JL (2008) Magmatism and tectonics in a tilted crustal section through a continental arc, eastern Transverse Ranges and southern Mojave Desert. Geol Soc Am Field Guide 11:1-18

Bateman PC, Dodge FCW, Kistler RW (1991) Magnetic susceptibility and relation to initial $^{87}Sr/^{86}Sr$ for granitoids of the central Sierra Nevada, California. J Geophys Res 96:19155-19568

Berman RG, Aranovich LY, Rancourt DG, Mercier PHJ (2007) Reversed phase equilibrium constraints on the stability of Mg-Fe-Al biotite. Am Mineral 92:139-150

Blundy J, Cashman K (2008) Petrologic reconstruction of magmatic system variables and processes. Rev Mineral Geochem 69:179-239

Bogaerts M, Scaillet B, Auwera JV (2006) Phase equilibria of the Lyngdal granodiorite (Norway): Implications for the origin of the metaluminous ferroan granitoids. J Petrol 47:2405-2431

Bucher K, Frost BR (2006) Fluid transfer in high-grade metamorphic terrains intruded by anorogenic granites: The Thor Range, Antarctica. J Petrol 47:567-593

Buddington AL, Lindsley DH (1964) Iron-titanium oxide minerals and synthetic equivalents: J Petrol 5:310-357

Carmichael ISE, Ghiorso MS (1990) The effect of oxygen fugacity on the redox state of natural liquids and their crystallizing phases. Rev Mineral Geochem 24:191-212

Claiborne LE, Miller CF, Walker BA, Wooden JL, Mazdab FK, Bea F (2006) Tracking magmatic processes through Zr/Hf ratios in rocks and Hf and Ti zoning in zircons: An example from the Spirit Mountain batholith, Nevada. Mineral Mag 70:517-543

Dale J, Holland T, Powell R (2000) Hornblende-garnet-plagioclase thermobarometry: a natural assemblage calibration on the thermodynamics of hornblende. Contrib Mineral Petrol 140:353-362

Dall'Agnol R, Oliveira DC (2007) Oxidized, magnetite-series, rapakivi granites of Carajas, Brazil: Implications for classification and petrogenesis of A-type granites. Lithos 93:215-233

Dall'Agnol R, Ramo OT, Magalhaes MS, Macambira MJB (1999) Petrology of the anorogenic, oxidised Jamon and Musa granites, Amazonian Craton: implications for the genesis of Proterozoic A-type granites. Lithos 46:431-462

Elliott BA, Ramo OT, Nironen M (1998) Mineral chemistry constraints on the evolution of the 1.88-1.87 Ga post-kinematic granite plutons in the Central Finland Granitoid Complex. Lithos 45:109-129

Ernst WG, Liu J (1998) Experimental phase-equilibrium study of Al- and Ti- contents of calcic amphibole in MORB - a semiquantitative thermobarometer. Am Mineral 83:952-969

Essene EJ (1982) Geologic thermometry and barometry. Rev Mineral 10:153-206

Essene EJ (1989) The current status of thermobarometry in metamorphic rocks. *In:* Evolution of Metamorphic Belts. Daly JS, Cliff RA, Yardley BWD (eds) Geological Society of London, p 1-44

Ferriss EDA, Essene EJ, Becker U (2008) Computational study of the effect of pressure on the Ti-in-zircon geothermometer. Eur J Mineral (Werner Schreyer Special Issue), (in press)

Ferry JM, Watson EB (2007) New thermodynamic models and revised calibrations for the Ti-in-zircon and Zr-in-rutile thermometers. Contrib Mineral Petrol 154:429-437

Foley BJ, Ball EN, Fischer GC, Thompson JM, Memeti V, Pignotta GS, Anderson JL, Paterson SR, Matzel J, Mundil R (2007) Downward ductile displacement of volcanic crust during pluton emplacement in the central Sierra Nevada: Undergraduate Team Research at USC. GSA Cordilleran Section meeting, Bellingham Paper #31-4

Frost BR (1991) Introduction to oxygen fugacity and its petrologic importance. Rev Mineral 25:1-9

Frost RB, Barnes CG, Collins WJ, Arculus RJ, Ellis DJ, Frost CR (2001) A geochemical classification of granitic rocks. J Petrol 42:2033-2048

Fu B, Page Z, Cavosie AJ, Fournelle J, Kita NT, Lackey JS, Wilde SA, Valley JW (2008) Ti-in-zircon thermometry: applications and limitations. Contrib Mineral Petrol 156:197-215

Ganguly J, Saxena S (1984) Mixing properties of aluminosilicate garnets: constraints from natural and experimental data, and applications to geothermo-barometry. Am Mineral 69:88-97

Gastil RG, Diamond J, Knaack C, Walawendar M, Marshall M, Boyles C, Chadwick B, Erskine B (1990) The problem of the magnetite-ilmenite boundary in southern and Baja California. *In:* The Nature and Origin of Cordilleran Magmatism. Anderson JL (ed) Geol Soc Am Memoir 174:19-32

Ghent ED, Stout MZ (1981) Geobarometry and geothermometry of plagioclase-biotite-garnet-muscovite assemblages. Contrib Mineral Petrol 76:92-97

Ghiorso MS, Sack RO (1991) Fe-Ti oxide thermometry: thermodynamic formulation and estimation of intensive variables in silicic magmas. Contrib Mineral Petrol 108:485-510

Green TH, Adam J (2002) Pressure effect on Ti- or P-rich accessory mineral saturation in evolved granitic melts with differing K_2O/Na_2O ratios. Lithos 61:271-282

Hammarstrom JM, Zen E (1986) Aluminum in hornblende, an empirical igneous geobarometer: Am Mineral 71:1297-1313

Harrison TM, Watson EB (1984) The behavior of apatite during crustal anatexis: equilibrium and kinetic considerations: Geochim Cosmochim Acta 48:1467–1477

Harrison TM, Watson EB, Aikman AB (2007) Temperature spectra of zircon crystallization in plutonic rocks. Geology 35:635-638

Hart CJR, Goldfarb RJ, Lewis LL, Mair JL (2004) The northern Cordilleran mid-Cretaceous plutonic province: Ilmenite/magnetite-series granitoids and intrusion-related mineralization. Resour Geol 54:253-280

Hayden LA, Watson EB (2007) Rutile saturation in hydrous siliceous melts and its bearing on Ti-thermometry of quartz and zircon. Earth Planet Sci Lett 258:561-568

Hayden LA, Watson EB, Wark DA (2007) A thermobarometer for sphene. Contrib Mineral Petrol Online 12 p. *http://dx.doi.org/10.1007/s00410-007-0256-y*

Hildreth W (1981) Gradients in silicic magma chambers: implications for lithospheric magmatism. J Geophys Res 86:10153-10192

Hodges KV, Crowley PD (1985) Error estimation and empirical geothermobarometry for pelitic systems. Am Mineral 70:702-709

Hoisch TD (1991) Equilibria with the mineral assemblage quartz = muscovite + biotite + garnet + plagioclase. Contrib Mineral Petrol 108:43-54

Holdaway MJ (2000) Application of new experimental and garnet Margules data to the garnet-biotite geothermometer. Am Mineral 85:881-892

Holdaway MJ (2001) Recalibration of the GASP geobarometer in light of recent garnet and plagioclase models and versions of the garnet-biotite geothermometer. Am Mineral 86:1117-1129

Holland T, Blundy J (1994) Non-ideal interactions in calcic amphiboles and their bearing on amphibole-plagioclase thermometry. Contrib Mineral Petrol 116:433-447

Hollister LS, Grissom GC, Peters EK, Stowell HH, Sisson VB (1987) Confirmation of the empirical correlation of Al in hornblende with pressure of solidification of calc-alkaline plutons. Am Mineral 72:231-239

Hoskin PWO, Kinney PD, Wyborn D, Chappell BW (2000) Identifying accessory mineral saturation during differentiation in granitoid magmas: an integrated approach. J Petrol 41:1365-1396

Hoskin PWO, Schaltegger U (2003) The composition of zircon and igneous and metamorphic petrogenesis. Rev Mineral Geochem 53:27-62

Ishihara S (1977) The magnetite-series and ilmenite-series granitic rocks. Mining Geol 27:293-305

Ishihara S, Hashimoto M, Machida M (2000) Magnetite/ilmenite-series classification and magnetic susceptibility of the Mesozoic-Cenozoic batholiths of Peru. Resour Geol 50:123-129

Janousek V, Braithwaite CJR, Bowes DR, Gerdes A (2004) Magma mixing in the genesis of Hercynian calc-alkaline granitoids: an integrated petographic and geochemical study of the Sazava intrusion, Central Bohemian Pluton, Czech Republic. Lithos 78:67-99

Johnson MC, Rutherford M J (1989) Experimental calibration of an aluminum-in-hornblende geobarometer with application to Long Valley caldera (California) volcanic rocks. Geology 17:837-841

Klimm K, Blundy JD, Green TH (2008) Trace element partitioning and accessory phase saturation during H2O-saturated melting of basalt with implications for subduction zone chemical fluxes. J Petrol 49:523-533

Kohn MJ, Spear FS (1990) Two new geobarometers for garnet amphibolites, with applications to southeastern Vermont. Am Mineral 75:89-96

Lindsay JM, Schmidtt AK, Trumbull RB, de Silva SL, Siebel W, Emmermann R (2001) Magmatic evolution of the La Pacana caldera system, central Andes, Chile: Compositional variation of two cogenetic, large volume felsic ignimbrites. J Petrol 42:459-486

Lopez-Moro FJ, Lopez-Plaza M (2004) Monzonitic series from the Variscan Tormes dome (Central Iberian Zone): petrogenetic evolution from monzogabbro to granitic magmas. Lithos 72:19-44

Manley CR, Bacon CR (2000) Rhyolite thermobarometry and the shallowing of the magmatic reservoir, Coso volcanic field, California. J Petrol 41:149-174

Manon MR, Essene EJ, Dachs E (2008) Low T heat capacity measurements and new entropy data for titanite: implications for thermobarometry of high pressure rocks. Contrib Mineral Petrol, in press. (UPDATE)

Matzel J, Bowring SA, Miller RB (2006) Timescales of pluton construction at differing crustal levels: examples from the Mount Stuart batholith and Ten Peak pluton, north Cascades, WA. Geol Soc Am Bull 118:1412-1430

Mazdab F, Wooden JL, Barth AP (2007) Trace element variability in titanite from diverse geologic environments: GSA Abstr Programs 39, no. 6, p. 406

Miller CF, McDowell SM, Mapes RW (2003) Hot and cold granites? Implications of zircon saturation temperatures and preservation of inheritance. Geology 31:529-532

Miller CF, Miller JS (2002) Contrasting stratified plutons exposed in tilt blocks, El Dorado Mountains, Colorado River rift, NV, USA. Lithos 61:209-224

Miller JS, Matzel JEP, Miller CF, Burgess SD, Miller RB (2007) Zircon growth and recycling during the assembly of large, composite plutons. J Volcanol Geotherm Res 167:282-299

Needy SK, Anderson JL, Wooden JL, Barth AP, Paterson SR, Memeti V, Pignotta GS (2008), Mesozoic magmatism in an upper- to middle-crustal section through the Cordilleran continental margin arc, eastern Transverse Ranges, California. In: Crustal cross-sections from the western North America Cordillera and elsewhere: Implications for tectonic and petrologic processes. Miller RB, Snoke AW (eds) Geol Soc Am Special Paper (accepted pending revisions)

Page F, Zeb F, Bin K, Noriko T, Fournelle J, Spicuzza, MJ, Schulze DJ, Viljoen F, Basei MAS; Valley JW (2007) Zircons from kimberlite; new insights from oxygen isotopes, trace elements, and Ti in zircon thermometry. Geochim Cosmochim Acta 71:3887-3903

Paterson SR, Memeti V, Anderson JL, Miller R, Zak J, Jacobs R, Seyum S, Shimono S, Wenrong C (2008) Transpression and downward crustal flow during the cretaceous high flux magmatic event in the Central Sierra Nevada, California: Abstract, GSA Cordilleran Section meeting, Las Vegas, Paper #34-7

Paterson SR, Miller RB, Anderson JL, Lund S, Bendixen J, Taylor N, Fink T (1994) Emplacement and evolution of the Mt. Stuart batholith. In: Geologic Field Trips in the Pacific Northwest Swanson DA, Haugerud RA (eds) Department of Geological Sciences, University of Washington in conjunction with the Geological Society of America, Seattle, 2:2F1-2F27

Percival JA, Mortensen JK (2002) Water deficient calc-alkaline plutonic rocks of northeastern Superior Province, Canada: significance of charnockitic magmatism: J Petrol 43:1617-1650

Piwinskii AJ (1968) Experimental studies of rock series, central Sierra Nevada batholith, California: J Geol 76:548-570

Putirka KD (2008) Thermometers and barometers for volcanic systems. Rev Mineral Geochem 69:61-120

Ren M, Omenda PA, Anthony EY, White JC, Macdonald R, Bailey DK (2006) Application of the QUILF thermbarometer to the peralkaline trachytes and pantellerites of the Eburru volcanic complex, East Africa Rift, Kenya. Lithos 9:109-124

Ribbe PH (1982) Titanite (Sphene). Rev Mineral 5:137-154

Rodriguez C, Selles D, Dungan M, Langmuir C, Leemanm W (2007) Adakitic dacites formed by intracrustal fractionation of water-rich parent magmas at Nevado de Longavi' Volcano (36.2°S) Andean Southern Volcanic Zone, Central Chile. J Petrol 48:2033-2061

Ryabchikov ID, Kogarko LN (2006) Magnetite compositions and oxygen fugacities of the Khibina magmatic system. Lithos 9:35-45

Schmidt MW (1992) Amphibole composition in tonalite as a function of pressure: an experimental calibration of the Al-in-hornblende barometer. Contrib MineralPetrol 110:304-310

Sial AN, Toselli AJ, Saavedra J, Parada MA, Ferreira VP (1999) Emplacement, petrological and magnetic susceptibility characteristics of diverse magmatic epidote-bearing granitoid rocks in Brazil, Argentina, and Chile. Lithos 46:367-392

Speer JA (1982) Zircon. Rev Mineral 5:67-112

Stein E, Dietl C (2001) Hornblende thermobarometry of granitoids from the Central Odenwald (Germany) and their implications for the geotectonic development of the Odenwald. Mineral Petrol 72:185-207

Stern CR, Huang W, Wyllie PJ (1975) Basalt-andesite-rhyolite-H2O: Crystallization intervals with excess H2O and H2O-undersaturated liquidus surface to 35 kilobars with implications for magma genesis. Earth Planet Sci Lett 28:189-196

Stone D (2000) Temperature and pressure variations in suites of Archean felsic plutonic rocks, Berens River area, northwest Superior Province, Ontario, Canada. Can Mineral 38:455-470

Takagi T (2004) Origin of magnetite- and ilmenite-series granitic rocks in the Japan arc. Am J Sci 304:169-202

Tollari N, Toplis MJ, Barnes S-J (2006) Predicting phosphate saturation in silicic magmas: An experimental study of the effects of melt composition and temperature. Geochim Cosmochim Acta 70:1518-1536

Velde B (1965) Phengite micas; synthesis, stability, and natural occurrence. Am J Sci 263:886-913

Walker BA Jr, Miller, CF, Claiborne, LE, Wooden JL, Miller JS (2007) Geology and geochronology of the Spirit Mountain batholith, southern Nevada: implications for timescales and physical processes of batholith construction. J Volcanol Geotherm Res 167:239-262

Wark DA, Hildreth W, Spear FS, Cherniak DJ, Watson EB (2007) Pre-eruption recharge of the Bishop magma system. Geology 35:235-238

Wark DA, Watson EB (2006) TitaniQ: a titanium-in-quartz geothermometer. Contrib Mineral Petrol 152:743-754

Watson EB, Harrison TM (1983) Zircon saturation revisited: temperature and composition effects in a variety of crustal magma types. Earth Planet Sci Lett 64:295-304

Watson EB, Liang Y (1995) A simple model for sector zoning in slowly grown crystals: Implications for growth rate and lattice diffusion, with emphasis on accessory minerals in crustal rocks. Am Mineral 80:1179-1187

Watson EB, Wark DA, Thomas JB (2006) Crystallization thermometers for zircon and rutile. Contrib Mineral Petrol 151:413-433

Wei C, Powell R (2004) Calculated phase relations in high pressure metapelites in the system NKFMASH ($Na_2O-K_2O-FeO-MgO-Al_2O_3-SiO_2-H_2O$). J Petrol 45:183-202

Wei C, Powell R (2006) Calculated phase relations in the system NCKFMASH ($Na_2O-CaO-K_2O-FeO-MgO-Al_2O_3-SiO_2-H_2O$) for high pressure metapelites. J Petrol 47:385-408

Wiebe RA, Wark DA, Hawkins DP (2007) Insights from quartz cathodoluminescence zoning into crystallization of the Vinalhaven granite, coastal Maine. Contrib Mineral Petrol 154:439-453

Wones DR (1981) Mafic silicates as indicators of intensive parameters in granitic magmas. Mining Geol 31:191-212

Wones DR (1989) Significance of the assemblage titanite + magnetite + quartz in granitic rocks. Am Mineral 74:744-49

Wones DR, Eugster HP (1965) Stability of biotite: experiment, theory, and application. Am Mineral 50:1228-1272

Wooden JL, Mazdab F, Claiborne LEL , Miller C.F, Barth AP (2006) Elemental analysis of zircon by High Mass Resolution USGS-Stanford SHRIMP-RG: Measuring and evaluating Ti-in-zircon temperatures and compositional characteristics. EOS 31:F-04

Wu CM, Zhang J, Ren L-D (2004a) Empirical garnet-biotite-plagioclase-quartz (GBPQ) geobarometry in medium- to high-grade metapelites. J Petrol 45:1907-1921

Wu CM, Zhang J, Ren L-D (2004a) Empirical garnet-muscovite-plagioclase-quartz geobarometry in medium- to high-grade metapelites. Lithos 78:319-332

Xirouchakis D, Lindsley DH, Frost BR (2001a) Assemblages with titanite ($CaTiOSiO_4$), Ca-Fe-Mg olivine and pyroxenes, Fe-Mg-Ti oxides, and quartz: Part I. Theory. Am Mineral 86:247-253

Xirouchakis D, Lindsley DH, Frost BR (2001b) Assemblages with titanite ($CaTiOSiO_4$), Ca-Fe-Mg olivine and pyroxenes, Fe-Mg-Ti oxides, and quartz: Part II. Application. Am Mineral 86:254-264

Zen E (1985) Implications of magmatic epidote-bearing plutons on crustal evolution in the accreted terranes of northwestern North America. Geology 13:266-269

Zen E (1989) Plumbing the depths of batholiths. Am J Sci 289:1137-1157

Zen E, Hammarstrom JM (1984) Magmatic epidote and its petrologic significance. Geology 12:515-518

Zhang SH, Zhao Y, Song B (2006) Hornblende thermobarometry of the Carboniferous granitoids from the Inner Mongolia paleo-uplift: implications for the tectonic evolution of the northern margin of the North China block. Mineral Petrol 87:123-141

Fluid Inclusion Thermobarometry as a Tracer for Magmatic Processes

Thor H. Hansteen

IFM-GEOMAR, Leibniz-Institute for Marine Sciences
Dynamics of the Ocean Floor
D-24148 Kiel, Germany

thansteen@ifm-geomar.de

Andreas Klügel

Fachbereich Geowissenschaften
Universität Bremen
D-28334 Bremen, Germany

akluegel@uni-bremen.de

INTRODUCTION

Fluid inclusions in minerals may form in any type of volcanic or plutonic rock ranging from mafic to silicic compositions. Because all igneous rocks reach fluid saturation at some stage during their evolution, fluids trapped as inclusions in magmatic minerals belong to a certain paragenesis or phase assemblage, which may include minerals, melts and one or more fluid phases. These fluid inclusions reflect one or more stages during rock evolution, and can be used to constrain multistage formation and evolution processes including ascent histories, magma chamber processes and crystallization behavior.

Fluid inclusions can provide thermobarometric data on various timescales. During ascent of mafic to intermediate magmas, fluid inclusions may form within hours to days and record transient magma stagnation levels, whereas chemical mineral-melt thermobarometry requires equilibrium mineral growth and thus typically reflects well-defined crystallization events (Roedder and Bodnar 1980; Wanamaker et al. 1990; Hansteen et al. 1998; Klügel et al. 2000; Frezzotti and Peccerillo 2004). Such data has only rarely been combined with melt inclusion investigations in order to depict detailed magma ascent histories (e.g., Bureau et al. 1998). During prolonged crystallization in magma chambers or in the plutonic environment, however, fluid inclusions reflect the equilibrium situation, and data from various thermobarometric methods should overlap.

This chapter focuses on the use of fluid inclusions as thermobarometers to constrain magmatic processes and timescales. After an introduction of basic principles and explanation of the most relevant fluid systems we show in a "cookbook" style how barometric data are derived, and discuss error magnitudes and pitfalls. Thermobarometric information attained from chemical mineral-melt equilibria are compared to that obtained from fluid inclusions in order to provide detailed accounts for magma ascent and crystallization. Although the term "fluid inclusion" is often used in a general way comprising all inclusions that were trapped in the liquid state (Roedder 1984), this chapter is restricted to CO_2- and H_2O-dominated inclusions and does not include sulfide- or silicate melt inclusions, which are covered in Kent (2008) and Metrich and Wallace (2008).

OCCURRENCE AND PETROGRAPHY OF FLUID INCLUSIONS

Every igneous mineral may contain fluid inclusions. Among pheno- and xenocrysts from volcanic rocks, a majority of inclusion investigations have been performed on the host minerals quartz, plagioclase, amphibole, pyroxene and olivine. In mantle xenoliths and in plutonic rocks the occurrence of thousands of fluid inclusions within a regular thin section is not uncommon (Roedder 1965). Inclusion sizes are one to three orders of magnitude smaller than the mineral in which they occur, typically <2 to 30 micrometers, the smaller ones being much more abundant than the larger ones. Depending on composition and density, fluid inclusions at room temperature contain one or more phases (Fig. 1). Because the lower refractive indices of most fluid inclusions compared to their host minerals cause internal reflections in the inclusion cavity, many inclusions appear dark in ordinary plane-polarized light under the petrographic microscope. Applying conoscopic light enhances the visibility of the inclusion contents, and is thus invaluable for fluid inclusion petrography.

Accurate observations of fluid inclusions rely on an optimal microscopic image. Three-dimensional petrography using 50 to 1000 µm thick doubly polished sections is essential for the successful determination of phase assemblages of inclusions and the chronology of trapping (e.g., Roedder 1984). Standard petrographic thin sections are 25 to 30 µm thick essentially providing a two-dimensional view only. As common inclusion sizes are comparable to the thickness of thin sections, most fluid inclusions are destroyed upon preparation and will simply not be discovered. The ideal slide thickness for petrographic inclusion investigations depends on the actual sizes of the fluid inclusions, and properties of the host mineral including color, optical anisotropy and cleavage. Because the visibility of inclusions within a mineral deteriorates quickly with increasing depth below the surface, the ideal sample thickness is often in the range between 60 and 120 µm. Minerals with a strong color require comparatively thin sections, while those with a good cleavage, like pyroxene, amphibole and plagioclase, may require comparatively thick sections for the inclusions to survive preparation.

INCLUSION CLASSIFICATION

Fluid inclusions may be classified according to their phases visible at room temperature, which may vary from single-phase (vapor, liquid or supercritical fluid), two-phase liquid-vapor, three-phase liquid-liquid-vapor (typical for mixed-system CO_2-H_2O inclusions at room temperature), and each of the above types may additionally contain one or more *daughter minerals* (defined as a mineral formed after sealing off of the inclusion from its surroundings; Roedder 1984) (Fig. 2). This descriptive classification may be expanded to classify complex inclusions containing several solid and fluid/liquid phases. The drawback of this classification is that inclusions with similar compositions and densities may be classified as distinct types. More often, inclusions in minerals are *chronologically* classified according to their time of trapping relative to host mineral growth (Sobolev and Kostyuk 1975; Roedder 1979, 1984). Generally, fluid inclusions may be trapped both during and after growth of the host crystal (Fig. 3).

Primary inclusions occur as isolated, single inclusions or in small non-oriented groups. They may also be aligned within and often parallel to mineral growth zones. Primary inclusions have been formed during, and as a direct result of, growth of the host crystal. Secondary inclusions occur as trails that extend to the surface of the host mineral and may crosscut several neighboring grains. They were thus formed after termination of crystal growth and represent fluids trapped during healing of fractures or microcracks. Pseudosecondary inclusions occur along healed microcracks which were formed during host mineral growth, and are thus *senso stricto* a subgroup of the primary inclusions. They are texturally distinguished by their occurrence along straight or curved trails or planes which do not coincide with growth zones

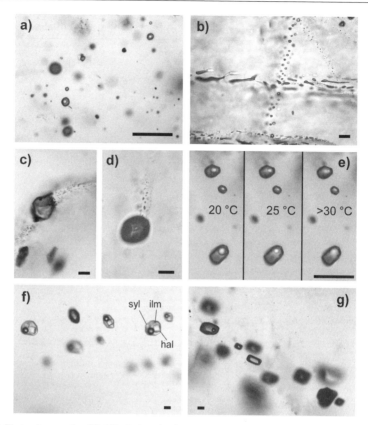

Figure 1. Photomicrographs of fluid inclusions in phenocrysts and xenoliths. Scale bars are 20 micrometers. (**a**) Texturally early CO_2-rich inclusions occurring as irregular groups in plagioclase from a magmatically overprinted tholeiitic gabbro xenolith (La Palma, Canary Islands). (**b**) Trails of mature secondary fluid inclusions cut by a young trail of immature inclusions with elongated shapes due to arrested necking-down processes (dunite xenolith, La Palma, Canary Islands). (**c**) Texturally early, leaked CO_2 inclusion with decrepitation trails extending from two edges; olivine phenocryst from a basanite lava (Gran Canaria, Canary Islands). (**d**) Decrepitated primary CO_2 fluid inclusion in olivine from a dunite xenolith (La Palma, Canary Islands). (**e**) Homogenization of two-phase CO_2 inclusions into the liquid phase (plagioclase from a tholeiitic gabbro xenolith). (**f**) Boiling assemblage comprising coexisting high-salinity liquid phase and low-salinity vapor phase inclusions. All inclusions homogenize at 512 ± 40 °C; the liquid phase has a salinity of 63 ± 3 wt% NaCl equivalents. The daughter crystals are halite (hal), sylvite (syl), and ilmenite (ilm). Miarolitic quartz in the Permian Eikeren alkali granite, Oslo Rift, Norway. (**g**) Boiling assemblage with coexisting liquid phase and low-salinity vapor phase inclusions. All inclusions homogenize at 351 ± 25 °C; the liquid phase has a salinity of 20 ± 2 wt% NaCl equivalents. Same locality as in f).

(Fig. 3a). Pseudosecondary trails may crosscut growth zones, but do not reach the grain boundaries of the host crystal.

The above nomenclature may be misleading if the primary mineralogy is not preserved, like in many plutonic rocks (Van den Kerkhof and Hein 2001), where fluid inclusion generations may be termed "early" and "late" only. In such cases, the relative chronology of the inclusion groups and trails can be constrained from their spatial distribution and crosscutting relationships (e.g., Touret 2001).

A *fluid inclusion assemblage* (FIA) describes a group of inclusions that were all trapped at the same time or during the same event (Goldstein and Reynolds 1994); the term "*group of*

Inclusion type	Abbreviation	Appearance	Essential phases
Monophase liquid	L	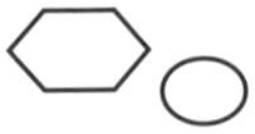	L = 100%
Monophase vapor	V		V = 100%
Liquid-rich, two-phase	L + V		L > 50%
Vapor-rich, two-phase	V + L		V = 50-80%
Multiphase solid	S + L ± V		L, S
Immiscible liquid	$L_1 + L_2 \pm V$		L_1, L_2

Figure 2. Classification scheme for fluid inclusions based upon inclusion contents observed at room temperature. L = liquid; V = vapor; S = solid. Modified after Shepherd et al. (1985).

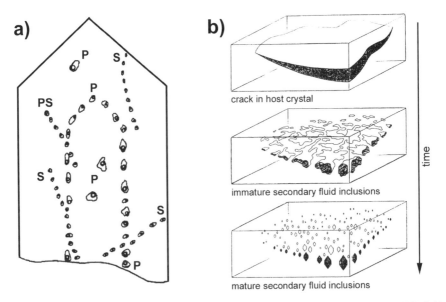

Figure 3. (a) Petrographic classification of primary (P), secondary (S) and pseudosecondary (PS) fluid inclusions based on their three-dimensional distribution within single crystals (see text for details). Modified after Shepherd et al. (1985). (b) Formation of secondary fluid inclusions through crack healing and consecutive "necking-down" processes. Modified from Roedder (1984).

synchronous inclusions" (GSI; Touret 2001) can be used as a synonym for FIA. Such a genetic, chronology-based classification is useful for rocks with clear-cut petrographic relationships, where in the ideal case each inclusion generation can be identified as part of a certain paragenesis. In many cases, however, a classification based on the chronology of isolated inclusions and healed inclusion-filled cracks may be ambiguous, and a classification scheme based on inclusion contents will be more practical. Thus, in order to successfully classify magmatic fluids, a combination of inclusion chronology, inclusion compositions and the assignment to a certain paragenesis in the rock should be used whenever possible. Inclusions trapped under liquidus conditions should be classified as parts of paragenesis including coexisting fluid and melt, e.g., occurring along single cracks or within primary groups.

FLUID INCLUSION FORMATION AND RE-EQUILIBRATION

Trapping during crystal growth

Most primary fluid inclusions represent fluid bubbles exsolved from a magma that adhered to the surface of an actively growing crystal, possibly at sites of increased dislocation density. Under favorable conditions such as periods of enhanced crystal growth these bubbles became subsequently overgrown and trapped (Roedder 1963, 1984). In subvolcanic magma chambers the rate of inclusion trapping can be strongly augmented by events that cause rapid degassing and thus increased liquidus temperature and increased crystallization rates. Such events may be reflected by increased occurrences of inclusions along growth zones of phenocrysts, and may generally be caused by changes in pressure, temperature and melt composition during magma movements or mixing (e.g., Pearce and Kolisnik 1990; Coombs and Gardner 1991; Troll and Schmincke 2002; Blundy and Cashman 2005). Primary fluid inclusions may also become trapped during recrystallization of minerals as a result of deformation or heating with partial dissolution, in which case they occur as irregular groups rather than singly or along growth zones (e.g., Hansteen et al. 1998; Fig. 1a).

Formation of secondary inclusions

Secondary and pseudosecondary inclusions originate as fluid-filled microcracks in the host mineral (Figs. 1b and 3b). Such cracks form after stress buildup in minerals upon pressure release, temperature change, and/or deformation. When an available fluid enters the crack due to capillary forces the crack tends to heal rapidly at magmatic temperatures (Roedder 1979, 1984). The healing process begins with formation of elongated, often wedge-shaped compartments enclosing portions of the fluid. By recrystallization processes the surface energy in the crack-fluid system, which is the product of surface tension and area, becomes minimized. If the host mineral has a finite solubility in the fluid, material of the host crystal dissolves from high-tension surfaces (small curve diameters) and re-precipitates on low-tension surfaces (greater curve diameters), a process termed *necking-down* (Fig. 3b). The minimum surface area is that of a regular arrangement of spheres within the former crack. The minimum surface energy is given by a partly or wholly faceted surface enclosing the same volume, referred to as a *negative crystal shape* of an inclusion.

Post-entrapment re-equilibration

Roedder's rules. An investigated primary or secondary fluid inclusion in principle represents the P-T-X conditions at the time of entrapment only if it satisfies the following criteria, often referred to as "Roedder's rules" (Roedder 1984): 1) a single homogeneous fluid phase was trapped, 2) the inclusion remained at a constant volume after trapping, and 3) nothing was added to or removed from the inclusion after trapping. Criterion 1) cannot be simply claimed but needs to be verified by careful microscopic investigations, although this is not easy for small inclusions. Criteria 2) and 3) address the basic assumption for the interpretation of fluid inclusions, namely, that they have behaved as an *isochoric system* having constant volume and mass under the conditions of interest. In most cases this assumption is a first approximation at best because, after inclusion trapping, the *P-T* path of the host mineral carried by an ascending magma differs strongly from the isochores of the trapped fluids. This results in the evolution of a pressure gradient between inclusions and surroundings of the host crystal. If the pressure differences become large, the host responds by plastic deformation (*stretching*) and/or brittle deformation (*decrepitation*), which is termed *volumetric re-equilibration* of inclusions (e.g., Bodnar 2003). Fluid inclusions in igneous rocks usually become over-pressurized, but at other *P-T* conditions an internal under-pressure may also develop.

Stretching. This type of volumetric re-equilibration describes permanent, plastic deformation (creep) of the enclosing crystal without loss of fluid from the inclusion. Stretching is not relevant at short time scales except for soft minerals (low Mohs hardness, e.g., fluorite and calcite) but is the typical mode of re-equilibration at low strain rates, often associated with long time scales. At high temperatures, however, stretching may also affect hard minerals and significantly reduce inclusion density in a short time. Experiments with synthetic $NaCl$-H_2O inclusions in quartz having an internal overpressure of about 210 MPa showed significant stretching after a few days at 625 °C and recovery to 150 MPa after 30 days (Vityk and Bodnar 1998). Re-equilibration experiments with olivine at 1400 °C and ambient pressure showed stretching of natural CO_2 inclusions and related density decrease from about 1.0 to 0.7 g/cm^3 within a few days (Wanamaker and Evans 1989). Stretching is problematic insofar as it affects some inclusions more than others, which cannot be recognized by optical microscopy (Vityk and Bodnar 1998). In addition, stretching may promote diffusive loss of fluid components through crystal defects involved in the deformation processes (Bakker and Jansen 1991). Stretching may however be inferred from different phases or volume proportions in inclusions belonging to the same fluid inclusion assemblage, and from the distribution of homogenization temperatures (and hence densities) of a fluid inclusion assemblage as is discussed below.

Compared to stretching, expansion of the host crystal due to decompression is generally of minor concern because most minerals in igneous rocks and xenoliths (olivine, pyroxene, feldspar, amphibole, quartz) have small compressibilities. The inclusion volume will merely increase by about 0.1% for every 100 MPa of pressure release, which can usually be ignored (Roedder and Bodnar 1980). Likewise, the thermal expansion of silicate minerals is also rather moderate (on the order of 0.2-0.5% per 100 °C; Robertson 1988).

Decrepitation. The typical mode of volumetric re-equilibration at high strain rates and for relatively hard minerals is (partial) decrepitation, sometimes also referred to as *leakage*. The host crystal fails by fracturing rather than creep and the inclusion leaks parts of its contents by advection along the rapidly propagating microcracks. If the fluid is completely lost after intense fracturing then the term *total decrepitation* is used. Decrepitation can be recognized in petrographic investigations by halos of minute fluid inclusions adjacent to inclusions and by microcracks ("whiskers") often extending from sharp corners (Fig. 1c,d). Leakage due to internal overpressure causes a drop in inclusion pressure and density to unpredictable values. If the cracks penetrate grain boundaries and become connected to the surrounding, the inclusion may adopt the actual ambient fluid pressure. In many cases the inclusion becomes sealed again by microcrack healing, which occurs in a short time at magmatic temperatures; e.g., in olivine within a few hours to days depending on crack size (Wanamaker et al. 1990). If leakage and subsequent sealing occur at identical *P-T* conditions such as during ponding of an ascending magma, then the inclusion may record the new ambient pressure. Thus the geological interpretation of partially decrepitated fluid inclusions can still be very important, as they often reflect a specific stage of host rock evolution.

Other re-equilibration processes. Necking-down during the evolution of secondary fluid inclusions can also be considered as re-equilibration (Fig. 3b). If formation and sealing of the individual daughter inclusions involves a homogeneous fluid then each inclusion has the same density as the entrapped fluid. However, if necking-down is accompanied by cooling and related phase changes then the densities of the new inclusions may differ strongly from each other (Roedder 1984). The composition and bulk density of inclusions may also change by *compositional re-equilibration* involving 1) diffusion of components through the host mineral and 2) reactions between fluid and inclusion walls that occur after trapping. These processes are discussed below in the section of missing H_2O. In general, experimental data suggest that various mechanisms of re-equilibration operate together in a specific environment and that extrapolation of data may be ambiguous. For a detailed treatment we refer to the review by Bodnar (2003) and references therein.

MICROTHERMOMETRY

Microthermometry is defined as the observation of phase changes occurring in a fluid inclusion upon heating. For comparatively well-known unary and binary fluid systems (i.e., CO_2, H_2O, CO_2-H_2O, H_2O-NaCl), such phase changes provide quantitative information on the composition and density of single fluid inclusions. For more complex systems involving three or more major components, microthermometry can provide semi-quantitative to qualitative data depending on the chemical system involved. Thus additional microanalytical methods are often needed to characterize the system, including non-destructive methods for major and minor elements, such as Laser Raman spectroscopy and Fourier Transform Infrared (FTIR) spectroscopy (e.g., Wopenka et al. 1990; Burke 2001). Raman and FTIR spectroscopy can provide species ratios (e.g., of CO_2, CH_4, N_2, CO, SO_2) in single fluid inclusions, and allow for identification of daughter minerals. Trace element analysis of single fluid inclusions can be performed *in-situ* with particle induced X-ray emission (PIXE) and synchrotron XRF, and by laser ablation ICP-MS by ablating the inclusion. A description of these methods is outside the scope of this chapter and the interested reader is referred to Samson et al. (2003), Cauzid et al. (2006), Simon et al. (2007) and references cited therein.

Microthermometry measurements are performed using a purpose-designed *heating-freezing stage* for the use with a petrographic microscope. Two designs are commercially available, namely the Reynolds gas flow stage (Fluid Inc., Denver, CO) (based on the USGS gas flow stage; Werre et al. 1979; Woods et al. 1981), and the Linkam stage (Shepherd 1981), which are built to operate in the temperature range between −196 °C (the temperature of the liquid nitrogen cooling agent) and about 600 °C. The petrographic microscope has to be fitted with long working distance lenses and condensor system. High-temperature heating stages for measurements above 600 °C include the commercially available Linkam TH 1500 (Oskarsson and Hansteen 1992) and the Vernadsky rapid-heating and quenching stage (designed at the Vernadsky Institute, Moscow; Zapunnyy et al. 1988). The latter attains heating and cooling rates up to about 100 °C/s.

Because metastability is a common problem in such microscopic systems as fluid inclusions, microthermometry measurements are always performed upon heating. The nucleation of new phases like a mineral (ice, solid CO_2, etc) or a vapor bubble requires significant undercooling, where the nucleation temperature is not reproducible. Thus microthermometry measurements comprise melting and dissolution reactions and homogenization of inclusions into liquid or vapor upon heating. The most common microthermometry data include a) temperature of first melting, denoted T_t (triple point temperature) for unary systems and T_m(eutectic) or T_m(initial) for binary and higher systems, respectively; b) final melting temperature T_m(final); c) homogenization temperature into the liquid (T_h(LV→L)) or into the vapor (T_h(LV→V)) phase, respectively; and d) T_{trap} (temperature of trapping) (cf. Roedder 1984). Partial homogenization, involving homogenization into the liquid or vapor phase in the presence of either daughter minerals or a second liquid phase, is still denoted T_h(LV→L) or T_h(LV→V), respectively.

Temperature standards for microthermometry include compounds with known melting points, and to an increasing extent comprise synthetic and natural fluid inclusions with known compositions and densities (e.g., Sterner and Bodnar 1991). For detailed discussions on lab techniques and interpretation of microthermometric data in general see Roedder (1984), Shepherd et al. (1985), and Samson et al. (2003).

FLUID SYSTEMS

In the following discussion of fluid phase systems, we focus on the use of microthermometry to obtain densities of magmatic fluids under *isochoric* conditions. When an isochoric system is heated, its internal pressure changes as a complex function of composition and density. Thus microthermometry observations are usually depicted in *isochoric phase diagrams*, rather than their isobaric or isothermal siblings used in petrology.

The most common fluid inclusions in magmatic systems belong to the pseudo-system H-O-C-N-S-NaCl, where NaCl represents highly-water-soluble salts including various halides and sulfates. The model systems briefly introduced here are CO_2 ($\pm CH_4$), and H_2O-NaCl, because these are most relevant to barometry and are good approximations of commonly occurring geological fluids. A more detailed description of these and other fluid systems is outside the scope of this chapter, and the reader is referred to Bowers and Helgeson (1983a,b), Van den Kerkhof (1990), Sterner and Bodnar (1991), Schmidt and Bodnar (2000), Diamond (2003) and references cited therein.

The CO_2 system

CO_2 is a comparatively simple unary system that can be used to explain basic features pertaining to most fluid systems. Depicted in a *P-T* projection, the system CO_2 comprises three primary phase fields solid (S), liquid (L) and vapor (V) (Fig. 4). Applying the Gibbs phase rule, the primary phase fields have a variance of 2 and are separated by univariant curves where S+L, S+V and, most commonly, L+V coexist. The L+V univariant curve ends at the *critical point* at 30.98 °C and 7.38 MPa (Span and Wagner 1996). The three phases S+L+V coexist only at the invariant *triple point*, which for pure CO_2 is at −56.56 °C and 0.52 MPa. The density ρ, which is essential for the interpretation of fluid phase systems, is not directly visible in the *P-T*-projection but can be visualized in a three-dimensional *P-T-*ρ diagram (Fig. 4a). Because each CO_2 inclusion can have only one unique density and hence one unique univariant two-phase curve, the entire array of these curves covers all possible densities.

Microthermometry analysis involves cooling the inclusion towards the temperature of liquid N_2 until its contents comprise solid and vapor phase CO_2. By heating up the inclusion liquid CO_2 appears at the invariant triple point at temperature T_t. Slight heating above T_t makes solid CO_2 melt completely, and the inclusion consists of liquid and vapor (Fig. 5). Depending on the inclusion density, further heating along the L+V curve produces one of the following reactions: (1) At densities below 0.468 g/cm³ (the critical density), the vapor bubble will expand at the expense of the liquid phase, until the inclusion *homogenizes into the vapor phase* at temperature T_h(LV→V). (2) At densities above 0.468 g/cm³, the liquid phase will expand at the expense of the vapor bubble, until the inclusion *homogenizes into the liquid phase* at temperature T_h(LV→L) (Figs. 1e and 5). (3) For inclusions having the critical density, the liquid phase and vapor bubble will maintain their original sizes until the inclusion homogenizes by fading and disappearance of the meniscus between both phases, termed *critical homogenization*. Inclusions having near-critical densities may also show critical behavior because the two-phase field is broad and flat near the critical temperature (Roedder 1965; Angus et al. 1976). Once homogenized, each inclusion consists of a *supercritical fluid* and must follow an isochore where further heating will not result in any additional phase changes.

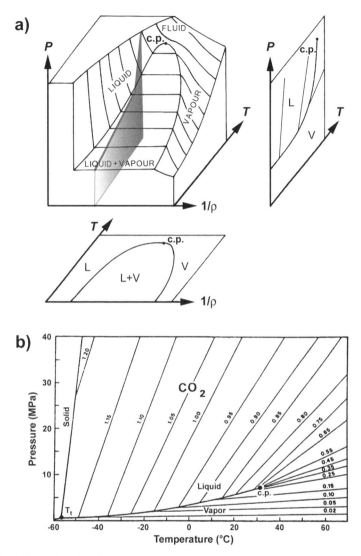

Figure 4. (a) Three-dimensional pressure (P) - temperature (T) - density (ρ) block diagram for the CO_2 system schematically showing the relationship between P-T and T-ρ projections. L= liquid; V= vapor; c.p. = critical point. The shaded section represents the density of an arbitrary inclusion homogenizing into the liquid phase. Modified from Roedder (1984) and Goldstein and Reynolds (1994). (b) P-T projection of the block diagram showing CO_2 isochores in the low-temperature range based on data from Angus et al. (1976); modified after Roedder (1984). Numbers on isochores are densities in g/cm³; T_t = triple point; c.p. = critical point.

To sum up, temperature measurements of *two phase transitions* are necessary and sufficient to characterize the properties of CO_2 or any other unary inclusions. Triple point melting is used to identify the chemical system, and homogenization into liquid or vapor is used for an accurate density determination. The volumetric and microthermometric behavior of fluids in the system CO_2-N_2-CH_4 has been summarized by Van den Kerkhof (1990), Kooi et al. (1998), and Van den Kerkhof and Thiéry (2001).

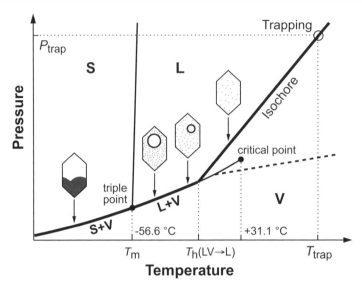

Figure 5. *P-T* projection illustrating the phase transitions that occur in an arbitrary pure CO_2 inclusion upon heating. The inclusion has a density above the critical density and thus homogenizes into the liquid. The homogenization temperature $T_h(LV \rightarrow L)$ defines the isochore, from which the pressure and temperature of trapping can be derived. Isochores of inclusions homogenizing into vapor run below the critical point as indicated by the dashed line. Modified after Van den Kerkhof (1988).

The H_2O-NaCl system

Overview. The low pressure part of the *P-T* phase diagram of pure H_2O has a topology similar to that of CO_2, but the absolute *P* and *T* values are different (Fig. 6). Phase transitions relevant to fluid inclusion investigations occur at comparatively higher temperatures, including the triple point at 0.15 °C and the critical point at 374 °C (representing a density of 0.32 g/cm³). Like for the CO_2 system, low-density inclusions homogenize into the vapor phase and high-density ones into the liquid phase. Isochores for high-density water inclusions are notably steeper than those for most CO_2 densities (Crawford 1981).

Adding NaCl to the H_2O system changes it from a unary to a binary, adding one degree of variance to each of the phase elements. At low temperatures, the triple point of H_2O is replaced by a binary melting curve in the H_2O-NaCl system (Fig. 7). The low-temperature part of the system is a *binary eutectic* with a minimum at −20.8 °C, and a eutectic composition of 23.3 wt% NaCl. Below 0.1 °C, the salt hydrate *hydrohalite* (HH; NaCl·$2H_2O$) is stable.

Phase changes observed during microthermometry. Upon heating of a low-salinity H_2O-NaCl inclusion, either hydrohalite or ice will melt last, where final ice melting reflects the lower salinities. The final melting temperatures of either ice (T_m(Ice)) or hydrohalite (T_m(HH)) can be used to estimate accurately the salinity and thus the composition of the inclusion. For inclusions containing halite as a daughter mineral, salinities are measured using the temperature of halite dissolution (see below).

Adding NaCl to the water system in amounts up to the room-temperature saturation level (26.3 wt% NaCl) has three major effects (Fig. 6): a) the bulk densities increase because salt dissolves readily in water with very little net volume expansion; b) the critical point moves to progressively higher temperatures with the amounts of salt added, leading to an extended two-phase L+V curve; and c) the isochores for a given liquid homogenization temperature become less steep.

Fluid Inclusion Thermobarometry: Tracer for Magmatic Processes 153

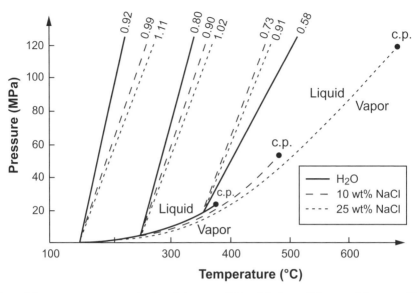

Figure 6. Pressure-temperature projection of the system H_2O and the low-salinity part of the system H_2O-NaCl at elevated temperatures. In the system H_2O, the L+V two-phase curve extends to the critical point at 374 °C. Addition of salt to the water system gradually increases the extent of the L+V two phase curve until nearly 700 °C for a salinity of 25 wt% NaCl. With increasing salinity the fluid densities increase and the isochores become somewhat less steep. Isochores are marked with densities in g/cm^3, and c.p. is the critical point for each respective composition. Modified after Crawford (1981).

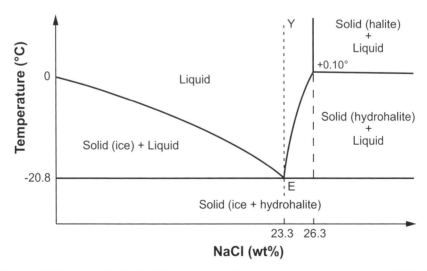

Figure 7. Melting properties for fluid inclusions in the low-temperature part of the system H_2O-NaCl. Salinities below the room-temperature saturation of 26.3 wt% NaCl can be estimated by either observing the temperature of final ice melting (for salinities below the eutectic composition of 23.3 wt% NaCl), or the temperature of final hydrohalite melting (for salinities between 23.3 and 26.3 wt%). Modified after Crawford (1981) and Hall et al. (1988).

High-salinity inclusions contain halite (H) as a daughter mineral at room temperature (Fig. 1f). Thus the two-phase L+V curve for the low-salinity compositions is replaced by a three-phase H+L+V curve (Fig. 8). In this case, the salinity of an inclusion can be accurately measured by observing the temperature of final halite dissolution (T_m(Halite)) at the appropriate *halite liquidus* (Bodnar et al. 1985; Chou 1987; Sterner et al. 1988, 1992). The inclusion density is measured by observing its homogenization into the liquid (T_h(LV→L)) or vapor (T_h(LV→V)) phase. Depending on bulk density, four different sequences of high-temperature phase transitions are possible for inclusions having the same salinity, i.e., for inclusions occurring along the same *isopleth* in the H_2O-NaCl system. This is illustrated by an example using four different inclusions (A to D) with compositions of 50 wt% NaCl and 50 wt% H_2O but four different densities (Fig. 8). A) Upon heating, the highest-density inclusion containing the smallest vapor bubble at room temperature will lose its bubble prior to halite dissolution. After homogenization into the liquid phase at temperature T_1 (Fig. 8), it will follow a path across the two-phase L+H field until final homogenization at the 50 wt% halite liquidus (T_m(Halite) at T_2). The resulting isochore is comparatively steep. B) Inclusion B having a lower density and thus a larger bubble at room temperature will homogenize into the liquid phase at T_2 and simultaneously dissolve its halite at the appropriate halite liquidus (T_m(Halite); Fig. 8). The isochore thus extends from the intersection point of the halite liquidus with the H+L+V three phase curve. C) Inclusion C has a still larger vapor bubble than inclusion B, and will homogenize into liquid after halite dissolution. It will thus cross the two-phase L+V field until homogenization (T_h(LV→L) at temperature T_3; Fig. 8). The vapor-rich inclusion D will also intersect the halite liquidus at T_2. Because it has a density lower than the critical density, it will homogenize into the vapor phase and must thus follow a low-pressure path across the two-phase L+V field until homogenization

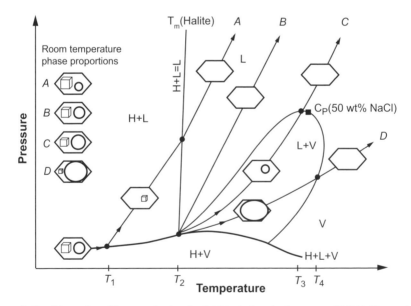

Figure 8. Possible modes of homogenization for fluid inclusions in the system H_2O-NaCl containing halite at room temperature. The example considers the 50 wt% NaCl isopleth, thus all inclusions have the same salinity but different densities decreasing from inclusion "A" to inclusion "D". The modes of final homogenization are (inclusion designation in parenthesis): (A) Halite melting T_m(Halite); (B) simultaneous halite melting and liquid phase homogenization T_h(LV→L); (C) liquid homogenization, and (D) vapor homogenization T_h(LV→V), respectively. c.p. is the critical point and T_m(H) the final melting temperature of halite. Modified from Konnerup-Madsen (1995).

into vapor ($T_h(LV{\rightarrow}V)$ at T_4; Fig. 8). The resulting isochore will be much less steep than for the higher-density inclusions with the same salinity.

Phase separation (boiling) in the H_2O-NaCl system. A simplified *P-T-X* diagram of the system H_2O-NaCl is shown in Figure 9. A large immiscibility field exisits within this system at elevated temperatures and low to moderate pressures (Sourirajan and Kennedy 1962; Bodnar et al. 1985; Chou 1987; Driesner and Heinrich 2007; Simon et al. 2007). For clarity, the low-temperature part of this field is shown as a pressure-composition projection in Figure 10. All saline aqueous fluids entering the immiscibility region will instantaneously separate into coexisting vapor and liquid phases. Taking the 800 °C isotherm at 100 MPa pressure as an example, the two equilibrium phases are a high-density liquid (brine) containing 70 wt% NaCl, and a low-density vapor containing about 2 wt% NaCl (points A and B in Figure 10). All fluids with compositions between 2 and 70 wt% NaCl at 800 °C and 100 MPa must instantaneously unmix to form such coexisting vapor and liquid phases. Generally, changes in each of the parameters *P*, *T* and *X* can result in boiling at elevated temperatures. A common mechanism for boiling of originally homogeneous fluids is pressure decrease, induced by fracturing of roof rocks above shallow intrusions or magma chambers (cf. Roedder and Bodnar 1980). Thus a rapid pressure drop corresponding to a change from lithostatic to hydrostatic conditions can be facilitated. Boiling of an originally homogeneous fluid phase may lead to volume expansion of the melt-fluid system and thus promote further wall-rock fracturing.

Other components. Most naturally occurring aqueous fluid inclusions contain major components in addition to H_2O and NaCl, the most important being $CaCl_2$ and $MgCl_2$ (Konnerup-Madsen 1979; Crawford 1981; Zhang and Frantz 1987). In order to relate natural samples to the H_2O-NaCl system, microthermometric measurements of aqueous inclusions are

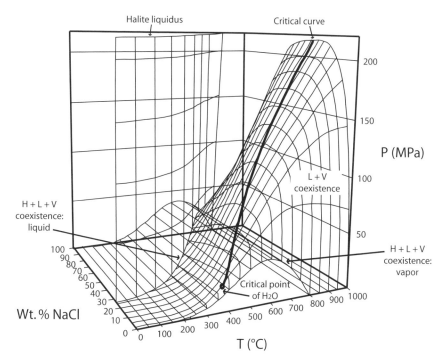

Figure 9. Topology of the system H_2O-NaCl at elevated temperatures and pressures. Modified after Driesner and Heinrich (2007).

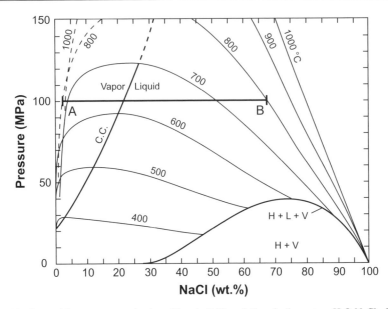

Figure 10. Composition-pressure projection of immiscibility relations in the system H_2O-NaCl, showing the boundaries for the L+V two-phase fields for selected temperatures. The compositions "A" and "B" connected with a horizontal tie line indicate the compositions of the coexisting vapor and liquid phases, respectively, at 800 °C and 100 MPa. Modified from figures in Bodnar et al. (1985) and Chou (1987).

generally treated as if they belong to this simplified system. Thus salinities measured in natural fluid inclusions are given as *wt% NaCl equivalents* (Clynne and Potter 1977).

Descriptions of the geologically relevant but complex ternary system H_2O-NaCl-CO_2 is beyond the scope of this paper. In general, however, adding CO_2 to the system H_2O-NaCl extends the two-phase (immiscibility) conditions to higher pressures and temperatures compared to either of the binary systems H_2O-NaCl or H_2O-CO_2 (Takenouchi and Kennedy 1965; Bowers and Helgeson 1983a,b; Roedder 1984; Brown and Lamb 1989; Labotka 1991; Schmidt and Bodnar 2000).

TRACKING VOLCANIC PLUMBING SYSTEMS USING CO_2 INCLUSIONS

Pure CO_2 or CO_2-dominated fluid inclusions occur ubiquitously in xenoliths and phenocrysts hosted by basaltic as well as evolved magmas (Roedder 1965, 1984). They provide an excellent geobarometer even though the method is based on some assumptions and has several sources of serious errors (Roedder and Bodnar 1980). This section is a "handbook" to show the usage of CO_2-dominated inclusions for barometry, to review the principles and limitations of the method, and to constrain sources and magnitudes of possible errors.

Examples. In a study of CO_2 inclusions in ultramafic xenoliths from Loihi seamount (Hawaii), Roedder (1983) was probably the first who related the depths recorded by primary and secondary inclusions to that of a magma chamber inferred from seismic data. Hansteen (1991) used the density distribution of CO_2-rich inclusions in olivine phenocrysts to show that a picritic magma from Iceland was trapped at 7-10 km depth near the crust-mantle boundary before eruption. Likewise, the density distributions of CO_2 inclusions in phenocrysts and xenoliths from the Canary Islands (Hansteen et al. 1991, 1998; Klügel et al. 1997, 2005; Galipp et al. 2006) and Madeira Archipelago (Schwarz et al. 2004; Klügel and Klein 2006) combined with clinopyroxene-melt barometry indicate the presence of well-defined levels of magma stagnation in both mantle (>15 km depth) and crust; a bimodal pressure distribution is not uncommon (Fig. 11). For Réunion Island, in contrast, fluid inclusion densities of olivine phenocrysts suggest that zones of magma crystallization and fluid trapping extend from near the surface to 15 km depth at least (Bureau et al. 1998). Pure CO_2 inclusions in quartz xenoliths from Vulcano Island yield two distinct density intervals, which suggests ponding of magmas first in a lower crustal reservoir (21-17 km) and then fractionation in a shallower reservoir (14-8 km; Zanon et al. 2003; Frezzotti and Peccerillo 2004); an additional zone of magma accumulation is indicated at 2-1.5 km depth (Clocchiatti et al. 1994) (Fig. 11). Belkin and De Vivo (1993) used CO_2 and H_2O-CO_2 inclusions in ultramafic cumulate xenoliths from plinian eruptions of Vesuvius to infer a depth range of crystallization of 4-10 km. Fulignati et al. (2004) interpreted CO_2-rich fluid inclusions in skarn xenoliths of the 1944 K-phonotephritic eruption of Vesuvius to indicate a pressure of <100 MPa for the shallow reservoir, which implies a multistage evolution before eruption. Looking at continental intraplate volcanoes, a change of magma ascent rates at Moho depths was inferred from the density distribution of two generations of CO_2 inclusions in mantle xenoliths from a volcanic field at the Hungary-Slovakia border (Szabó and Bodnar 1996). For the East Eifel volcanic field (Germany) overlapping pressures indicated by CO_2-inclusions in both cumulate and metamorphic xenoliths were taken as evidence for a main magma reservoir in the lower crust at 22-25 km depth, near the brittle-ductile transition (Sachs and Hansteen 2000).

These examples show that multi-stage ascent of magmas with stagnation at Moho or crustal depths is the rule rather than the exception, and that CO_2-dominated fluid inclusions played the major role in identifying these levels. Because the *P-T-V* properties of CO_2 are reasonably well known, the determination of inclusion densities and pressures is relatively straightforward. Nevertheless, some pitfalls and error sources do exist and need to be considered in any barometric study based on fluid inclusions.

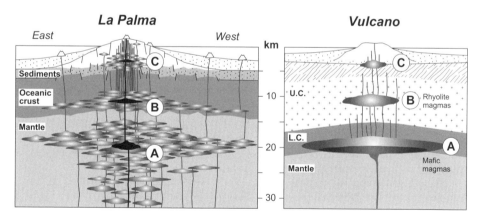

Figure 11. Models of the Recent magma plumbing systems beneath La Palma (Canary Islands) and Vulcano (Aeolian arc) based on data and illustrations from Hansteen et al. (1998); Zanon et al. (2003); Frezzotti and Peccerillo (2004); Klügel et al. (2005); and Peccerillo et al. (2006). For La Palma, the magma storage systems in the lower oceanic crust (B) and uppermost crust (C) were inferred by fluid inclusion barometry, whereas those in the mantle (A) reflect clinopyroxene-melt barometry data and subordinate fluid inclusion data. The active part of a single eruption is shown in black. For Vulcano, three main levels of magma storage have been identified by fluid inclusion barometry: (A) in the mafic granulite lower crust (L.C.) to upper mantle; (B) in the felsic granulitic upper crust (U.C.) where mafic to intermediate magmas rising from (A) fractionate to rhyolitic magmas; and (C) within the volcano-sedimentary terrain. The barometry-based models for both islands are consistent with petrological and geophysical data on the underlying lithological structures.

Rationale

In an ideal fluid inclusion pressure and temperature are related by an isochore described by an equation of state for the respective fluid enclosed. If the fluid composition X and the equation of state are known and T is determined by an independent method, then a pressure represented by the inclusion can be derived. By investigating a sufficiently large number of fluid inclusions, their density distributions can yield the depths of phenocryst crystallization and/or xenolith entrainment (which in many cases reflect a magma chamber) and the depths of temporary magma stagnation during ascent. This is highly valuable information on a volcanic plumbing system not easily available from other methods. We point out that the study of *both* primary and secondary inclusions is relevant and useful here because early as well as late processes characterize the nature of a plumbing system.

Applicability. Compared to barometers based on mineral-melt equilibria that are suitable for volcanic rocks (reviewed in Putirka 2008), fluid inclusion barometry offers two advantages: 1) it is not confined to certain mineral phases or phase assemblages but can be used for basically any mineral occurring as phenocryst phase or in a xenolith, and 2) it has the potential to reveal short-lived events during magma ascent since fluid inclusions can rapidly adapt to decreasing ambient pressures. On the other hand, processes occurring after trapping bias the results and limit the extent to which high pressures in the inclusions are preserved. In general, pressures recorded by both primary and secondary inclusions considered here always represent a *minimum* limit for their trapping pressure. Barometers using H_2O and CO_2 contents of silicate melt inclusions offer similar advantages and disadvantages as compared to barometers based on mineral-melt equilibria; see Metrich and Wallace (2008) for a discussion.

Fluid inclusion barometry is thus a tool complementary to mineral-melt barometry, and we strongly suggest using both methods wherever possible, combined with careful petrographic observations and additional petrological data, in order to obtain the most complete information for a reconstruction of the ascent history of magmas (e.g., Peccerillo et al. 2006). *Xenoliths* are particularly valuable in this respect as they are often derived from wall-rocks of a magma reservoir where melt accumulation and degassing cause high deviatoric stresses and fluid penetration. Many xenolith crystals become therefore fractured and rehealed under the presence of CO_2-rich fluids, which results in an extremely large number of secondary inclusions (Roedder 1965, 1984; Andersen and Neumann 2001) that may reflect the *P-T* conditions at the presumptive magma reservoir. Phenocrysts, in contrast, usually show far less secondary inclusions because of the small magnitude of deviatoric stresses exerted on them by the host melt.

Density distribution. The re-equilibration of primary and secondary fluid inclusions obviously results in a biased density distribution. The extent of re-equilibration and whether it occurs by either stretching or partial decrepitation depend on inclusion size, the rate at which *P-T* conditions change, and the deformation properties of the host mineral and hence its structure and composition (e.g., Bodnar 2003). The important point in fluid inclusion barometry is that data from natural systems may provide ample information on the magmatic plumbing system even if—or maybe because!—Roedder's rules 2) or 3) are not satisfied. The key to success is the *nature of the density distribution* of an investigated fluid inclusion suite (e.g., Hansteen et al. 1998; Touret 2001; Andersen and Neumann 2001).

This is schematically shown in Figure 12 for fluid inclusions in phenocrysts and xenoliths trapped at a deep level. If the host magma ascends from this level to the surface more or less continuously then the density distribution will be skewed and show a broad "re-equilibration tail" (Fig. 12a). Depending on factors such as the ascent velocity, the original density may still be recognizable as a frequency maximum or merely as an upper limit of measured values. In contrast, if the ascending magma stagnates temporarily at one or more levels such as the mantle-crust boundary, then the extent of re-equilibration is larger at these levels and additional

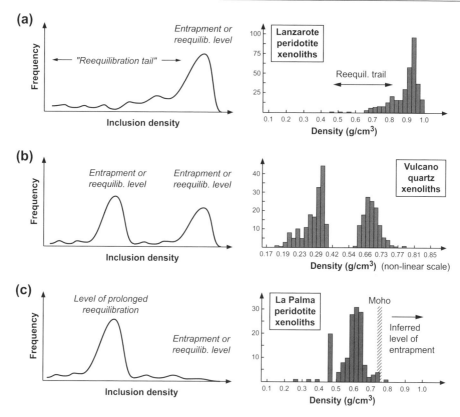

Figure 12. Sketch showing idealized (left) and measured (right) density distributions of fluid inclusion associations in phenocrysts or xenoliths hosted by magmas with different ascent histories. (a) Single-stage ascent from the depth of entrapment causes random re-equilibration of inclusions en route to the surface. This results in a single frequency maximum at the high-density end and a large spread and random distribution of data towards lower densities. The example for a suite of peridotite xenoliths from Lanzarote (Canary Islands) is based on data from Neumann et al. (1995). (b) Multi-stage ascent from depth with temporal ponding at a shallower level. This results in a large spread of inclusion densities with two distinct frequency maxima, as is exemplified by early and late fluid inclusions in quartz xenoliths from Vulcano Island where the second maximum corresponds to a shallow level of trapping and re-equilibration (data from first volcanic cycle; Zanon et al. 2003). (c) Multi-stage ascent from depth with prolonged ponding at a shallower level. This results in almost complete re-equilibration of inclusions with a single pronounced maximum and little spread of densities. The example for a suite of Recent peridotite xenoliths from La Palma (Canary Islands) is based on Hansteen et al. (1998).

frequency maxima become recognizable in the density distribution; new inclusions may also become trapped (Fig. 12b). An example is Vulcano Island where two distinct crustal reservoirs could be identified by distinct density intervals of two generations of CO_2-rich fluid inclusions in quartz xenoliths (Zanon et al. 2003). It is highly unlikely that such distributions are fortuitous if sufficient inclusions are analyzed. If magma ponds at a certain level for a long period of time (months to years), then almost complete re-equilibration may occur. The resulting density distribution shows a well-defined maximum at the shallower level and the memory of the former deep level is almost completely lost (Fig. 12c). This situation is well exemplified by the active Cumbre Vieja volcano on La Palma (Canary Islands) where fluid inclusions in xenoliths and phenocrysts show an almost Gaussian density distribution reflecting pressures within the lower crust but almost no memory of former mantle pressures (Klügel et al. 2005).

Density distributions broadly resembling those in Figure 12 have also been derived by decompression experiments with synthetic fluid inclusions in natural quartz (e.g., Bakker and Jansen 1991, Vityk and Bodnar 1998). Based on systematic variations of parameters such as decompression P-T path and duration of re-equilibration combined with statistical analyses of the results, comprehensive relations between P-T-t parameters, nature and extent of re-equilibration, and histogram properties could be shown. Although the experiments were restricted to aqueous inclusions in quartz and cannot be rigorously compared to other fluids and minerals, the results have general implications for the interpretation of density distributions. As a particular result, it was found that a small percentage of inclusions, that cannot be distinguished or recognized petrographically, maintained original densities even after extensive re-equilibration.

Determining inclusion density, pressure and depth

Composition of the inclusions. Most fluids enclosed in basalt-hosted xenoliths and phenocrysts consist of essentially pure CO_2 (Roedder 1965, 1984), but of course this needs to be verified for any fluid inclusion investigated for pressure determinations since exceptions do exist (e.g., Andersen and Neumann 2001, and references therein). Upon heating of pure CO_2 inclusions from ca. −190 °C, the first phase transition observed in microthermometric investigations is *instantaneous* melting of frozen CO_2 at −56.6 °C (T_m). If additional fluid components are present then the initial melting temperature will be lowered and a *temperature interval* will be noticeable between initial and final melting (T_m(initial) and T_m(final), respectively). Isochoric T-X sections of the CO_2-CH_4 and CO_2-N_2 systems show that T_m(initial) of CO_2 is depressed by about 0.13 °C per mol% of CH_4 or 0.24 °C per mol% of N_2 in an inclusion, and a melting interval is clearly observable for amounts >1 mol% of CH_4 or a few mol% of N_2 (Van den Kerkhof 1990). Likewise, an amount of 1-2 mol% of SO_2 lowers T_m(initial) by about 1.4 °C and causes a melting interval of about 0.8-1.3 °C (Frezzotti et al. 2002). A combination of microthermometric and Raman microspectrometric measurements of CO_2 inclusions in natural samples indicates the amounts of CH_4, N_2, SO_2, H_2S or CO to be less than 0.1 mol% if T_m(initial) is depressed by no more than 0.2 °C (Zanon et al. 2003). Although it is difficult to detect small amounts of additional components in CO_2-dominated inclusions by microthermometric investigations, it has been suggested that a few mol% do not significantly affect the interpretation of the trapping conditions of inclusions (Van den Kerkhof 1990; Frezzotti et al. 2002).

The presence of H_2O in CO_2-rich inclusions causes no depression of T_m(initial) because of its little solubility, but introduces additional phase transitions including formation of CO_2 clathrate hydrate at low T, its subsequent melting at $T \leq 10$ °C, and final homogenization of coexisting CO_2 and H_2O phases. Whereas amounts of <10 vol% of water are hard to detect in microscopic investigations (but are detectable by Raman and FTIR spectroscopy), clathrate formation in CO_2-rich inclusions is readily recognized (Roedder 1984). According to Burruss (1981) the bulk composition and density of such an inclusion can be obtained from the homogenization temperature of CO_2 liquid and vapor (yielding the density of the CO_2 phase as shown below) and from the relative volume proportions of both phases estimated at some temperature between that of CO_2 homogenization and 50 °C. The estimation of volume proportions, however, is not trivial and is subject to considerable error (Roedder 1984; Bakker and Diamond 2006).

Determination of inclusion density and error magnitude. The density of CO_2 inclusions is accurately derived from the temperature and nature of final homogenization. Once T_h(LV→L) or T_h(LV→V) have been established as shown above the bulk density can directly be derived from the relation shown in Figure 13 using e.g., the auxiliary equations (3.14) and (3.15) of Span and Wagner (1996). Since T_h(LV→L) can be determined with an accuracy and reproducibility of better than ±0.2 °C it follows from the slope of the saturated-liquid curve in Figure 13 that the densities are accurate to within 0.001-0.01 g/cm³ (0.1-2% relative

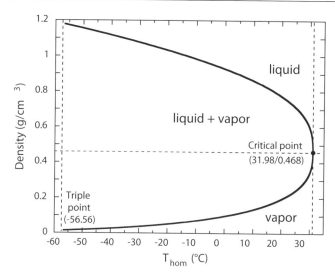

Figure 13. Isobaric section showing the relation between homogenization temperature and density of saturated liquid and vapor, respectively, for pure CO_2 inclusions calculated using the auxiliary equations (3.14) and (3.15) of Span and Wagner (1996); cf. Figure 4a.

uncertainty), except for near the critical temperature where $\Delta\rho/\Delta T$ becomes very high (2-8% relative uncertainty below 30.9 °C). The determination of $T_h(LV \rightarrow V)$ is less accurate than $T_h(LV \rightarrow L)$ because the complete evaporation of a liquid film is more difficult to recognize than elimination of a bubble that shows pseudo-Brownian motion. By assuming an accuracy of 1 °C for the determination of $T_h(LV \rightarrow V)$ the relative uncertainty for density estimates is 3-12% for values <0.35 g/cm³ (T_h < 30.2 °C). The microthermometric determination of densities of CO_2 inclusions is thus surprisingly accurate, except for near-critical densities, and is not a major error source for barometric investigations.

Which equation of state? A number of equations of state (EOS) for pure CO_2 have been published in the last decades based on different theoretical approaches, different P-T ranges of experimental data, and different adjustable parameters (see discussions in Mäder and Berman (1991); Seitz et al. (1994); Sterner and Pitzer (1994); and Span and Wagner (1996)). Igneous petrologists face the problem that their range of interest, about 100 to >1000 MPa at 800-1300 °C, is poorly represented by experiments and requires extrapolation beyond the data used to adjust the EOS parameters. Most commonly the Modified Redlich-Kwong-type EOS by Kerrick and Jacobs (1981) and Holloway (1981) (hereafter referred to as KJ81 and H81, respectively) were used in the past, assuming that their extrapolation to conditions of intratelluric magmas was adequate (e.g., Hansteen 1991; Szabó and Bodnar 1996; Hansteen et al. 1998; Zanon et al. 2003; Klügel et al. 2005).

Because of the paucity of experimental data for CO_2 at temperatures >800 °C it is difficult to judge the suitability of any EOS for magmatic systems. The calibration of the EOS of Span and Wagner (1996; SW96) comprises some 2800 experimental ρ-P-T data in the range from 0.1 MPa and −56 °C to 800 MPa and 830 °C with most P-T values falling within the technically important region of <30 MPa and <100 °C. This EOS includes the latest state-of-the-art data and is probably the most reliable equation for this range. In the high-T region the reported uncertainty of calculated densities is ±2% at 400-800 MPa, ±1% at 100-400 MPa, and ±0.2% at 10-100 MPa. Span and Wagner (1996) suggest that their equation should also yield reasonable data outside its validity range but the uncertainty cannot be estimated. Alternatively, for magmatic systems the EOS of Sterner and Pitzer (1994; SP94) is a good choice because 1) its calibration includes additional high *P-T* data and 2) with 28 parameters it is easier to implement in spreadsheets than the sophisticated SW96 equation. Sterner and Pitzer (1994) claim that their EOS provides valid isochores from 0.1 MPa and −57 °C to 10 GPa and >1700 °C, but because of lacking experimental data for the *P-T* conditions of magmatic systems the respective uncertainties were not stated. Based on new high-pressure experiments, Abramson and Brown (2004) suggest that pressures derived by the SW96 and SP94 equations need to be decreased by a few percent at *P-T* conditions of the uppermost mantle.

Figure 14 compares pressures calculated by SP94, deliberately used here as a reference, against those calculated by other common EOS using the software packages FLUIDS (Bakker 2003) and FLINCOR (Brown 1989) for a model temperature of 1200 °C. This comparison covers almost the entire density range of CO_2 inclusions reported so far. The frequently used KJ81 equation yields negligible deviations from the reference pressures for densities <0.8 g/cm^3 but increasingly overestimates pressures at higher densities (up to 11% at 1.1 g/cm^3). The H81 and Bottinga and Richet (1981; BR81) equations underestimate pressures for low densities by up to 10% and 7%, respectively, and overestimate for high densities by up to 10%. The SW96 equation shows only minor deviations from SP94 over the entire density range, with a maximum difference of −3% around the critical density of 0.466 g/cm^3. An example illustrates the effect of these differences: For an inclusion with a density of 1.1 g/cm^3 the different EOS yield a pressure range of 1032 to 1184 MPa at 1200 °C corresponding to a depth uncertainty of ca. 4-5 km; for 0.6 g/cm^3 the respective values are 269 to 303 MPa and ca. 1 km.

To summarize, the SP94 and SW96 equations are probably the best choice for calculating CO_2 isochores, with an advantage of SP94 being its easier implementation, but both equations

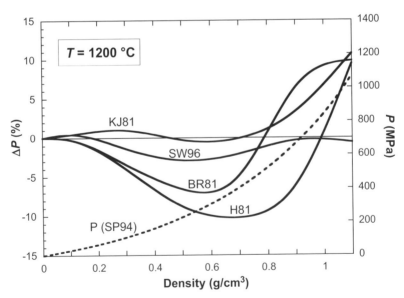

Figure 14. Plot of inclusion density against pressures calculated using different equations of state (EOS) for pure CO_2 and a trapping temperature of 1200 °C. Dashed line (right scale) shows the pressure given by the EOS of Sterner and Pitzer (1994) used here as a reference (SP94). Thick solid lines (left scale) show the relative deviation from the reference (ΔP in %) for pressures calculated using the following EOS: KJ81 (Kerrick and Jacobs 1981), H81 (Holloway 1981), BR81 (Bottinga and Richet 1981), and SW96 (Span and Wagner 1996). The calculations used the software packages FLUIDS 1 (Bakker 2003) and FLINCOR (Brown 1989).

still need to be tested by experimental data for the *P-T* conditions of magmatic systems. The KJ81 equation, often used in the past, retrieves pressures equally well for inclusions with densities below 0.8 g/cm^3 (Fig. 14). An electronic spreadsheet for the calculation of CO_2 densities and isochores from homogenization temperatures using SP94 is available from the authors upon request.

Trapping temperature. For a pressure to be calculated the temperature at the time of inclusion trapping (T_{trap}) must be known. For phenocrysts T_{trap} can be obtained by an independent method such as mineral-melt equilibrium (e.g., Putirka 2008 and references therein), homogenization temperature calculate of melt inclusions, or estimation based on melt composition. Eruption temperatures of lavas are generally a good proxy for T_{trap} since the magma temperature is unlikely to have changed significantly during the final ascent. Since ΔP/ΔT of CO_2 isochores is rather low at magmatic conditions and slightly increases with increasing fluid density (Fig. 15), calculated pressures are relatively insensitive to uncertainties in temperature determinations (Roedder 1965, 1983). For example, a choice of 1150 rather than 1200 °C results in a calculated pressure error of 2 MPa (3.8%) for an inclusion density of 0.2 g/cm^3, 12 MPa (4%) for 0.6 g/cm^3, and 34 MPa (3.1%) for 1.1 g/cm^3. Even a gross error in estimation of T_{trap} has a limited effect on calculated pressures as shown by a comparison of panels a) and c) in Figure 16. It can be concluded that, at least for phenocrysts, the determination of T_{trap} is not a critical issue in fluid inclusion barometry.

For crustal and mantle xenoliths the determination of T_{trap} is less straightforward because a xenolith's history before entrainment into the host magma, and the relation between xenolith entrainment and fluid entrapment, cannot always be established. Fluid and melt inclusions can be trapped long before xenolith entrainment, concomitant (and in some cases causally related) to entrainment, or during subsequent xenolith transport to the surface. In many cases the temperature of the host magma may be a good choice for T_{trap} because heating of a xenolith occurs rather rapidly: For a representative thermal diffusivity of 10^{-6} m^2/sec (Robertson 1988) the center of a spherical xenolith 10 cm in diameter reaches 90% of the surrounding melt

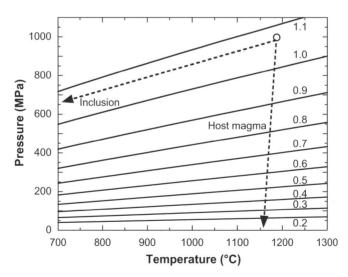

Figure 15. Isochores for pure CO_2 calculated using the EOS of Sterner and Pitzer (1994). Numbers at isochores indicate fluid density in g/cm^3. Dashed lines illustrate the strongly increasing pressure difference between an adiabatically ascending basaltic magma and an isochoric fluid inclusion trapped at *P-T* conditions indicated by the circle.

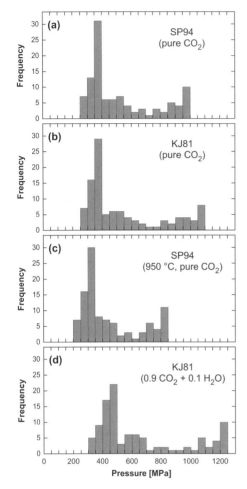

Figure 16. Effect of changing calculation parameters on pressure distributions inferred for CO_2 fluid inclusions in peridotite xenoliths from Gran Canaria. The calculations use measured densities from Hansteen et al. (1998), a model temperature of 1150 °C except for panel c), and the EOS of Sterner and Pitzer (1994) and Kerrick and Jacobs (1981) indicated as SP94 and KJ81, respectively. (a) Pressure distribution for SP94. (b) KJ81 yields a similar pressure distribution as SP94 except for higher pressures where KJ81 overestimates resulting in a larger data spread. (c) Decreasing the model temperature by as much as 200 °C causes a comparatively mild decrease of calculated pressures that increases with increasing inclusion density (compare with panel a). (d) Data for an assumed original fluid composition of 90 mol% CO_2 + 10 mol% H_2O, which was accounted for by increasing the measured density by 4.5% and by specifying the CO_2-H_2O system in KJ81 used to calculate the isochores. See text for details.

temperature within 12 minutes (cf. Fig. 29 of Carslaw and Jaeger 1959). Zanon et al. (2003) show that the combination of geothermometers for the host magmas and homogenization temperatures of melt inclusions in xenolith minerals yield a consistent and plausible range for T_{trap}.

For some xenolith types such as mantle peridotites the local ambient temperature can be determined by mineral geothermometers, which, however, are slow in adapting to temperature changes because chemical diffusivities in minerals are orders of magnitude smaller than thermal diffusivities. As a consequence, T_{trap} may actually be close to the host magma temperature even though xenolith minerals (and their cores in particular) indicate much lower values. The same holds for a regional geotherm when claimed to represent the P-T conditions of trapped fluids (e.g., Andersen and Neumann 2001). The actual T_{trap} can be anywhere between the regional geotherm and magma temperatures because xenoliths represent the vicinity of conduits and magma holding reservoirs and consequently were entrained at elevated temperatures. This view is based on the frequently observed abundance of deep-seated xenoliths in late rather than early lavas of an eruption, which plausibly reflects massive wall-rock collapse when magmatic overpressure declines (Cox et al. 1993; Klügel et al. 1999). The interpretation of xenoliths as biased samples overprinted by magmatism is also consistent with the penetration or infiltration of xenoliths by basaltic melts often observed or with the association of crustal or mantle xenoliths with cumulates (e.g., Kempton 1987; Wulff-Pedersen et al. 1996; Klügel 1998; Sachs and Hansteen 2000; Shaw et al. 2006). A regional geotherm is therefore not an adequate measure for T_{trap} of fluid inclusions in a volcanic region where local temperatures vary strongly on a kilometer- or meter-scale. Such local variations can also be reflected in a wide range of temperature estimates for mantle xenoliths from a single locality (e.g., Frezzotti et al. 2002), which together with the host magma temperature provide brackets for T_{trap} rather than a single value.

Calculation of depth. The derivation of depth from the calculated pressures appears to be a trivial issue at first sight, but it requires some caution because it is not always clear whether the derived pressure is *hydrostatic* or *lithostatic*. The local pressure at which a fluid is trapped is in principle a fluid pressure and hence not lithostatic, thus its relation to the height of the overlying rock mass is ambiguous. The fluid pressure during trapping can be equal to the lithostatic pressure or it can be lower if an overlying column of magma rather than rock needs to be considered. In this case further complications can arise because degassing and fragmentation of an erupting magma affect the fluid pressure at depth. The fluid pressure can also exceed the lithostatic pressure by up to a few tens of MPa, such as in the case of an overpressurized magma chamber. Roedder and Bodnar (1980) discuss a variety of pressure conditions for fluid entrapment and suggest that pressures are probably between lithostatic and hydrostatic in most natural situations. As an example, assumption of a lithostatic column (lithostatic pressure) with an average rock density $\rho = 3.3$ g/cm^3 as in the upper mantle will yield a 18% lower depth than assumption of a hydrostatic column (hydrostatic pressure) with a magma density $\rho_{melt} = 2.7$ g/cm^3 (Roedder 1983); for continental crust this difference becomes smaller. Remarkably, at high pressures, the uncertainty of the seemingly simple depth calculation may be larger than the uncertainties in determining density and pressure as discussed above!

The role of possibly missing H_2O

The predominance of pure CO_2 fluid inclusions in basalt-hosted xenoliths and phenocrysts does not necessarily preclude the earlier presence of H_2O in the trapped fluid. To the contrary, a significant amount of H_2O in the fluid phase at depth is indicated e.g., by degassing basalts (reviewed by e.g., Metrich and Wallace 2008) or by xenoliths with fluid inclusions being associated with hydrous mineral phases (Sachs and Hansteen 2000). Much of the trapped H_2O can be lost, however, by *compositional re-equilibration* (Bodnar 2003) involving 1) reactions and 2) preferred leakage and diffusion. First, H_2O in a mixed CO_2-H_2O inclusion may become removed through reaction between fluid and inclusion wall to produce hydrous minerals and carbonate (Andersen et al. 1984; Kleinefeld and Bakker 2002); in this case the observed pure CO_2 is actually a residual fluid. Occurrences of such reaction products in small amounts are often hard to recognize by visual observation. Second, and more significant, mixed CO_2-H_2O inclusions show preferential leakage of H_2O and thus a differentiation of supercritical fluid upon changing ambient conditions. Synthetic mixed inclusions in quartz that were re-equilibrated at reduced pressures and T between 560 and 825 °C showed a significant decrease in X_{H_2O} within days to weeks (Bakker and Jansen 1991; Hall and Sterner 1993). Here, non-decrepitative preferential H_2O leakage is thought to occur mainly along crystal defects and is probably enhanced by ductile deformation and recrystallization (Hollister 1990). Because of different molecule sizes and wetting properties, H_2O is more prone than CO_2 to migration along dislocations and planar defects whereas the latter is solely lost by hydrofractures (Bakker and Jansen 1994). In addition, hydrogen can diffuse out of a quartz-hosted inclusion within days (Mavrogenes and Bodnar 1994). In olivine, experiments revealed extremely fast diffusion of water-derived hydrogen and hydroxyl, which may leave an inclusion devoid of H_2O within hours at magmatic temperatures (Mackwell et al. 1985; Mackwell and Kohlstedt 1990). The rapidity of diffusion indicates that it is governed by the inherent Fe^{3+}/Fe^{2+} and point defect abundance in olivine. Likewise, the very fast transfer of oxygen through olivine as experimentally observed was explained by point-defect diffusion (Pasteris and Wanamaker 1988); therefore diffusion of water-derived species may be fast in other ferromagnesian minerals as well. Diffusive loss and preferential H_2O leakage are thus plausible explanations for the absence of H_2O in most fluid inclusions in mantle xenoliths.

For pressure determinations the problem of "missing H_2O" in fluid inclusions boils down to an estimation of their former H_2O content and its adequate correction (cf. Sachs and Hansteen

2000). If α is the molar H_2O/CO_2 ratio of a mixed CO_2-H_2O fluid trapped in an inclusion and all H_2O becomes subsequently lost, then for an isochoric system the original fluid density ρ_{orig} is simply derived from the measured density of the H_2O-free inclusion ρ_{meas} by: $\rho_{orig} = \rho_{meas} \times (1 + \alpha \times 18/44)$, i.e., corrected by the H_2O/CO_2 *weight* ratio. The isochore for the original fluid is then calculated using the corrected density and an equation of state for the CO_2-H_2O system (e.g., Kerrick and Jacobs 1981) with the respective H_2O content given by α. The estimation of the former H_2O content, however, is notoriously difficult and in most cases confined to an estimate of a *plausible upper limit*. Even for basalts where much data on CO_2 and H_2O solubilities exist, estimation of the CO_2/H_2O ratio in the fluid phase (vapor) is hampered by missing information on the H_2O and CO_2 contents in the melt at the time of fluid trapping and on the conditions of degassing such as open vs. closed system. In general, the vapor phase exsolved from basaltic melts at depth is strongly CO_2-dominated until most of the CO_2 is lost and significant H_2O exsolution occurs (Metrich and Wallace 2008). For MORB and intraplate basalts at depths exceeding a few kilometers, a maximum H_2O content in the fluid phase of 10 mol% is probably a conservative estimate in most cases (cf. Dixon and Stolper 1995); it may be higher at subduction zones. For 10 mol% of H_2O the correction factor for density is 1.045. Using the Kerrick and Jacobs (1981) EOS for the CO_2-H_2O system the resulting corrections for calculated pressures range from e.g., −7% for the 0.3 g/cm^3 isochore to +23% for the 0.8 g/cm^3 isochore at 1150 °C. These pressure corrections show a strong non-linear dependency on density and therefore strongly skewed calculated pressure distributions (Fig. 16d).

Volumetric re-equilibration during magma ascent: a case study

It has been discussed above that the healing of microcracks in basalt-hosted olivine occurs within a few hours to days (Wanamaker et al. 1990). Therefore, if a magma stagnates during its ascent for a period of more than about a day, even those inclusions showing petrographic evidence for leakage can under favorable conditions yield the pressure of this stagnation. We reiterate that it is not the density of a single decrepitated inclusion that provides the relevant barometric information, but the clustering of many inclusions around a certain density. However, the results can still be biased by stretching during subsequent magma ascent, as has been calculated for olivine-hosted CO_2 inclusions during isothermal decompression using a power-law creep model (Wanamaker and Evans 1989). An extreme decompression by 3000 MPa within 6 hours (magma ascent from 100 km depth to the surface at 5 m/s) would increase an inclusion's radius by about 0.8% at 1100 °C and 9% at 1200 °C resulting in a density decrease by 3% and 22%, respectively. Likewise, decompression by 1000 MPa within two days (magma ascent from 33 km depth at 0.2 m/s) would result in a density decrease by 8% at 1100 °C and 30% at 1300 °C. The stretching rate generally increases as the overpressure of inclusions increases, and the total amount of stretching for a fixed overpressure increases as the external pressure decreases (Wanamaker and Evans 1989). Clearly, stretching may significantly reduce inclusion densities in xenoliths and phenocrysts even during short ascent times, resulting in a wide range of homogenization temperatures as is often observed.

A case in point for volumetric re-equilibration is shown by CO_2 inclusions in basanite-hosted xenoliths and phenocrysts from the 1949 eruption on La Palma, Canary Islands (Hansteen et al. 1998; Klügel et al. 2000). Inclusions in tholeiitic gabbro xenoliths that show ample evidence for magmatic overprinting indicate a dominant pressure range of 270-330 MPa reflecting the lower oceanic crust where the gabbros originated. This density range overlaps perfectly with the ranges indicated by inclusions in ultramafic cumulates (90% of values between 190 and 340 MPa), peridotite mantle xenoliths (90% between 190 and 400 MPa), and clinopyroxene phenocrysts (Figs. 11 and 17). Remarkably, almost none of the investigated primary and secondary inclusions in mantle xenoliths showed mantle pressures. This was attributed to storage of the xenoliths in the lower crust for a long period of time, probably tens of years before eruption, where almost complete re-equilibration of inclusions and interactions between xenoliths and mafic magma occurred. Some re-equilibration did also occur during the final xenolith uplift, which

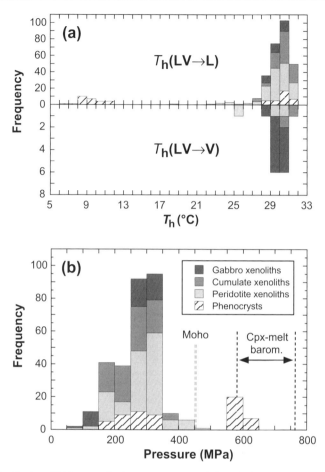

Figure 17. Distribution of (a) homogenization temperatures and (b) calculated pressures indicated by fluid inclusions in basanite-hosted xenoliths and phenocrysts from the 1949 eruption on La Palma (Canary Islands). The diagrams are based on data of Hansteen et al. (1998) recalculated with the EOS of Sterner and Pitzer (1994). A well-defined maximum between about 200 and 350 MPa is reflected in both crustal and mantle xenoliths as well as clinopyroxene phenocrysts, and is consistent with the depth of the lower oceanic crust where MORB gabbro xenoliths originate. Mantle pressures (above ca. 450 MPa) are indicated almost exclusively by olivine phenocrysts and less so by peridotites, which is taken as evidence for prolonged storage of mantle xenoliths in the lower crust. The pressure range indicated by fluid inclusions in olivine phenocrysts overlaps with that indicated by mineral-melt barometry of clinopyroxene phenocrysts.

took only a few hours (Klügel 1998), but its effect was limited because the frequency maximum of the pressure distribution shown in Figure 17 overlaps perfectly with the inferred pressure of origin of the gabbro xenoliths. The basanite carrying the xenoliths to the surface ascended from mantle depths and stagnated in the lower crust only briefly (Klügel et al. 2000) so that CO_2 inclusions in olivine phenocrysts could retain their original mantle pressures, which fit well with pressures indicated by clinopyroxene-melt barometry. CO_2 inclusions in clinopyroxene phenocrysts, however, became re-equilibrated in the lower crust because this mineral is more prone to leakage than olivine due to better cleavage. The results of this case study illustrate the usefulness of fluid inclusion barometry of xenoliths and phenocrysts especially when combined with mineral-melt barometry and geospeedometry.

PRESSURE INFORMATION FROM AQUEOUS FLUIDS IN SHALLOW MAGMA CHAMBERS AND PLUTONIC ROCKS

Background

Fluids coexisting with crystallizing magmas exhibit a wide range of chemical compositions, covering major parts of the ternary system H_2O-$NaCl$-CO_2 (Burnham 1979; Holloway and Blank 1994; Lowenstern 1994; Dixon et al. 1995; Schmidt and Bodnar 2000; Kamenetsky et al. 2002). Both magma composition and pressure exert major influence on fluid composition. Mafic intraplate magmas at depth typically contain comparable amounts of dissolved CO_2 and H_2O, which are exsolved upon ascent and crystallization (e.g., Metrich and Wallace 2008). Mafic subduction zone magmas are, however, H_2O dominated. Due to the higher solubility of H_2O than CO_2 in silicate melts at high pressures, the first fluid phase formed after initial fluid saturation in the uppermost mantle or lower to middle crust is CO_2-rich. If the magmatic fluids are progressively lost to the surroundings during crystallization and degassing in a shallow reservoir, the fluids continuously released from the magma will evolve towards H_2O-richer compositions. As discussed above, however, an ascending melt can produce CO_2-dominated fluids even in the upper crust, provided that either the initial water contents are low or that crustal ascent has been rapid and stagnation times limited.

In consequence, felsic magmas usually contain comparatively high amounts of dissolved water. At high temperatures and moderate to low pressures, both a high-salinity liquid and a vapor phase may coexist with silicate melt (this phenomenon is described above for the system H_2O-$NaCl$). Such fluid phase separation is common in the plutonic environment because a wide range of possible magmatic fluid compositions fall within the immiscibility fields for the systems H_2O-$NaCl$ and H_2O-$NaCl$-CO_2 at low pressures (Bodnar et al. 1985; Chou 1987). It should be stressed that the resulting major element compositions of the liquid and vapor phases are consequently governed by the boundaries of the immiscibility field rather than by the volatile contents of the melt.

In the following, we give examples of fluid inclusions representing the magmatic fluid phase in shallow magma chambers prior to or during volcanic eruptions, and those occurring in a variety of plutonic rocks. Both scenarios involve residual fluids exsolved from magmas at shallow levels, where the combined parameters pressure, temperature and fluid composition control whether fluid immiscibility ("boiling") occurs. The degassing rate, which is important for wall rock fracturing and eventually for the hazard potential of a shallow magma chamber, is intimately linked with the pressure-dependent degree of fluid oversaturation.

Fluids in shallow magma chambers

The number of fluid inclusion studies from volcanic rocks seems to correlate with magma chamber depth. The main reasons are probably the scarcity of fluid inclusions with adequate sizes in phenocrysts formed at shallow levels, and the relative difficulty of observing phase changes in low-density (i.e., vapor-rich) fluid inclusions.

Primary magmatic fluids coexisting with rhyolitic melt inclusions in quartz from igneous complexes in New Zealand and Chile have been interpreted as shallow-level magma chamber paragenesis (Davidson and Kamenetsky 2007). The saline fluids coexisting with the melts are comparatively enriched in halogens (chlorine and bromine). Lowenstern (1994) demonstrated immiscibility among vapor, highly saline liquid, and silicate melt during pre-eruptive crystallization of peralkaline rhyolites from Pantelleria, Italy. The Cl contents of Pantellerian rhyolites indicate equilibration at pressures between 50 and 100 MPa. Lava samples erupted from Mt. Unzen, Japan, in June 1991, contain abundant, complexly zoned plagioclase (Browne et al. 2006, and references therein), which contain coexisting rhyolitic melt inclusions and fluid inclusions (Hansteen, unpubl. data). Fluid inclusions are two-phase (about 3% liquid and

97% vapor) at room temperature, and are well represented by the system H_2O-NaCl-CO_2. The aqueous phase in the fluid inclusions has a salinity of 21.5 ± 1 wt% NaCl equivalents. The density of the CO_2-dominated phase was estimated at 0.02 g/cm^3, giving a total fluid inclusion density of about 0.05 g/cm^3 when a density of 1.15 g/cm^3 is assumed for the aqueous phase. Using a post-mixing magma temperature of 900 °C, the magmatic fluid inclusions were formed at a pressure of 35±15 MPa (1.5±0.5 km) (equation of state of Brown and Lamb 1989).

Xenoliths interpreted as wall-rock fragments from shallow magma reservoirs have successfully been used to estimate magma stagnation depths in both mafic (see above) and silicic systems. From the last La Fossa eruption of Vulcano Island (1888-90), Clocchiatti et al. (1994) described secondary H_2O-NaCl-CO_2 inclusions in metamorphic xenoliths giving equilibration pressures of 30-60 MPa, which translates into a location of the higher portions of the chamber at around 1500-2000 m depth (cf. Fig. 11). One cumulate xenolith from the primitive basaltic 1999 eruption of Cerro Negro, Nicaragua, contains abundant water-rich fluid inclusions with a salinity of 3.7 wt% NaCl equivalents (Hansteen and Weiss, unpubl. data). No CO_2 could be detected in the fluid inclusions, demonstrating the water-rich composition of the residual magmatic fluid. Inclusion densities correspond to a magma chamber pressure of 100 ±16 MPa. Compositional zoning on a cm-scale even within single glassy clasts demonstates late-stage magma mingling during this eruption. The two compositional endmembers give pressures of 275 ±25 MPa and 100 ± 20 MPa respectively, using the clinopyroxene-melt thermobarometer of Putirka et al. (1996), which testifies to a two-level magma chamber system at Cerro Negro. Based on volatile contents in melt inclusions, Roggensack et al. (1997) similarly estimated fractionation pressures of about 300 MPa for the 1992 eruption products, overlapping with our cpx-melt thermobarometry.

Summing up, fluid inclusions in phenocrysts and xenoliths have successfully been used to demonstrate shallow-level magma stagnation between 30 and 100 MPa for both basaltic and silicic volcanic systems. The composition of the fluids were found to be variable but well represented by the system H_2O-NaCl-CO_2.

Fluid inclusions in plutonic rocks

Plutonic rocks typically contain abundant fluid inclusions representing various stages of their evolution (e.g., Roedder and Coombs 1967; Konnerup-Madsen 1979; Weisbrod 1981). Given the time available for differentiation processes, *P-T* variations, interaction with external fluids of changing compositions and chemical re-equilibration of mineral paragenesis, fluid inclusion records in plutonic rocks may be hard to decipher. Early inclusion assemblages may also have partly or fully re-equilibrated at later stages (see above). Nevertheless, successful fluid inclusion studies have been performed on both xenolith samples and in-situ samples from a considerable number of plutonic rocks. Fluid inclusion studies are also an important tool in pegmatite research, a topic which is beyond the scope of this paper; the interested reader is referred to reviews by London (1996) and Thomas et al. (2006). Primary fluids exsolving from felsic magma and eventually evolving into hydrothermal fluids is a topic of long-standing interest, which has recently been highlighted by studies of subvolcanic and plutonic rocks (Kamenetsky et al. 2004; Davidson and Kamenetsky 2007), showing a continuous evolution from magmatic to hydrothermal fluids.

Fluid inclusions in deep-seated intrusions generally have a significant component of CO_2, probably related to its elevated solubility in silicate melts at high pressures. Konnerup-Madsen (1979) interpreted CO_2-dominated and CO_2-H_2O inclusions in quartz from the Precambrian Farsund plutonic complex, south Norway, to represent magmatic fluids indicating an origin at 0.5 to 0.6 GPa. Later inclusions related to uplift were saline (2 to 60 wt% NaCl equivalents), and contained significant amounts of the cations K^+, Ca^{2+} and Mg^{2+}. Frost and Touret (1989) investigated fluid inclusions in quartz from the Sybille Monzosyenite, Laramie Anorthosite

Complex, Wyoming, reporting that the primary magmatic fluid consisted predominantly of CO_2, which exsolved at 0.3 GPa and 950-1000 °C, confirming formation conditions inferred from mineralogic data. Halite and sylvite crystals, which occur as part of the magmatic mineral paragenesis trapped as inclusions in feldspar, were interpreted as remnants of immiscible hydrosaline melts, that may have coexisted with the CO_2-rich fluid. Frezzotti et al. (1994) also reported possible brine-CO_2 immiscibility at magmatic conditions of 0.3 GPa and 750 °C for peraluminous graphite-bearing leucogranites from Deep Freeze Range (northern Victoria Land, Antarctica).

In a classical study of peralkaline syenitic and granitic xenoliths from Ascencion Island, Harris (1986) reported coexisting silicate melt inclusions and low-salinity aqueous inclusions, and additionally silicate melt coexisting with both a high-salinity (> 40 wt% NaCl equivalents) aqueous liquid phase and a CO_2-rich vapor phase. Harris (1986) interpreted the homogeneous low-salinity fluids as the original magmatic composition, and the coexisting brines and CO_2-rich vapor inclusions to represent the conditions after phase separation at 100 to 200 MPa.

An example of repeated fluid boiling in a subvolcanic granite is represented by the Permian Eikeren alkali granite, Oslo rift (Hansteen and Burke 1990). Fluid evolution during cooling of the granite can be summarized as four stages: a) boiling during late crystallization at about 700 °C; b) near-isobaric cooling and closed system fluid evolution down to about 450 °C; c) repeated boiling at 412 ± 51°C (Fig. 18), leading to opening of the fluid system to external fluids, and d) rapid dilution by meteoric waters below 400 °C. Because both pressure and temperature can be obtained by microthermometry of boiling NaCl-H_2O solutions (see above; Figs. 1f and 1g), the two boiling events at about 700 and 412 °C act as reference points for the proposed cooling path. Boiling under magmatic conditions involved a liquid phase with a salinity of 48 ± 7 wt% NaCl equivalents, giving a pressure of about 95 MPa at 700 °C. Boiling at 412 °C represents a liquid phase with 47 ± 5 wt% NaCl equivalents, giving a pressure of 25 ± 6 MPa, interpreted as hydrostatic conditions. Sub-solidus fluids in the temperature interval between the two boiling events did not coexist with a vapor phase, indicating pressures well above hydrostatic conditions. Fluid immiscibility in magmatic systems has been described in detail by Roedder (1992).

Highly peralkaline (agpaitic) magmas have higher solubilities for water and halogens than other magmas, and thus exhibit different degassing behavior. Konnerup-Madsen and Rose-Hansen (1982) investigated magmatic H_2O-CO_2-CH_4 fluid inclusions in the agapaitic Ilímaussaq intrusion and alkaline Gardar granitic complexes (south Greenland), and found that the

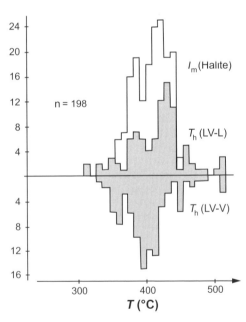

Figure 18. Boiling event at 412 ± 51 °C for the subvolcanic Eikeren Granite, Permian Oslo Rift, Norway. Pressure and temperature can be simultaneously estimated since both a highly saline liquid and a low-salinity vapor phase coexist as separate inclusions. Note that homogenization into liquid (T_h(LV→L)) and into vapor (T_h(LV→V)) occur at the same temperature, a necessary observation for the identification of boiling.

more alkalic magmas exsolved the more CH_4-rich fluids. Krumrei et al. (2007) demonstrated that hydrocarbons in fluid inclusions in the Ilímaussaq intrusion are of magmatic origin.

Summing up, fluid inclusions in a variety of mafic and silicic intrusive rocks indicate a pressure-dependent influence on fluid compositions, where the deeper magmas have elevated CO_2 contents. Agpaitic magmas may additionally contain hydrocarbons of magmatic origin. Fluid immiscibility resulting in the formation of a high-salinity liquid phase and a low-density vapor phase is common in plutonites covering pressures up to 0.6 GPa.

SUMMARY

Fluid inclusions can occur in any igneous mineral. In order to relate such inclusions to magmatic processes, their time of formation relative to host mineral growth must be considered. By petrographic investigations of doubly polished sections fluid inclusions assemblages are determined, and in the ideal case related to mineral paragenesis. Early and late generations of fluid inclusions can often be visually determined according to their occurrence in primary growth zones or in later, secondary trails.

In order to obtain thermobarometric information from fluid inclusions, their chemical composition and density have to be determined. Microthermometry is still a primary tool to provide the densities and also qualitative to semi-quantitative chemical information of single fluid inclusions, especially in the system H_2O-CO_2-NaCl. For more complex chemical compositions, additional methods like Laser Raman microprobe and micro-FTIR should be applied in order to characterize the systems.

Within a magma plumbing system of a volcanic rock, magma stagnation typically occurs at discrete levels. During ascent of a magma, increasing pressure gradients between fluid inclusions in phenocrysts and xenoliths and the host magma cause volumetric (leakage, stretching) and compositional (e.g., loss of H_2O) re-equilibration of the inclusions, and additionally new inclusions may form. The resulting density distribution of fluid inclusions depends on the ascent rates and stagnation periods, and may indicate the pressures of formation and/or re-equilibration as discrete frequency maxima. These pressures are readily determined from calculated isochores if the respective temperatures are known or can be estimated.

In shallow magma chambers and plutonic rocks, magmatic processes are often reflected by several generations of fluid inclusions, allowing for detailed investigations of fluid evolution. Phase separation involving immiscibility of a low-density vapor phase and a higher-density liquid phase at moderate pressures may lead to the formation of coexisting vapor-rich and water-rich fluid inclusions. By determining densities and compositions of both inclusion types, pressure and temperature can be simultaneously determined.

OUTLOOK

Integrated studies involving both silicate melt and fluid inclusions combined with additional petrological and geophysical data will increasingly be the way to go for igneous petrologists and volcanologists. For the former, a combination of melt and fluid inclusion geochemistry will eventually provide unprecedented insights into deep-seated magmatic processes, although the analytical effort may be high. Considering the limited number of microthermometric studies of volcanic rocks published so far, a wealth of basic information is still to be obtained on magma plumbing systems from all tectonic settings. For volcanologists, fluid inclusions are the prime source for barometric information pertaining to shallow magma reservoirs of active volcanoes and to short-duration stagnation of magma immediately prior to an eruption. Basic information like CO_2/H_2O ratios in the degassing fluids, important for volcano dynamics and eruptive behavior, are an intergral part of the microthermometric data. Fluid inclusion data are thus an invaluable tool to understand the past behavior of a volcano and, in conjunction with geophysical and geochemical monitoring, contribute to the understanding of a volcano's plumbing system, which is a prerequisite for hazard mitigation research (cf. Peccerillo et al. 2006). Thus the combination of efficient low-cost methods like detailed petrography and microthermometry is still highly relevant in volcanology and petrology.

In the plutonic environment, fluid inclusions are typically larger and more abundant than in their eruptive siblings, offering one explanation for the large number of microthermometric studies available. Sizeable inclusions and the common occurrence of high-salinity aqueous fluids with high trace element contents make such rocks ideal for microbeam analysis of single inclusions using e.g., laser ablation ICP-MS. In combination with detailed petrography and microthermometry, geochemical fluid evolution trends can be reconstructed.

ACKNOWLEDGMENTS

The research of the authors over the past years has been supported by the German Science Foundation (DFG) and the German Ministry for Education, Science, Research and Technology (BMBF). We gratefully acknowledge the constructive reviews from Ronald Bakker, Maria Luce Frezzotti, Keith Putirka and Paul Wallace that helped to improve the paper. This publication is contribution no. 165 of the Collaborative Research Center 574 "Volatiles and Hazards in Subduction Zones" at Kiel University.

REFERENCES

Abramson EH, Brown JM (2004) The melting curve of carbon dioxide and implication for the fluid. EOS Trans, Am Geophys Union 85(47):V41A-1358
Andersen T, Neumann ER (2001) Fluid inclusions in mantle xenoliths. Lithos 55:301 320
Andersen T, O'Reilly SY, Griffin WL (1984) The trapped fluid phase in upper mantle xenoliths from Victoria, Australia: implications for mantle metasomatism. Contrib Mineral Petrol 88:72-85
Angus S, Armstrong B, de Reuck KM, Altunin VV, Gadetskii OG, Chapela GA, Rowlinson JS (1976) International Tables of the Fluid State, Vol 3, Carbon dioxide. Pergamon Press, Oxford
Bakker RJ (2003) Package FLUIDS 1. Computer programs for analysis of fluid inclusion data and for modelling bulk fluid properties. Chem Geol 194:3-23
Bakker RJ, Diamond LW (2006) Estimation of volume fractions of liquid and vapor phases in fluid inclusions, and definition of inclusion shapes. Am Mineral 91:635-657
Bakker RJ, Jansen JBH (1991) Experimental post-entrapment water loss from synthetic CO_2-H_2O inclusions in natural quartz. Geochim Cosmochim Acta 55:2215-2230
Bakker RJ, Jansen JBH (1994) A mechanism for preferential H_2O leakage from fluid inclusions in quartz, based on TEM observations. Contrib Mineral Petrol 116:7-20
Belkin HE, DeVivo B (1993) Fluid inclusion studies of ejected nodules from plinian eruptions of Mt. Somma-Vesuvius. J Volcanol Geotherm Res 58:89-100
Blundy J, Cashman K (2005) Rapid decompression-driven crystallization recorded by melt inclusions from Mount St Helens volcano. Geology 33:793-796
Bodnar RJ (2003) Re-equilibration of fluid inclusions. In: Fluid Inclusions: Analysis and Interpretation, Short Course Series. Vol 32. Samson I, Anderson A, Marshall D (eds) Mineral Assoc Can, Vancouver, p 213-231
Bodnar RJ, Burnham CW, Sterner SM (1985) Synthetic fluid inclusions in natural quartz. III. Determination of phase equilibrium properties in the system H_2O-NaCl to 1000 °C and 1500 bars. Geochim Cosmochim Acta 49:1861-1873
Bottinga Y, Richet P (1981) High pressure and temperature equation of state and calculation of the thermodynamic properties of gaseous carbon dioxide. Am J Sci 281:615-660
Bowers TS, Helgeson HC (1983a) Calculation of the thermodynamic and geochemical consequences of nonideal mixing in the system H_2O-CO_2-NaCl on phase relations in geologic systems: Equation of state for H_2O-CO_2-NaCl fluids at high pressures and temperatures. Geochim Cosmochim Acta 47:1247-1275

Bowers TS, Helgeson HC (1983b) Calculation of the thermodynamic and geochemical consequences of nonideal mixing in the system H_2O-CO_2-NaCl on phase relations in geologic systems: Metamorphic equilibria at high pressures and temperatures. Am Mineral 68:1059-1075

Brown PE (1989) FLINCOR: A microcomputer program for the reduction and investigation of fluid-inclusion data. Am Mineral 74:1390-1393

Brown PE, Lamb WM (1989) P-V-T properties of fluids in the system $H_2O\pm CO_2\pm NaCl$: New graphical presentations and implications for fluid inclusion studies. Geochim Cosmochim Acta 53:1209-1221

Browne BL, Eichelberger JC, Patino LC, Vogel TA, Uto K, Hoshizumi H (2006) Magma mingling as indicated by texture and Sr/Ba ratios of plagioclase phenocrysts from Unzen volcano, SW Japan. J Volcanol Geotherm Res 154:103-116

Bureau H, Métrich N, Pineau F, Semet MP (1998) Magma-conduit interaction at Piton de la Fournaise volcano (Réunion Island): a melt and fluid inclusion study. J Volcanol Geotherm Res 84:39-60

Burke EAJ (2001) Raman microspectrometry of fluid inclusions. Lithos 55:139-158

Burnham CW (1979) Magmas and hydrothermal fluids. In: Geochemistry of hydrothermal ore deposits. 2nd ed. Barnes HL (ed) J. Wiley & Sons, New York, p. 71-136

Burruss RC (1981) Analysis of phase equilibria in C-O-H-S fluid inclusions. In: Short course in fluid inclusions: applications to petrology. Vol 6. Hollister LS, Crawford ML (eds), Mineral Assoc Can, Vancouver, p 101-137

Carslaw HS, Jaeger JC (1959) Conduction of Heat in Solids. Oxford University Press, Oxford

Cauzid J, Philippot P, Somogyi A, Ménez B, Simionovici A, Bleuet P (2006) Standardless quantification of single fluid inclusions using synchrotron radiation induced X-ray fluorescence. Chem Geol 227:165-183

Chou I-M (1987) Phase relations in the system NaCl-KCl-H_2O. III: Solubilities of halite in vapour saturated liquids above 445 C and redetermination of phase equilibrium properties in the system NaCl-H_2O to 1000 C and 1500 bars. Geochim Cosmochim Acta 51:1965-1975

Clocchiatti R, Del Moro A, Gioncada A, Joron JL, Mosbah M, Pinarelli L, Sbrana A (1994) Assessment of a shallow magmatic system: the 1888-90 eruption, Vulcano Island, Italy. Bull Volcanol 56:466-486

Clynne MA, Potter RW II (1977) Freezing point depression of synthetic brines (abst). Geol Soc Am Abstr Programs 9, 930

Coombs ML, Gardner JE (1991) Shallow-storage conditions for the rhyolite of the 1912 eruption at Novarupta, Alaska. Geology 29:775-778

Cox KG, Charnley N, Gill RCO, Parish KA (1993) Alkali basalts from Shuqra, Yemen: magmas generated in the crust-mantle transition zone? In: Magmatic Processes and Plate Tectonics, Geol Soc Spec Publ 76. Prichard HM, Alabaster T, Harris NBW, Neary CR (eds), p 443-453

Crawford ML (1981) Phase equilibria in aqueous fluid inclusions. In: Short course in fluid inclusions: applications to petrology. Vol 6. Hollister LS, Crawford ML (eds), Mineral Assoc Can, Vancouver, p 75-100

Davidson P, Kamenetsky VS (2007) Primary aqueous fluids in rhyolitic magmas: Melt inclusion evidence for pre- and post-trapping exsolution. Chem Geol 237:372-383

Diamond LW (2003) Systematics of H_2O inclusions. In: Fluid Inclusions: Analysis and Interpretation, Short Course Series. Vol 32. Samson I, Anderson A, Marshall D (eds) Mineral Assoc Can, Vancouver, p 55-78

Dixon JE, Stolper EM (1995) An experimental study of water and carbon dioxide solubilities in mid-ocean ridge basaltic liquids; Part II, Applications to degassing. J Petrol 36:1633-1646

Dixon JE, Stolper EM, Holloway JR (1995) An experimental study of water and carbon dioxide solubilities in mid-ocean ridge basaltic liquids. Part I: calibration and solubility models. J Petrol 36:1607-1631

Driesner T, Heinrich CA (2007) The system H_2O–NaCl. Part I: Correlation formulae for phase relations in temperature–pressure–composition space from 0 to 1000 °C, 0 to 5000 bar, and 0 to 1 X_{NaCl}. Geochim Cosmochim Acta 71:4880-4901

Frezzotti ML, Andersen T, Neumann ER, Simonsen SL (2002) Carbonatite melt-CO_2 fluid inclusions in mantle xenoliths from Tenerife, Canary Islands: a story of trapping, immiscibility and fluid–rock interaction in the upper mantle. Lithos 64:77-96

Frezzotti ML, Di Vincenzo G, Ghezzo C, Burke EAJ (1994) Evidence of magmatic CO_2-rich fluids in peraluminous graphite-bearing leucogranites from Deep Freeze Range (northern Victoria Land, Antarctica). Contrib Mineral Petrol 117:111-123

Frezzotti ML, Peccerillo A (2004) Fluid inclusion and petrological studies elucidate reconstruction of magma conduits. Eos Trans Am Geophys Union 85:157-163

Frost BR, Touret JLR (1989) Magmatic CO_2 and saline melts from the Sybille Monzosyenite, Laramie Anorthosite complex, Wyoming. Contrib Mineral Petrol 103:l78-186

Fulignati P, Marianelli P, Metrich N, Santacroce R, Sbrana A (2004) Towards a reconstruction of the magmatic feeding system of the 1944 eruption of Mt Vesuvius. J Volcanol Geotherm Res 133:13-22

Galipp K, Klügel A, Hansteen TH (2006) Changing depths of magma fractionation and stagnation during the evolution of an oceanic island volcano: La Palma (Canary Islands). J Volcanol Geotherm Res 155:285-306

Goldstein RH, Reynolds TJ (1994) Systematics of fluid inclusions in diagenetic minerals. SEPM Short Course 31, Tulsa, 199 pp

Hall DL, Sterner SM (1993) Preferential water loss from synthetic fluid inclusions. Contrib Mineral Petrol 114:489-500

Hall DL, Sterner SM, Bodnar RJ (1988) Freezing point depression of NaCl-KCl-H_2O solutions. Econ Geol 83:197-202

Hansteen TH (1991) Multi-stage evolution of the picritic Mælifell rocks, SW Iceland: A mineralogical and glass inclusion study. Contrib Mineral Petrol 109:225-239

Hansteen TH, Andersen T, Neumann ER, Jelsma H (1991) Fluid and silicate glass inclusions in ultramafic and mafic xenoliths from Hierro, Canary Islands: implications for mantle metasomatism. Contrib Mineral Petrol 107:242-254

Hansteen TH, Burke EAJ (1990) Melt-mineral-fluid interaction in peralkaline silicic intrusions in the Oslo rift, Southeast Norway. II: High-temperature fluid inclusions in the Eikeren-Skrim complex. Nor Geol Unders Bull 417:15-32

Hansteen TH, Klügel A, Schmincke HU (1998) Multi-stage magma ascent beneath the Canary Islands: evidence from fluid inclusions. Contrib Mineral Petrol 132:48-64

Harris C (1986) A quantitative study of magmatic inclusions in the plutonic ejecta of Ascension Island. J Petrol 27:251-276

Hollister LS (1990) Enrichment of CO_2 in fluid inclusions in quartz by removal of H_2O during crystal-plastic deformation. J Struct Geol 12:895-901

Holloway JR (1981) Compositions and volumes of supercritical fluids in the Earth"s crust. In: Short course in fluid inclusions: applications to petrology. Vol 6. Hollister LS, Crawford ML (eds), Mineral Assoc Can, Vancouver, p 13-38

Holloway JR, Blank JG (1994) Application of experimental results to C-O-H species in natural melts. Rev Mineral Geochem 30:187-230

Kamenetsky VS, Naumov VB, Davidson P, van Achterbergh E, Ryan CG (2004) Immiscibility between silicate magmas and aqueous fluids: a melt inclusion pursuit into the magmatic-hydrothermal transition in the Omsukchan Granite (NE Russia). Chem Geol 210:73-90

Kamenetsky VS, van Achterbergh E, Ryan CG, Naumov VB, Menagh TP, Davidson P (2002) Extreme chemical heterogeneity of granite-derived hydrothermal fluids: an example from inclusions in a single crystal of miarolitic quartz. Geology 30:459-462

Kempton PD (1987) Mineralogic and geochemical evidence for differing styles of metasomatism in spinel lherzolite xenoliths: Enriched mantle source regions of basalts? In: Mantle metasomatism. Vol 1. Menzies MA, Hawkesworth CJ (eds) Academic Press, London, p 45-89

Kent AJR (2008) Melt inclusions in basaltic and related volcanic rocks. Rev Mineral Geochem 69:273-331

Kerrick DM, Jacobs GK (1981) A modified Redlich-Kwong equation for H_2O, CO_2 and H_2O-CO_2 mixtures at elevated temperatures and pressures. Am J Sci 281:735-767

Kleinefeld B, Bakker RJ (2002) Fluid inclusions as microchemical systems: evidence and modelling of fluid-host interactions in plagioclase. J Metamorphic Geol 20:845-858

Klügel A (1998) Reactions between mantle xenoliths and host magma beneath La Palma (Canary Islands): constraints on magma ascent rates and crustal reservoirs. Contrib Mineral Petrol 131:237-257

Klügel A, Hansteen TH, Galipp K (2005) Magma storage and underplating beneath Cumbre Vieja volcano, La Palma (Canary Islands). Earth Planet Sci Lett 236:211-226

Klügel A, Hansteen TH, Schmincke HU (1997) Rates of magma ascent and depths of magma reservoirs beneath La Palma (Canary Islands). Terra Nova 9:117-121

Klügel A, Hoernle KA, Schmincke HU, White JDL (2000) The chemically zoned 1949 eruption on La Palma (Canary Islands): Petrologic evolution and magma supply dynamics of a rift-zone eruption. J Geophys Res 105(B3):5997-6016

Klügel A, Klein F (2006) Complex magma storage and ascent at embryonic submarine volcanoes from Madeira Archipelago. Geology 34:337-340

Klügel A, Schmincke HU, White JDL, Hoernle KA (1999) Chronology and volcanology of the 1949 multi-vent rift-zone eruption on La Palma (Canary Islands). J Volcanol Geotherm Res 94:267-282

Konnerup-Madsen J (1995) Basic microthermometric observations on fluid inclusions in minerals – a working manual. NorFA Short Course Notes, Copenhagen, pp 100, ISBN: 87-89980-12-3

Konnerup-Madsen J, Rose-Hansen J (1982) Volatiles associated with alkaline igneous rift activity: Fluid inclusions in the Ilímaussaq intrusion and the Gardar granitic complexes (south Greenland). Chem Geol 37:79-93

Konnerup-Madsen J (1979) Fluid inclusions in quartz from deep-seated granitic intrusions, South Norway. Lithos 12:13-23

Kooi ME, Schouten JA, Van den Kerkhof AM, Istrate G, Althaus E (1998) The system CO_2-N_2 at high pressure and applications to fluid inclusions. Geochim Cosmochim Acta 62:2837-2843

Krumrei TV, Pernicka E, Kaliwoda M, Markl G (2007) Volatiles in a peralkaline system: abiogenic hydrocarbons and F-Cl-Br systematics in the naujaite of the Ilímaussaq intrusion, South Greenland. Lithos 95:298-314

Labotka TC (1991) Chemical and physical properties of fluids. Rev Mineral 26:43-104

London D (1996) Granitic pegmatites. GSA Special Paper 315: The Third Hutton Symposium on the Origin of Granites and Related Rocks 315:305-319

Lowenstern JB (1994) Chlorine, fluid immiscibility and degassing in peralkaline magmas from Pantelleria, Italy. Am Mineral 79:353-370

Mackwell SJ, Kohlstedt DL (1990) Diffusion of hydrogen in olivine: Implications for water in the mantle. J Geophys Res 95:5079-5088

Mackwell SJ, Kohlstedt DL, Paterson MS (1985) The role of water in the deformation of olivine single crystals. J Geophys Res 90:11319-11333

Mäder UK, Berman RG (1991) An equation of state for carbon dioxide to high pressure and temperature. Am Mineral 76:1547-1559

Mavrogenes JA, Bodnar RJ (1994) Hydrogen movement into and out of fluid inclusions in quartz: Experimental evidence and geologic implications. Geochim Cosmochim Acta 58:141-148

Métrich N, Wallace PJ (2008) Volatile abundances in basaltic magmas and their degassing paths tracked by melt inclusions. Rev Mineral Geochem 69:363-402

Neumann ER, Wulff-Pedersen E, Johnsen K, Andersen T, Krogh E (1995) Petrogenesis of spinel harzburgite and dunite suite xenoliths from Lanzarote, eastern Canary Islands: Implications for the upper mantle. Lithos 35:83-107

Oskarsson N, Hansteen TH (1992) The use of graphite for the removal of oxygen from nitrogen puge-gas in high-temperature microthermometry using the Linkam TH1500 stage. Eur J Mineral 4:865-871

Pasteris JD, Wanamaker BJ (1988) Laser Raman microprobe analysis of experimentally re-equilibrated fluid inclusions in olivine: Some implications for mantle fluids. Am Mineral 73:1074-1088

Pearce TH, Kolisnik AM (1990) Observations of plagioclase zoning using interference imaging. Earth Sci Rev 29:9-26

Peccerillo A, Frezzotti ML, De Astis G, Ventura G (2006) Modeling the magma plumbing system of Vulcano (Aeolian Islands, Italy) by integrated fluid-inclusion geobarometry, petrology, and geophysics. Geology 34:17-20

Putirka KD (2008) Thermometers and barometers for volcanic systems. Rev Mineral Geochem 69:61-120

Robertson EC (1988) Thermal properties of rocks. Open file report 88-441, United States Geological Survey, Reston, VA

Roedder E (1963) Studies of fluid inclusions II: Freezing data and their interpretation. Econ Geol 58:167-211

Roedder E (1965) Liquid CO_2 inclusions in olivine-bearing nodules and phenocrysts from basalts. Am Mineral 50:1746-1782

Roedder E (1979) Origin and significance of magmatic inclusions. Bull Minéral 102:487-510

Roedder E (1983) Geobarometry of ultramafic xenoliths from Loihi Seamount, Hawaii, on the basis of CO_2 inclusions in olivine. Earth Planet Sci Lett 66:369-379

Roedder E (1984) Fluid inclusions. Reviews in Mineralogy, volume 12. Mineral Soc Am, Washington

Roedder E (1992) Fluid inclusion evidence for immiscibility in magmatic differentiation. Geochim Cosmochim Acta 56:5-20

Roedder E, Bodnar RJ (1980) geologic pressure determinations from fluid inclusion studies. Ann Rev Earth Planet Sci 8:263-301

Roedder E, Coombs DS (1967) Immiscibility in granitic melts, indicated by fluid inclusions in ejected granitic blocks from Ascension Island. J Petrol 8:417-451

Roggensack KL, Hervig RL, McKnight SB, Williams SN (1997) Explosive basaltic volcanism from Cerro Negro volcano: Influence of volatiles on eruption style. Science 277:1639-1642

Sachs PM, Hansteen TH (2000) Pleistocene underplating and metasomatism of the lower continental crust: a xenolith study. J Petrol 41:331-356

Samson I, Anderson A, Marshall D (2003) Fluid Inclusions: Analysis and Interpretation, Short Course Series, vol 32. Mineralogical Association of Canada, Vancouver

Schmidt C, Bodnar RJ (2000) Synthetic fluid inclusions: XVI. PVTX properties in the system H_2O-NaCl-CO_2 at elevated temperatures, pressures, and salinities. Geochim Cosmochim Acta 64:3853-3869

Schwarz S, Klügel A, Wohlgemuth-Ueberwasser C (2004) Melt extraction pathways and stagnation depths beneath the Madeira and Desertas rift zones (NE Atlantic) inferred from barometric studies. Contrib Mineral Petrol 147:228-240

Seitz JC, Blencoe JG, Joyce DB, Bodnar RJ (1994) Volumetric properties of CO_2-CH_4-N_2 fluids at 200°C and 1000 bars: A comparison of equations of state and experimental data. Geochim Cosmochim Acta 58:1065-1071

Shaw CSJ, Heidelbach F, Dingwell DB (2006) The origin of reaction textures in mantle peridotite xenoliths from Sal Island, Cape Verde: the case for "metasomatism" by the host lava. Contrib Mineral Petrol 151:681-697

Shepherd TJ (1981) Temperature-programmable, heating-freezing stage for microthermometric analysis of fluid inclusions. Econ Geol 76:1244-1247

Shepherd TJ, Rankin AH, Alderton DHM (1985) A practical guide to fluid inclusion studies. Chapman and Hall, New York, pp 239

Simon AC, Frank MR, Pettke T, Candela PA, Piccoli PM, Heinrich CA, Glascock M (2007) An evaluation of synthetic fluid inclusions for the purpose of trapping equilibrated, coexisting, immiscible fluid phases at magmatic conditions. Am Mineral 92:124-138

Sobolev VS, Kostyuk VP (1975) Magmatic crystallization based on a study of melt inclusions. "Nauka" Press, Novosibirsk, 182-253. (in Russian; Translated in part in Fluid Inclusion Research, Proc COFFI 9 (1976))

Sourirajan S, Kennedy GC (1962) The system H_2O-NaCl at elevated temperatures and pressures. Am J Sci 260:115-141

Span R, Wagner W (1996) A new equation of state for carbon dioxide covering the fluid region from the triple point temperature to 1100 K at pressures up to 800 MPa. J Phys Chem Ref Data 25:1509-1596

Sterner SM, Bodnar RJ (1991) Synthetic fluid inclusions. X: Experimental determination of P-V-T-X properties in the CO_2-H_2O system to 6 kb and 700°C. Am J Sci 291:1-54

Sterner SM, Chou I-M, Downs RT, Pitzer KS (1992) Phase relations in the system NaCl-KCl-H_2O. V: Thermodynamic-PTX analysis of solid-liquid equilibria at high temperatures and pressures. Geochim Cosmochim Acta 56: 2295-2309

Sterner SM, Hall DL, Bodnar RJ (1988) Synthetic fluid inclusions. V. Solubility relations in the system NaCl-KCl-H_2O under vapour-saturated conditions. Geochim Cosmochim Acta 52:989-1005

Sterner SM, Pitzer KS (1994) An equation of state for carbon dioxide valid from zero to extreme pressures. Contrib Mineral Petrol 117:362-374

Szabó CS, Bodnar RJ (1996) Changing magma ascent rates in the Nógrád-Gömör volcanic field, Northern Hungary / Southern Slovakia: Evidence from CO_2-rich fluid inclusions in metasomatized mantle xenoliths. Petrology 4:240-249

Takenouchi S, Kennedy GC (1965) The solubility of carbon dioxide in NaCl solutions at high temperatures and pressures. Am J Sci 263:445-454

Thomas R, Webster JD, Davidson P (2006) Understanding pegmatite formation: The melt and fluid inclusion approach. In: Melt inclusions in plutonic rocks. Webster JD (ed) Mineral Assoc Can, Vancouver, p 189-210

Touret JLR (2001) Fluids in metamorphic rocks. Lithos 55:1-25

Troll VR, Schmincke H-U (2002) Magma mixing and crustal recycling recorded in ternary feldspar from compositionally zoned peralkaline ignimbrite "A", Gran Canaria, Canary Islands. J Petrol 43:243-270

Van den Kerkhof AM (1988) The system CO_2-CH_4-N_2 in fluid inclusions: theoretical modelling and geological applications. Ph.D. Dissertation, Free University, Amsterdam

Van den Kerkhof AM (1990) Isochoric phase diagrams in the systems CO_2-CH_4 and CO_2-N_2: Application to fluid inclusions. Geochim Cosmochim Acta 54:621-629

Van den Kerkhof AM, Hein UF (2001) Fluid inclusion petrography. Lithos 55:27-47

Van den Kerkhof AM, Thiéry R (2001) Carbonic inclusions. Lithos 55:49-68

Vityk MO, Bodnar RJ (1998) Statistical microthermometry of synthetic fluid inclusions in quartz during decompression re-equilibration. Contrib Mineral Petrol 132:149-162

Wanamaker BJ, Evans B (1989) Mechanical re-equilibration of fluid inclusions in San Carlos olivine by power-law creep. Contrib Mineral Petrol 102:102-111

Wanamaker BJ, Wong TF, Evans B (1990) Decrepitation and crack healing of fluid inclusions in San Carlos olivine. J Geophys Res 95:15623-15641

Weisbrod A (1981) Fluid inclusions in shallow intrusives. In: Short course in fluid inclusions: applications to petrology. Vol 6. Hollister LS, Crawford ML (eds), Mineral Assoc Can, Vancouver, p 241-277

Werre RW Jr, Bodnar RJ, Bethke PM, Barton PB Jr (1979) A novel gas-flow fluid inclusion heating/freezing stage. Geol Soc Am Abstr Prog 11:539 (abstr)

Woods TL, Bethke PM, Bodnar RJ, Werne RW Jr (1981) Supplementary components and operation of the U.S. Geological Survey gas-flow heating/freezing stage. US Geol Survey Open File Report #81-954

Wopenka B, Pasteris JD, Freeman JJ (1990) Analysis of individual fluid inclusions by Fourier transform infrared and Raman microspectroscopy. Geochim Cosmochim Acta 54:519-533

Wulff-Pedersen E, Neumann ER, Jensen BJ (1996) The upper mantle under La Palma, Canary Islands: formation of Si-K-Na-rich melt and its importance as a metasomatic agent. Contrib Mineral Petrol 125:113-139

Zanon V, Frezzotti ML, Peccerillo A (2003) Magmatic feeding system and crustal magma accumulation beneath Vulcano Island (Italy): Evidence from CO_2 fluid inclusions in quartz xenoliths. J Geophys Res 108(B6, 2298): doi:10.1029/2002JB002140

Zapunnyy SA, Sobolev VS, Bogdanov AA, Slutskiy LV, Dmitriyev LV, Kunin LL (1988) An apparatus for hightemperature optical research with controlled fugacity. Geokhimiya 7:1044-1052

Zhang YG, Frantz JD (1987) Determination of homogenization temperatures and densities of supercritical fluids in the system $NaCl$-KCl-$CaCl_2$-H_2O using synthetic fluid inclusions. Chem Geol 64:335-350

Petrologic Reconstruction of Magmatic System Variables and Processes

Jon Blundy

Department of Earth Sciences
University of Bristol
Wills Memorial Building, Bristol BS8 1RJ, United Kingdom

Jon.Blundy@bristol.ac.uk

Kathy Cashman

Department of Geological Sciences
University of Oregon
Eugene, Oregon 97403-1272, U.S.A.

cashman@uoregon.edu

INTRODUCTION

Explosive volcanic eruptions constitute a major class of natural hazard with potentially profound economic and societal consequences. Although such eruptions cannot be prevented and only rarely may be anticipated with any degree of accuracy, better understanding of how explosive volcanoes work will lead to improved volcano monitoring and disaster mitigation. A major goal of modern volcanology is linking of surface-monitored signals from active volcanoes, such as seismicity, ground deformation and gas chemistry, to the subterranean processes that generate them. Because sub-volcanic systems cannot be accessed directly, most of what we know about these systems comes from studies of erupted products. Such studies shed light on what happens underground prior to and during eruptions, thereby providing an interpretative framework for *post hoc* evaluation of monitoring data. The aim of this review is to present some of the current petrological techniques that can be used for studying eruptive products and for constraining key magmatic variables such as pressure, temperature, and volatile content. We first review analytical techniques, paying particular attention to pitfalls and strategies for analyzing volcanic samples. We then examine commonly used geothermometry schemes, evaluating each by comparison with experimental data not used in the original geothermometer calibrations. As there are few mineral-based geobarometers applicable to magma storage regions, we review other methods used to determine pre-eruptive magma equilibration pressures. We then demonstrate how petrologically-constrained parameters can be compared to the contemporaneous monitoring record. These examples are drawn largely from Mount St. Helens volcano, for which there are abundant petrological and monitoring data. However, we emphasize that our approaches can be applied to any number of active volcanoes worldwide. Finally, we illustrate the application of these techniques to two different types of magmatic systems—large silicic magma chambers and small intermediate-composition magma storage regions—with particular focus on the combined evolution of melt volatiles and magmatic crystal contents in these two different scenarios.

MAGMATIC VARIABLES

Magmatic variables can be both intensive and extensive. Intensive variables are those whose values do not depend on the amount of magma or gas in the system, namely composition, crystallinity, temperature, storage pressure and oxidation state. By combining knowledge of these intensive variables with experimental measurements it is possible to constrain key additional variables that exert important controls on eruptive behavior, such as density and viscosity, or that can refine seismic or potential-field studies of volcanic regions, such as seismic wave velocity and electrical conductivity. Extensive magmatic variables are the counterpart of intensive variables—their value is proportional to the size of the system. Examples include the volume and depth of stored magma, the magmatic flux into and out of the sub-volcanic system, and various forms of heat and stored energy. Extensive variables are much harder to determine from petrology and require integration with other techniques such as seismic tomography, ground deformation and physical volcanology. Some extensive variables have complex relationships with intensive variables. For example, magma storage depth (an extensive variable) is related to storage pressure (an intensive variable) through the nature of the overburden, the magma flow velocity and the extent to which the system is over-pressured. In general, numerical modeling of the dynamics of volcanic systems requires knowledge of both intensive and extensive variables.

Magmatic variables evolve in space and time, reflecting the complexity of underground magmatic plumbing systems and processes by which those plumbing systems are recharged and emptied (e.g., Marsh 2006). Although no two volcanoes are alike, one can envisage a generic magmatic plumbing system that encapsulates the key variables (Fig. 1). Our generic volcano comprises a region of principal magma storage, typically referred to as a magma chamber or reservoir, and a plexus of fractures, which may or may not be magma-filled to form dykes, sills or conduits, which connect the chamber to the surface above and to the magma source region below. Intensive parameters can vary spatially within this system due to heat loss to (and physical exchange with) the wall rocks, heat input from new magma additions from below, post-eruptive changes in

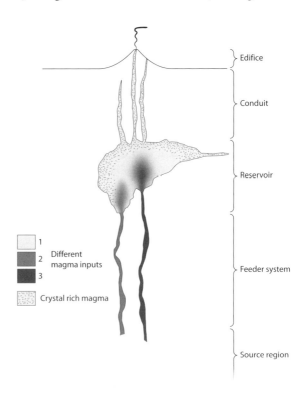

Figure 1. Schematic of a sub-volcanic magma system showing key features to be investigated using petrological techniques. The chamber is recharged via a feeder system from a deeper source region and discharged via a shallow conduit system connected to the surface. The magma reservoir is a zone of mixing and blending between different magma inputs from the source region. Crystallization and degassing occur in the conduit system and reservoir, which may extend laterally in the form of sills. The conduit connecting the magma reservoir to the surface may be either magma-filled or a fracture network.

the volcanic load, gas loss (or throughput) and associated redox reactions, and crystallization. Both intensive and extensive variables may show significant temporal variability that can serve to enhance or diminish the likelihood of eruption and that should be manifest as changes in the monitoring record.

Petrologically-determined intensive variables

We focus on those variables that can be deduced from petrology alone, i.e., those that can be determined from studying eruptive products in the laboratory. These include:

1. Temperature and pressure, determined by geothermometry and geobarometry using the chemistry of coexisting minerals and/or glasses.

2. The concentration, distribution and migration of volatile species determined by analysis of glassy melt inclusions.

3. The extent, distribution and timing of crystallization and degassing, as determined using textural and chemical constraints.

4. Experimental simulation of sub-volcanic systems involving matching of the phase assemblage and phase compositions of erupted rocks.

Linking intensive variables to the monitoring record

The chemistry and textural characteristics of erupted rocks contain a wealth of information about processes occurring within the sub-volcanic system prior to and during eruption. Magma that remained underground after one eruption may be discharged subsequently, so that some textural or chemical features may record information about previous eruptive episodes (e.g., Hammer et al. 1999; Cashman and McConnell 2005). Unraveling the complex testimony of volcanic rocks is a key objective of this review, and one that will be illustrated using selected examples. Linking this testimony to the contemporaneous monitoring record, however, is fraught with difficulties, even for the world's most extensively monitored volcanoes, such as Mount St. Helens (USA), Soufrière Hills (Montserrat), Unzen (Japan) and Mt. Etna (Italy). It is increasingly clear that only through a multidisciplinary effort involving volcano monitoring agencies, seismologists, petrologists, gas geochemists and numerical modelers will we be able to create an accurate picture of the underground geometry and dynamics of magmatic systems.

METHODOLOGY

Petrology relies to varying extents on chemical and textural characterization of eruptive products using microbeam techniques. Three commonly used analytical techniques that we review here are back-scattered electron microscopy, electron-microprobe analysis, and secondary ion mass spectrometry (or ion-microprobe analysis); other important techniques are discussed elsewhere in this volume (e.g., FTIR and micro-Raman in Metrich and Wallace 2008; LA-ICP-MS in Kent 2008; isotopic analysis in Ramos and Tepley 2008). We also briefly review new techniques for imaging and analysis (such as nanoSIMS and X-ray computer tomography) as well as methods of textural analysis.

Scanning electron microscopy (SEM)

SEMs can be used in several different modes for different applications. Most useful for petrologic applications is the acquisition of images from back-scattered electrons (BSEM) that originate from elastic collisions between the incident electron beam and atomic nuclei at or close to the sample surface. The intensity of the back-scattered signal depends primarily on the mean atomic number (Z) of the target. Because the back-scattered signal derives predominantly from the outermost few nanometers of the sample, BSEM does not suffer from significant blurring of chemical contrast due to vertical chemical heterogeneity, thus the technique provides an excellent map of Z variations at the sample surface. Lateral spatial resolution is controlled by the beam diameter, which is a function of the incident beam current used. Lower incident beam currents confer greater spatial resolution at the expense of reduced signal. The electron beam accelerating voltage is an important term in the back-scatter cross-section equation, however within the range commonly used for petrological work (5 to 20 kV), the influence of accelerating voltage is subordinate to that of target Z and beam current.

BSEM is most useful for elucidating fine-scale chemical zoning in minerals showing solid solution between end-members of contrasted Z, such as plagioclase and ferromagnesian minerals (Figs. 2 and 3), and for imaging crystals that have low Z contrast with the surrounding matrix glass (e.g., plagioclase in high-silica glass). When a sample has both low Z and high Z crystals (say, plagioclase and hornblende), high quality imaging usually requires separate image acquisition for the two phases using different analysis conditions (beam current and contrast settings). BSEM images can also be processed to generate mineralogical modes by binning the image for certain grayscale bands corresponding to particular minerals and then using image-processing software to calculate areas of the different bins (e.g., Hammer et al. 2000). However, care must be taken when minerals have similar and overlapping Z values, such as is the case for amphiboles and clinopyroxenes.

BSEM can also be used to generate qualitative diffusion profiles across individual chemical zones. In those cases where a single chemical component is giving rise to most or all of the backscatter contrast, quantitative diffusion profiles can be generated provided that independent analysis (e.g., by EPMA or SIMS) is used to constrain the relationship between backscatter intensity and composition (e.g., Ginibre et al. 2002; Morgan et al. 2004,

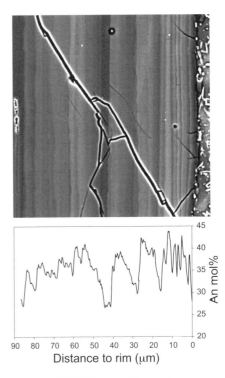

Figure 2. BSEM image of fine-scale zoning at the rim of a plagioclase crystal from Parinacota volcano, Chile. Below the BSEM image is a graph showing the compositional variation (as mol% anorthite) represented by that zoning. Image and data courtesy of C. Ginibre; further details can be found in Ginibre et al. (2002).

2006; Humphreys et al. 2008a). Examples include anorthite content of plagioclase (Fig. 2), Mg# of olivine (Fig. 3), H_2O content of matrix glass (Fig. 4) and Fe-Ti zonation in oxides (Devine et al. 2003). Quantitative diffusion profiles can be used to derive temporal information provided that diffusivity of the key chemical component at the conditions of interest is known (Costa et al. 2003; Costa and Chakraborty 2004; Humphreys et al. 2008a; Costa et al. 2008). As diffusivity is often anisotropic, some additional information on the orientation of the crystal across the diffusion interface is required. This can be obtained by conventional universal stage microscope techniques or from electron back-scatter diffraction analysis (EBSD, e.g., Schwartz et al. 2000; Randle and Engler 2000). Where multiple major components are diffusing, no simple relationship between backscatter intensity and composition can be obtained.

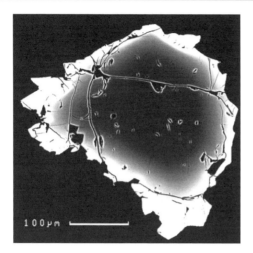

Figure 3. BSEM image of zoned olivine crystal from Santorini, Greece. The grayscale variation reflects variation from Mg-rich (dark) core to Fe-rich (light) rim. The gradation in grayscale across the core-rim boundary can be used to quantify the extent of Mg-Fe interdiffusion. Image courtesy of D. Morgan and V. Martin.

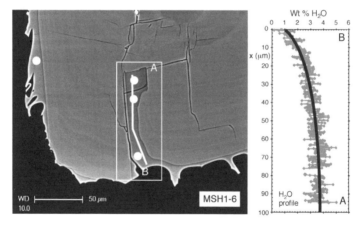

Figure 4. BSEM image of melt tube in plagioclase phenocryst from Mount St. Helens (left). The melt tube connects the interior of the crystal with the matrix glass and shows a grayscale variation due to an outwards decrease in dissolved H_2O. Ion-microprobe analyzes of H_2O in the melt tube and groundmass glasses (4 white circles on left hand image) are used to calibrate the relationship between dissolved H_2O and grayscale intensity to produce a diffusion profile (right) that can be modeled to estimate the time taken for H_2O to diffuse out of the interior of the crystal and into the matrix. Fit parameters give diffusion times (t) of 107-166 s for initial H_2O contents of 4.6-6.4 wt% in melt in the crystal interior. From Humphreys et al. (2008a).

Another type of SEM imaging uses cathodoluminescence (CL), an optical technique that captures photons generated by interaction of the electron beam with the sample surface. Although the origin of the CL signal is often unclear, as photons are emitted from defects in the crystal structure created either by vacancies or by the presence of trace elements, CL imaging can reveal complexities of crystal growth behavior that are not observable using other methods. Absolute CL intensity is difficult to quantify, however, because it is dependent on operating conditions; for this reason it is common to describe CL domains simply as "CL-dark" and "CL-bright". For petrologic applications, CL is a particularly effective technique for imaging zoning patterns in quartz (e.g., Peppard et al. 2001; Rusk and Reed 2002; Rusk et al. 2008) and zircon (e.g., Nemchin et al. 2001), where growth patterns revealed by CL imaging provide insight into both the mechanism (e.g., Ostwald ripening) and location (within the magma reservoir) of crystal growth. When the CL characteristics can be related to specific trace elements, such as Ti in quartz, and quantified using the microbeam techniques described above, spatial patterns of growth can be directly linked to intensive parameters responsible for crystallization through use of mineral equilibria, such as the titanium-in-quartz geothermometer of Wark and Watson (2006).

Electron microprobe analysis (EMPA)

EMPA uses a focused electron beam to generate X-rays via electron transfer between inner orbitals (K, L, M) in the target atoms. The energy (or wavelength) of the resulting X-radiation is chemically diagnostic and the intensity of a particular X-ray is a function of the concentration of the element that produced it within the target volume. X-rays can be detected as a function of their energy (energy-dispersive EMPA, or EDS) or their wavelength (wavelength-dispersive EMPA, or WDS). In EDS, the entire energy spectrum is collected simultaneously, while in WDS a series of moveable spectrometers are tuned to characteristic wavelengths for a particular length of time. In both cases, analysis time is several minutes, although for low abundance elements analysis times can be significantly longer. WDS involves more complex and expensive instrumentation but has lower detection limits than EDS, as well as less problematic calibration and data reduction. WDS is therefore considered the technique of choice for most petrological studies.

By scanning the beam across the sample it is possible to collect X-rays of a particular wavelength or energy and so generate a chemical map (Fig. 5) to complement a BSEM image

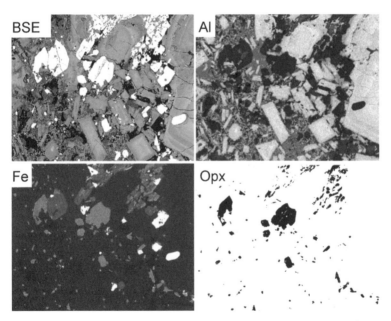

Figure 5. Maps of polished dacite sample (SH184, erupted September 1984) from Mount St. Helens. (a) BSEM map showing bright ferromagnesian mineral, mid-grey plagioclase and glass and black vesicles. (b) Al map allowing clear distinction between plagioclase (light) and glass (dark). (c) Fe map showing bright oxide minerals and distinguishes between pyroxene and hornblende. (d) Digitized binary image of orthopyroxene in sample. This image can be used to quantify the volume fraction of orthopyroxene in the sample and to derive crystal size distribution data. Similar maps can be generated for all other phases by combining individual chemical maps. Image is 1 mm across. Images courtesy of D. Muir.

of the same area. This offers the advantage of distinguishing between phases that cannot be readily distinguished by BSEM alone because they have similar Z values. In our example, while plagioclase crystals can be distinguished by BESM alone (confirmed by the Al map), distinguishing between the mafic phases (hornblende, orthopyroxene and Fe-Ti oxides) requires additional element maps, here illustrated by the identification of orthopyroxene using the Fe map. Chemical mapping is also a valuable technique when looking at chemical zonation of a component whose abundance is too low to affect BSEM intensity. A recent example is the mapping of phosphorous in igneous olivine to elucidate fine-scale zoning that is less sensitive to diffusive re-homogenization than Fe-Mg zoning (Milman-Barris et al. 2008).

As with BSEM, the intensity of the X-ray signal is a function of the input electron beam energy and flux: higher beam currents excite X-rays from lower concentrations of target elements at the expense of poorer spatial resolution, whereas higher accelerating voltages excite higher energy X-rays from heavier elements at the expense of greater sample penetration. Elements with Z greater than 4 (beryllium) can be analyzed by EMPA, although only elements with $Z > 8$ are routinely analyzed. With suitably intense electron beams and long counting times (several minutes) minimum detection limits of <100 ppm are attainable, although for some minor and trace elements considerable care must be taken to deconvolve interferences from minor X-ray lines of major elements, e.g., SiK_β on SrL_α, by selecting suitably clean background positions.

Analytical conditions are carefully selected for the type of materials being analyzed, particularly those that are, to variable degrees, unstable under the electron beam, e.g., hydrous silicate glasses, sodium-rich feldspars, carbonates, sulfates and phosphates. Most silicate minerals can be analyzed using 15-20 keV electrons, with a beam current of 5-15 nA, focused to a spot at the sample surface. Unlike backscatter electrons, X-rays are generated from a significant interaction volume below the sample surface, which limits the effective spatial resolution to 2-3 microns. Lower beam currents and accelerating voltages are recommended for microlites whose small sizes make them susceptible to X-ray excitation from surrounding and underlying glass. Microlite analyses should be carefully screened for glass contamination, manifest as stoichiometrically excess Si and/or anomalously high K.

Hydrous silicate glasses pose a particular analytical challenge because of alkali-migration during electron beam irradiation. The problem is most acute in rhyolites with >1 wt% dissolved H_2O and appears to be a consequence of electron implantation into the outer few microns of the sample. Cations within the sample are attracted electrostatically away from the surface towards the zone of highest electron density, moving at a rate dependent on their solid-state diffusivity. Conversely, H_2O concentrations increase towards the sample surface, possibly due to migration of OH^- away from the implanted electrons (Humphreys et al. 2006). The migration process is thermally mediated by heating at the sample surface due to phonon excitation by the electron beam. Those species with the highest diffusivities, i.e., Li, Na, K and H_2O, are the most strongly affected and H_2O-rich glasses are the most susceptible. SIMS depth-profiling of electron-beam irradiated regions (Humphreys et al. 2006) reveals that the process is a strong function of net electron beam energy and is irreversible. Alkali-migration can lead to low Na and correspondingly high Si and Al, as well as low analytical totals. Low totals are especially problematic when the shortfall from 100% is assumed to be due to H_2O as in the volatile-by-difference (VBD) technique (e.g., Devine et al. 1995; King et al. 2002).

The simplest way to mitigate beam damage is to reduce beam currents to 2-4 nA and defocus the electron spot over a 10-15 µm diameter. Na and K are counted first and for short periods (e.g., 6s peak, 3s background) calibrated using similar beam conditions, and run using a glass (or suite of glasses) of known H_2O content as a secondary standard. VBDs within 0.5 to 0.7 wt% of the true H_2O content should be attainable using this technique (Humphreys et al. 2006), assuming negligible concentrations of other unanalyzed species (e.g., CO_2, F, etc.). A drawback of this approach is that it requires relatively large pools of glass, which are not always available in highly vesicular pumices or where melt inclusions are small. Another method is to use conventional electron beam conditions, but to extrapolate the Na signal to zero time to compensate for the effects of migration. This approach has the advantage that smaller glass pools can be analyzed, but requires software capable of taking counts at regular intervals and requires some key assumptions about the shape of the Na decay curve with time. Agreement between VBD and known H_2O contents of rhyolite glasses to within 0.5 wt% is reported using this approach, particularly when combined with iterative addition of H_2O to the matrix correction (e.g., Devine et al. 1995; Roman et al. 2006). The final method is to freeze the sample within the analytical chamber to minimize thermal influences on alkali diffusion (Kearns et al. 2002). At liquid nitrogen temperatures a cold stage can effectively eliminate alkali migration. However, considerable technological modification to most electron-microprobes is required for cold-stage analysis and contamination build-up is a major problem.

Secondary ion mass spectrometry (SIMS)

SIMS is a microbeam technique in which a beam of focused ions (positively or negatively charged) is used to sputter particles, some of which are charged (secondary ions), from a sample surface under high vacuum (Shimizu and Hart 1982; Hinton 1995). The secondary ions are then accelerated into a double-focusing mass spectrometer where they are separated according to their energy and mass/charge ratio before being detected using an electron multiplier or Faraday cup. The most commonly used primary beams are O^- and Cs^+. SIMS analysis of geological samples is conventionally carried out on gold-coated polished thin sections or grain mounts.

Unlike EMPA the technique is destructive in that the sputtering process leaves a small hole at the sample surface. The size of the sputter crater depends on beam intensity and analytical duration, but for most petrological work is ~2 microns deep. SIMS affords several advantages over EMPA in that it boasts much greater sensitivity for many elements of interest and can be used to analyze light elements, such as Li, B, H and C. The latter can be readily converted to H_2O and CO_2. The ease with which elements can be analyzed depends primarily on their secondary ion-yield under the analytical conditions. For a given set of primary beam conditions, secondary ion-yields can vary by 4 orders of magnitude across the periodic table (Hinton 1990). Elements with favorable positive or negative secondary ion-yields (e.g., Li^+, B^+, Ti^+, REE^+, U^+, Th^+, Br^-, I^-) have detection limits of less than few 10s of ppb. Other elements, for example Group XII metals (Zn, Cd, Hg) and platinum group elements, have negligible high-energy secondary ion yields and cannot be analyzed at very low concentrations.

A further complication is the presence of molecular ion interferences. Sputtered ions include both monatomic and polyatomic species (molecular ions). Examples include the isobaric interferences of $^{24}Mg^{2+}$ on $^{12}C^+$, $^{29}Si^{16}O^+$ on $^{45}Sc^+$ and $^{137}Ba^{16}O^+$ on $^{153}Eu^+$. Low mass resolving power (~500) prevents direct resolution of these isobaric species on the basis of their small mass differences; for this reason, molecular ion interferences are commonly addressed using kinetic energy filtering, which relies on the narrower energy distribution of sputtered molecular ions compared to monatomic ions. Selective sampling of the high-energy tail of the energy distribution (by reducing the sample extraction voltage by 50-100 V) considerably reduces transmission of molecular ions, although in some cases additional conventional peak stripping is necessary. For example, to eliminate $^{29}Si^{16}O^+$ on $^{45}Sc^+$, one must first analyze $^{42}Ca^+$, from which $^{44}Ca^+$ can be calculated. The difference between the measured and calculated count rate at mass 44 is ascribed to $^{28}Si^{16}O^+$, from which $^{29}Si^{16}O^+$ can be calculated (assuming identical ion-yield for ^{28}Si and ^{29}Si) and subtracted from mass 45 to yield the corrected $^{45}Sc^+$.

In contrast to X-ray generation by electrons, the physics of sputtering is not well understood and SIMS relies heavily on calibration using standards of known composition whose matrix is closely matched to that of the target materials. It is possible to calibrate on a single multi-element standard, such as NIST glass SRM610, but careful consideration must be given to relative ion-yield difference between this glass and natural silicate glasses and minerals, which may be up to 25% relative. Quantitative SIMS analysis requires that the secondary ion count is referenced to that of a species whose concentration is known: conventionally this is ^{30}Si, whose concentration is determined by EMPA of the same spot.

Analysis of H_2O and CO_2. An important petrological application of SIMS is the analysis of H_2O (as $^1H^+$) and CO_2 (as $^{12}C^+$). Calibration for these volatile species requires use of a suite of glasses, similar in composition to those of interest, with known H_2O and CO_2 contents. As there is no empirical evidence that ^1H ion yields are affected by the presence of CO_2, and vice versa, it is possible to use calibrant glasses that contain both species to generate working curves that relate normalized ^1H and ^{12}C secondary-ion count rates to H_2O and CO_2, respectively, in the standards. In principal the secondary-ion counts should be corrected for variations in Si content (Shimizu 1997). In practice, the variation in relative ion yields is correlated with SiO_2 content, such that plotting $^1H/^{30}Si$ (or $^{12}C/^{30}Si$) versus H_2O (or CO_2) yields a single working curve for glasses of variable silica content (Fig. 6; Blundy and Cashman 2005). This (fortuitously) provides a convenient and robust means of determining H_2O and CO_2 that requires relatively few calibrant glasses.

The quality of the analysis depends critically on the instrumental background. Elevated background counts are a perennial problem for H_2O and CO_2 analysis because of their high concentration in organic materials used in sample preparation and mounting (especially epoxy resins used for grain mounts). The problem is greatest for CO_2 because of its low concentration in most geological materials and its inherently low positive secondary-ion yield. Surface contamination with H or C, especially from organic solvents or diamond polishing materials, can be minimized by allowing a few minutes "burn-in" of the beam before analysis and elimination of the first few analytical cycles. Surface contamination can be minimized further by chemical polishing with colloidal silica, by analyzing thin sections (rather than grain mounts), by using low volatility mounting resins (where necessary), and by working at high vacuum. The latter is facilitated on ion-microprobes fitted with multi-sample airlocks in which samples can be

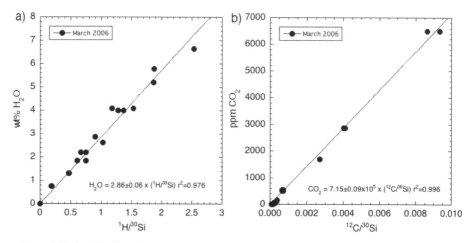

Figure 6. Typical SIMS working curves for (a) H_2O and (b) CO_2 derived from rhyolite, dacite and andesite glasses of known volatile content analyzed on a single day in March 2006 using the Edinburgh University Cameca ims4f ion microprobe. Abscissa denotes ratios of counts per second for $^1H^+$ and $^{12}C^+$, normalized to ^{30}Si. $^{12}C^+$ is corrected for $^{24}Mg^{2+}$ interference. H_2O based on 13 different glasses; CO_2 based on 10 different glasses. The equations of the working curves are given. Fully propagated uncertainties (1 s.d.), are approximately same size as symbol, or smaller.

pumped down to < 5×10^{-8} Torr prior to insertion into the analytical chamber, which is typically at < 2×10^{-9} Torr. Backgrounds can be monitored during analysis using a small chip of quartz mounted alongside the sample. Experience on the Cameca ims-4f at Edinburgh University using a 10kV (nominal) $^{16}O^-$ primary beam, 4.5 kV secondary accelerating 75±20 V offset, and 2-5 nA current at the sample surface, shows that ^{30}Si-normalised backgrounds measured on NIST28 quartz are typically 5×10^{-3} for $^1H^+$ and 1×10^{-4} for $^{12}C^+$ for samples pumped to 10^{-9} Torr or less in an airlock prior to analysis. This corresponds to minimum detection limits of the order 150 ppm or less for H_2O and 60 ppm or less for CO_2 in Mg-free materials (see below).

Alternatively, individual crystals or glass chips can be pressed into indium, thereby reducing the introduction of organics into the sample chamber. Using this technique coupled with a Cs^+ primary beam, high primary ion currents (200-300 nA) and an electron flood gun to reduce sample charging, detections limits for negative secondary ions equate to ≤3 ppm for C and ≤30 ppm for H_2O (Hauri et al. 2002). The improved detectability, however, comes at the expense of more time-consuming sample preparation, painstaking analytical set up, relatively large (15-30 micron) beam diameters and potential damage to the glass by the electron flood gun. For typical H_2O contents of silica-rich volcanic glasses (≥0.5 wt%) such enhanced detection limits are not warranted.

A further problem with CO_2 analysis is the isobaric interference of $^{24}Mg^{2+}$ on $^{12}C^+$. This interference is not removed by energy filtering as Mg^{2+}/Mg ratios do not significantly change with energy, and therefore must be resolved by peak stripping using $^{25}Mg^{2+}$ measured on mass 12.5. The small differences in ion-yield between ^{25}Mg and ^{24}Mg should be determined from analysis of an Mg-rich silicate glasses under the same analytical conditions as the ratio may differ slightly from the canonical value of 7.90. For Mg-poor rhyolite glasses with ~0.3 wt% MgO, the interference amounts to approximately 1000 ppm CO_2. Thus it is essential to determine the $^{24}Mg^+/^{25}Mg^+$ ion-yield ratio very precisely, to employ long count times on masses 12 and 12.5, and to measure background count rates for each sample analyzed to optimize determination of CO_2 concentrations of ≥100 ppm (e.g., Fig 6b). For more Mg-rich glasses the contribution from $^{24}Mg^{2+}$ is too large to be adequately removed by peak-stripping and higher mass resolving power (≥1500) must be used, at the expense of poorer secondary ion transmission.

NanoSIMS is a relatively new SIMS technique, increasingly used in cosmochemistry for the isotopic analysis of tiny interplanetary dust particles (Hoppe 2006). The technique has many similarities to conventional SIMS, although the coaxial ion optics of the primary and secondary ion beams enable the incident ion beam (O^- or Cs^+) to be focused to a much smaller spot at the sample surface (e.g., ~50 nm for Cs^+, 150 nm for O^-), while maintaining high secondary ion collection efficiency. Like conventional SIMS, NanoSIMS is not standardless and calibration against standards of known composition is required. There have been relatively few petrological applications of NanoSIMS to date (e.g., Hellebrand et al. 2005), and the considerable potential applications to trace element and volatile element variations in volcanic materials have yet to be explored. However, the high spatial resolution, linked to analytical sensitivity, makes NanoSIMS ideally suited to the study of fine-scale trace element zoning in minerals for use in diffusion geospeedometry, or chemical analysis of very small melt inclusions. We anticipate that NanoSIMS applications in volcanic petrology will increase substantially in the coming years.

Analysis of textures

The textures of volcanic samples provide fundamental information on conditions of magma storage and ascent. So integral are textures to our thinking about igneous rocks that they form the basis for much descriptive classification (e.g., obsidian vs. pumice; porphyritic vs. aphanitic; basalt vs. diabase vs. gabbro). Here we briefly review two aspects of textural analysis: data collection and data interpretation. More detailed discussions of crystal and vesicle size distribution analysis can be found elsewhere (e.g., Marsh 1988; Cashman and Marsh 1988; Cashman 1990; Higgins 1994, 2006; Kile et al. 2000; Klug et al. 2002; Madras and McCoy 2002; Gurioli et al. 2005; Hersum and Marsh 2006, 2007; Resmini 2007; Armienti 2008).

Image acquisition and processing. Modal analysis of phenocryst phases is traditionally performed by point counting using a mechanical stage mounted on a petrographic microscope (e.g., Chayes 1956). The uncertainty on point-counted modes can be estimated graphically using Van der Plas and Tobi (1965). The full range of textural features is best captured using a high-resolution slide scanner to obtain a digital image of a thin section (Fig. 7a), while groundmass phases are best imaged using BSEM or X-ray maps (see above). The resulting digital images can be analyzed for crystal (or bubble) size, shape, number density and orientation using automated image analysis programs available as free downloads (e.g., NIH Image, http://*rsb.info. nih.gov/nih-image/index.html*). Automated thresholding to produce the binary image required for analysis is easiest (and most accurate) when the contrast between the phase of interest and the background is large (e.g., plagioclase phenocrysts in a fine-grained matrix; Fig. 7a,b). When contrasts are not sufficient for automated analysis, individual phases may be distinguished using cross-polarized light, Z contrast variations in BSEM images or X-ray maps of selected elements (e.g., Fig. 5) and then outlined by hand.

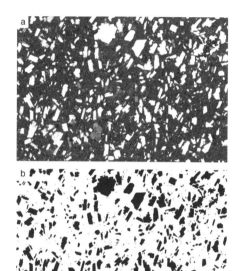

Figure 7. Example of a thin section scan of a 1956 cryptodome sample from Bezymianny Volcano, Kamchatka, showing white plagioclase crystals (a) and associated binary image of plagioclase (b) from which textural data can be easily extracted. The total image area is 107 mm^2, of which 21.9% is occupied by plagioclase phenocrysts. 1670 individual plagioclase crystals are identified, for an average of 15.57 crystals mm^{-2}.

Measurements from thin sections are two-dimensional. If bulk magma properties are to be investigated, measurements must be converted from 2-D to 3-D (e.g., Jerram and Higgins 2007); this conversion requires additional information about the crystal population (such as phase proportions and the distribution of crystal shapes). For this reason, there have been several techniques introduced to measure true 3-D crystal size data directly, including extraction of whole crystals from a sample (e.g., Dunbar et al. 1994; Bindeman 2005), optical focusing through a thick section with low crystallinity (e.g., Castro et al. 2003), statistical methods (e.g., Morgan and Jerram 2006; Armienti 2008) and X-ray computed tomography (XCT; e.g., Carlson 2006). Of these, XCT is potentially the most useful because images can be acquired rapidly, several phases can be isolated from a single image, spatial relations between individual bubbles or crystals are preserved, and images allow visualization of the phase topology that is not possible in 2D images.

XCT uses the attenuation of an X-ray beam passing through a specimen to image phases of different mass density or Z contrast. Measurements are typically made on a rotating specimen that allows the sample to be viewed from multiple angles; individual tomographic slices map attenuation within the plane of the X-ray beam. The resulting XCT image is a 3D array of voxels (volume elements), each of which is assigned an attenuation coefficient. Processing of these data requires smoothing and filtering algorithms that may limit resolution, particularly of thin bubble walls. 3D images are commonly presented as movies (e.g., Wright et al. 2006) or renderings of shaded relief (e.g., Gualda and Rivers 2006; Polacci et al. 2007). Current limitations of the technique relate to image resolution, which depends on both the sample size and average attenuation, and the need for large density contrasts between phases of interest. For this reason, volcanological applications of XCT have focused primarily on imaging vesicles (e.g., Lindquist and Venkatarangan 1999; Song et al. 2001; Shin et al. 2005; Wright et al. 2006; Polacci et al. 2007), although phenocrysts can also be imaged provided that there is sufficient contrast with matrix material (e.g., Gualda and Rivers 2006). Processed images can be used to test models of permeable flow through actual pore pathways (e.g., Wright et al. 2006) or for extraction of quantitative textural data (e.g., Song et al. 2001; Shin et al. 2005; Gualda 2006; Polacci et al. 2007).

Textural analysis. Textural data acquired by the methods reviewed above can be used to address both the pre-eruptive state of the magmatic system and the processes by which magma ascends and erupts. Both the bulk crystallinity and the proportions of phenocryst phases provide important information on magma storage conditions; these data typically provide baseline information for assessment of phase equilibria experiments designed to constrain intensive parameters of magma storage reservoirs (see below). In theory, measurement of the size distribution of different crystal phases could also provide information on time scales of crystallization, although this would require better constraints on crystal growth and nucleation rates than are currently available. Alternatively, time constraints provided by other methods such as short-lived isotopes (e.g., Gauthier and Condomines 1999) or diffusional profiles along phenocryst margins (e.g., Zellmer et al. 2003) can be used in conjunction with textural data to measure growth rates directly.

The size distributions of individual phases (crystals and bubbles) reflect the integrated crystallization and vesiculation history of a magma batch, and therefore provide critical information not only on the physical state of the magma at the time of eruption but also on changes in the physical state through space and time. Size analysis provides a particularly powerful tool when the emplacement history of the samples is well constrained. For example, lava samples collected along active flows through open channels (where cooling and crystallization are rapid) and lava tubes (where insulation slows cooling and crystallization) provide constraints on the syn-emplacement crystallization histories of these two lava types. Textural differences are evident in a simple comparison of the number of crystals per unit area (N_a) plotted as a function of crystallinity (ϕ), as shown in Figure 8. Rapidly cooled channel samples show a dramatic increase in crystal number (from 0 to 4000 mm^{-2}) with increasing crystallinity, indicating crystallization by the addition of crystals (*nucleation-dominated* crystallization) of average size (d):

$$d = \sqrt{\frac{\phi}{N_a}} \tag{1}$$

Envelopes enclosing the channel data show that the average crystal size in these samples ranges from 6 to 9 µm. In contrast, slow-cooling lava tube and lava lake samples have lower, and near-constant, crystal number densities (< 200-300 mm^{-2}) but larger crystal sizes (20-100 µm), indicating that in these cases crystallization was *growth-dominated*.

Average N_a data may be extrapolated to three dimensions by assuming that N_a is related to the volume-based number N_v as N_a/d (based on intersection probabilities; e.g., Underwood 1970; Hammer et al. 1999). A more detailed description of crystal textures, however, is provided by analysis of crystal size distributions (CSDs; e.g., Cashman and Marsh 1988). As

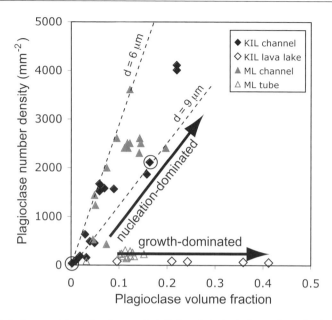

Figure 8. Plagioclase crystal number (per area) plotted against plagioclase volume fraction for samples collected both along active lava channels and from within lava tubes and lava lakes from Kilauea and Mauna Loa volcanoes, Hawaii. Dashed lines illustrate samples that are related by the addition of crystals of similar average size (labeled). We describe these crystallization trends as "nucleation-dominated", as compared to the "growth-dominated" trends where crystallinity increases are accomplished by changing the crystal size instead of the crystal number. Data from Folley (1999), Riker (2005) and Cashman (unpublished).

CSD theory is reviewed in detail by Armienti (2008), here we summarize techniques that are specifically relevant to the interpretation of volcanological processes. CSD theory utilizes the slope of the cumulative number distribution (the density distribution n, with units of number per volume per bin size) as a stable measure of the population distribution (e.g., Randolph and Larson 1971). Under steady state conditions, n is related to crystal size (L) as

$$n = n° \exp\left(-\frac{L}{L_d}\right) \qquad (2)$$

where $n°$, the nucleation density, is the intercept at $L = 0$, and L_d is the dominant size as measured by the slope of the distribution ($L_d = -1/\text{slope}$). Figure 9a shows example CSDs for the two (nucleation-dominated) Kilauea channel samples that are circled in Figure 8. These samples have linear CSDs (i.e., they are well described by Eqn. 2) with near parallel slopes (L_d = 3-4 µm). Note that L_d calculated from the CSDs is smaller than the average d of 9 µm estimated from the $N_a - \phi$ plot shown in Figure 8; this is because L_d is the mode of a distribution that is positively skewed (hence the mean derived from Fig. 8 is larger than the mode derived from the CSD). In fact, $3L_D$ (the mode of the volume-based distribution; Cashman 1992) is usually the closest approximation to the value d determined from $N_a - \phi$ plots.

The extrapolated number density of nuclei-sized crystals, $n°$, indicates the importance of nucleation in the crystallization process, while the area defined by the CSD curve provides a measure of the total number of crystals per unit volume: $N_T = n°L_d$. For example, the nucleation-dominated channel samples shown in Figure 9a show an increase in N_T over an order of magnitude (from 3×10^4 mm^{-3} to 7×10^5 mm^{-3}) as the plagioclase crystallinity increases from 1 to 16%. In contrast, time-sequential CSDs for the growth-dominated Makaopuhi lava

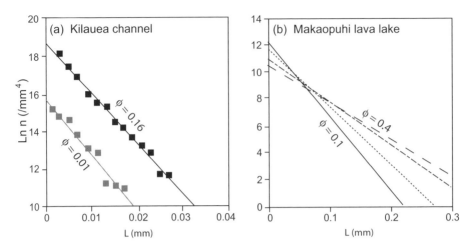

Figure 9. Plagioclase CSDs from time-sequential samples of basaltic lava. (a) Samples collected along a channelized lava flow from Kilauea (circled points in Fig. 8) that show a nucleation-dominated trend with similar slopes (crystal size) but different intercepts (crystal abundance). (b) Samples collected from Makaopuhi lava lake that show a growth-dominated trend controlled by diminishing cooling rates with depth in the lava lake (e.g., Cashman and Marsh 1988). Note scale differences between plots. Data from Folley (1999) and unpublished data.

lake samples (Fig. 9b) have decreasing slopes (L_d increases from 17 to 36 μm) and decreasing intercepts as plagioclase crystallinity increases from 10 to 41%. The net result is a slight decrease in crystal number density (from 4000 to 1500 mm^{-3}) as crystallization proceeds, probably the result of ripening (e.g., Cashman and Marsh 1988).

Parameters determined from a CSD plot may be related to average rates of crystal nucleation (J) and growth (G) by assuming that the dominant size (L_d) is a consequence of steady growth over an appropriate duration of time (τ): $L_d = G\tau$. In industrial crystallization systems, τ is the residence time of a system crystallizing at steady state. In geological applications, τ is the effective crystallization time, as determined by some independent measure. Nucleation rate ($J = dN_v/dt$) is related to growth rate as $J = n°G$, and thus can be determined for any CSD for which G is known. This relationship shows that $n°$ is a direct measure of the ratio between the rates of nucleation and growth (J/G) that are ultimately responsible for the final sample textures. Reported values of $n°$ vary over 11 orders of magnitude (from $<10^1$ to $>10^{12}$ mm^{-4}), illustrating the wide range in crystallization conditions experienced in magmatic systems. In theory, this ratio can be related directly to the undercooling, ΔT, if the appropriate thermodynamic parameters are known (Armienti et al. 1994). As J is directly proportional to $n°$, however, estimates of J are sensitive to the accuracy of crystal measurements at small sizes.

Finally, textural analysis also permits measurements of bubble and crystal orientations and shapes. Orientation can be used to infer the dynamics of magma flow and emplacement (e.g., Castro et al. 2002; Rust et al. 2003). Experimentally, it has been shown that crystal shape is closely related to the conditions of crystallization, with higher degrees of supersaturation leading to less equant shapes (more disparity between growth rates on different crystal faces, e.g., Donaldson 1976; Lofgren 1980; Hammer and Rutherford 2002). As most measurements are made in 2D, the biggest challenge to quantifying true crystal shapes lies in knowing the intersection probabilities for crystals of different morphology (e.g., Higgins 1994). Frequency distributions for different crystal shapes are presented by Morgan and Jerram (2006), who provide a simple computer program for fitting 2D aspect ratio distributions to 3D crystal shapes that can be used with *CSD Corrections* (http://geologie.uqac.ca/~mhiggins/csdcorrections.html), a computer program written by M. Higgins that calculates CSDs for different average particle shapes. A volcanological application of crystal shape analysis is illustrated by a study of tephra produced during early (pre-climactic) stages of the 1991 eruption of Mt. Pinatubo, Philippines (Hammer et al. 1999). Crystallization in these samples was driven by rapid decompression within the conduit followed by short (hours) periods of magma arrest at shallow levels prior to eruption. Longer magma residence times at shallow levels within the conduit resulted in both a change from nucleation-dominated to growth-dominated crystallization (as shown by CSDs) and an accompanying change in plagioclase crystal habit from tabular to prismatic (Fig. 10).

In summary, textural analysis can be used to track the evolution of the crystal (or bubble) populations in a volcanic sample. Measurement of crystal textures can then be combined with micro-analytical techniques (above) and temperature and pressure constraints (below) to determine not only the conditions of crystallization in magma storage regions but also phase changes resulting from decompression and/or cooling during magma ascent and eruption. The continued development of new imaging techniques (both physical – e.g., tomography – and chemical – e.g., BSEM or nanoSIMS imaging) suggests that the power of combined physical and chemical analysis of volcanic samples has only just begun to be realized.

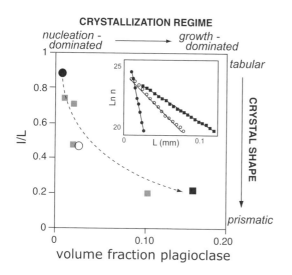

Figure 10. Variations in plagioclase shape (measured as I/L; ratio of intermediate:long axis) with increasing crystallization in pyroclasts from Pinatubo volcano, Philippines. Inset shows CSD plots for the samples identified by black symbols; comparison of the two plots shows that as crystallization becomes progressively dominated by crystal growth, crystal shapes become more equant. Data from Hammer et al. (1999).

GEOTHERMOMETRY

Geothermometers and geobarometers use equilibrium chemical reaction(s) between two or more phases (e.g., minerals, melt) in a rock. Because the extent of chemical equilibrium is determined by kinetic factors, such as solid-state diffusion, reactions become inhibited below a threshold (or "closure") temperature. In volcanic rocks this is often considered to be the eruption temperature, although care must be taken in cases where entrained wall rock or cumulate material has not fully equilibrated with the melt prior to eruption or where the component of interest, e.g., Al in clinopyroxene, has such low diffusivity that re-equilibration at any time after initial crystallization is unlikely. In all cases, textural evidence for equilibrium between the phases of interest must be examined before attempting thermobarometry.

The most widely used magmatic geothermometers involve exchange reactions between two components in coexisting mineral-mineral or mineral-melt pairs, or net transfer reactions in which one or more phases is consumed or produced. Thermobarometers require thermodynamic calibration of the equilibrium constant for a particular reaction or set of reactions. In general, geothermometers use reactions with large enthalpy changes, while barometers involve reactions with large volume changes. Thermodynamic data used to calculate the equilibrium constant may be derived calorimetrically or empirically. In either case a reliable description of activity-composition (a-X) relationships is required for determining the equilibrium constant from natural mineral or melt compositions.

Mineral-melt thermometry is reviewed by Putirka (2008) and a wide range of mineral-mineral thermometers and barometers is reviewed by Anderson (2008); here we limit our discussion to three commonly-used magmatic mineral-mineral thermometers:

1. Coexisting iron-titanium oxides (also used as an oxybarometer for the determination of magmatic oxygen fugacity)
2. Coexisting amphibole and plagioclase, with or without a coexisting silica phase
3. Coexisting orthopyroxene and clinopyroxene

Our primary objective is not to describe the thermodynamics of the thermometers, which is covered in detail elsewhere, but to assess their reliability in sub-volcanic magma systems by applying them to experimental data not included in the original calibrations.

Iron-titanium oxide thermometry and oxybarometry

Many intermediate and acid volcanic rocks contain two discrete Fe-Ti oxide phases: rhombohedral oxides close to the ilmenite ($FeTiO_3$)-hematite (Fe_2O_3) binary and cubic oxides close to the magnetite (Fe_3O_4)-ulvospinel (Fe_2TiO_4) binary. Minor components, notably Mg, Mn, Al and Zn, may substitute for Fe and Ti in these oxides. Minor components can create deviations from the binary join that, in some cases, lead to significant modifications to calculated temperatures. Bacon and Hirschmann (1988) use the Mg/Mn ratio of coexisting oxides as a test for equilibrium:

$$\log\left(\frac{Mg}{Mn}\right)_{mt} = 0.9317^{+0.0113}_{-0.0104} \times \log\left(\frac{Mg}{Mn}\right)_{ilm} - 0.0909^{+0.0785}_{-0.0787} \quad (3)$$

where Mg and Mn are atomic fractions in magnetite (mt) and ilmenite (ilm) and the error bounds denote the acceptable range for equilibrium.

The two key equilibria for iron-titanium oxide thermometry are:

$$FeTiO_3 + Fe_3O_4 = Fe_2TiO_4 + Fe_2O_3 \quad (4a)$$

$$4Fe_3O_4 + O_2 = 6Fe_2O_3 \quad (4b)$$

Reaction (4a) is an exchange reaction that can be used to estimate temperature; reaction (4b) is a redox reaction that can be used to estimate f_{O_2}. In practice the two reactions are combined to give a T-f_{O_2} estimate (Fig. 11). The reaction volumes of both (4a) and (4b) are sufficiently small that the effect of pressure on the T-f_{O_2} estimate is negligible for most sub-volcanic pressures.

Development of geothermometers requires accurate calibration of a-X relationships, which in the case of Fe-Ti oxides includes ordering in the spinel phase. Different formulations of a-X relationships provided by Andersen and Lindsley (1988), Andersen et al. (1993), Ghiorso and Sack (1991) and Ghiorso and Evans (2008) result in broadly similar T-f_{O_2} estimates except at very oxidizing conditions. A further complication is the calculation of oxide compositions using EMPA data, because only rarely are Fe^{2+} and Fe^{3+} contents determined independently (e.g., measuring oxygen by EMPA). Assumptions required about oxide stoichiometry, and hence mole fractions of the various components, have led to several different recalculation schemes. The simplest way to calculate T-f_{O_2} from Fe-Ti oxides is to use freeware programs: the Ghiorso and Evans (2008) formulation is available at *http://ctserver.ofm-research.org/*

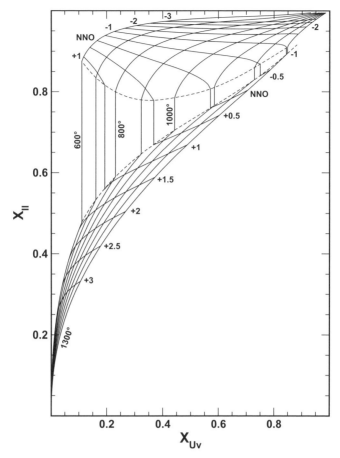

Figure 11. Plot of ilmenite mole fraction (X_{il}) in rhombohedral oxide versus ülvospinel mole fraction (X_{uv}) in coexisting cubic oxide, contoured for temperature (°C) and f_{O_2} ($\log_{10}$ units relative to NNO buffer) in the system Fe-Ti-O. The dashed lines denote the miscibility gap in the rhombohedral oxide phase; the presence of minor components profoundly affects the size and shape of this gap. Figure from Ghiorso and Evans (2008) kindly supplied by M. Ghiorso.

OxideGeothrm/OxideGeothrm.php; the formulation of Andersen and Lindsley is included in the QUILF package (Andersen et al. 1993). The reader is also referred to ILMAT, a useful downloadable Excel spreadsheet developed by LePage (2003) that calculates T-f_{O_2} from pairs of oxides using different thermometer formulations of *a-X* relationships and recalculations of oxide formulae from EMPA data.

Fe-Ti oxides have been shown experimentally to re-equilibrate quickly in response to changes in temperature and/or f_{O_2} (Hammond and Taylor 1982; Venezky and Rutherford 1999). The rate at which re-equilibration occurs depends on the size of the grains and their physical proximity. A common practice is to analyze several grains of titano-magnetite and several of ilmenite (typically much scarcer in relatively oxidized volcanic rocks) and use the averages in T-f_{O_2} calculations. Equilibrium between the two averaged compositions is normally checked using Equation (3). A problem with this approach is that, because the analyzed grains may vary in size, it is never clear whether *all* of the ilmenites are in equilibrium with *all* of the magnetites, particularly as their co-occurrence in a single volcanic sample is not in itself evidence for textural equilibrium. For example, eruptive processes can mix crystals that grew in different parts of a magma chamber.

A more reliable approach (Blundy et al. 2006) is to analyze only adjacent Fe-Ti oxide pairs that either share a common boundary (i.e., are in contact) or occur as inclusions in the same silicate phenocryst phase. In this way it is possible to ensure textural equilibrium, even in the case of zoned crystals (e.g., Devine et al. 2003), and to distinguish between crystals with different growth histories (e.g., phenocrysts, inclusions, microlites etc.). Each touching pair can then be checked for equilibrium using Equation (3) and used to calculate an individual T and f_{O_2}. This approach yields information on the temporal evolution of T and f_{O_2} that is lost through conventional averaging procedures. We assume that a homogeneous set of T-f_{O_2} estimates from all pairs will result only if the *entire* oxide population, including inclusions in phenocrysts, reached full chemical equilibrium prior to eruption. This is unlikely for most volcanic rocks, despite the relatively rapid rates of diffusive re-equilibration in oxides.

An example of this approach is provided by Fe-Ti oxides from Mount St. Helens volcano. Previous T-f_{O_2} estimates for the 1980-86 dacite magmas, based upon averaged analyzes, yielded temperatures that ranged from 860 to 1030 °C (Melson and Hopson 1981; Rutherford et al. 1985; Rutherford and Devine 1988; Rutherford and Hill 1993). Blundy et al. (2006) used touching (or co-included) pairs from samples spanning the 1980-81 phase of the eruption to demonstrate systematic variations between individual crystal types, both in a single sample and between different samples. Specifically, the lowest temperatures (880 °C) derived from inclusions in phenocrysts from the May 18th Plinian deposit, whilst microlites and phenocrysts from all subsequent eruptions showed elevated temperatures, which Blundy et al. (2006) attributed to magma heating by latent heat release during decompression crystallization. Figure 12 shows an expanded Mount St. Helens T-f_{O_2} dataset, extending from 1980 to 1986 (Blundy et al. 2008). Some of the lower temperatures recorded by cryptodome and May 1985 oxide pairs probably reflect entrainment of cooler material from the walls of the conduit. Significantly, oxide pairs from samples containing microlites formed during slow ascent and decompression crystallization showed systematically lower f_{O_2} than phenocrysts from the Plinian deposit (Fig. 12; see below). These features would not have been apparent if averaged oxide compositions were used.

How well does the thermometer work? To test the reliability of the different Fe-Ti oxide thermometers discussed above, we have applied them to the analyzed run products of experiments that (1) crystallized both oxides and (2) were not included in any of the calibrations. A total of 75 additional oxide pairs were taken from the following experimental studies: Venezky and Rutherford (1999), Beard and Lofgren (1991), Blatter and Carmichael (2001), Bogaerts et al. (2006), Dall'Agnol et al. (1999), Gardner et al. (1995), Martel et al. (1998), Pichavant et al. (2002), Rutherford and Devine (1988), Scaillet and Evans (1999). These data cover a wide range of pressure (50-690 MPa) and temperature (707-1000 °C). For the f_{O_2} calculations we have included only a subset ($n = 30$) of the data where the experimental f_{O_2} was buffered either by solid oxygen buffers (e.g., Ni-NiO or Re-ReO$_2$) or by controlling the partial pressure of H$_2$ using a Shaw membrane. The resulting range in experimental f_{O_2} is from NNO-0.1 to NNO+2.8 log units.

Temperature and f_{O_2} were calculated from the following formulations:

1. Andersen and Lindsley (1988) thermometer with Lindsley and Spencer (1982) oxide formula recalculation
2. Andersen and Lindsley (1988) thermometer with Stormer (1983) oxide formula recalculation
3. Ghiorso and Evans (2008)
4. QUILF (Andersen et al. 1993) using "FeTi" and "HM" reactions only.

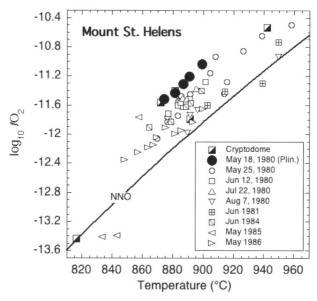

Figure 12. Temperature-f_{O_2} plot of Fe-Ti oxides in various eruptive phases of the 1980-86 eruption of Mount St. Helens. All data are taking from touching pairs of oxides or oxides co-included in the same silicate phenocryst phase. The NNO buffer is shown for reference. Only the plinian eruption of May 18th, 1980 (filled circles) lacks microlites; all other samples (open symbols) are microlite-bearing due to slower ascent and decompression crystallization. Oxide data are from Blundy et al. (2008).

For each formulation we calculated the average absolute deviation (*aad*) on T and f_{O_2} for the experimental dataset. For T, *aad* ranges from 33-60 °C (Fig. 13). The smallest *aad* comes from the thermometer of Andersen and Lindsley (1988) although this formulation fails to capture the full range in T, with systematic underestimates above 860°C and overestimates below. In contrast, Ghiorso and Evans (2008) and QUILF are more scattered, but reproduce the full range of T, although QUILF shows a systematic underestimate at all T. f_{O_2} estimates from Andersen and Lindsley (1988) and Ghiorso and Evans (2008) have similar *aad* of 0.25-0.34 log units; QUILF consistently underestimates f_{O_2} for this relatively oxidizing experimental dataset (*aad* = 1.09 log units).

Amphibole-Plagioclase thermometry

Many intermediate and acid volcanic rocks contain coexisting amphibole and plagioclase. The geothermometer of Holland and Blundy (1994) is based on two exchange equilibria involving these phases:

$$\text{edenite} + 4 \text{ quartz} = \text{tremolite} + \text{albite} \tag{5}$$

$$\text{edenite} + \text{albite} = \text{richterite} + \text{anorthite} \tag{6}$$

Equilibrium (5) is applicable to silica-saturated rocks, whereas (6) is applicable to silica-undersaturated rocks. Holland and Blundy (1994) developed *a-X* models for both amphibole and plagioclase and calibrated the thermometers (available at *http://rock.esc.cam.ac.uk/astaff/holland/hbplag.html*) on coexisting amphibole-plagioclase pairs from a large number of metamorphic rocks and experimental studies to cover a wide range of *P-T* and composition. As both thermometers have modest (±72 °C GPa^{-1}) pressure dependence at pressures of sub-volcanic magma chambers, a pressure of 200 or 300 MPa can be assumed in the absence of other constraints. Holland and Blundy (1994) report an *aad* of ±28 °C for thermometer (5) and

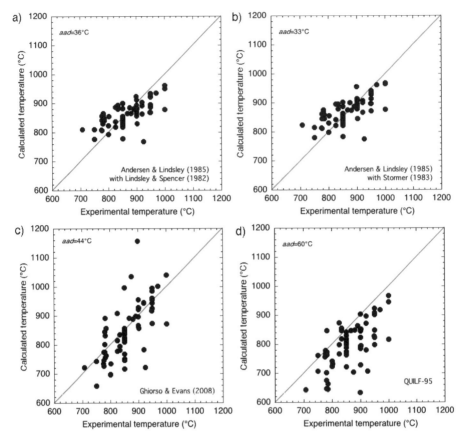

Figure 13. Comparison of calculated Fe-Ti temperatures for 75 experimental oxide pairs from studies not used in original thermometer calibrations. Four thermometers are shown: (a) Andersen and Lindsley (1988) thermometer with Lindsley and Spencer (1982) oxide formula recalculation; (b) Andersen and Lindsley (1988) thermometer with Stormer (1983) oxide formula recalculation; (c) Ghiorso and Evans (2008); and (d) QUILF-95 (Andersen et al. 1993). For each thermometer we show the number of points (*n*) and the average absolute deviation (*aad*). Data references are provided in the text. Note that we use *aad* in preference to Root Mean Square Error (*RMSE*) as *aad* gives a better impression of the average misfit of calculated and experimental temperatures for relatively small datasets with outliers. The latter can strongly skew the *RMSE* to high values.

±32 °C for thermometer (6) over a temperature range of 400-1000 °C and 100 to 1500 MPa pressure. Anderson et al. (2008) further consider the application of reactions (5) and (6) to granitic rocks and review the potential use of amphibole as a thermobarometer.

As with all thermometers, care must be taken to ensure that the amphibole-plagioclase pairs are in textural equilibrium. Both phases are commonly zoned, so it is essential to use the outermost rims of phenocrysts for thermometry, or touching pairs of crystals where possible. Many volcanic rocks contain amphibole phenocrysts with plagioclase inclusions and vice versa that can be used for thermometry provided that the immediately adjacent host mineral is analyzed (e.g., Bachmann and Dungan 2002; Couch et al. 2003a). As in the case of Fe-Ti oxides, we would caution against taking average compositions of amphibole and plagioclase from an entire thin section; instead we recommend that temperatures be calculated from a large number of (touching) pairs. The use of core-core pairs is only recommended where there is clear textural evidence that plagioclase cores and amphibole cores were precipitating simultaneously from the same melt; this is rarely easy to establish.

How well does the thermometer work? We have assembled a dataset of 294 amphibole-plagioclase pairs from experiments at magmatic temperatures published since 1994 and therefore not included in the original thermometer calibration. The dataset covers a pressure range of 42 to 1500 MPa and 640 to 1050 °C. The Holland and Blundy (1994) thermometers were used to calculate T for each experiment at the experimental P. As few experiments contain quartz in addition to amphibole and plagioclase, all temperatures were calculated with thermometer (6). The average absolute deviation between calculated and experimental temperature for the entire dataset is 61 °C (Fig. 14a). The deviation correlates negatively with the Mg# of the amphibole (Fig. 14b), suggesting that the *a-X* relationships of Holland and Blundy (1994) do not adequately address non-ideality between Fe^{2+} and Mg on the M1, M2 and M3 crystallographic sites. Evidently, in magmas with high Mg# amphibole-plagioclase thermometry offers a less accurate means of estimating magmatic temperatures than Fe-Ti oxide thermometry, but nonetheless a valuable one in rocks that lack two oxides.

Two-pyroxene thermometry

Many intermediate and acid volcanic rocks contain coexisting ortho- and clinopyroxene. The reader is referred to Putirka (2008) for a discussion of the various pyroxene equilibria upon which thermometry is based and the techniques used for recalculating pyroxene structural formulae.

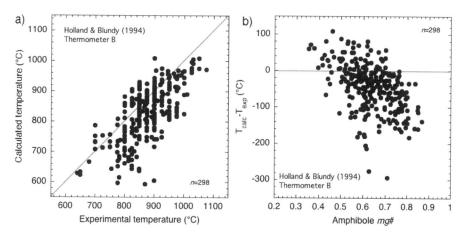

Figure 14. Application of Thermometer B of Holland and Blundy (1994) to 298 experimental amphibole-plagioclase pairs not included in original calibration dataset. (a) comparison of calculated and experimental temperatures; (b) difference between calculated and experimental temperature as a function of amphibole Mg#. The *aad* of the calculated temperature is 61 °C for the entire dataset. The negative correlation between calculated temperature and Mg# in (b) suggests that there is additional non-ideality on the amphibole M1-3 sites not fully addressed by the thermometer. Experimental data from Nakajima and Arima (1998), Bogaerts et al. (2006), Pichavant et al. (2002), Grove et al. (1997), Klimm et al. (2003), Rutherford and Devine (2008), Gardner et al. (1995), Donato et al. (2006), McCanta et al. (2007), Larsen (2006), Brooker and Blundy (unpublished data), Poli (1993), Rapp and Watson (1995), Ernst and Liu (1998), Sisson et al. (2005), Barclay and Carmichael (2004), Blatter and Carmichael (2001), Costa et al. (2004), Holtz et al. (2005), Kawamoto (1996), Martel et al. (1999), Moore and Carmichael (1998), Prouteau and Scaillet (2003), Sato et al. (1999), Scaillet and Evans (1999), Schmidt and Thompson (1996), Patino-Douce and Beard (1995), Alonso-Perez et al. (2008), López and Castro (2001), Nicholis and Rutherford (2004), Johannes and Koepke (2001), Molina and Poli 2000, Kaszuba and Wendlandt (2000), Couch (2002), Gardien et al. (2000), Sen and Dunn (1994), Dall'Agnol et al. (1999).

Three formulations of the two-pyroxene thermometer are in widespread usage:

1. Wells (1977)
2. Brey and Kohler (1990)
3. QUILF (Andersen et al. 1993)

How well do the thermometers work? We have assembled a dataset of 142 experimental two-pyroxene pairs from experiments at magmatic temperatures published since 1996. Data span a range of 0.1 to 1500 MPa and 819 to 1230 °C (Figure 15). The Wells (1977) thermometer (not shown), which has no pressure dependency, has an *aad* of 99 °C for the entire dataset. The *aad* of the Brey and Kohler (1990) thermometer is 66 °C (Fig. 15a) and comparable to that of the amphibole-plagioclase thermometer, although the errors are much larger than the ±30 °C (2 s.d.) quoted by Brey and Kohler (1990). The thermometer is much more reliable at temperatures above 1000 °C, and is therefore better suited to andesitic and basaltic magmas. QUILF (Fig 15b) systematically overestimates all experimental temperatures with an *aad* of 114 °C. However the uncertainty calculated QUILF is demonstrably more realistic and nearly 50% of the calculated temperatures fall within error of the experimental temperature.

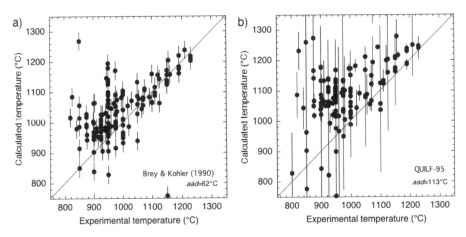

Figure 15. Comparison of calculated two-pyroxene and experimental temperature for 142 experimental pyroxene pairs using the thermometers of (a) Brey and Köhler (1990); and (b) QUILF-95 (Andersen et al. 1993). The *aad* for each thermometer is given. Error bars are those provided by original authors. Experimental data from Singh and Johannes (1996), Alonso-Perez et al. (2008), Moore and Carmichael (1998), Larsen (2006), Lopez and Castro 2001, Johannes and Koepke (2001), Grove et al. (2003), Sisson et al. (2005), Grove et al. (1997), Dall'agnol et al. (1999), Pichavant et al. (2002), Costa et al. (2004), Kawamoto (1996), Martel et al. (1999), Prouteau and Scaillet (2003), Sato et al. (1999), Blatter and Carmichael (1998, 2001), Pertermann and Lundstrom (2006), Pichavant and Macdonald (2007), Patino-Douce and Beard (1995), Müntener et al. (2001).

GEOBAROMETRY

In nature there are far fewer reactions suitable for magmatic geobarometry than for geothermometry at conditions appropriate for sub-volcanic systems. The aluminum-in-hornblende barometer—which relies on the presence of a low-variance mineral assemblage—is discussed by Anderson (2008), while Putirka (2008) discusses the barometric potential of some mineral-melt equilibria. Here we consider four alternative methods for estimating pressure in sub-volcanic systems:

1. Experimental reproduction of phase assemblages and compositions in natural volcanic rocks
2. The saturation pressure of volatiles in melt inclusions
3. The chemical composition of silica-rich glasses in equilibrium with quartz and feldspar
4. The proportion and number of plagioclase crystals.

Experimental reconstruction

For multiply-saturated volcanic rocks it is possible to reconstruct the pre-eruptive magma storage conditions using experiments designed to reproduce the phase assemblage and phase compositions of natural samples. This approach requires two key assumptions. The first is that the crystallizing assemblage represents equilibrium at the moment of eruption and the second is that the assemblage equates to equilibrium at a unique set of pressure, temperature and volatile conditions. The ubiquity of zoned crystals, especially plagioclase, in most volcanic rocks renders both of these assumptions critical when considering whole-rock samples. As crystallization over a range of pressures is probably the norm in many natural systems, experiments should be 'tuned' to examine specific parts of the system (such as the uppermost pre-eruptive storage region).

Pichavant et al. (2007) elegantly address experimental investigations of magma storage by examining the equilibration length-scales that operate in natural magmas, and the resulting effective composition, or "reactive magma," that can be studied experimentally. If the reactive magma at the pre-eruptive storage conditions is only a subset of the bulk rock volume, e.g., it excludes the cores of zoned crystals, then there is relatively little virtue in performing experiments on the bulk sample. As an illustration, Pichavant et al. (2007) demonstrate that starting experiments with a rock powder that contains xenocrystic calcic plagioclase will prevent determination of a unique set of pre-eruptive conditions because the starting material deviates substantially from the true reactive magma composition. "Partial equilibrium" experiments can be used to circumvent this problem, whereby natural samples are lightly crushed, thus chemically isolating crystal cores on the timescales of typical experimental runs. An alternative approach is to attempt to equilibrate the matrix glass composition with crystals of the same composition as the rims of the phenocrysts (e.g., Couch et al. 2003a). In practice this can be done either by using a glass separate from the natural sample as the experimental starting material, or by synthesizing a starting composition based on analysis of glass in the natural sample. If there are nucleation problems, the glass starting material may be seeded with tiny amounts (1-5%) of the desired crystallizing assemblage and checked for overgrowth or dissolution textures in the resultant run products. Implicit in this approach is the assumption that the reactive magma includes the groundmass glass and the phenocryst rims.

Typical pre-eruptive storage conditions of 0-500 MPa make externally-heated hydrothermal ("cold-seal") or gas-pressure (e.g., TZM) vessels, or internally-heated gas pressure (IHPV) vessels, the ideal experimental techniques for these studies. Pressure and temperature can be controlled precisely in both externally- and internally-heated apparatus. Temperature accuracy is considerably greater in IHPV because the thermocouple sits adjacent to the sample, in contrast to externally heated pressure vessels where the thermocouple is located inside the walls of the pressure vessel.

As f_{O_2} has a profound effect on phase relations and compositions, this intensive variable must also be controlled. In IHPV a small amount of H_2 can be added to the argon pressurizing gas and the fugacity of H_2 measured using a Shaw membrane (e.g., Martel et al. 1998, 1999; Scaillet and Evans 1999). f_{O_2} can then be calculated using the disproportionation reaction of H_2O,

$$2H_2O = O_2 + 2H_2 \tag{7}$$

$$\log f_{O_2} = \log K_w - 2\log f_{H_2} + 2\log f_{H_2O} \tag{8}$$

Here K_w is the disproportionation constant for H_2O (e.g., Robie et al. 1978) and f_{H_2O} is the fugacity of H_2O in the sample, which will be the fugacity of the pure gas in H_2O-saturated experiments. Provided that H_2 can diffuse through the noble metal sample capsule on the timescales of the experiment, f_{H_2} inside the capsule equates to that in the pressurizing gas. Although Au is relatively impermeable to H_2 (Chou 1987), it can be used as a capsule material in long duration experiments given a sustained source of H_2, e.g., Ar-H_2 gas mixtures.

In an externally-heated apparatus an (inner) sample capsule is embedded in an outer capsule containing a solid buffer (e.g., Ni and NiO, plus a small amount of H_2O; Chou 1987). f_{O_2} is buffered via exchange of H_2 between the sample and the buffer across the walls of the inner capsule. The sample capsule must be sufficiently permeable to H_2 (e.g., AgPd, AuPd, Pt) that the external buffer and sample can attain the same f_{H_2}, whereas the outer capsule should then be relatively impermeable to ensure that H_2 is not lost from the system during the experiment.

Where the sample is undersaturated in H_2O, the f_{O_2} is buffered at a value that deviates from that of the pure buffer by an amount controlled by the H_2O activity in the sample capsule (e.g., Klimm et al. 2003; Sisson et al. 2005). Specifically, as H_2O activity in the undersaturated sample is less than 1, the f_{O_2} in the sample is buffered at a value below that of the buffer (i.e., more reducing). Except for highly H_2O-undersaturated melt or H_2O-poor fluids, the deviation of the sample f_{O_2} from that of the buffer is quite small, e.g., 0.3 log units lower than the buffer for $a_{H_2O}^{melt} = 0.7$.

Starting conditions for equilibrium experiments typically use an initial temperature, f_{O_2} and/or pressure obtained from mineral thermobarometers to explore the neighbouring P-T-H_2O-f_{O_2} space for the best match to the natural assemblage. The extent to which this approach affords a unique solution depends on the number of degrees of freedom of the natural assemblage, the uncertainties in the petrologically-estimated P, T and f_{O_2}, and the variability in the natural phase compositions. The greater the number of equilibrium crystal phases, the fewer the degrees of freedom and the more unique the solution. Consider, for example, a volcanic rock containing phenocrysts of amphibole, plagioclase, orthopyroxene, ilmenite and magnetite in equilibrium with a vesicular (vapor-saturated) glass. The total number of major components is likely to be 11 (SiO_2, TiO_2, Al_2O_3, FeO, Fe_2O_3, MgO, CaO, Na_2O, K_2O, H_2O, CO_2). The total number of degrees of freedom in this 7-phase assemblage is 6. Thus even at fixed P, T, f_{O_2}, and vapor composition there are still 2 additional degrees of freedom. However, if the rock also contains clinopyroxene and alkali-feldspar, then the coexistence of these phases with glass would constitute a unique solution. The degrees of freedom may also be limited by independent compositional parameters, such as the anorthite content of the plagioclase, the Mg# of the pyroxene or amphibole, the Al_2O_3 content of the glass, or the proportion of a particular phase.

To illustrate the principles of the experimental approach, Figure 16 shows the results of a careful experimental investigation of pre-eruptive conditions for a Holocene dacite from Volcán San Pedro, Chile (Costa et al. 2004). Oxide-thermometry indicates a pre-eruptive temperature of 850 °C. For experiments conducted at this temperature and 200 MPa pressure we compare 7 independent parameters (crystallinity, modal plagioclase: amphibole ratio, SiO_2, Al_2O_3 and CaO in the glass, anorthite content (X_{An}) in plagioclase and Al_2O_3 content of amphibole) with those in the natural sample with respect to the bulk H_2O content of the magma. We note that all 7 parameters show a consistent value of 5.1±0.4 wt% H_2O, which overlaps with the H_2O content of glasses in experiments that yield the correct phase assemblage (Fig. 16, shaded region). These results show a clear example of a reactive magma composition that closely matches that of the bulk dacite used in the experiments. The small deviation of H_2O content estimated from glass CaO content, compared to other parameters, suggests that the natural

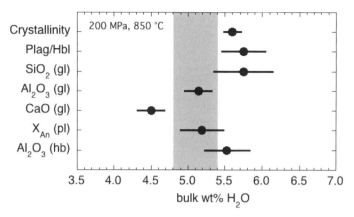

Figure 16. Summary of experimental phase relations for Volcán San Pedro, Chile, dacite from Costa et al. (2004). Results are expressed in terms of the bulk magmatic H_2O content that best matches the seven parameters from the natural sample listed on the ordinate axis: crystallinity; weight plagioclase:hornblende (Plag/Hbl) ratio; wt% SiO_2, Al_2O_3 and CaO in glass (gl); anorthite content of plagioclase (X_{An}) and Al_2O_3 content of hornblende. The data points are derived by linear regression of each parameter against H_2O for experiments performed at 850 °C and 200 MPa, the estimated pre-eruptive temperature and pressure, The error bars are based on uncertainty in the regressions and variability in the natural sample. The shaded bar denotes the range of H_2O contents that yield the correct phenocryst assemblage (±biotite). The convergence of all parameters on a pre-eruptive H_2O content of 5.1±0.4 wt% suggest that the bulk rock approximates the reactive magma at San Pedro, although the slight mismatch for CaO (gl) suggests the presence of small amounts of calcic plagioclase cores.

sample may contain a small quantity of calcic plagioclase cores that are not reactive, a finding consistent with the petrographic description of rare An-rich plagioclase cores in the San Pedro dacite (Costa et al. 2004). Pichavant et al. (2007) draw a similar conclusion about the Mont Pelée reactive magma composition based on the experiments of Martel et al. (1998, 1999).

How well does the barometer work? In Table 1 we list 40 published experimental studies designed to constrain pre-eruptive magma storage conditions at 25 different volcanoes. All of these studies provide a close match for many of the observed parameters in the natural samples. However, in most cases the match is not as close as observed for Volcan San Pedro (Fig. 16) and some parameters cannot be adequately matched at a single experimental condition.

A failure to match phase proportions and compositions at any reasonable permutation of P-T-H_2O-f_{O_2} conditions most likely reflects a failure of the natural sample (or the experiments) to attain equilibrium, omission of an important compositional variable such as CO_2, S or F, or violation of the reactive magma principle discussed above. For example, based on the experiments of Rutherford et al. (1985) and Rutherford and Devine (1988), Blundy and Cashman (2001) show that the experimentally determined P-T-H_2O-f_{O_2} conditions that best match the phase *compositions* of Mount St. Helens dacite, differ from the conditions that best match the phase *proportions*. One possible explanation for this discrepancy, as noted above, is that the experimental starting material contains calcic plagioclase cores that are relicts from earlier phases of the crystallization history of the magma (e.g., Cooper and Reid 2003; Berlo et al. 2007) and therefore not present in the reactive magma. Despite these caveats, however, careful experimental exploration of phase equilibrium space has been a cornerstone of magma petrology and continues to provide critical constraints on magma storage conditions (e.g., Hammer 2008).

Volatile saturation pressures

Another means of determining pressure is provided by the strong pressure dependence of the solubility of H_2O and CO_2, the dominant magmatic volatile species, in silicate melts.

Table 1. Experimental studies designed to reproduce pre-eruptive storage conditions beneath andesite and dacite volcanoes. Table lists the volcano studied and the P-T range investigated.

Source	Volcano	P range (MPa)	T range (°C)
Arce et al. (2006)	Nevado de Toluca, Mexico	100-250	780-880
Barclay and Carmichael (2004)	Jorullo, Mexico	40-300	950-1100
Barclay et al. (1998)	Soufrière Hills, Montserrat	50-200	800-940
Blatter and Carmichael (1998)	Zitácuaro, Mexico	0.1-292	950-1150
Blatter and Carmichael (2001)	Valle de Bravo, Mexico	52-301	909-1140
Browne and Gardner (2006)	Redoubt, AK	30-250	775-900
Coombs and Gardner (2001)	Novarupta, AK	20-200	760-900
Coombs et al (2000)	Southwest Trident, AK	50-200	1000-1100
Coombs et al (2002)	Southwest Trident, AK	90	890-1000
Costa et al. (2004)	San Pedro, Chile	54-405	800-950
Cottrell et al (1999)	Santorini, Greece	50-250	800-1050
Couch et al (2003a)	Soufrière Hills, Montserrat	5-225	825-1100
Donato et al. (2006)	Salina Island, Italy	200-300	755-825
Gardner et al. (1995)	Mount St. Helens, WA	100-350	850
Geschwind and Rutherford (1992)	Mount St. Helens, WA	100-390	760-930
Grove et al (2003)	Shasta, CA	0.1-200	940-1250
Grove et al. (1997)	Medicine Lake, CA	100-200	865-1050
Hammer et al. (2002)	Novarupta, AK	15-225	850-1050
Holtz et al (2005)	Unzen, Japan	100-300	775-875
Johnson and Rutherford (1989)	Fish Canyon, CO	200-500	740-930
Larsen (2006)	Aniakchak, AK	50-200	820-950
Luhr (1990)	El Chichón, Mexico	100-400	800-1000
Martel et al. (1998)	Mont Pelée, Martinique	200-311	850-925
Martel et al. (1999)	Mont Pelée, Martinique	200-400	850-1040
McCanta et al (2007)	Black Butte, CA	100-450	800-950
Merzbacher and Eggler (1984)	Mount St. Helens, WA	100-400	850-1000
Moore and Carmichael (1998)	Colima, Mexico	0.1-303	900-1251
Nicholls et al. (1992)	Taupo, New Zealand	200-500	700-800
Parat et al (2008)	Fish Canyon, CO	400	850-950
Pertermann and Lundstrom (2006)	Arenal, Costa Rica	500	1100-1300
Pichavant et al. (2002)	Mont Pelée, Martinique	400	950-1025
Prouteau and Scaillet (2003)	Pinatubo, Philippines	400-980	750-995
Rutherford and Devine (1988)	Mount St. Helens, WA	220-320	890-970
Rutherford and Devine (2003)	Soufrière Hills, Montserrat	10-250	780-950
Rutherford and Devine (2008)	Mount St. Helens, WA	70-260	840-940
Rutherford and Hill (1993)	Mount St. Helens, WA	120-220	790-940
Rutherford et al (1985)	Mount St. Helens, WA	0.1-320	847-1190
Sato et al (1999)	Unzen, Japan	29-196	800-1032
Scaillet & Evans (1999)	Piantubo, Philippines	220-390	750-900
Venezky and Rutherford (1997)	Mount Rainer, WA	25-200	900-945
Venezky and Rutherford (1999)	Unzen, Japan	40-200	750-1050

H_2O and CO_2 solubilities are dependent on melt and vapor compositions, for which various experimentally-calibrated solubility relationships are available. Perhaps the easiest relationship to apply is that of Newman and Lowenstern (2002), available as an Excel spreadsheet, *VolatileCalc*. This formulation has separate relationships for rhyolite and basalt liquids, but is not calibrated for andesite liquids, which lack experimental constraints. However, as most melt inclusions in andesites are rhyolite or dacite in composition, this is not a critical problem. Volatile solubility is weakly dependent on temperature, therefore an estimate of magmatic temperature is required to calculate the saturation pressure (see above). For a given concentration of H_2O and CO_2, an uncertainty of ±50 °C in temperature translates to an uncertainty in saturation pressure of approximately ±10 MPa at 200 MPa and ±5 MPa at 100 MPa. Alternative solubility models for mixed H_2O-CO_2 volatile phase include those of Tamic et al. (2001), which is only valid from 75-200 MPa, and Liu et al. (2005), valid from 0 to 500 MPa.

Magma storage pressures can be estimated from the volatile contents of melt inclusions (e.g., Wallace et al. 1999; Schmitt 2001; Liu et al. 2006) provided there is evidence that the melt was volatile-saturated at the time the melt inclusions became sealed. The only unequivocal evidence for volatile saturation in a magma containing H_2O and CO_2 is a decrease in CO_2 content of melt inclusions with increasing crystallization, as revealed, for example, by increasing contents of incompatible trace elements (e.g., Wallace et al. 1999). This is because gas loss, whether driven by isobaric volatile-saturated crystallization or by decompression of volatile-saturated magma, will always drive CO_2 preferentially into the vapor phase, leading to its preferential depletion, relative to H_2O, in the melt. In cases where the melt is not volatile-saturated, the estimated pressure based on volatile saturation will underestimate the true storage pressure. Note that the presence of gas bubbles in a melt inclusion does not necessarily indicate that the magma was gas-saturated as bubbles can also form by shrinkage during cooling.

Melt inclusions can provide information not only on the pressures at which magma was stored pre-eruptively, but also on the relationship between degassing and crystallization (Moore 2008). The different paths described by melt inclusions generated under a variety of different scenarios are described in detail in a later section. At active volcanoes, the calculated pressures from melt inclusions can be converted into depth, using a suitable density model, and used to corroborate independent constraints on magma storage depths, for example from seismology or geodetic measurements (see below).

Different pressure constraints are provided by groundmass glasses, which are invariably degassed such that saturation pressures, although accurate, pertain to the process of eruption (specifically, the pressure at which the magma passed through the glass transition temperature) rather than to conditions of magma storage. As kinetic factors may inhibit diffusion of H_2O and CO_2 into bubbles during eruption (e.g., Gonnermann and Manga 2005), pressures recorded in matrix glass may be over-estimates. Nonetheless, in the case of Mount St. Helens dacites, H_2O saturation pressures of 2 to 40 MPa (mean of 12 measurements = 16±12 MPa at an assumed eruption temperature of 900 °C) in the groundmass glass from the May 18th, 1980 plinian phase (Blundy and Cashman 2005; Humphreys et al. 2008a) agree well with the calculated pressure range over which the Mount St. Helens magma crossed the glass transition temperature (e.g., Carey and Sigurdsson 1985; Papale et al. 1998), suggesting that, in this case at least, kinetic effects played a minor role in limiting effective H_2O escape to bubbles.

How well does the barometer work? The principal limitation of using volatile solubility as a barometer is the obvious one that the magma must contain crystals with melt inclusions sufficiently large to analyze, as well as the need to distinguish equilibrium volatile saturation at the time of entrapment from post-entrapment processes, and the compositional dependence of volatile solubility. Some degree of disequilibrium may be inevitable for rapidly erupted volcanic liquids; where diffusion-limited disequilibrium has occurred, calculated saturation pressures will be overestimates. The compositional dependence of volatile solubility is well documented in simple systems (e.g., Johannes and Holtz 1996; Moore 2008), although it is rarely taken into account in solubility models, which tend to be based on generic melt compositions. Deviations in solubility from the models are likely to be greatest for peralkaline and peraluminous melts.

To gauge the reliability of saturation pressures calculated from dissolved volatile contents we have assembled a dataset of 118 published, experimentally-determined $H_2O\pm CO_2$ measurements (SIMS or FTIR analyzes only) on silicic to intermediate melt compositions, none of which were used in the original solubility model calibrations. We have used both static and decompression experiments to encompass the likely range of behaviors in natural volcanic rocks. We have not, however, included experiments that demonstrate homogeneous nucleation, where substantial supersaturations are possible (e.g., Mangan et al. 2004b). We have used the experimental temperature for our calculations and test the models of both Newman and Lowenstern (2002) and Liu et al. (2005); at all pressures below 300 MPa the two models agree to within < ±5 MPa.

The agreement between calculated and experimental pressures is remarkably good for all experiments at P < 400 MPa (Fig. 17). The large negative deviations from calculated pressures apparent in the 400 MPa experiments of Scaillet et al. (1995) are almost certainly the result of low quench rates in these IHPV experiments leading to some degassing during quenching. Ignoring those data, we calculate an *aad* between calculated and experimental pressures of 22 MPa for Newman and Lowenstern (2002) and 23 MPa for Liu et al. (2005). The *aad* would increase if we had to rely on alternative means of calculating temperature (e.g., from mineral thermometers). To assess the importance of melt composition on solubility we have also calculated saturation pressures for 17 glasses from the phonolite decompression experiments of Larsen and Gardner (2004). Despite the obvious compositional differences to the calibrant dataset, the calculated pressures are still good to ±35 MPa. Evidently small differences in melt composition from the rhyolites used in the solubility models propagate to relatively small uncertainties in calculated pressures.

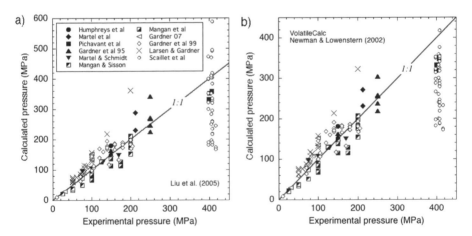

Figure 17. Calculated volatile solubility pressures. (a) Model of Liu et al. (2005); (b) *VolatileCalc* from Newman and Lowenstern (2002). Pressures calculated at experimental temperature. Data from Humphreys et al. (2008a), Martel et al. (1999) Pichavant et al. (2002), Gardner et al. (1995) Martel and Schmidt (2003), Mangan and Sisson (2000), Mangan et al. (2004a), Gardner (2007), Gardner et al. (1999), Larsen and Gardner (2004), Scaillet et al. (1995).

Haplogranite projection

An alternative means of estimating pressure, especially for rocks that lack suitable melt inclusions for volatile analysis, is to use the well known relationship between silica solubility in melts and pressure. Since the seminal work of Tuttle and Bowen (1958) it has been known that in the hydrous orthoclase-albite-silica (*Or-Ab-Qz*) "haplogranite" system, the silica content of melts saturated in feldspar(s) and quartz increases appreciably with decreasing pH_2O. Subsequent experimental studies, summarized in Johannes and Holtz (1996), have confirmed and refined Tuttle and Bowen's conclusions. At pressures of ≥ 500 MPa, there is a miscibility gap between K- and Na-feldspar and the ternary minimum becomes a eutectic connected to the *Ab-Or* binary by a two-feldspar cotectic. At pressures below 500 MPa, the Qz-Ab-Or-H_2O system is characterized by two primary phase volumes (quartz and feldspar) separated by a curved cotectic (Fig. 18). Because of complete solid solution between albite and orthoclase on the solidus at low pressure, the cotectic contains a minimum rather than a eutectic. For H_2O-saturated conditions, the silica phase volume decreases (i.e., silica solubility increases) with decreasing pressure, the minimum moves towards the Qz-Or join, and the temperature of the minimum (or eutectic) increases from 645 °C at 500 MPa, to 680 at 200 MPa, to ~950 °C at atmospheric pressure (Tuttle and Bowen 1958). At any given pressure, H_2O-undersaturation has little effect on the position of the cotectic, but shifts the minimum towards the Qz-Or join and increases its temperature. For example, at 200 MPa the minimum temperature changes from 764 °C at a water activity (a_{H_2O}) of 0.5 to 902 °C at a_{H_2O} = 0.1. The stable liquidus silica phase is high quartz, except at $P \leq 20$ MPa, where tridymite appears near the ternary minimum. At still lower pressures tridymite becomes stable along the entire cotectic (Tuttle and Bowen 1958). There are no reliable experimental data on the Qz-Ab-Or ternary at pressures below 50 MPa because of difficulties in attaining equilibrium on experimental time scales (e.g., Schairer 1950; Brugger et

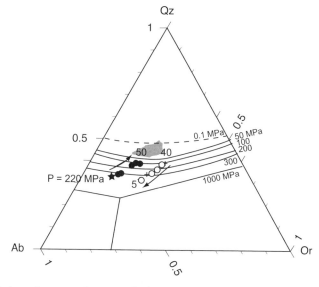

Figure 18. Haplogranite ternary (as summarized by Blundy and Cashman 2001), showing changes in experimental glass composition as a function of pressure, using projection scheme in equations (9a-c). Staring material for experiments was lightly crushed dacite pumice from the 1991 eruption of Mt. Pinatubo. Experimental charges were decompressed from 200 MPa to the final pressure and then allowed to equilibrate for one week. At pressures ≥ 50 MPa, experimental data lie close to calculated cotectic. However, at $P \leq$ 40 MPa, progression of melt compositions away from the Qz' cotectic probably reflects disequilibrium crystallization over the 7-day run duration. Redrafted from Hammer and Rutherford (2002).

al. 2003). A simple consequence of these observations is that silicate liquids with SiO_2 contents above ~75% can be generated only at low pressure, as saturation of the melt with a silica phase (typically quartz) at higher pressure limits the maximum possible SiO_2 content of the melt.

In natural silicic liquids, the relationship between SiO_2 and pressure is more complicated because of the presence of other components, notably Ca, which stabilizes plagioclase feldspar, and Mg+Fe, which stabilize amphibole and biotite. In an attempt to account for the effects of other components on SiO_2, Blundy and Cashman (2001) developed a projection scheme to convert natural glass compositions into the haplogranite ternary. Their projection scheme considered the effect of An on water-saturated phase relations in the system Qz-Ab-Or using experiments in the quinary system Qz-Ab-Or-An-H_2O and its constituent sub-systems (e.g., Johannes and Holtz 1996). Fit parameters were obtained by trial and error to obtain the following projection scheme:

$$Qz' = Qz_n \times (1 - 0.03An + 6 \times 10^{-5}[Or_n \times An] + 10^{-5}[Ab_n \times Or_n \times An]) \quad (9a)$$

$$Or' = Or_n \times (1 - 0.07An + 10^{-3}[Qz_n \times An]) \quad (9b)$$

$$Ab' = 100 - Qz' - Or' \quad (9c)$$

where the prime denotes the revised ternary co-ordinates and Qz, Ab, Or and An are the CIPW normative components. All units are expressed in weight percent, and the subscript n denotes that the normative components Qz, Ab and Or are first summed to 100%. As no attempt was made to account for the effect of normative corundum (Cor), the normalization scheme is not applicable to liquids with >1% Cor. For glasses saturated in a silica phase, the projection can be used to constrain the pressure of crystallization, while for glasses lacking a silica phase, minimum pressures and a liquid line of descent can be determined (Blundy and Cashman 2001).

How well does the barometer work? A principal limitation of this method is the lack of reliable experimental data on the haplogranite system at pressures below 50 MPa. Since Blundy and Cashman's (2001) original paper, a number of experimental studies have been published in which H_2O-saturated glasses containing feldspar and a silica phase (usually quartz) have been performed over a range of pressures, temperatures, and decompression paths (e.g., Hammer and Rutherford 2002; Couch et al. 2003b; Martel and Schmidt 2003). Only experiments performed at temperatures below 800 °C (Hammer and Rutherford 2002) contain a silica phase at pressures > 50 MPa; these experiments represent the most pertinent test of this technique. In experiments that involved rapid decompression to pressures between 200 and 50 MPa followed by equilibration times of one week, glass compositions provide a good estimate of relative pressure from the plotting position on the Qz'-Ab'-Or' plot, but only moderate accuracy in absolute pressure (e.g., Hammer and Rutherford 2003). At lower pressures, glass compositions diverge from the anticipated cotectic and move away from the Qz' apex (Fig. 18). We suspect that observed decreases in the Qz' component of the melt reflect disequilibrium crystallization because of the sluggish kinetics in these cool viscous melts (e.g., Cashman and Blundy 2000). Disequilibrium is suggested texturally by the observed isolated clusters of quartz-feldspar aggregates within apparently homogeneous matrix glass, a texture that is common in shallow plugs and domes (e.g., Cashman 1992; Hammer et al. 1999). These disequilibrium effects are probably transient, as suggested by broad beam analyzes of holocrystalline groundmass produced by shallow crystallization of feeder dikes below Unzen volcano (Noguchi et al. 2008a). These groundmass compositions plot between the 50 MPa and (approximate) 0.1 MPa cotectics in Figure 18, consistent with their crystallization of ~ 40 MPa.

We conclude that the haplogranite barometer can be used as a crude guide to pressure in the absence of other constraints. Evidence for disequilibrium precipitation of silica phases suggests that relative pressures are best established from the Ab' component of the glass (Fig. 19; see also Hammer and Rutherford 2003). The scheme would benefit from more equilibrium

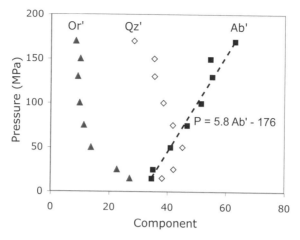

Figure 19. Individual haplogranite projection parameters (Qz', Ab', Or') plotted against pressure derived from decompression of H_2O-saturated magma of bulk composition representative of the reactive magma of the current eruption of Soufriere Hills, Montserrat (rhyolite). As seen in Figure 18, the Qz' component increases as P decreases to 50 MPa, then decreases at shallow pressure. Or' shows little change until P < 50 MPa, at which point it increases dramatically. Ab' shows the most consistent change with pressure, thereby exhibiting the greatest barometric potential for this bulk composition; regression equation shown on Figure. Data from Martel and Schmidt (2003).

experimental studies of low-P phase relations in haplogranites and natural granitic compositions, although there are substantial problems with obtaining equilibrium in such experiments (e.g., Schairer 1950; Brugger et al. 2003), problems that may also persist in nature (e.g., Cashman 1992; Cashman and Blundy 2000; Blundy and Cashman 2001). Simultaneous refinement of thermodynamic models of melts in the haplogranite system (e.g., Holland and Powell 2001) and their extrapolation to pressures below 100 MPa would greatly improve our understanding of kinetic influences on crystallization at shallow pressures.

Crystal textures

One final method of estimating pressure is to use crystal textures, where those textures can be reasonably assumed to result from decompression. In H_2O-saturated magmas, plagioclase ± orthopyroxene tend to be the dominant crystallizing phases at low pressure, because their stability is strongly dependent on pH_2O. This pressure-dependence can generate high supersaturations (effective undercooling > 200 °C) if H_2O-saturated magma is decompressed isothermally from ~ 200 MPa to near-surface conditions (Fig. 20). For this reason, phase proportions, alone, carry important information about equilibration conditions. However, as the kinetics of crystallization vary depending on decompression conditions, equilibrium may not be fully achieved, particularly at shallow pressures (e.g., Geschwind and Rutherford 1995; Hammer and Rutherford 2002; Couch et al. 2003b; Martel and Schmidt 2003; Larsen 2005; Suzuki et al. 2006). To infer magma storage and ascent conditions from textural data, therefore, we need to briefly review the kinetics of decompression-driven crystallization, as illuminated by these recent experiments (see Hammer 2008 for a more complete review).

The phase diagram shown in Figure 20 shows that the equilibrium abundance of both plagioclase and pyroxene increases dramatically with decreasing pressure, particularly at P < 50 MPa. This is demonstrated by experiments where H_2O-saturated silicic magma (typically at starting pressures of ~ 200 MPa) is decompressed rapidly and then allowed to equilibrate for at least one week (Fig. 21). Here plagioclase crystallinity increases gradually from 0 to 20%

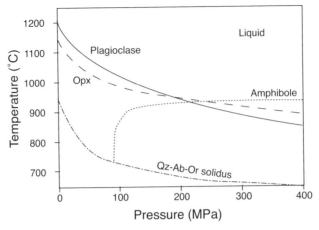

Figure 20. Phase diagram for H$_2$O-saturated ($P_{tot} = p$H$_2$O) Mount St. Helens dacite magma (based on compilation of Blundy and Cashman 2001). Note the strong sensitivity of phase boundaries to pH$_2$O. The low temperature stability of alkali feldspar + quartz have not been determined experimentally and are therefore not shown. Oxide phases are omitted for clarity; their composition and stability are sensitive to f_{O_2}.

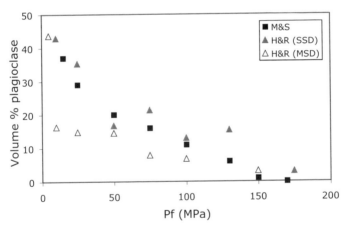

Figure 21. Variation in plagioclase abundance as a function of pressure for H$_2$O-saturated rhyolitic magmas. Filled symbols represent experiments where the magma was decompressed from 200 MPa to the final pressure and then allowed to equilibrate prior to quenching (Single Step Decompressions, SSD); open symbols represent decompression at a constant rate (1.2 MPa/hr) to the quench pressure (Multi-Step Decompressions, MSD). Plagioclase abundance increases rapidly at $P \leq 50$ MPa (also seen in natural samples, e.g., Blundy et al. 2006; Blundy and Cashman 2008). Data from Hammer and Rutherford (2002), Martel and Schmidt (2003).

between 200 and 50 MPa (0.13% MPa^{-1}), then rapidly from 20 to 45% between 50 and 5 MPa (0.56% MPa^{-1}). A similar pattern of crystallinity increase with decreasing pressure can be inferred from melt inclusion data obtained from pyroclasts erupted from Mount St. Helens in 1980 (Blundy et al. 2006) and from scoria and lapilli erupted from Mt. Etna in 2002 (Spilliaert et al. 2006). In contrast, experiments that simulate steady (~1.2 MPa hr^{-1}) decompression over ≤ 180 hours do not attain equilibrium plagioclase crystallinities, particularly at pressures between 5 and 50 MPa (Fig. 21). Thus while in theory plagioclase abundance should provide information on equilibration pressure, in practice this method can be used only where decompression conditions allow an equilibrium phase assemblage to be attained (either through temporary magma arrest at intermediate pressures for days to weeks or with magma ascent at rates $< \sim 1$ MPa hr^{-1}).

The kinetics of crystal nucleation may also provide information on equilibration pressures, especially at low pressures (high effective undercooling). Specifically, recent experiments suggest that rapid decompression to shallow pressures causes plagioclase crystallinity to increase rapidly by inducing very high rates of nucleation (Fig. 22). The plagioclase number density then stabilizes after 24-48 hours (shaded areas in Fig. 22b), while the abundance of plagioclase continues to increase slowly (crystallization becomes growth-dominated; Fig. 22a). These changes are illustrated in Figure 23a, which shows that plateau values of both number density and crystallinity increase with decreasing equilibration pressure. Because the data indicate that equilibrium crystal number densities are achieved more rapidly than equilibrium crystallinities, and as crystal number density varies by several orders of magnitude, it appears that crystal number density may provide the more reliable geobarometer (Fig. 23b).

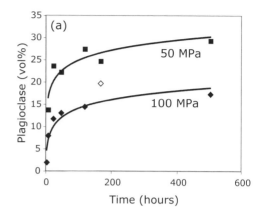

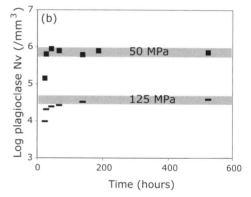

Figure 22. Temporal evolution of plagioclase (a) abundance and (b) number density resulting from instantaneous isothermal decompression from 160 MPa, 875 °C to different pressures (shown on curves). Time held at the lower pressure is given on the abscissa. Gray bars illustrate equilibrium plagioclase number achieved for each pressure. Experimental data from Couch et al. (2003b).

How well does the barometer work? Unfortunately, there are not sufficient experimental data to test the correlation shown in Figure 23b. We can, however, make some comparisons using data from natural samples. Episodic vulcanian/sub-plinian eruptions of Mount St. Helens during the summer of 1980 ejected magma that had been stored for weeks at different depths within the conduit (Cashman and McConnell 2005; Scandone et al. 2007). In most of these pyroclasts, log plagioclase number densities lie between 5.3 and 6.6 (Fig. 23c), which suggests equilibration pressures between 85 and 30 MPa using the correlation shown in Figure 23b. As these pressures are in good agreement with those estimated by Cashman and McConnell (2005) and Blundy et al. (2008) from the H$_2$O content of melt inclusions, in this example, at least, it appears that groundmass textures provide a fairly robust indicator of equilibration pressure (see also Clarke et al. 2007).

This calibration does not work under all conditions, however. First, slow and steady extrusion of dome lavas will cause crystal nucleation over a range of effective undercoolings, thus resulting number densities will not be representative of any single pressure (e.g., Couch et al. 2003b). Second, the calibration is for crystallization from melts of rhyolitic composition

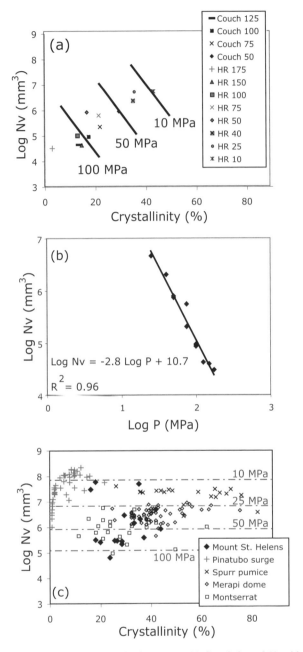

Figure 23. Number density (N_v) as a potential barometer. (a) Correlation of N_v with crystallinity for experiments of Couch et al. (2003) and Hammer and Rutherford (2002) for different equilibration pressures. Only plateau values are plotted, i.e., after N_v and crystallinity have attained equilibrium values (see Fig. 22). (b) Correlation of log number density with log equilibration pressure for experimental data in (a). (c) Variation in N_v with crystallinity for pumiceous samples of dacite from Pinatubo (Hammer et al. 1999) and Mount St. Helens (Cashman and McConnell 2005); andesite from Spurr (Gardner et al. 1998), and andesitic dome lavas from Montserrat (Clarke et al. 2007) and Merapi (Hammer et al. 2000). Dashed lines show pressures for rhyolitic compositions determined from the calibration shown in (b).

and cannot be applied directly to melts of other compositions (e.g, andesitic magmas of Merapi and Mt. Spurr, Fig. 23c). Third, silicic pyroclasts from the pre-climactic eruptions of Pinatubo (Hammer et al. 1999) and from the May 18th lateral blast of Mount St. Helens (Cashman and Hoblitt 2004) have extremely high plagioclase number densities (> 10^7 mm^{-3}) that have not been replicated in any decompression experiments (Fig. 23c). The volcanological context for these samples argues that their textures must result from rapid decompression to shallow levels (below ~15 MPa suggested by volatile contents of matrix glass and melt inclusions) followed by growth times of hours to weeks. As these pressures are consistent with those predicted from the calibration in Figure 23b, the inability to replicate these textures experimentally is puzzling, but may reflect degassing differences between melts saturated with H_2O alone and those that contain multiple volatile phases (see below).

Textures as geospeedometer? Textural characteristics, particularly formation of hornblende breakdown rims during decompression, have been used to monitor rates of magma ascent (Rutherford and Hill 1993; Rutherford and Devine 2003, 2008; Browne and Gardner 2006; Rutherford 2008) when calibrated with experimental data. As distinguishing between textures resulting from steady decompression versus temporary arrest may not be possible, application of textural geospeedometers requires field evidence for steady magma discharge (Q), such as continuous and long-lived effusion of lava domes.

Texturally, variations in effusion rate may be manifested as variations in both total crystal abundance and crystal number density (e.g., Cashman 1992; Geschwind and Rutherford 1995). Crystal abundance affects the magma rheology, which, in turn, controls the morphology of lava domes (e.g., Fink and Griffiths 1998). For example, slow effusion (< ~1-2 m^{3}s^{-1}) of H_2O-saturated andesitic magma allows complete crystallization during magma ascent to the Earth's surface and creates holocrystalline spines such as the famous spine of Mt. Pelée. In contrast, higher effusion rates (> 5-10 m^{3}s^{-1}) limit the extent of syn-ascent crystallization and produce flows with more fluid morphologies (e.g., Cashman et al. 2008). Crystal number density varies over orders of magnitude as a function of decompression rate (e.g., Couch et al. 2003) and therefore offers a potentially sensitive measure of both decompression rate (e.g., Fig. 24) and the rate of H_2O exsolution (Toramaru et al. 2008). Effusion rate control on plagioclase number density has been observed during effusive eruptions of Merapi (Hammer et al. 2000) and Unzen volcanoes (Noguchi et al. 2008b; Fig. 24b). Unfortunately, the natural data cannot

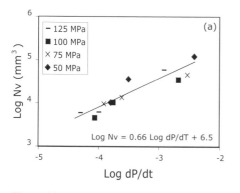
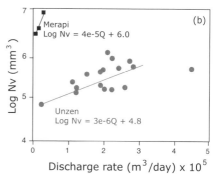

Figure 24. Textures as geospeedometers. (a) Variations in log number density (N_v) resulting from decompressions performed at different rates to the final pressures indicated in the legend. Note that in these experiments, the rate of decompression (dP/dt, in MPa/s) is more important in determining N_v than the final pressure. Data from Couch et al. (2003). (b) Relationship between plagioclase number density (per area) and magma flux (Q) at andesitic Merapi volcano, Indonesia (Hammer et al. 2000) and dacitic Unzen volcano, Japan (Noguchi et al. 2008b).

be directly compared with the experiments because the geometry of the Merapi and Unzen conduits is unknown (and thus discharge rate cannot be directly compared with decompression rate), although inferred ascent rates through the Unzen conduit of 0.008-0.05 ms^{-1} (~0.0002-0.001 MPas^{-1}; Noguchi et al. 2008b) yield an order of magnitude higher number density than predicted by the calibration shown in Figure 24a. In summary, textural geobarometers and speedometers require much more extensive calibration, but may provide extensive untapped potential for probing conditions of magma storage and ascent.

COMPARISON OF PETROLOGICAL DATA WITH MONITORING SIGNALS

A key objective of volcano petrology is to shed light on underground processes that precede and accompany eruptions and that are reflected in monitored signals, such as seismicity, gas chemistry, ground deformation and eruption dynamics. In this section we illustrate how the petrological techniques discussed above can be used, via reference to our ongoing research at Mount St. Helens. In particular, we show the utility of acquiring melt inclusion data sets that are both complete (i.e., including analyzes of major, trace, and volatile elements) and well constrained (i.e., are from samples for which both timing and eruption conditions are known).

Gas chemistry

Syn-eruptive and inter-eruptive fluxes of volcanic gas and its composition can be measured by a variety of indirect spectroscopic methods and direct analyzes of fumaroles. Both types of measurement are widely used as a volcano monitoring tool. The gases most widely measured by spectroscopic techniques are SO_2 and halogens. The more abundant volcanic gases, H_2O and CO_2, can also be determined by IR spectroscopy. However, their abundance in the atmosphere makes measurements of the volcanogenic flux subject to considerable uncertainty, although recent technical developments with unmanned vehicles have made direct CO_2 measurements in the plume possible (McGonigle et al. 2008). Alternatively, H_2O and CO_2 can be measured in enclosed volcanic craters where the beam path-length is short and atmospheric contamination kept to a minimum. To date such measurements have been confined to persistently degassing basaltic volcanoes (e.g., Burton et al. 2000; Aiuppa et al. 2007). At andesite and dacite volcanoes the only reliable means of measuring H_2O and CO_2 comes from analyzes of sampled fumarole gases. Fumarole gas chemistry can be used to determine the proportion and fugacity of the various H-C-O species, including f_{O_2}. In this section we focus on two types of volcanic gas measurements—fumarole gas determination of f_{O_2} and spectroscopic measurement of SO_2—and compare these data with petrologic information to gain insights into the degassing process.

Oxygen fugacity. Fifty high-temperature (650-830 °C) fumarole gas analyzes from Mount St. Helens in the period 1980-1982 are presented by Gerlach and Casadevall (1986). All gases are H_2O-rich with 91-99 mol% H_2O, as also observed during the 1991-1995 eruption of Unzen volcano (Ohba et al. 2008). CO_2 is the next most abundant species (1-8 mol%) with minor H_2, H_2S, CO and SO_2. Field chromatograph data were restored to equilibrium high-temperature values using thermodynamic data. The calculated f_{O_2} of the fumarole gases lies within NNO±0.3 at temperatures >700 °C but decreases to NNO-1.6 at lower temperatures (Fig. 25).

The fumarole f_{O_2} data can be compared to those calculated from coexisting Fe-Ti oxide pairs from magma erupted over the same time period (Fig. 12). Oxides from the May 18th Plinian eruption, which tapped magma stored at pressures of 160 to 320 MPa (Blundy and Cashman 2005; see above), define an f_{O_2} of NNO+0.8, which is significantly more oxidizing than the highest temperature fumarole gases. In contrast, oxide pairs from the cryptodome and post-May 18th eruptions show f_{O_2} down to NNO and slightly below, thereby overlapping the fumarole gases (Fig. 25). All of these samples contain microlites and show abundant evidence of shallow-level crystallization. Their apparent reduction is a consequence of degassing-related equilibria such as:

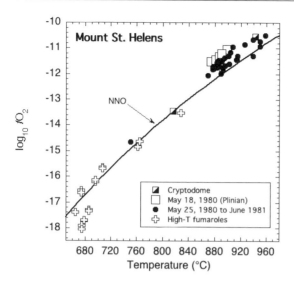

Figure 25. Comparison of temperature and oxygen fugacity (f_{O_2}) defined by oxide pairs with values from high-temperature fumaroles at Mount St. Helens. Conditions in the deep-seated magma chamber ($P \geq 160$ MPa), as sampled by the May 18th, 1980 plinian eruption, display higher f_{O_2} than the fumarole gases. In contrast, f_{O_2} from the cryptodome and post-May 18th eruptions, which show independent evidence for shallow ($P < 160$ MPa) crystallization, define lower f_{O_2} that extends down to the fumarole values. This is a consequence of degassing-related reduction, as demonstrated thermodynamically by Burgisser and Scaillet (2007). The NNO buffer is shown for reference. Oxide data from Blundy et al. (2008); fumarole data from Gerlach and Casadevall (1986).

$$Fe^{3+} \text{ (melt)} + S^{2-} \text{ (melt)} = Fe^{2+} \text{ (melt)} + S^{4+} \text{(vapor)}$$

Thermodynamic calculations by Burgisser and Scaillet (2007) demonstrate that such equilibria in the rhyolite-H-S-O system cause reduction in magmatic f_{O_2} by an amount related to the total S content of the magma, its initial f_{O_2}, and exsolved gas content, and the pressure at which gas and melt last equilibrate. Their calculations show that reduction in a rhyolitic melt with 200-250 ppm dissolved sulfur (a reasonable value for Mount St. Helens undegassed magma; Fig. 26a) at 200 MPa and NNO+1 and containing 0.1 wt% exsolved gas, is of the order 0.7 log units if equilibrium is maintained down to 10 MPa. This is in excellent agreement with the oxide data in Figure 25. The amount of reduction is only slightly greater for higher initial exsolved gas contents. These data show clearly that the f_{O_2} of fumarole gases corresponds not to the f_{O_2} of the deep-seated magma body, as tapped during the May 18th plinian phase, but to the shallow sub-volcanic conduit (Blundy et al. 2008). Clearly, as emphasized by Burgisser and Scaillet (2007), interpreting sub-volcanic magmatic processes in terms of fumarole gas chemistry must take into account gas-melt equilibrium occurring within the conduit.

Sulfur dioxide. UV spectroscopic (COSPEC) measurements of SO_2 emissions from volcanoes provide valuable constraints on the volume of magma degassed both during and between eruptions. The sulfur contents of melt inclusions can be combined with information about eruptive volume and duration to calculate the sulfur released during eruption, assuming complete degassing of the magma, to obtain the "petrological flux", F_p:

$$F_p = \frac{V}{t} \times \left(SO_2^{MI} - SO_2^{gm}\right)$$

where *MI* and *gm* denote SO_2 contents of melt inclusions and (degassed) groundmass glass, respectively, V is the erupted volume (in dense rock equivalent) and t is the eruptive duration. F_p can then be compared to the measured syn-eruptive flux. It has long been recognized that there is a substantial (one to two orders of magnitude) mismatch between the calculated petrological flux of SO_2 and the observed syn-eruptive flux for large silicic eruptions (e.g., Wallace 2001, 2005). This has led to the conclusion that large-scale eruptions involve gas loss from a much larger volume of magma than is actually erupted. Candidate processes include the addition of sulfur-rich magma to the base of the chamber shortly prior to eruption, convective

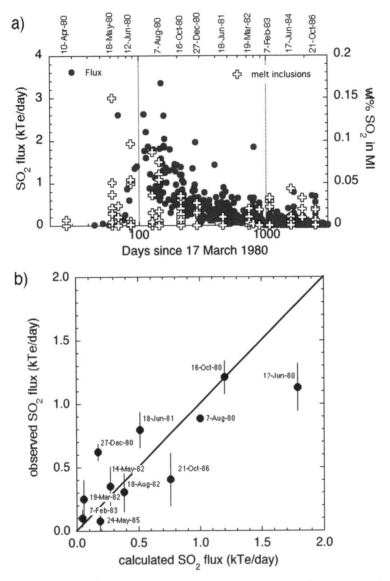

Figure 26. Observed and petrological fluxes at Mount St. Helens. (a) Temporal evolution of SO_2 in the eruption plume, as measured by airborne COSPEC (dots; McGee and Casadevall 1994), and SO_2 in melt inclusions (crosses; Blundy et al. 2008). Both parameters show a similar temporal pattern with a marked drop-off after May 18th 1980. The pre-eruptive SO_2 content of the melt is of the order 1000-1500 ppm (or 300-500 ppm S). Note the logarithmic abscissa; major eruptive events are labeled. Note also the significant inter-eruptive flux of SO_2. (b) Comparison of observed and calculated petrological SO_2 flux for post-May 18th dome-forming and dome-related eruptions. Calculated flux uses maximum SO_2 melt inclusion contents from (a) and the erupted volumes and durations of Swanson et al. (1987) and Swanson and Holcomb (1990). Observed fluxes are taken from McGee and Casadevall (1994), averaged over the duration of the eruptive period, with 1 s.d. error bars; it is not straightforward to estimate uncertainty on the petrological flux. Note the strong 1:1 correlation between the two parameters.

cycling of magma through the conduit system, or development of permeability throughout the magma reservoir during eruption.

The post-plinian eruptions of Mount St. Helens afford an excellent opportunity to compare petrological and observed fluxes because of the abundant gas flux data (McGee and Casadevall 1994) and the substantial melt inclusion database of Blundy et al. (2008). At Mount St. Helens, SO_2 fluxes generally decrease with time. This decrease is mirrored by the trend in melt inclusion SO_2 contents (Fig. 26a). We can calculate the maximum petrological flux based on the maximum SO_2 content in melt inclusions from any single eruptive phase and compare this with the observed syn-eruptive flux for 11 post-plinian episodes (Fig. 26b). The clear correlation of these data suggest that, in contrast to large plinian eruptions, smaller, dome-related events discharge a quantity of gas commensurate with the petrological data. This interpretation cannot, however, account for the inter-eruptive flux, which at times exceeds the subsequent syn-eruptive flux (Harris and Rose 1996; Fig. 26a). Either the inter-eruptive flux extracts gas from outside the small magma volumes that are subsequently erupted, or the upper reaches of the magma system are effectively replenished with gas on inter-eruptive timescales. Both interpretations require the development of permeability in dacitic magma. We consider that a better understanding of the physical mechanisms of gas transport in silicic magmas should be a key research priority.

Seismicity. A full discussion of the types and causes of sub-volcanic seismicity is beyond the scope of this review. There is, nonetheless, widespread consensus that the distribution of sub-volcanic earthquakes contains information about the underground storage and movement of magma and gas (see recent reviews of volcano seismicity by Neuberg 2000 and Chouet 2003). It is therefore useful to compare seismic information with melt inclusion data. For well-monitored volcanoes there are abundant data on the temporal evolution of seismic events that can be related to melt inclusion trapping pressures, as calculated from volatile solubility, provided suitable conversion is made between pressure and depth.

Converting pressure to depth beneath active volcanoes is not straightforward as the exact density structure beneath the volcano is rarely known. In the simplest case the system is assumed to be either lithostatic or "magmastatic", although it is likely that the dynamics of the system and the effect of the volcanic edifice on the underlying stress distribution make either assumption unreliable (e.g., Pinel and Jaupart 2000). Despite these limitations, useful inferences can be made from simple models. The difference between magma density and rock density is of the order 10%, if the presence of bubbles is ignored. We have already shown that the typical accuracy of melt inclusion pressure calculated from H_2O and CO_2 is ±22 MPa. In combination the likely uncertainty on calculated depth is of the order ±1.5 km for magma storage depths of ≤15 km.

In the case of Mount St. Helens, a density model is available from Williams et al. (1987). Assuming that the pressure in the system is lithostatic and taking into account the effect of the volcanic edifice, Blundy et al. (2008) arrive at the following conversion for total volatile pressure (P_{tot}) to depth (z) below sea-level:

$$z(km) = 0.029654 P_{tot} + 0.22704 P_{tot}^{0.5} - 2.95 \tag{10a}$$

This expression applies to the magma system prior to May 18th 1980. The removal of ~400 m of the edifice on May 18th, leads to a slightly revised version for subsequent eruptions:

$$z(km) = 0.03074 P_{tot} + 0.18334 P_{tot}^{0.5} - 2.55 \tag{10b}$$

Magma storage pressures can be calculated directly from saturation pressures for those melt inclusions where both H_2O and CO_2 data are available; where we lack CO_2 data, we calculate pH_2O from measured H_2O and then assume that $X_{H_2O} = 0.8$ in the coexisting vapor phase, leading to the correction $P_{tot} = 1.287 \, pH_2O$. Figure 27 compares the calculated melt inclusion

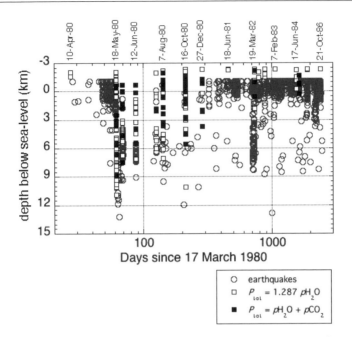

Figure 27. Comparison of melt inclusion trapping depths (squares), calculated from volatile saturation pressures (Newman and Lowenstern 2002) converted to depth (relative to sea-level) using equations (10a) and (10b), and earthquake depths (open grey circles) from the PNSN catalogue for the period 1980-1986. Note logarithmic abscissa. Selected eruptive episodes shown on top axis. The filled squares are for melt inclusions for which both H_2O and CO_2 were analyzed; open squares are for melt inclusions in which only H_2O was analyzed and XH_2O assumed to be 0.80, hence $P_{tot}=1.287 pH_2O$. Calculated depths assuming magmastatic pressure rather than lithostatic pressure would place the melt inclusion approximately 10% deeper.

trapping depths to the depths of contemporaneous earthquakes for the period 1980-1986 taken from the PNSN catalogue (*http://www.geophys.washington.edu/SEIS/PNSN/*). The close match between absolute depths and their temporal evolution strongly supports a link between seismicity and magma storage. Close inspection of Figure 27 reveals that in the months of 1980 following the plinian eruption of May 18th there were relatively few sub-volcanic earthquakes except those that followed magma withdrawal from depths of at least 6 km during individual explosive eruptions. After December 1980 shallow earthquakes appeared in the weeks prior to individual effusive eruptions, coinciding with a considerable reduction in eruptive vigor. The simplest interpretation of these data is that the magmatic system was relatively open to magma ascent during 1980, but subsequently became effectively plugged by viscous magma inhibiting further magma ascent (Scandone et al. 2007). Post-1980 melt inclusions testify to relatively shallow pre-eruptive magma storage during this time period, suggesting that magmas consistently paused, degassed, and crystallized at shallow levels prior to eruption.

Constraints on conduit dimensions. In the same way that melt inclusion SO_2 contents can be combined with eruptive volumes and durations to calculate a petrological sulfur flux, so the trapping depths of melt inclusions can be used to estimate conduit and chamber dimensions, particularly when melt inclusions are from relatively small, dome-related eruptions that derive predominantly from the conduit and uppermost reaches of the magma system. For example, if we assume that in a large population of melt inclusions the maximum and minimum trapping depths correspond to the top and bottom of the erupted magma volume, we can calculate the dimensions of a cylinder that encompasses this volume.

We have made calculations for the seven post-plinian eruptive episodes from Mount St. Helens for which we have more than 8 melt inclusions and for which the erupted volume is well constrained (e.g., Swanson and Holcomb 1990). In each case we assume that the shallowest erupted magma came from the very top of the conduit and use the melt inclusions to define the lowermost extent of the erupted magma, which we assume also to be within the conduit. In reality the shallowest erupted magma may derive from some distance below the edifice, while the deepest erupted magma may have been stored within the uppermost chamber rather than the conduit. Nonetheless, our calculations illustrate the potential of the approach.

We use both the deepest and second deepest melt inclusions for our calculations to safeguard against bias in the small datasets: for example, it is possible that some melt inclusions are a legacy from un-erupted magma of the plinian phase. We use the lithostatic model (Eqn. 10b) to calculate depths and assume that the conduit is a cylinder of radius, r, which can be obtained from:

$$r = \sqrt{\frac{V}{\pi z}} \tag{11}$$

where V is the erupted volume (in dense rock equivalent; taken from Rutherford and Hill 1993) and z is the depth of the deepest (or second deepest) melt inclusion. Note that using magma density to calculate pressure will increase z and reduce r slightly; the presence of bubbles will exacerbate this effect. However, because of the square root relationship of r to z, the consequences of adopting different density models are muted.

The calculated conduit radii are shown in Figure 28. They are in good agreement with those derived by Chadwick et al. (1988) from numerical modeling of crater-floor deformation that preceded four separate dome extrusion events in late 1981 and 1982 (Fig. 28, shaded area).

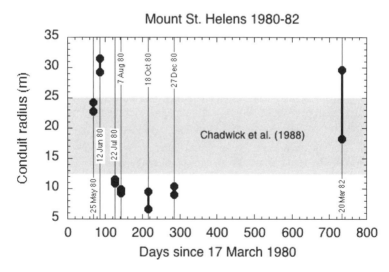

Figure 28. Temporal evolution of conduit radii for 7 post-May 18th 1980 eruptions of Mount St. Helens. Radii are calculated using equation (11). The two filled circles for each eruption denote the radii calculated from the greatest and second-greatest melt inclusion trapping depth. The shaded region is the range of conduit radii estimated by Chadwick et al. (1988) from crater-floor deformation data in 1981 and 1982. Note that changes in overburden density, for example due to magma vesiculation, cannot account for the observed variation in conduit radius; for example, a 30% reduction in overburden density result in only a 14% relative reduction in radius.

Through 1980 there is a marked decrease in r. This decrease accompanied a transition from explosive to effusive activity in late 1980 and has been interpreted to reflect plating of the shallow conduit walls with partially solidified magma (e.g., Geschwind and Rutherford 1995) or gradual collapse of the fully connected conduit (Scandone et al. 2007). After 1980, the diameter of the shallow conduit apparently stabilized until an explosive eruption in March 1982, when our data suggest an increase of the effective conduit radius, probably as the result of magma accumulation at intermediate depths combined with erosion of the walls of the shallow system during the eruption.

Melt inclusion data for the current eruption at Mount St. Helens are more limited and the maximum pressure recorded by any melt inclusion is 78 MPa (Blundy et al. 2008). Using a density of 2000 kg m^{-3} for the extruded spines and a total erupted volume of 73×10^6 km^3 we calculate $r = 78$ m for the current conduit. This is in good agreement with the value of 95 m independently derived by Iverson et al. (2006) on the basis of the extrusion flux, and substantially larger than the calculated values for 1980-1982 (Fig. 28).

Our calculations suggest that there is virtue in trying to use melt inclusion data to constrain conduit diameters for use in numerical modeling calculations, as conduit dimensions are notoriously hard to estimate by other means (see geospeedometry discussion above) and yet are a key input to any numerical model.

MAGMA ASCENT AND CRYSTALLIZATION

The previous section showed ways in which petrologic data can be integrated with volcano monitoring data to examine active systems. The power of petrologic studies, however, lies in their application to the generation and storage of magma between eruptive events. Here we illustrate ways in which intensive parameters derived from petrologic studies allow us to track the degassing and crystallization history of different magmatic systems.

Phenocryst abundance

The phenocryst abundance in volcanic samples provides information on conditions of magma storage, particularly in rapidly erupted pumice, where the rate of transport from the magma storage region to the surface is sufficiently fast to inhibit additional crystallization. In kinetic terms, the larger size and lower number density of phenocrysts reflects a dominance of crystal growth over crystal nucleation, a consequence of the relatively small undercoolings caused by either small degrees of cooling within magma storage regions or small reductions in magmatic pH_2O.

Although data on phenocryst phase proportions are surprisingly hard to find, some generalizations may be made. First, the abundance of plagioclase phenocrysts in samples of intermediate composition rocks erupted from stratovolcanoes is commonly 20-35%. This value appears to reflect magma residence times of years to centuries in subvolcanic magma storage regions. Explosively erupted samples with unusually high phenocryst contents (e.g., Pinatubo 1991; Pallister et al. 1996) typically have relatively low equilibration temperatures (< 800 °C) that suggest extensive cooling-related crystallization; basaltic magma inputs often trigger (or at least aid) eruption of these magmas. Second, the plagioclase phenocryst content of dome lavas is typically higher than that of pumice erupted explosively from the same storage region (that is, during the same eruption). This difference has been documented at Mount St. Helens (e.g., Cashman and Taggart 1983), Unzen (Nakada and Motomura 1999), Merapi (e.g., Hammer et al. 2000) and Mt. Pelée (Martel et al. 2000), and suggests that phenocrysts continue to grow during slow magma ascent from the storage region to the surface. Further evidence for continuous crystal growth during ascent is provided both by zoning patterns and compositional ranges in phenocryst rims (Cashman and Blundy 2000; Blundy and Cashman

2001; Berlo et al. 2007) and by equilibration pressures recorded in plagioclase-hosted melt inclusions (e.g., Blundy and Cashman 2005).

Melt inclusion compositions

The value of melt inclusion studies is well established for studying magmatic systems that crystallize olivine or quartz (e.g., Wallace 2005; Metrich and Wallace 2008). Here we review the use of plagioclase-hosted melt inclusions in studying conditions of magma storage and ascent. The ubiquity of plagioclase in many volcanic rocks, and the probability that plagioclase may record the entire crystallization and ascent history of the magma, make plagioclase melt inclusions an appropriate target of study. As it is unlikely that any single population of melt inclusions can be used to infer pre-eruptive magma storage conditions, we recommend using data from a large suite of melt inclusions with different textures and in different phenocryst phases to reconstruct an overall picture of the sub-volcanic system. A large sample size also allows the data to be screened for processes such as post-entrapment crystallization or diffusive volatile loss, or to subdivide the data according to host mineral and/or texture (e.g., Liu et al. 2006).

Once formed, plagioclase-hosted melt inclusions remain connected to the matrix melt via thin melt tubes (e.g., Stewart and Pearce 2004; Blundy and Cashman 2005; Humphreys et al. 2008a). Diffusion of chemical components along the melt tubes, coupled with crystallization of plagioclase around the walls of the melt inclusions, modifies the melt inclusion composition until the inclusion becomes isolated, or occluded, from the matrix melt. The extent to which diffusion maintains chemical equilibrium between inclusion and matrix melts depends on the timescale, temperature and diffusivities of the components of interest. Volatile elements (e.g., H_2O, CO_2, Li) will diffuse rapidly, while network-forming cations such as Si and Al are likely to be the rate-limiting step in controlling full chemical equilibration. Diffusive timescales based on the experimentally-determined interdiffusivity of Si between hydrous dacitic and rhyolite melts (Baker 1991) are ~7 hours for a (typical) 100 µm long melt tube at 900°C when the melt has 6 wt% H_2O and ~2 days with 3 wt% H_2O. Thus melt inclusions are likely to preserve the major element chemistry of the matrix melt on a timescale of days or less before their eventual occlusion except for the most slowly diffusing trace species, i.e., large, highly charged cations such as LREE, U and Th.

It is important to distinguish between this process and that of simple post-entrapment crystallization, which occurs under closed conditions. When the system is closed, elements that do not occur in the host crystal should increase with increasing post-entrapment crystallization, for example, Mg and Fe in plagioclase-hosted melt inclusions. If, instead, Mg and Fe show compatible behavior in both evolving plagioclase-hosted melt inclusions and in matrix glass, a ferromagnesian phase such as orthopyroxene is implicated. If no such daughter minerals are found in the melt inclusions then chemical exchange of Mg and Fe along the melt tubes must have maintained chemical coherence between the inclusion and matrix melts (e.g., Blundy et al. 2008).

The trace element chemistry of melt inclusions can be used to identify new inputs of magma into the system, provided that account is taken of the possible consequences of diffusive fractionation (Baker 2008). At Mount St. Helens, for example, REE patterns in melt inclusions from the current (2004-2008) eruption are quite distinct from those observed during the 1980-86 eruption, providing evidence for new melt being added to the sub-volcanic system since 1986 (Blundy et al. 2008). Similarly, Humphreys et al. (2008b) have described "exotic" melt inclusions from Shiveluch volcano, whose compositions suggest inputs of distinct melts into the sub-volcanic system. Melt inclusions therefore provide information on open-system processes for which there is abundant isotopic evidence from bulk rocks, single melt inclusions and individual zoned phenocrysts (Ramos and Tepley 2008).

Modeling volatile elements

Melt inclusions are most commonly used to interpret degassing paths followed by ascending and cooling magma. The systematics of H_2O and CO_2 in magmatic systems are discussed more fully by Moore (2008). Here we explore H_2O-CO_2 trajectories that would be followed by melts having different ascent and/or crystallization histories, as indicated by the textural data described above. We consider 5 distinct scenarios (Fig. 29a):

1. Decompression without crystallization
2. Rapid decompression crystallization
3. Slow decompression crystallization
4. Isobaric vapor-saturated crystallization
5. Isobaric crystallization initially vapor under-saturated

Our calculations consider only closed systems, where vapor bubbles and melt maintain chemical equilibrium and no vapor is lost from the system. Volatile solubility is taken from the model of Liu et al. (2005). Our model melt is initially at 900 °C and 325 MPa. For scenarios 1-4 the melt has initial contents of H_2O and CO_2 of 6.44 wt% and 500 ppm, respectively, consistent with an equilibrium vapor phase composition of 80 mol% H_2O. For the vapor-undersaturated scenario (5) the initial values are 4.34 wt% and 400 ppm. Isobaric crystallization is driven by cooling and occurs at 0.33 wt% crystals per °C for both vapor-saturated and undersaturated scenarios. Decompression crystallization occurs isothermally at rates of 0.2 wt% crystals per MPa ("rapid crystallization," scenario 2) and 0.1 wt% MPa^{-1} ("slow crystallization," scenario 3). Note that these rates bracket the value of 0.13% crystals per MPa derived from decompression experiments conducted between 200 and 50 MPa (e.g., Couch et al. 2003b). No account is taken of the release of latent heat of crystallization, which is of the order 2.5 °C for each wt% crystallized (Blundy et al. 2006).

Decompression without crystallization (scenario 1) leads to loss of both H_2O and CO_2 to the vapor phase (Fig. 29b). The greater solubility of H_2O compared to CO_2 means that the vapor becomes increasingly H_2O-rich as decompression occurs. The initial vapor comprises 80 mol% H_2O, while the final vapor is over 99.5 mol% H_2O, manifested as a sharp drop in CO_2 in the melt at near constant H_2O. Although not shown, open system decompression results in an even sharper fall in CO_2 at near constant H_2O (e.g., Papale 2005).

Decompression crystallization follows similar trajectories to decompression alone although there is a slight increase in H_2O contents during the earliest stages of crystallization. As both H_2O and CO_2 are considered perfectly incompatible in the crystallizing assemblage, the effect of crystallization is to increase both components in the melt. Because CO_2 is the less soluble it is preferentially driven off into the vapor phase, enriching the melt slightly in H_2O. The enrichment is greater in the case of fast decompression crystallization (scenario 2) than slow decompression crystallization (scenario 3); in both scenarios, decompression crystallization shows some similarities to open system behavior described above, although the increase in H_2O during the early stages of crystallization is quite distinct (Fig. 29d).

Isobaric vapor-saturated crystallization (scenario 4) increases H_2O in the melt throughout crystallization (see also Liu et al. 2006, Fig. 5b), due again to the preferential loss of CO_2 to the vapor and the (slight) increase in solubility with decreasing temperature; this is in contrast to the case for CO_2-free vapor-saturated systems, when H_2O contents remain approximately constant with crystallization (Blundy and Cashman 2005). The melt follows an isobar in CO_2-H_2O space (Fig. 29c), albeit polythermally, resulting in a trend distinct from that of decompression crystallization. CO_2 decreases with increasing crystallization (Fig. 29d), hence the previous assertion that the only unequivocal evidence for vapor saturation comes from a negative correlation between CO_2 and some index of crystallization (Wallace et al. 1999).

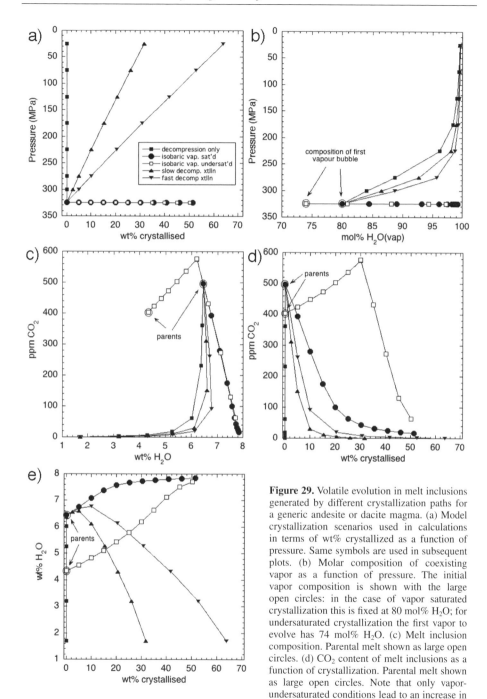

Figure 29. Volatile evolution in melt inclusions generated by different crystallization paths for a generic andesite or dacite magma. (a) Model crystallization scenarios used in calculations in terms of wt% crystallized as a function of pressure. Same symbols are used in subsequent plots. (b) Molar composition of coexisting vapor as a function of pressure. The initial vapor composition is shown with the large open circles: in the case of vapor saturated crystallization this is fixed at 80 mol% H_2O; for undersaturated crystallization the first vapor to evolve has 74 mol% H_2O. (c) Melt inclusion composition. Parental melt shown as large open circles. (d) CO_2 content of melt inclusions as a function of crystallization. Parental melt shown as large open circles. Note that only vapor-undersaturated conditions lead to an increase in CO_2 with crystallization, up to the point of saturation. (e) H_2O content of melt inclusions as a function of crystallization. Parental melt shown as large open circles. Note that even vapor-saturated conditions can lead to a small initial increase in H_2O with crystallization, up to the point of saturation. Isobaric conditions lead to a continual increase in H_2O during crystallization.

Finally, isobaric vapor-undersaturated crystallization (scenario 5) follows a unique trajectory on all diagrams in Figure 29. Initially crystallization increases both H_2O and CO_2 in the melt until vapor saturation is attained at ~30% crystallinity. At that point the equilibrium vapor phase is 74 mol% H_2O. Once vapor saturation occurs, CO_2 in the melt begins to fall rapidly. H_2O increases throughout crystallization. In Figure 29c, the vapor-undersaturated period of crystallization is marked by a positive trajectory, which turns sharply to follow the isobar (as per scenario 4) once saturation is reached.

The five scenarios above summarize simple, end-member degassing situations, as commonly invoked in the interpretation of melt inclusion H_2O and CO_2 data. However, because real magma chambers are thermally and spatially variable, we would expect different portions of the chamber to experience different degassing histories. For a large suite of melt inclusions it is unlikely, therefore, that any single degassing trajectory will be defined. In fact, one could envisage all five of the above scenarios occurring in different parts of a single large chamber. Moreover, magma recharge, convection, sidewall crystallization, eruptions from an overlying magma reservoir, volatile fluxing, and magma chamber over-pressurization can all serve to complicate the degassing signal. In the following section we consider how more realistic magma systems are likely to behave.

Combining degassing and crystallization

To illustrate the above complexities we have created a simple model magma chamber in which there is a vertical, lithostatic pressure gradient from the floor (325 MPa) to the roof (125 MPa) and a lateral thermal gradient from the walls (750 °C) to the interior (925 °C) that drives sidewall crystallization (Fig. 30). Magma entering the chamber at its base contains 5 wt% H_2O and 500 ppm CO_2, but is volatile-undersaturated. At 925 °C it is also crystal-free. The solubility of H_2O and CO_2 are taken from the expressions of Liu et al. (2005), and the weight fraction crystallinity (X_c) of the magma is taken to be a simple function of both pressure and temperature:

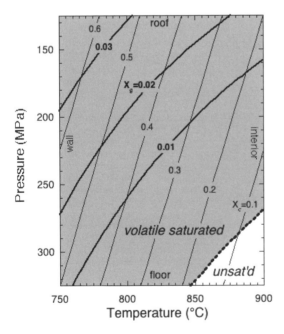

Figure 30. Model magma chamber in terms of pressure and temperature, contoured for crystal weight fraction (X_c) and exsolved gas weight fraction (X_g). The chamber is radially symmetrical about the right-hand axis. The magma initially enters the chamber at 900 °C under vapor-undersaturated conditions (lower right-hand corner) crystallizing in response to both decompression (towards the roof) and cooling (towards the wall). The relationship between P, T and X_c is given in Equation (10); the H_2O and CO_2 solubility is from Liu et al. (2005). The P-T conditions at which vapor-saturation is attained are shown with the bold dotted line. The magma is vapor saturated throughout the shaded portion of the figure.

$$X_c = \frac{(P-325)}{1000} + \frac{(900-T)}{300} \qquad (12)$$

where P is pressure in MPa and T is temperature in °C. This relationship is based crudely on experimental studies on andesite and dacite magmas. We assume that the magma chamber behaves as a closed system in which no gas is lost and we have not modeled the effects of vapor composition on liquidus temperatures (and hence crystallinity), or of latent heat release on magma temperature. Despite these simplifications, our model chamber shows many of the characteristics to be expected in natural systems, such as vertical and lateral gradients in crystallinity and bubble content (Fig. 30). Of course, magma chambers with different aspect ratios will differ in detail from the model in Figure 30, although the overall spatial variation of parameters will be similar. Note also that in the following discussion we ignore the effects of magma chamber convection on the thermal structure.

As formulated, magma attains volatile saturation at 270 MPa within the interior of the chamber, but at slightly higher pressure towards the cooler chamber walls. This divides the chamber into a thin lower layer (or lens) that is vapor-undersaturated and a dominant upper layer that is vapor-saturated (shaded region of Fig. 30). Decreasing pressure within the chamber leads to crystallization, as does cooling of the magma towards the walls. We have calculated the evolution of the melt composition along a series of isobaric sections (Fig. 31), which covers a substantial portion of the CO_2-H_2O-X_c plots (Fig. 30). Importantly, the entire population of equilibrium melts residing within the chamber (shown as shaded fields in Fig. 31) does not follow any of the simple end-member scenarios illustrated above.

Interpreting magmatic systems

We can use the insights gained from our model to interpret the behavior of the limited number of sub-volcanic systems for which substantial datasets of melt inclusion CO_2, H_2O and incompatible element analyzes exist. In particular, we compare the crystallization and degassing behavior of large silicic caldera-forming systems (Bishop Tuff, California; Oruanui, New Zealand) with that of smaller, less evolved and more frequently active systems (Mt. Etna, Italy; Popocatépetl, Mexico). For each example we have calculated the crystallinity (X_c) from incompatible elements measured within the inclusions by assuming perfect incompatibility (i.e., $D = 0$) and adopting the lowest incompatible element content of any melt inclusion as the parental melt. Ideally, these calculated crystallinities would be compared with actual crystal size distributions, but to date, no such complete study exists.

Large silicic systems. Although H_2O and CO_2 contents of rhyolitic ignimbrites produced during caldera-forming eruptions of the Bishop Tuff and Oruanui record evidence of both isobaric and polybaric crystallization (Fig. 32), the dominant signature is that of isobaric volatile-saturated crystallization. Evidence for isobaric crystallization in the Oruanui system includes the narrow (120-150 MPa) pressure range recorded by the melt inclusions (Fig. 32a) and the remarkable constancy of H_2O over a wide range of crystallinities (Fig. 32b). The low CO_2 contents of all Oruanui inclusions (Fig. 32c) suggest either some open system degassing early in the evolution of the system or a low CO_2 parent magma. Isobaric crystallization at about 200 MPa is also indicated by the range of crystallinities at near-constant H_2O shown by the early and middle Bishop Tuff samples (Fig. 32b). Late Bishop Tuff samples, however, show a pronounced decrease in CO_2 with increasing crystallinity (Fig. 32c) that provides evidence for vapor-saturation throughout the system (Wallace et al. 1999) and records crystallization over a wide pressure range (100-300 MPa). The nature of crystallization in the late Bishop Tuff has been the subject of detailed studies, some of which have suggested settling of shallow-formed crystals followed by growth at deeper levels (Anderson et al. 2000; Peppard et al. 2001). Alternatively, explaining this trend by the isobaric sections illustrated in Figure 31 would require crystallization across a large temperature interval (which is not observed).

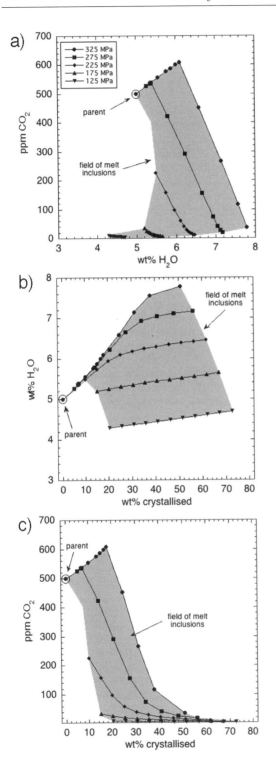

Figure 31. Calculated evolution of melt inclusions in the model magma chamber of Fig. 30 in terms of (a) CO_2 and H_2O contents; (b) H_2O as a function of wt% crystallized and (c) CO_2 as a function of wt% crystallized. The solid lines denote different isobaric section through the chamber and the large filled circle is the vapor undersaturated parental magma. The shaded region denotes the field of melt inclusions likely to be tapped during an eruption of the chamber. Note that they do not match any of the simple scenarios illustrated in Figure 29.

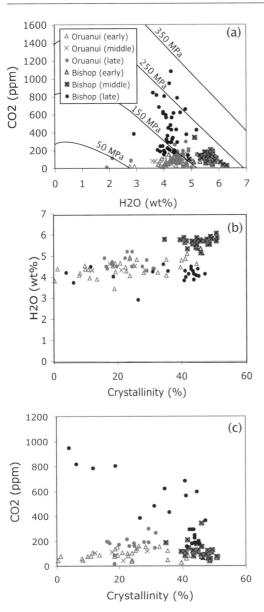

Figure 32. Volatile contents and crystallinity of early, middle and late Oruanui (New Zealand) and Bishop Tuff (USA) magmas as recorded by melt inclusions in terms of: (a) H_2O-CO_2, with isobars calculated at 760 °C from Liu et al. (2005); (b) H_2O versus wt% crystallized; (c) CO_2 versus wt% crystallized. Oruanui data are taken from Liu et al. (2006); crystallinity is calculated from Rb content of melt inclusions relative to parent Rb content of 89 ppm assuming $D_{Rb} = 0$. Bishop data are taken from Wallace et al. (1999); crystallinity is calculated from Th content of melt inclusions relative to parent Th content of 12.2 ppm assuming $D_{Th} = 0$.

Mt. Etna. Melt inclusions in samples from the 2001-2002 eruptions of Mt. Etna preserve a very different crystallization and degassing story to that seen in large rhyolitic systems. As indicated by contemporaneous monitoring data (e.g., Andronico et al. 2005), the basaltic magma that fed these eruptions ascended rapidly from intermediate crustal levels, particularly in 2001. This ascent path is nicely recorded by the volatiles trapped in melt inclusions and melt embayments, which indicate decompression and degassing from almost 500 MPa to the surface (Fig. 33a). Crystallization accompanying degassing is evinced by strong depletions in both H_2O (Fig. 33b) and CO_2 (Fig. 33c) with increasing crystallinity. The data also retain interesting details about the decompression path, as highlighted by the spread of H_2O and CO_2 data at pressures at and somewhat above 200 MPa, particularly in the 2002 samples (Fig. 33a). These data have been interpreted to indicate temporary magma storage at about 5 km, which appears to coincide with a stratigraphic boundary between underlying flysch and carbonate units (Spilliaert et al. 2006). The range in volatile compositions at this level is inferred to reflect variable addition of CO_2-rich gas to the ponded magma. An interesting consequence of this fluxing would have been dehydration of the magma (by addition of CO_2). Reduction of pH_2O, in turn, would promote crystallization, as seen in the 'blurring' of the CO_2-crystallinity data for the 2002 samples as compared to the simpler story of rapid CO_2 depletion in 2001 (Fig. 33c).

Small intermediate systems. There are only a few andesitic (and dacitic systems) for which complete (or near-complete) melt inclusion data sets have been published. Here we examine data from Popocatépetl (Mexico) and more limited data

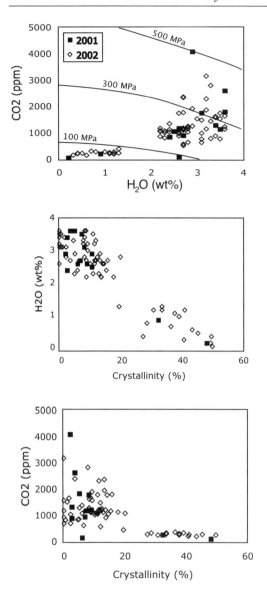

Figure 33. Volatile contents and crystallinity of basaltic tephra erupted from Mt. Etna (Italy) in 2001 and 2002 as recorded by melt inclusions in terms of: (a) H_2O-CO_2, with isobars from Metrich et al. (2004); (b) CO_2 versus wt% crystallized; (c) H_2O versus wt% crystallized. Data are taken from Metrich et al. (2004) and Spilliaert et al. (2006); crystallinity calculated from K_2O content of the melt inclusions relative to the parent concentration (from the appropriate bulk rock composition) assuming $D_K = 0$.

from Augustine (Alaska). Popocatépetl has been erupting intermittently since 1994, during which time it has emitted substantially more gas than magma (e.g., Delgados-Granados et al. 2001, 2008). The melt inclusion data are interesting in that they record of wide range of entrapment pressures (50-300 MPa) at near-constant H_2O (~ 2 wt%; Fig. 34a). The result is crystallization at constant H_2O (Fig. 34b) but a wide spread (and no obvious pattern) of CO_2 versus crystallinity (Atlas et al. 2006). These data are particularly intriguing when compared to those from Augustine, which have low CO_2 and well-developed degassing-crystallization trends (Fig. 34a,b). They also contrast with the strong degassing-crystallization trend preserved in H_2O-crystallinity data from Mount St. Helens (Blundy and Cashman 2005, 2008; Fig. 34b). We suspect that the range in melt inclusion compositions in Popocatépetl magma result from the strong gas fluxing observed in this system. In this way, Popocatépetl is similar to Etna, except that the overall crystallinity is higher (and average magma flux is lower). Like Etna, however, both the degassing and gas fluxing trends shown in Figure 34 record crystallization over a wide range in pressure that seems characteristic of small, rapidly replenished arc magma systems.

Implications for magma accumulation and eruption. If the conclusions about crystallization and degassing of magmatic systems drawn from the examples above can be generalized, they might provide important insights into conditions related to recharge and eruption patterns in different tectonic environments.

Bishop Tuff and Oruanui are large bodies of limited vertical extent where crystallization appears to be dominantly isobaric, volatile-saturated, and driven by cooling. They therefore resemble a version of our simple magma chamber (Fig. 30) with a high width:height ratio. The large size of these magma bodies requires that they form over long time

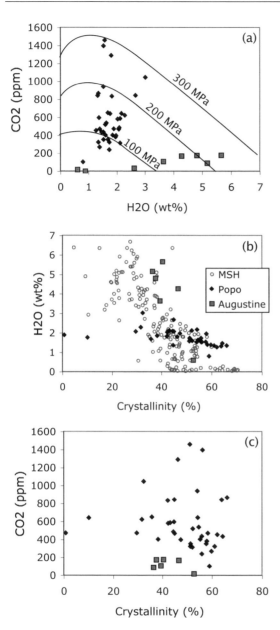

Figure 34. Volatile contents and crystallinity of Augustine (Alaska)1986 and Popocatépetl (Mexico)1994-1998 magma as recorded by melt inclusions in terms of: (a) H_2O-CO_2, with isobars calculated at 1050 °C from Liu et al. (2005); (b) H_2O versus wt% crystallized; (c) CO_2 versus wt% crystallized. Data are taken from Roman et al. (2006) and Atlas et al. (2006). Crystallinity is calculated from K_2O content of melt inclusions relative to parent K_2O content assuming $D_K = 0$. In (b), data from Mount St. Helens are shown for comparison (from Blundy and Cashman 2005).

periods, probably by periodic replenishment from below (e.g., Annen et al. 2006). One can envision large laterally extensive sheet-like bodies, similar to many granite batholiths, whose temperature wax and wane over time in response to episodic recharge events. Tectonic environments that promote large, shallow, periodically replenished magma bodies tend to be extensional. Eruptions are infrequent and involve wholesale roof collapse; it is not clear whether eruptions are triggered by external factors (e.g., basaltic inputs, roof collapse) or internal factors (e.g., overpressurization by gas accumulation during crystallization).

Etna, Augustine, Popocatépetl and Mount St. Helens are stratovolcanoes with small, vertically extensive magmatic systems over which substantial crystallization occurs due to the pressure drop alone. They therefore resemble our model magma chamber with high height:width ratio. While the pipe- or dyke-like shape of the magma reservoirs argues for crystallization controlled primarily by decompression, one might anticipate additional crystallization by cooling along the vertically extensive walls. However, not only does the steady fall in H_2O with increasing crystallinity argue against much cooling during crystallization, but combined melt inclusion, crystallinity and plagioclase-melt geothermometry data from Mount St. Helens show that magma undergoing decompression *heats up* due to the release of latent heat of crystallization (Blundy et al. 2006). Stratovolcanoes are common in arc environments. They tend to be characterized by small frequent eruptions that are often (although not always) triggered by mafic recharge events that mix incompletely with reservoir magma, as evidenced by the common presence of mafic enclaves (e.g., Montserrat, Unzen, Pinatubo). Absence of a cooling signature may derive from the relatively short repose intervals that characterize such eruptions.

SUMMARY

In summary, petrologic studies are so integral to the investigation of magmatic systems that they are often taken for granted by those who apply new geophysical and geochemical monitoring techniques to active volcanoes. Here we have shown ways in which the traditional strengths of petrologic studies of magma generation and storage can be integrated with real-time monitoring data to connect direct investigations of magmatic processes (particularly volatile exsolution and crystallization) with manifestations of those processes that are used in monitoring, such as seismicity and gas emissions.

"Traditional" petrologic methods used to study magma storage conditions focus on constraining intensive parameters such as temperature, pressure, oxygen fugacity and dissolved volatile contents. Our review of commonly-used mineral-mineral geothermometers shows that they are generally robust, with Fe-Ti oxides providing the best constraints. In contrast, geobarometry is notoriously difficult, particularly for upper crustal pressures. For this reason, we review alternative methods of estimating equilibration pressures, including phase equilibria experiments, volatile saturation models, matrix glass compositions and crystal textures. As no one method is perfect, we recommend using as many different methods as permitted by the sample set. Mineral-melt thermobarometry (Putirka 2008) offers significant advances in this regard.

We confine much of our comparison of petrologic and monitoring data to the well-characterized eruption of Mount St. Helens in the 1980s. This is not to suggest that Mount St. Helens should, in any way, be considered a reference standard, but simply because both the nature of the eruption (with styles ranging from plinian to sub-plinian, vulcanian, and effusive) and the wealth of monitoring and petrologic data allow us to compare these datasets for different eruptive conditions. We find an impressive correlation between the petrology and the monitoring data, which provides some interesting insights into the chemical and physical processes occurring during the 1980-1986 eruption. Specifically, comparison of f_{O_2} recorded in Fe-Ti oxides and crater fumaroles demonstrates that shallow degassing causes a reduction in magmatic f_{O_2}, as predicted by Burgisser and Scaillet (2007). Comparison of petrologic and measured SO_2 emissions shows these estimates to be similar for eruptive episodes, but implicates non-erupted magma in generating inter-eruptive gas fluxes, thereby raising questions about the physical mechanisms of gas transport in active, but non-erupting, magmatic systems. Comparison of petrologic estimates of magma storage pressures with seismic data provides important insights into the development of subvolcanic regions of magma storage and transport and raises the tantalizing possibility that the geometry of subvolcanic regions may control the style of ensuing eruptions. Where large numbers of melt inclusions from a single eruption are analyzed for H_2O and CO_2, there exists the possibility of constraining conduit diameter, an essential, but poorly-constrained, parameter in dynamical modeling. Improved geospeedometry techniques (either chemical or textural) would also improve geometric constraints.

In stepping back to examine the evolution of magma storage regions between eruptions, we demonstrate the importance of fully characterizing the phase proportions and composition of individual samples, as well as the full range of melt inclusion compositions contained within those phases. Simple models of degassing and crystallization illustrate the potential for comprehensive melt inclusion studies to elucidate magma decompression and crystallization histories, particularly when combined with trace element and isotopic analyzes of the same inclusions. Application of our simple models to examples of both large silicic and small intermediate systems demonstrates the profound difference in these systems, from their geometry and crystallization history to implied conditions of magma input and eruptive frequency. Generalization of the conclusions drawn from examination of the few systems for which complete data sets are available will require generation of similar (or improved, particularly by addition of textural data) datasets for more magmatic systems. As the tools for such analysis are now readily available, we hope this review will inspire collaborative efforts to fully characterize many different types of magmatic systems, and to integrate those observations with studies in physical volcanology, numerical modeling and volcano monitoring.

ACKNOWLEDGMENTS

JDB thanks NERC for the award of a Senior Research Fellowship (NER/K/S/2001/00771); KVC thanks NSF for funding support through award EAR0510437. We are grateful to M. Mangan and M. Humphreys for providing timely and constructive reviews of this manuscript, to B. Scaillet, J. Beard, M. Ghiorso, K. Klimm, V. Martin, D. Morgan, D. Muir, C. Ginibre, E. Gottesfeld, J. Riker, M. Folley and M. Humphreys for providing unpublished data or images, J. Craven, R. Hinton and S, Kearns for critical reading of the microbeam analytical sections and K. Putirka for his comments and editorial patience.

REFERENCES

Aiuppa A, Moretti R, Federico C, Giudice G, Gurrieri S, Liuzzo M, Papale P, Shinohara H, Valenza M (2007) Forecasting Etna eruptions by real-time observation of volcanic gas composition. Geology 35:1115-1118

Alonso-Perez R, Müntener O, Ulmer P (2008) Igneous garnet and amphibole fractionation in the roots of island arcs: experimental constraints on H_2O undersaturated andesitic liquids. Contrib Mineral Petrol (in press)

Andersen DJ, Lindsley DH (1988) Internally-consistent solution models for Fe-Ti-Mg-Mn oxides – Fe-Ti oxides. Am Mineral 73:714-726

Andersen DJ, Lindsley DH, Davidson PM (1993) QUILF: a Pascal program to assess equilibria among Fe-Mg-Mn-Ti oxides, pyroxenes, olivine, and quartz. Comp Geosci 19:1333-1350

Anderson AT Jr, Davis AM, Lu FQ (2000) Evolution of Bishop Tuff rhyolitic magma based on melt and magnetite inclusions and zoned phenocrysts. J Petrol 41:449-473

Anderson JL, Barth AP, Wooden JL, Mazdab F (2008) Thermometers and thermobarometers in granitic systems. Rev Mineral Geochem 69:121-142

Andronico D, Branca S, Calvari S, Burton M, Caltabiano T, Corsaro RA, Del Carlo P Garfì G, Lodato L, Miraglia L, Murè F, Neri M, Pecora E, Pompilio M, Salerno G, Spampinato L (2005) A multi-disciplinary study of the 2002-03 Etna eruption: Insights into a complex plumbing system. Bull Volcanol 67:314-330

Annen C, Blundy JD, Sparks RSJ (2006) The genesis of intermediate and silicic magmas in deep crustal hot zones. J Petrol 47:505-539

Arce JL, Macias JL, Gardner JE, Layer PW (2006) A 2.5 ka History of Dacitic Magmatism at Nevado de Toluca, Mexico: Petrological, $^{40}Ar/^{39}Ar$ Dating, and Experimental Constraints on Petrogenesis. J Petrol 47:457-479

Armienti P (2008) Decryption of igneous rock textures: crystal size distribution tools. Rev Mineral Geochem 69:623-649

Armienti P, Pareschi MT, Innocenti F, Pompilio M (1994) Effects of magma storage and ascent on the kinetics of crystal growth. Contrib Mineral Petrol 115:402-414

Atlas ZD, Dixon JE, Sen G, Finny M, Martin-Del Pozzo AL (2006) Melt inclusions from Volcan Popocate´petl and Volca´n de Colima, Mexico: Melt evolution due to vapor-saturated crystallization during ascent. J Volcanol Geotherm Res 153:221-240

Bachmann O, Dungan M (2002) Temperature-induced Al-zoning in hornblendes of the Fish Canyon magma, Colorado. Am Mineral 87:1062-1076

Bacon CR, Hirschmann MM (1988) Mg/Mn partitioning as a test for equilibrium between coexisting Fe-Ti oxides. Am Mineral 73:57-61

Baker DR (1991) Interdiffusion of hydrous dacitic and rhyolitic melts and the efficacy of rhyolite contamination of dacitic enclaves. Contrib Mineral Petrol 106:462-473

Baker DR (2008) The fidelity of melt inclusions as records of melt composition. Contrib Mineral Petrol doi10.1007/s00410-008-0291-3

Barclay J, Rutherford MJ, Carroll MR, Murphy MD, Devine JD, Gardner J, Sparks RSJ (1998) Experimental phase equilibria constraints on pre-eruptive storage conditions of the Soufriere Hills magma. Geophys Res Lett 25:3437-3440

Barclay J, Carmichael ISE (2004) A Hornblende Basalt from Western Mexico: Water-saturated Phase Relations Constrain a Pressure-Temperature Window of Eruptibility. J Petrol 45:485-506

Beard JS, Lofgren GE (1991) Dehydration melting and water-saturated melting of basaltic and andesitic greenstones and amphibolites at 1, 3 and 6.9 kb. J Petrol 32:365-401

Berlo K, Blundy J, Turner S, Hawkesworth C (2007) Textural and chemical variation in plagioclase phenocrysts from the 1980 eruptions of Mount St. Helens, USA. Contrib Mineral Petrol 154:291-308

Bindeman IN (2005) Fragmentation phenomena in populations of magmatic crystals. Am Mineral 90:1801-1815

Blatter DL, Carmichael ISE (1998) Plagioclase-free andesites from Zitacuro (Michoacan), Mexico: petrology and experimental constraints. Contrib Mineral Petrol 132: 121-138

Blatter DL, Carmichael ISE (2001) Hydrous phase equilibria of a Mexican high-silica andesite: A candidate for a mantle origin? Geochim Cosmochim Acta 65:4043–4065

Blundy J, Cashman K (2001) Magma ascent and crystallization at Mount St. Helens, 1980-1986. Contrib Mineral Petrol 140:631-650

Blundy J, Cashman KV (2005) Rapid decompression crystallization recorded by melt inclusions from Mount St. Helens volcano. Geology 33:793-796

Blundy J, Cashman K, Humphreys M (2006) Magma heating by decompression-driven crystallization beneath andesite volcanoes. Nature 443:76-80

Blundy J, Cashman K, Berlo K (2008) Evolving magma storage conditions beneath Mount St. Helens inferred from chemical variations in melt inclusions from the 1980-1986 and current eruptions. Sherrod DR, Scott WE, Stauffer PH eds, A volcano rekindled; the renewed eruption of Mount St. Helens, 2004–2006: U.S. Geological Survey Professional Paper 1750 (in press)

Bogaerts M, Scaillet B, van der Auwera J (2006) Phase equilibria of the Lyngdal granodiorite (Norway): Implications for the origin of metaluminous ferroan granitoids. J Petrol 47:2405-2431

Brey GP, Köhler T (1990) Geothermobarometry in Four-phase Lherzolites II. New Thermobarometers, and Practical Assessment of Existing Thermobarometers. J Petrol 31:1353-1378

Browne BL, Gardner JE (2006) The influence of magma ascent path on the texture, mineralogy, and formation of hornblende reaction rims. Earth Planet Sci Lett 246: 161-176

Brugger CR, Johnston AD, Cashman KV (2003) Phase equilibria in silicic systems at one-atmosphere pressure Contrib Mineral Petrol 146:356-369

Burgisser A, Scaillet B (2007) Redox evolution of a degassing magma rising to the surface. Nature 445:194-197

Burton MR, Oppenheimer C, Horrock LA, Francis PW (2000) Remote sensing of CO_2 and H_2O emission rates from Masaya volcano, Nicaragua. Geology 28:915-918

Carey S, Sigurdsson H (1985) The May 18, 1980 eruption of Mount St, Helens. 2. Modeling of dynamics of the plinian phase. J Geophys Res 90:2948-2958

Carlson WD (2006) Three-dimensional imaging of earth and planetary materials. Earth Planet Sci Lett 249:133-147

Cashman KV (1990) Textural constraints on the kinetics of crystallization of igneous rocks. Rev Mineral 24:259-314

Cashman KV (1992) Groundmass crystallization of Mount St. Helens dacite, 1980-1986: A tool for interpreting shallow magmatic processes. Contrib Mineral Petrol 109:431-449

Cashman K, Blundy J (2000) Degassing and crystallization of ascending andesite Phil Trans Roy Soc 358:1487-1513

Cashman KV, Hoblitt RP (2004) Magmatic precursors to the May 18, 1980 eruption of Mount St. Helens, WA. Geology 32:141-144

Cashman KV, Marsh BD (1988) Crystal size distribution (CSD) in rocks and the kinetics and dynamics of crystallization II. Makaopuhi lava lake. Contrib Mineral Petrol 99:292-305

Cashman KV, McConnell S (2005) Transitions from explosive to effusive activity – the summer 1980 eruptions of Mount St. Helens. Bull Volcanol 68:57-75

Cashman KV, Taggart JE (1983) Petrologic monitoring of 1981-1982 eruptive products from Mount St. Helens, Washington. Science 221:1385-1387

Cashman KV, Thornber CR, Kauahikaua JP (1999) Cooling and crystallization of lava in open channels, and the transition of pahoehoe lava to `a`a. Bull Volcanol 61:306-323

Cashman KV, Thornber CR, Pallister JS (2008) From dome to dust: shallow crystallization and fragmentation of conduit magma during the 2004-2006 dome extrusion of Mount St. Helens. *In* Sherrod DR, Scott WE, Stauffer PH, eds, A volcano rekindled: the renewed eruption of Mount St. Helens, 2004–2006. US Geol Surv Prof Pap 1750 (in press)

Castro JM, Cashman KV, Manga M (2003) A technique for measuring 3D crystal-size distributions of prismatic microlites in obsidian. Am Mineral 88:1230-1240

Castro JM, Manga M, Cashman KV (2002) Dynamics of obsidian flows inferred from microstructures: insights from microlite preferred orientations. Earth Planet Sci Lett 199:211-226

Chayes F (1956) Petrographic Modal Analysis. John Wiley & Sons, New York

Chou I M (1987) Oxygen buffer and hydrogen sensor techniques at elevated pressures and temperatures. *In*: Hydrothermal Experimental Techniques. Barnes HL, Ulme GC (eds) John Wiley, New York, p 69-99

Chouet BA (2003) Volcano seismology. Pure Appl Geophys 160:739-788

Clarke AB, Stephens S, Teasdale R, Sparks RSJ, Diller K (2007) Petrologic constraints on the decompression history of magma prior to Vulcanian explosions at the Souffriere Hills volcano, Montserrat. J Volc Geotherm Res 161:261-274

Coombs ML, Gardner JE (2001) Shallow-storage conditions for the rhyolite of the 1912 eruption at Novarupta, Alaska. Geology 29:775-778

Coombs ML, Eichelberger JC, Rutherford MJ (2000) Magma storage conditions for the 1953-1974 eruption of Southwest Trident volcano, Katmai National Park, Alaska. Contrib Mineral Petrol 140:99-118

Coombs ML, Eichelberger JC, Rutherford MJ (2002) Experimental and textural constraints on mafic enclave formation in volcanic rocks. J Volc Geotherm Res 119:125-144

Cooper KM, Reid MR (2003) Re-examination of crystal ages in recent Mount St. Helens lavas: Implications for magma reservoir processes. *Earth Planet Sci Lett* 213:149-167

Costa F, Chakraborty S, Dohmen R (2003) Diffusion coupling between trace and major elements and a model for calculation of magma residence times using plagioclase. Geochim Cosmochim Acta 67:2189-2200

Costa F, Chakraborty S (2004) Decadal time gaps between mafic intrusion and silicic eruption obtained from chemical zoning patterns in olivine. Earth Planet Sci Lett 227:517-530

Costa F, Scaillet B, Pichavant M (2004) Petrological and experimental constraints on the pre-eruption conditions of Holocene dacite from Volcán San Pedro (36°S, Chilean Andes) and the importance of sulfur in silicic subduction-related magmas. J Petrol 45:855–881

Costa F, Dohmen R, Chakraborty S (2008) Time scales of magmatic processes from modeling the zoning patterns of crystals. Rev Mineral Geochem 69:545-594

Cottrell E, Gardner JE, Rutherford MJ (1999) Petrologic and experimental evidence for the movement and heating of the pre-eruptive Minoan rhyodacite (Santorini, Greece). Contrib Mineral Petrol 135: 315-331

Couch S (2003) Experimental investigation of crystallization kinetics in a haplogranite system. Am Mineral 88:1471-1485

Couch S, Harford CL, Sparks RSJ, Carroll MJ (2003a) Experimental constraints on the conditions of formation of highly calcic plagioclase microlites at the Soufriere Hills Volcano, Montserrat. J Petrol 44:1455-1475

Couch S, Sparks RSJ, Carroll MR (2003b) The kinetics of degassing-induced crystallization at Soufriere Hills volcano, Montserrat. J Petrol 44:1477-1502

Dall'Agnol R, Scaillet B, Pichavant M (1999) An experimental study of a Lower Proterozoic A-type granite from the eastern Amazonian craton, Brazil. J Petrol 40:1673-1698

Davidson JP, Hora JM, Garrison J, Dungan MA (2005) Crustal forensics in arc magmas. J Volc Geotherm Res 140:157-170

Davidson JP, Morgan DJ, Charlier BLA, Harlou R, Hora JM (2007) Microsampling and isotopic analysis of igneous rocks: Implications for the study of magmatic systems. Ann Rev Earth Planet Sci 35:273-311

Devine JD, Gardner JE, Brack HP, Layne GD, Rutherford MJ (1995) Comparison of analytical methods for estimating H_2O contents of silicic volcanic glasses. Am Mineral 80:319-328

Devine JD, Rutherford, MJ, Norton GE, Young SR (2003) Magma storage region processes inferred from geochemistry of Fe-Ti oxides in andesitic magma, Soufriere Hills Volcano, Montserrat, WI. J Petrol 44:1375-1400

Donato P, Behrens H, De Rosa R, Holtz F, Parat F (2006) Crystallization conditions in the Upper Pollara magma chamber, Salina Island, Southern Tyrrhenian Sea. Mineral Petrol 86: 89–108

Donaldson CH (1976) An experimental investigation of olivine morphology. Contrib Mineral Petrol 57:187-213

Dunbar NW, Cashman KV, Dupre R, Kyle PR (1994) Crystallization processes of anorthoclase phenocrysts: Evidence from crystal compositions, crystal size distributions and volatile contents of melt inclusions. Antarctic Res Ser 66:129-146

Eberl DD, Kile DE, Drits VA (2002) On geological interpretations of crystal size distributions: Constant vs. proportionate growth. Am Mineral 87:1235-1241

Ernst WG, Liu J (1998) Experimental phase-equilibrium study of Al- and Ti-contents of calcic amphibole in MORB - A semiquantitative thermobarometer. Am Mineral 83:952-969

Folley M (1999) Crystallinity, rheology and surface morphology of basaltic lavas, Kilauea Volcano, Hawai`i. Unpublished MS Thesis, Univ Oregon

Fink JH, Griffiths (1998) Morphology, eruption rates, and rheology of lava domes: Insights from laboratory models. J Geophys Res 103: 527-545

Gardner JE (2007) Bubble coalescence in rhyolitic melts during decompression from high pressure. J Volcanol Geotherm Res 166:161–176

Gardner CA, Cashman KV, Neal CA (1998) Tephra-fall deposits from the 1992 eruption of Crater Peak, Alaska: implications of clast textures for eruptive processes Bull Volcanol 59: 537-555

Gardner JE, Hilton M, Carroll M (1999) Experimental constraints on degassing of magma: isothermal bubble growth during continuous decompression from high pressure. Earth Planet Sci Lett 168:201–218

Gardner JE, Rutherford M, Carey S, Sigurdsson H (1995) Experimental constraints on pre-eruptive water contens and changing magma storage prior to explosive eruptions of Mount St. Helens volcano. Bull Volc 57:1-17

Gauthier PJ, Condomines M (1999) ^{210}Pb-^{226}Ra radioactive disequilibria in recent lavas and radon degassing: inferences on the magma chamber dynamics at Stromboli and Merapi volcanoes. Earth Planet Sci Lett 172: 111-126

Gerlach TM, Casadevall TJ (1986) Evaluation of gas data from high-temperature fumaroles at Mount St. Helens, 1980-1982. J Volcanol Geotherm Res 28:107-140

Geschwind CH, Rutherford MJ (1992) Cummingtonite and the evolution of the Mount St. Helens (Washington) magma system: An experimental study. Geology 20:1011-1014

Ghiorso MS, Sack RO (1991) Fe-Ti oxide geothermometry – thermodynamic formulation and the estimation of intensive variables in silicic magmas. Contrib Mineral Petrol108: 485-510

Ghiorso MS, Evans BW (2008) Thermodynamics of rhombohedral oxide solid solutions and a revision of the Fe-Ti two-oxide geothermometer and oxygen-barometer. Am J Sci (in press)

Ginibre C, Kronz A, Woerner G (2002) High-resolution quantitative imaging of plagioclase composition using accumulated backscattered electron images; new constraints on oscillatory zoning. Contrib Mineral Petrol 142: 436-44

Gonnerman HM, Manga M (2005) Nonequilibrium magma degassing: Results from modeling of the ca. 1340 A.D. eruption of Mono Craters, California. Earth Planet Sci Lett 238:1-16

Grove TL, Donnelly-Nolan JM, Housch T (1997) Magmatic processes that generated the rhyolite of Glass Mountain, Medicine Lake volcano, N. California. Contrib Mineral Petrol 127: 205-223

Grove TL, Elkins-Tanton LT, Parman SW, Chatterjee N, Münetner O, Gaetani G (2003) Fractional crystallization and mantle-melting controls on calc-alkaline differentiation trends. Contrib Mineral Petrol 145: 515–533

Gualda GAR, Rivers M (2006) Quantitative 3D petrography using X-ray tomography: Application to Bishop Tuff pumice clasts. J Volcanol Geotherm Res154: 48-62

Gualda GAR (2006) Crystal size distributions derived from 3D datasets: sample size versus uncertainties. J Petrol 47: 1245-1254

Gardien V, Thompson AB, Ulmer P (2000) Melting of biotite + plagioclase + quartz gneisses: the role of H_2O in the stability of amphibole. J Petrol 41:651-666

Gurioli L, Houghton BF, Cashman KV, Cioni R (2005) Complex changes in eruption dynamics and the transition between Plinian and phreatomagmatic activity during the 79 AD eruption of Vesuvius Bull Volcanol 67:144-159

Halter WE, Pettke T, Heinrich CA, Rother-Rutishauser B (2002) Major to trace element analysis of melt inclusions by laser-ablation ICP-MS: methods of quantification. Chem Geol 183: 63-86

Hammer JE (2008) Experimental studies of the kinetics and energetic of magma crystallization. Rev Mineral Geochem 69:9-59

Hammer JE, Cashman KV, Voight B. Magmatic processes revealed by textural and compositional trends in Merapi dome lavas. J Volcanol Geotherm Res100:165–192

Hammer J, Rutherford M (2002) An experimental study of the kinetics of decompression-induced crystallization in silicic melt. J Geophys Res 107, doi:10.1029/2001JB000281

Hammer JE, Rutherford MJ, Hildreth W (2002) Magma storage prior to the 1912 eruption at Novarupta, Alaska. Contrib Mineral Petrol 144: 144-162

Hammer JE, Rutherford MJ (2003) Petrologic indicators of preeruption magma dynamics. Geology 31: 79-82

Hammer JE, Cashman KV, Voight B (2000) Magmatic processes revealed by textural and compositional trends in Merapi dome lavas. J Volcanol Geotherm Res 100:165-192

Hammer JE, Cashman KV, Hoblitt RP, Newman S (1999) Degassing and microlite crystallization during pre-climactic events of the 1991 eruption of Mt. Pinatubo, Phillipines. Bull Volcanol 60:355-380

Hammond PA, Taylor LA (1982) The ilmenite titano-magnetite assemblage – kinetics of reequilibration. Earth Planet Sci Lett 61:143-150

Harris DM, Rose WI (1996) Dynamics of carbon dioxide emissions, crystallization and magma ascent: hypotheses, theory, and applications to volcano monitoring at Mount St. Helens. Bull Volcanol 58:163-174

Hauri E, Wang J, Dixon JE, King PL, Mandeville C, Newman S (2002) SIMS analysis of volatiles in silicate glasses 1. Calibration, matrix effects and comparisons with FTIR. Chem Geol 183: 99-114

Hellebrand E, Snow J, Mostefaoui S, Hoppe P (2005) Trace element distribution between orthopyroxene and clinopyroxene in peridotites from the Gakkel Ridge: a SIMS and NanoSIMS study. Contrib Mineral Petrol 150: 486-504

Hersum TG, Marsh BD (2006) Igneous microstructures from kinetic models of crystallization. J Volcanol Geotherm Res 154: 34-47

Hersum TG, Marsh BD (2007) Igneous textures: on the kinetics behind the words. Elements 3: 247-252

Higgins M (1994) Numerical modeling of crystal shapes in thin sections: estimation of crystal habit and true size. Am Mineral 79: 113-119.

Higgins MD (2006) Quantitative Textural Measurements in Igneous and Metamorphic Petrology. Cambridge University Press, Cambridge, 276 pp

Hinton RW (1990) Ion microprobe trace-element analysis of silicates – measurement of multi-element glasses. Chem Geol 83:11-25

Hinton RW (1995) Ion microprobe analysis in geology. In: Potts PJ, Bowles JFW, Reed SJB, Cave MR (eds) Microprobe Techniques in the Earth Sciences, Chapman and Hall, London, pp 235-289

Holland T, Blundy J (1994) Non-ideal interactions in calcic amphiboles and their bearing on amphibole-plagioclase thermometry. Contrib Mineral Petrol 116(4): 433-447

Holland T, Powell R (2001) Calculation of phase relations involving haplogranitic melts using an internally consistent thermodynamic dataset. J Petrol 42:673-683

Holtz F, Sato H, Lewis J, Behrens H, Nakada S (2005) Experimental Petrology of the 1991–1995 Unzen Dacite, Japan. Part I: Phase Relations, Phase Composition and Pre eruptive Conditions. J Petrol 46:319-337

Hoppe P (2006) NanoSIMS: A new tool in cosmochemistry. Appl Surf Sci 252:7102-7106

Humphreys MCS, Kearns SL, Blundy JD (2006) SIMS investigation of electron-beam damage to hydrous, rhyolitic glasses: Implications for melt inclusion analysis. Am Mineral 91:667-679

Humphreys MCS, Menand T, Blundy JD, Klimm K (2008a) Magma ascent rates in explosive eruptions: Constraints from H_2O diffusion in melt inclusions. Earth Planet Sci Lett 270:25-40

Humphreys MCS, Blundy JD, Sparks RSJ (2008b) Shallow-level decompression crystallization and deep magma supply at Shiveluch Volcano. Contrib Mineral Petrol 155:45-61

Ihinger PD, Hervig RL, Macmillan PF (1994). Analytical methods for volatiles in glasses. Rev Mineral Geochem 30:67-121

Iverson RM, Dzurisin D, Gardner CA, Gerlach TM, LaHusen R, Lisowski M, Major JJ, Malone SD, Messerich JA, Moran SC, Pallister JS, Qamar AI, Schilling SP, Vallance JW (2006) Dynamics of seismogenic volcanic extrusion at Mount St. Helens in 2004–05. Nature 444:439-443

Jerram DA, Higgins MD (2007) 3D analysis of rock textures: quantifying igneous microstructures. Elements 3(4): 239-245

Johannes W, Holtz F (1996). Petrogenesis and Experimental Petrology of Granitic Rocks. Springer-Verlag, Berlin, 335 pp

Johannes W, Koepke J (2001) Incomplete reaction of plagioclase in experimental dehydration melting of amphibolite. Aust J Earth Sci 48:581-590

Johnson, M.C. and Rutherford, M.J. (1989) Experimentally determined conditions in the Fish Canyon Tuff, Colorado, Magma Chamber. J Petrol 30:711-737

Jurewicz SR, Watson EB (1985) The distribution of partial melt in a granitic system: the application of liquid phase sintering theory. Geochim Cosmochim Acta 49:1109-1121

Kaszuba JP, Wendlandt RK (2000) Effect of carbon dioxide on dehydration melting reactions and melt compositions in the lower crust and the origin of alkaline rocks. J Petrol 41:363-386

Kawamoto T (1996) Experimental constraints on differentiation and H_2O abundance of calc-alkaline magmas. Earth Planet Sci Lett 144:577-589

Kearns SL, Steen N, Erlund E (2002) Electronprobe microanalysis of volcanic glass at cryogenic temperatures. Microscop Microanal 8, Suppl 2:1562CD

Kent AJR (2008) Melt inclusions in basaltic and related volcanic rocks. Rev Mineral Geochem 69:273-331

Kile DE, Eberl DD, Hoch AR, Reddy MM (2000) An assessment of calcite crystal growth mechanisms based on crystal size distributions. Geochim Cosmochim Acta 64:2937-2950

King PL, Venneman, TW, Holloway JR, Hervig RL, Lowenstern JE and Forneris, JF (2002) Analytical techniques for volatiles: A case study using intermediate (andesitic) glasses. Am Mineral 87:1077-1089

Klimm K, Holtz F, Johannes W, King PL (2003) Fractionation of metaluminous A-type granites: an experimental study of the Wangrah Suite, Lachlan Fold Belt, Australia. Precamb Res 124:327–341

Klug C, Cashman KV, Bacon CR (2002) Structure and physical characteristics of pumice from the climactic eruption of Mt. Mazama (Crater Lake), Oregon. Bull Volcanol 64: 486-501

Larsen JF (2005) Experimental study of plagioclase rim growth around anorthite seed crystals in rhyodacitic melt. Am Mineral 90:417-427

Larsen JF (2006) Rhyodacite magma storage conditions prior to the 3430 yBP caldera-forming eruption of Aniakchak volcano, Alaska. Contrib Mineral Petrol 152:523–540

Larsen JF, Gardner JE (2004) Experimental study of water degassing from phonolite melts: implications for volatile oversaturation during magmatic ascent. J Volc Geotherm Res 134:109-124

LePage L (2003) ILMAT: an excel worksheet for ilmenite--magnetite geothermometry and geobarometry. Comput Geosci 29:673-678

Lindquist WB, Venkatarangan A (1999) Investigating 3D geometry of porous media from high resolution images. Phys Chem Earth 25:593-599

Lindsley DH, Spencer KJ (1982) Fe–Ti oxide geothermometry: Reducing analyzes of coexisting Ti-magnetite (Mt) and ilmenite (Ilm). Eos Trans, Am Geophys Union 63: 471

Liu Y, Zhang Y, Behrens H (2005) Solubility of H_2O in rhyolitic melts at low pressures and a new empirical model for mixed $H_2O–CO_2$ solubility in rhyolitic melts. J Volcanol Geotherm Res 143:219-235

Liu Y, Anderson AT, Wilson CJN, Davis AM, Steele IM (2006) Mixing and differentiation in the Oruanui rhyolitic magma, Taupo, New Zealand: evidence from volatiles and trace elements in melt inclusions. Contrib Mineral Petrol 151:71–87

Lofgren GE (1980) Experimental studies on the dynamic crystallization of silicate melts. *In*: Physics of Magmatic Processes. Hargraves RB (ed) Princeton University Press, Princeton, p 487-551

López S, Castro A (2001) Determination of the fluid–absent solidus and supersolidus phase relationships of MORB-derived amphibolites in the range 4–14 kbar. Am Mineral 86:1396-1403

Luhr JF (1990) Experimental phase relations of water- and sulfur-saturated arc magmas and the 1982 eruptions of El Chichon Volcano. J Petrol 31:1071-1114

Madras G, McCoy BJ (2002) Ostwald ripening with size-dependent rates: Similarity and power-law solutions. J Chem Phys 117: 8042-8049

Mangan MT (1990) Crystal size distribution systematics and the determination of magma storage times: The 1959 eruption of Kilauea volcano, Hawaii. J Volcanol Geotherm Res 44:295-302

Mangan M, Sisson T (2000) Delayed, disequilibrium degassing in rhyolitic magma: decompression experiments and implications for explosive volcanism. Earth Planet Sci Lett 183:441-455

Mangan M, Mastin L, Sisson T (2004a) Gas evolution in eruptive conduits: combining insights from high temperature and pressure decompression experiments with steady-state flow modeling. J Volcanol Geotherm Res 129:23-36

Mangan M, Sisson T, Hankins W (2004b) Decompression experiments identify kinetic controls on explosive silicic Eruptions. Geophys Res Lett 31, L08605, doi:10.1029/2004GL019509

Marsh BD (1988) Crystal size distributions (CSD) in rocks and the kinetics and dynamics of crystallization I. Theory. Contrib Mineral Petrol 99:277-291

Marsh BD (2006) Dynamics of magma chambers. Elements 2:287-292

Martel C, Pichavant M, Bourdier J-L, Traineau H, Holtz F, Scaillet B (1998). Magma storage conditions and control of eruption regime in silicic volcanoes: experimental evidence from Mt. Pelée. Earth Planet Sci Lett 156:89-99

Martel C, Pichavant M, Holtz F, Scaillet B (1999) Effects of f_{O_2} and H_2O on andesite phase relations between 2 and 4 kbar. J Geophys Res 104:29453-29470

Martel C, Bourdier JL, Pichavant M, Traineau H (2000) Textures, water content and degassing of silicic andesites from recent plinian and dome-forming eruptions at Mount Pelee volcano (Martinique, Lesser Antilles arc). J Volc Geotherm Res 96:191-206

Martel C, Schmidt B (2003) Decompression experiments as an insight into ascent rates of silicic magmas. Contrib Mineral Petrol 144: 397–415

McCanta MC, Rutherford MJ, Hammer JE (2007) Pre-eruptive and syn-eruptive conditions in the Black Butte, California dacite: Insight into crystallization kinetics in a silicic magma system. J Volcanol Geotherm Res 160:263–284

McGee KA, Casadevall TJ (1994) A Compilation of Sulfur Dioxide and Carbon Dioxide Emission-Rate Data from Mount St. Helens during 1980-88. US Geol Surv Open-File Rep 94-212

McGonigle AJS, Aiuppa A, Giudice G, Tamburello G, Hodson AJ, Gurrieri S (2008) Unmanned aerial vehicle measurements of volcanic carbon dioxide fluxes. Geophys Res Lett 35:L06303

Melson WG, Hopson CA (1981) Preeruption temperatures and oxygen fugacities in the 1980 eruptive sequence. USGS Prof Pap 1250:641-648

Métrich N, Wallace PJ (2008) Volatile abundances in basaltic magmas and their degassing paths tracked by melt inclusions. Rev Mineral Geochem 69:363-402

Metrich N, Allard P, Spilliaert N, Andronico D, Burton M (2004) 2001 flank eruption of the alkali- and volatile-rich primitive basalt responsible for Mount Etna's evolution in the last three decades. Earth Planet Sci Lett 228:1-17

Milman-Barris MS, Beckett JR, Baker MB, Hofmann AE, Morgan Z, Crowley MR, Vielzeuf D, Stolper E (2008) Zoning of phosphorus in igneous olivine. Contrib Mineral Petrol 155:739-765

Molina JF, Poli S (2000) Carbonate stability and fluid composition in subducted oceanic crust: an experimental study on H_2O-CO_2-bearing basalts. Earth Planet Sci Lett 176:295-310

Moore G (2008) Interpreting H2O and CO2 contents in melt inclusions: constraints from solubility experiments and modeling. Rev Mineral Geochem 69:333-361

Moore G, Carmichael ISE (1998) The hydrous phase equilibria (to 3 kbar) of an andesite and basaltic andesite from western Mexico: constraints on water content and conditions of phenocryst growth. Contrib Mineral Petrol 130:304-319

Morgan DJ, Blake SR, Rogers NW, DeVivo B, Rolandi G, Macdonald R, Hawkesworth CJ (2004) Time scales of crystal residence and magma chamber volume from modeling of diffusion profiles in phenocrysts: Vesuvius 1944. Earth Planet Sci Lett 222:933-946

Morgan DJ, Jerram DA (2006) On estimating crystal shape for crystal size distribution analysis. J Volcanol Geotherm Res154:1-7

Morgan DJ, Blake SR, Rogers NW, DeVivo B, Rolandi G, Davidson JP (2006) Magma chamber recharge at Vesuvius in the century prior to the eruption of A.D. 79. Geology 34: 845-848

Müntener O, Kelemen PB, Grove TL (2001) The role of H_2O during crystallization of primitive arc magmas under uppermost mantle conditions and genesis of igneous pyroxenites: an experimental study. Contrib Mineral Petrol 141:643-658

Nakada S, Motomura Y (1999) Petrology of the 1991-1995 eruption at Unzen: effusion pulsation and groundmass crystallization. J Volcanol Geotherm Res 89:173-196

Nakajima K, Arima M (1998) Melting experiments on hydrous low-K tholeiite: Implications for the genesis of tonalitic crust in the Izu-Bonin-Mariana arc. Island Arc 7:359-373

Nemchin AA, Giannini LM, Bodorkos S, Oliver NHS (2001 Ostwald ripening as a possible mechanism for zircon overgrowth formation during anatexis: theoretical constraints, a numerical model, and its application to pelitic migmatites of the Tickalara Metamorphics, northwestern Australia. Geochim Cosmochim Acta 65:2771-1788

Neuberg J (2000) Characteristics and causes of shallow seismicity in andesite volcanoes. Phil Trans R Soc 358:1533-1546

Newman S, Lowenstern JB (2002) VolatileCalc: a silicate melt-H_2O-CO_2 solution model written in Visual Basic for Excel. Comput Geosci 28:597-604

Nicholis MG, Rutherford MJ (2004) Experimental constraints on magma ascent rate for the Crater Flat volcanic zone hawaiite. Geology 32:489-492

Nicholls IA, Oba T, Conrad WK (1992) The nature of primary rhyolitic magmas involved in crustal evolution: Evidence from an experimental study of cummingtonite bearing rhyolites, Taupo Volcanic Zone, New Zealand. Geochim Cosmochim Acta 56:955-962

Noguchi S, Toramaru A, Nakada S (2008a) Groundmass crystallization in dacite dykes taken in Unzen Scientific Drilling Project (USDP-4) J Volcanol Geotherm Res doi:10.1016/j.jvolgeores.2008.03.037

Noguchi S, Toramaru A, Nakada S (2008b) Relation between microlite textures and discharge rate during the 1991-1995 eruptions at Unzen, Japan. J Volcanol Geotherm Res, doi:10.1016/j.jvolgeores.2008.03.025

Ohba T, Hirabayashi J-I, Nogami K, Kusakabe M, Yoshida M (2008) Magma degassing process during the eruption of Mt. Unzen, Japan in 1991 to 1995: Modeling with the chemical composition of volcanic gas. J Volcanol Geotherm Res doi:10.1016/j.jvolgeores.2008.03.040

Pallister JS, Hoblitt RP, Meeker GP, Knight RJ, Siems DF (1996) Magma mixing at Mount Pinatubo: Petrographic and chemical evidence from the 1991 deposits. *In:* Fire and Mud: Eruptions and Lahars of Mount Pinatubo, Philippines. Punongbayan RS, Newhall CG (eds) Univ Washington Press, Seattle, p 687-731

Papale P (2005) Determination of total H_2O and CO_2 budgets in evolving magmas from melt inclusion data. J Geophys Res 110, doi:10.1029/2004JB003033

Papale P, Neri A, Macedonio G (1998) The role of magma composition and water content in explosive Eruptions 1. Conduit ascent dynamics. J Volcanol Geotherm Res 87:75-93

Parat F, Holtz F, Feig S (2008) Pre-eruptive conditions of the Huerto Andesite (Fish Canyon system, San Juan volcanic field, Colorado): Influence of volatiles (COHS) on phase equilibria and mineral composition. J Petrol 49:911-935

Patiño-Douce AE, Beard BS (1995) Dehydration-melting of Biotite Gneiss and Quartz Amphibolite from 3 to 15 kbar. J Petrol 36:707-738

Peppard BT, Steele IM, Davis AM, Wallace PJ, Anderson AT (2001) Zoned quartz phenocrysts from the rhyolitic Bishop Tuff. Am Mineral 86:1034-1052

Pertermann M, Lundstrom C (2006) Phase equilibrium experiments at 0.5 GPa and 1100-1300 degrees C on a basaltic andesite from Arenal volcano, Costa Rica. J Volcanol Geotherm Res 157:222-235

Pichavant M, Martel C, Boudier J-L, Scaillet B (2002) Physical conditions, structure, and dynamics of a zoned magma chamber: Mount Pelée (Martinique, Lesser Antilles Arc). J Geophys Res 107, doi:10.1029/2001JB000315

Pichavant M, Macdonald R (2007) Crystallization of primitive basaltic magmas at crustal pressures and genesis of the calc-alkaline igneous suite: experimental evidence from St Vincent, Lesser Antilles arc. Contrib Mineral Petrol 154:535–558

Pichavant M, Costa F, Burgisser A, Scaillet B, Martel C, Poussineau S (2007) Equilibration Scales in Silicic to Intermediate Magmas Implications for Experimental Studies. J Petrol 48:1955-1972

Pinel V, Jaupart C (2000) The effect of edifice load on magma ascent beneath a volcano. Phil Trans Roy Soc 358:1515-1532

Polacci M, Baker DR, Bai L, Mancini L (2007) Large vesicles record pathways of degassing at basalt volcanoes. Bull Volcanol doi 10.1007/s00445-007-0184-8

Poli S (1993) The amphibolite-eclogite transition: an experimental study on basalt. Am J Sci 293:1061-1107

Pompilio M (1998) Melting experiments on Mt. Etna lavas; I, The calibration of an empirical geothermometer to estimate the eruptive temperature. Acta Vulcanol 10:1-9

Prouteau G, Scaillet B (2003) Experimental constraints on the origin of the 1991 Pinatubo dacite. J Petrol 44:2203-2241

Putirka KD (2008) Thermometers and barometers for volcanic systems. Rev Mineral Geochem 69:61-120

Ramos FC, Tepley FJ III (2008) Inter- and intracrystalline isotopic disequilibria: techniques and applications. Rev Mineral Geochem 69:403-443

Randle V, Engler O (2000) Texture Analysis: Macrotexture, Microtexture, and Orientation Mapping. CRC Press

Randolph AD, Larson MA (1971) Theory of Particulate Processes. Academic Press, New York, 251 pp

Rapp RP, Watson EB (1995) Dehydration melting of metabaslt at 8-32 kbar – implications for continental growth and crust-mantle recycling. J Petrol 36:891-931

Resmini RG (2007) Modeling of crystal size distributions (CSDs) in sills. J Volcanol Geotherm Res 161:118-130

Robie RA, BS Hemingway, JR Fisher (1978) Thermodynamic properties of minerals and related substances at 298.15 K and 1 bar (10^5 pascals) pressure and at higher temperature. USGS Bull 1452

Roman DC, Cashman KV, Gardner CA, Wallace PJ, Donovan JJ (2006) Storage, degassing and pre-eruption crystallization of compositionally heterogeneous magmas from the 1986 eruption of Augustine Volcano, Alaska. Bull Volcanol 68:240-254

Rusk BG, Reed MH (2002) Scanning electron microscope-cathodoluminescence analysis of quartz reveals complex growth histories in veins from the Butte porphyry copper deposit, Montana. Geology 30:727-730

Rusk BG, Lowers HA, Reed MH (2008) Trace elements in hydrothermal quartz: Relationships to cathodoluminescence textures and insights into vein formation. Geology 36:547-550

Rust A, Manga M, Cashman KV (2003) Determining flow type, shear rate and shear stress in magmas from bubble shapes and orientations J Volcanol Geotherm Res 122:111-132

Rutherford MJ (2008) Magma ascent rates. Rev Mineral Geochem 69:241-271

Rutherford MJ, Devine JD (1988) The May 18, 1980 eruption of Mount St. Helens 3. Stability and chemistry of amphibole in the magma chamber. J Geophys Res 93: 11949-11959

Rutherford MJ, Devine JD (2003) Magmatic conditions and magma ascent as indicated by hornblende phase equilibria and reactions in the 1995-2002 Soufriere Hills magma. J Petrol 44:1433-1454

Rutherford MJ, Devine JD (2008) Magmatic Conditions and Processes in the Storage Zone of the 2004–2006 Mount St. Helens Dacite. *In:* Sherrod DR, Scott WE, Stauffer PH (eds) A volcano rekindled: the renewed eruption of Mount St. Helens, 2004–2006. US Geol Surv Prof Pap 1750 (in press)

Rutherford MJ, Hill PM (1993) Magma ascent rates from amphibole breakdown: An experimental study applied to the 1980-1986 Mount St. Helens eruption. J Geophys Res 98:19667-19686

Rutherford MJ, Sigurdsson H, Carey S, Davis A (1985) The May 18, 1980 eruption of Mount St. Helens 1. Melt composition and experimental phase equilibria. J Geophys Res 90: 2929-2947

Sato H, Nakada S, Fujii T, Nakamura M, Suzuki-Kamata K (1999) Groundmass pargasite in the 1991–1995 dacite of Unzen volcano: phase stability experiments and volcanological implications. J Volc Geotherm Res 89:197-212

Scaillet B, Pichavant M, Roux J (1995) Experimental crystallization of leucogranite magmas. J Petrol 36:663-705

Scaillet B, Evans BW (1999) The 15 June 1991 eruption of Mount Pinatubo. I. phase equilibria and pre-eruption P-T-f_{O_2}-f_{H_2O} conditions of the dacite magma. J Petrol 40:381-411

Scandone R, Cashman KV, Malone SD (2007) Magma supply, magma ascent and the style of volcanic eruptions. Earth Planet Sci Lett 253:513-529

Schairer JF (1950) The alkali feldspar join in the system $NaAlSiO_4$-$KAlSiO_4$-SiO_2. J Geol 58: 512-517

Schmidt MW, Thompson AB (1996) Epidote in calc-alkaline magmas: An experimental study of stability, phase relationships, and the role of epidote in magmatic evolution. Am Mineral 81:462-474

Schmitt AK (2001) Gas-saturated crystallization and degassing in large-volume, crystal-rich dacitic magmas from the Altiplano-Puna, northern Chile. J Geophys Res 106:30,561-30,578

Schwartz AJ, Kumar M, Adams BL (2000) Electron Backscatter Diffraction in Materials Science. Kluwer Academic/Plenum Publishers, New York

Sen C, Dunn T (1994) Dehydration melting of a basaltic composition amphibolite at 1.5 and 2.0 GPa: implications for the origin of adakites. Contrib Mineral Petrol 117:394-409

Shimizu N, Hart SR (1982) Applications of the ion-microprobe to geochemistry and cosmochemistry. Ann Rev Earth Planet Sci 10:483-526

Shimizu N (1997) Principles of SIMS and modern ion microprobes. *In* Modern Analytical Geochemistry, Gill J (ed) Longman, p 235-242

Shin H, Lindquist WB, Sahagian DL, Song S-R (2005) Analysis of the vesicular structure of basalts. Comput Geosci 31:473-487

Silver LA, Ihinger PD, Stolper E (1990) The influence of bulk composition on the speciation of water in silicate glasses. Contrib Mineral Petrol 104:142-162

Singh J, Johannes W (1996) Dehydration melting of tonalites 2. Composition of melts and solids. Contrib Mineral Petrol 125:26-44

Sisson TW, Ratajeski K, Hankins WB, Glazner AF (2005) Voluminous granitic magmas from common basaltic sources. Contrib Mineral Petrol 148:635-661

Song S-R, Jones KW, Lindquist WB, Dowd BA, Sahagian DL (2001) Synchotron X-ray computed microtomography: studies on vesiculated basaltic rocks. Bull Volcanol 63:252-263

Spilliaert N, Allard P, Metrich N, Sobolev AV (2006) Melt inclusion record of the conditions of ascent, degassing, and extrusion of volatile-rich alkali basalt during the powerful 2002 flank eruption of Mount Etna (Italy). J Geophys Res 111 doi:10.1029/2005JB003934

Stewart M, Pearce TH (2004) Sieve-textured plagioclase in dacitic magma: Interference imaging results. Am Mineral 89:348-351

Stormer JC (1983) The effects of recalculation on estimates of temperature and oxygen fugacity from analyzes of multicomponent iron-titanium oxides. Am Mineral 66:586-594

Swanson DA, Holcomb RT (1990) Regularities in growth of the Mount St. Helens dacite dome, 1980–1986. *In* Lava flows and domes, emplacement mechanisms and hazard implications. IAVCEI Proceedings in Volcanology. Fink JH (ed) Springer-Verlag, Berlin 2:3–24

Tamic N, Behrens H, Holtz F (2001) The solubility of H_2O and CO_2 in rhyolitic melts in equilibrium with a mixed CO_2–H_2O fluid phase. Chem Geol 174:333–347

Thomas R (2000) Determination of water contents of granite melt inclusions by confocal laser Raman microprobe spectroscopy. Am Mineral 85:868-872

Thomas R, Davidson P (2007) Progress in the determination of water in glasses and melt inclusions with Raman spectroscopy: A short review. Acta Petrologica Sinica 23:15-20

Toramaru A, Noguchi S, Oyoshihara S, Tsune A (2008) MND (microlite number density)-water exsolution rate meter. J Volc Geotherm Res doi:10.1016/j.jvolgeores.2008.03.035

Tuttle OF, Bowen NL (1958) Origin of granite in the light of experimental studies in the system $NaAlSi_3O_8$-$KAlSi_3O_8$-SiO_2-H_2O. Geol Soc Am Mem 74: 153 pp

Underwood E (1970) Quantitative Stereology. Addison-Wesley, Massachusetts, 274 pp

Van der Plas L, Tobi AC (1965) A chart for judging reliability of point-counting results. Am J Sci 263:87-89

Venezky DY, Rutherford MJ (1997). Preeruption conditions and timing of dacite-andesite magma mixing in the 2.2 ka eruption at Mount Rainier. J Geophys Res 109:20069-20086

Venezky DY, Rutherford MJ (1999) Petrology and Fe-Ti oxide reequilibration of the 1991 Mount Unzen mixed magma. J Volcanol Geotherm Res 89:213-230

Wallace PJ (2001) Volcanic SO_2 emissions and the abundance and distribution of exsolved gas in magmatic bodies. J Volc Geotherm Res 108:85-106

Wallace PJ (2005) Volatiles in subduction zone magmas: concentrations and fluxes based on melt inclusion and volcanic gas data. J Volcanol Geotherm Res 140:217-240

Wallace PJ, Anderson AT Jr, Davis AM (1999) Gradients in H_2O, CO_2 and exsolved gas in a large-volume silicic magma system: Interpreting the record preserved in melt inclusions from the Bishop Tuff. J Geophys Res 104:20097-20122

Wark D, Watson E (2006) TitaniQ: a titanium-in-quartz geothermometer. Contrib Mineral Petrol 152:743-754

Watson EB (1982) Melt infiltration and magma evolution. Geology 10:236-240

Wells PRA (1977) Pyroxene thermometry in simple and complex systems. Contrib Mineral Petrol 62:129-139

Williams DL, Abrams G, Finn C, Dzurisin D, Johnson DJ, Denlinger R (1987) Evidence from gravity data for an intrusive complex beneath Mount St. Helens. J Geophys Res 92:10207-10222

Wright HMN, Roberts J, Cashman KV (2006) Permeability of anisotropic tube pumice – model calculations and measurements Geophys Res Lett 33, doi:10.1029/2006GL027224

Zellmer GF, Sparks RSJ, Hawkesworth CJ, Wiedenbeck M (2003) Magma Emplacement and Remobilization Timescales Beneath Montserrat: Insights from Sr and Ba Zonation in Plagioclase Phenocrysts. J Petrol 44:1413-1431

Magma Ascent Rates

Malcolm J. Rutherford

Department of Geological Sciences
Brown University
Providence, Rhode Island, 02912, U.S.A.

Malcolm_Rutherford@Brown.edu

INTRODUCTION

Volcanoes can erupt explosively, generating high columns of ash and occasional pyroclastic flows, or they can erupt slowly forming lava flows and domes. This variation reflects significant differences in mass eruption rate (e.g., Wilson et al. 1980). Mass eruption rates in turn are controlled by the rates of magma ascent through the volcanic conduit, and the conduit size. The magma ascent rate itself is a function of the pressure in the magma storage region, the physical properties of the magma, such as its density, viscosity and crystallinity, and the resistance to flow in the conduit that connects the magma storage zone to the surface (Papale and Dobran 1994; Mastin and Ghiorso 2001; Pinkerton et al. 2002; Sparks et al. 2006). A number of important characteristics of volcanic eruptions are affected or controlled by the rate at which magma ascends from depth. For example, the bubble content (i.e., vesicularity) and the degree of crystallization that develop in the melt phase can be significantly different in rapidly vs. slowly ascended magma. In fact, the inability of rapidly ascending magma to effectively lose exsolved gas may be one factor causing an eruption to change from effusive to explosive behavior, as recently documented for the Soufriere Hills eruption on Montserrat in the West Indies (Sparks et al. 1998). Therefore, in order to better understand the processes involved and the changes that occur in volcanic eruptions, it is important to quantify the rates at which different magmas rise to the surface.

The fact that the rate of magma ascent controls a number of reactions that occur in volatile-rich magmas suggests several ways to study magma ascent rates using reaction data and theoretical flow models. One reaction that is a direct result of magma ascent is the exsolution of volatiles from the melt to form gas bubbles (Sparks 1978). Water is the most abundant bubble-forming gas in intermediate and silica-rich composition magmas, but CO_2, SO_2, and Cl are also very important in some magma types. Available data suggest that these gases diffuse out of the melt at different rates as pressure exerted on an ascending magma decreases during ascent (Watson 1994; Baker and Rutherford 1996: Zhang and Behrens 2000). It is therefore theoretically possible to use the extent of degassing of each of these different species as an indicator of the time involved in magma ascent from depth. Degassing during ascent also causes the melt portion of the magma to crystallize. The rate and extent of this crystallization have been determined for some magma types and can be used to estimate magma ascent rates. Similarly, some crystals formed in magmas at depth can contain volatile species such as OH, CO_2 and S in their structure, e.g., biotite, hornblende, and Fe-sulfide, and these phenocrysts tend to become unstable and react with the surrounding melt during magma ascent. These reactions occur because the decrease in pressure during ascent causes a decrease in the volatile species dissolved in the melt that destabilizes the phenocryst. The amount of reaction between phenocryst and melt is controlled by the ascent rate and the kinetics of the mineral breakdown. These degassing-related reactions are particularly important in studying volatile-rich, subduction zone volcanoes. These reactions are less useful in studies of basaltic volcanic centers because of the generally lower volatile contents in these magmas, but examples have been described where such reactions occur (Vaniman et al. 1982; Corsaro and Pompillio 2004). Magma ascent rates can also be estimated using the rate of movement of seismic activity toward the surface (Endo and Murray 1992) and using theoretical analyses of conduit flow together with measurements of the mass eruption rate for a given event (Carey and Sigurdsson 1985).

Other methods of assessing magma ascent rates involve study of xenolith transport from depth by the magma, the extent of chemical reaction between these xenoliths and their host liquid in the magma, and evidence of U-Th-Ra isotopic disequilibrium in a magma (Condomines et al. 1988). The latter approach involves the identification of isotopic disequilibrium of the short-lived isotopes in the U and Th decay series; the disequilibrium indicates a maximum time of magma existence (4-5 half lives for the decay involved, i.e.; ~8000 years for ^{230}Th-^{226}Ra and 30 years for ^{232}Th-^{238}Ra) since the disequilibrium was generated. (This topic will be treated in other chapters in this volume, and therefore will not be considered further here.) The ability of magma to physically transport xenoliths to the surface depends on the ascent velocity and the rheology of the magma as well as the xenolith size and density (Spera 1984). Reactions between xenoliths and magma may occur where a xenocryst or xenolith and magma are in contact but not in equilibrium. In such cases, the crystal undergoes a time dependent diffusion exchange of atoms with the melt, allowing an assessment of the time involved in magma ascent from the point (depth) of crystal-melt contact which must be determined (Klugel 1998). The following sections describe some of the interesting results on magma ascent rate determinations using these different approaches, but first we consider the problem of the magma source, and where ascent begins.

MAGMA SOURCE AND MAGMA RESERVOIRS

One of the significant problems that must be addressed when attempting to determine magma ascent rates for volcanic eruptions is the question of where ascent was initiated. For many volcanoes there is good evidence for a magma storage region in the crust, a magma reservoir from which the final ascent began. Magma storage zones in the upper crust have been identified seismically for almost all well studied (recently active) volcanoes in subduction-zone environments, including Mount St. Helens (Scandone and Malone 1985), Mount Pinatubo in the Philippines (Pallister et al. 1996), Soufriere Hills on Montserrat (Aspinall et al. 1998), Unzen in Japan, and Redoubt (Power et al. 1994) in the Aleutians. The presence of these near surface (5 to 15 km depth) reservoirs is confirmed by petrological data (Fig. 1) that indicates the phenocryst assemblage equilibrated in the reservoir (Hammer and Rutherford 2003) if they did not form here initially (Rutherford and Devine 2008). Density contrasts between magma and surrounding rocks (Ryan 1993) also indicate that there are likely both deep and shallow magma reservoirs beneath many calc-alkaline volcanoes, and petrologic data indicates that two such reservoirs did exist at 3-5 and >8 km under Mount Rainier at the time of its most recent (2,200 years ago) explosive eruption (Venezky and Rutherford 1997). Using the methods outlined above, estimates of magma ascent rates for such volcanoes will generally refer to the transport from the higher, crustal-level storage chamber to the surface.

Eruption from a crustal reservoir has also generally been the case for basaltic volcanic centers such as Hawaii, Iceland, Etna, and the Canary Islands (Decker 1987; Klugel et al. 1997; Corsaro and Pompilio 2004). For eruptions of basaltic magma at mid-ocean ridges, available seismic evidence generally indicates a magma storage region in the crust or near the crust-mantle boundary. However, some magma, specifically kimberlite and more mafic magma erupting in mid-plate oceanic islands and in some island arcs (Carmichael 2002), appears to ascend directly from the mantle without encountering a crustal storage region. In such cases,

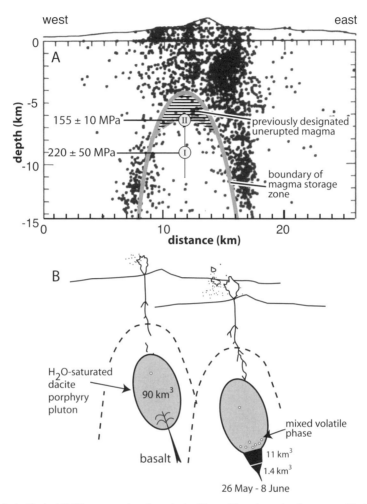

Figure 1. A. Vertical E-W cross-section through the Mount Pinatubo volcanic system. Black dots are earthquake foci mostly outside the magma volume. I and II show petrological constraints on phenocryst-melt equilibration depths. B. Sketch cross sections illustrating interaction between basalt and pre-existing dacite at Mount Pinatubo. [Used with permission of the Geological Society of America from Hammer and Rutherford (2003) *Geology*, Vol. 31, Fig. 3, p. 81.]

the magma ascent rate would refer to the time between separation of magma from its source in the Earth's mantle and its arrival at the surface. However, most of the magma ascent rate estimates are for volcanic centers where there is strong evidence of a crustal-level magma storage zone, and the ascent rate is for magma transport from this reservoir to the surface.

BUBBLE FORMATION AND EVOLUTION DURING MAGMA ASCENT

As magma rises from deep in the crust, volatiles dissolved in the melt will eventually tend to exsolve and form bubbles because of their lower solubility in melt at lower pressures. The depth at which the gas phase is generated depends on the CO_2 and H_2O dissolved in the magma, and the solubility of these species in the melt (e.g., Holloway and Blank 1994). Based on the abundances of CO_2 and H_2O determined for melt inclusions trapped in early formed phenocrysts (Fig. 2), the first gas phase generated in basalts and andesites is likely to be CO_2-rich (Roggensack et al. 1997; Metrich et al. 1999). The same is likely true for more evolved magmas as well, but they tend to also contain higher (3-7 wt%) dissolved H_2O (Cervantes and Wallace 2003; Mangiacapra et al. 2008). What is the rate of this gas exsolution process, and can measurements of the gas released be used to estimate the rates of magma ascent? Analyses of matrix glasses in both explosively erupted and extrusive magmas illustrate that most of the water originally in the melt at depth has been lost to the gas phase during ascent (Rutherford et al. 1985). In other words, even for the relatively rapid ascent involved in explosive eruptions of a volatile-rich magma, there appears to have been sufficient time during ascent for most of the original H_2O to escape from the melt. Other volatile species that have an affinity for the gas phase include CO_2, SO_2, and Cl, based on their common occurrence in volcanic gas emissions. The available diffusion data for silicate melts (Watson 1994) suggest that these species diffuse at somewhat slower rates than water. Thus, analytical data on the abundance of the different volatile species lost from the melt to the gas phase during an eruption may record the rate of magma ascent, although the rapid diffusion of H_2O may mean that it is nearly all transferred to the gas phase in most eruptions.

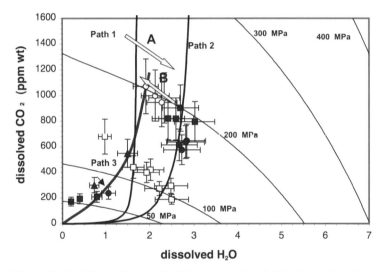

Figure 2. Plot of dissolved CO_2 vs. dissolved H_2O for melt inclusions in K-trachybasalt and latite magmas erupted at Campi Flegrei volcanic center, Italy. Paths 1, 2 and 3 show two possible decompression paths relating high- (200 MPa) and low-pressure inclusions. Arrows indicate how crystallization or CO_2 gas fluxing would move H_2O content at close to constant pressure. (Modified from Mangiocapra et al. 2008; Isobars are calculated from Papale et al. 2006.)

Experiments indicate that the loss of water from melt into gas bubbles during decompression (ascent) of a rhyolitic melt is generally very rapid (Hurwitz and Navon 1994; Gardner et al. 1999; Mangan and Sisson 2000; Gardner and Denis 2004; Baker et al. 2006; Gardner 2007). At pressure drops of 0.025 MPa/s (1 bar per 4 seconds) water is able diffuse out of the melt and into the bubbles rapidly enough to maintain equilibrium. At faster decompression rates of 0.25 and 1.0 MPa/s, water diffused too slowly to maintain equilibrium and the melts became supersaturated with water. These experiments indicate that a transition occurs between 0.025 and 0.25 MPa/s decompression rates, beyond which gas bubble-melt equilibria cannot be maintained. Given typical rhyolitic magma densities, these decompression rates equal ascent velocities of 0.7 and 5 m/s, rates much faster than are typically inferred for effusive lava and dome eruptions. In the high-silica rhyolite (77 wt% SiO_2) decompression experiments referenced above, the 825 °C temperature produces a magma that is relatively viscous, so water diffusion is relatively slow. In less viscous basaltic magmas, water diffusion is faster at the higher (~1200 °C) temperatures (Zhang and Stolper 1991). Numerical modeling of bubble growth in basaltic magmas suggests that for reasonable ascent rates, water diffusion and bubble growth are rapid enough to always maintain equilibrium between bubbles and melt (Toramaru 1995). This includes ascent rates up to 100 m/s.

The above discussion focuses on degassing of water, which is generally the fastest diffusing volatile species in melts. In contrast, the diffusivity of CO_2, S and Cl in rhyolitic melts is 10's to100's of times slower than water at a given temperature (Watson 1994: Baker and Rutherford 1996; Alletti et al. 2007). If magma ascends during explosive eruptions at rates controlled by water exsolution (the maximum at which water and melt can maintain equilibrium), then it is expected that slower diffusing gas species would not maintain equilibrium. Indeed, this appears to be the case, as matrix glasses in volcanic pumices typically contain much if not all of the pre-erupted contents of S and Cl. For example, the pre-eruptive concentrations of H_2O and Cl in the 1991 Pinatubo magma are about 6.4 wt% and 1200 ppm, respectively (Pallister et al. 1996), whereas their final concentrations dissolved in matrix glasses are about 0.4 wt% and 880 ppm. Degassing of Cl and S is complicated, however; since the partitioning of Cl and S between the melt and a water-rich gas phase is likely to vary strongly as a function of water content, pressure, melt composition, and temperature. Thus, in intermediate and mafic composition magmas, where there is significant dissolved CO_2 in the melt at depth, the initial gas phase will be CO_2-rich and only at much lower pressures (Fig. 2) will significant H_2O partition into the gas phase (Dixon and Stolper 1995; Wallace 2005). There is some preliminary data on how S, Cl and F partition into the CO_2-rich vs. the H_2O-rich gas phase (Nicholis 2006), but this is a relatively unstudied research area where additional work may be very productive in understanding magma ascent during volcanic eruptions. An important implication of the above findings is that water is able to rapidly diffuse out of silicate melts into bubbles during magma ascent such that the melt-bubble equilibrium is almost always continuously maintained (Gardner et al. 1999), and this is important because several of the reaction processes described below are driven by water degassing of the interstitial melt during magma ascent. Hence, the rates at which those reactions occur are not constrained by the exsolution of water from the melt, but by their own reaction kinetics.

There is one group of volatile studies where the diffusion rates of volatile species such as H_2O and CO_2 have been used with some success to determine the ascent rate of magmas during volcanic eruptions. Anderson (1991) initially illustrated the potential of this approach using gas bubbles in hourglass melt inclusions in quartz phenocrysts in the Bishop Tuff rhyolite. He showed that the bubble-melt volume ratios suggest a decompression time of ~2 minutes during the plinian phase of the eruption, i.e., an average ascent rate of 40 m/s for ascent from 180 MPa to 40 MPa where the magma was fragmented. Liu et al. (2007) refined this model by measuring the concentrations of both H_2O and CO_2 in quartz-hosted, bubble-free cylindrical melt inclusions that are connected to the surrounding bubble-rich melt. During magma ascent, bubbles form and grow outside the cylindrical inclusion creating a concentration gradient in the cylinder. Liu et al. (2007) used CO_2 and H_2O diffusion data to model the diffusion profiles that would be produced for different starting conditions and different length inclusions as a function of decompression rate. Fitting the measured melt diffusion profiles to the theoretical data suggests that average magma ascent rates were 0.05-0.35 m/s for the large phreato-magmatic Oruanui eruption of rhyolitic magma in New Zealand. This estimate is somewhat lower than the average velocity (~5-8 m/s) that is predicted for this magma by theoretical models (Papale et al. 1998), but the approach is promising because it makes use of both CO_2 and H_2O diffusive loss during decompression. Analyses of more complete profiles along a tubular inclusion may be possible in the future using an instrument like the NANOSIMS (Saal et al. 2008). Humphreys et al. (2008) used H_2O diffusion profiles in tubular glassy melt inclusions in plagioclase phenocrysts to estimate the decompression time (and average velocity) of the May 18 Mount St Helens plinian eruption. They obtained an average velocity of 37-64 m/s for the ascent of this magma from 7 km to the fragmentation depth (~500 m), significantly higher than estimates from mass eruption constraints (Scandone and Malone 1985; ~3 m/s, Geschwind and Rutherford 1995), and from numerical models (~4 m/s, Carey and Sigurdsson 1985; ~15 m/s, Papale et al. 1998). There are several possible explanations discussed for the high estimate from H_2O diffusion data, including the fact that at these ascent rates the decompression is too fast for H_2O diffusive equilibrium to be maintained; the lack of equilibrium would tend to give high ascent rate estimates as would the presence of CO_2 in the gas phase. Of course, the estimates from theoretical models may too low; the models continue to be revised as new data on critical parameters such as volatile solubility and melt viscosity becomes available.

GROUNDMASS CRYSTALLIZATION DURING MAGMA ASCENT

The degassing of the melt in ascending water-rich magma discussed in the previous section causes gas bubbles to form and/or grow, and also tends to cause the interstitial melt phase to crystallize, ascent time permitting. This crystallization occurs because any near adiabatic ascent path approaches and crosses the solidus for such magmas (Fig. 3). The process has been documented by the descriptions of numerous volcanic complexes (e.g., Mount St. Helens, etc.) where rapidly ascended magmas (explosive eruptions) erupt with a glassy groundmass but effusive eruptions of the same magma have experienced extensive groundmass

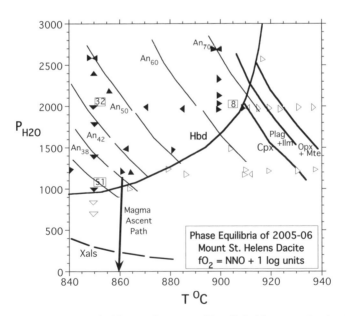

Figure 3. Phase equilibria of a typical intermediate composition, H_2O-rich magma showing the onset of crystallization of the major phenocryst phases plagioclse (Plag), orthopyroxene (Opx), Ca-pyroxene (Cpx), Amphibole (Amph), magnetite (Mte) and ilmenite (Ilm). The Mount St. Helens 2004-06 dacite ascended from a depth of approximately 4 km (120 MPa) where amphibole was stable and phenocryst content was ~50 vol%) at 860 °C in the final ascent. Numbers in boxes show the magma crystallinity at the P and T of the box. Curve labeled Xals is the approximate solidus for the dacite. (Modified after Rutherford and Devine 2008).

crystallization (Cashman 1992; Hammer et al. 1999; Couch et al. 2003). The process has also been demonstrated experimentally beginning as far back as the Tuttle and Bowen (1958) work on granite melts; these phase equilibrium studies show that melts with more than ~2 wt% dissolved water have anhydrous phenocryst-in curves that have a negative slope on a P-T diagram (Eggler and Burnham 1973). Although driven by the loss of volatiles from the melt with decreasing pressure, as discussed in the previous section, the crystallization of the groundmass melt does not necessarily occur at the same rate. For example, while the exsolution of water from a decompressing melt follows equilibrium solubility for decompression rates of < 0.25 MPa/s (magma ascent rates less than 5 m/s), a rhyolitic glass crystallizes much more slowly (Cashman 1988). The rate of crystallization will also differ for different temperatures and melt compositions, being slower for more silica-rich melts at a given water concentration.

A study of groundmass glass crystallization in the eruption products of the 1980-86 Mount St. Helens eruption products (Geschwind and Rutherford 1995) illustrated what can be done using the extent of groundmass crystallization to estimate magma ascent rates. Earlier work (Rutherford and Devine 1988) established that all of the erupted magmas came from a storage region at 8-11 km depth, and also established that the depth, temperature (880 °C) and dissolved water content (4.6 wt%) of the melt in that magma storage zone did not change over the time of these eruptions. In addition, the similarity of phenocryst and groundmass mineral compositions (Fe-Ti oxides) in these samples suggests that there was essentially no change in the temperature of any of these magmas during ascent. Geschwind and Rutherford (1995) concluded that differences observed in the groundmass crystallinity of these samples are the result of different ascent rates. Data on the extent of groundmass crystallization in natural Mount St. Helens samples were compared with the results of experiments on May 18 dacite samples decompressed from 220 to 2 MPa over different times, i.e., at different decompression rates (Fig. 4). These experiments demonstrate that nucleation and growth of microlites in the melt phase is detectable after a one-day constant-rate decompression from 220 to 2 MPa, and changes continuously for decompressions taking up to 8 days, as measured by

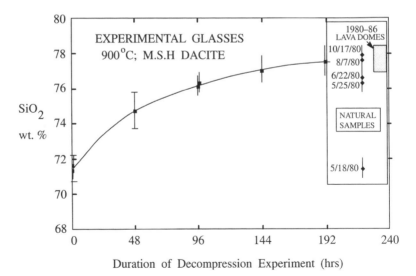

Figure 4. Plot of melt composition as a function of time in an H_2O-saturated dacite simulating adiabatic magma ascending from 220 MPa (8 km) to the surface at 900 °C. The compositions produced are compared with those in the natural dacite on the right. Ascent times of 100-200 hours are required to reproduce the natural groundmass glass of the post-May 18 Mount St. Helens lava dome samples. (Modified after Geschwind and Rutherford 1995.)

the change in melt (glass) composition. Using the composition of the remaining matrix glass in the experimental and natural samples, average magma ascent rates have been obtained which range from >0.2 m/s for the 1980 explosive plinian event to 0.01 to 0.02 m/s for the different dome forming eruptions of 1980-86 (Fig. 4). Nucleation is unlikely to have been a factor affecting groundmass crystallization in either the natural system or the experiments because the Mount St. Helens magma was crystal-rich prior to ascent. A similar study of a crystal-poor magma might show that growth of microlites from the groundmass melt takes much longer when phenocrysts are not present to serve as nucleation sites. Westrich et al. (1988) describe eruptions where the same phenocryst-poor, high-silica rhyolite erupted explosively and then effusively without groundmass crystallization in either case, even though the effusive magma must have lost the same amount of water over a much longer time frame.

Several recent studies have shown the difficulty in using groundmass crystallization to estimate magma ascent rates. Hammer and Rutherford (2002) demonstrated experimentally, that the crystal nucleation and growth rates in an ascending water-rich magma go through maxima at large and small degrees of under-cooling respectively. Thus, both the texture and the approach to equilibrium during crystallization depend on the ascent rate and also on whether the rate is steady or whether there are pauses in the rate of ascent. Although an adiabatic ascent is often assumed (e.g., Geschwind and Rutherford 1995), it is generally recognized that the ascending magma could experience either heating or cooling. Cooling could occur by loss of heat to the rocks surrounding the conduit or by gas bubble formation and expansion; heating would tend to occur from the release of the latent heat of crystallization as it occurred. Blundy et al. (2006) used plagioclase-melt geothermometry and pressures based on dissolved water in melt inclusions to identify an increase in temperature for final M.I. entrapment conditions during the 1980-82 Mount St. Helens eruptions (Fig. 5). The temperature increase, from 880 to as high as 940 °C for the post-May 18 eruptions, appears to confirm that release of latent heat of crystallization can raise the temperature of ascending magmas. A correlation of increased

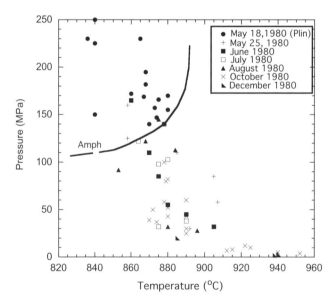

Figure 5. Plot of pressure of sample equilibration from volatiles in plagioclase melt inclusions vs. temperature from plagioclase-melt geothermometry for 1980 Mount St. Helens samples. Trend of the data shows an increase in temperature of more slowly ascended (and more crystalline) samples interpreted as the effect of latent heat build-up. (Modified after Blundy et al. 2006 to exclude May 18, 1980 and 1982 data).

crystallinity with increased M.I. sealing temperature was also determined in these same samples (Blundy et al. 2006). However, these eruptions occurred in a time period where the Mount St Helens eruption was alternating from periods of conduit-emptying explosions to effusive dome building (Swanson et al. 1987); thus the ascent path was definitely not constant-rate decompression, and some parts of each eruption may have spent days to weeks at a constant pressure. Furthermore, each hand sample from these eruptions contains amphiboles with a range of rim thicknesses that have been interpreted to indicate the samples are mixtures of magma newly ascended from depth with samples of the same magma that spent longer periods of time along the margins of the active conduit (Rutherford and Hill 1993). If the magma is a mixture, the determination of ascent rates from crystallinity arguments becomes very difficult unless the ascent path of the different magma components can be characterized.

The McCanta et al. (2007) study of the Black Butte dome near Mount Shasta (CA) indicates that this eruption involved an intermediate-composition, water-rich magma and it contains no evidence of magma mixing or interruptions in the magma ascent. It was determined experimentally that the unzoned plagioclase (Fig. 6), amphibole and titaniferous-magnetite phenocryst cores formed at a uniform set of conditions ($P = 350$ MPa, $T = 890$ °C). The melt interstitial to these phenocrysts was almost completely crystallized in the ascent of this magma to the surface as indicated by the flow alignment of the microphenocryts and microlites in the rock. Two plagioclase crystallization processes were used to assess the time involved in the magma ascent to the surface, and for comparison with the observed amphibole-melt reaction (next section). Normally zoned overgrowths (17±3 μm thick) on the uniform composition (An_{77}) cores range from the core composition to An_{40}, and represent essentially continuous crystallization during the magma ascent (assumed to be adiabatic). The plagioclase microphenocrysts (and microlites that make up much of the groundmass) have cores ≤ An_{65}, and represent a parallel crystallization process. According to the P-T ascent path the An_{60-65} plagioclase microphenocryst cores would have become stable at ~150 MPa. Crystal size distribution (CSD) analysis of the plagioclase crystal population (Fig. 7) verify that there were two separate plagioclase growth events; the phenocryst growth in the magma storage zone at ~300 MPa, and the microphenocrysts and phenocrysts that grew during the ascent. Using an ascent rate based on amphibole rim thicknesses, a growth rate for the plagioclase phenocryst rims and microphenocrysts ($G = 2.5$-8.7×10^{-8} mm/s) was obtained that is in the range ((1×10^{-7} to 2×10^{-11}) of previous determinations for experimental (Couch et al. 2003) and natural (Cashman 1992) samples; this agreement supports the model proposed for the Black Butte magma ascent.

Figure 6. Photomicrographs of plagioclase phenocrysts with relatively homogenous cores and thin (15-20 μm), normally zone rims in a groundmass rich in plagioclase microphenocrysts (50-150 μm) and microlites (<50 μm). Phenocryst cores are unzoned; thin (~20 μm) rims are normally zoned. (Modified after McCanta et al. 2007.)

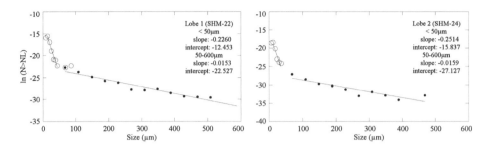

Figure 7. CSD plots of plagioclase for representative samples of the Black Butte (CA) dacite. Two discrete populations are present as indicated by the two line segments required to fit the data, one for phenocrysts (>150 μm) and microphenocrysts (50-150 μm) and one for the microlites (<50 μm). [Used with permission of Elsevier from McCanta et al. (2007), *J Volc Geotherm Res*, Vol. 160, Fig. 9, p. 276. *http://www.sciencedirect.com/science/journal/03770273*].

PHENOCRYST-MELT REACTIONS DURING MAGMA ASCENT

Reactions between volatile-bearing minerals and melt during magma ascent

Another reaction that is caused by the loss of volatiles from the melt as magma rises toward the surface is the breakdown of volatile-bearing phenocrysts, such as amphibole and biotite. Sulfur is also partitioned from the melt to the gas phase during magma ascent, and thus anhydrite and FeS, which occur in some volcanic rocks, would also tend to breakdown. Most of these reactions will be controlled by the kinetics of the reaction between the phenocrysts and the volatile-depleted melt. The rates of the phenocryst-melt reactions produced by decompression have generally not been determined because they are complex functions of many variables, including mineral and glass compositions, melt viscosity, and temperature. However, the rates of reaction between amphibole phenocrysts and the coexisting rhyolitic melt have been studied (Rutherford and Hill 1993; Browne and Gardner 2006), and these calibrations can be used to estimate ascent rates for dacite and andesite composition magmas.

The rims produced by the amphibole-melt reaction have been widely observed in volcanic rocks, particularly andesite to rhyolite composition effusive eruptions (MacGregor 1938; Garcia and Jacobson 1979). Textures suggest that the rim thickness can be accurately measured at least where they occur in more viscous dacite and andesite magmas. The textures of these reaction rims (Fig. 8) also indicate the reaction is between the amphibole and the melt and will begin when the melt no longer contains sufficient dissolved water to stabilize the amphibole (Rutherford and Hill 1993). Depending on the temperature of the ascending magma, the breakdown tends to begin when the pressure decreases below about 100 MPa (Fig. 3). The reaction of the amphibole with the water-deficient melt produces a fine-grained assemblage of Ca-pyroxene, low Ca-pyroxene, plagioclase and magnetite (Rutherford and Hill 1993). Buckley et al. (2006) re-examined this reaction, analyzing amphibole rims in the recent Soufriere Hills, Montserrat andesite; they emphasized that the reaction rim must be an open system, losing alkalis and H_2O to the surrounding groundmass.

Mount St. Helens sample studies show no detectable reaction between the amphibole phenocrysts and the groundmass glass (melt) in the May 18, 1980, pumice samples, even though the melt clearly lost all but about 0.3 wt% of the ~ 4.6 wt% dissolved water it contained before erupting (Rutherford et al. 1985). In almost all subsequently erupted samples, however, the amphibole phenocrysts have reaction rims at the contact of the crystal with the groundmass melt. The dome-forming eruption of October 1986 was an exception, containing only amphiboles without reaction rims. Each of the other post-May 18 samples had a main

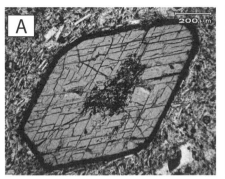 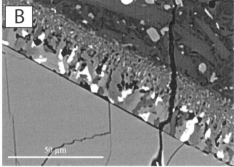

Figure 8. Photomicrograph showing a typical amphibole decompression-induced reaction rim on a 2 mm phenocryst with a hollow core (A) from dacite lava dome, and (B) a BSE close-up image of the same phenocryst (rim) showing coarse to fine texture outward from the amphibole. The phases in the rim, from white to black are magnetite, low- and high-Ca pyroxene, plagioclase, glass and vesicles. (Modified after McCanta et al. 2007.)

population of amphibole phenocrysts (30 to 100 vol% of the crystals present) with uniform thickness (0 to 50 μm range) reaction rims, and a second group with rims ranging to greater thicknesses. These amphiboles are interpreted to represent, respectively, a main batch of magma that ascended directly from the storage region at 8-11 km beneath the surface, and magma from previous eruptions left temporarily in the conduit before being mixed into new magma from depth. An experimental calibration of the amphibole breakdown reaction was made for a constant-rate ascent of the 1980 Mount St. Helens dacite magma at 900 °C (Fig. 9). Experiments that simulated ascent from 8 km depth to the surface in less than 4 days showed no reaction rims on amphibole crystals; 2 μm wide rims were produced in 7-day ascents, and 32 μm rims formed in 20 days. Using this calibration, the magma ascent rates during the lava

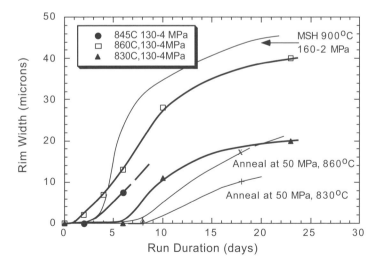

Figure 9. Plot of experimentally determined amphibole reaction-rim widths as a function of a constant rate ascent duration at temperatures of 830 to 900 °C. Two curves labeled "anneal" show rim development in experiments run at the indicated conditions using a 160 MPa assemblage as the starting material. (Modified after Rutherford and Devine 2003.)

dome phase of the Mount St. Helens eruption were determined to vary from as low as 0.004 m/s up to 0.015 m/s. This is in contrast with the 2 to 3 m/s ascent rate during the May 18, 1980 plinian phase of the eruption based on the mass eruption rate (Carey and Sigurdsson 1985; Geschwind and Rutherford 1995).

Rutherford and Devine (2003) estimated magma ascent rates by measuring the width of amphibole reaction rims for samples collected over time (1995-2002) as Soufriere Hills volcano on Montserrat erupted a phenocryst-rich siliceous andesite. This dome-forming lava experienced phenocryst-melt equilibration in a magma storage region at 5-6 km depth (130 MPa) and 830 °C. Reaction rims on amphibole crystals adjacent to groundmass melt showed significant changes in samples collected over the next several years. The first magma erupted contains amphiboles with 120 μm thick rims. The magma erupted two months later, in February 1996, has only 18 μm-thick rims, and the rims are only 10 μm wide in an April 1996 eruption. Most amphiboles in magma erupted in July through September 1996 contain no reaction rims. The phenocryst assemblage in these magmas is the same throughout this period, and thus the rims are not produced by changes in the magma storage region; they formed during magma ascent. Using an experimental calibration of rim width versus time at 830 °C (Fig. 9), absolute ascent rates and ascent rate changes can be calculated for the Montserrat eruption. The magma ascent rate increased from 0.001 to 0.012 m/s during the first 300 days of this eruption. Interestingly, there was an increase in the mass extrusion rate over this same time period (Fig. 10). The amphibole rim data show that this increase in extrusion rate was not produced simply by an increase in the size of the conduit; it required an increased magma ascent rate. Another significant observation for this eruption is the association of numerous explosive eruptions from the summit of the dome with periods of high magma ascent rate. These explosions indicate periods of less efficient separation of gas and magma, a logical consequence of higher magma ascent rates.

McCanta et al. (2007) saw evidence of a single continuous magma ascent at Black Butte Ca. based on the texture and composition of plagioclase of the dacite samples (Fig. 6). Looking

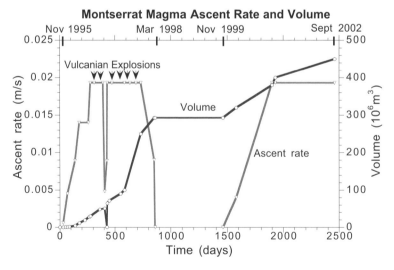

Figure 10. Ascent-rate and erupted volume vs. time for the 1995-2002 andesite eruptions at Soufriere Hills, Montserrat. Ascent rate is based on amphibole rim-thickness development (Figure 11), which goes to zero early and late in the time period shown. Note volcanian explosions associated with the high ascent rate eruptions in 1996-97. (Modified after Rutherford and Devine 2003.)

at the thickness of breakdown rims on amphiboles (Fig. 8) in the same samples, they found a peak at 34±10 μm in all four lobes of the dome. The small spread in the rim thicknesses observed confirms that the Black Butte eruption was a continuous event and there was no mingling of magma batches with different ascent histories as at Mount St Helens. Using the measured rim thickness range (24-44 μm) and the Rutherford and Hill (1993) calibration of rim growth rate, a constant rate magma ascent rate of 0.004-0.006 m/s was determined for the dacite magma ascent from ~200 MPa or ~8 km to the surface at 890 °C. This ascent rate is supported by the fact it yields a plagioclase growth rate (G = 2.5-8.7 × 10^{-8} mm/s; previous section) that is similar to other estimates of plagioclase growth in similar composition systems (Cashman 1992; Izbekov et al. 2002).

Several lines of evidence suggest that many intermediate and silicic composition eruptions involve magma ascent rates that are different in different parts of the conduit and also over time. The post-May 18 (1980) eruptions at Mount St. Helens, regularly involved a slow effusion of magma followed by an explosive eruption that rapidly removed significant amounts of magma that was in the conduit (Pallister et al. 1992). The sequence of explosions followed by additional slow lava effusion suggests that these eruptions involved magma batches that ascended at different rates. Similar eruptions occurred at Mount Pinatubo (Hoblitt et al. 1996), and at Soufriere Hills on Montserrat (Sparks et al. 1998). Browne and Gardner (2006) suggested that variations in the ascent rate occurred during the 1989-90 eruptions at Redoubt volcano. These variations are generally attributed to magma degassing which affects both the buoyancy of a rising magma and the physical properties of the magma as gas is lost to bubbles and bubbles may separate from the magma. By performing both single step and multi-step decompressions to a given final pressure, Browne and Gardner (2006) showed that rim growth on amphiboles in dacite magma is extremely slow just outside the stability field (at 840 °C) and also at very low pressures (Fig. 11). The highest rate of rim growth occurs at 30-40 MPa below the amphibole stability field in the single step decompressions, but the rim forms preferentially at lower pressures (10-50 MPa) in multi-step (~ constant-rate) decompression

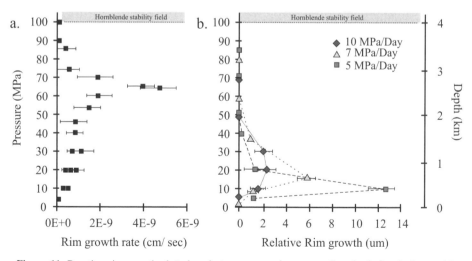

Figure 11. Reaction rim growth plotted against pressure and corresponding depth for single-step (a) and multi-step (b) decompression experiments on a 1989 Redoubt dacite. Reaction rim growth rate is equal to the reaction rim width/ total time below the amphibole stability boundary (from phase equilibria experiments). Relative rim growth is the difference in rim thickness from sequential experiments quenched at lower and lower pressures. [Published with permission of Elsevier from Browne and Gardner (2006), *Earth Planet Sci Lett*, Vol. 246, Fig. 5. p. 170. http://www.sciencedirect.com/science/journal/0012821X].

experiments. Clearly, the possibility of variations in ascent rate must be considered when using reaction rims on amphibole (or any other phenocryst-melt reaction driven by decompression) to determine magma ascent rates.

The above point about variations in ascent rate is further illustrated by the Nicholis and Rutherford (2004) study of amphibole rims developed in hawaiite (trachybasalt) that erupted as tephra and lava flows near Yucca Mountain (Nevada, USA). Amphiboles do not occur in all eruptions in the area, but when present, there are several phenocrysts in any sample collected, and many have a thick (~30 μm), fine-grained reaction rim. Nicholis and Rutherford (2004) determined the conditions where the amphibole would be stable ($P > 180$ MPa; $T \sim 900$ °C) and determined experimentally that no constant-rate decompression would produce a reaction rim like those in the natural sample. Amphiboles remaining in all but the very fast magma decompressions were invariably rounded and reduced in modal abundance, suggesting dissolution rather than crystallization at the rim (see Browne and Gardner 2006). The amphibole breaking down was replaced by additions to preexisting anhydrous phenocrysts in all experiments that produced groundmass crystallization comparable to that of the natural samples. It is speculated that these observations are due to the relatively low viscosity of the H_2O-rich trachybasalt magma. Comparison of plagioclase microlites in experimental samples with those in natural tephras (CSD analysis) indicated that magma ascent rates were >0.04 m/s for these hawaiites. The rimmed amphiboles were only produced when the ascending magma was additionally held for several days at ~ 20 MPa (~800 m depth), and Nicholis and Rutherford (2004) postulated that the final eruption product was produced by mixing samples of magma stored at the shallow depth into rapidly ascending magma of the same composition.

The 2004-08 eruption of Mount St. Helens consists of a series of domes or spines composed of phenocryst-rich dacite almost identical to the magma erupted there in 1980-86, but with some interesting differences. The recent eruption involves a degassed magma with low volatile emissions (Gerlach et al. 2008), has an Fe-Ti oxide temperature of 860±15 °C, is phenocryst rich (~40%), and has erupted with a largely crystalline groundmass (Pallister et al. 2008). The 2004-08 magma contains amphibole, sometimes rounded, with ~3 μm thick reaction rims in the first-erupted material and 5 μm thick rims in all subsequent eruptions (Thornber et al. 2008), somewhat less than the 10-20 μm thick rims in the 1980-85 lava domes. The thinner reaction rims in the recent magma could be the result of the lower temperature of the recent eruptions, but the amphiboles in magmas erupting in mid 2005 and beyond (when the mass eruption rate had dropped from ~ 6 to < 2 m^3/s (Schilling et al. 2008)) all have a thin outer zone of F-rich amphibole (Rutherford and Devine 2008). Thin breakdown rims are present on this F- and Si-rich amphibole. DeJesus and Rutherford (2007) investigated the origin and significance of the new-growth zone of F-rich amphibole. They suggest the factors involved in its formation are (1) increases in F and SiO_2 activity in the melt produced by crystallization during relatively slow magma ascent and (2) increased time at temperatures outside the stability field of OH-rich amphibole. A mechanism such as dissolution or abrasion of pre-existing amphibole is also required to get new amphibole growth in this magma since it has no modal or normative Ca-pyroxene. As shown in Figure 12, the F-, Si- and Fe-rich amphibole is stable to pressures as low as 50 MPa in dacitic magma at 860 °C. A decompression-induced reaction rim produced on this amphibole during magma ascent would only form after the pressure dropped below 50 MPa. The rim would be much thinner than rims developed on amphiboles without this outer F-rich layer and ascending at the same rate. Ascent rates of 0.005-0.015 m/s calculated for the 2004-06 Mount St. Helens eruption using the OH amphibole rim data (Thornber et al. 2008) appear applicable to the early 2005 spine, but give a maximum estimate of magma ascent for later 2005-06 eruptions. Although the presence of the F-, Fe-, and SiO_2-rich rims on amphibole phenocrysts is clearly important in assessing the significance of amphibole breakdown reactions, such rims have not yet been described for other eruptions (DeJesus and Rutherford 2007), and may be unique to the conditions and magma of this particular MSH eruption.

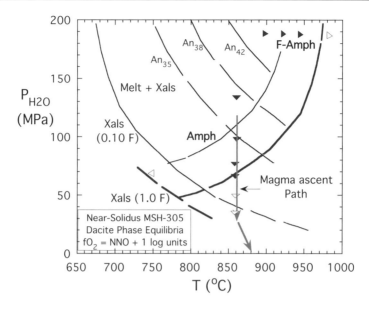

Figure 12. P_{H_2O}-Temperatue diagram showing the stability of F-rich amphibole relative to OH-rich amphibole from experiments on a 2004-06 Mount St. Helens dacite composition. F-Amph and Xals (1.0 F) show the change in amphibole stability and in the solidus with 1 wt% F in the melt; the light-weight lines show similar curves for the dacite with low F (1000 ppm) in amphibole and melt (Rutherford and Devine 2008). Adiabatic magma ascent line (with latent heat) shows that the decomposition of a F-rich amphibole will occur over a very short segment of the magma ascent path relative to the OH-rich phenocryst. (Modified after DeJesus and Rutherford 2007.)

Phenocryst-melt reactions driven by pre-eruption magma mixing

Evidence of phenocryst reactions that are the result of mingling and mixing of two contrasting magmas is almost ubiquitous in volcanic eruption products (Eichelberger 1978; Pallister et al. 1992; Nakada and Fuji 1993). In fact, zoning in phenocrysts such as plagioclase and amphibole and olivine often record the number of mixing events (Singer et al. 1995; Rutherford and Devine 2003; Ginibre et al. 2007). In most cases, it is not possible to use these disequilibria to assess magma ascent because the timing of the last mixing event relative to the onset of ascent and eruption is unknown, or because the depth of the mixing event is unknown. However, compositional zoning and disequilibrium in coexisting Ti-magnetite and ilmenite in magmas has proven to be useful in determining ascent rates because the time scales of re-equilibration are short (days to weeks). Nakamura (1995) extracted time information from zoned titanomagnetite in the 1991-95 Mount Unzen magma where a high-temperature basaltic andesite had mingled with a pre-existing dacite just prior to eruption. Enclaves of the phenocryst-poor mafic magma occur in the dacite and some of the basaltic andesite was also mixed into the dacite. Mingling is indicated by small blade-shaped Ca-rich plagioclase and pargasitic amphibole crystals characteristic of the enclaves in the groundmass of the mixed magma. A number of reactions began as a result of the magma mixing including embayment of quartz and biotite phenocrysts, growth of pargasitic amphibole rims on biotite phenocrysts and Ti-zoning in magnetite (Nakada and Fugii 1993). The thickness of the Ti-rich diffusion profile at the outer margin of magnetite phenocrysts is typically 20-30 µm for samples erupted over the 1991-93 time period. Nakamura (1995) measured and modeled the Ti-zoning in magnetite (Fig. 13) by assuming the temperature of the dacite was 800 °C, and concluded that the time from mixing to quench at the surface was a few months. Venezky and Rutherford

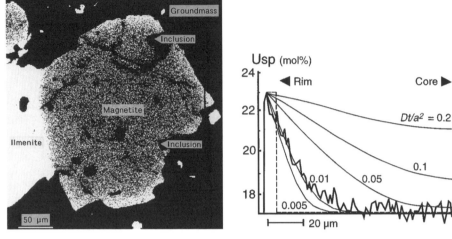

Figure 13. Ti X-ray image of zoned titanomagnetite and adjacent ilmenite phenocysts in the Mount Unzen 1991 dacite (left). Plot on the right shows the Ulvospinel component in the natural magnetite compared to a diffusion model of rim to core zoning produced at 800 °C. Dt/a2 = 0.1 yields a time of 2-4 months at 800 °C. (Published with permission of the Geological Society of America from Nakamura (1995), *Geology*, Vol. 23, Figs. 2, 3, p 808).

(1999) determined that the Mount Unzen mixed magma temperature was more likely 850-870 °C based on the composition of coexisting magnetite and ilmenite in the phenocryst outer margins and on coexisting pairs in the reaction rims around biotite. These temperatures were confirmed by the P-T-X_{H_2O} conditions (P = 160 MPa; T_1 = 790; T_2 = 1030 °C, $X_{H_2O,1}$ = 5 wt%; $X_{H_2O,2}$ = 3 wt%) determined from phase equilibria for the two end-member magmas involved in the mixing and their proportions. Using the Freer and Hauptman (1978) data for Ti diffusion in magnetite, Venezky and Rutherford (1999) determined that the 20-25 μm thick diffusion zone in the 1991-93 Mount Unzen magnetite phenocrysts would be produced in 11 to 30 days at 900 and 850 °C respectively. If ascent began at the time of mixing, the average ascent from 7 km to the surface would have been at the rate of 0.003 to 0.007 m/s. If the magma ascent began somewhat after mixing, the rate obviously would have been faster. The fact that decompression-induced amphibole reaction rims are not always present suggests that different batches of the erupted Mount Unzen magma represent a short range of times between mixing and ascent.

Devine et al. (2003) also used Ti zoning in titanomagnetite to estimate the rate of magma ascent for the andesite erupting at Soufriere Hills, Montserrat in 2001-02. As in the case of Mount Unzen, the zoning reflects the mixing of hot basaltic andesite into a preexisting storage zone. In this case the pre-mixing storage zone contained a siliceous andesite at ~830 °C at a pressure of 130 MPa (Barclay et al. 1998). The temperature of the hotter magma is not well constrained, but must have been >1050 °C because it contained no phenocrysts yet was able to crystallize pargasite when quenched against the cooler siliceous andesite. The mixing in of the mafic magma raised the temperature of the mixed magma to as high as 860 °C according to the phase equilibria (Rutherford and Devine 2003). If the temperature of the mixed magma were higher than 860 °C for as little as two days, a characteristic breakdown of the amphibole phenocrysts would have been produced and is not observed. Based on diffusion calculations and experiments at 860 °C, Devine et al. (2003) conclude that the Ti-rich rims on magnetite (Fig. 14) required a mixing and ascent time of ~20 days at this temperature. However, many batches of the magma that contain these zoned magnetite crystals also contain amphiboles with no decompression-induced reaction rims (Fig. 10). According to decompression experiments

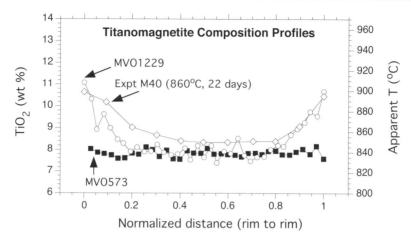

Figure 14. TiO$_2$ concentrations in profiles across titanomagnetite crystals in 1995-2002 Soufriere Hills volcano samples and one experimental sample. MVO573 (Sept. 1997) is from a rapidly ascended magma sample (no reaction on amphibole phenocrysts) and was used in experiments such as M40. The profile across the M40 titanomagnetite closely approximates those in natural samples like MVO1229 (Sept. 2001) and indicates the time required after mixing to create the zoning observed. Apparent temperature is calculated for each Ti-magnetite composition paired with the adjacent unzoned ilmenite. (Modified after Rutherford and Devine 2003.)

on samples of this material, the lack of rims on the amphibole requires ascent from 130 MPa (5-6 km) to the surface in < 4 days, a steady state ascent rate of > 0.017 m/s. Rutherford and Devine (2003) conclude that the < 4 day ascent began as many as 18 days after the heating and mixing event that affected the erupting magma in the period Sept. 2001-(MVO-1228-9), but at least some of the mixed magma reaching the surface in Sept. 2002 was heated for no more than 10 days (no Ti-zoning in magnetite) before ascending rapidly (< 4 days) to the surface.

MAGMA ASCENT RATES FROM MASS ERUPTION RATES

Measurements of the mass eruption rate of magma can be used to calculate conduit diameter and ascent velocity, if certain assumptions are made about an eruption (Wilson et al. 1980). This technique is illustrated by studies of the May 18, 1980, plinian Mount St. Helens eruption (Carey and Sigurdsson 1985; Geschwind and Rutherford 1995). Magma flowing through a conduit below the depth where it is saturated with a gas phase can be described as a one-phase, laminar flow of an essentially Newtonian fluid. For this case, magma ascent in the conduit can be modeled using the Hagen-Poiseuille law

$$M = (\Delta P/\Delta L) \cdot (\Pi \, \rho_F r^4)/8\eta$$

where M is mass flux (in kg/s), $\Delta P/\Delta L$ is the pressure gradient driving the flow, ρ_F is the fluid (magma) density, η is fluid viscosity, and r is conduit radius. Exsolution of a vapor phase changes the bulk density and viscosity of the magma and thus affects flow conditions, but exsolution is slow at first and does not significantly affect the flow until the magma is within a kilometer or two of the surface (Wilson et al. 1980). Calculations are made for depths greater than 1 km, assuming that the driving force for flow is due to the density difference between the surrounding rocks and the magma. For quantitative applications of the Hagen-Poiseuille law, numerical values for the magma mass flux, density, and viscosity of Mount St. Helens magmas are needed. Using recently revised data for these parameters (Geschwind and Rutherford 1995), and assuming the conduit radius remains constant at depth, the conduit radius during the explosive 1980 eruption of Mount St. Helens is calculated to be 33 m. For dome-forming eruptions in the period October 1980 to 1982, the calculated radius decreased to 7 m and remained essentially constant (Fig. 15).

Ascent velocities for the Mount St. Helens eruptions were calculated from the relationship expressing conservation of mass, $M = \Pi r^2 \rho_F u$, where u is the velocity. Given the conduit radii calculated above, conservation of mass requires that ascent velocity decreased from 2-3 m/s on May 18, 1980, to 0.08 m/s for the October 1980 and 1981 eruptions (Fig. 15). The former estimate is consistent with vent diameter estimates and the vent velocity is required to explain plinian ash columns (Carey and Sigurdsson 1985). However, the latter velocity leads to an ascent time of 30 hours from a depth of 8.5 km which is less than the approximately 4-8 days indicated by the petrological methods for the post-May 18, dome-forming eruptions (Rutherford and Hill 1993). The assumptions of constant magma viscosity, and even constant conduit radius are almost certainly wrong for the dome-forming eruptions because there were dome-destroying explosions followed by slow effusive dome growth during this period (Swanson et al. 1987). Excessive crystallization of microlites during slow ascent periods increases magma viscosity and decreases the magma velocity in the upper parts of the conduit. This could explain the differences between the calculated velocity and the velocity derived from hornblende breakdown for the 1980-82 dome lavas, and has also been modeled as significant for explaining the cycles of effusive dome growth and volcanian explosions that occurred at Soufriere Hills Montserrat in 1996-98 (Watts et al. 2002). Another complicating factor in calculating magma ascent rates of intermediate composition eruptions is the fact these magmas commonly contain evidence of a magma mixing event shortly before the eruption. This observation, which suggests that overpressures may have developed in the magma reservoir, is true for Soufriere Hills (Murphy et al. 2000; Rutherford and Devine 2003) and for the 1991 Mount Pinatubo eruptions (Pallister et al. 1996).

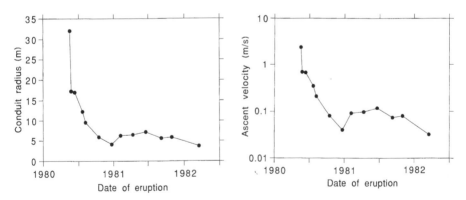

Figure 15. Plots showing change in conduit radius (left) and magma ascent velocity (right) with eruption date during the 1980-82 eruptions of Mount St. Helens as controlled by the observed Mass eruption rate and the Hagen-Poiseuille law (see text). Note the rapid decrease in both conduit radius and ascent rate during 1980. [Published with kind permission from Springer Science+Business Media from Geschwind and Rutherford (1995) *Bull Volc*, Vol. 57, Figs. 8, 9, p. 367.]

MAGMA ASCENT FROM SEISMIC DATA

Earthquakes are commonly associated with volcanic eruptions and volcano unrest and have been effectively used for monitoring and forecasting eruptions (e.g., Scandone and Malone 1985; Decker 1987; Endo et al. 1996). Volcanic earthquakes generally occur in four main types, high- and low-frequency earthquakes, explosion earthquakes, and tremor. The direct association of earthquakes with magma migration can be ambiguous, however, because earthquake generation requires stressed and brittle rock, and so aseismic subvolcanic regions may be unstressed and unfractured or hot and ductile. On the other hand, pressured magma may generate earthquakes in brittle rock adjacent to magma-bearing conduits. Hence, magma zones may be either seismic or aseismic.

One case of an association of earthquakes and possible magma movement occurred during the 1977 deflation of Krafla Volcano in Iceland. At the same time that the volcano began to deflate, earthquakes began under the central volcano and then migrated along the adjacent rift zone (Tryggvason 1994). When the hypocenters of this migrating tremor passed a geothermal drill-hole, fresh basaltic pumice erupted from the hole. The tremor hypocenters migrated along the rift system at a rate of about 0.5 m/s, but it is not clear whether magma moved at this rate or whether a pressure pulse moved through existing magma at this rate. Migration of tremor hypocenters like that at Krafla Volcano has also been observed in the East and Southwest Rift Zones of Kilauea Volcano, Hawaii (Aki and Koyanagi 1981; Klein et al. 1987). These swarms may define the motions of magma in the rift systems, as many of them are related to the eruption of basaltic magma. The rates of down-rift swarm movement span a range of about 0.005 to 8.5 m/s. In a few cases, upward movement of earthquake swarms in the rift is also identified, with rates on order of 0.03 to 1.7 m/s (Klein et al. 1987). Interestingly, this upward migration tends to slow as the magma approaches the surface.

Although it is expected that earthquake hypocenters should migrate as magma migrates beneath volcanoes, as in the above-described cases of Krafla and Hawaii, there are surprisingly few documented cases of such trends. A magma ascent velocity of 0.6-0.7 m/s is estimated for the explosive May 18, 1980 eruption of Mount St. Helens based on the arrival time of new magma at the surface as detected initially by a change in the seismic signals adjacent to the 8-11 km deep storage zone (Scandone and Malone 1985). Later in the 1980 phase of the eruption, seismic signals at depth and the following onset of an eruption at the surface indicate an average ascent velocity of 0.007-0.01 m/s, essentially identical to estimates from amphibole reaction rims (Rutherford and Hill 1993). Kagiyama et al. (1999) estimated the rise rate of magma at Mount Unzen (Japan) in 1991 as being 20 m/day (0.00023 m/s) using the depth and time of seismic events. This estimate is significantly lower than the rate obtained from Fe-Ti oxide zoning (above), possibly because it represents the average velocity of initial vent opening as compared to an ascent rate once flow has been established.

There have been two areas of volcano seismic study that have yielded new insight into magma ascent rates in the past 12 years. Endo et al. (1996) refined their modeling of pre-eruption real time seismic amplitude measurements (RSAM) for several volcanic eruptions including 1980-86 Mount St Helens, and 1991 Mount Pinatubo. They found that for many dome-building eruptions an exponential curve was a good fit to the build of RSAM counts prior to an eruption (Fig. 16). Using the deepest seismicity at the onset of the buildup, they determined that the 1986 dome magma at Mount St. Helens ascended from 1.4 km to the surface in 12 hours, i.e., magma ascended at an average rate of 0.032 m/s. This compares reasonable well with a minimum estimate of <0.018 m/s for the average ascent rate from 8 km depth to the surface based on the lack of amphibole reaction rims in this magma (Rutherford and Hill 1993). Another estimate of magma ascent rate comes from the work of White (1996) on the deep long period (DLP) earthquakes associated with the Mount Pinatubo eruptions of 1991. A swarm of DLP's were located 28-35 km beneath the Pinatubo summit between May 26 and 28, 1991, about 11.5 days prior to the arrival of a dome-forming mixed magma at the surface on June 7. The mixed magma consisted of 36 vol % basalt and 64% dacite (Pallister et al. 1996), the dacite having been stored at 7-11 km depth according to the dacite petrology. White (1996) interprets the DLP's to have been produced by the flow of deep-seated basaltic

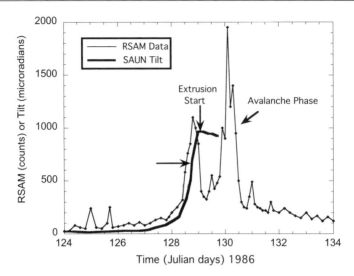

Figure 16. Real Time Seismic amplitude (RSAM) counts build up at station YEL at Mount St. Helens in May, 1986 compared to radial tilt at nearby station SAUN. RSAM for YEL peaked at 1400 UT, May 8 (Julian day 128), 1986. Tilt on the SAUN tiltmeter reversed direction several hours later; the extrusion of lava is assumed to have occurred at the tilt reversal. (Modified after Endo and Murray 1992.)

magma upward from near the base of the crust, a flow that created vibrating fluid-filled cracks adjacent to over-pressurized magma (Chouet et al. 1994). White (1996) speculates that the batch of basaltic magma that produced the May 26 DLP's may have been the same magma that reached the surface on June 7 having passed by and partially mingled with dacite. The average ascent rate from 35-40 km to the surface would be 0.031 to 0.035 m/s in such a case. Similar basaltic magma ascent rates were determined over this depth range at Kilauea (Decker 1987).

MAGMA ASCENT RATES BASED ON XENOLITH-MELT REACTIONS

Many alkali basalt magmas erupt at the surface carrying xenoliths with clear textural and compositional evidence to indicate that the inclusion was not in equilibrium with the host magma. Among these xenoliths, spinel and garnet peridotite are clearly derived from the earth's mantle based on their high pressure phase assemblage, and must have been transported from a source in the upper mantle (Carter 1970; Irving 1980). Reactions observed between the inclusion and the host magma include phenocryst dissolution (Brearley and Scarfe 1986; Edwards and Russell 1996), profiles of diffusive exchange (Klugel 1998), and zoning in inclusion glasses (Frey and Green 1974). The extent of reaction that takes place at the magma-inclusion contact can be used, along with available kinetic or diffusion data, to estimate the rate of magma ascent if certain constraints can be placed on the ascent process. For example, it must be possible to determine the depth at which the inclusion was incorporated into the magma, and it must also be possible to assess any changes in temperature during ascent. This section describes three examples where ascent-rate estimates have been made using such reactions.

Alkali basalt laden with inclusions of peridotite has erupted both historically (six eruptions in the past 500 years) and prehistorically on La Palma in the Canary Islands (Klugel et al. 1997). The inclusions are spinel-bearing peridotites ranging up to 15 cm in diameter, and they are commonly encased in a crystalline selvage (reaction) zone (< 1 mm thick) of titanium-rich augite, amphibole, and magnetite. The selvage minerals are fine-grained at the inclusion contact, and coarsen toward the host magma. Fluid inclusions in recrystallized grains and in the selvage minerals are essentially pure CO_2, and their densities indicate that the selvage minerals and recrystallized grains in the peridotite grew at pressures of 200 to 400 MPa (7-15 km depth). Olivine crystals in the peridotite adjacent to the selvage zone have developed diffusion zones that are 0.9 to 2.6 mm wide as a result of reaction with the magma (Fig. 17). This diffusion zone is interpreted to have developed in olivine at the xenolith margins while the magma was stored in a deep (7 to 15 km) crustal reservoir. Using diffusion data for Fe and Mg in olivine for a magma temperature of 1200 °C, the measured diffusion zone thicknesses yield development times of 8 to 110 years. These times are interpreted as indicating the length of time the inclusions spent in the magma storage zone and in ascent to the surface. Interestingly, there are some fractures and surfaces in the xenoliths in the Canary Island samples that have no selvage zones. The diffusion zones in olivine adjacent to the melt along these surfaces are 0 to 0.02 mm wide. The fractures with no selvage zone are interpreted to have formed by a fracturing event at the time of the final eruption (Klugel et al. 1997). The times calculated for development of the thin diffusion zones range from 0 to 4 days. If it is correct to assume the surfaces formed when the magma began to ascend, then average ascent rates of >0.06 m/s are calculated for the movement of this alkali basalt from 11 km to the surface. This is in the same general range as the >0.2 m/s magma ascent rates necessary to carry the xenoliths to the surface in this magma (Spera 1984). The lack of diffusion rims of intermediate thickness on olivine in the xenoliths is interpreted to mean that

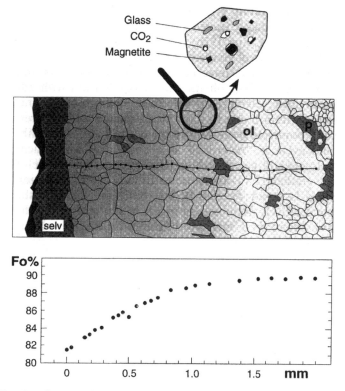

Figure 17. Drawing of a contact between an olivine-rich xenolith and host basalt with a thin selvage zone at the boundary as is typical of such occurrences in La Palma lavas (Canary Islands). Middle panel shows zoning profile (in olivine) along traverse in top view. Lowest panel shows diffusion-based solution for times required to form large 1-2 mm diffusion path (6-83 years) and a thin reaction zone formed on cracks in the olivine (<4 days). (Modified after Klugel 1998.)

the xenolith-bearing magma was stored in the lower crust and was not affected by a fracture-producing event in the 8 to 110 year period before the eruption began.

A second petrologic indicator of magma ascent rate in alkali basalt and kimberlite magmas is the amount of reaction that occurs between melt and garnet-bearing xenoliths; the garnet in these lithologies is unstable in contact with melt at depths <40 km in the earth's continental crust for temperatures >850 °C (Green and Ringwood 1967). Based on the rate of garnet conversion to spinel in experiments, O'Hara et al. (1971) concluded that well-preserved garnet lherzolite could not have been exposed to temperatures as high as 850 °C at pressures <1.2 GPa for more than a few hours. More recently, Canil and Fedortchouk (1999) have experimentally studied the rate of garnet xenocryst dissolution in kimberlitic melt at pressures of 1.5 to 2.5 GPa and 1400 to 1500 °C, and compared the results to dissolution and reaction rims seen in natural kimberlites. They found that the dissolution rate of garnet in H_2O-saturated kimberlite is very rapid, linear in time at a given T, and not significantly affected by differences in pressure. They also report dissolution features on garnet xenocrysts in a natural kimberlite ranging up to 25 μm in width on 1 to 8 mm diameter spherical crystals. Based on these studies, Canil and Fedortchouck (1999) conclude that the garnet xenocrysts spent <1 hour in 1200 °C magma once the melt and garnet were in contact (Fig. 18). Two factors prevent us from using this dissolution data to calculate other than minimum ascent rates. First, it is theorized that garnet xenocrysts are exposed to the kimberlitic melt when xenoliths are fragmented, but when this happens is not known; it may be at great depth, or it may be at the base of the diatreme zone at approximately 1-3 km (Mitchell et al. 1980; Sparks et al. 2006). Secondly, the data and calculations are done for a magma ascending adiabatically (Fig. 18), but if a gas phase is generated at significant depth, as seems likely based on the evidence for abundant CO_2 in kimberlite magmas, then a significant cooling is calculated from the work done by the expanding gas phase, and the temperature could drop from >1300 °C at 30 km (10 wt% CO_2) to ~1000 °C at the surface (Sparks et al. 2006). Canil and Fedortchouck (1999) consider the two above factors, but note that the 25 μm

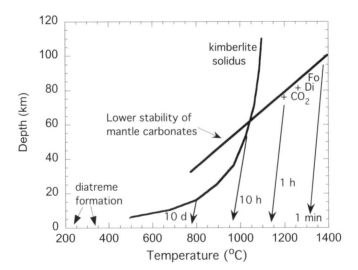

Figure 18. Schematic plot of depth vs. temperature showing possible ascent paths for the Grizzley (Canada) kimberlite magma based on garnet dissolution times. The arrows show simple adiabatic ascent paths possible at different temperatures and the times required for garnet preservation on that path. The kimberlite solidus is a minimum temperature constraint, and the lower pressure stability limit of carbonate in the mantle is used as a possible depth constraint for the release of garnet xenocrysts into kimberlitic magma. (Modified after Canil and Fedortchouk 1999.)

thick dissolution features occurring on fractures in garnet are likely to have developed during the explosive, late-stage emplacement of their kimberlite dike. They suggest that the explosive emplacement probably occurred in only seconds to minutes. This estimate is consistent with the Sparks et al. (2006) model for the final emplacement of kimberlite magmas.

A third example where reactions between xenoliths and the host magma have been used to estimate magma ascent rates has developed from analytical studies of H_2O in nominally anhydrous minerals like olivine in the xenoliths. The diffusion of H_2O in olivine varies with crystallographic orientation (Mackwell and Kohlstedt 1990) and so the concentration of H_2O must be determined in a profile with a known crystallographic orientation. Diffusion rates for H_2O in olivine vary with temperature and hydrogen fugacity, but are sufficiently fast that several mm diameter crystals will equilibrate in hours at >1000 °C (Mackell and Kohlstedt 1990). Recently, Peslier and Luhr (2006) described diffusion loss profiles in olivine crystals from spinel peridotite inclusions carried to the surface in alkali basalt magma. They previously found that associated pyroxenes contained significantly higher amounts of dissolved H_2O but there was no detectible H_2O loss from the pyroxene rims. Demouchy et al. (2006) measured similar H-loss profiles in olivine from a garnet lherzolite collected at a locality in Chile (Fig. 19). In both of these studies it was assumed that the diffusive loss of water began when the peridotite sample left the mantle source region at 40 and 60-70 km depth respectively, and the Mackwell and Kohlstedt (1990) experimental data were used to estimate a time for the profiles to develop. Peslier and Luhr (2006) derive times of 18-65 hours which correspond to magma ascent rates of 0.2 to 0.5 m/s; Demouchy et al. (2006) determine times of 1.9 to 6.3 hours for temperatures of 1200 to 1290 °C which is equivalent to a magma ascent rate of 6±3 m/s from 60 km. Both of these estimates are very similar to rates of alkali basalt ascent determined from

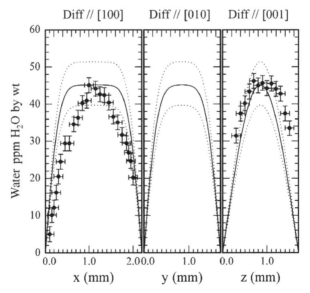

Figure 19. Water content profiles in a representative olivine crystal in a Garnet peridotite Xenolith from Pali-Aike, Chile (Demouchy et al. 2006). Individual data points are plotted for two crystallographic directions in the crystal. Solid lines are calculated diffusion profiles based on previous experimental work for different directions at 1200-1290 °C with an initial water content of 312 ppm H_2O and a final 0 ppm at the rim. Calculated profiles yield ascent rates of 1.9 h at 1290 °C and 6.3 h at 1200 °C for ascent from 70 km. [Published with permission of the Geological Society of America from Demouchcy et al. (2006), *Geology*, Vol. 34, Fig. 4, p. 431.]

dissolutions rates of garnet and from theoretical models of magma flow (Sparks et al. 2006), but the 6 m/s determination is at high end of the range. The modeling of H-loss profiles (or other fast-diffusing elements) shows significant promise for assessing magma ascent rates, but again there is a significant need for additional work to determine if the diffusive loss of H from olivine begins in the mantle as assumed. A related problem is the amount of water dissolved in the surrounding alkali basalt, and the effect it has on H-loss from olivine xenocrysts.

XENOLITH TRANSPORT, FLOW MODELING AND MAGMA ASCENT RATES

The presence of dense mantle xenoliths in tephra and lava flow eruptions of alkali basalt and kimberlite indicates that the magma carried them from depth (e.g., Irving 1980). An average settling rate for xenoliths (minimum magma ascent rate) can be calculated by balancing the different forces on the xenoliths during ascent, and by making certain assumptions about the rheological properties of the magma (McGetchin and Ulrich 1973; Mitchell et al. 1980; Spera 1984; Sparks et al. 2006). Figure 20 shows the results of settling calculations for spherical xenoliths, a density difference of 500 kg/m^3 between the xenoliths and the melt (melt density = 2,800 kg/m^3), and an assumption of Bingham plastic rheology for the xenolith-laden magma. As a Bingham fluid, the magma has a yield strength ($\sigma°$) that correlates well with the apparent viscosity. Yield strength of 100 N/m^2 corresponds to a phenocryst content of ~25 vol% in a basalt. The calculation indicates that a 20-cm-wide xenolith will settle at a rate of 0.1 m/s. Obviously, the ascent rate of a magma carrying this xenolith would have to be >0.1 m/s in order to carry the xenolith to the surface; the xenolith settling velocity gives a minimum estimate of magma ascent rate. Published minimum estimates for alkali basalt ascent rates range from 0.1 to 2 m/s based on a lack of chemical zonation in phenocrysts, or a preservation of chemical zonation in melt pools (Brearley and Scarfe 1986). Two factors that

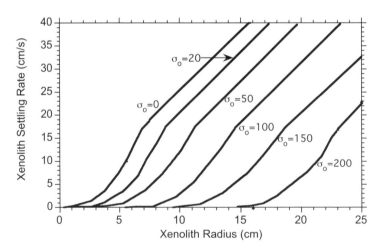

Figure 20. Plot showing calculated xenolith settling velocity as a function of xenolith radius for different magma yield strengths (σ_0). A density difference between the xenolith and melt of 500 kg/m^3 is assumed; the melt density is 2,800 kg/m^3 and a Bingham rheology (viscosity of 350 poise) is assumed. Units of yield strength are N/m^2. A yield strength of 100 N/m^2 corresponds approximately to a crystallinity of 25 vol%. (Adapted from Spera 1984). The slight change in slope on the yield strength curves is due to the increased importance of inertia relative to friction and buoyancy forces at high xenolith Reynolds numbers. The calculations were made using Stokes law {$U = 2gr2(dr)/9 \eta$} modified to accommodate variations in the properties of the melt and nodules (Modified after Spera 1984.)

have not been considered in the settling calculation are the effect of nearby xenoliths and the effect of bubbles in the magma. If significant, both of these would decrease the settling rate of xenoliths, and therefore reduce the minimum ascent rate required to carry a xenolith of a given size to the surface. Based on the relative abundance of vesicles and xenoliths in alkali basalt, however, it is possible that neither has a significant effect on calculated ascent rates since there is no obvious correlation between crystal content in the alkali basalt and xenolith abundance (Spera 1984). Spera (1984) also argues that while the volatile abundance in such magmas is likely to significantly affect the initial velocity of the xenolith-carrying magmas, increases in ascent rate at depths >3-5 km will not result from volatile exsolution or expansion because the solubility of the CO_2-rich gas exsolved at depth changes slowly with pressure, and the gas is relatively incompressible at >200 MPa.

Kimberlite, another mantle xenolith-bearing magma, is known to be very gas-rich and also commonly has high phenocryst and/or xenolith contents (Dawson 1971). Spera (1984) did not use xenolith settling rates to constrain magma ascent rate for kimberlite magmas, possibly because the model (Fig. 20) depends strongly on good estimates for the magma viscosity (yield strength). However, theoretical models of magma flow have been developed for evaluating ascent of magmas that range from basalt to kimbelite in composition (McGetchin and Ulrich 1973: Mitchell et al. 1980; Spera 1984; Sparks et al. 2006). Following McGetchin and Ulrich (1973), Spera (1984) estimates that the ascent rate of kimberlitic magma is in the range of 10 to 30 m/s, higher than the ascent rate of alkali basalt by a factor of 10. However, Mitchell et al. (1980) consider that the ascent of such gas-rich magmas is probably best considered as taking place in three stages. Stage 1 (100-10 km) may be thought of as flow of a Newtonian fluid in a pipe, although the flow will not be Newtonian if the magma contains a significant load of xenoliths and xenocrysts. Additionally, there now seems to be general acceptance that kimberlitic magma flow is in thin (0.3-1 m) dikes for stage 1 depths (Mitchell et al. 1980; Sparks et al. 2006). Stage 2 (10-1 km) is where the gas phase becomes important in determining magma properties and ascent rates; the time spent in this stage is 3-4 % of the total ~100 km ascent time in most models. Stage 3 is the explosive diatreme stage (1-0 km) and lasts for an estimated 7 s out of a total ascent time of 130 minutes (Mitchell et al. 1980). The velocity of the magma in the final two stages is very high as required to explain the diatreme deposit characteristics. The important point however is that the most recent theoretical models (Table 1) suggest that kimberlitic magma velocity in the dike stage (1) averages from 2.9-16.8 m/s (Sparks et al. 2006), very much overlapping with the 10-30 m/s estimates from the earliest models.

Table 1. Ascent rates for dike transported kimberlite magma calculated using $u = 7.7[w^5/\{\mu(\rho g \Delta r)^3\}]^{1/7} g \Delta \rho$ (Sparks et al. 2006).

Dike width (2w) (m)	Velocity $\Delta\rho = 100$ kg/m^3	Velocity $\Delta\rho = 300$ kg/m^3
0.2	2.9	5.4
0.3	3.8	7.2
0.6	6.3	11.7
1.0	9.1	16.8

SUMMARY OF MAGMA ASCENT RATE ESTIMATES

The estimates of average magma ascent rates for silicic to intermediate composition lava extrusions are fairly well constrained in the range of 0.001 m/s for very slowly ascending magmas, to 0.015 m/s for more rapidly ascending extrusive magmas (Table 2). These

Table 2. Magma ascent rate estimates from different observations.*

Volcano	Observation*	Explosive Ascent rate (m/s)	Extrusive Ascent rate (m/s)
Mount St. Helens	groundmass crystallization	>> 0.2	0.01 - 0.02
Mount St. Helens	hornblende rims	> 0.18	0.04 - 0.15
Mount St. Helens	calculation from mass eruption rate	1 - 2	0.01 - 0.1
Mount St. Helens	seismicity	0.6	0.007 - 0.01
Soufriere Hills, Mont.	amphibole rims	> 0.2	0.001 - 0.012
Soufriere Hills, Mont.	magnetite	> 0.2	0.003 - 0.015
Mount Unzen	magnetite zonation	not present	0.002
Black Butte CA	amphibole rims & plagioclase growth	not present	0.004-0.006
Hualali, HI alk basalt	xenolith transport	not present	> 0.1
La Palma, Canary Is.	olivine zonation	not present	> 0.06
Xenocrysts in alk basalt	hydrogen zoning in olv	0.2 to 0.5 m/s	
Kimberlite	theoretical modeling	> 4 -16	
Kimberlite	garnet dissolution	1.1 to 30 m/s (final 2 km)	

* Different types of observations are discussed in various sections of the paper.

estimates come from a variety of observations made at different volcanoes, and it is important to remember that they are average rates. Ascent rates are likely to vary over the vertical extent of the conduit carrying the magma, because the density and viscosity of the magma will vary considerably over this length, particularly in water-rich magmas undergoing crystallization and volatile exsolution. Additionally, the effective cross section area of the conduit at depth is generally not known. One topic for future study is the nature of changes in flow within volcanic conduits. It has also been suggested that convection may occur in conduits (Witter et al. 2005); in such a case, the magma ascent would be very difficult to assess, but a record should be preserved in some of the phenocryst and groundmass minerals such as the Fe-Ti oxides. It is rather interesting that observations made for basaltic composition magma extrusions give essentially the same range of average ascent rates as obtained for more silica-rich magmas.

The assessment of magma ascent rates during explosive eruptions is more complex because there is almost certainly a much greater variation in the ascent rate from the base of the conduit to the surface. This is particularly true for eruptions where there is evidence for alternating effusive dome-forming eruptions and conduit-clearing explosive events (e.g., 1980 Mount St. Helens and 1996-2000 Soufriere Hills, Montserrat). The magma leaves the top of the conduit at essentially sonic velocities for truly explosive eruptions (Papale et al. 1998; Sparks et al. 2006). This high rate of ascent is achieved as a result of the low density and rapid expansion of the gas phase in these gas-rich magmas. Deep in the conduit, very little of the gas is exsolved, the magma is dense, and depending on the size and shape of the conduit, flow rates are calculated to be much lower. In our view, some of the more profitable areas for study of magma ascent lie with investigations of the behavior of the various magmatic volatiles; the rates of their diffusion in melts, and the processes by which gas can be lost from magmas during ascent. Additionally, since so many eruptions appear to be staged from a magma storage zone in the earth's crust, a better understanding of storage zone processes (e.g., magma injection, convection and mixing) would greatly improve our understanding of the magma ascent process.

REFERENCES

Alletti M, Baker DR, Feda C (2007) Halogen diffusion in a basaltic melt. Geochim Cosmochim Acta 71:3570-3580

Aki K, Koyanagi RY, (1981) Deep volcanic tremor and magma ascent under Kilauea, Hawaii. J Geophys Res 86:7095-7109

Anderson AT (1991) Hourglass inclusions: Theory and application to the Bishop Rhyolitic Tuff. Am Mineral 76:530-547

Aspinall WP, Miller AD, Lynch LL, Latchman JL, Stewart RC, White RA, Power JA (1998) Soufriere Hills eruption, Montserrat: Volcanic earthquake locations and fault plane solutions. Geophys Res Lett 25:3397-3401

Baker, DR, Lang P, Robert G, Bergevin J-F, Allard E, Bai L (2006) Bubble growth in slightly supersaturated albite melt at constant pressure. Geochim Cosmochim Acta 70:1821-1838

Baker LL, Rutherford M (1996) Sulfur diffusion in rhyolite melts. Contrib Mineral Petrol 123:335-344

Barclay J, Rutherford MJ, Carroll MR, Murphy MD, Devine JD, Gardner J, Sparks RSJ (1998) Experimental phase equilibria constraints on pre-eruptive conditions of the Soufriere Hills magma. Geophys Res Lett 25:3437-3440

Blundy J, Cashman K, Humphreys M (2006) Magma heating by decompression-driven crystallization beneath andesite volcanoes. Nature 44:76-80

Brearley M, Scarfe CM (1986) Dissolution rates of upper mantle minerals in an alkali basalt melt at high pressure: an experimental study and implications for ultramafic xenolith survival. J Petrol 27:1157-1182

Browne BL, Gardner JE (2006) The influence of magma ascent path on the texture, mineralogy, and formation of Hornblende reaction rims. Earth Planet Sci Lett 246:161-176

Buckley VJE, Sparks RSJ, Wood BJ (2006) Hornblende dehydration reactions during magma ascent at Soufriere Hills Volcano, Montserrat. Contrib Mineral Petrol 151:121-140

Canil D, Fedortchouk Y (1999) Garnet dissolution and the emplacement of kimberlites. Earth Planet Sci Lett 167:227-237

Carey S, Sigurdsson H (1985) The May 18, 1980, eruption of Mount St. Helens 2: Modeling of dynamics of the Plinian phase. J Geophys Res 90:2948-2958

Carmichael ISE (2002) The andesite aqueduct: perspectives on the evolution of intermediate magmatism in west central (105-99°W) Mexico. Contrib Mineral Petrol 143:641-663

Carter JL (1970) Mineralogy and chemistry of the Earth's upper mantle based on the partial fusion-partial crystallization model. Bull Geol Soc Am 81:2021-2034

Cashman KV (1988) Crystallization of Mount St. Helens 1980-1986 dacite: a quantitative textural approach. Bull Volcanol 50:194-209

Cashman KV (1992) Groundmass crystallization of Mount St. Helens dacite, 1980-1986: a tool for interpreting shallow magmatic processes. Contrib Mineral Petrol 109:431-449

Cervantes P, Wallace P (2003) The role of water in subduction zone volcanism: New insights from melt inclusions in high-Mg basalts from central Mexico. Geology 31:235-238.

Chouet BA, Page RA, Stephens CD, Lahr JC, Power JA (1994) Precursory swarms of long-period events at Redoubt Volcano (1989-1990), Alaska: their origin and use as a forecasting tool. J Volc Geotherm Res 62:95-135

Condomines M, Hemond C, Allegre CJ (1988) U-Th-Ra radioactive disequilibria and magmatic processes. Earth Planet Sci Lett 90:243-262

Corsaro R, Pompillio M (2004) Dynamics of magmas at Mount Etna. *In:* Mount Etna: Volcano Laboratory. Bonaccorso A, Calvari S, Colteli M, Del Negro C, Falsaperla S (eds) AGU Geophysical Monograph 143:91-110

Couch S (2003) Experimental investigation of crystallization kinetics in a haplogranitic system. Am Mineral 88:1471-1485

Couch S, Sparks RSJ, Carroll MR (2003) The kinetics of degassing-induced crystallization at Soufrierre Hills Volcano, Montserrat. J Petrol 44:1477-1502

Dawson JB (1971) Advances in kimberlite geology. Earth Sci Rev 7:182-214

Decker RW (1987) Dynamics of Hawaiian volcanoes: an overview. *In:* Volcanism in Hawaii. Decker RW, Wright TL Staufer PH (eds) U.S. Geol Survey Prof Paper 1350:997-1018

Demouchy S, Jacobsen SD, Gaillard F, Stern CR (2006) Rapid magma ascent recorded by water diffusion profiles in mantle olivine. Geology 34:429-432

Dejesus S, Rutherford MJ (2007) The origin and significance of F-rich amphibole in the 2004-6 Mount St. Helens magma. EOS Trans, AGU 88(52) Fall Meet Suppl Abstract V41B-0603

Devine JD, Rutherford MJ, Norton GE, Young SR (2003) Magma storage zone processes inferred from geochemistry of Fe-Ti oxides in andesitic magma, Soufriere Hills volcano, Montserrat. J Petrol 44:1375-1400

Dixon JE, Stolper EM (1995) An experimental study of water and carbon dioxide solubilities in mid-ocean ridge basaltic liquids: Part II: Applications to degassing. J Petrol 36:627-654

Edwards BR, Russell JK (1996) A review and analysis of silicate mineral dissolution experiments in natural silicate melts. Chem Geol 130:233-245

Eichelberger JC (1978) Andesitic volcanism and crustal evolution. Science 275:21-27

Eggler DH, Burnham CW (1973) Crystallization and fractionation trends in the system andesite-H_2O-CO_2-O_2 at pressures to 10 kb. Bull Geol Soc Am 84:2517-2532

Endo ET, Murray T (1992) Real-time seismic amplitude measurement (RSAM): a volcano monitoring and prediction tool. Bull Volc 53:533-545

Endo ET, Murray TL, Power JA (1996) A comparison of pre-eruption, real-time seismic amplitude measurements for eruptions at Mount St. Helens, Redoubt Volcano, Mount Spurr, and Mount Pinatubo. *In:* Fire & Mud: eruptions and lahars of Mount Pinatubo, Philippines. Newhall CG, Punongbayan RS (eds) University of Washington Press, Seattle, p 233-246

Freer R, Hauptman Z (1978) An experimental study of magnetite-titanomagnetite interdiffusion. Phys Earth Planet Inter 16:223-231

Frey FA, Green DH (1974) The mineralogy, geochemistry and origin of lherzolitic inclusions in Victorian basanites. Geochim Cosmochim Acta 38:1023-1059

Garcia MO, Jacobsen SS (1979) Crystal clots amphibole fraction and the evolution of calc-alkaline magmas. Contrib Mineral Petrol 69:319-327

Gardner JE (2007) Heterogeneous bubble nucleation in highly viscous silicate melts during instantaneous decompression from high pressure. Chem Geol 236:1-12

Gardner JE, Hilton M, Carroll MR (1999) experimental constraints on degassing of magma: Isothermal bubble growth during continuous decompression from high pressure. Earth Planet Sci Lett 168:201-218

Gardner JE, Denis M-H (2004) Heterogeneous bubble nucleation on Fe-Ti oxide crystals in high-silica rhyolitic melts. Geochim Cosmochim Acta 68:3587-3597

Gerlach TM, McGee KA, Doukas MP (2008) Emission rates of CO_2, SO_2, H_2S, scrubbing, and pre-eruption excess volatiles at Mount St. Helens, 2004-2005. *In:* A volcano rekindled: the first year of renewed eruption at Mount St. Helens, 2004-2006. Sherrod, DR, Scott WE, Stauffer PH (eds) US Geol Survey Prof Pap 1750, Chpt 26 (in press)

Geschwind C-H, Rutherford MJ (1995) Crystallization of microlites during magma ascent: the fluid mechanics of 1980-86 eruptions at Mount St. Helens. Bull Volc 57:356-370

Ginibre C, Worner G, Kronz A (2007) Crystal zoning as an archive for magma evolution. Elements 3:261-266

Green DH, Ringwood AE (1967) An experimental investigation of the gabbro to eclogite transformation and its petrological applications. Geochim Cosmochim Acta 81:767-833

Hammer JE, Cashman K, Hoblitt RP (1999) Degassing and microlite crystallization during pre-climatic events of the 1991 eruption of Mount Pinatubo, Philippines. Bull Volc 60:355-380

Hammer JE, Rutherford MJ (2002) Kinetics of decompression-induced crystallization in silicic melt. J Geophys Res 107, doi:10.1029/2001JB000281

Hammer, JE Rutherford MJ (2003) Glass Composition geobarometry: a petrologic indicator of pre-eruption Pinatubo dacite magma dynamics. Geology 31:79-82

Hoblitt RP, Wolfe EW, Scott WE, Couchman MR, Pallister JS, Javier D (1996) The preclimatic eruptions of Mount Pinatubo, June 1991. *In:* Fire & Mud: eruptions and lahars of Mount Pinatubo, Philippines. Newhall CG, Punonngbayan RS (eds) University of Washington Press, Seattle, p 457-512

Holloway JR, Blank JG (1994) Application of experimental results to C-O-H species in natural melts. Rev Mineral 30:187-230

Humphreys MCS, Menad T, Blundy JD, Klimm K (2008) Magma ascent rates in explosive eruptions: Constraints from H_2O diffusion in melt inclusions. Earth Planet Sci Lett 270:25-40

Hurwitz S, Navon O (1994) Bubble nucleation in rhyolitic melts: Experiments at high pressure, temperature and water content. Earth Planet Sci Lett 122:267-280

Irving AJ (1980) Petrology and Geochemistry of composite ultramafic xenoliths in alkali basalts and implications for magmatic processes in the mantle. Am J Sci 280A:389-426

Izbekov PE, Eichelberger JC, Patino LC, Vogel TA, Ivanov BA (2002) Calcic cores of plagioclase phenocrysts in andesite from Karyminski volcano: evidence for rapid introduction by basaltic replenishment. Geology 80:799-802

Kagiyama T, Utada H, Yamamoto T (1999) Magma ascent beneath Unzen volcano, SW Japan, deduced from the electrical resistivity structure. J Volc Geotherm Res 89:35-42

Klein FW, Koyanagi RY, Nakata JS, Tanigawa WR (1987) The seismicity of Kilauea's magma system. *In:* Volcanism in Hawaii Decker RW, Wright TL Staufer PH (eds) U.S. Geol Survey Prof Paper 1350:1019-1185

Klugel A, Hansteen TH, Schminke H-U (1997) Rates of magma ascent and depths of magma reservoirs beneath La Palma (Canary Islands). Terranova 9:117-121

Klugel A (1998) Reactions between mantle xenoliths and host magma beneath La Palma (Canary Islands): constraints on magma ascent rates and crustal reservoirs. Cont Mineral Petrol 131:237-257

Liu Y, Anderson AT, Wilson CJN (2007) Melt pockets in phenocrysts and decompression rates of silicic magmas before fragmentation. J Geophys Res 112:B06204 doi:10.1029/2006JB004500

MacGregor AG (1938) The volcanic history and Petrology of Montserrat with observations on Mt. Pele in Martinique: Royal Society expedition to Montserrat BWI. Phil Trans Roy Soc Ser B 229:1-90

Mackwell SJ, Kohlstedt DL (1990) Diffusion of hydrogen in olivine: implications for water in the mantle. J Geophys Res 95:5079-5088

Mangan M, Sission T (2000) Delayed, disequilibrium degassing in rhyolite magma: decompression experiments and implications for explosive volcanism. Earth Planet Sci Lett 183:441-455

Mangiacapra A, Moretti R, Rutherford MJ, Civetta L, Orsi G, Papale P (2008) The deep magma system at Campi Flegrei caldera (Italy). Geophys Res Lett (in review)

Mastin LG, Ghiorso MS (2001) Adiabatic temperature changes of magma-gas mixtures during ascent and eruption. Contrib Mineral Petrol 141:307-321

McCanta MC, Hammer JE, Rutherford MJ (2007) Pre-eruptive and syn-eruptive conditions in the Black Butte, CA dacite: Insight into crystallization kinetics in a silicic magma system. J Volc Geotherm Res 160:263-284

McGetchin TR, Ullrich GW (1973) Xenoliths in Maars and diatremes with inferences for the Moon, Mars, and Venus. J Geophys Res 78:1833-1853

Metrich N, Schiano P, Clocchiati R, Maury RC (1999) Transfer of Sulfur in subduction settings: an example from Batan Island (Luzon volcanic arc, Philippines). Earth Planet Sci Lett 167:1-14

Mitchell RH, Carswell DA, Clarke DB (1980) Geological implications and validity of calculated equilibrium conditions for ultramafic xenoliths from the pipe 200 kimberlite, Northern Lesotho. Contrib Mineral Petrol 72:205-217

Murphy MD, Sparks RSJ, Barclay J, Carroll MR, Brewer TS (2000) Remobilization of andesitic magma by intrusion of mafic magma at the Soufriere Hills volcano, Montserrat, West Indies. J Petrol 41:21-42

Nakada S, Fujii T (1993) Preliminary report on the activity at Unzen volcano (Japan) November 1990- November 1991: dacite lava domes and pyroclastic flows. J Volc Geotherm Res 54:319-333

Nakamura M (1995) Continuous mixing of crystal mush and replenished magma in the ongoing Unzen eruption. Geology 23:807-810

Nicholis MG, Rutherford MJ (2004) 32 Experimental constraints on magma ascent rate for the Crater Flat volcanic zone hawaiite. Geology 32:489-492

Nicholis MG (2006) Degassing processes in basaltic magmas: terrestrial and lunar applications. PhD dissertation, Brown University, Providence RI

O'Hara MJ, Richardson SW, Wilson G (1971) Garnet peridotite stability and occurrence in crust and mantle. Contrib Mineral Petrol 32:48-68

Pallister JS, Hoblitt, RP, Crandell DTR, Mullineaux DR (1992) Mount St Helens a decade after the 1980 eruptions: Magmatic models, chemical cycles, and a revised hazards assessment. Bull Volc 54:126-146

Pallister JS, Hoblitt RP, Meeker GP, Knoght RJ, Siems DF (1996) Magma mixing at Mount Pinatubo: Petrographic and chemical evidence from the 1991 deposits. In: Fire & Mud: eruptions and lahars of Mount Pinatubo, Philippines. Newhall CG, Punonngbayan RS (eds) University of Washington Press, Seattle, p 687-732

Pallister JS, Thornber CR, Cashman KV, Clynne MA, Lowers HA, Brownfield IA, Meeker GP (2008) Petrology of the 2004-2006 Mount St. Helens lava dome – implications for magmatic plumbing and eruption triggering. In: A volcano rekindled: the first year of renewed eruption at Mount St. Helens, 2004-2006. Sherrod, DR, Scott WE, Stauffer PH (eds) US Geol Survey Prof Paper 1750, chpt 19 (in press)

Papale P, Dobran, F (1994) Magma flow along the volcanic conduit during the plinian and pyroclastic flow phases of the May 18, 1980, Mount St. Helens eruption. J Geophys Res 99:4355-4373

Papale P, Neri A, Macedonio G (1998) The role of magma composition and water content in explosive eruptions: 1. Conduit ascent dynamics. J Volc Geotherm Res 87:75-93

Papale P, Moretti R, Barbato D (2006) The compositional dependence of the saturation surface of H_2O+CO_2 fluids in silicate melts. Chem Geol 229:78-95

Peslier AH, Luhr JF (2006) Hydrogen loss from olivines in mantle xenoliths from Simcoe (USA) and Mexico: mafic alkalic magma ascent rates and water budget of the sub-continental litosphere. Earth Planet Sci Lett 242:302-319

Pinkerton H, Wilson L, Macdonald R (2002) The transport and eruption of magma from volcanoes: a review. Contemp Phys 43:197-210

Power JA, Lahr JC, Page RA, Chouet BA, Stephens CD, Harlow DH, Murray TL, Davies JN (1994) Seismic evolution iof the 1989-90 eruption sequence of redoubt volcano, Alaska. J Volc Geotherm Res 62:69-94

Roggensack K, Hervig RL, McNiight SB, Williams SN (1997) Explosive basaltic volcanism from Cerro Negro volcano: influence of volatiles on eruptive style. Science 277:1639-1642

Rutherford MJ, Sigurdsson H, Carey S (1985) The May 18, 1980 eruption of Mount St. Helens, 1: Melt compositions and experimental phase equilibria. J Geophys Res 90:2929-2947

Rutherford MJ, Devine JD (1988) The May 18, 1980 eruption of Mount St. Helens III: Stability and chemistry of amphibole in the magma chamber. J Geophys Res 93:11,949-11,959

Rutherford MJ, Hill PM (1993) Magma ascent rates from amphibole breakdown: Experiments and the 1980-1986 Mount St. Helens eruptions. J Geophys Res 98:19667-19685

Rutherford MJ, Devine JD (2003) Magmatic conditions and magma ascent as indicated by hornblende phase equilibria and reactions in the 1995-2001 Soufriere Hills magma. J Petrol 44:1433-1454

Rutherford MJ, Devine J (2008) Magmatic conditions and processes in the Storage Zone of the 2004-06 Mount St. Helens Eruption: The record in Amphibole and Plagioclase phenocrysts. *In:* A volcano rekindled: the first year of renewed eruption at Mount St. Helens, 2004-2006. Sherrod, DR, Scott WE, Stauffer PH (eds) US Geol Survey Prof Pap 1750, chpt 31 (in press)

Ryan M (1993) Neutral buoyancy and the structure of mid-ocean ridge magma reservoirs. J Geophys Res 98:22321-22338

Saal A, Hauri E, LoCasio M, Van Orman J, Rutherford M, Cooper R (2008). The volatile content of lunar volcanic glasses: Evidence for the presence of water in the lunar interior. Nature 454:192-195

Scandone RS, Malone S (1985) Magma supply, magma discharge and readjustment of the feeding system of Mount St. Helens during 1980. J Volc Geotherm Res 23:239-262

Schilling SP, Thompson RA, Messerich JA, Iwatsubo EY (2008) Mount St. Helens lava dome growth and surface modeling, 2004-2005. *In:* A volcano rekindled: the first year of renewed eruption at Mount St. Helens, 2004-2006. Sherrod, DR, Scott WE, Stauffer PH (eds) US Geol Survey Prof Pap 1750, chpt 8 (in press)

Singer BS, Dungan MA, Layne GD (1995) Textures and Sr, Ba, Mg, Fe, K, and Ti compositional profiles in volcanic plagioclase: Clues to the dynamics of calc-alkaline magma chambers. Am Mineral 80:776-798

Sparks RSJ (1978) the dynamics of bubble formation and growth in magmas: a review and analysis. J Volc Geotherm Res 3:1-37

Sparks RSJ, Young SR, Barclay J, Calder ES, Cole P, Daroux B, Davies MA, Druitt TH, Harford C, Herd R, James M, Lejeune AM, Loughlin S, Norton G, Skerrit G, Stasiuk MV, Stevens NS, Toothill J, Wage G, Watts R (1998) Magma production and growth of the lava dome of the Soufriere Hills volcano, Montserrat: November 1995 to December 1997. Geophys Res Lett 25:3421-3425

Sparks RSJ, Baker L, Brown RJ, Field M, Schumacher J, Stripp G, Walters A (2006) Dynamical constraints on kimberlitic volcanism. J Volc Geotherm Res 155:18-48

Spera FJ (1984) Carbon dioxide in petrogenesis III: role of volatiles in the ascent of alkaline magma with special reference to xenolith-bearing magmas. Contrib Mineral Petrol 88:217-232

Swanson DA, Dzurisin D, Holcomb TR, Iwatsubo EY, Chadwick WW Jr., Casadeval TJ, Ewert JW, Heliker CC (1987) Growth of lava domes at Mount St. Helens, Washington. Geol Soc Am Spec Pap 212: 1-16

Thornber C, Pallister JS, Lowers H, Rowe M, Mandeville C, Meeker G (2008) The chemistry, mineralogy and Petrology of amphibole in Mount St Helens 2004-2006 dacite. *In:* A volcano rekindled: the first year of renewed eruption at Mount St. Helens, 2004-2006. Sherrod, DR, Scott WE, Stauffer PH (eds) US Geol Survey Prof Pap 1750, chpt 32 (PAGE RANGE)

Toramaru A (1995) Numerical study of nucleation and growth of bubbles in viscous magma. J Geophys Res 104:20097-20122

Tryggvason E (1994) Surface deformation at the Krafla volcano, North Iceland, 1982-1992. Bull Volc 56:98-107

Tuttle OF, Bowen NL (1958) Origin of granite in the light of experimental studies in the system $NaAlSi_3O_8$-$KAlSi_3O_8$-SiO_2-H_2O. Geol Soc Am Mem 74

Vaniman DT, Crowe BM, Gladney ES (1982) Petrology and geochemistry of Hawaiitic lavas from Crater Flat, Nevada. Contrib Mineral Petrol 80:341-357

Venezky D, Rutherford MJ (1997) Pre-eruption conditions and timing of magma mixing in the 2.2 ka C-layer, Mount Rainier. J Geophys Res 102:20,069-20,086

Venezky D, Rutherford MJ (1999) Petrology and Fe-Ti oxide reequilibration of the 1991 Mount Unzen mixed magma. J Volc Geotherm Res 89:213-230

Wallace PJ (2005) Volatiles in subduction zone magmas: concentrations and fluxes based on melt inclusion and volcanic gas data. J Volc Geotherm Res 140:217-240

Watson EB (1994) Diffusion in volatile-bearing magmas. Rev Mineral 30:371-411

Watts RB, Herd RA, Sparks RSJ, Young SR (2002) Growth patterns and emplacement of the andesitic lava dome at Soufriere Hills Volcano, Montserrat. *In:* The Eruption of Soufriere Hills volcano Montserrat from 1995 to 1999. Druitt TH, Kokelar BP (eds) Geo Soc London Memoir 21:115-152

Westrich HR, Stockman HW, Eichelberger JC (1988) Degassing of rhyolitic magma during ascent and emplacement. J Geophys Res. 93:6503-6511

White RA (1996) Precursory deep long-period earthquakes at Mount Pinatubo spatio-temporal link to a basalt trigger. *In:* Fire & Mud: Eruptions and Lahars of Mount Pinatubo, Philippines. Newhall CG, Punonngbayan RS (eds) University of Washington Press, Seattle, p 233-246

Witter JB, Kress VG, Newhall CG (2005) Volcan Popocatepetl, Mexico. Petrology, magma mixing, and intermediate sources for volatiles for the 1994-present eruption. J Petrol 46:2337-2366

Wilson L, Sparks RSJ, Walker GPL (1980) Explosive volcanic eruptions IV: the control of magma properties and conduit geometry on eruption column behavior. Geophys J Roy Astron Soc 63:117-148

Zhang Y Stolper EM (1991) Water diffusion in basaltic melt. Science 351:306-309

Zhang Y Behrens H (2000) H_2O diffusion in rhyolite glasses. Geochim Cosmochim Acta 55:441-456

Melt Inclusions in Basaltic and Related Volcanic Rocks

Adam J.R. Kent

Department of Geosciences
Oregon State University
Corvallis, Oregon, 97331, U.S.A.

adam.kent@geo.oregonstate.edu

INTRODUCTION

Melt inclusions are small parcels of melt trapped in crystals within magmatic systems, and are analogous to fluid inclusions formed by trapping of hydrothermal and other fluids during mineral growth in fluid-mineral systems (Sorby 1858; Roedder 1979, 1984). After trapping, melt inclusions are potentially isolated from external melt and thus provide a way to investigate melts trapped during magmatic evolution—driven by processes such as crystal-liquid separation, vapor saturation and degassing, magma mixing and assimilation—which can dramatically alter the compositions of the eventual erupted (or intruded) magmatic end products. Melt inclusions are a powerful tool for the study of basaltic magma systems and their mantle source regions, and are widely used to study the origin and evolution of mantle-derived magmas. Melt inclusions have specific uses in the study of volatile elements (see chapters by Metrich and Wallace 2008, Moore 2008, and Blundy and Cashman 2008), but also provide unique information about the range of melt compositions present within basaltic magmatic systems, and how these reflect mantle sources and the processes that occur during melt generation, evolution, transport and eruption. This review outlines techniques used to obtain chemical and other information from melt inclusions, discusses the processes which lead to melt inclusion trapping in phenocryst minerals, examines the possible means by which melt inclusion compositions might be fractionated during trapping or during subsequent re-equilibration with the host mineral or external melt, and discusses some implications of melt inclusion compositions for the nature of basaltic melt generation and transport systems.

This review is largely restricted in scope to studies of volcanic rocks of basaltic and related composition. This refers to rocks erupted as lavas or tephra with broadly basaltic compositions: SiO_2 ~45-52 wt%, relatively high MgO and FeO, and typically containing one or more of the following minerals as phenocrysts: forsterite-rich olivine, anorthite-rich plagioclase, orthopyroxene, clinopyroxene and a spinel phase (aphyric basalts also occur but are less useful for melt inclusion studies). These are magmas that ultimately derive from melting of the mantle, but most have also experienced some modification during ascent and magma storage within the oceanic or continental crust. Melt inclusions are found in all terrestrial environments in which basaltic and related rocks occur, as well as in rocks of basaltic composition from lunar, martian and other meteorite samples (e.g., Bombardieri et al. 2005; Nekvasil et al. 2007). The ubiquity of melt inclusions in basaltic rocks emphasizes that the conditions amenable to trapping melt inclusions occur during the evolution of almost all basaltic magmas, regardless of environment of formation. It is also worth noting that there is a significant body of literature devoted to the study of melt inclusions in rocks of other compositions—particularly those related to silicic volcanic and plutonic rocks, but also carbonatites, kimberlites and other rarer magma types (e.g., Nielsen and Veksler 2002; Kamenetsky et al. 2004), and inclusions also occur within mafic cumulates and peridotitic residues from mantle melting (e.g., Schiano and Bourdon 1999; Schiano et al. 2004) and in layered mafic intrusions (e.g., Jakobsen et al. 2005). These studies are largely beyond the scope of this review.

There are three primary advantages that the study of melt inclusions, in comparison to studies based on whole rock or matrix glass samples of erupted basaltic rocks, can confer—of which the first two are the most important in the literature. Firstly, melt inclusions trapped at pressures that are higher than those of eventual eruption can preserve abundances of volatile elements (H, C, Cl, S, F; see in this volume Metrich and Wallace 2008; Moore 2008, and Blundy and Cashman 2008) and these may be directly measured within trapped glasses. In many cases the abundances of volatile species can be inferred or demonstrated to be largely unaffected by shallow-pressure degassing (e.g., Sisson and Bronto 1998; Newman et al. 2000; Saal et al. 2002; Cervantes and Wallace 2003a) and approach the compositions of undegassed magmas. Melt inclusions also provide the opportunity to measure the isotopic composition (H, C, S, Cl) and speciation (H, S) of volatile elements and to examine the relationship between volatiles and other compositional or petrological features. This has been particularly beneficial for efforts to understand the role of water in formation of primitive basalts in arc and back arc environments (e.g., Sisson and Layne 1993; Sisson and Bronto 1998; Newman et al. 2000; Cervantes and Wallace 2003a; Walker et al. 2003), studies of the interplay between volcanic eruption and volatile degassing (e.g., Anderson 1975; Wallace and Anderson 1998; Cervantes and Wallace 2003b; Gurenko et al. 2005) and for efforts to study the volatile structure of the terrestrial mantle (e.g., Sobolev and Chaussidon 1996; Lassiter et al. 2002; Saal et al. 2002). In many cases this information cannot be gleaned from study of the host lavas, as these have degassed or are altered. Even lavas erupted under significant confining pressures of water often have melt inclusions that preserve volatile abundances that appear appreciably less degassed than those measured in host matrix glasses (Sobolev 1996; Newman et al. 2000; Saal et al. 2002).

Secondly melt inclusions typically host a wider diversity of melt compositions than are represented by the bulk or matrix glass compositions of the host and associated lavas or tephras from a given lava suite or location. This is an almost ubiquitous observation in basaltic rocks, and is commonly interpreted to indicate that magmatic diversity in most basaltic systems is significantly greater in extent than the range of lava compositions that are erupted (e.g., Nielsen et al. 1995; Sobolev 1996; Kent et al. 1999a; Slater et al. 2001), although alternate interpretations are possible (e.g., Danyushevsky et al. 2004). The best explanation for this phenomenon is that inclusion-bearing magmas represent mixtures of smaller volumes of melts of variable compositions that are present within basaltic melt generation and melt transport systems. Mixing and blending to produce the erupted bulk compositions reduces compositional diversity. Melts trapped within melt inclusions that formed prior to or during mixing preserve a greater range of compositions (Sobolev and Shimizu 1993; Nielsen et al. 1995; Sobolev 1996; Saal et al. 1998, 2005; Kent et al. 1999a, 2002a; Sours-Page et al. 1999; Norman et al. 2002). This makes intuitive sense when considering the volumes that individual lavas (up to many cubic kilometers) and melt inclusions (typically a few hundred cubic microns) represent. Importantly, melt inclusions may also trap compositions that are more primitive than erupted magma or liquid compositions, and in some cases may provide access to melts that approach primary magma compositions.

Variations in magma compositions within and between comagmatic suites are a primary source of information used to study the origin and evolution of basaltic and related rocks. Thus the inherently greater diversity of compositions evident in melt inclusions often allow the processes that drive magmatic diversity to be examined with greater resolution than studies based on bulk rock compositions, and melt inclusions may also preserve melts present in low volumetric proportions that are not evident amongst the erupted bulk compositions. Melt inclusion suites have been described which exhibit compositional variations that reflect all of the major drivers of compositional change in basaltic and related rocks, including variations in mantle source composition and mineralogy and differences in the style and degree of melting, mantle-melt interaction, crystal fractionation, mixing and assimilation, and degassing.

A third advantage is that melt inclusions trapped within resistant phenocryst phases in rocks that have been altered by weathering, hydrothermal alteration or low temperature metamorphism, may provide the *only* means to directly establish the composition of magmatic liquids (e.g., McDonough and Ireland 1993; Shimizu et al. 2001; Kamenetsky et al. 2002), particularly for elements that are mobile during alteration (e.g., volatiles, alkalis and alkali earths). This is particularly valuable for study of ancient and altered lavas.

Despite these clear advantages, caution must also be exercised when interpreting data from melt inclusion suites. Melt inclusions are typically small, with a restricted volume available for analysis, and specialized techniques are required to obtain chemical information. Measured compositions, particularly trace element abundances and isotopic compositions, often have higher uncertainties than conventional analyses based on relatively large amounts of bulk rock powder or glass, although this may be offset to some degree by greater chemical and isotopic diversity. Of additional major concern for studies of melt inclusions are processes that result in chemical modification of the inclusion away from the original trapped composition. In many cases this may produce measured compositions that were never present as a discrete melt within the host magma system. In fact it is emphasized that melt inclusions very rarely, if ever, preserve the exact composition of the trapped melt. This is largely due to the effects of crystallization of the host mineral from the melt after trapping (e.g., Roedder 1979; Danyushevsky et al. 2000; Kress and Ghiorso 2004), but also because the potential exists for inclusions to interact with their host minerals and external melts through diffusive exchange (e.g., Qin et al. 1992; Danyushevsky et al. 2000; Gaetani and Watson 2000, 2002; Cottrell et al. 2002; Spandler et al. 2007; Portnyagin et al. 2008) and/or may reflect chemical fractionation during the formation of the inclusion (e.g., Faure and Schiano 2005; Baker 2008). In some instances the effects of post-entrapment modification may be identified and reversed using laboratory heating or numerical treatments. In other cases use of specific chemical parameters, such as incompatible element abundances and ratios can minimize the effects of modification. However there is also the possibility that irreversible changes to composition may occur in a manner that cannot be determined or subsequently reversed. Thus the challenge is determining whether the compositional variations evident in a particular suite of melt inclusions reflect primary petrological variations or inclusion-specific processes. In the latter case the danger is that inclusion compositions may be interpreted as petrological in origin and provide misleading insight into the processes of mama generation and evolution.

Another question to be carefully considered is the degree to which melt inclusions are representative of the magmatic systems in which they occur. The use of data from melt inclusions to draw conclusions about the host magma requires an extrapolation across spatial scales of $\sim 10^{10}$ or more. The processes which lead to inclusion formation may not produce inclusion populations that directly reflect the ranges of melt compositions present at the time of trapping, and some compositions may be trapped in preference to others. Melt inclusions may also trap melts of anomalous compositions produced by boundary layer trapping, mineral dissolution and reaction and other localized processes, and that never existed in significant quantities within the magmatic system (e.g., Danyushevsky et al. 2003, 2004; Yaxley et al. 2004; Faure and Schiano 2005; Baker 2008). There is also the possibility that sample selection and preparation itself may promote a bias, unconscious or otherwise, towards specific compositions—the most obvious being selection of inclusions large enough for chemical analysis. Thus careful consideration needs to be given to establishing petrological and chemical relations between the rocks and minerals that host inclusions and the inclusions themselves. An important source of information concerning the origin and significance of melt inclusions is the comparison between inclusion populations and the bulk or matrix glass composition of the lava or tephra which host inclusions, as well as other associated lavas.

Previous reviews of melt inclusions in basaltic and related rocks

A number of reviews concerning aspects of melt inclusion studies are already available in the literature (Sobolev and Kostyuk 1975; Roedder et al. 1979, 1984; Lowenstern 1995, 2003; Sobolev 1996; Frezzotti 2001; Danyushevsky et al. 2002a; Schiano 2003) and it is not the intention of this study to repeat in detail material available in these. In particular, the reader is referred to these works for more detailed information on the use of melt inclusions to obtain estimates of the conditions of melt inclusion formation via physical measurements and experimental reheating (Sobolev and Kostyuk 1975; Roedder 1979, 1984; Danyushevsky et al. 2002a; Schiano 2003). Discussions of the measurement and use of volatile abundances in melt inclusions and applications of melt inclusions in silicic systems are available in companion articles by Metrich and Wallace (2008), Moore (2008) and Blundy and Cashman (2008). Tepley and Ramos (2008) also deal with measurement of radiogenic isotope compositions in melt inclusions.

TECHNIQUES USED FOR STUDY OF MELT INCLUSIONS

Petrographic examination of melt inclusions

Petrographic examination of melt inclusions can provide considerable information and is required before commencing any subsequent analytical protocol. In many cases where the host mineral is translucent and relatively clear melt inclusions may be observed via examination of the host mineral grains under a petrographic microscope at medium to high magnification (typically ≥ 10× objective) in transmitted light. Every effort should be made to examine melt inclusions within the host mineral, although this is not always possible. Refractive index oil or ethanol may be used to increase the amount of light passing through the host mineral. Observation of inclusions at adequate magnification to distinguish key petrographic features, and distinction between melt inclusions and small mineral inclusions, is generally not possible using binocular microscopes, even ones of high quality. In cases where the surface of the host mineral is covered by thin layers of glass, groundmass or alteration products then use of double polished thick sections (similar to those used for fluid inclusion studies) or polished grain mounts can improve viewing by allowing more light to pass into the crystal interior. An alternate approach is to lightly leach the outside of crystals with acid to remove surface coatings that impede transmitted light and several workers use cold leaching in a commercially available ~50% aqueous solution of fluoroboric (HBF_4) acid for this purpose.

The use of regular 30 μm thick petrographic thin sections is not encouraged for identification and documentation of melt inclusions, as any inclusion that is large enough for quantitative analysis (≥ 30 μm) will intersect one or more surfaces of the section and thus will not be fully available for viewing in three dimensions. Features of importance to interpretation may be removed by polishing, and in some cases features which appear to be melt inclusions in thin section or in a polished surface may turn out instead to be truncated apophyses, regions of crystal intersection, or zones of crystal resorbtion. In addition, fractures that indicate connection between inclusion and the external environment on the outside of the host crystal may also have been removed.

The common petrographic features of melt inclusions have been discussed in previous reviews (e.g., Sobolev and Kostyuk 1975; Roedder 1979, 1984; Frezzotti 2001; Danyushevsky et al. 2002a; Schiano 2003). The primary influence on the petrographic appearance of melt inclusions is the thermal history experienced by the host minerals, together with the dissolved volatile contents and the relative timing of trapping and vapor exsolution. Figure 1 shows examples of common petrographic features of melt inclusions from basaltic and related

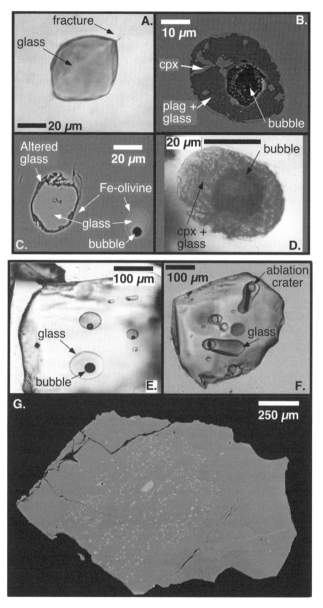

Figure 1. Photographs of melt inclusions. (A) Transmitted light photo of glassy melt inclusion in olivine without bubble. A small fracture has nucleated at one end. (B) Backscattered electron (BSE) image of olivine hosted melt inclusion containing clinopyroxene, plagioclase and glass as well as a bubble. (C) BSE photo showing two olivine-hosted melt inclusions. One (to the right) consists of glass and a bubble and shows a well-developed Fe-rich rim on olivine surrounding the inclusion, the inclusion on the left shows alteration of glass around the rim with remnant glass in the center. (D) Olivine hosted inclusion in transmitted light containing a mixture of fine dendritic clinopyroxene and glass and a bubble. (E) Polyhedral olivine crystal in transmitted light showing multiple ovoid and elongate melt inclusions containing glass and shrinkage bubbles. (F) Polyhedral olivine crystal in transmitted light showing multiple glassy and bubble free elongate to round melt inclusions. Craters remaining after LA-ICP-MS analyses are also evident (G) BSE photo of an anorthite-rich MORB plagioclase phenocryst showing large numbers of light colored melt inclusions, many occurring within inclusion rich bands.

volcanic rocks. The most common of these are glass (the result of quenching of trapped silicate liquid), daughter minerals (that nucleate and grow after trapping), minerals that are co-trapped with the inclusion (these do not nucleate within the trapped liquid but may grow further after trapping), and vapor saturation or shrinkage bubbles.

In rapidly cooled rocks, such as tephras or glassy submarine pillow rims, melt inclusions can consist entirely of glass ± vapor or shrinkage bubble ± co-trapped mineral grains, and daughter minerals will be absent (Fig. 1; e.g., Roedder 1979). However it is important to note that even completely glassy inclusions have almost always still experienced crystallization of the host mineral on the wall of the inclusion (the interface between inclusion and host mineral). As the new mineral growth occurs in thin shells on the inclusion wall, mineral growth may be apparent on compositional profiles or backscattered electron images (Fig. 1).

Inclusions that have experienced relatively slow cooling rates may contain daughter mineral phases. These minerals often have skeletal or dendritic textures indicative of rapid mineral growth (Fig 1), indicating that significant undercooling or supersaturation is required for nucleation. Formation of daughter crystals will result in closed system chemical fractionation between mineral phases and residual liquid and may result in stabilization of mineral phases and one or more residual liquid compositions that are not found elsewhere in the magmatic system under investigation (Frezotti 2001). This phenomenon has been used to study crystallization and immiscibility within trapped liquids (e.g., Clocchiatti and Massare 1985; Jambon et al. 1992; Nielsen and Veksler 2002). The formation of glassy or microcrystalline inclusions in any given sample during cooling may in part also reflect inclusion size, composition and volatile abundances (e.g., Lowenstern 2003) and thus may vary significantly between inclusions, even within the same sample.

Bubbles, typically representing small fractions of the volume of the inclusion are also often found within glassy and crystalline melt inclusions (Fig. 1). Although often referred to as "shrinkage bubbles" bubbles represent a separate phase formed by vapor exsolution (Roedder 1984), although the greater thermal expansion of glass relative to crystalline host may assist in bubble formation during cooling. Studies, largely of silicic melt inclusions, suggest that significant underpressure may be required to nucleate a bubble phase (e.g., Mangan and Sisson 2000), and bubble formation and relative size is also a function of inclusion size and cooling history: smaller inclusions and/or those forming at faster cooling rates (Tait 1992; Lowenstern 1994) may not nucleate bubbles. In basaltic and related rocks where high-pressure vapor exsolution produces a CO_2-rich vapor phase (e.g., Dixon et al. 1995) bubbles in inclusions trapped at higher pressures may only contain CO_2 in significant amounts (e.g., Hauri 2002; Metrich and Wallace 2008), but might also be potentially important hosts for noble gases.

Preparation of melt inclusions for analysis

Petrological and geochemical information from melt inclusions comes from a number of sources, including petrographic and experimental observations and micro chemical analyses (e.g., Roedder 1979, 1984; Sobolev 1996; Danyushevsky et al. 2002a). Many earlier studies of melt inclusions focused on studying melt inclusion crystallization or obtaining estimates of the conditions of trapping through heating experiments, and although not the focus of this review (see Sobolev and Kostyuk 1975; Roedder 1979, 1984; Danyushevsky et al. 2002a; Schiano 2003), this approach continues to provide valuable information. Recent work has increasingly focused the use of microanalytical techniques to determine elemental abundances and isotopic compositions of material trapped within inclusions.

Experimental rehomogenization of melt inclusions. For determination of the chemical or isotopic composition of the trapped melt it is almost always preferable to analyze a homogeneous glass rather than a mixture of various crystalline phases and residual glass, and slowly cooled melt inclusions containing mixtures of crystals and glass require reheating and quenching prior to analysis. Most applicable analytical techniques are based on calibration of a specific response, such as emission of characteristic X-rays, infrared absorption, or ion yields produced by ion beam sputtering or laser ablation, and this response may vary between different crystalline and glass phases. There may also be specific problems with analysis of individual phases, such as the rapid Na loss reported by Nielsen et al. (1995) in albitic quench crystals in plagioclase-hosted MORB melt inclusions. In addition, for surface analysis techniques, the proportions of solid phases evident at the sample surface may not be representative of the three dimensional volume of the inclusion, and the same applies to non-glassy inclusions where some portion may have been removed by sample preparation prior to analysis.

Homogenization of melt inclusions is generally accompanied in one of two general ways (Nielsen et al. 1998; Danyushevsky et al. 2002a). Inclusions can be heated individually and observed using a microscope heating stage (Roedder 1979, 1984; Danyushevsky et al. 2002a), with specially designed "Vernadsky" style stages also allowing rapid quenching (e.g., Danyushevsky et al. 2002). In this manner the temperatures of disappearance of various mineral phases and vapor bubbles can be directly measured. Alternatively, inclusion-bearing crystals may be heated in a furnace—typically a 1-atmosphere redox controlled furnace (e.g., Nielsen et al. 1995, 1998), although high-pressure apparatus has also been used—to a pre-determined temperature and quenched. Nielsen et al. (1998) and Danyushevsky et al. (2002a) discuss the advantages and disadvantages of each approach. The use of a heating stage provides a direct estimate of the trapping temperature of an inclusion (although this may be difficult to interpret; Danyushevsky et al. 2002a; Massare et al. 2002) but is hampered by slower throughput, greater sample preparation and increased potential for fracturing of the host crystal (Nielsen et al. 1998). Homogenization within a gas-mixing furnace provides no direct measurement of trapping temperatures although an iterative approach combined with calculation of phase equilibria may allow these to be constrained (Nielsen et al. 1995, 1998). In many cases the period of time that inclusions are held at the estimated trapping temperature is kept to a minimum to avoid diffusive loss of water during heating (see below), and if this is significant it can also have an effect on the estimated temperature of trapping and measured redox state (Danyushevsky et al. 2002a; Rowe et al. 2007). Alternatively, Gaetani and Watson (2000) advocate a longer period of isothermal heating (up to 72 hours) following homogenization and characterization of compositional gradients adjacent to inclusions to assess the degree of inclusion-host equilibration. Where compositional gradients are absent Gaetani and Watson (2000) estimate that the trapping temperature is constrained to within ~±20 °C. Due to rapid loss of water via diffusive exchange this approach would only work for relatively dry melts.

Both gas mixing furnace and heating stage homogenization techniques also allow for rapid quenching of heated inclusions, which is required to reduce quench growth of the host mineral during cooling. Nielsen et al. (1995) showed that quench crystallization with plagioclase-hosted melt inclusions resulted in lower measured Ca and Al contents at quench times of 3s, compared to the faster quenches (~0.5 s) provided by the 1-atm furnace. Overall both homogenization approaches are widely used for study of melt inclusions, although direct comparisons between the two are relatively rare. Norman et al. (2002) treated inclusions from Koolau, Hawaii, using both methods and report essentially unchanged results.

For any rehomogenized inclusion it is important to ensure that analyzed melt inclusions have not "breached" or vented through cracks during heating. Contraction during cooling often results in the development of fractures at the margins of melt inclusions (e.g., Tait 1992) and if these extend to the outside of the host crystal they can provide a pathway for loss of volatile elements during reheating. In addition, fractures can also allow low temperature hydrous fluids to devitrify and chemically alter inclusions (Fig 1). Prior to reheating affected inclusions can be recognized by the presence of alteration within the inclusion—often concentrated at the margins of the inclusion or adjacent to fractures (e.g., Fig. 1C, 2). However, where inclusions have been homogenized experimentally, alteration minerals are also remelted and incorporated into the inclusion composition resulting in an apparently glassy inclusion of modified composition (Nielsen et al. 1998). This is particularly a problem in samples that have experienced significant submarine or subaerial weathering, and one example is also shown in Figure 2. In addition, for species that are undersaturated in the inclusion, rehomogenization can also result in incorporation of material from within the fracture and outside the host crystal into the inclusion via diffusive transport along melt-filled fractures. This phenomenon has been documented in MORB melt inclusions where MnO contents in excess of 10 wt% have been measured and related to Mn derived from Mn-rich coatings formed on fractures and on the external grain boundaries of the host crystals (Nielsen et al. 1998). Homogenized inclusions that have been modified by alteration prior to rehomogenization can sometimes be recognized petrographically as they may show cuspate or irregular margins resulting from reaction between inclusion and host olivine (e.g., Nielsen et al. 1998; Danyushevsky et al. 2002a). Nielsen et al. (1998) also emphasized that anomalously low S and/or Cl occur in inclusions that have vented during heating due to the low solubilities of these elements in basaltic melts at atmospheric pressure. These inclusions are often disregarded (Fig. 2; Nielsen et al. 1998, 2000; Sour-Page et al. 2002), although it should be noted that this might also result

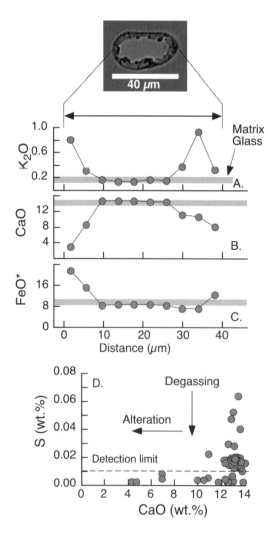

Figure 2. (A), (B) and (C) Selected oxide compositions measured across an altered and unhomogenized melt inclusion (see inset BSE photo) from Baffin Island, Canada (Kent, A.J.R. unpublished data). The composition of matrix glass (sample 10021; Kent et al. 2004b) is also shown. Melt inclusions show significant alteration and compositional change within altered material adjacent to inclusion margins, but no apparent deviation from expected composition in the visually unaltered central portion of the inclusion. (D) Composition of rehomogenized melt inclusions from Baffin Island show low CaO contents in some inclusions with low S contents, suggesting they incorporate low Ca alteration products. The range of S contents above detection limits is probably the result of shallow pressure degassing.

in removal of pristine melt inclusion compositions trapped at very shallow pressures (Fig. 2). Sulfur is particularly useful as it is typically present at relatively high concentrations in basaltic melts and in many cases the pre-degassing minimum S contents can be estimated based on measured FeO contents (e.g., Mathez 1976, Wallace and Carmichael 1992).

Mounting and polishing melt inclusions. Preparation of melt inclusions for geochemical analysis typically requires mounting inclusion-bearing mineral grains in epoxy or other mounting media and then polishing to expose inclusions. Some techniques require more specialized techniques, such as the preparation of double polished wafers for some spectroscopic analyses, although measurement of major and trace elements by a combination of EMP, SIMS and/or LA-ICP-MS requires only that the inclusion be polished on one side to expose the inclusion. Two approaches can be taken. Firstly workers can prepare grain mounts containing a number of inclusion-bearing crystals and polish to an arbitrary level. All inclusions exposed can then be documented and analyzed, and the process repeated if necessary. This is a relatively inefficient use of material, as many inclusions hosted by the mounted grains may not be exposed at the level of polishing, but where crystalline material is relatively plentiful and inclusions abundant it is often sufficient. This approach is also less labor intensive and has the beneficial effect of providing an essentially random sample of the available inclusions. The alternate approach is to select specific grains that contain melt inclusions prior to polishing through inspection with the petrographic microscope, and then each crystal is polished individually to expose the inclusion. This is considerably more labor intensive but is a more efficient use of material where sample and or inclusions are scarce. However there is also the question of bias, unconscious or otherwise, by selection of individual inclusions and this may lead to overemphasis on certain trapped compositions within a dataset (e.g., Danyushevsky et al. 2004).

Several methods for mounting individual inclusions are available, including the published techniques of Thomas and Bodnar (2002) and Hauri et al. (2002). In addition, polishing using crystalbond™ or other thermoplastic cement or mounting of individual crystals using epoxy in small tubes cut from ¼″ metal pipe are also used. More recently workers planning to use SIMS for volatile and other analyses are mounting crystals using indium metal in preference to epoxy to produce mounts with low vacuum backgrounds (e.g., Saal et al. 2002).

Chemical analysis of melt inclusions

A primary requirement for analysis of individual melt inclusions is that analytical techniques are capable of providing information from small sample volumes. This largely restricts analysis of melt inclusions to microbeam-based approaches using focused beams of ions, photons, electrons, protons or X-rays, although dissolution of individual crystals has also been used for measurement of radiogenic isotope compositions within melt inclusions (Tepley and Ramos 2008). Because of these limitations, interest in geochemical studies of melt inclusions has been strongly tied to technical advances in microchemical analysis techniques. Development and widespread application of electron microprobe techniques enabled earlier studies (e.g., Anderson and Wright 1972; Anderson 1974; Sobolev and Kostyuk 1975; Sobolev et al. 1983a,b and others), and application of spectroscopic techniques for measurement of water and CO_2 allowed the first measurements of H_2O and CO_2 contents (e.g., Anderson 1974). Subsequent development of the ion microprobe for measuring volatiles and trace elements, and more recently, laser ablation ICP-MS for *in situ* trace element analyses, allowed application of trace element techniques to the study of melt inclusions, and the integration of melt inclusion data sets with the large amount of data already available for whole rock and glass samples. These approaches also now offer the means to make *in situ* measurements of some stable and radiogenic isotope systems.

In the following we briefly discuss the most common of these: electron microprobe (EMP) analysis, Secondary Ion Mass Spectrometry (SIMS) and laser ablation inductively coupled mass spectrometry (LA-ICP-MS). Fourier Transform Infrared spectroscopy (FTIR) is also commonly used to determine H_2O and CO_2 abundance and speciation, but is discussed elsewhere in this volume (Metrich and Wallace 2008; Moore 2008), and other spectroscopic techniques have also been used to measure composition or speciation. The purpose is not to exhaustively review these techniques but to focus on the specific application to melt inclusions. In addition a number of other techniques are not yet in wide usage, but hold considerable promise for future basaltic melt inclusions studies. Examples are synchrotron-based methods to determine element speciation and volatile contents (e.g., Bonnin-Mosbah et al. 2002; Sutton et al. 2005; Metrich et al. 2006; Seaman et al. 2006), and Laser Raman spectroscopy to investigate the volatile contents of unexposed melt inclusions (Thomas et al. 2006).

An important point to note is that melt inclusions require analysis of a restricted volume with limited number of atoms. For trace elements the total amount of material available for analysis within individual inclusions is very limited. A single 100 µm diameter melt inclusion contains just a few picograms of a trace element present at 1 µg/g concentration. Even with sensitive analytical techniques the uncertainties associated with melt inclusion analyses are often worse than that obtained from conventional analyses of bulk rock samples—where the number of atoms available for analysis is typically much greater. However, in accord with the general inverse relation between compositional diversity and analytical volume evident in many natural materials, the poorer precision of measurements is often compensated for by the greater range of measured element and isotopic compositions observed in melt inclusion suites.

Electron Microprobe Analysis (EMP). The most widely used technique for study of chemical composition of melt inclusions is the electron microprobe, which provides measurements of major and minor element compositions (typically Si, Ti, Al, Cr, Fe, Mn, Mg, Ca, Ni, Na, K, P) as well as the volatile elements F, Cl and S (e.g., Nielsen et al. 1998; Davis et al. 2003; Straub and Layne 2003; Blundy and Cashman 2008), and may also be used to measure the oxidation state of S through the SK_α peak scan technique (e.g., Wallace and Carmichael 1994; Rowe et al. 2007). The electron microprobe also provides the user with additional means to examine inclusions using backscattered electron and compositional imaging and is indispensable as a means to determine the composition of the mineral host using standard mineral analysis protocols.

The spatial resolution of most EMP analyses of basaltic glass is < 10-20 µm, and so the relatively small size of most melt inclusions typically place no extra restrictions on analytical protocols, and for many melt inclusions even allows for multiple measurements of each inclusion to test homogeneity and analytical reproducibility (Nielsen et al. 1998). Electron probe measurements of melt inclusions are thus directly comparable to routine measurements of basaltic glass and precision and accuracy is limited by the quality of the instrumental calibration, control of sample devolatilization, the duration of analysis, and the abundance of each element. Significant redistribution of volatile elements (H, Li, Na) in glass during EMP analysis using standard beam conditions has been shown to occur in felsic melt inclusions (Humphreys et al. 2006). Sodium redistribution is also known to occur in basaltic glasses although mobility of Li and H has not been demonstrated in mafic samples, which are generally less prone to electron beam damage. Advances in techniques for direct measurement of oxygen and correction for the effects of electron beam damage also promise improvements in the ability to estimate H_2O contents in melt inclusions and other glasses by difference (Metrich and Wallace 2008).

Most elements present at concentrations of $>\sim 1$ wt% can be measured with accuracy and precision $\ll 5\%$. Elements typically present in lower abundances (Cr, Mn, Ni, K, P, Cl, S, F) may be measured less precisely, with uncertainties of up to ±10-50% depending on abundance. Analysis of these minor species may also require longer count times, which slows analytical throughout. Analysis of Cl and S, often present at abundances of $\ll 0.1\%$ may require total analysis times as long as 10 minutes (Nielsen et al. 1998; Kent et al. 1999a; Danyushevsky et al. 2003; Davis et al. 2003) to obtain sufficient precision. Analysis of F may also require corrections for spectral overlap from Fe (e.g., Davis et al. 2003; Straub and Layne 2003).

Several workers have also exploited the ability of EMP analysis to study the speciation of sulfur trapped within melt inclusions from basaltic rocks. This technique has been widely applied to volcanic glasses (Wallace and Carmichael 1994) but is also suited to melt inclusions (Rowe et al. 2007). Further discussion of the technique may be found in the chapter by Metrich and Wallace in this volume (2008). Fine shifts in the wavelength of the SK_α X-ray are known to correlate with S speciation, generally characterized as $S^{6+}/\Sigma S$, which in turn is a function of the oxygen fugacity, composition and pressure and temperature of entrapment (Wallace and Carmichael 1994). Rowe et al. (2007) demonstrated that melt inclusions could be used to provide an estimate of oxidation state of basaltic melt trapped in inclusions, although the effects of natural degassing, host mineral crystallization, overheating during re-homogenization and natural Fe-loss (Danyushevsky et al. 2000) can also promote significant changes in oxygen fugacity and sulfur speciation. Perhaps the most important potential change is that related to diffusive loss (or gain) of H^+ via equilibration between trapped melt and the host or external ambient melt. Basaltic melts are potentially very sensitive to this process (see below).

Secondary Ion Mass Spectrometry (SIMS). Analyses of lithophile trace elements (typically one or more of the following elements: Li, Be, B, Sc, Ti, V, Mn, Rb, Sr, Y, Zr, Nb, Cs, Ba, rare earth elements (REE), Hf, Ta, Pb, Th, U), present in concentrations of <1 to > 1000 µg/g, in melt inclusions from silicic and basaltic rocks were first conducted by secondary ion mass spectrometry (SIMS) from the 1980's and early-mid 1990's (e.g., Dunbar et al. 1992; McDonough and Ireland 1993; Sobolev and Shimizu 1993; Sobolev and Danyushevsky 1994; Gurenko and Chaussidon 1995; Nielsen et al. 1995; Sobolev 1996; Shimizu 1998). The advent of reliable techniques for measurement of trace element abundances using SIMS and other techniques was an important driving force for expanded interest in the study of melt inclusions during this time. Although laser ablation ICP-MS (LA-ICP-MS) has also become a widespread method of trace element analyses of inclusions (see below) SIMS analysis is still routinely used for melt inclusion studies and in addition to trace elements also offers the possibility for measurements of volatile element (H, C, F, S, Cl), stable isotope (H, Li, B, O, C, S, Cl) signatures and Pb isotope compositions.

General reviews and discussion of SIMS analysis, including applications of SIMS to analysis of trace element abundances in silicate minerals and glasses, can be found in Shimizu and Hart (1982) and Hinton (1995). SIMS analysis uses a focused primary ion beam, generally $^{16}O^-$, for determination of lithophile trace element abundances as positive secondary ions or $^{133}Cs^+$ for analysis of volatile elements and some light stable isotopes as negative secondary ions. The primary beam, focused to a spot size of 15-30 µm, is used to sputter material from the melt inclusion or other region of analytical interest. A small proportion (generally < 1%) of the sputtered atoms are ionized during the collision process and accelerated into a double-focusing mass spectrometer where count rates at mass peaks corresponding to ions of interest can be measured using electron multipliers or Faraday Cup detectors. The physics of the ion sputtering and extraction process is not well understood and elemental concentrations are typically calculated by reference to matrix matched standards of known composition analyzed during the same analytical session (Shimizu and Hart 1982; Hinton 1995).

The concentration of unknown elements can be calculated from the following:

$$C_i^U = C_s^U \frac{\left(C_i^R / C_S^R \right)}{\left(I_{ij}^R / I_{Sk}^R \right)} \left(I_{ij}^U / I_{Sk}^U \right) \tag{1}$$

Where:

C^U_i = the concentration of trace element i in unknown material U

C^U_S = the concentration of the internal standard element S in the unknown material U.

C^R_i / C^R_S = the ratio of the known concentrations of trace element i to the internal standard element S in reference standard R.

I^R_{ij}/I^R_{Sk} = normalized ion yield (the ratio of measured ion beam intensities) for isotope j of trace element i and isotope k of the internal standard element S in the reference standard

I^U_{ij}/I^U_{Sk} = normalized ion yield for isotope j of trace element i and isotope k of the internal standard element S in the unknown material

If multiple standards are used to generate a calibration then the term $(C^R_i/C^R_S)/(I^R_{ij}/I^R_{Sk})$ can be replaced by the slope of the line defined by multiple standards on an x-y plot of I^R_{ij}/I^R_{Sk} vs. C^R_i/C^R_S (as long as this line projects through the origin).

The internal standard element is typically a major element that is present in known quantity (generally taken from electron microprobe measurements) and that has a minor isotope suitable for analysis. Examples include ^{29}Si, ^{30}Si, ^{42}Ca, ^{43}Ca, and ^{47}Ti. The use of an internal standard corrects for changes in ion yield related to instrumental variations and also due to the geometry of analyzed inclusions. Application of Equation (1) requires the following condition be met: (i) that the variation between measured normalized ion intensities (I^U_{ij}/I^U_{Sk}) and C^U_i/C^U_S is linear and that when C^U_i is zero that I^U_{ij}/I^U_{Sk} is zero (i.e., that there is no background or that this has been corrected for prior to calculation of I^U_{ij}/I^U_{Sk}); (ii) a well-characterized homogenous standard material is available for calibration; and (iii) that there are no differences in relative ion yield between the isotopes of interest and the internal standard between the standard and unknown materials. The latter phenomenon is generally termed a *matrix effect*. The adherence to these requirements can be tested using secondary standards and fortunately for basaltic melt inclusions there are a number of well-characterized basaltic composition glass standards available (see below).

Secondary ions produced by the sputtering process often have a large range of initial kinetic energies, and an electrostatic analyzer (ESA) and slit located between the ESA and magnet is used to preclude entrance of ions with aberrant energies or trajectories into the magnet. In addition, a small offset in the accelerating voltage is also used to select ions with a specified energy range. Known as *energy filtering*, for trace element analysis this technique involves selecting ions with an energy range centered on +60-100 eV with a slit width equivalent to ±10-30 eV. Although this reduces ion beam intensities the technique is critical for trace element analyses using the smaller Cameca f-series ion probes, as selection of secondary ions with significant initial kinetic energy both diminishes peak interferences from ionized molecular species (particularly metal oxides $M^{16}O^+$) without resorting to higher mass resolving power (and concomitant loss in transmission) and minimizes possible matrix effects. In some cases energy filtering is not sufficient to reduce molecular interferences and this can limit the analysis of certain elements or requires additional correction techniques such as peak stripping. For this reason SIMS analysis of some transition metals (Sc, V, Ni, Cu, Zn) and Rb can be problematic (Hervig et al. 2006). Another important example is the requirement for a matrix-based correction procedure for removing Ba and light REEO$^+$ peaks from middle REE and middle REEO$^+$ contributions from heavy REE (Zinner and Crozaz 1986; Fahey et al. 1987).

One important feature of SIMS analysis is that ion sputtering rates are relatively low and as a result, trace element measurements by SIMS typically takes 30-90 minutes per analysis. However, in combination with the relatively high useful yield of the technique (defined as the proportion of atoms of a given isotope removed from the analytical volume that produce detectable ions during analysis; Fig. 3; Hervig et al. 2006) slow sputter rates result in relatively good preservation of inclusion material and for larger inclusions there is often sufficient material remaining for additional analyses. This can be a tremendous advantage when multiple analyses are planned for individual inclusions—such as measurements of trace elements, volatile abundances and/or stable isotope compositions, and is one of the primary distinctions between SIMS and LA-ICP-MS (see below). Detection limits for SIMS trace element analyses depend on the instrument used, the size and location of the energy bandwidth chosen for filtering (Fig. 3A) and the primary beam intensity, but are typically low, in the ng/g to µg/g range and sufficient for trace element measurements even in incompatible element depleted basalt compositions (Fig. 3A; Hervig et al. 2006). Analytical precision is largely a function of signal intensity via the Poisson relationship (the standard deviation $s = 1/\sqrt{n}$, where n is the number of counts obtained for a given isotope) and thus varies with elemental and isotopic abundance. Accuracy is controlled by the quality of the calibration procedure, including the degree to which standards are homogenous and well characterized. Overall, both precision and accuracy is generally 5-10% relative for elements present at > 1-10 µg/g abundance and 10-40% for elements present at lower abundances (e.g., Sobolev 1996; Shimizu 1998; Slater et al. 2001; Kent and Elliott 2002; Kent et al. 2002a; Portnyagin et al. 2007; Sadofsky et al. 2008).

SIMS can also be used for measuring abundance of volatile elements in melt inclusions and other volcanic glasses (e.g., Sobolev 1996; Sobolev and Chaussidon 1996; Kent et al. 1999a; Hauri et al. 2002; Lassiter et al. 2002; Gurenko et al. 2005; Benjamin et al. 2007; Portnyagin et al. 2007; Sadofsky et al. 2008; Blundy and Cashman 2008). Recent developments of volatile measurements using negative secondary ions have lowered detection limits and enabled precise measurement of a full suite of volatile elements (H, C, F, S, Cl) in a single analysis (Hauri 2002; Hauri et al. 2002; Saal et al. 2002). Further discussion of volatile measurement techniques may be found in the chapter by Blundy and Cashman (2008).

Laser Ablation Inductively-Coupled Mass Spectrometry. Laser ablation has become a powerful and widely used technique for *in situ* microsampling of solid materials in the Earth and other sciences (Durrant 1999), and provides a sensitive means for measurement of trace element abundances in small sample volumes in silicate minerals and glasses. Reviews and discussion of relevant aspects of the LA-ICP-MS technique and ICP-MS mass spectrometry may be found in Perkins and Pearce (1995), Niu and Houk (1996), Eggins et al. (1998), Russo et al. (2002), Jackson and Gunther (2003); Hattendorf et al. (2003) and Heinrich et al. (2003). Applications of LA-ICP-MS for analysis of silicate melt inclusion has also been discussed by Taylor et al. (1997) and for fluid and melt inclusions by Halter et al. (2002), Heinrich et al. (2003), and Pettke et al. (2004), and is now a widely applied technique (e.g., Kamenetsky et al. 1997; Norman et al. 2002; Danyushevsky et al. 2003; Yaxley et al. 2004; Rowe et al. 2006; Kohut et al. 2006; Benjmain et al. 2007; Elburg et al. 2007; Wade et al. 2007; Johnson et al. 2008; Vigouroux et al. 2008). LA-ICP-MS can provide information on a similar suite of trace elements as SIMS, including all commonly used lithophile trace elements, but provides increased access to transition metals and also, via specialized analyses, to select platinum group and chalcophile elements (e.g., Sun et al. 2003). LA-ICP-MS has also been used to measure Sr and Pb isotope compositions (see below).

Analyses of trace element abundances by LA-ICP-MS in melt inclusions and other materials use a pulsed laser to ablate selected sample regions. Modern laser ablation systems of 266 nm, or more commonly, 213 or 193 nm wavelength allow focusing of the laser to a small spot (generally from 10-100 µm) on a selected location with a homogenous energy profile across the spot, enabling ablation of a flat bottomed circular crater (e.g., Eggins et al. 1998). Interaction between each laser pulse and the sample surface results in localized destruction of chemical bonds, either via thermal interaction or more advantageously via a "phase explosion" and ejection of material as a jet of plasma and particulates above the sample surface (Russo et

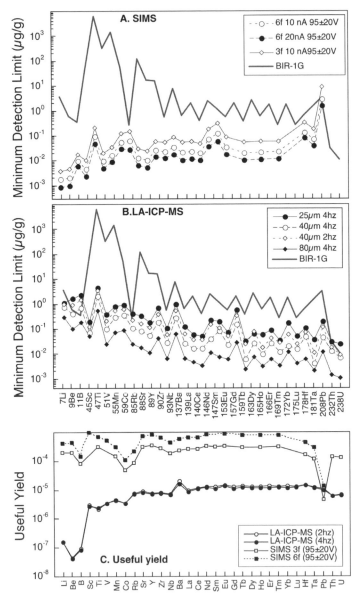

Figure 3. Detection limits and useful yields (defined as the proportion of atoms within a given analyzed volume that produce detectable ions during analysis) detected during analysis) for trace elements measured by SIMS and LA-ICP-MS analyses. Detection limits and useful yields for SIMS are from Hervig et al. (2006) and for LA-ICP-MS from Kent, A.J.R. unpublished data, calculated from the methods of Perkins and Pearce (1995). (A) Detection limits for SIMS, showing data for both Cameca 6f and 3f instruments using a 10 or 20 nA $^{16}O^-$ primary beam and secondary energy offset of 95 ± 20V (Hervig et al. 2006). (B) Detection limits for LA-ICP-MS using three spot sizes (25, 40 and 80 μm diameter) and two pulse rates (2 and 4 Hz). The composition of depleted basalt standard glass BIR-1G is shown as a representative depleted basaltic composition for comparison. In general both SIMS and LA-ICP-MS have sufficiently low detection limits for trace element analyses in melt inclusions, even in depleted melt compositions, although specific elements can be problematic (Pb for SIMS; Be, B, Rb for LA-ICP-MS). (C) Calculated useful yields for selected trace elements for SIMS using energy filtering (for Cameca 3f and 6f) and LA-ICP-MS.

al. 2002). UV laser systems, particularly those operating at 213 or, even better, at 193 nm are preferred as they produce finer particulate, resulting in higher sensitivity and lesser elemental fractionation associated with incomplete breakdown of ablated particulate within the plasma furnace (e.g., Eggins et al. 1998; Gunther and Heinrich 1999; Jackson and Gunther 2003). Ablation is carried out within a closed cell with continuous flow of inert gas (Ar, or more commonly He; Eggins et al. 1998) sweeping ablated particulates into the ICP furnace where they are rapidly atomized and ionized prior to entering the high vacuum mass analyzer (Niu and Houk 1996; Gunther and Heinrich 1999). Rapidly scanning quadrupole or sector field ICP-MS mass analyzers are used to serially monitor selected isotopes from the elements of interest with short dwell times for individual masses (Longerich et al. 1996). ICP ion sources are capable of efficiently ionizing the majority of elements in the periodic table (with the exception of some halogens; e.g., Perkins and Pearce 1995) and thus a large number of elements are potentially available for analysis.

As with SIMS, the physics of laser-solid interaction and particulate transport are incompletely understood (e.g., Russo et al. 2002) and elemental abundances are calculated relative to analyses of well characterized standard glasses, using internally-standardized techniques similar to those described above for SIMS. The same Equation (1) is used to calculated unknown concentrations, together with analysis of standard glasses for calibration, with similar assumptions as described for SIMS. The most common choice of internal standard for LA-ICP-MS analysis of basaltic melt inclusions is Ca, although Ti, Mg and Si have also been used. Calibration of LA-ICP-MS analyses are best when the internal standard and analyzed trace elements have similar volatility, and Ca, Mg, Ti and to a lesser extent Si, are relatively refractory, as are most common lithophile trace elements (e.g., Li, Sr, Y, Zr, Nb, Ba, REE, Hf, Th and U). More volatile elements calibrated using these internal standards may be more problematic and include Pb, Cs, Rb, Zn, P and B. However it is also emphasized that use of He as the sweep gas, along with deep UV wavelength ablation systems with homogenized energy distributions, minimizes elemental fractionation, often to levels that are similar or less than overall measurement precision (<5%; e.g., Eggins et al. 1998; Heinrich et al. 2003). Production of complex molecular species including metal oxides is also more limited (e.g., Kent and Ungerer 2006) and can be largely controlled by adjustment of plasma conditions. Background count rates are higher for LA-ICP-MS and are generally corrected for before to calculation of normalized ion ratios by subtraction of count rates measured before ablation starts.

Overall internal precision for LA-ICP-MS analyses of trace elements is again largely a product of the number of counts recorded for each isotope during analysis via the Poisson relationship and precision is inversely proportional to elemental and isotopic abundances. For larger inclusions and analyses of matrix glasses it is often possible to use a larger spot size and further improve precision. Ablation rates depend on the bulk composition of the material and ablation parameters, but are generally in the order of 0.1-0.2 μm/pulse in most silicates (e.g., Eggins et al. 1998; Paul et al. 2005), allowing most inclusions of ≥30 μm diameter or more to be analyzed in ~30-60 seconds of ablation at pulse rates of 2-5 Hz. At these conditions detection limits are typically in the low μg/g to ng/g range and are typically sufficient for measurement of depleted basalt compositions (Fig. 3A).

Most analyses reported in the literature are done on inclusions that are glassy (either through natural quenching or rehomogenization) and polished to expose the inclusion, and where major element compositions have been determined via prior electron microprobe analysis. In this way the protocols used for analysis of inclusions are similar to those used for other silicate glasses and minerals, except the size of the inclusion may limit the diameter of the laser spot used. The smallest inclusions analyzed by LA-ICP-MS are typically in the 30-50 μm range. Analyses of this type typically report precisions and accuracy that are better than 10-15% at the 2s limit (e.g., Fig. 4), with detection limits in the low μg/g to ng/g levels (e.g., Kent and Ungerer 2005).

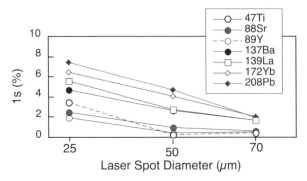

Figure 4. Variations in observed standard deviations calculated from three repeat measurements of selected trace element abundances in NIST 612 glass (measured using the isotopes shown in the legend) as a function of laser spot size. Measurements used a laser pulse rate of 3 Hz.

Halter et al. (2002), Heinrich et al. (2003), and Pettke et al. (2004) describe an alternate method, adapted from techniques developed for analysis of fluid inclusions, where the laser is used to drill vertically through the host crystal to completely remove unexposed inclusions in a single ablated volume. This approach involves ablating considerable amounts of the host mineral in addition to the inclusion, and contributions from this must be accounted for (Halter et al. 2002). Isotopes corresponding to all major and trace elements of interest are monitored and LA-ICP-MS is thus used to determine both major element and trace element concentrations. Although useful for reconnaissance, this approach is not widely used for petrological studies of melt inclusions, as it results in complete destruction of the inclusion, and an alternate means are required to estimate the composition of the internal standardizing element, which can introduce additional uncertainty (Halter et al. 2002). There is also no provision for determination of the abundances of volatile elements (Cl, S, H, C, F) in the destroyed inclusions, and although measured incompatible trace element abundances by this technique can have precision and accuracy that are similar to conventional LA-ICP-MS analyses (Pettke et al. 2003), analyses of compatible elements will generally contain an additional component of uncertainty related to the correction for host mineral within the ablated volume. Measurements reported to date of major element abundances (Heinrich et al. 2003; Pettke et al. 2004) are also less precise than those routinely obtained by electron microprobe and also limit petrological applications of the obtained data.

Comparison between SIMS and LA-ICP-MS for trace element analysis. SIMS and LA-ICP-MS are currently the two most commonly applied techniques for measuring trace element abundances in melt inclusions. The two techniques are based on quite different physical principles and instrumentation but rely on almost identical methods for calibration and calculation of trace element abundances. The accuracy of both techniques is limited by the quality of this calibration and adherence to the assumptions listed above. Precision is largely a factor of signal size during analysis—which primarily reflects element and isotopic abundances. At present the same standard glasses are widely used in SIMS and LA-ICP-MS laboratories (examples include the NIST 610 and 612 glasses, MPI-DING glass set and the BCR-2G, BHVO-2G and BIR-2G glasses from the USGS; e.g., Pearce et al. 1997; Rocholl 1998; Rocholl et al. 2000; Raczek et al. 2001; Jochum et al. 2000, 2005a, 2006; Kent et al. 2004a). For this reason trace element data produced by SIMS and LA-ICP-MS should be comparable, and this is borne out by the few comparative studies available (e.g., Kent et al. 2004b; Pettke et al. 2005). An additional comparison is shown in Figure 5 where we show the results of analysis of a suite of melt inclusions hosted in olivine by SIMS (data from Kent et al. 2002a) and LA-ICP-MS. The results are similar to other studies where direct comparisons have been made between SIMS

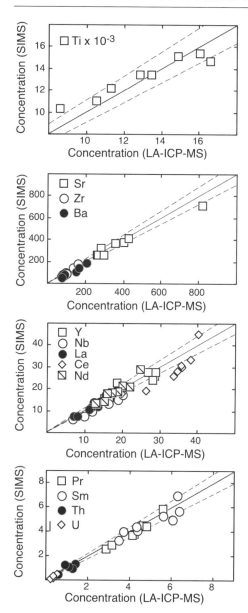

Figure 5. Comparison between trace element abundances measured in the same olivine hosted melt inclusions by SIMS and LA-ICP-MS. SIMS data are from Kent et al. (2002a) and subsequent LA-ICP-MS analyses followed techniques outlined in Kent et al. (2004b). Solid line shows 1:1 relation and dashed lines show ±10% deviation from this.

and LA-ICP-MS trace element analysis of volcanic glasses, and in general there is good agreement within ± 10-15% across a wide range of concentrations. In most cases these differences are well within the combined uncertainties of the two techniques.

The primary advantage of SIMS over LA-ICP-MS for trace element analyses is that the higher useful yield of the technique results in removal of relatively small amounts of material per analysis (Fig. 3; Hervig et al. 2006). SIMS useful yields for lithophile trace elements range between $\sim 10^{-3}$-10^{-4} for f-series Cameca ion probes using energy filtering (Hervig et al. 2006). A typical analysis using a 10 nA primary $^{16}O^-$ beam focused to a 25 μm diameter spot will sputter a crater from 2-5 μm deep during a typical 30-90 minute analysis. Except where inclusions are at the smaller end of the analyzable range (20-30 μm), this can leave material remaining for further work—although repolishing may be required. Useful yields for LA-ICP-MS are typically 1-2 orders of magnitude lower than SIMS (Fig. 4). Coupled with faster sample removal rates during ablation (Paul et al. 2005; Kent and Ungerer 2006), this results in removal of significantly more material during LA-ICP-MS trace element analyses, often limiting options for further analysis. For example a 30-50 second ablation analysis using a laser pulsed at 3-4 Hz will excavate a crater of ~15-40 μm depth and can often completely obliterate smaller inclusions (< ~50 μm).

The primary advantage of LA-ICP-MS over SIMS is that analysis times are significantly shorter—typically 2 minutes per individual analysis, allowing for significantly higher throughput during laboratory analysis. Production of molecular species are also much lower and rates of ionization are higher during LA-ICP-MS analysis (e.g., Kent and Ungerer 2005) and thus no special processing techniques are required for REE analysis (c.f. Zinner and Crozaz 1986; Fahey et al. 1987) and many transition and chalcophile metals are simpler to measure.

It is also worth noting that analyses of incompatible elements in melt inclusions by SIMS or LA-ICP-MS are both relatively insensitive to accidental incorporation of the host mineral into the analyzed volume. This is particularly the case where the internal standardizing isotope is also much higher in concentration in the inclusions than in the host mineral (e.g., ^{42}Ca, ^{43}Ca or ^{47}Ti in olivine hosted inclusions). As shown in Figure 6 incorporation of up to 60-80% or more of the host mineral into the volume sputtered or abated for analysis will still produce less than 10-15% change in the measured concentration of moderately to highly incompatible elements ($K_D \leq 0.1$-0.01). The effect is understandably the reverse for compatible elements, which are very sensitive to incorporation of the host—addition of only a few percent will cause large increases in the concentrations measured in the analyzed volume. Where contributions from the host mineral are significant they can be subtracted from the measured total via comparison to an analysis of the host mineral and calculation of relative ablated masses of host and inclusion (e.g., Halter et al. 2002), although this may introduce additional uncertainties.

Isotopic compositions of melt inclusions. Both SIMS and LA-ICP-MS offer the ability to determine some radiogenic isotope compositions of melt inclusions, as well as for SIMS measurement of some stable isotope systems (H, Li, B, C, O, S, Cl). As with trace element measurements *in situ* isotopic analyses generally have worse precision than for accepted solution-based isotopic measurement techniques, but this is again is often compensated for by relatively large isotopic variations observed within individual inclusion suites. Work on radiogenic isotope systems to date has largely concentrated on measurements of Pb isotope composition. For SIMS these measurements are made using large magnet instruments capable of high transmission such as the Cameca 1270/1280 series. Saal et al. (1998, 2005) reported measurements of ^{208}Pb/^{206}Pb and ^{207}Pb/^{206}Pb in relatively Pb-rich melt inclusions from several OIB suites and other studies have also used a similar approach to measure Pb isotope compositions from other localities (e.g., Kobayashi et al. 2004; Yurimoto et al. 2004). Uncertainties in ^{208}Pb/^{206}Pb and ^{207}Pb/^{206}Pb ratios are generally in the range of 0.2-1.2%, depending on signal intensity. These studies have documented large variations in the Pb composition of melt inclusions from single lavas and lava suites, and have been interpreted to represent highly heterogeneous Pb isotope compositions within the mantle

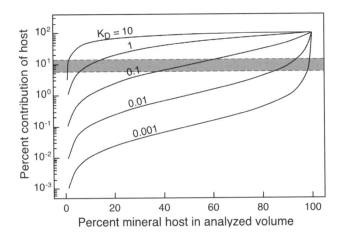

Figure 6. Calculated percent contributions to measured trace element concentrations in melt inclusions resulting from incorporation of the mineral host into the analyzed volume. Contributions are shown for a range of mineral-melt K_D. Grey dashed box shows the range of 5-15% uncertainties typical for trace element analyses by SIMS or LA-ICP-MS.

source of individual lavas, or to record interaction between melts and mantle lithosphere (Saal et al. 1998, 2005; Kobayashi et al. 2004; Yurimoto et al. 2004). Analyses of the low abundance isotope, ^{204}Pb may only be feasible in inclusions with sufficiently high Pb contents (e.g., Saal et al. 2005), and so are rarely reported in the literature. Much of the variation in the Pb isotope system is described by ratios of radiogenic isotopes to ^{204}Pb, and thus this remains a limitation. Application of multicollector SIMS instruments and improvements in sensitivity promise improvements in measurement of ^{204}Pb. LA-ICP-MS instruments using sector field magnets have also been used to measure Pb isotope compositions of melt inclusions, either using multicollector arrays (Paul et al. 2006) or via rapid electrostatic sequential scanning of ^{206}Pb-^{207}Pb-^{208}Pb abundances measured by electron multiplier to overcome the relatively short signal duration during ablation of individual inclusions (Jochum et al. 2006). For the latter approach the low abundance of ^{204}Pb as well as dynamic range issues precludes measurement of this isotope. Paul (2006) reports measurements made by laser ablation multicollector ICP-MS of Pb isotope composition, including ^{204}Pb, from melt inclusions from ocean island basalt samples from Mangaia and Pitcairn islands as well as a subduction-related basalt from Vanuatu. The range of reported Pb isotope compositions is significantly less than that reported for Mangaia inclusions by Saal et al. (1998) using SIMS.

LA-ICP-MS has also been used to measure Sr isotope compositions of individual melt inclusions, again using multicollector instruments (Jackson and Hart 2006). This measurement requires large spot sizes, and correspondingly large melt inclusions in order to generate sufficient signal during ablation. In addition, as melt inclusions typically include significant Rb a correction is required to remove the interference of ^{87}Rb from ^{87}Sr, which cannot be resolved using current instruments. The best precisions are generally achieved in inclusions with the lowest Rb/Sr. As a result reported ^{87}Sr/^{86}Sr ratios have greater uncertainty than those made by solution-based methods with Jackson and Hart (2006) reporting a typical uncertainty of ± 320 parts per million, although reported variations from Samoan olivine-hosted inclusions are considerably larger than this. An alternative that is also promising is to use the microdrilling, chemical processing and solution analysis techniques routinely used for study of isotopic variations in phenocrysts (e.g., Ramos and Tepley 2008) to study individual melt inclusions (e.g., Harlou et al. 2005).

SIMS instruments have also been used for measurements of stable isotope compositions of melt inclusions. Measurements of δD, δ^7Li, δ^{11}B, δ^{13}C, δ^{18}O and δ^{34}S, have been reported in basaltic melt inclusions, with correction for instrumental mass fractionation typically made by comparison to analyses of basaltic glass standards of known isotopic composition (e.g., Hauri 2002; Hauri et al. 2002; Kobayashi et al. 2004). Measurements of δ^{37}Cl have also been made by SIMS and are applicable to melt inclusions (e.g., Layne et al. 2004). Both large and small format SIMS instruments have been used for stable isotope measurements in melt inclusions, although the greater degree of transmission in the large format instruments is an inherent advantage where high mass resolving power is required and/or overall precision is limited by counting statistics on minor isotopes. Analyses of stable isotope compositions in melt inclusions using SIMS are again more uncertain than those made by conventional mass spectrometric techniques, with internal precisions of ~1-5 per mil reported for H, Li, B, C, S and Cl (e.g., Gurenko and Chaussidon 1997, 2001; Rose et al. 2001; Hauri et al. 2002; Kobayashi et al. 2004) and 0.1-1.5 per mil for O (Gurenko et al. 2001; Gurenko and Chaussidon 2002), with precision dependent on abundance of individual elements in specific inclusions (e.g., Hauri et al. 2002). Stable isotope analyses of melt inclusions often show variations well in excess of these uncertainties, and consequently provide important information about the origin of basaltic melt inclusions, including insight into the roles of the role of degassing, assimilation and mantle source composition in basaltic magma systems (Gurenko and Chaussidon 1997, 2002; Gurenko et al. 2001; Rose et al. 2001; Hauri 2002; Kobayashi et al. 2004).

FORMATION AND EVOLUTION OF MELT INCLUSIONS

Melt inclusion host minerals

The most common mineral hosts used for the study of melt inclusions, in approximate order of importance in basaltic and related rocks, are olivine, plagioclase, spinel, clinopyroxene and orthopyroxene. Olivine is often preferred as it is an early-formed mineral, together with spinel in many basaltic and related magmas, preserving inclusions from the relatively early stages of magmatic evolution, and readily traps melt inclusions in many volcanic rocks. Olivine also has the advantage of being translucent and relatively incompatible with respect to most commonly used lithophile trace elements (the exception being transition metals such as Ni, Mn, Co, Cr, V). This is advantageous for three reasons (i) incorporation of a small amount of the host mineral in the volume of inclusion analyzed for trace elements has little effect on measured trace element abundances (Fig. 6); (ii) the rate at which diffusive equilibration occurs between a trapped inclusion and the mineral host or external melt is lower for elements with lower mineral-melt distribution coefficients (Qin et al. 1992); and (iii) continued crystallization of the host mineral on the interior of the inclusion does not fractionate incompatible element abundances and has little overall effect on measured incompatible trace element abundances beyond the effect of concentration due to olivine removal. Olivine-hosted inclusions can also be simpler to analyze (as analysis of a small proportion of the host mineral does not significantly affect measured incompatible trace element abundances; Fig. 6). Plagioclase is also a commonly used mineral for melt inclusions, as it can contain very large numbers of melt inclusions (e.g., Fig. 1). In many basaltic rocks, including MORB and related rocks, it is often the dominant phenocryst phase (Sinton et al. 1993; Nielsen et al. 1995, 2001; Hansen et al. 2000; Sours-Page et al. 1999, 2002; Font et al. 2007). Questions remain about the ability of plagioclase to preserve trace element signatures over the timescales equivalent to magma ascent and storage within the crust (Cottrell et al. 2002; Danyushevsky et al. 2002a), although variations which match those of the host lavas are evident in a number of MORB suites (Sinton et al. 1993; Nielsen et al. 1995, 1998; Sours-Page et al. 1999, 2002). Melt inclusions also commonly occur in pyroxene, although studies of melt inclusions hosted in pyroxenes are less common in basaltic and related rocks, and are often made in concert with studies of inclusions in co-existing olivine (e.g., Sobolev and Danyushevsky 1994; Kent et al. 2002a).

Melt inclusions also occur within spinel, including Cr-rich spinel formed during the early stages of crystallization of primitive basaltic magmas (Kamenetsky 1996, 2002; Schiano et al. 1997; Shimizu et al. 2001) providing an alternate mineral host for studies of melt inclusions trapped in primitive basalts. However studies of spinel are limited in many cases by the small size of spinel phenocrysts and contained inclusions in most basaltic rocks (e.g., Shimizu et al. 2001). In addition, inclusions within spinel can only be examined in reflected light, placing limitations on the petrographic examination of inclusions. Nevertheless several studies demonstrate that studies of melt inclusions in Cr-spinel can provide useful information (Kamenetsky 1995, 1996; Shimizu et al. 2001; Kamenetsky et al. 2002). One specific advantage of spinel may be its resistance to subsequent weathering or hydrothermal alteration, allowing melt inclusions to provide a means to estimate the compositions of trapped melts in rocks where the whole rock composition and other phenocryst phases have been heavily modified by weathering or alteration (e.g., Shimizu et al. 2001; Kamenetsky et al. 2002).

Formation of melt inclusions

The common, near ubiquitous, presence of melt inclusions in phenocrysts phases in basaltic rocks demonstrates that the processes that lead to trapping and formation of inclusions occur during the evolution of almost all basaltic magmas, although abundance and size of inclusions can vary between individual samples. In general terms melt inclusions form at the interface between melt and crystalline phases in regions of relatively slow crystal growth and/or mineral dissolution or in regions isolated by the growth of intersecting crystal faces. Reviews, such as those by Roedder (1979, 1984) and Sobolev and Kostyuk (1975), on the basis of petrographic observations and analogies to fluid inclusions, suggest a number of general mechanisms of melt inclusion formation, summarized in Figure 7. Of these, crystallization and textural equilibration following rapid skeletal, dendritic or hopper growth (Fig. 7D,E) or differential dissolution (Fig. 7A), and localized slow crystal growth rates associated with defects or the presence of other phases at the crystal-liquid interface (Fig. 7B,C) are probably the most relevant for basaltic magmas (Kohut and Nielsen 2004; Faure and Schiano 2005; Goldstein and Luth 2006; Baker 2008). Small secondary inclusions may also form in healed fractures (Fig. 7F), but are often too small for analysis. Growth following rapid differential dissolution may also be important for plagioclase (Nakamura and Shimikata 1998; Michael et al. 2002).

Experimental studies aimed at understanding melt inclusion formation in basaltic and related rocks are relatively few (Nakamura and Shimikata 1998; Kohut and Nielsen 2004; Faure and Schiano 2005; Goldstein and Luth 2006; Baker 2008) but informative, and emphasize the importance of the thermal history and relations between formation mechanisms and inclusion composition. Experiments to date have largely been aimed at olivine, plagioclase and to a lesser degree clinopyroxene, but inclusions in other minerals probably form through similar processes. Faure and Schiano (2005) identified two different inclusion trapping mechanisms in forsterite in the CMAS system. In rapidly growing skeletal or dendritic crystals, where crystal growth is controlled by diffusion at the crystal-liquid interface, inclusions were trapped by intersection or encirclement of melt by hopper, swallowtail or rod crystal morphologies. Inclusions trapped in this way were typically small, irregular and showed orientations and

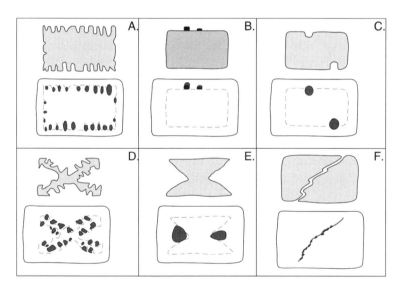

Figure 7. Schematic illustrations of common melt inclusion formation mechanisms relevant to basaltic rocks. Top and bottom images in each panel represent early and later phases in crystal growth. (A) crystal growth following a period of rapid mineral dissolution. (B) Inclusions forming where other minerals abut the host-melt interface, inclusions are associated with mineral inclusions that did not crystallize from the trapped inclusion. (C) Defects at the crystal interface limit localized growth rates. Inclusions are randomly distributed within the host. (D) Textural equilibration and overgrowth following rapid dendritic growth, inclusions are controlled by crystallographic orientations. (E) Overgrowth of skeletal or hopper crystals, inclusions form in symmetric locations. (F.) Healing of melt-filled fractures, inclusions are typically small and define a surface. These may be referred to as secondary inclusions, following common fluid inclusion nomenclature. Modified from Roedder (1979).

distributions influenced by crystallographic orientation. In polyhedral crystals, where growth rates are slower and growth is controlled by interface attachment, inclusions form at the site of defects at the crystal-liquid interface (in natural systems this probably includes the presence of other mineral phases, although Faure and Schiano [2005] did not investigate this directly). Inclusions formed in this way in polyhedral olivine were larger with more equant form, and were randomly distributed within olivine crystals.

Other experimental studies show the importance of isothermal growth and annealing following rapid crystal growth. Kohut and Nielsen (2004) were able to form melt inclusions in forsteritic olivine and anorthitic plagioclase in a primitive MORB melt through rapid cooling followed by isothermal crystal growth. In this case isothermal conditions promoted slower crystal growth and trapped inclusions during formation of tabular overgrowths over dendritic or skeletal crystals. Goldstein and Luth (2006) formed melt inclusions in forsterite crystallizing from a haplobasaltic composition over a range of cooling rates of 7-250 °C/hour (but not at a cooling rate of 1 °C/hour). Inclusions were significantly more abundant, and resulted in multiple inclusions forming within individual crystals, when cooling was followed by a six-hour isothermal period prior to final quenching. Goldstein and Luth (2006) also suggested that inclusions formed by trapping melt within interstices formed by agglomeration or simultaneous growth of individual platy olivine crystals. Baker (2008) was also able to form melt inclusions ranging from ~8-200 μm diameter in plagioclase and clinopyroxene at 1 GPa using a rapid cooling (50°/minute) to a temperature 75 °C below the liquidus (~1225 °C) followed by isothermal crystallization for periods ranging 36 minute to 14 hours. Inclusions were more common in the longer duration experiments, consistent with the importance of isothermal crystal growth and textural equilibration following rapid cooling (e.g., Kohut and Nielsen 2004), although embayments in grain boundaries (interpreted by the author as the initial stages of inclusion formation) were common in all experiments.

Overall, the range of cooling histories used in experimental studies of melt inclusion formation do not deviate significantly from those expected during the formation, transport and eruption of basaltic magmas and thus there is good reason to believe that the mechanisms of inclusion formation identified in these experimental studies are relevant to natural samples. Nakamura and Shimikata (1998) were also able to produce inclusions in plagioclase by promoting rapid dissolution followed by subsequent slow crystal growth. In this case they reproduced textures similar to the "spongy" zones commonly observed in plagioclase from convergent margin lavas and although these textures are observed in basaltic rocks this trapping mechanism may be more relevant to intermediate and evolved compositions (Michael et al. 2002).

It is also important to note that the smooth walled equant or ovoid geometries commonly observed in melt inclusions (Fig. 1) probably largely result from dissolution and reprecipitation of olivine after trapping, as the inclusion seeks to minimize surface energy, and thus may not provide petrographic information on the mechanism of inclusion trapping. Experimental studies by Schiano et al. (2006) shows that such textural re-equilibration processes appear isochemical, and may be relatively fast at the temperatures of basaltic magmas. Under thermal gradients textural re-equilibration may also be accompanied by migration of inclusions within the individual host mineral grains.

Changes in composition during inclusion formation. A key question for petrological studies of melt inclusions is whether inclusion formation also promotes changes in composition of the trapped melt. The primary concern is whether the preferential incorporation of boundary layers that form at growing crystal-liquid interfaces result in trapped compositions that differ from the composition of melt present at the time of trapping, and thus lie off the liquid line of descent. Crystals growing rapidly with respect to the rates of component diffusion in silicate melts depletes the adjacent liquid in compatible elements, and enriches it in incompatible elements, forming a compositional boundary layer. Once crystallization ceases or slows then boundary layers will disperse at a rate controlled by diffusion within silicate liquids. If boundary layers are relatively thick (>> ~10 μm) and inclusion trapping is rapid compared to the time scales required for boundary layers to re-equilibrate with external melt following crystal growth then trapped melt compositions may lie off the liquid line of descent. Examples of this are evident from experimental studies. Faure and Schiano (2005) show that inclusions hosted with rapidly growing dendritic and skeletal crystals have marked enrichments of Al_2O_3 over CaO, reflecting the slower diffusion rate of Al in silicate melts (Liang et al. 1996; Fig. 8A), whereas inclusions in slower growing polyhedral crystals lie close to the predicted liquid line of descent for the experimental composition. Thus the rate of crystal growth and mechanism of inclusion trapping may be important to understanding the composition of natural melt inclusions, and interpretation of inclusions may benefit from careful petrography and documentation of crystal growth mechanisms. This may be difficult to discern in many olivine crystals due to textural equilibration and further growth. However mapping of P_2O_5 zoning in olivine may aid in recognition of rapidly grown portions of crystals (Milman-Barris et al. 2008), and studies to date suggests that inclusions in olivine often occur in P_2O_5-rich zones interpreted to form during rapid crystal growth.

Experimental studies also show that the extent to which boundary layers are incorporated into inclusions is also dependent on the thermal history of samples. The experiments of Kohut and Nielsen (2004) and Goldstein and Luth (2006), where isothermal crystallization times are relatively long, show little deviation of trapped inclusion compositions away from the liquid line of descent—even where inclusions occur in skeletal and dendritic crystals. Isothermal growth (e.g., Fig. 8), was probably sufficient to allow boundary layers formed during rapid crystal growth to dissipate prior to melt isolation. The importance of thermal history and isothermal cooling is also highlighted in recent work by Baker (2008), who showed that boundary layers enriched in S, Cl, P_2O_5, as well as in major elements such as Al_2O_3 and FeO formed at the margins of rapidly growing plagioclase and pyroxene. Boundary layers (defined by zones of enrichment in compatible elements and depletion in compatible elements) were narrower (but showed greater changes in concentration) where isothermal crystallization times were shortest. Longer isothermal periods revealed broader boundary layers with less enrichment or depletion, consistent with collapse of steep gradients after the cessation of rapid crystal growth. Baker (2008) modeled these variations in terms of two-stage cooling and diffusive equilibration. For basaltic melts at 1200 °C, the elements that diffuse relatively slowly (P, D ≈ 2×10^{-13} m^2/sec) require a period of ~100 hours to drop to a value within ~10 % of that of the of the external melt, whereas more rapidly diffusing elements (Baker 2008, uses the example of S, D ≈ 2×10^{-12} m^2/sec; Freda et al. 2005) require much shorter time periods—generally << 10 hours (Fig. 9).

Despite experimental demonstration of the potential for boundary layer contributions to trapped melt inclusions (e.g., Faure and Schiano 2005; Baker et al. 2008), there is less evidence from natural basaltic inclusion suites to suggest that this is a concern—particularly where studies concentrate on inclusions that are sufficiently large for routine geochemical analysis (typically ≥ ~30 μm in diameter). Efforts to recognize the effects of trapping boundary layers in natural samples, by comparing the size of inclusions with elements with different diffusivities (e.g., Nielsen et al. 1995, 1998; Kuzmin and Sobolev 2004) show little to no evidence for preferential incorporation of boundary layers in smaller inclusions. A study of Icelandic picrites showed that fractionation of CaO/Al_2O_3 attributable to boundary layer incorporation into melt inclusions only occurred in inclusion 15 μm in diameter or less (Kuzmin and Sobolev 2004). A summary of data from olivine-hosted melt inclusions from a number of primitive magmas shown in Figure 8B-E also suggests fractionation of CaO/Al_2O_3 in response to boundary layer trapping may be less common in natural basaltic melt inclusions. In particular there is little evidence for large relative enrichments in Al evident in fast-trapped inclusions seen by Faure and Schiano (2005). Where large deviations in CaO/Al_2O_3 are observed in inclusion suites these typically extend to higher CaO/Al_2O_3 in inclusions than in the host glass or lavas (Kamenetsky

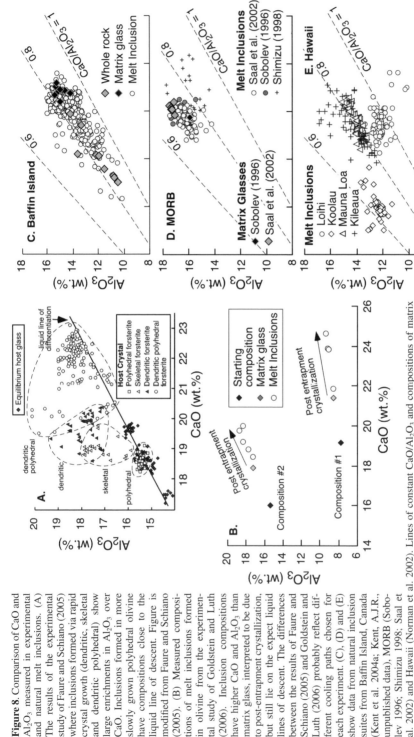

Figure 8. Comparison of CaO and Al_2O_3 measured in experimental and natural melt inclusions. (A) The results of the experimental study of Faure and Schiano (2005) where inclusions formed via rapid crystal growth (dendritic, skeletal and dendritic polyhedral) show large enrichments in Al_2O_3 over CaO. Inclusions formed in more slowly grown polyhedral olivine have compositions close to the liquid line of descent. Figure is modified from Faure and Schiano (2005). (B) Measured compositions of melt inclusions formed in olivine from the experimental study of Goldstein and Luth (2006). Inclusion compositions have higher CaO and Al_2O_3 than matrix glass, interpreted to be due to post-entrapment crystallization, but still lie on the expect liquid lines of descent. The differences between the results of Faure and Schiano (2005) and Goldstein and Luth (2006) probably reflect different cooling paths chosen for each experiment. (C), (D) and (E) show data from natural inclusion suites from Baffin Island, Canada (Kent et al. 2004a; Kent, A.J.R., unpublished data), MORB (Sobolev 1996; Shimizu 1998; Saal et al. 2002) and Hawaii (Norman et al. 2002). Lines of constant CaO/Al_2O_3 and compositions of matrix glasses are also shown. Although there are distinct differences in CaO/Al_2O_3 between suites there are no large departures from the expected liquid line of descent for inclusions within individual suites, similar to those demonstrated by Faure and Schiano (2005), implying that little fractionation of this ratio occurs during natural inclusion formation (see text).

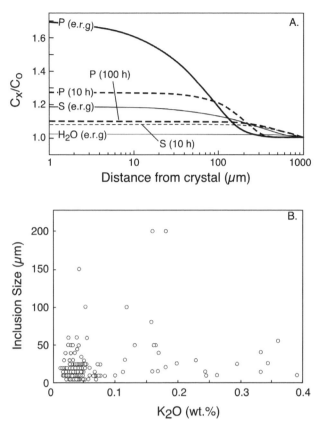

Figure 9. (A) Modeled changes in concentration of P, S and H_2O (given as C_x/C_o, the ratio of concentration at distance x to that at infinite distance from the crystal) adjacent to a growing crystal immediately following the end of rapid crystal growth (e.r.g.), and after 10 or 100 hours of isothermal crystal growth. Calculations are for a basaltic melt at 1200 °C. Modified from Baker (2008). (B) Relation between inclusion size for rehomogenized inclusions from Endeavour segment of the Juan de Fuca Ridge (Sours Page et al. 1999; Nielsen et al. 1998). Nielsen et al. (1998) interpreted the lack of covariation as evidence against equilibration between host and inclusion, but also argues against trapping of K_2O-enriched boundary layers.

et al. 1998; Shimizu 1998; Schiano et al. 2000; Norman et al. 2002; Danyushevsky et al. 2004; Ehlburg et al. 2007; Laubier et al. 2007), and are proscribed to petrological variations such as differences in the degree of melting and amount of clinopyroxene present within the mantle source (e.g., Schiano et al. 2000; Norman et al. 2002), interaction between mantle melts and clinopyroxene during melt transport (Shimizu 1998) or melts produced by localized dissolution and reaction (Danyushevsky et al. 2004). This potential for petrological variation limits the use of the CaO/Al_2O_3 ratio as an unambiguous means of determining the presence of boundary layer trapping, unless petrological variations in CaO/Al_2O_3 in host and associated lavas are well defined (e.g., Vigouroux et al. 2008). Baker (2008) also suggest that Cl/P_2O_5 or S/P_2O_5 ratios be used for detecting boundary layer effects, although the application of these ratios is also limited by the possibility of Cl and S loss via degassing prior to trapping, the complex solubility of S during changes in melt composition and redox state (e.g., Mathez 1976; Wallace and Carmichael 1992; Rowe et al. 2007) and difficulty obtaining sufficiently precise P_2O_5 measurements to recognize enrichment in MORB and other basalts with naturally low P_2O_5 contents.

The approach of Michael et al. (2002), where compositional parameters are compared to diffusivity (or proxies of diffusivity) for individual elements, can be modified and extended to other suites, and provides an alternate means to investigate the role of boundary-layer driven diffusion in natural melt inclusion suites. An example is shown in Figure 10 as a comparison between composition and the known diffusivity of these elements in basaltic melts for olivine-hosted melt inclusions from Baffin Island, Canada and Theistareykir, Iceland. In Figure 10A, where the standard deviation calculated for individual elements is compared to diffusivity, correlations are poor to non-existent, suggesting a minimal role for diffusion-driven elemental fractionation in controlling the overall trace element variability evident in these melt inclusion suites. There are also poor correlations evident between the degree of enrichment or depletion relative to primitive mantle and diffusivity in representative individual melt inclusions (Fig. 10B), suggesting that the lithophile trace element composition of individual inclusions are not controlled by boundary layer trapping.

Although further investigation of the role of boundary layer trapping is required, the limited evidence for variations in natural inclusion suites to date may reflect the fact that isothermal crystallization following rapid crystal growth plays an important role in forming melt inclusions within natural magmatic systems. Further, even where trapping of boundary layers occurs, inclusions that are small enough to record the effects are also probably too small for routine analysis (< ~30 μm). Selection of larger inclusions for analysis may preferentially select those that formed from defects at the crystal-liquid interface during slower growth, rather than through rapid crystal growth (Faure and Schiano 2005). It is also possible that future studies could reveal fractionations related to boundary layer trapping in elements that diffuse more slowly in melts, such as Zr and P_2O_5 (Baker 2008). Overall boundary layer trapping is probably of more concern for study of lower temperature and more silicic magmas, where lower temperatures and higher degree of polymerization promote slower melt diffusion rates. In silicic melts the lower limit of inclusion

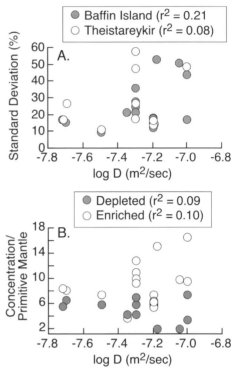

Figure 10. (A) Comparison between diffusivity in basaltic melt at 1200 °C with the standard deviation calculated from individual trace element abundances in 25 olivine-hosted melt inclusions from Baffin Island (Kent, A.J.R. unpublished data) and 47 olivine-hosted inclusions from Theistareykir, Iceland (Slater et al. 2001). Diffusion data are from LaTourette et al. (1996, 1997) and Richter et al. (2003). Note that LREE have all been assigned the same diffusivity as Nd and heavy HREE as Yb. Data from Richter et al. (2003) were calculated at 1200 °C using the same activation energy as Ba (181 kJ/mol; LaTourette et al. 1996). (B) Comparison of diffusivity with primitive-mantle-normalized concentration for representative enriched and depleted melt inclusions from Baffin Island, Canada (Kent, A.J.R. unpublished data). The poor correlations in (B) and (C) suggest that diffusivity in basaltic melts does not control the variations in melt inclusion compositions evident in these suites. As shown in Figure 17 the standard deviation and mineral-melt partition coefficients are more strongly correlated.

size, below which inclusions are effected by boundary layer effects is estimated at ~50 µm (e.g., Lu et al. 1995).

Evolution of melt inclusions after trapping

Considerable effort has gone into understanding the processes that occur in melt inclusions after trapping, and how these might effect the eventual measured inclusion compositions (e.g., Qin et al. 1992; Tait 1992; Sobolev and Danyushevsky 1994; Sobolev 1996; Nielsen et al. 1998; Schiano and Bourdon 1999; Danyushevsky et al. 2000, 2002a,b; Kress and Ghiorso 2004; Gaetani and Watson 2000, 2002; Cottrell et al. 2002; Spandler et al. 2007; Portnyagin et al. 2008).

The compositional evolution of a melt inclusion following trapping is related to the pressure and temperature path followed by the individual inclusion, and thus may diverge significantly from that experienced by the surrounding melt. At the time of formation the temperature and pressure of the inclusion will exactly match that of the surrounding melt, and as the volumes of melt inclusions are small, temperature within an inclusion will remain linked to that of the surrounding melt. However, as the bulk modulus of silicate liquids (1.2×10^{10} Pa; Murase and McBirney 1973) are about an order of magnitude less than the elastic moduli of silicate minerals (1.97×10^{11} Pa; Gebrande 1982), and because dissolved volatile components exsolve in response to decompression, pressure changes within trapped inclusions are relatively small during magmatic decompression (Tait 1992; Schiano and Bourdon 1999). Quantitative models show that the pressure change experienced by a melt inclusion during decompression will only be a small fraction (at least < 30%, and probably below 10%) of the total pressure change experienced by the host crystal and magma (Fig. 11; e.g., Tait 1992; Schiano and Bourdon 1999). Moreover, if the degree of expansion of an inclusion exceeds the tensile strength of the host mineral then cracking and decrepitation (loss of pressure and volatile components) of the inclusion can also occur. The threshold pressure required to rupture the mineral host decreases with inclusion radius (Fig. 11B) consistent with the common observation that large melt inclusions often have cracks within the adjacent mineral host. Fracturing can lead to changes in pressure within the inclusion and exsolution and loss of a vapor phase (either during initial decompression or laboratory rehomogenization), as well as providing a pathway for subaerial or submarine weathering to alter inclusion compositions (Nielsen et al. 1998).

Post entrapment crystallization (PEC). Once an inclusion has been trapped and the temperature decreases, crystallization of the host phase on the inclusion wall will commence (Sobolev and Kostyuk 1975; Roedder 1984; Sobolev 1996; Kress and Ghiorso 2004). This is an inevitable consequence of cooling the melt-host system and will occur in all trapped inclusions. As nucleation sites are readily available at the margin of the inclusion, newly crystallized host will almost always occur as concentric shells at the inclusion wall. This may not be detectable using optical microscopy, but is often apparent in BSE or compositional images from the electron microprobe (Fig. 1). In the absence of significant re-equilibration between the newly grown crystal and the mineral host, new mineral growth will be near fractional. For olivine this results in crystallization of an increment of olivine that is more magnesian than the trapped melt, following established Fe^{2+}/Mg partitioning relationships, where K_D (FeO/MgO) olivine-melt = 0.30 ± 0.04 (Roedder and Emslie 1970; Ford et al. 1983; Falloon et al. 2007; Putirka et al. 2007). This will deplete the trapped melt in Mg relative to Fe^{2+} and the next increment of olivine crystallized will be less Mg-rich (but still more Mg rich than coexisting melt). This produces a zone of Fe-rich olivine surrounding the melt inclusion, depletes the melt inclusion in Mg (Fig. 12) and other compatible elements, and increases concentrations of olivine-incompatible elements. Crystallization in plagioclase-hosted inclusions likewise results in progressive depletion of Ca over Na and formation of more albitic rims around inclusions (Nielsen et al. 1995).

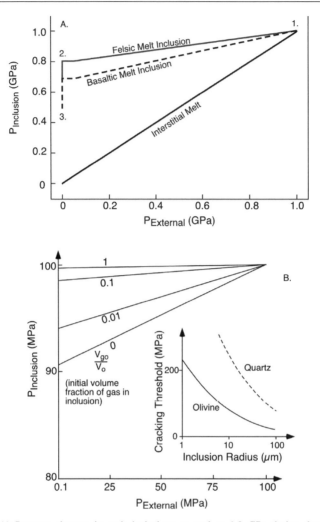

Figure 11. (A) Pressure changes in melt inclusions trapped at 1.0 GPa during decompression to atmospheric pressure. Decompression paths are shown for basaltic and felsic composition melt inclusions. The decompression of interstitial glass is shown for comparison. Decompression between points 1. and 2. represent adiabatic decompression from 1 GPa to 1 atm and then cooling from 1200 °C to 25 °C (point 3). Felsic and basaltic trajectories differ slightly due to differences in the bulk moduli of these liquids. Modified from Schiano and Bourdon (1999). (B) Calculated decrease in pressure within an olivine-hosted melt inclusion, as a function of external pressure and different volumes of CO_2 gas present at the starting pressure. The presence of an initial gas phase leads to higher pressures within the inclusion following decompression (Tait 1992). Inset shows increase in the cracking threshold pressure of olivine-hosted inclusion as a function of inclusion radius. The curve for quartz-hosted inclusions is also shown for reference. Calculated differences in cracking thresholds arise due to differences in surface energy (Tait 1992). Larger melt inclusions are more likely to crack their mineral host. Modified from Tait (1992).

In cases where there has been minimal diffusive re-equilibration between inclusion and host after trapping (see 3.4.1 below), corrections for post entrapment crystallization in olivine can be made via experimental reheating to the temperature of trapping (Sobolev and Kostyuk 1975; Nielsen et al. 1995, 1998; Danyushevsky et al. 2002a; Schiano 2003) or by numerical

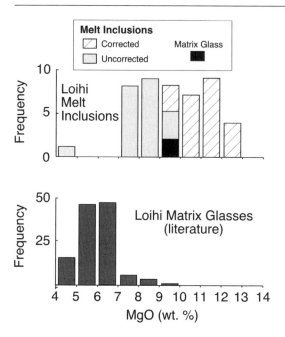

Figure 12. Measured and corrected MgO contents for melt inclusions and matrix glasses from two samples from Loihi seamount. Also shown are MgO contents for matrix glasses from Loihi reported in the literature. Corrected MgO contents in melt inclusions, calculated using the numerical approach outlined in the text, are greater than those in associated matrix glasses, and are also greater than the majority of glasses reported from Loihi from the literature (includes both tholeiitic and alkaline compositions). Corrected MgO contents approach those estimated for primary Hawaiian melts. Modified from Kent et al. (1999a).

correction schemes. For the latter, olivine of a composition calculated to be in equilibrium with the trapped melt is added iteratively to the measured trapped composition until equilibrium between inclusion and host is obtained (Sobolev and Chaussidon 1996; Sobolev 1996). In most instances cases this requires addition of less than 10-15% (by weight) olivine (e.g., Sobolev and Chaussidon 1996; Kent et al. 1999a; Lassiter et al. 2002) and thus has little effect on olivine incompatible elements beyond the uncertainties associated with measurements. This procedure works well for olivine as FeO/MgO partitioning is relatively insensitive to pressure and temperature (e.g., Roedder and Emslie 1979; Putirka et al. 2007) and so is robust over the crystallization conditions of basaltic magmas, although if pressure of crystallization is known independently then this can also be incorporated into the calculations (Falloon et al. 2007). This approach does require additional constraints on the Fe^{2+}/Fe^{3+} ratio. For MORB most workers follow the formulations of Kilinc et al. (1983) and Kress and Carmichael (1991) for a given oxygen fugacity.

Similar approaches have not been developed for plagioclase, as cation partitioning relationships are more dependent on pressure and temperature. Nielsen et al. (1995) emphasized the use of phase equilibria modeling to determine if experimentally reheated inclusions were in equilibrium with their host. Kress and Ghiorso (2004) also present a more thermodynamically rigorous procedure to correct for post entrapment crystallization using the MELTS algorithm that takes into account changes in oxygen fugacity, temperature and pressure during crystallization.

It should also be noted that the elements that are most sensitive to correction for post entrapment crystallization are those that are compatible within the host mineral, and these may thus be less robust in corrected melt inclusion compositions. A common example is the MgO content of melt inclusions in basaltic rocks, which, because of the high MgO content of forsterite-rich olivine, is often strongly depleted during post entrapment crystallization (e.g., Fig. 12). For example a melt inclusion with an uncorrected MgO content of 5.1 wt% (and FeO/MgO ~2) has a corrected MgO content of 9.1 wt% (an increase of 180%) when calculated to be in equilibrium with Fo$_{89}$ olivine. Variation of just ± 1 Fo unit in the composition of the host olivine is equivalent to a range of corrected MgO contents between 8.1-10.3 wt%. In comparison, for the same correction to Fo$_{89}$, incompatible element contents decrease by only ~15% due to the addition of olivine, and ratios of incompatible elements are unaffected by the correction. For this reason incompatible elements and their ratios are often considered more robust sources of information in melt inclusions than compatible elements.

Re-equilibration with host and or external melt

The potential for melt inclusions to preserve magmatic compositions and remain isolated during magmatic evolution has also lead to scrutiny of the potential for diffusive exchange between the inclusion and the host mineral and/or the external melt (Qin et al. 1992; Danyushevsky et al. 2000, 2002a,b; Gaetani and Watson 2000, 2002; Cottrell et al. 2002; Spandler et al. 2007). The central question is whether the composition of melt inclusions may be altered significantly by post-entrapment re-equilibration, and if so whether this effect can be reversed by numerical schemes or neutralized by restricting interpretations to elements or indices that are less sensitive to re-equilibration. If an inclusion exchanges completely with external melt then the trapped melt composition will only reflect the composition of the most recent external melt the host crystal has encountered, if partial equilibration occurs then there is the opportunity for elements or species with different equilibration rates to be fractionated from one another (Qin et al. 1992; Cottrell et al. 2002) and thus to induce chemical signals that do not reflect real magmatic variations.

Qin et al. (1992) laid the theoretical basis for understanding diffusive equilibration of melt inclusions by developing analytic solutions to the diffusion equation for isothermal equilibration of a spherical inclusion located concentrically within a spherical mineral host. This approach has subsequently been further addressed using theoretical, experimental and/or observational approaches (Danyushevsky et al. 2000, 2002a,b; Gaetani and Watson 2000, 2002; Cottrell et al. 2002; and Spandler et al. 2007). Inclusion equilibration rate is a function of diffusivity of the species of interest within the host crystal, relative size of inclusion and host (or proximity of the nearest crystal face), partition coefficient for species where $K_D < 0.1$, cooling rate, and degree of initial disequilibrium between the host and inclusion (e.g., Fig. 13). These studies show that more rapid re-equilibration is expected in inclusions that are smaller, have larger ratios of inclusion radius to host mineral radius, and that cool more slowly. Species that are more compatible and/or have higher diffusivities, and that are further from initial equilibrium will also equilibrate more rapidly. The rate of equilibration scales proportionately with the diffusivity and with the inverse square of inclusion size—thus an inclusion that is twice as large will equilibrate four times more slowly (Cottrell et al. 2002).

Iron loss in olivine-hosted inclusions. One specific concern is the effect of re-equilibration between inclusion and host and/or external melt on the Fe^{2+} content and Fe/Mg ratio of trapped melts within olivine-hosted melt inclusions (Danyushevsky et al. 2000; Gaetani and Watson 2000, 2002). Iron and Mg are compatible within olivine and Fe-Mg interdiffusion is relatively rapid at temperatures and oxygen fugacities relevant to basaltic melts (at 1200 °C D = $10^{-15.0 \pm 0.6}$ m^2/sec; Buening and Buseck 1973; Misener 1974; Jurewicz and Watson 1988). During eruption and in other cases where crystallization and cooling rates are relatively rapid compared to Fe-Mg interdiffusion, post entrapment crystallization occurs in a near fractional manner and can be readily reversed by experimental heating or numerical methods described above (e.g., Fig. 12; Sobolev 1996; Danyushevsky et al. 2000).

Where cooling rates are slower relative to crystallization, then diffusive equilibration between the inclusion and host olivine and/or external melt can become significant (Danyushevsky et al. 2000, 2002b; Gaetani and Watson 2000, 2002). The net effect of this is Fe loss. This is because, as Fe-rich olivine grown at the inclusion host interface equilibrates with more magnesian olivine in the remainder of the host mineral, the inclusion also seeks to re-equilibrate

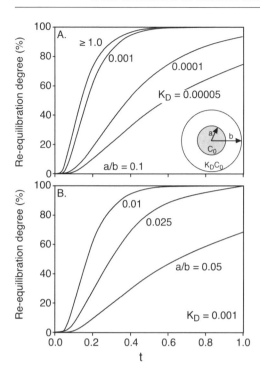

Figure 13. Re-equilibration between melt inclusions and ambient melt. Modified from Qin et al. (1992). Axes show equilibration percent versus τ, a dimensionless time parameter defined as $\tau = Dt/b^2$. Where D is diffusivity, t is time and b is host mineral radius (see inset). (A) The effect of variation in mineral melt partition coefficient (K_D) showing that equilibration is faster where an element is more compatible within the host. (B) The effect of variation in the ratio of inclusion radius to host radius (see inset in A), equilibration is faster when this ratio is larger. C_0 refers to the concentration of a given element within the melt inclusions, K_D is the mineral melt partition coefficient.

with the rim by adding Fe to maintain Fe/Mg equilibrium. Such Fe-loss is generally recognized by Fe contents of inclusions that are below those of the liquid line of descent followed by whole rock and matrix glass compositions (Fig. 14B-D; Danyushevsky et al. 2000, 2002b). Further, as inclusions hosted in more magnesian olivines are subject to a greater degree of Fe-loss, effected inclusions will also often define a correlation between the forsterite content of the host olivine and measured Fe contents within melt inclusions (Fig 14A; e.g., Danyushevsky et al. 2000, 2002b; Yaxley et al. 2004).

The effects of Fe loss are schematically illustrated in Figure 15. Due to volume considerations Fe loss has little net effect on the bulk composition of the larger host crystal or external melt, but will deplete the inclusion in Fe as it seeks to maintain Fe-Mg exchange equilibrium with the host mineral (Danyushevsky et al. 2000, 2002b; Gaetani and Watson 2000, 2002). Iron-loss has been documented in olivine-hosted inclusions in a number of basaltic and associated suites, including those from subduction zones (where the effect is more common; Sobolev and Danyushevsky 1994; Danyushevsky et al. 2000), and in komatiite, ocean island basalt and continental flood basalt suites (Danyushevsky and Sobolev 1994; Danyushevsky et al. 2000, 2002b; Norman et al. 2002; Yaxley et al. 2004). An analogous process has also been suggested to occur in slowly cooled olivine-hosted spinel mineral inclusions (Scowen et al. 1991).

Gaetani and Watson (2000) emphasize that Fe loss is irreversible through experimental reheating of inclusions or other technique involving simple addition of olivine. Iron loss can result in Fe^{2+}/Mg ratios that are low, and if these are corrected by following normal olivine addition procedures or experimental reheating then this will lead to overestimation of Mg# contents of the trapped inclusion compositions. Inclusions may also be more oxidized through loss of Fe^{2+} (Rowe et al. 2007). In addition, experimental reheating may overestimate trapping temperatures by as much as 50 °C in cases where complete re-equilibration occurs over a large interval of temperature (Danyushevsky et al. 2000). Numerical techniques are available to estimate

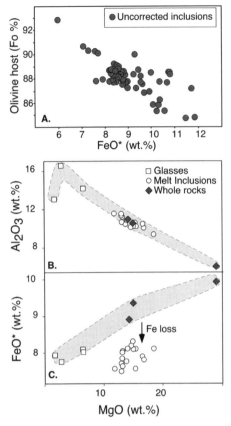

Figure 14. Indications of Fe loss in melt inclusions. (A) Negative correlation between forsterite content of the host olivine and Fe content of an inclusion as the result of Fe-loss during re-equilibration between olivine and melt (see text). Modified from Yaxley et al. (2005). The extent of Fe loss is greater in the more magnesian olivine host phenocrysts (Danyushevsky et al. 2000). (B) Anomalously low FeO contents in melt inclusions in olivine-hosted melt inclusions from Tonga boninites compared to the liquid line of descent defined by host glasses and whole rock compositions. (C) Al_2O_3 contents lie on the estimated liquid line of descent. Modified from Danyushevsky et al. (2000).

the amount of Fe lost. Danyushevsky et al. (2000) presented an iterative correction scheme to reverse the effects of Fe loss in homogenized inclusions which enables calculation of homogenization temperatures, although this requires an independent estimate of the FeO content of the trapped liquid and thus may be subject to additional uncertainty. Danyushevsky et al. (2002b) also present an alternate correction scheme based on measurement and modeling of Fe diffusion profiles in the host mineral surrounding the inclusion. Vigouroux et al. (2008) corrected for Fe-loss in naturally quenched and un-rehomogenized inclusions by adding Fe^{2+} back into each melt inclusion until the FeO* of the inclusion was restored to the MgO vs. FeO* trend defined by whole rock and groundmass glass compositions. Another promising approach may be to use parameterizations for MgO partitioning between olivine and basaltic liquid (Putirka 2005; Falloon et al. 2007; Putirka et al. 2007) to determine the original MgO content of the trapped liquid and using Fe^{2+}/Mg partitioning relations to establish the FeO content.

Danyushevsky et al. (2000) considered the case where equilibration occurred between a melt inclusion and "infinite" host olivine, disregarding the possibility of continued equilibration between the host and the external melt during continued magmatic evolution. Gaetani and Watson (2000, 2002) used numerical modeling to study equilibration with external melt and demonstrated that where cooling rates were sufficiently slow equilibration with external melt may also result in loss of Fe relative to Mg. Cooling rates as rapid as 1-2 °C/year are sufficient to maintain exchange equilibrium between an inclusion and its host melt during magmatic evolution, and these are considerably faster than estimates of cooling rates within crustal basaltic magma chambers (e.g., Cooper et al. 2001).

The effect of re-equilibration on major element contents in other mineral hosts remains to be explored. Diffusion rates of Ca and Na in plagioclase are relatively slow (Grove et al. 1984), as they require coupled diffusion with Al, and thus might not be expected to re-equilibrate with basaltic melt inclusions over timescales relevant to igneous systems. Danyushevsky et al. (2000) also suggested that equilibration within plagioclase hosted inclusions in MORB may produce lower FeO and TiO_2 contents, based on melt inclusions from ODP Hole 896A, although they note that there is no clear physical model apparent for this.

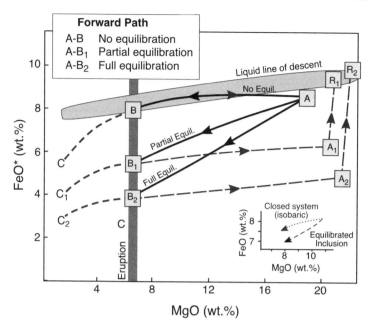

Figure 15. The schematic effects of equilibration between melt inclusions and host olivine (modified from Danyushevsky et al. 2000). Shaded field represents an inferred liquid line of descent. Path A-B represents the effect of crystallization of olivine on the walls of the host mineral from a starting composition of A during cooling, but without equilibration between the new olivine and the remainder of the host crystal. Path A-B is reversible via rehomogenization or numerical treatments for olivine crystallization after entrapment (see text). Path A-B$_2$ represents full equilibration between new olivine grown on the interior of the inclusion and the host olivine via Fe-Mg interdiffusion, path A-B$_1$ represent an arbitrary partial re-equilibration path. Paths B-C, B$_1$-C$_1$ and B$_2$-C$_2$ represent the paths followed by melt inclusions during rapid cooling after eruption, where crystallization rate is rapid relative to the rate of Fe-Mg diffusion in olivine and near-fractional growth of olivine occurs. From C, C$_1$ or C$_2$ post-eruptive olivine growth can be corrected for to return inclusions to B, B$_1$,or B$_2$, but continued correction of partially or fully equilibrated inclusions moves inclusions further along B$_1$-A$_1$ or B$_2$-A$_2$. The homogenized compositions A$_1$, A$_2$ have lower Fe and higher Mg# than the original trapped inclusion and if fully equilibrated with host olivine (R$_1$, R$_2$) also have higher MgO contents and Mg#'s.

The inset (bottom right) shows the MgO and FeO variations expected for a 100 μm diameter melt inclusion in a 2 mm diameter olivine trapped at 15 kb and transported to 5 kb with ~200 °C temperature decrease (0.5 °C/year; Gaetani and Watson 2002). The entrapment pressure of the inclusion is preserved during decompression. The dotted line shows the trend followed with no Fe-Mg equilibration (dotted line) and the dashed line reflects full equilibration between inclusion host and external melt. Modified from Gaetani and Watson (2002), axes have been adjusted to be the same scale as the larger plot.

Modification of incompatible trace elements by diffusive equilibration. As noted above the rate of diffusive equilibration of a given species between a melt inclusion and the host crystal or external melt depends on the diffusivity, inclusion and host size and mineral melt partition coefficients (Fig. 13; Qin et al. 1992). Although low mineral melt partition coefficients should restrict the rate of re-equilibration for many incompatible lithophile trace elements, if significant equilibration does occur then variations in diffusivities and partition coefficients could result in changes in relative abundances—often a valuable source of petrological information. Although many studies of melt inclusions have explicitly or implicitly assumed that negligible diffusive modification of inclusions may have occurred after trapping, recent work have demonstrated that changes in some compatible (e.g., Fe in olivine) and/or rapidly diffusing components (e.g., H$^+$, H$_2$O) occur in some natural inclusion suites (e.g., Danyushevsky et al. 2000; Cottrell et al. 2000; Hauri 2002; Rowe et al. 2007). Every inclusion suite studied should be assessed on its own merits for possible diffusive modification using the approaches detailed below.

Cottrell et al. (2002) developed a numerical model to study the effect of re-equilibration between host and inclusion on incompatible element abundances, and applied insight from their results to interpret data from olivine and plagioclase hosted trace element abundances in MORB suites. Although knowledge of diffusivity of some trace elements limited interpretations, Cottrell et al. (2002) suggested that significant re-equilibration could result in fractation of REE contents within some melt inclusions, particularly in plagioclase where elements such as Sr, Eu and Ba can be compatible or slightly incompatible (Fig. 16). Comparison of model

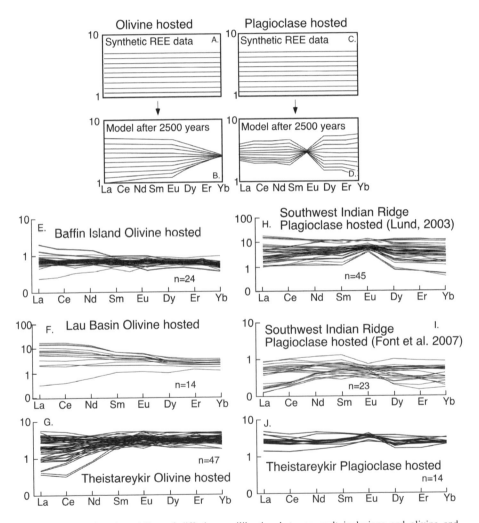

Figure 16. Results of modeling of diffusive equilibration between melt inclusions and olivine and plagioclase hosts, modified from Cottrell et al. (2002). Panels A-D show the calculated effects of diffusive re-equilibration for olivine and plagioclase hosted compositions with parallel primitive mantle normalized patterns for REE (A and C). Panels B and D show the calculated effect after 2500 years. Panels E-J show compiled data for olivine and plagioclase hosted melt inclusions from the literature. Data sources: Kamenetsky et al. (1997); Slater et al. (2001); Kent, A.J.R. unpublished data.

results for REE with compiled data sets appears to indicate that Eu, and possible the HREE contents in plagioclase-hosted melt inclusions are effected by re-equilibration with the host for number for plagioclase-hosted melt inclusion suites (Fig. 16). In contrast, predicted patterns for REE re-equilibration in olivine-hosted melt inclusion suites do not resemble equilibrated compositions and probably reflect magmatic diversity.

Cottrell et al. (2002) also emphasized comparisons of the standard deviation, calculated for individual elements within suites of melt inclusions from the same samples, and mineral-melt or peridotite-melt partition coefficients. The rationale is that bulk partition coefficients (for mantle melting) and mineral melt partition coefficients (for crystal fractionation) drive variations among incompatible trace elements in primitive melts. It has been recognized for some time that standard deviations for trace elements in suites of MORB matrix glasses correlate well with bulk partition coefficient for peridotite melting (Hofmann 1988). Thus if melt inclusions sample melt diversity and preserve it unaltered by re-equilibration or other inclusion-specific processes then the variance exhibited by melt inclusions should correlate with bulk or mineral-melt partition coefficients. If partitioning also drives compositional variations for the lavas that host melt inclusions via the same set of magmatic processes then variances shown by trace elements in melt inclusion populations should correlate with those of the host and associated lava or glasses, although this might not be a 1:1 correlation.

It is clear from the work of Cottrell et al. (2002) and from the additional comparisons shown in Fig. 17 that plagioclase-hosted melt inclusion suites can show anomalous standard deviations compared to partition coefficients or host lava compositions. Standard deviations calculated for host glass or lavas suites are well correlated with bulk partition coefficients for peridotite melting (the correlations with log D shown in Fig. 17 have $r^2 > 0.85$) emphasizing the role of partitioning in compositional variation. Standard deviations calculated for olivine hosted melt inclusion suites also correlate well with both $D^{peridotite-melt}$ ($r^2 > 0.8$) and with standard deviations for the same elements in host lava suites (not shown but r^2 are > 0.8). However correlations between $D^{peridotite-melt}$ and plagioclase hosted inclusion suites are considerably worse ($r^2 < 0.5$). Moreover Sr and Eu, and Ba appear to have standard deviations in most plagioclase hosted inclusion suites that are lower than elements of similar $D^{peridotite-melt}$, whereas Zr has elevated standard deviation (Fig. 17). Overall this suggests that careful examination of trace element abundances in plagioclase for the effects of re-equilibration is required before compositional variations are interpreted as the result of magmatic processes.

The above analysis and the work of Cottrell et al. (2002) suggests that trace element contents of olivine-hosted inclusions remain relatively unaffected by re-equilibration after trapping, at least for the timescale relevant for transport and storage of basaltic magmas, and this appears consistent with observations from olivine-hosted melt inclusion suites. Cottrell et al. (2002) used estimates for diffusivities of REE and other trace elements in olivine, as these had not been measured. Recent experimental determinations of diffusivities of trace elements in olivine show that they may be considerably faster than previously thought, with D_{REE} at 1300 °C $\sim 10^{-15}$ m^{2}sec^{-1} (Spandler et al. 2007). On this basis Spandler et al. (2007) argued that diffusive re-equilibration of REE and other trace elements was sufficiently fast, occurring on timescales of years, relative to the longer timescales of ascent of basaltic magma and storage within the crust (e.g., Faul 2001; Cooper et al. 2001; Condomines et al. 2003), to preclude preservation of mantle-derived compositions in olivine and spinel hosted basaltic melt inclusions.

It is currently unclear whether the rapid equilibration envisaged by Spandler et al. (2007) is relevant to natural melt inclusion suites. More recent measurements of the diffusion rate of REE in olivine record values that are 3-4 orders of magnitude slower (Cherniak et al. 2007) than those of Spandler et al. (2007), and if the diffusivities of Spandler et al. (2007) are used for the model of Cottrell et al. (2002) then the pattern predicted by re-equilibration shown in Fig 16 would occur within just 2-3 years (Cottrell, E. *personal communication*). This finding

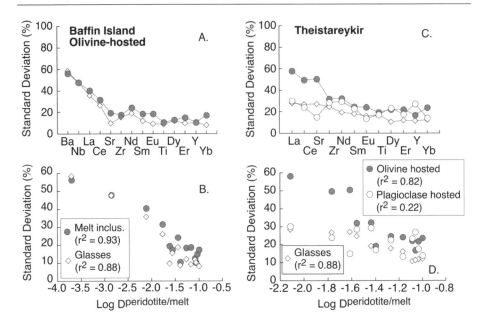

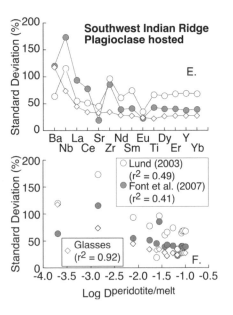

Figure 17. Comparison of standard deviations calculated for trace element abundances measured in plagioclase and olivine hosted melt inclusion suites and matrix glass compositions from basaltic samples from the Southwest Indian Ridge (SWIR), Baffin Island and Theistareykir, Iceland. Panels (A), (C), and (E) show comparisons between percent standard deviations for melt inclusion suites compared to those for host glass or lava other associated glasses or lavas from each location. Panels (B), (D) and (F) show percent standard deviations for melt inclusions and host and associated glasses or lavas compared to bulk D for peridotite melting (from Cottrell et al. 2002, and sources therein). Data sources are Slater et al. (2001), Lund (2003), Meyzen et al. (2003), Kent et al. (2004b), and Kent, A.J.R. unpublished data.

is at odds with the compositions of natural melt inclusion suites, where, as discussed above, REE profiles and standard deviations do not match those expected from equilibration, but are often closely matched to those of the host and associated lava compositions. Moreover compositional variations amongst REE and other lithophile trace elements in olivine-hosted melt inclusions that are directly attributable to mantle variations are relatively common. One clear example are the melt inclusion suites from subduction zone lavas that show trace element signatures that relate to subduction fluxes introduced to the mantle wedge (e.g., Sobolev and

Danyushevsky 1994; Kamenetsky et al. 1997; Sisson and Bronto 1998; Kent and Elliott 2002; Cervantes and Wallace 2003a; Walker et al. 2003) and relate water contents (itself a rapidly diffusing component within olivine; Mackwell and Kohlstead 1990; Portnyagin et al. 2008) to mantle melting (e.g., Cervantes and Wallace 2003a). In other cases chemical variations relating to mantle melting or mantle source variations can be preserved alongside chemical variations relating to crustal residence of magmas (Kent et al. 2002a; Yaxley et al. 2004; Kamenetsky and Gurenko 2007).

Improved analytical capabilities, specifically increased analytical sensitivity and access to greater spatial resolution for *in situ* analyses, offer additional means to identify and constrain the role of diffusive equilibration in altering the composition of trace element signatures within melt inclusions. Although these approaches have not yet been widely applied, spatially-resolved measurements of the compositions of the minerals that host inclusions can be used to search for trace elements diffusion gradients adjacent to melt inclusions, as has been done for Fe/Mg diffusion (e.g., Danyushevsky et al. 2000). Such gradients could be useful in establishing the presence and extent of diffusive exchange. In addition, analyses of trace element abundances in host minerals adjacent to melt inclusions, can provide a direct indication of whether trace element distribution between host and inclusion conform to expected equilibrium partitioning relationships. Mineral-melt partition coefficients are generally well documented for most of the common phenocryst phases in basaltic melts (e.g., Nielsen et al. 1992; Rollinson 1993; Bindemann et al. 1998), although abundances of highly incompatible elements in some host minerals may be more difficult to measure. A recent study by Weinsteiger et al. (2007) shows that Ti partitioning between plagioclase phenocrysts and contained melt inclusions conformed to expected K_D (plagioclase-melt) values (0.03 ± 0.01) over a large range of individual inclusion TiO_2 contents. This suggests that Ti variations in inclusions accurately reflect the compositions of trapped liquids, and not chemical fractionation related to inclusion trapping or re-equilibration.

A final important point is that regardless of the rate of diffusive equilibration of trace elements with the external melt, significant equilibration will only occur in the presence of a gradient in chemical potential. For melt inclusions trapped within crystals that remain within an external melt of similar composition there is little to drive changes in trapped melt compositions, unless magmatic evolution significantly modifies the composition of the host melt, or partitioning changes dramatically with changes in pressure or temperature. Gaetani and Watson (2002) argue that polybaric magmatic evolution will provide a gradient for re-equilibration of Fe/Mg within olivine-hosted inclusions, but that alone would provide little chemical gradient in incompatible element abundances. Thus it would be possible for an inclusion to preserve its trapped composition almost indefinitely—even if the rate of equilibration were rapid. It is argued below that in the majority of cases the diversity of melt inclusions evident in individual lavas is a reflection of mixing of magmas and melts of different compositions to produce the eventual erupted lava compositions. If this is the case, and if final mixing and blending occurs relatively late in the processes of melting, melt transport and crustal storage prior to eruption, then melt inclusions could easily preserve a range of near-pristine mantle-derived compositions. Differences in gradients in chemical potential may also prove an important difference between experimental results, where gradients are generally steep by design and natural samples, where the concentration differences between inclusion and external melt are typically much lower.

Diffusive loss or gain of H_2O. A special case when considering diffusive re-equilibration is the loss or gain of water by melt inclusions after trapping. For melt inclusions trapped at relatively high temperatures, and or with slow cooling rates the diffusion rate of H_2O as H^+ in olivine is relatively rapid (e.g., Mackwell and Kohlstead 1990). There are also recent experimental results that suggest that diffusion of molecular water may also be rapid enough to promote changes in melt inclusion water contents (Portnyagin et al. 2008).

If external hydrogen fugacity differs from that within the inclusion, then H$_2$O loss or gain via diffusion can occur during slow cooling or laboratory reheating (Roedder 1979; Sobolev and Danyushevsky 1994; Danyushevsky et al. 2000; Massare et al. 2002; Hauri et al. 2002; Rowe et al. 2007). This may be especially germane to studies where samples come from the massive interiors of thick lava flows, where cooling rates may be relatively slow relative to flow tops or tephra. For this reason many studies, particularly for subduction-related volcanism, have successfully focused on studies of scoria and other rapidly-cooled tephra (e.g., Anderson 1975; Roggensack et al. 1997; Cervantes and Wallace 2003a; Benjamin et al. 2007; Wade et al. 2007) and have measured high water contents. Unfortunately for studies based on older portions of the geologic record, tephra and other rapidly cooled materials such as flow tops are often poorly preserved and the massive portions of flows may be the only sample materials available.

Danyushevsky et al. (2002a) argued that loss or gain of H$_2$O via H$^+$ diffusion requires dissociation of H$_2$O and is limited by the amount of Fe^{2+} available, because few other species are available to accommodate the dissociation. Thus the potential H$_2$O loss is limited by (Danyushevsky et al. 2002a):

$$H_2O_{loss} (wt\%) = 0.125 \times FeO (wt\%) \qquad (2)$$

and is typically < 1 wt% in most basaltic and related rocks (Danyushevsky et al. 2002a). Further H loss is possible if Fe^{2+} is available in the host crystal, as in olivine, but this is marked by the presence of fine magnetite "dust" adjacent to the inclusion wall which is readily identifiable in reflected light or using BSE imagery (Nielsen et al. 1998; Danyushevsky et al. 2002a). Gain of water via hydrogen diffusion is likewise limited by the amount of Fe^{3+} available to reduce via (Danyushevsky et al. 2002a):

$$H_2O_{gain} (wt\%) = 0.113 \times Fe_2O_3 (wt\%) \qquad (3)$$

Given the lower Fe$_2$O$_3$ contents in most basaltic magmas potential H$_2$O gain in this manner is probably more limited (Danyushevsky et al. 2002a).

From these constraints H$_2$O loss in this manner is probably proportionally less significant in volatile rich magmas such as those found in subduction zones (e.g., Sisson and Layne 1993; Cervantes and Wallace 2003a; Walker et al. 2003; Gurenko et al. 2005), but could be more significant for inherently drier MORB or OIB magmas (e.g., Dixon et al. 1991, 2002; Kent et al. 1999a; Saal et al. 2002).

Recent experimental work by Portnyagin et al. (2008) suggest that diffusion of molecular water through olivine may also be sufficiently rapid (similar to H$^+$ diffusion) to modify the water contents of trapped melt inclusions, although the actual diffusion mechanism remains unclear. Portnyagin et al. (2008) showed that "dry" melt inclusions (H$_2$O < 0.5 wt%) in olivine from the Galapagos Plateau were able to acquire as much as 2.5 wt% H$_2$O after two days in water-bearing melt at 200 MPa and 1140 °C. This water gain is very rapid, and suggests that it could easily occur within natural inclusion suites if water contents of the external melt were modified by crystal fractionation, magma mixing, degassing or other petrological processes. This observation may also explain decoupling between water and components of similar bulk partition coefficient evident in some MORB and subduction related melts (e.g., Sobolev and Chaussidon 1996; Portnyagin et al. 2008), although in some cases this could also be caused by diffusive decoupling of water and lithophile trace elements (e.g., REE) during long term mantle storage prior to melting (Workman et al. 2006).

Many workers have used the covariation between water and other elements of either similar volatility or mantle-melting behavior to investigate the veracity of water contents measured in basaltic melt inclusions. Many studies show that systematic variations, related either to coherent trends among volatile elements (CO$_2$, Cl, S) related to degassing (e.g., Roggensack 1997; Cervantes and Wallace 2003a; Kamenetsky et al. 2007) or consistent covariation between H$_2$O and elements with similar bulk partition coefficients during mantle melting or crystal fractionation (e.g., K$_2$O, La, Ce, Nd, Ba; e.g., Cervantes and Wallace 2003b; Saal et al. 2005; Wade et al. 2006; Sadofsky et al. 2008).

Other chemical parameters can also be used to recognize H$_2$O mobility. Diffusion of H will result in changes in hydrogen isotope composition due to the slower diffusivities of D relative to H. Loss of H$_2$O equates with higher δD as H diffuses more rapidly than D. Hauri (2002) reported results of a study of olivine hosted melt inclusions from Hawaiian subaerial lavas and demonstrated that inclusions from the massive portions of lavas could record unmodified H$_2$O contents, but differentiation of these from those that had lost H$_2$O via diffusion required careful evaluation of H$_2$O and δD, and moreover samples that preserved unmodified H$_2$O contents were not consistently related to flow morphology of other geologic relationships that could be used in a predictive sense.

The oxidation state of sulfur, as determined by SKα peak shift methods is also a sensitive indicator of redox changes in trapped melts related to water loss or gain via H$^+$ diffusion—particularly when H$_2$O is considered in units of weight percent. This can potentially be used to determine whether inclusions have lost water, particularly in companion with other indices of redox state, such as V/Sc ratios (e.g., Canil 1997; Li and Lee 2004). Rowe et al. (2007) show that loss or gain of just 0.1 wt% water would be associated with f_{O_2} change of ~ 1 log unit, which is well within the uncertainty of the SKα technique for most basaltic melt compositions (Wallace and Carmichael 1994; Rowe et al. 2007).

ORIGIN OF COMPOSITIONAL VARIATIONS IN MELT INCLUSIONS

A common and key observation concerning the chemical compositions of melt inclusions, and one that has already been emphasized, is that they exhibit a high degree of compositional diversity, even after correction for post-entrapment crystallization and diffusive Fe-loss. This diversity is expressed in major and trace element abundances, in isotopic compositions, or a combination of all of these, but in many cases variations in incompatible lithophile elements or isotopic composition provide the most robust indications as they are less sensitive to the effects of postentrapment crystallization and diffusive re-equilibration (see above). In many cases the range of chemical compositions evident in melt inclusions from a single sample may rival or exceed that present in the entire suite of host and associated lava or tephra samples (in this instance the latter refers to lavas or tephra which are demonstrably products of the same magmatic system as the host sample, although not necessarily part of a consanguinous series). Examples of this phenomenon are shown in Figures 16-21 and are apparent from studies of all environments where basaltic magma commonly erupt, including MORB (e.g., Sobolev and Shimizu 1993; Shimizu 1998; Nielsen et al. 1995; Sours-Page et al. 1999, 2002; Slater et al. 2001; Font et al. 2007; Laubier et al. 2007); subduction zones (e.g., Kamenetsky et al. 1997; Kent and Elliott 2002; Walker et al. 2003; Wallace and Cervantes 2003a; Rowe et al. 2008), Ocean Islands (e.g., Saal et al. 1998; Kent et al. 1999a; Sobolev et al. 2000; Norman et al. 2002) and Continental Flood Basalts (e.g., Kent et al. 2002a; Yaxley et al. 2005).

Despite the potential for modification of trapped melt compositions during entrapment or post trapping equilibration, it is probable that the majority of compositional variations observed in melt inclusion suites relate to directly variations in the compositions of the trapped melts themselves. Evidence for this assertion come from a number of sources, with one of the most compelling being that melt inclusion compositions often define compositional arrays that overlie and/or extend those evident in host and associated lavas (e.g., Anderson and Brown 1972; Sobolev et al. 1983a,b, 1991; Sinton 1993; Sobolev and Danyushevsky 1994; Gurenko and Chaussidon 1995, 1997; Kamenetsky et al. 1995, 1997; Kamenetsky 1995; Nielsen et al. 1995; Sobolev 1996; Portnyagin et al. 1997; Saal et al. 1998, 2002, 2005; Shimizu 1998; Kent et al. 1999a, 2002a; Sours-Page 1999; Danyushevsky et al. 2000; Shimizu et al. 2001; Hansen and Gronvold 2000; Slater et al. 2001; Kent and Elliott 2002; Norman et al. 2002; Cervantes and Wallace 2003a,b; Gurenko et al. 2005; Jackson and Hart 2006; Benjamin et al. 2007; Kohut et al. 2006; Ehlburg et al. 2007; Font et al. 2007; Kamenetsky and Gurenko 2007; Laubier et al. 2007; Portnyagin et al. 2007; Rowe et al. 2007; Sadofsky et al. 2008; Johnson et al. 2008; Vigouroux et al. 2008; Rowe et al. 2008 and many others). In addition melt inclusions may also correlate with host mineral compositions (e.g., Kamenetsky et al. 2002; Kamenetsky and Gurenko 2007; Elburg et al. 2007; Weinsteiger et al. 2007) and/or have phase equilibria that infer that melt inclusion compositions are in equilibrium with the same source or phenocryst mineralogy as the host lavas (Nielsen et al. 1995, 1998; Kamenetsky 1995; Sobolev 1996; Shimizu 1998; Slater et al. 2001; Yaxley et al. 2005; Vigouroux et al. 2008).

If melt inclusions do inherit the bulk of their compositional diversity from the melts they trap then two important questions are: (1) what is the ultimate source of this diversity, and (2) why are inclusions more variable than the bulk compositions of the rocks that host them? A view that is widely expressed in the literature is that the range of melt inclusions derives directly from the diversity of melt compositions present within the host magmatic system, and reflect the same range of well-documented petrological processes known to drive variations within lavas, glasses and other bulk samples, including variations in mantle source composition and mineralogy, differences in the degree and style of melting, melt mixing and variable melt aggregation, crystallization and crystal-liquid separation, and assimilation and degassing. The volumes of individual eruptive units in basaltic volcanoes vary, but even the smallest eruptions are many, many orders of magnitude greater in volume than individual melt inclusions, which rarely exceed a few thousand cubic microns. Erupted lava or tephra compositions are thus considered to represent blended mixtures of a number of different melt compositions, whereas melt inclusions sample this diversity at a smaller scale, consistent with their smaller volumes and protect trapped melts from subsequent mixing and blending (e.g., Dungan and Rhodes 1978; Sinton et al. 1993; Sobolev and Shimizu 1993; Nielsen et al. 1995, 2000; Sobolev 1996; Kent et al. 1999a, 2002a; Slater et al. 2001; Norman et al. 2002; Maclennan et al. 2003). In this sense melt inclusions can be broadly considered the unmixed (or less-mixed) equivalents of erupted lavas or tephra (e.g., Sobolev and Shimizu 1993; Sobolev 1996; Shimizu 1998; Kent et al. 1999a; Slater et al. 2001; Maclennan et al. 2003). It is also important to note that although the above model considers composition variations as spatial features of magma systems, melt inclusions are also trapped at discrete times and thus may likewise preserve temporal variations. Thus melt inclusions may preserve magmatic compositions from earlier stages in the liquid line of descent relative to the bulk or glass compositions of erupted materials. It is also probable that the analysis techniques used for bulk samples, which generally involve fine crushing or dissolution of rock promote artificial "averaging" of material that may be more heterogeneous at finer scales relevant to trapped melt inclusions. However, analyses of matrix glasses in submarine basalts made using the same microbeam techniques used for melt inclusions rarely shows compositional diversity at the 10-100 μm scale relevant to melt inclusions (e.g., Kent et al. 1999a,b, 2002b, 2004b; Norman et al. 2002; Workman et al. 2006), and glasses typically appear homogeneous within analytical uncertainties. A comparison of the chemical variations evident in lava and melt inclusion samples from the Borgarhraun flow of Theistareykir volcano, Iceland by Maclennan et al. (2003) showed that both showed compositional variations related to incomplete mixing of the products of a single melting column together with crystal liquid separation. The degree of variability in inclusions was greater than that in whole rock lavas samples, consistent with inclusions sampling a volume of material ~30 times less than lava samples. Variation in compatible element appeared to record short length scale variation in crystal-liquid separation, presumably occurring within shallow magma chambers, whereas incompatible element variations related to longer length scales controlled by the outputs of the melting column.

A general model where melt inclusions provide a greater sample of magmatic diversity is supported by a number of observations. Firstly, as noted above, trends defined by melt inclusion suites often extend and or overlie trends shown by bulk lava, tephra or matrix glass samples, and compositional variations in individual melt inclusion suites have been related to almost all of the various igneous processes that drive compositional variations in basaltic systems. This includes variations in mantle source composition (e.g., Sinton et al. 1993; Nielsen et al. 1995; Sobolev and Chaussidon 1996; Kamenetsky et al. 1997; Saal et al. 1998; Sours-Page et al. 1999, 2002; Sobolev et al. 2000; Kent and Elliott 2002; Wallace and Cervantes 2003a; Elburg et al. 2007), variations in the degree and style of melting and in the mineralogy of the mantle source (e.g., Sobolev and Shimizu 1993; Nielsen et al. 1995; Gurenko and Chaussidon 1995; Sobolev 1996; Shimizu 1998; Slater et al. 2001; Maclennan et al. 2003; Ehlburg et al. 2007; Lubier et al. 2007); mantle-melt interaction during melt transport (e.g., Shimizu et al. 1998), and crystal fractionation and assimilation (Kent et al. 1999a, 2002a; Lassiter et al. 2002; Danyushevsky et al. 2003, 2004; Yaxley et al. 2004; Kamenetsky and Gurenko 2007). Secondly studies have also documented close correspondences between the calculated average composition of melt inclusions from a given sample and that of their host lavas (e.g., Nielsen et al. 1995; Kent et al. 1999a, 2002a; Kent and Elliot 2002; Slater et al. 2001; Norman et al. 2002; Maclennan et al. 2003), which directly implies that the range of compositions sampled by melt inclusions are also those that blend to produce individual lavas in which each inclusion suite occurs (and where the correspondence is 1:1 that the proportion of each composition sampled by inclusions is also approximately the same as its contribution to the host lava). An example is shown in Figure 18 where K_2O contents for lava compositions in MORB from the Endeavour segment of the Juan de Fuca ridge correlate directly with the average K_2O contents of melt inclusions from each sample, despite the large range in K_2O in inclusions from each individual samples (Sours-Page 1999). Figure 19 shows trace element abundances in olivine hosted melt inclusions from two crustally contaminated basaltic lavas from Yemen (Kent et al. 2002a). Although trace elements, particularly those like Ba, K, U that are enriched in continental crust, vary widely within individual inclusions, the average inclusion composition (after correction for fractionation of olivine) is a close match to the host lava in each instance. In each of these

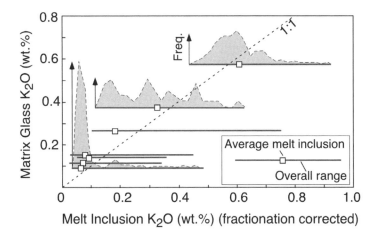

Figure 18. Comparison between K_2O contents in host lavas versus those measured in plagioclase-hosted melt inclusions for MORB from the Endeavour segment of the Juan de Fuca Ridge (modified from Sours-Page et al. 1999). The average melt inclusion composition and overall range are shown as well as representative histograms (dashed lines and grey fields) showing the relative proportions of K_2O found in melt inclusions from three different samples. Despite the large range of inclusion compositions there is a close correspondence between the average composition and that of the host.

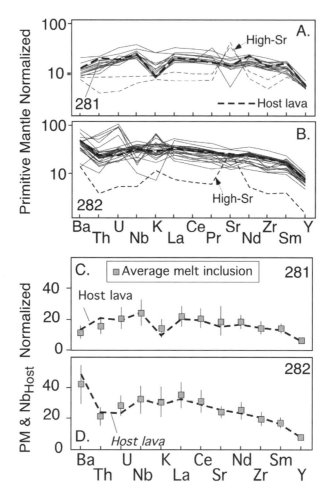

Figure 19. Trace element abundances measured in olivine hosted melt inclusions from two crustally contaminated lavas (samples 281, 282) from Yemen. As shown in (A) and (B) inclusions show large variations in primitive mantle normalized trace element abundances, including a small number of high-Sr inclusions. (C) and (D) show average melt inclusion abundances normalized to primitive mantle and also to Nb contents (the latter to correct for differences in crystal fractionation). Error bars reflect one standard deviation in inclusion compositions. Despite the wide variation in trace element abundances the average composition is a close match to the host lava. Modified from Kent et al. (2002a).

two examples the similarity of average inclusions to the host lava, even for elements that vary widely in concentration, strongly suggest that inclusions and host sample the same range of compositions, albeit at different scales. Although it has not yet been widely used within the literature, comparisons between the variance or standard deviations of melt inclusion suites and those determined for host and associated lavas or tephras can form an additional basis for investigating the relationship of inclusions and bulk rock or matrix glass compositions (e.g., Fig. 17; Cottrell et al. 2002).

Some melt inclusion suites also appear to show changes in melt diversity that is consistent with the above model and in so doing can also provide insight into the architecture and behavior of basaltic magma systems. An example is given in Figure 20(A) where K_2O/TiO_2 and Zr/Y

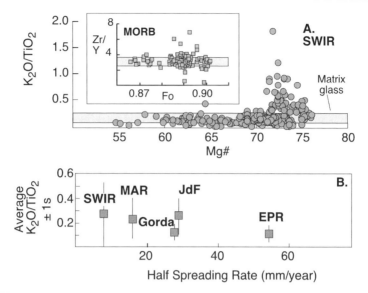

Figure 20. (A) Comparison between K_2O/TiO_2 and Mg# ($100X_{Mg}/[X_{Mg} + X_{Fe2+}]$ where X denotes mol fraction) measured in plagioclase-hosted melt inclusions from the Southwest Indian Ridge (SWIR; Lund 2003). The inset shows Zr/Y in MORB melt inclusions versus the forsterite content (Fo) of the host olivine (modified from Sobolev 1996). (B) Average melt inclusion K_2O/TiO_2 ± 1s versus representative half spreading rates for compiled MORB melt inclusion data sets for the Southwest Indian Ridge, Mid Atlantic Ridge (MAR), Juan de Fuca Ridge (JdF), Gorda Ridge and East Pacific Rise (EPR). Data from Nielsen et al. (1995); Kamenetsky 1996; Shimizu 1998; Sours-Page (1999, 2002); Lund (2003) and Nielsen, R.L. (unpublished data).

(these elemental ratios are used here as indices of melt variation) are compared to measured Mg#, or host olivine composition. Both examples show greater degrees of magmatic diversity in the more primitive melts and reduced diversity in more evolved compositions. This is consistent with a magmatic system where a range of diverse primitive melts are introduced and then mix and homogenize as they differentiate to less evolved compositions. Not all melt inclusion suites show this behavior, and this contrast may reflect differences in the timing of magmatic inputs, mixing and eruption between different basaltic magma storage and transport systems. In other cases melt inclusion diversity may also reflect the effect of external forcing factors. As an example of the latter Figure 20(B) shows the apparent decrease in melt inclusion diversity (as reflected by the standard deviation calculated for individual melt inclusion suites) with increasing spreading rate for compiled MORB melt inclusion suites. The most diverse inclusion suites, at least in terms of K_2O/TiO_2 come from the slower spreading centers. Although this might also reflect variations in mantle source character between ridges, the difference is consistent with the idea that established magma chambers in higher melt flux medium and fast spreading centers promote greater degrees of magmatic processing and mixing. In slower spreading ridges more transient magma supply limits the presence of permanent magma chambers, and thus inclusions preserve more variable compositions (Michael and Cornell 1998; Sours-Page et al. 2002). Similar observations and arguments have been made for the comparison between melt inclusion suites from ridge and off axis seamounts (Sours-Page et al. 2002).

From the above discussion it is now relatively straightforward to address the question of what the diversity evident melt inclusions actually represents: in the context of the model described above inclusion simply sample the compositional variations produced by the same sets of petrological processes that drive compositional variations within erupted lavas, tephras or matrix glasses. Melt inclusions sample this diversity at a finer scale, and may also sample some compositions preferentially. Survival of melt inclusions with variable compositions within crystals from the same hand sample, even with inclusions from the same phenocryst (e.g., Sobolev et al. 2000; Kent et al. 2002a) shows that the rates of crystal transport within a basaltic magmatic system (or the rates at which crystals from one melt are introduced into another) are faster than the rates are which melts can mix and homogenize, or that inclusions can re-equilibrate with external melt once trapped. It is interesting to note that an analogous conclusion has also recently been made concerning the rates of crystal transport being considerably more rapid than rates of crystallization and crystal re-equilibration in silicic melts (Ruprecht et al. 2008). In addition, although we have argued that magmatic systems are diverse and that individual erupted units may represent aggregation of a number of melts or magmas of different composition, the relative homogeneity of matrix glasses in many samples with variable melt inclusions shows that the mixing and blending of multiple melts *does* occur efficiently prior to final quenching. This may occur relatively late in magmatic evolution (e.g., Kent et al. 1999a), and could well be related to final eruption and emplacement.

Finally, in the context of the above discussion it is worth restating some of the advantages of using melt inclusions for studies of basaltic and related volcanic rocks. The greater degree of variation evident in inclusion suites offers greater scope for using measured compositional to characterize petrological processes and to test hypotheses. Inclusions may also provide access to melts produced during stages of magmatic evolution that are not well represented in bulk rocks samples—generally the initial stages of the liquid lines of descent, and to volumetrically small or depleted melts that are likewise poorly represented in bulk compositions. This is not to say that there will always be clear agreement as to the causes of melt inclusion variations—even in apparently simple magmatic systems petrologists often argue over the causes of compositional variations. There is no reason to expect this would not continue!

How representative are trapped melts?

The volumes of melt inclusions are exceedingly small compared to their host lavas or tephras, and even smaller in comparison to overall size of the magmatic systems which produce them. Thus extrapolation across many orders of magnitude is implicit when applying petrologic information obtained from inclusions to entire magmatic systems.

One of the most difficult relationships to establish is that between host lavas and so called "anomalous melt inclusions". In contrast to previous definitions (e.g., Spandler et al. 2007) which consider any inclusions of different composition to the host rock as anomalous, herein anomalous inclusion are defined as those which do not lie on compositional trends defined by the majority of contained melt inclusions together with host and associated lavas and tephra samples (e.g., Danyushevsky et al. 2003, 2004) and that can be reasonably be ascribed to a common set of petrologic processes that drive both inclusions and whole rock or glass compositions. Anomalous melt inclusions occur in many different melt inclusion suites (e.g., Sobolev and Shimizu 1998; Sobolev et al. 2000; Kent et al. 2002a; Danyushevsky et al. 2003, 2004) although they are subordinate within the majority of melt inclusion suites. There are a number of possible origins for anomalous inclusions. They have been interpreted as compositions that are present at volumetrically low fractions within the host magmatic systems, and thus are effectively swamped by other compositions within bulk lava or tephras, especially if the anomalous compositions have low trace element abundances (e.g., Sobolev and Shimizu 1993; Sobolev et al. 2000; Danyushevsky et al. 2003, 2004). In this case inclusions may represent discrete melt compositions that are petrologically significant, but volumetrically minor. Alternatively anomalous inclusions may have compositions that have been strongly altered by inclusion-specific processes during entrapment and subsequent diffusive equilibration and thus may not reflect any real melt composition present within the host magmatic system (e.g., Michael et al. 2002). Finally anomalous melt inclusions may represent real melt compositions (i.e., melts that exist discretely within the magmatic system) that are preferentially sampled by melt inclusions, and thus over-represented within inclusion populations compared to their contributions to bulk magma compositions. One example of the uncertainty concerning interpretation of the so-called "ultra-depleted melt" (UDM) and "Sr-rich" melt inclusions found in some basaltic rocks. Sobolev and Shimizu (1993) reported the first example of an UDM inclusion, characterized by extreme depletion in LREE and other incompatible elements, from 9°N on the Mid Atlantic Ridge and interpreted this as evidence for near-fractional melting of melt-depleted material from the top of a mantle melting column, consistent with the compositions of clinopyroxene in abyssal peridotites (Johnson et al. 1990). Subsequent studies have suggested that these compositions could also relate to diffusion-driven fractionation during melt transport (Van Orman et al. 2002) or channelized melt transport (Spiegelman and Kelemen 2003). Strontium-rich inclusions were reported from Hawaiian lavas (Sobolev et al. 2000) but also occur in lavas from MORB, Icelandic basalts and continental flood basalts (e.g., Figure 19, 21; Gurenko and Chaussidon 1995; Kamenetsky et al. 1998; Kent et al. 2002a; Danyushevsky et al. 2004). Sobolev et al. (2000) interpreted Sr-rich inclusions from Mauna Loa, Hawaii, to represent melts from gabbroic components recycled within mantle plumes. The Sr-rich trace element signature is only apparent in small numbers

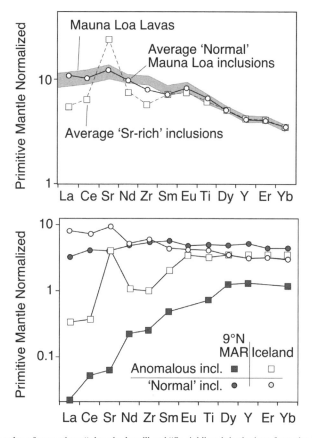

Figure 21. Examples of anomalous "ultra depleted" and "Sr-rich" melt inclusions from the (A) Mauna Loa, Hawaii, and (B) the Mid Atlantic Ridge, and Iceland (Gurenko and Chaussidon 1995; Sobolev et al. 2000; Danyushevsky et al. 2003). Normal composition melt inclusions are shown for comparison in both panels and the field of Mauna Loa lava compositions is also shown in (A).

of inclusions, and for Hawaii is absent in whole rock compositions, although some MORB and Icelandic basalts have Sr-rich characteristics (Fig. 21; Danyushevsky et al. 2003, 2004). For example, only six out of 160 analyzed inclusions from Mauna Loa analyzed by Sobolev et al. (2000) contain the Sr-rich signature, and a subsequent study by Norman et al. (2002) found no Sr-rich compositions among an additional 80 analyzed inclusions from Mauna Loa. The low abundance of Sr-rich and ultradepleted compositions complicates the interpretation of these as mantle-derived compositions, and subsequent workers have suggested that they instead result from interaction of basaltic melts and plagioclase-bearing crustal lithologies (Michael et al. 2002; Danyushevsky et al. 2003, 2004).

A final question to consider is whether the process of melt inclusion formation itself can create a bias in the compositions of melts that are preserved. Crystallization is tied to the thermal evolution of magmas, such that cooling promotes mineral growth, and melt inclusion formation, and thus could also lead to preferential trapping and preservation of melts formed in the cooler portions of magmatic systems within melt inclusion populations. Danyushevsky et al. (2003, 2004) and Yaxley et al. (2004) have argued that melt compositions reflecting mineral dissolution and assimilation of enriched melts may be preferentially trapped within melt inclusions, as mineral dissolution, reaction and assimilation occurs together with cooling and phenocryst growth in boundary layers where hot basaltic melts come into contact with cooler wallrocks or crystal mushes and undergo rapid compositional change and crystallization (e.g., Kent et al. 2002a). Importantly this could promote a bias where compositions reflect interaction with wallrocks or crystal much zones being trapped in the most mafic phenocrysts. This bias may also be exacerbated by selection of larger inclusions as more amenable to microanalytical techniques (Danyushevsky et al. 2004).

MELT INCLUSIONS IN BASALTIC AND RELATED ROCKS

The following is a brief summary of some the applications of melt inclusions to studies of basaltic and related volcanic rocks, focused on the main geodynamic environments in which basalts are found. The goal is not to provide a systematic or exhaustive survey, but to give the reader a sense of the important contributions that melt inclusions have made towards understanding of basaltic magmatism.

Mid-ocean ridges

Much of the early work on melt inclusions in basaltic rocks was focused on studies of MORB and mid ocean ridge magma systems (e.g., Dungan and Rhodes 1978; Sinton et al. 1993), and melt inclusions continue to play an important role in our understanding of the most voluminous magmatic systems on earth. Sobolev and Shimizu (1993) in an influential paper identified the presence of an "ultra-depleted" composition in an olivine-hosted melt inclusion (Fig. 21) from the Mid Atlantic Ridge, and subsequently from other ridge localities (Sobolev et al. 1994). These compositions were consistent with the highly depleted nature of clinopyroxenes from abyssal peridotites (Johnson et al. 1990), and were interpreted to be depleted melts produced during the final stages of near-fractional melting of adiabatically upwelling mantle beneath ridges. Similar arguments suggesting that melt inclusions sample incompletely aggregated melts have been made in other studies of MORB (Shimizu 1998; Laubier et al. 2007) and for Iceland (Gurenko and Chaussidon 1995; Slater et al. 2001; Maclennan et al. 2003), and thus melt inclusions provide important constraints on mantle melting and transport beneath ridges. Subsequent studies have suggested alternate mechanisms for formation of ultra depleted melt compositions, including fluxed melting of mantle harzburgite (Nielsen et al. 2000), diffusive fractionation during melt transport (Van Orman et al. 2002), dissolution and reaction within magma mush zones (Danyushevsky et al. 2003, 2004) or diffusional equilibration between inclusions and trapped melts (Michael et al. 2002). However the results of Sobolev and Shimizu

(1993) were also an early indication of the ability of basaltic melt inclusions to contain highly variable trace element compositions, and provided a spur for subsequent work. Since then, studies of MORB melt inclusions have focused on the nature of primitive MORB melts, and their interaction with residual mantle and crystal-rich cumulates (e.g., Kamenetsky et al. 1998; Kamenetsky and Crawford 1998; Shimizu 1998; Danyushevsky et al. 2003, 2004; Kamenetsky and Gurenko 2006; Laubier et al. 2007; Font et al. 2007). In addition to MORB, melt inclusions have also proven valuable to the study of ophiolite sequences (e.g., Sobolev et al. 1983b, 1996; Portnyagin et al. 1997).

A number of studies of MORB focus on inclusions trapped within anorthite-rich plagioclase phenocrysts and megacrysts (e.g., Sinton et al. 1993; Nielsen et al. 1995; Sours-Page et al. 1999, 2002; Hansen and Gronvold 2000; Lund 2003; Font et al. 2007) (e.g., Fig. 1G). Although, as noted, questions remain about the ability of plagioclase to preserve some trace element signatures (Cottrell et al. 2002; Danyushevsky et al. 2002a), studies of plagioclase provide a tremendous amount of information concerning the diversity of compositions present within MORB magma systems. Nielsen et al. (1995) exploited this diversity, together with the fact that large numbers of inclusions could be analyzed, to devise statistical tests of models for melt production and sampling beneath mid ocean ridges. This approach has not been widely employed; however it should prove useful where data from large numbers of inclusions are available.

Melt inclusions have also contributed to an understanding of volatile contents in MORB magmas and their depleted mantle source (Sobolev 1996; Sobolev and Chaussidon 1996; Saal et al. 2002), demonstrating that MORB magmas are relatively dry. Saal et al. (2002) analyzed melt inclusions from depleted magmas from the Siqueiros fracture zone and argued for constancy of volatile/non volatile element ratios within magmas derived from the depleted mantle (e.g., CO_2/Nb, H_2O/Ce, F/P, Cl/K and S/Dy). The constancy of these ratios is of assistance when studying degassed samples as it allows estimation of primary volatile abundances using the compositions of non-volatile trace elements, although it has subsequently been argued that CO_2/Nb ratios are more variable (Cartigny et al. 2008) than considered by Saal et al. (2002).

Subduction zones

Melt inclusion studies have had a large impact on understanding magma genesis in subduction zones, as they provide a link between melt compositions and the abundances of H_2O and other volatile elements that are thought to drive partial melting (e.g., Stolper and Newman 1994; Kelley et al. 2006). Anderson (1973, 1974) was the first to demonstrate high water contents in melt inclusions from subduction-related basalts, and a large number of studies have now used melt inclusions to constrain the pre-degassing contents of volatile elements in basaltic and related melts within subduction zones (e.g., Sisson and Layne 1993; Sobolev and Chaussidon 1996; Sisson and Bronto 1998; Newman et al. 2000; Cervantes and Wallace 2003a; Walker et al. 2003; Wallace 2005; Wade et al. 2006; Benjamin et al. 2007; Portnyagin et al. 2007; Kamenetsky et al. 2007; Johnson et al. 2008; Sadofsky et al. 2008; Vigouroux et al. 2008). These studies provide convincing evidence that basaltic magmas from subduction zones have significantly higher water contents (at least up to 5 wt% water) compared to basalts from other geodynamic environments (e.g., Kovalenko et al. 2000; Wallace 2005)—a key observation supporting models that consider that hydrous slab-derived fluxes to arc and back-arc magmas responsible for initiating mantle partial melting. In addition, melt inclusion studies have also demonstrated the presence of relatively dry magmas in some subduction zones (e.g., Sisson and Layne 1993; Sisson and Bronto 1998), which are interpreted to result from decompression melting related to flow within the mantle wedge.

Melt inclusions have been used to study changes in subduction related fluxes across or along arc segments (e.g., Walker et al. 2003; Rowe et al. 2008) and in some cases show that water contents may be decoupled from supposedly fluid-mobile elements (Walker et al. 2003), challenging established views of element transport. Inclusions also provide evidence for the complexity of subduction-related magma systems, highlighting variations produced during differentiation (e.g., Sobolev and Danyushevsky 1994; Lee and Stern 1998; Elburg et al. 2007) and demonstrating that significant variations in slab-derived trace element fluxes may occur within, and between, individual arc melting systems (Kamenetsky et al. 1997; Kent and Elliott 2002; Elburg et al. 2007). Finally melt inclusions also provide a means to relate volatile abundances to eruption style (e.g., Roggensack et al. 1997; Cervantes and Wallace 2003a; Gurenko et al. 2005; Kamenetsky et al. 2007), and so can be used to evaluate volcanic hazards.

Oceanic islands

Studies of melt inclusions in ocean island basalts (OIB) and related rocks are fewer than those from MORB or subduction environments, but nevertheless are important for understanding the nature and origin of mantle heterogeneity sampled by OIB magmas. Saal et al. (1998), in one of the first applications of *in situ* Pb isotope studies in melt inclusions, demonstrated that OIB source regions of individual basalts may sample material with highly diverse isotopic compositions. Their results suggest tremendous heterogeneity in Pb isotopic composition (comparable to the range shown by all terrestrial basalts in terms of ^{208}Pb-^{207}Pb-^{206}Pb), at the scales sampled by individual lavas. This is an important insight given that the range of isotopic compositions of OIB lavas largely defines the range of terrestrial basaltic rocks, and is used to constrain the composition and evolution of the mantle (e.g., Hofmann 1997). Together with subsequent work (Saal et al. 2005) these measurements show that large variations exist in melt inclusions from lavas representing the HIMU, EM-I and EM-II mantle isotopic endmembers, suggesting that all lavas also probably contain a common Pb isotope contribution from overlying depleted oceanic lithosphere. Subsequent studies of Pb isotopes in OIB melt inclusions further emphasize the variability of mantle source regions that contribute to individual lavas (Kobayashi et al. 2004; Yurimoto et al. 2004), and it has also been shown that melt inclusions from individual OIB have measurable variations in H, Sr, B and Li isotope signatures (Hauri 2002; Kobayashi et al. 2004; Jackson and Hart 2006).

Melt inclusions also provide a means to assess the chemical heterogeneity of OIB magmas in more detail than possible using lava compositions alone. Norman et al. (2002) used a large data set for olivine-hosted melt inclusions from Hawaiian picrites to argue for the effects of both variations in degree of melting and source composition in controlling the composition of melts erupted from Hawaiian volcanoes. They suggested that these inclusions sampled the Mauna Loa isotopic component at larger degrees of melting. Sobolev et al. (2000) identified rare Sr-rich melt inclusions from Mauna Loa lavas (e.g., Fig. 21) and argued that these represented recycled gabbroic components within the Hawaiian plume—although subsequent studies have argued that these may instead result from reaction or mineral dissolution within magma transport systems (Danyushevsky et al. 2004).

Studies of OIB and related rocks also indicate contamination by assimilation of brine or altered oceanic crustal components could be relatively common (Kent et al. 1999a; Lassiter et al. 2002; Hauri 2002), and may play an important role in determining the eventual composition of erupted OIB magmas. In addition, melt inclusions provide important constraints on the volatile contents of primary OIB magmas (Wallace 1998; Lassiter et al. 2002; Davis et al. 2003), and show that degassing, crystal fractionation and contamination, as well as mantle source variations, may all affect the contents of volatile elements. Some studies also show a relationship between eruptive behavior and magmatic volatile contents in Hawaiian volcanoes, and melt inclusions have been used to estimate pre-eruptive volatile contents and fluxes (e.g., Wallace and Anderson 1998).

Continental flood basalts

Even fewer studies of continental flood basalts (CFB) have been made using melt inclusions. This may in part reflect the relatively limited appearance of olivine and other phenocryst phases in some CFB lavas sequences, although there is clear promise for future studies. Workers from the former Soviet Union conducted studies on the Siberian Traps (e.g., Sobolev et al. 1981; Ryabchikov et al. 2001a,b, 2002) in an effort to constrain primary magma compositions and conditions of magma formation. Elsewhere, studies have emphasized the role of crustal interactions in controlling melt inclusion compositions (e.g., Kent et al. 2002a; Yaxley et al. 2004; Nielsen et al. 2006), although it is also clear that melt inclusions in CFB may preserve primitive melts and mantle-derived compositional variations (e.g., Larsen and Pedersen 2000; Kent et al. 2002a; Nielsen et al. 2006).

Melt inclusions have also been used to study the volatile contents of CFB melts. Thordarson and Self (1996a) used measurements of S, Cl and F contents of melt inclusions in olivine and plagioclase to estimate pre-degassing magmatic volatile contents, and together with a volatile budget model (the so called "petrological method") estimated the atmospheric release of volatile elements during eruption of the voluminous Roza flow of the Columbia River Flood Basalts. These authors applied this same melt-inclusion-based approach to estimate atmospheric loading of volatile elements during other large basaltic fissure eruptions (e.g., Thordarson and Self 1996b).

CONCLUDING REMARKS

Melt inclusions constitute a powerful tool for studying basaltic magmas, and represent a rich source of information about basaltic magma systems. Analysis of multiple inclusion from just a single sample (or even in some cases a single crystal) can produce an array of melt compositions that allow petrologists to study magmatic volatile contents, to better gauge the range of compositions present within magmatic systems, to gain access to relatively primitive melt compositions trapped in early-formed phenocrysts, and to study melt compositions in altered rocks. In concert with careful petrological and geochemical study of the rocks and minerals that host melt inclusions, as well as others from the same magmatic systems, melt inclusions can provide unique information about the nature of magma generation and transport and the compositions of produced melts and magmas.

Formation of melt inclusions appears to be a normal part of the evolution of most basaltic magmas, with trapping probably occurring during slow crystal growth immediately after episodes of rapid cooling and crystallization. Concerns exist that processes that lead to trapping of melt inclusions may also alter the compositions of trapped melts and that inclusions may subsequently equilibrate with their host mineral or external melts. A number of methods are available for assessing the possibility and extent of modification of trapped melt inclusion compositions, and the basis for most of these is a careful comparison between inclusions and the compositions of rocks that host them. Such comparisons show that much of the variability evident in most melt inclusion suites represents variations in the composition of trapped melts themselves. The range of melt inclusion compositions observed in basaltic volcanic rocks tell us that considerable compositional diversity exists within many basaltic magma systems, and that this diversity is relatively poorly sampled by the bulk compositions of erupted lavas or tephras, or by analyses of matrix glasses. Although it is often possible to reconstruct the major and trace element compositions of inclusions after post entrapment crystallization and diffusive modification, the most robust sources of information from melt inclusions are incompatible elements that diffuse relatively slowly, and for most inclusions and mineral hosts this includes most of the commonly used lithophile trace elements.

To date interest in the study of melt inclusions in basaltic rocks has been linked to technical advances in the ability to make accurate chemical measurements in small sample volumes. This should continue for the foreseeable future, in part driven by increasing access to *in situ* trace element measurement techniques as LA-ICP-MS and (to a lesser extent) SIMS instrumentation become more common. In addition, improvements and increased application of techniques for more precise stable and radiogenic isotope measurements will provide new dimensions for the study of melt inclusions.

Improvements in instrumentation and access are also leading to the generation of increasing numbers of large multi-element (and in some cases isotopic) datasets from melt inclusions. In addition online database efforts within the petrological community are making compilation of large published datasets easier. This is creating a tremendous amount of data and a companion effort to develop alternate techniques for analysis and synthesis of such large data sets is also warranted. One option is expansion of the multivariate analysis techniques used by Slater et al. (2001), although few other workers have explored this approach to date.

Finally, there is also an important role for continued studies of the processes of inclusion formation and evolution using natural and experimental samples. Recent experiment studies have raised important questions about the effect of trapping and diffusional re-equilibration on melt inclusion compositions. However, to date there have not been equivalent attempts to relate these observations to the compositional variations evident in natural inclusion suites. Reproduction of complex natural kinetic phenomenon, such as those involved in formation and evolution of melt inclusions, is difficult within the laboratory environment and on non-geologic timescales. Thus, although continued experimental studies of melt inclusion formation and evolution processes are vital, they are more valuable when accompanied by attempts to reconcile their results with observations of the natural melt inclusion record.

ACKNOWLEDGMENTS

The author would like to thank colleagues and students past and present for discussions concerning melt inclusions and related subjects, particularly Paul Wallace, Michael Wiedenbeck, Doug Phinney, Michael Rowe, Marc Norman, Ian Hutcheon, Joel Baker, David Peate, Ingrid Peate, Roger Nielsen, Ed Kohut, John Eiler, Jon Woodhead, Janet Hergt, Steve Eggins, Alex Sobolev, Glenn Gaetani, Nobu Shimizu, Greg Yaxley, Erik Hauri and David Graham. The author would also particularly like to thank Ed Stolper for providing early encouragement to an economic geologist interested in petrology. The author also gratefully acknowledges the ongoing support of the National Science Foundation.

Several reviewers, including Katie, Kelley, Liz Cottrell, Leonid Danyushevsky, Keith Putirka and Benedetto de Vivo also provided valuable feedback on early versions of this chapter.

REFERENCES

Anderson AT (1973) The before-eruption H_2O content of some high alumina magmas. B Volcanol 37:530-552
Anderson AT (1974) Evidence for a picritic, volatile-rich magma beneath Mt Shasta, California. J Petrol 15:243-267
Anderson AT, Wright TL (1972) Phenocrysts and glass inclusions and their bearing on oxidation and mixing of basaltic magmas, Kilauea Volcano, Hawaii. Am Mineral 57:188-216
Anderson AT, Brown GG (1975) CO_2 and formation pressures of some Kilauean melt inclusions. Am Mineral 78:794-803
Baker DR (2008) The fidelity of melt inclusions as records of melt composition. Contrib Mineral Petrol doi: 10.1007/s00410-008-0291-3
Benjamin ER, Plank T, Wade JA, Kelley KA, Haun EH, Alvarado GE (2007) High water contents in basaltic magmas from Irazu Volcano, Costa Rica. J Volcanol Geotherm Res 168:68-92
Bindeman IN, Davis AM, Drake MJ (1998) Ion microprobe study of plagioclase-basalt partition experiments at natural concentration levels of trace elements. Geochim Cosmochim Acta 62:1175-1193
Blundy J, Cashman K (2008) Petrologic reconstruction of magmatic system variables and processes. Rev Mineral Geochem 69:179-239

Bombardieri DJ, Norman MD, Kamenetsky VS, Danyushevsky LV (2005) Major element and primary sulfur concentrations in Apollo 12 mare basalts: The view from melt inclusions. Meteorit Planet Sci 40:679-693

Bonnin-Mosbah M, Metrich N, Susini J, Salome M, Massare D, Menez B (2002) Micro X-ray absorption near edge structure at the sulfur and iron K-edges in natural silicate glasses. Spectrochim Acta B 57:711-725

Buening DK, Buseck PR (1973) Fe-Mg lattice diffusion in olivine. J Geophys Res 78:6852-6862

Canil D (1997) Vanadium partitioning and the oxidation state of Archaean komatiite magmas. Nature 389:842-845

Cartigny P, Pineau F, Aubaud C, Javoy M (2008) Towards a consistent mantle carbon flux estimate: Insights from volatile systematics (H_2O/Ce, delta D, CO_2/Nb) in the North Atlantic mantle (14 degrees N and 34 degrees N). Earth Planet Sci Lett 265:672-685

Cervantes P, Wallace P (2003a) Magma degassing and basaltic eruption styles: a case study of similar to 2000 year BP Xitle volcano in central Mexico. J Volcanol Geotherm Res 120:249-270

Cervantes P, Wallace PJ (2003b) Role of H_2O in subduction-zone magmatism: New insights from melt inclusions in high-Mg basalts from central Mexico. Geology 31:235-238

Cherniak DJ (2007) REE diffusion in olivine. Eos Trans, Am Geophys Union 88:MR13C-1397

Clocchiatti R, Massare D (1985) Experimental crystal-growth in glass inclusions - the possibilities and limits of the method. Contrib Mineral Petrol 89:193-204

Cooper KM, Reid MR, Murrell MT, Clague DA (2001) Crystal and magma residence at Kilauea Volcano, Hawaii: Th-230-Ra-226 dating of the 1955 east rift eruption. Earth Planet Sci Lett 184:703-718

Cottrell E, Spiegelman M, Langmuir CH (2002) Consequences of diffusive re-equilibration for the interpretation of melt inclusions. Geochem Geophys Geosys 3: doi: 000175371900001

Danyushevsky LV, Della-Pasqua FN, Sokolov S (2000) Re-equilibration of melt inclusions trapped by magnesian olivine phenocrysts from subduction-related magmas: petrological implications. Contrib Mineral Petrol 138:68-83

Danyushevsky LV, Leslie RAJ, Crawford AJ, Durance P (2004) Melt inclusions in primitive olivine phenocrysts: The role of localized reaction processes in the origin of anomalous compositions. J Petrol 45:2531-2553

Danyushevsky LV, McNeill AW, Sobolev AV (2002a) Experimental and petrological studies of melt inclusions in phenocrysts from mantle-derived magmas: an overview of techniques, advantages and complications. Chem Geol 183:5-24

Danyushevsky LV, Perfit MR, Eggins SM, Falloon TJ (2003) Crustal origin for coupled 'ultra-depleted' and 'plagioclase' signatures in MORB olivine-hosted melt inclusions: evidence from the Siqueiros Transform Fault, East Pacific Rise. Contrib Mineral Petrol 144:619-637

Danyushevsky LV, Sokolov S, Falloon TJ (2002b) Melt inclusions in olivine phenocrysts: Using diffusive re-equilibration to determine the cooling history of a crystal, with implications for the origin of olivine-phyric volcanic rocks. J Petrol 43:1651-1671

Davis MG, Garcia MO, Wallace P (2003) Volatiles in glasses from Mauna Loa Volcano, Hawai'i: implications for magma degassing and contamination, and growth of Hawaiian volcanoes. Contrib Mineral Petrol 144:570-591

Dixon JE, Clague DA, Stolper EM (1991) Degassing history of water, Sulfur, and carbon in submarine lavas from Kilauea Volcano, Hawaii. J Geol 99:371-394

Dixon JE, Stolper EM, Holloway JR (1995) An experimental study of water and carbon dioxide solubilities in mid ocean ridge basaltic liquids .1. Calibration and solubility models. J Petrol 36:1607-1631

Dixon JE, Leist L, Langmuir C, Schilling JG (2002) Recycled dehydrated lithosphere observed in plume-influenced mid-ocean-ridge basalt. Nature 420:385-389

Dunbar NW, Hervig RL (1992) Volatile and trace element composition of melt inclusions from the lower Bandelier Tuff: Implications for magma chamber processes and eruptive style. J Geophys Res 97:15151-15170

Dungan MA, Rhodes JM (1978) Residual glasses and melt inclusions in basalts from Dsdp Legs 45 and 46 - evidence for magma mixing. Contrib Mineral Petrol 67:417-431

Durrant SF (1999) Laser ablation inductively coupled plasma mass spectrometry: achievements, problems, prospects. J Anal At Spectrom 14:1385-1404

Eggins SM, Kinsley LK, Shelley JMG (1998) Deposition and elemental fractionation processes during atmospheric pressure laser sampling for analysis by ICP-MS. Appl Surf Sci 127-129:278-286

Elburg MA, Kamenetsky VS, Foden JD, Sobolev A (2007) The origin of medium-K ankaramitic arc magmas from Lombok (Sunda arc, Indonesia): Mineral and melt inclusion evidence. Chem Geol 240:260-279

Fahey AJ, Goswami JN, McKeegan KD, Zinner E (1987) ^{26}Al, ^{244}Pu, ^{50}Ti and trace element abundances in hibonite grains from CM and CV meteorites. Geochim Cosmochim Acta 51:329-350

Falloon TJ, Danyushevsky LV, Ariskin A, Green DH, Ford CE (2007) The application of olivine geothermometry to infer crystallization temperatures of parental liquids: implications for the temperature of MORB magmas. Chem Geol 241:207-233

Faure F, Schiano P (2005) Experimental investigation of equilibration conditions during forsterite growth and melt inclusion formation. Earth Planet Sci Lett 236:882-898

Font L, Murton BJ, Roberts S, Tindle AG (2007) Variations in melt productivity and melting conditions along SWIR (70 degrees E-49 degrees E): evidence from olivine-hosted and plagioclase-hosted melt inclusions. J Petrol 48:1471-1494

Ford CE, Russell DG, Craven JA, Fisk MR (1983) Olivine liquid equilibria - temperature, pressure and composition dependence of the crystal liquid cation partition-coefficients for Mg, Fe^{2+}, Ca and Mn. J Petrol 24:256-265

Freda C, Baker DR, Scarlato P (2005) Sulfur diffusion in basaltic melts. Geochim Cosmochim Acta 69:5061-5069

Frezzotti ML (2001) Silicate-melt inclusions in magmatic rocks: applications to petrology. Lithos 55:273-299

Gaetani GA, Watson EB (2000) Open-system behaviour of olivine-hosted melt inclusions. Earth Planet Sci Lett 183:27-41

Gaetani GA, Watson EB (2002) Modeling the major-element evolution of olivine-hosted melt inclusions. Chem Geol 183:25-41

Gebrande H (1982) Elastic wave velocities and constant of elasticity of rock and rock forming minerals. In: Physical properties of rocks. G Angenheister (ed) Springer Verlag, Berlin, Lanalt-Bornsetin series, group V, vol. 1b, 3.1.2, p 8-34

Goldstein SB, Luth RW (2006) The importance of cooling regime in the formation of melt inclusions in olivine crystals in haplobasaltic melts. Can Mineral 44:1543-1555

Grove TL, Baker MB, Kinzler RJ (1984) Coupled CaAl-NaSi diffusion in plagioclase feldspar - experiments and applications to cooling rate speedometry. Geochim Cosmochim Acta 48:2113-2121

Gunther D, Heinrich CA (1999) Comparison of the ablation behaviour of 266 nm Nd : YAG and 193 nm ArF excimer lasers for LA-ICP-MS analysis. J Anal At Spectrom 14:1369-1374

Gurenko AA, Belousov AB, Trumbull RB, Sobolev AV (2005) Explosive basaltic volcanism of the Chikurachki Volcano (Kurile arc, Russia): Insights on pre-eruptive magmatic conditions and volatile budget revealed from phenocryst-hosted melt inclusions and groundmass glasses. J Volcanol Geotherm Res 147:203-232

Gurenko AA, Chaussidon M (1995) Enriched and depleted primitive melts included in olivine from Icelandic tholeiites: Origin by continuous melting of a single mantle column. Geochim Cosmochim Acta 59:2905-2917

Gurenko AA, Chaussidon M (1997) Boron concentrations and isotopic composition of the Icelandic mantle: Evidence from glass inclusions in olivine. Chem Geol 135:21-34

Gurenko AA, Chaussidon M (2002) Oxygen isotope variations in primitive tholeiites of Iceland: evidence from a SIMS study of glass inclusions, olivine phenocrysts and pillow rim glasses. Earth Planet Sci Lett 205:63-79

Gurenko AA, Chaussidon M, Schmincke HU (2001) Magma ascent and contamination beneath one intraplate volcano: Evidence from S and O isotopes in glass inclusions and their host clinopyroxenes from Miocene basaltic hyaloclastites southwest of Gran Canaria (Canary Islands). Geochim Cosmochim Acta 65:4359-4374

Halter WE, Pettke T, Heinrich CA, Rothen-Rutishauser B (2002) Major to trace element analysis of melt inclusions by laser-ablation ICP-MS: methods of quantification. Chem Geol 183:63-86

Hansen H, Gronvold K (2000) Plagioclase ultraphyric basalts in Iceland: the mush of the rift. J Volcanol Geotherm Res 98:1-32

Hattendorf B, Latkoczy C, Günther D (2003) Laser ablation ICP-MS. Anal Chem A 75:341A-347A

Harlou R, Pearson DG, Nowell GM, Davidson JP, Kent AJR (2005) Sr isotope studies of melt inclusions by TIMS. Geochim Cosmochim Acta 69:A380

Hauri E (2002) SIMS analysis of volatiles in silicate glasses, 2: isotopes and abundances in Hawaiian melt inclusions. Chem Geol 183:115-141

Hauri E, Wang JH, Dixon JE, King PL, Mandeville C, Newman S (2002) SIMS analysis of volatiles in silicate glasses 1. Calibration, matrix effects and comparisons with FTIR. Chem Geol 183:99-114

Heinrich CA, Pettke T, Halter WE, Aigner-Torres M, Audetat A, Gunther D, Hattendorf B, Bleiner D, Guillong M, Horn I (2003) Quantitative multi-element analysis of minerals, fluid and melt inclusions by laser-ablation inductively-coupled-plasma mass-spectrometry. Geochim Cosmochim Acta 67:3473-3497

Hervig RL, Mazdab FK, Williams P, Guan Y, Huss GR (2006) Limits of quantitative analysis by secondary ion mass spectrometry. Chem Geol 227:83-99

Hinton RW (1995) Ion microprobe analysis in geology. In: Microprobe techniques in the earth sciences. Potts PJ, Bowles JFW, Reed SJB, Cave MR (eds) Chapman and Hall, London, p 235-290

Hofmann AW (1988) Chemical differentiation of the Earth: The relationship between mantle, continental-crust, and oceanic crust. Earth Planet Sci Lett 90:297-314

Hofmann AW (1997) Mantle geochemistry: The message from oceanic volcanism. Nature 385:219-229

Humphreys MCS, Kearns SL, Blundy JD (2006) SIMS investigation of electron-beam damage to hydrous, rhyolitic glasses: Implications for melt inclusion analysis. Am Mineral 91:667-679

Jackson MG, Hart SR (2006) Strontium isotopes in melt inclusions from Samoan basalts: Implications for heterogeneity in the Samoan plume. Earth Planet Sci Lett 245:260-277

Jackson SE, Gunther D (2003) The nature and sources of laser induced isotopic fractionation in laser ablation-multicollector-inductively coupled plasma-mass spectrometry. J Anal At Spectrom 18:205-212

Jakobsen JK, Veksler IV, Tegner C, Brooks CK (2005) Immiscible iron- and silica-rich melts in basalt petrogenesis documented in the Skaergaard intrusion. Geology 33:885-888

Jambon A, Lussiez P, Clocchiatti R, Weisz J, Hernandez J (1992) Olivine growth-rates in a tholeiitic basalt - an experimental-study of melt inclusions in plagioclase. Chem Geol 96:277-287

Jochum KP, Dingwell DB, Rocholl A, Stoll B, Hofmann AW, Becker S, Besmehn A, Bessette D, Dietze H-J, Dulski P, Erzinger J, Hellebrand E, Hoppe P, Horn I, Janssens K, Jenner GA, Klein M, McDonough WF, Maetz M, Mezger K, Müker C, Nikogosian IK, Pickhardt C, Raczek I, Rhede D, Seufert HM, Simakin HG, Sobolev AV, Spettel B, Straub S, Vincze L, Wallianos A, Weckwerth G, Weyer S, Wolf D, Zimmer M (2000) The preparation and preliminary characterisation of eight geological MPI-DING reference glasses for in-situ microanalysis. Geostandard Geoanal Res 24:87-133

Jochum KP, Stoll B, Herwig K, Willbold M, Amini M, Hofmann AW, Aarburg S, Abouchami W, Alard O, Bouman C, Becker ST, Brätz H, de Bruin D, Canil D, Cornell D, Dalpé C, Danyushevsky LV, Dücking M, Groschopf N, Günther D, Guillong M, de Hoog C-J, Höfer H, Horz K, Jacob D, Kasemann SA, Kent AJR, Klemd R, Lahaye Y, Latkoczy C, Ludwig T, Mason P, Meixner A, Misawa K, Nash BP, Pfänder J, Premo WR, Raczek I, Rosner M, Stracke A, Sun W, Tiepolo M, Vannuci R, Vennemann T, Wayne D, Woodhead JD, Zack T (2006) MPI-DING reference glasses for in-situ microanalysis: New reference values for element concentrations and isotope ratios. Geochem Geophys Geosys 7, Q02008, doi:10.1029/2005GC001060

Jochum KP, Willbold M, Raczek I, Stoll B, Herwig K (2005b) Chemical characterisation of the USGS reference glasses GSA-1G, GSC-1G, GSD-1G, GSE-1G, BCR-2G, BHVO-2G and BIR-1G using EPMA, ID-TIMS, ID-ICP-MS and LA-ICP-MS. Geostandard Geoanal Res 29:285-302

Jochum KP, Stoll B, Herwig K, Amini A, Abouchami W, Hofmann AW (2005b) Lead isotope ratio measurements in geological glasses by laser ablation-sector field-ICP mass spectrometry (LA-SF-ICPMS). Int J Mass Spectrom 242:281-289

Johnson J, Nielsen RL, Fisk M (1996) Plagioclase-hosted melt inclusions in the Steens basalt, southeastern Oregon. Petrology 4:247-254

Johnson ER, Wallace PJ, Cashman KV, Granados HD, Kent AJR (2008) Magmatic volatile contents and degassing-induced crystallization at Volcán Jorullo, Mexico: Implications for melt evolution and the plumbing systems of monogenetic volcanoes. Earth Planet Sci Lett 269:477-486

Johnson KTM, Dick HJB, Shimizu N (1990) Melting in the oceanic upper mantle - an ion microprobe study of diopsides in abyssal peridotites. J Geophys Res-Solid Earth Planets 95:2661-2678

Jurewicz AJG, Watson EB (1988) Cations in olivine. 2. Diffusion in olivine xenocrysts, with applications to petrology and mineral physics. Contrib Mineral Petrol 99:186-201

Kamenetsky MB, Sobolev AV, Kamenetsky VS, Maas R, Danyushevsky LV, Thomas R, Pokhilenko NP, Sobolev NV (2004) Kimberlite melts rich in alkali chlorides and carbonates: A potent metasomatic agent in the mantle. Geology 32:845-848

Kamenetsky V (1996) Methodology for the study of melt inclusions in Cr-Spinel, and implications for parental melts of MORB from Famous areas. Earth Planet Sci Lett 142:479-486

Kamenetsky VS (1995) New method for determining primitive melts - example of melt inclusions in spinel from tholeiites of the famous area, Mid-Atlantic Ridge. Petrology 3:473-477

Kamenetsky VS, Crawford AJ, Eggins S, Muhe R (1997) Phenocryst and melt inclusion chemistry of near-axis seamounts, Valu Fa Ridge, Lau Basin: insight into mantle wedge melting and the addition of subduction components. Earth Planet Sci Lett 151:205-223

Kamenetsky VS, Eggins SM, Crawford AJ, Green DH, Gasparon M, Falloon TJ (1998) Calcic melt inclusions in primitive olivine at 43°N MAR: evidence for melt-rock reaction/melting involving clinopyroxene-rich lithologies during MORB generation. Earth Planet Sci Lett 160:115-132

Kamenetsky VS, Gurenko AA (2007) Cryptic crustal contamination of MORB primitive melts recorded in olivine-hosted glass and mineral inclusions. Contrib Mineral Petrol 153:465-481

Kamenetsky VS, Pompilio M, Metrich N, Sobolev AV, Kuzmin DV, Thomas R (2007) Arrival of extremely volatile-rich high-Mg magmas changes explosivity of Mount Etna. Geology 35:255-258

Kamenetsky VS, Sobolev AV, Eggins SM, Crawford AJ, Arculus RJ (2002) Olivine-enriched melt inclusions in chromites from low-Ca boninites, Cape Vogel, Papua New Guinea: evidence for ultramafic primary magma, refractory mantle source and enriched components. Chem Geol 183:287-303

Kamenetsky VS, Sobolev AV, Joron JL, Semet MP (1995) Petrology and geochemistry of Cretaceous ultramafic volcanics from eastern Kamchatka. J Petrol 36:637-662

Kelley KA, Plank T, Grove TL, Stolper EM, Newman S, Hauri E (2006) Mantle melting as a function of water content beneath back-arc basins. J Geophys Res-Solid Earth 111: doi: 000240948900002

Kent AJR, Clague DA, Honda M, Stolper EM, Hutcheon ID, Norman MD (1999b) Widespread assimilation of a seawater-derived component at Loihi Seamount, Hawaii. Geochim Cosmochim Acta 63:2749-2761

Kent AJR, Peate DW, Newman S, Stolper EM, Pearce JA (2002b) Chlorine in submarine glasses from the Lau Basin: seawater contamination and constraints on the composition of slab-derived fluids. Earth Planet Sci Lett 202:361-377

Kent AJR, Baker JA, Wiedenbeck M (2002a) Contamination and melt aggregation processes in continental flood basalts: constraints from melt inclusions in Oligocene basalts from Yemen. Earth Planet Sci Lett 202:577-594

Kent AJR, Clague DA, Honda M, Stolper EM, Hutcheon ID, Norman MD (1999b) Widespread assimilation of a seawater-derived component at Loihi Seamount, Hawaii. Geochim Cosmochim Acta 63:2749-2761

Kent AJR, Elliott TR (2002) Melt inclusions in Mariana arc lavas: Implications for the formation and evolution of arc magmas. Chem Geol 183:265-288

Kent AJR, Jacobsen B, Peate DW, Waight TE, Baker JA (2004a) Isotope dilution MC-ICP-MS rare earth element analysis of geochemical reference materials NIST SRM 610, NIST SRM 612, NIST SRM 614, BHVO-2G, BHVO-2, BCR-2G, JB-2, WS-E, W-2, AGV-1 and AGV-2. Geostandard Geoanal Res 28:417-429

Kent AJR, Norman MD, Hutcheon ID, Stolper EM (1999a) Assimilation of seawater-derived components in an oceanic volcano: evidence from matrix glasses and glass inclusions from Loihi seamount, Hawaii. Chem Geol 156:299-319

Kent AJR, Stolper EM, Francis D, Woodhead J, Frei R, Eiler J (2004b) Mantle heterogeneity during the formation of the North Atlantic Tertiary Province: Constraints from trace element and Sr-Nd-Os-O isotope systematics of Baffin Island picrites. Geochem Geophys Geosys 5: doi 000225030200004

Kent AJR, Ungerer CA (2005) Production of barium and light rare earth element oxides during LA-ICP-MS microanalysis. J Anal At Spectrom 20:1256-1262

Kent AJR, Ungerer CA (2006) Analysis of light lithophile elements (Li, Be, B) by laser ablation ICP-MS: Comparison between magnetic sector and quadrupole ICP-MS. Am Mineral 91:1401-1411

Kilinc A, Carmichael ISE, Rivers ML, Sack RO (1983) The ferric-ferrous ratio of natural silicate liquids equilibrated in air. Contrib Mineral Petrol 83:136-140

Kobayashi K, Tanaka R, Moriguti T, Shimizu K, Nakamura E (2004) Lithium, boron, and lead isotope systematics of glass inclusions in olivines from Hawaiian lavas: evidence for recycled components in the Hawaiian plume. Chem Geol 212:143-161

Kohut E, Nielsen RL (2004) Melt inclusion formation mechanisms and compositional effects in high-An feldspar and high-Fo olivine in anhydrous mafic silicate liquids. Contrib Mineral Petrol 147:684-704

Kohut EJ, Stern RJ, Kent AJR, Nielsen RL, Bloomer SH, Leybourne M (2006) Evidence for adiabatic decompression melting in the Southern Mariana arc from high-Mg lavas and melt inclusions. Contrib Mineral Petrol 152:201-221

Kovalenko VI, Naumov VB, Yarmolyuk VV, Dorofeeva VA (2000) Volatile components (H_2O, CO_2, Cl, F, and S) in basic magmas of various geodynamic settings: Data on melt inclusions and quenched glasses. Petrology 8:113-144

Kress VC, Carmichael ISE (1991) The compressibility of silicate liquids containing Fe_2O_3 and the effect of composition, temperature, oxygen fugacity and pressure on their redox states. Contrib Mineral Petrol 108:82-92

Kress VC, Ghiorso MS (2004) Thermodynamic modeling of post-entrapment crystallization in igneous phases. J Volcanol Geotherm Res 137:247-260

Kuzmin DV, Sobolev AV (2004) Boundary layer contribution to the composition of melt inclusions in olivine. Geochim Cosmochim Acta 68:A544-A544

Larsen LM, Pedersen AK (2000) Processes in high-Mg, high-T magmas: Evidence from olivine, chromite and glass in palaeogene picrites from West Greenland. J Petrol 41:1071-1098

Lassiter JC, Hauri EH, Nikogosian IK, Barsczus HG (2002) Chlorine-potassium variations in melt inclusions from Raivavae and Rapa, Austral Islands: constraints on chlorine recycling in the mantle and evidence for brine-induced melting of oceanic crust. Earth Planet Sci Lett 202:525-540

LaTourrette T, Wasserburg GJ (1997) Self diffusion of europium, neodymium, thorium, and uranium in haplobasaltic melt: The effect of oxygen fugacity and the relationship to melt structure. Geochim Cosmochim Acta 61:755-764

LaTourrette T, Wasserburg GJ, Fahey AJ (1996) Self diffusion of Mg, Ca, Ba, Nd, Yb, Ti, Zr, and U in haplobasaltic melt. Geochim Cosmochim Acta 60:1329-1340

Laubier M, Schiano P, Doucelance R, Ottolini L, Laporte D (2007) Olivine-hosted melt inclusions and melting processes beneath the FAMOUS zone (Mid-Atlantic Ridge). Chem Geol 240:129-150

Layne GD, Godon A, Webster JD, Bach W (2004) Secondary ion mass spectrometry for the determination of delta Cl-37 Part I. Ion microprobe analysis of glasses and fluids. Chem Geol 207:277-289

Lee J, Stern RJ (1998) Glass inclusions in Mariana arc phenocrysts: A new perspective on magmatic evolution in a typical intra-oceanic arc. J Geol 106:19-33

Li ZXA, Lee CTA (2004) The constancy of upper mantle fO_2 through time inferred from V/Sc ratios in basalts. Earth Planet Sci Lett 228:483-493

Liang Y, Richter FM, Davis AM, Watson EB (1996) Diffusion in silicate melts .1. Self diffusion in $CaO-Al_2O_3-SiO_2$ at 1500 degrees C and 1 GPa. Geochim Cosmochim Acta 60:4353-4367

Longerich HP, Jackson SE, Gunther D (1996) Laser ablation inductively coupled plasma mass spectrometric transient signal data acquisition and analyte concentration calculation. J Anal At Spectrom 11:899-904

Lowenstern JB (1994) Chlorine, Fluid immiscibility, and degassing in peralkaline magmas from Pantelleria, Italy. Am Mineral 79:353-369

Lowenstern JB (1995) Application of silicate melt-inclusions to the study of magmatic volatiles. In: Magmas, Fluids and Ore Deposits. Thompson J (ed) Mineral Assoc Canada Short Course, 23, p 71-100

Lowenstern JB (2003) Melt inclusions come of age: volatiles, volcanoes and Sorby's legacy. In: Melt inclusions in volcanic systems. de Vivo B, Bodnar RJ (eds) Elsevier, Amsterdam, p 258

Lu FQ, Anderson AT, Davis AM (1995) Diffusional gradients at the crystal/melt interface and their effect on the composition of melt inclusions. J Geol 103:591-597

Lund MJ (2003) Plagioclase-hosted melt inclusion along the Southwest Indian Ridge. MS Thesis, University of Copenhagen, Copenhagen

Mackwell SJ, Kohlstedt DL (1990) Diffusion of hydrogen in olivine: Implications for water in the mantle. J Geophys Res 95:5079-5088

Maclennan J, McKenzie D, Hilton F, Gronvold K, Shimizu N (2003) Geochemical variability in a single flow from northern Iceland. J Geophys Res- Solid Earth 108: doi: 000181501800001

Mangan M, Sisson T (2000) Delayed, disequilibrium degassing in rhyolite magma: decompression experiments and implications for explosive volcanism. Earth Planet Sci Lett 183:441-455

Massare D, Metrich N, Clocchiatti R (2002) High-temperature experiments on silicate melt inclusions in olivine at 1 atm: inference on temperatures of homogenization and H_2O concentrations. Chem Geol 183:87-98

Mathez EA (1976) Sulfur solubility and magmatic sulfides in submarine basalt glass. J Geophys Res 81:4269-4276

McDonough WF, Ireland TR (1993) Intraplate origin of komatiites inferred from trace elements in glass inclusions. Nature 365:432-434

Métrich N, Wallace PJ (2008) Volatile abundances in basaltic magmas and their degassing paths tracked by melt inclusions. Rev Mineral Geochem 69:363-402

Metrich N, Susini J, Foy E, Farges F, Massare D, Sylla L, Lequien S, Bonnin-Mosbah M (2006) Redox state of iron in peralkaline rhyolitic glass/melt: X-ray absorption micro-spectroscopy experiments at high temperature. Chem Geol 231:350-363

Meyzen CM, Toplis MJ, Humler E, Ludden JN, Mevel C (2003) A discontinuity in mantle composition beneath the southwest Indian ridge. Nature 421:731-733

Michael PJ, Cornell WC (1998) Influence of spreading rate and magma supply on crystallization and assimilation beneath mid-ocean ridges: Evidence from chlorine and major element chemistry of mid-ocean ridge basalts. J Geophys Res-Solid Earth and Planets 103:18325-18356

Michael PJ, McDonough WF, Nielsen RL, Cornell WC (2002) Depleted melt inclusions in MORB plagioclase: messages from the mantle or mirages from the magma chamber? Chem Geol 183:43-61

Milman-Barris MS, Beckett JR, Baker MB, Hofmann AE, Morgan Z, Crowley MR, Vielzeuf D, Stolper E (2008) Zoning of phosphorus in igneous olivine. Contrib Mineral Petrol 155:739-765

Misener DJ (1974) Cation Diffusion in Olivine. In: Geochemical Transport and Kinetics. Hofman AW, Giletti BJ, Yoder HS, Yund RA (eds) Carnegie Institution of Washington, Washington, 53, p 117-129

Moore G (2008) Interpreting H_2O and CO_2 contents in melt inclusions: constraints from solubility experiments and modeling. Rev Mineral Geochem 69:333-361

Murase T, Mcbirney AR (1973) Properties of Some Common Igneous Rocks and Their Melts at High-Temperatures. Geol Soc Am Bull 84:3563-3592

Nakamura M, Shimakita S (1998) Dissolution origin and syn-entrapment compositional change of melt inclusion in plagioclase. Earth Planet Sci Lett 161:119-133

Nekvasil H, Filiberto J, McCubbin FM, Lindsley DH (2007) Alkalic parental magmas for chassignites? Meteorit Planet Sci 42:979-992

Newman S, Stolper EM, Stern RJ (2000) H_2O and CO_2 from the Mariana arc and backarc systems. Geochem Geophys Geosys 1: doi:1999GC000027

Nielsen RL, Crum J, Bourgeois R, Hascall K, Forsythe LM, Fisk MR, Christie DM (1995) Melt inclusions in high-An plagioclase from the Gorda Ridge: an example of the local diversity of MORB parent magmas. Contrib Mineral Petrol 122:34-50

Nielsen RL, Gallahan WE, Newberger F (1992) Experimentally determined mineral-melt partition-coefficients for Sc, Y and Ree for olivine, ortho-pyroxene, pigeonite, magnetite and ilmenite. Contrib Mineral Petrol 110:488-499

Nielsen RL, Michael PJ, Sours-Page R (1998) Chemical and physical indicators of compromised melt inclusions. Geochim Cosmochim Acta 62:831-839

Nielsen RL, Sours-Page R, Harpp K (2000) Role of a Cl-bearing flux in the origin of depleted ocean floor magmas. Geochem Geophys Geosys 1, doi:1999GC000017

Nielsen TFD, Veksler IV (2002) Is natrocarbonatite a cognate fluid condensate? Contrib Mineral Petrol 142:425-435

Nielsen TFD, Turkov VA, Solovova IP, Kogarko LN, Ryabchikov ID (2006) A Hawaiian beginning for the Iceland plume: Modelling of reconnaissance data for olivine-hosted melt inclusions in Palaeogene picrite lavas from East Greenland. Lithos 92:83-104

Niu HS, Houk RS (1996) Fundmental aspects of ion extraction in ICP-MS. Spectrochim Acta B 51:779-815

Norman MN, Garcia MO, Kamenetsky VS, Nielsen RL (2002) Olivine-hosted melt inclusions in Hawaiian picrites: Equilibration, melting and plume source characteristics. Chem Geol 183:143-168

Paul B (2006) A new perspective on melt inclusions: Development of novel in-situ analytical protocols. Ph.D. dissertation, University of Melbourne, Melbourne

Paul B, Woodhead JD, Hergt J (2005) Improved in situ isotope analysis of low-Pb materials using LA-MC-ICP-MS with parallel ion counter and Faraday detection. J Anal At Spectrom 20:1350-1357

Pearce NJG, Perkins WT, Westgate JA, Gorton MP, Jackson SE, Neal CR, Chenery SP (1997) A compilation of new and published major and trace element data for NIST SRM 610 and NIST SRM 612 glass reference materials. Geostandard Newslett 21:115-144

Perkins WT, Pearce NJG (1995) Mineral microanalysis by laserprobe inductively coupled plasma mass spectrometry. *In*: Microprobe Techniques in the Earth Sciences. Potts PJ, Bowles JFW, Reed SJB, Cave MR (eds) Chapman and Hall, London, p 291-325

Pettke T, Halter WE, Webster JD, Aigner-Torres M, Heinrich CA (2004) Accurate quantification of melt inclusion chemistry by LA-ICPMS: a comparison with EMP and SIMS and advantages and possible limitations of these methods. Lithos 78:333-361

Portnyagin M, Almeev R, Matveev S, Holtz F (2008) Experimental evidence for rapid water exchange between melt inclusions in olivine and host magmas. Earth Planet Sci Lett 272:541-552

Portnyagin M, Hoernle K, Plechov P, Mironov N, Khubunaya S (2007) Constraints on mantle melting and composition and nature of slab components in volcanic arcs from volatiles (H_2O, S, Cl, F) and trace elements in melt inclusions from the Kamchatka Arc. Earth Planet Sci Lett 255:53-69.

Portnyagin MV, Danyushevsky LV, Kamenetsky VS (1997) Coexistence of two distinct mantle sources during formation of ophiolites: a case study of primitive pillow-lavas from the lowest part of the volcanic section of the Troodos Ophiolite, Cyprus. Contrib Mineral Petrol 128:287-301

Putirka KD (2005) Mantle potential temperatures at Hawaii, Iceland, and the mid-ocean ridge system, as inferred from olivine phenocrysts: Evidence for thermally driven mantle plumes. Geochem Geophy Geosy 6, doi: 000229144800002

Putirka KD, Perfit M, Ryerson FJ, Jackson MG (2007) Ambient and excess mantle temperatures, olivine thermometry, and active vs. passive upwelling. Chem Geol 241:177-206

Qin Z, Lu F, Anderson ATJ (1992) Diffusive re-equilibration of melt and fluid inclusions. Am Mineral 77:565-576

Raczek I, Stoll B, Hofmann AW, Jochum KP (2001) High-precision trace element data for the USGS reference materials BCR-1, BCR-2, BHVO-1, BHVO-2, AGV-1, AGV-2, DTS-1, DTS-2, GSP-1 and GSP-2 by ID-TIMS and MIC-SSMS. Geostandard Newslett 25:77-86

Ramos FC, Tepley FJ III (2008) Inter- and intracrystalline isotopic disequilibria: techniques and applications. Rev Mineral Geochem 69:403-443

Richter FM, Davis AM, DePaolo DJ, Watson EB (2003) Isotope fractionation by chemical diffusion between molten basalt and rhyolite. Geochim Cosmochim Acta 67:3905-3923

Rocholl A (1998) Major and trace element composition and homogeneity of microbeam reference material: Basalt glass USGS BCR-2G. Geostand Newslett - J Geostand Geoanal 22:33-45

Rocholl A, Dulski P, Raczek I (2000) New ID-TIMS, ICP-MS and SIMS data on the trace element composition and homogeneity of NIST certified reference material SRM 610-611. Geostand Newslett - J Geostand Geoanal 24:261-274

Roedder E (1979) Origin and significance of magmatic inclusions. Bull Mineral 102:487-510

Roedder E (1984) Fluid Inclusions. Rev Mineral 12:1-644

Roedder E, Emslie RF (1970) Olivine-liquid equilibrium. Contrib Mineral Petrol 29:275-289

Roggensack K, Hervig RL, McKnight SB, Williams SN (1997) Explosive basaltic volcanism from Cerro Negro volcano: Influence of volatiles on eruptive style. Science 277:1639-1642

Rollinson HR (1993) Using Geochemical Data: Evaluation, Presentation, Interpretation. Longman, Singapore

Rose EF, Shimizu N, Layne GD, Grove TL (2001) Melt production beneath Mt. Shasta from boron data in primitive melt inclusions. Science 293:281-283

Rowe MC, Kent AJR, Nielsen RL (2007) Determination of sulfur speciation and oxidation state of olivine hosted melt inclusions. Chem Geol 236:303-322

Rowe MC, Kent AJR, Nielsen RL (2008) Subduction influence on basaltic oxygen fugacity and trace- and volatile-elements across the Cascade Volcanic Arc. J Petrol (in press)

Rowe MC, Nielsen RL, Kent AJR (2006) Anomalously high Fe contents in rehomogenized olivine-hosted melt inclusions from oxidized magmas. Am Mineral 91:82-91

Ruprecht P, Bergantz GW, Dufek J (2008) Modeling of gas-driven magmatic overturn: Tracking of phenocryst dispersal and gathering during magma mixing. Geochem Geophys Geosys doi:10.1029/2008GC002022

Russo RE, Mao XL, Liu HC, Gonzalez J, Mao SS (2002) Laser ablation in analytical chemistry - a review. Talanta 57:425-451

Ryabchikov ID, Solovova IP, Ntaflos T, Buchl A, Tikhonenkov PI (2001) Subalkaline picrobasalts and plateau basalts from the Putorana plateau (Siberian continental flood basalt province): II. Melt inclusion chemistry, composition of "primary" magmas and P-T regime at the base of the superplume. Geochem Int 39:432-446

Ryabchikov IA, Solovova IP, Kogarko LN, Bray GP, Ntaflos T, Simakin SG (2002) Thermodynamic parameters of generation of meymechites and alkaline picrites in the Maimecha-Kotui province: Evidence from melt inclusions. Geochem Int 40:1031-1041

Ryabchikov ID, Ntaflos T, Buchl A, Solovova IP (2001) Subalkaline picrobasalts and plateau basalts from the Putorana plateau (Siberian continental flood basalt province): I. Mineral compositions and geochemistry of major and trace elements. Geochem Int 39:415-431

Saal AE, Hart SR, Shimizu N, Hauri EH, Layne GD (1998) Pb isotopic variability in melt inclusions from oceanic island basalts, Polynesia. Science 282:481-484

Saal AE, Hart SR, Shimizu N, Hauri EH, Layne GD, Eiler JM (2005) Pb isotopic variability in melt inclusions from the EMI-EMII-HIMU mantle end-members and the role of the oceanic lithosphere. Earth Planet Sci Lett 240:605-620

Saal AE, Hauri EH, Langmuir CH, Perfit MR (2002) Vapour undersaturation in primitive mid-ocean-ridge basalt and the volatile content of Earth's upper mantle. Nature 419:451-455

Sadofsky SJ, Portnyagin M, Hoernle K, van den Bogaard P (2008) Subduction cycling of volatiles and trace elements through the Central American volcanic arc: evidence from melt inclusions. Contrib Mineral Petrol 155:433-456

Schiano P (2003) Primitive mantle magmas recorded as silicate melt inclusions in igneous minerals. Earth Sci Rev 63:121-144

Schiano P, Bourdon B (1999) On the preservation of mantle information in ultramafic nodules: glass inclusions within minerals versus interstitial glasses. Earth Planet Sci Lett 169:173-188

Schiano P, Clocchiatti R, Boivin P, Medard E (2004) The nature of melt inclusions inside minerals in an ultramafic cumulate from Adak volcanic center, Aleutian arc: implications for the origin of high-Al basalts. Chem Geol 203:169-179

Schiano P, Clocchiatti R, Lorand JP, Massare D, Deloule E, Chaussidon M (1997) Primitive basaltic melts included in podiform chromites from the Oman Ophiolite. Earth Planet Sci Lett 146:489-497

Schiano P, Eiler JM, Hutcheon ID, Stolper EM (2000) Primitive CaO-rich, silica-undersaturated melts in island arcs: Evidence for the involvement of clinopyroxene-rich lithologies in the petrogenesis of arc magmas. Geochem Geophys Geosys 2, doi: 1999GC00032

Schiano P, Provost A, Clocchiatti R, Faure F (2006) Transcrystalline melt migration and Earth's mantle. Science 314:970-974

Scowen PAH, Roeder PL, Helz RT (1991) Reequilibration of chromite within Kilauea Iki Lava Lake, Hawaii. Contrib Mineral Petrol 107:8-20

Seaman SJ, Dyar MD, Marinkovic N, Dunbar NW (2006) An FTIR study of hydrogen in anorthoclase and associated melt inclusions. Am Mineral 91:12-20

Shimizu K, Komiya T, Hirose K, Shimizu N, Maruyama S (2001) Cr-spinel, an excellent micro-container for retaining primitive melts - implications for a hydrous plume origin for komatiites. Earth Planet Sci Lett 189:177-188

Shimizu N (1998) The geochemistry of olivine-hosted melt inclusions in a FAMOUS basalt ALV519-4-1. Phys Earth Planet Int 107:183-201

Shimizu N, Hart SR (1982) Applications of the ion micro-probe to geochemistry and cosmochemistry. Annu Rev Earth Planet Sci 10:483-526

Sinton CW, Christie DM, Coombs VL, Nielsen RL, Fisk MR (1993) Near-primary melt inclusions in anorthite phenocrysts from the Galapagos Platform. Earth Planet Sci Lett 119:527-537

Sisson TW, Bronto S (1998) Evidence for pressure-release melting beneath magmatic arcs from basalt at Galunggung, Indonesia. Nature 391:883-886

Sisson TW, Layne GD (1993) H_2O in basaltic and basaltic andesite glass inclusions from four subduction-related volcanoes. Earth Planet Sci Lett 117:619-637

Slater L, McKenzie D, Gronvold K, Shimizu N (2001) Melt generation and movement beneath Theistareykir, NE Iceland. J Petrol 42:321-354

Sobolev AV, Clocchiatti R, Dhamelincourt P (1983a) Variations of the temperature, Melt composition and water-pressure during olivine crystallization in oceanitic rocks from the Piton Fournaise volcano (Reunion island, 1966 eruption). C R Acad Sci Ii 296:275-281

Sobolev AV (1996) Melt inclusions as a source of primary petrographic information. Petrology 4:209-220
Sobolev AV, Hofmann AW, Nikogosian IK (2000) Recycled oceanic crust observed in 'ghost plagioclase' within the source of Mauna Loa lavas. Nature 404:986-990
Sobolev AV, Chaussidon M (1996) H_2O concentrations in primary melts from suprasubduction zones and mid-ocean ridges: Implications for H_2O storage and recycling in the mantle. Earth Planet Sci Lett 137:45-55
Sobolev AV, Danyushevsky LV (1994) Petrology and geochemistry of boninites from the north termination of the Tonga trench - constraints on the generation conditions of primary high-Ca boninite magmas. J Petrol 35:1183-1211
Sobolev AV, Gurenko AA, Shimizu N (1994) Ultra-depleted melts from Iceland: data from melt inclusion studies. Mineral Mag 58A:860-861
Sobolev AV, Kamenetsky VS, Kononkova NN (1991) New data on petrology of Siberia Meymechites. Geokhimiya 1084-1095
Sobolev AV, Migdisov AA, Portnyagin MV (1996) Incompatible element partitioning between clinopyroxene and basalt liquid revealed by the study of melt inclusions in minerals from Troodos lavas, Cyprus. Petrology 4:307-317
Sobolev AV, Shimizu N (1993) Ultra-depleted primary melt included in an olivine from the Mid-Atlantic ridge. Nature 363:151-154
Sobolev AV, Tsamerian OP, Zakariadze GS, Shcherbovskii AJ (1983b) Compositions and conditions of the crystallization of the lesser Caucasus Ophiolite Volcanogenic Complex melts according to the data of melt inclusion study. Doklady Akademii Nauk Sssr 272:464-468
Sobolev VS, Kostyuk VP (1975) Magmatic crystallization based on a study of melt inclusions. Fluid Incl Res 9:182– 235
Sorby HC (1858) On the microscopic structures of crystals, indicating the origin of minerals and rocks. Geol Soc London Q J 14:453-500
Sours-Page R, Nielsen RL, Batiza R (2002) Melt inclusions as indicators of parental magma diversity on the northern East Pacific Rise. Chem Geol 183:237-261
Sours-Page RL, Johnson KTM, Nielsen RL, Karsten JL (1999) Local and regional variation of MORB parent magmas: Evidence from melt inclusions from the Endeavour segment of the Juan de Fuca ridge. Contrib Mineral Petrol 134:342-363
Spandler C, O'Neill HSC, Kamenetsky VS (2007) Survival times of anomalous melt inclusions from element diffusion in olivine and chromite. Nature 447:303-306
Spiegelman M, Kelemen PB (2003) Extreme chemical variability as a consequence of channelized melt transport. Geochem Geophys Geosys 4: doi 000184337400001
Stolper E, Newman S (1994) The role of water in the petrogenesis of Mariana Trough magmas. Earth Planet Sci Lett 121:293-325
Straub SM, Layne GD (2003) The systematics of chlorine, fluorine, and water in Izu arc front volcanic rocks: Implications for volatile recycling in subduction zones. Geochim Cosmochim Acta 67:4179-4203
Sun W, Bennet VC, Eggins SM, Arculus RJ, Perfit MR (2003) Rhenium systematics in submarine MORB and back-arc basin glasses: laser ablation ICP-MS results. Chem Geol 196:259-281
Sutton SR, Karner J, Papike J, Delaney JS, Shearer C, Newville M, Eng P, Rivers M, Dyar MD (2005) Vanadium K edge XANES of synthetic and natural basaltic glasses and application to microscale oxygen barometry. Geochim Cosmochim Acta 69:2333-2348
Tait S (1992) Selective preservation of melt inclusions in igneous phenocrysts. Am Mineral 77:146-155
Taylor RP, Jackson SE, Longerich HP, Webster JD (1997) In situ trace-element analysis of individual silicate melt inclusions by laser ablation microprobe inductively coupled plasma-mass spectrometry (LAM-ICP-MS). Geochim Cosmochim Acta 61:2559-2567
Thomas JB, Bodnar RJ (2002) A technique for mounting and polishing melt inclusions in small (< 1 mm) crystals. Am Mineral 87:1505-1508
Thomas R, Kamenetsky VS, Davidson P (2006) Laser Raman spectroscopic measurements of water in unexposed glass inclusions. Am Mineral 91:467-470
Thordarson T, Self S (1996) Sulfur, chlorine and fluorine degassing and atmospheric loading by the Roza eruption, Columbia River Basalt Group, Washington, USA. J Volcanol Geotherm Res 74:49-73
Thordarson T, Self S, Oskarsson N, Hulsebosch T (1996) Sulfur, chlorine, and fluorine degassing and atmospheric loading by the 1783-1784 AD Laki (Skaftar fires) eruption in Iceland. Bull Volcanol 58:205-225
Van Orman JA, Grove TL, Shimizu N (2002) Diffusive fractionation of trace elements during production and transport of melt in Earth's upper mantle. Earth Planet Sci Lett 198:93-112
Vigouroux N, Wallace PJ, Kent AJR (2008) Volatiles in high-K magmas from the western Trans-Mexican Volcanic Belt: evidence for fluid-flux melting and extreme enrichment of the mantle wedge by subduction processes. J Petrol 49:1589-1618
Wade JA, Plank T, Melson WG, Soto GJ, Hauri EH (2006) The volatile content of magmas from Arenal volcano, Costa Rica. J Volcanol Geotherm Res 157:94-120

Walker JA, Roggensack K, Patino LC, Cameron BI, Matias O (2003) The water and trace element contents of melt inclusions across an active subduction zone. Contrib Mineral Petrol 146:62-77

Wallace P, Carmichael ISE (1992) Sulfur in basaltic magmas. Geochim Cosmochim Acta 56:1863-1874

Wallace PJ (1998) Water and partial melting in mantle plumes: Inferences from the dissolved H_2O concentrations of Hawaiian basaltic magmas. Geophys Res Lett 25:3639-3642

Wallace PJ, Anderson AT (1998) Effects of eruption and lava drainback on the H_2O content of basaltic magmas at Kilauea volcano. Bull Volcanol 59:327-344

Wallace PJ, Carmichael ISE (1994) S-speciation in submarine basaltic glasses as determined by measurements of S Kα X-ray wavelength shifts. Am Mineral 79:161-167

Weinsteiger AB, Kent AJR, Nielsen RL, Tepley FJ (2007) Are variations in TiO_2 contents in anorthite-hosted MORB melt inclusions controlled by magma compositions or boundary layer phenomena? Eos Trans 88(52): V11B-058.

Workman RK, Hauri E, Hart SR, Wang J, Blusztajn J (2006) Volatile and trace elements in basaltic glasses from Samoa: Implications for water distribution in the mantle. Earth Planet Sci Lett 241:932-951

Yaxley GM, Kamenetsky VS, Kamenetsky M, Norman MD, Francis D (2004) Origins of compositional heterogeneity in olivine-hosted melt inclusions from the Baffin Island picrites. Contrib Mineral Petrol 148:426-442

Yurimoto H, Kogiso T, Abe K, Barsczus HG, Utsunomiya A, Maruyama S (2004) Lead isotopic compositions in olivine-hosted melt inclusions from HIMU basalts and possible link to sulfide components. Phys Earth Planet Interior 146:231-242

Zinner E, Crozaz G (1986) A method for the quantitative measurement of rare-earth elements in the ion microprobe. Int J Mass Spec 69:17-38

# Interpreting H_2O and CO_2 Contents in Melt Inclusions: Constraints from Solubility Experiments and Modeling

Gordon Moore

Department of Chemistry & Biochemistry
Arizona State University
Tempe, Arizona, 85287-1604, U.S.A.

gordon.moore@asu.edu

INTRODUCTION

Due to their volatile nature and low solubility in silicate melts at the surface of the Earth, the *direct* measurement of the volatile components H_2O and CO_2 in magmatic systems is dependent on the presence of glass inclusions trapped in a crystal host prior to eruption. These inclusions, along with the glassy rinds of submarine pillow lavas, represent one of the few windows researchers have into the pre-eruptive chemical characteristics of magmatic systems (e.g., Anderson et al. 2000; Roggensack 2001a; Metrich and Wallace 2008). This information is critically important, as it is volatile exsolution and expansion that provides much of the energy for explosive eruptions, plus it informs our broader understanding of geochemical fluxes in igneous environments (Wallace 2005), and helps us understand specific magmatic behavior such as pre-eruptive phase equilibria (Moore and Carmichael 1998). The dissolved volatile concentration in magmas also strongly influences their physical properties such as density and viscosity (Lange 1994; Ochs and Lange 1999), which in turn affect volcanological behavior such as eruption style (Sparks et al. 1994; Zhang et al. 2007).

In order to use melt inclusion volatile content measurements for petrologic interpretation, experimental volatile solubility data are critically necessary. Experimental solubility constraints allow the measured volatile contents in melt inclusions to be used to estimate intensive properties such as a minimum depth of entrapment of the inclusion (i.e., a calculated fluid saturation pressure), as well as to indicate the type of degassing process (i.e., open or closed system) that occurred during the emplacement and eruption of the magma (Fig. 1). These estimates assume, of course, that the magma is saturated in a fluid containing H_2O and/or CO_2, and are not appropriate for a magma that is fluid-undersaturated. With the assumption of fluid saturation however (see in this volume Blundy and Cashman 2008 and Metrich and Wallace 2008 for discussion of evidence for mid- to lower-crustal fluid saturation in magmas), solubility models place mass balance constraints on magmas, lead to estimates of fluid/magma mass ratios, and to models for their evolution during crystallization and degassing (Wallace et al. 1995; Papale 2005). Determining all of these characteristics is crucial for assessing trends in pre-eruptive and eruptive behavior, and is essential information for volcano hazard monitoring and prediction at any specific volcano.

The importance of volatiles such as H_2O in influencing igneous processes was recognized as early as Bowen (1928). Quantification of their properties however did not significantly begin until the late 1950's with the efforts of researchers like C.W. Burnham and his pioneering measurements of the high *P-T* intensive parameters of H_2O (Burnham et al. 1969), including its solubility in silicate melts (Burnham and Davis 1971, 1974). This laid the groundwork for

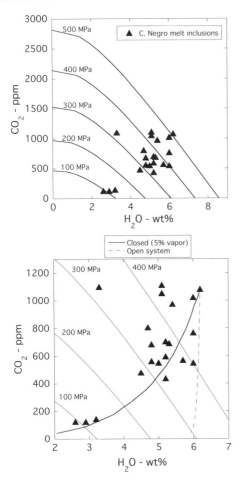

Figure 1. A. An example of volatile concentrations of CO_2 and H_2O in calc-alkaline melt inclusions from a single eruption of Cerro Negro volcano (data from Roggensack 2001a). Isobaric concentration lines calculated using the VolatileCalc solubility model for basalts (Newman and Lowenstern 2002). **B.** Magnified view of same Cerro Negro melt inclusion data set showing calculated degassing paths calculated using VolatileCalc. The bold solid line represents closed system degassing assuming the presence initially of 5% fluid. The dashed line is the calculated open-system degassing path. Note that while the general trend of the data is explained by closed system degassing, the variation in the data remains difficult to interpret by any single degassing process.

further quantification and the development of new measurement techniques in the 1980's. In particular, Fourier transform infrared spectroscopy (FTIR) (Stolper 1982; Stolper et al. 1987) and secondary ion mass spectrometry (SIMS) (Hervig and Williams 1988; Hervig 1992) revolutionized both field and experimental igneous petrology by allowing precise determination of H_2O and CO_2 contents in glassy samples at a spatial resolution of tens of microns. These analytical improvements have also facilitated the development of models for solubility behavior in melts as a function of pressure, temperature, and melt composition, and contributed significantly to our understanding of volatile contents in magmas.

This review chapter will focus on the significant amount of new experimental and modeling work over a broad range of melt and fluid compositions that has been performed since Carroll and Holloway (1994) published their important RiMG volume entitled Volatiles in Magmas. Although a significant amount of work on synthetic melts has also been conducted as part of a successful effort to elucidate the various solubility mechanisms and dependencies at work in silicate melts (e.g., Holloway et al. 1992a; Kohn and Brooker 1994; Behrens et al. 1996; Miyagi et al. 1997; Brooker et al. 1999), this data has proved to be difficult to incorporate into general solubility models for natural melts (Moore et al. 1998; Papale 1999; Papale et al. 2006). Therefore, this chapter will only include the recent work (i.e., since 1994) on pure and mixed ($H_2O + CO_2$) solubility in *natural* melt compositions, and its application to interpreting pre-eruption volatile contents measured in melt inclusions and submarine glasses.

EXPERIMENTAL VOLATILE SOLUBILITY STUDIES IN NATURAL MELTS

The "good" solubility experiment

Our ability to extract useful information from observed melt inclusion volatile contents is certainly dependent on our understanding of the solubilities of those volatiles as a function of pressure (P), temperature (T), and melt/fluid composition. This understanding comes only from experimental measurement of the solubility behavior in various melt compositions. Once obtained, the resulting experimental dataset can be used to develop models that allow interpolation and extrapolation of the data to conditions and compositions that may or may not have been investigated or be experimentally possible, but that are important nonetheless to understanding and predicting the character of volcanic systems (e.g., VolatileCalc; Fig. 1). In order to build the most precise and stable solubility models however, the best possible attainable experimental data are required. The data must therefore meet several requirements in order to be suitable for use in creating any empirical model, and these are discussed in the following sections.

Experimental apparatus for volatile solubility studies. The solubility of H_2O and/or CO_2 in silicate liquids are experimentally determined by equilibrating a known melt composition under fluid-saturated conditions at known P and T, and then quenching the liquid to a glass. These studies can be conducted in a variety of high P-T apparatus that meet the requirements specific to this type of experiment. These requirements include: 1) a relatively large sample volume (0.3 to 1 cc) to accommodate the inherently large molar volume of fluids, and the subsequent volume change associated with dissolution of fluid into the silicate liquid, especially at lower (< 300 MPa) P conditions, 2) a rapid quench from super-liquidus T to below the glass transition T, particularly for mafic samples and any experimental condition resulting in high dissolved H_2O content (> 4 wt% H_2O) in the liquid, and 3) near-hydrostatic P conditions to minimize capsule failure due to shear, as well as to minimize error in estimating the experimental P.

The high P-T apparatus that meet these requirements include the rapid-quench cold seal vessel (Ihinger 1991), the rapid-quench pressure vessel (Holloway et al. 1992b), and the large volume piston cylinder (Baker 2004; Moore et al. 2008a). Each of these techniques has advantages and disadvantages for the solubility experiment. For example, the cold seal (CS) and pressure vessel (PV) offer hydrostatic and precise P control with large sample volumes, unlike the relatively small volumes of the piston cylinder (PC). The problem of their relatively slow quench rates has been overcome by the development of rapid-quench modifications (e.g., Ihinger 1991; Holloway et al. 1992b). The upper P-T ranges for the CS and PV methods are constrained however (~1150 °C, ~300 MPa; Holloway and Wood 1988) by the material strength of the vessel and other considerations such as short furnace life at high T, and the occurrence of large T gradients at higher P. Thus, they are typically used for more silicic (i.e., at lower P and T) compositions. While being able to achieve the higher T and extremely fast quench rates (> 150 °C/s) needed for mafic compositions, the PC suffers from a disadvantage in that its solid pressure media (as opposed to fluid or gas media in CS or PV) requires a minimum force to overcome internal friction that leads to non-hydrostatic conditions, and thus is normally run at pressures greater than 500 MPa. The PC also has relatively smaller sample volumes, making fluid-saturated experiments at lower P difficult. These factors have resulted in an experimental "gap" in P space for solubility studies, particularly for the more mafic compositions, between 300 MPa (upper region for the CS and PV) and 500 MPa (minimum P for PC). Recent work with the large volume piston cylinder (Baker 2004; Moore et al. 2008a) has shown that with a modified sample assembly (Fig. 2), this apparatus can be used for solubility determinations

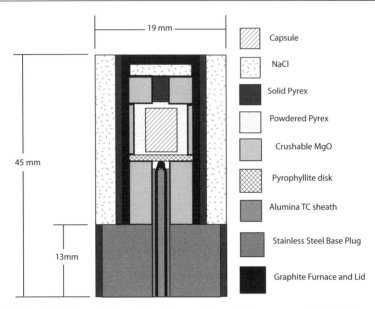

Figure 2. Schematic of piston cylinder solid media assembly used for low pressure (< 500 MPa) experiments, including H_2O-CO_2 fluid solubility and volatile-bearing phase equilibria experiments. From Moore et al. (2008a).

at pressures lower than previously thought (down to 300 MPa), and presents the capability to bridge this experimental *P-T* region for volatile solubility studies.

Analytical techniques for determining melt and fluid compositions. Once a solubility experiment has been successfully conducted, it is necessary to determine, at a minimum, the dissolved volatile content of the glassy run product. If the experiment involves mixed fluid components (e.g., H_2O + CO_2), it is critically important to also determine the composition of the fluid (i.e., the H_2O/CO_2 ratio) that was in equilibrium with the melt. The analysis of the glassy solid portion of the run product can be conducted using several methods, including the same micro-beam techniques used for melt inclusion and submarine glass determinations (e.g., SIMS and FTIR; see Ihinger et al. 1994; Metrich and Wallace 2008). For those experimentalists with the luxury of large sample volumes, using more precise bulk techniques such as high T vacuum manometry and Karl-Fischer titration is preferred due to their first-order nature (i.e., the calibrations for these techniques do not depend on secondary standards; see Ihinger et al. 1994).

To determine the fluid composition in mixed volatile experiments, two techniques are generally used; a mass balance/gravimetry technique and a low T vacuum manometry method. The mass balance technique involves weighing the sample capsule, then lowering the T of the capsule such that the H_2O freezes but the CO_2 remains a vapor (usually using liquid nitrogen). The capsule is then punctured, releasing only the CO_2 and then re-weighed. The capsule is then heated above 100 °C to release all the H_2O, and then weighed a final time. While quite a simple technique, it requires a large enough amount of fluid to overcome any weighing errors, and thus success is also dependent on the precision of the balance used. Unfortunately, the large volume necessary to equilibrate a melt with such a large amount of fluid can be prohibitive, and some studies using this technique report errors in the fluid composition between 20 and 100% (e.g., Kadik and Lukanin 1973; Tamic et al. 2001; Botcharnikov et al. 2006). The low T vacuum manometry method consists of puncturing the sample capsule under vacuum, and then collecting and separating the resulting fluid in the vacuum line. The amounts of the H_2O

and CO_2 are then measured using manometric techniques precise to ± 10 micromoles of fluid (Moore et al. 2008a), enabling a far more precise determination of the fluid composition in equilibrium with the melt, particularly for mixed fluid compositions near $X^f_{H_2O} \sim 0.5$. The error in the determination increases however as the pure end member fluid compositions are approached (i.e., as the amount of the minor fluid component approaches 10 micromoles).

Thus, a "good" solubility experiment is conducted in an appropriate high P-T apparatus that allows stable equilibration of the volatiles with the silicate melt. It also employs a fast enough quench from super-liquidus T to form a glass and prevent any retrograde equilibration of the volatiles with the silicate portion of the sample. Once a successful sample is retrieved, precise characterization of both the fluid (if mixed volatiles are involved) and the melt composition is critical. For researchers who have apparatus with large sample volumes at their disposal, bulk analytical techniques for the dissolved volatile contents are desirable for their precision and their first-order nature (Ihinger et al. 1994). Their disadvantage is that they usually destroy a significant amount of the sample, particularly if duplicate measurements are desired. Non-destructive micro-analytical techniques for dissolved H_2O and CO_2 in glasses such as FTIR and SIMS can be successfully used, but must be used in conjunction with well-characterized standards of similar composition to the sample being investigated (Hauri et al. 2002; King et al. 2002; Ihinger et al. 1994; Moore et al. 2008b).

The solubility of pure H_2O and CO_2 in natural melts

Almost all of the early solubility experimental work was done on pure, single component H_2O or CO_2 fluids in mafic to silicic melt compositions (e.g., Goranson 1931; Hamilton et al. 1964; Burnham and Davis 1971). Because H_2O was recognized to have a significant impact on the chemical and physical behavior of silicate melts, and because CO_2 was discovered to have a relatively small solubility (Wyllie and Tuttle 1959), much of the early work focused on H_2O rather than CO_2. This focus on H_2O was also due to the fact that dissolved volatile contents were first determined using a mass balance technique that involved puncturing the capsule and measuring the weight loss due to the evaporation of the undissolved fluid component. This required a significant amount of the volatile component to be dissolved in order to minimize the associated weighing errors. Another early approach was to measure the size of the "dimples" in the capsule walls, thereby estimating the amount of excess fluid present in vesicles during the experiment. This technique, used mostly in PV work, required not only the presence of a large amount of fluid, but also required a capsule material ductile enough to easily occupy the fluid space during quench and the associated decrease in volume (e.g., gold or silver) of the fluid (Burnham and Jahns 1962). The disadvantage of these capsule materials however, is their low melting point that limits the upper T of the experiments, and thus the composition of the melt that can be explored.

Spectroscopic techniques such as FTIR and SIMS have since supplanted these earlier methods. These newer techniques have resulted in an increase in the precision of the volatile content measurements, which is critical for volatile components with low solubility (e.g., CO_2). And in the case of FTIR, added information regarding the speciation of the dissolved volatile in the melt is also obtained (e.g., Zhang and Xu 2007; Hui et al. 2008). The following section will briefly discuss the results of the early single component fluid solubility measurements, and then focus on the more recent determinations and their implications for understanding natural magmatic systems.

Pure H_2O. Despite the analytical constraints, early measurements on pure H_2O solubilities laid the groundwork for our current understanding of volatile behavior in melts. For H_2O, many solubility measurements were made in a broad variety of synthetic and natural compositions (e.g., Hamilton et al. 1964; see also McMillan 1994 for references). From these measurements, it became possible to define the P and T dependence of H_2O solubility in mafic to silicic melts, which led to predictions of the speciation of the dissolution reaction that were later generally confirmed by spectroscopic measurements (Stolper 1982; McMillan 1994). The development of the FTIR technique in the 1980's and 1990's for measuring dissolved H_2O content and speciation in silicate glasses was also accompanied by solubility studies on natural compositions such as rhyolites (Silver et al. 1990; Ihinger 1991). These data were used to develop new solubility models based on the behavior of volatile species (e.g., molecular H_2O) observed spectroscopically (Fig. 3). This work has ultimately led to a geothermometer/speedometer based on the speciation of H_2O in silicic glass (Newman et al. 1988; Stolper 1989; Zhang et al. 2000; Zhang and Xu 2007), and has improved our understanding of the diffusion properties of H components in melts and glasses (Zhang 1999; Zhang 2007).

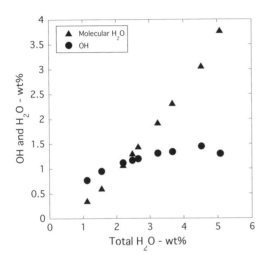

Figure 3. Speciation of dissolved H_2O versus total dissolved H_2O in rhyolite as measured using FTIR spectroscopy. Note the crossover point at ~2.5 wt% total H_2O where the concentration of molecular H_2O equals that of the OH species. Data from Silver et al. (1990) equilibrated at 850 °C and quenched at 200 °C/min. See Zhang et al. (1997) and Zhang (1999) for discussion of the dependence of speciation of H_2O on cooling rate.

Recent work measuring H_2O solubility in natural compositions has consisted of significantly expanding the compositional range of the melts investigated, as well as detailed studies at lower P (Table 1). For example, Carroll and Blank (1997) have measured the solubility of H_2O in a peralkaline phonolite, as well as calibrated its infrared absorptivity coefficients. This study found that the phonolite composition behaves similarly to jadeitic glass in terms of its H_2O solubility. Di Matteo et al. (2004) explored the H_2O solubility of a similar, but more trachytic, composition and found similar behavior as the phonolite of Carroll and Blank (1997). A very detailed, pure H_2O solubility study on calc-alkaline rhyolite at low P (0.098-25 MPa) was conducted by Liu et al. (2005), yielding very important data for late stage degassing of hydrous rhyolites. Behrens and Jantos (2001) have also conducted a detailed study on the effect of compositional variation in rhyolites on the H_2O solubility across the peralkaline-peraluminous join.

The determinations of H_2O solubility in mafic compositions are unfortunately much more limited than those for silicic melts. This is partially due to the fact that H_2O-rich mafic melts are famously difficult to quench to a glass, as well as that most mafic igneous systems generally contain significant amounts of CO_2, making the pure H_2O solubility behavior somewhat less applicable to natural systems. Nevertheless, several workers have tackled the necessary experimental difficulties and made solubility measurements on mafic compositions. These include the several pure H_2O runs in the low P study of dissolved volatiles in a MORB tholeiite by Dixon et al. (1995), as well as the study of Pineau et al. (1998) on a basaltic andesite. Other measurements have been made in the reconnaissance studies of Moore et al. (1995, 1998) in the effort to develop a broad compositional model for H_2O solubility. Intermediate

Table 1. Recent experimental studies containing data on the solubility of pure H_2O and CO_2 in natural melts.

Reference	Melt Composition	P - MPa	T - °C	Apparatus[1]	Analytical[2] Methods
Pure H_2O					
Behrens and Jantos 2001	rhyolite	50-500	800	EHPV	KF
Botcharnikov et al. 2004*	rhyodacite	200	850	CS	ED
Botcharnikov et al. 2006	andesite	500	1200	RQ-IHPV	FTIR
Botcharnikov et al. 2007*	andesite	200	1050, 1200	RQ-IHPV	KF, FTIR
Carroll and Blank 1997	phonolite	19-150	850-973	RQ-CS	FTIR, Man
Di Matteo et al. 2004	trachyte	20-200	850	CS	FTIR, KF
Dixon et al. 1995	MORB[3]	18-72	1200	RQ-IHPV	FTIR, Man
Larsen and Gardner 2004	phonolite	50-200	825-850	CS	FTIR
Liu et al. 2005	rhyolite	0.1-25	700-1200	IHPV	FTIR
Moore et al. 1995, 1998	various	65-311	900-1130	RQ-IHPV	Man
Pineau et al. 1998	basaltic andesite	50-300	1200-1250	IHPV	Man
Tamic et al. 2001	rhyolite	200, 500	800, 1100	CS, IHPV	FTIR, KF
Webster et al. 1999*	basalt, andesite	200	1100-1170	IHPV	SIMS
Webster and Rebbert 1998*	rhyolite	50-200	700-975	IHPV	SIMS
Yamashita 1999	rhyolite	22-100	850-1200	RQ-IHPV	FTIR
Pure CO_2					
Botcharnikov et al. 2006	andesite	200, 500	1200, 1300	RQ-IHPV	FTIR
Brooker et al. 2001	melilitite	1200-2700	1300-1600	PC	FTIR, EA
Jendrzejewski et al. 1997	MORB[3]	1200, 1300	25-195	IHPV	FTIR, Man
Thibault and Holloway 1994	leucitite	100-2000	1200-1600	IHPV, PC	EA, FTIR, SIMS

[1] Apparatus abbreviations are: CS – cold seal vessel; EHPV – externally heated pressure vessel; IHPV – internally heated pressure vessel; RQ – rapid quench.
[2] Methods include: EA – COH elemental analyzer; ED – electron probe difference; FTIR – Fourier transmission infrared spectroscopy; KF – Karl Fischer titration; Man – high temperature vacuum manometry; SIMS – secondary ion microprobe spectrometry.
[3] Mid-ocean ridge tholeiitic basalt.
*– may contain significant amounts of halogens and/or sulfur.

compositions (i.e., andesitic melts) are also relatively lacking in terms of pure H_2O solubility data for similar reasons as the mafic melts (see Table 1 for recent references), but include measurements by Botcharnikov et al. (2006, 2007), and Moore et al. (1998).

General H_2O solubility behavior in natural melts. A plot of the H_2O solubility for basalt and rhyolite as a function of increasing P is shown in Figure 4. It can be seen that at low P (>70 MPa) and their appropriate T, the pure H_2O solubility of the two compositions are very similar in terms of weight percent of the melt. To some this is often somewhat surprising, as it is commonly understood that H_2O solubility increases with silica content of the melt. It is, however, purely a function of the inverse T dependence of H_2O solubility, and the ~350 °C difference between a basalt and rhyolite. At higher P however, the greater P dependence of the H_2O solubility in the rhyolite dominates, and the H_2O contents of the two compositions diverge quickly. The addition of certain components, such as alkalies, have also been shown to greatly increase the H_2O solubility (Fig. 5; Carroll and Blank 1997; Behrens and Jantos 2001; Di Matteo et al. 2004; Larsen and Gardner 2004) in natural melts, and highlight the necessity of measuring solubility behavior across a broad range of conditions and melt compositions. Another important characteristic of the H_2O component and its solubility that should not be overlooked is its absolute concentration value. It has been measured in melt inclusions from mafic arc magmas at amounts greater than 6 weight percent (e.g., Sisson and Layne 1993; Roggensack 2001a; Wallace 2005). Because H_2O is such a light molecule, at these concentrations the H_2O component comprises 15-20 mol percent of these magmas (calculated on an oxide basis), superseded only by SiO_2 and Al_2O_3, and therefore becomes one of the dominant components of the melt on a molecular basis.

Pure CO_2. Investigations into the solubility of pure CO_2 in natural silicate melts have followed a similar track to that of H_2O, with the added complication of its low (generally 100-1000's of ppm) dissolved concentration making it inherently difficult to precisely measure. Wyllie and Tuttle (1959) made some of the first experiments on CO_2 in geologically relevant systems (granitic) and recognized that the amount of CO_2 dissolved in the melts was small relative to H_2O. This fact, along with the imprecise measurement techniques used at the time (e.g., gravimetric, beta-track radiography, and electron microprobe deficit; Blank and Brooker 1994) led to much debate in the 1970's and early 1980's as to what the solubility of CO_2 and its dependencies in natural melts were. Since then, significant and successful CO_2 solubility experimental studies, particularly of dissolution mechanisms and speciation in synthetic melts

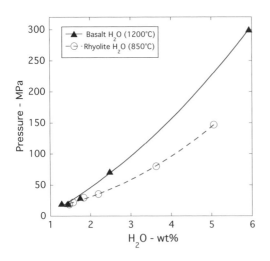

Figure 4. Dissolved H_2O concentrations in pure H_2O-saturated basalt and rhyolite as a function of pressure. The basalt data is at 1200 °C (Hamilton et al. 1964; Dixon et al. 1995) and the rhyolite is at 850 °C (Silver et al. 1990). Note the similar concentrations of H_2O at low pressure (i.e., < 2 wt% H_2O).

(e.g., Kohn et al. 1991; Brooker et al. 2001), were done. More recently however, there have been only a few studies conducted on the solubility of pure CO_2 in natural melts. Similar to H_2O, the work that has been done in this area has focused on increasing the range of melt compositions, as well as the P and T conditions studied (Table 2).

General solubility behavior of pure CO_2 in natural melts. In general, but to varying degrees, the concentration of total dissolved CO_2 in natural melts increases with increasing P, and decreases with increasing T. These dependencies have been systematically measured in rhyolitic (Fogel and Rutherford 1990) and mafic (Brooker et al. 2001; Jendrzejewski et al. 1997; Shilobreyeva and Kadik 1990; Pan et al. 1991; Thibault and Holloway 1994) compositions. Other studies have determined pure or near pure CO_2 solubilities in other compositions as well, but they are often a small part of a larger study involving mixed volatiles (e.g., Dixon et al. 1995; Botcharnikov et al. 2006; King and Holloway 2002). The total solubility and speciation

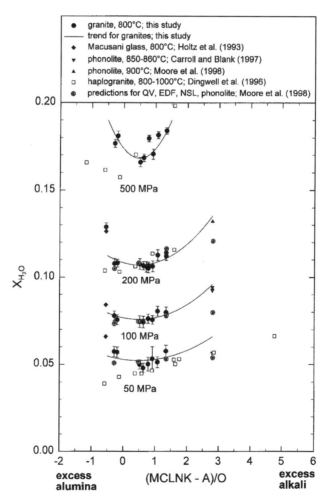

Figure 5. Summary diagram taken from Behrens and Jantos (2001) showing the effect of compositional changes in rhyolites on the H_2O solubility at various pressures. X_{H_2O} is the mole fraction of H_2O dissolved in the melt. See text for definition of the compositional parameter (MCLNK-A)/O.

Table 2. Recent experimental studies containing data on the solubility of mixed H_2O-CO_2 fluid in natural melts.

Reference	Melt Composition	P - MPa	T - °C	Apparatus[1]	Analytical[1] Methods
Behrens et al. 2004a	dacite	100-500	1250	RQ-IHPV	KF, SIMS, FTIR
Botcharnikov et al. 2006	andesite	500	1200	RQ-IHPV	FTIR
Botcharnikov et al. 2007*	andesite	200	1050, 1200	RQ-IHPV	KF, FTIR
Dixon et al. 1995	MORB[3]	18-72	1200	RQ-IHPV	FTIR, Man
Jakobsson 1997	icelandite	1000	1400	PC	FTIR, EA
King and Holloway 2002	andesite	1000	1300	PC	FTIR, SIMS
Moore et al. 2008b	basalt	300-700	1200	PC	FTIR, Man, SIMS
Tamic et al. 2001	rhyolite	200, 500	800, 1100	CS, IHPV	FTIR, KF

1 - See Table 1 for abbreviations.
* - may contain significant amounts of halogens and/or sulfur.

of dissolved CO_2 are both a strong function of melt composition, with silicic melts containing only molecular CO_2 (Behrens et al. 2004b), and mafic melts having only the carbonate complex (Blank and Brooker 1994). The intermediate compositions are more complicated in that they can contain both the molecular and carbonate species of dissolved CO_2 (Behrens et al. 2004a; King et al. 2002). Despite it being well established that CO_2 solubility is strongly dependent on composition, little experimental work has been done to constrain this dependence in natural melts. Some experimental work has explored the compositional dependence of pure CO_2 on tholeiitic and alkalic basalts (Pan et al. 1991; Thibault and Holloway 1994), while Brooker et al. (2001) investigated silica-undersaturated alkalic melts at high P. Unfortunately, no study exists that explores the effect of composition for mafic or silicic compositions in a systematic way similar to the detailed study for H_2O solubility dependence in rhyolites of Behrens and Jantos (2001).

The solubility of mixed fluids (H_2O + CO_2) in natural melts

Perhaps the most important experimental solubility studies for understanding and interpreting melt inclusion data are those that involve mixed volatiles (H_2O + CO_2). This is true because melt inclusions typically contain measurable amounts of both components. And, because of the interplay between the mixing properties of a binary fluid and intensive parameters such as P, T, and melt composition, experimental constraints on mixed solubility behavior are critical to properly modeling the system. Unfortunately, a survey of the literature shows that there is a relative dearth of these experiments, and not nearly enough to precisely understand melt inclusion volatile contents given the number of variables, particularly for mafic and intermediate compositions (Table 2). The situation is beginning to be recognized, however, as more and more melt inclusion measurements are made, and the need grows to precisely interpret them, particularly for subduction-related volcanoes (e.g., Roggensack 2001a,b; Cervantes and Wallace 2003a; Wade et al. 2006; Johnson et al. 2008; Metrich and Wallace 2008). Several important experimental studies have also recently highlighted the current inadequacies of the existing data and the resulting effect on the available mixed volatile solubility models (Moore et al. 2005, 2008c).

The earliest mixed volatile experiments on natural melts were conducted by Wyllie and Tuttle (1959), and pioneering work was also performed in the 1970's and early 1980's by several groups (Kadik et al. 1972; Brey and Green 1973, 1975; Mysen et al. 1975; Jakobsson and Holloway 1986). All of these researchers were stymied however by their inability to precisely measure the dissolved components (particularly CO_2) in the melts. As was mentioned before, mixed volatile studies also require precise fluid composition determination, adding another complication to these already difficult experiments. The modern analytical techniques of SIMS and FTIR have been as important to advances in mixed volatile studies as they have been for the pure end-member solubility studies. But, because of the complication of the presence of both CO_2 and H_2O species dissolved in the melts, calibration of these techniques has proven more difficult than for the pure component studies. This is especially true for the use of FTIR on intermediate compositions, where CO_2 is present in the melt as both a carbonate and molecular species (Mandeville et al. 2002; King et al. 2002). Recent work using the SIMS technique has also shown the possibility of a matrix effect of H_2O that suppresses the ionization of carbon in mafic glasses (Moore et al. 2008b), complicating the SIMS determination of CO_2 in mafic glasses containing greater than 3 wt% dissolved H_2O. All of these complications arise from the fact that there are not enough well-characterized glass standards available that contain significant amounts of both H_2O and CO_2, and point out that much work remains to be done in this area.

General behavior of mixed volatile solubility in natural melts. Despite the complicated nature of mixed solubility experiments, there are several excellent studies that have elucidated the solubility behavior of mixed volatiles in natural melts. In particular is the work of Tamic et al. (2001) who conducted a detailed investigation of the H_2O-CO_2 solubility in a rhyolite up to 500 MPa (Table 2). Also included are the study of Behrens et al. (2004b) from 100-500 MPa on a dacitic liquid, and the low-P study of a mid-ocean ridge basalt of Dixon et al. (1995). The results of Dixon et al. (1995) at high T (1200 °C) and low P (<100 MPa) are suitably modeled assuming that the solubility of both CO_2 (as a carbonate ion) and molecular H_2O behave in a Henrian manner, and that the two volatile components mix linearly in the fluid. This linear behavior breaks down however for both components at higher P (>150 MPa). This observed behavior in basalt is similar to that seen in the high-P study on rhyolite of Tamic et al. (2001) and the dacite of Behrens et al. (2004a), where linear solubility dependence for H_2O is seen at 200 MPa. At higher P (500 MPa) however, both of these studies report a greater dependence on the fluid composition (i.e., on H_2O fugacity). CO_2 solubility is also non-linear at all conditions in these studies, in contrast to the linear behavior reported by Blank et al. (1993) at lower P. Behrens et al. (2004a) also report a marked increase in CO_2 solubility (as well as increasing carbonate/molecular species ratio) with increasing H_2O content, indicating that H_2O stabilizes CO_2 in dacitic melts. All of these observations point to the conclusion that the influence of H_2O on CO_2 solubility in natural melts is complicated, particularly at higher P (>100 MPa). This conclusion is significant, particularly for mafic and intermediate magmas that are thought to have relatively deep origins, and to contain significant amounts of H_2O and CO_2 prior to eruption.

Other recent studies (Moore et al. 2005, 2008c; Roggensack and Moore 2008) have investigated the influence of compositional variation on H_2O-CO_2 solubility in calcic and calc-alkaline basalts, with the goal of directly determining solubility behavior in melt compositions similar to those found in arc-related melt inclusions. One of the striking results from these studies is the strong positive dependence of the CO_2 solubility of these hydrous arc basalts on their calcium content (Fig. 6). Unfortunately, many existing solubility models cannot account for this dependence, leading to a significant overestimation of the minimum saturation P and an incorrect estimate of the equilibrium fluid composition for calc-alkaline basaltic melt inclusions. These discrepancies are due again to the lack of experimental data available for a broad range of melt compositions on which to base solubility models, particularly for calcic to calc-alkaline compositions. This topic will be discussed in more detail in the following sections on solubility models and their application to melt inclusion data.

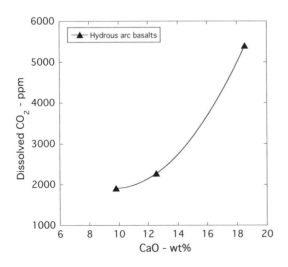

Figure 6. The influence of CaO content on the CO_2 solubility in calcic to calc-alkaline basalt compositions. Saturated H_2O-CO_2 experimental data at 400 MPa and 1200 °C from Roggensack and Moore (2008) containing 3-5 wt% H_2O.

VOLATILE SOLUBILITY MODELS FOR NATURAL SILICATE MELTS

Many diverse mathematical approaches have been used to model the solubility of H_2O and/or CO_2 in silicate melts (see Table 3). Most models are based on thermodynamic relationships, and range from simple descriptions of P and T dependencies to those that make use of certain solubility characteristics (e.g., linear Henrian behavior). Some are based on melt composition dependencies that are inferred to be related to melt structure (e.g., the NBO/T polymerization parameter of Mysen 1987). Others are simply *ad hoc* mathematical expressions (e.g., Taylor series expansions) that were chosen only because they fit the data well. Each model type has its particular strengths and weaknesses, and care must be taken with all of them so that the user is not misled into believing a prediction that could be in serious error, or over-interpreting the physical significance of a statistical fit parameter.

Table 3. Volatile solubility models.

Reference	Melt Composition	P - MPa	T - °C	Model Type[1]
H_2O solubility				
Behrens and Jantos 2001	rhyolitic	0.1-200; 500	800	comp
Carroll and Blank 1997	phonolitic	0.1-191	850	reg sol
Di Matteo et al. 2004	trachytic	20-200	850	comp
Dixon et al. 1995	basaltic	0.1-98	1200	reg sol
Liu et al. 2005	rhyolitic	0.1-500	700-1200	emp
Moore et al. 1998b	various	0.1-300	700-1200	comp
Papale 1997	various	0.1-1000	>730	reg sol
Yamashita 1999	rhyolitic	0.1-100	700-1200	reg sol
Zhang 1999	rhyolitic	0.1-800	500-1350	
CO_2 solubility				
Dixon et al. 1995	basaltic	0.1-98	1200	reg sol
Dixon et al. 1997	alkali basalt	0.1-500	1200	comp
Papale 1997	various	0.1-1000	>730	reg sol
Mixed volatile solubility				
Behrens et al. 2004a[2]	dacitic-rhyolitic	100-500	850-1250	reg sol
Liu et al. 2005	rhyolitic	0.1-500	700-1200	emp
Newman and Lowenstern 2002	basalt, rhyolite			various
Papale 1999	various	<1000	>730	comp reg sol
Papale et al. 2006	various			comp reg sol
Tamic et al. 2001[3]	rhyolitic	75-500	800-1100	emp

[1] - Model types include: comp – compositional, accounts for compositional variation, but may include an ad hoc fit equation form; comp reg sol – compositional regular solution model; emp – empirical, generally an ad hoc form of fit equation; reg sol – regular solution model, no compositional dependence; various – uses many different models for its calculations.
[2] - CO_2 solubility for H_2O-CO_2 bearing silicic compositions.
[3] – H_2O solubility for H_2O–CO_2 bearing rhyolite.

To simplify the discussion of the recent modeling work, each model is grouped in terms of the volatile component that is modeled (Table 3). Although grouped in this way, any given model may or may not be applicable to calculating solubility for a pure or mixed volatile fluid, depending on how the model is constructed. For example, the general compositional model for H_2O by Moore et al. (1998) is intended for calculating the solubility of pure H_2O. But because it uses H_2O fugacity as a parameter, it can also be used successfully to calculate H_2O solubility for mixed volatile cases (i.e., by assuming an linear mixing model for CO_2 and H_2O in the fluid, where the fugacity of H_2O in the fluid mixture is equal to the fugacity of pure H_2O at the P and T of interest times the mole fraction of H_2O in the mixture), as long as the final H_2O fugacity value does not significantly exceed that used to calibrate the model. Other models, such as the rhyolite/dacite-CO_2 solubility model of Behrens et al. (2004a), are based on mixed hydrous-CO_2 bearing datasets and thus are only correctly used in the mixed volatile case.

H_2O solubility models

Silicic models (>65 wt% SiO_2). A majority of H_2O solubility studies from the literature prior to 1994 were conducted on silicic compositions (McMillan 1994). This is also true for the recent work reviewed here (Table 1). There are also many other studies that fall under the silicic classification that are not listed in Table 1 because they are synthetic or haploid compositions (e.g., Holtz et al. 1995, 2000; Romano et al. 1996; Dingwell et al. 1997; Schmidt et al. 1999; Behrens et al. 2001). All of this experimental attention has resulted in a good understanding of the mechanisms of H_2O dissolution in rhyolites, as well as many models for H_2O solubility in silicic melts. And, for rhyolitic melts specifically, it has resulted in two types of models that are currently in general use. The first type is based on the regular solution model developed first for the system albite-diopside-silica-H_2O by Silver and Stolper (1985), and then later applied to rhyolite melts (Silver et al. 1990). The second is a model that accounts for changes in H_2O solubility due to compositional variation in rhyolites by using a metal cation/total oxygen ratio (Behrens and Jantos 2001).

The regular solution model for H_2O solubility in rhyolites of Silver and Stolper (1985) and Silver et al. (1990) is described and discussed by Holloway and Blank (1994). This type of model has the characteristic that its fit parameters can be equated with thermodynamic parameters (e.g., partial molar volume) for the particular component being dissolved in the melt of interest. This type of interpretation has been shown by Ochs and Lange (1999) to be incorrect however, as the regressed values are subject to significant error stemming from the low P nature of the solubility data, and uncertainty in the activity model for H_2O in the melt. By using direct density determinations on hydrous melts, Ochs and Lange (1999) have convincingly shown that there is only one value for the partial molar volume of H_2O (22.89 ±0.55 cc/mol) that is independent of melt composition, as well as H_2O concentration.

Nevertheless, the formalism of Silver et al. (1990) has been applied to other solubility data sets, and represent precise solubility models for various melt compositions. For example, Carroll and Blank (1997) use the model successfully to develop an expression for the solubility of H_2O in a silicic phonolite. The H_2O solubility model of Yamashita (1999) for rhyolite incorporates the measured partial molar volume value for H_2O of Ochs and Lange (1999) and compositional independence into his regular solution model (i.e., it is not a variable parameter) with great success at P up to 100 MPa (Fig. 7). Zhang (1999) also addresses the H_2O partial molar volume inconsistency directly, and shows the difficulties in deriving partial molar parameters using solubility data, as well as developing an excellent H_2O solubility model for rhyolites (Fig. 7) that can convincingly be extrapolated to high P and T. Despite all of the complications over the interpretation of the fit parameters, the regular solution model formalism of Silver and Stolper (1985) provides an H_2O solubility model that has been well used to recover the experimental solubility values and to precisely estimate H_2O solubility contents for particular rhyolitic compositions.

H_2O and CO_2 Content in Melt Inclusions: Experiments & Modeling 347

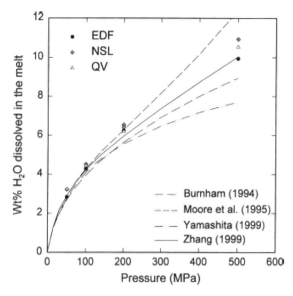

Figure 7. H_2O concentration versus pressure for peralkaline (NSL), meta-aluminous (QV), and peraluminous (EDF) rhyolitic compositions, including various predicted solubility curves. Note the stability of the Zhang (1999) model up to 500 MPa, demonstrating the utility of accounting for H_2O speciation. Figure taken from Behrens and Jantos (2001).

The second type of model for estimating the H_2O solubility of rhyolites is that developed by Behrens and Jantos (2001) that accounts for solubility changes due to variation in the bulk melt composition of the rhyolite. This model is based on a series of solubility experiments conducted over a range of P on a broad range of rhyolites (Table 1), including peralkaline, peraluminous, and subaluminous silicic melts. This study convincingly shows that H_2O solubility has a minimum for subaluminous compositions, with an increase in solubility in both the peralkaline and peraluminous direction (Fig. 5). This variation in H_2O solubility can be modeled using a compositional parameter (MCLNK-A/O = $100\times(2Mg + 2Ca + Li + Na + K - Al)/O$), although the degree of scatter in their data led them to conclude that the model is only semi-quantitative. This is attributed by the authors to the fact that the MCLNK-A/O parameter accounts for neither the concentration nor oxidation state of iron in their experiments, implying a significant role of iron in determining H_2O solubility in silicic melts.

Mafic models (< 52 wt% SiO_2). As there are significantly less data for H_2O solubility in mafic melts, there are subsequently far fewer models that have been developed to calculate it. In fact, the only model specifically developed for estimating H_2O solubility in basalts is that of Dixon et al. (1995). This model uses the regular solution formalism discussed above for rhyolitic compositions, but derives different values for the fit parameters. Other models available for mafic melt calculations are those that fall into the general compositional category discussed below.

General compositional models for H_2O solubility. The effort of Burnham and workers on developing a general H_2O solubility model followed their measurements in silicic compositions (Burnham and Davis 1971; Burnham and Davis 1974), as well as their unique P-V-T measurements of supercritical H_2O fluid at high P and T (Burnham et al. 1969). This work culminated in a thermodynamic-based model that assumed a linear relationship between the activity of the H_2O component and the square of the mole fraction dissolved in the melt

(Burnham 1981), and equilibrium constants were developed to calculate H_2O solubility for a broad range of melt compositions (Burnham 1994) based on feldspar-like melt components. A corollary of the main assumption of their model is that OH is the dominant dissolving species, and that molecular H_2O is not present in significant amounts. This assumption has been shown to be true at lower P, but at higher P and H_2O fugacities molecular H_2O becomes the dominant dissolving species present (Fig. 3; Stolper 1982; Silver et al. 1990). Nevertheless the Burnham model, as it has come to be known, has been used productively to predict pure H_2O solubility in a broad range of natural melt compositions, and certainly provided a useful framework for understanding H_2O solubility.

The empirical model of Moore et al. (1998) represents a successful attempt to create a simple model that defines the P, T, and compositional dependence of H_2O solubility for a very broad range of melts. Using a series of forty-one H_2O solubility determinations on fourteen melts chosen to independently maximize compositional differences, they fit the data to a thermodynamically-based equation of the form:

$$2\ln X_{H_2O}^{melt} = \frac{a}{T} + \sum_i b_i X_i \left(\frac{P}{T}\right) + c \ln f_{H_2O}^{fluid} + d \tag{1}$$

where $X_{H_2O}^{melt}$ = the mole fraction of H_2O dissolved in the melt; $f_{H_2O}^{fluid}$ = the fugacity of H_2O in the fluid; T = temperature in Kelvin; P = pressure in bars; and X_i = the anhydrous mole fractions of the oxide components. The fit parameters for Equation (1) are: a, b_i (i = Na_2O, FeOt, Al_2O_3), c, and d, and the fugacity of H_2O is calculated using the modified Redlich Kwong equation of state (Holloway and Blank 1994). This model has proven successful in estimating H_2O solubility across a spectrum of natural melt compositions, and is available as a spreadsheet that facilitates calculation and use (Moore et al. 1998). Behrens and Jantos (2001) correctly point out however that this model does not capture the compositional dependence they observed in rhyolites, and suggest that another compositional parameter similar to theirs could be incorporated as more data become available. Di Matteo et al. (2004) go a step further and re-regress the Moore et al. model, finding a single compositional parameter that reproduces their data in trachytic compositions. They find however that the resulting expression is useful only for trachytic melts. Also, as is described by Zhang (1999) and the authors themselves, the Moore et al. (1998) model is strictly empirical in nature, and thus should not be extrapolated beyond the maximum H_2O fugacity for which it is calibrated (~3000 bars).

In perhaps the most ambitious effort to date to develop a compositionally dependent model for H_2O, Papale (1997) combined the regular solution mixing model for the anhydrous melt components of Ghiorso et al. (1983) with an exhaustive database of pure H_2O and CO_2 solubility measurements (640 for H_2O, 263 for CO_2) to derive the regular solution mixing parameters for hydrous and CO_2-bearing natural melts. One significant result from this work is the observation that the CO_2 solubility dataset suffers from significant inconsistency in the measurements, and this leads to difficulty in creating a model that recovers all the experimental observations. The H_2O dataset however appears to be internally consistent, although there are inconsistencies with the synthetic and mono-mineralic compositions. The resulting expression for H_2O solubility has ten interaction parameters between water and the melt components, and reproduces the dataset to within ±10% of most of the data. The Papale (1997) model also predicted the observation of a compositionally independent partial molar volume for the H_2O component that was subsequently confirmed by the work of Ochs and Lange (1999). Unfortunately, because of the heavy computational nature of this model, it has not been used to its fullest potential by other workers. It has also since been improved and modified to include mixed fluid behavior by Papale (1999) and Papale et al. (2006), which will be discussed in more detail below.

CO_2 solubility models

As the amount of recent work determining the pure CO_2 solubility in natural melts is relatively small (Table 1), so are the number of studies that attempt to model its behavior (Table 3). Much of the earlier work, particularly on silicic compositions, has been reviewed by Blank and Brooker (1994) and Holloway and Blank (1994). Significant for the behavior of dissolved CO_2 in basalts though, are the more recent solubility studies of Dixon et al. (1995) and Dixon (1997). Using their low-P determinations of CO_2 solubility, Dixon et al. (1995) developed a regular solution model for CO_2 in a mid-ocean ridge tholeiite. In order to account for the significant CO_2 solubility changes due to compositional variation in basalts, Dixon (1997) developed a compositional parameter, PI, (where PI= $-6.50[Si^{+4} + Al^{+3}] + 20.17[Ca^{+2} + 0.8K^+] + 0.7Na^+ + 0.4Mg^{+2} + 0.4Fe^{+2}$) based on cation proportions and the solubility determinations on an alkalic leucitite (Thibault and Holloway 1994) and basanite (Dixon and Pan 1995). This allowed the estimation of CO_2 solubility from tholeiitic to strongly alkalic basalts as a function of composition. The compositional dependence was further simplified with the observation that the PI compositional parameter varied linearly as a function of the SiO_2 content of the basalts, thus facilitating a simple estimate based on P, T, and SiO_2 concentration (Dixon 1997). It is this parameterization that is incorporated into the mixed volatile solubility basalt model of VolatileCalc (Newman and Lowenstern 2002; see further discussion below). While this model works well for tholeiitic to alkalic basalts, it is not able to precisely estimate the CO_2 solubility of other compositions such as calcic and calc-alkaline basalts (Moore et al. 2008b). One cause of this is that the PI compositional parameter becomes negative for certain basalt compositions (typically, high Al_2O_3 and high CaO; see Fig. 8), leading to nonsensical negative CO_2 solubility concentrations. Unfortunately, these types of basaltic compositions commonly occur in subduction-related volcanic arcs, where there are many ongoing melt inclusion studies that require precise estimation of solubility properties for CO_2.

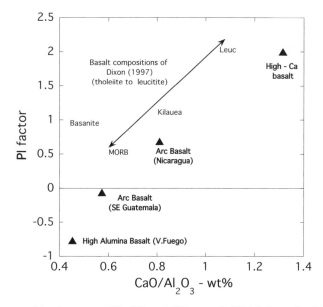

Figure 8. The compositional parameter (PI) of Dixon (1997) versus CaO/Al_2O_3 for arc basaltic compositions (triangles). Arrow shows the range of PI for the tholeiitic and alkali-rich basalts used to calibrate the model. Note the negative values of PI for Volcan Fuego and the Guatemalan basalt.

Papale (1997) also applied the approach of combining the anhydrous mixing properties of melts with solubility data to modeling the behavior pure CO_2 solubility in a wide range of natural melts. As was noted in the previous H_2O solubility modeling section, much of the pure CO_2 solubility data in the literature used by Papale (1997) at the time was inconsistent due to the lack of precise measurement techniques. Therefore, the dataset was difficult to model with this approach, and required the elimination of many outliers (Papale 1997). Perhaps because of the scatter in the data used, the model required twice as many mixing parameters (twenty, as opposed to ten for H_2O) between CO_2 and the melt components to fully describe the solubility variation.

Mixed ($H_2O + CO_2$) solubility models

Modeling the dissolution of an H_2O-CO_2 fluid in natural silicate melts is a complicated endeavor that requires a significant amount of precise experimental data. As was discussed earlier, the dataset must not only include the precise measurement of the H_2O and CO_2 content of the fluid-saturated glass, but also the composition of the H_2O-CO_2 fluid that was in equilibrium with the melt during the experiment.

There are several approaches that have been taken in developing the current mixed volatile solubility models. One is to use fugacity-based, pure fluid solubility models and make assumptions about the mixing behavior of the H_2O and CO_2 components in the fluid and melt. Generally, these are simplifying assumptions such as linear mixing in the fluid, which leads to dissolved solubility values that are calculated using a fugacity value equal to the pure fluid fugacity multiplied by the mole fraction of the component of interest in the fluid ($f_{H_2O} = X_{H_2O} f_{H_2O}$). This is the approach taken by Newman and Lowenstern (2002) in their mixed volatile solubility algorithm for basalts and rhyolites known as VolatileCalc. This mode of calculation assumes however that there is no influence of one component on the solubility of the other, which we have seen earlier, is valid at low P (< 100 MPa) but is certainly incorrect for silicic melts at higher P (Tamic et al. 2001; Behrens et al. 2004a). The lack of experimental data at higher P for mafic compositions leaves these assumptions essentially untested for basaltic melts, although they appear reasonable for low-P calculations (Dixon et al. 1995). Another approach is that of Behrens et al. (2004a), Tamic et al. (2001), and Liu et al. (2005), all of whom based their individual solubility expressions for H_2O and CO_2 in silicic melts on a dataset that included mixed volatile measurements. A significantly more comprehensive approach has been taken by Papale (1999) and Papale et al. (2006), who combined the available pure and mixed volatile solubility data with their previous work using the regular solution model for the anhydrous melt components of Ghiorso et al. (1983) to obtain the fluid/melt interaction parameters for both the volatile components.

Models specific to silicic compositions. For the solubility of CO_2 in hydrous silicic liquids, the model of Behrens et al. (2004a) explores the behavior of dacitic and rhyolitic liquids. Interestingly, they note the presence of detectable amounts of the dissolved carbonate species present in dacite, the stabilization of which they attribute to the high concentration of H_2O in their experiments. This observation is supported by a similar one made in andesites and rhyolites at higher P (King and Holloway 2002; Moore et al. 2006), and significantly complicates the determination of total CO_2 in these melts using FTIR by increasing the number of peaks that must be identified and calibrated. As a result, Behrens et al. (2004a) resorted to the use of SIMS for their CO_2 measurements and report a significant increase of CO_2 solubility with H_2O content. They also observe a significant change in CO_2 solubility behavior of a high P non-linear to low P linear dependence on fluid composition. The resulting regular solution model from this work is therefore only useful in hydrous silicic magmas (although they present a model for anhydrous melts as well), from 100-500 MPa, and over a range mole fraction of H_2O in the fluid of 0.1-0.9. The study of Liu et al. (2005) explores the apparently non-ideal behavior that occurs in H_2O-CO_2 bearing rhyolites at high P in the extreme ends of the H_2O-CO_2 fluid composition space, and develop an ad hoc empirical expression that can be used for pure H_2O and CO_2 fluids, and reproduces the non-linear behavior in the mixed fluid compositional regions not well constrained by the model of Behrens et al. (2004a). Although excellent at recovering the experimental values from the dataset (± 20%), care must be taken when extrapolating the CO_2 solubility expression of Liu et al. (2005) to pressures or conditions that are outside of the experimental conditions of the dataset because of its exponential nature.

The H_2O solubility models of Tamic et al. (2001) and Liu et al. (2005) for mixed volatile-bearing rhyolites take a very similar approach in that they both use series type equations to model their H_2O-CO_2 solubility data. Tamic et al. (2001) chose a truncated Taylor series expansion equation type that relates the H_2O solubility to P, T, and fluid composition. This model has eighteen fit parameters that reproduce a broad range of solubility data from the literature to within ±5.5% relative. The expression for H_2O solubility of Liu et al. (2005) is similarly dependent on the fluid composition, but has fewer fit parameters (six), and reports a 2σ error of 15%. As mentioned above, caution must be used with these types of empirical models, particularly when extrapolating them beyond the bound of their data set (i.e., $P > 500$ MPa).

The rhyolite model in VolatileCalc (Newman and Lowenstern 2002) uses the regular solution parameters for H_2O from Silver et al. (1990), and those for CO_2 are from Blank et al. (1993). The fugacities of the two fluid components are assumed to mix linearly, and their pure, end-member fugacities are calculated using the modified Redlich-Kwong equation of state (Holloway 1977; Flowers 1979). The solubilities of each component in the melt are also assumed to be independent of each other, and to behave linearly (Henrian) with respect to changes in fluid composition. As the rhyolite model is based essentially on pure and mixed fluid solubility data for meta-aluminous rhyolite at low P (<150 MPa), it is capable of generating precise estimates for the solubility parameters, and essentially matches other models that only calculate the pure component solubilities. There are, however, several caveats that must be taken into account. First, there is no compositional dependence built in to the calculation for rhyolite that can account for the H_2O solubility behavior observed by Behrens and Jantos (2001) across the peralkaline-peraluminous compositional spectrum of silicic melts. At the same time, the compositional effect on CO_2 in silicic magmas is also known to be important (Stolper et al. 1987), but remains unconstrained. Thus, the bulk composition of any silicic melt inclusion suite must be shown to be close to meta-aluminous and relatively constant in order to properly apply VolatileCalc to it. Secondly, we have also seen in the work of Tamic et al. (2001) and Behrens et al. (2004a) that CO_2 solubility appears to be dependent on H_2O content at P equal to and greater than 200 MPa, which should cause users of VolatileCalc to consider the output at these higher P carefully, particularly for values at the ends of the fluid composition spectrum.

Mixed solubility models for basalts. The mixed volatile solubility calculation package of Newman and Lowenstern (2002) can also be used to estimate minimum saturation P and degassing trends for basaltic melt inclusion studies. VolatileCalc incorporates the regular solution parameters for H_2O solubility determined by Dixon et al. (1995), and uses the CO_2 solubility parameters from Dixon et al. (1995) and Pan et al. (1991). It also uses the compositional parameterization of Dixon (1997) discussed in the earlier section on CO_2 solubility. This results in a flexible model that allows users to calculate solubility parameters from nephelinite to tholeiite by inputting only the SiO_2 content of the melt. As this compositional dependence is constructed using solubility data from tholeiitic and alkali-rich basalts, it is capable of precisely estimating the solubility parameters within this compositional range. Unfortunately, the model is not capable of correctly estimating the mixed solubility behavior of calcic or calc-alkaline compositions, as these basalts are outside of the compositional parameterization of Dixon (1997). In particular, the effect of increasing calcium content has been shown to have an especially strong positive effect on the CO_2 solubility in basalts (Fig. 6; Moore et al. 2008b). Because of the strong P dependence of CO_2, applying VolatileCalc to these compositions results in an overestimation of the minimum saturation P by as much as

35-50% relative, and typically outputs results that are beyond the range of P calibration of the model itself. Similarly, the fluid compositions estimated by VolatileCalc are also significantly incorrect compared to experimental values (see discussion below).

General compositional models for mixed volatile solubility. Currently, the only model available that accounts for compositional variation in H_2O-CO_2 solubility from basalts to rhyolites is that of Papale (1999), which was updated by Papale et al. (2006). This model uses the entire available dataset for C-O-H-silicate liquid solubility to calibrate its multi-component regular solution model. There are 865 measurements of H_2O, 173 for CO_2, and 84 measurements for mixed volatiles, giving an indication of the breadth of the database used. One improvement of this work over the earlier studies is that the number of recent CO_2 measurements is now large enough that the older (pre-1980's), inconsistent determinations could be thrown out. This has resulted in a more stable and systematic dataset to model, although they report that the larger number of H_2O determinations has resulted in higher confidence for the calculations of that component. Nevertheless, the ambitious model of Papale et al. (2006) has reached the point where its authors can begin contemplating effects such as how the oxidation state of iron in Fe-rich melts might affect the solubility values.

The compositional variation in melt inclusions and its implications for interpreting volatile contents

We have seen from the previous discussions that melt composition plays an important role in determining the solubility of H_2O and CO_2 in magmas. Therefore, it must be accounted for in any solubility model being used to interpret melt inclusion volatile contents. Unfortunately, accounting for compositional effects on solubility is rarely performed in most melt inclusion studies, due to the fact that there are no models readily available that have this capability (The model of Papale et al. 2006 is an obvious exception, but has only recently become broadly available due to the efforts of Dr. Mark Ghiorso of OFM Research who has created a website calculator that allows several types of calculations using the model, as well as to Dr. Papale's generosity in making the code available.). While the basalt model of VolatileCalc can account for compositional variation in alkali-rich basalts, it is not able to precisely calculate solubility behavior for subduction-related basalts. The rhyolite model, as we have seen, also does not have the ability to account for solubility changes due to compositional variation.

This section will examine the implications of compositional effects for interpreting melt inclusion volatile contents using both VolatileCalc and the Papale et al. (2006) models. In both cases, an attempt is made to test the model output for rhyolitic and basaltic compositions with experimental data not used in their respective databases. This is a difficult task, as the number of mixed volatile solubility measurements since the publication of Papale et al. (2006) is limited. As a result, the data of Tamic et al. (2001) is used for rhyolites, even though it is contained within the Papale et al. (2006) dataset. Plus, selected data on calcic and calc-alkaline basalts from Moore et al. (2008) are used to benchmark the basalt models. We also ask the question whether the scatter that is difficult to explain using reasonable degassing histories in many melt inclusion suites (e.g., Fig. 1), could be due to compositional effects on the melt solubility.

Compositional variation in silicic magmas. The effect of compositional variation on the two solubility models (VolatileCalc; Newman and Lowenstern 2002 and Papale et al. 2006) will be assessed here using the mixed solubility data of Tamic et al. (2001) and Behrens et al. (2004a), representing a meta-aluminous rhyolite and dacite respectively. The test was approached as if the experimental data were melt inclusion measurements, and the models used accordingly to calculate minimum saturation P and fluid compositions. The resulting values for the fluid composition and P are plotted against the measured experimental values in Figure 9. It can be seen that both VolatileCalc and Papale do an excellent job of recovering the

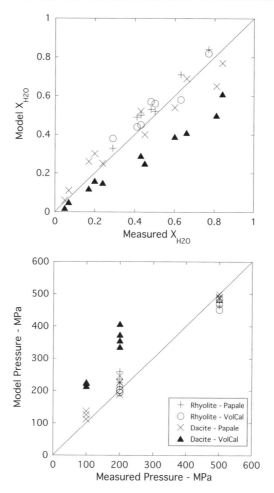

Figure 9. Model versus measured values for pressure and fluid compositions (X_{H_2O}) in a rhyolite and dacite using VolatileCalc and Papale et al. (2006). Rhyolite data taken from Tamic et al. (2001) and the dacite values are from Behrens et al. (2004a). Neither dataset is used to calibrate VolatileCalc, while the Papale et al. model does contain the Tamic et al. data.

200 and 500 MPa rhyolitic data of Tamic et al. (2001). VolatileCalc reproduces the 200 MPa data better by a few percent, while Papale is slightly better at 500 MPa. The major distinction between the two models occurs when the dacite data of Behrens et al. (2004a) is considered. The Papale model performs as well with the dacite data as it does for the rhyolite, while VolatileCalc overestimates the saturation P by a factor of two, and underestimates the H_2O content of the fluid by similar amounts. Of course, Newman and Lowenstern (2002) did not design the rhyolite model in VolatileCalc to be applied to a dacite composition (66 wt% SiO_2), and the estimates for the appropriate rhyolite are quite good. Although somewhat extreme, this exercise highlights the potential dangers inherent in using a model based on a single composition to calculate solubilities for a range of melt compositions.

Compositional variation in arc basaltic melt inclusions. The compositional variation found in melt inclusions from basalts and basaltic andesites in arc settings is represented in Figure 10 (for further discussion of compositional variation in basaltic melt inclusions, see

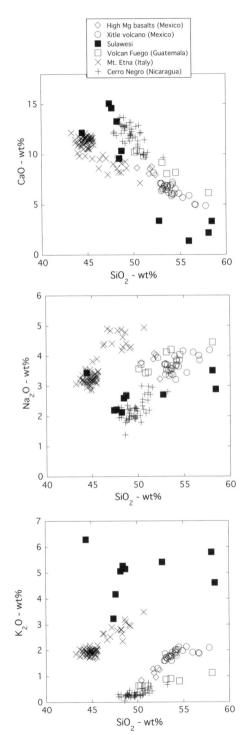

Kent 2008 in this volume). Even within single eruptive suites (e.g., Cerro Negro, Guatemala) significant total variation is found in both the calcium and total alkali content, both of which influence the total CO_2 solubility in mafic melts. And, these values can vary as much as 50% for any given silica content, which is an important observation given that the compositional parameterization of Dixon (1997) for basalts in VolatileCalc assumes a linear dependence on silica content. The significance of this is that any compositional variation at a single value for silica is not accounted for (except for the tholeiite to alkali-rich basalt trend) in the Dixon (1997) model. Thus, before using these models to calculate solubility parameters for melt inclusions found in arcs, they should be benchmarked against experimental solubility data conducted on compositions similar to the melt inclusions. Unfortunately, there are very few studies on arc compositions, and as a result, select data from Moore et al. (2008c) and Botcharnikov et al. (2007) will be used here. As the Botcharnikov et al. (2007) data is andesitic, it is not appropriate to use VolatileCalc for it, and only the Papale et al. (2006) model will be used.

Again approaching this test as if the experimental data were melt inclusion measurements, we calculated minimum saturation P and fluid compositions for the mafic compositions of Moore et al. (2008) using the basalt module of VolatileCalc. The results are plotted versus the experimental values in Figure 11. Because the experiments were conducted at 400, 500 and 600 MPa, some of the dataset is beyond the stated capability of VolatileCalc of 500 MPa. Considering

Figure 10. Harker diagrams for melt inclusion compositions taken from arc-related basalts. Note the broad range of compositions within single suites of inclusions. Data from: high-Mg basalts and Xitle (Cervantes and Wallace 2003a,b); Sulawesi (Elburg et al. 2007); Volcan Fuego (Roggensack 2001b); Mt. Etna (Spilliaert et al. 2006); Cerro Negro (Roggensack 2001a).

only the 400 and 500 MPa data however, we see that except for one point the predicted saturation P are overestimated by a minimum of 25%, extending up to 50%. Similar deviations from the experimental values are seen in the fluid compositions as well.

The results of the same test for the Papale model, adding the andesite composition of Botcharnikov et al. (2007), are shown in Figure 12. The Papale model appears to recover the low-P andesite data and the high P calc-alkaline basalt values quite well. The other high-P estimates suffer from significant scatter (average 22% residual) however. For the most part the scatter in the P values are not systematic, except for that of the high calcium basalt which are systematically overestimated (and may be outside the compositional range of the model). The fluid composition estimates for the high calcium basalt experiments are also anomalously and significantly low, while those for the andesite appear systematically high. Taking into account the error in the measured experimental parameters for mixed solubility data (particularly that for fluid composition measurements), and given the broad compositional range that this model covers, these results are certainly acceptable and represent a significant and important improvement in our ability to model the behavior of C-O-H fluids in volcanic systems where magma compositions may vary significantly.

Implications of compositional effects on observed melt inclusion variation. Melt inclusion volatile contents are often presented on a CO_2 versus H_2O plot similar to Figure 1. While this is a simple way to represent the volatile content variation found in a given suite of melt inclusions, what cannot be shown on this plot is the bulk compositional variation of the melt inclusions themselves. Because of the observed variation in melt inclusion bulk compositions (Fig. 10; Kent 2008), and the strong effect of melt composition on solubility, which has been discussed above, modeled solubility curves such as CO_2-H_2O isobars, isopleths (variation at constant fluid composition), and degassing trends for a single composition on a CO_2-H_2O diagram are meaningless unless it can be shown that the compositional variation of the melt inclusion suite is small, and similar to that used in the model.

Melt inclusion volatile content data may be better examined if the compositional effects on the volatile solubility are accounted for. By plotting the calculated minimum saturation P using a compositionally dependent model such as Papale et al. (2006) against the measured values for CO_2 and H_2O in the melt inclusions (Fig. 13) it becomes possible to see the "P trajectory" of any given suite of inclusions, and to locate depth regions important to the fluid and melt evolution of the magma. This is difficult to accurately envision using the CO_2-H_2O plot, as the saturation P of each inclusion is dependent on its individual composition. With the melt inclusion data of Roggensack (2001a) for Cerro Negro volcano recast in this way, it becomes possible to identify important pressures where a majority of the sampled melt inclusions were trapped (assuming the system was saturated), and a more isobaric character (between 150-250 MPa) to the fluid-melt evolutionary history becomes apparent. Future work in this area will be to define the complicated behavior of degassing trends (open vs. closed system, etc) using compositionally dependent models such as Papale et al. (2006).

ACKNOWLEDGMENTS

This work was supported by N.S.F. grants EAR-0409863 and EAR-0610005. The thoughtful reviews of Rebecca Lange, Keith Putirka, Paul Wallace, and an anonymous reviewer greatly improved the manuscript. I would also like to thank Kurt Roggensack for many animated discussions about volatiles, melt inclusions, experiments, and what it all means. And much credit goes to John R. Holloway for introducing me to volatile solubility experiments so many years ago, and to Ian S.E. Carmichael who taught me that models can only be as good as the data they are based on.

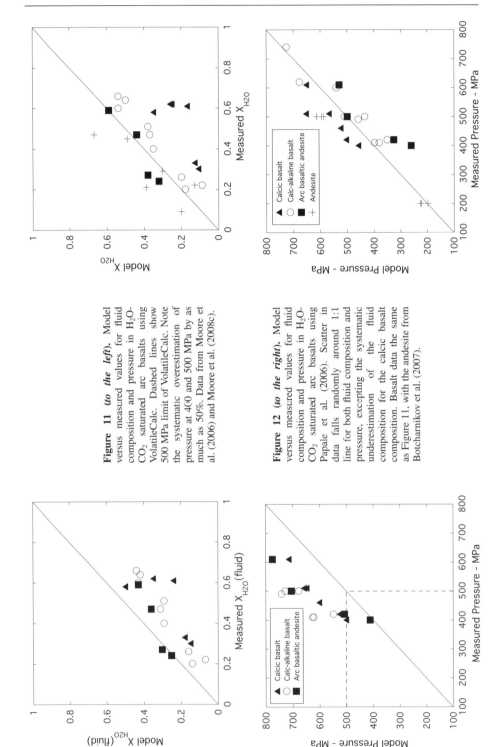

Figure 11 (*to the left*). Model versus measured values for fluid composition and pressure in H_2O-CO_2 saturated arc basalts using VolatileCalc. Dashed lines show 500 MPa limit of VolatileCalc. Note the systematic overestimation of pressure at 400 and 500 MPa by as much as 50%. Data from Moore et al. (2006) and Moore et al. (2008c).

Figure 12 (*to the right*). Model versus measured values for fluid composition and pressure in H_2O-CO_2 saturated arc basalts using Papale et al. (2006). Scatter in data falls randomly around 1:1 line for both fluid composition and pressure, excepting the systematic underestimation of the fluid composition for the calcic basalt composition. Basalt data the same as Figure 11, with the andesite from Botcharnikov et al. (2007).

H_2O and CO_2 Content in Melt Inclusions: Experiments & Modeling 357

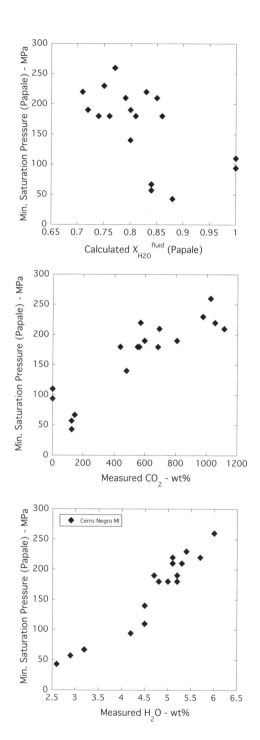

Figure 13. Calculated minimum saturation pressure vs. calculated fluid composition, measured CO_2, and H_2O content for melt inclusions from Cerro Negro (Roggensack 2001a). Saturation pressures calculated using Papale et al. (2006). By recasting the melt inclusion bulk compositions and volatile contents into saturation pressures, it becomes possible to more precisely identify pressure regions critically important to the fluid/melt evolution of the magma. For this case, note the essentially isobaric evolution of both the fluid and melt volatile composition between 250 and 150 MPa. Compare to the oft-used CO_2 vs. H_2O plot (e.g., Fig. 1A and 1B) where such behavior is difficult to discern.

REFERENCES

Anderson AT, Davis AM, Lu F (2000) Evolution of Bishop Tuff rhyolitic magma based on melt and magnetite inclusions, and zoned phenocrysts. J Petrol 41:449-473

Baker DR (2004) Piston-cylinder calibration at 400 to 500 MPa: A comparison of using water solubility in albite melt and NaCl melting. Am Mineral 89:1553-1556

Behrens H, Jantos N (2001) The effect of anhydrous composition on water solubility in granitic melts. Am Mineral 86:14-20

Behrens H, Meyer M, Holtz F, Benne D, Nowak M (2001) The effect of alkali ionic radius, temperature, and pressure on the solubility of water in $MAlSi_3O_8$ melts (M = Li, Na, K, Rb). Chem Geol 174:275-289

Behrens H, Ohlhorst S, Holtz F, Champenois M (2004a) CO_2 solubility in dacitic melts equilibrated with H_2O-CO_2 fluids: Implications for modeling the solubility of CO_2 in silicic melts. Geochim Cosmochim Acta 68:4687-4703

Behrens H, Romano C, Nowak M, Holtz F (1996) Near-infrared spectroscopic determination of water species in glasses of the system MalSi3O8 (M=Li, Na, K): an interlaboratory study. Chem Geol 128:41-63

Behrens H, Tamic N, Holtz F (2004b) Determination of the molar absorption coefficient for the infrared absorption band of CO_2 in rhyolitic glasses. Am Mineral 89:301-306

Blank JG, Brooker RA (1994) Experimental studies of carbon dioxide in silicate melts: solubility, speciation, and stable carbon isotope behavior. Rev Mineral 30:157-186

Blank JG, Stolper EM, Carroll MR (1993) Solubilities of carbon dioxide and water in rhyolitic melt at 850 °C and 750 bars. Earth Planet Sci Lett 119:27-36

Blundy J, Cashman K (2008) Petrologic reconstruction of magmatic system variables and processes. Rev Mineral Geochem 69:179-239

Botcharnikov RE, Behrens H, Holtz F (2006) Solubility and speciation of C-O-H fluids in andesitic melt at T=1100-1300 °C and P=200 and 500 MPa. Chem Geol 229:125-143

Botcharnikov RE, Holtz F, Behrens H (2007) the effect of CO_2 on the solubility of H_2O-Cl fluids in andesitic melt. Eur J Mineral 19:671-680

Bowen NL (1928) The evolution of igneous rocks. Princeton Univ Press, New Jersey

Brey G, Green D (1975) The role of CO_2 in the genesis of olivine melilitite. Contrib Mineral Petrol 49:93-103

Brey GP, Green DH (1973) CO_2 solubility and solubility mechanisms in silicate melts at high pressures. Contrib Mineral Petrol 49:93-103

Brooker RA, Kohn SC, Holloway JR, McMillan PF (2001) Structural controls on the solubility of CO_2 in silicate melts Part I: bulk solubility data. Chem Geol 174:225-239

Brooker RA, Kohn SC, Holloway JR, McMillan PF, Carroll MR (1999) Solubility, speciation, and dissolution mechanisms of CO_2 in melts along the $NaAlO_2$-SiO_2 join. Geochim Cosmochim Acta 63:3549-3565

Burnham (1994) Development of the Burnham model for prediction of H_2O solubility in magmas. Rev Mineral 30:122-129

Burnham CW (1981) The nature of multicomponent aluminosilicate melts. In: Chemistry and Geochemistry of solutions at high temperatures and pressures. Phys Chem Earth 13&14:191-227

Burnham CW, Davis NF (1971) The role of H_2O in silicate melts: I. P-V-T relations in the system $NaAlSi_3O_8$-H_2O to 10 kilobars, 700-1100 °C. Am J Sci 274:902-940

Burnham CW, Davis NF (1974) The role of H_2O in silicate melts: II. Thermodynamic and phase relations in the system $NaAlSi_3O_8$-H_2O to 10 kilobars and 1000 °C. Am J Sci 274:902-940

Burnham CW, Holloway JR, Davis NF (1969) Thermodynamic properties of water to 1000 °C and 10000 bars. Geol Soc Am Spec Paper 132:1-96

Burnham CW, Jahns RH (1962) A method for determining the solubility of water in silicate melts. Am J Sci 260:721-745

Carroll MR, Blank JG (1997) The solubility of H_2O in phonolitic melts. Am Mineral 82:549-556

Carroll MR, Holloway JR (eds) (1994) Volatiles in Magmas. Reviews in Mineralogy. Vol 30, Min Soc America, Fredericksburg, Virginia

Cervantes P, Wallace P (2003b) Magma degassing and basaltic eruption styles: a case study of ~2000 year BP Xitle volcano in central Mexico. J Volcanol Geotherm Res 120:249-270

Cervantes P, Wallace PJ (2003a) Role of H_2O in subduction-zone magmatism: New insights from melt inclusions in high-Mg basalts from central Mexico. Geology 31:235-238

Di Matteo V, Carroll MR, Behrens H, Vetere F, Brooker RA (2004) Water solubility in trachytic melts. Chem Geol 213:187-196

Dingwell DB, Holtz F, Behrens H (1997) The solubility of H_2O in peralkaline and peraluminous granitic melts. Am Mineral 82:434-437

Dixon JE (1997) Degassing of alkali basalts. Am Mineral 82:368-378

Dixon JE, Pan V (1995) Determination of the molar absorptivity of dissolved carbonate in basanitic glass. Am Mineral 80:1339-1342

Dixon JE, Stolper EM, Holloway JR (1995) An experimental study of water and carbon dioxide solubilities in mid-ocean ridge basaltic liquids. Part I: Calibration and solubility models. J Petrol 36:1607-1631

Elburg M, Kamenetsky V, Nikogosian I, Foden J, Sobolev A (2006) Coexisting high- and low-calcium melts identified by mineral and melt inclusion studies of a subduction-influenced syn-collisional magma from South Sulawesi, Indonesia. J Petrol 47:2433-2462

Flowers GC (1979) Correction of Holloway's (1977) adaptation of the modified Redlich-Kwong equation of state for calculation of the fugacities of molecular species in supercritical fluids of geologic interest. Contrib Mineral Petrol 69:315-318

Fogel RA, Rutherford MJ (1990) The solubility of carbon dioxide in rhyolitic melts: a quantitative FTIR study. Am Mineral 75:1311-1326

Ghiorso MS, Carmichael ISE, Rivers ML, Sack RO (1983) The Gibbs free energy of mixing of natural silicate liquids; an expanded regular solution approximation for the calculation of magmatic intensive variables. Contrib Mineral Petrol 84:107-145

Goranson RW (1931) The solubility of water in granitic magmas. Am J Sci 22:481-502

Hamilton DL, Burnham CW, Osborn EF (1964) Solubility of water and effects of oxygen fugacity and water content on crystallization in mafic magmas. J Petrol 5:21-39

Hauri E, Wang J, Dixon JE, King PL, Mandeville C, Newman S (2002) SIMS analysis of volatiles in silicate glasses 1. Calibration, matrix effects and comparisons with FTIR. Chem Geol 183:99-114

Hervig RL (1992) Ion probe microanalyses for volatile elements in melt inclusions. EOS Trans Am Geophys Union 73:367

Hervig RL, Williams P (1988) SIMS microanalyses of minerals and glasses for H and D. *In:* Secondary Ion Mass Spectrometry, SIMS VI. Benninghoven A, Huber AM, Werner HW (eds) J. Wiley & Sons, New York, p 961-964

Holloway JR (1977) Fugacity and activity of molecular species in supercritical fluids. *In:* Thermodynamics in Geology. Fraser D (ed) Reidel, Boston Mass, p 161-181

Holloway JR, Behrens H, Dingwell DB, Taylor RP (1992a) Water solubility in aluminosilicate melts of haplogranitic composition at 2 kbar. Chem Geol 96:289-302.

Holloway JR, Blank J (1994) Application of experimental results to C-O-H species in natural melts. Rev Mineral 30:187-230

Holloway JR, Dixon JE, Pawley AR (1992b) An internally heated, rapid-quench, high-pressure vessel. Am Mineral 77:643-646

Holloway JR, Wood BJ (1988) Simulating the Earth: Experimental Geochemistry. Unwin Hyman, Winchester Mass

Holtz F, Behrens H, Dingwell DB, Johannes W (1995) H_2O solubility in haplogranitic melts: Compositional, pressure, and temperature dependence. Am Mineral 80:94-108

Holtz F, Roux J, Behrens H, Pichavant M (2000) Water solubility in silica and quartzofeldspathis melts. Am Mineral 85:682-686

Hui H, Zhang Y, Xu Z, Behrens H (2008) Pressure dependence of the speciation of dissolved water in rhyolitic melts. Geochim Cosmochim Acta 72:3229-3240

Ihinger PD (1991) An experimental study of the intersection of water with granitic melt. PhD thesis, California Institute of Technology

Ihinger PD, Hervig RL, McMillan PF (1994) Analytical methods for volatiles in glasses. Rev Mineral 30:67-121

Jakobsson S (1997) Solubility of H_2O and CO_2 in an icelandite at 1400 °C and 10 kilobars. Contrib Mineral Petrol 127:129-135

Jakobsson S, Holloway JR (1986) Crystal-liquid experiments in the presence of a C-O-H fluid buffered by graphite + iron + wustite: Experimental method and near-liquidus relations in basanite. J Volcanol Geotherm Res 59:265-291

Jendrzejewski N, Trull TW, Pineau F, Javoy M (1997) Carbon solubility in Mid-Ocean Ridge basaltic melt at low pressures (250-1950 bar). Chem Geol 138:81-92

Johnson ER, Wallace P, Cashman KV, Delgado Granados H, Kent, AJR (2008) Magmatic volatile contents and degassing-induced crystallization at Volcan Jorullo, Mexico: Implications for melt evolution and the plumbing systems of monogenetic volcanoes. Earth Planet Sci Lett 269:477-486

Kadik AA, Lukanin OA (1973) The solubility-dependent behavior of water and carbon dioxide in magmatic processes. Geochem Int 10:115-129

Kadik AA, Lukanin OA, Lebedev YB, Kolovushkina EY (1972) Solubility of H_2O and CO_2 in granite and basalt at high pressures. Geochem Int 9:1041-1050

Kent AJR (2008) Melt inclusions in basaltic and related volcanic rocks. Rev Mineral Geochem 69:273-331

King PL, Holloway JR (2002) CO_2 solubility and speciation in intermediate (andesitic) melts: The role of H_2O and composition. Geochim Cosmochim Acta 66:1627-1640

King PL, Vennemann TW, Holloway JR, Hervig RL, Lowenstern JB, Forneris JF (2002) Analytical techniques for volatiles: A case study using intermediate (andesitic) glasses. Am Mineral 87:1077-1089

Kohn SC, Brooker RA (1994) The effect of water on the solubility and speciation of CO_2 in aluminosilicate glasses along the join SiO_2-$NaAlO_2$. Mineral Mag 58:489-490.

Kohn SC, Brooker RA, Dupree R (1991) ^{13}C MAS NMR: A method for studying CO_2 speciation in glasses. Geochim Cosmochim Acta 55:3879-3884

Lange RA (1994) The effect of H_2O, CO_2, and F on the density and viscosity of silicate melts. Rev Mineral 30:331-370

Larsen J, Gardner J (2004) Experimental study of water degassing from phonolite melts: implications for volatile oversaturation during magmatic ascent. J Volcanol Geotherm Res 134:109-124

Liu Y, Zhang Y, Behrens H (2005) Solubility of H_2O in rhyolitic melts at low pressures and a new empirical model for mixed H_2O-CO_2 solubility in rhyolitic melts. J Volcanol Geotherm Res 143:219-225

Mandeville CW, Webster JD, Rutherford MJ, Taylor BE, Timbal A, Faure, K (2002) Determination of molar absorptivities for infrared absorption bands of H_2O in andesitic glasses. Am Mineral 87:813-821

McMillan PF (1994) Water solubility and speciation models. Rev Mineral 30:131-156

Métrich N, Wallace PJ (2008) Volatile abundances in basaltic magmas and their degassing paths tracked by melt inclusions. Rev Mineral Geochem 69:363-402

Miyagi I, Yurimoto H, Takahashi E (1997) Water solubility in albite-orthoclase join and JR-1 rhyolite melts at 1000 °C and 500 to 2000 bars, determined by micro-analysis with SIMS. Geochem J 31:57-61.

Moore G, Carmichael ISE (1998) The hydrous phase equilibria (to 3 kbar) of an andesite and basaltic andesite from western Mexico: constraints on water content and conditions of phenocryst growth. Contrib Mineral Petrol 130:304-319

Moore G, Roggensack K, Hervig RL, Vennemann T (2008b) Effect of variable H_2O contents on calibrations of carbon measurements in basaltic glass using SIMS and FTIR. Geochem Geophys Geosyst (in prep)

Moore G, Roggensack K, Holloway J (2006) Dissolved carbonate species in mixed-volatile rhyolitic melts: carbon speciation correlates with dissolved H_2O content. EOS Trans AGU Fall Meet 87:52

Moore G, Roggensack K, Klonowski S (2008a) A low-pressure – high-temperature technique for the piston-cylinder. Am Mineral 93:48-52

Moore G, Roggensack K, Vennemann T (2005) Compositional variation in arc lavas and their associated melt inclusions: Implications for magmatic volatile contents and degassing behavior. EOS Trans AGU Fall Meet 86:52

Moore G, Roggensack K, Vennemann T, Hervig RL (2008c), Mixed (H_2O + CO_2) volatile solubility in calcic to calc-alkaline basalts from 400 to 700 MPa. Contrib Mineral Petrol (in prep)

Moore G, Vennemann T, Carmichael ISE (1995) Solubility of water in magmas to 2 kbar. Geology 23:1099-1102

Moore G, Vennemann T, Carmichael ISE (1998) An empirical model for the solubility of H_2O in magmas to 3 kilobars. Am Mineral 83:36-42

Mysen B, Arculus RJ, Eggler DH (1975) Solubility of carbon dioxide in natural nephelenite, tholeiite and andesite melts to 30 kbar pressure. Contrib Mineral Petrol 53:227-239

Mysen BO (1987) Magmatic silicate melts: Relations between bulk composition, structure, and proppertites. *In*: Magmatic Processes: Physicochemical Principles. Mysen BO (ed) Geochemical Society, University Park, Pennsylvania, 375-399

Newman S, Epstein S, Stolper E (1988) Water, carbon dioxide, and hydrogen isotopes in glasses from the ca. 1340 A.D. eruption of the Mono craters, California: constraints on degassing phenomena and initial volatile contents. J Volcanol Geotherm Res 35:75-96

Newman S, Lowenstern JB (2002) VolatileCalc: a silicate melt-H_2O-CO_2 solution model written in Visual Basic for excel. Comp Geosci 28:597-604

Ochs FA, Lange RA (1999) The density of hydrous magmatic liquids. Science 283:1314-1317

Pan V, Holloway JR, Hervig RL (1991) The pressure and temperature dependence of carbon dioxide solubility in tholeiitic basalt melts. Geochim Cosmochim Acta 55:1587-1595

Papale P (1997) Modeling of the solubility of a one-component H_2O or CO_2 fluid in silicate liquids. Contrib Mineral Petrol 126:237-251

Papale P (1999) Modeling of the solubility of a two-component H_2O + CO_2 fluid in silicate liquids. Am Mineral 84:477-492

Papale P (2005) Determination of total H_2O and CO_2 budgets in evolving magmas from melt inclusion data. J Geophys Res 110:B03208

Papale P, Moretti R, Barbato D (2006) The compositional dependence of the saturation surface of H_2O + CO_2 fluids in silicate melts. Chem Geol 229:78-95

Pineau F, Shilobreeva S, Kadik A, Javoy M (1998) Water solubility and D/H fractionation in the system basaltic andesite-H_2O at 1250 °C and between 0.5 and 3 kbars. Chem Geol 147:173-184

Roggensack K (2001a) Sizing up crystals and their melt inclusions: a new approach to crystallization studies. Earth Planet Sci Lett 187:221-237

Roggensack K (2001b) Unraveling the 1974 eruption of Fuego volcano (Guatemala) with small crystals and their young melt inclusions. Geology 29:911-914

Roggensack K, Moore G (2008) Explosive arc basalts: H2O-CO2 solubility experiments and melt inclusion evidence for a possible magmatic gas origin of the Tiscapa maar, Nicaragua. Geology (in prep)

Romano C, Dingwell DB, Behrens H, Dolfi D (1996) Compositional dependence of H_2O solubility along the joins $NaAlSi_3O_8$-$KAlSi_3O_8$, $NaAlSi_3O_8$-$LiAlSi_3O_8$, and $KAlSi_3O_8$-$LiAlSi_3O_8$. Am Mineral 81:452-461

Schmidt BC, Holtz F, Pichavant M (1999) Water solubility in haplogranitic melts coexisting with H_2O-H_2 fluids Contrib Mineral Petrol 136:213-224

Shilobreyeva SN, Kadik AA (1990) Solubility of CO_2 in magmatic melts at high temperatures and pressures. Geochem Int 27:31-41

Silver L, Stolper E (1985) A thermodynamic model for hydrous silicate melts. J Geol 93:161-178

Silver LA, Ihinger PD, Stolper EM (1990) The influence of bulk composition on the speciation of water in silicate glasses. Contrib Mineral Petrol 104:142-162

Sisson TW, Layne GD (1993) H_2O in basalt and basaltic andesite glass inclusions from four subduction-related volcanoes. Earth Planet Sci Lett 117:619-635

Sparks RSJ, Barclay J, Jaupart C, Mader HM, Phillips JC (1994) Physical aspects of magmatic degassing I. Experimental and theoretical constraints on vesiculation. Rev Mineral 30:413-446

Spilliaert N, Allard P, Métrich N, Sobolev A (2006) Melt inclusion record of the conditions of ascent, degassing, and extrusion of volatile-rich alkali basalt during the powerful 2002 flank eruption of Mount Etna (Italy). J Geophys Res 111:B04203

Stolper E (1982) The speciation of water in silicate melts. Geochim Cosmochim Acta 46:2609-2620

Stolper E (1989) Temperature dependence of the speciation of water in rhyolitic melts and glasses. Am Mineral 74:1247-1257

Stolper E, Fine G, Johnson T, Newman S (1987) Solubility of carbon dioxide in albitic melt. Am Mineral 72:1071-1085

Tamic N, Behrens H, Holtz F (2001) The solubility of H_2O and CO_2 in rhyolitic melts in equilibrium with a mixed CO_2-H_2O fluid. Chem Geol 174:333-347

Thibault Y, Holloway JR (1994) solubility of CO_2 in a Ca-rich leucitite: effects of pressure, temperature and oxygen fugacity. Contrib Mineral Petrol 116:216-224

Wade JA, Plank T, Melson WG, Soto GJ, Hauri E (2006) The volatile content of magmas from Arenal volcano, Costa Rica. J Volcanol Geotherm Res 157:94-120

Wallace P (2005) Volatiles in subduction zon magmas: concentrations and fluxes based on melt inclusion and volcanic gas data. J Volcanol Geotherm Res 140:217-240

Wallace PJ, Anderson AT, Davis AM (1995) Quantification of pre-eruptive exsolved gas contents in silicic magmas. Nature 377:612-616.

Wyllie PJ, Tuttle OF (1959) Effect of carbon dioxide on the melting of granite and feldspars. Am J Sci 257:548-655

Yamashita S (1999) Experimental study of the effect of temperature on water solubility in natural rhyolite melt to 100 MPa. J Petrol 40:1497-1507

Zhang Y (1999) H_2O in rhyolitic glasses and melts: measurement, speciation, solubility, and diffusion. Rev Geophys 37:493-516

Zhang Y (2007) Silicate melt properties and volcanic eruptions. Rev Geophys 45:RG4004

Zhang Y, Jenkins J, Xu Z (1997) Kinetics of the reaction $H_2O + O = 2OH$ in rhyolitic glasses upon cooling: geospeedometry and comparison with glass transition. Geochim Cosmochim Acta 61:2167-2173

Zhang Y, Xu Z (2007) A long-duration experiment on hydrous species geospeedometer and hydrous melt viscosity. Geochim Cosmochim Acta 71:5226-5232

Zhang Y, Xu Z, Behrens H (2000) Hydrous species geospeedometer in rhyolite: Improved calibration and application. Geochim Cosmochim Acta 64:3347-3355

Volatile Abundances in Basaltic Magmas and Their Degassing Paths Tracked by Melt Inclusions

Nicole Métrich

*Laboratoire Pierre Sue
CNRS-CEA, CE-Saclay
Gif sur Yvette, 91191, France*

nicole.metrich@cea.fr

Paul J. Wallace

*Department of Geological Sciences
University of Oregon
Eugene, Oregon, 97403-1272, U.S.A.*

pwallace@uoregon.edu

INTRODUCTION

The abundances of CO_2, H_2O, S and halogens dissolved in basaltic magmas are strongly variable because their solubilities and ability to be fractionated in the vapor phase depend on several parameters such as pressure, temperature, melt composition and redox state. Experimental and analytical studies show that CO_2 is much less soluble in silicate melts compared to H_2O (e.g., Javoy and Pineau 1991; Dixon et al. 1995). As much as 90% of the initial CO_2 dissolved in basaltic melts may be already degassed at crustal depths, whereas H_2O remains dissolved because of its higher solubility such that H_2O contents of basaltic magmas at crustal depths may reach a few percents. Most subduction-related basaltic magmas are rich in H_2O (up to 6-8 wt%; Sisson and Grove 1993; Roggensack et al. 1997; Newman et al. 2000; Pichavant et al. 2002; Grove et al. 2005) compared to mid-ocean ridge basalts (<1 wt%; Sobolev and Chaussidon 1996; Fischer and Marty 2005; Wallace 2005).

During magma movement towards the surface, exsolution of major volatile constituents (CO_2, H_2O) causes gas bubble nucleation, growth, and possible coalescence that exert a strong control on the dynamics of magma ascent and eruption (Anderson 1975; Sparks 1978; Tait et al. 1989). Gas bubbles have the ability to move faster than magma (Sparks 1978), particularly in low viscosity basaltic magmas. Bubble accumulation, coalescence and foam collapse give rise to differential transfer of gas slugs and periodic gas bursting (Strombolian activity; Jaupart and Vergniolle 1988, 1989) or periodic lava fountains (Vergniolle and Jaupart 1990; Philips and Wood 2001) depending on magma physical properties and ascent rate. It is also thought that strombolian and lava fountain activities critically depend on the magma rise speed, the transition between the two styles being controlled by the bubble diameter for the same magma ascent rate (Wilson and Head 1981; Parfitt and Wilson 1995; Parfitt 2004). A model of gas segregation derived from laboratory experiments (<10 vol% gas fraction) was recently applied to Stromboli suggesting that the transition between episodic explosive eruption of gas-poor, relatively dense magma and more explosive, gas-rich, low viscosity magma would take place for critical values of the magma discharge rate of 0.1-1 $m^3 s^{-1}$ (Menand and Philips 2007).

Whether the magma rises with its gas bubbles entrained (closed-system degassing) or the gas bubbles are able to segregate from the melt (open-system degassing) further affects gas compositions and dissolved volatile contents of ascending melts. Accordingly, the abundances of H_2O, S, and halogens in the melt and the composition of the exsolving gas phase will change in response to the pressure-dependent solubility and diffusivity of each volatile component.

Water is commonly the most abundant volatile constituent and has a direct influence on magma properties and rheology. It drastically affects the silicate melt density because of its relatively large partial molar volume (17-25 cm^3/mol) when dissolved (between 750 and 1250 °C, 0-1 GPa) compared to other silicate melt components (Ochs and Lange 1999). The effect of 1 wt% dissolved H_2O on the density of basaltic melt would be equivalent to increasing the temperature by ~400 °C or decreasing the pressure by ~500 MPa (Ochs and Lange 1999). Dissolved H_2O also has a strong effect on melt viscosity (e.g., Shaw 1972; Richet et al. 1996; Giordano and Dingwell 2003), such that loss of H_2O during ascent causes melts to become increasingly viscous. Water exsolution and loss also exert a main control on the melt crystallization (temperature and sequence of mineral phase appearance). The effect of H_2O on phase equilibria is such that H_2O loss enhances crystallization, particularly of plagioclase, whereas adding H_2O to a melt depresses plagioclase crystallization in basaltic systems (e.g., Sisson and Grove 1993; Métrich and Rutherford 1998; Grove et al. 2003; Di Carlo et al. 2006; Pichavant and Macdonald 2007). Finally, increasing the H_2O content of a melt also increases the diffusivities of other volatiles such as S (Freda et al. 2005) and halogens (Aletti et al. 2007). Thus the amount of H_2O dissolved in magmas is one of the key parameters that must be known to model physical properties and crystallization paths during ascent.

Assessing the volatile abundances in basaltic magmas, their degassing paths during ascent, and the magmatic volatile budgets of active volcanoes has been a major objective of many studies in the last 15 years. Because of strong pressure-dependent solubilities, the volatile components are almost totally lost from erupted volcanic products, and therefore melt droplets (melt or glass inclusions) entrapped during crystal growth have been widely used for addressing these questions. Recent developments in microanalyses of volatile elements in melt inclusions and glasses have led to large data sets on CO_2 and H_2O, which complement data for other volatile species (Cl, S, F), trace elements and isotopes.

In this chapter we first review information on how melt inclusions form, the analytical techniques for quantitative determination of volatile concentrations, and the results of experimental solubility studies of volatiles. Second, we provide a general summary of volatile abundances in basaltic magmas, volatile behavior during magma ascent, and vapor exsolution pressures on the basis of melt inclusions hosted in olivine crystals from natural basaltic samples. Third, we examine the relationships between melt inclusion data, eruption styles, and volcanic gas fluxes. Finally, we discuss systematics of the primary or initial volatile contents in basaltic magmas from different tectonic environments. Based on our review of the extensive literature on these topics, we conclude by summarizing a few key questions for future research.

WHAT ARE MELT INCLUSIONS AND HOW DO THEY FORM?

Primary melt inclusions form in crystals when some process interferes with the growth of a perfect crystal, causing a small volume of melt to become encased in the growing crystal. This can occur from a variety of mechanisms, including (1) skeletal or other irregular growth forms due to strong undercooling or non-uniform supply of nutrients, (2) formation of reentrants by resorption followed by additional crystallization, and (3) wetting of the crystal by an immiscible phase (e.g., sulfide melt or vapor bubble) or attachment of another small crystal (e.g., spinel on olivine) resulting in irregular crystal growth and entrapment of that phase along with silicate melt (Sobolev and Kostyuk 1975; Roedder 1984; Lowenstern 1995). Most melt inclusions contain only silicate melt at the time of trapping, but this single phase may exsolve one or more vapor bubbles and/or crystallize daughter crystals post-entrapment as a result of cooling

(Lowenstern 1995). Melt inclusions formed by the third mechanism described above contain multiple phases at the time of trapping, though additional vapor exsolution or crystallization may occur post-entrapment (Roedder 1984). Magma mixing may cause multiple populations of primary melt inclusions to be present in phenocrysts from a single volcanic rock or tephra layer (Anderson 1976).

Because mafic melt inclusions hosted in olivine phenocrysts are the main focus of this review, we summarize recent experimental data that bear on the formation of inclusions in this mineral. Having constraints on the olivine texture and melt inclusion formation are suitable before an accurate interpretation of volatile evolution. Building on an extensive experimental literature on the morphology of olivine (e.g., Donaldson 1976), there have been several recent studies (Faure et al. 2003; Faure and Schiano 2004, 2005) in which dynamic crystallization experiments involving forsterite in the CMAS system have been compared with natural olivine and melt inclusion morphologies. These studies found 5 different crystal morphologies at increasing degrees of undercooling: polyhedral, tabular, skeletal (hopper), dendritic (swallowtail), and feather shape. More than one morphological type was commonly produced in a single experiment. The summary below is based on these studies.

Polyhedral crystals form at relatively low cooling rates (~1 to 2 °C/h) and have well defined crystal faces. Experimentally grown polyhedral crystals commonly contained spherical or elongated melt inclusions (Fig. 1A). Natural melt inclusion-bearing olivines commonly show similar morphology, providing evidence that at least some inclusions form by slow, near-equilibrium growth (Fig. 1B). Slowly grown, polyhedral olivines also show embayed textures (Faure and Schiano 2005); such textures in natural olivines are commonly interpreted to be the result of disequilibrium caused by resorption, but the experimental results show that these form during near-equilibrium growth.

Tabular shape, in which crystals are rectangular in sections parallel to [010], are a less common form in dynamic crystallization experiments, and the crystals do not contain trapped melt inclusions. Olivine evolves from tabular to skeletal (hopper) to dendritic (swallowtail) with increasing degree of undercooling (Faure and Schiano 2005). Evolution from tabular to skeletal (hopper) is particularly important in forming the U-shaped cavities in some hopper crystals, as described below (Faure et al. 2003).

Skeletal crystals (hopper morphology) form at higher cooling rates and degrees of undercooling and appear either as hexagonal or hourglass shapes depending on the plane of orientation. A characteristic feature of hopper crystals is that the ends of the prismatic crystal created in the experiments have large glass-filled cavities. In larger crystals, these cavities have a U-shaped profile, whereas in smaller crystals they are V-shaped (Faure et al. 2003). These cavities can become sealed by additional olivine growth if such crystals formed by rapid cooling are subjected to short (~5 min.), isothermal annealing. This results in a closed, skeletal olivine in which two melt inclusions are symmetrically distributed with regard to the center of the crystal (Fig. 2A). Longer isothermal annealing times (several hours) cause the cavities to become completely filled by the growth of olivine (Faure et al. 2003; Faure and Schiano 2005). Olivine crystals with pairs of melt inclusions similar in morphology to the experimentally grown, closed skeletal crystals with U-shaped cavities occur in Kilauea olivines (Fig. 2B). Melt inclusions with an hourglass morphology were described from Paricutin volcano (Luhr 2001). This morphology is similar to closed skeletal crystals with V-shaped cavities (Fig. 2B), but in some cases growth was sufficiently rapid to leave a central channel in the crystal unfilled such that the two cavities at the ends of the prism are connected (Luhr 2001).

More complex olivine and inclusion textures are created in closed dendritic olivine, which are formed when initial rapid cooling forms a swallowtail olivine followed by isothermal annealing to form a final morphology that appears polyhedral (Fig. 3A; Faure and Schiano 2005). This produces a well faceted crystal with abundant melt inclusions showing preferred orienta-

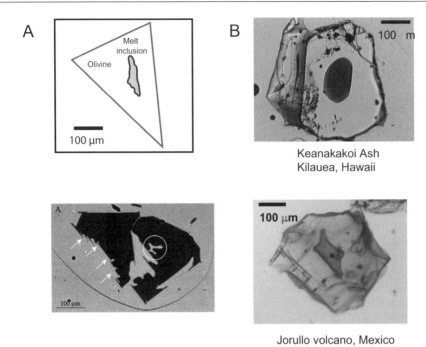

Figure 1. Comparison of (A) experimentally produced and (B) natural polyhedral olivines and their melt inclusions. (A) Upper left: sketch showing a polydedral forsterite crystal with a large rounded glass inclusion near the center (based on SEM image of an experimental olivine produced at a cooling rate 2 °C/hr; Faure et al. 2003). Lower left: backscattered electron image of forsterite crystal showing polyhedral morphology with large embayment and skeletal (stair-like) overgrowths (marked by arrows) produced in a second stage of more rapid cooling. [Photo used by permission of Elsevier Limited from Faure and Schiano (2005) *Earth Planet Sci Lett*, Vol. 236, Fig. 8A, p. 890.] (B) Upper right: photomicrograph of small polyhedral olivine with large, central, partly faceted melt inclusion from the Keanakakoi Ash Member, Kilauea Volcano, Hawaii. Lower right: similar to above, but with large, central irregular inclusion. Olivine is from Jorullo Volcano, Mexico.

tion. Natural examples of such textures are an indication of the complex cooling histories by which some olivines form (Figs. 3B, C).

Compositional boundary layers can form adjacent to growing crystals due to competition between diffusion in the silicate melt and crystal growth. A primary concern in melt inclusion studies is whether the trapped melt accurately represents the bulk melt surrounding the crystal at the time of entrapment or has been affected by boundary layer enrichment or depletion (Roedder 1984; Baker 2008). In the dynamic crystallization experiments described above (Faure and Schiano 2005), polyhedral olivines contained melt inclusions that were similar to the parental melt regardless of inclusion size (Fig. 4). For such crystals, there were no compositional gradients in the glass surrounding the olivine, suggesting that growth was controlled by interface attachment rather than diffusion processes. In contrast, more rapid growth morphologies (skeletal, dendritic) contain inclusions that are moderately to strongly enriched in Al_2O_3 compared to the parental melt (Fig. 4). Because both CaO and Al_2O_3 are incompatible in olivine, *in situ* crystallization in the experiments should cause correlated increases in both elements. The anomalously high Al_2O_3 contents in melt inclusions in rapidly grown olivines and the presence of diffusional gradients in glass adjacent to the experimental crystals suggests that Al_2O_3 builds up in the boundary layer because it has a much lower

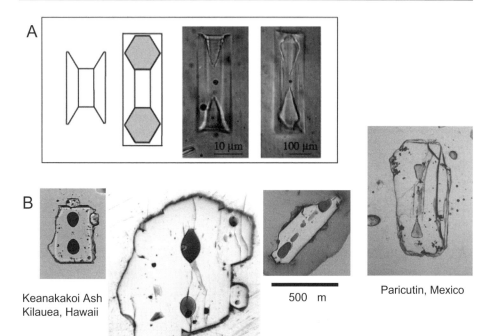

Figure 2. Comparison of (A) experimentally produced and (B) natural hopper (skeletal morphology) olivines and their melt inclusions. (A) Formation of closed skeletal olivine with symmetrically distributed melt inclusions. Sections are parallel to the (010) plane. The two photomicrographs on the upper right show evolution from hopper (cooling rate 422 °C/hr) to closed hopper crystal (cooling rate 437 °C/hr; then charge was reheated for 5 min. below the liquidus temperature. [Photos used by permission of Elsevier Limited from Faure and Schiano (2005) *Earth Planet Sci Lett*, Vol. 236, Fig. 11A, p. 895.] (B) Photomicrographs of closed-hopper morphology in olivine phenocrysts from the Keanakakoi Ash Member, Kilauea Volcano (3 photos on left), and Paricutin Volcano, Mexico (photo on right).

diffusivity than CaO in silicate melts. Baker (2008) demonstrates that P_2O_5 has very low diffusivity in silicate melts as well, and shows that the ratio Cl/P_2O_5 is useful as an indicator of whether melt inclusions have been affected by boundary layer enrichment processes.

The dynamic crystallization experiments provide a framework for testing whether natural melt inclusions have been affected by boundary layer enrichment processes by comparing CaO and Al_2O_3 in inclusions to bulk rock or tephra and matrix glass. Shown in Figure 5A are CaO and Al_2O_3 data for olivine-hosted melt inclusions from high-K minettes and basanites from the Colima Graben in western Mexico (Vigouroux et al. 2008). The olivine hosts are mostly polyhedral, and the inclusions vary from ellipsoidal to irregular in shape. Most of the melt inclusions have CaO/Al_2O_3 ratios that are similar to values for the bulk tephra (Fig. 5A). Although a few inclusions have anomalous CaO/Al_2O_3 ratios, these are more likely the result of augite fractionation. The data show no evidence for the strongly enriched Al_2O_3 values seen in melt inclusions in rapidly grown experimental forsterite crystals in the CMAS system (cf. Fig. 4). Similarly, melt inclusions from Etna scoriae show no evidence of anomalously high Al_2O_3 (Spilliaert et al. 2006a). Interestingly, olivine-hosted melt inclusions from Paricutin Volcano, many of which have closed skeletal forms indicating rapid crystal growth (Luhr 2001), also do not show any anomalous behavior of Al_2O_3 (Fig. 5B).

The results summarized above show that compositions of natural olivine-hosted melt inclusions are partly at odds with the results of dynamic crystallization experiments in that rapid

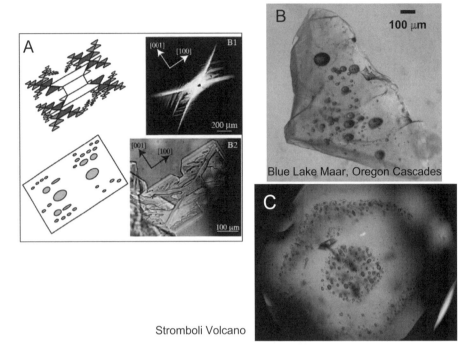

Figure 3. Comparison of (A) experimentally produced and (B) natural dendritic polyhedral olivines and their melt inclusions. (A) Swallowtail olivine that is reheated below the liquidus temperature develops a morphology similar to polyhedral olivine, but with numerous melt inclusions showing preferential orientations. [Photo used by permission of Elsevier Limited from Faure and Schiano (2005) *Earth Planet Sci Lett*, Vol. 236, Fig. 11B, p. 895.] (B) Photomicrograph of olivine crystal from Oregon Cascades showing numerous melt inclusions with preferred orientation, similar to experimental textures in (A). (C) Olivine crystal from Stromboli showing concentric zones rich in small melt inclusions showing preferred orientations. This likely represents rapid cooling or H_2O-loss induced pulses of swallowtail growth followed by slow crystallization.

growth morphologies do not show any Al_2O_3 enrichment compared to bulk sample values. One possibility is that the presence of several wt% H_2O and/or additional melt components such as Fe increase the diffusivity of components such as Al_2O_3 in the melt. While we suggest that all melt inclusion data sets should be scrutinized in terms of elements that are incompatible in olivine, the lack of Al_2O_3 enrichment in natural olivine-hosted melt inclusions suggests that H_2O, which diffuses faster than Al_2O_3, should not have been affected by boundary layer enrichment processes during melt inclusion formation. This interpretation is consistent with recent modeling results (Baker 2008). Diffusivities of H_2O and CO_2 in mafic melts have only been measured at a limited set of compositions and temperatures (Baker et al. 2005; Zhang et al. 2007). Recent experimental studies of S and Cl diffusion in silicate melts show that in nominally anhydrous basaltic melts, Cl diffusivity is about 10 times lower than that of H_2O, similar to that of CO_2, and a factor of 5 higher than that of S (Freda et al. 2005; Alletti et al. 2007). The presence of H_2O increases the diffusivity of Cl by no more than a factor of 3 (Alletti et al. 2007). Based on available data, models of incompatible element build-up in a boundary layer adjacent to a growing crystal suggest that trapped basaltic melt inclusions may, if formed by sufficiently rapid growth, be enriched in S, Cl, and CO_2 as a result of boundary-layer effects, but the enrichments are likely to be on the order of 10-30% or less (Baker 2008).

Different diffusivities for S and Cl should result in some scattering of S/Cl ratios in melt inclusions. However, many basaltic melt inclusion data sets show no significant variations in

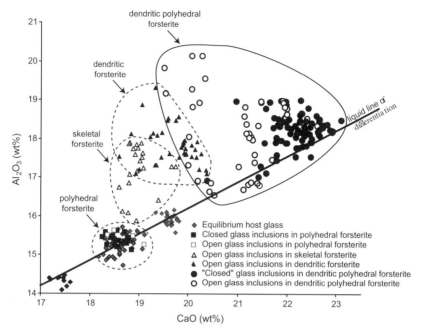

Figure 4. CaO vs. Al_2O_3 showing compositions of melt inclusions in polyhedral, dendritic polyhedral, skeletal and dendritic forsterite produced experimentally in the CMAS system. Both CaO vs. Al_2O_3 are incompatible during forsterite growth. The initial melt composition is shown by the solid diamonds in the lower left, and the final quenched residual melts are shown by the gray diamonds. [Figure used by permission of Elsevier Limited from Faure and Schiano (2005), *Earth Planet Sci Lett*, Vol. 236, Fig. 4, p. 887.]

the S/Cl ratio (before S degassing or sulfide immiscibility) for inclusions having contrasting morphologies indicative of different crystallization rates for their host olivines. This is illustrated by average values of the S/Cl ratios ($\pm 1\sigma$) in H_2O-rich, oxidized basaltic inclusions of Vesuvius (S/Cl = 0.34±0.05; Marianelli et al. 2005); Stromboli (S/Cl =1.1±0.1; Métrich et al. 2005), and Etna (2.0±0.2; Spilliaert et al. 2006a). Instead, diffusion might significantly affect the S/Cl ratio in melt and gas during dynamic degassing controlled by bubble growth.

POST-ENTRAPMENT MODIFICATION OF MELT INCLUSIONS

A number of processes can potentially modify the compositions of melt inclusions after they are trapped inside their crystal host (e.g., Danyushevsky et al. 2002). Most published studies on post-entrapment modification are based on melt inclusions in olivine, and we therefore review mechanisms of post-entrapment change and the ensuing limitations to data interpretation (see also Kent 2008). During post-entrapment cooling, crystallization of the included melt occurs along the melt–crystal interface, depleting the melt in constituents that enter the crystalline phase and enriching it in elements incompatible in the crystal. For olivine-hosted melt inclusions, post-entrapment crystallization of olivine causes a strong depletion in MgO of the included melt, but other major elements are much less affected (Danyushevsky et al. 2000). Diffusive exchange can also occur between the melt and crystal. This particularly affects the FeO^T contents of many melt inclusions in olivine (Danyushevsky 2001), but both CaO (Gaetani and Watson 2000, 2002) and trace element abundances (Cottrell et al. 2002) can also be modified.

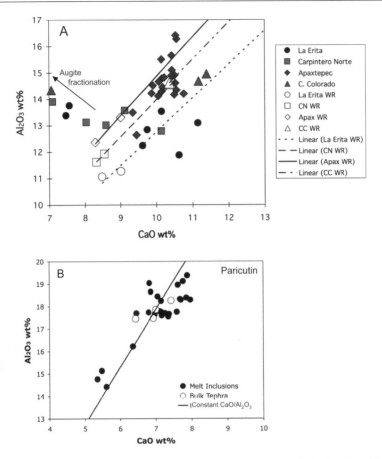

Figure 5. CaO vs. Al_2O_3 showing compositions of natural olivine-hosted melt inclusions from (A) high-K minette and basanite tephra from cinder cones near Volcan Colima in the western Trans-Mexican Belt (data from Vigouroux et al. 2008) and (B) basaltic andesite tephra from Paricutin volcano, Mexico (data from Luhr 2001). Solid and dashed lines show constant CaO/ Al_2O_3 ratios for the bulk tephra samples corresponding to each of the melt inclusion data sets. Some of the inclusions in (A) show evidence for augite fractionation, but none of the inclusions show strong enrichments like those found in many of the CMAS melt inclusions in Figure 4.

Of primary concern for studies of volatiles in basaltic systems is the potential for post-entrapment diffusive loss of volatiles from melt inclusions through the olivine host. During 1-atm heating experiments on olivine-hosted melt inclusions, systematic increases in homogenization temperature occur over time (Sobolev et al. 1983) in response to the deformation of the host crystal (Massare et al. 2002). FTIR measurements have verified that this is coupled with a decrease in hydroxyl and molecular H_2O concentrations in the inclusions, and almost total dehydration of inclusions can occur within a few hours at high temperature (Massare et al. 2002). It is likely that the dehydration is caused by rapid proton diffusion through the olivine due to the very low H_2 fugacity of the environment surrounding the olivine in the heating experiments (Danyushevsky et al. 2002), or preferential diffusion of water-derived species (OH, molecular H_2O) through olivine dislocations and lattice defects (Massare et al. 2002). Hydrogen loss causes oxidation of Fe in the melt inclusion that results in precipitation of magnetite in a characteristic magnetite dust texture (Danyushevsky et al. 2002). Loss of water from olivine-hosted melt inclusions by proton diffusion either during experimental heating studies or in slowly cooled lava flows causes anomalously high D/H ratios because of the more rapid diffusion of H relative to D (Hauri 2002).

The amount of H_2O that can be lost from an inclusion by hydrogen diffusion was suggested to be limited to ≤ 1 wt% H_2O by the amount of FeO^T in the inclusion and the stoichiometry of the oxidation reaction (Danyushevsky et al. 2002), but larger H_2O losses have been documented experimentally, suggesting that other mechanisms may also operate (Massare et al. 2002). Recent experiments have shown that melt inclusions with initially low H_2O concentrations (<1 wt%) can gain substantial H_2O in as little as 2 days at magmatic temperature when surrounded by more H_2O-rich melt (Portnyagin et al. 2008). In these experiments, the addition of H_2O to the system changes the mineral-melt equilibria leading to dissolution of pre-existing olivine and MgO enrichment in the melt. These results have been suggested to indicate that molecular H_2O is a rapidly diffusing species through the olivine and that melt inclusions can either gain or lose H_2O depending on changes in the magmatic environment surrounding the crystal (e.g., loss of H_2O during ascent and degassing or increases in H_2O during isobaric fractional crystallization).

There is no evidence, as far as we are aware, for diffusive loss of CO_2, Cl, or S from melt inclusions (e.g., Portnyagin et al. 2008). However, diffusive loss of Fe from melt inclusions can lead to exsolution of a sulfide globule in the inclusion and consequent decrease in dissolved S content of the silicate melt (Danyushevsky et al. 2002).

Another important post-entrapment modification affecting melt inclusions is the formation of shrinkage bubbles (Roedder 1984). During post-entrapment (but pre-eruptive) cooling, the much greater thermal contraction of the melt relative to the crystal host results in formation of a vapor bubble (shrinkage bubble) that can significantly deplete the melt in CO_2 because of its low solubility (Anderson and Brown 1993; Cervantes et al. 2002). For olivine-hosted melt inclusions, a decrease in temperature of 100 °C is estimated to create a shrinkage bubble that is 0.9 to 1.6 vol% of the inclusion (Anderson and Brown 1993; Riker 2005). Further expansion of the bubble occurs during eruptive cooling, but this happens over such a short timescale that little to no additional volatiles are likely to diffuse from the melt to the bubble (Anderson and Brown 1993). In experiments on olivines (Fo_{88}) from a Mauna Loa picrite, reheating to 1400 °C for <10 minutes rehomogenized the shrinkage bubble (≤3 vol%) into the melt, but even with rapid quenching a vapor bubble commonly formed during quench (Cervantes et al. 2002). Analyses of the rehomogenized melt inclusions by FTIR showed that ~80% of the total CO_2 that was dissolved in the melt at the time of trapping had been lost to the shrinkage bubbles during an ~100 °C cooling interval between entrapment and eruption. Smaller cooling intervals would result in less loss of CO_2 to shrinkage bubbles. Basaltic melt inclusions in Hawaiian olivines that experience ≤40 °C cooling between entrapment and eruption generally do not contain shrinkage bubbles (Riker 2005).

During ascent and eruption, volatiles can be lost if the host crystal ruptures around an inclusion or if an inclusion is not fully sealed in the host. The resulting depressurization in the inclusion forms one or more vapor bubbles, and such inclusions can typically be recognized by thin films of glass that form where melt is injected into the cracks. Slow cooling of melt inclusions in lava flows or domes causes crystallization of the trapped melt, and such inclusions must be experimentally reheated for analysis (Danyushevsky et al. 2002). However, additional volatiles can be lost if the inclusion host is cracked or if there is diffusion of H or molecular H_2O through the host, and these lost volatiles cannot be restored to the melt through reheating and rehomogenization.

Given the potential for volatile loss from partially to fully crystallized melt inclusions from slowly cooled lava flows and domes, it is preferable to analyze naturally glassy melt inclusions from rapidly cooled tephra deposits (e.g., Anderson and Brown 1993; Roggensack et al. 1997; Métrich et al. 2001) or submarine samples. When analysis of rehomogenized melt inclusions is necessary, short heating times (preferably ≤ 15 min) are required to prevent H_2O loss by diffusion (Danyushevsky et al. 2002; Massare et al. 2002) or decreases in melt S and changes in S^{6+}/S_{total} due to oxidation of the inclusion (Danyushevsky et al. 2002; Rowe et al. 2007). Finally, it should be remembered that even naturally glass melt inclusions from tephra or submarine samples have the potential to have lost or gained some H_2O by diffusion through the olivine host during pre-eruptive storage or ascent. Cases where this has been unequivocally documented are rare (e.g., Portnyagin et al. 2008).

ANALYTICAL TECHNIQUES FOR VOLATILES

The main microanalytical techniques that are now routinely used to analyze the major volatiles in silicate melt inclusions are secondary ion mass spectrometry (SIMS or ion microprobe; H_2O, CO_2, S, Cl, F), Fourier Transform Infrared (FTIR) Spectroscopy (H_2O, CO_2), electron microprobe (S, Cl, F, S^{6+}/S^{2-} ratio), and confocal micro-Raman spectroscopy (H_2O). In addition, a nuclear microprobe technique has been used to analyze C in glasses.

A thorough review of the theory and application of SIMS and FTIR for analysis of volatiles can be found in Ihinger et al. (1994). More recent discussions of calibration issues, matrix effects, and comparison of SIMS and FTIR for analysis of glasses is given by Hauri et al. (2002) and Blundy and Cashman (2008). An advantage of SIMS relative to FTIR is that melt inclusions to be analyzed only need to be exposed on one side of the host crystal, but the SIMS technique is destructive whereas FTIR is not. Another advantage of SIMS for volatile analysis is that all major volatiles (H_2O, CO_2, S, Cl, F) together with trace elements can be analyzed on the same instrument. A comparison of SIMS, FTIR, and bulk extraction by hydrogen manometry for andesitic glasses can be found in King et al. (2002). A summary of sample preparation issues and the applications of SIMS and FTIR to volatile analysis in melt inclusions can be found in Nichols and Wysoczanski (2007).

An essential part of analysis of H_2O and CO_2 by FTIR is determination of absorption (or extinction) coefficients for the species of interest. The fundamental OH stretching vibration that forms an assymetric band with maximum around 3570 cm^{-1} is commonly used to measure total H_2O in a sample. This band has an absorption coefficient that is only weakly dependent on major element composition (Dixon et al. 1995), in contrast to the bands for molecular H_2O (5200 and 1630 cm^{-1}) and hydroxyl (4500 cm^{-1}; Dixon et al. 1995; King et al. 2002; Mandeville et al. 2002), which have absorption coefficients that more strongly depend on bulk composition. For basaltic melt inclusions with less than ~4 wt% dissolved H_2O, using the band at 3570 cm^{-1} insures better precision on H_2O measurements. The absorption coefficient for carbonate ion in mafic glasses is also strongly dependent on major element glass composition (Dixon and Pan 1997).

For quantitative determination of H_2O and CO_2 by transmission FTIR, the thickness of the melt inclusion-bearing wafer must be measured precisely, and this is normally done either with a high-precision digital micrometer or by optically measuring the thickness under a microscope (e.g., Wallace et al. 1999). A microprofilometer can also be used for this (Roggensack 2001). Spectroscopic methods for determining the wafer thickness include a method based on ratioing the absorbances at different regions of the background (Agrinier and Jendrzejewski 2000) and using the spacing of interference fringes in reflectance spectra taken on the host crystal (Wysoczanski and Tani 2006). Recently, a technique has been developed for analyzing small melt inclusions that are fully enclosed inside a wafer but not actually intersected at the wafer surface (Nichols and Wysoczanski 2007).

A technique for using confocal micro-Raman spectroscopy to measure H_2O in melt inclusions has recently been developed and calibrated (Thomas 2000; Behrens et al. 2006;

Di Muro et al. 2006). Besides being non-destructive, like FTIR, a major advantage of this technique is that inclusions do not have to be double-face polished and in transparent crystals they can be analyzed *in situ* without exposure to surface (Thomas et al. 2006). This feature and the very small spatial resolution (1-2 μm) make it possible to analyze a large number of inclusions within a single host crystal. With proper calibration, this technique is very promising for analyzing H_2O in glasses ranging from basaltic to rhyolitic in composition (Mercier et al. 2008). For most compositions the minimum detection limit is assessed to be ~0.1 wt% of H_2O_T, or even lower in Fe-poor glasses.

Carbon concentrations can be determined using the $^{12}C(d, p)^{13}C$ nuclear reaction (e.g., Varela et al. 2000; Spilliaert et al. 2006a). The samples are irradiated with an incident deuteron beam of 1.45 MeV and the protons detected between 2.520 and 3.036 MeV behind a 9 μm thick Al screen with a surface barrier detector (1500 μm depleted depth). Any surface contamination is identified by a high energy peak. At 1.45 MeV incident energy the main interferences are due to (d, p) nuclear reactions with the major constituents of the matrix such as ^{24}Mg, ^{28}Si, ^{27}Al. In order to take into account the matrix effect, the background is measured on C-free silicate glasses. In silicate glasses (density = 2.6 to 2.7 g/cm^3), the analyzed depth is 8-9 μm. The minimum detection limit for carbon varies from 40 to 15 ppm as the integrated charge increases from 1 to 3 °C. The uncertainty (1σ) using reference basaltic glasses containing 300 and 100 ppm of carbon is 10%.

The electron microprobe is commonly used to analyze Cl, S and F in melt inclusions, as well as the major element composition of inclusions and their host crystals. Technical details regarding analysis of F and Cl in glasses by electron microprobe are reviewed by Carroll and Webster (1994). Sulfur occurs in multiple valence states (S^{2-}, S^{6+}) in glasses, and the wavelength of S Kα radiation changes as a function of the mean oxidation state (Carroll and Rutherford 1988; Wallace and Carmichael 1994; Jugo et al. 2005a). Therefore it is important to measure the appropriate peak position to use during the electron microprobe analysis of S in silicate glasses, and the peak position used for glass samples should be different from that used to analyze sulfide or sulfate standards used for calibration. More details regarding the S Kα peak shift technique and interpretation of the results in terms of oxygen fugacity are discussed by Kent (2008).

Major element data for melt inclusions are often used to estimate the amount of H_2O in the inclusion by the difference of the major element total (including S, Cl and F) from 100%. However, electron beam radiation causes permanent damage to hydrous glasses during electron microprobe analysis because alkali ions migrate away from the surface of the irradiated sample (Humphreys et al. 2006). If not properly corrected for, electron microprobe analyses of hydrous glasses will result in values of Na_2O that are far too low, and this effect seriously degrade estimates of H_2O by difference. Methods for accurate analysis of Na_2O and H_2O by difference in hydrous glasses are discussed in Blundy and Cashman (2008).

VOLATILE SOLUBILITIES AND VAPOR MELT PARTITIONING

Volatile components occur as dissolved species in silicate melts, but they can also be present in an exsolved vapor phase if a melt is vapor saturated. In laboratory experiments it is possible to saturate melts with a nearly pure vapor phase (e.g., H_2O saturated), though the vapor always contains at least a small amount of dissolved solute. In natural systems, however, multiple volatile components are always present, and data from volcanic gases, melt inclusions, and submarine pillow rim glasses show that the major volatiles present in magma are H_2O, CO_2, S, Cl and F. When the sum of the partial pressures of all dissolved volatiles in a silicate melt equals the confining pressure, the melt becomes saturated with a multicomponent (C-O-H-S-Cl) vapor phase (Verhoogen 1949). Referring to natural magmas as being H_2O

saturated or CO_2 saturated is thus, strictly speaking, incorrect because the vapor phase is never pure and always contains more than one volatile component.

Earlier experimental studies on volatile solubility in silicate melts generally focused on melts saturated with a single volatile component, but more recent studies on basaltic to andesitic melts have examined solubility relations when multiple volatile components are present (Luhr 1992; Dixon et al. 1995; Webster et al. 1999; King and Holloway 2002; Botcharnikov et al. 2007; Moore 2008). Though experimentally difficult, it is particularly useful to measure the composition of the vapor phase coexisting with vapor-saturated melt so that vapor-melt partitioning of volatiles can be determined as a function of temperature, pressure, melt composition, and oxygen fugacity. When compared with experimental solubility studies, the data on dissolved H_2O and CO_2 concentrations in olivine-hosted melt inclusions that we review in this chapter provide strong evidence that basaltic magmas are vapor-saturated in the middle to upper crust. In this section, we review recent experimental solubility data and thermodynamic models useful for understanding vapor saturation and degassing in basaltic systems.

For H_2O and CO_2, the maximum solubility in a silicate melt is controlled by saturation with the pure vapor phase. In contrast, for S, Cl, and F, the maximum solubility is limited by saturation with a solid or immiscible liquid phase such as anhydrite, Fe-S-O liquid, alkali chloride melt, or alkali fluoride melt (e.g., Luhr 1990; Wallace and Carmichael 1992; Carroll and Webster 1994; Webster et al. 1999; Jugo et al. 2005a).

Because H_2O is typically the most abundant volatile component in natural melts, and H_2O and CO_2 together exert the largest contribution towards the total vapor pressure, many experimental studies have focused on the solubility behavior of these two volatiles. H_2O dissolves in silicate melts to form both OH^- and molecular H_2O species (Silver et al. 1990), and CO_2 dissolves in basaltic melts as carbonate ions (Fine and Stolper 1986). At pressures up to 1 kbar (100 MPa), both dissolved carbonate and molecular H_2O obey Henry's Law solubility behavior such that vapor saturation isobars on a diagram of H_2O vs. CO_2 show a negative diagonal trend (Fig. 6; Dixon and Stolper 1995). This makes it possible to estimate the vapor saturation pressures for melt inclusions using H_2O-CO_2 data (assuming the partial pressures of the other volatile components are small in comparison), and these values can be equated to the trapping pressure of the inclusion if it formed from vapor-saturated melt. However, the solubilities of both H_2O (Moore et al. 1998) and CO_2 (Dixon 1997) in silicate melts are also dependent on melt composition as well as temperature, so these effects must also be taken into account in making vapor saturation pressure calculations. Solubility, degassing and vapor-melt partitioning of H_2O and CO_2 can be calculated for nephelinitic to basaltic and rhyolitic melts with the VolatileCalc program (Newman and Lowenstern 2002), which uses thermodynamic solubility models calibrated with experimental data. Software for performing similar calculations using a different solubility model (Papale et al. 2006), is now available (*http://ctserver.ofm-research.org/Papale/Papale.php*). The accuracy of these models is evaluated in light of new H_2O-CO_2 solubility data by Moore (2008).

For basaltic melts with oxygen fugacities near or below the NNO buffer, S saturation is controlled by an immiscible Fe-S-O liquid phase (Wallace and Carmichael 1992). Dissolved S concentrations of sulfide-liquid-saturated melts show a strong positive dependence on melt FeO and temperature and a negative dependence on pressure (Wallace and Carmichael 1992; Mavrogenes and O'Neill 1999; O'Neill and Mavrogenes 2002; Liu et al. 2007). In these systems, the saturation concentration of S in the melt is also a strong function of the ratio of network modifying to network forming cations (Liu et al. 2007). There has been little experimental investigation of the effect of dissolved H_2O on S concentration of sulfide-liquid-saturated melts, but Liu et al. (2007) report experimental data on an Etna basalt at 1 GPa in which increasing H_2O from ~0 to 4 wt% caused a decrease in melt S content by about 50%.

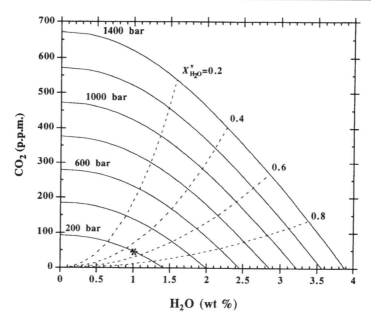

Figure 6. Inverse correlation of CO_2 and H_2O in vapor-saturated basaltic melt at various constant pressures and 1200 °C. Solid curves are each for a different constant pressure (vapor saturation isobars). Dashed curves show isopleths of constant vapor composition. Both the isopleths and isobars are calculated using a thermodynamic solubility model calibrated with experimental data (Dixon and Stolper 1995). This figure can be used to determine the pressure and composition of the vapor phase in equilibrium with a basaltic melt that is vapor saturated. [Figure used by permission of Oxford University Press, from Dixon and Stolper (1995), *J Petrol*, Vol. 36, Fig. 1, p. 1635.]

In more oxidized melts, maximum dissolved S is controlled by saturation with either anhydrite (Carroll and Rutherford 1987; Luhr 1990) or immiscible sulfate melt (Jugo et al. 2005a). Melt S concentrations are much higher for sulfate-saturated melts than for sulfide-liquid-saturated melts (Fig. 7), and dissolved S shows a strong positive dependence on both temperature and pressure but much less dependence on melt composition. Measurements of S Kα wavelength in glasses show that the dramatic increase of S solubility with oxygen fugacity reflects a change from dominantly S^{2-} in the melt to dominantly S^{6+} (Carroll and Rutherford 1988; Métrich and Clocchiatti 1996; Jugo et al. 2005b). It has been shown that the presence of S^{4+} species in natural oxidized glass inclusions and low pressure silicate glasses is an analytical artifact caused by electron microprobe beam damage (Métrich et al. 2003) or by irradiation with the synchrotron X-ray beam during analysis (Wilke et al. 2008). Recent X-ray absorption spectroscopy determination of the sulfur oxidation state in experimental glasses and inclusions points to the fact that sulfur in reduced basaltic magmas is possibly oxidized upon cooling (Métrich et al. 2008).

Empirical and thermodynamic models for prediction of S concentrations in sulfide-saturated melts include models calibrated for anhydrous basaltic melts (Holzeid and Grove 2002; O'Neill and Mavrogenes 2002; Li and Ripley 2005), and anhydrous to hydrous basaltic to rhyolitic melts (Moretti and Ottonello 2005; Scaillet and Pichavant 2005; Liu et al. 2007). These models were compared by Liu et al. (2007), who concluded that their model predicted S concentrations at sulfide saturation in hydrous melts better than the models calibrated for anhydrous systems. They also found that the Scaillet and Pichavant (2005) model was more successful at predicting S concentrations at sulfide saturation in Fe-poor rhyolitic melts, but

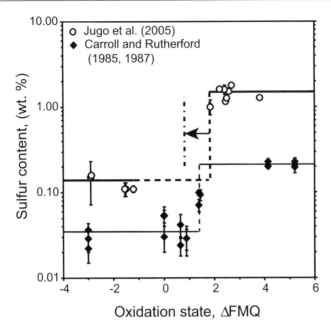

Figure 7. Experimentally determined S concentration as a function of oxygen fugacity in basaltic melts at 1300 °C and 1 GPa and trachyandesite melts (with <17 wt% FeO) at 200 MPa and 927-1027 °C. The arrow indicates the expected shift in the stability of sulfate with decreasing silica activity (Sisson 2003). [Figure used by permission of Oxford University Press, from Jugo et al. (2005b), *J Petrol*, Vol. 46, Fig. 3, p. 794.]

less successful for other melt compositions. One advantage of the Scaillet and Pichavant (2005) model for use in melt inclusion studies is that it explicitly includes the relationship between melt S content and f_{S_2}. This makes it possible to calculate the mole fractions of H_2S and SO_2 in the vapor phase in equilibrium with melt if the oxygen fugacity, temperature, pressure and H_2O content are known.

Experimental studies show that Cl is highly soluble in basaltic melts, with maximum dissolved values as high as 2.9 wt% for H_2O-poor melts saturated with molten NaCl at 2 kbar (200 MPa) and 1080-1210 °C (Webster et al. 1999). With increasing H_2O in the system, the molten NaCl contains some dissolved H_2O and is therefore best described as a hydrosaline melt. The dissolved Cl content does not vary much for melts containing as much as ~4 wt% H_2O (Fig. 8). At dissolved H_2O contents of 5 to 6 wt%, the melt is in equilibrium with an H_2O-Cl vapor phase rather than hydrosaline melt. Within the resolution of the data, it appears that there is no invariant point where hydrosaline melt and H_2O-Cl vapor coexist, but rather that there is complete miscibility between the two phases. Olivine-hosted basaltic melt inclusions generally have Cl concentrations << 1 wt%, indicating that the melts would have been saturated with Cl-bearing vapor rather than dense hydrosaline melt during storage, ascent and degassing.

The maximum Cl contents of andesitic melts determined experimentally under similar conditions as those described above are slightly lower (2.5 wt% Cl) than for basaltic melt (Botcharnikov et al. 2007). In contrast to the basalt experiments, however, the H_2O-Cl data for andesitic melts suggests coexistence of hydrosaline melt and H_2O-Cl vapor for melts with high dissolved H_2O and Cl, and the addition of CO_2 to the vapor appears to increase the extent of the immiscibility gap between the two phases (Botcharnikov et al. 2007). This behavior, though, is unlikely to affect natural andesitic melts because, like their basaltic counterparts, they generally do not contain high enough Cl to reach saturation with hydrosaline melt.

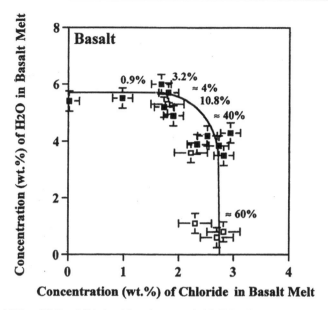

Figure 8. Solubilities of H_2O and Cl in basaltic melt saturated with Cl-bearing aqueous vapor or hydrosaline melt at 200 MPa and 1040-1170 °C. Numbers next to data points give the measured wt% Cl in the vapor or hydrosaline melt; numbers preceded by ≈ represent calculated values that may have very large errors. Filled squares are melts with 9-10.9 wt% FeO and open squares are melts with 6-8.1 wt% FeO. Silicate melt + vapor are stable along the horizontal part of the curve, and silicate melt + hydrosaline melt are stable along the vertical part. Vapor and hydrosaline melt are inferred to be completely miscible along the curved portion. [Figure used by permission of Elsevier Limited, from Webster et al. (1999), *Geochim Cosmochim Acta*, Vol. 63, Fig. 2, p. 733.]

A detailed review of experimental solubility studies of F in silicate melts can be found in Carroll and Webster (1994).

H_2O AND CO_2 ABUNDANCES IN BASALTIC MAGMAS: CRYSTALLIZATION DEPTHS, MAGMA ASCENT AND DEGASSING PROCESSES

H_2O-CO_2 abundances in basaltic magmas

Direct assessment of the dissolved amounts of H_2O and CO_2 in magmas comes from melt inclusions, although experimental phase equilibria also provide strong constraints on magmatic H_2O contents. A large literature is dedicated to basaltic melt inclusions, specifically hosted in olivine, because this phase is usually present along the overall crystallization paths of mafic magmas. The possible post-trapping evolution of olivine-hosted melt inclusions is also relatively well studied (see section on post-entrapment modifications). In contrast, systematic studies on basaltic melt inclusions in clinopyroxene are scarce.

Available data on CO_2 contents of basaltic melt inclusions from arc settings indicate a limited range of concentrations (0-≤0.25 wt%; Fig. 9A). Dissolved CO_2 contents >0.3 wt% have only rarely been reported in basaltic melt inclusions, whereas primary basaltic melts are inferred to have high CO_2 contents up to 0.6 wt% in MORB (Fisher and Marty 2005), 0.7 wt% at Kilauea (Gerlach et al. 2002), and possibly as high as 0.6-1.3 wt% in arcs (Wallace 2005). Similarly, olivine-hosted melt inclusions from Etna contain ≤0.4 wt% CO_2 (Métrich et al. 2004), whereas an original CO_2 content of ~1.5 wt% for Etna primary magma (Allard

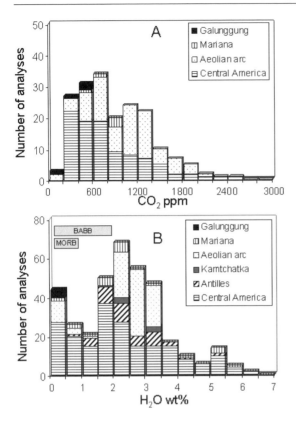

Figure 9. Compilation of CO_2 (A) and H_2O (B) concentrations in melt inclusions from subduction-related basalts and basaltic andesites. Note that CO_2 concentrations <2000 ppm and most of the low H_2O concentrations are likely the result of partial degassing before inclusion entrapment (see text). Data from Central America [Fuego (Roggensack 2001; Sisson and Layne 1993); Popocatepelt and Colima (Atlas et al. 2006); Cerro Negro (Roggensack 2001); Irazù (Benjamin et al. 2007); Arenal (Wade et al. 2006); BVF and Pacaya (Walker et al. 2003); Paricutin (Lurh 2001); Mexico (Cervantes and Wallace 2003a,b; Johnson et al. 2008)]; Lesser Antilles (Bouvier et al. 2008); Kamchatlka (Portnyagin et al. 2007); Aeolian arc [Stromboli (Métrich et al. 2001; Bertagnini et al. 2003); Vulcano (Gioncada et al. 1998)]; Mariana (Newman et al. 2000); and Galunggung (Sisson and Bronto 1998).

et al. 1997) was inferred from the time-averaged CO_2/S ratio in summit plume emissions and the S content of the most primitive melt inclusions. Olivine-hosted melt inclusions from high-K minettes and basanites from the Colima Graben in western Mexico contain ≤0.53 wt% CO_2 (Vigouroux et al. 2008). Primary magmas may therefore be vapor saturated at crustal depths (Anderson 1995; Wallace 2005), and exsolved CO_2-rich gas bubbles would enhance the magma buoyancy and thus its eruptive style as discussed below.

Water concentrations recorded by melt inclusions in subduction settings show a broad range from 0.2 to 6 wt% (Fig. 9B) that overlaps with values for back arc basin basalts (BABB). In comparison, basaltic melt inclusions from non-arc settings average 0.12 and 0.5 wt% H_2O, respectively, for N-MORB and E-MORB (Sobolev and Chaussidon 1996), and vary mostly between 0.2 and 0.8 wt% in Hawaiian basalts (Anderson and Brown 1993; Wallace and Anderson 1998). Some lower values for Hawaiian basalts have probably been affected by H diffusive loss from melt inclusions in slowly cooled lava flows (Hauri 2002).

The large range of H_2O contents in arc basaltic melt inclusions results from several factors: variable degassing at middle to upper crustal depths before inclusion entrapment, H_2O increase during magma differentiation, and ultimately, variable H_2O content in the magma source region. As an example of the effects of differentiation, basaltic andesite melt inclusions from Fuego volcano (Sisson and Layne 1993) contain some of the highest H_2O concentrations (~6 wt%) in arc magmas but have less than 5 wt% of MgO and are trapped in Fe-rich olivine (Fo~74). An example of the effects of low pressure degassing before inclusion entrapment comes from Stromboli scoriae, in which olivine-hosted melt inclusions contain low H_2O (<0.5 wt%; Métrich et al. 2001).

Focusing on melt inclusions that best represent magmatic values prior to degassing, the H_2O contents of subduction-related basaltic magmas varies from 0.2-0.3 wt% (Galunggung, Indonesia arc; Sisson and Bronto 1998) to 5.2 wt% (Central Mexico; Cervantes and Wallace 2003a). These inclusions were trapped in Mg-rich olivines (Fo≥88) and contain high concentrations of S (~0.2 and 0.6 wt%, respectively) and CO_2 (0.075 and 0.2 wt%, respectively). They likely record the primitive signature of magma prior to significant degassing of H_2O and S. Additional arguments support this conclusion. In Galunggung samples, melt inclusions have an H_2O content that remains identical and low (0.2-0.3 wt%) while their CO_2 content decreases from 0.075 wt% down to a few ppm upon decompression. Moreover, their basaltic composition is very similar to that of the host magmas, even though melt inclusions suffered some post-entrapment crystallization and Fe-loss (Sisson and Bronto 1998). Low H_2O concentrations (0.2 wt%) have also been reported in high-alumina basaltic melt inclusions hosted in Fo~88 olivine from Black Crater basalt in the Cascade arc (Medicine Lake volcano; Sisson and Layne 1993). In the latter case there is also agreement between low H_2O content of melt inclusion and their MORB-like composition. Similarly, low H_2O contents (0.1-0.6 wt%) associated with low Cl (0.02-0.03 wt%) have been reported in picritic to basaltic melt inclusions in Southern Mariana arc olivines (Fo_{89-91}). These low values have been interpreted as evidence for adiabatic decompression melting of a relatively anhydrous but metasomatized mantle wedge (Kohut et al. 2006).

In contrast, inclusions recording high H_2O concentrations from Central Mexico basalts show the strongest enrichments of large ion lithophile elements (LILE) relative to high field strength elements (HFSE), which is a characteristic of subduction-related magmas worldwide. This pattern is consistent an origin related to mantle fluxing by an H_2O-rich component derived from the subducted slab (Cervantes and Wallace 2003a). Hence there is consistency between the H_2O content and the magma geochemical signature as discussed in the final section below.

A large range in H_2O concentrations is exemplified at the scale of the Mariana arc where primitive magmas are believed to contain 1-3 wt% H_2O prior to degassing (Newman et al. 2000). A similar range is also illustrated at the scale of a single volcano at Mt. Shasta, where the pre-eruptive water contents of magmas range from <1 to possibly as high as 8-10 wt% based on melt inclusion studies and experimental petrology (Grove et al. 2002, 2005).

As deduced from melt inclusion data reported in the literature, the H_2O contents of subduction-related basaltic magmas, prior to extensive degassing, is between <1 and 5.2 wt%, after data filtering (excluding MgO < 5 wt%). It includes boninites (1.7 wt% on average) and island arc tholeiites (1.6 wt% on average; Sobolev and Chaussidon 1996).

Pressures of crystallization and magma degassing

Estimation of vapor saturation pressures. Because of the low solubility of CO_2 in basaltic magmas (e.g., Dixon et al. 1995), most of them are saturated with CO_2-rich vapor at high pressure. This is in agreement with the larger CO_2 concentrations in the magma sources as deduced from volcanic and hydrothermal CO_2 emissions and fluxes compared to estimates based on CO_2 dissolved in melt inclusions (e.g., Fisher and Marty 2005; Wallace 2005). Accordingly, most olivine-hosted melt inclusions are probably trapped under vapor-saturated conditions. Evidence of vapor saturation can come from the occurrence of heterogeneous melt/gas trapping in olivines, the presence of CO_2-vapor bubbles and the inability of melt inclusions to be experimentally homogenized. Assuming vapor saturation and limited loss of CO_2 to shrinkage bubbles, the total vapor (or fluid) pressure ($P_{CO_2}+P_{H_2O}$) at the time of melt entrapment and, therefore, the corresponding depth of magma ascent and degassing, can be assessed from the CO_2 and H_2O contents dissolved in melt inclusions. These pressures are commonly calculated using the program VolatileCalc (Newman and Lowenstern 2002). However, the vapor saturation pressures calculated for H_2O-rich basalts can be overestimated by using this program at higher pressures because high H_2O content strongly enhances CO_2 solubility in basaltic melt by stabilizing carbonate species at pressure ≥400 MPa (Botcharnikov et al. 2005; Moore 2008). In fact, only a few pressure estimates for melt inclusions apparently reach 500 MPa (as reported by Roggensack et al. 1997; Walker et al. 2003 for Central America; Cervantes and Wallace 2003a and Vigouroux et al. 2008, for Mexico). Instead, most of the vapor saturation pressures for basaltic melt entrapment in olivine range from 300-400 to a few MPa, at which pressures the effect of water on CO_2 solubility would be less significant. An evaluation of the accuracy of VolatileCalc and the method of Papale et al. (2006) is given by Moore (2008).

Open-system vs. closed-system degassing at basaltic volcanoes. Models for degassing of magmas saturated with H_2O+CO_2 vapor predict that the CO_2/H_2O ratio in the vapor phase is higher than that of the coexisting melt, and this ratio in the melt decreases rapidly during degassing (Newman et al. 2000).

Under open-system conditions, when the vapor is extracted from the melt (Rayleigh distillation), the CO_2 concentration in the melt decreases with negligible loss of H_2O until the melt reaches the vapor saturation pressure for pure H_2O. An example of open-system degassing is illustrated by Mariana Trough samples (Newman et al. 2000). Basaltic melt inclusions hosted in olivine contain up to 0.09 wt% CO_2 and their host glasses have as little as 0.02 wt% CO_2 whereas the water contents of inclusions and host glasses are similar (2.23±0.7 vs. 2.14±0.13 wt%) as are their chemical compositions.

However, closed-system degassing—where the melt remains in equilibrium with its coexisting gas phase (Dixon et al. 1995)—is suggested by melt inclusion data from several volcanoes. In such a situation, exsolution of H_2O into the gas phase is more significant and ultimately induces magma crystallization, as illustrated by the following examples. An example of H_2O loss and crystallization during magma ascent under closed-system conditions is given by the basaltic-andesite scoria from the 1723 eruption of Irazù in Costa Rica (Benjamin et al. 2007). In that case, the H_2O-CO_2 degassing path is calculated under closed-system conditions assuming that the magma was coexisting with 2 wt% (~10 vol%) exsolved gas at ~150 MPa (Fig. 10A). As the pressure decreased down to <20 MPa, and the temperature decreased by 30 °C, the olivines became richer in FeO and their melt inclusions in K_2O. The overall story of crystallization and magma degassing seems rather simple, although the olivine compositional range (Fo_{87-79}) is quite large and brackets the equilibrium olivine composition calculated for the host lava (Fo_{83}). All olivines have normal zoning and rims of Fo_{78}. The most primitive melt inclusions trapped in olivine Fo_{87} record the highest CO_2 (445 ppm) and H_2O (3.1 wt%) concentrations and thus the highest vapor saturation pressure of melt entrapment (125-150 MPa).

Samples from prehistoric activity of Arenal in Costa Rica also illustrate coupled magma degassing and polybaric crystal fractionation (Wade et al. 2006). The most mafic inclusions have a high alumina basaltic composition considered to be the closest to parental Arenal magmas. They show the highest H_2O (3.9 wt%), CO_2 (<0.03 wt%) and sulfur (0.3 wt%) contents and the highest entrapment pressures (~200 MPa). Decreasing H_2O contents (down to 1.1 wt%) correlate broadly with host olivine composition (Fo_{79-76}) and extent of crystallization, as indicated by increasing SiO_2 and K_2O in the melt inclusions. Overall, they record mafic magma ascent and degassing from 200 to 20 MPa, under apparent closed-system conditions that were simulated with 1 wt% exsolved gas and crystal fractionation.

Despite the good fit of the Irazu and Arenal melt inclusion data to closed-system degassing models, we note that the required initial exsolved gas contents of 2 and 1 wt%, respectively, require either a much higher initial CO_2 content for the magma (on the order of 1 to 2 wt% CO_2) or some open-system process of gas enrichment. In the following section, we discuss the growing body of evidence for open-system addition of gas to ascending magmas.

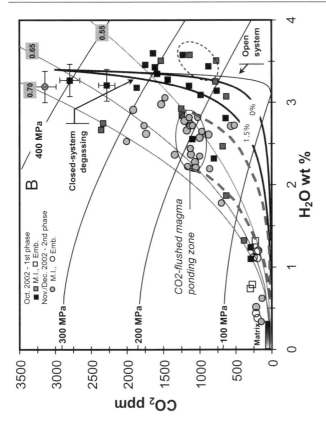
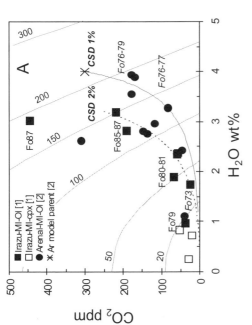

Figure 10. Diagram showing the variation of the CO_2 and H_2O concentrations in melt inclusions in (A) olivine and clinopyroxene from, respectively, the 1763 and 1963-5 eruptions of Irazù volcano ([1] Benjamin et al. 2007) and olivine from prehistoric tephra layers of Arenal ([2] Wade et al. 2006). The variations observed in the diagrams are by far larger than the analytical error (see [1] and [2] for details on error and accuracy of SIMS analyses). For each sample set the composition of the host olivine is indicated as Fo mol%. The closed-system degassing curves were calculated using VolatileCalc (Newman and Lowenstern 2002), assuming starting conditions of 49 wt% SiO_2, 1100 °C and 2 wt% exsolved vapor for Irazù samples (Benjamin et al. 2007) and 1 wt% exsolved vapor phase for Arenal samples (Wade et al. 2006); (B) from the 2002 flank eruption of Mt Etna (Spilliaert et al. 2006a). Degassing path of the basaltic melt containing 3.4 ± 0.2 wt % dissolved H_2O at 400 MPa is calculated in either open system (OSD) or closed system (CSD) with 0 and 1.5 wt % initial gas phase. Isobars, degassing trends and curves of equilibrium between the melt and a vapor phase of fixed composition (isopleths) are calculated using VolatileCalc (Newman and Lowenstern 2002). Data points for the first stage, in particular samples emitted during lava fountains at the onset of the eruption (black squares), fit with closed system magma degassing with 1.5 wt % initial gas phase. Melt inclusions having a large bubble (dashed contour domain) suffered CO_2-diffusion from the melt towards the bubble during cooling. Most of the erupted magma is represented by the cluster of inclusions (second phase) trapped at between 220 and 170 MPa that depart from the CSD trend. Variable magma dehydration was regarded as the possible result of deep CO_2 flushing of magma during its ponding at an average pressure of 200 MPa (~ 8 km depth). This latter process can also account for a water depleted inclusions trapped at pressures of 300 MPa. Abbreviations are: M.I. for melt inclusions and Emb. for embayments.

Deviation from CO_2-H_2O equilibrium degassing trends. Melt inclusion H_2O and CO_2 data from some volcanoes show variations that cannot be explained by either open- or closed-system degassing. Such features are exemplified by melt inclusions hosted in olivine (Fo_{80-82}) from scoriae emitted during explosive activity of the 2002 flank eruption at Mt Etna (Spilliaert et al. 2006a). Most of the melt inclusions were trapped at pressures ranging from ~400 to 200 MPa (Fig. 10B). In a first population of melt inclusions, the CO_2 and H_2O variations follow a closed-system degassing path that was described above as being the result of decompression of the primitive basalt already coexisting with 1.5 wt% vapor phase at ~400 MPa. Closed-system ascent of this basaltic magma gave rise to lava fountains at the onset of the eruption. In contrast, melt inclusions from samples representative of the largest volume of trachybasaltic magma delivered during the second phase of the 2002 eruption were trapped in olivine within a narrow pressure range from 220 to 170 MPa. The inclusions delineate a "water-depletion" trend where H_2O decreases (from 2.9 to 2.2 wt%) while CO_2 weakly varies (1140-1050 ppm). The range of H_2O concentrations is larger than the analytical error (6-8% relative). This trend deviates from that of closed-system degassing unless the vapor amount reaches 15 to 20 wt%, which is physically implausible. A process of mixing between the H_2O-rich primitive ascending melt that was extruded previously, at the onset of the eruption, and already degassed magma was rejected because of their high sulfur (0.29-0.32 wt%) and Cl (0.15-0.16 wt%). The selective water-depletion in Etna basalt at high pressure was interpreted as resulting from the re-equilibration of the magma with a deeper-derived CO_2-rich gas phase, during ponding and crystallization of magma at ~200±20 MPa (equivalent to 8 km depth). In order to buffer a gas phase with 65 mol% of CO_2 in equilibrium with the melt inclusions, either 5 wt% of pure CO_2 or ≥8 wt% of an H_2O-CO_2 mixture with ≤0.85 mol% CO_2 would have to have been added to the ponding magma (Spilliaert et al. 2006a). Continuous CO_2 flushing of Etna's plumbing system is consistent with the high CO_2 content (~1.5 wt%) inferred in primary Etna magma (e.g., Allard et al. 1997). It was proposed that early CO_2-melt separation may thus be a common process at Mt Etna, except in the case of very fast magma ascent, which allows closed-system rise of bubble-melt mixtures.

Another example where fluxing of CO_2-rich gas appears to have played an important role is the historic eruption (1759-1774) of Jorullo volcano in central Mexico (Johnson et al. 2008). Jorullo melt inclusions trapped some of the most volatile-rich (≤5.3 wt% H_2O, ≤0.1 wt% CO_2), primitive (≤10.5 wt% MgO; olivine up to $Fo_{90.6}$) melts yet measured in an arc setting, as well as more degassed, evolved compositions. About half of the analyzed melt inclusions deviate from either open- or closed-system degassing trends, but these inclusions span a wider range of pressures (150-350 MPa) than in the case of Etna. These differences are consistent with differences in the plumbing systems. Jorullo was a short-lived, monogenetic volcano, in which olivine crystallized over a wide range of depths from the middle to upper crust, probably in a network of dikes and sills, rather than mostly crystallizing in a single, larger magma ponding zone as occurs at Etna at ~8 km depth.

Deviation from calculated closed-system degassing trends was also reported at Mt Vesuvius, where melt inclusions show H_2O depletion coupled with limited variations in melt CO_2 contents (Marianelli et al. 2005). Note that the compositional variation of melt inclusions also affects the degassing trend because the bulk composition can have a strong effect on CO_2 solubility (Moore 2008). The evolution of CO_2 vs. H_2O in Arenal and Fuego samples reported in Wade et al. (2006), and shown in Figure 10A, could also indicate some deviation - larger than the analytical error - from closed-system conditions.

Gas fluxing related to deeply-derived magmatic CO_2 could be a common process at basaltic volcanoes. However, further evaluation of this hypothesis requires large systematic data sets on melt inclusions from individual volcanic systems and assessment of the effects of disequilibrium degassing (e.g., Gonnermann and Manga 2005) in such highly dynamic basaltic systems.

Gas fluxing, H_2O loss, and crystallization. We described previously how closed-system degassing can cause H_2O loss from melt and thereby result in crystallization during ascent. The process of gas fluxing involving CO_2-rich gas also causes H_2O loss from melt and can therefore be effective in driving crystallization.

At Jorullo volcano, Mexico, the high MgO content of the early erupted magmas combined with initially high H_2O and the effects of gas fluxing caused olivine-only crystallization over a wide range of depths (10-400 MPa), as indicated by melt inclusion data (Johnson et al. 2008). This effect can be seen by examining phase relations for both H_2O-saturated and H_2O-undersaturated conditions (Fig. 11). Melt inclusion data demonstrate that olivine crystals formed at a maximum pressure of 400 MPa under H_2O-undersaturated conditions, although the magma was vapor saturated because of the presence of CO_2. Fluxing of CO_2-rich gas derived from deeper within the plumbing system would remove H_2O from the melt, even though the melts were H_2O undersaturated. Thus the melts would have been below their relevant H_2O undersaturated liquidii (Fig. 11), forcing small quantities of olivine to crystallize during ascent. Larger amounts of olivine would have crystallized once the melts crossed the H_2O-saturated liquidus (at ~150-200 MPa). Increases in melt inclusion K_2O with decreasing pressure and host olivine Fo content for early erupted Jorullo melt inclusions are consistent with ≤14% crystallization, which matches the prediction based on the phase diagram. Later in the eruption, low pressure degassing caused extensive shallow crystallization of olivine, plagioclase, and minor clinopyroxene. The K_2O increases in late erupted melt inclusions require up to 29-36% crystallization before melt inclusion entrapment, again consistent with phase equilibria.

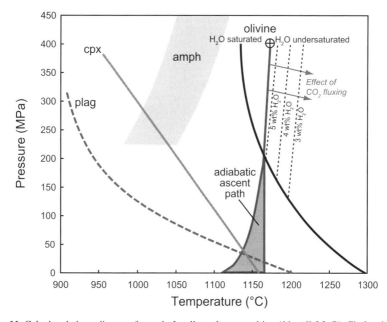

Figure 11. Calculated phase diagram for early Jorullo melt composition (11 wt% MgO). Circle with cross indicates hypothetical starting melt based on the highest melt inclusion entrapment pressure of 400 MPa with 5.3 wt% H_2O. The adiabatic ascent path includes the effects of crystallization, gas exsolution, and gas expansion. See Johnson et al. (2008) for a complete description of how the phase diagram was calculated. [Figure used by permission of Elsevier Limited from Johnson et al. (2008), *Earth Planet Sci Lett*, Vol. 236, Fig. 8, p. 483.]

Whether H_2O loss occurs by open-system gas fluxing or closed-system degassing, in either case it causes crystallization of magma during ascent or storage. As shown by some of the examples given above, degassing-induced crystallization during basaltic magma ascent is a common process. However, recognizing the effects of degassing-induced crystallization can be complicated by several additional processes. These include: (i) magma mixing involving degassed and undegassed magmas, as exemplified by Popocatépetl and Colima (Atlas et al. 2006); (ii) hybridization (mingling) in which inclusions and their host crystals are derived from different magmas distributed vertically in the crust and thus record different entrapment pressures (e.g., from 100 to 440 MPa in the case of the 1974 vulcanian eruption at Fuego, Guatemala; Roggensack 2001); and (iii) assimilation at upper crustal levels as reported at Paricutin (Mexico; Luhr 2001) and Jorullo (Johnson et al. 2008).

Sulfur and halogen degassing

Bulk degassing rates and exsolution pressures. The degassed fraction of a volatile species can be derived from the difference between the volatile content of undegassed melt inclusions and that of the degassed residual melt corrected for crystallization (Sisson and Layne 1993).

As with H_2O and CO_2, the initial dissolved S in a magma is almost totally lost prior to or upon eruption. Residual matrix glasses commonly contain <50 to ~150 ppm S (e.g., Luhr 2001; Métrich et al. 2001; Cervantes and Wallace 2003b; Witter et al. 2004, 2005; Spilliaert et al. 2006b), except in the case of products emitted during highly explosive subaerial and submarine eruptions. Progressive exsolution of S from melt into the gas phase during decompression has been widely described, and S generally decreases with decreasing H_2O, as shown for Central American samples (Fig. 12A). Sulfur starts degassing at rather high pressure in oxidized, H_2O-rich magmas in which S is dominantly dissolved as sulfate, as calculated from electron microprobe S $K\alpha$ wavelength shifts (Wade et al. 2006; Benjamin et al. 2007). Pressure of 140 MPa is reported for efficient degassing of S that is prevalently dissolved as sulfate in the H_2O-rich Etna basalt (Spilliaert et al. 2006b). In that case, ~80% of the initial S is lost in a pressure

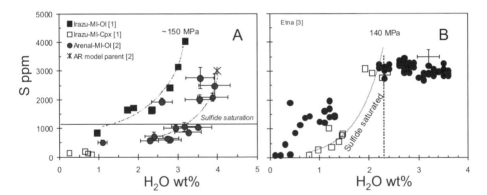

Figure 12. Variation of H_2O and S concentrations in melts inclusions from (**A**) Irazù samples ([1] Benjamin et al. 2007) and Arenal ([2] Wade et al. (2006). The authors calculated the S degassing paths based on the approach in Sisson and Layne (1993), using the K_2O-H_2O relationship to calculate the proportion of H_2O in the separating assemblage of crystals and bubbles and assuming no fractionation of K_2O. Sulfur degassing path was calculated from water mass balance and $D^{vap/melt}$ = 110 and 70 for S in, respectively, Irazù (Benjamin et al. 2007) and Arenal (Wade et al. 2006) samples; (B) from Etna olivine hosted melt inclusions from samples of the 2002 flank eruption (filled circle) and of three lava fountains in 2000 (open square). The pressure of S degassing is assessed from the CO_2 and H_2O content of melt inclusions. In a CSD process, the S decrease is fitted with $D_S = \alpha D^*_S$ values from 0 to 60, where D^*_S is the instantaneous gas–melt partition coefficient and α the mass fraction of total vapor that is accumulated at each step of degassing upon ascent ([3] Spilliaert et al. 2006b).

range between 140 and 10 MPa. This range of pressures is far higher than the low pressure of S exsolution (≤10) MPa reported for submarine basalts (e.g., Dixon et al. 1991).

The evolution of S and H_2O during decompression depends on magma ascent rate and degassing conditions (open vs. closed system). Changes in redox state towards more reduced conditions can drive the melt to sulfide saturation during late stage crystallization at shallow depth, and this affects the S degassing path. Faster magma ascent inhibits these processes until low pressure, causing S degassing to be controlled by vapor-melt partitioning. In cogenetic basaltic and trachybasaltic magmas at Etna, the S content of sulfide-saturated melt inclusions trapped in Fo<75 olivine representative of the shallow, partly degassed magma are only half that of sulfide-free inclusions in Fo~80 olivine that crystallized during fast ascent of basaltic magma under dominantly closed-system conditions (Fig. 12B). In contrast to the differences in S content, the H_2O contents of the different inclusions are comparable (~1.5 wt%).

The amount of initial Cl lost by degassing is in the range of 10-50% (e.g., Hawaii – Gerlach and Graeber 1985; Etna – Spillaert et al. 2006b; Laki – Thordarson et al. 1996; Stromboli – Métrich et al. 2001). Fluorine is much less affected by degassing, with typically <15% of the initial amount lost (references as above). The lower extent of Cl and F degassing from magma causes their concentrations to increase in melt inclusions with increasing differentiation, in contrast to S, in agreement with the higher solubilities of Cl and F in mafic melts (Webster 2004). As an example, Cl positively correlates with U and F in Arenal melt inclusions (Wade et al. 2006). Chlorine has been found to significantly degas at rather low pressure (<20-10 MPa) at Etna even though it starts to be slightly exsolved into the gas phase at pressure ≤100 MPa (Spilliaert et al. 2006b). Moreover, the rather large range of Cl concentrations demonstrates that Cl degassing strongly depends on the dynamics of magma ascent and extrusion and therefore on kinetic effects, in the absence of Cl-bearing mineral phases. Kinetic effects have been demonstrated during decompression experiments on H_2O-rich rhyolitic melts where the amount of Cl retained in matrix glass depends on the size of erupted pumice clasts (Gardner et al. 2006). Similar effects may occur in basaltic melts in which Cl diffusion would be faster because the diffusion in H_2O-rich basaltic melts is significantly higher than that of hydrous rhyolite (Aletti et al. 2007). Actually, Cl loss is enhanced in highly fragmented clasts from lava fountaining activity (Spilliaert et al. 2006b).

ERUPTION STYLES AND PROCESSES – HOW DO THEY RELATE TO MELT INCLUSION DATA?

Rapid ascent of volatile-rich magma under closed-system conditions results in highly explosive eruptions (possibly plinian) because there is no time for gas to escape (Cashman 2004). Large plinian eruptions of silicic magmas are common, whereas basaltic magmas generally give rise to less explosive eruptions because of their lower viscosity, although plinian basaltic eruptions, typical of a dispersed flow, can occur (Vergniolle and Mangan 2000). The explosivity of basaltic eruptions, the transition between explosive and effusive style or the transition from lava fountaining to strombolian activity depends on the regime of gas bubble transfer and the magma supply and ascent rates (e.g., Pioli et al. 2008). Combining the compositions of melt inclusions and their host crystals, dissolved volatile contents, and host magma compositions sheds light on the processes that control magma eruptive style.

A rare example of a sub-plinian eruption involving deeply derived picritic magma has been reported at Mt Etna (Coltelli et al. 2005). The olivine (Fo_{90-91}) of the explosive products contain melt inclusions rich in MgO (12.6 wt% on average) that are very similar in major (including FeO) and trace elements to the bulk rocks after correction for post-entrapment crystallization of the host olivine on the inclusion walls (Kamenetsky et al. 2007). The H_2O

and CO_2 contents of the melt inclusions average 3.2±0.3 wt% and 0.29±0.2 wt%, respectively, excluding a few values affected by CO_2 loss. The inclusion entrapment pressures—estimated to be >400 MPa—are most likely underestimated because the entrapped glass coexists with gaseous CO_2 and carbonates present in the vapor bubble. The high explosivity of the eruption has been attributed to the rapid ascent of H_2O and CO_2-rich picritic magma under dominantly closed-system degassing conditions (Coltelli et al. 2005).

More commonly, basaltic volcanoes are characterized by vigorous lava fountains and/or strombolian and violent strombolian activity. The Kilauea Iki (Hawaii) eruption of 1959 is a well studied example of lava fountaining, but it is somewhat unusual in that it involved picritic magma (Anderson and Brown 1993; Anderson 1995). In this example, the olivine (Fo_{87-88}) in equilibrium with the picritic magma mostly crystallized at relatively low pressure (<~100 MPa), as inferred from their melt inclusions. These data have been interpreted as resulting from crystallization of a newly intruded, hot, MgO-rich magma that rose through and mixed with cooler degassed magma stored at shallow depths (Anderson and Brown 1993). It has been proposed that gassy picritic magma (15 wt% MgO in melt + 0.3 wt% total of exsolved and dissolved CO_2) becomes buoyant relative to the stored, degassed tholeiitic magma only at pressures less than 200 MPa (Anderson 1995).

More typical, differentiated Kilauea magma erupted at Pu'u O'o along the east rift zone giving rise to high lava fountains (up to 400 m high, Parfitt 2004) and less powerful spattering activity. Melt inclusions in the former case have preserved relatively high H_2O contents (0.4-0.5 wt%), whereas inclusions from the spattering activity contain lower values of 0.1-0.3 wt% H_2O. These lower H_2O values require degassing at pressures significantly lower than those in Kilauea's summit magma reservoir (Wallace and Anderson 1998). The low H_2O values have been interpreted as indicating that the magma was already degassed prior to olivine crystallization and melt inclusion entrapment.

Closed-system ascent and degassing of Etna basaltic magma was inferred for the 100-300 m high lava fountaining that occurred at the onset of the 2002 flank eruption of Mt Etna, on the basis of the CO_2 and H_2O degassing path recorded in melt inclusions (Fig. 10B). In this case undegassed, primitive melt inclusions are representative of the erupted basaltic magma for major elements and trace element patterns (Spilliaert et al. 2006a; Métrich et al. 2007). This interpretation is consistent with the hypothesis that lava fountaining events are controlled by magma speed rise, such that rapid ascent does not allow time for gas transfer or bubble loss from rising magmas (e.g., Wilson and Head 1981; Parfitt and Wilson 1995).

Similarly, fast ascent of deep-seated, volatile-rich magma under closed-system conditions gives rise to the most explosive activity and the emission of basaltic pumice at Stromboli (Aeolian islands, Italy). These products contain MgO-rich olivines with undegassed melt inclusions (on average: H_2O = 2.8±0.2 wt%; S = 0.22±0.02 wt%; CO_2 = 0.1-0.2 wt%) that were trapped at pressures of 200-300 MPa (Métrich et al. 2001; Bertagnini et al. 2003). In contrast, gas jets and mild explosions lasting a few seconds propel lava lumps and dense scoriae during the rythmic, strombolian activity at the craters. These scoriae carry low pressure crystals with melt inclusions (<1.2 wt% H_2O, <0.01 wt% S, low CO_2) that mirror the extensively degassed and viscous magma stored in the uppermost parts of the conduits, capping the volcanic system (Landi et al. 2004).

Mafic cinder cones commonly exhibit strongly pulsatory explosive activity described as violent strombolian that produces both moderately high eruption columns (2-6 km) with abundant fine ash and simultaneous lava effusion from lateral flank vents (Pioli et al. 2008). Examples of such activity are the historic eruptions of Paricutin and Jorullo in Mexico, where melt inclusion data show that basaltic to basaltic andesite magmas had initial H_2O contents of 3.5-5.3 wt% and CO_2 ≤ 1000 ppm (Luhr 2001; Johnson et al. 2008; Pioli et al. 2008).

VOLCANIC DEGASSING: INTEGRATION OF VOLCANIC GAS COMPOSITIONS AND GAS FLUXES WITH MELT INCLUSIONS

Syn-eruptive magma degassing and volatile degassing budgets

The mass of volatiles released by syn-eruptive magma degassing can be estimated from the initial dissolved volatile contents recorded by undegassed melt inclusions, the residual volatile contents in bulk rock or matrix glass samples, and the volume of magma erupted. This approach, known as the "petrologic method", has been widely applied and debated for assessing the S emissions from volcanic eruptions because S (primarily as SO_2 and H_2S species) injected into the stratosphere is converted into H_2SO_4 aerosols that have an effect on climate and the Earth's atmosphere (see Robock and Oppenheimer 2003 for a review). The petrological estimates of SO_2 degassing have been compared with independent ground-based, airborne and satellite measurements of SO_2 flux at active volcanoes (see Wallace 2001 and Self et al. 2006 for reviews).

The mass of SO_2 (M_{SO_2}) released from a volcanic eruption is estimated using the petrologic method according to the relation: $M_{SO_2} = 2\ M_{magma}\ (C_{MI}-C_{WR})$, where M_{magma} is the mass of bulk magma derived from the volume of erupted products corrected for their vesicularity, C_{MI} is the S concentration in undegassed melt inclusions representative of the bulk erupted magma, C_{WR} is the S concentration in bulk rock, or in matrix glass that is corrected for the extent of crystallization. The factor of 2 is included because the mass of SO_2 is two times that of S (SO_2/S mass ratio = 2).

Large basaltic eruptions can deliver huge masses of volcanic gas into the atmosphere during the emission of voluminous lava flows, along 10 to 100 km long fissures, at high effusion rates, with occasionally violent lava fountain activity (Self et al. 2006). The 8 month long 1783-1784 Laki flood basalt eruption in Iceland released 14.7 km³ of quartz tholeiitic magma and is a well studied example of syn-eruptive degassing. The large volume and homogeneous composition of Laki products imply that they were erupted from a compositionally uniform magma body in a deep-seated (>10 km) reservoir (Thordarson et al. 1996). The major and trace element compositions of melt inclusions from the eruptive products are very similar to those of the bulk erupted magma, but the recent discovery of heterogeneous oxygen isotope values for individual olivine crystals from this eruption has called into question the origin of the crystals (Bindeman 2008). The eruption released 122 Mt of SO_2, 15 Mt of Cl and 7 Mt of F as assessed from petrologic estimates (Thordarson et al. 1996). These authors concluded that the estimates for the Laki eruption were reliable because similar values for H_2SO_4 (~200 Mt) were provided by independent estimates. By integrating data on the melt inclusions, lavas, tephra and deposits, it is shown that 80% of the S mass was released during the explosive eruptive phases at the vents and carried by eruption columns, which reached heights of 9-13 km, to the lower stratosphere (Thordarson and Self 2003). Less than 20% of the sulfur mass was released from the lava to form localized haze within the lower troposphere.

Similarly, good agreement has been obtained between the petrologic estimates of SO_2 released during the 94-day 2002 flank eruption at Etna and the total SO_2 emissions derived from optical correlation spectrometer (COSPEC) measurements during the eruption (Spilliaert et al. 2006a), considering the uncertainties in the erupted product volumes (±50%) and COSPEC data (±20-25%). This agreement implies that there was no excess release of SO_2 during the eruption; the SO_2 emissions were essentially supplied by syn-eruptive bulk degassing of the extruded magma. Good agreement between petrologic and spectroscopic estimates of SO_2 output has also been demonstrated at Arenal by considering melt inclusions that are most representative of undegassed basaltic andesite magma (Wade et al. 2007).

The consistency between petrologic and other independent estimates of SO_2 output at some basaltic volcanoes demonstrates that the petrologic method is useful when the only source of

S comes from syn-eruptive magma degassing. As shown by the example of Laki above and also by cases discussed in Wallace (2001), this is most commonly the case in non-subduction related volcanoes. Petrologic estimates of SO_2 emissions from ancient flood basalts are likely to provide realistic values (Sharma et al. 2004).

Excessive degassing and the role of degassed but unerupted magma

The S contents of undegassed melt inclusions have been widely used to assess the mass and volume of degassed magma that is necessary to sustain measured SO_2 fluxes during quiescent degassing (only gas emissions) at active basaltic volcanoes. This assessment is based on the relationship: $M_{magma} = \Phi_{SO_2}/2X_S$, where M is the mass of degassed magma, Φ_{SO_2} is the time-averaged flux of SO_2, and X_S is the mass fraction of S degassed from magma based on melt inclusions and bulk rock or matrix glass samples, as described above for the petrologic method.

As an example, only a small fraction (<10%) of the degassed magma is actually erupted at Stromboli (Allard et al. 2008). This interpretation is derived from the S concentrations (0.22 wt%) in high pressure melt inclusions trapped in Fo_{88-89} olivine (Bertagnini et al. 2003), the time averaged flux of SO_2 (250 tons d^{-1}), and the bulk magma extrusion rate (0.018 m^3 s^{-1}) over the last 30 years.

Quiescent and intrusive degassing—where the amount of SO_2 released is in excess compared to estimates based on syn-eruptive degassing of erupted magma—is commonly reported at open conduit basaltic volcanoes such as Etna (Allard 1997), Villarica (Witter et al. 2004), and Popocatepetl (Witter et al. 2005), amongst others.

Differential gas bubble transfer

Because gas bubbles can segregate from the melt at different depths, the volatile components fractionate according to their solubilities. The composition of the exsolved gas phase will thus change as a function of the pressure-related solubility of each volatile species. Knowing the abundance of S, Cl, F, H_2O and CO_2 dissolved in melt inclusions and the pressure interval over which degassing has occurred, as derived from the H_2O and CO_2 content, it is possible to calculate the pressure-related evolution of S and halogens and the S/Cl and S/F ratios in the gas phase as the pressure decreases.

At Mt Etna, the S/Cl mole ratio (1.09 × weight ratio) in the gas phase has been estimated from melt inclusion data as function of pressure, assuming closed- and open-system degassing, with and without sulfide immiscibility (Spilliaert et al. 2006b; Fig. 13). The S/Cl ratio of the gas phase is controlled by S outgassing until about 10-20 MPa and becomes influenced by Cl exsolution only at lower pressures, when the melt is already depleted in S. Pure open-system degassing fails to explain the S/Cl ratios (≤0.2) commonly measured in gas emissions during Etna eruptions. Under closed-system conditions, the S/Cl ratio of the gas phase evolves more progressively down to 3.7 (molar) in gas emitted at the surface. Any gas bubbles segregated from the melt at depth will have a high S/Cl ratio. As an example, gas bubbles segregated from ~1.8 km have a high S/Cl value of 10 (Fig. 13), as measured in the gas emission during one of the cyclic lava fountains on June 2000 at Etna crater (Allard et al. 2005). The lava fountaining has been interpreted—on the basis of this high S/Cl ratio in gas emissions—as deriving from bubble accumulation prior to eruption (Allard et al. 2005) according to the bubble foam model (Vergniolle and Jaupart 1990).

GEOCHEMICAL SYSTEMATICS OF THE PRIMARY VOLATILE CONTENTS OF MAFIC MAGMAS

One of the most important conclusions that can be drawn from the data reviewed in this chapter is that most of the observed phenocrysts in basaltic magmas crystallize during ascent,

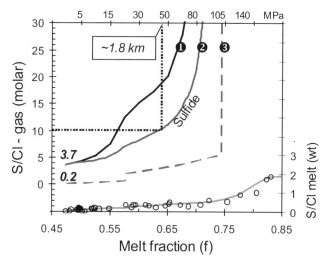

Figure 13. Pressure-related evolution of the S/Cl ratio in the melt and the gas phase of Mt Etna, under closed-system conditions (curves 1 and 2) and open-system conditions (curve 3), modified from Spilliaert et al. (2006b). The S/Cl ratio in the gas phase is calculated following the relationship $(X_S/X_{Cl})_{gas} = (D_S/D_{Cl}) \times (X_S/X_{Cl})_{melt}$; where X and D are, respectively, the concentrations and the bulk partition coefficients of S and Cl as determined from the melt inclusion data. The S/Cl ratios in melt are given in weight units, whereas the calculated gas ratios are given in molar units in order to have direct comparison with values measured in gas emissions. CSD curves and the final S/Cl ratio in surface gas are calculated with an average bulk Cl loss of 55%. Pure OSD conditions produce much lower S/Cl ratio (<0.2) in surface gases than measured in emissions of both eruption types (Spilliaert et al. 2006b). The CSD curves illustrate the significant influence of sulfide immiscibility on the S/Cl degassing path. The lithostatic depth of ~*1.8* km below vents is computed from rock densities in the sedimentary basement of Etna. The main interpretation that gas bubbles segregated from rather low pressure fits with a process of differential gas transfer (see text).

degassing, and shallow storage in the crust. As such, melt inclusions in crystals preserve a record of these processes, but the inclusions represent melts that had variably degassed prior to entrapment. This variable degassing hinders our ability to determine the primary volatile contents of basaltic melts formed by partial melting of the Earth's upper mantle. The extent to which this variable degassing hides the primary volatile contents is inversely proportional to the solubility of the component in question. Thus for low solubility components like CO_2 (and He), melt inclusions formed in the middle to upper crust may never record values that are close to primary values (Wallace 2005). The major volatile components that degas the least are F and Cl, and they are least affected by this problem. H_2O and S lie between the two extremes and show variable degrees of degassing in melt inclusion data sets.

A standard approach has been to analyze as large a number of inclusions as is practical and to then assume that the highest pressure inclusions are best representative of the undegassed magma. In the case of H_2O, an independent check on this can be made using experimental phase equilibria (e.g., Di Carlo et al. 2006; Johnson et al. 2008), mineral-melt hygrometers (Housh and Luhr 1991; Hamada and Fuji 2007), and partitioning of OH between melt and nominally anhydrous minerals (e.g., Hauri et al. 2006). However, agreement between these various approaches only confirms the melt H_2O content during crystallization, but it does not preclude higher primary magmatic H_2O contents when melts are in the upper mantle or lower crust. In this section, we review the systematics of the primary volatile contents of basaltic magmas from arcs, as inferred from melt inclusion data, and compare them to what is known from submarine glasses from mid-ocean ridges and oceanic islands, with the goal of better understanding volatile recycling and magma formation in subduction zones.

Several recent studies have used melt inclusions to examine volatile variations either with distance from the trench or with position along the arc where there are important along-arc variations in subduction parameters. Variations in volatile content with distance from the trench were studied in Guatamala by Walker et al. (2003). For arc front volcanoes, they found maximum H_2O contents of 6.2 wt% (Fuego) and 2 wt% (Pacaya). However, the Pacaya inclusions have CO_2 below detection limit, so the low H_2O values are likely the result of partial degassing before inclusion entrapment. Monogenetic basaltic cinder cones located up to 80 km behind the volcanic front have maximum H_2O contents similar to Pacaya (~2 wt%). Overall, the results for Guatamala show no consistent across arc-variation, but the highest H_2O contents (Fuego) are found only at the arc front. Distinctly higher CO_2 contents were found in some of the monogenetic cones behind the front. Walker et al. (2003) speculated that this could be the result of devolatilization of carbonate sediments with increasing slab depth, but given how strongly CO_2 degasses at crustal pressures, this could also be caused by greater average crystallization depths for the back-arc cones.

Complementing this is a recent study of along-arc variations in volatiles in Central America based on data from 11 volcanic centers in Guatamala, Nicaragua and Costa Rica (Sadofsky et al. 2008). The thickness of the downgoing oceanic crust, composition of subducted sediments, and slab dip angle all vary along this arc. For the northern part of the arc, the downgoing plate is normal-thickness, 25 Ma oceanic crust, overlain by ~450 m of carbonate ooze and hemipelagic clay, and probably underlain by strongly serpentinized mantle. In contrast, beneath central Costa Rica the downgoing plate is relatively thick, 15-20 Ma oceanic crust formed at the Galapagos spreading ridge and overprinted by the Galapagos hotspot. Estimated parental magma H_2O contents (for melts in equilibrium with Fo_{91} olivine) vary from 2.7 to 5.5 wt%, but show no consistent variation along the arc despite the large differences in the nature of the subducting slab (Sadofsky et al. 2008). The highest values were for Cerro Negro in Costa Rica and Irazu in central Costa Rica. Parental magma S, Cl and F also show no systematic variations along the arc, though Irazù volcano at the southern end of the arc contains much higher values than the other volcanoes (Benjamin et al. 2007). Sadofsky et al. (2008) concluded that melting beneath all parts of the Central American arc is triggered by addition of H_2O-rich components from the downgoing slab.

In an across arc study of olivine-hosted (Fo_{73-91}) melt inclusions from the Kamchatka and northern Kurile arcs, Portnyagin et al. (2007) calculated mean parental H_2O concentrations (melts in equilibrium with Fo_{90} olivine) for the different volcanoes studied ranging from 1.8 to 2.6 wt%, with values generally decreasing with increasing distance from the trench. Interestingly, the parental H_2O estimates based on the highest measured melt inclusion H_2O contents (rather than averages) at each volcano show no such pattern but are instead constant at 2.5-2.8 wt% H_2O. However, about two-thirds of the data set are for olivines from lava flows that were experimentally rehomogenized, and the H_2O values are lower than those found for olivine-hosted melt inclusions from rapidly cooled tephra at Kluchevskoy ($\leq$3.9 wt% H_2O in Fo_{88} olivine and $\leq$7.1 wt% H_2O in Fo_{80} olivine; Auer et al. 2008). Mean parental S, Cl, and F based on the melt inclusion data all show increases with increasing distance from the trench.

In a similar across arc study of olivine-hosted (Fo_{82-91}) melt inclusions from nine cinder cones in the central Trans-Mexican Volcanic Belt (TMVB), Johnson et al. (2008.) found parental magma H_2O contents (for melts in equilibrium with Fo_{90} olivine) that are high at the volcanic front (3.5-5.2 wt% H_2O), but remain high (3.0-4.5 wt%) in the cones over 130 km behind the front (3.0-4.5 wt% H_2O). The center farthest from the front, a tuff ring, has much lower water (~1 wt%). Parental magma S contents are variable across the arc, ranging from 850-2150 ppm, and parental Cl concentrations remain high over most of the arc (~860-1360 ppm), decreasing to 390 ppm in the tuff ring far behind the front. The high volatile concentrations across the central TMVB are surprising given that the subducting Cocos plate is young (~11-15 Ma at the trench), and therefore relatively hot, and thus significant devolatilization should occur beneath the forearc region. Low B concentrations in whole rock samples are consistent with shallow slab dehydration (Hochstaedter et al. 1996), so the data from the TMVB show that H_2O can be strongly decoupled from highly fluid-mobile elements such as B.

To better understand the recycling of volatiles in subduction zones and their role in mantle melting, it is necessary to compare volatile concentrations to non-volatile trace elements of both similar and contrasting behavior. For example, H_2O and Ce have similar bulk partition coefficients during partial melting of mantle peridotite, making the H_2O/Ce ratio a valuable indicator of sources that have become enriched in H_2O (e.g., Michael 1995). Some incompatible trace elements in arc magmas, such as Y and the high field strength elements (HFSE), Ti, Zr, Nb, Ta, have similar abundances to values for depleted and enriched MORB, leading to the interpretation that these elements are supplied primarily by the mantle wedge (Pearce and Peate 1995). This suggests that these elements are generally not mobile in the H_2O-rich component(s) transferred from the slab to the wedge, though they can be mobile under certain circumstances (Kessel et al. 2005; Barry et al. 2006). Comparison of HFSE and Y systematics for arc and non-arc magmas shows that the mantle wedges beneath arcs are variable in composition, with compositions ranging from more depleted than the depleted MORB mantle (DMM) to values requiring a source similar to that for EMORB magmas (Pearce and Peate 1995; Wallace and Cervantes 2003; Portnyagin et al. 2007). The more depleted sources form as a result of previous partial melting and melt extraction, and heterogeneities in the wedge can exist even on relatively small spatial scales (e.g., Wallace and Carmichael 1999). The enriched vs. depleted character of the mantle source is well illustrated by the Nb/Y ratio because both these elements are relatively fluid immobile, but Nb is more highly incompatible and thus is depleted during partial melting (Pearce and Peate 1995).

Volatiles and fluid-mobile elements (e.g., K, Ba, Sr, U, Pb) transferred from the slab to the wedge overprint and selectively enrich the variably depleted mantle wedge (Fig. 14A). Partial melting of DMM to form depleted MORB results in magmas with low H_2O/Y ($\leq$100; le Roux et al. 2006) whereas partial melts from an OIB-type mantle source yield higher values ($\leq$1000) as well as much higher Nb/Y. Data from olivine-hosted melt inclusions from Kamchatka, Central America, and the TMVB have H_2O/Y varying from ~1000 to 5400. The mantle wedge for Kamchatka appears to be as or more depleted than DMM (Portnyagin et al. 2007) whereas the wedges beneath Central America and the TMVB are less depleted and more heterogeneous (Fig. 14A). Interestingly, the data for Kamchatka suggest that the mantle wedge is less strongly enriched in H_2O than beneath Central America and the TMVB, despite the fact that the plate subducting beneath Kamchatka is much older.

The high values of H_2O/Y for basaltic arc magmas require mantle source H_2O concentrations in the range from 0.3 to nearly 1 wt%, compared to 70-160 ppm that is estimated for DMM (Workman and Hart 2005). Grove et al. (2006) presented a model for hydrous flux melting in arcs. H_2O-rich fluids rising from the subducted slab or from breakdown of chlorite in the mantle wedge just above the slab rise through an inverted thermal gradient in the wedge caused by corner flow of hot asthenospheric mantle. When the fluid encounters mantle at the temperature of the wet peridotite solidus, a small mass fraction of H_2O-saturated melt forms, with ~20-30 wt% dissolved H_2O. As these melts continue to rise through the inverted thermal gradient, they react with the peridotite, extracting basaltic components and diluting the initially very high H_2O contents. By the time the melts reach the hottest part of the wedge, at temperatures ~1400 °C, the partial melts contain ~5 wt% H_2O, similar to the highest values for primitive arc magmas.

Some arc basaltic magmas are relatively H_2O-poor. In the TMVB, the H_2O-poor types have high Nb/Y (Fig. 14A; Cervantes and Wallace 2003(a or b)), and their isotopic and trace element characteristics are similar to basalts erupted in the Mexican Basin and Range province

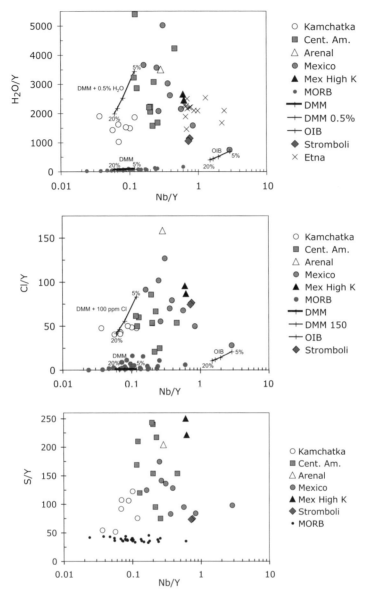

Figure 14. Relationship between major volatiles and "conservative" trace elements, which are not likely to be mobile in fluids released from subducted slabs (Pearce and Peate 1995). (A) Comparison of Nb/Y vs. H_2O/Y in arc basaltic magmas and MORB. Red line shows 5 to 20% batch partial melts of DMM (Workman and Hart 2005) containing 110 ppm H_2O calculated using partition coefficients from Kelley et al. (2006). Black line near middle of figure shows 5 to 20% batch partial melts of DMM with 0.5 wt% H_2O. Black line in lower right shows 5 to 20% batch partial melts of an OIB source with 0.2 wt% H_2O. Data are shown for Kamchatka (Portnyagin et al. 2007), Central America (Sadofsky et al. 2008), Arenal volcano, Costa Rica (Wade et al. 2006), central Mexico (Johnson et al. 2008), high-K minettes and basanites in western Mexico (Vigouroux et al. 2008), Stromboli (Bertagnini et al. 2003), Etna (Kamenetsky et al. 2007), and East Pacific Rise MORB (le Roux et al. 2006). (B) Nb/Y vs. Cl/Y. Partial melting curves as in (A), with 0.38 ppm Cl in DMM, a fluid enriched DMM source with 100 ppm Cl, and an OIB source with 50 ppm Cl. (C) Nb/Y vs. S/Y.

well behind the arc (Luhr 1997). The H_2O/Y values of these intraplate-type alkali basalts are similar to other OIB (Fig. 14A), indicating that the mantle source region for these magmas has not been significantly enriched in H_2O and trace elements by subduction processes. In the Cascades of Northern California, the H_2O-poor types are high-alumina olivine tholeiites with trace element characteristics more similar to MORB (Sisson and Layne 1993; Grove et al. 2002). Another well documented example of H_2O-poor basaltic magma in arcs comes from the 1982-1983 eruptions of Galunggung volcano, Indonesia, where basaltic melt inclusions contain 0.2 to 0.3 wt% H_2O (Sisson and Bronto 1998). In general, H_2O-poor arc magmas are inferred to be the result of decompression melting caused by upwelling in the mantle wedge (Sisson and Layne 1993), probably caused by advection of hot mantle from behind the arc due to corner flow (Grove et al. 2006).

The Cl concentrations of basaltic arc magmas are also much higher than those of depleted MORB, EMORB, and most OIB-type magmas (Fig. 14B). Cl/Y ratios for arc magmas show similar patterns as H_2O/Y, and the Cl concentrations require ~100 to 400 ppm Cl in the mantle source compared to 0.38 ± 0.25 ppm that is present in DMM (Workman and Hart 2005). The wide range of Cl/H_2O ratios for arc and back-arc magmas requires that Cl and H_2O are strongly fractionated from one another either during devolatilization of the slab or during migration of slab-derived fluids through the mantle wedge (Fig. 15; Kent et al. 2002; Wallace 2005). Fluid inclusions in eclogites from exhumed subduction zone complexes can be used to understand the compositions of fluids released during dehydration of subducted oceanic crust. Philippot et al. (1998) used such data to suggest that fluids released from the upper oceanic crust, which is much more strongly altered during hydrothermal alteration near spreading ridges, have salinities of 3.1-4.0 wt% NaCl. In contrast, fluids derived from deeper in the oceanic crust, which has been hydrothermally altered at higher temperatures than the upper crust, are more saline, with 17-45 wt% NaCl. Thus the salinities of fluids released from the dehydrating oceanic crust appear to bracket most values for arc and back-arc magmas, though the data from Cerro Negro in Nicaragua appear somewhat anomalous in requiring low salinity fluids. Fluids released during breakdown of serpentinite in the downgoing slab are estimated to have 4-8 wt% NaCl (Scambelluri et al. 2004), also similar to many values for arc and back-arc magmas. However, fluids derived from serpentinite dehydration in the downgoing slab have light $\delta^{18}O$ values, and evidence for strong involvement of such fluids has only been found in the northern Central American arc (Eiler et al. 2005).

Olivine-hosted melt inclusions from basaltic arc magmas typically have 900-2500 ppm S, but values extend as high as 4000-6000 ppm (Wallace 2005; Benjamin et al. 2007). These values are higher than for MORB magmas of the same FeO^T but are comparable to some OIB (Wallace and Carmichael 1992). As with H_2O and Cl, arc magmas also have higher S/Y than MORB and OIB (Fig. 14C). The higher S/Y ratios may reflect addition of subduction-recycled S from the slab to the wedge, oxidation of the mantle wedge by fluids such that residual sulfide is no longer stable during melting, or large degrees of melting that exhaust residual sulfide (Wallace 2005; Righter et al. 2008). However, the higher S contents relative to MORB of comparable FeO^T cannot be derived from low–f_{O_2} DMM because of solubility constraints imposed by sulfide saturation (Mavrogenes and O'Neil 1999), so breakdown of sulfide by oxidation or large degrees of mantle melting is required. Estimated mantle source concentrations for basaltic arc magmas are 250-500 ppm S (Métrich et al. 1999; de Hoog et al. 2001), higher than DMM estimates of 80-300 ppm S (Chaussidon et al. 1989).

CONCLUDING REMARKS

We have provided here a review of published literature showing the relevance of studies on olivine-hosted melt inclusions for determining the volatiles dissolved in magma and their

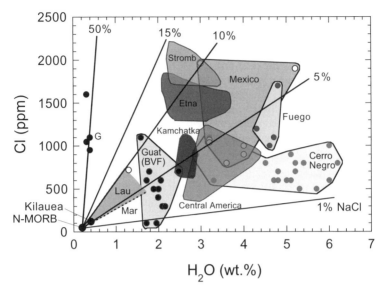

Figure 15. H_2O vs. Cl for olivine-hosted melt inclusions from arc basalts. Data are shown for Cerro Negro (Roggensack et al. 1997), Fuego (Roggensack 2001), Guatamalan volcanoes behind the volcanic front (Guat BVF; Walker et al. 2003), the Trans-Mexican Volcanic Belt (Luhr 2001; Cervantes and Wallace 2003a), Stromboli (Metrich et al. 2001; Bertagnini et al. 2003), Galunggung volcano, Indonesia (G; Sisson and Bronto 1998), parental melt compositions from Kamchatka (Portnyagin et al. 2007), parental melt compositions for Central America (Sadofsky et al. 2008), and undegassed Etna inclusions (Spilliaert et al. 2006a). For Cerro Negro, only inclusions from the 1992 explosive eruption are plotted; partially degassed inclusions from the effusive 1995 eruption are not shown. Note that a Mexican sample with OIB-like composition has low H_2O (open circle with 1.3 wt% H_2O). Field for Lau Basin back-arc basalts is from Kent et al. (2002). The average Cl/H_2O ratio for Mariana Trough basalts (Stolper and Newman 1994) is shown as a dashed line and coincides with the low end of values for the Lau Basin. Lines show basaltic melt compositions produced by addition of fluid with 1, 5, 10, 15, and 50 equivalent wt% NaCl to an N-MORB mantle source. Most of the data have not been corrected for varying degrees of olivine fractionation, and thus do not represent primary magma values, but such correction should only affect the absolute concentrations and not the Cl/H_2O ratios.

evolution during magma ascent and decompression. Based on this, we summarize the following main points and raise some questions for future work:

1. Olivine-hosted melt inclusions lead to pressure estimates that only rarely exceed 300-400 MPa, and provide a powerful tool for tracking degassing between these pressure and surface. Such a range of pressures is generally lower than those deduced from CO_2-rich fluid inclusions (Hansteen and Klügel 2008). This discrepancy raises an important question for future research. Presently, the volatile data for melt inclusions hosted in high pressure clinopyroxenes are very scarce, and these should be investigated further. Regarding olivine, its crystallization in basaltic magmas depends on temperature and the melt water content. Its growth rate is strongly influenced by the rate of decompression and undercooling and thus the conditions of magma ascent. Melt inclusion formation is favored by temperature and pressure gradients and undercooling. Olivine-hosted melt inclusions from rapidly cooled explosive products are the most suitable for avoiding extensive post-entrapment interaction between the inclusions and their host crystals. Data from such inclusions provide a rich record of dynamic processes. They allow us to infer the conditions of

magma ascent, crystallization, and degassing. For future work, combining textural studies of natural and experimental olivines, their melt inclusion morphologies, and the dissolved volatile contents of the inclusions would allow a better estimate of the timescales of magma crystallization and ascent.

2. Quantifying the H_2O and CO_2 contents of melt inclusions is a difficult task. The inclusions must have suffered limited post-entrapment evolution. Their interpretation must be tightly associated with a precise study of both the inclusions and their host mineral composition and also that of the bulk eruptive products. The efforts dedicated in the last 15 years to developing analytical techniques to get quantitative analyses of both the major (CO_2, H_2O) and minor (S, halogens) volatile components, with uncertainties that on average are ≤10%, have been very successful. These efforts have led to a rapid increase in the quality and quantity of melt inclusion volatile data that in turn have been integrated with trace element and isotope geochemistry, surface gas emission measurements, and geophysical data relevant to understanding depths of magma storage.

3. Because CO_2 is extensively lost to the gas phase at high pressure, melt inclusions generally are not representative of the CO_2 concentrations of magmas at high pressure. However, a new idea that has emerged is that magma flushing by CO_2-rich gas has the potential to cause magma dehydration and affect the degassing path of H_2O. Efforts must be made to improve the modeling of CO_2-H_2O evolution during decompression. Improvements to models of magma degassing require experimental and thermodynamic data on the solubility of CO_2 in H_2O-bearing basaltic melts.

4. Melt inclusion data can be used to model degassing of S and halogens from melt as a function of pressure. Segregation of gas bubbles from the melt at depth will result in high ratios of S/Cl in the gas phase released by erupting magmas.

5. The petrologic method based on melt inclusions has proven to be reliable for estimating the SO_2 output from syn-eruptive degassing of large flood basalts and basaltic magmas ascending in closed system conditions, provided there is no differential gas transfer (gas loss) prior to eruption.

6. For volcanic systems underdoing quiescent degassing, the S contents of undegassed melt inclusions can be combined with SO_2 fluxes measured in surface gas emissions to estimate the volume of underlying, non-erupted magma that has degassed.

7. The effects of shallow degassing processes must be considered in studies that attempt to elucidate the primary volatile contents of magmas based on olivine-hosted melt inclusions. Systematic volatile and trace element analyses in subduction-related magmas make it possible to (i) distinguish magmas formed by decompression melting of MORB- or OIB-type mantle sources from those formed by partial melting of a mantle source fluxed by slab-derived H_2O-rich components, (ii) address specific questions on the relative fractionation of H_2O and Cl during slab dehydration, and (iii) investigate the origins of the high S concentrations in some arc basalts. Future progress in understanding volatile recycling in arcs requires integration of melt inclusion data, geodynamic models for the thermal structure in the downgoing plate and mantle wedge, thermodynamic and experimental data on H_2O storage and its effect on partial melting in the wedge, and models of fluid and melt transport through the mantle.

Finally, one of the main points that arises here is that combining melt inclusion data with (i) accurate studies of their host olivines and the mineralogy of the host magmas, (ii) experimental data, and (iii) field work and gas measurements is a necessity and represents a main challenge for the next few years.

ACKNOWLEGMENTS

We are grateful to K. Putirka, J. Lowenstern, P. Papale and G. Moore for their comments and thorough reviews that strongly contributed to improving this manuscript. We would also like to thank Emily Johnson for providing several of the photomicrographs of melt inclusions.

REFERENCES

Agrinier P, Jendrzejewski N (2000) Overcoming problems of density and thickness measurements in FTIR volatile determinations: a spectroscopic approach. Contrib Mineral Petrol 139:265-272
Alletti M, Baker DR, Freda C (2007) Halogen diffusion in a basaltic melt. Geochim Cosmochim Acta 71:3570-3580
Allard P, Jean-Baptiste P, D'Alessandro W, Parello F, Parisi B, Flehoc C (1997) Mantle-derived helium and carbon in groundwaters and gases of Mount Etna, Italy. Earth Planet Sci Lett 148:501-516
Allard P (1997) Endogenous magma degassing and storage at Mount Etna. Geophys Res Lett 24:2219-2222
Allard P, Burton M, Muré F (2005) Spectroscopic evidence for a lava fountain driven by previously accumulated magmatic gas. Nature 43:407-410
Allard P, Aiuppa A, Burton M, Caltabiano T, Federico C, Salerno G, La Spina A. (2008) Crater gas emissions and the magma feeding system of Stromboli volcano. Learning from Stromboli. AGU Geophys Monograph (in press)
Anderson AT (1975) Some basaltic and andesitic gases. Rev Geophys Space Phys 13:37-55
Anderson AT (1976) Magma mixing: Petrological process and volcanological tool. J. Volcanol Geotherm Res 1:3-33
Anderson AT (1995) CO_2 and the eruptibility of picrite and komatiite. Lithos 34:19-25
Anderson AT, Brown GG (1993) CO_2 contents and formation pressures of some Kilauean melt inclusions. Am Mineral 78:794-803
Atlas ZD, Dixon JE, Sen G, Finny M, Martin-Del Pozzo AL (2006) Melt inclusions from volcan Popocatépelt and volcan Colima, Mexico: melt evolution due to vapor-saturated crystallization during ascent. J Volcanol Geotherm Res 153:221-240
Auer S, Bindeman I, Wallace P, Ponomareva V, Portnyagin M (2008) The origin of hydrous, high $\delta^{18}O$ voluminous volcanism: Diverse oxygen isotope values and high magmatic water contents within the volcanic record of Klyuchevskoy Volcano, Kamchatka, Russia. Contrib Mineral Petrol, doi:10.1007/s00410-008-0330-0
Baker DR, Freda C, Brooker RA, Scarlato P (2005) Volatile diffusion in silicate melts and its effects on melt inclusions. Ann Geophys 48:699-717
Baker DR (2008) The fidelity of melt inclusions as records of melt composition. Contrib Mineral Petrol, doi:10.1007/s00410-008-0291-3
Barry TL, Pearce JA, Leat PT, Millar IL, le Roex AP (2006) Hf isotope evidence for selective mobility of high-field-strength elements in a subduction setting: South Sandwich Islands. Earth Planet Sci Lett 252:223-244
Benjamin ER, Plank T, Wade JA, Kelley KA, Hauri EH, Alvarado GE (2007) High water contents in basaltic magmas from Irazù volcano, Costa Rica. J Volcanol Geotherm Res 168:68-92
Behrens H, Roux J, Neuville DR, Siemann M (2006) Quantification of dissolved H_2O in silicate glasses using confocal microRaman spectroscopy. Chem Geol 229:96-112
Bertagnini A, Métrich N, Landi P, Rosi M (2003) Stromboli volcano (Aeolian Archipelago, Italy): An open window on the deep-feeding system of a steady state basaltic volcano. J Geoph Res 108(B7):2336, doi:10.1029/2002JB002146
Bindeman I (2008) Oxygen isotopes in mantle and crustal magmas as revealed by single crystal analysis. Rev Mineral Geochem 69:445-478
Blundy J, Cashman K (2008) Petrologic reconstruction of magmatic system variables and processes. Rev Mineral Geochem 69:179-239
Botcharnikov RE, Holtz F, Behrens H (2007) The effect of CO_2 on the solubility of H_2O-Cl fluids in andesitic melt. Eur J Mineral 19:671-680
Botcharnikov R, Freise M, Holtz F, Berhens H (2005) Solubility of C-O-H mixtures in natural melts: new experimental data and application range of recent models. Ann Geophys 48:633-646
Bouvier AS, Métrich N, Deloule E (2008) Slab-derived fluids in magma sources of St Vincent (Lesser Antilles Arc): volatiles and light elements imprints. J Petrol 49:1427-1448, doi:10.1093/petrology/egn031
Carroll MR, Rutherford MJ (1987) The stability of igneous anhydrite: experimental results and applications for sulfur behaviour in the 1982 El Chichon trachyandesite and other evolved magmas. J Petrol 28:781-801

Carroll MR, Rutherford MJ (1988) Sulphur speciation in hydrous experimental glasses of varying oxidation states: results from measured wavelength shifts of sulphur X-ray. Am Mineral 73:845-849

Carroll MR, Webster JD (1994) Solubilities of sulfur, noble gases, nitrogen, chlorine and fluorine in magmas. Rev Mineral Geochem 30:231-279

Cashman KV (2004) Volatile controls on magma ascent and eruption. *In*: The State of the Planet: Frontiers and Challenges in Geophysics. AGU Geophys Monogr Series 150. Sparks RSJ, Hawkesworth CJ (eds), Washington DC, p 109-124

Cervantes P, Kamenetsky V, Wallace P (2002) Melt Inclusion Volatile Contents, Pressures of Crystallization for Hawaiian Picrites, and the Problem of Shrinkage Bubbles. EOS Trans, AGU, 83(47), Fall Meet Suppl Abstract V22A-1217

Cervantes P, Wallace P (2003a) Role of H_2O in subduction-zone magmatism: New insights from melt inclusions in high-Mg basalts from central Mexico. Geology 31:235-238

Cervantes P, Wallace P (2003b) Magma degassing and basaltic eruption styles: a case study of 2000 year BP Xitle volcano in Central Mexico. J Volcanol Geotherm Res 120:249-270

Chaussidon M, Albarède F, Sheppard SM (1989) Sulphur isotope variations in the mantle from ion microprobe analyses of micro-sulphide inclusions. Earth Planet Sci Lett 92:144-156

Coltelli M, Del Carlo P, Pompilio M, Vezzoli L (2005) Explosive eruption of a picrite: The 3930 BP subplinian eruption of Etna volcano (Italy). Geophys Res Lett 32:L23307, doi:10.1029/2005GL024271

Cottrell E, Spiegelman M, Langmuir CH (2002) Consequences of diffusive re-equilibration for the interpretation of melt inclusions. Geochem Geophys Geosys 3, doi:10.1029/2001GC000205

Danyushevsky LV, Della-Pasqua FN, Sokolov S (2000) Re-equilibration of melt inclusions trapped by magnesian olivine phenocrysts from subduction-related magmas: petrological implications. Contrib Mineral Petrol 138:68-83

Danyushevsky LV (2001) The effect of small amounts of H_2O on crystallization of mid-ocean ridge and bark-arc basin magmas. J Volcanol Geotherm Res 110:265-280

Danyushevsky LV, McNeill AW, Sobolev AV (2002) Experimental and petrological studies of melt inclusions in phenocrysts from mantle-derived magmas: an overview of techniques, advantages and complications. Chem Geol 183:5-24

de Hoog JCM, Taylor BE, van Bergen MJ (2001) Sulfur isotope systematics of basaltic lavas from Indonesia: implications for the sulfur cycle in subduction zones. Earth Planet Sci Lett 189:237-252

Di Carlo I, Pichavant M, Rotolo S, Scaillet B (2006) Experimental Crystallization of a High-K Arc Basalt: the Golden Pumice, Stromboli Volcano (Italy), J Petrol 1-27 doi:10.1093/petrology/egl011

Di Muro A, Villemant B, Montagnac G, Scaillet B, Reynard B (2006) Quantification of water content and speciation in natural silicic glasses (phonolite, dacite, rhyolite) by confocal MicroRaman spectroscopy. Geochim Cosmochim Acta 70:2868-2884

Dixon JE (1997) Degassing of alkalic basalts. Am Mineral 82:368-378

Dixon JE, Clague DA, Stolper EM (1991) Degassing history of water, sulfur, and carbon in submarine lavas from Kilauea volcano, Hawaii. J Geol 99:371-394

Dixon JE, Stolper EM (1995) An experimental study of water and carbon dioxide solubilities in mid-ocean ridge basaltic liquids. Part II: Applications to degassing. J Petrol 36:1633-1646

Dixon JE, Stolper EM, Holloway JR (1995) An experimental study of water and carbon dioxide solubilities in mid-ocean ridge basaltic liquids. Part I: calibration and solubility models. J Petrol 36:1607-1631

Dixon JE, Pan V (1997) determination of the molar absorptivity of dissolved carbonate in basaltic glass. Am Mineral 80:1339-1342

Donaldson CH (1976) An experimental investigation of olivine morphology. Contrib Mineral Petrol 57:187-213

Eiler JM, Carr MJ, Reagan M, Stolper E (2005) Oxygen isotope constraints on the sources of Central American arc lavas. Geochem Geophys Geosys 6: Q07007, doi:10.1029/2004GC000804

Faure F, Trolliard G, Nicollet C, Montel J-M (2003) A developmental model of olivine morphology as a function of the cooling rate and the degree of undercooling. Contrib Mineral Petrol 145:251-263

Faure F, Schiano P (2004) Crystal morphologies in pillow basalts: implications for mid-ocean ridge processes. Earth Planet Sci Lett 220:331-344

Faure F, Schiano P (2005) Experimental investigation of equilibration conditions during forsterite growth and melt inclusion formation. Earth Planet Sci Lett 236:882-898

Fine GJ, Stolper EM (1986) Carbon dioxide in basaltic glasses: concentrations and speciation. Earth Planet Sci Lett 76:263-278

Fisher T, Marty B (2005) Volatile abundances in the sub-arc mantle: insights from volcanic and hydrothermal gas discharges. J Volcanol Geotherm Res 140:205-216

Freda C, Baker DR, Scarlato P (2005) Sulfur diffusion in basaltic melts. Geochim Cosmochim Acta 69:5061-5069

Gaetani GA, Watson EB (2000) Open system behavior of olivine hosted melt inclusions. Earth Planet Sci Lett 83:27-41

Gaetani GA, Watson EB (2002) Modeling the major-element evolution of olivine-hosted melt inclusions. Chem Geol 183:25-41

Gardner JE, Burgisser A, Hort M, Rutherford M, (2006) Experimental and model constraints on degassing of magma during ascent and eruption. *In:* Neogene-Quaternary continental margin volcanism: A perspective from Mexico. Siebe C, Macias JL, Aguirre-Diaz GJ (eds) Geol Soc Am Spec Paper 402:99-113

Gerlach TM, Graeber EJ (1985) Volatile budget of Kilauea volcano. Nature 313:273-277

Gerlach TM, McGee KA, Elias T, Sutton AJ, Doukas MP (2002) The carbon dioxide emission rate of Kïlauea Volcano; implications for primary magma CO_2 content and summit reservoir dynamics. J Geophys Res 107:2189

Gioncada A, Clocchiatti R, Sbrana A, Bottazzi P, Massare D, Ottolini L (1998) A study of melt inclusions at Vulcano (Aeolian islands, Italy): insights on the primitive magmas and on the volcanic feeding system. Bull Volcanol 60:286-306

Giordano D, Dingwell DB (2003) Viscosity of hydrous Etna basalt: implications for Plinian-style basaltic eruptions. Bull Volcanol 65:8-14

Gonnermann HM, Manga M (2005) Nonequilibrium magma degassing: Results from modeling of the ca. 1340 A.D. eruption of Mono Craters, California. Earth Planet Sci Lett 238:1-16

Grove TL, Chatterjee N, Parman SW, Medard E (2006) The influence of H_2O on mantle wedge melting. Earth Planet Sci Lett 249:74-89

Grove TL, Baker M, Price RC, Parman SW, Elkins-Tanton LT, Chatterjee N, Muentener O (2005) Magnesian andesite and dacite lavas from Mt. Shasta, northern California: products of fractional crystallization of H_2O-rich mantle melts. Contrib Mineral Petrol 148:542-565

Grove TL, Elkins LT, Parman SW, Chatterjee N, Müntener O, Gaetani GA (2003) Fractional crystallization and mantle-melting controls on calc-alkaline differentiation trends. Contrib Mineral Petrol 145:515-533

Grove TL, Parman SW, Bowring SA, Price RC, Baker MB (2002) The role of an H_2O-rich fluid component in the generation of primitive basaltic andesites and andesites from Mt. Shasta region, N. California. Contrib Mineral Petrol 142:375-396

Hansteen TH, Klügel A (2008) Fluid inclusion thermobarometry as a tracer for magmatic processes. Rev Mineral Geochem 69:143-177

Hamada M, Fujii T (2007) H_2O-rich island arc low-K tholeiite magma inferred from Ca-rich plagioclase-melt inclusion equilibria. Geochem J 41:437-461

Hauri E (2002) SIMS analysis of volatiles in silicate glasses, 2: isotopes and abundances in Hawaiian melt inclusions. Chem Geol 183:15-141

Hauri E, Wang J, Dixon JE, King PL, Mandeville C, Newman S (2002) SIMS analysis of volatiles in silicate glasses, 1: Calibration, matrix effects and comparisons with FTIR. Chem Geol 183:99-114

Hauri EH, Gaetani GA, Green TH (2006) Partitioning of water during melting of the Earth's upper mantle at H_2O-undersaturated conditions. Earth Planet Sci Lett 24:715-734

Hochstaedter AG, Ryan JG, Luhr JF, Hasenaka T (1996) On B/Be ratios in the Mexican Volcanic belt. Geochim Cosmochim Acta 60:613-628

Holzeid A, Grove TL (2002) Sulfur saturation limits in silicate melts and their implications for core formation scenarios for terrestrial planets. Am Mineral 87:227-237

Housh TB, Luhr JF (1991) Plagioclase-melt equilibria in hydrous systems. Am Mineral 76:477-492

Humphreys MCS, Kearns SL, Blundy JD (2006) SIMS investigation of electron-beam damage to hydrous, rhyolitic glasses: Implications for melt inclusion analysis. Am Mineral 91:667-679

Ihinger PD, Hervig RL, McMillan PF (1994) Analytical methods for volatiles in glasses. Rev Mineral 30:67-121

Jaupart C, Vergniolle S (1988) Laboratory models of Hawaiian and Strombolian eruptions. Nature 331:58-60

Jaupart C, Vergniolle S (1989) The generation and collapse of a foam layer at the roof of a basaltic magma chamber. J Fluid Mechanic 203:347-380

Javoy M, Pineau F (1991) The volatiles record of a "popping' rock from the Mid-Atlantic Ridge at 14°N: chemical and isotopic composition of gas trapped in the vesicles. Earth Planet Sci Lett 107:598-611

Johnson EJ, Wallace PJ, Cashman KV, Delgado Granados H, Kent A (2008) Magmatic volatile contents and degassing-induced crystallization at Volcan Jorullo, Mexico: Implications for melt evolution and the plumbing systems of monogenetic volcanoes. Earth Planet Sci Lett 269:477-486

Johnson EJ, Wallace PJ, Delgado Granados H, Manea VC, Kent A, Bindeman I, Donegan C (2008) The origin of H_2O-rich subduction components beneath the Michoacán-Guanajuato Volcanic Field, Mexico: insights from magmatic volatile contents, oxygen isotopes, and 2-D thermal models for the subducted slab and mantle wedge. J Petrol (in review)

Jugo PJ, Luth RW, Richards JP (2005a) Experimental data on the speciation of sulfur as a function of oxygen fugacity in basaltic melts. Geochim Cosmochim Acta 69:477-503

Jugo PJ, Luth RW, Richards JP (2005b) An experimental study of the sulfur content in basaltic melts saturated with immiscible sulphide or sulfate liquids at 1300 °C and 1.0 GPa. J Petrol 46:783-798

Kamenetsky V, Pompilio M, Métrich N, Sobolev AV, Kuzmin DV, Thomas R (2007) Arrival of extremely volatile-rich high-Mg magmas changes explosivity of Mount Etna. Geology 35:255-258, doi:10.1130/G23163A.1

Kelley KA, Plank T, Grove TL, Stolper EM (2006) Mantle melting as a function of water content beneath back-arc basins. J Geophys Res 111, B09208, doi:10.1029/ 2005JB003732

Kent AJR (2008) Melt inclusions in basaltic and related volcanic rocks. Rev Mineral Geochem 69:273-331

Kent AJR, Peate DW, Newman S, Stolper EM, Pearce JA (2002) Chlorine in submarine glasses from the Lau Basin: seawater contamination and constraints on the composition of slab-derived fluids. Earth Planet Sci Lett 202:361-377

Kessel R, Schmidt MW, Ulmer P, Pettke T (2005) Trace element signature of subduction-zone fluids, melts and supercritical liquids at 120-180 km depth. Nature 437: doi:10.1038/nature03971

King PL, Vennemann TW, Holloway JR, Hervig RL, Lowenstern JB, Forneris JF (2002) Analytical techniques for volatiles: A case study using intermediate (andesitic) glasses. Am Mineral 87:1077-1089

King PL, Holloway JR (2002) CO_2 solubility and speciation in intermediaite (andesitic) melts: the role of H_2O and composition. Geochim Cosmochim Acta 66:1627-1640

Kohut EJ, Stern R, Kent AJR., Nielsen RL, Bloomer SH, Leybourne M (2006) Evidence of adiabatic decompression melting in the Southern Marianna Arc from high-Mg lavas and melt inclusions. Contrib Mineral Petrol, doi:10.1007/s00410-006-0102-7

Landi P, Métrich N, Bertagnini A, Rosi M (2004) Dynamics of magma mixing and degassing recorded in plagioclase at Stromboli (Aeolian Archipelago, Italy). Contrib Mineral Petrol 147:213-237

Le Roux PJ, Shirey SB, Hauri EH, Perfit MR, Bender JF (2006) The effects of variable sources, processes and contaminants on the composition of northern EPR MORB (8-10°N and 12-14°N): Evidence for volatiles (H_2O, CO_2, S) and halogens (F, Cl). Earth Planet Sci Lett 251:209-231

Li C, Ripley EM (2005) Empirical equations to predict the sulfur content of mafic magmas at sulfide saturation and applications to magmatic sulfide deposits. Mineral Deposit 40:18-230

Liu Y, Samaha N-T, Baker DR (2007) Sulfur concentration at sulfide saturation (SCSS) in magmatic silicate melts. Geochim Cosmochim Acta 71:1783-1799

Lowenstern JB (1995) Application of silicate-melt inclusions to the study of magmatic volatiles. *In:* Magmas, Fluids and Ore Deposits. Thompson JFH (ed) Mineral Assoc Canada Short Course Series 23:71-100

Luhr JF (1990) Experimental phase relations of water and sulfur saturated arc magmas and the 1982 eruptions of El Chichon volcano. J Petrol 31:1071-1114

Luhr JF (1992) Slab-derived fluids and partial melting in subduction zones - insights from 2 contrasting Mexican volcanoes (Colima and Ceboruco). J Volcanol Geothem Res 54:1-18

Luhr JF (1997) Extensional tectonics and the diverse primitive volcanic rocks in the western Mexican Volcanic Belt. Can Mineral 35:473-500

Luhr JF (2001) Glass inclusions and melt volatile contents at Paricutin volcano, Mexico. Contrib Petrol Mineral 142:261-283

Mandeville CW, Webster JD, Rutherford MJ, Taylor BE, Timbal A, Faure K (2002) Determination of molar absorptivities in infrared absorption bands of H_2O in andesitic glasses. Am Mineral 87:813-821

Marianelli P, Sbrana A, Métrich N, Cecchetti A (2005) The deep feeding system of Vesuvius involved in recent violent strombolian eruptions. Geophys Res Lett 32:L02306, doi:10.1029/2004GL021667

Massare D, Métrich N, Clocchiatti R (2002) High-temperature experiments on silicate melt inclusions in olivine at one atmosphere. Inference on temperatures of homogenization and H_2O concentrations. Chem Geol 183:87-98

Mavrogenes JA, O'Neill HC (1999) The relative effects of pressure, temperature and oxygen fugacity on the solubility of sulfide in mafic magmas. Geochim Cosmochim Acta 63:1173-1180

Menand T, Philips JC (2007) Gas segregation in dykes and sills. J Volcanol Geotherm Res 159:393-408

Mercier M et al. (2008) Influence of glass polymerization and oxidation on microRaman water analysis in alumino-silicate glasses. Geochim Cosmochim Acta (accepted)

Métrich N, Clocchiatti R (1996) Sulfur abundance and its speciation in oxidized alkaline melts. Geochim Cosmochim Acta 60:4151-4160

Métrich N, Ruhterford MJ (1998) Low pressure crystallization paths of H_2O-saturated basaltic-hawaiitic melts from Mt Etna: Implications for open-system degassing of basaltic volcanoes. Geochim Cosmochim Acta 62:1195-1205

Métrich N, Schiano P, Clocchiatti R, Maury RC (1999) Transfer of sulfur in subduction settings. An example from Batan island (Luzon volcanic arc, Philippines). Earth Planet Sci Lett 67:1-14

Métrich N, Bertagnini A, Landi P, Rosi M (2001) Crystallization driven by decompression and water loss at Stromboli volcano (Aeolian island) Italy. J Petrol 42:1471-1490

Métrich N, Susini, Galoisy L, Calas G, M. Bonnin-Mosbah·M., Menez B (2003) X-ray microspectroscopy of sulphur in basaltic glass inclusions. Inference on the volcanic sulphur emissions. J Phys 4 104:393-398

Métrich N, Allard P, Spilliaert N, Andronico D, Burton M (2004) 2001 flank eruption of the alkali- and volatile-rich primitive basalt responsible for Mount Etna's evolution in last three decades. Earth Planet Sci Lett 228:1-17

Métrich N, Berry AJ, O'Neill H St C, Susini J (2008) The oxidation state of sulfur in synthetic and natural glasses determined by X-ray absorption spectroscopy. Geochim Cosmochim Acta (submitted)

Métrich N, Bertagnini A, Landi P, Rosi M, Belhadj O (2005) Triggering mechanism at the origin of paroxysms at Stromboli (Aeolian archipelago, Italy): the 5 April 2003 eruption. Geophys Res Lett 32:L103056, doi:10.1029/2004GL022257

Métrich N, Kamenetsky V, Allard P, Pompilio M (2007) Trace element and volatile signature of Etna magma source(s): a melt inclusion approach. 2007 Goldschmidt Conference Cologne. Geochim Cosmochim Acta Suppl 71(15):A658

Michael PJ (1995) Regionally distinctive sources of depleted MORB: evidence from trace elements and H_2O. Earth Planet Sci Lett 131:301-320

Moore G, Vennemann T, Carmichael ISE (1998) An empirical model for the solubility of H_2O in magmas to 3 kilobars. Am Mineral 83:36-42

Moore G (2008) Interpreting H_2O and CO_2 contents in melt inclusions: constraints from solubility experiments and modeling. Rev Mineral Geochem 69:333-361

Moretti R, Ottonello G (2005) Solubility and speciation of sulfur in silicate melts: The conjugated Toop-Samis-Flood-grjotheim (CTSFG) model. Geochim Cosmochim Acta 69:801-823

Newman S, Lowenstern JB (2002) VolatileCalc: a silicate melt-H_2O-CO_2 solution model written in Visual Basic Excel. Comp Geosci 2:597-604

Newman S, Stolper EM, Stern RJ (2000) H_2O and CO_2 in magmas from Mariania arc and back arc systems. Geochem Geophys Geosyst 1:1999GC000027

Nichols ARL, Wysoczanski RJ (2007) Using micro-FTIR spectroscopy to measure volatile contents in small and unexposed inclusions hosted in olivine crystals. Chem Geol 242:371-384

Ochs FA, Lange RA (1999) The density of hydrous magmatic liquids. Science 283:1314-1317

O'Neill HSC, Mavrogenes JA (2002) The sulfide capacity and sulfur content at sulfite saturation of silicate melts at 1400 °C and 1b. J Petrol 43:1049-1087

Papale P, Moretti R, Barbato D (2006) The compositional dependence of the saturation surface of H_2O + CO_2 fluids in silicate melts. Chem Geol 229:78-95

Parfitt EA (2004) A discussion of the mechanisms of explosive basaltic eruptions. J Volcanol Geotherm Res 134:77-107

Parfitt EA, Wilson L (1995) Explosive volcanic eruptions - IX. The transition between Hawaiian-style lava fountaining and strombolian explosive activity. Geophys J Int 121:226-232

Pearce JA, Peate DW (1995) Tectonic implications of the composition of volcanic arc magmas. Annu Rev Earth Planet Sci 23:251-285

Pichavant M, Mysen BO, Macdonald R (2002) Source and H_2O content of high-MgO magmas in island arc settings: an experimental study of a primitive calc-alkaline basalt from St. Vincent, Lesser Antilles arc. Geochim Cosmochim Acta 66:2193-2209

Pichavant M, Macdonald R (2007) Crystallization of primitive basaltic magmas at crustal pressures and genesis of the calc-alkaline igneous suite: experimental evidence from St Vincent, Lesser Antilles arc. Contrib Mineral Petrol 2007, doi10.1007/s00410-007-0208-6

Philippot P, Agrinier P, Scambelluri M (1998) Chlorine cycling during subduction of altered oceanic crust. Earth Planet Sci Lett 161:33-44

Philips JC, Woods AW (2001) Bubble plume generated during recharge of basaltic magma reservoirs. Earth Planet Sci Lett 186:297-309

Pioli L, Erlund E, Johnson E, Cashman K, Wallace P, Rosi M, Delgado Granados H (2008) Explosive dynamics of violent Strombolian eruptions: The eruption of Parícutin Volcano 1943-1952 (Mexico). Earth Planet Sci Lett 271:359-368

Portnyagin M, Hoernle K, Plechov P, Mironov N, Khununaya S (2007) Constraints on mantle melting and composition and nature of slab components in volcanic arcs from volatiles (H_2O, S, Cl, F) and trace elements in melt inclusions from the Kamchatka Arc. Earth Planet Sci Lett 255:53-69

Portnyagin M, Almeev R, Matveev S, Holtz F (2008) Experimental evidence for rapid water exchange between melt inclusions in olivine and host magma. Earth Planet Sci Lett 272:541-552

Richet P, Lejeune AM, Holtz F, Roux J (1996) Water and viscosity of andesite melts. Chem Geol 128:185-197

Righter K, Chesley JT, Calazza CM, Gibson EK, Ruiz J (2008) Re and Os concentrations in arc basalts: The roles of volatility and source region fO_2 variations. Geochim Cosmochim Acta 72:926-947

Riker J (2005) The 1859 eruption of Mauna Loa Volcano, Hawaii: Controls on the development of long lava channels. Unpublished M.S. Thesis, University of Oregon

Robock A, Oppenheimer C (2003) Volcanism and the Earth's Atmosphere. Geophys Monograph 139, Washington DC

Roedder E (1984) Fluid Inclusions, Am Mineral Soc, Chelsea, Michigan

Roggensack K, Hervig RL, McKnight SB, Williams SN (1997) Explosive basaltic volcanism from Cerro Negro volcano: influence of volatiles on eruptive style. Science 277:1639-1642

Roggensack K (2001) Unraveling the 1974 eruption of Fuego volcano (Guatemala) with small crystals and their young melt inclusions. Geology 29:911-914

Rowe MC, Kent AJR, Nielsen RL (2007) Determination of sulfur speciation and oxidation of olivine-hosted melt inclusions. Chem Geol 236:303-322

Sadofsky SJ, Portnyagin M, Hoernle K, van den Bogaard P (2008) Subduction cycling of volatiles and trace elements through the Central American volcanic arc: evidence from melt inclusions. Contrib Mineral Petrol 155, doi:10.1007/s00410-007-0251-3

Scaillet B, Pichavant M (2005) A model of sulphur solubility for hydrous mafic melts: application to the determination of magmatic fluid compositions of Italian volcanoes. Ann Geophys 48:671–698

Scambelluri M, Fiebig J, Malaspina N, Muntener O, Pettke T (2004) Serpentinite subduction: Implications for fluid processes and trace-element recycling. Int Geol Rev 46:595-613

Self S, Widdowson M, Thordarson T, Jay AE (2006) Volatile fluxes during flood basalt eruptions and potential effects on global environment: A Deccan perspective. Earth Planet Sci Lett 248:517-531

Sharma K, Blake S, Self S, Krueger AJ (2004) SO_2 emissions from basaltic eruptions, and excess sulfur issue. Geophys Res Lett 31:L13612, doi:10.1029/2004GL0.19688

Shaw HR (1972) Viscosities of magmatic liquids: an empirical method of prediction. Am J Sci 272:870-893

Silver LA, Ihinger PD, Stolper E (1990) The influence of bulk composition on the speciation of water in silicate glasses. Contrib Mineral Petrol 104:142-162

Sisson TW (2003) Native gold in a Hawaiian alkalic magma. Econ Geol 95:643-648

Sisson TW, Bronto S (1998) Evidence for pressure-release melting beneath magmatic arcs from basalt at Galunggung, Indonesia. Nature 391:883-885

Sisson TW, Layne GD (1993) H_2O in basalt and basaltic andesite glass inclusions from 4 subduction-related volcanoes. Earth Planet Sci Lett 117:619-635

Sisson TW, Grove TL (1993) Temperatures and H_2O contents of low-MgO high-Alumina basalts. Contrib Mineral Petrol 113:167-184

Sobolev AV, Clocchiatti R, Dhamelincourt P (1983) Les variations de la température, de la composition du magma et l'estimation de la pression partielle d'eau pendant la cristallisation de l'olivine dans les océanites du Piton de la Fournaise (Réunion, éruption de 1966). C R Acad Sci Paris 296:275-280

Sobolev AV, Chaussidon M (1996) H_2O concentrations in primary melts from supra-subduction zones and mid-oceanic ridges: Implications for H_2O storage and recycling in the mantle. Earth Planet Sci Lett 137:45-55

Sobolev VS, Kostyuk VP (1975) Magmatic Crystallisation as Based on the Study of Melt Inclusions. "Nauka", Novosibirsk, pp 232

Sparks RSJ (1978) The dynamics of bubble formation and growth in magmas. J Volcanol Geotherm Res 3:1-37

Spilliaert N, Allard P, Métrich N, Sobolev A (2006a) Melt inclusion record of the conditions of ascent, degassing and extrusion of volatile-rich alkali basalt during the powerful 2002 flank eruption of Mount Etna (Italy). J Geophys Res 111:B04203, doi:10.1029/2005/JB003934

Spilliaert N, Métrich N, Allard P (2006b) S-Cl-F degassing pattern of water-rich alkali basalt: modelling and relationship with eruption styles on Mount Etna volcano. Earth Planet Sci Lett 248:772-786

Stolper EM, Newman S (1994) The role of water in the petrogenesis of Mariana trough magmas. Earth Planet Sci Lett 121:293-325

Tait S, Jaupart C, Vergniolle S (1989) Pressure, gas content and eruption, periodicity in a shallow crystallizing magma chamber. Earth Planet Sci Lett 92:107-123

Thomas R (2000) Determination of water contents of granite melt inclusions by confocal laser Raman microprobe spectroscopy. Am Mineral 85:868-872

Thomas R, Kamenetsky VS, Davidson P (2006) Laser Raman spectroscopic measurements of water in unexposed glass inclusions. Am Mineral 91:467-470

Thordarson T, Self S (2003) Atmospheric and environmental effects of the 1783-1784 Laki eruption: A review and reassessment. J Geophys Res 108(D1):4011, doi:10.1029/2001JD002042

Thordarson T, Self S, Oskarsson N, Hulsechosch T (1996) Sulfur, chlorine and fluorine degassing and atmospheric loading by the 1783-1784 AD (Skaftar Fires) eruption in Iceland. Bull Volcanol 58:205-225

Vergniolle S, Jaupart C (1990) Dynamics of degassing at Kilauea Volcano, Hawaii. J Geophys Res 95:2793-2809

Vergniolle S, Mangan S (2000) Hawaian and strombolian eruption. *In*: Encyclopedia of Volcanoes. Sigurdsson H (ed) Academic Press, p 447-461

Verhoogen J (1949) Thermodynamics of a magmatic gas phase. Univ Calif Bull Dept Geol Sci 28:91-136

Varela ME, Métrich N, Bonnin-Mosbah M, Kurat G (2000) Carbon in glass inclusions of Allende, Vigarano, Bali and Kaba (CV3) olivines. Geochim Cosmochim Acta 64:3923-3930

Vigouroux N, Wallace PJ, Kent AJR (2008) Volatiles in high-K magmas from the western Trans-Mexican Volcanic Belt: Evidence for fluid-flux melting and extreme enrichment of the mantle wedge by subduction processes. J Petrol, doi:10.1093/petrology/egn039

Wade JA, Plank T, Melson WG, Soto GJ, Hauri AH (2006) Volatile content of magmas from Arenal volcano, Costa Rica. J Volcanol Geotherm Res 157:94-120

Walker J, Roggensack K, Patino LC, Cameron BI, Matias O (2003) The water and trace element contents of melt inclusions accross an active subduction zone. Contrib Mineral Petrol 146:62-77

Wallace PJ (2001) Volcanic SO_2 emissions and the abundance and distribution of exsolved gas in magma bodies. J Volcanol Geotherm Res 108:85-106

Wallace P (2005) Volatiles in subduction zone magmas: concentrations and fluxes based on melt inclusion and volcanic gas data. J Volcano Geotherm Res 140:217-240

Wallace P, Carmichael ISE (1992) Sulfur in basaltic magmas. Geochim Cosmochim Acta 56:1863-1874

Wallace PJ, Carmichael ISE (1994) S speciation in submarine basaltic glasses as determined by measurements of $SK\alpha$ X-ray wavelength shifts. Am Mineral 79:161-167

Wallace P, Anderson AT (1998) Effects of eruption and lava drainback on the H_2O contents of basaltic magmas at Kilauea volcano. Bull Volcanol 59:327-344

Wallace PJ, Carmichael ISE (1999) Quaternary volcanism near the Valley of Mexico: Implications for subduction zone magmatism and the effects of crustal thickness variations on primitive magma compositions. Contrib Mineral Petrol 135:291-314

Wallace P, Anderson AT, Davis AM (1999) Gradients in H_2O, CO_2 and exsolved gas in a large-volume silicic magma system: interpreting the record preserved in melt inclusions from the Bishop Tuff. J Geophys Res 104:20,097-20,122

Webster JD, Kinzler RJ, Mathez EA (1999) Chloride and water solubility in basalt and andesite melts and implication for magmatic degassing. Geochim Cosmochim Acta 63:729-738

Webster JD (2004) The exsolution of magmatic hydrosaline chloride liquids. Chem Geol 210:33-48

Wilke M, Jugo PJ, Klimm K, Susini J, Botcharnikov R, Kohn SC, Janousch M (2008) The origin of S^{4+} detected in silicate glasses by XANES. Am Mineral 93:235-240

Wilson L, Head JW (1981) Ascent and eruption of basaltic magma on the Earth and Moon. J Geophys Res 86:2971-3001

Witter JB, Kress VC, Delmelle P, Stix J (2004) Volatile degassing, petrology, and magma dynamics of the Villarica lava lake, Southern Chile. J Volcanol Geotherm Res 134:303-337

Witter JB, Kress VC, Newhall CG (2005) Volcan Popocatépelt, Mexico. Petrology, Magma mixing and immediate sources of volatiles for the 1994-present eruption. J Petrol 46:2337-2366

Workman RK, Hart SR (2005) Major and trace element composition of the depleted MORB mantle (DMM). Earth Planet Sci Lett 231:53-72

Wysoczanski RJ, Tani K (2006) Spectroscopic FTIR imaging of water species in silicic volcanic glasses and melt inclusions: an example from the Izu-Bonin arc. J Volcanol Geotherm Res 156:302-314

Zhang YX, Xu XJ, Zhu MF (2007) Silicate melt properties and volcanic eruptions. Rev Geophys 45:RG4004

Inter- and Intracrystalline Isotopic Disequilibria: Techniques and Applications

Frank C. Ramos
Department of Geological Sciences
New Mexico State University
Las Cruces, New Mexico, 88003-8001, U.S.A.

framos@nmsu.edu

Frank J. Tepley III
College of Oceanic and Atmospheric Sciences
Oregon State University
Corvallis, Oregon, 97331-5503, U.S.A.

ftepley@coas.oregonstate.edu

INTRODUCTION

In the last ~20 years, we have seen a significant expansion of techniques related to geochemical and isotopic microsampling of materials in the earth sciences. From constraining pre-eruptive histories of flood basalt magmas to identifying the natal rivers of origin of anadromous fishes, these techniques have had significant impacts in a wide variety of scientific fields. Nowhere has the impact been greater than in identifying the sources, processes, and timing of processes involved in igneous magmatic systems. Both technique refinements and the development of new technologies have aided in advancing microsampling applications, thus allowing for a better understanding of the sources and mechanisms responsible for changing geochemical and isotopic signatures in natural systems. In this chapter, we focus on the techniques and technologies associated with radiogenic isotope microsampling and review applications of these techniques as utilized in scientific investigations.

Isotope microsamping is a logical extension of earlier studies that evaluated individual components of magmas and magma systems, including melts and minerals. From the use of petrographic microscopes and the later introduction of the electron microprobe, the focus on internal chemical variations in melts and minerals is critical to assessing the petrogenetic histories of igneous rocks. Even today, these technologies are used to ensure that further trace element and isotopic analyses are undertaken in a textural and major element context. For trace elements and isotopes, early studies (e.g., Cortini and van Calsteren 1985) confirmed variations in the melt and mineral components of many igneous rocks but focused on mineral or glass separates. Potential information associated with isotopic variations retained by individual crystals or internal variations within individual crystals was lost. Later studies such as Geist et al. (1988) focused on isotope variations within single megacrysts to constrain mixing scenarios between basaltic, andesitic, and rhyolitic magmas, and Davidson et al. (1990) used microdrilling to document isotopic variations across an enclave-host magma interface and to identify xenocrystic crystals from within enclaves. These represent early studies that truly focused on "internal" isotopic variations of magmatic components from igneous systems and foreshadowed the potential of future studies utilizing isotopes at the microscale to identify magmatic processes and constrain their impacts on magmatic systems.

Isotope microsampling can be separated into two broad categories: 1) microspatial analyses and 2) microanalyte analyses. For microspatial analyses, samples commonly originate from micromilling, laser ablation sampling, or secondary ionization mass spectrometry (SIMS). For microanalyte analyses, samples are commonly obtained from single crystals or micromilling. As implied by the name, microspatial analyses target a limited area/volume of material such as a limited number of rings of a fish otolith or a single growth layer of a magmatic phenocryst, with the ultimate intent of constraining isotopic variability in the growth history of the target material. Constraining magmatic growth histories of phenocrysts exemplifies the utility of microspatial analyses. Phenocrysts grow from core to rim by the addition of sequential layers of material. Plagioclase, for example, typically contains high concentrations of Sr that allow for small portions of mineral to be successfully analyzed for $^{87}Sr/^{86}Sr$ ratios. By analyzing sequential layers, changes in the isotopic character of the magma in which the crystal grew, imposed by open-system processes, can be constrained (Tepley et al. 2000; Ramos et al. 2005). Thus, microspatial $^{87}Sr/^{86}Sr$ isotope analyses are well-suited to identify and track the effects of open-system processes occurring during mineral growth.

With microanalyte techniques, spatial resolution may be sacrificed to obtain the minimum required amount of analyte needed for a successful analysis. An example is that of Sr or Pb in quartz. The mineral structure and composition of quartz (SiO_2) does not readily allow for Sr and Pb elemental substitution, two elements commonly used as radiogenic tracers. As a result, Sr and Pb concentrations in quartz are low and typically result from melt or mineral inclusions captured during growth of individual crystals. Low Sr and Pb concentrations and the small size of quartz-hosted inclusions usually preclude them from being individually analyzed by microspatial techniques. However, these crystals can be approached from a microanalyte perspective using single grain analyses where all analyte is assumed to result from inclusions rather than the quartz host. Such single crystals must however be large enough to contain enough inclusions to yield the minimum required amount of analyte for successful analyses. Thus, microspatial and microanalyte techniques overlap to some degree but they also diverge so as to become very different in nature and, thus, scientific interpretation.

To comprehensively address isotope microanalysis techniques, we introduce the types of isotope analyses undertaken, the technologies involved in these isotope analyses, the critical aspects of these isotope analyses, and review a range of applications which have been undertaken. Our intent is to offer a broad review of successful applications, not an exhaustive review of all studies.

MICROSPATIAL ISOTOPE ANALYSIS

Early petrographic studies focusing on mineral phenocrysts from volcanic rocks documented textural evidence for highly variable magmatic histories recorded during mineral growth. Assessments of these varied histories were initially evaluated by measuring *in situ* major element variations (e.g., core to rim transects) using the electron microprobe targeted toward individual mineral phenocrysts. Such measurements were critical in assessing major element chemical changes of magmas occurring during mineral crystallization. These changes could result from a myriad of magmatic processes including, but not limited to, crystal fractionation, magma mixing, magma recharge, and/or processes associated with assimilation/fractional crystallization (AFC). Major element variations, however, could only account for broad assessments supporting major element modification of magmas in which crystals grew. Major elements alone could not, for example, confirm the presence of external inputs into magmatic systems such as required by AFC processes. Thus, identifying endmembers responsible for such external changes was more problematic.

In contrast to major elements, radiogenic isotopes offer an intriguing alternative in assessing the influence of either internal (closed-system) or external (open-system) inputs into magmatic systems and constraining the isotopic composition of the endmembers involved. Combining major element and isotopes has vast potential in understanding igneous petrogenesis in addition to other scientific fields. Initially, microdrilling (referred to as micromilling in this chapter) using diamond-embedded bits designed for the dental and engineering industries was used to obtain small samples in a microspatial context. Early attempts obtained samples on scales of 300 μm diameter by 300 μm depth and typically targeted Sr isotopes (e.g., Tepley et al. 1999). Fortunately, for most systems, even this level of microspatial sampling documented large isotopic variations in magmatic phenocrysts resulting from open-system processes affecting phenocryst-related host magmas. Early targets included minerals such as plagioclase and calcite, minerals generally characterized by high Sr contents (≥500 ppm). After milling, samples were dissolved and purified using chromatographic techniques and analyzed using thermal ionization mass spectrometry (TIMS). Analyte amounts were large (≥100 ng Sr) and required purification to removed interfering elements (e.g., Rb) prior to analysis. Partly as a result, micromilling became accepted as a tool for better understanding processes associated with magmagenesis and the popularity of the technique increased to the point that commercial micromills were built.

Since the mid- to late-1990s, micromilling has been a relatively common technique for acquiring samples, especially for materials requiring purification using column chromatography (e.g., Rb-rich potassium feldspar, Knesel et al. 1999). In addition, bits are now either diamond-embedded or made from highly pure tungsten carbide. Both are commonly used for silicate and carbonate applications that target a range of problems in igneous petrology. In addition, applications associated with biology and ecology have used isotopes to trace mammal movements (e.g., Hoppe et al. 1999) and identify the natal rivers of origin of marine caught salmon (e.g., Barnett-Johnson et al. 2005). Even mineral formation rates in meteorites have been examined using micromilling (Bizzarro et al. 2004; Thrane et al. 2006).

In addition to micromilling, two *in situ* techniques are commonly utilized for microspatial isotope analyses. The first uses secondary ionization mass spectrometry (SIMS) to analyze material that is removed from the surface of samples bombarded by oxygen ions, cesium ions, or electrons. Such analyses are typically used to measure radiogenic Pb isotope ratios that do not require high measurement precisions in minerals such as zircon (Davis et al. 2003; Ireland and Williams 2003). More recently, common Pb isotope ratios have been measured to variable precisions in U- and Th-poor materials such as melt inclusions. These measurements are generally limited to ratios involving the more abundant ^{206}Pb, ^{207}Pb, and ^{208}Pb isotopes and have been useful in identifying mantle endmember components involved in ocean island basalt (OIB) volcanism (e.g., Saal et al. 1998). In addition to Pb isotopes, recent work by Weber et al. (2005) measured Sr isotopes in aragonite otoliths to evaluate natal river sources of salmon (Bacon et al. 2004). As such, expansion of SIMS-related techniques have also allowed for a broader spectrum of microspatial studies.

Microspatial analyses have also benefited from the attainment of increased accuracies and greater measurement precisions using laser ablation sampling. Advances in laser sampling systems and multi-collector inductively coupled mass spectrometry have combined to greatly expand potential targets for Sr and Pb isotope analyses to include highly elementally complex materials such as clinopyroxene and volcanic rock groundmass. In the last 5-7 years, great strides have been made in obtaining highly accurate Sr and Pb isotope measurements with variable but consistently greater measurement precisions which on occasion rival those typical of TIMS (e.g., Ramos et al. 2004; Kent 2008). Below, we review microspatial techniques and applications.

Micromilling

Technique. Historically, the concept of removing growth zones or intracrystalline subsamples from larger samples for chemical analyses was governed by the scientific benefit of such analyses. All sub-disciplines in geology have gained some advantage from applying such techniques from evaluating sulfur isotopes in sulfur-bearing phases in breccias (Lambert et al. 1982), to obtaining X-ray diffraction data from micromilled minerals in calc-silicate granulites (Maaskant et al. 1980), to identifying fish provenance by micromilling otoliths for Sr isotopic analyses (Kennedy et al. 2002). Applying micromilling in igneous petrology for the purposes of evaluating isotopes in magma-related materials was first accomplished by Davidson et al. (1990) in which the cores of plagioclase feldspar crystals, and a small droplet of chilled mafic melt in the same hand sample, were milled using small bits and a basic drill press. Since then, micromilling has been and continues to be used successfully in a range of igneous-related applications.

Applications to igneous petrogenesis, however, are limited by the capability to physically sample intended target materials and to efficiently capture micromilled materials for chemical processing. As early as 1945, an experimental motorized microdrill was developed for extracting hard minerals from polished surfaces of large samples (Wagner 1945). In 1977, a device using a series of levers and joints that allowed for fine movements of a probe or hypodermic needle called a micromanipulator was used for extracting small crystals, 10-20 μm in diameter, from thin sections for further chemical analysis (Rickwood 1977). It was also common to use either hand-held drills (e.g., Dremel™) or mounted drill presses (e.g., Sherline™ vertical mill) assisted with either binocular or petrographic microscopes to mill material from a variety of target materials as long as spatial resolution was not critical. Advances in micrometer stages, optical systems, and finally, computer-assisted or -driven milling devices allowed for sampling on fine spatial scales that are used today (e.g., NewWave™ MicroMill™ or New Hermes™ Vanguard VMC).

Davidson et al. (1990), and many subsequent microdrilling studies, successfully analyzed Sr isotopes in plagioclase. Plagioclase, by far the most common mineral in arc-related volcanic rocks, has great potential to record temporal chemical and isotopic changes associated with magma chamber dynamics (Anderson 1983; Stamatelopoulou-Seymour et al. 1990; Blundy and Shimizu 1991; Singer et al. 1995). It is usually one of the first minerals on the liquidus and commonly crystallizes during magmatic differentiation of basalts and andesites. Compositional zoning in plagioclase typically reflects primary growth owing to slow CaAl-NaSi diffusive exchange within the crystal structure (Grove et al. 1984), and textural features preserved in phenocrysts help identify changing physical and chemical parameters in the host magma from which it crystallizes (e.g., Ginibre et al. 2002, 2004). Combined chemical and textural zoning may be used to reconstruct the conditions in which plagioclase grew, and by proxy, the chemical and physical evolution of the magma itself. Lastly, plagioclase usually contains high concentrations of Sr and relatively low concentrations of Rb which make it ideally suited for Sr isotope evaluations targeting initial Sr isotope ratios (i.e., $^{87}Sr/^{86}Sr$ ratios that have not been measurably influenced by radiogenic ingrowth). In addition to plagioclase, phases such as potassium feldspar, pyroxene, hornblende, apatite, glass, etc., may also be analyzed for Sr isotopes or integrated with additional constraints such as Pb or Nd isotopes from these same phases.

Ultimately an optimal mineral phase and isotope system must be chosen that will yield the most productive results for addressing the specific scientific problem of interest. To accomplish this goal, a balance between elemental concentrations, the precision of the milling technique, the ability to remove milled material from the crystal surface, and the ability of the analytical technique (TIMS or solution MC-ICPMS) to analyze the chemically processed sample accurately and precisely (as the precision required may be a function of the specific scientific application) must be pursued. As noted by Davidson et al. (2007), there is a trade-off between the concentration of the target element and the required minimum amount of material milled and processed such that ample analyte must be acquired to provide for precise and accurate results. As Figure 1 illustrates, higher concentrations of the target element require less overall amounts of material to be sampled, which offers the potential for finer spatial resolution. Alternatively, lower concentrations require greater amounts of milled sample, and thus, coarser spatial resolution.

For example, Sr readily substitutes for Ca in plagioclase feldspars, thus concentrations typically range from a few hundred to a few thousand ppm. Approximately 1-3 ng of Sr are needed for a highly precise TIMS analysis (Charlier et al. 2006), although processing such small samples requires exceptional care. Therefore, a plagioclase with a relatively low Sr concentration of 100 ppm would require ~0.01 mg of sample to yield 1 ng Sr. Alternatively, for a plagioclase with 1000 ppm Sr, only 0.001 mg of sample is required. Clearly the limiting step may be the ability to recover, handle, and process such small amounts of material for analysis. To do so requires having a consistent blank and good knowledge of the blank concentration and isotopic composition (at such low analyte levels, blank corrections become increasingly important). The ideal circumstance is to have a well-characterized sample-blank relationship where blank corrections are minimal.

Micromilling techniques vary between 1) milling individual holes in crystal growth zones to 2) milling shallow troughs along growth zones, in each case generating enough material to undertake appropriate isotopic analyses. Early milling efforts used individual holes in which several small (<300 µm diameter x <300 µm depth) holes were drilled into a phenocryst with overlapping holes used to obtain finer spatial resolution (e.g., Siebel et al. 2005; Tepley et al. 1999, 2000). One disadvantage associated with this method is the lack of lateral or depth-related spatial resolution. Knesel et al. (1999) used a variant of this technique by utilizing a single, large-diameter hole (~600 µm) drilled into sanidine phenocrysts with material removed

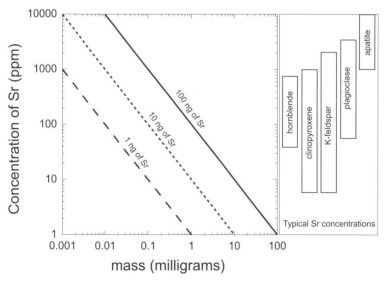

Figure 1. Diagram of concentration versus mass associated with different mineral phases and the masses of material needed to produce 1 ng, 10 ng and 100 ng of target analyte for analysis. Also shown are ranges of Sr concentrations in selected minerals found in typical volcanic rocks. [Used by permission of Annual Reviews, from Davidson et al. (2007), *Annual Review of Earth and Planetary Sciences*, Vol. 35, Supplemental Material, Fig. 1, p. 1].

at various controlled depths. So, although surficial lateral resolution was lost, depth resolution was maintained by carefully controlling milling depths. The shallow-trough method of milling, described in Charlier et al. (2006), uses a computer-controlled milling machine equipped with a binocular microscope, an adjustable speed motor, and a computer controlled X-Y stage reproducible to ±1 micron. The computer-controlled milling machine allows for sampling multiple holes within growth zones and for designating the milling depth of each hole, such that sufficient material is removed from the sample using multiple holes rather than a single deep hole. The advantage is that lateral- and depth-related spatial resolution is increased in comparison to milling single holes.

Depending on the milling device, the user will have to determine the desired milling location. Hand-operated milling machines are best used with a micrometer stage so as to maximize lateral spatial resolution. Previous workers used a detachable micrometer stage (Fig. 2) that was secured to both the vertical mill and a petrographic microscope (e.g., Tepley et al. 1999, 2000). The advantage with this technique is the ability to precisely determine locations of drill sites under the microscope, record locations, re-fit the stage to a vertical mill, and then precisely return to pre-determined drilling sites. Vertical spatial resolution, however, suffers using this method because of the imprecision of determining when the bit is at the sample surface. Computer-controlled milling machines are more precise for both lateral and vertical spatial resolutions, and software accompanying milling machines helps to determine drill tip location, sample surface location, and milling depth (Charlier et al. 2006). Options such as milling depth, location, holes versus troughs, and rate of milling are controlled using accompanying software.

Following milling, recovering material from the sample surface can be accomplished either by the dry method, collecting material with a small plastic scoop, or the wet method, collecting material using water. In the wet method, a drop of ultra-pure (18.2 MΩ) water is used as both a cooling agent for the bit and sample collection medium. As material is milled from the sample surface, milled particulates collect in the water creating a slurry. Charlier et al. (2006) describe the added technique of placing a warmed square of Parafilm™ with a small ~4 mm hole cut into it on the sample surface. The hole in the Parafilm™ is placed over the desired milling site and aids in slurry collection by confining the water to the drilling area and preventing it from draining into possible surface cracks of the polished thick section. Both methods use a micro-pipette with 0.5-10 μL tip to suction/collect the slurry. Suctioning additional water drops from the sample surface may be needed to ensure greater recovery of milled material.

If determining the isotopic compositions of samples is the primary goal, samples can be placed directly into digestion beakers and be dissolved. However, if isotope dilution is required, samples must be dried, weighed, and spiked. In this case, a sample weighing method must be devised. Early studies (Tepley et al. 1999, 2000) use a small square piece of clean aluminum foil, shaped around a circular mold, to make a flat-bottomed, aluminum collection vessel into which the sample is pipetted and dried. Once dry, the aluminum vessel containing the sample is weighed and the weight recorded. Sample is then carefully scraped into a digestion beaker. The aluminum vessel, less sample, is then re-weighed; the difference between the two is the sample weight. In general, this method results in relatively poor sample recoveries as a result of remaining sample adhering to the aluminum vessel or loss resulting from static electricity. Errors in sample weight affect element concentrations determined by isotope dilution but will not compromise the isotopic composition of the sample.

Charlier et al. (2006) developed a more precise method of weighing micromilled samples. These authors used a small receptacle made of gold foil that was weighed, loaded with sample, dried, re-weighed, and placed into the Teflon digestion beaker along with digestion acids. The gold foil was impervious to digestion acids and was removed and re-weighed after digestion where the difference in weight was the sample weight. Similarly, Ramos et al. (2005) used pre-

(a) Sherline microdrill.

Petrographic microscope with
micrometer stage attached.

Microdrill with micrometer stage
attached.

Microdrill with manual Z-direction control and
plate for micrometer attachment.

(b) New Wave Instruments Micromill.

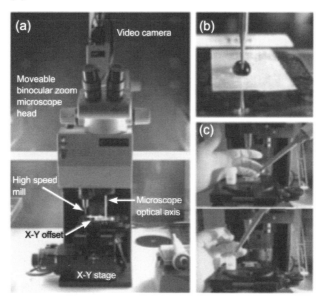

Figure 2. Illustrations of different milling devices. a) Sherline™ vertical mill shown with removable micrometer stage attached to petrographic microscope and to the mill. The sample is secured in a removable upper portion of the micrometer stage. This allows the operator to maintain a centered stage while being able to remove the sample for washing between milling events. The upper portion can then be re-attached to the stage in the same orientation and settings. b) Images illustrating various components of the NewWave™ MicroMill™ including the moveable stage, binocular microscope and video camera, tungsten carbide mill bit in sample collection water slurry, and illustrations of how the sample is pipetted into a beaker for chemical analysis. [Used by permission of Elsevier from Charlier et al. (2006), *Chemical Geology*, Vol. 232, Fig. 1, p. 117].

weighed, pre-cleaned Teflon disks on which samples were dried and re-weighed. The samples plus Teflon disks were placed into digestion beakers. The advantage of these methods was that the user could theoretically recover the entire dried sample for a fairly accurate sample weight.

Samples sizes vary according to bit diameters and milling depths and, most importantly, are controlled by how much material is needed for high precision analysis. Because samples are generally small, additional precautions must be taken to ensure cleanliness of beakers and reagents used in the dissolution process. Using multiply-distilled reagents and freshly distilled acids during dissolution is recommended. The amount of reagents used in the digestion process is limited by 1) the minimal amount of reagent used to maintain low blanks, and 2) the minimum amount of reagents required to prevent samples from drying during the digestion process. If too small a volume of reagent is used, heating will vaporize the reagent and require an extra cooling stage for re-condensing the sample or require re-digestion. Approximately 0.3-0.5 ml of reagent is minimally required to meet these conditions. Typical beakers used in this process are Savillex™ 3 ml Teflon screw-top beakers which are cleaned in a thorough fashion. Further sample processing and column chromatography is described in the Single Crystal Analyses section below.

Standard dissolution techniques for most rock digestions can also be used for micromilled samples: HF and concentrated HNO_3 for the first stage, 7N HNO_3 for the second stage, and 6N HCl for the third stage (Ramos 1992). Samples can then be re-dissolved in the acid required for chromatography. Samples are generally spiked prior to digestion unless sample splitting is required. In general, splitting while samples are dissolved in 6N HCl is best. Mixed spikes (e.g., Rb/Sr, Sm/Nd) are typically used to ensure that critical element ratios are preserved even if weighing errors have occurred. Required spike amounts can be calculated using a spreadsheet and estimated concentrations of elements of interest in the sample.

Applications to volcanic rocks. Whole-rock isotopic measurements are important in understanding broad petrogenetic histories of volcanic systems. However, whole-rock sampling produces average chemical and isotopic characteristics that either obscure or completely eliminate potential variations in mineral and melt components of rocks both isotopically and geochemically. Micromilling techniques were developed to extract isotopic and geochemical traits that whole-rock processing and whole-crystal analyses cannot obtain, thus allowing researchers to more thoroughly constrain the petrogenetic histories of volcanic and plutonic systems. Here, we review micromilling studies that have aided and, in some cases, complicated our understanding of crustal-level volcanic processes.

Micromilling, sample processing, and undertaking TIMS analyses are time consuming, therefore a thorough sampling methodology is required prior to initiating studies. Micromilling can be combined with a range of additional techniques including Nomarski Differential Interference Contrast (NDIC), a reflected light technique (Anderson 1983; Pearce and Clark 1989) used to enhance compositionally dependent textural features, and electron microprobe analysis to elucidate potential compositional variations in target phenocrysts. The combined use of Nomarski interferometry, electron microprobe analysis, and Sr-isotopic microanalysis to provide textural, chemical and isotopic information is termed Crystal Isotope Stratigraphy (CIS; Davidson and Tepley 1997; Davidson et al. 1998). These techniques help define a more thorough sampling strategy and, once accomplished, provide information that would have otherwise been lost as a result of the averaging effects of whole-rock techniques.

One of the first studies to employ these techniques was that of Tepley et al. (2000) who investigated the 1982 eruption of El Chichón Volcano in Mexico. Tilling and Arth (1994) had previously found that mineral separates (plagioclase, clinopyroxene, hornblende, anhydrite, apatite and groundmass) from the same hand sample of the 1982 eruption were isotopically heterogeneous. Tepley et al. (2000) found texturally complex plagioclase phenocrysts characterized by multiple dissolution zones, large variations in An contents associated with dissolution surfaces, and monotonic decreases in $^{87}Sr/^{86}Sr$ from core to rim (Fig. 3). They interpreted these data as reflecting crystal growth in an increasingly contaminated magma reservoir that was recharged with similar trachyandesitic magma with lower $^{87}Sr/^{86}Sr$. The lower $^{87}Sr/^{86}Sr$ recharge magma presumably fluxed the reservoir with hotter and more volatile-saturated melt causing textural discontinuities in the plagioclase, with consequent large magnitude changes in An contents, and eventual eruption. Decreasing $^{87}Sr/^{86}Sr$ ratios in plagioclase resulted from crystal growth following re-equilibration in newly mixed magma. Thus, utilizing crystal isotope stratigraphy, Tepley et al. (2000) were able to elucidate a more detailed picture of the petrogenetic evolution of El Chichón Volcano that otherwise would have been overlooked as a result of whole-rock chemical and isotope averaging.

Numerous other studies have used crystal isotope stratigraphy to evaluate the petrogenetic histories of a range of different volcanic rocks. Tepley et al. (1999) investigated mixing

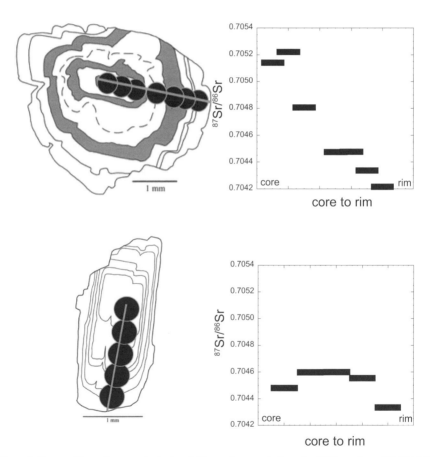

Figure 3. Cartoon illustrating two plagioclase feldspar phenocrysts from the 1982 eruption of El Chichón volcano (Mexico). The drawings show dissolution zones (stippled areas), dissolution boundary (dashed line), micromilling holes, and electron microprobe traverse locations. Graphs to the right show changing $^{87}Sr/^{86}Sr$ ratios from cores to rims for each crystal with filled rectangles representing results from individual micromill holes. In both cases, $^{87}Sr/^{86}Sr$ ratios decrease from cores to rims resulting from crystal growth in magma experiencing recharge with a lower $^{87}Sr/^{86}Sr$ melt. [Used by permission of Oxford University Press, from Tepley et al. (2000), *Journal of Petrology*, Vol. 41, Fig. 3, p. 1402].

relationships and mafic inclusion disaggregation at Chaos Crags, a series of rhyodacite and dacite domes containing variably quenched basaltic andesite inclusions on the flank of Lassen Peak in Lassen Volcanic National Park, California (USA). The principal findings from this study were that three styles of mixing occurred in the system: 1) limited mixing at the interface between rhyodacite magma and a ponded basalt layer to form a basaltic andesite, 2) mechanical dispersal of blobs of this material into the rhyodacite magma, and 3) disaggregation of the inclusions through shear forces in the resident magma. In the initial mixing episode, rhyodacitic magma containing large plagioclase phenocysts mixed into the basalt. These plagioclase phenocrysts reacted in the hotter and more mafic liquid and formed reaction textures and upon cooling, grew rims in equilibrium with the predominant magma composition. These rims had higher An contents and lower $^{87}Sr/^{86}Sr$ ratios than cores. Disaggregation of the basaltic andesite inclusions dispersed the contents of the inclusions into the more felsic magma in a later mixing event. In this study, the complex pathways of mixing were easily discerned through combined textural analyses, crystal chemistry variations, and isotopic analyses using micromilling. It was suggested that these types of mixing processes were more common than previously thought and that they may be responsible for the creation of intermediate lavas in arc environments.

Hora (2003) and Davidson et al. (2007) also documented the complex mixing of magmas and movement of crystals within a volcanic system through use of single crystal isotope analyses and intra-crystalline micromilling and isotopic analyses at Ngauruhoe Volcano (New Zealand). Micromilled plagioclase crystals in basaltic andesite lavas showed increases in $^{87}Sr/^{86}Sr$ from cores to rims which were consistent with crystals growing in magmas that became progressively more contaminated with higher $^{87}Sr/^{86}Sr$ melts. Accompanying clinopyroxene phenocrysts with low $^{87}Sr/^{86}Sr$ ratios suggested that they crystallized in a low $^{87}Sr/^{86}Sr$ melt prior to plagioclase crystallization. Subsequent remobilization, prior to eruption, combined the two minerals in later-erupted volcanic rocks. Just as in Tepley et al. (1999), micromilling allowed researchers to document changes in isotopic values within crystals and, when integrated with textural and compositional information and single crystal isotopic analyses, to trace the pathways of interaction of magmas in evolving volcanic systems.

Similar magma/crystal dynamics were illustrated by Perini et al. (2003). These authors showed that potassium feldspar megacrysts hosted in mafic alkaline potassic, ultrapotassic, and differentiated rocks retained both initial Sr isotopic homogeneity and heterogeneity and could be seen in eruptive products at nearby volcanoes in central Italy. At Monte Cimino, isotopic results from micromilling potassium feldspar megacryst cores and rims suggested that they nucleated and grew in a trachytic magma but were subsequently mixed into latite and olivine latite melts. At Vico volcano, micromilling and isotopic analyses of potassium feldspar megacrysts and isotopic analyses of whole-rock hosts showed a complex relationship. Three scenarios were observed; 1) a megacryst was isotopically homogenous from core to rim and in isotopic equilibrium with its host trachyte; 2) a megacryst was isotopically heterogeneous from core to rim and not in isotopic equilibrium with its host tephri-phonolite; and 3) a megacryst was isotopically homogenous from core to rim but not in isotopic equilibrium with its host olivine latite. As a result, Perini et al. (2003) created a complex model in which some crystals were derived from an older trachytic melt similar in composition to the host trachytic melt, some crystals were co-genetic with their current host, and some crystals were xenocrysts.

As these studies have illustrated, crystals commonly record complex textural, compositional, and isotopic characteristics that result from volcanic systems that recycled previously crystallized phenocrysts, commonly referred to as antecrysts, from crystal-rich mush piles genetically related to the system but not actively crystallizing from the prevalent magma (e.g., Jerram and Davidson 2007). An excellent example that illustrated these complex interactions was Armienti et al. (2007). This study compared whole-rock $^{87}Sr/^{86}Sr$ ratios with those of rims and cores of clinopyroxene "phenocrysts" and a hornblende megacryst from an historic eruption of Mount Etna (Italy). Results showed that pyroxenes and hornblende were in isotopic disequilibrium with their hosts, except for a very thin rim on a single clinopyroxene crystal that was in ^{87}Sr/^{86}Sr equilibrium with the host rock. To further justify an antecrystic origin, these authors evaluated crystal size distributions (CSDs) (see Armienti 2008 in this volume) of the mineral phases hosted in lavas erupted in 2001. Clinopyroxenes with disequilibrium isotopic ratios were part of a group of larger crystals that defined a separate trend on the CSD diagram which supported a different growth history in comparison to smaller phenocrystic minerals in the lavas. Together, evidence suggested that crystals in isotopic disequilibrium with their host rocks grew in a different, but related, magma compared to the prevalent magma feeding the current eruption. The authors suggested that these "disequilibrium" crystals originated from a cumulate mush that existed at depth, and that some were entrained in recently erupted magma.

Chadwick et al. (2007) also applied crystal isotope stratigraphy to plagioclase-bearing basaltic andesite lavas from Merapi volcano. They documented significant Sr isotope variations in cognate plagioclase phenocrysts. Chadwick et al. (2007) attributed the isotopic variability in phenocrysts to extensive magma-crust interaction in two forms: 1) direct incorporation of high-An content, high-^{87}Sr/^{86}Sr plagioclase xenocrysts originally from a sedimentary protolith which were present as crystal cores that were subsequently overgrown by plagioclase in equilibrium with resident magma, and 2) plagioclase crystal growth in resident magma that was progressively contaminated by calcareous and volcaniclastic xenoliths and carbonate crust with concurrent crystal fractionation. Constraining the exchange of mass between various reservoirs in this magmatic system through micromilling allowed Chadwick et al. (2007) to speculate that incorporation of the proposed crustal material (limestone) may have major effects on the volatile budget of the volcano. They suggested that this interaction, which is more significant than previously thought, may have significant consequences on the eruptive behavior at Merapi Volcano and other volcanoes built on carbonate platforms.

Lastly, Morgan et al. (2007) combined textural analysis in the form of crystal size distributions (CSD) and microsampling techniques to understand magma supply and mixing processes at Stromboli Volcano (Italy). Plagioclase phenocrysts from a 26,000 year-old sample were targeted in which time and growth rate data were determined through CSD, and changing isotopic information of the supply magma was determined through micromilling and TIMS. The authors constructed a model in which isotopic micromilling results and respective sampling positions were overlaid and then stretched and shifted to generate an isotopic timeline. The end result was an isotopic evolution line that represented, by proxy, magma evolution of the evolving Stromboli magmatic system. Combining the information, Morgan et al. (2007) were able to determine the timescales of mixing, crystallization and contamination prior to eruption of Stromboli. In this study, as with Waight et al. (2000a), it was shown that micromilling was a powerful tool for determining magma chamber processes. However, when combined with other process-related techniques or contrasting *in situ* isotopic analyses, the technique became even more useful.

Applications to plutonic rocks. Unlike volcanic rocks, plutonic rocks cool slowly and are thus affected by diffusive re-equilibration associated with long-term high-temperature environments, or by variable isotopic exchange associated with hydrothermal fluids. Additionally, plutonic age determinations and high parent/daughter ratios in potential target isotopic systems can yield sources of error that must be evaluated. Discussions of these problems, some of which are outlined here, can be found in Davidson et al. (2008) and Davidson et al. (2005). Given these caveats, we discuss notable studies that apply micromilling to plutonic rocks and host minerals that show a range of alteration from little or no diffusive re-equilibration to full isotopic resetting.

One of the first to utilize intracrystalline isotopic micromilling on plutonic rock components was Waight et al. (2000a) who determined initial ^{87}Sr/^{86}Sr and $\varepsilon Nd_{(i)}$ on cm-scale feldspars in a 395 Ma S-type granite from the Wilson's Promontory Batholith of the

Lachlan Fold Belt (SE Australia). This research tested whether isotope variations could be preserved in plutonic rocks with obvious magma mixing/mingling textures that remained at moderate temperatures for millions of years. Additionally, this study was an excellent plutonic analogue to volcanic equivalents in which the interactions between different magmas and crystals were explored (e.g., Tepley et al. 1999). Evidence for magma mingling included numerous mafic microgranular enclaves hosted in a monzogranite, and feldspar megacrysts in both the host and enclaves. Micromilling of megacrysts revealed significant decreases in initial $^{87}Sr/^{86}Sr$ and complementary increases in $\varepsilon Nd_{(i)}$ from cores to rims in both alkali and plagioclase feldspars (Fig. 4). High-temperature diffusive or hydrothermal re-equilibration would effectively produce homogeneous isotope signatures rather than a complementary divergent trend, therefore such processes could be ruled out. Thus, this research verified not only that micromilling techniques could be applied to slow-cooling, old plutonic systems when evaluated for the possibility of hydrothermal alteration, but also that application of these techniques to petrogenetic problems could trace the movement of crystals between plutonic systems with different isotope characteristics.

Not all plutonic systems are immune to alteration processes however. Waight et al. (2000b) documented both initial Sr and Nd isotopic heterogeneity and homogeneity in plagioclase megacrysts in comparison with their respective host magmas residing in enclaves in two I-type plutons from the Lachlan Fold Belt (SE Australia). In the Swifts Creek pluton, $^{87}Sr/^{86}Sr$ and $^{143}Nd/^{144}Nd$ isotope ratios of plagioclase megacrysts were distinct from their host enclaves which confirmed results from earlier studies that concluded an admixed nature to the enclaves (Eberz and Nicholls 1988; Eberz et al. 1990). In this case, Waight et al. (2000b) modeled feldspars as having crystallized in a primitive granitoid which were then transferred to their current host. In the Bridle Track pluton, $^{87}Sr/^{86}Sr$ ratios of feldspars, the enclaves in which they resided, and the granitic host of the enclaves were indistinguishable. $^{143}Nd/^{144}Nd$ ratios of the same crystals and hosts were slightly different which suggested a mingled magma origin, as did field evidence, petrology and whole-rock data. Waight et al. (2000b) concluded that a

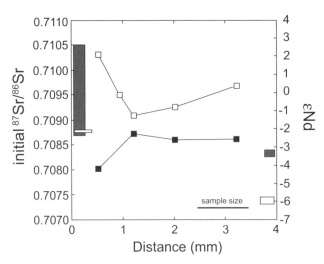

Figure 4. Sr and Nd isotope traverses from a potassium feldspar megacryst hosted in a mafic enclave from Wilson's Promontory Batholith (Australia). Unfilled squares represent initial $^{87}Sr/^{86}Sr$ ratios and filled squares represent initial εNd values for the same micromilled holes. Filled boxes represent initial isotope range of host granite and unfilled boxes represent initial isotope ratio of enclave for $^{87}Sr/^{86}Sr$ (left) and εNd (right), respectively. Modified after Waight et al. (2000a).

significant Sr isotope equilibration process occurred to homogenize Sr isotopes of megacrysts, enclaves, and granitoid host, but was unable to homogenize Nd isotopes completely. In these cases, the use of multiple isotope systems constrained the altered nature of the samples and allowed for appropriate interpretation.

Siebel et al. (2005) determined that sub-solidus thermo-metamorphic or hydrothermal alteration affected $^{87}Sr/^{86}Sr$ ratios in potassium feldspars residing in various Late Paleozoic plutonic rocks of the Bavarian Forest. In some cases, potassium feldspars were homogenized along with bulk groundmass via a post-crystallization fluid phase. In other cases, interaction of post-magmatic hydrothermal fluids produced isotopic heterogeneities. In all cases, the propagation of errors associated with assumed crystallization ages, parent/daughter ratio measurements, and isotopic ratio measurements were evaluated to ensure accurate initial age calculations.

In many of these studies, the use of micromilling allowed workers to follow the paths of crystal migration and growth in various magmas. Halama et al. (2002) focused on plagioclase megacrysts from gabbroic dikes in the Gardar Province (South Greenland) to trace the path of plagioclase-rich mushes, postulated to migrate from the crust-mantle interface, through the crust eventually ponding within the crust and forming massive anorthosites. Halama et al. (2002) used major element zoning patterns, *in situ* trace element variations, and $^{87}Sr/^{86}Sr$ ratio variations in the plagioclase megacrysts to formulate a model for anorthosite formation.

Another plutonic micromilling study that elucidated processes similar to magma interaction processes in volcanic environments was Perlroth (2000), documented in Davidson et al. (2005, 2008). This study explored Sr isotope contrasts in potassium feldspar megacrysts in the ~400 Ma Shap Granite (NW England). The Shap Granite has abundant potassium feldspar megacrysts and mafic inclusions, many of which contain potassium feldspar megacrysts rimmed by plagioclase. A leading hypothesis was that potassium feldspar megacrysts grew in the Shap Granite and were then transferred into a boundary layer mixture along an interface between the two magmas which eventually produced mafic enclaves of a mixed nature. Thermal and chemical equilibration of the potassium feldspar with the inclusion mafic magma allowed the plagioclase rims to grow. There is some debate (Pidgeon and Aftalion 1978; Rundle 1992) about the age of the Shap Granite, and this led to equivocation regarding age-corrected initial Sr isotope ratios of samples micromilled from a potassium feldspar megacryst. However, if a younger age were accepted for the Shap Granite, then the isotopic data became consistent with megacryst growth in the granite followed by crystal transfer and rim re-equilibration within mafic enclaves, thus mimicking processes involved in volcanic systems.

The significance of these plutonic studies is threefold: 1) original isotopic disparities in minerals of old plutonic rocks can persist for extended time periods, and can be measured using micromilling, however this is not always the case, 2) multiple isotopic systems (Sr, Nd, Pb) can be used in a complementary fashion in some cases to rule out the effects of both slow cooling, high-temperature alteration and/or hydrothermal isotopic resetting, and 3) interpretations based on micromilling techniques require precise and accurate measurements of the age of the plutonic body, the parent/daughter ratio and isotopic ratio of the sample in question, and appropriate error propagation of all measurements.

In long-lived volcanic and plutonic systems, sub-solidus isotopic diffusive re-equilibration is a process that warrants careful evaluation. For example, Gagnevin et al. (2005) not only documented Sr isotopic zoning in potassium feldspar megacrysts in the 7 Ma Monte Capanne monzogranite (Elba, Italy), but also evaluated whether diffusive re-equilibration may have affected the system. Micromilling transects from core to rim of several potassium feldspar megacrysts revealed high to low initial $^{87}Sr/^{86}Sr$ values. These observations suggested crystal growth in a crustally contaminated magma with high initial $^{87}Sr/^{86}Sr$ ratios followed by growth in magma recharged with mafic melt with lower initial $^{87}Sr/^{86}Sr$. However, to test whether changes in $^{87}Sr/^{86}Sr$ ratios were due to isotope re-equilibration, Gagnevin et al. (2005a) modeled the isotopic data to determine timescales required for diffusive exchange. Results showed that at least 20 m.y. were needed to produce the isotopic profiles of the megacrysts. Thus, they considered this amount of time unrealistic given other published results for similar granite crystallization histories.

Isotopic variations in plutonic minerals also allowed researchers to constrain cooling times based on the *lack* of diffusive equilibration in minerals with intracrystalline isotope contrasts. An example was the Rum Intrusion (NW Scotland) where Tepley and Davidson (2003) demonstrated that select plagioclases crystallized in liquids contaminated by isotopically diverse wallrocks whereas other plagioclases crystallized in homogeneous uncontaminated magmas. Both types of crystals were then transported to the floor of the magma reservoir where they cooled together. Preservation of isotopic heterogeneity within some of these crystals allowed Tepley and Davidson (2003) to calculate a maximum amount of time that the Rum magma chamber was held at high temperature based on the diffusivity of Sr in plagioclase feldspar. Figure 5 illustrates how cooling rate (linear or exponential) affects the isotopic equilibration of crystals of various lengths. Given that isotopic disequilibrium (regardless of the magnitude) in the Rum plagioclases existed on length scales of 1-2 mm, only cooling rates below the 2

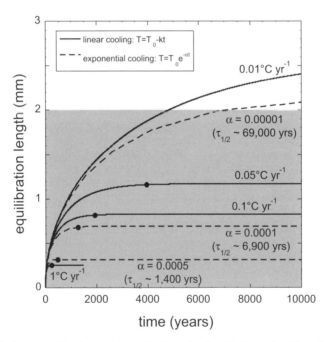

Figure 5. Diffusive re-equilibration models used to determine the effective scale of Sr exchange owing to cooling from an initial temperature T_0. Both linear and exponential models are considered. In the linear cooling models, k represents linear cooling rates of 0.01 to 1°C yr^{-1} and is represented in the diagram by solid lines. In the exponential cooling models, represented in the diagram by dashed lines, α is the effective temperature decay constant and $\tau_{1/2}$ is the time taken for the temperature to drop from T_0 to $T_0/2$ (see Reddy et al. (1996) for mathematic formulations of these equations). Diffusive exchange is dependent on plagioclase composition. In this case, the diagram illustrates labradorite (An$_{50-70}$) at 1200°C. Since all plagioclase crystals evaluated are ≤2 mm, only the modeled cooling rates within the shaded box are acceptable solutions. Note that diffusive exchange effectively ceases when temperature cools to below 1000 °C. This is indicated by the solid dot on each cooling curve. [Used by permission of Springer-Verlag, from Tepley and Davidson (2003), *Contributions to Mineralogy and Petrology*, Vol. 145, Fig. 11, p. 639].

mm equilibration length line (shaded area) were acceptable solutions to the cooling rate determination. Ultimately, the modeling indicated that the cooling rate for the Rum Intrusion was ~0.1 °C yr^{-1} or faster and Sr isotope equilibration essentially ceased at $T = 1000$ °C (circle on cooling path line) after which there could be no further appreciable diffusion or addition to the equilibration length.

The issue of *in situ* crystallization versus crystal transport was revisited by Davidson et al. (2008) who documented isotopically distinct initial ^{87}Sr/^{86}Sr ratios in plagioclase and pyroxene phenocrysts in the Dias Intrusion (Dry Valleys, Antarctica). This work evaluated whether minerals crystallized *in situ* (in which case the assumption is that different crystals share the same isotopic composition owing to growth in the same magma) or in magmas with different isotopic compositions and were transported to their current hosts. Scatter in age-corrected isotopic results suggested that plagioclase and pyroxene crystallized in different environments and were transported to their current position in the Dias Intrusion.

As these studies demonstrate, the possibility of extracting petrogenetic information in plutonic rock mineral phases through micromilling is possible, however, several issues must be evaluated. As Davidson et al. (2008) point out, one issue is whether the system under study can support mineral-scale isotopic heterogeneity. Small, upper crustal intrusions tend to cool more quickly than large, high-temperature, slow-cooling magmas. The added time associated with slow-cooling magmas tends to force isotopic equilibration. Another issue is confidence in age corrections of the magmatic body. Measurement errors associated with the age of the body and parent/daughter ratios (e.g., Rb/Sr, Sm/Nd) magnify resulting errors associated with determining whether initial, mineral-scale isotopic heterogeneity existed. Lastly, effects owing to post-emplacement alteration need to be evaluated. As Waight et al. (2000a,b) have shown, using multiple isotopic systems may alleviate the uncertainty associated with data obtained through singular isotopic system analyses.

Secondary ionization mass spectrometry (SIMS)

Technique. Microspatial measurements of Pb and Sr isotopes have been undertaken using secondary ionization mass spectrometry (SIMS). SIMS based analyses can be targeted to micron-length sample sites although analyses can vary widely in measurement precisions as a result of relatively low ion beam intensities (e.g., compared to TIMS). For Pb isotope measurements using SIMS, analyses typically focus on the more abundant ^{206}Pb, ^{207}Pb, and ^{208}Pb masses and are only minimally affected by potential interfering elements/molecules (e.g., HfSi$^+$) as a result of the use of variable mass resolutions (e.g., 1800-2000 for feldspars and 3000 for glasses; Layne and Shimizu 1997) which still allow for high ion beam transmissions. For melt inclusions, a ^{16}O$^-$ ion beam and high resolution (~3500) is used to ensure minimal influence of potential interfering elements at the Pb isotope masses (Saal et al. 1998, 2005). In addition, typical spot sizes are ~20-30 microns. And although precisions are generally poor (0.2 to 1.2%), Pb isotope ratios have proven to vary widely in target materials, such that SIMS analyses have successfully identified large Pb isotope variations in, for example, melt inclusions.

Applications. Saal et al. (1998) first used SIMS to measure Pb isotope ratios in olivine-hosted melt inclusions to identify an isotopically diverse suite of magmas involved in ocean island volcanic plumbing systems that could not be resolved by conventional whole-rock analyses. These authors found that melt inclusions retained a range of EMII and HIMU mantle endmember components that spanned 50% of the worldwide Pb isotope variation observed in ocean island basalts in only a few sampled basalts. This study also demonstrated that large ranges of Pb isotope signatures were retained by olivine-hosted melt inclusions in individual basalt flows. These signatures were only retained in melt inclusions because melt aggregation had homogenized variable melt fractions prior to eruption. Saal et al. (2005) expanded this investigation to include three endmember mantle compositions (HIMU, EMI, and EMII) and presented a model in which melts of an ocean lithospheric component (i.e., Pacific MORB) were thought to mix with wallrock components originating from either HIMU, EMI, and EMII magmas as reflected by linear Pb isotope trends.

Kobayashi et al. (2004) also analyzed melt inclusions for Pb isotopes in conjunction with lithium and boron isotopes at Hawaii and concluded that recycled materials from ancient subducted crust were incorporated into the mantle sources generating Hawaiian basaltic volcanism. Results again confirmed that melt inclusions retained greater isotopic diversity, as compared to whole rocks, and successfully demonstrated the utility of combining SIMS-measured Pb isotope ratios with additional isotopic constraints to identify and constrain the chemical nature of the sources contributing to ocean island volcanism. And although these studies focused on melt inclusions, Layne and Shimizu (1997) demonstrated that Pb isotopes could also be successfully measured in plagioclase, potassium feldspar, pyrite, and chalcopyrite using SIMS.

In addition to Pb isotopes, Weber et al. (2005) developed a procedure to measure Sr isotopes in fish otoliths using the Sensitive High Resolution Ion Microprobe (SHRIMP). Compared to laser ablation and TIMS, SHRIMP generated $^{87}Sr/^{86}Sr$ ratios had relatively poor precisions ($^{87}Sr/^{86}Sr$ ±0.001 to 0.0001) but even at these levels Bacon et al. (2004) effectively identified the natal rivers of salmon originating from North American cratonic regions and western North American Paleozoic accreted terrains. SIMS Sr isotope measurements usually had greater amounts of interfering elements and molecules, however, and beam intensities were generally lower than TIMS due the limited volume sampled.

Laser ablation sampling

Sr isotope technique. Christensen et al. (1995) first demonstrated the utility of measuring Sr isotopes in carbonate and plagioclase using laser ablation (LA) sampling in conjunction with multi-collector inductively coupled plasma mass spectrometry (MC-ICPMS). The advantages of laser sampling were clear: samples were easily prepared and analyses could be obtained rapidly, bypassing time-intensive digestion and purification procedures. Although true for some minerals such as carbonate, accurate analyses of other materials still offer significant challenges. Since the Christensen et al. (1995) evaluation, refinements in the utility and longevity of commercial laser ablation systems and advances in multi-collector inductively coupled mass spectrometers have combined to greatly expand the use of laser sampling associated with microspatial isotope analyses. Further studies by Davidson et al. (2001) evaluated plagioclase glasses with variable Rb/Sr ratios, while Waight et al. (2002), Bizzarro et al. (2003), and Schmidberger et al. (2003) expanded the range of analyzed materials to include apatite, clinopyroxene, and sphene. Ramos et al. (2004) also demonstrated that highly accurate Sr isotope measurements could be obtained on less Sr enriched carbonates (≤1000 ppm Sr), plagioclases (≤750 ppm Sr), and clinopyroxenes (≤50 ppm Sr), in addition to elementally complex materials such as basaltic and basaltic andesite fine-grain groundmass (Figs. 6, 7 and 8). Additionally, accurate assessments of Sr isotopes using laser ablation sampling of aragonite otoliths have greatly benefited ecological studies of fishes (e.g., Thorrold and Shuttleworth 2000; Woodhead et al. 2005; Barnett-Johnson et al. 2005).

Two requirements are generally necessary for successful laser ablation isotope measurements: 1) analyte abundances must be relatively high, although precisions associated with materials characterized by low analyte abundances may still be sufficient to address investigations involving large dynamic isotope ranges, and 2) potential interfering elements and molecules must be minimal or their respective impact on measured ratios must be accurately removed. If high analytical precisions are not required (e.g., $^{87}Sr/^{86}Sr$ of ≥0.00004), laser sampling can be a very efficient means of acquiring data from a time perspective, and thus, is highly advantageous as compared to undertaking micromilling and purification using column chromatography.

Since laser ablation analyses involve direct sampling of unpurified materials, measurement accuracies may be highly dependent on the presence of interfering elements, thus target materials with low interfering element abundances are generally preferred. Such effects are

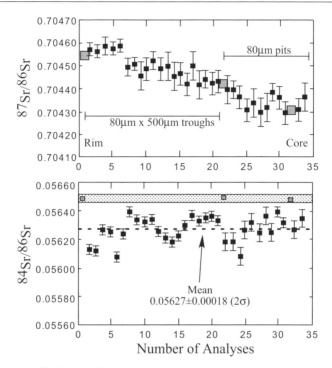

Figure 6. Core to rim $^{87}Sr/^{86}Sr$ and $^{84}Sr/^{86}Sr$ variations of a single plagioclase crystal from Pisgah Crater, California (USA) determined using laser ablation sampling. Gray filled squares and stippled field indicate ratios measured using micromilling and TIMS. Thirty-four $^{87}Sr/^{86}Sr$ ratios measured using laser ablation sampling (filled squares with 2σ error bars in upper diagram) using troughs and pits follow the trend defined by micromilled samples and confirm variations resulting from assimilation of materials in crustal-level magma chambers (Ramos and Reid 2005). In contrast to $^{87}Sr/^{86}Sr$ ratios, $^{84}Sr/^{86}Sr$ ratios are consistently lower than TIMS determined values with a mean of 0.05627±0.00018. This may result from potential interfering elements or molecules, or inaccurate Kr corrections. [Used by permission of Elsevier, from Ramos et al. (2004), *Chemical Geology*, Vol. 211, Fig. 8, p. 149].

best exemplified in Sr isotope measurements. Sr isotope ratios can be influenced by a variety of interfering elements including ^{87}Rb, ^{84}Kr, ^{86}Kr, $^{168}Er^{2+}$(mass/charge = 84), $^{168}Yb^{2+}$, $^{172}Yb^{2+}$, $^{174}Yb^{2+}$, and $^{176}Yb^{2+}$ (Davidson et al. 2001; Ramos et al. 2004). Many potential molecular species such as Ca dimers (e.g., $^{44}Ca^{43}Ca$, Waight et al. 2002), Ca argides (e.g., $^{44}Ca^{40}Ar$, Woodhead et al. 2005), and Fe oxides (e.g., $^{56}Fe^{16}O_2^+$, Schmidberger et al. 2003) may also influence Sr isotope ratios (Fig. 9). In addition, isotopes such as ^{85}Rb, which is monitored for ^{87}Rb corrections, may also be influenced by interfering elements such as $^{170}Yb^{2+}$ and $^{170}Er^{2+}$. Thus, even isotopes that are measured for interference corrections may be affected, potentially compromising accurate interference corrections. For the VG Element Axiom™ and Nu Plasma™ mass spectrometers, Ca-dimers and Ca-argides pose significant challenges to obtaining accurate Sr isotope measurements in Ca-rich minerals such as carbonate and plagioclase. Careful interference correction procedures, however, have been successfully used to obtain accurate Sr isotope measurements (e.g., Woodhead et al. 2005). In contrast, Sr isotope measurements on Ca-doped NBS 987 standard solutions and Ca-rich minerals, including carbonate and plagioclase, are only minimally affected (Hart et al. 2005) or not affected (Ramos et al. 2004) by Ca dimers and Ca argides when using the ThermoFinnigan Neptune™. Although it is unclear why Ca-dimers and Ca-argides are either present or absent, differences in the design of plasma interfaces or the standard running parameters used (per

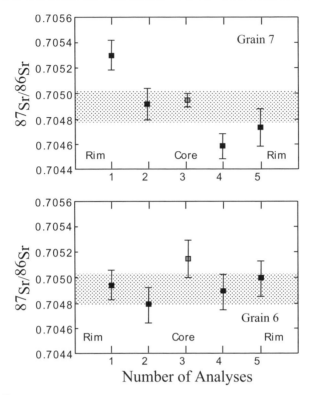

Figure 7. $^{87}Sr/^{86}Sr$ ratios of two single clinopyroxene crystals from a primitive Grand Canyon basalt (MgO wt% of ~10%) from the western United States determined using laser ablation sampling. Filled squares with 2σ error bars represent multiple laser ablation trough analyses and gray squares with 2σ error bars represent micromilled samples from the same crystals. Stippled fields reflect $^{87}Sr/^{86}Sr$ ratios measured from four separate single crystals using column chromatography and TIMS. $^{87}Sr/^{86}Sr$ ratios measured using 160x500 μm laser ablation troughs in grain 6 overlap those of single crystals (lower diagram) confirming accuracy of measured $^{87}Sr/^{86}Sr$ ratios. Results for grain 7, however, reflect variable $^{87}Sr/^{86}Sr$ ratios that lie outside the field of single crystals illustrating the capability of laser ablation sampling to identify within crystal variations that are not detectable using single crystal analyses. [Used by permission of Elsevier, from Ramos et al. (2004), *Chemical Geology*, Vol. 211, Fig. 9, p. 151].

machine/manufacturer) may result in variations in ionization potentials and in the interfering molecules and ions present. And although not directly evaluated, similarities between Sr isotope ratios obtained on plagioclase using laser sampling in conjunction with the GV IsoProbe™ compared to those obtained using micromilling and TIMS (Davidson et al. 2001), also allow for only a minimal influence of Ca dimers and Ca argides during analyses. Tables 1a and 1b show two different collector configurations for Sr isotope measurements and relevant first order interferences for the ThermoFinnigan Neptune™ and Nu Plasma™, respectively. Select interfering elements may be present during all analyses, Rb and Kr for instance, but the presence of others may depend on the specific mass spectrometer being used. Great care must be taken to identify and constrain all potential interferences. Ultimately, correcting for interfering elements affecting Sr isotope masses, and relevant corrections, is both the most challenging and most critical aspect of obtaining accurate Sr isotope ratios using laser ablation. Thus, it is beneficial to confirm the accuracy of results using alternative, proven methods such as micromilling/TIMS.

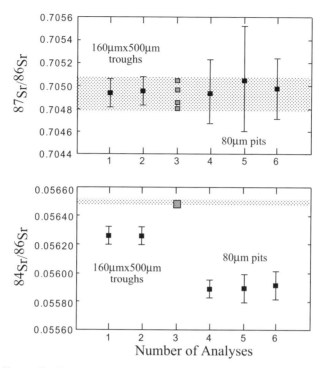

Figure 8. $^{87}Sr/^{86}Sr$ and $^{84}Sr/^{86}Sr$ variations of fine-grained Pisgah Crater (USA) basaltic groundmass. Filled squares with 2σ error bars result from laser ablation sampling using 160×500 μm troughs and 80 μm diameter pits. Gray squares and stippled field represent results from four separate micromilled holes of fine-grained groundmass. $^{87}Sr/^{86}Sr$ results reflect variable errors but results overlap those from micromilled samples and attest to the accuracy of $^{87}Sr/^{86}Sr$ obtained using laser ablation sampling. $^{84}Sr/^{86}Sr$ variations deviate from micromilled samples analyzed using TIMS and partly reflect variations presumably resulting from pit-induced fractionation effects. [Used by permission of Elsevier, from Ramos et al. (2004), *Chemical Geology*, Vol. 211, Fig. 8, p. 149].

To offer a general starting point in which to undertake Sr isotope analyses using laser ablation, Tables 2a and 2b are presented to illustrate machine and laser settings. These parameters are used in conjunction with the ThermoFinnigan Neptune™/UP213 New Wave™ laser and the Nu Plasma™/Lambda Physik Compex 110ARF™ excimer laser, respectively, and will change depending on the specific mass spectrometer and laser being used. Many of the previously mentioned studies utilize different mass spectrometers and lasers and will include basic information for machine and laser settings. Additionally, the Faraday collector configurations and major interfering elements described above can be used initially and refined as necessary. Further evaluations should be undertaken to ensure that additional interferences are not present. This can be accomplished in many ways, but Ramos et al. (2004) had success in using NBS987 standard solutions doped with potential interferences such that specific elements or molecules could be individually evaluated as potential interferences.

Once interferences are identified, a correction procedure must be designed and applied to remove their impact from the desired ratios of interest. This may be complex. For the ThermoFinnigan Neptune™, all Faraday collectors are used to measure the isotopes of Sr and interfering elements and molecules. If more isotopes were needed to be measured, a peak-hopping routine would be required. It is critical that ion intensities from at least one isotope per interfering species is measured, independent of the effects of additional interfering elements

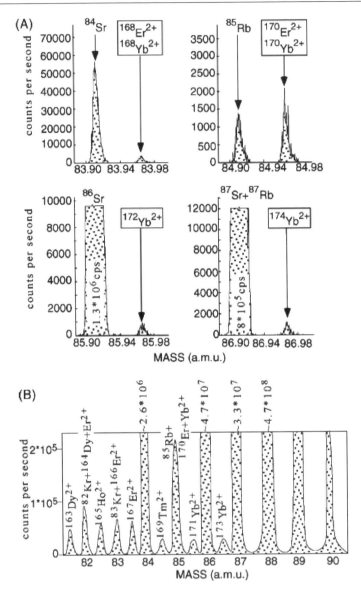

Figure 9. A) Mass scans at resolving power of 3000 illustrating the presence and magnitude of a range of interfering elements on the VG Axiom MC-ICPMS at or near Sr and Rb masses of interest. B) Mass scan at normal resolving power illustrating the presence and magnitude of a range of half-mass and whole-mass interfering elements. [Used by permission of Elsevier, from Waight et al. (2002), International *Journal of Mass Spectrometry*, Vol. 221, Fig. 4, p. 234].

or molecules. Ramos et al. (2004) offered a detailed description of the off-line correction sequence utilized for Sr isotope analyses on the ThermoFinnigan Neptune™.

Firstly, Kr contributions are removed by measuring on-peak baselines during normal gas flow through the ablation cell without firing the laser. Background Kr signals, present as a result of Kr impurities in the He and Ar gas supplies, are directly measured as elevated ion

Table 1a. Collector block configuration of the ThermoFinnigan Neptune™ MC-ICPMS (modified after Ramos et al. 2004) with potential interferences illustrated.

Collector	L4	L3	L2	L1	C	H1	H2	H3	H4
Mass	83	83.5	84	85	85.5	86	86.5	87	88
Isotope of interest	$^{83}Kr_{11.5\%}$	$^{167}Er^{2+}$	$^{84}Sr_{0.56\%}$	$^{85}Rb_{72.2\%}$	$^{171}Yb^{2+}$	$^{86}Sr_{9.86\%}$	$^{173}Yb^{2+}$	$^{87}Sr_{7.00\%}$	$^{88}Sr_{82.6\%}$
Isobaric interferences			$^{84}Kr_{57.0\%}$			$^{86}Kr_{17.3\%}$	+	$^{87}Rb_{27.8\%}$	
Er^{2+}	$^{166}Er^{2+}$		$^{168}Er^{2+}$	$^{170}Er^{2+}$					
Yb^{2+}			$^{168}Yb^{2+}$	$^{170}Yb^{2+}$		$^{172}Yb^{2+}$		$^{174}Yb^{2+}$	$^{176}Yb^{2+}$

Table 1b. Collector block configuration of the Nu Plasma™ MC-ICPMS (modified after Woodhead et al. 2005) with potential interferences illustrated.

Collector	L4	L3	L2	L1	C	H1	H2	H3	H4	H5	H6
Mass	82	83	84		85		86		87	88	89
Isotope of interest	$^{82}Kr_{11.5\%}$	^{83}Kr	$^{84}Sr_{0.56\%}$		$^{85}Rb_{72.2\%}$		$^{86}Sr_{9.86\%}$		$^{87}Sr_{7.00\%}$	$^{88}Sr_{82.6\%}$	$^{89}Y_{100\%}$
Isobaric interferences			$^{84}Kr_{57.0\%}$				$^{86}Kr_{17.3\%}$		$^{87}Rb_{27.8\%}$		
Argides	$^{42}Ca^{40}Ar$	$^{43}Ca^{40}Ar$	$^{44}Ca^{40}Ar$				$^{46}Ca^{40}Ar$			$^{48}Ca^{40}Ar$	
Ca Dimers	$^{42}Ca^{40}Ca$	$^{43}Ca^{40}Ca$	$^{44}Ca^{40}Ca$				$^{46}Ca^{40}Ca$			$^{48}Ca^{40}Ca$	

Table 2a. Operating parameters of the ThermoFinnigan NeptuneTM MC-ICPMS and Nu PlasmaTM MC-ICPMS (modified from Ramos et al. 2004 and Woodhead et al. 2005).

	Neptune	**Nu Plasma**
RF power	1200 W	1350 W
Argon cooling gas flow rate	15 L/min	13 L/min
Auxiliary gas flow rate	0.8 L/min	0.9 L/min
Interface cones	Nickel	Nickel
Acceleration voltage	10 kV	4 kV
Mass resolution	400	400
Mass analyzer pressure	$5\text{-}8 \times 10^{-9}$ mbar	4×10^{-9} mbar
Detection system	Nine Faraday collectors	Eight Faraday collectors
Sampling mode	Three or ten blocks of 10×8 s integrations for laser and solution analysis, respectively	
Background/baseline	3 min on peak in 2.0% HNO_3	on peak
Nebulizer	Glass cyclonic spray chamber fitted with a Micromist PFA nebulizer	?
Uptake mode	Free aspiration	Free aspiration
Sample uptake rate	50 µL/min	?
Typical sensitivity on ^{88}Sr	55-60 V/ppm (10-11 ohm resistors)	
Ar sample gas flow rate	0.60-1.05 L/min, optimized to maximize ^{88}Sr signal	0.85 L/min, optimized to maximize ^{88}Sr signal
Beam dispersion	(Dispersion Quad)	−25

Table 2b. Operating parameters of UP213 New WaveTM and Helex/ Lambda Physik CompexTM 110ArF excimer laser (modified from Ramos et al. 2004 and Woodhead et al. 2005).

Laser	**UP213 New Wave™ Laser**	**Helex Lambda Physik Compex 110ArF excimer**
Wavelength	213 nm	193 nm
Spot dimensions	160 µm × 500 µm trough or 80 µm × 500 µm trough or 80 µm diameter spot	105 µm diameter spot
Energy density	~7-10 J	
Pulse rate	10 Hz	5 Hz
Carrier gas	Helium	Helium
He auxiliary gas flow rate (for analysis of Sr isotopes in plagioclase)	0.85 to 0.95 L/min, optimized every run session	0.2 L/min, optimized every run session

signals (compared to electronic baselines) at the 83, 84, and 86 mass positions and removed during baseline correction. Typical signals were always less than 3 mV of ^{83}Kr. Although this methodology worked for Ramos et al. (2004), Hart et al. (2005) questioned whether on-peak baselines accurately account for Kr contributions and instead used the ^{84}Kr intensity corrected for ^{84}Sr contributions to calculate ^{86}Kr contributions. Thus, as is typical for laser ablation analyses, the user will have to evaluate which procedure works more effectively in this regard.

After Kr, REE^{2+} corrections are undertaken. The ^{167}Er^{2+} and ^{171}Yb^{2+} ion intensities can be measured in the low 3 and center Faraday collectors, respectively, which are set at half-mass positions (Tables 1a and 1b). From the ^{167}Er^{2+} ion intensity, interfering ^{166}Er^{2+}, ^{168}Er^{2+} and ^{170}Er^{2+} intensities can be calculated using the ^{166}Er^{2+}/^{167}Er^{2+}, ^{168}Er^{2+}/^{167}Er^{2+}, and ^{170}Er^{2+}/^{167}Er^{2+} ratios measured from reference solutions. These contributions can then be mathematically removed from the mass 83, 84, and 85 ion intensities prior to correcting for the next interfering element/molecule. Yb interference corrections can then be pursued in the same fashion.

After REE^{2+}, ^{87}Rb corrections can then be undertaken. Ramos et al. (2004) applied a procedure that used the interference corrected ^{85}Rb intensity and the mass bias factor determined from the interference corrected ^{86}Sr/^{88}Sr ratio to determine the ^{87}Rb contribution to the ion intensity at mass 87. In contrast, Hart et al. (2005), which intended to measure glasses with significantly higher Rb/Sr ratios, determined that ^{87}Rb contributions could not be accurately assessed to the degree needed for samples with high Rb/Sr by using ^{86}Sr/^{88}Sr determined mass bias corrections. As a result, these authors applied Rb corrections that incorporated ^{85}Rb/^{87}Rb ratios determined on an external set of glasses in which respective ^{87}Sr/^{86}Sr ratios had previously been measured.

Although a complete set of correction procedures are reviewed here, selected materials may not require correction for all potential interferences, thus appropriate correction procedures will need to be determined for each type of material analyzed. In addition, if analyses are undertaken at poor vacuum conditions, peak-tailing corrections may also need to be applied.

After correcting for Rb, standard normalization of Sr isotopes to ^{86}Sr/^{88}Sr = 0.1194 can be undertaken to correct for mass spectrometer mass bias/mass fractionation. Note that this normalization is ~10 times larger than that for TIMS, thus inaccurate normalization can have a much greater impact on resulting normalized Sr isotope ratios. Ultimately, this correction procedure proved effective in generating accurate ^{87}Sr/^{86}Sr isotope ratios but only variably accurate ^{84}Sr/^{86}Sr ratios in a broad array of target materials (Ramos et al. 2004). For analyses using the Nu Plasma™, it is unclear how separation between Ca dimers, Ca argides, and REE^{2+} would be accomplished, and as of yet, successful Sr isotope analyses of REE-rich minerals such as clinopyroxene are absent. For the GV Isoprobe™, no detailed evaluation of interferences has been undertaken probably because most are removed by use of a collision cell. It has been demonstrated, however, that relatively accurate Sr isotope ratios can be obtained on REE-rich minerals such as clinopyroxene (Schmidberger et al. 2003), so it is assumed that REE interfering effects are generally minor. Vroon et al. (2008) offers a review of correction and reduction procedures from a range of studies using laser ablation for further reference.

To analyze Sr isotopes using laser ablation, we suggest starting with previously determined mass spectrometer settings, laser settings, and collector configurations. A program should then be initiated to evaluate all potential elemental and molecular interferences and create appropriate correction procedures. Once done, users should analyze "known" samples with previously determined Sr isotope ratios. Modern marine carbonates are commonly used to evaluate measurement accuracies as modern seawater ^{87}Sr/^{86}Sr is well-known. Overall however, there is a dearth of reference materials for laser ablation applications, thus it is commonly easiest to analyze potential target materials using micromilling, chromatography, and TIMS or purified solution MC-ICPMS to obtain accurate Sr isotope ratios of reference materials. Reproducing "known" results will give users a means in which to evaluate measurement accuracy and reproducibility and an opportunity to test correction procedures for a range of materials. Such tests should be done for each type of material to be analyzed (i.e., carbonate, plagioclase, clinopyroxene, apatite, etc) and will be useful for evaluating at which concentration levels potential interferences can be accurately corrected (e.g., Rb/Sr ≤ 0.05).

Sr isotope applications. Studies measuring $^{87}Sr/^{86}Sr$ ratios of minerals using laser ablation sampling encompass a wide range of applications. Carbonates, for example, offer a generally Sr-rich mineral with few interferences. For evaluating analytical accuracies and precisions associated with carbonates, shells characterized by modern marine $^{87}Sr/^{86}Sr$ ratios are commonly measured (e.g., Christensen et al. 1995; Outridge et al. 2002; Ramos et al. 2004; Woodhead et al. 2005). Alternative methods include directly comparing $^{87}Sr/^{86}Sr$ ratios measured using laser ablation sampling to results determined on purified samples, including freshwater otoliths, acquired using micromilling and TIMS (Barnett-Johnson et al. 2005). For otolith studies, spot and raster sampling have been used to document sequential changes in $^{87}Sr/^{86}Sr$ ratios that correlate to fish movements from freshwater to marine environments (Fig. 10; e.g., Woodhead et al. 2005), between freshwater environments (e.g., Outridge et al. 2002), and within natal rivers of origin (e.g., Barnett-Johnson et al. 2008).

Studies identifying the natal rivers of fishes include Atlantic croaker (Thorrold and Shuttleworth 2000), tropical shad (Milton and Chenery 2003), barramundi (McCulloch et al. 2005), and California salmon (Barnett-Johnson et al. 2005). Barnett-Johnson et al. (2008) also correlated the spatial distribution of $^{87}Sr/^{86}Sr$ ratios with distinctive microtextural features indicative of wild or hatchery origins (Barnett-Johnson et al. 2007) to evaluate wild and hatchery salmon population dynamics in California offshore marine fisheries. This study demonstrated that natal rivers associated with both wild and hatchery fish could be accurately determined using $^{87}Sr/^{86}Sr$ ratios measured by laser ablation sampling. The combination of otolith studies to date offers a foundation in which individual otolith growth rings can be used to estimate times associated with fish movements through different freshwater environments prior to outmigration to marine environments. These time constraints may aid in evaluating ecological and biological factors to ensure the greatest chance for successful marine migration and future reproduction of endangered anadromous fish species.

In contrast to otoliths, Bizzarro et al. (2003) focused on analyses of magmatic apatite and carbonate phenocrysts to identify open-system processes affecting alkaline igneous rocks of the Sarfartoq province (Greenland). This study successfully combined variable textural and mineral $^{87}Sr/^{86}Sr$ isotopic compositions, and concluded that similar $^{87}Sr/^{86}Sr$ variations in both apatite and carbonate crystals result from coeval crystallization occurring during continuous modification of the host magma as a result of magmatic recharge. Magmas acquired variable $^{87}Sr/^{86}Sr$ ratios that originated from melts added to the magma reservoir from a range of subcontinental lithospheric- and plume-related carbonate-rich mantle components. Similar processes occurred throughout the history of the Sarfartoq province.

For plagioclase, early studies focused on identifying and tracking open-system processes reflected in $^{87}Sr/^{86}Sr$ variations retained in phenocrysts from basalts at Long Valley, California (USA) (Christensen et al. 1995) and El Chichón (Mexico) (Davidson et al. 2001). Christensen et al. (1995) used $^{87}Sr/^{86}Sr$ ratios acquired by laser ablation sampling to demonstrate that plagioclase phenocrysts, which otherwise were characterized by uniform major element compositions, retained variable $^{87}Sr/^{86}Sr$ ratios resulting from magma mixing associated with post-caldera basaltic magmatism at Long Valley. Additionally, Davidson et al. (2001) used laser rastering to measure $^{87}Sr/^{86}Sr$ profiles along traverses across plagioclase phenocrysts from El Chichón, an arc-related volcano. These traverses were shown to be generally symmetric, were correlated to textural features revealed by Nomarsky imaging, and demonstrated differential rim and core $^{87}Sr/^{86}Sr$ signatures confirmed by micromilling and TIMS. El Chichón variations resulted from sequential growth of plagioclase cores in magma with higher $^{87}Sr/^{86}Sr$ and growth

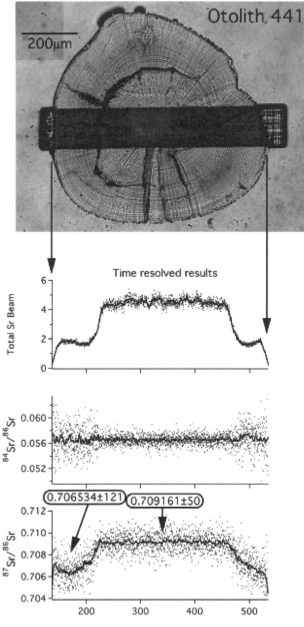

Figure 10. Sr ion beam intensity, $^{84}Sr/^{86}Sr$, and $^{87}Sr/^{86}Sr$ transect profiles of an otolith. Profiles reflect marine (0.70916) and freshwater (0.70653) $^{87}Sr/^{86}Sr$ signatures. Such profiles can be used to identify the $^{87}Sr/^{86}Sr$ signatures of natal rivers and track fish outmigration to marine environments, inmigration to freshwater environments, or migration between freshwater environments. [Reproduced by permission of The Royal Society of Chemistry, from Woodhead et al. (2005), *Journal of Analytical Atomic Spectrometry*, Vol. 20, Fig. 2, p. 26].

of rims in magma with lower $^{87}Sr/^{86}Sr$. For both studies, not only were the effects of open-system processes such as magma mixing identified, but the isotope compositions of the different mixed magmas were constrained.

In contrast to smaller volcanic systems, $^{87}Sr/^{86}Sr$ ratios of plagioclase obtained using laser ablation sampling were also used to 1) confirm the presence of a wide range of $^{87}Sr/^{86}Sr$ variations in all major formational units of flood basalts erupted on the Columbia Plateau (USA) and 2) constrain the processes associated with continental, plume-related volcanism. Early-erupted Imnaha basalts, previously thought to result from melting of Yellowstone plume mantle without significant external inputs, exhibited a wide range of $^{87}Sr/^{86}Sr$ ratios in plagioclase phenocrysts while host basalts exhibited only subtle variations (Eckberg et al. 2006). The overall range of $^{87}Sr/^{86}Sr$ variations, best observed in plagioclase, originated from plume-related magma interacting with either accreted fluid-fluxed subcontinental mantle/crust or cratonic crust encountered during ascent (Wolff et al. 2008). Where magmas interacted with fluid-fluxed materials, plagioclase $^{87}Sr/^{86}Sr$ ratios were lower than plume-related signatures. Where magmas interacted with cratonic crust, presumably while residing in crustal chambers, plagioclase crystals retained highly elevated $^{87}Sr/^{86}Sr$ ratios.

Further plume-related magmatism on the Columbia River plateau generated Picture Gorge basalts and the volumetrically immense Grande Ronde basalts. Both whole-rocks and plagioclase crystals from these formations retained wider ranging $^{87}Sr/^{86}Sr$ signatures in comparison to earlier-erupted Imnaha basalts. These variations resulted from more extensive interactions with either fluid-fluxed accreted materials (Picture Gorge basalts) or greater amounts of assimilation of cratonic crust (Grande Ronde basalts) and reflected the same processes affecting Imnaha basalts, but on a massive scale. Interestingly, laser ablation analyses of plagioclase crystals were key in identifying the processes that influenced isotopic and trace element signatures of early erupted magmas, processes that would greatly influence the geochemical and isotopic nature of future eruptions of large-volume flood basalts such as those comprising the Grande Ronde formation.

In addition to tracing the sources and processes generating Columbia River flood basalts, $^{87}Sr/^{86}Sr$ ratios measured using laser ablation sampling were also used to evaluate crystal residence times for all major Columbia River basalt formations (Ramos et al. 2005). Results indicated that most plagioclase phenocrysts, and where available clinopyroxene, reflected 1) disequilibrium with host melts and 2) phenocryst incorporation into large-volume flood basalt magmas at varied times prior to eruption, some less than five years prior (Fig. 11).

Open-system processes influencing the products of the 1915 Lassen volcano (USA) eruption were also evaluated by Salisbury et al. (2008). This study pursued an integrated approach using crystal size distributions (CSDs), textural, and *in situ* analyses including $^{87}Sr/^{86}Sr$ ratios measured by laser ablation sampling. Andesite- and dacite-hosted plagioclase crystals with $^{87}Rb/^{86}Sr$ ratios as high as 0.20 were successfully analyzed. Sequential trough analyses documented subtle core to rim $^{87}Sr/^{86}Sr$ variations that resulted from the complex interplay of a range of dacite, andesite, and basaltic andesite magmas. Phenocrysts of plagioclase, with uniform $^{87}Sr/^{86}Sr$ ratios, were thought to have experienced simple crystallization histories over 100s to 1000s of years in a dacitic magma chamber. Microphenocrysts, however, grew over months in an environment in which andesite and basaltic andesite were mixed, while microlites reflected equilibrium growth during ascent and eruption from the resulting hybrid magma. This study exemplified the utility of integrating a range of chemical and physical parameters to understand the histories of individual eruptive episodes of arc volcanoes.

For clinoproxene, Schmidberger et al. (2003) targeted individual crystals from a single mantle xenolith and documented a large range in $^{87}Sr/^{86}Sr$ ratios. This range resulted, at least partially, from the infiltration and reaction with kimberlitic host melt during magmatic ascent but may have also been largely present before magmatism. If so, highly variable $^{87}Sr/^{86}Sr$ ratios

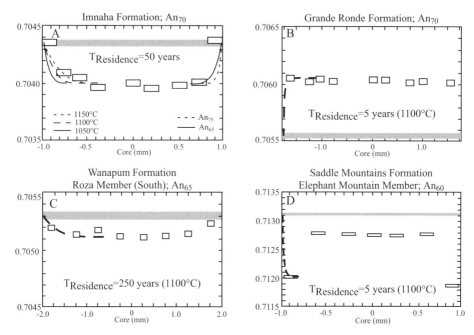

Figure 11. Selected plagioclase phenocryst $^{87}Sr/^{86}Sr$ profiles from Columbia River flood basalts (modified from Ramos et al. 2005) compared to host rock $^{87}Sr/^{86}Sr$ (gray fields). Phenocrysts in Columbia River flood basalts generally reflect isotope disequilibrium with host magmas. Profiles are from A) early Imnaha basalt, B) Grande Ronde basalt, C) Roza basalt, and D) late-erupted Saddle Mountain basalt. These profiles are modeled using simple one-dimensional diffusion to determine maximum phenocryst residence ages. Curves in A) reflect temperatures of 1150°C, 1100°C, and 1050°C (left side) and curves for An_{75} and An_{65} (right side). Dashed curves in B), C) and D) are for 1100°C and An_{75}. Modeling suggests that plagioclase phenocrysts reside in Columbia River flood basalts for only short times (<250 years), indicating that mineral and melt components of flood basalt magmas were undergoing assembly immediately before or during eruption.

present at the single xenolith scale would indicate direct evidence for mantle heterogeneity at small scales.

Although not all-inclusive, these studies reflect the broad utility of $^{87}Sr/^{86}Sr$ isotope analyses obtained using laser ablation samples in a microspatial isotope context. The potential for these analyses to address new questions associated with biological and petrological fields is extensive and future studies will undoubtedly use such analyses in even more creative ways.

Pb isotope technique. In contrast to Sr, Pb isotopes measured using laser ablation have far fewer apparent interfering elements or molecules. By introducing thallium into the sample stream via a nebulizing spray chamber prior to the sample entering the torch of the mass spectrometer, analytical precisions associated with Pb isotope ratios are increased substantially as a result of mass fractionation correction using $^{203}Tl/^{205}Tl$ normalization (Horn et al. 2000). As a result, Tl normalized Pb isotope ratios sampled using laser ablation attain precisions and accuracies that approach those of double-spike TIMS measurements although some variances have been documented (Thirwall 2002). Alternatively, sample-standard bracketing can be effectively used, commonly using NBS610 glass as a standard. Thus, laser ablation sampling offers an opportunity to acquire accurate and precise Pb isotope ratios in a highly time-efficient fashion.

Figure 12 illustrates results from six separate 50×250 μm troughs analyzed from a single Valles caldera potassium feldspar crystal using the ThermoFinnigan Neptune™ (Wolff

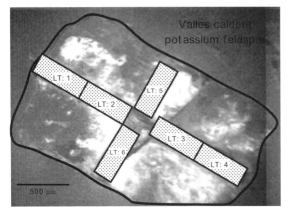

	Single Crystal Analysis
$^{208}Pb/^{204}Pb$:	37.555 (5)
$^{207}Pb/^{204}Pb$:	15.489 (2)
$^{206}Pb/^{204}Pb$:	17.802 (2)
$^{208}Pb/^{206}Pb$:	2.1096 (2)
$^{207}Pb/^{206}Pb$:	0.8701 (1)

Laser Trough:	1 (LT: 1)	2 (LT: 2)	3 (LT: 3)	4 (LT: 4)	5 (LT: 5)	6 (LT: 6)
$^{208}Pb/^{204}Pb$:	37.59 (8)	37.46 (5)	37.56 (7)	37.67 (8)	37.53 (6)	37.47 (7)
$^{207}Pb/^{204}Pb$:	15.50 (3)	15.45 (2)	15.49 (3)	15.54 (3)	15.48 (3)	15.46 (3)
$^{206}Pb/^{204}Pb$:	17.81 (4)	17.75 (2)	17.80 (3)	17.86 (4)	17.79 (3)	17.76 (3)
$^{208}Pb/^{206}Pb$:	2.1099 (5)	2.1098 (4)	2.1094 (5)	2.1089 (7)	2.1098 (4)	2.1094 (4)
$^{207}Pb/^{206}Pb$:	0.8703 (2)	0.8703 (2)	0.8700 (2)	0.8697 (5)	0.8701 (1)	0.8701 (1)
^{208}Pb:	141 mV	157 mV	162 mV	142 mV	150 mV	172 mV

Figure 12. Image of a Valles caldera potassium feldspar crystal sampled for Pb isotopes using laser ablation. Six 250 μm troughs were sampled with results of each analysis illustrated (LT 1-6). After sampling, the single crystal was dissolved, purified, and analyzed using the procedures of Wolff and Ramos (2003). Results attest to the accuracy of Pb isotope ratios measured using laser ablation sampling in respect to ratios of both the more abundant Pb masses and ratios normalized to the least abundant ^{204}Pb mass.

and Ramos 2004). Analyses of isotope ratios involving the more abundant ^{206}Pb, ^{207}Pb, and ^{208}Pb isotopes average 2.1095 and 0.8701 for $^{208}Pb/^{206}Pb$ and $^{207}Pb/^{206}Pb$ respectively, with all individual analyses lying within in-run error of these averages. To evaluate accuracy, the crystal was then dissolved, Pb was purified using anion exchange chromatography, and the sample was analyzed using Tl-doping following the procedures of Wolff and Ramos (2003). Results for this crystal are 2.1096±0.00002 and 0.8701±0.0001 for $^{208}Pb/^{206}Pb$ and $^{207}Pb/^{206}Pb$, respectively. For potassium feldspar, laser and single crystal results yield identical $^{208}Pb/^{206}Pb$ and $^{207}Pb/^{206}Pb$ ratios and attest to the accuracy of laser ablation, Tl-normalized Pb isotope measurements.

To evaluate ratios normalized to less abundant ^{204}Pb, $^{206}Pb/^{204}Pb$, $^{207}Pb/^{204}Pb$, and $^{208}Pb/^{204}Pb$ average 17.80, 15.49, and 37.55 respectively for the same six laser ablation analyses. These ratios are characterized by poorer measurement precisions, as reflected by two decimal place precisions, as a result of much lower ^{204}Pb ion intensities. And unlike larger peaks, ^{204}Pb is corrected for interfering ^{204}Hg which is present in minor amounts presumably as impurities in He or Ar carrier gases. In addition, very minor signals originate from the potassium feldspar crystal itself, potentially near the crystal surface. To minimize such surface contributions for both Hg and Pb, ~3μm of the surface was ablated along the targeted sampling area prior to analysis. To correct for Hg, ^{202}Hg is monitored and used to calculate ^{204}Hg contributions using the $^{202}Hg/^{204}Hg$ ratio measured in separate Hg solutions. Results for the single crystal analysis are 17.802±0.002, 17.489±0.002, and 37.555±0.005 for $^{206}Pb/^{204}Pb$, $^{207}Pb/^{204}Pb$, and $^{208}Pb/^{204}Pb$ respectively. Thus, laser ablation measurements relative to ^{204}Pb have poorer overall analytical precisions but still lie within in-run error of the single crystal results, again demonstrating measurement accuracy for potassium feldspar.

In contrast to Sr, Pb contents of most magmatic materials are substantially lower. In this case, the Pb content of the potassium feldspar is generally between 10-15 ppm, a concentration level that still yields highly accurate and precise ^{208}Pb/^{206}Pb and ^{207}Pb/^{206}Pb ratios and highly accurate and relatively precise ^{206}Pb/^{204}Pb, ^{207}Pb/^{204}Pb, and ^{208}Pb/^{204}Pb ratios. In contrast to the Tl normalized analytical procedure described above, Simon et al. (2007) pursue a sample/standard bracketing methodology that normalizes laser ablation results to nebulized NBS981 solutions to correct for mass bias. This procedure is reserved, however, for measuring ratios involving the larger ^{206}Pb, ^{207}Pb, and ^{208}Pb isotopes from potassium feldspars and related glasses from Long Valley rhyolites where interferences are presumed to be minor. Further studies by Kent (2008) evaluate Pb isotope accuracy, reproducibility, and interference corrections using both Faraday and ion multipliers, and sample/standard bracketing using standard/reference glasses.

Pb isotope applications. There are a range of studies that focus on Pb isotopes analyzed using laser ablation. Below, selected studies are reviewed that either focus on uncommon radiogenic materials or materials characterized by common Pb isotope signatures. Studies targeting highly radiogenic materials such as zircon are not included. For example, Willigers et al. (2002) obtained Pb ages using laser ablation sampling, including ^{204}Pb results using an ion multiplier, for apatite, monazite, and sphene. While sphene ages diverged from those determined by TIMS, apatite and monazite ages were within error of TIMS results for equivalent mineral separates. These authors attributed this sphene age discrepancy to the presence of interferences across the Pb mass spectrum, although individual interferences were not identified.

For materials characterized by common Pb isotope ratios, Mathez and Waight (2003) used laser ablation sampling to determine ^{207}Pb/^{206}Pb and ^{208}Pb/^{206}Pb ratios of sulfide and plagioclase crystals to evaluate the evolution of the Bushveld Complex (South Africa), a large layered mafic intrusion. These authors found Pb isotope disequilibrium between sulfides and plagioclase, and between plagioclases from different layers of the intrusion. For sulfide-plagioclase pairs, sulfides typically had lower ^{208}Pb/^{206}Pb and higher ^{207}Pb/^{206}Pb ratios than plagioclases that were attributed to the addition of Pb from surrounding country rocks. Pb isotope variations between plagioclase crystals from different layers of the mafic intrusion resulted from the inputs of multiple sources of Pb during crystallization. Different Pb isotope ratios between sulfides and plagioclase however, resulted from Pb addition to sulfides after temperatures dropped below those needed for Pb exchange in plagioclase. And, these different Pb sources were thought to have obliterated any potential isotopic signatures of the mantle sources associated with the original magma.

Pb isotope ratios of Bishop Tuff-hosted potassium feldspars obtained by Simon et al. (2007) were used to confirm that feldspar crystals grew from magmas associated with the Bishop Tuff, not magmas related to precaldera glass mountain lavas. Similar to Wolff and Ramos (2003), Simon et al. (2007) used Pb isotopes of potassium feldspar to discount the possibility that the Bishop Tuff resulted from long-lived rhyolitic magmatism and instead concluded that the Bishop Tuff more-likely resulted from the accumulation of many small, different magmas. In addition, Pb isotope ratios defined linear trends between young Long Valley basalts and local evolved crust undermining any potential for accurate age determinations using Rb/Sr isotopes.

In addition to igneous applications, Tyrrell et al. (2007) utilized the efficiency of laser ablation sampling to analyze a large suite of potassium feldspar crystals to trace sediment provenance associated with the breakup of Pangea. Potassium feldspar crystals in selected Triassic and Jurassic sandstones were targeted to identify crystal populations based on Pb isotopes. Both sandstones had two main populations of crystals with the Jurassic sandstone also having one additional outlier population. Crystals from each population were found together independent of facies and stratigraphic position. Isotope ratios from these distinct populations were compared with potential source rocks found throughout the North Atlantic region. Sources were then either discounted or included and paleodepositional directions were determined for sediments contributing to these sandstones. In addition to constraining sediment transport directions, dispersal distances were also determined, as well as identification of different sediment sources associated with different periods of North Atlantic rifting.

Pb isotopes analyzed using laser ablation sampling also aided geologic structure-based studies. Connelly and Thrane (2005) utilized laser ablation sampling to rapidly measure a large suite of potassium feldspar crystals from currently adjoining early Proterozoic orthogneiss terranes in west Greenland to evaluate a potential suture zone between them. These authors focused on initial Pb isotope ratios of potassium feldspars to distinguish whether these terranes shared similar magmatic sources and were thus related. Two populations of feldspars were present, each with distinctive initial Pb isotope ratios, which suggested that these orthogneiss bodies did not originate from similar mantle/crustal sources. As a result, these authors concluded that a cryptic suture must be present between the two west Greenland orthogneiss terranes.

Overall, Pb isotopes sampled using laser ablation offer an accurate means in which to acquire data in a time-efficient manner. Current applications attest to the potential utility of Pb isotopes results measured using laser ablation sampling for a wide variety of scientific applications in the earth sciences.

MICROANALYTE ANALYSIS

Single crystal analyses

Technique. Together with micromilling, single crystal isotope analyses target a limited sample microspatially (i.e., a single crystal) but unlike micromilling, single crystals usually contain low overall amounts of analyte because either the crystal is small or the crystal is characterized by low analyte concentrations. In addition to single crystals, multi-crystal samples may also be targeted using the same technique because many multi-crystal samples (e.g., quartz) also have low analyte abundances. For Sr and Pb isotopes, olivine and quartz best demonstrate the utility of single crystal analyses.

Olivine, $(Mg,Fe)_2SiO_4$, and quartz, SiO_2, are found in a wide variety of igneous rocks and have little tolerance for incorporating Sr or Pb into their respective mineral structures as a result of elemental size and charge considerations. As a result, both typically have very low Sr and Pb concentrations. In contrast, mineral or melt inclusions captured during olivine or quartz growth may contain significant amounts of Sr and/or Pb. The generally small size of these inclusions, however, may require large amounts of host mineral to obtain minimal analyte abundances. For olivine, inclusions of basaltic melt can contain Sr concentrations ≥1000 ppm. If, for example, melt inclusions account for 1% of the total weight of an olivine crystal and contain 1000 ppm Sr, a 1 mg (0.001 g) single crystal would yield ~10 ng of Sr, an amount easily analyzed using TIMS. If, however, the crystal contained only 0.1% melt inclusions, a 1 mg crystal would yield ~1 ng of Sr. A ~1 ng Sr sample can still be analyzed to a reasonable precision ($^{87}Sr/^{86}Sr$ ≤0.0001) using TIMS but uncertainty of whether the measured isotope ratio reflects that of melt inclusions magnifies as the amount of Sr contamination resulting from the digestion and purification process becomes a greater percentage of the total amount of analyte analyzed. This "processing" contamination is measured as a total process blank. If such blanks can be limited to ≤10 pg Sr, small (~1 ng) samples can be analyzed successfully while maintaining blank levels of ≤1%. Maintaining low blank levels is the most critical aspect of accurately analyzing single crystals, or small micromilled samples, characterized by low analyte abundances. While Sr samples as small as 0.5 ng can be successfully analyzed (e.g., Wolff et al. 1999), uncertainties in blank corrected ratios increase dramatically when processing blanks exceed 1% of total analyte.

For quartz crystals from high-silica rhyolites, captured melt inclusions may have low Sr but high Pb concentrations. Current studies targeting comendites and pantellerites from Baitoushan volcano (Ramos et al. 2007, 2008) determine quartz-hosted melt inclusion Sr concentrations of ≤1 ppm while Pb concentrations are ≥40 ppm. These quartz crystals are characterized by high melt inclusion contents, typically >20%. For a 3 mg quartz crystal, the Sr yield would be <0.6 ng while the Pb yield would be >24 ng. While analyzing 0.6 ng of Sr is challenging, especially given that associated blank would have to be ≤6 pg to maintain a 1% blank contribution, analyzing 24 ng of Pb is easily accomplished using solution MC-ICPMS in conjunction with Tl-normalization (Wolff and Ramos 2003). Thus, single quartz crystals could be analyzed for Pb isotopes but multiple crystals would be required for Sr isotope analyses.

To undertake low blank single crystal or micromilled Sr isotope measurements, a method of column chromatography must be pursued which offers an opportunity to attain total process blanks of generally less than ~20 pg. Two methods are prevalent: 1) a stripping method using Sr spec^{TM} resin and 2) a chromatographic method using standard cation exchange resin. The first uses a small volume of Sr spec^{TM} resin loaded onto ~1 ml shrink-fit Teflon columns with a frit inserted at the column tip (e.g., Davidson et al. 1998; Charlier et al. 2006). This method removes unwanted ions by flushing the resin with ~3N HNO_3 and removing Sr with pure ultrafiltered water. Total process blanks using this method can be <10 pg (e.g., Harlou et al. 2005).

During the Wolff et al. (1999) study, the presence of organic ion signals that appeared at both Sr and Rb masses was identified in samples processed using batches of Sr spec^{TM} resin made prior to 1998. As a result, pyrex columns with high aspect ratios (height/width >25) were designed that use ~0.5 ml of 200-400 mesh cation exchange resin to undertake low blank Sr chromatography. These small-volume, high-aspect ratio columns were fitted with microporous frits and minimized blank while maximizing Rb/Sr chemical separations. The procedure for Rb and Sr sequestration was essentially the same as larger cation exchange columns commonly used for whole-rock purifications, but the resin and eluant volumes were dramatically reduced. These small-volume columns were used to undertake column purifications of quartz and potassium feldspar crystals (Wolff et al. 1999; Knesel et al. 1999) and plagioclase, clinopyroxene, amphibole and olivine crystals (Ramos and Reid 2005) and maintained blanks that were typically <20 pg and commonly <10 pg.

To obtain low blanks using either Sr spec^{TM} or cation exchange resins, the resins must be extensively pre-cleaned prior to use and all acid components used in the digestion and chromatography procedures must be fresh and doubly distilled. Normally these acids are distilled the night before use and extracted directly from the distillery. In addition, 3 ml Teflon digestion vessels are used and must be cleaned extensively. All should undergo cleaning in Teflon bottles containing distilled aqua regia that sit on a warm hot plate for more than a week. All pipette tips and bottles containing distilled acids should be cleaned prior to undertaking chemistry. For pre-cleaning cation exchange resin, a ~20 ml disposable Bio Rad Econo-column^{TM} is filled with resin and flushed with freshly distilled 2.5N HCl, 6.0N HCl, and ultrafiltered H_2O for at least two days prior to use. For Sr spec^{TM} resin, hot water is commonly flushed through the resin in a similar fashion.

To determine the total amount of Sr in the samples to be analyzed, isotope dilution using a high-purity Rb/Sr spike is required. Spikes should be added to samples and blanks prior to undergoing digestion. A high Rb/Sr spike and a low Rb/Sr spike are commonly used depending on the Rb/Sr ratio of the target material (e.g., plagioclase would require low Rb/Sr spike and potassium feldspar would require high Rb/Sr spike). For crystals from high-silica rhyolites (e.g., potassium feldspar and quartz), a spike with very high Rb/Sr may be required. Isotope dilution results will 1) yield Rb and Sr concentrations and $^{87}Rb/^{86}Sr$ ratios for age dating or age correction if required and 2) allow for the determination of total analyte abundances so that samples can be stripped of blank contributions. Such contributions will be minor with blanks that are <1% of the analyte but may become more significant when blanks exceed 1%. For actual blank measurements, minimal amounts of spike should be used (commonly ~1 μL) so that pre-spiked $^{87}Sr/^{86}Sr$ ratios of blank can be determined and samples can be corrected for the effects of blank. It may also be useful to determine blank $^{87}Sr/^{86}Sr$ ratios from samples collected from flushing greater amounts of acid (i.e., greater than used in normal column chromatography) through columns without adding spike. Accurately constraining the $^{87}Sr/^{86}Sr$ ratio of the blank will be critical for correcting samples in which blank exceeds 1% of the total analyte.

For analyses of small samples, thermal ionization mass spectrometry (TIMS) is usually preferred to MC-ICPMS. Loading of samples requires TaO_2 and dilute phosphoric acid that must also be devoid of Sr (see Charlier et al. 2006). Standard loading procedures include adding 1 μL of TaO_2 suspended in a solution of 5% nitric acid and 1 μL of 5% phosphoric acid. Filaments for analyses should be made of rhenium and brought to temperatures that exceed normal Sr analytical conditions and be manually checked for Sr and Rb signals in the mass spectrometer prior to loading. Commonly, filaments have to set at high temperatures to allow for all Sr and especially Rb to be depleted. Care should be used for analyzing small Sr samples and NBS987 Sr standards should be analyzed at similar concentration levels to evaluate mass spectrometer analytical accuracy and reproducibility at low analyte levels.

For Pb purifications, anion exchange resin and 1N HBr can be used in similarly small columns as that used for Sr microchemistry. Resin again needs to be pre-cleaned prior to use and HBr needs to be doubly distilled to maintain low blanks. Teflon cleaning procedures should be followed as described for Sr and columns should be cleaned accordingly. Additional chromatography procedures may also be used. Studies associated with single zircons have also been successful in purifying Pb while minimizing blank, thus select procedures from these studies could also be pursued. Ultimately, Pb isotope analyses should be undertaken using either double-spiking and TIMS or MC-ICPMS. Alternatively, Tl-doping or standard sampling bracketing can be used with MC-ICPMS.

Sr and Pb applications. Single crystal isotope analyses were used in a range of applications. The earliest focused on a wide variety of minerals to identify and constrain the effects of open-system processes and discriminate between competing models describing the petrogenesis of S-type rhyolites from Italy (Feldstein et al. 1994). Two separate rhyolites were involved: a more radiogenic $^{87}Sr/^{86}Sr$ endmember and a less radiogenic $^{87}Sr/^{86}Sr$ endmember. Single crystal analyses included plagioclase, potassium feldspar, biotite, and cordierite, in addition to glass. For plagioclase, textural evidence was also integrated with isotope results. Coarse and medium, highly sieved plagioclase crystals with low initial $^{87}Sr/^{86}Sr$ ratios were determined to be associated with clinopyroxene megacrysts and mafic enclaves, materials not directly related to host rhyolites. All other plagioclase crystals had higher initial $^{87}Sr/^{86}Sr$ ratios and were considered phenocrystic. These phenocrystic plagioclase $^{87}Sr/^{86}Sr$ ratios were generally lower than those of phenocrystic potassium feldspar crystals but higher than accompanying glass (assumed to reflect the melt component). In addition, individual biotite crystals retained initial $^{87}Sr/^{86}Sr$ ratios that spanned most of the total observed range in both rhyolitic glasses and crystals. Thus, a complex petrogenetic history was involved. Feldstein et al. (1994) concluded that this history included crystallization of phenocrysts occurring at a mixing interface between two layers of rhyolitic magma in which minerals grew in an environment that offered a range of $^{87}Sr/^{86}Sr$ ratios. Additionally, diffusion modeling was used to constrain possible crystal residence times and suggested that this process occurred relatively quickly (<10,000 years) as much of the observed isotopic heterogeneity would have been eliminated if time periods were longer. This study exemplified the use of single crystal isotope analyses to evaluate the petrogenetic history of highly complicated rocks involving a range of melt and mineral components derived from multiple sources.

More recent studies targeting large rhyolitic magma systems included Wolff et al. (1999) and Knesel et al. (1999). These studies used Sr isotopes of single potassium feldspar crystals to evaluate open-system processes affecting Bandelier Tuff and Taylor Creek (New Mexico, USA) rhyolites. Wolff et al. (1999) focused on linear trends defined by potassium feldspar crystals in the Otowi member of the Bandelier Tuff (Valles caldera). These trends could either have been Rb/Sr isochrons or mixing lines (line "A" in Fig. 13 defined by potassium feldspars). If isochronous, results suggested extensive residence of Bandelier high-silica rhyolitic magma for ~270,000 years prior to eruption. Alternatively, linear trends could have also resulted from two-component mixing without age significance.

To further discriminate between these possibilities, Wolff and Ramos (2003) evaluated similar potassium feldspar crystals using single crystal Pb isotope analyses. Single crystal Pb isotope results also defined a linear array that, if isochronous, yielded an age in excess of 10 Ma, an age that far exceeded any reasonable magma residence time. Thus, Pb isotope variations must have resulted from mixing. Sr was highly depleted (~3 ppm) in Valles magmas compared to Pb (~40 ppm), therefore Sr isotopes must have been modified long before Pb isotopes given any reasonable contaminant (e.g., local crustal lithologies), and both must then have resulted from mixing not aging. Thus, results from combined single crystal Sr and Pb isotopes undermined the possibility of long-lived magma residence at Valles caldera and called into question similar assessments for other large rhyolitic magma systems. Additional work focused on $^{87}Sr/^{86}Sr$ variations in glasses from glomerocrysts and melt inclusions in quartz. Overall, results suggested mixing (Fig. 13) between three magmatic components: 1) magmas with crustal Sr and Pb isotope signatures, 2) magmas with more-mantle like Sr and Pb isotope signatures, and 3) minor amounts of local Proterozoic crust. Mixing between these components resulted in a complex array of Sr and Pb isotope variations that may be typical for large rhyolitic magma systems.

Knesel et al. (1999) also focused on $^{87}Sr/^{86}Sr$ ratios of potassium feldspar crystals derived by both single crystal and micromilling analyses. This study integrated crystal size parameters from Taylor Creek rhyolites to constrain the evolutionary history of a single dome from another

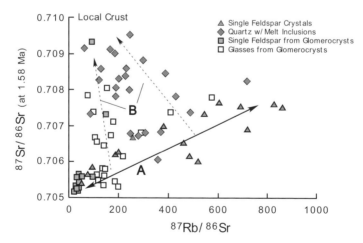

Figure 13. Single crystal feldspar, quartz separates, and glass $^{87}Rb/^{86}Sr$ and $^{87}Sr/^{86}Sr$ results from pumice and pumice hosted glomerocrysts of the Bandelier Tuff. Variations result from mixing 1) highly-evolved, rhyolitic magmas with crustal isotope characteristics with 2) highly-evolved, magmas with more mantle-like $^{87}Rb/^{86}Sr$ and $^{87}Sr/^{86}Sr$ characteristics (double-headed, bold arrow indicated by "A"). In addition, crystals and glasses that diverge to the upper left of this diagram (dashed arrows indicated by "B") originate from melt, present as glass in glomerocrysts or melt inclusions in quartz, that has also incorporated local Proterozoic crust after mixing (Wolff and Ramos 2004).

large rhyolitic magma system. Figure 14a shows that as individual crystal size increased, initial $^{87}Sr/^{86}Sr$ values increased to a maximum and then decreased to a minimum. Interestingly, initial $^{87}Sr/^{86}Sr$ profiles of variably sized sanidine crystals showed the same pattern, in this case core-to-rim, of increasing to a maximum then decreasing to a minimum with the added observation that the smallest crystals lacked the initial increasing trend (Fig. 13b). Knesel et al. (1999) postulated that these trends resulting from crystal growth initially in a magma chamber contaminated by incongruent melts of crustal components with higher $^{87}Sr/^{86}Sr$ (initial increasing $^{87}Sr/^{86}Sr$ trend) followed by progressive recharge of a similar composition magma with lower $^{87}Sr/^{86}Sr$ (decreasing $^{87}Sr/^{86}Sr$ trend). Larger, presumably older crystals found throughout the chamber grew during contamination and magma recharge events and recorded sequentially increasing and then decreasing initial $^{87}Sr/^{86}Sr$ ratios from core to rim, whereas smaller, presumably younger crystals grew during the magma recharge phase only and recorded only decreasing initial $^{87}Sr/^{86}Sr$ ratios. In this study, the combined use of micromilling, single crystal, and crystal size techniques proved to be critical to identifying a complex petrogenetic history.

In contrast to successful studies focused primarily on single potassium feldspar crystals, additional mineral and melt components were used by Ramos et al. (2007) to evaluate magma generation processes at Baitoushan volcano (China/North Korea). At Baitoushan, minerals and melt obtained from at least three recent (<10 ka) comendite and pantellerite eruptions reflected disequilibrium at many levels. Figure 15 illustrates potassium feldspar and pumice fragment $^{87}Sr/^{86}Sr$ variations found in highly-evolved, alkali-rich comendite and trachyte components comprising the 980 AD eruption which emitted ~100 km³ of volcanic material. Initially the

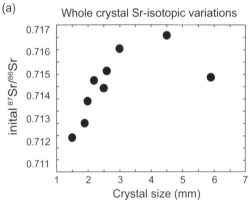
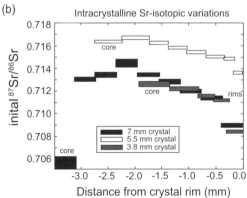

Figure 14. a) Initial $^{87}Sr/^{86}Sr$ versus crystal size in sanidines from the Taylor Creek Rhyolite complex. This diagram illustrates increasing then decreasing initial $^{87}Sr/^{86}Sr$ ratios in crystals with decreasing size. Smaller crystals do not record increasing initial $^{87}Sr/^{86}Sr$ ratios because they reflect later growth in a magma chamber that was initially contaminated by wallrock melt with high $^{87}Sr/^{86}Sr$ and then progressively fluxed with rhyolitic magma with low $^{87}Sr/^{86}Sr$. b) Diagram of changing $^{87}Sr/^{86}Sr$ values in three micromilled sanidine phenocrysts of different sizes. The largest crystal showed increasing $^{87}Sr/^{86}Sr$ near the core to a maximum $^{87}Sr/^{86}Sr$, and then decreasing $^{87}Sr/^{86}Sr$ from the middle of the crystal to the rim. Smaller crystals recorded only portions of this trend owing to later growth in the magma chamber. [Used by permission of Oxford University Press, from Knesel et al. (1999), *Journal of Petrology*, Vol. 40, Fig. 4 and 7, p. 779 and 781].

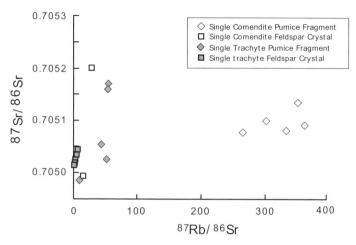

Figure 15. Feldspar and pumice $^{87}Sr/^{86}Sr$ and $^{87}Rb/^{86}Sr$ results from the ~980 AD Baitoushan comendite eruption (Ramos et al. 2007, 2008). The eruption involved two distinct magma components: 1) a early-erupted, highly-evolved comendite and 2) a later-erupted, less-evolved trachyte. Unfilled squares and diamonds represent results from individual comendite potassium feldspar crystals and pumice fragments, respectively. Filled squares and diamonds represent results from individual trachyte potassium feldspar crystals and pumice fragments, respectively. Comendite and trachyte crystals and pumices retain distinct isotope characteristics. Trachyte potassium feldspar crystals define a linear trend (also in $^{87}Sr/^{86}Sr$ versus 1/Sr) that results from 2-component mixing that does not involve comendite as a mixing endmember. Accompanying trachyte pumices are characterized by greater $^{87}Sr/^{86}Sr$ variations imposed after feldspar fractionation. In contrast, comendites have potassium feldspar crystals that retain even greater $^{87}Sr/^{86}Sr$ variations than trachytes that are both higher and lower that accompanying comendite pumice fragments. Results reflect open-system magmatic processes most-likely involving a range of highly-evolved, variably radiogenic magmas rather than variably enriched crustal materials.

eruption produced comendite, while the final stages were comprised of trachyte. For comendites, potassium feldspar crystals retained $^{87}Sr/^{86}Sr$ ratios which were both higher and lower than accompanying pumices while the reverse was seen in trachyte components where potassium feldspar crystals retained a limited range in $^{87}Sr/^{86}Sr$ while trachyte pumices spanned most of the range of comendite potassium feldspars. These relationships reflected the complex interplay of highly-evolved magmas and most-likely resulted from the mixing of a range of melt and mineral components which were not directly identified in the eruptive products.

In addition to the 980 AD eruption, a pantellerite was erupted from Baitoushan at ~0 AD. This pantellerite retained a range of mineral components including potassium feldspar, clinopyroxene, and quartz. These different components retained a range of $^{87}Sr/^{86}Sr$ ratios that were different than accompanying ratios of pumices (Fig. 16A). Melt, as reflected by pumice, generally defined a ~400 ka isochron that suggested an extended magma residence age. Mineral components, however, did not lie along this trend and reflected more consistent, less radiogenic $^{87}Sr/^{86}Sr$ ratios. In addition, Pb isotope ratios of the same crystals defined negative trends which undermined any age relevance for Rb/Sr isotopes and must have resulted from the mixing of magmas with highly variable Pb isotope signatures that were potentially related to the incorporation of local Archean lithologies assimilated at deep crustal levels. These variations are different than those reflected in the 980 AD eruption and must have resulted from either 1) a range of magmas whose presence was limited in time such that magmas with similar signatures were not seen in later eruptions or 2) that a range of magmas existed but did not communicate potentially as a result of the involvement of a complex magma plumbing system that prevented potential interactions.

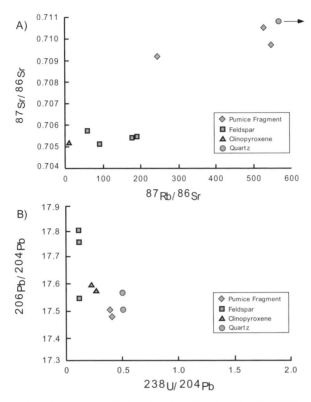

Figure 16. Rb-Sr and U-Pb systematics of minerals and pumices from the ~0 AD Baitoushan pantellerite eruption (Ramos et al. 2007). A) Illustrates differences in $^{87}Sr/^{86}Sr$ ratios of individual potassium feldspar crystals, individual clinopyroxene crystals, and a quartz separate (~20 crystals) as compared to accompanying pumice fragments. Quartz has high $^{87}Sr/^{86}Sr$ and $^{87}Rb/^{86}Sr$ (>1000) and lies to the left of this graph (as indicated by the arrow). Pumice fragments generally define a linear trend that, if isochronous, suggests ~400,000 years of magma residence prior to eruption. Potassium feldspar and clinopyroxene crystals, however, have relatively consistent $^{87}Sr/^{86}Sr$ ratios that are significantly lower than pumice fragments. B) Pb isotope results from the same crystals (except quartz symbols represent two individual crystals) that define a negative trend with all minerals and pumice fragments having variable signatures that cannot result from extended magma residence. Results are consistent with Pb addition after the bulk of crystallization of clinopyroxene and potassium feldspar, or the scavenging of crystals from unrelated magmas by the magma represented by pantellerite pumice fragments.

In contrast to rhyolites, Ramos and Reid (2005) identified assimilation processes occurring during basaltic magmagenesis at Pisgah Crater, California (USA) using single olivine crystals, in addition to plagioclase, clinopyroxene, and amphibole. This study clearly demonstrated that melt inclusions in olivine crystals from alkalic, continental basalts reflected both the 1) greatest overall Sr isotope variations of all magmatic components present and 2) sequentially increasing Sr isotope ratios in progressively erupted lavas. Results confirmed that basalts acquired a range of isotope characteristics that resulted from the incorporation of materials during residence at crustal depths (Glazner and Farmer 1992) and not as a result of the mixing of mantle-derived magmas (Reiners 2002). Further work by Mickiewicz (2007) combined crystal size distributions (CSDs) and $^{87}Sr/^{86}Sr$ ratios obtained by micromilling to evaluate the crustal residence history and magma chamber dynamics associated with these basalts. This work also clearly identified individual sieve-textured plagioclase xenocrysts which retained highly radiogenic $^{87}Sr/^{86}Sr$ ratios that confirmed an origin involving crustal assimilation.

In contrast in igneous applications, Muller et al. (2000) used micromilling to mill out small pieces of synkinematc minerals (not milling them directly) from mylonites and dykes from the Eastern Alps. Using this technique, samples could be inspected prior to milling thick sections to ensure exclusion of mineral inclusions that could potentially impact Rb and Sr budgets. Although this study mainly targeted white micas that grew in pressure shadows of feldspar and andalusite during deformation, potassium feldspar, plagioclase, biotite, and andalusite were also included. Multiple white mica samples, and some additional minerals, were used to generate isochrons dating 1) mica formation resulting from mylonitization affecting their respective host rocks and 2) mica formation ages associated with small local dikes in the area. Most mylonitization ages were similar to nearby intrusion ages, suggesting that most white micas, and thus mylonitization, resulted from deformation due to pluton emplacement. In addition, dikes yielded similar ages as both intrusion and mylonitization ages. Thus, micromilling was successfully used to obtain portions of single crystals that required low blank chromatography to obtain dates of deformation related to local intrusive rocks.

Overall, single crystal isotope studies have proven valuable in addressing questions related to igneous and metamorphic petrogenesis and discriminating between potential models describing how volcanic and metamorphic rocks and minerals obtain their chemical and isotopic characteristics. Further studies will undoubtedly integrate even more chemical, textural, and size parameters to better identify and constrain the effects of processes involved in magmagenesis and metamorphism.

SUMMARY

As reviewed above, the utility of microscale isotope sampling is only beginning to be appreciated. Whether micromilling, single crystal analyses, or *in situ* analyses using SIMS or laser ablation sampling, these techniques offer vast potential for understanding a range of natural systems. Although this manuscript mainly focuses on applications targeting magmatic minerals and melt components, investigations of biological materials have also benefited greatly. As such, additional fields will also benefit from such isotope sampling techniques in the future. As has been the case for the last ~20 years, advances in both sampling systems and measuring technologies will continue, increasing the possibilities and range of opportunities to apply multiple isotope systems in scientific applications. Ultimately, these techniques will allow evaluations of natural systems in new and innovative ways, ways that are bound to expand our understanding of natural processes in revolutionary ways.

ACKNOWLEDGMENTS

The ideas and techniques illustrated in this chapter have been developed over time by a number of researchers and collaborators. We would like to express our gratitude to these people, too numerous to list, who aided our understanding of systems reviewed in this manuscript. For discussions about microanalytical topics, we would like to acknowledge J. Wolff, J. Gill, J. Davidson, B. Charlier, P. Holden, D. Tollstrup, W. Bohrson, S. Hart, K. Knesel, and R. Barnett-Johnson. In addition, J. Wolff, J. Gill, and A. Kent, allowed access to unpublished data and S. Rodgers, A. Kinch, B. Shurtleff and R. Wilson assisted with analyses. We would also like to thank K. Putirka, J. Wolff, and T. Waight for thorough reviews of the manuscript. Investigations of and unpublished data for Baitoushan and Valles caldera high-silica rhyolites were supported by National Science Foundation grants to Ramos (EAR-0538214) and Wolff (EAR-001013).

REFERENCES

Anderson AT (1983) Oscillatory zoning of plagioclase: Nomarski interference contrast microscopy of etched polished thin sections. Am Mineral 68:125-129

Armienti P (2008) Decryption of igneous rock textures: crystal size distribution tools. Rev Mineral Geochem 69:623-649

Armienti P, Tonarini S, Innocenti F, D'Orazio M (2007) Mount Etna pyroxene as a tracer of petrogentic processes and dynamics of the feeding system. In: Cenozoic Volcanism in the Mediterranean Area. Beccaluva L, Bianchini G, Wilson M (eds) Geol Soc Am Spec Paper 418:265-276

Bacon CR, Weber PK, Larsen KA, Reisenbichler R, Fitzpatrick JA, Wooden JL (2004) Migration and rearing histories of chinook salmon (Oncorhynchus tshawytscha) determined by ion microprobe Sr isotope and Sr/Ca transects of otoliths. Can J Fish Aquat Sci 61:2425-2439

Barnett-Johnson R, Grimes CB, Royer CF, Donohoe C (2007) Identifying the contribution of wild or hatchery Chinook salmon to the ocean fishery using otolith microstructure as natural tags. Can J Fish Aquat Sci 64:1-10

Barnett-Johnson R, Pearson, TE, Ramos FC, Grimes CB, MacFarlane RB (2008) Tracking natal origins of salmon using isotopes, otoliths, and landscape geology. Limno Oceanograph 53:1633-1642

Barnett-Johnson R, Ramos FC, Grimes CB, MacFarlane RB (2005) Validation of Sr isotopes in otoliths by laser ablation multicollector inductively coupled plasma mass spectrometry (LA-MC-ICPMS): opening avenues in fisheries science applications. Can J Fish Aquat Sci 62:2425-2430

Bizzarro M, Baker JA, Haack, H (2004) Mg isotope evidence for contemporaneous formation of chondrules and refractory inclusions. Nature 431: 275-278

Bizzarro M, Simonetti A, Stevenson RK, Kurszlaukis S (2003) In situ $^{87}Sr/^{86}Sr$ investigation of igneous apatites and carbonates using laser-ablation MC-ICP-MS. Geochim Cosmochim Acta 67:289-302

Blundy JD, Shimizu N (1991) Trace element evidence for plagioclase recycling in calc-alkaline magmas. Earth Planet Sci Lett 102:178-19

Chadwick JP, Troll VR, Ginibre C, Morgan D, Gertisser R, Waight TE, Davison JP (2007) Carbonate assimilation at Merapi Volcano, Java, Indonesia: insights from crystal isotope stratigraphy. J Petrol 48:1793-1812

Charlier BLA, Ginibre C, Morgan D, Nowell GM, Pearson DG, Davidson JP, Ottley CJ (2006) Methods for the microsampling and high-precision analysis of strontium and rubidium isotopes at single crystal scale of petrological and geochronological applications. Chem Geol 232:114-133

Christensen JN, Halliday AN, Lee D, Hall CM (1995) In situ Sr isotopic analysis by laser ablation. Earth Planet Sci Let 136:79-85

Connelly JN, Thrane K (2005) Rapid determination of Pb isotopes to define Precambrian allochthonous domains: An example from west Greenland. Geology 33:953-956

Cortini M, van Calsteren PWC (1985) Lead isotope differences between whole-rock and phenocrysts in recent lavas from southern Italy. Nature 314: 343-345

Davidson JP, Charlier BLA, Hora JM, Perlroth R (2005) Mineral isochrons and isotopic fingerprinting: pitfalls and promises. Geology 33:29-32

Davidson JP, De Silva SL, Holden P, Halliday A (1990) Small-scale disequilibrium in a magmatic inclusion and its more silicic host. J Geophys Res 95:17,661-17,675

Davidson JP, Font L, Charlier BLA, Tepley FJ III (2008) Mineral-scale Sr isotope variation in plutonic rocks-a tool for unraveling the evolution of magma systems. Proc R Soc Edinb 97:357-367

Davidson JP, Morgan D, Charlier BLA, Harlou R, Hora JM (2007) Microsampling and isotopic analysis of igneous rocks: Implications for the study of magmatic systems. Annu Rev Earth Sci 35:273-311

Davidson JP, Tepley FJ III (1997) Recharge in volcanic systems: Evidence from isotopic profiles of phenocysts. Science 275:826-829

Davidson JP, Tepley FJ III, Knesel KM (1998) Isotopic fingerprinting may provide insights into evolution of magmatic systems. EOS Trans Am Geophys Union 79:185, 189, 193

Davidson JP, Tepley FJ III, Palacz Z, Meffan-Main S (2001) Magma recharge, contamination and residence times revealed by in situ laser ablation isotopic analysis of feldspar in volcanic rocks. Earth Planet Sci Let 184:427-442

Davis DW, Williams IS, Krogh TE (2003) Historical development of zircon geochronology. Rev Mineral Geochem 53:145-181

Eberz GW, Nicholls IA (1988) Microgranitoid enclaves from Swifts Creek Pluton SE-Australia: textural and physical constraints on the nature of magma mingling processes in the plutonic environment. Geol Rundsch 77:713-736

Eberz GW, Nicholls IA, Maas R, McCulloch MT, Whitford DJ (1990) The Nd- and Sr-isotopic composition of I-type microgranitoid enclaves and their host rocks from the Swifts Creek Pluton, southeast Australia. Chem Geol 85:119-134

Eckberg A, Wolff JA, Ramos FC, Hart GL, Tollstrup DL (2006) The purity of the COSYMA in the Imnaha basalt: Strontium isotope ratio variations in plagioclase phenocrysts. EOS Transactions 52: Abstract V51D-1703

Feldstein SN, Halliday AN, Davies GR, Hall CM (1994) Isotope and chemical microsampling: Constraints on the history of an S-type rhyolite, San Vincenzo Tuscany, Italy. Geochim Cosmochim Acta 58:943-958

Gagnevin D, Daly JS, Poli G, Morgan D (2005) Microchemical and Sr isotopic investigation of zoned K-feldspar megacrysts: insights into the petrogenesis of a plutonic system and disequilibrium processes during crystal growth. J Petrol 46:1689-1724

Geist DJ, Myers JD, Frost CD (1988) Megacryst-bulk rock isotopic disequilibrium as an indicator of the contamination processes: The Edgecumbe Volcanic Field, SE Alaska. Contrib Mineral Petrol 99:105-112

Ginibre C, Kronz A, Worner G (2002) High-resolution quantitative imaging of plagioclase compositions using accumulated Back-Scattered Electron images: new constraints on oscillatory zoning. Contrib Mineral Petrol 142:436-448

Ginibre C, Wörner G, Kronz A (2004) Structure and dynamics of the Laacher See magma chamber (Eifel, Germany) from major and trace element zoning in sanidine: a catholuminescence and electron microprobe study. J Petrol 45:2197-2123

Glazner AF, Farmer GL (1992) Production of isotopic variability in continental basalts by cryptic crustal contamination. Science 55:72-74

Grove TH, Baker MB, Kinzler RJ (1984) Coupled CaAl-NaSi diffusion in plagioclase feldspar: experiments and applications to cooling rate speedometry. Geochim Cosmochim Acta 48:2113-2121.

Halama R, Waight T, Markl G (2002) Geochemical and isotopic zoning patterns of plagioclase megacrysts in gabbroic dykes from the Gardar Province, South Greenland: implications for crystallization processes in anorthositic magmas. Contrib Mineral Petrol 144:109-127

Harlou R, Pearson DG, Nowel GM, Davidson JP, Kent AJR (2005) Sr studies of melt inclusions by TIMS. Geochim Cosmochim Acta 69/10S:A380

Hart SR, Ball L, Jackson M (2005) Sr isotopes by laser ablation PIMMS: Applications to cpx from samoan peridotite xenoliths. WHOI Plasma Facility Open File Technical Report 11

Hoppe KA, Koch PL, Carlson RW, Webb SD (1999) Tracking mammoth and mastodons: Reconstruction of migratory behavior using strontium isotope ratios. Geology 27:439-442

Hora JM (2003) Magmatic differentiation processes at Ngauruhoe Volcano, New Zealand: constraints from chemical, isotopic and textural analysis of plagioclase crystal zoning. MSc Dissertation, University of California, Los Angeles, Los Angeles, California

Horn I, Rudnick RL, McDonough WF (2000) Precise elemental and isotope ratio determination by simultaneous solution nebulization and laser ablation-ICP-MS: application to U-Pb geochronology. Chem Geol 164:281-301

Ireland TR, Williams IS (2003) Considerations in zircon geochronology. Rev Mineral Geochem 53:215-241

Jerram DA, Davidson JP (2007) Frontiers in textural and microgeochemical analysis. Elements 3:235-238

Kennedy BP, Klaue A, Blum JD, Folt CL, Nislow KH (2002) Reconstructing the lives of fish using Sr isotopes in otoliths. Can J Fish Aquat Sci 57:2280-2292

Kent AJR (2008) *In situ* analysis of Pb isotope ratios using laser ablation MC-ICP-MS: Controls on precision and accuracy and comparison between Faraday cup and ion counting systems. JAAS, doi: 10.1039/b801046c

Knesel KM, Davidson JP, Duffield WA (1999) Evolution of silicic magma through assimilation and subsequent recharge: Evidence from Sr-isotopes in sanidine phenocrysts, Taylor Creek Rhyolite, NM. J Petrol 40:773-786

Kobayashi K, Tanaka R, Moriguti T, Shimizu K, Nakamura E (2004) Lithium boron and lead isotope systematics of glass inclusions in olivines from Hawaiian lavas: evidence for recycled components in Hawaiian plume. Chem Geol 212:143-161

Lambert IB, Drexel JF, Donnelly TH, Knutson J (1982) Origin of breccias in the Mount Painter area, South Australia. J Geol Soc Austral 29:115-125

Layne GD, Shimizu N (1997) Measurement of lead isotope ratios in common silicate and sulfide phases using the CAMECA IMS 1270 Ion Microprobe. *In:* Secondary Ion Mass Spectrometry, SIMS XI. Gillen G et al. (eds) John Wiley, p 63-65

Maaskant P, Coolen JJMMM, Burke EAJ (1980) Hibonite and coexisting zoisite and clinozoisite in a calc-silicate granulite form southern Tanzania. Mineral Mag 43:995-1003

Mathez EA, Waight TE (2003) Lead isotope disequilibrium between sulfide and plagioclase in the Bushveld Complex and the chemical evolution of large layered intrusions. Geochim Cosmochim Acta 67:1875-1888

McCulloch M, Cappo M, Aumend J, Muller W (2005) Tracing the life history of individual barramundi using laser ablation MC-ICP-MS Sr-isotopic and Sr-Ba ratios in otoliths. Mar Freshwater Res 56:637-644

Mickiewicz S (2007) Constraining continental basaltic magma chambers processes: Textural and *in situ* geochemical investigation of plagioclase from Pisgah Crater, California. MSc Dissertation, Central Washington University, Washington

Milton DA, Chenery SR (2003) Movement patterns of the tropical shad (Tenualosa ilisha) inferred from transects of 87Sr/86Sr isotope ratios in their otoliths. Can J Fish Aquat Sci 60:1376-1385

Morgan DJ, Jerram DA, Chetkoff DG, Davidson JP, Pearson DG, Kronz A, Nowell GM (2007) Combining CSD and isotopic microanalysis: magma supply and mixing processes at Stromboli volcano, Aeolian Islands, Italy. Earth Planet Sci Lett 260:419-431

Muller W, Mancktelow NS, Meier, M (2000) Rb-Sr microchrons of synkinematic mica in mylonites: an example from the DAV fault of the Eastern Alp. Earth Planet Sci Lett 180:385-397

Outridge PM, Chenery SR, Babaluk JA, Reist JD (2002) Analysis of geological Sr isotope markers in fish otoliths with subannual resolution using laser ablation-multicollector-ICP-mass spectrometry. Environ. Geology 42:891-899

Pearce TH, Clark AH (1989) Nomarski interference contrast observations of textural details in volcanic rocks. Geology 17:757-759

Perini G, Tepley FJ III, Davidson JP, Conticelli S (2003) The origin of K-feldspar megacrysts hosted in alkaline potassic rocks from central Italy: a track for low-pressure processes in mafic magmas. Lithos 66:223-240

Perlroth R (2000) An investigation of crystal transfer between melts. MSc Dissertation, University of California, Los Angeles, Los Angeles, California

Pidgeon RT, Aftalion M (1978) Co-genetic and inherited zircon U-Pb systems in granites. Palaeozoic granites of Scotland and England. Geol J Special Issue 10:183-220

Ramos FC, Reid MR (2005) Distinguishing melting of heterogeneous mantle sources from crustal contamination: Insights from Sr isotopes at the phenocryst scale, Pisgah Crater California. J Petrol 46:999-1012

Ramos FC, Rodgers SL, Gill JB (2007) Evaluating the timing of volcanism at Baitoushan volcano (North Korea/China) in the context of open-system effects: Insights from Sr, Nd, and Pb isotopes at the single grain scale. AGU Fall Meeting abs. #V51C-0712

Ramos FC, Rodgers, SL, Wolff JW, Gill JB (2008) Connecting high silica rhyolite-hosted phenocrysts to potential crystal mush intrusive complements: Insights from Baitoushan volcano and Valles caldera. GSA Cordilleran/Rocky Mountains Section, Abs. w/ Programs, p18-7 p 84

Ramos FC, Wolff JA, Tollstrup DL (2004) Measuring $^{87}Sr/^{86}Sr$ variations in minerals and groundmass from basalts using LA-MC-ICPMS. Chem Geol 211:135-158

Ramos FC, Wolff JA, Tollstrup DL (2005) Sr isotope disequilibrium in Columbia River flood basalts: Evidence for rapid shallow-level open-system processes. Geology 33:457-460

Ramos FC (1992) Isotope geology of the Central grouse Creek Mountains, Box Elder County, Utah. MS Thesis University of California, Los Angeles, Los Angeles, California

Reddy SM, Kelley SP, Wheeler J (1996) A $^{40}Ar/^{39}Ar$ laser probe study of micas from the Seisa zone, Italian Alps: Implications for metamorphic and deformation histories. J Metamorph Geol 14:493-508

Reiners P (2002) Temporal-compositional trends in intraplate eruptions: Implications for mantle heterogeneity and melting processes. Geochem Geophys Geosys 10.1029/2001GC000250

Rickwood PC (1977) A technique for extracting small crystals from thin sections. Am Mineral 62:382-384

Rundle CC (1992) Review and assessment of isotopic ages from the English Lake District. British Geological Survey, Onshore Geology Series, Technical Report WA/92/38

Saal AE, Hart SR, Shimizu N, Hauri EH, Layne GD (1998) Pb isotopic variability in melt inclusions from oceanic island basalts, Polynesia. Science 282(5393):1481-1484

Saal AE, Hart SR, Shimizu N, Hauri EH, Layne GD, Eiler JM (2005) Pb isotopic variability in melt inclusions from the EMI-EMII-HIMU mantle end-members and the role of oceanic lithosphere. Earth Planet Sci Lett 240:605-620

Salisbury MJ, Bohrson WA, Clynne M, Ramos FC, Hoskin P (2008) Plagioclase crystal populations identified by crystal size distribution and *in situ* chemical data: implications for timescales of magma chamber processes associated with the 1915 Eruption of Lassen Peak, CA. J Petrol 2008; doi: 10.1093/petrology/egn045

Schmidberger SS, Simoneti A, Francis D (2003) Small-scale Sr isotope investigations of clinopyroxenes from peridotite xenoliths by laser ablation MC-ICP-MS-implications for mantle metasomatism. Chem Geol 199:317-329

Siebel W, Reitter E, Wenzel T, Blaha U (2005) Sr isotope systematics of K-feldspars in plutonic rocks revealed by the Rb-Sr microdrilling technique. Chem Geol 222:183-199

Simon JI, Reid MR, Young ED (2007) Lead isotopes from LA-MC-ICPMS: tracking the emergence of mantle signatures in an evolving silicic magma system. Geochim Cosmochim Acta 71:2014-2035

Singer BS, Dungan MA, Layne GD (1995) Textures and Sr, Ba, Mg, Fe, K, and Ti compositional profiles in volcanic plagioclase: clues to the dynamics of calc-alkaline magmas chambers. Am Mineral 80:776-798

Stamatelopoulou-Seymour K, Vlassopoulos D, Pearce TH, Rice C (1990) The record of magma chamber processes in plagioclase phenocrysts at Thera volcano, Aegean volcanic arc, Greece. Contrib Mineral Petrol 104:73-84

Tepley FJ III, Davidson JP (2003) Mineral-scale Sr-isotope constraints on magma evolution and chamber dynamics in the Rum layered intrusion, Scotland. Contrib Mineral Petrol 145:628-641

Tepley FJ III, Davidson JP, Clynne MA (1999) Magmatic interactions as recorded in plagioclase phenocrysts of Chaos Crags, Lassen Volcanic Center, California. J Petrol 40:787-806

Tepley FJ III, Davidson JP, Tilling RI, Arth JG (2000) Magma mixing, recharge and eruption histories recorded in plagioclase phenocrysts from El Chichón volcano, Mexico. J Petrol 41:1397-1411

Thirwall M (2002) Multicollector-ICPMS analysis of Pb isotopes using a ^{207}Pb-^{204}Pb double spike demonstrates up to 400ppm/amu systematic errors in Tl normalization. Chem Geol 184:255-279

Thorrold SR, Shuttleworth S (2000) *In situ* analysis of trace elements and isotope ratios in fish otoliths using laser ablation sector field inductively coupled plasma mass spectrometry. Can J Fish Aquat Sci 57:1232-1242

Thrane K, Bizzarro M, Baker JA (2006) Extremely brief formation interval for refractory inclusions and uniform distributions of 26 Al in the early solar system. Astrophys J 646:L159-L162

Tilling RI, Arth JG (1994) Sr and Nd isotopic compositions of sulfur-rich magmas of El Chichón volcano, Mexico. IAVCEI abstract

Tyrrell S, Haughton PD, Daly JS (2007) Drainage reorganization during the breakup of Pangea revealed by in-situ Pb isotopic analysis of detrital k-feldspar. Geology 35:971-974

Vroon PZ, Wagt BV, Koornneef JM, Davies GR (2008) Problems in obtaining precise and accurate Sr isotope analysis from geological material using laser ablation MC-ICPMS. Anal Bioanal Chem 390:465-476

Wagner WR (1945) An experimental micro-drill. Science 101:127-128

Waight TE, Baker J, Peate D (2002) Sr isotope ratio measurements by double-focusing MC-ICPMS: techniques, observations and pitfalls. Int J Mass Spectrom 221:229-244

Waight TE, Dean AA, Maas R, Nicholls IA (2000b) Sr and Nd isotopic investigations towards the origin of feldspar megacrysts in microgranular enclaves in two I-type plutons of the Lachlan Fold Belt, southeast Australia. Austral J Earth Sci 47:1105-1112

Waight TE, Maas R, Nicholls IA (2000a) Fingerprinting feldspar phenocrysts using crystal isotopic composition stratigraphy: implications for crystal transfer and magma mingling in S-type granites. Contrib Mineral Petrol 139:227-239

Weber PK, Bacon CR, Hutcheon ID, Ingram BL, Wooden JL (2005) Ion microprobe measurement of strontium isotopes in calcium carbonate with application to salmon otoliths. Geochim Cosmochim Acta 69:1225-1239

Willigers BJA, Baker JA, Krogstad EJ, Peate DW (2002) Precise and accurate *in situ* Pb-Pb dating of apatite, monazite, and sphene by laser ablation multi-collector ICP-MS. Geochim Cosmochim Acta 66 6:1051-1066

Wolff JA, Ramos FC (2003) Pb isotope variations among Bandelier Tuff feldspars: No evidence for a long-lived silicic magma chamber. Geology 31:533-5363

Wolff JA, Ramos FC (2004) Sr and Pb in high-silica rhyolites: Lessons from the Bandelier Tuff. EOS Trans. AGU 85(17), Jt. Assem. Suppl. Abstract V12A-02

Wolff JA, Ramos FC, Davidson JP (1999) Sr isotope disequilibrium during differentiation of the Bandelier Tuff: Constraints on the crystallization of a large rhyolitic magma chamber. Geolology 27:495-498

Wolff JA, Ramos FC, Hart GL, Patterson JD, Brandon AD (2008) Columbia River flood basalts from a centralized crustal magmatic system. Nature Geosci 1:163-167

Woodhead J, Swearer S, Hergt J, Maas Roland (2005) *In situ* Sr-isotope analysis of carbonates by LA-MC-ICP-MS: interference corrections, high spatial resolution and an examples from otolith studies. J Anal At Spectrom 20:22-27

Oxygen Isotopes in Mantle and Crustal Magmas as Revealed by Single Crystal Analysis

Ilya Bindeman

Department of Geological Sciences
University of Oregon
Eugene, Oregon, 97403-1272, U.S.A.

bindeman@uoregon.edu

PART I: INTRODUCTION

Oxygen is the most abundant element in the Earth's crust, mantle, and fluids and therefore its isotopic composition provides robust constraints on magma genesis. Application of oxygen isotope geochemistry to volcanology and igneous petrology provides a much needed foundation for radiogenic isotope and trace element approaches. Since isotope fractionations at high temperature are small, there is a demand for high analytical precision in order to recognize and interpret small (tenths of permil) variations in isotopic composition. Recently improved analytical techniques involving lasers and ion microprobes, and reduction in sample and spot size, has painted a picture of isotope complexity on a single crystal scale that is helpful in interpreting magma genesis and evolution. In this chapter a review is provided for several classic examples of silicic and basic magmatism, including Yellowstone and Iceland, that shows isotope zoning and heterogeneity reaching several permil. Isotope heterogeneity fingerprints crystal sources and provides constraints on diffusive and recrystallizational timescales. These new lines of evidence reveal that magma genesis happens rapidly, at shallow depths, and through batch assembly processes.

Oxygen isotope geochemistry spans more than 50 years of investigation and is the most developed among other traditional (e.g., C, N, H, S, Li, B) and less-traditional (e.g., Fe, Mo, Cu) stable isotope systems. While we provide basic concepts of isotope fractionation below, the reader is referred to three prior RiMG volumes on stable isotopes (Valley, Taylor, and O'Neil 1986 – volume 16; Valley and Cole 2001 – volume 43; Johnson et al. 2004 – volume 55), and the Hoefs (2005) and Sharp (2006) textbooks for greater treatment and historic perspective. Finally this Chapter does not deal with oxygen isotopic variations in meteorites and planetary igneous materials, and interested readers are referred to RiMG volume 68, "Oxygen in the Solar System" (MacPherson et al. 2008).

The remaining novel aspects of oxygen isotope geochemistry of igneous and metamorphic rocks are related to continuing application of laser fluorination analysis to phenocrysts that provide superior analytical precision to other methods. In particular, 1) Many topical studies should include major element oxygen measured at new analytical levels as a component in multi-isotope and trace elemental investigation, 2) Older studies that relied on whole-rock methods should be reassessed, 3) Many igneous systems such as continental and oceanic flood basalts, large silicic igneous provinces, and island arc magmas should be reevaluated, 4) Oxygen isotopes should be studied on a single crystal level. This Chapter outlines the perspective that magmatic crystals should be studied individually when possible, or by size fractions, to demonstrate whether crystals are isotopically homogeneous and is in equilibrium with other minerals within common host magma. As is demonstrated below, in many classic examples of igneous systems around the world, magmatic crystals are not in equilibrium and thus bulk phenocryst oxygen isotope analysis does not reveal the full details of petrogenesis, nor do they provide a proxy for magmatic values.

Basics of oxygen isotope variations in nature and their causes

Oxygen consists of three isotopes: ^{16}O (99.76%), ^{17}O (0.04%), and ^{18}O (0.2%). Mass-dependent, oxygen isotopic variations are described through a ratio of ^{18}O to ^{16}O and the delta notation that is traditionally used in stable isotope geochemistry:

$$\delta^{18}O = (R_{SA}/R_{ST} - 1) \times 1000 \tag{1}$$

where R_{SA} and R_{ST} are absolute ratios of $^{18}O/^{16}O$ in sample and standard and the standard is Vienna Mean Standard Ocean Water (VSMOW) with $^{18}O/^{16}O$ absolute ratios of 0.020052 (Baertschi 1976). A $\Delta^{18}O$ measure of variation between the sample and the standard is in permil (‰), or part per thousand. It is not recommended to use any other standard (e.g., PDB) when describing oxygen isotopic variations in nature since it only causes confusion. The multiplication factor of 1000 is used to display natural isotope variations in whole numbers and not fractions; it is used because isotope variations due to chemical or physical processes in nature are small, in the third or second decimal place of the $^{18}O/^{16}O$ ratio.

Oxygen isotopic variations on Earth span about 100‰ (Fig. 1). Half of these variations, and nearly all negative $\delta^{18}O$ values are occupied by meteoric waters that are isotopically-light as a result of Rayleigh distillation upon vapor transport and precipitation. Silicate rocks and magmas occupy the positive part of this diagram with the absolute majority of mantle rocks, basaltic magmas and chondritic meteorites plotting in a narrow range of 5.5 to 5.9‰. Silicate, oxide, carbonate, and phosphate minerals and igneous, sedimentary, and metamorphic rocks such as carbonates and especially diatoms that precipitate from water at low temperatures are the highest $\delta^{18}O$ materials because of large positive isotope fractionation factors between silica, carbonates and water at low temperatures (e.g., Friedman and O'Neil 1977; Chacko et al. 2001; Hoefs 2005). Metasedimentary rocks and igneous rocks such as S-type granites inherit high-$\delta^{18}O$ supracrustal signature from the source, while rocks that represent remelting of hydrothermally altered rocks that interacted with low-$\delta^{18}O$ meteoric water at high-temperature are low-$\delta^{18}O$ (see below).

Equilibrium isotope fractionation factors between minerals and melts. The equilibrium isotope fractionation factor is defined as:

$$\alpha = R_A/R_B \tag{2}$$

where R_A and R_B are absolute isotope ratios of individual, coexisting minerals at equilibrium. Since α variations are typically in the second or third decimal point, a more convenient parameter, $1000\ln\alpha$ is used.

Because of the difference in chemical bonds affecting vibrational frequencies of oxygen in minerals, a heavier isotope of oxygen partitions itself into a mineral with stronger (more covalent) Si-O-M bonds. As a general rule, isothermal high-T distribution of ^{18}O among coexisting igneous phases is explained by the proportion of Si-O and M-O bonds (Taylor 1968; Zheng et al. 1993a,b; Chiba et al. 1989; Hoefs 2005), while the cation identity in homovalent substitution (e.g., Fe vs. Mg) plays an insignificant third order role. Thus the common igneous minerals will become progressively lighter from quartz (pure silicate) to magnetite (pure M-O oxide). For example, a polymineralic granite at 850 °C with the whole-rock value of 7.8‰ will have decreasing $\delta^{18}O$ values of its constituent minerals in this sequence: quartz (8.2‰) > albite ≈ K-Fsp (7.5‰) > anorthite (6.6‰) > zircon (6.4‰) ≥ pyroxene (6.3‰) ≈ amphibole ≥ biotite ≥ garnet ≈ olivine (6.1‰) > sphene (5.4‰) ≥ ilmenite (4.9‰) > apatite ≥ magnetite (3.5‰). The quoted $\delta^{18}O$ values were calculated using experimental and empirical fractionation factors (e.g., Taylor and Sheppard 1986; Chiba et al. 1989; Zheng 1993a,b; Chacko et al. 2001; Valley

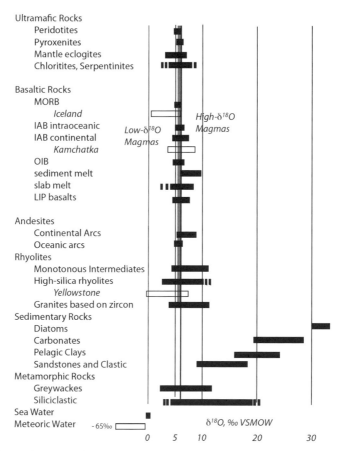

Figure 1. Oxygen isotopic variations in magmas and rocks on Earth. Gray band between 5.5 and 5.9 represents "normal-δ^{18}O magmas" for ultramafic and basaltic rocks that constitute predominant mantle and products of its melting; closed system differentiation of such magma into andesite and rhyolite that show normal δ^{18}O values of 5.8 to 6.5 permil (Fig. 4). High-δ^{18}O rocks and magmas result from low-temperature precipitation or interaction with sea water involving large positive 1000ln$\alpha_{mineral-water}$ isotope fractionation (Fig. 2a). Low-δ^{18}O rocks and magmas result from high-temperature interaction with, or subsequent re-melting of, materials that interacted with meteoric waters at high-temperatures involving small isotope fractionations. Data are from both original sources and compilations by Muehlenbachs (1998); Eiler (2001), Valley et al. (2005); Hoefs (2005); Sharp (2006). Selected areas emphasized in this review are shown as open boxes.

et al. 2003). Establishing the equilibrium fractionation factors for igneous minerals is still an active area of experimental research, especially for refractory accessory minerals, for which direct experiments are difficult (e.g., Krylov et al. 2002; Valley et al. 2003; Trail et al. 2008).

For isotopic fractionation between minerals and melt, most variation is caused by the changing composition of melt. In an attempt to quantify the mineral-melt fractionation factors, the easiest approach is to treat the melt as a mixture of normative minerals and then calculate the weighted average of individual mineral-melt (normative sum) fractionation factors (Matthews et al. 1994; Palin et al. 1996; Eiler 2001; Appora et al. 2002). Although the natural silicate melts are more complicated mixtures of various components those chemical variations are often treated by a regular solution model (Ghiorso and Sack 1995), for the purpose of isotope fractionation, non idealities of melt mixing functions play a minor role. The normative mineral approach assumes that a normative mineral in melt has the same partition function ratios for isotope distribution; it has been demonstrated to be largely correct assumption for a variety of quartzofeldspathic mixtures, despite a few tenths of one permil disagreement between crystalline quartz and quartz-melt as measured in experiments (Matthews et al. 1998).

When $\delta^{18}O$ values are relatively similar and not extremely low or high in $\delta^{18}O$, the fractionation factor $1000\ln\alpha$ is numerically very close to the simple difference between measured $\delta^{18}O$ values, or $\Delta^{18}O$ (e.g., Hoefs 2005; Sharp 2006):

$$1000\ln\alpha_{Quartz\text{-}Zircon} \approx \Delta^{18}O_{Quartz\text{-}Zircon} = \delta^{18}O_{Quartz} - \delta^{18}O_{Zircon} \qquad (3)$$

Isotope fractionation of oxygen between coexisting minerals, like any other stable isotope, is a strong function of temperature; for solid substances at temperatures higher than ~50-100 °C, the isotope fractionation factor $1000\ln\alpha$ is a linear function of $1/T^2$ (Biegielsen and Meyer 1947; Urey 1947). The reverse quadratic rather than $1/T$ dependence that is common in thermodynamics of element partitioning is explained by the fact that the factor that causes isotope fractionations-isotope vibrational frequencies, are themselves increasing with temperature. Figure 2 presents isotope fractionations between major igneous minerals and water as a function of T. Note that isotope fractionations are largely pressure-independent for common upper mantle-crustal pressures because isotope substitutions do not have significant volume effects. In $1000\ln\alpha$ vs. $1/T^2$, isotope exchange between minerals, and most likely minerals and silicate melt, can be described by a single A factor:

$$1000\ln\alpha = 10^6 A/T^2 \qquad (4)$$

defining linear dependence passing through the origin at infinite temperature at which isotope fractionations are zero.

Because of the reverse quadratic relationship between $1000\ln\alpha$ and temperature (Fig. 2) isotope fractionations between coexisting minerals, melts and fluids are small at high magmatic temperatures, typically less than 2-3‰. It underlies the importance of high precision, better than 0.1‰, to resolve subtle isotopic differences between different sources for magmas. Small (1-2‰) isotope fractionation between minerals were traditionally used to predict that isotope fractionation are consistent with magmatic temperatures and not with subsolidus exchange (e.g., Taylor 1986), meaning that they reflect primary sources rather than secondary isotope effects.

Isotope thermometry. An important aspect of oxygen isotope analysis of coexisting mineral assemblages is the ability to use the experimentally-determined A factors and measured $\Delta^{18}O$ as input parameters to calculate temperature in Equation (4). Temperature estimates are more easily determined for mineral pairs with large A factors such as quartz and magnetite (Fig. 2B). Several assumptions should be evaluated in order for isotope thermometry to reflect primary magmatic values. First, minerals should be fresh and unaltered, as can be demonstrated under an electron microscope. Second, the rock or magma should have cooled rapidly so that the measured isotope temperature reflects quenched pre-eruptive conditions and does not reflect the effects of slow plutonic cooling with sequential mineral closure to isotope exchange (e.g., Eiler et al. 1993; Farquhar et al. 1993). Third, minerals should be phenocrysts and not isotopically-distinct xenocrysts.

Because the first two conditions are rarely met for metamorphic and likewise plutonic rocks, the initial optimism to be able to use pressure-independent oxygen isotope thermometry (where P and T can be correlative based on chemical geothermometers and geobarometers) nearly disappeared, except for simplest bimineralic assemblages involving refractory accessory phases where possibility for exchange is limited (e.g., Valley 2001). However, for fresh volcanic rocks that quenched rapidly, measured $\Delta^{18}O$ values can be used to determine preeruptive temperature. Young pyroclastic volcanic rocks such as single pumice or tephra

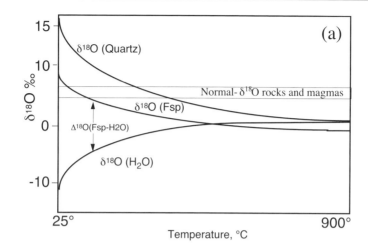

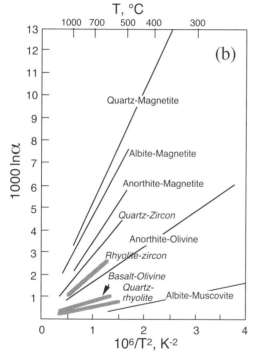

Figure 2. Oxygen isotope equilibrium between minerals, melts and water as a function of temperature. a) Schematic diagram illustrating $\delta^{18}O$ as a function of temperature for a closed-system exchange between minerals and water. Isotope variations at high temperature are smaller, and water becomes isotopically heavier than rocks ($\Delta^{18}O_{Fsp-H2O} < 0$) after a crossover temperature of 400-550 °C, and that high-T water-rock interaction will lead to depletion of rocks with respect to ^{18}O (Friedman and O'Neil 1977; Chacko et al. 2001). When sea water with 0‰ is considered, low-T hydrothermal alteration may lead to higher-$\delta^{18}O$ values while interaction with meteoric water will lead to rock $\delta^{18}O$ depletion at nearly every temperature. Hydrothermal alteration will lead to highly heterogeneous distribution of oxygen isotopes in rocks depending on local (i.e., mm-scale) water-rock ratios. b) Mineral-mineral and mineral-melt isotope fractionation as a function of $1/T^2$. Data sources: Thin lines are experimental data (Chiba et al. 1989). Thicker lines represent empirical measurements (Eiler 2001; Bindeman and Valley 2002; Valley et al. 2003; Bindeman et al. 2004).

clasts perhaps present the best examples of quenched eruptive products that preserve magmatic $\Delta^{18}O$ values, and thus temperatures (e.g., Bindeman and Valley 2002, Fig. 3).

Mineral vs. whole rock analysis. Oxygen isotope ratios are used in much the same way as radiogenic isotopes or trace elemental ratios to define or identify different geochemical reservoirs or sources of magmas. In this sense, whole rock oxygen isotope analysis should theoretically represent magma and its source, and such analysis was a common practice in early studies. The early investigation of oxygen isotopes in igneous rocks relied on whole rock analysis by conventional methods; however, large oxygen isotopic variations put forward by

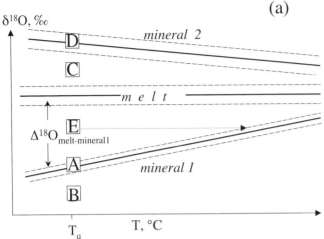

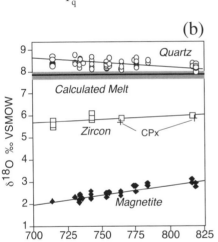

Figure 3. a) Recognition of $\Delta^{18}O_{\text{melt-mineral}}$ isotope equilibria and disequilibria. With decreasing temperature, isotope fractionations (divergent lines) between minerals (letter boxes) and melt increase up until the temperature of eruptive quench, T_q, oxygen diffusion is expected to operate without closure. Crystal A (mineral 1) and crystal D (mineral 2) are in isotope equilibrium among themselves and the melt, while individual crystals E, B and C of mineral 1 record disequilibrium $\Delta^{18}O_{\text{melt-mineral}}$ values. Crystal B comes from a lower-$\delta^{18}O$ source, while crystal C comes from a higher-$\delta^{18}O$ source, and they are therefore xenocrysts. Crystal E is only subtly, by a few tenths of one permil, heavier than the equilibrium value, and can be interpreted to represent crystallization at higher-temperatures and then capture into a lower-temperature magma differentiate similar to cumulate entrapment. Crystal E could be a "proto"- or "ante" cryst. b) An example of isotope equilibria among minerals and melt in Bishop tuff, a large volume rhyolite deposit in which individual pumice clasts record equilibrium $\Delta^{18}O$ melt-mineral isotope fractionations, divergent with decreasing temperature (from Bindeman and Valley 2002). Each vertical column of symbols represent analyses from a single pumice clast erupted from different depth of the thermally-zoned magma chamber ranging from 6 to 11 km (Wallace et al. 1999) thus quenching specific magmatic $\Delta^{18}O_{\text{melt-mineral}}$ values. The obtained isotopic temperatures agree very well with Fe-Ti oxide temperatures in the Bishop tuff clasts.

these early researchers were later reinterpreted as being mostly due to secondary alteration effects with surface waters (Taylor 1986). Studies of minerals instead of rocks, and in particular refractory phenocrysts such as quartz, zircon and olivine, yielded a picture of subtle subpermil variations in $\delta^{18}O$ within the mantle (Valley et al. 1998; Eiler 2001) and narrower ranges in many crustal rocks (e.g., Valley et al. 2005). Fresh glass of modern or recent volcanic rocks that is unaltered can be used to directly infer primary magmatic $\delta^{18}O$ values. However, volcanic glass in old rocks is metastable and exchanges oxygen with waters even before there is clear petrographic evidence of devitrification or development of microscopic clays (e.g., Taylor 1986). Iridescence in sanidine and glass, which may signal the appearance of microscopic clays can be used as an indicator of $\delta^{18}O$ modification.

Unaltered phenocrysts in igneous rocks provide a better view of true magmatic $\delta^{18}O$ value, but they are different from the $\delta^{18}O$ value of the magma by a fractionation factor $\Delta^{18}O_{\text{mineral-melt}}$ that is a function of temperature, melt composition, and, for the case of plagioclase and similar heterovalent solid solution, their Ca/Na and Al/Si ratios. In studies involving comparison of $\delta^{18}O$ between different magmas or sections in a drillcore, it is better to present data for a single mineral if possible (e.g., Wang et al. 2003). In some other applications, a calculated magma value can be generated and compared, especially if $\delta^{18}O$ values of multiple phenocrysts are obtained. Additionally, when relying on $\delta^{18}O$ values of crystals in old volcanic or plutonic rocks, it is better to avoid feldspars and micas, and rely on "refractory" phases that are resistant to secondary isotope exchange, and do not have solid solution that fractionates oxygen isotopes. For that matter, quartz and zircon in silicic rocks, and olivine in basic rocks, provide reliable proxies for their parental melts, while plagioclase and K-feldspar are less reliable.

The use of phenocrysts requires extra time for mineral separation—a task that was tedious for conventional resistance-furnace fluorination that necessitates 5-25 mg of monomineralic separate. With the reduction of sample size by ten-fold, mineral separation time is now trivial or minimal for laser fluorination and ion microprobe methods. This translates into better quality, inclusion-free, alteration-free separate. Most importantly, the 0.5-2 mg typical sample size for the laser fluorination analysis overlaps with sizes of typical phenocrysts of most rock-forming minerals, thus allowing them to be studied individually, an approach utilized in this Chapter.

Oxygen isotopes in mantle-derived rocks, normal-$\delta^{18}O$, high-$\delta^{18}O$, and low-$\delta^{18}O$ magmas

Basaltic magmas of mid-ocean ridges and most common island arc basalt are characterized by a relatively narrow 5.7±0.2‰ range of $\delta^{18}O$ values calculated from phenocrysts or measured directly in fresh glasses (Fig. 1). The OIB basalts document greater range of +4 to +6‰ and reflect recycling of materials that interacted with surface waters at low or high temperature into the mantle (Fig. 1). Eclogite nodules and serpentinized ultramafic rocks exposed on the surface document greater $\delta^{18}O$ ranges (Kyser et al. 1982; Harmon and Hoefs 1995). High-$\delta^{18}O$ magmatic values are seen in intermediate and silicic rocks in continental arcs and collision zones, in magmas that were derived from or interacted with metasedimentary protoliths; low-$\delta^{18}O$ values characterize caldera settings and rift zones that involve shallow crustal recycling of rocks interacted with meteoric waters at high-T. Reviews by Taylor (1986) and recent reviews by Eiler (2001) and Valley et al. (2005) provide a comprehensive account of whole-rock and phenocryst-based oxygen isotope geochemistry of the mantle and the crust. This chapter challenges these studies somewhat by documenting new complexity revealed by individual phenocryst studies, which shows far greater ranges in individual phenocryst $\delta^{18}O$ values that reflect crystallization from or exchange with the diverse melts (open boxes in Fig. 1).

It is convenient to define most common basic magma as "normal-$\delta^{18}O$" and consider products of its differentiation-crystallization as a "normal-$\delta^{18}O$ differentiation array" (Fig. 4). It has been noted in many natural examples of closed-system differentiation series that differentiation of basalt leads to a small subpermil increase in $\delta^{18}O$ melt value, with a particular magnitude and trajectory of increase weakly dependent on the sequence of phase appearances that affect $\Delta^{18}O_{mineral-melt}$ (Anderson et al. 1971; Taylor and Sheppard 1986; Eiler 2001).

In order to better understand mineral-melt oxygen isotope partitioning, and calculate basaltic magma differentiation trends based on experimental mineral-mineral and mineral-melt partitioning, a computational approach can be taken. The rich existing database on mineral-mineral isotope fractionation (Friedman and O'Neil 1977; Chiba et al. 1989; Chacko et al. 2001, and others), is now complemented by isotope exchange experiments involving CO_2 gas as an exchange medium and silicate melts of variable composition (see summary in Eiler 2001). We recommend continuing with the practice of treating melt as a mixture of normative mineral components and calculate $\Delta^{18}O_{phenocryst-melt}$ isotope fractionation as a weighed sum of $\Delta^{18}O_{phenocryst-normative\ mineral}$ in melt between the phenocryst and each normative component of the melt for a given temperature. This approach has been used by Eiler (2001) to demonstrate small (<0.2-0.3‰) isotopic differences between variably differentiated basic rocks as a function of basalt MgO content, by Zhao and Zheng (2003) for rocks of variable composition, and by Bindeman et al. (2004) for extended island arc series as a function of SiO_2. The latter study

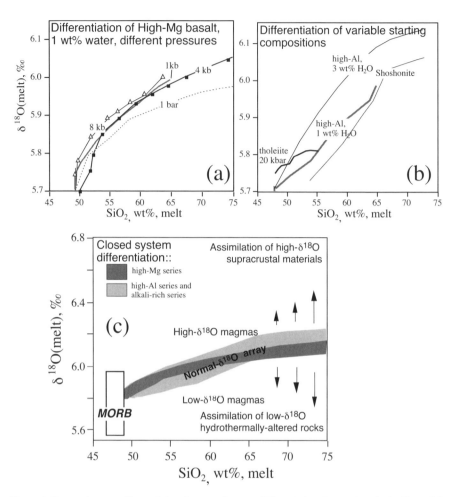

Figure 4. Oxygen isotope effects of closed system igneous differentiation that produce insignificant 0.2-0.5‰ increase and concave-down trend in $\delta^{18}O$-SiO_2 coordinates. a) Differentiation of high-Mg basalts in MELTS program at different pressures; b) differentiation of various starting compositions; c) results of multiple numerical crystallization experiments for the two most common island arc series, labeled as "normal-$\delta^{18}O$" magma differentiation array, that separates fields of high-$\delta^{18}O$ magmas and low-$\delta^{18}O$ magmas, whose genesis requires open system behavior. After Bindeman et al. (2004).

demonstrated that for a typical igneous rock, a smooth, concave downward trajectory, and an increase of 0.3-0.4‰ is expected in $\delta^{18}O_{melt}$-SiO_2 coordinates from basalt to rhyolite (Fig. 4). Only some rare magmatic series involving unusual Fe-enrichment in the residual melt such as the Skaergaard intrusion yield a nearly flat trajectory (e.g., Anderson et al. 1971; Kalamarides 1984; Bindeman et al. 2008a). A trajectory of $\delta^{18}O$ increase with SiO_2 is steeper from basalt to andesite because low-$\delta^{18}O$ minerals such as olivine and pyroxene fractionate early, driving melt toward higher $\delta^{18}O$ values (Fig. 4). With the appearance of feldspar, the trend flattens while crystallization of near-eutectic quartz+feldspar leads to a nearly constant $\delta^{18}O$ value of melt and separating cumulate assemblage. The latter is important for high-silica rhyolites that are near-eutectic systems; since the chemical composition of bulk cumulates is the same as the residual melt, the $\Delta^{18}O_{cumulate-melt}$ fractionation is zero.

High-δ¹⁸O magmas. The $\delta^{18}O$ value of crustal rocks is diverse (Fig. 1, 4), and the majority of crustal igneous rocks have $\delta^{18}O$ values higher than the mantle magma differentiation array. The majority of cases of isotopically-heavy magma values are related to the exchange, or derivation from, a high-$\delta^{18}O$ metasedimentary silicate rock (such as metapelites, carbonates, or metagreywackes), that originally crystallized from, or exchanged with, sea water at surface temperatures. Oxygen isotopes serve as an important parameter of crustal assimilation and derivation, especially when coupled with other monitors of these processes such as $^{87}Sr/^{86}Sr$ isotopes (Taylor 1980, 1986), U-series, and trace elements (Finney et al. 2008).

Low-δ¹⁸O magmas. A subset of igneous rocks, silicic and basic, that represent remelting, assimilation or exchange with hydrothermally-altered rocks (altered by heated meteoric waters) represent $\delta^{18}O$ magmas. The low-$\delta^{18}O$ rocks were earlier known to occur in only a few locations such as Yellowstone and Iceland (e.g., Taylor 1986). The author of this Chapter is a strong believer of far greater abundance of low-$\delta^{18}O$ magmas than is currently thought both in terms of volume and the number of volcanic units; however, low-$\delta^{18}O$ magmas are likely underrepresented in pre-Tertiary geologic record because being of shallow genesis they are eroded away.

PART II: SINGLE PHENOCRYST ISOTOPE STUDIES

A recent upsurge in interest in single crystal studies reflects the advent of trace elemental and isotopic microbeam and microdrilling techniques that are suitable for dating and fingerprinting single crystals. These approaches have led to the realization that many, if not most, igneous systems contain isotopically diverse and chemically distinct crystal populations that are "heterogeneous on all scales" (e.g., Dungan and Davidson 2004). Minerals that are found in magmas as "pheno"crysts may not necessarily have crystallized from their host melt and may be "proto"crysts or cumulates that are entrained into the more differentiated host product, "ante"crysts capured from chamber walls that represent prior episode(s) of magmatism in the same place, or "xeno"crysts captured from rocks that are much older and are not related to the current cycle of magmatism. For review of these topics see Ramos and Tepley (2008) of this volume. Collectively, there is a growing consensus that histories of crystals and crystal populations ("crystal cargo") and their host melt may be decoupled and thus isotope disequilibria may provide insights into the origin of both crystals and magma.

Many examples of isotopic zoning and disequilibria in volcanic phenocrysts, with emphasis on volcanic arcs, have been described, and a small level of residual zoning may in fact characterize the majority of igneous rocks, even in slowly cooled plutonic examples (Tepley and Davidson 2003). Therefore, the presence of isotopically-zoned phenocrysts provides "blessing rather than a curse" into a potentially important record of magma sources and timescales of magmatic processes.

Modern methods of oxygen isotope analysis of crystals

Sizes of typical "pheno"crysts in volcanic rocks are in the 0.5-2 mm scale, with their relative abundance, expressed through crystal size distribution (Fig. 5), an important textural characterization of a rock that is pertinent to the kinetics of its crystallization (Streck 2008; Armenti 2008 and references therein) and crystal inheritance. A certain mass of material is required for a precise single crystal oxygen isotope analysis, and as seen on Figure 6b, the precision of an isotopic measurement decreases with the method and the amount of material available. For example, the quoted 0.5-2mm range corresponds to 0.35 to 22 mg of crystal mass with density 2.8 g/cm³.

The most precise method for oxygen isotope analysis—laser fluorination—was developed in the 1990s (Sharp 1990; Valley et al. 1995). This method relies on fluorination of ~0.3-2 mg material followed by the multiple-cycle analysis of generated micromol quantities of O_2 or

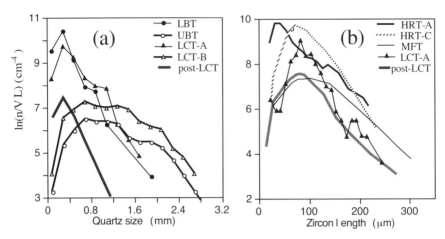

Figure 5. Crystal Size distributions of quartz and zircons in rapidly quenched pumice clasts in large volume tuffs and smaller volume lavas. There are significant differences in crystal sizes and CSD even within the same tuff units. The abundant concave down, lognormal CSDs suggests solution and re-precipitation recycled smaller crystals and the outermost rims of large crystals (e.g., Simakin and Bindeman 2009). CSDs can potentially be used to correlate different magma batches with different crystallization conditions. Abbreviations: HRT- Huckleberry Ridge tuff, LCT- Lava Creet tuff, MFT- Mesa Falls tuff, LBT and UBT are Lower and Upper Bandelier tuffs.

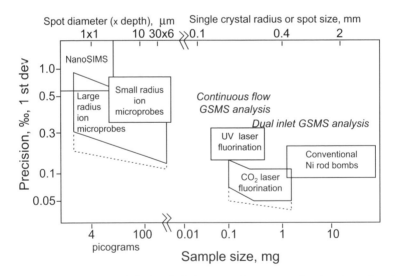

Figure 6. Precision of various analytical methods for oxygen isotopes as a function of sample weight or spot size; precision ranges are current and are based on informed prejudices of the author. Dashed lines represent "best day" precisions.

CO_2 gas on a mass spectrometer against a standard gas of known isotopic composition (dual inlet on Fig. 6). The dual inlet mode of sufficiently large micromol quantities of gas remains the base method of stable isotope mass spectrometry (as TIMS for radiogenic isotopes) analysis with precision improving as a square root of the number of unknown gas-reference gas comparison cycles, commonly better than 0.01-0.02‰ for a typical 6-10 cycle analysis. Most analytical error in oxygen isotope analysis is related to the chemistry of oxygen extraction by fluorination.

Overall external precision in CO_2 laser fluorination labs in Wisconsin, Caltech, and Oregon is "better than 0.1‰" 1 standard deviation, with the uncertainty on standards of ±0.05-0.07‰ common for many analytical sessions (Valley et al. 1995; Eiler 2001; Bindeman et al. 2008c).

The advantage of CO_2 laser fluorination over other methods include: 1) quantitative, 100% reaction of silicate material at temperatures approaching 2000 °C; at such temperatures isotope fractionations between generated gas and potential O-bearing residue are much smaller (ca Fig. 3) than that at 500 °C conventional resistance furnace analysis. 2) High temperature provides the ability to heat and react even the most refractory minerals such as olivine and zircon; 3) reduction of sample size to a single phenocryst that generate enough gas for a multiple cycle, dual inlet analysis. 4) Mineral separation is much less tedious and the quality control is better since it is possible to consistently rely on the highest purity concentrate.

The most important current *in situ* analytical technique for light stable isotopes is ion microprobe analysis. This method uses secondary ions generated by the collision of a primary beam with a target; secondary ions are then analyzed on a mass spectrometer using a variety of secondary ion collection and filtering techniques (Hinton 1995; Ireland 1995; Riciputi et al. 1998; Kita et al. 2004). Spot size can vary from ~2 to 20 μm on large multicollector (e.g., Cameca 1270/80, Page et al. 2007) and small (Cameca 7f) radius ion microprobes, and the analysis time can be remarkably short (6-10 min per spot). However, achieving an appropriate flat polish of the surface to be analyzed and other aspects of sample preparation requirements can be time consuming and challenging. The author's own use of several large-radius, dual collector instruments yielded external precision of 0.15-0.24‰ on a single spot analysis. NanoSims, an instrument that is used chiefly for nanoscale imaging (e.g., Badro et al. 2007), provides permil precision of submicron particles and areas.

The UV laser fluorination methods (Rumble et al. 1997; Young et al. 1998, 2008) have similar to the ion microprobe precision but greater spot sizes; the advantage of UV laser fluorination is the ability to precisely measure ^{17}O within single minerals that is important for extraterrestrial applications (e.g., Young et al. 2008 and references therein).

While CO2 laser fluorination remains the most precise technique for single crystal isotope studies, the improved techniques of large radius ion microprobe analysis for single spots makes the latter appropriate for studying 1-2‰ level isotope variations required for high-temperature igneous environments. These two methods are likely to complement each other for quite some time in the future.

$\delta^{18}O$ heterogeneity and $\Delta^{18}O_{crystal-melt}$ disequilibria

At equilibrium (Figs. 2-3), a volcanic rock should have consistent $\delta^{18}O$ values of minerals and $\Delta^{18}O_{mineral-mineral}$ and $\Delta^{18}O_{mineral-melt}$ isotope fractionations. However, this chapter documents that examples of isotope disequilibria recognized by single phenocryst isotope studies are more common than previously thought.

Oxygen isotope disequilibria may characterize selected xenocrystic crystals in a population or be the property of the majority of crystals in a rock. The second case clearly requires interpretation of melt and crystal origin in the course of a single petrogenetic process. It is also possible that one mineral may have disequilibrium $\Delta^{18}O$ values while others are in high-temperature equilibrium.

The sign and magnitude of $\Delta^{18}O_{crystal-melt}$ disequilibria may be variable, i.e., magma may be lighter or heavier than the measured or computed values of magma required for equilibrium with each mineral that occurs in it (Fig. 2-3). If a low-$\delta^{18}O$ magma mixes with a high-$\delta^{18}O$ magma, or by assimilating high-$\delta^{18}O$ country rock, the $\Delta^{18}O_{crystal-melt}$ may be small or negative. As an example, magmatic values of refractory minerals that survive slow plutonic cooling, such as zircon and garnet (Lackey et al. 2006), may have a variable sign for the $\Delta^{18}O_{mineral-mineral}$ fractionation.

Petrogenetic processes leading to diverse $\delta^{18}O$ values and $\Delta^{18}O_{crystal\text{-}melt}$ disequilibria. Unlike many radiogenic isotope ratios that can suffer significant changes at <1% addition of isotopically-contrasting component, oxygen is a major element. Modification of $\delta^{18}O$ value by more than several tenths of 1 permil requires volumetrically-significant mass transformations, by many percents to tens of percent. Such proportions bring mass and heat balance problems to the forefront of the discussion of crystal and magma origins. The following processes are proposed to affect both $\delta^{18}O$ values, and cause $\Delta^{18}O_{crystal\text{-}melt}$ disequilibria.

(i) Interaction with waters or brines derived from country rocks or stoped blocks (e.g., Friedman et al. 1974; Muehlenbachs et al. 1974; Hildreth et al. 1984; Taylor 1986)

(ii) Rapid assimilation of rocks significantly different in $\delta^{18}O$ that affects magma and not protocrysts (Taylor 1986; Balsley and Gregory 1998; Spera and Bohrson 2004).

(iii) Partial melting of rocks with groundmass that has suffered hydrothermal alteration followed by mixing with more normal magma (e.g., Bacon et al. 1989).

(iv) Complete or bulk melting of shallow hydrothermally-altered rocks followed by an eruption of magma at the surface (e.g., Bindeman and Valley 2001).

It should be noted that sometimes oxygen isotopic values are the only evidence of petrogenetic processes; this is particularly true when magma exchanges with predecessor rocks of the same chemical or isotopic values (e.g., Bindeman et al. 2008b). Lack of radiogenic ingrowth often renders radiogenic systems to be unable to resolve sources that were in supracrustal environments before being melted (e.g., Lackey et al. 2005). In order to be preserved, transformation of melt $\delta^{18}O$ that leads to $\Delta^{18}O_{crystal\text{-}melt}$ disequilibria should proceed faster relative to oxygen diffusive timescales for the largest crystal size, so that reequilibration processes by intracrystalline oxygen diffusion and/or solution-reprecipitation have not erased the memory of the processes described above. For that matter, volcanic rocks and especially pyroclastic igneous rocks preserve more evidence of disequilibria since volcanic eruption quenches their isotope values (Auer et al. 2008). Fresh tephra, scoria, and pumice clasts may provide the best material to study since they undergo the most rapid cooling, while other volcanic products can suffer posteruptive growth and reequilibration.

Toward isotopic equilibrium: diffusion vs. solution reprecipitation. It is likely that both intracrystalline diffusion and solution reprecipitation play a role in isotope equilibria. Figures 7 and 11 present two contrasting examples on how to recognize these two processes given intracrystalline zoning patterns. Element mapping, backscatter-electron, and cathodoluminescence imaging are appropriate tools to recognize one from another in the course of isotope investigation. Furthermore, crystal size distribution of minerals aids in deciding the presence or absence of solution-reprecipitation episodes and on quantifying the amount of material redeposited (Simakin and Bindeman 2008). In particular, abundant lognormal crystal size distributions of zircon and quartz (Fig. 5) suggests that tens of percent of zircon mass was dissolved and reprecipitated, and that the smallest crystals, as well as the rims of the larger crystals, were recycled more than once. The reprecipitation of the rim will not affect the cores of the largest crystals that should retain distinct isotopic values, which are only possible to anneal by intracrystalline diffusion.

Rates of intracrystalline oxygen diffusion vary by 4 orders of magnitude for different minerals (Fig. 8). Increases in oxygen and water fugacity typically decreases oxygen diffusion rates for zircon, olivine, and quartz (Ryerson et al. 1989; Farver and Yund 1991; Watson and Cherniak 1997). Rates of solution reprecipitation are highly variable and depend strongly on the driving forces—undersaturation caused by changes in temperature or composition (e.g., Watson 1996), and are kinetically-limited by the diffusion of the components such as silica through the melt (Zhang et al. 1989). When driving forces for dissolution are great (e.g., at large undersaturations) dissolution may erase evidence of early intracrystalline isotope diffusion and

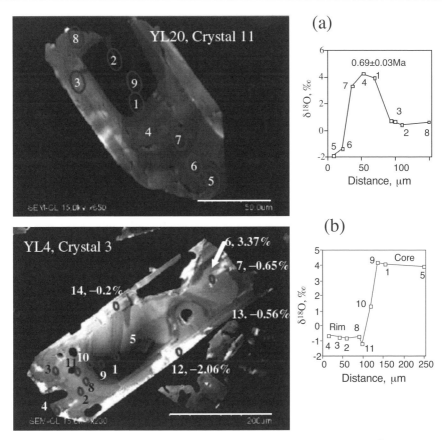

Figure 7. Cathodoluminescence image of isotopically zoned, older zircons from low-δ^{18}O rhyolites of Yellowstone and ion microprobe profiles. Sharp boundaries in isotopic composition indicate that solution and re-precipitation in (b) played more important role than intracrystalline diffusion (e.g., bell-shaped profile) in (a). The solution and re-precipitation generated step-function isotope profiles, with higher δ^{18}O cores surrounded by the low-δ^{18}O rims; faster rates of solution and re-precipitation are capable of erasing the evidence of intracrystalline diffusion. Modified after Bindeman et al. (2008b).

leave sharp boundaries of oxygen isotope distribution within zircon (Fig. 7).

The amount of δ^{18}O heterogeneity and the $\Delta^{18}O_{mineral-melt}$ disequilibria can be used to predict mineral diffusive timescales if it is assumed that exchange between mineral and melt is rate limited by slow intracrystalline oxygen diffusion. If there is textural evidence that solution-reprecipitation played a more important role in isotope reequilibration, then timescales obtained using diffusion coefficients and intracrystalline diffusion provides the *maximum* time the crystals could have resided in isotopically distinct melt.

PART III: CASE STUDIES OF OXYGEN ISOTOPE DISEQUILIBRIA IN IGNEOUS ROCKS BASED ON SINGLE CRYSTAL ISOTOPE ANALYSIS

Below a review is provided on the examples of oxygen isotope disequilibria found in archetypal examples of basic and silicic magmatism around the world, with emphasize on olivine-basalt, plagioclase-basalt, zircon-rhyolite, and quartz-rhyolite disequilibria that

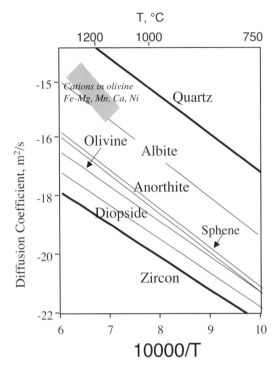

Figure 8. Experimental studies of diffusion with common igneous minerals show four orders of magnitude difference in oxygen diffusion coefficients, and that in olivine cation diffusion is faster than oxygen diffusion. As a result, isotopically-zoned xenocrysts with annealed compositional zoning should be common. Data sources: Zircon, Watson and Cherniak (1997), wet; Sphene, Morishita et al. (1996); Quartz, Farver and Yund (1991), wet; Olivine, Ryerson et al. (1989), QFM; Diopside, Farver (1989); Anorthite and albite, Elphick et al. (1988, 1986) wet.

pertain to the origin of mafic and silicic magmas and their crystals. The examples given here clearly demonstrate the need to explore many more igneous systems for disequilibria relationships, and that single phenocryst isotopic studies of refractory minerals such as olivine and zircon, that can only crystallize from, or exchange with, the high-temperature *melt*, document the tremendous variety of $\delta^{18}O$ melts that are present in the crust and the mantle.

Basaltic igneous systems: olivine in basalts

Iceland. The most extreme example of heterogeneous olivine $\delta^{18}O$ values (3‰ range), and $\Delta^{18}O_{melt\text{-}olivine}$ and $\Delta^{18}O_{melt\text{-}plagioclase}$ disequilibria is within the large-volume basalts of Iceland, as described by Bindeman et al. (2006b; 2008c). The greatest 2.2-5.2‰ range of olivine $\delta^{18}O$ values is measured in the products of 1783-1784AD, 15 km³ Laki fissure eruption (Fig. 9) and individual plagioclase phenocrysts in the same samples also exhibit disequilibrium relations with olivine and melt. Furthermore, the basaltic tephra erupted over the last 8 centuries and as late as in November 2004 from the Grímsvötn central volcano, which together with Laki are a part of a single volcanic system, is indistinguishable in $\delta^{18}O_{melt}$ from Laki glass. All basalts have exceptionally low-$\delta^{18}O$ melt value of 3.1 ± 0.1‰, homogeneous among 15 km³ of Laki basalt. This suggests that they tapped a homogeneous and relatively long-lived, well mixed low-$\delta^{18}O$ magma reservoir that retained its distinct $\delta^{18}O$ as well as ($^{226}Ra/^{230}Th$), ($^{230}Th/^{232}Th$), $^{87}Sr/^{86}Sr$ and trace elemental values (Sigmarsson et al. 1991). In order to generate such low-$\delta^{18}O$ value, digestion of tens of percent of low-$\delta^{18}O$ crust is required.

The preservation of oxygen isotope disequilibria between olivine, melt and plagioclase (Figs. 9, 10) suggests that no more than a hundred years has elapsed since incorporation of plagioclase into basaltic melt because plagioclase has relatively fast oxygen isotope diffusion (Elphick et al. 1986, 1988; Fig. 8). Likewise, time estimates based on mineral diffusive timescales of oxygen and cations in olivine crystals suggests crystal residence for hundreds of years in a well-mixed reservoir under Grimsvotn (e.g., Bindeman et al. 2006b). Similarly short (<10² yr) time estimates were obtained by Gurenko and Chaussidon (2002) and Gurenko and Sobolev (2006) in their oxygen isotope study of isotopically-zoned olivines from Midfell area.

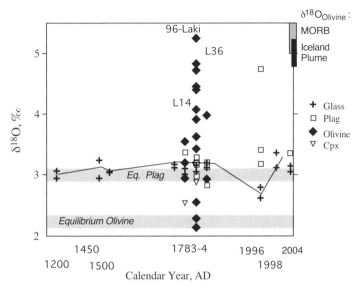

Figure 9. Extreme isotope heterogeneity and disequilibria of individual olivine and plagioclase crystals in the Laki–Grímsvötn magma system based on historical tephra and products of the Laki eruption. There have been relatively constant $\delta^{18}O_{melt}$ values for eight centuries and extreme diversity in $\delta^{18}O$ values of "pheno"crysts. The $\delta^{18}O$ values of An_{75} plagioclase and olivine in equilibrium with 3.1±0.1‰ melt are shown. In the 1996 eruption, plagioclase spans a far wider range than in prior events, which is interpreted here to indicate shorter residence time and later entrainment of feldspars in magma (Bindeman et al. 2006b).

Olivines in other large-volume Holocene basalts of Iceland also display remarkable variety in their $\delta^{18}O$ and Fe-Mg zoning patterns (Fig. 11). The preservation of Fe-Mg and Ni zoning in some of those olivines suggests that these grains spend less than 100 years in the magma prior to eruption (e.g., Costa and Dungan 2005). Ion microprobe oxygen isotope profiling of single olivine crystals has revealed the presence of crystals that are not zoned but are surrounded by a thin rim (preeruptive xenocrysts); crystals that are zoned with respect to Fe-Mg and $\delta^{18}O$ (zoned crystals); and crystals that are only zoned with respect to $\delta^{18}O$ but not composition (compositionally equilibrated, annealed crystals). Moreover, while the majority of crystals have high-$\delta^{18}O$ cores and low-$\delta^{18}O$ melt-equilibrated rims, there were a few crystals with the reverse relationship- crystals with high-$\delta^{18}O$ cores that are diffusively grading into a lower $\delta^{18}O$ rim, and crystals with relatively abrupt compositional and $\delta^{18}O$ overgrowth boundaries (Fig. 11). The proportion of "short-residence" zoned crystals with thin overgrowth rims is variable from unit to unit. The majority of crystals in Laki with relatively constant Fo_{75} composition and subtle residual Ni, Ca, Mn zoning requires longer residence (100s of years).

It appears that the best model to explain isotope disequilibria and heterogeneous $\delta^{18}O$ values in minerals in large volume Holocene basalts from Iceland include magmatic digestion and erosion of Pleistocene hyaloclastites (Bindeman et al. 2008c). Mechanical and thermal magmatic erosion characterizes many lava tubes (Williams et al. 2001, 2004) and it is likely to occur upon basaltic magma flow in the complex network of dikes and sills of the Icelandic rift-related magma plumbing system.

Hawaii. Interesting evidence on $\Delta^{18}O_{melt\text{-}olivine}$ disequilibria is coming from another classic example of basaltic volcanism—Hawaii. Garcia et al. (1998) analyzed coexisting bulk olivine and matrix glass from the 1983-1997 Puu Oo eruption episode of Kilauea volcano in Hawaii

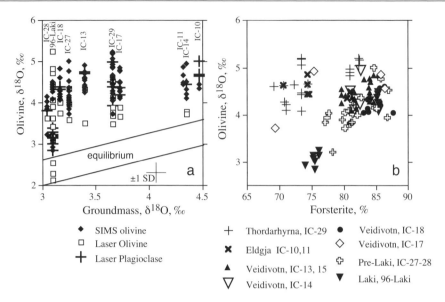

Figure 10. Widespread isotopic heterogeneity of olivine and plagioclase crystals in basalts from Iceland analyzed by laser fluorination and ion microprobe (SIMS). Sample number and corresponding source center are shown; each symbol corresponds to separate lava flow. a) $\delta^{18}O$ values of glass and olivines vs. $\delta^{18}O$ value of their host groundmass (measured by laser) arranged by sample (unit) number. Olivines in each sample are lower than the mantle olivine by more than 1‰, and are higher than the expected equilibrium value of olivines with their host groundmass. The field labeled equilibrium denotes $\delta^{18}O_{olivine}$ values that would be in equilibrium with their host melt; calculated equilibrium fractionation between olivine and these Icelandic melt compositions of 0.7±0.3‰, defining the range of 0.4-1‰ were obtained using the parameterization of Bindeman et al. (2004). b) $\delta^{18}O$ values vs. forsterite content of olivines analyzed by ion microprobe and electron microprobe in the same spot. Average standard deviation (±1 SD) on single spot analysis by the ion microprobe are ±0.24‰, and ±0.1‰ by the laser fluorination. There is a positive correlation of $\delta^{18}O$ vs. %Fo for some, while in others there is a distinct bimodal distribution. From Bindeman et al. (2008c).

and observed moderate $\Delta^{18}O_{melt-olivine}$ disequilibria deviating by up to 0.8‰ from the estimated equilibria $\Delta^{18}O_{melt-olivine} = 0.7\pm0.1$‰ (Fig. 12a). The majority of variation characterizes glass that is too light for rather constant $\delta^{18}O_{olivine}$ values. Both olivine and matrix glass in this prolonged eruption are variably depleted by 0.3-0.8‰ relative to the normal mantle and thus require assimilation or exchange with low-$\delta^{18}O$ hydrothermally-altered rocks inside the Kilauea edifice. It appears that little or no correlation exists between radiogenic isotopes, trace element ratios and oxygen isotope parameters with the exception that late erupted, more magnesian basalts of the Puu Oo eruption were closer to mantle $\delta^{18}O$ values and had smaller $\Delta^{18}O_{melt-olivine}$ disequilibria. Garcia et al. (1998) noticed that the lavas that traveled in the shallow, near horizontal, magma pathways for longer periods of time (Fig. 12b) exhibited greater levels of depletion and $\Delta^{18}O_{melt-olivine}$ disequilibria. Although these authors had difficulties finding the right mass balance or a particular mechanism of $^{18}O/^{16}O$ preruptive exchange, the fact that variable $\Delta^{18}O_{melt-olivine}$ and $\delta^{18}O_{melt}$ values occurred in the same eruption limits the timescale of magma-rock interaction to a duration of preeruptive residence, of months to years.

In a subsequent study of the Kilauea volcanic record, Garcia et al. (2008) described even larger variability in $\Delta^{18}O_{melt-olivine}$ and $\delta^{18}O_{melt}$ values in products of 1650-2000AD historic eruptions from Kilaeua summit. In contrast to the Grimsvotn-Laki system described by Bindeman et al. (2006b), the Hawaiian record indicated variable $\delta^{18}O_{melt}$ values and relatively constant $\delta^{18}O_{olivine}$ values that led Garcia et al. (2008) to suggest that melt depletion happened

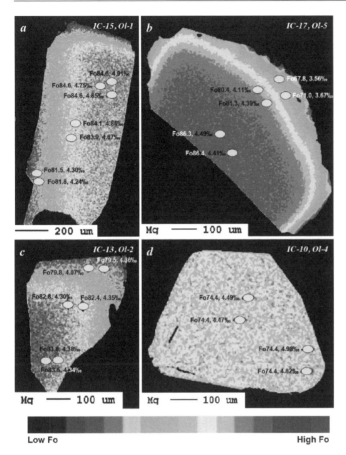

Figure 11. Mg- elemental maps showing various types of elemental and isotopic zoning in olivines from large-volume basalts from Iceland (Bindeman et al. 2008c). Ovals are pits left after ion microprobe analyses of O isotopes. (a-b) grains with normal compositional and isotopic zoning with decreasing Fo and $\delta^{18}O$ toward the rim; (c) grain with compositional zoning but without isotope zoning; (d) compositionally unzoned grain with subtle increase in $\delta^{18}O$ toward the rim. [Used with permission of Oxford University Press from Garcia et al. (1998) *Journal of Petrology*, Vol. 39, p. 803-817.]

after olivine grew, perhaps shortly pre- or syneeruptively (Fig. 12b), and that the difference between Kilauea and Grimsvotn is related to the much smaller magma chamber under the former. There appears to be a correlation of $\delta^{18}O_{melt}$ with Pb and Sr isotopes of hydrothermally-altered more radiogenic Mauna Loa lavas, and thus they may serve as a viable assimilant for historic Kilauea lavas.

Isotopically homogenous olivines in Kilauea provide a nice natural analogue to constrain rates of oxygen exchange in olivine. This example suggests that days to months of residence of originally normal-$\delta^{18}O$ olivines in low-$\delta^{18}O$ melt was insufficient to cause heterogeneous $\delta^{18}O$ values in olivine. The examples of large volume basalts from Iceland excluding Laki (Figs. 10-11) continue the trend of exchange and demonstrate that longer olivine residence time in these basalt has led to olivines acquiring low-$\delta^{18}O$ values and preserving $\delta^{18}O$ heterogeneity. In Laki, the maximum $\delta^{18}O_{olivine}$ heterogeneity and $\Delta^{18}O_{olivine-melt}$ disequilibria indicate the longest residence of hundreds of years (e.g., Bindeman et al. 2006b).

Kamchatka. In a study of olivine-matrix relationship of Klyuchevskoy volcano in Kamchatka, the largest in Eurasia, Auer et al. (2008) found a large $\Delta^{18}O_{melt\text{-}olivine}$ range (Fig. 13) with both positive and negative deviations from equilibrium. Unlike Hawaii and Iceland with low-$\delta^{18}O$ values, the prolific and tall Klyuchevskoy volcano displays the highest known $\delta^{18}O_{olivine}$ and $\delta^{18}O_{melt}$ values (Dorendorf et al. 2000; Auer et al. 2008). While most magmas in Klyuchevskoy and the surrounding Central Kamchatka Depression area are high-

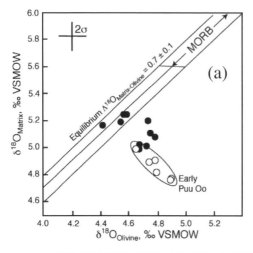

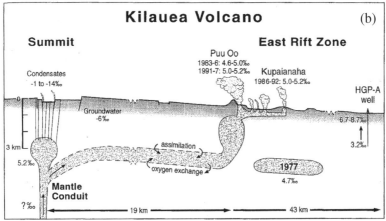

Figure 12. Olivine-Matrix disequilibria in products of Kilaeua eruptions. a) Disequilibria in 1983-1997 Puu Oo eruption. b) Magma pathways within Kilauea volcano. [Used by permission of Oxford University Press from Garcia et al. (1998), *J Petrol*, Vol. 38, Fig. 1, p. 805.]

$\delta^{18}O$ (Portnyagin et al. 2007), the Mg rich component is more normal-$\delta^{18}O$, while high-Al basalts exhibit the highest $\delta^{18}O_{olivine}$ values of 7.6‰. Both high-Mg and high-Al basalt are hydrous with up to 7 wt% water in Fo$_{80}$ melt inclusions (Auer et al. 2008).

Given that variable in $\delta^{18}O$ olivines also exhibit Fe-Mg disequilibria, and given the correlation of $\delta^{18}O_{olivine}$ with Al_2O_3 and MgO content of the parental melt, the best explanation for Klyuchevskoy is pre-eruptive mixing between variable $\delta^{18}O$ cumulates and high-Al and high-Mg basalts, as well as direct magma mixing between high-$\delta^{18}O$, high-Al, and lower-$\delta^{18}O$, high-Mg basalts that are observed in several historic eruptions. Isotopic and compositional heterogeneity of Klyuchevskoy basalts (Fig. 13) is explained by the short time of weeks these olivines spend in magma.

Summary. A histogram of $\delta^{18}O_{olivine}$ values measured in the samples discussed above (single crystals and bulk of several or size fraction crystals) are plotted in Fig. 14 and compares them with the two previously published compilations of bulk analysis: $\delta^{18}O_{whole\ rock}$ (Harmon and Hoefs 1995, conventional methods), and $\delta^{18}O_{bulk\ olivine}$ values (Eiler 2001, laser fluorination). It can be seen that the single crystal based studies document far greater $\delta^{18}O$ range than could have possibly been anticipated by bulk analysis of either whole rocks or multi-crystal concentrates. Olivine is a refractory, early-crystallizing, high-temperature phenocryst, and thus the existence

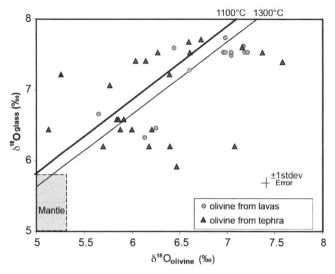

Figure 13. Oxygen isotopic heterogeneity of olivine phenocrysts and their host groundmass for historic or recent (<7 ka) tephra and lava samples from Klyuchevskoy volcano, Kamchatka, Russia. Notice that olivines in a single sample (defined by horizontal arrays) spans up to a few permil to the left and right of equilibrium $\Delta^{18}O_{\text{olivine-groundmass}}$ values shown at respective temperatures by black lines. This extreme $\delta^{18}O_{\text{olivine}}$ diversity is accompanied by significant Fe-Mg disequilibria between olivine and groundmass. The disequilibria and heterogeneity in this case is explained by short-term pre-eruptive mixing between variable-$\delta^{18}O$ magma with variable $\delta^{18}O$ olivines, and by preeruptive mixing between cumulate and magma. From Auer et al. (2008).

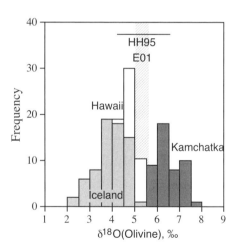

Figure 14. Histogram of $\delta^{18}O$ values of forsteritic olivine measured largely as single phenocrysts in selected examples of basaltic igneous rocks discussed in this work. Hatched pattern marked E01 indicates the majority of olivine values in mantle derived magmas based largely on bulk monomineralic separates summarized by Eiler (2001), while thin horizontal line indicates $\delta^{18}O$ values of basalt (-0.5 permil for basalt-olivine fractionation) from Harmon and Hoefs (1995). Notice that refractory single olivine records great variety of $\delta^{18}O$ values of melt from which it crystallized or exchanged with exceeding either laser-based (E01) or conventional, whole-rock values (HH95).

of such heterogeneity demonstrates the existence of basaltic melts either in the mantle or the crust, that span 6‰.

Silicic igneous systems

Isotopic analysis of individual phenocrysts or intra-crystalline domains is a novel tool to fingerprint crystal sources and to recognize separate magma batches of silicic magmas or solid sources from which the crystals once crystallized or were trapped (Tepley et al. 1999; Bindeman and Valley 2001; Wolff and Ramos 2003). Below, we discuss oxygen isotope disequilibria with the examples of quartz and zircon, two common phases of constant composition and thus simple $\Delta^{18}O_{mineral-melt}$ fractionation (e.g., Fig. 3). Zircon has a particular advantage for oxygen isotope studies: it is one of the most refractory, alteration-resistant minerals that retains oxygen isotopic values (Watson 1996; Valley 2003), it is a prime mineral for U-Th-Pb geochronology, and it is a prime mineral for other isotopic (e.g., Hf) and trace elemental studies (Hanchar and Hoskin 2003).

Isotope zoning and Residual memory in Zircon. U-Pb ion microprobe dating of zircon crystallization in volcanic samples demonstrates that zircon crystallization (and closure to subsequent U-Pb exchange) commonly predates the eruption of host lavas as determined by Ar-Ar dating. This has been documented for a range of different sized magma systems. Miller and Wooden (2004) found that zircons in Devils Kitchen rhyolite (Coso volcanic center, California) are up to 200 k.y. older than corresponding K-Ar ages, and several zircon populations with different pre-eruptive histories could be identified, signifying an origin from different magma batches. Simon and Reid (2005) suggested that zircons in the Glass Mountain rhyolites (Long Valley, California) record episodes of punctuated and independent evolution rather than the periodic tapping of a long-lived magma chamber. U-Pb zircon ages in combination with Ar-Ar potassium feldspar ages for the Geysers pluton (California Coast Ranges) indicate that the shallow portions of the pluton cooled to <350 °C within ~200 ka, whereas at the same time zircons from just solidified granitoids at deeper levels became remobilized by the heat of newly intruded magma (Schmitt et al. 2003a,b). Charlier et al. (2005) found zircon ages spanning 100 k.y. in the Taupo volcanic zone in New Zealand, and interpreted zircons to be derived by bulk remobilization of crystal mush and assimilation of metasediment and/or silicic plutonic basement rocks. In their study of Crater Lake volcanic rocks, Bacon and Lowenstern (2005) identified parental rocks as a source for antecrystic zircons and plagioclase in the form of co-erupted granodioritic blocks and magmas. These studies did not reach a universal conclusion on the state of silicic magma bodies in the crust, but favored variably "long" zircon residence in "mushy" upper crustal magma chambers (e.g., Vazquez and Reid 2002; Reid 2003).

Oxygen isotope analysis of U-Pb dated zircon populations provides an important additional insight into the origin of zircons and their host magmas. Many large silicic magma systems display remarkably different degrees of oxygen isotope disequilibria and diversity in their phenocryst populations.

Case studies of large silicic magma systems

Bishop tuff. On one extreme, there is a nearly homogeneous $\delta^{18}O$ magma body parental to Bishop tuff (Fig. 3b, Bindeman and Valley 2002) in which zircon appears to be in perfect $\Delta^{18}O$ equilibrium with other minerals; on the other extreme there are rhyolites from Yellowstone, with tremendous diversity in age, $\delta^{18}O$ and $\Delta^{18}O_{Zircon-melt}$ values (Fig. 15). The erupted Bishop tuff magma body was equilibrated with respect to the major element oxygen (Fig. 3b) and did not retain zircon or other phenocrystal evidence of inheritance from the Glass Mountain magmas (Simon and Reid 2005). The pre-Bishop Tuff, Glass Mountain rhyolites, erupted over a time-span exceeding 1 Ma, exhibit heterogeneity with respect to Sr, Nd, Pb, and O isotopes between different domes (Davies and Halliday 1998; Bindeman and Valley 2002; Simon and Reid 2005). Subsequent accretion of these magma batches led to the formation of the Bishop Tuff magma body, which averaged isotopic differences in the melt, equilibrated $\delta^{18}O$ in minerals, and rejuvenated U-Pb ages of zircons (Reid and Coath 2000). Several other large volume rhyolites: Cerro Galan, Toba, Fish Canyon, both monotonous-intermediates, and high-silica rhyolites, exhibit relative isotope homogeneity between early and late eruption products (Bindeman and Valley 2002).

Yellowstone. On the other end of the spectrum, Yellowstone intracaldera volcanic rocks belonging to the Upper Basin lavas are an example where almost the entire population of zircons is inherited from precaldera source rocks spanning 2 m.y., based on both U-Pb zircon ages and

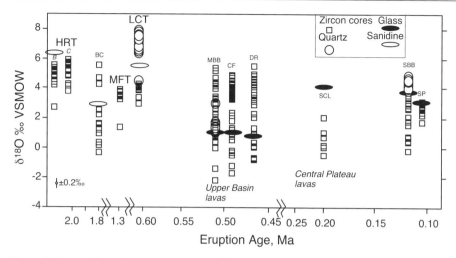

Figure 15. Isotope disequilibria and diversity of $\delta^{18}O$ values of zircons and quartz from selected rhyolitic lavas and tuffs from Yellowstone Plateau volcanic field, Wyoming, USA measured by the large radius ion microprobe. Analyses of $\delta^{18}O$ are plotted vs. Ar-Ar eruptive age (Gansecki et al. 1996; Lanphere et al. 2002); the age of the SBB is by U/Th dating of zircon rims, (Watts et al., unpublished data); BC age is by K-Ar Obradovich (1992). HRT, MFT, and LCT are Huckleberry Ridge (Members B and C), Mesa Falls, and Lava Creek tuff caldera forming eruptions; Unit abbreviations: BC - Blue Creek, MBB and SBB are Middle and South Biscuit Basin, DR- Dunraven Road, SCL- Scaup Lake, SP- Solfatara Plateau. Modified after Bindeman et al. (2008b).

$\delta^{18}O$ values (Bindeman et al. 2001; 2008b). Tremendous ranges of $\delta^{18}O$ values of phenocrysts and variable $\Delta^{18}O_{mineral-melt}$ and $\Delta^{18}O_{quartz-zircon}$ disequilibria in the majority of 6000 km³ volcanic rocks in Yellowstone (Fig. 15) are revealed by individual quartz phenocryst analysis by laser fluorination, and individual zircon analysis by ion microprobe. These rhyolites are the products of nearly wholesale remelting and recycling of hydrothermally altered materials from earlier eruptive cycles, and they preserve extreme oxygen isotopic variability and zoning in phenocrysts, including quartz (Bindeman and Valley 2001).

Timber Mt. Caldera Complex, Nevada. Intermediate between Yellowstone and Long Valley, are the large volume low-$\delta^{18}O$ Ammonia tanks rhyolites of Timber Mountain caldera in Nevada (Fig. 16) that exhibit $\Delta^{18}O_{zircon-melt}$ disequilibria but demonstrate $\Delta^{18}O_{Quartz-melt}$ and $\Delta^{18}O_{Sphene-melt}$ equilibria. Neither different parts of the large Ammonia Tanks tuff magma body, exemplified by the study of individual pumice clasts dispersed by the caldera-forming eruptions (Mills at al. 1997; Tefend et al. 2007), nor crystal populations, exemplified by the study of phenocrysts (Bindeman et al. 2006a), were unable to achieve whole-rock isotopic and chemical equilibration. Diversity of $\delta^{18}O$ zircon values in Ammonia Tanks rhyolites is supported by the diversity of their ages that spans the entire history of Timber Mountain caldera complex (Bindeman et al. 2006a).

Recognizing magma batches using individual phenocrysts. At Yellowstone and Timber Mt. caldera complexes, lavas and tuffs with poly-age and diverse $\delta^{18}O$ zircon populations represent accretion of independent homogenized magma batches that were generated rapidly by remelting of source rocks of various ages and $\delta^{18}O$ values. Sufficient magma residence time may allow annealing $\delta^{18}O$ and $\Delta^{18}O$ values of quartz, then sphene, and then zircon. However, isotope zoning in zircon in these two caldera complexes within even large-volume rhyolites suggest that "residual memory" of accretion and magma derivation from low-$\delta^{18}O$ hydrothermally-altered rocks has not yet been erased (Fig. 17). Large volume tuffs of Yellowstone and Timber

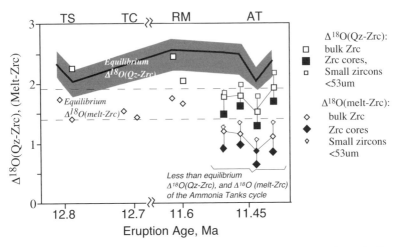

Figure 16. Zircon-quartz and zircon-melt oxygen isotope disequilibria in large volume rhyolites from Timber Mt/Oasis Valley caldera complex, Nevada, using size fraction analysis of zircons by laser fluorination. Large volume units Topopah Spring (TS), Tiva Canyon (TC), and Rainier Mesa (RM) show equilibrium relations, while zircons in all four of Ammonia Tanks (AT) cycle samples have smaller than equilibrium fractionations vs. quartz, due to the inherited high-δ^{18}O cores which are present even in smaller zircons; Smaller zircons show closer equilibrium with quartz and melt than larger zircons, because the latter have higher- δ^{18}O cores. Sphene and quartz in these samples are in isotopic equilibrium. Four samples in the Ammonia Tanks cycle are plotted in accordance with their relative eruption sequence at around 11.45 Ma. Modified after Bindeman et al. (2006a).

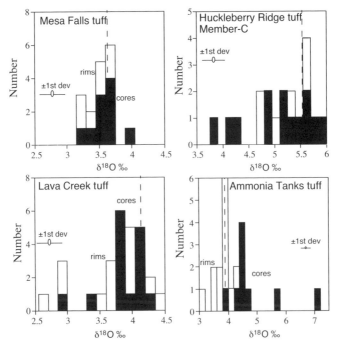

Figure 17. Oxygen isotope diversity and zoning of zircons from major tuff units from Yellowstone and Ammonia Tanks tuff of Timber Mt. caldera complex. Dashed vertical line represents equilibrium value with the host glass. Data from Bindeman et al.. (2006a, 2008b).

Mt. exhibit the $\Delta^{18}O_{Quartz\text{-}melt}$ equilibria, small $\Delta^{18}O_{zircon\text{-}melt}$ disequilibria, and diverse $\delta^{18}O_{zircon}$ values (Figs. 15-16, Bindeman et al. 2006a, 2008b).

When $\delta^{18}O_{zircon}$ values are plotted against the eruptive volume in three large silicic systems (Fig. 18) there appears to be an empirical correlation that suggests decreasing $\Delta^{18}O_{zircon\text{-}melt}$ disequilibria with increasing volume to ~100-300 km³. This correlation may indicate that magma bodies of ~100-300 km³ in size represent well-mixed reservoirs which exist for a long enough time to anneal isotope disequilibria in zircons and other minerals. However, greater heterogeneity that appears in larger, supervolcanic volumes of magma, may reflect late stage or preeruptive magma batch addition or stratification. The only exception from this rule is provided by the Kilgore tuff, 1800 km³ eruptive unit in the Heise volcanic field in Idaho (Morgan and McIntosh 2005; Bindeman et al. 2007), which appears to have greater homogeneity in zircon age and $\delta^{18}O$ values.

Phenocryst heterogeneity in plutonic rocks

Plutonic rocks have undergone prolonged cooling that should in theory anneal isotope disequilibria, and evidence of batch segregation given diffusion coefficients and times involved (Fig. 8). However, zircon, a mineral with the slowest oxygen diffusion, demonstrates $\Delta^{18}O_{zircon\text{-}other\ mineral}$ disequilibria in studied examples of plutonic rocks. King and Valley (2001) and Lackey et al. (2006) (Fig. 19) studied the oxygen isotopic composition of minerals in the Idaho Batholith and Sierra Nevada Batholith respectively, using two refractory minerals with slow oxygen diffusion—garnet and zircon. At equilibrium, both orthosilicates should have nearly identical $\delta^{18}O$ values and $\Delta^{18}O_{zircon\text{-}garnet} = 0\pm0.1‰$ (Valley et al. 2003). King and Valley (2001) and Lackey et al. (2006) observed that near the contact with high-$\delta^{18}O$ country rocks, zircon retained its average intraplutonic $\delta^{18}O$ values, while other minerals, including garnet, increased their ^{18}O from the assimilation of high-$\delta^{18}O$ country rocks. In both examples, lower than equilibrium $\delta^{18}O$ zircon values, and correspondingly higher than zero $\Delta^{18}O_{garnet\text{-}zircon}$ values indicated late stage contamination by a high-$\delta^{18}O$ wallrock that did not affect the previously crystallized zircon. However, it is also possible that their data indicate batch

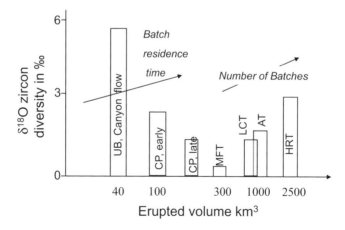

Figure 18. Oxygen isotope diversity of zircons from major tuff units from Yellowstone and Timber Mt. Calderas, and smaller volume rhyolitic lavas as determined by ion microprobe analysis of individual zircons. Achieving isotope equilibrium is interpreted to represent batch residence time that increases with unit size and seems to reach least heterogeneity for units several hundred cubic kilometers erupted as a single cooling unit such as Mesa Falls tuff. More voluminous units develop greater heterogeneity due to the number of erupted Members. Unit abbreviations are the same as in Figures 15 and 16; CP and UB are Central Plateau and Upper Basin post Lava Creek tuff intra-caldera lavas of Yellowstone.

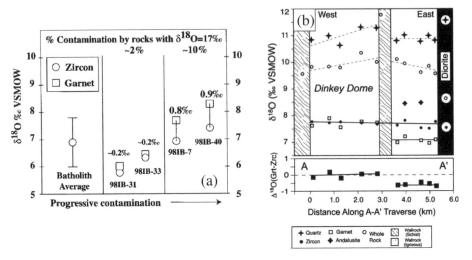

Figure 19. Two examples of plutonic oxygen isotope disequilibria between zircon and other minerals. a) Idaho Batholith (King and Valley 2001) showing progressively divergent $\Delta^{18}O_{garnet\text{-}zircon}$ fractionation closer to the contact with high $\delta^{18}O$ country rock; b) Variations in $\delta^{18}O$ and $\Delta^{18}O_{garnet\text{-}zircon}$ across Dinkey Dome pluton, Sierra Nevada Batholith. In both examples lower than equilibrium $\delta^{18}O$ zircon values, and correspondingly greater than zero $\Delta^{18}O_{garnet\text{-}zircon}$ values indicate late stage contamination by a high-$\delta^{18}O$ wall-rock that did not affect already crystallized zircon. Equilibrium $\Delta^{18}O_{garnet\text{-}zircon} = 0$ value is expected to be ~0‰ (Valley et al. 2003). [Used with kind permission of Springer Science+Business Media from Lackey et al. (2005) *Contrib Mineral Petrol*, Vol. 151, p. 20-44, and King and Valley (2001) *Contrib Mineral Petrol*, Vol. 142, p. 72-88.]

accretion similar to the volcanic examples outlined above: it is noteworthy that $\Delta^{18}O_{zircon\text{-}garnet}$ disequilibria was only preserved near intrusive contacts because these likely involved intraplutonic quench. Both King and Valley (2001) and Lackey et al. (2006) emphasized slow oxygen diffusion and high closure temperature for zircon, and thus the ability of zircon to retain its original crystallized magmatic value.

In their ion microprobe study of zircons in plutonic rocks, Bolz (2001), Appleby (2008), and Appleby et al. (2008) documented further examples of $\delta^{18}O$ heterogeneity. The latter paper presents detailed ion microprobe data for zircons in two diorites associated with the Scottish Caledonian Lochnagar pluton. The diorites are very similar with respect to whole-rock chemical composition and age but have very different zircon oxygen isotope compositions. One diorite sample forms a homogeneous zircon population but the other covers a large range of $\delta^{18}O$ values of several permil. Variations mostly occur between individual zircon crystals, but $\delta^{18}O$ zoning is also present within a few crystals. While in the majority of cases $\delta^{18}O$ increases with zircon growth, as is found in the studies of Lackey et al. (2006) and King and Valley (2001), there are individual zircon crystals with the reverse relationship in the same sample, suggesting that several mixing events of variable $\delta^{18}O$ batches occurred. The same diversity and zoning pattern is also found in a larger sample suite from Lochnagar and Etive plutons (Appleby 2008), and $\delta^{18}O$ zoning and disequilibria, supported by U-Pb age, differences in Hf isotopic values of zircons, and their REE values, appears to be a common phenomenon in these two plutons.

Summary. Highly diverse $\delta^{18}O$ zircon values of rhyolites in the examples discussed in this Chapter (Fig. 15 and 17) can be complemented by analyses of $\delta^{18}O$ values in single zircon and other refractory igneous minerals (see Bolz 2001; Bindeman et al. 2008a,c; Appleby et al. 2008) that demonstrate a picture of diverse silicic magma sources. Since zircons can only exchange oxygen upon crystallization from melt or through a prolonged exchange with melt, such diverse $\delta^{18}O_{zircon}$ values record the diversity of crustal melts that are coming from high-$\delta^{18}O$ and low-$\delta^{18}O$ sources and then mix effectively. It is likely that plutons and large volume batholiths are assemble by pulses (Grunder 1995; Davidson et al. 2001; Coleman et al. 2004; Glazner et al. 2004; de Silva and Gosnold 2007) and we now find single crystal isotopic evidence of these assembly processes. The lack of isotope zoning and disequilibria in some large-volume rhyolites such as Bishop tuff (Fig. 3b) suggest convection and annealing due to magma residence and these topics are discussed below.

PART IV: ISOTOPE DISEQUILIBRIA: WHAT HAVE WE LEARNED ABOUT MAGMA GENESIS?

Mineral-diffusive timescales and magma residence times

At high magmatic temperatures, timescales of diffusive equilibration of isotopes and trace elements between crystals and magma are highly variable but are usually short (from <1 to 100s years) relative to magma generation and segregation timescales (e.g., Costa et al. 2008). In this sense, they provide a chronological resolution that is unattainable by most other methods, especially for older rocks in which U-series methods cannot be used (e.g., Bindeman and Valley 2001; Costa et al. 2008). For the youngest volcanic rocks, the age of the melt could be independently constrained by the short-lived U series isotopes, such as ($^{226}Ra/^{230}Th$) and ($^{210}Pb/^{226}Ra$) activity ratios those half-lives of 1600 yr and 22 yr respectively are comparable with the cation and oxygen diffusion times (Bindeman et al. 2006b; Cooper and Reid 2008). Accessory minerals with slow diffusion coefficients, and particularly zircon (Fig. 8, Watson and Cherniak 1997; Valley 2003) are valuable probes of transient changes by diffusion and solution reprecipitation at high magmatic temperatures, although there appears to be growing evidence that solution-reprecipitation plays a more important role (e.g., Page et al. 2007) and thus diffusion timescales provide maximum time for isotope transformation.

Rapid magma genesis and high magma production rates

The existence of isotopically-zoned and disequilibrium crystals that constitute the majority of crystal populations (Figs. 9, 10, 15, 17) suggests rapid magma genesis. Hundreds to thousands of years time seems reasonable to generate small volume igneous bodies of a few cubic kilometers or less, given average magma production rates in arcs and plumes of 0.001-0.01 km^3/yr (Dufek and Bergantz 2005; McBirney 2006). The remarkable observation which is coming from single crystal isotope studies is that large volume (40-1000 km^3) rhyolites (Figs. 15, 17) and large volume (3-20 km^3) basalts (Figs. 9-13), all have isotope disequilbria for the majority of their crystals. For example, the voluminous 1000 km^3 Ammonia Tanks rhyolite from the Timber Mt. caldera complex demonstrate that individual, isotopically-distinct melt batches generated by reheating of hydrothermally-altered rocks were able to coalesce over time-scales of <150 ky, or likely <10 ka (Fig. 20) into a ~1,000 km^3 size magma body (Mills et al. 1997; Tefend et al. 2007; Bindeman et al. 2006a). In this period of time, hundreds of cubic kilometers of hydrothermally-altered low-$\delta^{18}O$ protolith were melted and digested, and inherited zircons with older ages and higher-$\delta^{18}O$ cores survived the hydrothermal alteration and melting. High magma production rates of ~0.01 km^3/yr are thus estimated, which require intrusion of substantial volumes of basaltic magmas on the order of many hundreds of km^3, that preceded the melting process and may require ignimbrite "flare-up" (e.g., de Silva and Gosnold 2007).

Likewise, in order to generate a Laki-size basaltic flow of 15 km^3 in 1000 to 8000 years based on mineral-diffusive timescales of isotopically-zoned crystal populations, and excess of ($^{226}Ra/^{230}Th$), high magma production and accumulation rates of 0.002-0.02 km^3/yr are required. It should also be noted that these are minimum estimates based on the erupted volumes and

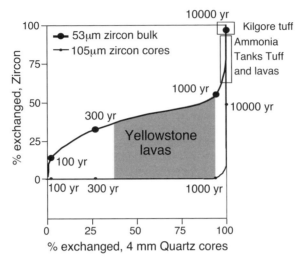

Figure 20. Timescale of magma genesis based on isotope re-equilibration between quartz in zircon within rhyolitic melt at 850 °C using diffusion coefficients from Watson and Cherniak (1997) and Farver et al. (1991), and diffusion in a sphere equation from Crank (1975). A mineral core constitutes the inner 50% of diameter. The field for isotopically-diverse Yellowstone zircons and quartz is shown and requires a 500-5000 yr residence time of xenocrysts in magma. In Ammonia Tanks tuff, quartz $\delta^{18}O$ zoning is totally annealed, while the remaining subtle 1.5‰ zoning in zircons is consistent with 15,000-20,000 yr of residence at the same conditions. This model predicts that disequilibria due to unequal exchange between quartz and zircon would persist for up to 2000 yr; after that only zircon would show internal zoning lasting for up to ~30,000 yr. Diffusion is assumed to become important when temperatures increase during melting and to stop during eruption. The range of zircon zoning in Ammonia Tanks tuff is consistent with ~10,000 yr from melting to eruption, but for the Kilgore ignimbrite (Heise volcanic field) is longer (Bindeman et al. 2007). The range of zircon and quartz zoning in Yellowstone's low-$\delta^{18}O$ rhyolites (Bindeman and Valley 2001) and from Timber Mt. (Bindeman et al. 2006b) are consistent with shorter residence times in smaller-volume low-$\delta^{18}O$ lavas. While this model is based on diffusion, this solution provide maximum timescales if solution and re-precipitation played more important role.

assuming complete evacuation of magma reservoirs and thus could be several times greater if the erupted proportion is only a fraction of the total magma generated. The quoted magma production rates are high but lower than Hawaii's hot spot eruption rates of ~ 0.01-0.1 km³/yr.

Very shallow petrogenesis and the role of hydrothermal carapaces

Abundant low-$\delta^{18}O$ signatures of crystals that we found in great abundance uniquely fingerprint shallow magma petrogenesis since meteoric water does not penetrate deeper than ~10 km, where the closed porosity in rocks no longer allows low-$\delta^{18}O$ waters to circulate (e.g., Taylor and Shappard 1986). Caldera settings and rift zone environments facilitate hydrothermal fluid flow and shallow petrogenesis through assimilation of hydrothermally-altered carapace around the magma chambers, which we call "crustal cannibalization." Documented examples of shallow level magma genesis presents significant challenges for heat and mass balance. Mechanically, shallow crustal cannibalization processes may involve reactive bulk assimilation (Dungan and Davidson 2004; Beard et al. 2005) which starts when interstitial melt forms interconnected networks, causing the roofrock to collapse and to disintegrate into individual phenocrysts or crystal clusters. The resulting crystal mush then becomes homogenized by convection. Isotopic diffusion and re-equilibration of crystals will last until the eruptive quench. In this model, magma cannibalizes earlier erupted rocks and the old hydrothermal system that penetrated through them. Self-cannibalization generates a low-$\delta^{18}O$ component that lowers the $\delta^{18}O$ value of the magma. At Yellowstone and Crater Lake, large scale, nearly wholesale melting of hydrothermally altered crust occurred rapidly and effectively enough that a significant portion of the old crust with abundant inherited "antecrysts" of zircons, quartz, and plagioclase (Bacon and Lowenstern 2005; Bindeman et al. 2008b) are preserved in the erupted magmas. Likewise, oxygen isotopic heterogeneity of olivine and plagioclase crystals, and disequilibrium relations between crystals and melt in large volume basaltic eruptions in Iceland and basalts of Kilauea (Figs. 11-12) suggest shallow, sometimes preeruptive modification of basalts in the upper few kilometers of crust.

Abundant evidence of shallow magma petrogenesis contradicts common knowledge that it is easy to generate magma in the middle or lower crust where ambient heat balance around magma bodies prevents rapid heat dissipation. Magma genesis by remelting in shallow environments is aided by: 1) rocks in caldera and rift environments that are pre-heated by prior long-term volcanic activity and subsidence that brings surface-altered rocks down to the underlying heat sources (Fig. 21); 2) erupted rocks are porous, fractured and thus relatively easy to disintegrate by a reactive assimilation processes compared to their plutonic equivalent; 3) shallowly emplaced eruptive products contain abundant volcanic glass which means that no latent heat of fusion is required for their remelting (i.e., melting efficiency is greater for volcanic rocks than for plutonic rocks); 4) shallow volcanic rocks in hydrothermal settings are pre-conditioned and saturated with water that promotes flux melting.

Accretion of large silicic magma bodies

The generation and eruption of silicic magmas continues to present challenges in answering questions about the size and longevity of crustal magma bodies, their physical state as either stagnant, near-solidus cumulate mushes, near-liquidus convecting liquids, or initially solid rocks, and the time and depth of their segregation (e.g., Annen and Sparks 2002; Bachmann and Bergantz 2003, 2004; Dufek and Bergantz 2005; Bindeman et al. 2008b).

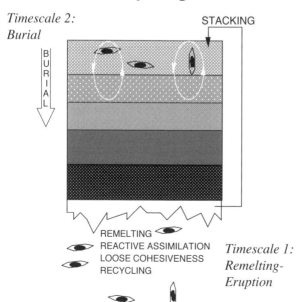

Figure 21. Magmatic cannibalization in shallow crustal environments and timescales of crystal recycling. Progressive burial by caldera collapses, rifting and/or overloading brings erupted volcanic rocks and their sub-volcanic equivalent down deeper to the melting zone. Hydrothermal alteration serves as a flux to cause melting, and glassy, porous nature of these rocks makes re-melting and disaggregation mechanically easier. Timescale 2 is the age between the core and the rim of zircon, while timescale 1 is the crystal diffusive/recrystallization age.

The common theme of documented examples of isotope disequilibria presented in this chapter is batch assembly of magma bodies with initially diverse phenocryst populations, followed by increased mixing and annealing with increased residence time. There is plenty of radiogenic isotope evidence that the accretion of large silicic magma bodies includes batches formed by differentiation of basaltic magmas in the crust, and by partial melting of the radiogenic crust due to heat transport by basaltic intrusions (e.g., see Grunder 1995; Annen and Sparks 2002, and Dufek and Bergantz 2005 for a review of the current literature). Partial melting of the variable-$\delta^{18}O$ crust will very often generate isotopically-distinct individual magma batches, and isotopes of major element oxygen nicely reflect these processes, especially when constrained by zircons. High-$\delta^{18}O$ batches reflect supracrustal materials, while low-$\delta^{18}O$ values document batches that are generated by melting of hydrothermally-altered crustal carapace. Oxygen isotope evidence from the Yellowstone and Timber Mt calderas (Figs. 15-17) indicates that source rocks were cooled below the solidus, altered by heated meteoric waters that imprinted the rocks with low $\delta^{18}O$ signatures, and then remelted in distinct pockets by intrusion of mafic magmas. Each pocket of new melt was variable in $\delta^{18}O$, but inherited zircons retained earlier age and $\delta^{18}O$ values.

The model of remelting and batch accretion presented in this chapter contradicts the commonly held model of a single, large-volume, mushy-state magma chamber that is periodically reactivated and produces rhyolitic off-springs.

The origin of "phenocrysts" and interpretation of their melt inclusions

The presence of $\delta^{18}O$ zoning and variability among phenocryst populations, as discussed in numerous examples above, suggest that crystals are captured from different $\delta^{18}O$ sources and thus record different pre-eruptive residence times. For example, phenocryst studies of Icelandic basalts (Figs. 11-12) reveal crystal-scale complexity that requires olivine recycling at different times prior to eruption. The evidence presented above suggests that prolonged annealing and isotopic homogenization of xenocrystic or antecrystic crystal populations is capable of erasing magma memory of crystal origin. Two views on crystal origin are possible: (1) crystals grew from the melt in which they occur, (2) crystals are "annealed beyond recognition" and mixed in with phenocrystic populations that grew from the melt. From the point of view of an isotope geochemist, it is still possible to recognize diverse crystal origins, their fragmentation and recrystallization histories (Fig. 22), and therefore fingerprint magma batches. These annealed crystals may or may not be in chemical equilibria with the host melt and exhibit subtle isotope evidence of diverse sources.

Heterogeneous magmatic origins of olivines in basalts clearly requires the need for reassessment of their melt inclusions, including volatile content (e.g., Thordarson and Self 2003). The same applies to melt inclusion in isotopically annealed quartz or zircon xenocrysts in silicic rocks, which may record melt compositions from previous eruption episodes that predate the host magma by many hundreds thousands years. Melt inclusion compositional studies, combined with identification of melt $\delta^{18}O$ sources and ages (e.g., melt inclusions in zircon), would provide powerful insights into the chemistry of assembled magma batches.

In this chapter, we presented evidence that annealed crystals are more common than phenocrysts in many basaltic and silicic magma systems around the world. This means that magmas and their crystals cargo may be difficult to produce, but are easy to recycle. Taken as a whole, these results beg the question: are there any "pheno"crysts left?

ACKNOWLEDGMENTS

Author's prior work with Alfred Anderson, John Valley, John Eiler, and prior and ongoing collaborations with Leonid Perchuk, Olgeir Sigmarsson, Paul Wallace, Vera Ponomareva,

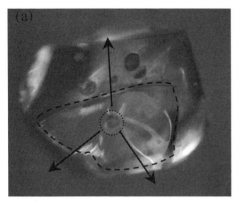 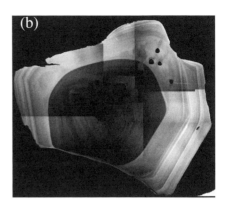

1 mm

Figure 22. Melt Inclusions in pheno(xeno)crysts: reheated, re-homogenized, decrepitated, and exchanged with the external melt. a) Fragmented host quartz crystal from the Bishop tuff after experimental reheating that lead to melt inclusion decrepitation and fragmentation of crystal. b) Cathodoluminescence image Late Bishop tuff quartz showing sharp boundaries between the dark-CL core and bright-CL rim with melt inclusions in both places (see Wark et al. 2007). Bright CL rims and dark CL cores are nearly identical isotopically (Bindeman and Valley 2002). Image courtesy: Julie Roberge

Axel Schmitt, Andrey Gurenko, and Maxim Portnyagin has led to formulation and maturation of ideas presented here. Informal reviews by Jim Palandri, Erwan Martin, Kathryn Watts, comments by John Valley, and formal reviews by Mike Garcia, Jade Star Lackey, Keith Putirka and an anonymous reviewer have dramatically improved presentation of this chapter. Supported by NSF (EAR0537872) and the University of Oregon.

REFERENCES CITED

Anderson AT Jr., Clayton RN, Mayeda TK (1971) Oxygen isotope geothermometry of mafic igneous rocks. J Geol 79:715-729
Annen C, Sparks RSJ (2002) Effects of repetitive emplacement of basaltic intrusions on thermal evolution and melt generation in the crust. Earth Planet Sci Lett 203:937-955
Appleby SK (2008) The origin and evolution of granites: an in-situ study of zircons from Scottish Caledonian granites, PhD Dissertation, University of Edinburgh, Scotland
Appleby SK, Graham CM, Gillespie MR, Hinton RW, Oliver GJH, EIMF (2008) A cryptic record of magma mixing in diorites revealed by high-precision SIMS oxygen isotope analysis of zircons. Earth Planet Sci Lett 269:105-117
Armienti P (2008) Decryption of igneous rock textures: crystal size distribution tools. Rev Mineral Geochem 69:623-649
Auer S, Bindeman I, Wallace P, Ponomareva V, Portnyagin M (2008) The origin of hydrous, high-$\delta^{18}O$ voluminous volcanism: diverse oxygen isotope values and high magmatic water contents within the volcanic record of Klyuchevskoy volcano, Kamchatka, Russia. Contrib Mineral Petrol DOI 10.1007/s00410-008-0330-0
Bachmann O, Bergantz GW (2003) Rejuvenation of the Fish Canyon magma body: a window into the evolution of a large-volume silicic magma systems. Geology 71:789-792
Bachmann O, Bergantz GW (2004) On the origin of crystal-poor rhyolites: extracted from batholithic mushes J Petrology 45:1565-1582
Bacon CR, Adami LH, Lanphere MA (1989) Direct evidence for the origin of low-$\delta^{18}O$ silicic magmas: quenched samples of a magma chamber's partially-fused granitoid walls, Crater Lake, Oregon. Earth Planet Sci Lett 96:199-208
Bacon CR, Lowenstern JB (2005) Late Pleistocene granodiorite source for recycled zircon and phenocrysts in rhyodacite lava at Crater Lake Oregon. Earth Planet Sci Lett 233:277-293

Badro J, Ryerson FJ, Weber PK, Ricolleau A, Fallon SJ, Hutcheon ID (2007) Chemical imaging with NanoSIMS: A window into deep-Earth geochemistry, Earth Planet Sci Lett: 262:543-551

Baertschi P (1976) Absolute ^{18}O Content of standard mean ocean water. Earth Planet Sci Lett 31:341-344

Balsley SD, Gregory RT (1998) Low-δ^{18}O magmas: why they are so rare? Earth Planet Sci Lett 162:123-136

Beard JS, Ragland PC, Crawford ML (2005) Reactive bulk assimilation: A model for crust-mantle mixing in silicic magmas. Geology 33:681-684

Biegelsen J, Mayer MG (1947) Calculation of equilibrium constants for isotope exchange reactions. J Chem Phys 15:261-267

Bindeman IN, Brooks CK, McBirney AR, Taylor HP (2008a) The low-δ^{18}O, late-stage ferrodiorite magmas in the Skaergaard Intrusion: Result of liquid immiscibility, thermal metamorphism, or meteoric water incorporation into magma? J Geology, (in press)

Bindeman IN, Fu B, Kita N, Valley JW (2008b) Origin and evolution of Yellowstone silicic magmatism based on ion microprobe analysis of isotopically-zoned zircons. J Petrol 49:163-193

Bindeman IN, Gurenko AA, Sigmarsson O, Chaussidon M (2008c) Oxygen isotope heterogeneity and disequilibria of olivine phenocrysts in large volume basalts from Iceland: Evidence for magmatic digestion and erosion of Pleistocene hyaloclastites. Geochim Cosmochim Acta doi: 10.1016/j.gca.2008.06.010

Bindeman IN, Ponomareva VV, Bailey JC, Valley JW (2004) Volcanic arc of Kamchatka: a province with high-δ^{18}O magma sources and large-scale ^{18}O/^{16}O depletion of the upper crust. Geochim Cosmochim Acta 68: 841-865

Bindeman IN, Schmitt AK, Valley JW (2006a) U-Pb zircon geochronology of silicic tuffs from Timber Mt Caldera complex, Nevada: rapid generation of large volume magmas by shallow-level remelting. Contrib Mineral Petrol 152:649-665

Bindeman IN, Sigmarsson O, Eiler JM (2006b) Time constraints on the origin of large volume basalts derived from O-isotope and trace element mineral zoning and U-series disequilibria in the Laki and Grímsvötn volcanic system. Earth Planet Sci Lett 245:245-259

Bindeman IN, Valley JW (2001) Low-δ^{18}O rhyolites from Yellowstone: Magmatic evolution based on analyses of zircon and individual phenocrysts. J Petrol 42:1491-1517

Bindeman IN, Valley JW (2002) Oxygen isotope study of Long-Valley magma system: Isotope thermometry and role of convection. Contrib Mineral Petrol 144:185-205

Bindeman IN, Valley JW, Wooden JL, Persing HM (2001) Post-caldera volcanism: In situ measurement of U-Pb age and oxygen isotope ratio in Pleistocene zircons from Yellowstone caldera. Earth Planet Sci Lett 189:197-206

Bindeman IN, Watts KE, Schmitt AK, Morgan LA, Shanks PWC (2007) Voluminous low-δ^{18}O magmas in the late Miocene Heise volcanic field, Idaho: Implications for the fate of Yellowstone hotspot calderas. Geology 35:1019-1022

Bolz V (2001) The oxygen isotope geochemistry of zircon as a petrogenetic tracer in high temperature contact metamorphic and granitic rocks. MS Thesis, University of Edinburgh, Scotland.

Chacko T, Cole DR, Horita J (2001) Equilibrium oxygen, hydrogen and carbon isotope fractionation factors applicable to geologic systems. Rev Mineral Geochem 43:1-81

Charlier BLA, Wilson CJN, Lowenstern JB, Blake S, Van Calsteren PW, Davidson JP (2005) Magma generation at a large hyperactive silicic volcano (Taupo New Zealand) revealed by U-Th and U-Pb systematics in zircons. J Petrol 46:3-32

Chiba H, Chacko T, Clayton RN, Goldsmith JR (1989) Oxygen isotope fractionations involving diopside, forsterite, magnetite, and calcite: Application to geothermometry, Geochim Cosmochim Acta 53:2985-2995

Coleman DS, Gray W, Glazner AF (2004) Rethinking the emplacement and evolution of zoned plutons: Geochronologic evidence for incremental assembly of the Tuolumne Intrusive Suite, California. Geology 32:433-436

Cooper KM, Reid MR (2008) Uranium-series crystal ages. Rev Mineral Geochem 69:479-554

Costa F, Dohmen R, Chakraborty S (2008) Time scales of magmatic processes from modeling the zoning patterns of crystals. Rev Mineral Geochem 69:545-594

Costa F, Dungan M (2005) Short time scales of magmatic assimilation from diffusion modeling of multiple elements in olivine. Geology 33:837-840

Crank J (1975) The Mathematics of Diffusion, 2nd Edition. Oxford: Oxford University Press

Davidson JP, Tepley F III, Palacz Z, Meffan-Main S (2001) Magma recharge, contamination and residence times revealed by in situ laser ablation isotopic analysis of feldspar in volcanic rocks. Earth Planet Sci Lett 184:427-442

Davies GR, Halliday AN (1998) Development of the Long Valley rhyolitic magma system: Sr and Nd isotope evidence from glasses and individual phenocrysts. Geochim Cosmochim Acta 62:3561-3574

de Silva SL, Gosnold WD (2007) Episodic construction of batholiths: insights from the spatiotemporal development of an ignimbrite flare-up. J Volcanol Geotherm Res 167:320-335

Dorendorf F, Wiechert U, Worner G (2000) Hydrated sub-arc mantle: A source for the Klyuchevskoy volcano, Kamchatka/Russia. Earth Planet Sci Lett 175:69-86

Dufek J, Bergantz GW (2005) Lower crustal magma genesis and preservation: A stochastic framework for the evaluation of basalt-crust interaction. J Petrology 46: 2167-2195

Dungan MA, Davidson J (2004) Partial assimilative recycling of the mafic plutonic roots of arc volcanoes: An example from the Chilean Andes. Geology 32:773-776

Eiler JM (2001) Oxygen isotope variations in basaltic lavas and upper mantle rocks. Rev Mineral Geochem 43:319-364

Eiler JM, Valley JW, Baumgartner LP (1993) A new look at stable isotope thermometry. Geochim Cosmochim Acta 57:2571-2583

Elphick SC, Dennis PF, Graham CM (1986) An experimental-study of the diffusion of oxygen in quartz and albite using an overgrowth technique. Contrib Mineral Petrol 92:322-330

Elphick SC, Graham CM, Dennis PF (1988) An ion microprobe study of anhydrous oxygen diffusion in anorthite - a comparison with hydrothermal data and some geological implications. Contrib Mineral Petrol 100:490-495

Farquhar J, Chacko T, Frost BR (1993) Strategies for high-temperature oxygen-isotope thermometry - a worked example from the Laramie anorthosite complex, Wyoming, USA. Earth Planet Sci Lett 117:407-422

Farver JR (1989) Oxygen self-diffusion in diopside with application to cooling rate determinations. Earth Planet Sci Lett 92:386-396

Farver JR, Yund RA (1991) Oxygen diffusion in quartz: dependence on temperature and water fugacity. Chem Geol 90:55-70

Finney B, Turner S, Hawkesworth C, Larsen J, Nye C, George R, Bindeman I, Eichelberger J (2008) Magmatic differentiation at an island-arc caldera: Okmok volcano, Aleutian Islands, Alaska. J Petrol doi:10.1093/petrology/egn008

Friedman I, Lipman PW, Obradovich JD, Gleason JD, Christiansen RL (1974) Meteoric water in magmas. Science 184:1069-1072

Friedman I, O'Neil JR (1977) Compilation of stable isotope fractionation factors of geochemical interest (Geological Survey Professional Paper 440-KK. In: Data of Geochemistry, 6th Ed. Fleischer M (ed) Washington, D.C., U.S. Gov Print Off p. 1-12

Gansecki CA, Mahood GA, McWilliams MO (1996) ^{40}Ar/^{39}Ar geochronology of rhyolites erupted following collapse of the Yellowstone caldera, Yellowstone Plateau volcanic field: implications for crustal contamination. Earth Planet Sci Lett 142:91-107

Garcia MO, Ito E, Eiler JM (2008) Oxygen isotope evidence for chemical interaction of Kilauea historical magmas with basement rocks. J Petrol 49:757-769

Garcia MO, Ito E, Eiler JM, Pietruszka AJ (1998) Crustal contamination of Kilauea Volcano magmas revealed by oxygen isotope analyses of glass and olivine from Puu Oo eruption lavas. J Petrol 39:803-817

Ghiorso MS, Sack RO (1995) Chemical mass transfer in magmatic processes. 4. A revised and internally consistent thermodynamic model for the interpolation and extrapolation of liquid-solid equilibria in magmatic systems at elevated temperatures and pressures. Contrib Mineral Petrol 119:197-212

Glazner AF, Bartley JM, Coleman DS, Gray W, Taylor RZ (2004) Are plutons assembled over millions of years by amalgamation from small magma chambers? GSA Today 14:4-11

Grunder AL (1995) Material and thermal roles of basalt in crustal magmatism: A case study from eastern Nevada. Geology 23:952-956

Gurenko AA, Chaussidon M (2002) Oxygen isotope variations in primitive tholeiites of Iceland: evidence from a SIMS study of glass inclusions, olivine phenocrysts and pillow rim glasses. Earth Planet Sci Lett 205:63-79

Gurenko AA, Sobolev AV (2006) Crust-primitive magma interaction beneath neovolcanic rift zone of Iceland recorded in gabbro xenoliths from Midfell, SW Iceland. Contrib Mineral Petrol 151:495-520

Hanchar JM, Hoskin PWO (eds) (2003) Zircon. Reviews in Mineralogy and Geochemistry Vol. 53, Mineralogical Society of America, Washington, D.C.

Harmon RS, Hoefs J (1995) Oxygen-isotope heterogeneity of the mantle deduced from global O-18 systematics of basalts from different geotectonic settings. Contrib Mineral Petrol 120:95-114

Hildreth W, Christiansen RL, O'Neil JR (1984) Catastrophic isotopic modification of rhyolitic magma at times of caldera subsidence, Yellowstone Plateau Volcanic Field. J Geophys Res 89:8339-8369

Hildreth W, Halliday AN, Christiansen RL (1991) Isotopic and chemical evidence concerning the genesis and contamination of basaltic and rhyolitic magmas beneath the Yellowstone Plateau Volcanic Field. J Petrol 32:63-138

Hinton RW (1995) Ion microprobe analysis in geology. In: Microprobe Techniques in the Earth Sciences. PJ Potts, JFW. Bowlcs, SJB. Reed and MR Cave (eds). London, Chapman and Hall: 235-289

Hoefs J (2005) Stable Isotope Geochemistry, Springer, 5th edition

Ireland TR (1995) Ion microprobe mass spectrometry: techniques and applications in cosmochemistry, geochemistry, and geochronology. *In:* Advances in Analytical Geochemistry. Hyman M, Rowe MW (eds). London, JAI Press Inc. 2:1-118

Johnson CM, Beard BL, Albarede F (eds) (2004) Geochemistry of Non-Traditional Stable Isotopes. Rev Mineral Geochem, volume 55. Mineralogical Society of America

Kalamarides RI (1984) Kiglapait geochemistry VI: Oxygen isotopes. Geochim Cosmochim Acta 48:1827-1836

King EM, Valley JW (2001) The source, magmatic contamination, and alteration of the Idaho batholith. Contrib Mineral Petrol 142:72-88

Kita NT, Ikeda Y, Togashi S, Liu YZ, Morishita Y, Weisberg MK (2004) Origin of ureilites inferred from a SIMS oxygen isotopic and trace element study of clasts in the Dar al Gani 319 polymict ureilite. Geochim Cosmochim Acta 68: 213-4235

Krylov DP, Zagnitko VN, Hoernes S, Lugovaja IP, Hoffbauer R (2002) Oxygen isotope fractionations between zircon and water: Experimental determination and comparison with quartz-zircon calibrations. Eur J Mineral 14:849-853

Kyser TK, O'Neil JR, Carmichael ISE (1982) Genetic relations among basic lavas and ultramafic nodules: Evidence from oxygen isotope compositions. Contrib Mineral Petrol 81:88-102

Lackey JS, Valley JW, Hinke HJ (2006) Deciphering the source and contamination history of peraluminous magmas using delta 18-O of accessory minerals: examples from garnet-bearing plutons of the Sierra Nevada batholith. Contrib Mineral Petrol 151:20-44

Lackey JS, Valley JW, Saleeby JB (2005) Supracrustal input to magmas in the deep crust of Sierra Nevada batholith: Evidence from high-delta O-18 zircon. Earth Planet Sci Lett 235:315-330

Lanphere MA, Champion DE, Christiansen RL, Izett GA, Obradovich JD (2002) Revised ages for tuffs of the Yellowstone Plateau volcanic field: Assignment of the Huckleberry Ridge Tuff to a new geomagnetic polarity event. Geol Soc Am Bull 114:559-568

MacPherson GJ, Mittlefehldt DW, Jones JH, Simon SB, Papike JJ, Mackwell S (eds) (2008) Oxygen in the Solar System. Rev Mineral Geochem, volume 68. Mineralogical Society of America

Marsh BD (1998) On the interpretation of crystal size distributions in magmatic systems. J Petrol 39:553-599

Matthews A, Palin JM, Epstein S, Stolper EM (1994) Experimental study of $^{18}O/^{16}O$ partitioning between crystalline albite, albitic glass and CO_2 gas. Geochim Cosmochim Acta 58:5255-5266

Matthews A, Stolper EM, Eiler JM, Epstein S (1998) Oxygen isotope fractionation among melts, minerals and rocks. *In*: Proceedings of the 1998 Goldsmith Conference, Toulouse, pp. 971–972. London Mineralogical Society.

McBirney AR (2006) Igneous Petrology, 3rd Edition, Johns and Bartlett

Miller JS, Wooden JL (2004) Residence resorption and recycling of zircons in Devils Kitchen rhyolite Coso Volcanic Field, California. J Petrol 45:2155-2170

Mills JG Jr., Saltoun BW, Vogel TA (1997) Magma batches in the Timber Mountain magmatic system SW Nevada volcanic field Nevada, USA. J Volcanol Geotherm Res 78:185-208

Morgan LA, McIntosh WC (2005) Timing and development of the Heise volcanic field, Snake River Plain, Idaho, western USA. Geol Soc Am Bull 117:288-306

Morishita Y, Giletti BJ, Farver JR (1996) Volume self-diffusion of oxygen in titanite: Geochem J 30:71-79

Muehlenbachs K, Anderson AT, Sigvaldasson GE (1974) Low-$\delta^{18}O$ basalts from Iceland. Geochim Cosmochim Acta 38:577-588

Muehlenbachs K (1998) The oxygen isotopic composition of the oceans, sediments and the seafloor. Chem Geol 145:263-273

Obradovich JD (1992) K-Ar ages of Yellowstone Volcanic Field volcanic rocks. US Geol Surv Open File Rep 92408

Page FZ, Ushikubo T, Kita NY, Riciputi LR, Valley JW (2007) High precision oxygen isotope analysis of picogram samples reveals 2-µm gradients and slow diffusion in zircon. Am Mineral 92:1772-1775

Palin JM, Stolper EM, Epstein S (1996) Oxygen isotope partitioning between rhyolitic glass/melt and CO_2: an experimental study at 550–950 degrees C and 1 bar. Geochim Cosmochim Acta 60:1963-1973

Portnyagin M, Bindeman IN, Hoernle K, Hauff F (2007) Geochemistry of primitive lavas of the Central Kamchatka Depression. *In*: Magma Generation at the Edge of the Pacific Plate. AGU Monograph series 172 "Volcanism and Subduction: the Kamchatka Region" p. 199-239

Ramos FC, Tepley FJ III (2008) Inter- and intracrystalline isotopic disequilibria: techniques and applications. Rev Mineral Geochem 69:403-443

Reid MR (2003) Timescales of magma transfer and storage in the crust. *In:* The Crust, Treatise on Geochemistry, Vol. 3. Rudnick RL (ed) Elsevier, Oxford, UK. p 167-193

Reid MR Coath CD (2000) In situ U-Pb ages of zircons from the Bishop Tuff: No evidence for long crystal residence times. Geology 28:443-446

Riciputi LR, Paterson BA, Ripperdan RL (1998) Measurement of light stable isotope ratios by SIMS: Matrix effects for oxygen, carbon, and sulfur isotopes in minerals. Int J Mass Spectrom 178:81-112

Rumble D, Farquhar J, Young ED, Christensen CP (1997) In situ oxygen isotope analysis with an excimer laser using F2 and BrF5 reagents and O-2 gas as analyte. Geochim Cosmochim Acta: 61:4229-4234

Ryerson FJ, Durham WD, Cherniak DJ (1989) Oxygen diffusion in olivine- effect of oxygen fugacity and implications for creep. J Geophys Res Solid Earth 94:4105-4118

Schmitt AK, Grove M, Harrison TM, Lovera O, Hulen JB, Walters M (2003a) The Geysers - Cobb Mountain Magma System California (Part 1): U-Pb zircon ages of volcanic rocks conditions of zircon crystallization and magma residence times. Geochim Cosmochim Acta 67:3423-3442

Schmitt AK, Grove M, Harrison TM, Lovera O, Hulen JB, Walters M (2003b) The Geysers-Cobb Mountain Magma System California (Part 2): Timescales of pluton emplacement and implications for its thermal history. Geochim Cosmochim Acta 67:3443-3458

Sharp Z (2006) Principles of Stable isotope Geochemistry. Prentice Hall

Sharp ZD (1990) A laser-based microanalytical method for the in situ determination of oxygen isotope ratios of silicates and oxides. Geochim Cosmochim Acta 54:1353-1357

Sigmarsson O, Condomines M, Gronvold K, Thordarson T (1991) Extreme magma homogeneity in the 1783-84 Lakagigar eruption: origin of a large volume of evolved basalt in Iceland. Geophys Res Lett 18:2229-2232

Simakin AG, Bindeman IN (2008) Evolution of crystal sizes in the series of dissolution and precipitation events in open magma systems. J Volcanol Geotherm Res doi: 10.1016/j.jvolgeores.2008.07.012

Simon JI, Reid MR (2005) The pace of rhyolite differentiation and storage in an 'archetypical' silicic magma system Long Valley. Earth Planet Sci Lett 235:123-140

Spera FJ, Bohrson WA (2004) Open-system magma chamber evolution: an energy-constrained geochemical model incorporating the effects of concurrent eruption, recharge, variable assimilation and fractional crystallization. J Petrol 45:2459-2480

Streck MJ (2008) Mineral textures and zoning as evidence for open system processes. Rev Mineral Geochem 69:595-622

Taylor HP (1968) The oxygen isotope geochemistry of igneous rocks. Contrib Mineral Petrol 19:1-71

Taylor HP Jr. (1980) The effects of assimilation of country rocks by magmas on O^{18}-O^{16} and Sr^{87}/Sr^{86} systematics in igneous rocks. Earth Planet Sci Lett 47:243-254

Taylor HP Jr. (1986) Igneous rocks: II. Isotopic case studies of circumpacific magmatism. Rev Mineral 16:273-316

Taylor HP Jr., Sheppard SMF (1986) Igneous rocks: I. Processes of isotopic fractionation and isotopic systematics. Rev Mineral 16:227-272

Tefend KS, Vogel TA, Flood TP, Ehrlich R (2007) identifying relationships among silicic magma batches by polytopic vector analysis: A study of the Topopah Spring and Pah Canyon Ash-flow Sheets of the Southwest Nevada Volcanic Field. J Volcanol Geotherm Res 167:198-211

Tepley FJ, Davidson JP (2003) Mineral-scale Sr-isotope constraints on magma evolution and chamber dynamics in the Rum layered intrusion, Scotland. Contrib Mineral Petrol 145:628-641

Tepley FJ, Davidson JP, Clynne MA (1999) Magmatic interactions as recorded in plagioclase phenocrysts of Chaos Crags, Lassen Volcanic Center, California. J Petrol 40:787-806

Thordarson T, Self S (2003) Atmospheric and environmental effects of the 1783-1784 Laki eruption: A review and reassessment. J Geophys Res.-Atm D1 Article Number: 4011

Trail D, Bindeman IN, Watson EB (2008) Experimental determination of quartz-zircon oxygen isotope fractionation. (in prep)

Urey HC (1947) The thermodynamic properties of isotopic substances. J Chem Soc (Lond) 1947:562-581

Valley JW (2001) Stable isotope thermometry at high temperatures. Rev Mineral Geochem 43:365-413

Valley JW (2003) Oxygen isotopes in zircon. Rev Mineral Geochem 53:343-385

Valley JW, Bindeman IN, Peck WH (2003) Calibration of zircon-quartz oxygen isotope fractionation. Geochim Cosmochim Acta 67:3257-3266

Valley JW, Cole D (eds) (2001) Stable Isotope Chemistry. Rev Geochem Mineral, volume 43. Mineralogical Society of America

Valley JW, Kinny PD, Schulze DJ, Spicuzza MJ (1998) Zircon megacrysts from kimberlite: oxygen isotope variability among mantle melts. Contrib Mineral Petrol 133:1-11

Valley JW, Kitchen N, Kohn MJ, Niendorf CR, Spicuzza MJ (1995) UWG-2, a garnet standard for oxygen isotope ratio: strategies for high precision and accuracy with laser heating. Geochim Cosmochim Acta. 59:5223-5231

Valley JW, Lackey JS, Cavosie AJ, Clechenko C, Spicuzza MJ, Basei MAS, Bindeman IN, Ferreira VP, Sial AN, King EM, Peck WH, Sinha AK, Wei CS (2005) 4.4 billion years of crustal maturation: oxygen isotope ratios of magmatic zircon. Contrib Mineral Petrol 150:561-580

Valley JW, Taylor HP Jr., O'Neil JR (eds) (1986) Stable Isotopes in High Temperature Geological Processes. Rev Geochem Mineral, volume 16. Mineralogical Society of America

Vazquez JA, Reid MR (2002). Time scales of magma storage and differentiation of voluminous high-silica rhyolites at Yellowstone caldera, Wyoming. Contrib Mineral Petrol 144:274-285

Wallace PJ, Anderson AT, Davis AM (1999) Gradients in H_2O, CO_2, and exsolved gas in a large-volume silicic magma system: Interpreting the record preserved in melt inclusions from the Bishop Tuff. J Geophys Res 104:20097-20122

Wang ZG, Kitchen NE, Eiler JM (2003) Oxygen isotope geochemistry of the second HSDP core. Geochim Geophys Geosystems 4 Article Number: 8712

Wark DA, Spear FS, Cherniak DJ, Watson EB (2007) Pre-eruption recharge of the Bishop magma system. Geology 35:235-238

Wark DA, Watson EB (2006) TitaniQ: A titanium-in-quartz geothermometer. Contrib Mineral Petrol 152:743-754

Watson EB (1996) Dissolution, growth and survival of zircons during crustal fusion: Kinetic principles, geological models and implications for isotopic inheritance. Trans R Soc Edinburgh: Earth Sci 87:43-56

Watson EB, Cherniak DJ (1997) Oxygen diffusion in zircon. Earth Planet Sci Lett 148:527-544

Williams DA, Kadel SD, Greeley R, Lesher CM, Clynne MA (2004) Erosion by flowing lava: geochemical evidence in the Cave Basalt, Mount St. Helens, Washington. Bull Volcanol 66:168-181

Williams DA, Kerr RC, Lesher CM, Barnes SJ (2001) Analytical/numerical modeling of komatiite lava emplacement and thermal erosion at Perseverance, Western Australia. J Volcanol Geotherm Res 110:27-55.

Wolff JA, Ramos FC (2003) Pb isotope variations among Bandelier Tuff feldspars: no evidence for a long-lived silicic magma chamber. Geology 31:533-536

Young ED, Fogel ML, Rumble D, Hoering TC (1998) Isotope-ratio-monitoring of O_2 for microanalysis of O^{18}/O^{16} and O^{17}/O^{16} in geological materials. Geochim Cosmochim Acta 62:3087-3094

Young ED, Kuramoto K, Marcus RA, Yurimoto H, Jacobsen SB (2008) Mass-independent oxygen isotope variation in the solar nebula. Oxygen in the solar system: Rev Mineral Geochem 68:187-218

Zhang YX, Walker D, Lesher CE (1989) Diffusive crystal dissolution. Contrib Mineral Petrol 102:492-513

Zhao Z, Zheng Y-F (2003) Calculation of oxygen isotope fractionation in magmatic rocks. Chem Geol 193:59-80

Zheng YF (1993a) Calculation of oxygen isotope fractionation in anhydrous silicate minerals. Geochim Cosmochim Acta 57:1079-1091

Zheng YF (1993b) Calculation of oxygen isotope fractionation in hydroxyl-bearing silicates. Earth Planet Sci Lett 120:247-263

Uranium-series Crystal Ages

Kari M. Cooper

Department of Geology
University of California
Davis, California, 95616, U.S.A.

kmcooper@geology.ucdavis.edu

Mary R. Reid

Department of Geology
Northern Arizona University
Flagstaff, Arizona, 86011, U.S.A.

mary.reid@nau.edu

INTRODUCTION

As the collected papers in this volume aptly demonstrate, chemical and textural records contained within crystals have become increasingly useful tools for understanding the evolution of magmatic systems. An essential component of studies of the dynamics of magmatic systems is the ability to place the thermal, physical, and compositional evolution of magmas in a temporal context. Studies of the kinetics of crystal growth and reaction, and of diffusive re-equilibration of crystals and melts, can provide information about the duration of magmatic processes such as magma ascent (Rutherford 2008), crystal growth (Hammer 2008), and residence of crystals in melts other than the ones in which they crystallized (Costa 2008). However, the only method of extracting information from young crystals about the absolute age of magmatic processes is isotopic dating. The timescales of many sub-volcanic magmatic processes such as magma transport, differentiation, crystallization, and storage within the crust, appear to be commensurate with the half-lives of uranium-series (U-series) nuclides, and therefore this isotopic system offers the greatest potential to quantify timescales and rates of magmatic processes. We focus here on crystal dating using the decay products of ^{238}U and ^{235}U, including ^{238}U-^{230}Th-^{226}Ra-^{210}Pb disequilibria, ^{235}U-^{231}Pa disequilibria, and U-Pb dating of young crystals. In this chapter, we review the principles of U-series dating of crystals, the types of information that have been gained so far from U-series dating, and the potential for combining U-series crystal ages with other crystal-scale and magma-scale information to unravel the dynamics of magmatic systems.

Principles of U-series dating

U-Pb dating is based on the accumulation of ^{206}Pb and ^{207}Pb over time due to decay of ^{238}U and ^{235}U, respectively, and is conceptually similar to most isotopic dating techniques. U-Pb dating of zircon, both by TIMS and *in situ* techniques, has been reviewed recently (Parrish and Noble 2003; Ireland and Williams 2003; Kosler and Sylvester 2003), and we refer readers to these papers for fundamentals of the techniques. Here, we are concerned primarily with U-Pb dating of young zircon (<1 Ma), and the complications inherent in dating young crystals and techniques to address these factors are detailed in the second section. U-series disequilibrium dating is less widely-applied and often is less intuitive; therefore, we briefly review here the principles relevant to U-series dating of crystals in magmatic systems. For a comprehensive view of U-series geochemistry, we refer the reader to two earlier reviews of the principles of U-series geochemistry and its application to geological problems (Ivanovich and Harmon 1992; Bourdon et al. 2003a).

^{238}U decays to ^{206}Pb through a series of intermediate daughter nuclides, including ^{234}U, ^{230}Th, ^{226}Ra, and ^{210}Pb; similarly, ^{235}U decays to ^{207}Pb through a series of intermediate nuclides, including ^{231}Pa (Fig. 1). Each of the intermediate daughter nuclides is both radiogenic and radioactive, and half-lives of the intermediate daughters range from seconds to 245 k.y. (Fig. 1). The currently-accepted values for half-lives of the daughters discussed in this chapter are compiled in Table 1. Left undisturbed for a sufficient period of time, any U-bearing system will attain a state known as secular or radioactive equilibrium, where the number of decays per unit time (i.e., the activity) of each of the intermediate daughters is equal to that of the parent isotope of uranium. Secular equilibrium is in fact a dynamic steady state, where the declining activity of each daughter nuclide tracks the slowly decreasing activity of ^{238}U or ^{235}U. Once attained, a system will remain in secular equilibrium until some process fractionates one or more of the daughter nuclides from their parents. Because the intermediate daughters have a wide range of geochemical behaviors in natural systems, geological events will, more often than not, produce disequilibrium in at least some parent-daughter pairs in the decay chain. In the case of magmatic systems, processes such as partial melting, assimilation of wall rocks, crystallization of minerals, and degassing have all been shown to produce disequilibria in the various phases present. After disequilibria have been produced, the system (and each sub-system; e.g., individual phases) will

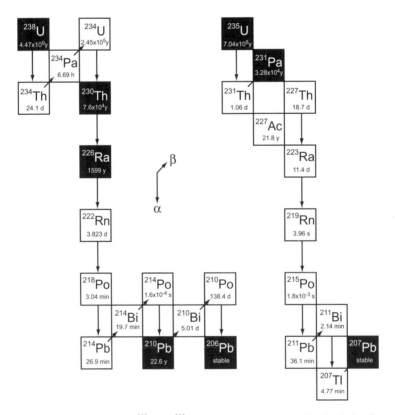

Figure 1. Schematic decay chains for ^{238}U and ^{235}U. The isotopes used in the U-series dating discussed in this chapter are shown as black boxes. The half-lives shown are those compiled in Bourdon et al. (2003).

Table 1. Half-lives of U-series nuclides discussed in this chapter.

Parent-daughter pair	Half-life of daughter (yr)	Approximate useful timescale (yr)
^{238}U-^{230}Th	75,690 ± 230	1000-400,000
^{230}Th-^{226}Ra	1599 ± 4	100-10,000
^{226}Ra-^{210}Pb	22.6 ± 0.1	1-100
^{235}U-^{231}Pa	32,760 ± 220	500-175,000

Half-lives are those compiled in Bourdon et al. (2003b). Note that useful timescale for each parent-daughter pair is only a broad estimate and the accessible timeframe can vary depending on the initial disequilibria and parent/daughter ratio.

return to secular equilibrium through decay of excess daughter or ingrowth of daughter that is in deficit relative to the activity of the parent. In all of the parent-daughter pairs discussed here, the half-life of the daughter is significantly shorter than that of the parent; in this case, secular equilibrium will be attained at a rate governed by the half-life of the daughter product. If the initial abundance of the daughter after a disturbance is known or can be estimated, the degree of disequilibrium measured at present can be used to calculate the time since disequilibrium was generated. The timescale over which disequilibrium can be detected—and therefore the useful time range for each parent-daughter pair for dating—depends on the initial degree of disequilibrium and the accuracy of the measurement of the abundances of parent and daughter. A useful rule of thumb is that disequilibria can be detected for ~5 times the half-life of the daughter (Table 1); however, with the increased precision of modern mass spectrometric analyses, and when analyzing phases such as zircon which are expected to have very large parent/daughter ratios and therefore extreme disequilibria at the time of crystallization, dating in favorable cases can be extended to significantly longer timescales. A lower limit to the range of time accessible with each parent-daughter pair is given by the time required for daughter abundances to deviate measurably from the initial conditions, and therefore is also a function of the analytical uncertainty and the confidence with which the initial daughter abundance can be estimated. Approximate time ranges accessible with each parent-daughter pair discussed here are compiled in Table 1.

The mathematics governing parent-daughter abundances in a decay chain are presented in some detail in Ivanovich and Harmon (1992), and we will not reproduce the full derivation of the equations here. For decay of a parent to daughter where the daughter is also radioactive, the general equation governing the abundance of the daughter product is given by

$$N_2 = \left(\frac{\lambda_1}{\lambda_2 - \lambda_1}\right) N_1^0 \left(e^{-\lambda_1 t} - e^{-\lambda_2 t}\right) + N_2^0 e^{-\lambda_2 t} \qquad (1)$$

where N_2 is the number of atoms of the daughter present at time t, N_2^0 is the number of atoms of the daughter present initially (t_0), N_1^0 is the number of atoms of the parent present initially, and λ_1 and λ_2 are the decay constants of the parent and daughter, respectively. Assuming that the time of interest is short with respect to the half-life of the parent and that the half-life of the parent is significantly longer than the half-life of the daughter, and making use of the relation that the decay rate (activity) for a given nuclide is equal to $N\lambda$, Equation (1) can be simplified to give

$$(N_2) = (N_1)^0 \left(1 - e^{-\lambda_2 t}\right) + (N_2)^0 e^{-\lambda_2 t} \qquad (2)$$

where parentheses around the chemical symbols, by convention, denote activities (the number of decay events per unit time).

U-series dating of crystals

U-series measurements can thus be used to date crystals in igneous systems, and can provide information about absolute ages of events or processes that occurred within the past few years to decades (e.g., ^{226}Ra-^{210}Pb disequilibria) to thousands of years to hundreds of thousands of years (^{238}U-^{230}Th disequilibria). U-series dating of crystals can be applied through bulk mineral separates for major phases, which include material from hundreds to thousands of grains, or through *in situ* analyses of sub-grain domains in accessory minerals with high concentrations of U or Th. We present information about the details of requirements for sample size, other analytical considerations, and techniques for calculating ages in the second section (*"Analytical and practical considerations"*). Once an analysis has been obtained, the interpretation of a U-series date for a crystal or bulk mineral aggregate depends on the geologic context, including the extent to which the analyzed grains represent a simple crystallization history or more complex histories involving protracted crystal growth, recycling of material from earlier phases of magmatic activity and/or incorporation of crustal material during magma storage. U-series crystal ages have been used thus far to examine the timescales and dynamics of a wide variety of magmatic processes, including magma storage within the crust, crystallization and differentiation, recycling of cogenetic crystal material from young plutonic bodies or crystal mush zones, thermal and chemical evolution of magmas during storage, and dating eruptions, among other applications. We discuss these various interpretations of U-series crystal ages, and the information that can be gained in different scenarios of varying complexity, in the following sections. As is increasingly obvious from studies such as those reviewed in the third section (*"What do crystal ages tell us?"*), coupling U-series crystal ages with other crystal-scale information has the potential to provide unique insights into the evolution of magmatic systems, and in the fourth section (*"Resolving ages from complex crystal populations: Incorporating other observations"*) we discuss various combinations of types of information and how they may provide insights well beyond what is available from crystal ages or crystal-scale chemical information in isolation. In the final section (*"Key questions and future directions"*), we discuss the potential to capitalize on and expand on these types of combined studies, and propose some areas that are likely to be fruitful avenues for further research.

ANALYTICAL AND PRACTICAL CONSIDERATIONS

Introduction

The ability to measure U-series ages of minerals depends ultimately on the number of atoms of each nuclide present in the analyzed volume relative to the sensitivity of the analytical technique, which in turn reflects a combination of the concentration of U-series elements within the mineral of interest, the abundance of that mineral in a rock (and therefore the ability to purify a mineral separate of sufficient size for analysis), and the analytical technique used. Measurements of many of the U-series nuclides in whole-rock samples are now routine, but our ability to measure crystal ages is still largely limited by analytical considerations. The analytical challenges of U-series mineral analyses largely result from two factors. First, the U-series elements are incompatible in most major minerals in igneous rocks, requiring significantly larger sample sizes for analysis of minerals relative to whole-rocks or glass. Second, the abundance of each nuclide at radioactive equilibrium is inversely proportional to its half-life, which means that the nuclides with shorter half-lives (e.g., ^{226}Ra or ^{210}Pb) are orders of magnitude less abundant in magmas than are ^{238}U or ^{230}Th. The result in practice is that while *in situ* measurements of U-series nuclides are now possible by secondary ionization mass spectrometry (SIMS) and potentially by laser ablation- inductively coupled mass spectrometry

(LA-ICP-MS; Stirling et al. 2000; Bernal et al. 2005), they are limited to trace phases with high concentrations of U and/or Th. Measurement of ^{238}U-^{230}Th-^{226}Ra-^{210}Pb disequilibria in major phases requires preparation of a bulk mineral separate and measurement by thermal ionization mass spectrometry (TIMS), or multicollector inductively coupled mass spectrometry (MC-ICP-MS; also known as plasma ionization multicollector mass spectrometry or PIMMS), or, in the case of ^{210}Pb, alpha spectrometry.

This section covers the practical side of U-series crystal dating; here, we review the requirements for successful analysis in terms of sample size and preparation for both bulk separate and *in situ* analyses. We then review the methods (and time investment required) for analysis by TIMS, MC-ICP-MS, and SIMS, followed by a discussion of methods for calculating ages from the different types of data, along with some of the complications that can arise and methods for addressing them.

Practical limitations: sample preparation and analysis

Sample size requirements. Figure 2 illustrates concentrations of U, Th, and Ra, and ^{210}Pb activities expected for different minerals precipitated from magmas with a range of U concentrations. Many variables influence the exact abundances of parent and daughter nuclides present in a mineral and therefore the amount required for analysis (e.g., age of the mineral separate, the degree of U-Th or Th-Ra disequilibria in the magma, the mineral major-element composition and temperature of crystallization, which in turn affect the partitioning behavior, etc.), but this figure can be used to estimate in a broad sense how large a mineral sample would be required for analysis of a given element. Partition coefficients used to construct Figure 2 are given in Table 2. *In situ* measurements require the highest concentrations, and are only possible for trace phases where U and/or Th are compatible (e.g., zircon, allanite). Although concentrations of U and Th in apatite may be high enough to analyze ^{238}U-^{230}Th disequilibria by SIMS, uranium and thorium will not be greatly fractionated during apatite crystallization, which would lead to difficulties in correcting for initial ^{230}Th present in the crystals and therefore imprecise ages. ^{238}U-^{230}Th dating of bulk mineral separates requires ~10 ng of Th and U, and is feasible for most of the minerals shown.

Current *in situ* techniques are not sensitive enough to allow analysis of radium; even analysis of ^{230}Th by *in situ* techniques is only feasible for high-U phases such as zircon, where ^{230}Th abundances will be high. Radium abundances in igneous rocks are generally 6-7 orders of magnitude lower than elemental abundances of Th and U. However, given the relatively high ionization efficiency of Ra by TIMS (and similar although somewhat lower total efficiency by MC-ICP-MS), it is possible to measure radium concentrations in samples containing as little as only a few femtograms of Ra, and measurements of ~10 fg generally produce good results (albeit with somewhat larger uncertainties than is typical for whole-rock samples, where a total of tens to hundreds of femtograms of Ra are analyzed). The calculations of concentrations necessary for 0.1 g-5 g mineral separates shown in Figure 2c assume that a total of 10 fg of radium is analyzed. Minerals where radium is relatively less incompatible (plagioclase, amphibole, and phlogopite) or even compatible (K-feldspar) are attractive targets for ^{230}Th-^{226}Ra dating and feldspar was one of the first minerals dated using Ra-Th (e.g., Reagan et al. 1992). Relatively few analyses of amphibole have been reported in the literature (e.g., Turner et al. 2003a,b), and to our knowledge no analyses of mica, but these minerals have significant potential in future dating applications. As is apparent from Figure 2c, concentrations of Ra in olivine, clinopyroxene, orthopyroxene, and zircon are expected to be extremely low at the time of crystallization, and Ra in pure separates of very young minerals could not reliably be measured. However, because of the high Th/Ra ratios in these minerals, ^{226}Ra ingrowth will be rapid, and concentrations of Ra expected in equilibrium with Th in the same minerals are several orders of magnitude higher (Fig. 2). Thus the age of the mineral being analyzed will have a significant effect on the expected concentration and therefore on the amount required

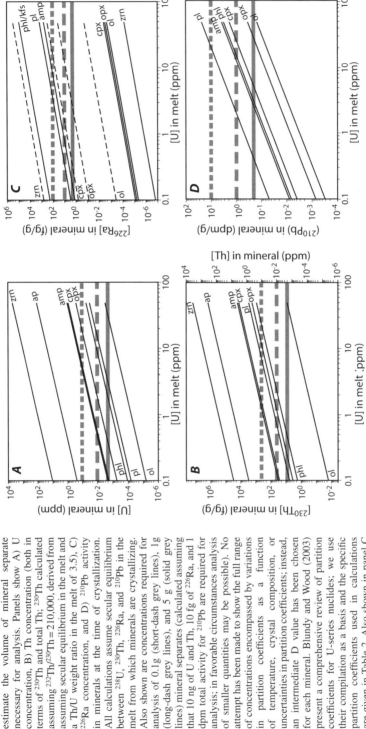

Figure 2. Plots of U concentration vs. daughter concentration in various minerals, intended to be used as a way to broadly estimate the volume of mineral separate necessary for analysis. Panels show A) U concentration, B) Th concentration (both in terms of ^{230}Th and total Th; ^{230}Th calculated assuming ^{232}Th/^{230}Th = 210,000, derived from a Th/U weight ratio in the melt of 3.5), C) ^{226}Ra concentration, and D) ^{210}Pb activity in minerals at the time of crystallization. All calculations assume secular equilibrium between ^{238}U, ^{230}Th, ^{226}Ra, and ^{210}Pb in the melt from which minerals are crystallizing. Also shown are concentrations required for analysis of 0.1g (short-dash grey lines), 1g (long-dash grey lines), and 5 g (solid grey lines) mineral separates (calculated assuming that 10 ng of U and Th, 10 fg of ^{226}Ra, and 1 dpm total activity for ^{210}Pb are required for analysis; in favorable circumstances analysis of smaller quantities may be possible). No attempt has been made to show the full range of concentrations encompassed by variations in partition coefficients as a function of temperature, crystal composition, or uncertainties in partition coefficients; instead, an intermediate D value has been chosen for each mineral. Blundy and Wood (2003) present a comprehensive review of partition coefficients for U-series nuclides; we use their compilation as a basis and the specific partition coefficients used in calculations are given in Table 2. Also shown in panel C (dashed lines with labels) are concentrations of Ra in olivine, opx, cpx, and zircon at equilibrium with ^{230}Th in the minerals.

Table 2. Partition coefficients used to construct Figure 2.

Mineral	D_U	D_{Th}	D_{Ba}	D_{Ra}	D_{Ra}/D_{Ba}	D_{Pb}
plagioclase	6.0×10^{-4}	0.003	0.42	0.079	0.19	0.61
K-feldspar, Or_{50}	—	—	13	5	0.38	—
amphibole	0.019	0.018	0.41	0.033	0.08	0.10
phlogopite	1×10^{-3}	1×10^{-3}	5	5	1	0.1
clinopyroxene	0.017	0.016	1.5×10^{-5}	2.0×10^{-7}	0.01	0.05
orthopyroxene	0.005	0.002	1.2×10^{-5}	1.8×10^{-7}	0.015	0.009
olivine	6.0×10^{-5}	9.5×10^{-6}	1.0×10^{-5}	1.2×10^{-7}	0.02	0.004
apatite	2.21	1.68	0.27	—	—	—
zircon	100	16.7	0.004	4.0×10^{-9}	1.0×10^{-6}	—

All partition coefficients derived from values or equations in Blundy and Wood (2003) unless otherwise noted; Ra partition coefficients calculated using Ba as a proxy as described for the different minerals in Blundy and Wood (2003). Dashes where no data are available. Calculations for Ba, Ra, and Pb in plagioclase assume An_{50} at 950 °C. U, Th, and Pb partitioning data for amphibole from Tiepolo et al. (2000). U, Th, and Pb in phlogopite from LaTourrette et al. (1995). D_{Ba} in phlogopite from Blundy and Wood (2003) Figure 22, corresponding to ~900 °C. D_{Ba} in clinopyroxene from calculations in Cooper et al. (2001). D_{Ba} in orthopyroxene estimated using experimental data for Sr reported in Blundy and Wood (2003) and their estimate that D_{Ba} is 50-100 times lower. D_{Ba} for olivine chosen to be within the range reported in the GERM database, yet not higher than D_{Ba} for pyroxene. D_U, D_{Th}, and D_{Ba} for apatite are averages of the data from empirical studies of Mahood and Stimac (1990) and Luhr et al. (1984); we did not use the Ds of Bea et al. (1994) in the calculations for U and Th because they are an order of magnitude higher than the other two estimates.

for analysis. Several studies have successfully analyzed ^{230}Th-^{226}Ra disequilibria in pyroxene (e.g., Volpe and Hammond 1991; Cooper et al. 2001; Turner et al. 2003a,b). When analyzing a given mineral separate for ^{238}U-^{230}Th-^{226}Ra disequilibria, the concentration of radium is generally the limiting factor dictating sample size.

The very short half-life of ^{210}Pb (22.6 yr) and incompatible behavior of Pb in most rock-forming minerals combine to produce extremely low abundances of ^{210}Pb in most rock-forming minerals. Therefore, ^{210}Pb in minerals remains one of the most challenging of the measurements of U-series nuclides, and generally requires preparation of larger samples than is required for other analyses. Analysis of ^{210}Pb generally requires ~1-5 g of mineral separate, in addition to that required for U-Th-Ra analyses. Nevertheless, with alpha counting by ^{210}Pb proxy, ^{210}Pb activity can be analyzed to better than 1% accuracy (1 sigma; e.g., Reagan et al. 2006).

Sample preparation for SIMS. It is possible to analyze accessory phases directly in rocks and thin sections ("in situ" *sensu stricto*) but it is more efficient to cluster a number of grains in a single mount. Accessory phases like zircon can be concentrated from crushed rock material using conventional heavy liquid and magnetic separation techniques. Alternatively, an HF-resistant mineral like zircon can be isolated by selective dissolution of the host rocks. If applied exclusively to glass shards, the latter method is a potential means of isolating only zircons that are not included within phenocrysts (e.g., Bindeman et al. 2006). Grains can be handpicked from the concentrate, mounted onto a surface (typically a piece of sticky tape) along with standards, and fixed in epoxy. Most workers then polish the mount using Al_2O_3 or diamond paste (0.25 to 1 micrometer diameter) so that grain interiors are exposed. Analyses of unpolished zircon rims can be performed by placing the faces of zircon grains flush to the mount surface (Reid and Coath 2000; Bindeman et al. 2006). This permits a spatial resolution of near-rim heterogeneity that is an order of magnitude lower than conventional spot analyses.

Depth profiling either by continuous analysis (Vazquez 2004; Reid 2008) or by incrementally polishing samples to discrete depth intervals (Bindeman et al. 2006) can also be performed.

Once the grain mount is prepared, the grains are mapped in reflected light and/or by backscattered electron (BSE) imaging. It is advantageous to also characterize internal compositional heterogeneities and/or included mineral/melts before analysis, using either cathodoluminesence (CL; zircon) or BSE (allanite, chevkinite). Either before or after grain mapping, the mount is ultrasonically cleaned—sometimes in HCl—to remove surface Pb contamination and a thin coat of gold (10-100 nm) is then applied to minimize charging of the sample surface during ion beam sputtering. Some studies (e.g., Bacon et al. 2000; Bacon and Lowenstern 2005; Bindeman et al. 2006) focused their studies on U-rich zircons that had been pre-selected using BSE, CL, or scanning ion imaging. Others focused on grains that were relatively small and homogeneous (Sano et al. 2002). These approaches maximize the age resolution by minimizing the uncertainties for ^{206}Pb and ^{230}Th and the chances that the ion beam averages multiple age domains, but could favor a specific zircon population to the exclusion of others.

Preparation of mineral separates for bulk analyses and chemical separation. Preparation of mineral separates of 0.1-5 g is generally sufficient for U-series dating of minerals, with larger mineral separates required for ^{226}Ra and ^{210}Pb analysis and/or for mafic samples and smaller mineral separates required for ^{238}U-^{230}Th analysis and/or silicic samples. In practice, the amount of time required to produce the mineral separate will also depend on the purity required for the mineral separate. Most isochron dating of minerals (including ^{238}U-^{230}Th dating) can tolerate the inclusion of a significant fraction (~10%) of impurities in the mineral separates, because the effect will be to move data points closer together along the same isochron line while keeping the slope and therefore the calculated age the same. However, as detailed below, accurate ^{230}Th-^{226}Ra and ^{226}Ra-^{210}Pb dating requires correction of the data for the effects of impurities in the mineral separate, and therefore it is generally necessary to have mineral separates that contain at most a few percent impurities. Therefore, in addition to the requirement of larger sample sizes, ^{230}Th-^{226}Ra and ^{226}Ra-^{210}Pb dating require mineral separates of higher purity, and are thus quite time-consuming.

Once mineral separates have been prepared, chemical separation of the elements of interest requires a significant effort, again scaled by the technique and specific nuclides analyzed. Chemical separation and mass spectrometric techniques for U-series nuclides have recently been reviewed by Goldstein and Stirling (2003), and we refer readers to that paper for analytical details. Analytical procedures vary from one lab to another; we summarize here information from several labs which have produced data on bulk mineral separates (Turner et al. 2003b; Rogers et al. 2004; Tepley et al. 2006; Cooper and Donnelly 2008). In brief, ^{238}U-^{230}Th analyses alone of mineral separates generally require dissolution, sample spiking and equilibration, followed by 1-4 ion-exchange column steps; all told, dissolution, chemistry and preparation for mass spectrometry requires on the order of two to three weeks per batch of 8-10 samples. Radium and ^{210}Pb analyses require additional effort in chemical separation. However, most labs have adopted techniques that allow separation of U, Th, and Ra from the same sample, which increases efficiency in chemical processing (e.g., Goldstein and Stirling 2003; Turner et al. 2003a,b; Rogers et al. 2004; Tepley et al. 2006; Cooper and Donnelly 2008). Radium separation for mineral separates requires an additional 4-5 ion-exchange column steps beyond what is required for U-Th preparation, and given the increased sample size requirements (which lead to large columns for the initial steps), the total process for sample dissolution, spiking and equilibration, and column chemistry, for U-Th-Ra analyses requires approximately 4-6 weeks for a batch of samples.

To date, no techniques have been developed to allow separation of ^{210}Pb from the same aliquot used for U-Th-Ra measurements, so analysis of ^{226}Ra-^{210}Pb disequilibria requires a mineral separate that is approximately twice as large as analysis of ^{238}U-^{230}Th-^{226}Ra alone.

Column chemistry for Pb requires one ion-exchange column and preparation for analysis by alpha spectrometry (autoplating) requires approximately 6-12 hours per sample, which totals approximately 1-2 weeks for sample dissolution, chemistry, autoplating, and counting per batch of samples.

Mass spectrometry: in situ analyses. Many accessory phases have sufficiently high U and/or Th contents so that domains within individual minerals can be dated by *in situ* methods, especially those of secondary ion mass spectrometry. Ages may be based on the magnitude of ^{238}U-^{230}Th disequilibrium or on the amount of in-grown ^{206}Pb, the ultimate daughter of ^{238}U. Recently, protocols for ^{235}U-^{231}Pa dating of zircon have been developed (Schmitt 2007). *In situ* dating of young accessory phases has mainly focused on zircon and has been performed at UCLA (e.g., Reid et al. 1997; Dalrymple et al. 1999), University of Western Australia (e.g., Brown and Fletcher 1999), Stanford University (e.g., Bacon et al. 2000; Lowenstern et al. 2000), and Hiroshima University (Sano et al. 2002). Relevant analytical conditions for U-Pb and Th-U SIMS dating of zircon in the various laboratories are described by Reid et al. (1997), Brown and Fletcher (1999), Sano et al. (2002), Schmitt et al. (2003a), Bacon and Lowenstern (2005), and Schmitt (2006). Dating of magmatic allanite and chevkinite is described by Vazquez and Reid (2004) and Vazquez (2008).

In the ion microprobe, an accelerated (primary) beam of ions is focused onto a sample surface in order to eject atomic and molecular species from the target minerals (Fig. 3). The resulting secondary ions are then accelerated into the mass spectrometer. For U-Pb analyses, flooding of the sample surface with O_2 can enhance Pb ion yields under favorable circumstances

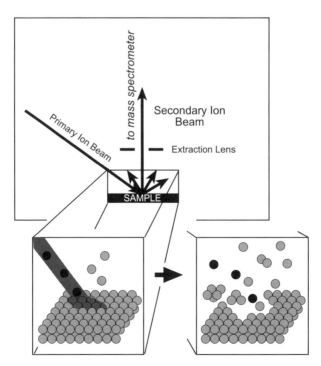

Figure 3. Schematic illustration of the approach of secondary ionization mass spectrometry. A primary ion beam (black circles) bombards the sample, generating secondary species from a sample's surface (gray circles). The secondary ions are extracted into the mass spectrometer where the relative mass abundances are measured.

by factors of ~2 (zircon; Ireland and Williams 2003) to 10 (baddelyite; Chamberlain et al. 2007) in certain instruments (Cameca ims 1270/1280). Areas sputtered from the sample surfaces typically range from 25-45 micrometers in diameter (Fig. 4), although results for beam spots as small as 10 micrometers have been reported (e.g., Brown and Fletcher 1999). The depth resolution of SIMS spots is, as noted above, more than an order of magnitude smaller than the lateral dimensions of the spot and this effect can be exploited to advantage (e.g., Reid and Coath 2000).

Before data acquisition can begin, sample surfaces are typically sputtered or rastered for 2-4 minutes to remove the gold coat and possible surface contamination in the analysis area. This means that the outermost few tenths of a micrometer of a mineral rim cannot be analyzed. Even with this precautionary measure, the more slowly sputtered margins of the analysis pit may contribute common Pb as the analysis proceeds. One solution to this is to reduce an aperture diameter in the secondary ion path so that ions only from the central portion of the analysis pit are emitted into the mass spectrometer. However, the reduction in the common Pb correction may not compensate for the resulting loss in precision arising from the attendant reduction in overall secondary beam intensity.

For U-Pb dating, secondary ion beam intensities are measured for Zr_2O_4 ($^{90}Zr_2O_4$, $^{92}Zr^{94}ZrO_4$, or $^{94}Zr_2O_4$), ^{206}Pb, ^{207}Pb, ^{238}U, ^{232}ThO, and ^{238}UO (with "O" referring to ^{16}O), along with a background analysis near mass ^{204}Pb. Collection of ^{204}Pb, ^{208}Pb, and ^{232}Th beam intensities may also be included. For U-Th dating, a Zr_2O_4 peak, one or more backgrounds (e.g., at mass/charge = 246.3), and ^{230}ThO, ^{232}ThO, and ^{238}UO are measured. Tuning the spectrometer to a mass resolution of >4000 (U-Th) to >4800 (U-Pb) ensures that molecular interferences (HfO_2 and REE oxides) on the masses of interest are well-resolved when dating of zircons is being performed. Under these conditions, ^{206}Pb count rates in young zircons range from 0.5 to 16 counts per second (cps); ^{230}ThO count rates range from 0.4-5 cps. For allanite, a mass resolution of ~9000 is required for U-Th analyses but even at this resolution ^{230}ThO intensities in allanite can still be as high as several cps (Vazquez and Reid 2004). Counts

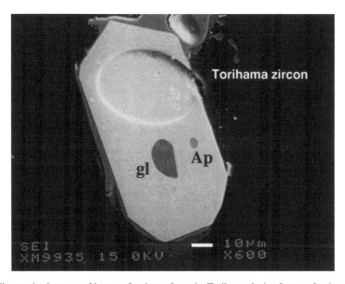

Figure 4. Electron back-scattered image of a zircon from the Torihama dacite, Japan, after ion microprobe analysis. Note the well-defined and flat-bottomed crater created by ion sputtering and the presence of glass and apatite inclusions and glass selvages on the grain. [Used with permission of Elsevier from Sano et al. (2002) *Journal of Volcanology and Geothermal Research*, Vol. 117, p. 288.]

for key peaks are typically integrated over several data cycles totaling hundreds of seconds. Background noise count rates are less than 0.05 cps (except where unusually high primary beam currents have been applied), negligible for most species of interest except for ^{230}ThO.

U, Th, Pb and their oxides have different ionization yields during sputtering and so a correction factor must be applied to measured ratios to obtain true ratios of ^{20x}Pb/^{23x}U and ^{230}Th/^{238}U. The relative sensitivities of Pb, Th, and U in zircon have been determined using various standards, as extensively discussed by Ireland and Williams (2003). For U-Th dating, ThO/UO relative sensitivity factors can be obtained from the difference between the ^{230}ThO/^{238}U ratio measured in "ancient" ($t > 400$ k.y.) zircons and that expected at secular equilibrium (e.g., Reid 1997; Charlier et al. 2005). Recent studies (Bacon and Lowenstern 2005; Charlier et al. 2005) have propagated the uncertainty in this fractionation factor into the reported age uncertainty. A somewhat more convoluted but more precise approach is to obtain relative sensitivity factors by comparing the ^{230}ThO/^{238}U ratio measured to time-integrated Th/U of zircon standards based on ^{208}Pb/^{206}Pb (Reid et al. 1997).

Uranium contents are typically estimated by dividing the measured ratios of a uranium species to a zirconium oxide species (e.g., ^{238}UO/^{94}Zr$_2$O$_4$) on the unknown by the same ratio obtained on the reference standard and multiplying that quotient by the standard's U concentration. Standards sufficiently homogeneous in U abundance include SL13 (238 ppm U), CZ3 (550 ppm U), and 91500 (81.2 ppm U). An analogous approach may be applied to determine Th concentrations or the Th concentrations can be determined directly from the U concentration by multiplying the latter by the measured Th/U.

Mass spectrometry: bulk mineral separates. Techniques for analysis of U-series nuclides by alpha spectrometry, TIMS, MC-ICP-MS, or single-collector, high-resolution ICP-MS, have been reviewed elsewhere (Goldstein and Stirling 2003), and we focus this discussion on the practical considerations of time and effort required to perform these analyses using different techniques. The time required for analysis of bulk mineral separates after chemical separation depends greatly on the technique used, and also can vary significantly from one sample to the next. TIMS analyses require approximately two hours each for U and Th concentration measurements by isotope dilution with an additional 4-8 hours for measurement of Th isotopic compositions (if run separately from the Th ID measurement) and an additional 2-8 hours for radium concentration measurement. MC-ICP-MS offers a significant advantage in sample throughput, with for example a single measurement of thorium isotopic composition in a sample requiring as little as 15-20 minutes. However, this is a somewhat misleading comparison, as instrument calibration, tuning, and analysis of standards (especially in the case of sample-standard bracketing) all require significant amounts of time. Perhaps a more appropriate comparison is that, for example, one can expect to complete analysis of Th isotopic compositions of 1-2 samples per day of instrument time by TIMS, and 8-10 samples per day by MC-ICP-MS.

Analysis of radium in mineral separates can be done by TIMS (see refs in Goldstein and Stirling 2003), by MC-ICP-MS (Pietruszka et al. 2002; Ball et al. 2008; Cooper and Donnelly 2008; Sims et al. 2008) or by single-collector, high-resolution ICP-MS (Sims et al. 2003; Cooper et al. 2005). Radium measurements by TIMS require ~4-8 hours of instrument time per sample, whereas radium measurements by MC-ICP-MS require ~1 hour per sample (when time for washout, calibration, and standards is included). The lower ionization efficiency of radium by TIMS compared to ICP-MS is counterbalanced by the lower transmission efficiency of ICP-MS, leading to a somewhat smaller effective ion yield by ICP-MS (0.5%, compared to 1-10% by TIMS; Goldstein and Stirling 2003). Because of the tradeoff between time required for analysis and effective ion yield, it is our experience that with small samples (10's of femtograms of Ra) more precise results are obtained by TIMS than by MC-ICP-MS, whereas with larger samples (~100 fg), results with good precision (~1% relative errors, 2-sigma) can be obtained by using MC-ICP-MS, with the advantage that the analyses are much more rapid than by TIMS.

Alpha spectrometry. Analysis of ^{210}Pb by alpha spectrometry is the most time-consuming of the ^{238}U-^{230}Th-^{226}Ra-^{210}Pb measurements. ^{210}Pb activity is measured by alpha spectrometry using ^{210}Po as a proxy. Samples are spiked with ^{209}Pb and counted for approximately one week. ^{210}Pb activities are determined either by assuming secular equilibrium between ^{210}Pb and ^{210}Po for samples older than ~2 years (with a half-life of 138 days, ^{210}Po will be in equilibrium with its parent ^{210}Pb within two years)(Turner et al. 2004; Reagan et al. 2006) or, for younger samples, by repeated measurements of ^{210}Po over a period of months at various stages of ^{210}Po ingrowth toward secular equilibrium, which are extrapolated to equilibrium with ^{210}Pb (Reagan et al. 2005, 2006). Despite the analytical challenges, a few ^{226}Ra-^{210}Pb data for plagioclase have been published (Reagan et al. 2006, 2008) and further measurements hold promise to unravel the most recent crystallization history of volcanic minerals.

Summary. The net result is that the amount of time and effort required for preparation of a mineral separate, preparation of the separate for analysis (mounting and polishing for SIMS analysis or chemical separation of the elements of interest for TIMS/MC-ICP-MS), and analysis time, vary widely between the different parent-daughter pairs, different compositions of rocks, and different analytical techniques. However, even the most straightforward of these analyses requires a significant effort in terms of time, effort, and funding, which provides an incentive for careful consideration during sample selection. Nevertheless, with well-chosen samples, the investment is worthwhile considering that the U-series system provides information about timing and duration of magmatic events that is simply not possible to gain by other techniques and that contributes a unique perspective on magmatic processes.

Age calculations

U-Th-Pb ages by SIMS. For young accessory phases, ^{207}Pb/^{206}Pb deviations from the modern-day production ratio of 0.04605 reflect contributions from sources other than *in situ* U decay. An important source of this deviation can be contributions from "common Pb". For U-bearing minerals (e.g., zircons) of identical age, variable mixtures of radiogenic and common Pb will define an array on a plot of ^{207}Pb/^{206}Pb versus ^{238}U/^{206}Pb (Fig. 5). The intercept with concordia (the locus of points defining the covariation in ^{207}Pb/^{206}Pb versus ^{238}U/^{206}Pb over time; Tera and Wasserburg 1974) defines the age of the zircons and the ordinate intercept defines the common ^{207}Pb/^{206}Pb (e.g., Baldwin and Ireland 1995). If zircons are of various ages, it may be more meaningful to calculate model ages by assuming a common Pb value, e.g., the (initial) isotopic composition of the host rock. In this case, the common Pb correction equates to extrapolating from the common Pb value on the ordinate through the data point to the concordia curve (Fig. 5). In some cases, rather than a magmatic value for common Pb, corrections or allowances for a contaminant Pb may be justified (Schmitt 2003b). Analyses with a large common Pb component could be the result of the primary beam overlapping or intercepting inclusions (glass or mineral) and rejection of these results may be justified.

A second source of deviation of ^{207}Pb/^{206}Pb from the modern-day production ratio arises from initial disequilibria in the U-series decay chains. Of these disequilibria, only ^{238}U-^{230}Th disequilibrium is likely to have a detectable effect on the U-Pb age but the superimposed effect of radioactive decay means that the initial ^{230}Th/^{238}U of accessory phases cannot be measured directly. One approach for estimating the initial ^{230}Th/^{238}U fractionation is to compare the ^{232}Th/^{238}U for each ionprobe spot to that of the host or, alternatively, to a putative parental melt composition. In instances where a grain has returned to secular equilibrium, the true age (T) can be calculated from the measured ^{206}Pb*/^{238}U from the following relationship (Schärer 1984):

$$^{206}Pb^* = {}^{238}U\left[\left(e^{\lambda_{238}T}-1\right)+\frac{\lambda_{238}}{\lambda_{230}}(f-1)\right] \quad (3)$$

where $f = [\text{Th/U}]_{\text{crystal}}/[\text{Th/U}]_{\text{melt}}$ and ^{230}Th is assumed to fractionate relative to ^{238}U to the same

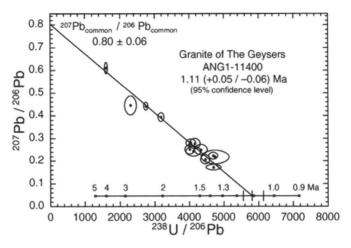

Figure 5. Tera-Wasserburg concordia diagram. Shown are 13 analyses of zircon from the Geysers granite, California. Line fit to the data gives a y-intercept 0.80 ± 0.06, the common $^{207}Pb/^{206}Pb$ ratio for the zircons if they are isochronous in age. The isochron intersects concordia (horizontal line with tic marks at discrete age increments) at 1.11 +0.05/−0.06 (2σ) Ma. This age agrees with an age of 1.13 ± 0.04 Ma obtained by Dalrymple et al. (1999) for the same zircons. Note that neither age is corrected for an initial, crystallization-induced ^{230}Th deficit in the zircons. [Used with permission of the Geological Society of America from Bacon et al. (2000) *Geology*, Vol. 28, p. 468.]

degree as ^{232}Th. For zircon, a Th/U fractionation factor predicted from Th and U zircon:melt partition coefficients, $f = 0.17$ (Blundy and Wood 2003), equates to a maximum (at secular equilibrium) deficit between measured and true ages of 90 k.y. while average f values of 0.19-0.26 estimated from a large number of igneous rocks (Bindeman et al. 2006) equate to maximum age deficits of 80-88 k.y. Younger crystals and more complex disequilibrium scenarios are addressed by Ludwig (1977).

U-Th isochron and model ages for ion microprobe spot analyses can be calculated from the activities of ^{230}Th, ^{232}Th, and ^{238}U. Activities can be calculated from ion ratios using the decay constants in Table 1. The calculation of U-Th isochron ages is described in the following section and is warranted when minerals form at the same time with the same initial Th isotope ratio. Individual spot analyses can alternatively be expressed as model ages by correcting the accessory phase activity ratios for an initial Th isotope composition. One approach for this is to calculate two-point mineral-melt isochrons using the present-day $(^{230}Th/^{232}Th)$-$(^{238}U/^{232}Th)$ characteristics of the melt. This assumes, in effect, that the minerals either crystallized from their host melt or from melts very similar to it. Ratios for the melts may be obtained by TIMS or MC-ICP-MS following the procedures already outlined. Ten-fold and greater internal variations in U/Th are often observed in and between zircons (e.g., Brown and Smith 2004; Miller and Wooden 2004; Charlier et al. 2005; Simon and Reid 2005) which suggests that zircons grow under variable conditions involving widely varying melts. Under favorable circumstances it may be possible to determine U/Th ratios in melt (now glass) entrapped in the zircons. An alternative approach is to fix the initial $^{230}Th/^{232}Th$ and assume that there is no ^{238}U-^{230}Th disequilibrium in the melt (i.e., assuming $(^{230}Th/^{232}Th) = (^{238}U/^{232}Th)$). The typical zircon has a large enough U/Th ratio so that uncertainty in the initial ratio is a modest contribution to the overall model age uncertainty except for zircon analyses with low $(^{238}U/^{232}Th)$. At the opposite extreme, the U/Th ratio of allanite is so low that the variation in $(^{230}Th/^{232}Th)$ depends only on its initial Th isotope ratio and the amount of time elapsed (i.e. the second term on the right in Equation (4) below can be neglected).

We are aware of two interlaboratory comparisons of SIMS ages on young accessory phases. Bacon and Lowenstern (2000) used a SHRIMP-RG ion microprobe to analyze zircons from the Geysers granite that had previously been analyzed using a CAMECA ims 1270 by Dalrymple et al. (1999). They obtained a ^{238}U/^{206}Pb age of 1.11 ± 0.06 Ma (95% confidence level; MSWD = 0.60) and ^{207}Pb/^{206}Pb ratio of 0.80 ± 0.06 for common Pb (Fig. 5), compared to a ^{238}U/^{206}Pb age of 1.13 ± 0.04 Ma (MSWD = 1.1) and a ^{207}Pb/^{206}Pb ratio of 0.91 ± 0.09 for common Pb obtained previously. Schmitt and Vazquez (2006) used a CAMECA ims 1270 ion microprobe to analyze zircons from Salton Buttes that had previously been analyzed by Brown et al. (2004) using a SHRIMP-RG and found closely overlapping U-Th ages between ~10 and ~18 ka.

^{238}U-^{230}Th-^{226}Ra-^{210}Pb ages of bulk separates. The classical approach to U-series dating using mineral separates is an isochron approach (e.g., Kigoshi 1967; Allegre 1968). The most common form of the isochron diagram used for the ^{238}U-^{230}Th parent-daughter pair is shown in Figure 6a. As with all isochron diagrams, this is a plot of parent vs. daughter (in this case expressed as activities), both normalized to an isotope of the daughter whose abundance does not change over the time of interest (in this case, ^{232}Th, which, while not stable, has a half-life that is long enough that decay can be neglected on the timescales of interest). The form of Equation (2) relevant to calculating ^{238}U-^{230}Th isochron ages is:

$$\frac{(^{230}\text{Th})}{(^{232}\text{Th})} = \frac{(^{230}\text{Th})_0}{(^{232}\text{Th})} e^{-\lambda_{230}t} + \frac{(^{238}\text{U})_0}{(^{232}\text{Th})}\left(1 - e^{-\lambda_{230}t}\right) \quad (4)$$

where, because uranium activity will not change measurably over the timescales of interest, U_o can be assumed to be equal to $U_{measured}$.

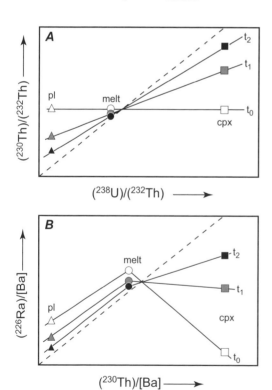

Figure 6. Schematic isochron diagrams for A) ^{238}U-^{230}Th and B) ^{230}Th-^{226}Ra. Shown are plagioclase (triangles), melt (circles) and clino-pyroxene (squares); the sense of U/Th, Ra/Ba and Th/Ba fractionation in minerals is consistent with known partitioning behavior, but the extent of fractionation is schematic. Melt was assumed to start with modest ^{230}Th-excess and ^{226}Ra-excess, but the overall behavior for minerals will be the same in the case of the opposite sense of disequilibria in the melt. Tie lines between symbols are isochrons representing the time of initial crystallization (t_0) and two subsequent times (t_1 and t_2) corresponding to approximately one and two half-lives of the daughter isotope, respectively. Note that isochrons on Ba-normalized ^{230}Th-^{226}Ra isochrons (B) are in general kinked lines, although they will approach a straight line as the phases approach secular equilibrium.

The typical assumptions for age calculations using an isochron approach are required: 1) the duration of crystallization must be short relative to the half-life of the daughter isotope; 2) all phases used to calculate ages must have precipitated at the same time from the same magma; and 3) each phase must have remained a closed system with respect to parent and daughter since the time of crystallization. Minerals with the same crystallization age will form an inclined array with slope proportional to the crystallization age. The ^{238}U-^{230}Th isochron has a number of distinctive features in comparison to 'conventional' isochron diagrams (such as Rb-Sr). For example, radioactive equilibrium is represented on the ^{238}U-^{230}Th isochron diagram by a line of unity slope (known as the equiline), and any phase with initial disequilibrium will move toward the equiline via ingrowth or decay of the daughter ^{230}Th. This means that phases can move vertically upward or downward, rather than the continual increase in daughter isotope abundance (and corresponding decrease in parent isotope abundance) shown by 'conventional' systems. These trajectories will be near-vertical on a U-Th isochron diagram, because decay of ^{238}U and ^{232}Th are slow enough that their activities will not change measurably over the timescales of interest. Any phases that have aged enough to reach radioactive equilibrium will remain on the equiline until some additional geologic event fractionates parent from daughter; this means that a mineral array that coincides with the equiline has an indeterminate age. In addition, as each phase exponentially approaches equilibrium the rate of vertical movement on an isochron diagram will slow; this leads to asymmetric age uncertainties, as a symmetric analytical uncertainty in ^{230}Th/^{232}Th ratio will represent a longer period of time on the side of the data point closest to the equiline than on the side of the data point farther from the equiline. Finally, the initial (^{230}Th)/(^{232}Th) ratio of the array is given by the intersection of the mineral array with the equiline, rather than the intersection with the y-axis. The ^{238}U-^{230}Th isochron approach has been used successfully to date bulk mineral separates in a large number of cases (e.g., Allegre 1968; Volpe and Hammond 1991; Reagan et al. 1992; Volpe 1992; Schaefer et al. 1993; Bourdon et al. 1994; Black et al. 1997, 1998a,b; Heath et al. 1998; Bourdon et al. 2000; Zellmer et al. 2000; Cooper et al. 2001; Heumann and Davies 2002; Charlier et al. 2003; Turner et al. 2003a,b; Tepley et al. 2006; Snyder et al. 2007), suggesting that the assumptions required for its use are often justified. We discuss in later sections cases where one or more of these assumptions may have been violated, and the information that may still be derived from an isochron approach even in these more complex scenarios.

An analogous diagram can be constructed for ^{230}Th-^{226}Ra parent-daughter pair, with the additional complication that Ba is used as a normalizing element because no longer-lived isotope of radium exists (Fig. 6b). In the case of the Ba-normalized ^{230}Th-^{226}Ra isochron diagram, an additional assumption that Ba is an exact chemical analog for Ra is required (e.g., Volpe and Hammond 1991; Reagan et al. 1992). As has been predicted for most minerals (Blundy and Wood 1994, 2003) and recently shown experimentally in the case of plagioclase (Miller et al. 2007a), there are in fact significant differences between the partitioning behavior of radium and barium in most rock-forming minerals. In most cases, radium is predicted to behave more incompatibly than barium (or, in the rare cases like K-feldspar where Ra and Ba are both compatible, less compatibly), leading to the preferential exclusion of Ra from the crystal lattice and to a correspondingly low Ra/Ba ratio in the mineral at the time of crystallization relative to Ra/Ba in the liquid (Blundy and Wood 2003). This effect will lead to deviations of co-precipitating phases from a straight-line array on a Ba-normalized ^{230}Th-^{226}Ra isochron diagram (Fig. 6b), and age calculations assuming two-point straight-line isochrons can be either older or younger than the true crystallization age. In order to circumvent this problem, Cooper and Reid (2003) developed the use of a radium evolution diagram to provide a method of graphically estimating ages (Fig. 7). This diagram plots (^{226}Ra)/[Ba] vs. time in mineral phases and groundmass/whole rock/glass. (^{226}Ra)/[Ba] at the time of measurement plots along the y-axis, and curves representing the evolution of (^{226}Ra)/[Ba] over time are calculated based on measured (^{230}Th)/[Ba] ratios. To this point, the treatment of the data is

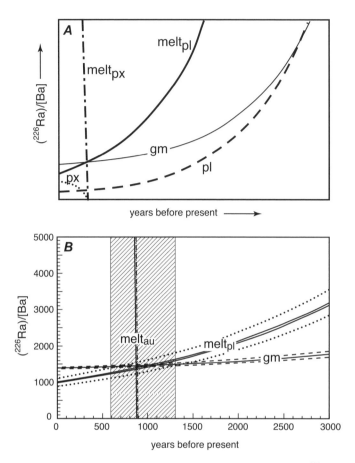

Figure 7. Ra evolution diagrams. A) Schematic Ra evolution diagram showing (^{226}Ra)/[Ba] ratios in minerals and melt projected back in time as predicted by their measured (^{230}Th)/[Ba] ratios. The time of measurement is at the y-axis. Shown are curves for plagioclase (dashed line), pyroxene (dotted line) and groundmass (assumed to represent the melt composition; thin solid line); compositions of pyroxene and plagioclase must be corrected for the effects of impurities before plotting. The intersection of the curves for pyroxene and plagioclase with the curve for groundmass represent the conventional two-point isochron ages uncorrected for Ra/Ba fractionation. Also shown are curves for melt in equilibrium with plagioclase and pyroxene (labeled melt$_{pl}$ and melt$_{px}$, respectively), calculated at each point in time by dividing (^{226}Ra)/[Ba] in each mineral at that point in time by the ratio D_{Ra}/D_{Ba} for that mineral. The age of crystallization is given by the point in time where the composition of the melt was equal to that in melt in equilibrium with the minerals, shown graphically by the intersection of curves for melt$_{pl}$ and melt$_{px}$ with the curve for groundmass. Note that the two-point mineral-melt isochron ages uncorrected for Ra/Ba fractionation can be either too old (plagioclase) or too young (pyroxene) depending on the initial (^{226}Ra)/(^{230}Th) ratio and Ra/Ba ratio in the minerals. B) Example evolution diagram constructed using data for a sample of the 1955 eruption of Kilauea volcano, Hawaii (modified from Cooper et al. 2001). Solid curves for melt in equilibrium with plagioclase correspond to An$_{70}$ at 1160 °C (upper curve) and An$_{60}$ at 1100 °C (lower curve); the dotted curves allow for 10% uncertainty in calculated D_{Ra}/D_{Ba} for plagioclase. Dashed lines for groundmass evolution through time indicate propagation of measurement uncertainties. Box with diagonal ruling represents age range of intersection of curves for groundmass and melt in equilibrium with plagioclase. Curves for melt in equilibrium with augite (melt$_{au}$) are also shown; solid and dot-dash lines delimit range of curves permitted by ± 35% uncertainty in partition coefficients and variations due to temperature. Data are consistent with co-precipitation of pyroxene and plagioclase from a melt with composition of the groundmass at an age of 1000 +300/−400 y (Cooper et al. 2001).

conceptually the same (although graphically different), from an isochron approach, and the intersection of curves for mineral and liquid (e.g., plagioclase or pyroxene and groundmass in Fig. 7) represents the conventional two-point isochron age. However, the true age must take into account differential partitioning of Ra and Ba, and therefore at each point in time, (^{226}Ra)/[Ba] in the mineral is corrected for the effects of fractionation by dividing by the ratio D_{Ra}/D_{Ba}. This new curve represents the composition of a melt in chemical equilibrium with the mineral (as defined by the partition coefficients) at any given point in time, and the model age of crystallization is given by the intersection of this curve with the curve for evolution of (^{226}Ra)/[Ba] in the melt (represented by whole-rock, glass, or groundmass data, whichever is most appropriate in a given case). We consider this to be a model age because it is subject to uncertainties in the partition coefficients for Ra and Ba. At this point, only one experimental determination of radium partition coefficient (in anorthite) has been published (Miller et al. 2007a). Therefore, we rely on calculations of radium partition coefficients based on an elastic-strain model using barium as the proxy element. Some confidence in this approach is given by the fact that the elastic-strain model agrees with the experimental data for anorthite (Miller et al. 2007a). A full discussion of calculation of these partition coefficients is beyond the scope of this paper, and we refer the reader to Blundy and Wood (2003) for a complete review of partitioning behavior of U-series elements and discussion of how best to estimate radium partition coefficients for minerals.

A further complication of calculating ^{230}Th-^{226}Ra ages—relative to calculating ^{238}U-^{230}Th or other ages where an isochron approach is feasible—is that impurities in the mineral separates will result in differences in the calculated ages. As is apparent from Figure 6b, mixtures of different phases will generally move data points away from a tie-line connecting pure minerals and melt, and will therefore affect the calculated ages. In effect, the model age calculations assume that the mineral separates are pure because they rely on partition coefficients to correct for Ra/Ba fractionation during crystallization; therefore, the composition of the pure mineral phases must be estimated in order to calculate ages. Two approaches can be used to improve estimates of Ra, Th and Ba concentrations in the pure minerals: one method is to prepare mineral separates that are as pure as possible, which will minimize the necessity for correction. However, due to the common presence of inclusions of other minerals and melt within phases such as plagioclase, K-feldspar, pyroxene, or amphibole, it is generally not practical to prepare separates that are 100% pure. Instead, it is necessary to correct the measured U-series element concentrations for the effects of impurities. This can be accomplished by taking a split of the dissolved mineral separate prior to U-Th-Ra analysis and analyzing its trace-element composition by ICP-MS, and comparing this to *in situ* measurements of trace elements in the pure mineral phase by SIMS or LA-ICP-MS. The trace-element concentrations of the bulk mineral solution can then be compared to trace-element data for other phases which may be present (e.g., melt, other mineral phases) in order to correct the measured U, Th, and Ra concentrations for the effects of these impurities. To date, this approach has generally relied on only a few elements (e.g., Cooper et al. 2001; Cooper and Reid 2003; Turner et al. 2003a,b), but by incorporating a suite of elements (e.g., Tepley et al. 2006) and by taking a least-squares approach to fitting the data, these uncertainties could be further minimized. Even with the relatively crude approach of using only a few elements for correction of impurities, the magnitude of the uncertainty on the ages contributed by the correction for impurities is of the same order as the uncertainty due to the choice of pre-eruptive temperature and average crystal composition used when calculating partition coefficients to use for correction for initial Ra/Ba fractionation. The age calculations are sensitive only to the ratio D_{Ra}/D_{Ba}, not to the absolute magnitude of D_{Ra} and D_{Ba}, both of which change in concert with changes in the partitioning parameters (site radius and Young's modulus), and therefore uncertainties in these variables affect the absolute values of the partition coefficients much more than the ratio (as discussed in more detail in Cooper et al. 2001). As a result, the uncertainties in ^{230}Th-^{226}Ra model ages are dominated by uncertainties in the pure mineral compositions and in the pre-eruptive temperature, with measurement uncertainties for the U-series nuclides and uncertainties in the partitioning parameters playing a relatively minor role. This offers the potential that by improving our ability to quantitatively account for impurities in the future, we can significantly improve the uncertainty in model ages.

There are few analyses of ^{226}Ra-^{210}Pb disequilibria in mineral separates reported in the literature; two examples are analyses of plagioclase separates in lavas from Arenal (Reagan et al. 2006) and Mount St. Helens (Reagan et al. 2008). An isochron approach has not been applied to ^{226}Ra-^{210}Pb crystal ages, in part because the ages of crystallization (for at least some part of crystals) in systems analyzed to date are not short with respect to the half-life of ^{210}Pb (e.g., Reagan et al. 2006, 2008). Lead is incompatible in calcic plagioclase, becoming compatible in sodic plagioclase (Bindeman et al. 1998), and ratios of D_{Pb}/D_{Ra} will be greater than unity in all plagioclase compositions (Blundy and Wood 2003). Using the method of Blundy and Wood (2003 and references therein) to calculate partition coefficients for Ra, and comparing to those for Pb calculated using the data of Bindeman et al. (1998), initial (^{210}Pb)/(^{226}Ra) ratios in plagioclase can be estimated and ^{226}Ra-^{210}Pb model ages calculated from the time required for decay from estimated initial ratios to the measured (^{210}Pb)/(^{226}Ra) ratios (Reagan et al. 2006, 2008). As discussed in more detail below, this approach captures an average crystallization age of the crystals (albeit weighted toward young crystal growth) and therefore can be used to estimate the proportion of crystal growth that occurred within decades of eruption.

Effect of mixing on mineral ages. Mixing arrays will produce straight lines on isochron diagrams which can be mistaken for arrays with age significance, therefore it is useful here to assess the effects of mixing on mineral ages. Mixing in this context can refer to several different scenarios: 1) mixing between different phases of the same age in the same rock; 2) mixing between old and young components of a given mineral in the same rock, where all mineral fractions precipitated from the same liquid; 3) mixing between different components of a given mineral which may have the same or different ages but which precipitated from different liquids. Scenarios 2 and 3, and the implications for age interpretations, are discussed below; here we briefly discuss the effects on calculated ages of mixing between different phases within the same system. In the case of the ^{238}U-^{230}Th system, minerals of the same age define straight lines on an isochron diagram. Mixing between any two phases also results in a straight-line array where the mixed data lie between the two end members. Therefore, mixing between phases in the same rock will not affect the slope of the isochron array (or the calculated age). However, mixing will generally decrease the spread of the array in terms of U/Th ratios, which will lead to more uncertainty on the slope of the line for a given analytical uncertainty, and therefore will increase the uncertainty of the age.

^{230}Th-^{226}Ra and ^{226}Ra-^{210}Pb ages are both model age calculations, dependent on the assumption that calculated partition coefficients applied to the measured ratios will accurately reflect interelement fractionation during mineral growth. This will only be true if the measured ratios reflect the composition of the mineral in question, rather than a mixture of the mineral being analyzed and any impurities in the bulk separate. As shown on Figure 6, mixtures between different phases of the same age can significantly change both (^{226}Ra)/[Ba] and (^{230}Th)/[Ba] in the mineral separate. Depending on the mineral in question and the nature of the impurities, impurities can have the effect of either increasing or decreasing the model ages compared to the true crystallization age. This effect is exacerbated in the case of minerals where radium, thorium and barium are highly incompatible (e.g., pyroxene), in which case even small mass fractions of impurities can dominate the U-series budgets of the mineral separate. For example, Cooper et al., (2001) calculated that of the 1.37 fg/g Ra in the pyroxene separate, the pure pyroxene contains only ~0.21 fg/g, indicating that ~85% of the radium present in the pyroxene separate originated from impurities, even though the same calculations estimated <1% impurities (melt+plagioclase) in the separate. As discussed above, this highlights the need for taking care to minimize impurities during preparation of mineral separates, and the need to develop more robust methods of estimating the composition and percentage of impurities in order to correct the bulk separate data for these effects. Exactly how best to go about this will depend on the details of the sample and must ultimately be treated on a case-by-case basis. Even in the absence of mass-balance calculations using many elements in multiple phases, in most cases maximum and minimum concentrations can be estimated by identifying the phases that will have the most leverage on the system and by quantifying their contributions using mass-balance calculations (e.g., Cooper et al. 2001). Similar arguments apply to ^{226}Ra-^{210}Pb ages.

Effect of diffusion on crystal ages

A chemical potential for U-series elements could potentially be imposed between crystals and liquid if crystal cores are isolated by continued crystal growth as the melt continues to differentiate, or if crystals are entrained by a new host melt with different composition from the original melt. In addition, chemical potentials could develop in minerals simply due to changes in concentration due to decay. Whatever the mechanism producing chemical gradients, they could drive diffusion of parent and daughter nuclides into or out of crystals, which in turn could affect U-series age calculations and could either increase or decrease apparent ages. The effect of diffusive re-equilibration on crystal ages depends on the time that a crystal spends in contact with the melt and on the diffusivity of the parent and daughter elements, which generally are slow at magmatic temperatures. For example, diffusion of U, Th, and Pb in zircon at magmatic temperatures is slow enough that diffusive mobility is negligible over the timescales discussed here (Cherniak and Watson 2003). Temperatures high enough to allow significant mobility of U and Th in zircon would be above the zircon saturation temperature for most silicic magmas, and therefore zircon would likely dissolve before diffusively re-equilibrating with a new host liquid. Vazquez and Reid (2004) estimated that Th diffusion in allanite would only affect a few micrometers in allanite over a period of 100,000 years at temperatures that characterize allanite-bearing rhyolites (<800 °C).

Diffusion of U and Th in plagioclase is very slow at magmatic temperatures, based on diffusion parameters calculated using the model of Van Orman et al. (2001) which predicts diffusion coefficients of order 10^{-21} m^2/sec at 1200 °C (Saal and Van Orman 2004; Van Orman et al. 2006). Diffusivity of radium in plagioclase has not been measured, but can also be estimated using experimentally-derived Ba diffusion parameters (Cherniak 2002) and the elastic strain model of Van Orman et al. (2001, 2006); this estimate is of the same order of magnitude as Ba (Cherniak 2002), which is approximately three orders of magnitude faster than U and Th (Van Orman et al. 2006). Given these estimates of diffusion parameters for U, Th, and Ra in plagioclase, U-Th crystal ages are not likely to be affected by diffusion during tens to hundreds of thousands of years of residence at magmatic temperatures. Ra-Th ages will be robust for tens to hundreds of thousands of years at temperatures less than ~1000 °C, but ages could potentially have been affected by thousands of years of residence at temperatures >1100 °C (Cooper and Reid 2003).

Diffusion of U and Th in diopside is also very slow (diffusion coefficients on the order of 10^{-20} m^2/s at 1200 °C; Van Orman et al. 1998, 2006), and U-Th ages of clinopyroxene are therefore likely to be robust at magmatic temperatures for the timescales of interest for U-series dating. Radium diffusion parameters for clinopyroxene have been estimated by Van Orman et al. (2006), and are expected to be 1-3 orders of magnitude faster than U and Th, suggesting that (as with Ra in plagioclase) protracted residence at relatively high magmatic temperatures (>1000 °C) could result in some diffusive effects on Ra-Th ages for pyroxene.

Although the diffusivity of Ra and Ba is predicted to be faster than that of U and Th in plagioclase, it is still relatively slow (diffusion coefficients on the order of 10^{-20} m^2/s at 1000 °C; Cherniak 2002). Watson (1996a) and Watson and Liang (1995) proposed a model for trace-element enrichment in a crystal lattice due to surface effects when crystal growth rates are rapid relative to diffusion of the element of interest. Growth entrapment refers to the incorporation of concentrations of incompatible elements in a crystal that are higher than concentrations predicted by bulk crystal-liquid equilibrium, due to the differences in crystal lattice sites at the surface monolayer compared to sites in the interior of the crystal. These higher surface concentrations can be trapped within a growing crystal if the element of interest cannot diffuse outward faster than the crystal-liquid interface grows outward (Watson and Liang 1995; Watson 1996a). Given estimates for Ra and Ba diffusion rates compared to estimates of plagioclase crystal growth rates, Ra/Ba ratios (and therefore calculated ages) could be affected by growth entrapment during crystallization at temperatures less than 1000 °C (Fig. 8; Cooper and Reid 2003). The relative effects of such growth entrapment on different elements are not well known, but, given that both D's will be closer to unity, it is likely that Ra/Ba ratios would be increased relative to bulk chemical equilibrium due to a decreased ability of the crystal to discriminate between the two elements, which would lead to a decrease in apparent crystal age (or even an undefined crystal age; see Cooper and Reid 2003). The concordance between U-Th and Ra-Th ages of plagioclase in most cases, together with the concordance of Ra-Th ages calculated from different phases, suggests that this is not a general phenomenon, but it may be important to consider in the case of rapid decompression-induced crystallization of plagioclase, which may be orders of magnitude faster than cooling-induced crystallization.

In summary, diffusive re-equilibration between crystals and liquids is unlikely to affect U-series crystal ages, except in the case of ^{230}Th-^{226}Ra ages of plagioclase and pyroxene

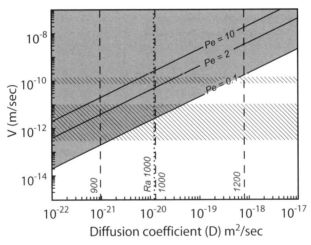

Figure 8. Plot of diffusivity vs. growth rate illustrating conditions of potential growth entrapment (modified after Cooper and Reid 2003). Solid lines represent contours of constant growth Peclet number $Pe = Vl/D_i$, where V is the crystal growth rate, l is the half-thickness of the enriched surface layer (assumed to be 50 angstroms), and D_i is the lattice diffusivity of element i (Watson and Liang 1995). Growth entrapment (i.e., concentrations higher than bulk crystal-liquid equilibrium) of element i will be measurable at $Pe = 0.1$, significant for $Pe > 2$, and ~100% effective (i.e., concentrations on the interior will be equal to those in the surface layer) at $Pe = 10$ (Watson and Liang 1995). Therefore, these values delimit regions of inefficient, partial, and efficient growth entrapment, respectively; the region where growth entrapment is possible is shaded. Estimates of plagioclase growth rates for the 1980-1986 dacite dome are shown by ruled fields. Experimentally-determined diffusion coefficients for Ba in labradorite (An$_{67}$) from Cherniak (2002) are shown by vertical dashed lines (labeled with temperatures in °C). Also shown is diffusion coefficient for Ra at 1000 °C (dotted line) calculated by Cherniak using the method of Van Orman and others (2001). Based on these diffusivities and growth rates, growth entrapment is expected for crystallization at temperatures less than 1000 °C, and could occur to some degree at temperatures up to 1200 °C if growth rates were similar to maximum estimates. See text for discussion.

held at high temperatures (>1000 °C) for thousands to tens of thousands of years. Growth entrapment of high concentrations of incompatible trace elements, which will be controlled by the rate of diffusion of cations, is unlikely to affect ^{230}Th-^{226}Ra ages in the case of relatively slow, thermal-driven crystallization. In contrast, this may be an important factor to consider in the case of rapid, decompression-induced crystallization of plagioclase, and would have the effect of making calculated plagioclase model ages too young.

WHAT DO CRYSTAL AGES TELL US?

Introduction

Early analyses of U-series disequilibria in minerals were performed in an attempt to date the eruption age of a lava (Kigoshi 1967; Taddeucci et al. 1967); some examples of successful dating of eruptions using this method are presented below. However, it was also recognized early on that crystallization of minerals in a volcanic rock could pre-date eruption by a significant time period (e.g., Allegre 1968). The number of U-series analyses of minerals in volcanic rocks has increased dramatically over the past two decades, primarily driven by analytical advances which have decreased the sample size requirements and made analysis of minerals more practical. This applies both to bulk mineral analyses, where multicollector mass spectrometry (TIMS or MC-ICP-MS) have allowed analysis of smaller samples at higher precision than was possible by alpha spectrometry (e.g., Goldstein and Stirling 2003), and to the development of *in situ* dating of accessory minerals with relatively high U and Th contents (Reid et al. 1997; Vazquez and Reid 2004). The greater analytical and interpretive challenges of ^{230}Th-^{226}Ra and ^{226}Ra-^{210}Pb mineral data have limited the number of ^{230}Th-^{226}Ra-^{210}Pb mineral data compared to ^{238}U-^{230}Th analyses, but the database of ^{230}Th-^{226}Ra and even ^{226}Ra-^{210}Pb crystal ages has also grown significantly over the past decade.

The recognition that many, if not most, crystals in volcanic rocks pre-date eruption by significant time periods (thousands to hundreds of thousands of years; Reid 2003), coupled with increasing evidence of mineral-melt disequilibrium from techniques such as single-crystal or sub-crystal trace-element and isotopic analyses, textural information, and diffusion studies (e.g., Ramos and Tepley 2008; Streck 2008; Costa 2008), indicates that many crystals contain records of magmatic processes that pre-date their residence in the host magma. However, given the complexity of magmatic processes recorded in crystals, the traditional distinctions between "phenocrysts" precipitated from the host magma and "xenocrysts" incorporated from wall rocks unrelated to the magmatic system are insufficient to describe the diversity of origins of crystal material within a magma. To distinguish phenocrysts from cogenetic crystals that are recycled from earlier phases of magmatism within the same system, Wes Hildreth (presentation at the Penrose Conference on "Longevity and Dynamics of Rhyolitic Magma Systems", 2001) proposed adoption of the term "antecrysts," which we use here. Miller et al. (2007b) further proposed that the term "phenocryst" is a textural term that does not apply to, for example, zircon crystals that may have grown from the host liquid but which are nevertheless quite small relative to other crystals in the rock. They propose the term "autocryst" to identify crystals grown from the host liquid, and we adopt that terminology here.

Especially in the case of bulk mineral separates, where ages are averaged over many hundreds or thousands of crystals, U-series ages may reflect contributions from crystals with different origins ranging from phenocrysts *sensu stricto* (i.e., autocrysts) to antecrysts scavenged from mush zones or young solidified intrusions produced by earlier magma batches that are part of the magmatic activity at the same center, to true xenocrysts incorporated from wall rocks that are unrelated to the magmatic activity at a given center. Interpretation of crystal ages in terms of the thermal and chemical history of a given magmatic system thus depends to a large degree on whether the crystals analyzed show evidence of antecrystic or xenocrystic components, or whether they are dominated by autocrysts. In the latter case, crystal ages can provide information on magma residence times, average crystal growth rates, and differentiation timescales of magmas (Reid 2003). In more complex scenarios, where there is independent evidence of antecrystic or xenocrystic components in the minerals analyzed, crystal ages can still provide information about the age and proportions of various crystal components in a magma, and, when linked with crystal thermometry, barometry, textural information, and trace-element or isotopic information, can provide unique insights toward unraveling magmatic histories. In particular, U-series ages provide time information over timescales that are not accessible through any other technique, and therefore can be used to place other chemical and textural information in a temporal context. For example, as discussed in more detail below, U-series data have been used to identify different crystal components that are present in related magmas; to identify magmas within the same system that were present at the same time but did not apparently interact with each other chemically; to quantify the age and proportion of old and young crystal populations present in a magma; to identify young plutonic material that is related to erupted magmas and to identify crystals originating from these young plutons which were incorporated into young magmas; to track the thermal history of a magma body through time; and to identify ages of prominent differentiation events, among other applications. Crystal ages are thus crucial in understanding the rates and dynamics of magmatic processes such as crystal growth, storage, recycling, magma differentiation and mixing.

Using U-Pb and decay series ages to date eruptions

U-Pb analyses, especially of zircon, are a longstanding and accurate means of dating igneous rocks (see Davis et al. 2005, for a historical overview); regular improvements in analytical sensitivity continue to reduce age uncertainties and refine interpretations (e.g., Crowley et al. 2007). Early on, U-series analyses of igneous rocks likewise focused on its potential for dating volcanic rocks, albeit necessarily restricted to eruptions that occurred in the past <350 k.y. Even so, this method provides a means of directly dating volcanic rocks that is complementary or an alternative to other methods. The latter includes $^{40}Ar/^{39}Ar$ dating (which may be subject to inheritance, especially in mafic rocks) and the more indirect approaches of ^{14}C-dating and thermoluminescence-dating, and cosmogenic exposure age dating (which depend on recovery of organic materials or soils, or minimization of the effects of erosion, respectively). As described above, the age sensitivity of dating volcanic rocks using U-series disequilibrium depends on magnitude of parent-daughter fractionation and on the duration of return to secular equilibrium.

In a series of recent studies, Jicha and others (2005, 2007, 2008) obtained internal U-Th mineral isochrons for major and minor minerals (e.g., plagioclase, clinopyroxene, orthopyroxene, magnetite) and for groundmass separated from lava flows and pyroclastic rocks in arc settings (Seguam Island, Aleutians; Puyehue–Cordón Caulle volcanic complex, Chile; and Mount Adams volcanic field, Washington). In virtually every case, the U-Th isochron ages are indistinguishable from high precision $^{40}Ar/^{39}Ar$ ages and, by extension, from the time of eruption. An important conclusion to be drawn from these studies is that the major mineral phases must have grown shortly (<1000 yr, based on analytical uncertainties) before eruption. Dating of accessory phases from young volcanic systems has also yielded ages in some cases that are concordant with those obtained by other methods. In their study of the Bishop Tuff, Crowley et al. (2007) chemically processed and analyzed individual zircons and obtained ^{238}U-^{206}Pb ages for most that, at 767.1 ± 0.9 ka, are largely concordant with the 770.4 ± 3.6 ka eruption age obtained by $^{40}Ar/^{39}Ar$ dating (Sarna-Wojcicki et al. 2000, corrected to Taylor Creek sanidine neutron fluence monitor). ^{235}U-^{207}Pb ages are, however, ~52 k.y. older and further work is needed to resolve this difference. A U-Th isochron defined by zircon and apatite aliquots from Puy du Dome, France (Condomines et al. 1997) also passes through data points for potassium feldspar and biotite and the derivative age is in excellent agreement with thermoluminescence and ^{14}C age dating. Most crystallization must have occurred within a thousand years before eruption.

As this review explores in greater detail below, the timing of phenocryst growth can be resolvably older than that of eruption. Accordingly, a better approach to dating volcanic eruptions would seem to be to date groundmass phases. This presents the difficult challenge of separating sufficient mineral quantities for precise analysis while avoiding the contaminating effects of phenocryst fragments and/or glass (which can easily swamp the U-series nuclide signatures of the mineral phases). Nonetheless, Sims et al. (2007) used the internal isochron defined by groundmass and groundmass plagioclase to estimate an eruption age of 68 +24/−20 ka (2σ) for the Bluewater Flow, New Mexico. Discrepancies between this age and ages from $^{40}Ar/^{39}Ar$ (2.3-5.6 Ma) and cosmogenic nuclide abundances (39-56 ka) can be explained by the effects of inherited Ar and of erosion, respectively.

Pre-eruptive crystal growth—'simple' scenarios

In situ ages. Over the past decade, a number of studies have dated individual accessory phases, and subdomains within those phases, from young (most <2 Ma), mainly silicic, systems. Instead of being eruption-aged, most zircon and allanites give crystallization ages— and attendant closure to Th and Pb isotope exchange (Cherniak and Watson 2003; Vazquez and Reid 2004)—in excess of those of eruption, as gauged mainly by the difference between the crystallization ages and Ar-Ar and/or K-Ar eruption ages. Thus rather than focusing on dating eruption, the attention has shifted to using accessory phase ages to gain insights into the magma evolution. Simon et al. (2008) recently summarized existing results for U/Th and U/Pb ion microprobe analyses of young zircons in terms of 1) the oldest zircon age and 2) mean pre-eruption ages, calculated as the difference between mean zircon age (t_{zir}) and the eruption age (t_{erup}) (reproduced here as Table 3). More detailed information about many of these results is provided later in this chapter.

Volumes of erupted magma that have been studied range from <1 km^3 (e.g., Reid et al. 1997; Simon and Reid 2005) to ≥1000 km^3 (e.g., Bindeman et al. 2001; Bindeman et al. 2006). Many studies find zircon ages that are several hundreds of k.y. older than the eruption age and mean pre-eruption intervals, $\Delta\ t_{zir}$-t_{erup}, for virtually all eruptions lie in the range of several tens to a few hundreds of k.y. (Table 3; Reed 2003; Simon et al. 2008). Smaller volume effusive eruptions tend to have longer pre-eruption intervals than larger explosive eruptions but there is considerable overlap in this aspect of the systems. A straightforward interpretation of these results is that magmas are stored for tens to hundreds of thousands of years before erupting. On the other hand, age distributions commonly show peaks in the timing of crystallization (e.g., Reid et al. 1997; Brown and Fletcher 1999; Reid and Coath 2000; Vazquez and Reid 2002; Miller and Wooden 2004; Charlier et al. 2005). Sequential eruptions from a given center may exhibit common age populations (e.g., Reid et al. 1997; Vazquez and Reid 2002; Charlier et al. 2005), suggestive of repeated remobilization of the same magma reservoir or subvolcanic intrusion. In other cases, temporally and geographically proximal eruption centers may emit eruptions with distinct zircon age populations (e.g., Charlier et al. 2003). U-Pb dating of igneous xenoliths (e.g., Bacon and Lowenstern 2005; Schmitt 2006) and of shallow plutons (Schmitt et al. 2003b) reveal zones in which solidification was complete or nearly complete at the same time as inferred magma storage. Such results beg questions about the geologic significance of the long pre-eruption intervals. Do large pre-eruption ages reflect long storage intervals of crystals in liquid-rich systems and therefore the residence time of what might conventionally be considered "magma"? Or do crystals become remobilized and/or recycled from crystalline mush or subvolcanic plutons by new introductions of melt, in which the ages reflect a mean crystal age for the magma and the evolution of the subvolcanic system? Considered in this light, the apparently "long" mean pre-eruption ages of most silicic systems (Table 3; Simon et al. 2008) could integrate an evolutionary history during which the accessory phases may have been immersed in liquid for only a fraction of this time.

Table 3. Summary of age characteristics of silicic extrusions (from Simon et al. 2008). [Used with permission of Elsevier from Simon et al. (2008) *Earth and Planetary Science Letters,* Vol. 266, p. 190.]

Sample	Locality	Eruption age[a] (ka)	(ka)	Oldest zircon age[b] (ka)	(se)	Mean pre-eruption interval (ka) Calculated[c]	2se	Modeled	sd	Dating method
Explosive large volume										
Oruanui	Taupo, NZ	27	27	178 (218)	4 (125)	18	5	52	3	U/Th U/Th
Las Tres	Baja, MX	36	36	168 (289)	48 (100)	85	12	74	9	U/Th U/Th
Tihoi	Taupo, NZ	45	45	166 (227)	58 (280)	27	8	62	5	U/Th U/Th
Earthquake	Taupo, NZ	62	63	151 (258)	71 (98)	47	19	62	9	U/Th U/Th
Rotoiti	Taupo, NZ	62	64	229	89	15	15	60	8	U/Th
Okaia	Taupo, NZ	29	29	158	48	27	15	71	6	U/Th
Kos Plateau	Aegean Arc	161	164	287 479	40 74	64	14	80	3	U/Th U/Pb
Whakamaru	Taupo, NZ	340	338	608	20	93	22	92	7	U/Pb
Rockland Tuff	Lassen Peak, CA	576	584	814	69	70	20	134	13	U/Pb
Lava Creek	Yellowstone, WY	639	655	710	30	8	23	62	8	U/Pb
Early Bishop	Long Valley, CA	760	777	926	36	64	8	104	6	U/Pb
Late Bishop	Long Valley, CA	760	777	926	25	34	7	84	9	U/Pb
Effusive small volume										
South Deadman	Long Valley, CA	0.6	0.6	271	79	43	29	184	14	U/Th
Crater Lake	Crater Lake, OR	27	27	175	79	7	5	52	7	U/Th
Pitchstone Plateau[d]	Yellowstone, WY	60	61	233	47	40	11	33	5	U/Th
West Yellowstone[d]	Yellowstone, WY	109	110	306	21	53	20	66	8	U/Th and U/Pb
Deer Mountain	Long Valley, CA	115	118	285	89	38	25	90	9	U/Th
Solfatara Plateau[d]	Yellowstone, WY	116	117	199	30	50	12	55	7	U/Th
Dry Creek"	Yellowstone, WY	159	161	200	70	6	26	40	8	U/Th
Devils Kitchen	Coso, CA	587	602	800	20	84	22	123	8	U/Pb
Tyler Valley Gey	Geysers, CA	670	689	1148	48	261	30	181	18	U/Pb
YA Glass Mountain	Long Valley, CA	885	894	1119	64	87	18	94	17	U/Pb
YG Glass Mountain	Long Valley, CA	900	916	1215	91	17	17	130	13	U/Pb
CM0002 ACR Gey	Geysers, CA	1150	1169	1488	36	151	20	93	10	U/Pb
ACR Gey	Geysers, CA	1150	1169	1566	96	191	20	124	13	U/Pb
CM0004 rhyodacite Gey	Geysers, CA	1100	1128	1499	39	182	20	143	16	U/Pb
K3154 ACR Gey	Geysers, CA	1180	1205	1628	32	155	20	153	11	U/Pb
CM0003 dacite Gey	Geysers, CA	1000	1017	1408	39	243	20	216	18	U/Pb
OD Glass Mountain	Long Valley, CA	1686	1712	2031	64	141	13	223	13	U/Pb
OL Glass Mountain	Long Valley, CA	1926	1945	2284	55	154	33	119	16	U/Pb
OC Glass Mountain	Long Valley, CA	2045	2066	2463	61	166	29	247	20	U/Pb
Pine Mt Gey	Geysers, CA	2170	2208	2672	105	252	30	145	23	U/Pb

Data sources included in Simon et al. 2008
[a] Eruption ages listed first as originally reported and then after correction for K-Ar decay constant error and choice and ascribed age of neutron influence monitor (cf. Renne et al. 1998); [b] Parenthetical dates reflect zircon growth that is probably older than the oldest reported zircon dates because they are within secular equilibrium and imprecise.; [c] Weighted mean pre-emption ages calculated from U/Th data will be bias towards a younger age because U/Th ages have larger age uncertainty as dates approach secular equilibrium.; [d] Not small volume

Bulk separate analyses. Mineral ages determined from bulk analyses represent an average of the ages of all crystals/zones within the population analyzed, weighted by volume of the components of different ages and by the concentration of parent in each of the components. Interpretations of the geologic significance of such an age depend on exactly what scenario one envisions for crystal growth and the relation between crystals and host liquid. In the simplest case, the assumption is that phenocryst-sized crystals precipitated from the host liquid during a short time interval relative to the half-lives of the isotopes of interest. The extent to which a given system conforms to this simple scenario depends in part on the half-life of the daughter isotope in a given parent-daughter pair: a 'short' crystallization interval may be a more accurate approximation for U-Th ages than for Ra-Th or ^{226}Ra-^{210}Pb ages. Documenting bulk chemical equilibrium between crystals and liquid is a necessary (though not sufficient) condition for demonstrating crystallization from the host liquid, and therefore for interpreting crystal ages in terms of magma residence. In addition, a simple and rapid crystallization history will lead to concordant ages between different parent-daughter pairs and between phases expected to be crystallizing at the same time; thus, as discussed in more detail below, discordant ages are themselves an indication of a more complex history recorded in the crystals. One somewhat more complex scenario is that of progressive crystal growth from the same magma, in which case the crystal age will represent an average of the duration of crystallization (albeit weighted by growth rate and by the parent concentration in different zones). Bulk sample crystallization ages in these kinds of relatively simple scenarios can be interpreted in terms of different geologic processes such as crystal and magma residence times or differentiation ages.

For example, whether interpreted as a 'burst' of crystallization of short duration or as an average age of progressive crystal growth, if the crystals precipitated in equilibrium with the host liquid (i.e., autocrysts), the U-series crystal age can provide a minimum residence time for the host magma (e.g., Reagan et al. 1992; Schaefer et al. 1993, Heath et al. 1998; Bourdon et al. 2000; Cooper et al. 2001; Heumann et al. 2002; Tepley et al. 2006). For example, in the case of the 1952 eruption of Kilauea, the case for interpreting plagioclase and pyroxene ages in terms of minimum magma residence time is bolstered by (1) concordant ^{230}Th-^{226}Ra model ages for pyroxene and plagioclase (after correction for the effects of Ra-Ba fractionation) because the two phases are expected to co-precipitate in evolved Hawaiian magmas, (2) concordant ^{238}U-^{230}Th and ^{230}Th-^{226}Ra ages for each of the bulk separates, and (3) evidence for major-element chemical equilibrium between crystals and liquid (Cooper et al. 2001). Mixed crystal ages have been observed in bulk separates of a trace phase as well: Charlier and Zellmer (2000) measured ^{238}U-^{230}Th ages of bulk separates of zircon of different sizes, finding systematically older ages for larger size fractions. They interpreted this observation to reflect episodic or continuous crystallization, emphasizing that the age represents an average crystallization age rather than the onset of crystallization.

Bulk crystal ages for crystals with relatively simple histories can also be interpreted in terms of differentiation ages—i.e. the average crystallization age records the age of a dominant interval of crystallization/differentiation. For example, Turner et al. (2003b) measured ^{238}U-^{230}Th-^{226}Ra disequilibria in whole-rock samples and mineral separates from lavas and cogenetic cumulate xenoliths collected from Sangeang Api, Indonesia. Ra-Th ages (corrected for the effects of Ra-Ba fractionation) are equivalent for cumulates and minerals separated from the lavas, which Turner et al. (2003b) interpret to reflect the timescale (~2000 yr) of the dominant crystallization/differentiation interval.

Crystal recycling—autocrysts vs. xenocrysts. vs. antecrysts

With increasing evidence for complex interactions between different magmas, or between magmas and crystal mush or solidified melts from a prior injection of magma in a given system (for example, see other chapters in this volume: Bindeman 2008; Costa 2008; Kent 2008; Streck 2008; Ramos and Tepley 2008), it is possible—or even likely—that simple crystallization from a single magma, as described in the previous section, is the exception rather than the rule. Especially in the case of bulk mineral separates where ages are averaged over many hundreds or thousands of crystals, crystal ages may reflect contributions from crystals with different origins ranging from phenocrysts *sensu stricto* (autocrysts) to antecrysts scavenged from remnants of earlier phases of the magmatic activity at the same center, to true xenocrysts. Although the diversity of origins of minerals within bulk separates may complicate interpretation of U-series data in terms of ages, the data can provide unique constraints on the origin and age of different components contributing to the crystal population in a given magma.

Xenocryst identification by in situ dating. An outstanding feature of *in situ* mineral dating is its ability to isolate contributions from xenocrysts—or xenocrystic cores—from those of autocrysts. Restitic or recycled crystals could skew the results of age dating of mineral separates. Potential contaminants acquired during sample or in laboratory preparations, while not strictly xenocrystic in the normal sense, might also be identified. The presence of xenocrysts could also signal a role for crustal melting and/or assimilation and can be a means of fingerprinting the sources of those inputs. For example, a potential contribution from greywacke sources to the generation of the Taupo volcanics, New Zealand, was provided by discovery of Cretaceous-aged zircons (Charlier et al. 2005). Alternatively, if independent evidence for assimilation exists, the absence of xenocrysts could provide insights into the conditions of magma evolution by requiring that xenocrysts are either completely resorbed during melting and/or are very effectively isolated from the melt.

When considered in total, a notable conclusion from the work to date is that zircons that predate the onset of a given center's magmatic activity are remarkably rare in volcanic rocks. Less than 1% (10 of >200 zircons) analyzed in nine units erupted over the past 45 k.y. from Taupo volcano, New Zealand, were found to be older than Quaternary in age (Charlier et al. 2005). Brown and Smith (2004), in particular, targeted their efforts at recognizing xenocrystic material in the 1.21 Ma Ongatiti ignimbrite of the Taupo Volcanic Zone. Cathodoluminescence imaging revealed only four resorbed cores—out of 300 zircons studied—and these yielded 119 to 340 Ma U-Pb ages that link the xenocrysts to Mesozoic and Paleozoic metasedimentary rocks (Fig. 9). Young, 15-30-μm-thick rims on those zircons are inferred to represent growth in the thousands of years before eruption, indicating that the cores are not simply accidental material eroded from the conduit during or immediately prior to eruption (Brown and Smith 2004). Bindeman et al. (2006) noted that, of more than 200 individual zircons analyzed in studies of Yellowstone volcanism (Bindeman et al. 2001; Vazquez and Reid 2002), only three Cretaceous and Triassic—and no Archean basement—xenocrysts were found. Similarly, fewer than 2% of the >120 Long Valley zircons studied so far (Reid and Coath 2000; Simon and Reid 2005; Simon et al. 2007) are xenocrysts and these are from underlying Mesozoic crust. Many other studies report no xenocrysts (e.g., Lowenstern et al. 2000; Charlier et al. 2003; Bacon and Lowenstern 2005; Bachmann et al. 2007).

Perhaps less surprisingly, settings in which silicic magmas are normally in the minority may have relatively few xenocrysts. Lowenstern et al. (2006) found a single Pan-African basement zircon amongst the otherwise young zircons that characterize granophyre xenoliths entrained in rift-related basalts from the Alid Volcanic Center, Eritrea (Fig. 10). Schmitt and Vazquez (2006) dated zircons in rhyolites and magmatic xenoliths associated with continental rupture and development of nascent oceanic crust in the Salton Trough area of southern California. Two of 44 zircons studied yielded Proterozoic ages; a single xenocryst of Jurassic age has also been found (Fig. 9; Brown et al. 2004). The Proterozoic zircons are interpreted to be detrital grains incorporated during the eruption since visible overgrowths on these zircons are absent and sedimentary fragments in the lavas are commonly smeared out and intermingled with the obsidian. Pilot et al. (1998) reported 330 and 1,600 Ma old zircons from gabbros exposed near the Kane fracture zone of the Mid-Atlantic Ridge. In contrast, Moeller et al. (2006) obtained U-Pb dates on zircons found within polished petrographic thin sections of gabbroic rocks from similar samples from the same locations. They found only ~1 Ma and younger zircons and attributed the previously reported Paleozoic-Proterozoic ages to contamination during sample preparation.

Only a few studies report substantial evidence for xenocrysts. More than two thirds (fourteen of 20) of analyzed zircons from the ~13 k.y. lower Laacher See Tuff yielded pre-Quaternary ages ranging in age from ~390 Ma to 2.7 Ga (Schmitt 2006). While some are magma-hosted xenocrysts, most lack adhering glass and are inferred to be crystals derived

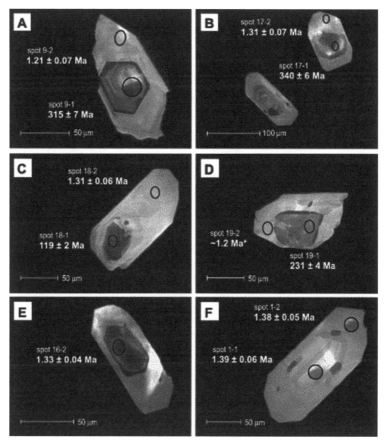

Figure 9. Cathodoluminescence images of selected Ongatiti zircon grains showing spot ages for xenocrystic cores surrounded by magmatic-aged rims. Black ellipses show approximate analysis areas. [Used with permission of Elsevier from Brown and Smith (2004) *Journal of Volcanology and Geothermal Research*, Vol. 135, p. 252.]

from disseminated country rock fragments. Three of 7 zircon grains dated from the 7 k.y. old rhyolite eruption of Cotopaxi volcano, Ecuador, gave Proterozoic ages (0.8-1.2 Ga; Garrison et al. 2006). In that case, younger ages on the remaining zircons (18-30 Ma) are inferred to represent hybrid grains wherein young rims mantle old cores. Five of 14 zircons in the ~36 ka La Virgen tephra from Las Tres Vírgenes composite cone in Baja, California, yielded ages >25 Ma (Schmitt et al. 2006). In contrast, none of the 30 ignimbrite zircons from two adjacent calderas yields an age that deviates from mean zircon ages of <1.4 Ma.

In summary, while it is difficult to generalize, it nonetheless appears that xenocrystic zircon, and especially xenocryst-cored zircons, are an uncommon feature of many volcanic rocks.

Autocrysts versus antecrysts: in situ dating. Accessory phase dating of young plutonic inclusions shows that individual intrusions can have a range of crystallization ages. Xenocrysts notwithstanding, the age distributions may be continuous (e.g., Charlier et al. 2003; Bachmann et al. 2007) or discontinuous (e.g., Lowenstern et al. 2000; Vazquez et al. 2007) but generally overlap the duration of magmatic activity at a given volcanic center. As with xenoliths, volcanic

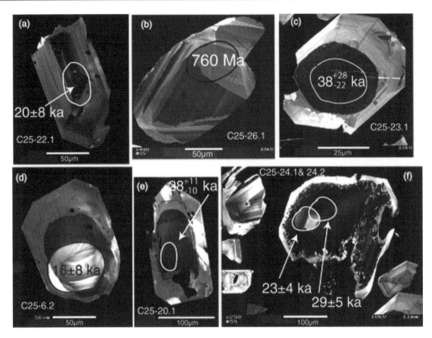

Figure 10. Cathodoluminescence images for zircons from a granitic xenolith from a 15.2 rhyolite pumice deposit at Alid volcanic center, Eritrea. Note the generally euhedral grain shapes and the growth-related oscillatory zoning represented by the dark and light bands throughout some grains (e.g., A) or mantling rounded cores (e.g., C). White and black ellipses show approximate analysis areas. [Used with permission of Oxford University Press from Lowenstern et al. (2006) *Journal of Petrology*, Vol. 47, p. 2114.]

rocks exhibit a variety of age distributions, from unimodal to multimodal. In some instances, the ages correspond to known periods of earlier activity (Bacon and Lowenstern 2005). Disequilibrium mineral assemblages (e.g., Nakada et al. 1994; Gardner et al. 2002; Bachmann et al. 2002; Schwartz et al. 2005) have been used to suggest that so-called "phenocrysts" are actually antecrysts, crystals disaggregated or otherwise remobilized from young crystal-rich domains, whether from earlier crystallized magmas or cumulate mush at the margins of magma bodies (e.g., Mahood 1990; Nakada et al. 1994; Harford and Sparks 2001). Recycling may occur by assimilation or rejuvenation of these subvolcanic intrusions. While age distributions alone are usually insufficient to distinguish antecrysts from phenocrysts (other relevant observations are described in the next section), they are often key arbiters of such a distinction.

One line of evidence for the presence of antecrysts has been identification of U-Th isotopic affinities between melt- and xenolith-hosted zircons. Ca. 27 ka rhyodacite tephra and lavas, interpreted to be early leaks from the magma chamber responsible for the climatic eruption of Mount Mazama, yield an approximately unimodal zircon age distribution that peaks at 45 k.y. (~20 k.y. before eruption; Fig. 11) and matches the zircon age distribution of a granodiorite block (Fig. 12; Bacon and Lowenstern 2005). Tailing of the age distribution to >200 ka and a possible clustering of ages around ~110 k.y. show that older zircons are also present. Exclusive of xenocrysts, U-Th isotope characteristics of zircons in the Lower Laacher See phonolite, Germany, are isochronous with those of syenite xenoliths and define a single age population that crystallized within ~5 k.y. of eruption (Fig. 12; Schmitt 2006). The older peak in a bimodal zircon age distribution for a Salton Sea obsidian matches a discrete ~18 ka zircon population in a felsite xenolith, suggesting episodic growth of zircon and possible recycling of pre-existing zircon-bearing rocks (Schmitt and Vazquez 2006).

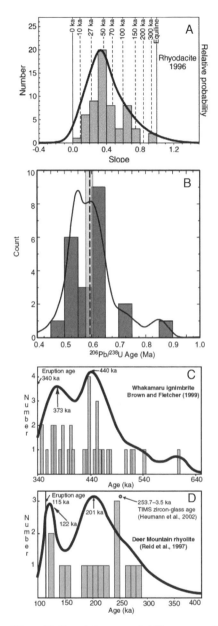
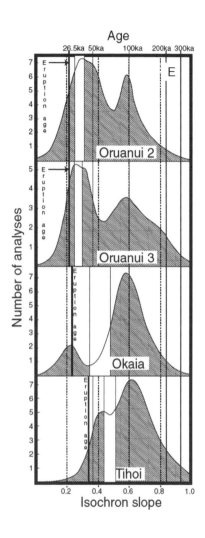

Figure 11. Cumulative age probability curves and, for most, histograms for U-Pb or U-Th age distributions obtained by SIMS dating of domains within zircons. Distributions in Figures 11A and 11E are plotted with respect to U-Th isochron slopes; corresponding ages are shown by dashed vertical lines. Age distributions are unimodal, bimodal, and multimodal. A) U-Th age distribution for zircons from an ~27 ka Crater Lake rhyodacite (Bacon and Lowenstern 2005); B) U-Pb age distribution for zircon rims from the ~600 ka Devils Kitchen rhyolite (Miller and Wooden 2004); C) U-Pb age distribution for zircons from the ~340 ka Whakamaru Ignimbrite (from Charlier et al. 2005 based on data for Brown and Fletcher 1999); D) U-Th age distribution for zircons from the ~115 ka Deer Mountain rhyolite (from Charlier et al. 2005 based on data for Reid et al. 1997); and E) U-Th age distributions in various Taupo rhyolites (eruption ages shown by dark vertical bar; from Charlier et al. 2005). Also shown for reference by vertical white bars are TIMS ages.

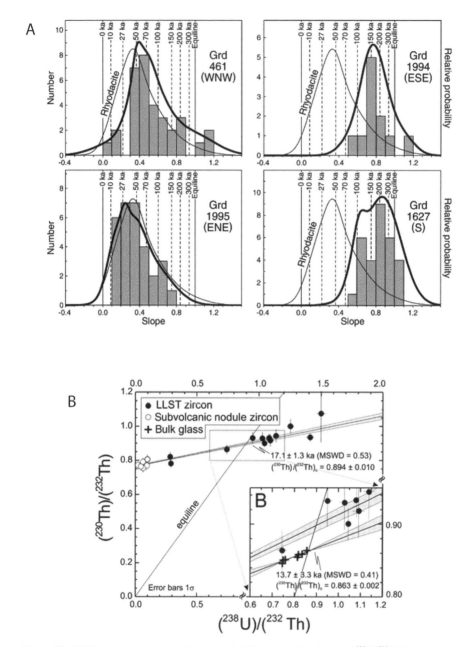

Figure 12. A) Histograms and cumulative age probability curves for slopes of ^{238}U–^{230}Th isochrons for SIMS zircon data from 4 granodiorite samples (heavy lines) and a rhyodacite (thin line) from Mt. Mazama. Vertical dashed lines are reference ages. ENE, etc., in parentheses give approximate azimuths of collection sites relative to center of Crater Lake caldera. B) (^{230}Th)/(^{232}Th) versus (^{238}U)/(^{232}Th) diagram for Laacher See zircons including best-fit ages and initial (^{230}Th)/(^{232}Th)$_0$ values. A xenocryst (excluded from the regression) and another spot were omitted from this plot for clarity. Inset shows close-up with bulk glass separate data for Laacher See phonolitic pumice (Bourdon et al. 1994), recalculated ages, and recalculated (^{230}Th)/(^{232}Th)$_0$. LLST—Lower Laacher See tephra. MSWD—mean square of weighted deviates. Modified after Bacon and Lowenstern (2005) and Schmitt (2006).

Three of the younger and more voluminous Central Plateau Member lavas of Yellowstone caldera, Wyoming, have nearly identical unimodal zircon age populations even though they erupted at times separated by as much as ~50 k.y. (from 160 to 110 ka). Vazquez and Reid (2002) conclude that these lavas tapped a new magma reservoir that differentiated approximately 40 k.y. before the first eruption. In contrast, over half of the zircons in the somewhat older (0.4-0.5 Ma) post-caldera rhyolites are similar to those in older Yellowstone volcanic rocks and could represent crystals recycled from earlier volcanic and/or plutonic rocks (Bindeman et al. 2001). Zircons in the pre-, syn-, and postcaldera eruptions at Ammonia Tanks, Nevada, are similarly interpreted to be recycled from older eruptive cycles (Bindeman et al. 2006).

Antecrysts have also been identified on the basis of considering zircon age distributions in the context of textural observations. Based on the crystal-rich nature of and presence of crystal clots within the ~10 km^3 Earthquake Flat rhyolite, New Zealand, a unimodal zircon population that crystallized ~60 k.y. before eruption is thought to represent antecrysts remobilized not long before eruption from a largely frozen crystalline pluton (Charlier et al. 2003). Similarly, textural characteristics of the 160 ka Kos Plateau Tuff, Greece, were taken to indicate that a broadly unimodal zircon age population that peaks at ~230 ka (70 k.y. before eruption) but tails to significantly older ages (~500 ka; Bachmann et al. 2007) reflects rejuvenation of crystal mush zircons (Fig. 13). The general concordance in age between these zircons and pumice-hosted granitic clasts led Bachmann et al. (2007) to further suggest that the latter were excavated either from the marginal "rind" of a magma chamber or from co-genetic plutonic rocks.

Multimodal age distributions have also been taken as evidence that crystal populations represent mixtures of antecrysts and autocrysts. Even though U-Pb ages for nearly half of dated zircon cores and rims from the 587 ka (K-Ar age) Devils Kitchen rhyolite of the Coso Volcanic Field, California, cluster around 600 ka, the rim ages span a minimum of 200 k.y. (Fig. 11; Miller and Wooden 2004). Peaks in the zircon age distributions are present at 620 and 670 ka, as are subordinate peaks at 730 and 800 ka. Miller and Wooden infer that zircons with relatively old rims (>700 ka) were probably recycled from earlier intrusions whereas the ~670 ka cores that are rimmed by younger growth probably represent zircons incorporated into younger, zircon-saturated melts. Zircon age distributions for the various dacites, rhyodacites and rhyolites associated with the Taupo Volcanic Field, New Zealand, range from unimodal to multimodal (Fig. 11; Brown and Fletcher 1999; Charlier et al. 2003; Charlier et al. 2005). Core to rim age variations in excess of 250 k.y. in some Whakamaru zircons were attributed by Brown and Fletcher (1999) to long magma residence times, with a role for crystal recycling. Charlier et al. (2005) noted the clustering of ages in the Whakamaru results and in the Taupo volcanics they studied and attributed the episodic nature of zircon crystallization to remobilization of zircons from mush instead of prolonged residence in a liquid-dominated magma body. In particular, a prominent cluster of ~100 ka model ages in the pre-26.5 ka Oruanui eruptives corresponds to an earlier phase of modest volume dome-building and is interpreted to reflect crystals derived from partly to wholly solidified plutonic bodies.

Complex crystal populations and bulk mineral ages. The presence of crystals with multiple origins within a bulk mineral separate precludes a simple interpretation of the bulk crystal age in terms of magma residence or average crystal growth rate. However, U-series ages can provide unique insights about the origin and especially the age of different components. U-series age data are particularly useful when combined with other types of chemical or textural information about the crystals, which can in turn provide independent information about the origin and abundance of different components. Measuring ages of mineral separates may be one of the only ways to recognize antecrysts in the case of a magmatic system with limited compositional range. The isotopic compositions of U-series nuclides and, in some cases, the crystal ages themselves can be used as tracers of different populations within a given magmatic system. The effect on U-series ages of mixtures of different minerals of the

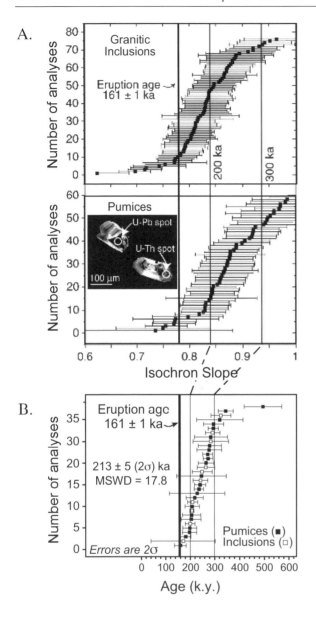

Figure 13. Isochron slopes and age distributions of zircons from Kos Plateau Tuff, Greece, showing presence of continuous age distributions. A) Model isochron slopes and corresponding ^{238}U-^{230}Th age distributions for granitoid xenoliths (top) and pumices (bottom). B) ^{238}U-^{206}Pb age distributions for granitoid xenoliths. Weighted U-Th mean ages are 228±14 ka (1σ; n = 70; MSWD = 1.08) and 194±10 ka (1σ; n = 74; MSWD = 1.6) for pumice and granitic clasts, respectively. Weighted U-Pb age is 213±5 (2σ; n = 33; MSWD = 17.8) for pumice and clasts combined. Eruption age of 161 ± 1 ka shown for reference. Based on Bachmann et al. (2007).

same age, and of progressive crystal growth for a given mineral from the same magma, was discussed earlier; here, we address somewhat more complex scenarios of mixing of minerals and the effects on ages calculated for bulk separates.

Where multiple mineral phases define an isochron on a ^{238}U-^{230}Th isochron diagram, it is generally difficult to maintain a straight line array by mixing of different mineral populations. For example, Heath et al. (1998) present several scenarios of mixing between phyric magmas, between phyric and aphyric magmas, and between a cumulate and a magma (Fig. 14). In general, producing a linear array with appropriate slope requires a fortuitous balance between the age, proportion, and initial ^{230}Th/^{232}Th of each component, and where multiple phases are present mixing of mineral populations is unlikely to explain a linear array. However, in cases where a mineral array can be dominated by one component, such a model could be appropriate, and in some cases the combination of ^{230}Th-^{226}Ra and ^{238}U-^{230}Th data are easiest to explain by mixing scenarios (see below).

In the case of accessory phases that are inclusions in major phases, the U-Th budget of the mineral separate may be so dominated by even small amounts of trace phases that the bulk mineral separate age essentially reflects accessory phase crystallization (e.g., Heumann et al. 2002; Charlier et al. 2003). However, it is still possible to derive some age information from the bulk separate in some of these cases. For example, plagioclase separates from rhyolite pumice and dome samples of the 1305 AD Kaharoa eruption of Tarawera Volcano, New Zealand, plot on a zero-age isochron but to the right of the equiline on a ^{238}U-^{230}Th isochron diagram, which is interpreted to reflect the presence of young zircon inclusions within the bulk separate (observed in thin section); nevertheless, the plagioclase must still be contemporaneous with or younger than the zircon inclusions, requiring a young age for plagioclase as well (Klemetti and Cooper 2007).

In the case of ^{230}Th-^{226}Ra ages, mixtures of different minerals may have more profound effects on calculated ages than in the case of ^{238}U-^{230}Th ages. As discussed earlier, the ^{230}Th-^{226}Ra data for bulk mineral separates must be corrected for the presence of other phases in order to allow calculation of model ages. It is still possible, however, that even a mineral separate containing only a single phase may represent a mixture of crystals of different ages and origins. The extent to which model ages calculated from such mixed populations are meaningful depends on how much factors such as composition (in particular, (^{226}Ra)/[Ba] and (^{230}Th)/[Ba]) of the parent liquid and the crystallization temperature may have varied between the different crystal

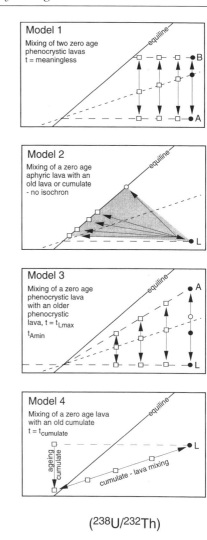

Figure 14. Schematic ^{238}U-^{230}Th isochron diagrams illustrating consequences of mixing different crystal populations in several scenarios. Model 1: mixing of two populations of zero-age minerals crystallized from different magmas. In this case, it would require a fortuitous combination of proportions of mixing for each mineral phase to produce an inclined linear array. Model 2: Mixing of an aphyric lava with old crystals; no isochron will be produced. Model 3: Mixing of zero-age crystals with older crystals formed from melt of similar composition. Like Model 1, this requires a fortuitous proportion of old and young crystals to produce a linear array. Model 4: mixing of a zero-age lava with old cumulate minerals via overgrowths containing melt inclusions. In this case the apparent age is meaningless. [Used with permission of Elsevier from Heath et al. (1998) *Earth and Planetary Science Letters*, Vol. 160, p. 58.]

populations. In the case where crystals of different ages precipitate from the same magma, the model age will represent the average age of crystallization. In somewhat more complex cases where there is evidence for contributions of crystals from multiple magmas, it can still be possible to calculate average ages by assuming a limited range for (^{226}Ra)/[Ba] in the liquids from which the minerals crystallized; this assumption can be tested in cases where ^{238}U-^{230}Th-^{226}Ra disequilibria in multiple lavas of diverse ages have been analyzed (although there is always some degree of uncertainty about whether erupted lavas are representative of the diversity of compositions sub-surface). However, model age calculations in many cases may be relatively robust to assumptions about liquid (^{226}Ra)/[Ba], within the range measured for a given volcanic system (e.g., Cooper and Reid 2003; Tepley et al. 2006). Overall, model ages that represent averages of mixed crystal populations can still be estimated, albeit with somewhat larger uncertainties due to the uncertainty in (^{226}Ra)/[Ba] of the magma(s) from which the minerals crystallized.

Xenocrysts in bulk mineral separates. A unique feature of U-series ages compared to other geochronologic techniques is that once a given parent-daughter pair has attained secular equilibrium, the ratio of parent to daughter remains constant. Therefore, very old crystals have no more leverage on the average age of a bulk population than do crystals that have just recently attained secular equilibrium. This is in contrast to systems where the parent/daughter ratio continually decreases over time (e.g., U-Pb), and it means that xenocrysts with ages of millions or hundreds of millions of years have relatively little influence on the average age of a mineral separate. However, it also means that other techniques (e.g., U-Pb dating or crystal-scale measurements of long-lived isotopic tracers such as Sr or Nd) must be used in order to definitively distinguish xenocrysts from antecrysts.

Discordant ages and antecrysts in bulk mineral separates. In the simple scenario of crystallization over a short duration followed by closed system aging, dates of the same samples using different parent-daughter pairs should yield the same age. Relatively few studies exist where multiple parent-daughter pairs have been measured for the same samples, although the data set is growing. In many cases where ^{238}U-^{230}Th and ^{230}Th-^{226}Ra ages have been measured on the same mineral separates, the two systems yield different apparent ages (e.g., Volpe and Hammond 1991; Volpe 1992; Cooper and Reid 2003; Turner et al. 2003a). Furthermore, in all cases where ^{226}Ra-^{210}Pb disequilibria have been measured on plagioclase separates, measurable disequilibrium yields model crystallization ages of decades (Reagan et al. 2006, 2008), even though ^{238}U-^{230}Th and ^{230}Th-^{226}Ra model ages for plagioclase in the same samples are thousands to tens of thousands of years old (Tepley et al. 2006; Cooper and Donnelly 2008). Similarly, anorthoclase separates from Mt. Erebus show ^{228}Th-^{232}Th disequilibria, indicating that some crystal growth occurred within 30 yr of eruption (Reagan et al. 1992), despite ^{230}Th-^{226}Ra ages of ~2500 years (note that although these isochron ages were not corrected for the effects of Ra-Ba fractionation, and therefore the true ages are likely somewhat younger than the quoted ages, the crystal ages are not likely to be within error of eruption age).

Although some cases of discordant ages may be explained as mixing arrays with old apparent ages, a more general explanation is that the discordant ages reflect protracted or episodic crystal growth spanning thousands to tens of thousands of years, where the different parent-daughter pairs will record different average ages for the same proportion of old and young material (Cooper and Reid 2003). An effect of the unchanging parent/daughter ratio at secular equilibrium is that all crystals in secular equilibrium, whatever their true age, appear to be the same 'age', as determined by the time required for a given parent-daughter pair to reach secular equilibrium—approximately 350 k.y. for ^{238}U-^{230}Th, 10 k.y. for ^{230}Th-^{226}Ra , and 100 yr for ^{226}Ra-^{210}Pb. One implication of this effect is that the average age of a crystal population is always biased toward young crystal growth, but that the apparent age of a mixture will vary with the age of the old component and between different parent-daughter pairs. Mixtures of crystals where the old component is not in secular equilibrium for either parent-daughter pair would be expected to produce concordant apparent ages (unless the concentration of parent varies in a different way for each parent-daughter pair, or the liquid from which the minerals crystallized had a different composition, between the two components). If the older component is in secular equilibrium for the shorter-lived daughter, discordant apparent ages between different parent-daughter pairs are an expected result of mixing crystal populations of different ages (Fig. 15).

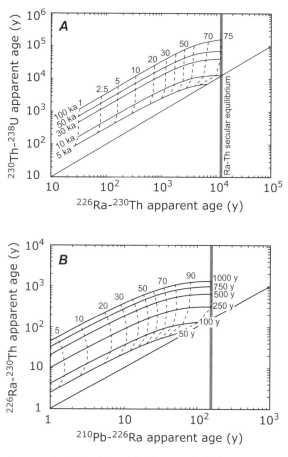

Figure 15. Apparent ages produced by mixing old and young plagioclase components in a bulk mineral separate. A) ^{230}Th-^{226}Ra apparent age vs. ^{230}Th-^{238}U apparent age (modified from Cooper and Reid 2003). B) ^{226}Ra-^{210}Pb apparent age vs. ^{230}Th-^{226}Ra apparent age. In each panel, fine straight line highlights concordant ages, and thick grey line represents secular equilibrium for the shorter-lived parent-daughter pair. Solid curves are constructed by calculating apparent ages (with reference to the young liquid) of a mixed crystal population, where the young component is assumed to be zero age and where the age of the old component is varied to produce the different solid curves on the diagrams. Dashed lines represent contours of a constant percentage of old plagioclase, where the percentage is labeled. The scenario modeled is that some plagioclase crystals ('old' crystals) precipitated from a melt, aged, and then were incorporated into a new liquid with similar composition to the original magma, followed by a second phase of crystallization that occurred in equilibrium with the new magma immediately prior to eruption and analysis. Both magmas are assumed to have (^{230}Th)/(^{238}U) = 1.10, (^{226}Ra)/(^{230}Th) = 1.30, and (^{210}Pb)/(^{226}Ra) = 1.0 at the time of crystal growth and the same concentrations of U, Th, Ra, Ba, and ^{210}Pb, leading to the equal concentrations of U, Th, and Ba in old and young plagioclase. Abundances of Ra and ^{210}Pb, and ^{230}Th/^{232}Th of old plagioclase are calculated by accounting for U, Th, Ra, and ^{210}Pb decay since the time of crystallization.

There is a tradeoff between the age of the old component and the percentage of old material that can account for any given combination of two apparent ages (Fig. 15). However, if independent evidence of the proportion of old and young crystal growth exists (e.g., textural information or other chemical evidence fingerprinting different components), the age of the older component can be estimated. In the case where ^{238}U-^{230}Th, ^{230}Th-^{226}Ra, and ^{226}Ra-^{210}Pb data all exist for the same samples, the combination of apparent ages for the three systems may be enough to uniquely estimate the age and proportion of old and young components (although it should be noted that this would not be possible in the case of more than two growth episodes). Thus there is additional information to be gained from analysis of multiple parent-daughter pairs in the same samples.

For example, discordant ^{238}U-^{230}Th and ^{230}Th-^{226}Ra ages in plagioclase have been measured in two samples of the ~2000-ybp Castle Creek eruptive period at Mount St. Helens (Fig. 16; Volpe and Hammond 1991; Cooper and Reid 2003). Textural evidence for disequilibrium between plagioclase and liquids, combined with major-element and trace-element zoning in plagioclase beyond that expected for crystallization from a single magma, suggest that these discordant ages reflect mixing of multiple crystal populations where the older component is tens of thousands of years old and makes up on the order of 40-50% of the bulk mineral separate (Cooper and Reid 2003). It is interesting in this context that the four other Mount St. Helens samples analyzed in the same study (with eruption ages between 2000 ybp and 1982) also show textural and trace-element evidence for mixed crystal populations, but do not have discordant ^{238}U-^{230}Th and ^{230}Th-^{226}Ra ages (Cooper and Reid 2003). One possible interpretation is that the older component in the case of the four younger samples is not old enough to create discordant ages. Furthermore, all of the average plagioclase ^{230}Th-^{226}Ra model ages overlap, regardless of whether the Ra-Th and U-Th ages are discordant, indicating that some crystal component in each magma must have been present beneath the volcano at the same time (Fig. 16).

Another example of discordant ages is seen in samples from Tonga and from Soufriere, Lesser Antilles, where ^{238}U-^{230}Th plagioclase ages are 30-50 ka whereas the same mineral separates show ^{230}Th-^{226}Ra disequilibria (Fig. 17; Turner et al. 2003a). An initial ^{238}U-^{230}Th study of minerals from Soufriere lavas and cumulates concluded that, whereas they could not rule out mixing producing the linear array of ^{238}U-^{230}Th data that they measured in samples from Soufriere, they preferred a model where the crystals represented phenocrysts precipitated from the magma (i.e., autocrysts) and therefore magma residence times on the order of tens of thousands of years (Heath et al. 1998). Turner et al. (2003a) present additional analyses of samples from Soufriere, including ^{230}Th-^{226}Ra data for minerals separated from a cumulate xenolith. They found that the cumulate minerals defined a linear array on a ^{238}U-^{230}Th isochron with the same apparent age as the earlier samples (40-50 ka), yet all of the mineral separates preserved ^{230}Th-^{226}Ra disequilibria. Based on these discordant ages and on crystal size distribution (CSD) evidence for complex crystal populations, Turner et al. (2003a) concluded that the data best fit a model where the cumulate minerals have old cores (tens of thousands of years) with young overgrowths produced during entrainment in a new magma. Turner et al. (2003a) also present ^{238}U-^{230}Th-^{226}Ra data for Tonga lavas and one cumulate xenolith, in addition to crystal size distribution data which shows evidence for multiple crystal populations, and they interpret the combined data set to reflect storage of plagioclase as cumulates, which were remobilized and incorporated into subsequent magmas passing through the same system. In addition, Turner et al. (2003a) also present calculations of apparent ages of mixtures, finding that discordant apparent ages are expected in the case of old cores mixed with young rims.

A similar pattern of discordant data is also shown in data for the 2004-2008 eruption of Mount St. Helens (Cooper and Donnelly 2008). Whole-rock samples of dacite erupted from October 2004-April 2005 overlap with samples erupted in the 1980's on a ^{238}U-^{230}Th isochron diagram. However, plagioclase separates in two samples erupted only a month apart (October 2004 and November 2004) have ^{230}Th/^{232}Th ratios that are significantly different from each other

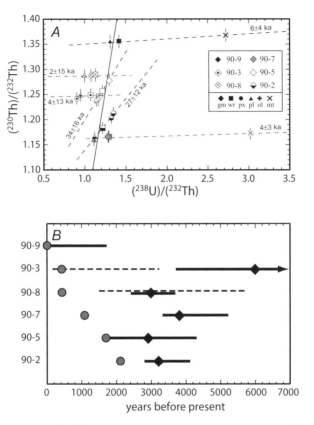

Figure 16. Comparison of ^{238}U-^{230}Th isochron ages and fractionation-corrected ^{230}Th-^{226}Ra ages for minerals separated from Mount St. Helens samples (modified after Cooper and Reid 2003). A) ^{238}U-^{230}Th isochron; data from Volpe and Hammond (1991). Apparent ages are indicated by numbers along dashed isochron arrays. Abbreviations as follows: groundmass (gm), whole rock (wr), pyroxene (px), plagioclase (pl), olivine (ol), magnetite (mt). Solid line represents the equiline. B) Summary of fractionation-corrected ^{226}Ra-^{230}Th ages for plagioclase (solid lines) and pyroxene (dashed lines). Filled diamonds represent "best-estimate" plagioclase model ages using mean plagioclase compositions and pre-eruptive temperatures based on bulk composition. Lines indicate range of plagioclase model ages allowing for variations in An (± An$_{10}$) and temperature (900-1000 °C for dacites, 1000-1100 °C for andesites, and 1000-1200 °C for the basalt). Arrow pointing to right for MSH 90-3 plagioclase age indicates that the upper limit is >8 ka. Range of pyroxene model ages are shown by fine dashed lines; this range primarily reflects propagation of uncertainties through mass-balance calculations of Ra in pure pyroxene; uncertainties in model D$_{Ra}$/D$_{Ba}$ have a very minor effect. Grey circles represent eruption ages for each sample (sample numbers along vertical axis). Note that apparent U-Th ages are within error of Ra-Th ages for all samples except 90-5 and 90-2, likely indicating an old crystal component in these two samples. See text for discussion.

and from plagioclase in a sample of 1982 dacite (Volpe and Hammond 1991). The apparent whole-rock/plagioclase age for one sample is ~40 ka, yet plagioclase in this sample also has significant ^{230}Th-^{226}Ra disequilibria (Cooper and Donnelly 2008) and ^{226}Ra-^{210}Pb disequilibria (Reagan et al. 2008). These data are most consistent with mixing of plagioclase with diverse ^{230}Th/^{232}Th ratios and different ages, ranging from tens of thousands of years to a few decades prior to eruption.

The presence of ^{226}Ra-^{210}Pb disequilibria or ^{228}Th-^{232}Th disequilibria in other feldspar separates with ^{230}Th-^{226}Ra apparent ages of thousands of years (Reagan et al. 1992, 2006,

2008) suggests that a very young crystal component, perhaps reflecting crystallization during ascent and eruption, may be ubiquitous in volcanic rocks but is rarely recognized when using other mineral dating techniques.

Overall, there are many cases where discordant ages and/or other textural and chemical information about the crystal populations indicate recycling of crystals within the same magmatic system. Especially in cases where there is limited compositional variability between the erupted products (e.g., Mount St. Helens 2004-2008 eruption compared to the 1980-86 eruption), age information may be the most effective way not only to recognize crystal recycling, but also to quantify the proportion and ages of various components. Collectively, these cases where discordant ages are observed suggest that by combining age information with textural and crystal-chemical information it may be possible to reconstruct a significant part of a volcano's history from the crystal record.

Insights from dating of subvolcanic plutons

Many lavas and pyroclastic deposits contain plutonic blocks that could represent magmas or crystal mush that solidified in subvolcanic domains or, in the case of anatexis, could be the remnants of magma sources. Dating of these blocks is a means of distinguishing between these possibilities and in the case of the former, providing insights into the emplacement and growth of subvolcanic intrusions. The more complete volcano-plutonic record provided by considering the crystallization history of plutonic as well as volcanic rocks can better illuminate the dynamical evolution of a volcanic system.

Plutonic zircons studied to date have come mainly from dioritic to granodioritic blocks but zircons have been obtained

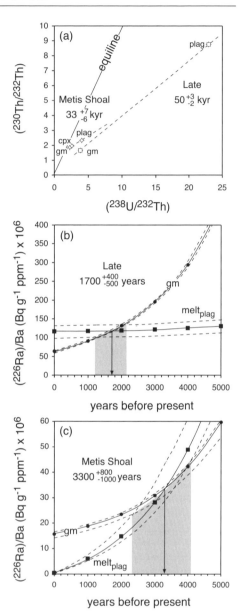

Figure 17. Diagrams showing discordant ^{238}U-^{230}Th (A) and fractionation-corrected ^{230}Th-^{226}Ra ages (B, C) for samples from Tonga. [Used with permission of Elsevier from Turner et al. (2003) *Earth and Planetary Science Letters*, Vol. 214, p. 284.]

from syenitic and gabbroic inclusions as well (e.g., Schmitt and Vazquez 2006; Bacon et al. 2007; Schmitt 2007). The presence of interstitial melt in plutonic blocks has been attributed to partial melting (Schmitt and Vazquez 2006; Bachmann et al. 2007) and to partial solidification (Charlier et al. 2003). Granophyric intergrowths of quartz and feldspar are common in granitoid xenoliths, as are intercrystalline voids and miarolitic cavities (Charlier et al. 2003; Lowenstern et al. 2006; Schmitt 2006; Bacon et al. 2007). Small to millimeter-sized zircon crystals may occur in and adjacent to the inclusion vesicles (Fig. 18). Micrometer-sized cavities and tunnels in some zircons suggest phase changes (solidification or resorption) in the presence of aqueous fluid (Bacon et al. 2007) while other zircons exhibit evidence of sub-solidus exsolution (Fig. 18; Schmitt 2006). Ages for zircon populations dominated by these crystals may be biased towards the latest stages of crystallization.

Figure 18. Backscatter electron images of Lower Laacher See Tuff zircon (Zrn) with A) ThSiO$_4$ (Tho) rods present versus B) homogeneous zircon. Dotted ellipses indicate U-Th ion microprobe spot locations. C) Secondary electron image of vesicle-grown zircon in a syenitic subvolcanic nodule. [Used with permission of the Geological Society of America from Schmitt (2006) *Geology*, Vol. 34, p. 598.]

Zircons and major phases in nodules associated with dominantly mafic volcanic centers provide little evidence for protracted crystallization histories. For example, well-resolved U–Th zircon ages for Salton Buttes granophyre xenoliths indicate crystallization within a few 1000 yr (Schmitt and Vazquez 2006). Batches of dioritic magma may have evolved for ≤10^4 years beneath Hawaiian volcanoes based on the similarity between core and rim ages on individual crystals (Vazquez et al. 2007). Low-U zircons in the Alid, Eritrea, granophyre may have grown from less differentiated magma <15,000 years prior to final solidification (Lowenstern et al. 2006). Differences between ages of growth zones or between individual grains in individual nodules are not statistically significant in granitoid xenoliths from Medicine Lake, California (Lowenstern et al. 2000), except for a single Medicine Lake crystal with a core that crystallized before all others and ~50 k.y. before its rim. The same generalities apply to Crater Lake xenoliths (Bacon and Lowenstern 2005) where multiple analyses of zircon grains mostly yield similar ages on the same grains. Some of the granodiorite zircons yield younger ages at their ends or margins than nearer to or in the center and one grain with a core in U-series secular equilibrium has multiple younger ages towards rims. U-Th-Ra dating of minerals in cumulate mafic and ultramafic xenoliths entrained in potassic lavas from Sangeang Api, Indonesia, yield ages that are both young (<2 ka) and concordant with mineral ages for one of the lavas, suggesting that a dominant interval of differentiation produced subvolcanic cumulates which were entrained in the magmas during eruption (Turner et al. 2003b). In addition, as discussed in a previous section, crystal size distributions together with discordant U-Th and Th-Ra ages of minerals in cumulate xenoliths and lavas from Tonga and Soufriere suggest multiple stages of crystal growth and storage as cumulate or plutonic bodies, followed by entrainment of crystals and additional crystal growth within the host magma (Turner et al. 2003a).

Some comagmatic (*sensu lato*) blocks yield relatively simple zircon age distributions (e.g., Alid granophyres, Eritrea; Lowenstern et al. 2006). Others, such as dioritic to leucocratic xenoliths from Hawaii (Vazquez et al. 2007), yield bimodal age populations with peaks that are separated by 60 to 200 k.y. Zircons from four granodiorite blocks from Crater Lake yield a broad range in crystallization ages (between ~20 ka and ~300 ka) that cluster at 50–70, ~110, and ~200 ka (Bacon and Lowenstern 2005). The older core in the Medicine Lake granitoid grain noted above may also be derived from an earlier intrusion. Thus, magmas that solidify beneath young volcanic centers apparently recycle zircons from earlier intrusions.

Comparisons between accessory phase ages for volcanic rocks and the plutonic blocks they host provide a means of linking volcano-plutonic records, as in the case of zircons in a ~27 ka Crater Lake rhyodacite lava that apparently crystallized at the same time as zircons in an associated granodiorite block (Bacon and Lowenstern 2005). U-Th zircon ages corroborate evidence from crystal sizes and cathodoluminescence patterns for a common zircon population between an erupted Salton Sea rhyolite and a possible shallow intrusive (Schmitt and Vazquez 2006). The age concordance between zircons extracted from Kos, Greece, granitic clasts and pumice zircons shows that the granitic inclusions were derived from co-genetic intrusive rocks (Fig. 13; Bachmann et al. 2007). Relatively uniform U-Th isotope characteristics for zircon crystals present in the vesicles of juvenile syenitic nodules from Laacher See cluster close to the isochron defined by zircons from the Lower Laacher See tephra and similarly suggest a cogenetic origin for plutonic and volcanic zircon populations (Fig. 12; Schmitt 2007).

In some cases, the timing of zircon crystallization in xenoliths correlates with episodes of volcanic activity rather than with volcanic zircon populations. Petrographic, chemical, and isotopic similarities between the Alid, Eritrea, granophyres and host tephras suggest that the granophyres may be wall fragments from the shallow magma chamber responsible for the Alid tephra. Zircons apparently grew late in the crystallization history of granophyre blocks (and are absent in the tephra) and virtually all crystallized within the same time period as doming and volcanism at the Alid volcanic center (36–15 ka; Fig. 10). The timing of zircon crystallization in Crater Lake granodiorite blocks also clusters at intervals that correspond to known periods of dacitic volcanism (Bacon and Lowenstern 2005).

The more complete volcano-plutonic record provided by dating plutonic nodules can better reveal the complexity of subvolcanic plumbing systems. For example, some of the glass-bearing granitoid fragments from Rotoiti, New Zealand, had been interpreted on mineralogical and chemical grounds to be the incompletely crystallized portion of a magma system that erupted >200 k.y. earlier (Charlier et al. 2003). Model ^{238}U-^{230}Th zircon age distributions and means that are very similar to those found in a host ignimbrite pumice and by TIMS analyses of two pumice zircon mineral separates, respectively, are inferred to contradict this. The ~55 ka age yielded by another mineralogically distinct granitoid is essentially eruption-aged and dissimilar to the main Rotoiti zircon population. Thus the Rotoiti granitoid clasts apparently crystallized synchronously with the main Rotoiti magma body but in separate pockets of magma.

Before they were dated, melt-bearing granitic inclusions had been taken as evidence for an anatectic origin for the Kos Plateau Tuff (Keller 1969). However, even the anhedral cores observed in both pumice and clast zircons proved to be no older than 500 k.y., negating the possibility that the clasts are restitic from melting of pre-Kos granites (Fig. 13). In contrast, ~30 and ~9 ka zircons in inferred partial melt pockets in mafic xenoliths attest to melting and recycling of preexisting basaltic crust associated with extension of the Salton Trough (Schmitt and Vazquez 2006). Similarly, based on zircon dating of xenoliths, two brief but voluminous episodes of mafic volcanism at Medicine Lake volcano, California, at approximately 90 and 30 ka were accompanied by intrusion of intermediate to silicic granitoids thought to represent melts of young crust formed in association with the high-heat flux (Lowenstern et al. 2000). Lastly, zircon ages for highly fractionated nodules provide evidence for a more complex interplay between alkaline and subalkaline magmas beneath Hawaii than previously inferred (Vazquez et al. 2007).

RESOLVING AGES FROM COMPLEX CRYSTAL POPULATIONS: INCORPORATING OTHER OBSERVATIONS

Introduction

Taken in isolation, it can be nearly impossible to insightfully interpret U-Pb or U-series ages in terms of magma evolution because of the possible effects of crystal inheritance or recycling and mixing of different crystal populations. In some instances, it may be possible to resolve questions about the extent to which a particular date represents a single geologic event simply by applying more than one age dating technique. Discordant ages from different parent-daughter pairs, for example, can themselves be an indication that crystals of more than one age are present in a bulk mineral separate (see previous section). More often, petrologic and geochemical observations at both the rock and mineral scale will be informative. It is becoming increasingly commonplace to couple trace element, oxygen isotope, and thermometry analysis with *in situ* crystal ages. These help to subdivide crystal populations and/or fractions of crystals according to the conditions under which they may have grown. It is less common to couple other observations of *in situ* crystal chemical or isotopic data with bulk mineral crystal ages, but other textural (e.g., CSD or petrographic studies), chemical, or isotopic information about the magma as a whole has been used to provide interpretive context for bulk mineral ages.

In this section, we review studies that have combined multiple approaches to understanding the implications of crystal ages in terms of magma evolution, including combining multiple dating techniques in the same sample, dating multiple size fractions of the same mineral, combining U-series ages with trace-element or isotopic data from the same crystals, combining age information and crystal thermometry, and incorporating textural information. Each of these approaches offers different perspectives on the processes affecting the chemical and thermal evolution of magmas and, using crystal ages to provide a temporal context, these types of combined studies can be very powerful tools for studying the dynamics of magmatic systems.

Combining U-Th and U-Pb SIMS dating

Accessory phases that give ^{238}U-^{230}Th results within error of secular equilibrium could be no more than a few hundreds of thousands of years old or could be Archean in age. If the former, minerals may have too little ^{206}Pb in-growth for acquisition of a precise ^{238}U-^{206}Pb age. As an illustration of this, results for a zircon from La Vírgen volcano, Mexico, are within error of possible secular equilibrium (289+/−100 ka; 1σ) and give a consistent but poorly constrained U-Pb age of 127±81 ka (Schmitt et al. 2006). Nonetheless, reconnaissance U-Pb surveys of zircons on the ion microprobe can be used to show whether there is little ^{206}Pb, supporting inferences that the zircons are all young and that xenocrysts are absent (Lowenstern et al. 2000; Charlier et al. 2003; Bacon and Lowenstern 2005; Bacon et al. 2007). A particularly instructive age comparison that emphasizes the importance of identifying the age of old zircon is provided by the discordant U-Pb vs. U-Th results of Garrison et al. (2006) for a rhyolite from Cotopaxi volcano, Ecuador. Four zircons that are not in ^{238}U-^{230}Th equilibrium at the 2σ level give ^{238}U -^{206}Pb model ages of 18 to 30 Ma, considerably younger than the 0.8-1.2 Ga zircons also detected in the same rhyolite. This example demonstrates that young rims on old zircons can produce ^{238}U-^{230}Th disequilibrium while preserving old, if geologically meaningless, U-Pb ages.

In those cases where xenocrysts are identified, U-Pb dating can identify possible sources and provide insights into the role of crustal recycling and the duration of zircon dissolution (Vazquez and Reid 2002; Lowenstern et al. 2006; Schmitt 2006; Schmitt and Vazquez 2006; Schmitt et al. 2006). Cores of West Yellowstone zircons that give ^{238}U-^{230}Th close to or within error of secular equilibrium yield three ^{207}Pb-corrected ^{238}U-^{206}Pb isotope ages between ~200-350 ka and two older ages of 770±109 ka and 1192±88 ka (Table 3). In especially favorable circumstances, it can be possible to obtain either model or isochron ages using both U-Th and U-Pb systems. For example, Vazquez et al. (2007) obtained a ^{238}U-^{206}Pb isochron age of 261±28 ka (2σ; n = 24; MSWD = 1.8) that is concordant with a ^{238}U-^{230}Th isochron age of 257 +60/-46 ka (2σ; n = 24; MSWD = 0.69), both on zircons from Hualalai leucocratic xenolith (Fig. 19). Bachmann et al. (2007) obtained continuous U-Th and U-Pb age distributions for the Kos Plateau rhyolite zircons, from eruption to 340 ka, with concordant weighted mean ages: 228±14 ka (1σ; n = 70; MSWD = 1.08) and 194±10 ka (1σ; n = 74; MSWD = 1.6) for U-Th on pumice and granitic clasts, respectively, and 213±5 ka (2σ; n = 33; MSWD = 17.8) for U-Pb on pumice and clasts combined (Fig. 13).

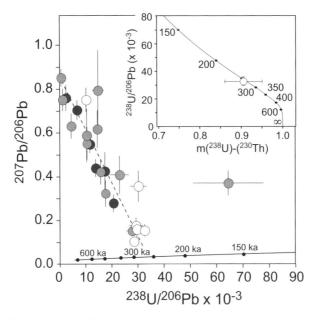

Figure 19. Tera-Wasserburg concordia diagram for zircons from Hualalai, Hawaii, leucocratic xenoliths. Data have not been corrected for common Pb. Concordia curve has been modified to account for the initial ^{230}Th deficit of zircon assuming f = 0.28 (see Section on 'Analytical and Practical Considerations' for further details). Inset shows that U-Pb ages are concordant with the U-Th age (m(^{238}U)-(^{230}Th)) signifies model age isochron slope). Modified after Vazquez et al. (2007).

U-Th/U-Pb spots versus mineral separates

Zircon spot analyses versus TIMS analyses of zircon aliquots. In cases of unimodal age populations, fitting of *in situ* data to a single isochron may be warranted. Thermal ionization mass spectrometric analyses of zircon populations may be justified as well, and can provide additional information on whether the distribution of spot ages is representative of the overall zircon population (e.g., Charlier and Zellmer 2000; Charlier et al. 2005). The TIMS data can, in turn, help to assess the match between the initial Th or Pb isotope ratio for the zircons to that for the host melt (e.g., Schmitt 2006). However, care must be taken to ensure that there is not a subtle age variation with parent-daughter ratio which could produce an array with a slope that is different from the crystallization age and thus yield an erroneous estimate of the initial ratio.

Zircon spot analyses combined with bulk analyses of major phases. In a few cases, zircon spot ages and ^{238}U-^{230}Th dating of bulk mineral separates (major phases +/− zircon) have been done on the same samples, providing additional information for interpretation. For example, U-series disequilibria have been used to examine processes beneath Laacher See volcano, Germany, both through analysis of bulk mineral separates (Bourdon et al. 1994) and *in situ* analyses of zircon (Schmitt 2006). Bulk mineral-glass U-series ages for minerals separated from the Lower Laacher See tephra and from syenite cumulate nodules are either significantly younger (by ca. 5 k.y.) or older (by ca. 12, ca. 22, and ca. 400 k.y.) than the eruption age (Bourdon et al. 1994). The older ages are interpreted as contamination of the mineral separates by secular equilibrium xenocrysts that shifted bulk $(^{230}$Th$)/(^{238}$U$)$ closer to the equiline (Bourdon et al. 1994), whereas heterogeneous initial $(^{230}$Th$)/(^{232}$Th$)$ for melt and crystals could account for apparent crystallization ages younger than eruption (Schmitt 2006). Schmitt (2006) recalculated ages for the samples analyzed by Bourdon et al. (1994) using an estimated $(^{230}$Th$)/(^{232}$Th$)$ in melt derived from the initial $(^{230}$Th$)/(^{232}$Th$)$ inferred from zircon analyses. These recalculated ages are in agreement with the eruption age and are also close to zircon crystallization ages. The new zircon ages, along with recalculated mineral-glass isochron ages, lead to a significant revision of the interpretation of magma storage and differentiation times beneath Laacher See, from tens of thousands of years (Bourdon et al. 1994) to only a few thousand years prior to eruption (Schmitt 2006).

A second example comes from Long Valley Caldera, California. Long Valley Caldera formed during the 760-770 ka eruption responsible for the Bishop Tuff; this eruption was both preceded by and followed by eruption of rhyolitic domes and lava flows (the precaldera Gass Mountain rhyolites and the postcaldera Inyo and Mono rhyolites). Zircon U-Th ages by SIMS (Reid 1997) and bulk mineral ages (Heumann et al. 2002) have been measured for postcaldera rhyolites. Zircon in the Bishop Tuff and the precaldera Glass Mountain rhyolites has been dated by using SIMS U-Pb analyses (Reid and Coath 2000: Simon and Reid 2005; Simon et al. 2007) and TIMS U-Pb analyses (Crowley et al. 2007). Individual feldspar and melt inclusion-bearing quartz from the Bishop Tuff and Glass Mountain rhyolites have been analyzed for their Sr (by TIMS) and Pb (by laser-ablation MC-ICPMS) isotope signatures (Christensen and DePaolo 1993; Christensen and Halliday 1996; Davies and Halliday 1998; Simon et al. 2007), and Rb-Sr model ages have been reported.

Heumann et al. (2002) present bulk ^{238}U-^{230}Th mineral separate ages for combinations of glass, sanidine, plagioclase, and amphibole separated from post-caldera rhyolite domes and lava flows at Long Valley. One sample produces a mineral array with slope corresponding to an age that is 135 k.y. before eruption, similar to a clustering in zircon spot ages (Fig. 11; Reid et al. 1997) for the same eruption. However, the glass and whole-rock analyses are not collinear with the other mineral data, suggesting that the minerals did not have a simple crystallization history. The glass and mineral data for other samples in their study do not in general form linear arrays, and high concentrations of U and Th in some mineral separates suggest that the U-Th systematics of mineral separates for major phases are dominated by the presence of zircon and/or allanite inclusions. They conclude that their data are best modeled by multiple crystallization events for trace phases (zircon and allanite), some of which were captured in zircon single-crystal analyses and spot ages, and some of which are reflected in mineral-glass bulk analyses which are dominated by zircon or allanite inclusions. Rb-Sr dating of feldspars in the same samples suggests feldspar fractionation at ~250 ka, followed by zircon crystallization from ~250 ka to ~140 ka.

Reid and Coath (2000) and Simon and Reid (2005) presented *in situ* U-Pb dating of zircon from the Bishop Tuff and pre-caldera Glass Mountain rhyolites. Zircon from the Glass Mountain rhyolites yielded ages that cluster between 2.0 and 1.7 Ma and between 1.1 and 0.85 Ma and support differentiation and crystallization in these magmas hundreds of k.y. before eruption. They found that Bishop Tuff zircons had mean crystallization ages that preceded eruption (at 760-770 ka) by 90 k.y. and concluded that the Glass Mountain and Bishop Tuff eruptions tapped different magma bodies. This conclusion is supported by additional Pb isotope data for feldspars separated from Glass Mountain and Bishop Tuff rhyolites (Simon et al. 2007) which show distinct Pb isotope compositions for feldspars from precaldera and caldera-related rhyolites. In contrast, Crowley et al. (2007) found ^{238}U-^{206}Pb crystallization ages for the Bishop Tuff that are essentially concordant with the ^{40}Ar/^{39}Ar eruption age, as described earlier in this chapter. Additional work is needed to resolve the discrepancies between the SIMS and TIMS age results.

A similar conclusion, that eruptions from the same system tapped distinct magmas, was shown for New Zealand rhyolites by Charlier et al. (2003). Charlier et al. (2003) present ^{238}U-^{230}Th data for major phase separates (magnetic separate, plagioclase, and biotite) along with ^{238}U-^{230}Th age of zircon separates by TIMS and *in situ* zircon ^{238}U-^{230}Th ages by SIMS from two rhyolitic eruptions from the Okataina volcanic center that likely occurred within months of each other (the Rotoiti and Earthquake Flat eruptions). Differences in zircon age spectra and mean zircon ages for the two eruptions indicate that they tapped distinct magma bodies (Charlier et al. 2003). In this case, the U-Th systematics for major phases appear to be dominated by inclusions of accessory minerals (monazite or zircon), such that isochron ages essentially give average accessory phase crystallization; these bulk mineral ages agree with TIMS ages for zircon separates and with the mean of zircon spot ages (Charlier et al. 2003), and thus appear not to provide information about the evolution of the magmatic system beyond what was available from the zircon analyses.

However, in other cases, comparisons of bulk mineral separate ages with *in situ* ages can provide a more complete view of the geochemical evolution of a magma. For example, Charlier and coworkers (Charlier and Zellmer 2000; Charlier et al. 2005) analyzed ^{238}U-^{230}Th disequilibria in bulk separates (plagioclase, orthopyroxene, hornblende, magnetite, and zircon separates of different size fractions), as well as *in situ* ^{238}U-^{230}Th data for individual zircon grains separated from samples of the Oruanui eruption of Taupo, New Zealand. The zircon separates analyzed by TIMS give average ages that vary systematically with size fraction, which is interpreted to reflect progressive crystallization where the young growth is preferentially sampled in the smaller size fractions (Charlier and Zellmer 2000). A ^{238}U-^{230}Th isochron derived from glass and major phases is permissive of co-crystallization of the major phases with the youngest zircons at ~33 ka (Charlier et al. 2005), yet the zircon spot ages extend to much older ages, suggesting that many of the zircons are antecrystic and record a longer period of magmatic evolution than the major phases.

Recently, Klemetti and Cooper (2007) present *in situ* ^{238}U-^{230}Th data for zircons and ^{238}U-^{230}Th-^{226}Ra data for bulk plagioclase separated from the ca. 1300AD Kaharoa eruption of Tarawera volcano in the Okataina Volcanic Center, New Zealand. Zircon spot ages range from within error of the eruption age to >300 ka (within error of secular equilibrium), though the majority of ages fall in the range of 20-50 ka. Plagioclase ^{238}U-^{230}Th data yield a two-point plagioclase-glass isochron within error of the eruption age. High U/Th and high Zr measured in splits of the dissolved bulk plagioclase separate, together with petrographic observation of zircon inclusions in plagioclase, suggest that the ^{238}U-^{230}Th data for the bulk plagioclase primarily reflect the presence of zircon inclusions. However, the plagioclase itself must still be dominantly young in order to host young zircon inclusions. Further evidence for youthful plagioclase comes from ^{230}Th-^{226}Ra disequilibrium measured in the plagioclase separates, which requires that the plagioclase and any included zircon must be less than a few thousand years old. Thus, the combined data require that the plagioclase (plus zircon inclusions) crystallized shortly before eruption, whereas the bulk of the zircon grains measured by SIMS have cores that are tens of thousands of years old, suggesting that most of the zircons were antecrystic, whereas the plagioclase appears likely to have grown from the host liquid.

In other cases, the bulk mineral ages may be dominated by mixing signatures, where the zircon ages provide age information. Garrison et al. (2006) analyzed ^{238}U-^{230}Th disequilibria in bulk mineral separates from two andesites and two rhyolites from Cotopaxi Volcano, Ecuador, in addition to the U-Th spot ages for seven at least partially xenocrystic zircons described previously in this section. The bulk mineral age for one of the andesites is undefined, with the other yielding an apparent age of 28 ka; however, based on textural evidence, Garrison et al. (2006) interpret this array to reflect mixing rather than having age significance. ^{238}U-^{230}Th isochron ages for the rhyolites are 76 and 112 ka, compared to eruption ages of 7.2 and 6.3 ka. Collectively, these data suggest melt generation in multiple stages (including melting of crust adding xenocrysts to the rhyolites), mixing, and extended residence (75-100 ka) of silicic magmas within the crust prior to eruption.

In summary, zircon spot ages combined with bulk mineral ages for major phases and zircon can provide significantly more insights about the evolution of magmas than can either technique alone. Although the details of the information each provides may vary from one magmatic system to another, the combination of ages for major phases and on trace phases allows a more complete history of the magma to be determined, including the temporal relationship between crystallization of major and trace phases.

Multiple size fractions

As discussed previously, Charlier and Zellmer (200) and Charlier et al. (2005) analyzed bulk zircon separates of multiple size fractions, and found in some cases that there were systematic differences in age between different sizes of crystals. This type of study suggests that, similar to the additional information about magma evolution provided by dating multiple phases within a given system, multiple size fractions of a given phase can provide additional insights about the duration of crystallization for a given phase, or perhaps the proportion of xenocrysts and antecrysts that may be present in a magma. Few studies to date have attempted U-series dating of multiple size fractions of major phases. Donnelly and Cooper (2006) present U-Th data for three size fractions of plagioclase in a sample erupted from Mount St. Helens in 2004 (Fig. 20). All size fractions have ^{230}Th/^{232}Th much lower than that in the host whole-rock, suggesting either that the plagioclase is tens of thousands of years old or, more likely, that the bulk separate represents a mix of older plagioclase (which might also have crystallized from a lower-^{230}Th/^{232}Th magma) with young plagioclase in equilibrium with the host magma (Donnelly and Cooper 2006; Cooper and Donnelly 2008). The three size fractions define a linear array, with the largest (250-500 µm) size fraction having the lowest ^{230}Th/^{232}Th ratio, but the intersection of the array with the ^{230}Th/^{232}Th ratio in the melt also has higher U/Th than any of the measured plagioclase or than partitioning would predict for zero-age plagioclase (Fig. 20). One possibility is that the different size fractions represent mixing between an old, large, low-^{230}Th/^{232}Th antecrystic plagioclase population and a second end-member that is itself a mix between zero-age plagioclase and whole-rock material adhering to the outside of grains or melt inclusions. The U/Th ratio of the largest size fraction is similar to that predicted from partitioning (if the crystals precipitated from a magma with similar U/Th to the host whole rock), suggesting that this fraction may contain a large percentage of old plagioclase. The addition of ^{230}Th-^{226}Ra data for the separates (in progress) will test whether the larger size fraction does indeed contain a smaller proportion of zero-age plagioclase, which will then allow an assessment of the characteristics and likely origin of the old plagioclase population.

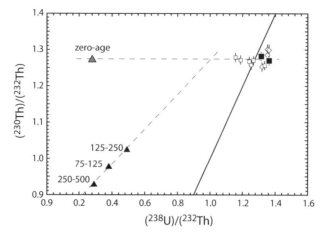

Figure 20. ^{238}U-^{230}Th isochron showing data for whole rocks and multiple size fractions of plagioclase in Mount St. Helens dacite. Shown are analyses of two splits of SH304-2A1 whole-rock (black squares), three size fractions of plagioclase separated from SH304-2A1 (black triangles; labeled with size fraction in micrometers), additional whole-rock analyses of other dacites erupted from Mt St. Helens during 2004-2006 (open squares), whole-rock analyses of samples erupted from 1980-1986 (open diamonds), and calculated composition of zero-age plagioclase in equilibrium with SH304-2A1 whole rock (grey triangle). All whole-rock data from Cooper and Donnelly (2008); analyses of plagioclase size fractions reported in Cooper and Donnelly (2008; 125-250 μm size fraction) and Donnelly and Cooper (2006). Solid line is the equiline, dashed lines represent mixing lines between zero-age plagioclase and whole-rock, and a mixing line defined by the three size fractions of plagioclase. See text for discussion.

Textural perspectives

Textural features of crystals can provide insights into the pre-eruption paragenesis of the minerals. These include internal chemical heterogeneities within individual crystals and the distribution of grains, grain sizes, and grain shapes in a crystal population. One important distinction is whether grains have selvages of glass and therefore were probably immersed in liquid rather than being isolated from liquid as inclusions in other phenocrysts or in xenoliths before eruption and mineral separation. For zircon, instances have been reported where the absence of adhering glass correlates to pre-Quaternary grains and where these grains are interpreted to be accidental crystals derived from disseminated country rock fragments (Schmitt et al. 2006). The presence of glass selvages on other zircon xenocrysts shows that those grains were magma-hosted.

Grain shapes and textural features of zircons and/or associated "phenocrysts" can help to elucidate the paragenesis and evolution of a crystal population. Variable crystal shapes, from euhedral to anhedral crystal shapes, have been suggested to indicate growth in conditions in that ranged from melt-rich to melt-poor (Vazquez et al. 2007). Some studies have noted evidence of rounding of zircons, indicative of partial resorption (e.g., Schmitt et al. 2003b; Bacon and Lowenstern 2005; Schmitt 2006). These are generally taken as evidence of near-eruption injection of heat into a system, ostensibly by mafic recharge, and possible derivation of zircon from more crystalline portions of a system. For example, Bacon and Lownstern (2005) found that zircons from granodiorite blocks are similar in size, shape, and CL zoning to zircons from the rhyodacite that entrained the blocks, suggesting a strong affinity between the two. Plagioclase crystals in the granodiorites exhibit evidence of subsolidus exsolution in the form of crystallographically oriented Fe-oxide needles. Identification of these same features in at least 80% of plagioclase crystals in the rhyodacite suggests that they are dominantly antecrysts derived from plutonic rocks. Schmitt (2006) correlated the presence of ThSiO$_4$

rods in Laacher See tephra zircon (Fig. 18), tentatively interpreted as exsolution features, to the same features in associated syenite subvolcanic nodules. Finally, studies have noted that granophyre zircons have unusual "spongy" textures (Schmitt and Vazquez 2006; Bacon et al. 2007), inferred to reflect zircon growth in the presence of a volatile-rich melt or fluid.

Crystal Size Distributions (CSD). A sensitive test of whether the distributions of crystal sizes reflect those of growth and not selective accumulation and/or fractionation is to compare the size distribution to the crystal population densities (see review by Armienti 2008). If estimates for crystal growth rates and its temporal dependence (e.g., linear) are also available, such crystal size distributions (CSD) could be used to fingerprint crystal populations and changes in the populations (if sampled by multiple eruptions), to identify the presence of mixed or inherited crystal populations (where breaks in the CSD slope are found), and to estimate the residence time of crystal populations at nominal growth rates.

As discussed above, Turner et al. (2003a) presented crystal size distributions in lavas and cumulate xenoliths from Tonga and from Soufriere volcano, Lesser Antilles, combined with ^{238}U-^{230}Th-^{226}Ra data for bulk mineral separates from the same samples. The CSDs for these lavas are kinked, indicating more than one crystal population within the lavas, consistent with discordant ^{238}U-^{230}Th-^{226}Ra data suggesting that the crystals in Tonga and Soufriere lavas contain more than one component (Turner et al. 2003a).

Zircon dissolution experiments suggest growth rates on the order of 10^{-15}-10^{-17} cm/s for silicic rocks (Watson 1996b). Residence times for zircon populations calculated using these growth rates are of the order of 100–200 k.y. (Bindeman et al. 2001; Bindeman and Valley 2002). Both conventional separation procedures and extraction of zircons by selective HF-dissolution have been applied, with no apparent differences. However, a key element of uncertainty in these interpretations is the extent to which progressive recrystallization has modified the CSD (Fig. 21; Bindeman et al. 2001, 2006; Bindeman and Valley 2002, 2003; Bindeman 2003).

Cathodoluminescence. Back-scattered, secondary electron, and cathodoluminescence images of polished grain mounts can reveal chemical variations within minerals. Of these, cathodoluminescence (CL), the detection of light emitted in the visible portion of the spectrum when a sample is bombarded with electrons, provides the finest-scale detail for zircon (Fig. 10). Differences in shading reflect relative differences in REE and U concentrations; accordingly, CL can be used to detect chemically distinct zones within individual grains and so is often used to guide the choice of analytical spots. CL can also supplement the optical identification of glass selvages on crystals (Fig. 4), thus providing evidence that the grains were surrounded by melt rather than having been occluded in other minerals.

Within any given zircon populations, the CL record of crystal growth may range from simple to complex. In most cases, CL imaging of young zircon provides evidence for fine-scale oscillatory (magmatic) zoning (e.g., Bacon et al. 2000, Vazquez and Reid 2002, Schmitt et al. 2003b, Miller and Wooden 2004; Schmitt 2006, Schmitt and Vazquez 2006). Sector zoning in zircon is also relatively common, resulting from significant differences in the trace element contents of the sectors and identified by patches of relative CL brightness. Fine-scale oscillatory zoning may be superimposed on an overall pattern of U and Th growth zoning (e.g., Miller and Wooden 2004). More rarely zircons may contain internal boundaries where zoning is truncated, indicative of periods of where crystals experienced episode(s) of resorption followed by subsequent growth. In more pronounced cases, distinct cores, often anhedral and embayed and therefore indicative of more pronounced dissolution, are present. These cores may be associated with ages that are demonstrably older than the rims (Vazquez and Reid 2002; Schmitt et al. 2003b,c; Brown and Smith 2004; Miller and Wooden 2004). Bindeman et al. (2001) reported that ~5-20% of zircons from the 0.4-0.5 Ma post-caldera rhyolites at Yellowstone caldera had resorbed cores whereas Brown and Smith (2004) found

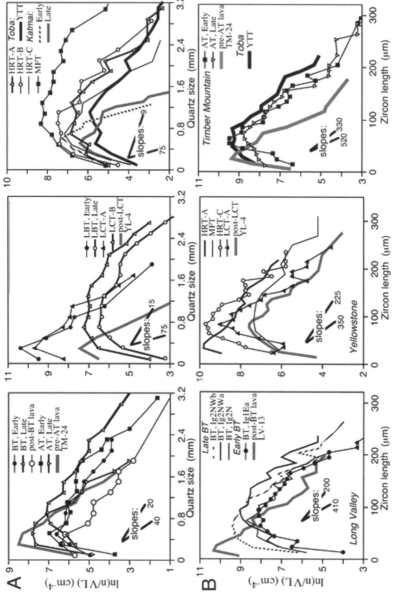

Figure 21. Quartz (A) and zircon (B) size distributions in various caldera-related lavas. Abbreviations are LBT: Lower Bandelier Tuff; BT: Bishop Tuff; LCT: Lava Creek Tuff; HRT: Huckleberry Ridge Tuff; MFT: Mesa Falls Tuff; AT: Ammonia Tanks Tuff; YTT: Youngest Toba Tuff. 150-400 quartz and 250-700 zircon crystals were measured for each CSD. Y-axis shows population density, expressed in terms of number of nuclei (#) of a characteristic length (L) in a sample volume (V). CSD slopes are also given and these can be used to derive residence times if a constant growth rate is assumed. See Armienti (2008) for further discussion of CSD. Modified after Bindeman (2003).

only four crystals with resorbed cores among more than 300 zircons from the 1.21 Ma Ongatiti ignimbrite, Taupo Volcanic Zone (Fig. 9), and Bacon and Lowenstern (2005) found no CL evidence for inherited zircon cores in a Crater Lake granodiorite block despite searching hundreds of grains. A relative lack of clearly distinct cores would seem to be consistent with the general lack of xenocrysts in most volcanic systems studied to date. On the other hand, more subtle internal resorption boundaries may separate episodes of growth where multimodal age distributions are found (e.g., Miller et al. 2004; Charlier et al. 2005; Schmitt et al. 2006).

Using a combination of CL imaging and age results, Bindeman et al. (2001) identified four populations (excluding xenocrysts) in the aforementioned post-caldera rhyolites at Yellowstone. Schmitt and Vazquez (2006) noted that the patchy CL images of Salton Buttes rhyolite zircons were similar to those observed in zircons from associated and similarly-aged granophyre zircons. Brown and Fletcher (2004) found that many of the larger zircons in the Ongatiti ignimbrite have a common internal structure consisting of featureless rims of 10-50 µm surrounding finely zoned internals, often separated by a dark CL zone with sparse melt inclusions. Thus CL patterns can enable episodes of growth to be recognized and may help to identify different zircon populations.

Crystal-scale major element/trace element variations

Major elements. U-Th dating of the allanite can be especially informative since this mineral, an accessory epidote found in some silicic magmas, exhibits diagnostic interelement substitutions. Vazquez and Reid (2004) found that variations in MnO/MgO and La/Nd in allanite are sensitive indicators of magmatic differentiation and that the complexly zoned allanites from the Youngest Toba Tuff, Indonesia, revealed crystallization environments that alternated between less and more differentiated magmas. Euhedral to subeuhedral intragrain zone boundaries suggest unrestricted growth at conditions that were more liquid than solid. Minerals may have been exchanged and mixed between magmas of different compositions. By pairing these chemical variations with U-Th ages, Vazquez and Reid found a pronounced increase in the chemical heterogeneity of the grains that suggested strong chemical diversification of the magma within a few tens of k.y. of eruption.

U, Th, and Th/U. Zircons from volcanic and nonmiarolitic plutonic rocks typically have a few tens to several thousand of parts per million (ppm) each of U and Th (Belousova et al. 2002). In their recent compilation of magmatic zircon Th/U (Fig. 22), Bindeman et al. (2006) corroborated the earlier characterization of Hoskin and Schaltegger (2003): most igneous zircons have Th/U ~ 0.5, with a range from ~0.2 to ~0.9. Exceptions include carbonatitic zircons (Th/U up to 9,000; Hoskin and Schaltegger 2003) and, as described further here, some granophyre zircons (Th/U up to 150; Schmitt 2006).

Many workers have compared Th/U and/or U and Th concentrations within and between zircons, and between zircons and their host melts to assess the affinities amongst them (Schmitt 2006; Bacon et al. 2007). For example, Timber Mountain zircons have comparatively anomalous Th/U ratios (in excess of unity and up to 4.7) but most can broadly be explained by normal Th/U partitioning as they crystallized from their high Th/U host magmas (Bindeman et al. 2006). Commonly, several-fold to more than an order of magnitude variations in U contents within and between zircons from a given sample are observed (Brown and Smith 2004; Miller and Wooden 2004; Charlier et al. 2005; Simon and Reid 2005). Interface kinetics and the compositional- and temperature-dependences of partition coefficients will tend to amplify the variations in zircon relative to the chemical variations in melts from which it grows. Additionally, it can be difficult to assign a normal trend to the variation in U and Th/U because of the heterogeneity—sometimes local—in accessory phase assemblages. U-rich cores may be statistically older than their U-poor rims (Bindeman et al. 2001; Miller and Wooden 2004; Schwartz et al. 2005) or the opposite trend in U concentration may occur. Extreme variations in U concentration may characterize zircons that show evidence of multiple periods of growth

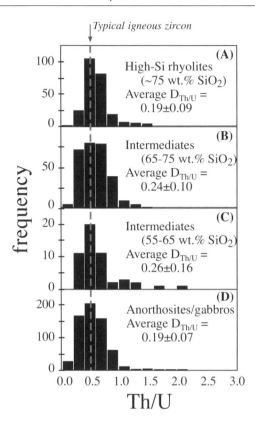

Figure 22. Histograms showing Th/U ratios in zircons from a variety of host rocks. Data for high-silica rhyolites and intermediate rocks are dominated by results for Cenozoic igneous rocks from the western U.S. Low silica rocks include granodiorites, gabbroic anorthosites, and gabbros from MORB and intracontinental rift settings. Modified after Bindeman et al. (2006).

and resorption, where magmatic zoning is clearly truncated by multiple internal dissolution surfaces (e.g., Brown and Fletcher 1999; Miller and Wooden 2004), and these variations enhance the chances of recognizing resorption surfaces. Most workers conclude that a given zircon population can grow in a variety of chemical environments, including ones representing a spectrum of melt fractions (Brown and Fletcher 1999; Schmitt et al. 2003c; Miller and Wooden 2004; Charlier et al. 2005).

High Th (up to ~5 wt%) and Th/U (up to 150) characterizes zircons from miarolitic plutonic rocks from some volcanic centers (Schmitt 2006; Bacon et al. 2007). Other granophyre zircons that apparently grew in the presence of hydrothermal fluids (Schmitt and Vazquez 2006) have moderately high Th but the average to low Th/U observed in hydrothermal veins (Hoskin 2005). The former may be related, in part, to microinclusions of monazite (Schmitt and Vazquez 2006) or to advanced degrees of melt differentiation in the presence of a highly oxidizing vapor phase that reduces the tendency of U to partition into zircon (Bacon et al. 2007).

Other trace elements. As with Th and U, there can be significant variations in the other trace element contents in zircons and particularly in the rare earth elements (REE). Yttrium concentration variations of up to an order of magnitude are correlated with Th and U variations in Yellowstone zircons (Bindeman et al. 2001) and with zones that are dark and light under cathodoluminescence. Such large variations cannot be explained by equilibrium crystal growth from a single, mostly liquid-rich domain and support inferences from U and Th concentrations and from Th/U variations for growth of zircons in environments that range from melt-rich to melt-poor magma, including in plutonic (subvolcanic) rocks (e.g., Bacon et al. 2000, 2007; Bacon and Lowenstern 2005; Schmitt 2006). A factor of two variation in trace elements, especially Y, concentrations was also found between different crystal sectors of single Yellowstone crystals, showing that, as noted above, crystal growth kinetics are important in determining absolute abundances (Watson and Liang 1995; Watson 1996a). These observations suggest that trace element concentrations in zircon are dependent on local variations in concentration in the source and on the kinetics of zircon crystallization (e.g., Hoskin and Ireland 2000).

The effects of kinetics and local trace element *concentrations* notwithstanding, the relative trace element characteristics of zircon may help to assess the affinity of a zircon to its host and, as warranted, to fingerprint possible sources of zircon inheritance (e.g., Belousova et al. 2002; Hoskin 2005). Diagnostic features may include HfO_2 contents, the ratio of LREE:HREE (zircons are generally strongly HREE-enriched), and the presence and magnitude of anomalies in Ce (Ce/Ce*, related to the preferential incorporation in zircon of Ce^{4+} compared to Ce^{3+}) and Eu (Eu/Eu*, mirroring the influence of feldspar fractionation on the host melts; e.g., Bacon et al. 2007; Vazquez et al. 2007). Vazquez et al. (2007) found that REE concentrations in Hualalai zircons are typical of zircon from evolved alkalic magmas. Schmitt (2006) matched REE patterns of inherited Laacher See zircons (Fig. 23) to syenite (negative Eu anomalies) rather than carbonatite (little or no Eu anomalies). Elevated HfO_2 concentrations, relatively high Th/U and LREE/HREE (and weak or absent positive Ce anomalies) may be diagnostic of crystallization in the presence of an oxidizing fluid (Hoskin 2005). These same features are observed in the spongy textured granophyric zircons and suggests late-stage crystallization of these zircons in the presence of an aqueous fluid (Fig. 23; Schmitt and Vazquez 2006; Bacon et al. 2007). As noted above, Schmitt and Vazquez (2006) attribute these characteristics at least in part to the presence of small (~5 μm) monazite inclusions whereas Bacon et al. (2007) suggest that oxidation of late-stage melts causes zircons to exclude U^{5+} and U^{6+}, leading to high Th/U. Ongoing investigations aimed at linking volcanoes and plutons will undoubtedly lead to additional means of fingerprinting zircon origins.

Trace elements in major minerals. Compared to zircon, relatively little effort has focused on coupling trace-element variations with ages of major phases. This likely reflects in part the fact that it is easier to make connections between information from trace elements at the sub-crystal scale with ages at a sub-crystal scale than it is to link sub-crystal trace-element information with bulk separate ages. Despite this difference in spatial scale sampled, it is still possible to make some broad inferences about the timescales of the magmatic processes that are reflected in trace-element zoning. For example, as discussed earlier, sub-crystal zoning in barium in plagioclase from Mount St. Helens—beyond that predicted from major-element variations and partitioning behavior—provides evidence that the cores of many crystals are antecrystic and, by extension, that crystals have not been held at high temperatures long enough for diffusion to equilibrate crystals with the host liquid. One further implication is that ^{230}Th -^{226}Ra crystal ages have not been affected by diffusive exchange with the host liquid. Similarly, ^{238}U-^{230}Th data for minerals separated from dacite at Kameni, Santorini yield a U-Th isochron age of 18 +19/−16 ka (Zellmer et al. 2000), whereas Sr diffusion profiles in Kameni lavas indicate crystal residence times of less than 100-450 years (Zellmer 1999). These data collectively suggest that a relatively small proportion of the overall history of antecrysts was spent suspended in the host magma at high temperatures.

Crystal ages and thermometry

Experimental determinations of zircon solubility make it possible to bracket the conditions under which zircon will be stable in silicic magmas (Watson and Harrison 1983).

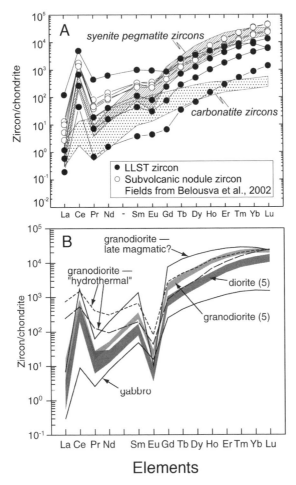

Figure 23. Chondrite-normalized rare earth element patterns for zircons from: A) Lower Laacher See tephra (LLST) and subvolcanic nodules and B) Veniaminoff volcano xenoliths. LLST tephra and syenite zircons both share REE patterns characterized by heavy rare earth element (HREE) enrichment, prominent positive Ce anomalies, and significant negative Eu anomalies. These patterns are similar to those for evolved syenitic zircon but differ from those of carbonatitic zircons where negative Eu anomalies caused by feldspar fractionation are typically absent (e.g., Belousva et al. 2002). Most Veniaminof zircons also have chondrite-normalized rare earth element (REE) patterns characteristic of igneous zircon (Belousova et al. 2002), with increasingly negative Eu anomalies being consistent with crystallization differentiation involving plagioclase. Bacon et al. (2007) attribute 1) high REE concentrations and large Eu anomalies to crystallization from highly differentiated interstitial melts; 2) unusually large Ce/Ce* in diorite and granodiorite zircon to crystallization in the presence of an oxidizing fluid; and 3) elevated light REE concentrations and REE patterns in two spongy-textured granodiorite zircons to hydrothermal zircons (Hoskin 2005). After Schmitt et al. (2006) and Bacon et al. (2007).

Zircon saturation mainly depends on the Zr content and temperature of a magma but the bulk chemical composition of a magma (especially the relative proportion of network modifiers to network formers) also affects zircon solubility. Pressure, including water pressure, does not seem to directly affect the stability of zircon to an appreciable degree (Watson and Harrison 1983) but may have an indirect effect by influencing the stability of other phases. Saturation temperatures obtained from the glass or groundmass compositions of a volcanic rock may be those closest to the pre-eruptive conditions but estimates based on whole-rock compositions are usually not significantly different at low degrees of crystallinity.

Some studies have compared the zircon saturation temperature (T_{zirc}) to estimated magma temperatures (T_{mag}), particularly those from Fe-Ti oxides, in order to gain insights in thermal history in the lead up to eruption (Fig. 24). For those cases where $T_{mag} > T_{zirc}$, zircons will be dissolving and the model of Watson (1996b) can be used to calculate the time necessary for zircon dissolution to occur. The model assumes that Zr diffusion is the rate-limiting process for zircon dissolution and thus that there is a static chemical boundary layer in melt around a zircon. Based on a preponderance of ~200 ka zircons in the coarsely porphyritic rhyolites of Deer Mountain and Inyo Domes, Long Valley, California, Reid et al. (1997) argue that the magma responsible for them could not have been at $T > T_{zirc}$ for more than 1 k.y. since the zircon crystallized. For those domes, iron–titanium oxide temperatures (809 ±4 °C) are similar to those for zircon saturation (795-805 °C; Fig. 24). In contrast, iron–titanium oxide temperatures (~850–880 °C) for the 27 ka rhyodacite eruption at Mt. Mazama exceed the zircon saturation temperature by ~50 °C (Bacon and Lowenstern 2005). Rhyodacite zircons, inferred to be antecrysts based on age and phase similarities to granodiorite blocks, must have been dissolving. A modest amount of dissolution (5%) could explain the subtle rounding of zircon that is observed and requires dispersal of the granodiorite into the magma within a few

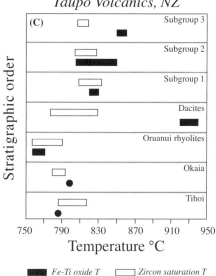

Figure 24. Comparison between Fe-Ti oxide temperatures (closed bars and circles) and zircon saturation temperatures (open bars) in Long Valley western moat rhyolites, Mt. Mazama rhyodacites and granodiorites, and Taupo volcanics. Instances where paired temperatures overlap and where they are discrepant are found. See text for discussion. Modified after Charlier et al. (2005) with data from Reid et al. (1997), Bacon and Lowenstern (2005), and references therein.

tens of years of eruption. Even after that eruption, dissolution would have continued in the reservoir so that by the time the climactic eruption of Mount Mazama responsible for the Crater Lake caldera occurred, no zircons remained. Complete dissolution would have only required ~700 yr.

Among the Taupo Volcanic Field dacites to rhyolites, the unit Ω dacite has $T_{mag} > T_{zirc}$ (Fig. 24; Charlier et al. 2005), indicating that zircons should have been dissolving in the lead up to eruption. Unit Ω is one of the only Taupo magmas in which pre-350 ka Quaternary and older inherited zircons are found. Calculations based on Watson (1996b) show that the largest zircons (250 μm) will dissolve in <100 yr in the Unit Ω dacite. This and the subhedral to euhedral morphologies of the zircons imply that the Unit Ω dacite must have been erupted within tens of years of the zircons being introduced into the dacite magma.

Somewhat different variants on the case of $T_{mag} > T_{zirc}$ are those where, rather than immediately before eruption, some earlier interval in the magmatic history is inferred to have been at temperatures greater than T_{zirc}. In one instance, quartz-hosted glass (melt) inclusions are significantly undersaturated in zircon at the equilibration temperatures obtained for the major mineral phases (Schmitt et al. 2003a). Thus, zircon growth must postdate melt-entrapment in quartz. Zircon saturation temperatures similar or less than those inferred for magma liquidus temperatures have also been used to explain the absence of zircon xenocrysts (e.g., Schmitt et al. 2003c).

Other studied volcanic eruptions have $T_{mag} \leq T_{zir}$. In these cases, zircons can reasonably be assumed to preserve the last time that temperatures passed through the zircon saturation temperature, except where thermal excursions over the zircon liquidus have been modest and/or short-lived. Cases where zircon saturation temperatures are greater than or are similar to Fe-Ti oxide temperatures have been used to suggest that 1) zircons date the attainment of conditions that correspond to zircon saturation (e.g., Reid et al. 1997; Reid and Coath 2000; Vazquez and Reid 2002); 2) magmas fluctuated between zircon saturated and undersaturated conditions as they cycled between cooling, crystallization, and magma mixing or recharge events (e.g., Brown and Smith 2004; Miller and Wooden 2004); and/or 3) transient heat pulses could incorporate zircons that then persist, resulting in zircon inheritance and Zr excesses (Bindeman et al. 2006). Magmatic fluctuations could account for growth that was episodic rather than continuous and thus explain multimodal age populations.

The recent development of a Ti-in-zircon geothermometer (Watson et al. 2006) means that temperatures can be obtained for individual zircons and, significantly, for zones within zircons. Activities for TiO_2 and SiO_2 in the melt must be obtained or assumed and, if temperatures have varied, may include allowances for activity variations in response to changing accessory phases and mineral proportion. Uncertainties in the Ti activity notwithstanding, Schmitt and Vazquez (2006) could use the Ti contents of the Salton Trough zircons to estimate that partial melting of hydrothermally altered basalts produced rhyolites at temperatures between ~770 and 835 °C.

Vazquez and Reid (2004) also obtained crystal-scale temperature information, in this case for allanite, by correlating the chemical variations within individual allanites to those expected for equilibrium with experimental melts produced at temperatures that ranged from 700 to 770 °C (Fig. 25). In this way, they were able to show that the thermochemical evolution of individual allanites from Toba caldera, Indonesia, can be complex and reflect crystallization under conditions that could have included mingling between recharge and resident magmas and/or during intrareservoir self-mixing.

Single-crystal or sub-crystal oxygen isotopic data

Zircons are resistant to weathering and to isotopic exchange at surface and near-surface conditions, and therefore can preserve oxygen-isotopic compositions imparted at the time of crystallization. Coupling of *in situ* oxygen isotope analyses with U-Pb radiometric age determinations has proven to be a potent means for investigating the crustal evolution of Earth (e.g., Valley 2003; Kemp et al. 2006). Specific to the dynamics of magmatic processes, the oxygen isotope compositions of zircon mirror those of the melts from which they crystallize. It is therefore possible to use zircon-melt isotopic fractionation factors to determine whether the isotopic composition of a zircon's host is different from the one that was responsible for zircon growth. In particularly amenable cases, this can be used to distinguish antecrysts from true xenocrysts. Where isotopic disequilibrium between a zircon and its hosts pertains, it is additionally possible to use the intragrain O isotope gradient in zircon to estimate the residence time of zircons in the new host. Core-rim variations in the oxygen isotope values of zircons can be obtained by successively abrading and analyzing zircons grains or by ion microprobe analyses (e.g., Bindeman and Valley 2000).

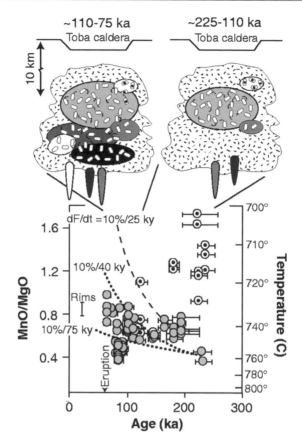

Figure 25. Temporal changes in the crystal growth environment of the Younger Toba Tuff, Indonesia, as inferred from internal covariations between MnO/MgO and U-Th ages in allanites. Predicted variations in temperature (right-hand axis) associated with changes in MnO/MgO are based on experimental crystallization of Younger Toba Tuff magma at 100-200 MPa (Gardner et al. 2002). Also shown are trends of fractional crystallization for different fractionation rates, using two starting compositions. Increase in the diversity of allanite composition between 110 and 75 ka is inferred to reflect interaction between and mingling of variably fractionated magma bodies with temperatures between ~760 and 715 °C. MnO/MgO obtained by electron microprobe analyses; U-Th ages obtained by ion microprobe analyses. [Used with permission from The American Association for the Advancement of Science from Vazquez and Reid (2004) *Science*, Vol. 305, p. 993.]

Some eruptions that occurred in the aftermath of caldera collapses are characterized by low-$\delta^{18}O$ lavas and tephras (e.g., the 0.4 -0.5 Ma postcaldera eruptions at Yellowstone caldera complex, Wyoming, and the Ammonia Tanks pre- to post-caldera cycle that followed in the wake of the Rainier Mesa caldera eruption, Timber Mountain/Oasis Valley, Nevada), with $\delta^{18}O$ ranging to more than 5‰ lower than mantle values. These ^{18}O-depleted melts are generally inferred to be derived by remelting of older, hydrothermally altered, but otherwise chemically similar, rhyolites (e.g., Bacon et al. 1989; Balsley and Gregory 1998; Bindeman et al. 2001b). Cores of zircons associated with these eruptions have normal rather than low $\delta^{18}O$ values and their ages overlap the timing of earlier eruptions, suggesting that the zircons are inherited in some fashion (Bindeman and Valley 2000, 2001; Bindeman et al. 2001, 2006, 2008). Core-to-rim decreases in $\delta^{18}O$ (by as much as 5‰; Fig. 26; Bindeman et al. 2001) indicate that the

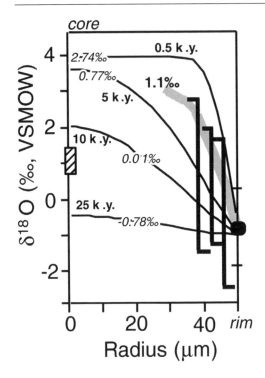

Figure 26. Model isotope profiles (thin curves) for 52.5 μm radius zircons compared to $\delta^{18}O$ values obtained by successive air abrasion of zircons from Middle Biscuit Basin, Yellowstone (bold vertical bars). Isotope values relative to VSMOW (Vienna standard mean ocean water). Durations of diffusional re-equilibration at 850 °C between cores with $\delta^{18}O = +4‰$ and rims with $\delta^{18}O \sim -1‰$ indicated; associated whole zircon $\delta^{18}O$ values shown by italicized numbers. Bold vertical bars connect measured core (left-oriented horizontal marks) and rim (right-oriented horizontal marks) $\delta^{18}O$ values; rim compositions obtained by difference between whole zircon and core values. The sharp contrast between the core and rim compositions of the zircons suggests residence times of $<10^2$-10^3 yr at 850 °C or 10^3-10^4 yr at 750 °C. An inferred zoning profile, shown by the thick grey line, yields a weighted $\delta^{18}O$ that corresponds to the measured $\delta^{18}O$ of whole grains. Also shown for reference are the isotopic composition of host obsidian (striped box) and the rim isotopic composition of four zircons obtained by ion microprobe analyses (large filled circle). See Bindeman and Valley (2000) for additional details.

zircons grew from and/or partially exchanged oxygen with ^{18}O-depleted melts. The failure of zircon to equilibrate with its host limits the residence time of zircons in the melt: the duration of melting must be on the order of 10 k.y. or less at Yellowstone, based on solid-state diffusion modeling, or less than a few thousand years at Ammonia Tanks Tuff, Nevada, based on the dissolution calculations by Watson (1996). Lower Laacher See Tuff zircons have $\delta^{18}O$ values up to 1.3‰ lower than expected for equilibrium with glass in their host phonolite (Schmitt 2006); as noted elsewhere, extrapolated initial Th isotope ratios for the zircons are also out of equilibrium with their host. Oxygen (and extrapolated initial Th) isotope similarity between zircons in the tephra and in syenite nodules lead to the conclusion that the zircons date partial to complete crystallization of subvolcanic apophyses. The absolute values of the thorium and oxygen isotope ratios in zircon also support evidence from the presence of abundant crustal xenocrysts for extensive crustal contamination.

Melts with $\delta^{18}O$ values of +5.4 to +6.6‰ are estimated to be responsible for the Salton Trough zircons based on their oxygen isotope compositions and these match values obtained for Salton Buttes obsidians. Schmitt and Vazquez (2006) show that these values are too low for silicic magma genesis by melting of continental crust or by fractional crystallization of tholeiitic basalt and conclude that melting of basalts in the presence of geothermal fluids could be responsible for generation of silicic magmas in the Salton Trough (Schmitt and Vazquez 2006). In support of this interpretation, zircons located within plagioclase-rich domains in the basaltic xenoliths have $\delta^{18}O$ values that overlap the range of rhyolite zircons.

Summary

Various combinations of crystal age information and crystal-scale chemical or isotopic data can thus serve several purposes: first, the addition of other information can help interpret age data, and in particular can help distinguish antecrysts from xenocrysts and identify mixed crystal populations that may have distinct ages. Second, the addition of age information to studies involving thermometry (whether by using zircon saturation temperatures or Ti-in-zircon thermometry) can replace qualitative thermal histories of magmas with quantitative thermal histories. In cases where sub-crystal chemical information is also available, the thermal and chemical evolution of magmas can be quantified. Finally, in cases where isotopic and/or chemical disequilibrium between crystals and melts has been preserved, crystal ages and diffusion modeling can be combined to delimit the duration of crystal residence at high temperatures and/or within a given melt. Thus, the combination of crystal age information with other textural, chemical, and isotopic data represents one of the most promising pathways toward furthering our understanding of magma dynamics.

KEY QUESTIONS AND FUTURE DIRECTIONS

What can crystal ages tell us about magmatic processes?

The unique contribution of U-series crystal ages to studies of magmatic processes is the ability to quantify absolute ages of crystal populations or even individual zones within crystals, on timescales commensurate with many magmatic processes. What is simultaneously the great strength and the interpretive challenge of using crystal age information to understand magmatic processes is that what is dated is the record of magma evolution contained within the crystals, not the magma as a whole. This has the advantage that the crystals record processes and events that are averaged and therefore obscured in the liquid fraction of the magma, but the weakness is that the crystal only contains a record of what was happening while it was growing. For example, there could be substantial periods of time that are represented only by discontinuities in the crystal zoning record. In cases where crystal size and age have been combined to calculate apparent average crystal growth rates, these rates are much slower than experimentally-determined crystal growth rates, suggesting significant periods of time where the crystals were not growing or were even resorbed. In the case of bulk crystal aggregates, any interpretation must take into account the fact that analyses are averaging over many grains and may also reflect the influence of melt inclusions or accessory phase inclusions, so the crystal age may not correspond to the age of the major phase constituting the bulk of the separate. We can evaluate the extent to which this is the case by *in situ* (e.g., LA-ICP-MS or SIMS) measurements of trace elements in the crystal compared to trace-element contents of the dissolved bulk separates.

These interpretive challenges notwithstanding, the studies described in earlier sections demonstrate the key role that crystal ages can play in understanding processes operating beneath the surface in magmatic systems. For example, crystal ages provide some of the strongest evidence for the longevity of magmatic systems, which have forced a re-evaluation of the physical nature of magmas when stored for long periods of time, the interactions between different magmas passing through the same system, and the implications of common (if not ubiquitous) crystal recycling (see section "*Crystal recycling*" above). Crystal ages have also provided evidence that magma bodies erupted close together in space and time have distinct crystal ages, which requires that the magmas were not interconnected—or at the least, not well-mixed (e.g., Charlier et al. 2003). Conversely, in different settings, crystal ages have shown that magmas which erupted from the same volcanic system, thousands of years apart, contain crystals with the same age spectra and therefore likely were tapping the same magma bodies and/or cumulate bodies (e.g., Reid et al. 1997). Interpretations such as these would simply not have been possible without age information.

Although the foregoing examples show that crystal ages even in (relative) isolation can provide insights into magmatic processes, crystal ages are most powerful as tools for unraveling magma dynamics when combined with other information about the dated crystals and their relation to host melts. For example, even in simple cases, crystal ages may be related to magma residence times or to differentiation timescales, but these interpretations rely on having corroborating evidence for chemical equilibrium between crystals and host melts. Independent sources of information at the crystal scale are even more important to understanding crystal age information in cases where the crystal history may be more complex; additional information is necessary in order to test multiple interpretations that are permitted by the age data alone. As demonstrated by the examples presented above, the addition of age information from multiple parent-daughter pairs and/or multiple size fractions, and the addition of textural and crystal-scale geochemical data (trace-element or isotopic data) to age information, can provide the means to take the crystal ages beyond simply understanding when crystals grew to understanding the rates and timescales of processes such as the thermal and chemical evolution of magmas, and both temporal and trace-element or isotopic evidence for recycling of crystalline material from earlier phases of magmatism.

Most U-series ages, however, are not suited to capturing some very rapid or short-lived magmatic processes. The most commonly-used parent-daughter pairs (^{238}U-^{230}Th and ^{230}Th-^{226}Ra) yield ages that capture magmatic events on timescales of centuries to hundreds of thousands of years, and thus do not record processes such as re-equilibration between crystals and new host liquids (diffusion ages of residence) or degassing-induced crystallization (one caveat being that anomalously high Ra/Ba ratios may provide a clue that crystallization was too rapid to allow trace-element equilibration). An exception, however, is ^{226}Ra-^{210}Pb crystal ages, which yield information about crystal growth on years to decades timeframe. This timescale is similar to the timescales of magma storage given by diffusion studies (e.g., Costa 2008), and there is a great deal of promise in combining ^{226}Ra-^{210}Pb data with diffusion modeling to access the youngest part of a magma's crystallization and crystal recycling history.

What do we get from the different scales of analyses?

The strength of *in situ* analysis is the ability to focus in on a relatively restricted period of crystal growth, and potentially to identify ages of more than one growth episode in a single crystal. *In situ* ages also give information about the distribution of ages within a sample, rather than a single age for a large number of crystals. On the other hand, a large number of analyses is required in order to characterize the age of a population within a sample, and this is especially true if there is more than one population of crystals. With our current analytical capabilities, *in situ* analyses are also restricted to accessory phases. This can be an advantage in that the conditions of zircon growth are more restricted than many of the major phases, so that relating ages to parameters such as temperature may be more straightforward than with major phases. However, it can also be a disadvantage in that crystal growth may be more subject to kinetic limitations on diffusion of trace elements through silicic melts, and the trace-element compositions of zircon or allanite may therefore be more likely to reflect kinetic effects than are major phases.

In contrast, bulk mineral separates—especially of major rock-forming minerals—require analysis of hundreds to thousands of crystals, and crystal ages thus derived are necessarily averaging over the entire population of crystals sampled. The disadvantage is that it is more difficult to identify multiple phases of crystal growth or multiple crystal populations, and an average age derived from a mixed-age mineral separate may not correspond to any geological event. On the other hand, one advantage is that the influence of a few crystals of distinct age (for example, xenocrysts in secular equilibrium) on the average age is relatively limited. In addition, bulk crystal data are likely to be representative of the overall crystal population(s) within a given magma, as long as an appropriate size range is sampled. Furthermore, fractional crystallization of major minerals is one of the main controls on the chemical evolution of magmas; therefore, bulk crystal ages may be more easily related to processes that operate at the magma reservoir scale. *In situ* ages and bulk mineral separates thus offer complementary information, and it can be particularly powerful to combine the two techniques.

Future directions

Crystal ages are clearly an important part of the picture in understanding the dynamics of magmatic systems. Equally clearly, these data are most powerful when combined with other types of information about the crystals that are dated, and the links between the crystals and both the host magmas and any other magmas from which at least some of the crystals precipitated. Progress has been made in combining age information with other textural and chemical information, but the potential for combined studies is only starting to be tapped. In particular, what is necessary in order to interpret crystal ages in the context of the dynamics and evolution of a given magmatic system is additional information with which to link the crystal ages to the chemical and textural evidence for magma evolution. Some examples (though not a complete list, by any means) of potentially fruitful directions for future work follow.

Crystal-scale trace-element and isotopic data. Additional measurements combining trace-element information at the sub-crystal scale with age information could be particularly useful in tracking mixed magmas and crystal recycling. For example, trace element data could potentially be used to identify the overall degree of evolution of a magma during crystallization, to identify any discontinuities in trace-element zoning that could reflect magma recharge or incorporation of crystals into a new magma, and in general to assess the extent to which crystals have attained chemical equilibrium with their host melts. In combination with absolute age data, studies of trace-element diffusion can provide information as to the duration of storage of crystals at high temperatures (i.e., in magmas as opposed to storage in mush zones or as young plutonic bodies). Sub-crystal isotopic data in combination with age data could be especially useful for identifying antecrystic or xenocrystic components to crystal populations, and to quantifying the proportion of crystals that represent different populations; in the case where discordant ages are obtained from different parent-daughter pairs, this kind of information could be used to quantify the age of old and young components of crystallization, and therefore the timescales over which crystals are stored and recycled.

Thermometry. A number of new types of thermometers have recently been developed, and their potential in combination with crystal age data to develop time-temperature (+/− composition) paths recorded in crystals has only started to be explored. In particular, Ti-in-zircon thermometry together with spot ages in the same crystals can allow the use of zircon crystals as tracers of the thermal state of a magma over time. There is also great potential in combining these temperature-time studies with sub-crystal trace-element data to obtain information about the chemical state of the magma over time.

Melt inclusions. Little has been done to date in combining information from melt inclusions with age data for the same crystals. Study of melt inclusions in dated crystals could be particularly useful for constraining the volatile history of magmas (especially for putting the volatile contents of melt inclusions within a temporal framework) and for examining the isotopic and trace-element evolution of magmas.

Numerical modeling. Numerical modeling could provide a conceptual framework for interpreting crystal-scale age data as well as trace-element data. For example, numerical models have recently been developed that allow tracking of crystals that are treated individually rather than as part of a continuum fluid (Ruprecht et al. 2008). These models can allow crystal growth and interactions with the host magma through kinetic and trace-element partitioning laws, and have the potential to provide synthetic crystal zoning profiles that are produced over a known timeframe, which can be compared to data from real samples in order to understand what scale of trace-element variation reflects magma-wide properties and what scale may primarily reflect the local chemical environment of each crystal. By combining this information with age data, we could potentially examine the timescales over which different kinds of processes may be recorded in crystals.

Kinetics of crystal growth. The effects of rapid crystallization may affect the way that parent and daughter are incorporated into crystals, and thus the calculated ages. It is important to understand these effects, and especially the circumstances under which these effects are likely to be important, in order to have confidence in interpretations based on age data and trace-element data.

Final thoughts

Overall, the geochemical/petrological community is moving toward a more integrated approach to the study of crystals in magmas, and the records of magmatic processes contained within them. This volume summarizes a wide variety of approaches to understanding crystal records, and each chapter contains examples of ways in which multiple data types may be combined, yet these integrated studies are just beginning to show their full potential. Magmatic systems are themselves complex, and this complexity is reflected in the crystals contained within them. Such complexity demands a comprehensive approach in order to make progress, and the next decade promises to yield many more insights into the dynamics of magmatic systems.

ACKNOWLEDGMENTS

This work was supported in part by NSF grants EAR 0714455 and EAR 0649295 to KMC and EAR 0538309 to MRR. We thank Mark Reagan, Axel Schmitt, and Simon Turner for constructive reviews which improved the manuscript. We also thank editors Keith Putirka and Frank Tepley for further editorial comments and suggestions, and for their efforts in putting together this volume.

REFERENCES

Allègre CL (1968) ^{230}Th dating of volcanic rocks: a comment. Earth Planet Sci Lett 5:209-210
Armienti P (2008) Decryption of igneous rock textures: crystal size distribution tools. Rev Mineral Geochem 69:623-649
Bachmann O, Charlier BLA, Lowenstern JB (2007) Zircon crystallization and recycling in the magma chamber of the rhyolitic Kos Plateau Tuff (Aegean arc). Geology 35:73-76
Bachmann O, Dungan MA, Lipman PW (2002) The Fish Canyon magma body, San Juan volcanic field, Colorado: rejuvenation and eruption of an upper-crustal batholith. J Petrol 43:1469-1503
Bacon CR, Adami LH, Lanphere MA (1989) Direct evidence for the origin of low-^{18}O silicic magmas: quenched samples of a magma chamber's partially-fused granitoid walls, Crater Lake, Oregon. Earth Planet Sci Lett 96:199-208
Bacon CR, Lowenstern JB (2005) Late Pleistocene granodiorite source for recycled zircon and phenocrysts in rhyodacite lava at Crater Lake, Oregon. Earth Planet Sci Lett 233:277-293
Bacon CR, Persing HM, Wooden JL, Ireland TR (2000) Late Pleistocene granodiorite beneath Crater Lake caldera, Oregon, dated by ion microprobe. Geology 28:467-470
Bacon CR, Sisson TW, Mazdab FK (2007) Young cumulate complex beneath Veniaminof caldera, Aleutian arc, dated by zircon in erupted plutonic blocks. Geology 35:491-494
Baldwin SL, Ireland TR (1995) A tale of two eras; Pliocene-Pleistocene unroofing of Cenozoic and late Archean zircons from active metamorphic core complexes, Solomon Sea, Papua New Guinea. Geology 23:1023-1026
Ball L, Sims KWW, Schwieters J (2008) Measurement of ^{234}U/^{238}U and ^{230}Th/^{232}Th in volcanic rocks using the Neptune MC-ICP-MS. J Anal At Spectrom 23:173-180
Balsley SD, Gregory RT (1998) Low-^{18}O silicic magmas: why are they so rare? 162:123-136
Bea F, Pereira MD, Stroh A (1994) Mineral/leucosome trace-element partitioning in a peraluminous migmatite (a laser-ablation-ICP-MS study). Chemical Geology 117:291-312
Belousova EA, Griffin WL, O'Reilly SY, Fisher NI (2002) Igneous zircon: trace element composition as an indicator of source rock type. Contrib Mineral Petrol 143:602-622
Bernal J-P, Eggins SM, McCulloch MT (2005) Accurate in situ ^{238}U-^{234}U-^{234}Th-^{230}Th analysis of silicate glasses and iron oxides by laser-ablation MC-ICP-MS. J Anal At Spectrom 20:1240-1249
Bindeman I (2008) Oxygen isotopes in mantle and crustal magmas as revealed by single crystal analysis. Rev Mineral Geochem 69:445-478

Bindeman IN (2003) Crystal sizes in evolving silicic magma chambers. Geology 31:367-370
Bindeman IN, Davis AM, Drake MJ (1998) Ion microprobe study of plagioclase-basalt partition experiments at natural concentration levels of trace elements. Geochim Cosmochim Acta 62:1175-1193
Bindeman IN, Schmitt AK, Valley JW (2006) U–Pb zircon geochronology of silicic tuffs from the Timber Mountain/Oasis Valley caldera complex, Nevada: rapid generation of large volume magmas by shallow-level remelting. Contrib Mineral Petrol 152:649-665
Bindeman IN, Valley JW (2000) Formation of low-^{18}O rhyolites after caldera collapse at Yellowstone, Wyoming, USA. Geology 28:719-722
Bindeman IN, Valley JW (2001) Low-δ^{18}O rhyolites from Yellowstone: magmatic evolution based on analyses of zircons and individual phenocrysts. J Petrol 42:1491-1517
Bindeman IN, Valley JW, Wooden JL, Persing HM (2001) Postcaldera volcanism: in situ measurement of U–Pb age and oxygen isotope ratio in Pleistocene zircons from Yellowstone caldera. Earth Planet Sci Lett 189:197-206
Bindeman IN, Valley JW (2002) Oxygen isotope study of the Long Valley magma system, California: isotope thermometry and convection in large silicic magma bodies. Contrib Mineral Petrol 144:185-205
Bindeman IN, Valley JW (2003) Rapid generation of both high and low-delta O-18 large-volume silicic magmas at the Timber Mountain/Oasis Valley caldera complex Nevada. Geol Soc Am Bull 115:581-595
Bindeman IN, Fu B, Kita N, Valley JW (2008) Origin and evolution of Yellowstone silicic magmatism based on ion microprobe analysis of isotopically-zoned zircons. J Petrol 49:163-193
Black S, Macdonald R, Barreiro B, Dunkley PN, Smith M (1998a) Open system alkaline magmatism in northern Kenya: evidence from U-series disequilibria and radiogenic isotopes. Contrib Mineral Petrol 131:364-378
Black S, Macdonald R, DeVivo B, Kilburn CRJ, Rolandi G (1998b) U-series disequilibria in young (A.D. 1944) Vesuvius rocks: Preliminary implications for magma residence times and volatile addition. J Volcanol Geotherm Res 82:97-111
Black S, Macdonald R, Kelly MR (1997) Crustal origin for peralkaline rhyolites from Kenya: Evidence from U-series disequilibria and Th-isotopes. J Petrol 38:277-297
Blundy J, Wood B (1994) Prediction of crystal-melt partition coefficients from elastic moduli. Nature 372:452-454
Blundy J, Wood B (2003) Mineral-melt partitioning of uranium, thorium and their daughters. Rev Mineral Geochem 52:59-123
Bourdon B, Henderson GM, Lundstrom CC, Turner SP (eds) (2003a) Uranium-Series Geochemistry. Reviews in Mineralogy and Geochemistry Vol. 52. Mineralogical Society of America. Washington, DC
Bourdon B, Turner S, Henderson GM, Lundstrom CC (2003b) Introduction to U-series geochemistry. Rev Mineral Geochem 52:59-123
Bourdon B, Worner G, Zindler A (2000) U-series evidence for crustal involvement and magma residence times in the petrogenesis of Parinacota volcano, Chile. Contrib Mineral Petrol 139:458-469
Bourdon B, Zindler A, Worner G (1994) Evolution of the Laacher See magma chamber: Evidence from SIMS and TIMS measurements of U-Th disequilibria in minerals and glasses. Earth Planet Science Lett 126:75-90
Brown SJA, Fletcher IR (1999) SHRIMP U–Pb dating of the preeruption growth history of zircons from the 340 ka Whakamaru ignimbrite, New Zealand: evidence for >250 ky magma residence times. Geology 27:1035-1038
Brown K, Carter C, Fohey N, Wooden J, Yi K, Barth A (2004) A study of the origin of rhyolite at mid-ocean ridges; geochronology and petrology of trachydacite and rhyolite from Salton Sea, California, and Torfajokull, Iceland. Geol Soc Am Abstr Programs 36:79
Brown SJA, Smith RT (2004) Crystallisation history and crustal inheritance in a large silicic magma system: ^{206}Pb/^{238}U ion probe dating of zircons from the 1.2 Ma Ongatiti ignimbrite, Taupo Volcanic Zone. J Volcanol Geotherm Res 135:247-257
Chamberlain KR, Harrison TM, Schmitt AK, Heaman LM, Swapp SM, Khudoley AK, Sears JW, Prokopiev AV (2007) In situ micro-baddeleyite U-Pb dating method: an age from any dike. Geol Soc Am Abstr Programs 39:97
Charlier B, Zellmer G (2000) Some remarks on U-Th mineral ages from igneous rocks with prolonged crystallization histories. Earth Planet Sci Lett 183:457-469
Charlier BLA, Peate DW, Wilson CJN, Lowenstern JB, Storey M, Brown SJA (2003) Crystallization ages in coeval silicic magma bodies: ^{238}U-^{230}Th disequilibrium evidence from the Rotoiti and Earthquake Flat eruption deposits, Taupo Volcanic Zone, New Zealand. Earth Planet Sci Lett 206:441-457
Charlier BLA, Wilson CJN, Lowenstern JB, Blake S, Van Calsteren PW, Davidson JP (2005) Magma generation at a large, hyperactive silicic volcano (Taupo, New Zealand) revealed by U-Th and U-Pb systematics in zircons. J Petrol 46:3-32
Cherniak DJ, Watson EB (2003) Diffusion in zircon. Rev Mineral Geochem 53:113-143
Cherniak DJ (2002) Ba diffusion in feldspar. Geochim Cosmochim Acta 66:1641-1650

Christensen JN, DePaolo DJ (1993) Time scales of large volume silicic magma systems: Sr isotopic systematics of phenocrysts and glass from the Bishop Tuff, Long Valley, California. Contrib Mineral Petrol 113:100-114

Christensen JN, Halliday AN (1996) Rb-Sr ages and Nd isotopic compositions of melt inclusions from the Bishop Tuff and the generation of silicic magma. Earth Planet Sci Lett 144:547-561

Condomines M (1997) Dating recent volcanic rocks through ^{230}Th-^{238}U disequilibrium in accessory minerals: Example of the Puy de Dôme (French Massif Central). Geology 25:375-378

Cooper KM, Donnelly CT (2008) ^{238}U-^{230}Th-^{226}Ra disequilibria in dacite and plagioclase from the 2004-2005 eruption of Mount St. Helens *In*: A Volcano Rekindled: The First Year of Renewed Eruption at Mount St. Helens, 2004-2006. Sherrod DR, Scott WE, Stauffer PH (eds) US Geological Survey Professional Paper 1750 (in press)

Cooper KM, Reid MR (2003) Re-examination of crystal ages in recent Mount St. Helens lavas: Implications for magma reservoir processes. Earth Planet Sci Lett 213:149-167

Cooper KM, Reid MR, Murrell MT, Clague DA (2001) Crystal and magma residence at Kilauea Volcano, Hawaii: ^{230}Th-^{226}Ra dating of the 1955 east rift eruption. Earth Planet Sci Lett 184:703-718

Cooper KM, Sims KWW, Eiler JM (2005) Time scales of assimilation and mixing at Krafla volcano, Iceland. EOS: Trans AGU v. 86: Fall Meet Suppl, Abstract V13B-0539

Costa F, Dohmen R, Chakraborty S (2008) Time scales of magmatic processes from modeling the zoning patterns of crystals. Rev Mineral Geochem 69:545-594

Crowley JL, Schoene B, Bowring SA (2007) U-Pb dating of zircon in the Bishop Tuff at the millennial scale. Geology 35:1123-1126

Dalrymple GB, Grove M, Lovera OM, Harrison TM, Hulen JB, Lanphere MA (1999) Age and thermal history of The Geysers plutonic complex (felsite unit), Geysers geothermal field, California; a ^{40}Ar/^{39}Ar and U-Pb study. Earth Planet Sci Lett 173:285-298

Davis DW, Williams IS, Krogh TE (2003) Historical Development of Zircon Geochronology. Rev Mineral Geochem 53:145-181

Davies GR, Halliday AN (1998) Development of the Long Valley rhyolitic magma system: strontium and neodymium isotope evidence from glasses and individual phenocrysts. Geochimica Cosmochimica Acta 62:3561-3574

Donnelly CT, Cooper KM (2006) Comparison of U-Th-Ra disequilibria in multiple crystal populations in lava from the current eruption of Mt. St. Helens. EOS: Trans AGU v. 87: Fall Meet Suppl, Abstract V54B-03

Gardner JE, Layer PW, Rutherford MJ (2002) Phenocrysts versus xenocrysts in the youngest Toba Tuff: Implications for the petrogenesis of 2800 km^3 of magma. Geology 30:347-350

Garrison J, Davidson J, Reid M, Turner S (2006) Source versus differentiation controls on U-series disequilibria: Insights from Cotopaxi Volcano, Ecuador. Earth Planet Sci Lett 244:548-565

Goldstein SJ, Stirling CH (2003) Techniques for measuring uranium-series nuclides: 1992-2002. Rev Mineral Geochem 52:23-57

Hammer JE (2008) Experimental studies of the kinetics and energetic of magma crystallization. Rev Mineral Geochem 69:9-59

Harford CL, Sparks RSJ (2001) Recent remobilisation of shallow-level intrusions on Montserrat revealed by hydrogen isotope composition of amphiboles. Earth Planet Sci Lett 185:285-297

Heath E, Turner SP, Macdonald R, Hawkesworth CJ, van Calsteren P (1998) Long magma residence times at an island arc volcano (Soufriere, St. Vincent) in the Lesser Antilles: evidence from ^{238}U-^{230}Th isochron dating. Earth Planet Sci Lett 160:49-63

Heumann A, Davies GR (2002) U-Th disequilibrium and Rb-Sr age constraints on the magmatic evolution of peralkaline rhyolites from Kenya. J Petrol 43:557-577

Hoskin PWO (2005) Trace element composition of hydrothermal zircon and the alteration of Hadean zircon from the Jack Hills, Australia. Geochim Cosmochim Acta 69:637-648

Hoskin PWO, Ireland TR (2000) Rare earth element chemistry of zircon and its use as a provenance indicator. Geology 28:627-630

Hoskin PWO, Schaltegger U (2003) The composition of zircon and igneous and metamorphic petrogenesis. Rev Mineral Geochem 53:27-62

Ireland TR, Williams IS (2003) Considerations in zircon geochronology by SIMS. Rev Mineral Geochem 53:215-241

Ivanovich M, Harmon RS (eds) (1992) Uranium-series Disequilibria: Applications to Earth, Marine, and Environmental Sciences. Oxford University Press, Oxford

Jicha BR, Singer BS, Beard BL, Johnson CM (2005) Contrasting timescales of crystallization and magma storage beneath the Aleutian Island arc. Earth Planet Sci Lett 236:195-210

Jicha BR, Singer BS, Beard BL, Johnson CM, Moreno-Roa H, Naranjo JA (2007) Rapid magma ascent and generation of ^{230}Th excesses in the lower crust at Puyehue–Cordón Caulle, Southern Volcanic Zone, Chile. Earth Planet Sci Lett 255:229-242

Jicha BR, Johnson CM, Hildreth W, Beard BL, Hart GL, Shirey SB, Singer BS (2008) Deciphering crust vs. mantle inputs and the timescales of magma genesis at Mount Adams using ^{238}U-^{230}Th disequilibria and Os isotopes. Earth Planet Sci Lett (in press)

Keller J (1969) Origin of rhyolites by anatectic melting of granitic crustal rocks; the example of rhyolitic pumice from the island of Kos (Aegean Sea). Bull Volcanol 33:924-959

Kemp AIS, Hawkesworth CJ, Paterson BA, Kinny PD (2006) Episodic growth of the Gondwana supercontinent from hafnium and oxygen isotopes in zircon. Nature 439:580-583

Kent AJR (2008) Melt inclusions in basaltic and related volcanic rocks. Rev Mineral Geochem 69:273-331

Kigoshi K (1967) Ionium dating of igneous rocks. Science 156:932-934

Klemetti EW, Cooper KM (2007) Cryptic young zircon and young plagioclase in the Kaharoa Rhyolite, Tarawera, New Zealand: Implications for crystal recycling in magmatic systems. EOS Transactions AGU, Fall Meet Suppl, Abstract V41F-03

Kosler J, Sylvester PJ (2003) Present trends and the future of zircon in geochronology: laser ablation ICPMS. Rev Mineral Geochem 53:243-275

LaTourrette T, Hervig RL, Holloway JR (1995) Trace element partitioning between amphibole, phlogopite, and basanite melt. Earth Planet Sci Lett 135:13-30

Lowenstern JB, Charlier BLA, Clynne MA, Wooden JL (2006) Extreme U–Th disequilibrium in rift-related basalts, rhyolites and granophyric granite and the timescale of rhyolite generation, intrusion and crystallization at Alid Volcanic Center, Eritrea. J Petrol 47:2105–2122

Lowenstern JB, Persing HM, Wooden JL, Lanphere M, Donnelly-Nolan J, Grove TL (2000) U-Th dating of singe zircons from young granitoid xenoliths: new tools for understanding volcanic processes. Earth Planet Sci Lett 183:291-302

Ludwig KR (1977) Effect of initial radioactive-daughter disequilibrium on U-Pb isotope apparent ages of young minerals. J Res U. S. Geol Survey 5:663–667

Luhr JF, Carmichael ISE, Varekamp JC (1984) The 1982 eruptions of El Chichon Volcano, Chiapas, Mexico: mineralogy and petrology of the anhydrite-bearing pumices. J Volcanol Geotherm Res 23:69-108

Mahood GA (1990) Reply to comment on "evidence for long residence times of rhyolitic magma in the Long Valley magmatic system: the isotopic record in precaldera lavas of Glass Mountain". Earth Planet Sci Lett 99:395-399

Mahood GA, Stimac JA (1990) Trace-element partitioning in panterllerites and trachytes. Geochim Cosmochim Acta 52:2257-2276

Miller SA, Burnett DS, Asimow PD, Phinney DL, Hutcheon ID (2007a) Experimental study of radium partitioning between anorthite and melt at 1 atm. Am Mineral 92:1535-1538

Miller JS, Matzel JEP, Miller CF, Burgess SD, Miller RB (2007b) Zircon growth and recycling during the assembly of large, composite arc plutons. J Volcanol Geotherm Res 167:282-299

Miller JS, Wooden JL (2004) Residence resorption and recycling of zircons in Devils Kitchen Rhyolite Coso Volcanic Field, California. J Petrol 45:2155-2170

Moeller A, Hellebrand E, Whitehouse M, Cannat M (2006) Trace elements in young oceanic zircons. EOS Transactions, American Geophysical Union 87: Suppl. 26, Dec. 2006

Nakada S, Bacon CR, Gartner AE (1994) Origin of phenocrysts and compositional diversity in pre-Mazama rhyodacite lavas, Crater Lake, Oregon. J Petrol 35:127-162

Parrish RR, Noble SR (2003) Zircon U-Th-Pb geochronology by isotope dilution-thermal ionization mass spectrometry (ID-TIMS). Rev Mineral Geochem 53:183-213

Pietruszka AJ, Carlson RW, Hauri EH (2002) Precise and accurate measurement of ^{226}Ra–^{230}Th–^{238}U disequilibria in volcanic rocks using plasma ionization multicollector mass spectrometry. Chem Geol 188:171-191

Pilot J, Werner C-D, Haubrich F, Baumann N (1998) Palaeozoic and Proterozoic zircons from the Mid-Atlantic Ridge. Nature 393:676-679

Ramos FC, Tepley FJ III (2008) Inter- and intracrystalline isotopic disequilibria: techniques and applications. Rev Mineral Geochem 69:403-443

Reagan M, Tepley III FJ, Gill J, Wortel M, Hartman B (2005) Rapid time scales of basalt to andesite differentiation at Anatahan volcano, Mariana Islands. J Volcanol Geotherm Res 146:171-183

Reagan MK, Cooper KM, Pallister JS, Thornber CR, Wortel M (2008) Timing of degassing and plagioclase growth in lavas erupted from Mount St. Helens, 2004-2005, from ^{210}Po-^{210}Pb-^{226}Ra disequilibria. In: A volcano rekindled: the first year of renewed eruption at Mount St. Helens, 2004–2006. Sherrod DR, Scott WE, Stauffer PH (eds) USGS Professional Paper 1750 (in press)

Reagan MK, Tepley III FJ, Gill JB, Wortel M, Garrison J (2006) Timescales of degassing and crystallization implied by ^{210}Po-^{210}Pb-^{226}Ra disequilibria for andesitic lavas erupted from Arenal volcano. J Volcanol Geotherm Res 157:135-146

Reagan MK, Volpe AM, Cashman KV (1992) ^{238}U- and ^{232}Th-series chronology of phonolite fractionation at Mount Erebus, Antarctica. Geochim Cosmochim Acta 56:1401-1407

Reid MR (2003) Timescales of magma transfer and storage in the crust *In*: The Crust. Rudnick RL (ed) Elsevier, Oxford, pp 167-193

Reid MR (2008) How long does it take to supersize an eruption? Elements 4:23-28

Reid MR, Coath CD (2000) In situ U-Pb ages from zircons from the Bishop Tuff: No evidence for long crystal residence times. Geology 28:443-446

Reid MR, Coath CD, Harrison TM, McKeegan KD (1997) Prolonged residence times for the youngest rhyolites associated with Long Valley Caldera: ^{230}Th-^{238}U ion microprobe dating of young zircons. Earth Planet Sci Lett 150:27-39

Renne PR, Swisher CC, Deino AL, Karner DB, Owens T, DePaolo DJ (1998) Intercalibration of standards, absolute ages and uncertainties in ^{40}Ar/^{39}Ar dating. Chem Geol 145:117–152

Rogers NW, Evans PJ, Blake S, Scott SC, Hawkesworth CJ (2004) Rates and timescales of fractional crystallization from U-238-Th-230-Ra-226 disequilibria in trachyte lavas from Longonot volcano, Kenya. J Petrol 45:1747-1776

Ruprecht P, Bergantz GW, Dufek J (2008) Modeling of gas-driven magmatic overturn: Tracking of phenocryst dispersal and gathering during magma mixing. Geochemistry Geophysics Geosystems 9: doi 10.1029/2008GC002022

Rutherford MJ (2008) Magma ascent rates. Rev Mineral Geochem 69:241-271

Saal AE, Van Orman JA (2004) The Ra-226 enrichment in oceanic basalts: Evidence for melt-cumulate diffusive interaction processes within the oceanic lithosphere. Geochem Geophys Geosys 5:2003GC000620

Sano Y, Tsutsumi Y, Terada K, Kaneoka I (2002) Ion microprobe U-Pb dating of Quaternary zircon: implication for magma cooling and residence time. J Volcanol Geotherm Res 117:285-296

Sarna-Wojcicki AM, Pringle MS, Wijbrans J (2000) New ^{40}Ar/^{39}Ar age of the Bishop Tuff from multiple sites and sediment rate calibration for the Matuyama-Brunhes boundary. J Geophys Res 105:21,431–21,443

Schaefer SJ, Sturchio NC, Murrell MT, Williams SN (1993) Internal ^{238}U-series systematics of pumice from the November 13, 1985, eruption of Nevado del Ruiz, Colombia. Geochim Cosmochim Acta 577:1215-1219

Scharer U (1984) The effect of initial ^{230}Th disequilibrium on young U-Pb ages: the Makalu case, Himalaya. Earth Planet Sci Lett 67:191-204

Schmitt A (2006) Laacher See revisited: High-spatial-resolution zircon dating indicates rapid formation of a zoned magma chamber. Geology 34:597-600

Schmitt AK (2007) Ion microprobe analysis of (^{231}Pa)/(^{235}U) and an appraisal of protactinium partitioning in igneous zircon. Am Mineral 92:691-694

Schmitt AK, Grove M, Harrison TM, Lovera O, Hulen JB, Walters M (2003a) The Geysers-Cobb mountain magma system California (Part 2): timescales of pluton emplacement and implications for its thermal history. Geochim Cosmochim Acta 67:3443-3458

Schmitt AK, Grove M, Harrison TM, Lovera OM, Hulen J, Walters M (2003b) The Geysers–Cobb Mountain Magma System, California (Part 1): U–Pb zircon ages of volcanic rocks, conditions of zircon crystallization and magma residence times. Geochim Cosmochim Acta 67:3423-3442

Schmitt AK, Lindsay JM, de Silva S, Trumbull RB (2003c) U-Pb zircon chronostratigraphy of early-Pliocene ignimbrites from La Pacana, north Chile: implications for the formation of stratified magma chambers. J Volcanol Geotherm Res 120:43-53

Schmitt AK, Stockli DF, Hausback BP (2006) Eruption and magma crystallization ages of Las Tres Virgenes (Baja California) constrained by combined ^{230}Th/^{238}U and (U–Th)/He dating of zircon. J Volcanol Geotherm Res 158:281–295

Schmitt AK, Vazquez JA (2006) Alteration and remelting of nascent oceanic crust during continental rupture: Evidence from zircon geochemistry of rhyolites and xenoliths from the Salton Trough, California. Earth Planet Sci Lett 252:260-274

Schwartz JJ, John BE, Cheadle MJ, Miranda EA, Grimes GB, Wooden JL, Dick HJB (2005) Dating the growth of oceanic crust at a slow-spreading ridge. Science 310:654-657

Simon JI, Reid MR (2005) The pace of rhyolite differentiation and storage in an 'archetypical' silicic magma system, Long Valley, California. Earth Planet Sci Lett 235:123-140

Simon JI, Reid MR, Young ED (2007) Lead isotopes by LA-MCICPMS: tracking the emergence of mantle signatures in an evolving silicic magma system. Geochim Cosmochim Acta 71:2014-2035

Simon JI, Renne PR, Mundil R (2008) Implications of pre-eruptive magmatic histories of zircons for U–Pb geochronology of silicic extrusions. Earth Planet Sci Lett 266:182-194

Sims KWW, Blichert-Toft J, Fornari D, Perfit MR, Goldstein S, Johnson P, DePaolo DJ, Hart SR, Murrell MT, Michael P, Layne G, Ball L (2003) Aberrant youth: Chemical and isotopic constraints on the young off-axis lavas of the East Pacific Rise. Geochem Geophys Geosys 4:8621, doi: 8610.1029/2002GC000443

Sims KWW, Ackert Jr. RP, Ramos FC, Sohn RA, Murrell MT, DePaolo DJ (2007) Determining eruption ages and erosion rates of Quaternary basaltic volcanism from combined U-series disequilibria and cosmogenic exposure ages. Geology 35:471-474

Sims KWW, Gill JB, Dosseto A, Hoffman DL, Lundstrom CC, Williams RW, Ball L, Tollstrup D, Turner S, Prytulak J, Glessner JJG, Standish JJ, Elliott T (2008) An inter-laboratory assessment of the thorium isotopic composition of synthetic and rock reference materials. Geostand Geoanal Res 32:65-91

Snyder DC, Widom E, Pietruszka AJ, Carlson RW, Schmincke H-U (2007) Time scales of formation of zoned magma chambers: U-series disequilibria in the Fogo A and 1563 A.D. trachyte deposits, Sao Miguel, Azores. Chem Geol 239:138-155

Streck MJ (2008) Mineral textures and zoning as evidence for open system processes. Rev Mineral Geochem 69:595-622

Stirling CH, Lee D-C, Christensen JN, Halliday AN (2000) High-precision in situ $^{238}U-^{234}U-^{230}Th$ isotopic analysis using laser ablation multiple-collector ICPMS. Geochim Cosmochim Acta 64:3737-3750

Taddeucci A, Broecker WS, Thurber DL (1967) ^{230}Th dating of volcanic rocks. Earth Planet Sci Lett 3:338-342

Tepley III FJ, Lundstrom CC, Gill JB, Williams RW (2006) U-Th-Ra disequilibria and the time scale of fluid transfer and andesite differentiation at Arenal volcano, Costa Rica (1968-2003). J Volcanol Geotherm Res 157:147-165

Tera F, Wasserburg GJ (1974) U–Th–Pb systematics on lunar rocks and inferences about lunar evolution and the age of the moon, Proceedings from the 5th Lunar Science Conference (Supplement 5). Geochim Cosmochim Acta 2:1571-1599

Tiepolo M, Vannucci R, Bottazzi P, Oberti R, Zanetti A (2000) Partitioning of rare earth elements, Y, Th, U, and Pb between pargasite, kaersutite, and basanite to trachyte melts: Implications for percolated and veined mantle. Geochemistry Geophysics Geosystems 1: 2000GC000064

Turner S, Black S, Berlo K (2004) Pb^{210}-Ra^{226} and Ra^{228}-Th^{232} systematics in young arc lavas: implications for magma degassing and ascent rates. Earth Planet Sci Lett 227:1-16

Turner S, George R, Jerram DA, Carpenter N, Hawkesworth C (2003a) Case studies of plagioclase growth and residence times in island arc lavas from Tonga and the Lesser Antilles, and a model to reconcile discordant age information. Earth Planet Sci Lett 214:279-294

Turner S, Foden J, George R, Evans P, Varne R, Elburg M, Jenner G (2003b) Rates and processes of potassic magma evolution beneath Sangeang Api Volcano, East Sunda Arc, Indonesia. J Petrol 44:491-515

Valley JW (2003) Oxygen isotopes in zircon. Rev Mineral Geochem 53:343-385

Van Orman JA, Grove TL, Shimizu N (1998) Uranium and thorium diffusion in diopside. Earth Planet Sci Lett 160:505-519

Van Orman JA, Grove TL, Shimizu N (2001) Rare earth element diffusion in diopside: Influence of temperature, pressure and ionic radius, and an elastic model for diffusion in silicates. Contrib Mineral Petrol 141:687-703

Van Orman JA, Saal AE, Bourdon B, Hauri EH (2006) Diffusive fractionation of U-series radionuclides during mantle melting and shallow-level melt-cumulate interaction. Geochim Cosmochim Acta 70:4797-4812

Vazquez JA (2004) Time scales of silicic magma storage and differentiation beneath caldera volcanoes from uranium-238-thorium-230 disequilibrium dating of zircon and allanite. PhD. thesis (unpubl), Earth and Space Sciences, UCLA, pp 284

Vazquez, JA (2008) Allanite and chevkinite as absolute chronometers of rhyolite differentiation. Geol Soc Am Abstr Programs 40:74

Vazquez JA, Reid MR (2002) Time scales of magma storage and differentiation of voluminous high-silica rhyolites at Yellowstone caldera, Wyoming. Contrib Mineral Petrol 144:274-285

Vazquez JA, Reid MR (2004) Probing the accumulation history of the voluminous Toba magma. Science 305:991-994

Vazquez JA, Shamberger PJ, Hammer JE (2007) Plutonic xenoliths reveal the timing of magma evolution at Hualalai and Mauna Kea, Hawaii. Geology 35:695–698

Volpe AM (1992) ^{238}U-^{230}Th-^{226}Ra disequilibrium in young Mt. Shasta andesites and dacites. J Volcanol Geotherm Res 53:227-238

Volpe AM, Hammond PE (1991) ^{238}U-^{230}Th-^{226}Ra disequilibria in young Mount St. Helens rocks: time constraint for magma formation and crystallization. Earth Planet Sci Lett 107:475-486

Watson EB (1996a) Surface enrichment and trace-element uptake during crystal growth. Geochim Cosmochim Acta 60:5013-5020

Watson EB (1996b) Dissolution, growth and survival of zircons during crustal fusion: kinetic principles, geological models and implications for isotopic inheritance. Trans Royal Soc Edinburgh Earth Sci 87:43-56

Watson EB, Harrison TM (1983) Zircon saturation revisited: temperature and compositional effects in a variety of crustal magma types. Earth Planet Sci Lett 64:295-304

Watson EB, Liang Y (1995) A simple model for sector zoning in slowly grown crystals; implications for growth rate and lattice diffusion, with emphasis on accessory minerals in crustal rocks. Am Mineral 80:1179-1187

Watson EB, Wark DA, Thomas JB (2006) Crystallization thermometers for zircon and rutile. Contrib Mineral Petrol 151:413-433

Zellmer G, Turner S, Hawkesworth C (2000) Timescales of destructive plate margin magmatism: new insights from Santorini, Aegean volcanic arc. Earth Planet Sci Lett 174:265-281

Zellmer GF, Blake S, Vance D, Hawkesworth C, Turner S (1999) Plagioclase residence times at two island arc volcanoes (Kameni Islands, Santorini, and Soufriere, St. Vincent) determined by Sr diffusion systematics. Contrib Mineral Petrol 136:345-357

Time Scales of Magmatic Processes from Modeling the Zoning Patterns of Crystals

Fidel Costa

Institut de Ciencies de la Terra Jaume Almera, CSIC
c/ LLuis Sole i Sabaris s/n
Barcelona 08028, Spain

fcosta@ija.csic.es

Ralf Dohmen, Sumit Chakraborty

Institut für Geologie, Mineralogie und Geophysik
Ruhr Universität Bochum
D-44780 Bochum, Germany

Ralf.Dohmen@rub.de; Sumit.Chakraborty@rub.de

INTRODUCTION

The advent of polarized light microscopy in the middle of the 19th century allowed mineralogists and petrologists interested in igneous rocks to recognize the widespread occurrence of fine-scale heterogeneities in the optical properties of minerals (e.g., see Young 2003 for details). The interpretation of mineral zoning patterns as archives of magmatic processes has been with us for some time (e.g., Larsen et al. 1938; Tomkeieff 1939). The development of the electron microprobe in the 1960's allowed mineral zoning profiles to be quantitatively analyzed and modeled (e.g., Bottinga et al. 1966; Moore and Evans 1967). Enhanced textural observations permitted recognition of many kinds of detailed structures during the growth and dissolution of magmatic minerals (Fig. 1; e.g., Anderson 1983; Pearce and Kolisnik 1990), and these processes were also explored using experimental and numerical models (e.g., Albarède and Bottinga 1972; Lofgren 1972; Kirkpatrick et al. 1976; Loomis 1982). Major element zoning patterns are routinely measured, and with the arrival of the ion microprobe, the identification of heterogeneous distribution of trace elements opened the window for more realistic and sophisticated scenarios and models (Fig. 1; e.g., Kohn et al. 1989; Blundy and Shimizu 1991; Singer et al. 1995). We now measure abundances of naturally occurring isotopes at the scale of tens of micrometers (e.g., Davidson et al. 2007a; Ramos and Tepley 2008) and this allows *in situ* dating of crystals (e.g., Cooper and Reid 2008).

Major and trace element or isotopic zoning in igneous minerals has been used to unravel the details of a large range of processes: from magma mixing, to magma degassing, magma transport, and fractionation (e.g., Blundy and Shimizu 1991; Singer et al. 1995; Umino and Horio 1998; Ginibre et al. 2007; Davidson et al. 2007b; Streck 2008). Crystal stratigraphy has also been used to construct lineages and genealogies between different crystals from the same or from different geological units (Wallace and Bergantz 2002). Zoning in crystals is also important because it may record the paths to irreversible magma mixing which cannot be inferred from studies of glasses alone (but see for example Perugini et al. 2004; Petrelli et al. 2006).

The advances in the spatial resolution and precision of *in situ* measurements of mineral compositions have also been beneficial to the experimental determination of the rates at which elements diffuse through crystals. The availability of large data sets of zoning patterns of

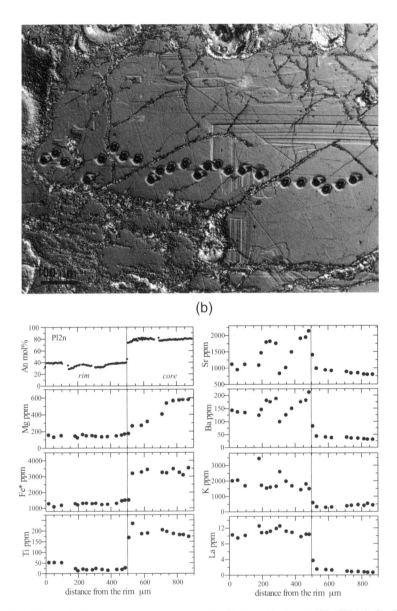

Figure 1. (a) Nomarski DIC image of a plagioclase crystal from a gabbroic xenolith of Volcán San Pedro (Costa et al. 2002). The crystal shows a well-defined core and a rim separated by an oscillatory-zoned interval. Ion microprobe analysis pits and the trace of the electron microprobe traverse between the pits are evident. (b) Major (An = anorthite mol%) and trace element profiles across this crystal. Note the abrupt change in the An mol% which coincides with the position of the core-rim boundary (location marked by a thin vertical line). Similar profiles are recorded by all trace elements except Mg. These relations were interpreted by Costa et al. (2003) in other crystals of the same rock suite to indicate that Mg was partially-equilibrated by diffusion between the core and the rim. Measurement and modeling of these trace element profiles provides information about magmatic processes and their time scales. The data are from Costa (2000), where more details can be found.

various elements in natural crystals combined with reliable diffusion coefficients for these same minerals have shown that diffusion within initially zoned minerals may partially or completely erase such zoning. This has led to a new perspective wherein the zoning of crystals is a record of magmatic variables and processes (e.g., Ginibre et al. 2007; Streck 2008), and the extent of erasure is a measure of the time during which changes in these variables occur. The first applications were on meteorites (e.g., Wood 1964; Goldstein and Short 1967; Short and Goldstein 1967) and extraterrestrial igneous rocks from the Moon or Mars (e.g., Taylor et al. 1973; Takeda et al. 1975; Walker et al. 1977), primarily to determine cooling rates of these magmas. However, the fundamental temporal information that can be extracted from compositional zoning is the duration of a particular process (e.g., see Turner and Costa 2007; Chakraborty 2008 for recent reviews). Compositional zoning in a crystal may be used to obtain time if: (1) one or more of the magmatic environmental variables such as temperature (T), pressure (P), volatile fugacities, or the composition of the liquid change (creating zoning), and (2) the diffusion of the elements of interest in the mineral is fast enough to partially erase the zoning, but slow enough for the crystal to not fully equilibrate to the new set of conditions (Fig. 2). Once a zoning pattern is established in a crystal it is possible under certain conditions to obtain a time constraint and correlate this directly to the process that is recorded in the crystal. Such constraints are most reliable when they are based on multiple determinations of durations from a single traverse by modeling the diffusion of multiple elements (e.g., Costa and Dungan 2005; Morgan and Blake 2006). As different elements diffuse at different rates, it is possible to obtain time scales for more than one process from a single thin section. Using multiple elements in a given crystal, multiple crystals of the same mineral from a single section, and different minerals from the same thin section makes obtaining large numbers of time determinations in a single thin section an opportunity that still needs to be fully exploited.

This chapter is designed to provide the reader with the knowledge necessary to go from measuring a concentration profile in a mineral to obtaining the duration of a process by applying the diffusion equations. This includes (i) presenting the basics of the equations that govern diffusion, (ii) clarifying the steps involved in extracting temporal information using

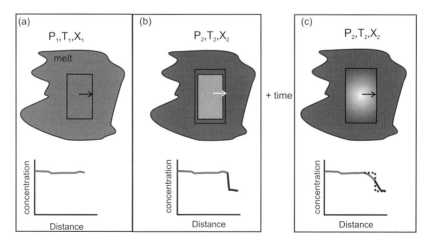

Figure 2. (a) Schematic representation of a crystal growing from a melt with a given set of variables (pressure: P, temperature: T, composition: X). A possible compositional traverse direction through the crystal is marked with an arrow and the corresponding concentration profile is shown below. (b) A change in one of the environmental variables is recorded in this case as a zone of different composition. (c) If diffusion in the crystal occurs fast enough, the original zoning is modified with time. One can use such information to constraint the processes and their durations.

these equations (including providing a brief introduction to numerical methods and some computer programs that use the method of finite differences to do such calculations), (iii) describing the parameters that control diffusion coefficients (with an Appendix that explains how the dependencies arise), (iv) considering how crystal growth and diffusion are combined to create observed profiles, and (v) discussing to what extent the necessary information can be obtained from nature. We have also compiled information on magmatic rates using the diffusion modeling approach and we compare these to the information from radiogenic isotopes.

EQUATIONS TO DESCRIBE DIFFUSION PROCESSES

In this section we introduce some basic concepts related to the physical and mathematical description of diffusion. Our presentation is oriented towards the application of the equations to obtaining time scales of magmatic processes and in some cases we have preferred to discuss the material from an intuitive perspective rather than in a rigorous manner. The reader is referred to Watson (1994), Chakraborty (1995, 2008), Lasaga (1998), Ganguly (2002) and Watson and Baxter (2007) for general aspects of diffusion in the Earth sciences, and to the textbooks of Manning (1968), Flynn (1972), Philibert (1991), Allnatt and Lidiard (1993), and Glicksman (2000) for a rigorous treatment of diffusion in solids.

Diffusion, flux and constitutive laws

Diffusion can be defined, following Onsager (1945), as the relative motion of one or more particles of a system relative to other particles of the same system. Random motions result from the thermal energy contained in a system, and they occur in all materials at all times at temperatures above the absolute zero. The diffusive motion of particles in air around us is perceptible and large (we can smell things pretty quickly), whereas the diffusive motion of Mg atoms in an olivine crystal sitting on our desk is imperceptibly slow. The important point is that the existence of a concentration gradient or driving force of any kind is not necessary for diffusion. What we are often interested in, however, is the quantity called flux, J. It is defined as the amount of material (expressed as the number of moles, mass of a substance, volume of a substance, and so forth) that crosses a plane of unit surface area in a given direction per unit time. This is a vector quantity and has units such as mol·m^{-2}·s^{-1}. To obtain a net flux or directed motion due to diffusion, a driving force that coaxes particles (e.g., atoms, molecules, ions) to preferentially move in one direction is required. In transport of matter, the commonly applicable driving force is a chemical potential gradient—matter flows in the direction of decreasing chemical potential. Onsager (1945) assumed that the flux, J_i, of a diffusing component i, is proportional to the gradient of the corresponding chemical potential, $\partial \mu_i/\partial x$ (diffusion in one dimension), and the two are related to each other via a proportionality constant, L_i:

$$J_i = -L_i \frac{\partial \mu_i}{\partial x} \qquad (1)$$

This relationship is intuitive, but it is not very practical because chemical potential is not an easily observable or measurable quantity. Chemical potentials can be related to measurable concentrations by the definition of a reference state and its associated reference chemical potential, μ°: $\mu_i = \mu^\circ + RT \ln(\gamma_i C_i)$, where γ_i is the activity coefficient. In the case of an ideal solution or a diluted component (for exceptions see below), where the derivative $\partial \ln(\gamma_i)/\partial x$ is zero, this relationship can be shown to be equivalent (e.g., Schmalzried 1981; Ganguly 2002) to the empirical relationship observed by A. Fick about 100 years earlier (Fick 1855). It connects the flux, J_i, to the concentration gradient of the species, $(\partial C_i/\partial x)$:

$$J_i = -D_i \frac{\partial C_i}{\partial x} \qquad (2)$$

The proportionality constant, D_i, is the diffusion coefficient of component i. This relationship, often referred to as Fick's first law, is analogous to other laws relating fluxes of different kinds to relevant driving forces (e.g., electrical current to electrical potential gradient: Ohm's Law; heat flux to the gradient of temperature: Fourier's Law) and is of practical use because concentration is an easily measurable quantity. The connection to the chemical potential gradient is the reason that Fick's law is applicable to diffusion in such disparate media as gases, liquids, and solids, even though the mechanisms of diffusion in these media are entirely different. The homogenization of a concentration gradient, a process that is a major concern of this chapter, is usually the result of a directed flux and not just random diffusion. Diffusion is ubiquitous, directed flux occurs only in the presence of a suitable driving force.

It is possible to obtain other kinds of mathematical relationships between directed flux and the measurable quantity, concentration. Some examples that will concern us in this chapter are the following:

(i) Diffusion in an anisotropic medium (e.g., a non-cubic crystal):

$$J_i^x = -D_i^{xx}\frac{\partial C_i}{\partial x} - D_i^{xy}\frac{\partial C_i}{\partial y} - D_i^{xz}\frac{\partial C_i}{\partial z}$$
$$J_i^y = -D_i^{yx}\frac{\partial C_i}{\partial x} - D_i^{yy}\frac{\partial C_i}{\partial y} - D_i^{yz}\frac{\partial C_i}{\partial z} \quad (3)$$
$$J_i^z = -D_i^{zx}\frac{\partial C_i}{\partial x} - D_i^{zy}\frac{\partial C_i}{\partial y} - D_i^{zz}\frac{\partial C_i}{\partial z}$$

Total flux, J_i^x, along a direction x is not just proportional to the concentration gradient along this direction. It is the sum of three terms, each of which has the form of Fick's law. This set of equations can be simplified if the axes are transformed into the principal axis system of the diffusion tensor (e.g., see Ganguly 2002). These principal axes may coincide with the crystallographic symmetry axes in some crystal classes, as is the case for orthorhombic olivine. Equation (3) can be expressed in matrix notation by the compact form $\mathbf{J}_i = -\mathbf{D}_i \cdot \nabla C_i$.

(ii) Diffusion accompanied by growth or dissolution—the moving boundary problem:

$$J_i = -D_i\frac{\partial C_i}{\partial x} + v \cdot C_i \quad (4)$$

Here the total flux, J_i, is the result of flux due to diffusion plus another flux due to the net growth or dissolution of a crystal in the surrounding medium (e.g., melt) at a rate, v. This equation describes flux of matter in a growing / dissolving crystal if the x-axis is defined to be such that the origin ($x = 0$) is always at the surface of the growing / dissolving crystal.

(iii) Flux due to a combination of forces; e.g., chemical and electrical forces acting on charged particles (e.g., ions) placed in a chemical as well as an electrical potential gradient:

$$J_i = -L_i\left[\frac{\partial \mu_i}{\partial x} - z_i \cdot e \cdot \frac{\partial \psi}{\partial x}\right] \quad (5)$$

where z_i is the charge of the diffusing component i, $\partial\psi/\partial x$ is the gradient of the mean electrical potential and e is the elemental charge. This is particularly important with respect to diffusion in ionic crystals, such as silicates, because it ensures that a charge imbalance does not develop due to diffusion of ions. Such constraints do not apply to metallic substances. Care should be taken when extrapolating the analysis of diffusion phenomena from such systems to silicates.

(iv) Diffusion in a one-dimensional, multicomponent system with n components, wherein there is coupling between the fluxes of different species (e.g., when charge balance constraints need to be fulfilled in a system containing ionic particles). The coupling of charged particles can be formally considered with the electric field imposed by the flux of the charged particles as in Equation (5) (for more details see Lasaga 1979; Schmalzried 1981):

$$J_1 = -D_{11}\frac{\partial C_1}{\partial x} - D_{12}\frac{\partial C_2}{\partial x} \ldots - D_{1n-1}\frac{\partial C_{n-1}}{\partial x}$$

$$J_2 = -D_{21}\frac{\partial C_1}{\partial x} - D_{22}\frac{\partial C_2}{\partial x} \ldots - D_{2n-1}\frac{\partial C_{n-1}}{\partial x} \quad (6)$$

$$\vdots$$

$$J_{n-1} = -D_{n-11}\frac{\partial C_1}{\partial x} - D_{n-12}\frac{\partial C_2}{\partial x} \ldots - D_{n-1n-1}\frac{\partial C_{n-1}}{\partial x}$$

This kind of expression was developed intuitively by Onsager as an analogy to the expression derived by Fick for simple binary systems. This set of equations can be written in matrix notation as $\mathbf{J} = -\mathbf{D}\cdot\partial\mathbf{C}/\partial x$, where $\mathbf{J} = \{J_1, J_2, \ldots, J_{n-1}\}$ and $\mathbf{C} = \{C_1, C_2, \ldots, C_{n-1}\}$. The flux of the n-th component directly follows from the flux of the other components by an additional constraint such as mass balance in a closed system. Onsager demonstrated that this form of the equation fulfils some basic requirements of thermodynamics, particularly when stated in terms of the more physically reasonable (but practically difficult to measure) chemical potential gradients. Experimental studies in different kinds of systems (ranging from aqueous solutions through silicate crystals and melts) have shown that such functions are an adequate description of diffusion in multi-component systems (e.g., Cooper 1965; Chakraborty et al. 1995; Liang et al. 1997).

Expressions that relate fluxes to various measurable parameters (concentration in the most common cases of interest to us) are known as *constitutive relations*. Generally speaking, these are relationships between fluxes and the forces that cause them, the connection between the two being made through parameters that are characteristic of specific media, and which are generally termed material constants. The diffusion coefficient, D, (Eqn. 2) and the constant L (Eqn. 1) are examples of material constants. The mathematical form of these equations describes the nature of the process; the values of the material constants are the causes of different behavior in different systems. This chapter is concerned with the material constant termed diffusion coefficient. It is necessary to ensure that D has the dimensions $[L]^2 \cdot [T]^{-1}$ (e.g., $m^2 \cdot s^{-1}$) irrespective of the choice of units for measuring concentration, flux etc.

Time dependence: the continuity equation

The reader would be left wondering why the variable time, t, does not appear explicitly in any of the equations shown above. Time is introduced by employing one of two universal constraints that any process must follow—the requirement of mass balance (the 2nd being the requirement that total entropy must increase in any natural process). In terms of fluxes, this is stated most simply as: when a specified system is observed over any given length of time, the difference between what comes in and what goes out is the net change in the amount contained in that system. This is a loose statement of a rigorous mathematical law called the Gauss Divergence Theorem (Fig. 3). The principle is the same that we use to balance our bank accounts (Input − Output = Change of balance).

Consider the volume of a rectangular parallelepiped where each side has incrementally small lengths e.g., dx, dy, dz. We observe the number of particles entering and leaving it through its faces. To simplify matters, we can specify that the flux occurs only in the x-direction. If the change of flux, J, in the x-direction is $(\partial J/\partial x)$, then we can describe the flux at position $x + dx$ to the first order as:

$$J(x+dx) = J(x) + \frac{\partial J}{\partial x}\cdot dx \quad (7)$$

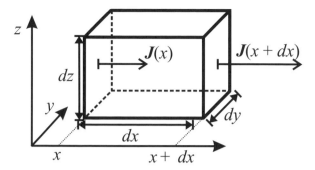

Figure 3. Nomenclature for the mass balance of fluxes in a rectangular solid. Volume of a rectangular parallelepiped with each side has incrementally small lengths e.g., dx, dy, dz. $J(x)$ and $J(x+dx)$ are the flux parallel to the x-direction at positions x, and $x+dx$, respectively.

The total number of particles (atoms, ions) entering and leaving through the sides of the parallelepiped is:

$$(J(x+dx) - J(x)) \cdot dy \cdot dz = \frac{\partial J}{\partial x} \cdot dx \cdot dy \cdot dz \tag{8}$$

The number of particles in the incremental volume element $dx \cdot dy \cdot dz$, defined as $C(x,t)$, changes to $C(x,t+dt)$ over an infinitesimal time interval, dt (short time where the flux is effectively constant) due to the flux imbalance:

$$C(x, t+dt) - C(x,t) = -\frac{\partial J}{\partial x} \cdot dt \tag{9}$$

The minus sign follows from the fact that the concentration has to increase when $J(x+dx) < J(x)$. The time derivative of C is defined mathematically as the limit of $(C(x,t+dt) - C(x,t))/dt$ when dt goes to zero and Equation (9) can be also expressed in the following differential form:

$$\frac{\partial C(x,t)}{\partial t} = -\frac{\partial J}{\partial x} \tag{10}$$

This derivation, based only on mass balance, is totally independent of the form of the constitutive equation we have chosen. Equation (10) is the continuity equation and replacing J in this equation using one of the expressions from the previous section allows us to obtain the corresponding time dependent equation. The simplest form of this relationship, for a binary exchange with a constant diffusion coefficient, is obtained by substituting Equation (2) on Equation (10) yields:

$$\frac{\partial C(x,t)}{\partial t} = D \frac{\partial^2 C(x,t)}{\partial x^2} \tag{11a}$$

This is often described as the diffusion equation, or Fick's second Law. Equation (11a) is valid only when D is independent of C or x; otherwise, the diffusion equation takes the following (expanded) form:

$$\frac{\partial C(x,t)}{\partial t} = \frac{\partial D}{\partial x} \cdot \frac{\partial C(x,t)}{\partial x} + D \cdot \frac{\partial^2 C(x,t)}{\partial x^2} \tag{11b}$$

The combination of a constitutive equation related to fluxes with the continuity equation leads to an equation such as (11a), which relates the change of concentrations in time to the concentration gradients in the form of a partial differential equation. It describes the evolution of concentration, C, as a function of time, t, at different spatial coordinates, $C(x,t)$. To obtain the specific concentration distribution (= concentration profile), $C(x)$, at any point of time t after the initiation of diffusion, we need three additional equations for each medium (e.g., crystal, melt) in which diffusion occurs. The first of these describes the initial condition (= concentration profile in our case) from which diffusion commences. The remaining two equations provide additional constraints that allow the concentrations at the boundaries of the diffusion system (= boundary conditions) to be specified. The boundary conditions may be formulated by specifying either the concentration or the flux at the boundaries of the system. If the system of interest is a crystal, a boundary condition is needed to specify the concentration or flux evolution at its surface. If this crystal is in equilibrium with a large body of melt at constant temperature, the concentration at the surface of the crystal would remain constant and obey the conditions of equilibrium, at the same time that diffusion strives to homogenize the concentration gradients in its interior. This condition may be expressed as $C_s = K$ where C_s is the concentration at the surface and K is a constant. Various forms of initial and boundary conditions are discussed below.

Once the initial and boundary conditions have been specified, an analytical solution may be derived in the form of a formula: $C(x,t) = $ ....with the diffusion coefficient, features of the initial concentration distribution, and the variables x and t appearing in some form on the right hand side. If the diffusion coefficient and initial conditions are known, the concentration at any position (x) and time (t) can be calculated. The analytical solutions for many different initial and boundary conditions for different geometries can be found in Crank (1975) and Carslaw and Jaeger (1986). These solutions are exact and calculations using these are fast, but they are limited to simple shapes for the diffusion medium (e.g., sphere, slab) and relatively simple boundary conditions (e.g., constant composition, variation according to some prescribed function). These simplifications introduce approximations when modeling shapes and concentration variations in real systems.

An alternative to analytical solutions (Appendix I) is to discretize the continuity equations using numerical methods. This chapter is accompanied by two computer programs that illustrate the implementation of these methods for a natural case: an Excel© spreadsheet (olivine_diffu_ excel.xls), and a Mathematica© notebook (olivine_diffu_mathema.nb). These can be found on the internet at *http://www.minsocam.org/MSA/RIM*. For most cases of interest, speed is not a problem for typical desktop computers, and the loss of accuracy (compared to analytical solutions) is inconsequential. Major benefits of this approach are that realistic shapes of crystals, variations in initial and boundary conditions, and the form of the continuity equations may be implemented with a minimum demand on mathematical abilities. Numerical methods are becoming the method of choice for modeling natural systems due to these advantages.

MODELING NATURAL CRYSTALS

We follow a forward modeling approach to obtain information on time (= duration) by modeling concentration gradients observed in natural crystals. Petrological constraints and intuition are applied to infer an initial concentration distribution, and the suitable form of the continuity equation. Equations (11a or 11b) are two possibilities, but others may be obtained by combining Equation (10) with any one of Equations (3, 4, 5, or 6). Petrological analysis should be employed to formulate suitable boundary conditions. The relevant questions are of the type: Was the crystal closed to element exchange with its surroundings during diffusion? And was the composition at its surface held constant during diffusion? A discretized form of the suitable continuity equation (Appendix I), in combination with these initial and boundary conditions, allows concentration profiles to be calculated for different durations of diffusion if diffusion coefficients are known. The progressive evolution of the concentration profile from the chosen initial condition is tracked, until a profile is obtained that matches the observed concentration profile or two-dimensional element map. The duration for which the calculated profile matches the observed profile is the length of time during which the natural crystal experienced diffusion, and this is the time scale of interest. Uncertainties arise from the choice of various models and parameters as they do in any modeling approach. These include the choice of the form of the continuity equation, initial and boundary conditions, and diffusion coefficients (see section about uncertainties below for more details). The diffusion equations formulated above typically hold for cases where temperature remains constant during diffusion, but non-isothermal diffusion is common in nature. Some of the aspects discussed here have been addressed by Ganguly (2002) and Chakraborty (2006) in the context of similar modeling in metamorphic rocks.

The key parameter in such models is the diffusion coefficient. A detailed discussion of this quantity follows below but for purposes of illustration in the examples that follow immediately we use the following expression (Dohmen and Chakraborty 2007a,b) for the calculation of the diffusion coefficient (in m²·s⁻¹) for Fe-Mg diffusion in olivine parallel to [001]:

$$D_{\text{Fe-Mg}} = 10^{-9.21} \cdot \left(\frac{f_{O_2}}{10^{-7}}\right)^{1/6} \cdot 10^{3(X_{Fe}-0.1)} \exp\left(-\frac{201000 + (P-10^5)\cdot 7\cdot 10^{-6}}{RT}\right)$$

where T is in Kelvin, P and f_{O_2} (pressure and oxygen fugacity, respectively) are in Pascals, X_{Fe} is the mole fraction of the fayalite component, and R is the gas constant in J·mol⁻¹·K⁻¹. Note that the diffusion coefficient depends (a) exponentially on temperature and (b) on intensive thermodynamic variables such as f_{O_2}. After discussing the modeling approaches, we discuss how such dependencies arise in the section on *Diffusion Coefficients*. It is shown that such expressions are now based on a microscopic understanding of diffusion mechanisms so that uncertainties in our knowledge of diffusion coefficients have been considerably reduced and time scales retrieved from diffusion modeling are more robust.

Initial and boundary conditions

There is no unique way of estimating the initial conditions, as this depends on the mineral and the nature of the problem. Figure 4 shows some possible initial distributions that may be associated with commonly observed chemical zoning in crystals: gradual (Fig. 4a), abrupt (Fig. 4b), homogeneous (Fig. 4c), and oscillatory (Fig. 4d). In the figure are also shown how such initial distributions evolve by diffusion (one profile at an arbitrary point of time in each

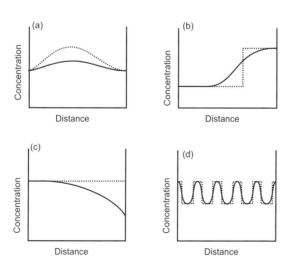

Figure 4. End-member types of initial zoning patterns used for obtaining time scales (dotted lines) and distribution after a given diffusion time (solid lines). (a) Smooth zoning that might be created during magma differentiation. (b) Abrupt zoning could be created during magma mixing. (c) Homogenous initial zoning where diffusion is driven by later change of concentrations at the boundary. (d) Oscillatory zoning.

case). The shapes of these profiles are good indicators of different initial distributions, although inferences concerning initial conditions are not always straightforward. Some strategies that have been employed are:

(1) Comparisons of zoning profiles for elements with very different diffusivities. Unmodified or less modified profiles of slowly diffusing elements can give clues to the initial distributions of rapidly diffusing elements. The profile shapes of the very slowly diffusing major element compositions CaAl-NaSi (Grove et al. 1984) in zoned plagioclase may be used to constrain the initial concentration profile (Costa et al. 2003) of a rapidly diffusing element such as Mg (LaTourrette and Wasserburg 1998). This approach also works in sanidine (Morgan and Blake 2006), where the slowly diffusing Ba may allow the initial distribution of faster Sr to be constrained and used to obtain time scales. A similar framework is applicable to olivine if zoning of slowly diffusing elements like P (Millman-Barris et al. 2008) can be used as a proxy for the initial profile shapes of the much faster diffusing divalent cations (e.g., Fe-Mg, Ni, Mn). This approach is powerful for distinguishing among the types of initial profiles (Fig. 4), provided it can be established that (i) the slower and faster diffusing elements had similar pre-diffusion distributions (not always the case, e.g., see the distribution of REE vs. Fe, Mg etc. in garnets, Denison and Carlson 1997), (ii) diffusive coupling (see below) has not modified the profile of the slower diffusing species.

(2) Using an arbitrary maximum concentration range for natural samples as an extreme initial concentration to obtain a maximum time for the diffusion process (e.g., Zellmer et al. 1999). A variant of this approach is to start with the arbitrary initial case of oscillatory zoning and track the progressive elimination of short-wavelength features by diffusion. The smallest surviving wavelength in the measured profile is an indication of the upper limit of diffusion-duration that the crystal may have experienced (Trepmann et al. 2004). A related method has been to use an extremely sharp step-like change and assume that the measured gradient is due entirely to diffusion (e.g., Fig. 4b; Morgan et al. 2004, 2006; Wark et al. 2007). Although all of these choices of initial conditions would yield maximum times, it is particularly useful for short diffusion profiles. These approaches have been used to model zoning gradients in garnet, clinopyroxene, feldspars, and quartz.

(3) An initially homogeneous concentration profile is often a good assumption for modeling olivine (e.g., Fig. 4c; Costa and Chakraborty 2004; Costa and Dungan 2005). The fact that robust time estimates have been obtained may reflect the growth of unzoned crystals from a large reservoir of liquid, or that they equilibrated at a high temperature before the final magmatic event that drove diffusion. The assumption of simple initial profiles, with *a posteriori* verification through the robustness of retrieved time scales, is therefore a useful variant.

(4) Smooth zoning profiles may be produced by magma fractionation during crystal growth (Fig. 4a), with or without accompanying diffusion. Algorithms such as those based on minimization of free energy allow such growth profiles to be simulated and used as initial conditions for diffusion. This has been done for metamorphic minerals (e.g., Loomis 1986; Spear et al. 1990) and we illustrate it below in an igneous system using the MELTS algorithm (Ghiorso and Sack 1995).

All other factors remaining the same, the use of different initial profiles gives different equilibration times. This effect is not always large. We have illustrated this with a series of numerical models that use two extreme possibilities for the shape of the initial profile: one is a homogenous crystal (e.g., Fig. 5a) and the other is a core with a rim overgrowth (e.g., Fig. 5b). The differences in retrieved time scales in these cases are less than a factor of 1.5 as can be seen in Figure 5. Initial conditions should be constrained as well as possible, but the uncertainties related to a lack of knowledge of this condition are not critical.

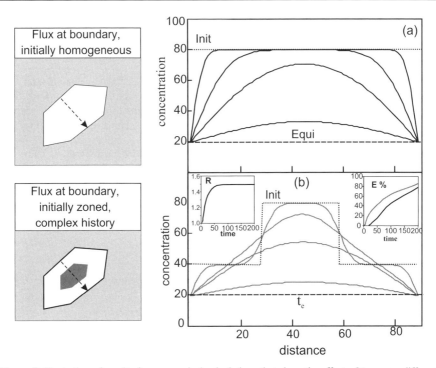

Figure 5. Illustration of results from numerical calculations that show the effect of two very different initial conditions on calculated times. Left panels show a hypothetical crystal surrounded by a matrix. The arrow marks the position of the compositional traverse. The right panel shows the time evolution of diffusion calculations for two crystals with different initial profiles at the same diffusion times. The initial profile (Init) is shown as a dotted line, and the equilibrium concentration (Equi) in shown as dashed line. (a) The initial concentration is homogeneous and diffusion is driven by the change in composition at the boundary. (b) The initial concentration profile is defined by step-function zoning and diffusion is driven a change in the boundary and by the original zoning profile. The inset at the right of the figure compares the time evolution for the two initial profiles as measured by the percentage of equilibration E% (= 100 − [100×{C-Equi}/{Init-Equi}]) at the center of the crystal. The crystal with the high homogenous concentration (black line) takes longer to equilibrate than the one with an initial step (grey line). The inset in the left side of the panel shows that the E% difference between the two initial profiles is at most a factor of 1.5 as measured by the value of R (= [E% of crystal b]/ [E% of crystal a]). This means that if we are able to constrain the maximum initial concentration, the precise shape of the initial profile introduces a limited uncertainty. The composition at the boundary is considered to be constant with time (for cases with changing conditions at the boundary see figures in next section).

The *boundary conditions* arise from the nature of exchange of the elements of interest at the boundary of the crystal with its surrounding matrix. Concentration at the boundary is defined if local and instantaneous equilibrium between the mineral surface and matrix occurs, which commonly appears to be the case (but not always; e.g., Dohmen and Chakraborty 2003). The boundary may be closed to exchange of matter (no flux) if the mineral is surrounded by a phase where the element of interest does not partition significantly or if the diffusion rate of the element in the surrounding matrix is much slower than in the mineral. Open boundaries are expected to be far more common in magmatic systems where minerals are in diffusive communication via melts. Open boundaries may result in the concentration at the rim of a crystal remaining fixed (e.g., when the crystal sits in a large volume of melt in which the concentration and diffusion rate of the element of interest is significant) or variable (as when the melt reservoir has limited volume, or when the ambient conditions such as temperature, oxygen fugacity, etc. change on the relevant time scale). It is important to differentiate between these scenarios because the times obtained by modeling a system with closed boundaries are much shorter than those obtained with open boundaries (Fig. 6; see also Chakraborty and Ganguly 1991 for some analytical solutions).

These considerations allow isothermal diffusion models to be evaluated, and these are often adequate for volcanic systems. Such models yield the duration of time that a heterogeneous crystal experienced high temperatures; i.e., prior to quenching by eruption. Most studies on volcanic systems have used isothermal models. However, crystals may have experienced thermal oscillations, such as those that may result from convection in a magma reservoir or due to multiple thermal pulses (e.g., Singer et al. 1995). As long as there is no overall heating or cooling trend, the results of modeling the duration of diffusion at a constant intermediate temperature are the same as that when oscillations are taken into account (e.g., Lasaga and Jiang 1995). The next level of sophistication is to incorporate the effect of changing temperature during diffusion. This is crucial for modeling plutonic systems with protracted thermal histories.

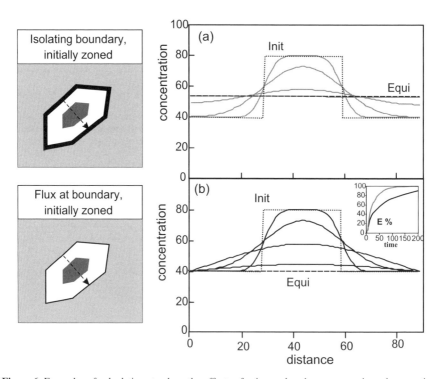

Figure 6. Examples of calculations to show the effects of using a closed vs. an open boundary, on time scales that are obtained. The main abbreviations and symbols are those used in Figure 5. The evolution of the concentration profiles for the two crystals with different boundary conditions but identical initial profiles are shown for the same diffusion times. (a) The crystal is closed to flux with the surrounding environment. (b) The crystal is open to flux and has a constant composition at the boundary. The inset at the right of the figure compares the time evolution for the two initial profiles as measured by the percentage of equilibration E% at the center of the crystal. A closed-system model reaches equilibrium much faster. Incorrectly applying this case to nature could underestimate the time scales by more than one order of magnitude, as shown in the inset where the percentage equilibration (%E) is plotted against time.

Non-isothermal diffusion

The diffusion models outlined above change in two important ways when non-isothermal diffusion is considered. "Non-isothermal" refers to systems undergoing a progressive change in temperature (e.g., heating, cooling) rather than modest oscillations about a mean temperature (thermal cycling). To model diffusion in a non-isothermal system we need to consider that: (i) diffusion coefficients change because they are strong functions of temperature, and (ii) the boundary conditions (e.g., the concentrations at the rims of crystals) change as a consequence of changing temperature. The magnitude of effect (i) is determined by the temperature dependence of diffusion coefficients (activation energy, see below) whereas the magnitude of effect (ii) is determined primarily by the enthalpy change of the exchange reaction between the mineral and its matrix for the element of interest. The modal proportions of mineral and matrix and the diffusion rates of the element in the two media influence the effect of (ii) as well.

The non-isothermal diffusion problem commonly may be reduced to an isothermal diffusion problem. This is accomplished by identifying a "constant," characteristic temperature of diffusion, T_{Ch}, such that the duration obtained by modeling isothermal diffusion at this temperature is the same as the one that would be obtained from a non-isothermal model. Chakraborty and Ganguly (1991) show that for diffusion processes with activation energies on the order of 200 kJ·mol^{-1}, there is a simple relationship between the peak temperature attained in a variety of non-isothermal histories, T_{Pk}, and the characteristic temperature: $T_{Pk} \sim 0.95\ T_{Ch}$ in Kelvin. If it is known that a pluton cooled slowly from a peak temperature of 1100 °C, a diffusion calculation carried out at the constant characteristic temperature of 1030 °C would yield a very good approximation of the duration of cooling to the temperature where diffusion effectively froze. This simplification is a consequence of the fact that diffusion rates depend exponentially on temperature. Much of the total diffusion during any thermal history occurs at temperatures very close to the thermal peak.

A second consequence of this exponential dependence allows us to use diffusion as a tool to measure time scales; diffusion effectively stops and the system (= concentration profiles in our case) 'freezes' with falling temperature. Analytical solutions were derived for systems undergoing cooling histories in a number of early works that made use of this behavior (Dodson 1973, 1976, 1986; Lasaga et al. 1977; Lasaga 1983). Dodson (1973) provided a formulation for the closure temperature (T_c); i.e., the temperature at which a system 'freezes' with respect to diffusion:

$$\frac{Q}{RT_c} = \ln\left(-\frac{AD_o RT_c^2}{Qsa^2}\right) \qquad (12)$$

Q is the activation energy for diffusion, D_o is the pre-exponential factor, A is a geometric factor, s is the cooling rate in the vicinity of the closure temperature, and a is the diffusing distance. This equation applies to the mean concentration in a zoned crystal (e.g., T_c is really a mean closure temperature), it only uses a single cooling rate, the reservoir of the crystal is infinite in size, and diffusion within it is taken to be infinitely fast. It is also assumed that sufficient diffusion has occurred to completely erase the initial composition. Ganguly and Tirone (1999, 2001) present formulations in which the mean closure temperature can be calculated for minerals that still record their initial concentrations and these should be of wide applicability to igneous systems. Comparable formulations are used to determine the closure temperatures of radiogenic isotope systems, and thereby to determine to which thermal events the dates obtained from these systems pertain. In the context of this manuscript it is relevant to point out that one can also calculate closure temperature profiles (Dodson 1986; Ganguly and Tirone 1999, 2001), where the system closes earlier and at higher temperatures in the crystal core than in the rim. Thus, there is not a unique closure temperature of a mineral and radiogenic system.

It depends on the position and also on the precision of the measurements of the elements or isotopes of interest (e.g., Charlier et al. 2007). These statements can be understood by considering the diffusion equation that describes the in-growth of the radiogenic daughter isotope (in one dimension; e.g., Albarède 1996):

$$\frac{\partial C_{DA}}{\partial t} = D(t)_{DA} \frac{\partial^2 C_{DA}}{\partial x^2} + \lambda N_o \exp(-\lambda t)$$

where C_{DA} is the concentration of the daughter, N_o is the number of parent atoms present initially, $D(t)_{DA}$ is the diffusion coefficient which depends on time through the thermal history, but not on composition or position, λ is the decay constant, and x is distance.

Numerical methods allow us to explore variations that are free from the restrictions noted above. To illustrate the effects of a cooling history on the calculated diffusion durations and the types of profiles that should result, we have performed a series of numerical calculations. The thermal history and cooling rates at four different positions within a spherical 1 km³ of magma intrusion (for details about the parameters used see caption of Fig. 7) are used to calculated how the diffusion coefficients change with time at each point. We have used the relation $x = 4\sqrt{Dt}$ for characterizing and comparing the diffusion distance x at different locations and for different elements (see Fig. 7). The diffusion distance varies depending on the thermal history and on position of the crystal within the pluton. Longer histories are recorded in the interior of the pluton. These numerical simulations illustrate that: (1) the use of a single temperature and crystal zoning data from a single location within the pluton is likely to give ambiguous results. The best location for using a single temperature might be close to the pluton rims because these undergo faster cooling, analogous to the volcanic systems considered above (this is only valid for a single intrusion event), and, (2) the zoning patterns of multiple types of crystals at multiple locations within the pluton should provide a picture of the temperature time evolution of the plutons, and thus allow discrimination between different scenarios of pluton growth and formation (e.g., Glazner et al. 2004). Mineral zoning has rarely been applied to obtain time scales for magmatic or hydrothermal processes in plutonic rocks (e.g., Coogan et al. 2002, 2005a, 2007; Tepley and Davidson 2003; Gagnevin et al. 2005; Davidson et al. in press), but it seems that such modeling could provide interesting insights into the evolution of plutons.

The results shown in Figure 7 also illustrate the effect of closure temperature on diffusion length scales. After a given time and temperature, the subsequent cooling history does not have any effect on the concentrations (e.g., diffusion distance vs. time relations are horizontal, as in Fig. 7e to f). In Figure 8 we show calculated closure temperatures for a number of elements in different minerals. These may be used as guidelines for first order estimates of how much diffusion to expect (length of diffusion zones). The plot can help to evaluate which elements in which minerals are expected to show significant diffusion zoning or erasure of magmatic zoning in different volcanic and plutonic settings.

Variable boundary conditions – a numerical simulation using MELTS

The preceding section focuses on the first of two complications that are introduced by non-isothermal diffusion—the variation of diffusion coefficient with changing temperature. These considerations allow first order estimates of length and time scales during non-isothermal

---

Figure 7. (*caption continued from facing page*)

using diffusion data of Dohmen and Chakraborty (2007a) and Dohmen et al. (2007). The relation $x = 4\sqrt{Dt}$ describes the penetration distance of diffusion for a semi-infinite medium from an infinite source, e.g., in the analytical solution: $(C_s - C)/(C_s - C_\infty) = \mathrm{erf}[(x)/(2\sqrt{Dt})]$, with C_s= concentration at the boundary, C_∞= concentration at the infinite medium. This would correspond to the profile schematically shown in Figure 4c.

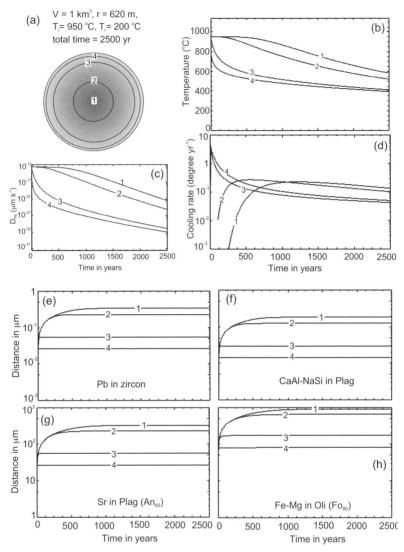

Figure 7. Numerical simulations illustrating the effect of thermal history on the diffusion distances and on the time scales that may be retrieved from crystal zoning patterns of plutonic rocks. (a) Parameters used in the calculations (T_i = initial temperature of the magma, T_r = initial temperature of the country rock, r = radius of the sphere, V = volume of the sphere) and illustration of the sphere and locations at which the temperatures were retrieved. We used a solution for a conductive cooling of a sphere with a thermal diffusivity of 10^{-6} m$^2 \cdot$s^{-1}. Position 1 is at the centre of the sphere, position 2 is at half the radius of the sphere (310 m from boundary), position 3 is 62 m from wall-rock boundary, and position 4 is 25 m from the boundary. The calculation was run for a total of 2500 years. (b) Temperature-time plot for the 4 positions. Note how the temperature close to the wall-rock boundary (positions 3 and 4) changes much faster than those at the interior. (c) Diffusion rate–time evolution for Pb in zircon (data of Cherniak and Watson 2001) to illustrate the effect of temperature-time path on diffusion rates. (d) Cooling rate evolution with time at the four different positions. The amount of diffusion that will occur also depends on the cooling rate, which is very variable depending on the position. (e) to (f) amount of diffusion distance for selected elements or components in different minerals. CaAl-NaSi in plagioclase (= Plag) using diffusion data of Grove et al. (1984), Sr in plagioclase using diffusion data of Cherniak and Watson (1994), Fe-Mg in olivine (= Oli)

(continued on facing page)

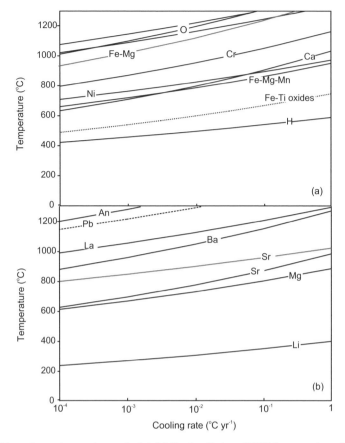

Figure 8. Mean closure temperatures calculated following Dodson (1973) for a number of elements in olivine, plagioclase, clinopyroxene, Fe-Ti oxides, and zircon. (a) Labels are elemental diffusion in olivine of composition Fo_{90} at 1 bar, NNO oxygen buffer, and parallel to [001] when applicable. Grey line for clinopyroxene (diffusion data from Dimanov and Sautter 2000), and dotted line for Fe-Ti oxides (diffusion data from Freer and Hauptman 1978). Data sources for olivine: Fe-Mg from Dohmen and Chakraborty (2007a, b), Ca from Coogan et al. (2005b), Mn, Ni from Petry et al. (2004), H from Demouchy and Mackwell (2006), Cr from Ito and Ganguly (2006), O from Gérard and Jaoul (1989), Ryerson et al. (1989), and Dohmen et al. (2002b). (b) Labels are diffusion of elements or anorthite in plagioclase (~ An_{70}) and also of Pb in zircon (dashed line; diffusion data from Cherniak and Watson 2001), and Sr in sanidine (grey line; diffusion data from Cherniak 1996). Diffusion data sources for plagioclase: Sr from Cherniak and Watson (1994), Ba from Cherniak (2002), La from Cherniak (2003), Mg from LaTourrette and Wasserburg (1998) extrapolated to An_{67} using Costa et al. (2003) equation, Li from Giletti and Shanahan (1997), and anorthite from Grove et al. (1984). All calculations for a sphere of 500 μm radius.

diffusion. Detailed descriptions of concentration profile shapes require adequate treatment of varying boundary conditions due to temperature changes. The early analytical models cited in the last section, as well as the work in the 1970's and 80's on extraterrestrial igneous processes cited in the introduction, considered variation of boundary conditions with temperature. However, the simplified thermal histories (e.g., linear or exponential cooling rates) and variations in boundary conditions considered in these analytical models are not always capable of capturing the complexities of real systems. Complications in natural systems may include non-linear changes of temperature with time, and changes in the coexisting phase assemblage during the course of magmatic evolution. The availability of software that allows the total free energy of a system to be minimized (e.g., MELTS: Ghiorso and Sack 1995; PERPLEX: Connolly 1990; DOMINO-THERIAK: de Capitani and Brown 1987) has opened a window of opportunity for diffusion modeling. These thermodynamic calculations yield the modal abundances and compositions of all phases in assemblages as they evolve along selected P-T-f_{O_2}-f_{H_2O} paths. Such information may be stored in discrete, digitized forms and implemented as boundary conditions in numerical models of diffusion. Diffusion modeling carried out in combination with such thermodynamic modules has potential for accurately describing profile shapes that develop during non-isothermal diffusion.

We have performed a series of numerical simulations wherein we have used MELTS (Ghiorso and Sack 1995) to calculate the equilibrium mineral-melt assemblage in a basaltic composition. We have focused on how the composition of olivine changes with decreasing T (from near liquidus at 1220 °C to 1170 °C) at constant P (100 MPa). We have chosen a crystal size of ~500 µm and a total diffusion time of 6 months (Fig. 9). These conditions have been used to generate three types of diffusion simulations: (i) constant temperature and composition at the boundary, (ii) cooling history but constant composition at the boundary, and (iii) cooling history and variable composition at the boundary (Fig. 9). We highlight the following observations:

(1) The profile shapes for cases (i) and (ii) are very similar. The only difference is that in case (ii) the extent of diffusion is smaller (shorter diffusion profiles) because the effect of decreasing diffusion coefficient due to decreasing temperature has been ignored in case (i). If we model the profile obtained in case (ii) using a single temperature as in case (i), we obtain very similar durations of diffusion; e.g., 5 months instead of 6. This shows that even if a volcanic rock has experienced a cooling history, the errors are typically not large if we use only a single temperature to model the process. Moreover, in a natural case we would obtain the temperature from a geothermometer that is also subject to the closure condition (Eqn. 12) and may not necessarily record the highest T, and therefore in practice the time difference would probably be even smaller. Such an effect could also be accounted for by using the characteristic temperature concept discussed above. The extent of this mismatch increases with increasing duration and/or extent of diffusion.

(2) The profile shapes produced in case (iii) are different from those produced in the other two cases. The shape of the profile for Fo mol%, for example, cannot be fitted in detail by any model in which the composition at the boundary is held constant. The Ca profile shape is impossible to obtain using any model similar to those considered in cases (i) and (ii). This means that detailed modeling of natural crystals provide hints on the importance or necessity of addressing variations in boundary conditions. The time scales obtained from case (iii) and case (i) are different, the latter being a factor of 4 shorter. This is mainly because the concentration gradient that is observed is a product of diffusion as well as variations in boundary conditions, and the latter effect is not accounted for in case (i). However, the shapes of the calculated profiles would provide a poor description of the observed profile shapes and this mismatch should be taken as an indication of the inadequacy of a constant-composition boundary condition model.

(3) All the three diffusion models lead to similar and linear relations when the concentrations of the different endmembers of the olivine are plotted against each other (Fo vs. Ni, Fo vs. Ca, and Fo vs. Mn; Fig. 9i-j). These linear trends are very different from the kinks shown by some of the equilibrium compositions. These plots show that combination of thermodynamic packages and diffusion models permit us to distinguish between olivine compositions that have been largely affected by diffusion and those that are unmodified after the creation of growth zoning. The specific relations may vary from one natural case to another and to be meaningful the calculations should be tailored (e.g., particular P, T, composition) to the magmatic system under study.

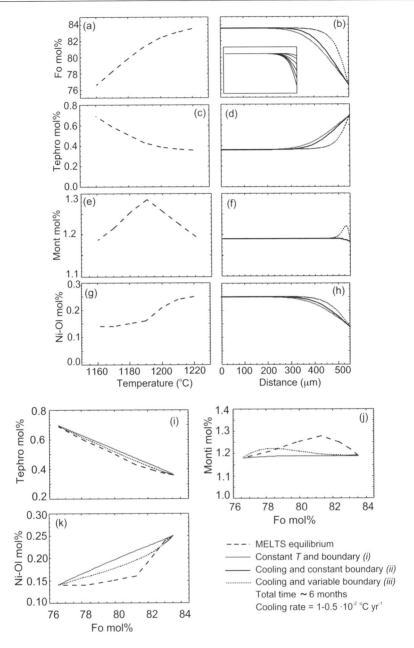

Figure 9. Numerical simulations to show the effects of cooling history and change in boundary compositions on the shapes of diffusion profiles. The temperature vs. composition of the olivine were calculated using MELTS (Ghiorso and Sack 1995) and correspond to the fractionation mode (equilibrium mode gave very similar results) of a basaltic composition at the quartz-fayalite-magnetite oxygen buffer (QFM) and 100 MPa (panels a, c, e, g of the figure). The olivine end members are Fo mol% = 100· Mg/(Mg+Fe+Ca+Mn+Ni), Tephro mol% = 100·Mn/(Mg+Fe+Ca+Mn+Ni), Mont = 100·Ca/(Mg+Fe+Ca+Mn+Ni), Ni-Ol mol% = 100·Ni/(Mg+Fe+Ca+Mn+Ni). Fractionating minerals were olivine, clinopyroxene and plagioclase. The changes in slope in the T vs. olivine composition plots occur where the two Ca-rich minerals start to co-

(continued on facing page)

Diffusion and crystal growth – the moving boundary problem

A scenario that probably occurs commonly in nature is diffusion concurrent with growth or dissolution of a crystal. This case belongs to the category of moving boundary problems, and in pure mathematical terms, it simply involves using constitutive Equation (4) and the continuity equation for describing how the concentration changes with time. If we focus on using numerical methods, then the difficulty of obtaining analytical solutions for different cases is of little consequence. However, major problems arise from (i) ascertaining how much growth or dissolution occurs at what stage of the evolution of a crystal, and (ii) the rate law that applies to growth or dissolution. The numerical implementation of a dissolution + diffusion model is more problematic than the implementation of a growth + diffusion model despite their mathematical similarities. The diffusion process itself may be modified in different ways during dissolution, but it is still a poorly understood topic. Consequently, we will focus here on the combination of growth and diffusion. Until now, applications of diffusion with growth or dissolution models have been rather limited (Nakamura 1995b: olivine growth; Watson 1996: zircon growth; Ozawa and Nagahara 2000: olivine dissolution).

The thermodynamic free energy minimization algorithms help to alleviate some of the problems associated with moving boundaries. With a given P-T-f_{O_2}-f_{H_2O} path for a bulk composition, the conditions at which a particular phase is produced can be obtained. If growth is instantaneous (i.e., there is no kinetic delay) and is uniformly distributed over all available crystals in a system (constrained, for example, by a crystal size distribution), the change in size of a crystal can be tracked numerically along with changes in modal abundances and composition. The impact of open-system behavior and resulting changes in bulk composition (e.g., due to magma mixing) are easily incorporated in stepwise-discretized numerical systems.

To illustrate the process, we have made a series of numerical simulations using the MELTS algorithm. The change in modal abundance of olivine indicated by the MELTS simulation is translated into growth of olivine and this adds layers of different compositions over the calculated range of P-T conditions. The size of each layer is determined by the amount of olivine that crystallized in the thermodynamic model; this olivine was plated onto a pre-existing olivine of spherical shape and size as in the simulations considered in the last section. We varied the growth rates from $\sim 10^{-8}$ to 10^{-10} cm·s^{-1}. These are within the limits that have been estimated from natural examples where the crystallization times are known or can be estimated (Cashman 1990). We highlight the following observations from the results shown in Figure 10:

(1) The profiles produced by only growth or by diffusion plus growth, with growth rates between 10^{-8} to 10^{-9} cm·s^{-1}, have shapes that are quite different from those produced

---

Figure 9. (*caption continued from facing page*)

crystallize. The length of the crystal and total time (6 months) were chosen arbitrarily but are similar to those relevant for magmatic systems. Cooling rates are 1-0.5×10^{-2} °C·yr^{-1}. The three diffusion models give different profiles (panels b, d, f, h). Single temperature and boundary composition produces the longest diffusion distance because we have used the initial temperature (grey line). The model with a cooling history (black dotted line) and constant boundary gives a similar shape, only shorter diffusion distance due to the lower temperatures (solid black line). The model that incorporates cooling and change at the boundary (about 12 mol% per year) shows a smaller diffusion distance yet, and a different profile shape which should make such profiles recognizable. Inset in panel (b) shows the detailed evolution of the non-isothermal and variable boundary model [referred to case (iii) in the main text]. Also shown are binary plots of olivine composition using the end-members (i-k) where it can be seen that the kinked trends generated from magma fractionation are different from the linear ones produced by diffusion alone. The symbols (i), (ii) and (iii) refer to the different cases explained in the text. Diffusion data for olivine come from: Fe-Mg from Dohmen and Chakraborty (2007a, b), Ca from Coogan et al. (2005b), Ni-Mn from Petry et al. (2004) all parallel to [001], and at QFM buffer and 1 atmosphere.

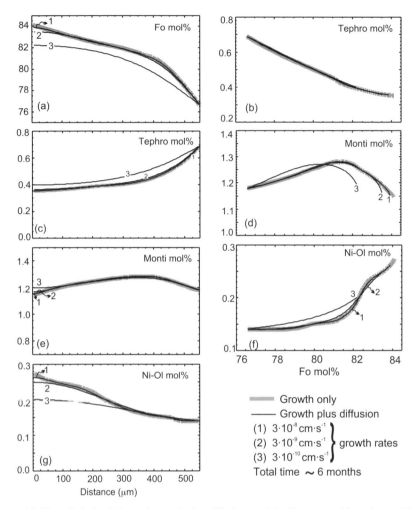

Figure 10. Numerical simulations of growth plus diffusion models. The composition of crystallizing olivine is that calculated by the MELTS thermodynamic algorithm (Ghiorso and Sack 1995) under the conditions explained in the caption of Figure 9 and in the text. The concentration profile without diffusion is shown as a thick grey line. The other lines are for growth plus diffusion at various growth rates. The same diffusion data and conditions were used as in the previous section.

by a diffusion only model in our sample calculation (Fig. 9). This means that they should be easily recognizable in that such shapes, if observed in nature, cannot be fitted by a diffusion-only model. It should be possible to carry out a generalized analysis based on non-dimensional parameters such as Peclet numbers to evaluate the nature of interaction between growth rates and length of diffusion profiles. Such an analysis geared explicitly to the determination of time scales from diffusion profiles in magmatic systems does not exist yet, but see Ozawa and Nagahara (2001) for a related example.

(2) As the growth rate decreases diffusion has a larger impact on the shape of the compositional profile. This departure occurs at a growth rate of ca. 10^{-10} cm·s^{-1} or less. The growth plus diffusion profiles of Fo, Ni and Mn, can be fitted with a diffusion-only

equation if we use a homogeneous starting profile (with the homogeneous concentration equivalent to the maximum measured concentration) and a constant-composition boundary condition at the rim (e.g., like model (i) shown in Fig. 9). The times we obtain are about twice as long as those used to produce the profiles in the diffusion plus growth models. Although this is a significant difference, it is reasonable considering the fact that we have used a simple model for a complex situation involving growth plus diffusion with changes in temperature and boundary conditions with time.

The growth plus diffusion profile for Ca cannot be fitted with a diffusion-only model without invoking a more complex initial or boundary condition than that assumed for the other elements. Moreover, we would find a very large difference in retrieved times compared to that obtained by fitting profiles of the other elements. These observations show that careful study of multi-element profiles permits an assessment of the assumed boundary conditions.

(3) The effects of concurrent diffusion and growth can also be recognized in binary plots of olivine composition as noted in the previous section. The Fo vs. Mn plot does not discriminate between growth and diffusion because Fe-Mg and Mn have similar geochemical behaviors during magma fractionation and their diffusion rates are also practically identical. The Fo vs. Ca plot is much more complex, and although there is a change in its pattern that depends on the growth and diffusion rates it is not clear that this should be easily recognizable in natural crystals. The most useful plot is that of Fo vs. Ni. The relation between these two appears to be non-linear in a fractionating magma, but with increasing diffusive modification the trend becomes more linear.

The proposed relations above that should allow one to distinguish between zoning produced by growth and diffusion are illustrated with examples from some natural olivine crystals (Fig. 11). One is from the Upper Placeta San Pedro sequence of the Tatara San Pedro volcanic complex. The compositional zoning in this crystal was shown by Costa and Dungan (2005) to have been the result of diffusion (Fig. 11a). The other crystal is from a lava of the Middle Estero Molino sequence, also from the Tatara San Pedro volcanic complex (QCNE.1; M. Dungan unpublished data). This crystal shows a non-linear relationship between Fo and Ni (Fig. 11b), indicating that there is a significant component of growth zoning retained in this crystal. Other criteria to distinguish between growth and diffusion zoning are based on the anisotropy of diffusion, which is discussed in more detail below. Linear relations between Fo and Ni are a necessary but not unique condition for identifying zoning produced by diffusion. Linear relations can also be obtained from growth alone, in particular for the low-Fo part of the trend (Fig. 10f). We note that these models are only illustrative and for application to a specific case the particular *P-T*-composition relations of the system should be used. If a diffusion-only calculation is used to model a concentration profile produced by growth + diffusion, time scales are overestimated. The effect of dissolution, to the extent that it can be modeled at all (see above), will be to underestimate time scales.

DIFFUSION COEFFICIENTS

The use of the diffusion equation to obtain time scales of magmatic processes is critically dependent on the accuracy of the relevant diffusion coefficients and whether experimentally determined diffusion coefficients apply to the natural processes that we want to model. Although here we do not review the experimental approach to obtaining diffusion data (see Baker 1993; Watson 1994; Chakraborty 1995, 2008), we want to make two observations. The first is that the advancements in experimental and analytical equipment based on thin film technology and nanometer-sized diffusion zones allow the determination of diffusion coefficients in many cases directly at the relevant P, T, and other environmental variables

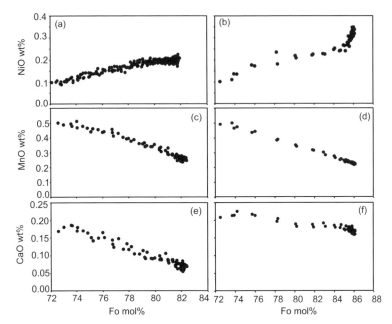

Figure 11. Plots of data from natural crystals to show effect of diffusion vs. growth. (a-c) Plot of crystal Oll_T6 from Upper Placeta San Pedro lava from the Tatara-San Pedro volcanic complex. The crystal is normally zoned (i.e., decreasing core to rim concentrations) of Fo, Ni, and Mn (traverses found in Costa and Dungan 2005). They show linear relations between Fo and other elements, which is indicative (but not proof) of profiles controlled mainly by diffusion (e.g., compare with Fig. 10f). This is in accord with the fact that Costa and Dungan (2005) found almost identical times when they modeled the profiles of Fo, Ni, Ca and Mn as being produced only by diffusion. (d-f) Crystal QCNE.1 Oliv-15 from the package G of Middle Estero Molino Sequence of the Tatara San Pedro volcanic complex (unpublished data of M. Dungan). This crystal is also normally zoned in Fo, Ni, and Mn. This crystal shows kinked relations between Fo and Ni and Fo and Ca. These indicate that diffusion has not been able to erase the growth zoning, particularly in the Fo-rich core (e.g., compare with plot Fig. 10f). If one would fit the zoning profiles of this crystal using diffusion models, different times would be obtained for different elements which would also indicate the presence of a significant component of growth zoning.

of interest for magmatic processes (e.g., Dohmen et al. 2002a). This removes the need for extrapolation from experimental data to the natural conditions in many cases. The second is that the number and the quality (e.g., exploring the full range of environmental variables) of diffusion coefficients has increased dramatically, and we understand the mechanisms of diffusion better. It is currently possible to use almost any igneous rock to obtain time scales of magmatic processes with a level of certainty that was not possible a couple of decades ago.

We first present the different types of diffusion coefficients, beginning with a discussion of atomistic mechanisms, and present examples to illustrate which diffusion coefficients to use in different situations. We discuss the intensive variables and other parameters that control the diffusion coefficients with a detailed analysis of Fe-Mg diffusion in olivine for the purpose of clarifying why (and how) the diffusion coefficients depend on specific thermodynamic variables (Appendix II). This also allows us to explore what is buried within the activation energies and pre-exponential factors in Arrhenius-type descriptions of the diffusion coefficient. We hope that these sections will allow the user of diffusion coefficients to use them with more confidence, with an understanding of their limitations in the applications to natural situations. Other publications that have described some of these aspects in the geological literature include Watson (1994), Chakraborty (1995, 2008), Ganguly (2002), Watson and Baxter (2007).

Types of diffusion coefficients

Diffusion in single crystals requires point defects. The most common diffusion mechanism involves vacancies but other possibilities are the interstitial mechanism, direct exchange, ring mechanism, the interstitialcy mechanism (see e.g., Philibert 1991). The vacancy mechanism implies that each atom surrounding a vacant crystallographic site in an otherwise homogeneous medium has a probability to jump into it, thereby leaving a new vacant site behind. This process is self diffusion and does not require the presence of a concentration gradient on the macroscopic scale. The atomic jumps can be treated statistically. For a particular atom, i, sitting on a particular crystallographic site a theoretical diffusion coefficient can be defined (Flynn 1972; Philibert 1991) as:

$$D_i = w \cdot l^2 \tag{13}$$

where w is the probability per unit time (or frequency) of a jump to any of the adjacent crystallographic sites at a distance l. This diffusion coefficient is called the **self diffusion coefficient** (or random walk diffusion coefficient). Equation (13) relates a macroscopically measurable property (D_i) and parameters that characterize the diffusion mechanism on an atomistic scale.

The most common experimental methods used to study self diffusion employ isotopic tracers to make the diffusion process measurable. An isotopic concentration gradient is artificially created in an experimental charge and the concentration distribution evolves with time as predicted by Fick's 2nd law (Eqn. 11a,b). The diffusion coefficient of an isotope j of an element i that describe its flux solely in response to the isotopic concentration gradient in a chemically homogeneous medium will be called the **tracer diffusion coefficient** of the element, $D_j^{*(i)} = f \cdot D_i$. The factor f considers effects of the geometry (coordination number, symmetry) and the diffusion mechanism (e.g., probability that the atom jumps back without contributing to a net flux; for details see Flynn 1972; Schmalzried 1981; Philibert 1991); it is effectively a constant with a value close to unity in most cases (see examples in Schmalzried 1981; Philibert 1991). The terms self- and tracer diffusion coefficients have been used interchangeably in the literature. The tracer diffusion coefficient is not necessarily the diffusion coefficient of a trace element, although it might be justified to use a tracer diffusion coefficient to describe the re-equilibration of a trace element concentration gradient (see below). The important point from the perspective of a modeler is that the tracer diffusion coefficients can be, and often are, functions of composition; i.e., the tracer diffusion coefficient of Mg in a Fo$_{90}$ olivine is different from that in a Fo$_{75}$ olivine.

Tracer diffusivities may apply strictly to only few natural examples of magmatic systems, such as those involving zoning in oxygen (e.g., Bindeman 2008) and Sr isotopes in crystals (although in this case the absence of an overall Sr concentration gradient needs to be demonstrated; Ramos and Tepley 2008). A concentration gradient of a trace or minor element, or a major element, in a solid solution will be involved in most other applications, and these gradients have two consequences: First, this implies the existence of a chemical potential gradient which must be used to describe the flux of the element (Eqn. 1). This process is called **chemical diffusion** and chemical diffusion coefficients (interdiffusion in binary, or multi-component systems) are used to describe it. The numerical magnitudes of chemical and tracer diffusion coefficients can be very similar, but they are different physical quantities. Secondly, as various crystallographic sites are occupied by more than one element, the flux of the element of interest cannot be treated independently from those of the other elements occupying the same site and formalisms to handle multi-component diffusion are required.

Multi-component diffusion in silicates

In an ionic crystal or solution the flux of an ion implies a flux of charge that creates an electrochemical potential gradient (Eqn. 5). This potential must be compensated by a corresponding counter-flux in the absence of an external electrical field in order to maintain charge neutrality for the bulk crystal. The diffusion of an element in a mineral is an interdiffusion process in which at least two elements (or other structural units of a single crystal, e.g., electrons, vacancies, see below) are involved. The formulation of Lasaga (1979) accounts for the coupling of diffusion in an ionic crystal. This framework specifies the coupled fluxes of cations in a silicate wherein the Si sub-lattice remains fixed, providing a fixed reference frame for the description of the fluxes of other ions. The constitutive equation describing the fluxes of n diffusing components or elements in one dimension is given by Equation (6). The individual elements of the diffusion matrix (D_{ij}) are related to the tracer diffusion coefficients of the components and the thermodynamic properties of the silicate (Lasaga 1979) by:

$$D_{ij} = D_i^* \delta_{ij} + D_i^* C_i \frac{z_j}{z_n} \left(\frac{\partial \ln \gamma_i}{\partial C_j} - \frac{\partial \ln \gamma_i}{\partial C_n} \right)$$
$$- \left[\left(D_i^* z_i C_i \right) \bigg/ \sum_{k=1}^{n} z_k^2 C_k D_k^* \right] \times \left[z_j \left(D_j^* - D_n^* \right) + \sum_{k=1}^{n} z_k C_k D_k^* \left(\frac{\partial \ln \gamma_k}{\partial C_j} - \frac{z_j}{z_n} \frac{\partial \ln \gamma_k}{\partial C_n} \right) \right] \quad (14)$$

Here i and j are indices for the component, z_i – charge of component i, D_i^* – tracer diffusion coefficient of component i, C_i – concentration of component i, δ_{ij} – the Kronecker delta symbol (1 for $i = j$, 0 otherwise). The concentration gradient of the nth component is determined from that of the other components according to the equation:

$$\frac{\partial C_n}{\partial x} = -1/z_n \cdot \sum_{j=1}^{n-1} z_j \cdot \frac{\partial C_j}{\partial x} \quad (15)$$

Garnet was one of the first silicates or minerals where multi-component diffusion was experimentally studied (Elphick et al. 1985; Loomis et al. 1985; Chakraborty and Ganguly 1992) and satisfactorily described by the application of the Lasaga (1979) model.

Example 1: Chemical interdiffusion of divalent cations. Let us consider diffusion of divalent Fe and Mg in a binary solid solution such as olivine. In the absence of any other mechanism of charge compensation, the net flux of the two ions at any given point must be zero (i.e., there must be equal number of Fe^{2+} and Mg^{2+} crossing a point in opposing directions per unit area per unit time). A single equation can describe the flux (consider Eqn. 6) where Mg has been chosen as the dependent component:

$$J_{Fe} = -D_{FeMg} \cdot \frac{\partial C_{Fe}}{\partial x} = -J_{Mg} \quad (16)$$

The interdiffusion coefficient D_{FeMg} can be derived from Equation (14) to be:

$$D_{FeMg} = \frac{D_{Fe}^* \cdot D_{Mg}^*}{X_{Fe} \cdot D_{Fe}^* + X_{Mg} \cdot D_{Mg}^*} \cdot \left(1 + \frac{\partial \ln \gamma_{Fe}}{X_{Mg}} \right) \quad (17)$$

Where X_{Fe} was introduced as the molar fraction of the Fe-endmember defined as $C_{Fe}/(C_{Fe} + C_{Mg})$ with an analogous definition for the Mg-endmember, X_{Mg} [= (1 − X_{Fe}) in this binary case]. The term in parenthesis in Equation (17) is called the thermodynamic factor and is equal to one if the solid solution behaves ideally, which is a good first approximation for olivine. Tracer diffusion coefficients (D_i^*) may depend on the major element composition, X_{Fe}. This must be considered when Equation (14) and its form for the binary system (Eqn. 17) are used. The interdiffusion coefficient in a binary system approaches the tracer diffusion coefficient of the dilute component if one component is very diluted (e.g., X_{Fe} approaches 0); it follows that $D_{FeMg} \sim D_{Fe}^*$ (Fig. 12). If a major component has a much smaller tracer diffusion coefficient than the dilute component, then the interdiffusion coefficient deviates significantly from the

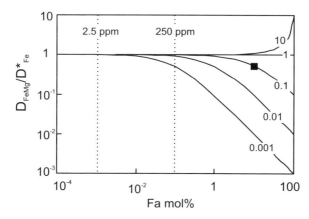

Figure 12. Plot of the chemical diffusion coefficient for FeMg interdiffusion (D_{FeMg}) normalized to the tracer diffusion coefficient of Fe (D_{Fe}^*) against the molar fraction of Fe for various ratios of D_{Mg}^*/D_{Fe}^*, which are indicated by the numbers on the curves. The chemical diffusion coefficient becomes effectively equal to D_{Fe}^* if X_{Fe} is below a critical value for all ratios. The critical value becomes smaller for decreasing D_{Mg}^*/D_{Fe}^*. If D_{Mg}^* would be even larger than D_{Fe}^* the chemical diffusion process could be modeled by D_{Fe}^* for binary solid solutions with $X_{Fe} < 0.1$. The solid square is a value calculated with experimental Mg tracer diffusion and Fe-Mg diffusion data for olivine with $X_{Fe} = 0.1$ (Chakraborty et al. 1994 and Chakraborty 1997, respectively), which indicates that D_{Fe}^* is ten times larger than D_{Mg}^*.

tracer diffusion coefficient even for relatively small concentrations of the dilute component (Fig. 12). Therefore, it is not always justified to use the tracer diffusion coefficient for a trace element, and some estimation of expected diffusion rates can help to decide whether tracer diffusion coefficients are applicable. Experimental diffusion data for ~Fo$_{90}$-olivine show that the tracer diffusion coefficients of Fe and Mg differ roughly by a factor of 10 (Fig. 12), which implies that for $X_{Fe} < 0.01$ the approximation $D_{FeMg} = D_{Fe}^*$ is justified.

These considerations can be extended to multi-component systems by imposing the condition that C_i tends to zero for any given i in Equation (14). The algebra is more involved, but the basic principle is the same; i.e., the process of diffusion of a trace element in a multi-component system is decoupled from the diffusion and concentration gradients of the other components as long as the diffusion rate of one other (preferably major) component is fast enough to ensure charge balance, and the transport rate is well described by the tracer diffusion coefficient. The tracer diffusion coefficient of some trace elements may also depend on the composition of the multi-component solution and therefore depend on the instantaneous concentrations and gradients of the other components in the system.

Natural olivine containing Mn, Ni and Ca as trace elements is a multi-component phase which can be described as (Fe,Mg,Mn,Ni,Ca)$_2$SiO$_4$. The transport of trace elements that occur as divalent cations in the same octahedral sites as the major elements Fe and Mg can be modeled (Costa and Dungan 2005) using their respective tracer diffusion coefficients (e.g., Ni, Mn: Petry et al. 2004; Ca: Coogan et al. 2005b), without consideration of diffusive coupling (note the contrast with Fe-Mg diffusion). Nonetheless, the values of these tracer diffusion coefficients depend on the composition (and hence on instantaneous concentration gradients) in the olivine; e.g., D_{Ni}^* in an olivine of composition Fo$_{90}$ is ~ 4.0×10^{-17} m^2·s^{-1} at 1000 °C, whereas the same tracer diffusion coefficient is ~ 4.0×10^{-16} m^2·s^{-1} in an olivine of composition Fo$_{40}$.

Example 2: Chemical diffusion of Sr^{2+} in plagioclase. Sr occurs at trace element concentration levels in most natural plagioclases and its diffusivity has been experimentally investigated in the presence of concentration (Cherniak and Watson 1994; Giletti and Casserly

1994) and isotopic gradients (LaTourette and Wasserburg 1998). Cherniak and Watson (1994) demonstrated that chemical diffusion of Sr occurs mainly by an exchange with Ca in anorthite-rich plagioclase. Equation (14) for the binary form with the two components Ca^{2+} and Sr^{2+} can be used (analogous to Eqn. 17) to describe the chemical diffusion coefficient of Sr^{2+}:

$$D_{Sr}^{ch} = \frac{D_{Sr}^* \cdot D_{Ca}^*}{X_{Sr} \cdot D_{Sr}^* + X_{Ca} \cdot D_{Ca}^*} \cdot \left(1 + \frac{\partial \ln \gamma_{Sr}}{X_{Sr}}\right) \quad (18)$$

It is valid to assume that the thermodynamic factor is 1 for low concentrations of Sr in an unzoned plagioclase (i.e., constant anorthite content), because the activity coefficient is constant (Henry's Law limit). LaTourette and Wasserburg (1998) determined that $D_{Ca}^* \sim 4.6 \times 10^{-17}$ $m^2 \cdot s^{-1}$ and $D_{Sr}^* \sim 7.1 \times 10^{-17}$ $m^2 \cdot s^{-1}$ at 1400 °C and $X_{An} = 0.95$. As the ratio of these tracer diffusion coefficients is close to 1, we can directly conclude from Figure 12 that $D_{Sr}^{ch} = D_{Sr}^*$ for molar fractions of Sr smaller than 0.01. The chemical diffusion coefficient of Sr varies from 7.1×10^{-17} $m^2 \cdot s^{-1}$ to 6.7×10^{-17} $m^2 \cdot s^{-1}$ for a variation of 3 orders of magnitude of Sr/(Sr+Ca) from 0.001 to 0.1 at 1400 °C. Given that typical Sr contents in plagioclase are 500-1000 wt ppm, chemical diffusion of Sr in anorthite-rich plagioclase can be modeled using its tracer diffusion coefficient.

The situation described above will change when the anorthite content of the plagioclase changes: not only does the chemical diffusion coefficient of Sr in plagioclase depend strongly on the anorthite content, but the mechanism of exchange which accounts for the charge balance in the crystal (as shown by Cherniak and Watson 1994) changes as well. In addition to the dependence of D_{Sr}^* on anorthite content, a more complicated situation arises if the plagioclase is zoned in the major components (e.g., anorthite content). The activity coefficient γ_i of a trace element i is in most cases constant for a fixed major element composition, but this may vary within a single crystal if it is zoned with respect to major elements. This means that the term $\partial \ln(\gamma_i)/\partial x$ is non-zero, and Equation (1) cannot be transformed to Equation (2). An extension of Equation (2) which considers this additional derivative has been proposed by Costa et al. (2003) and they have applied this approach to the modeling of Mg gradients in plagioclase.

$$J_i = -D_i^* \cdot \frac{\partial C_i}{\partial x} + D_i^* \cdot C_i \cdot \frac{A_i}{RT} \cdot \frac{\partial X_{An}}{\partial x} \quad (19)$$

A formulation such as this is applicable to any other trace element diffusing in a crystal that is zoned with respect to major elements. The additional driving force for the diffusion flux is related to the concentration gradient of the anorthite content, which has been treated as frozen for the duration being modeled because diffusivity of the anorthite component is much slower compared to that of trace elements (e.g., Grove et al. 1984). The factor A_i is a thermodynamic parameter related to activity coefficients. Its value can be empirically approximated to a first degree from the compositional dependence of the trace element partitioning between plagioclase and melt (i.e., dependence on anorthite content; Costa et al. 2003).

In summary, the tracer diffusion coefficients of all diffusing components plus the thermodynamic mixing properties of the solid solution must be known to model concentration profiles in crystals via the diffusion matrix (Eqn. 14). Some simplifications are possible in the absence of the full body of required information. The assumptions that mixing is ideal or that some components are effectively immobile considerably simplifies the calculations without introducing significant errors in time scales retrieved from modeling.

Parameters that determine tracer diffusion coefficients

Any factor that affects tracer or self diffusion coefficients affects all of the other diffusion coefficients as well. The values of diffusion coefficients in common silicate systems are mainly controlled by temperature, pressure, water fugacity, oxygen fugacity, composition, and crystallographic direction. The types and numbers of variables that need to be used to describe diffusion coefficients are determined by the Gibbs phase rule; i.e., there is a connection between transport properties and an equation from equilibrium thermodynamics. Temperature always has a strong effect because diffusion is a thermally activated process. Pressure and composition normally affect diffusion coefficients, although the consequences may be undetectably small in many systems. The activities of components and volatile fugacities may, but do not always, influence diffusion. An Arrhenius equation that describes the temperature and pressure dependencies is commonly used to calculate diffusion coefficients from experimental data:

$$D = D_o \cdot \exp\left(\frac{-Q - \Delta V(P - 10^5)}{R \cdot T}\right) \tag{20}$$

Here Q is the activation energy at 10^5 Pa, ΔV is the activation volume, P is pressure in Pascals, R is the gas constant, and D_o the pre-exponential factor.

The magnitude of the dependencies of the diffusion coefficient on T and P depend on the size of Q and ΔV. The lowest activation energies for elements diffusing in silicates are those for H and Li diffusion in olivine, pyroxenes, and plagioclase, which are around 100-200 kJ·mol^{-1} (e.g., Giletti and Shanahan 1997; Coogan et al. 2005a; Ingrin and Blanchard 2006). Divalent cations for the same silicates tend to have activation energies between 200-300 kJ·mol^{-1} (e.g., Giletti and Casserly 1994; Cherniak and Watson 1994; LaTourrette and Wasserburg 1998; Dimanov and Sautter 1999; Petry et al. 2004; Coogan et al. 2005b; Dohmen and Chakraborty 2007a; Dohmen et al. 2007). Increasingly larger activation energies in the ranges of 300 to 500 kJ·mol^{-1} have been found for trivalent and tetravalent cations (e.g, van Orman et al. 2001; Dohmen et al. 2002b; Cherniak 2003; Ito and Ganguly 2006; Costa and Chakraborty 2008). Knowing the activation energy, the effect of changing temperature at a given P can be estimated using $\ln D(T_2) - \ln D(T_1) = -Q\,[R(T_2 - T_1)]^{-1}$. For an activation energy of 200 kJ·mol^{-1}, a decrease of temperature from 1200 °C to 1000 °C lowers the diffusion coefficient by a factor of ~50. The effect of a change of 200 °C is much larger at lower T due to the exponential function, and a drop from 800 to 600 °C varies D by about 4 orders of magnitude.

The effect of pressure on diffusion rates has been investigated experimentally to a lesser degree, and so the values of the activation volumes are less constrained. The size of the activation volume is commonly relatively small, so experimentally challenging measurements of diffusion coefficients over a large pressures ranges are required. Most activation volumes for silicates are positive and within the ranges of 0 to 16×10^{-6} m^3·mol^{-1} (e.g., review in Béjina et al. 2003; Holzapfel et al. 2007). Positive activation volumes imply that diffusion rates become slower at higher pressures at a given temperature. The effect of pressure alone at a given temperature can be obtained using $\ln D(P_2) - \ln D(P_1) = -\Delta V\,(P_2 - P_1)\,[RT]^{-1}$. Using a ΔV of 7×10^{-6} m^3·mol^{-1} (Fe-Mg in olivine; Holzapfel et al. 2007) at 1200 °C, an increase of pressure from 1 atmosphere to 1 GPa decreases the Fe-Mg diffusion by less than a factor of 2. The implication is that pressure dependence may be ignored for modeling many crustal processes, but it is crucial to consider these effects when dealing with mantle processes. It is apparent that the effect of pressure on diffusion rates is much stronger at lower temperatures because of the smaller contribution from the T term in the denominator.

The influence of other variables such as f_{O_2}, f_{H_2O}, and composition are commonly included in the pre-exponential factor (e.g., Hier-Majumder et al. 2005; Dohmen and Chakraborty 2007a; Costa and Chakraborty 2008). The importance of volatile fugacities and compositions vary from case to case and it is not possible to generalize, but in some situations their effects can be more significant than the pressure effect noted above for crustal processes. Appendix II includes a derivation to demonstrate how these variables influence diffusion coefficients. This analysis, based on point defect thermodynamics, shows that a generalized form of a diffusion coefficient is

$$D_i^* = D° \cdot \exp\left(-\frac{\Delta H_f + \Delta H_m + P \cdot (\Delta V_f + \Delta V_m)}{RT}\right) \cdot a_1^{m_1} \cdot a_2^{m_2} \cdot \ldots \cdot a_{n-1}^{m_{n-1}} \quad (21)$$

where the various terms are defined in detail in Appendix II, but by comparing this equation to Equation (20) we observe that multiple energies are involved in the exponential, and that the pre-exponential includes the activities or fugacities of components (the a terms in equation above). We discuss below the effects of impurities (e.g., trace elements), oxygen fugacity, and water fugacity on diffusion rates, using Fe-Mg diffusion in olivine as an example.

Effect of water and impurities (trace elements). A fluid, or any impurity, can influence diffusion in a medium only if it affects the properties of that medium in some manner. Consideration of point defects and how they influence diffusion rates tells us that the two possibilities are if fluids or impurities alter (1) the concentration and/or (2) the mobility of point defects responsible for diffusion. Recalling that point defects in silicates are charged entities (a missing Mg^{2+} ion in the structure is effectively a doubly negatively charged vacancy), we recognize that incorporation of fluid species or other impurities has the potential to balance or create charge imbalances. This will affect diffusion rates by altering the concentrations of point defects (e.g., vacancies).

We have developed a very simplified point defect model for olivine considering only two point defects for the charge neutrality condition; vacancies, and Fe^{3+} on the metal site (Appendix II). This constraint can be easily extended to the case of natural olivine where several impurities are present:

$$[Al'_{Si}] + [Li'_M] + [Na'_M] + 2 \cdot [V''_M] = [Fe^{\bullet}_M] + [Cr^{\bullet}_M] + [Al^{\bullet}_M] + [Sc^{\bullet}_M] + [OH^{\bullet}_O] \quad (22)$$

The parentheses [...] denote the number of defects per formula unit of olivine: $[V''_M]$ is a vacancy on a M site (e.g., missing Fe^{2+} or Mg^{2+}); $[Fe^{\bullet}_M]$ is Fe^{3+} on a M site; $[Al'_{Si}]$ is Al^{3+} on the Si site; $[Li'_M]$ is Li^+ on the metal site; $[Na'_M]$ is Na^+ on the metal site; $[Cr^{\bullet}_M]$ is Cr^{3+} on the metal site; $[Al^{\bullet}_M]$ is Al^{3+} on the metal site. Negatively charged defects must exactly balance positively charged defects (Eqn. 22). The concentrations of such defects are thermodynamically defined by chemical potentials or fugacities of the chemical components in the system or environment of the mineral. Only the vacancies on the metal site are relevant for Fe-Mg diffusion. Following Dohmen and Chakraborty (2007a), the effect of impurities on the concentration of these metal vacancies can be evaluated by defining the equivalent charge of the impurities, $[IP3^{\bullet}_M]$, as follows:

$$[IP3^{\bullet}_M] = [Cr^{\bullet}_M] + [Al^{\bullet}_M] + [Sc^{\bullet}_M] + [OH^{\bullet}_O] - [Al'_{Si}] - [Li'_M] - [Na'_M] \quad (23)$$

where some impurities are mutually charge balanced. The effective concentrations of the impurities, as far as charge balance of vacant metal sites is concerned, are much lower than the absolute concentrations of each of the impurities. The charge balance condition can be expressed here in the reduced form:

$$2 \cdot [V''_M] = [IP3^{\bullet}_M] + [Fe^{\bullet}_M] \quad (24)$$

Considering the formation reaction of metal vacancies and Fe^{3+} (Eqn. A5), we can calculate (equilibrium constant and activity coefficients from Tsai and Dieckmann 2002) the concentration of these two defects for a given value of $[IP3^{\bullet}_M]$. Figure 13a shows that for "pure" olivine the logarithm of V''_M increases with the logarithm of f_{O_2} with a slope of 1/6 (compare with Eqn. A11). This dependence becomes weaker with increasing $[IP3^{\bullet}_M]$ until it vanishes and the metal vacancies become directly proportional to $[IP3^{\bullet}_M]$ with a constant of proportionality of 1/2 (compare with Eqn. 24). In the case of San Carlos olivine, such a transition from f_{O_2}-controlled defect chemistry to impurity-controlled defect chemistry occurs at ~800 °C (Dohmen and Chakraborty 2007a). The degree to which the trace element composition may influence the diffusion rate of the major elements can be calculated and predicted by the point defect model.

This is important when considering whether experimentally determined diffusion coefficients apply to the natural case to be modeled. The effect of water on diffusion rates can be evaluated using the expressions shown above, provided that the amount of water incorporated in olivine (as some form of defect species) is known as a function of water fugacity (Fig. 13b,c). If sufficiently high concentrations of H are incorporated in a crystal with a given type of impurity, the charge neutrality condition may change (vacancies may be charge balanced by H species rather than by Fe^{3+}). This would lead to a change in diffusion behavior that would manifest itself macroscopically as a "dry" to "wet" transition (e.g., Costa and Chakraborty 2008). **Other factors** that are not treated here can influence the value of D. These include the effects of non-hydrostatic pressure and transport along fast diffusion paths (e.g., dislocations). Some discussion of these aspects can be found in Chakraborty (2008).

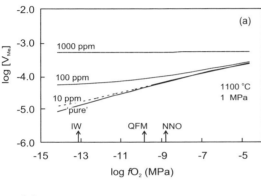

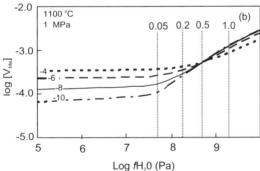

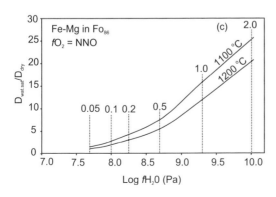

Figure 13. (a) Concentrations of the metal vacancies in units of atoms per formula unit of olivine (M_2SiO_4) as calculated from the point defect thermodynamic data of Tsai and Dieckmann (2002). Concentration in ppm for three different values of the equivalent charge of impurities are shown as a function of f_{O_2}. Arrows mark the position of different solid oxygen fugacity buffers: NNO = f_{O_2} at the Ni-NiO buffer, QFM = f_{O_2} at the quartz-fayalite-magnetite buffer, IW = f_{O_2} at the iron-wüstite buffer. Equations of buffers were taken from Huebner and Sato (1970), and Eugster and Wones (1962), (b) Concentration as a function of f_{H_2O} (equivalent to Psat.) for various f_{O_2}. The numbers at the individual curves denote the log of the f_{O_2} (in bar). Solubility of H_2O in olivine is calculated after Zhao et al. (2004). The numbers on the dotted lines are pressures in GPa and indicate the water fugacity at water saturation (water fugacity calculated after Pitzer and Sterner 1994). The calculation was done for 1 MPa and 1100 °C. For the calculation procedure see Dohmen and Chakraborty (2007a). (c) Ratio of the diffusion coefficient obtained under water-bearing conditions (data of Hier-Majumder et al. 2005) and water-absent conditions (data of Dohmen and Chakraborty 2007 a,b). The numbers on the dotted lines are pressures in GPa and indicate the water fugacity at water saturation. This corresponds to the maximum effect of water for the given total pressure. The effect of T on f_{H_2O} is not shown but it is small for the T range used here.

DIFFUSION MODELING – POTENTIAL PITFALLS AND ERRORS

We now have a complete toolbox of modeling methods and an understanding of the diffusion coefficients that is necessary to make use of these methods. The application of these methods is, however, subject to potential but avoidable pitfalls. It is also necessary to determine the magnitude of uncertainties from various sources to which the modeling results are subject.

Multiple dimensions, sectioning effects and anisotropy

Compositional data available for diffusion modeling are typically obtained from thin sections of rocks, thereby requiring the use of one-dimensional concentration profiles, or two-dimensional images of concentration distribution. One could access the three dimensional zoning distribution by repeated polishing and analysis but this is extremely time consuming. Extracting time scales from one-dimensional or two-dimensional models of a process that occurred in 3-D can introduce errors because the diffusive flux from dimensions that are not being modeled are neglected (e.g., Eqn. 3). This leads to an overestimate of the time required to obtain a given extent of diffusive modification of a concentration distribution. It is difficult to generalize because the effects depend on the shape and size of the crystal, on the diffusion coefficient, and on the duration of the diffusion process (Fig. 14). Smaller errors arise from the neglect of additional dimensions if the amount of diffusion that has occurred is small (e.g., shorter the diffusion profiles) relative to the size of the crystals. It is possible to overcome this effect by using an analytical solution for a three-dimensional geometry to model the concentration profiles observed in one or two dimensions. The standard geometries (e.g., sphere, cylinder) that are typically considered in these solutions constitute an over simplification of real crystal morphologies. The numerical methods we are discussing in this work make it possible to extract time scales from fitting two-dimensional composition maps rather than one-dimensional profiles. Such applications would improve the robustness of the modeling results (Fig. 15).

Another related effect is that of the orientation of the crystal and the zoning with respect to the plane of the thin section. The ideal sections in which to study zoning are those that go through the centre of the crystal and are also perpendicular to one or more crystal faces (Pearce 1984). Mineral grains exposed in thin sections rarely conform to both of these requirements. Compositional traverses should be taken perpendicular to crystal margins because inclined traverses or sections that are not perpendicular to the diffusion front yield artificially longer diffusion profiles (= longer time scales from modeling these profiles; e.g, Fig. 14; Ganguly et al. 2000; Costa et al. 2003; Costa and Chakraborty 2004). Criteria that may be used to identify suitable sections or traverse directions include: (i) Choosing the strongest gradient available in a thin section for modeling diffusion of a given element in a given mineral (e.g., Fig. 1, An profile), (ii) If the zoning is normal or reverse, then the maximum (minimum for reverse) concentration found at the core of a crystal is likely to be from the most central section of the mineral available in the thin section, (iii) Using the normalization criteria of Pearce (1984) where the concentration is plotted against the diameter raised to the power of three. This method applies only to cubic or spherical crystals.

Diffusion in non-cubic crystal classes is anisotropic in the same way that optical properties are. Therefore, the diffusion coefficient along an arbitrary direction can be calculated if the diffusion coefficients along the three principal axes are known (Eqn. 3; which may or may not coincide with crystallographic axes, just as in the case of optic axes):

$$D_i^{traverse} = D_i^a (\cos\alpha)^2 + D_i^b (\cos\beta)^2 + D_i^c (\cos\gamma)^2$$

Here $D_i^{traverse}$ is the diffusion coefficient parallel to the traverse direction in which composition is measured, and D_i^a, D_i^b, and D_i^c are the diffusion coefficients parallel to the a, b, and c principal axes; and α, β, and γ are the angles between the traverse and each axis, respectively. Although

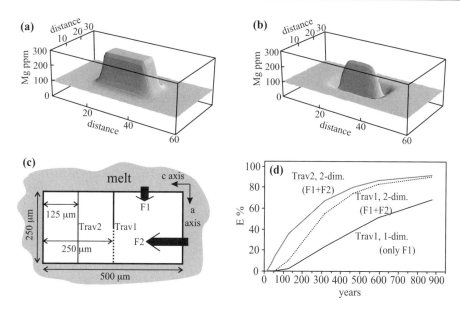

Figure 14. Illustration of numerical calculations that show that two-dimensional effects on diffusion lengths are important for elongate crystals (such as plagioclase) or for elements that show a marked anisotropy of diffusion (such as Fe-Mg in olivine). (a) initial and (b) diffused two-dimensional zoning distribution of the plagioclase (Costa et al. 2003) Ignoring the elongated shape of plagioclase tends to overestimate the time scales by more than an order of magnitude. (c) Olivine crystal with the horizontal dimension parallel to the c axis and the vertical dimension parallel to the a axis. Also shown are the positions of two traverses (Trav1, solid and dotted lines; Trav2, gray line). Note that in a real situation the crystal would exchange matter with the surrounding melt (shown as gray zone outside the crystal) in all three dimensions. The contributions of the fluxes (J) from two of these dimensions are shown as large black arrows ($F1$, $F2$). $F1$ represents the flux from the direction parallel to the traverses and $F2$ is the flux from a direction perpendicular to them. Diffusion parallel to the c axis is 6 times faster than that parallel to a or b, and this leads to a contribution from $F2$ that is 6 times greater than that of $F1$. (d) Plot of the time evolution of the concentration at the center of the crystal E% showing the effects two dimensions and that of the position of the traverse (compare Trav1 with Trav2). Neglecting the flux from the horizontal direction can lead to an overestimate of the time by an order of magnitude or more.

this anisotropy has not always been characterized, it is known for many important mineral groups such as olivine, clinopyroxene, and apatite (see Brady 1995). Modeling concentration profiles with diffusion coefficients that are not appropriate for the direction in which the compositions were measured will yield incorrect time scales. However, it is relatively straightforward to characterize the orientation of a crystal in a thin section with electron backscatter diffraction (EBSD). If anisotropic diffusion data are available, compositional data may be combined with EBSD observations to calculate the appropriate diffusion coefficient for modeling a profile measured in an arbitrary direction (but perpendicular to a crystal margin – see above).

Aside from reducing artifacts that result when anisotropy of diffusion is not considered in the course of modeling, diffusion anisotropy is also an excellent tool for identifying compositional gradients that have been produced by diffusion. Two-dimensional element maps reveal the differences in width of the compositional zoning in different directions. If anisotropic diffusion data are available, the expected difference in diffusion widths along different directions may be calculated and compared with the observed differences in the element map. Using the relationship $x^2 \sim Dt$, one obtains that $x^a/x^c \approx \sqrt{D^a/D^c}$. If zoning has been produced by diffusion, the observed ratio of widths of zoning along different directions

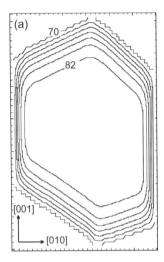
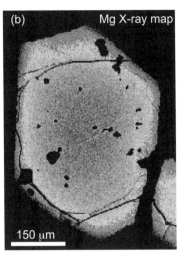

Figure 15. In addition to acquiring multiple one-dimensional traverses in a single crystal it is also useful to obtain and model two-dimensional X-ray distribution maps. This can be done using the method of finite differences, although the complex geometry of many crystals can be better modeled with finite elements rather than finite differences. (a) Simplified two-dimensional model of the olivine crystal shown in (b). Note the asymmetry of the zoning in the modeled and natural crystal. This is due to the fact that anisotropy of diffusion of Fe-Mg is such that diffusion parallel to [001] is about 6 times faster than that parallel to [010] or [100]. The numbers in (a) indicate the Fo content and increase by 2%, with the lowest value at 70 and up to 82.

should be close to that calculated from the diffusion data, once sectioning effects have been accounted for (see above). Alternately, if the zoning is produced by growth or dissolution, the widths are more likely to be independent of crystallographic direction, or at least the ratio would not correspond to that predicted from diffusion data. This method was applied by Costa and Dungan (2005) to establish that the compositional profiles they modeled were produced by diffusion with little or no contribution from growth or dissolution.

Uncertainties in the calculated times from the diffusion models

Uncertainties associated with calculated time scales are mainly of two types: (a) those linked to the parameters that enter into the calculations (e.g., diffusion coefficients, temperature), and (b) those related to the inferred conditions or model at which diffusion took place (e.g., initial or boundary conditions).

Most geothermometers have an uncertainty of ± 30 °C; if results of direct phase equilibria experiments are available for the studied system then the uncertainty can be reduced to ± 10 °C. How uncertainty in the determination of temperature translates into uncertainties in the calculated time scales depends on the activation energy and the specific temperature. Errors increase with increasing Q and decreasing temperature. At magmatic temperatures (e.g., $T >$ 900 °C): for Q between 200 and 300 kJ·mol^{-1}, a 50 °C uncertainty changes the calculated times by a factor of ~2; for 500 kJ·mol^{-1} by as much as a factor of ~4. For 100 kJ·mol^{-1}, this uncertainty reduces to a factor of 1.3. These factors are all <1.5 for uncertainties of 20 °C in cases where the temperature can be experimentally determined (e.g., Costa and Chakraborty 2004). If D also depends strongly on other parameters (e.g., P or f_{O_2}), the related uncertainties should also be considered. These are typically much smaller than the contribution related to temperature. Uncertainties in the diffusion coefficients themselves are variable. Recently published data have uncertainties for activation energies of about 10%, and experimentally determined diffusion coefficients are within a factor of two at any given temperature. Another potential source of uncertainty is related to the precision of the compositional data in the traverse that is modeled. This could take the form of spatial resolution or profiles affected by convolution effects (e.g., Ganguly et al. 1988), or simply compositional data with low precision (e.g., for some trace elements from electron microprobe analyses). Profiles fit to even scattered data yield uncertainties in retrieved time scales of less than a factor of 2. A factor of 2 to 4 could be considered to be the overall uncertainty for the determination of time scales from diffusion modeling in magmatic systems.

Other uncertainties may arise from the choice of the form of the constitutive equation, estimation of initial and boundary conditions, from sectioning, and from ignoring additional dimensions. These uncertainties are variable and need to be evaluated in each case, but if maximum care is taken during each step described above, they can be considerably reduced. We favor modeling multiple elements in single crystals, acquiring data from multiple crystals in a single sample, and examining multiple samples from a given system. This approach should limit the uncertainty in retrieved time scales by minimizing artifacts. Such datasets also provide a test for reproducibility and allow for a statistical treatment of the results.

A CHECKLIST FOR OBTAINING TIME SCALES OF MAGMATIC PROCESSES BY DIFFUSION MODELING

(1) It is important to develop independent constraints on the nature of the magmatic process whose duration or time scale one hopes to obtain from diffusion modeling. A working hypothesis of what caused the zoning (magma mixing?, crystal fractionation?) and the physical situation under which the zoning developed (in a large melt reservoir?, surrounded by which specific minerals?) is useful for identifying suitable constitutive equations and initial and boundary conditions to be used in modeling. For example, if field or microscopic evidence indicates that magma mixing took place, the nature of the mixing end-members may help to define initial conditions.

(2) Expected time scales and knowledge of diffusion coefficients (Fig. 8, Fig. 13) may be used to obtain a preliminary estimate of which elements in which minerals to focus on. Once a set of elements and minerals are identified, it is necessary to evaluate the availability of diffusion coefficients. Diffusion coefficients that have been measured using well-defined experimental protocols at conditions close to those to be modeled (T, f_{O_2}, P, composition of the crystal) are ideal. Once the diffusion coefficients have been selected, it is necessary to identify: (a) on which parameters they depend, on the basis of experimental data, and (b) on which parameters they should depend, according to theoretical considerations related to point defects. A series of tables with diffusion coefficients that have been used to obtain time scales can be found at www.ruhr-uni-bochum.de/planet-diffusion.

(3) A petrological evaluation should be carried out to determine the relevant parameters that influence the diffusion coefficients e.g., T, f_{O_2}.

(4) Careful selection of the locations and orientations of concentration profiles or element maps is essential to avoid sectioning effects and to ensure that concentration profiles are measured perpendicular to crystal margins in two dimensions. If diffusion in the crystal is anisotropic, then it is necessary to obtain orientation data (e.g., EBSD) and to calculate diffusion coefficients in the directions of the compositional traverses. If possible, (a) two-dimensional element maps should be acquired to make use of diffusion anisotropy and to identify true diffusion-related zoning, (b) concentration profiles of multiple elements should be measured to verify the validity of initial and boundary conditions and to evaluate reproducibility, and (c) compositional traverses

in at least two directions at high angles to each other should be measured to make use of diffusion anisotropy as a crosscheck on calculations.

(5) Identify the form of the constitutive equation appropriate for modeling and formulate the relevant continuity equation. The appropriate type of diffusion coefficient (tracer-, chemical-, multi-component- etc.) also needs to be chosen. Even if trace element concentration profiles are to be modeled, it is essential to determine if diffusive coupling may be important (e.g., plagioclase) or if tracer diffusion coefficients can be used (see discussion of Sr diffusion in plagioclase earlier).

(6) Determine the best estimate of the nature of the initial concentration distribution and flux at the boundary on the basis of the physical scenario being visualized. Insert these as initial and boundary conditions in the continuity equation. Various approaches for identifying initial and boundary conditions have been discussed. At this stage, it is useful to discretize the equations (continuity equation, initial condition, boundary condition) and write out the appropriate form for the numerical method of choice (e.g., explicit finite difference, implicit methods).

(7) Decide on isothermal or non-isothermal conditions for constructing the model. A preliminary estimate for non-isothermal conditions may be obtained by using an isothermal model calculated at the characteristic temperature. Preliminary estimates of time scales may be obtained at this stage, and checks for consistency may be made (e.g., between traverses in multiple directions, traverses of multiple elements, and results from one-dimensional vs. two-dimensional models etc.). The shapes of calculated profiles should closely match those of the measured profiles.

(8) Diffusion models may be combined with thermodynamic free energy minimization programs, if more details are desired (e.g., concentration profile shapes not well described by model calculations although the zoning is shown to be due to diffusion).

(9) These steps need to be applied to multiple crystals from single and multiple thin sections in order to obtain a suitable statistics and confidence in the retrieved time scales. Crosschecks are provided by the consistency of retrieved time scales from (a) one-dimensional vs. two-dimensional models, (b) data from multiple traverses and multiple crystals, taking anisotropy into account, (c) data from multiple elements in the same crystal, and (d) data from multiple minerals, if possible. In view of the various possible sources of artifacts, time scales calculated from only one or two concentration profiles may be spurious.

TIME SCALES OF MAGMATIC PROCESSES THAT HAVE BEEN OBTAINED FROM MODELING THE ZONING OF CRYSTALS AND THEIR RELATION TO OTHER INFORMATION ON TIME SCALES OF PROCESSES

A number of recent publications have used diffusion modeling to understand magmatic processes, but only a few studies have specifically addressed the duration over which a process has operated. Most studies have focused on modeling the zoning patterns in crystals from volcanic rocks, but some studies have examined plutonic processes (Coogan et al. 2002, 2005a, 2007; Tepley and Davidson 2003; Gagnevin et al. 2005; Davidson et al. in press). We have summarized below some studies that have used diffusion modeling of different minerals and elements to determine the durations of magmatic processes. We have concentrated on relatively recent works wherein determination of time scales was a significant focus. The times obtained for a variety of processes range from a few days to a few years, and rarely extend to more than about a hundred years. Following this compilation, we compare these time scales with those obtained using radioactive isotopes. We note a few examples where both methods have been applied and discuss whether we should expect the same answer from these different methods. Recent reviews of the time scales of magmatic processes can be found in Condomines et al. (2003), Reid (2003), Hawkesworth et al. (2004), Peate and Hawkesworth (2005), Turner and Costa (2007) and Costa (2008). The two last references contain the most information about time scales obtained from diffusion modeling.

Time scales from diffusion modeling

Open-system magmatic processes include magma assimilation, metasomatism, magma mixing or mingling, and the time between mafic melt intrusion into a reservoir and the subsequent eruption. The relatively large number of modeling studies devoted to this subject probably reflects the importance of such processes for understanding magma differentiation (e.g., Eichelberger 1978) and their potential for triggering eruption and volcanic hazards (e.g., Sparks et al. 1977). This focus also reflects the fact that such magmatic events typically leave a clearly identifiable record in the crystals (e.g., abrupt zoning patterns; e.g., Fig. 4b). This means that in these cases diffusion modeling with reliable initial and boundary conditions is possible through identification of initial conditions via characterization of the end members involved in mixing or assimilation. Many studies have used Fe-Ti zoning in magnetite (e.g., Nakamura 1995a; Venezky and Rutherford 1999; Coombs et al. 2000; Chertkoff and Gardner 2004). Although compositional traverses in magnetite might be used for interpreting some processes (e.g., Devine et al. 2003), there is only very limited diffusion data (e.g., Freer and Hauptman 1978); the retrieved time scales should be used with some caution until all the parameters that control diffusion are fully investigated. Other studies have used major and minor element zoning in olivine (e.g., Nakamura 1995b; Coombs et al. 2000; Pan and Batiza 2002; Costa and Chakraborty 2004; Costa and Dungan 2005). Diffusion in olivine is well constrained (e.g., see caption of Fig. 10) and diffusion data for many elements, including the effects of different environmental variables are known. Fe-Mg zoning in clinopyroxene has been used for determining the time between magma intrusion and eruption (e.g., Costa and Streck 2004; Morgan et al. 2004, 2006), although the diffusion data for these elements is also limited (e.g., Brady and McCallister 1983; Dimanov and Sautter 2000). Zoning of trace elements (Sr, Mg, Ba) in plagioclase has also been applied (e.g., Costa et al. 2003, 2008) and there exists a number of recent determinations of diffusion coefficients which have been reported for a large range of thermodynamic variables and numbers of elements (e.g., caption of Fig 10). Wolff et al. (2002) and Bindeman et al. (2006) used the disequilibrium between oxygen isotope composition of feldspars and quartz or olivines to obtain time scales for assimilation. There is a significant amount of experimental data available for O diffusion (see review by Cole and Chakraborty 2001), but the fact that it is still not possible to measure detailed profiles of the relevant oxygen isotopes in natural crystals (e.g., only core and rim separates) make this type of time determination more prone to uncertainties (e.g., there is no fitting of complete profiles). A recent development is the use of Ti zoning in quartz (Wark et al. 2007), which is also based on new determinations of Ti diffusion coefficients (Cherniak et al. 2007), and the development of a new geothermometer (Wark and Watson 2006). This should open new avenues for understanding the time-temperature histories of silica-rich magmas. The studies that we have highlighted include a variety of compositions and geological settings from mafic-silicic mixing in arc volcanoes to mixing between different types of basalts in mid-ocean ridges. Despite this variety of process and magma types, the majority of time scales obtained lie between a few days to a few years and they rarely exceed 100 years.

Another rate that has been investigated is the time of magma transport from the mantle to the surface, via studies of mantle xenoliths and xenocrysts. Klügel (2001), Shaw (2004), Shaw et al. (2006), and Costa et al. (2005), used Fe-Mg zoning in olivine to obtain transport times that are variable, from some months to a few hundred years. Such large ranges of times appear to reflect rapid and direct transport of some mantle xenoliths to the surface, and the residence of others in intermediate crustal reservoirs before reaching the surface (e.g, Klügel 2001). A recent new finding is that H is zoned in mantle olivines (Demouchy et al. 2006; Peslier and Luhr 2006). Such crystals also have been used to infer mantle-to-surface transport rates of several hours to days. Although there are some determinations of H diffusion in natural olivine (e.g., Kohlstedt and Mackwell 1998; Demouchy and Mackwell 2006), the experimental studies have identified multiple diffusion mechanisms that operate at different rates and it is still not clear which of them applies to natural crystals. More investigations of H diffusion are necessary before it can be used confidently for obtaining time scales.

Time scale for the generation of silica-rich magmas by partial melting or remobilization of crustal rocks have also been estimated with oxygen isotopes in quartz and zircon (Bindeman and Valley 2001), and Sr and Ba zoning in plagioclase (Zellmer et al. 1999, 2003). These studies found time scales somewhat longer than those noted up to now, with a range between few decades to a few thousand years. This is much longer than the days to several weeks recorded by the Fe-Ti oxides for the same processes (Devine et al. 2003). This could reflect that the Fe-Ti oxides record only the last thermal pulse involved in the process. The evidence for previous temperature perturbations in the Fe-Ti zoning may have been erased due to the very fast diffusion rates (e.g., Freer and Hauptman 1978). Morgan and Blake (2006) and Zellmer and Clavero (2006) have used Sr and Ba zoning in sanidine. Morgan and Blake (2006) modeled magma residence and differentiation times for a large silicic magma to be ca. 100 k.y. There is significant diffusion data on sanidine (e.g., Cherniak 1996 and 2002), but the difficulty is to model a suitable differentiation history. This requires a model that considers changing composition at the crystal boundary with time, as well as growth during diffusion (e.g., see previous sections). Morgan and Blake (2006) exploited the different diffusion rates of Sr and Ba to partly overcome such complexities.

Relation of time scales obtained from modeling crystal zoning to radioactive isotopes. The time scales calculated with diffusion modeling are short in geological terms and are rarely more than a few hundred years. This is in contrast with the time determinations from radioactive isotopes obtained from bulk-rocks, mineral separates, or multiple age determination in a single crystal, which are in the range of a few 10's to 100's of kyr (e.g., Turner and Costa 2007). This difference is probably due in part to the fact that even using U-series disequilibria it is difficult to access the days to a hundred years range of magmatic processes, except for the rare cases when ^{210}Pb-^{226}Ra disequilibria can be used (e.g., Sigmarsson 1996; Condomines et al. 2003). We are not aware of any study that has applied diffusion modeling and radiogenic isotope dating to exactly the same crystals, so we cannot quantitatively evaluate differences between the two clocks. Such exercises have been undertaken on metamorphic rocks and the calculated durations of processes of interest are the same using both methods (e.g., Ducea et al. 2003; Spear 2004; Hauzenberger et al. 2005). However, these long-lived radiogenic isotopes and long-time scales processes are quite different from the ones we have been discussing.

A magmatic system to which many radiogenic isotopic dating methods and diffusion studies have been applied is the Bishop Tuff of Long Valley, California (e.g., Hildreth and Wilson 2007; for a review see Costa 2008). If we leave aside age determinations with the Rb-Sr system, dating of zircons indicates residence times on the order of 50 to 100 kyr (Reid and Coath 2000; Simon and Reid 2005). This would correspond to the time between zircon crystallization and a much later eruption. Morgan and Blake (2006) obtained similar times by modeling the Sr and Ba zoning patterns in sanidine (~100 kyr). This is in agreement with the higher end of the spectra of residence times for zircon and would suggest that zircon and sanidine are coeval, that they crystallized during the development of the zonation of the Bishop Tuff, and that the times obtained from diffusion zoning and radioactive isotopes are compatible. Notwithstanding this apparent coincidence, recent investigations and diffusion modeling of the Ti zoning patterns in quartz by Wark et al. (2007) indicate maximum time gap of about 100 years between the creation of the Ti rich rims and eruption. This high-Ti zone can be interpreted as a high temperature event (Wark et al. 2007) and the short calculated times may correspond to the timing of the last intrusion that triggered the eruption. This example illustrates that even where different times are obtained by different methods it is not necessary that one of them is wrong. For the case of the Bishop Tuff it seems plausible that the accumulation of several hundreds of km^3 of erupted magma, took the tens of thousands of years suggested by the zircon data and the sanidine diffusion model, whereas the Ti-zoning in quartz rims records only the last, short-lived event immediately prior to eruption. Detailed petrographic analysis is crucial to interpret the different durations because different minerals or even different parts of some crystals may record different aspects of a magma's history.

SUMMARY AND PROSPECTS

(1) The zoning patterns of crystals are a valuable source of information that can be used to reconstruct the processes which have occurred in magmatic reservoirs and volcanic conduits. We stress that detailed analyses of the mineral zoning and modeling based on chemical diffusion laws also allows the determination of the durations and rates of some magmatic processes. This potential has been exploited by relatively few studies, but it is capable of providing data that contribute to progress in the understanding and quantification of magmatic systems.

(2) Modeling and interpretation of zoning patterns can be anchored within the existing theoretical framework and experimental data for diffusion in solid silicates. There are still many unknowns with respect to the atomistic mechanisms of diffusion and the influences of some experimental variables. Nonetheless, the available information allows robust time estimates to be obtained for some systems (olivine and plagioclase mainly) and certain magmatic processes that characterize open systems.

(3) The ranges of time scales that can be accessed from modeling mineral zoning patterns depend on the diffusion rates and on our capacity to resolve concentration gradients. As the ranges of diffusion coefficients may vary by more than 13 orders of magnitude in a single mineral (e.g., from CaAl – NaSi exchange in anorthite to Li diffusion in plagioclase) and we can measure concentration gradients in crystal zones ranging over about 6 orders of magnitude in length (some tens of nm to cm), the ranges of time scales that can be accessed is almost unlimited. The combination of measuring multiple elements in a single crystal, and in multiple crystals and minerals, offers the opportunity to retrieve a large number of time determinations from a single thin section. The availability and precision of diffusion coefficients will improve in the future, as will our capacity to analyze increasingly low concentrations and resolve increasingly smaller gradients. An increasing array of processes and their time scales will become accessible in the future.

(4) The time scale determinations that have been performed so far on igneous systems focus on volcanic rocks: i.e., much interesting research remains to be done in modeling zoning patterns in plutonic crystals. Processes such as magma mixing or assimilation occur rapidly and may predate eruption by a few months to about 100 years. These times are ~ two to three orders of magnitude shorter than those obtained from radioactive isotopes. However, detailed analysis of the two data sets shows that the two methods are retrieving times associated with different magmatic processes or events. These results are more complementary than contradictory. More studies focussing on the same types of materials are needed to test the degree of difference and complementariness between diverse approaches. This will require precise determination of diffusion coefficients plus isotopic and element abundances on short length scales combined with insightful petrographic scrutiny.

ACKNOWLEDGMENTS

FC thanks Michael Dungan and Brad Singer for introducing him to the fun and interesting science of looking at the zoning patterns of crystals. M. Dungan is also thanked for many discussions, the data shown for the QCNE1 crystal, and a careful review of this manuscript. Reviews by Dan Morgan and Keith Putirka were useful to improve the clarity of the manuscript. The sections about crystal growth and diffusion were motivated by discussions with the participants of the Tatara-San Pedro volcanic complex GSA Field Forum, especially by the criticisms of Jon Blundy. Many thanks to Keith Putirka and Frank Tepley for making this short course happen and their careful editing. FC is supported by a Ramon y Cajal Fellowship from the Spanish MEC. RD and SC acknowledge generous support of their research from the German Science Foundation (DFG) over the years. Most recently they have been funded by the SFB 526 program of the DFG.

REFERENCES

Albarède F (1996) Introduction to geochemical modeling. Cambridge University Press
Albarède F, Bottinga Y (1972) Kinetic disequilibrium in trace element partitioning between phenocryst and host lava. Geochim Cosmochim Acta 35:141-256
Allnatt AR, Lidiard AB (1993) Atomic Transport in Solids. Cambridge University Press, Cambridge
Anderson AT (1983) Oscillatory zoning of plagioclase: Nomarski interference contrast microscopy of etched polished section. Am Mineral 68:125-129
Baker DR (1993) Measurement of diffusion at high temperatures and pressures in silicate systems. In: Experiments at high pressure and applications to the earth's mantle. Mineral Assoc Canada Short Course 21:305-355
Béjina F, Jaoul O, Liebermann RC (2003) Diffusion in minerals at high pressure: a review. Phys Earth Planet Inter 139:3-20
Bindeman I (2008) Oxygen isotopes in mantle and crustal magmas as revealed by single crystal analysis. Rev Mineral Geochem 69:445-478
Bindeman IN, Sigmarsson O, Eiler J (2006) Time constraints on the origin of large volume basalts derived from O-isotope and trace element mineral zoning and U-series disequilibria in the Laki and Grímsvötn volcanic system. Earth Planet Sci Lett 245:245-259
Bindeman IN, Valley JW (2001) Low $\delta^{18}O$ rhyolites from Yellowstone: magmatic evolution based on analyses of zircons and individual phenocrysts. J Petrol 42:1491-1517
Blundy JD, Shimizu N (1991) Trace element evidence for plagioclase recycling in calc-alkaline magmas. Earth Planet Sci Lett 102:178-197
Bottinga Y, Kudo A, Weill D (1966) Some observations on oscillatory zoning and crystallization of magmatic plagioclase. Am Mineral 51:792-806
Brady JB (1995) Diffusion data for silicate minerals, glasses, and liquids. In: Mineral Physics and Crystallography: A Handbook of Physical Constants. AGU Reference Shelf 2. Ahrens TH (ed) Am Geophys Union, Washington, D.C., p 269-290
Brady JB, McCallister RH (1983) Diffusion data for clinopyroxenes from homogenization and self-diffusion experiments. Am Mineral 68:95-105
Carslaw HS, Jaeger JC (1986) Conduction of Heat in Solids. Oxford University Press, New York
Cashman K (1990) Textural constraints on the kinetics of crystallization of igneous rocks. Rev Mineral 24:259-314
Chakraborty S (1995) Diffusion in silicate melts. Rev Mineral Geochem 32:411-504
Chakraborty S (1997) Rates and mechanisms of Fe-Mg interdiffusion in olivine at 980 °C – 1300 °C. J Geophy Res 102:12317-12331
Chakraborty S (2006) Diffusion modelling as a tool for constructing timescales of evolution of metamorphic rocks. Mineral Petrol 88:7-27
Chakraborty S (2008) Diffusion in solid silicates: a tool to track timescales of processes comes of age. Ann Rev Earth Planet Sci 36:153-190
Chakraborty S, Dingwell DB, Rubie DC (1995) Multicomponent diffusion in ternary silicate melts in the system K20-A1203-Si02: II. Mechanisms, systematics, and geological applications. Geochim Cosmochim Acta 59:265-277
Chakraborty S, Farver JR, Yund RA, Rubie DC (1994) Mg tracer diffusion in synthetic forsterite and San Carlos olivine as a function of P, T, and fO2. Phys Chem Mineral 21:489-500

Chakraborty S, Ganguly J (1991) Compositional zoning and cation diffusion in garnets. In: Diffusion, Atomic Ordering and Mass Transport. Advances in Physical Geochemistry, Vol. 8. Ganguly J (ed) Springer-Verlag, New York, p 120-175

Chakraborty S, Ganguly J (1992) Cation diffusion in aluminosilicate garnets: experimental determination in spessartine-almandine diffusion couples, evaluation of effective binary diffusion coefficients, and applications. Contrib Mineral Petrol 111:74-86

Charlier BLA, Bachmann O, Davidson JP, Dungan MA, Morgan DJ (2007) The upper crustal evolution of a large silicic magma body: Evidence from crystal-scale Rb–Sr isotopic heterogeneities in the Fish Canyon magmatic system, Colorado. J Petrol 48:1875-1894

Cherniak DJ (1996) Strontium diffusion in sanidine and albite, and general comments on strontium diffusion in alkali feldspars. Geochim Cosmochim Acta 60:5037-5043

Cherniak DJ (2002) Ba diffusion in feldspar. Geochim Cosmochim Acta 66:1641-1650

Cherniak DJ (2003) REE diffusion in feldspar. Chem Geol 193:25-41

Cherniak DJ, Watson EB (1994) A study of strontium diffusion in plagioclase using Rutherford backscattering spectroscopy. Geochim Cosmochim Acta 58:5179-5190

Cherniak DJ, Watson EB (2001) Pb diffusion in zircon. Chem Geol 172:5-24

Cherniak DJ, Watson EB, Wark DA (2007) Ti diffusion in quartz. Chem Geol 236:65-74

Chertkoff DG, Gardner JE (2004) Nature and timing of magma interactions before, during, and after the caldera-forming eruption of Volcán Ceboruco, Mexico. Contrib Mineral Petrol 146:715-735

Cole DR, Chakraborty S (2001) Rates and mechanisms of isotope exchange. Rev Mineral Geochem 43:83-223

Condomines M, Gauthier P-J, Sigmarsson O (2003) Timescales of magma chamber processes and dating of young volcanic rocks. Rev Mineral Geochem 52:125-174

Connolly JAD (1990) Multivariable phase-diagrams - an algorithm based on generalized thermodynamics. Am J Sci 290:666-718

Coogan LA, Hain A, Stahl S, Chakraborty S (2005b) Experimental determination of the diffusion coefficient for calcium in olivine between 900 °C and 1500 °C. Geochim Cosmochim Acta 69:3683-3694

Coogan LA, Jenkin GRT, Wilson RN (2002) Constraining the cooling rate of the lower oceanic crust: a new approach applied to the Oman ophiolite. Earth Planet Sci Lett 199:127-146

Coogan LA, Jenkin GRT, Wilson RN (2007) Contrasting cooling rates in the lower oceanic crust at fast- and slow-spreading ridges revealed by geospeedometry. J Petrol 48:2211-2231

Coogan LA, Kasemann SA, Chakraborty S (2005a) Rates of hydrothermal cooling of new oceanic upper crust derived from lithium-geospeedometry. Earth Planet Sci Lett 240:415-424

Coombs ML, Eichelberger JC, Rutherford MJ (2000) Magma storage and mixing conditions for the 1953-1974 eruptions of Southwest Trident volcano, Katmai National Park, Alaska. Contrib Mineral Petrol 140:99-118

Cooper AR (1965) Model for multi-component diffusion. Phys Chem Glasses 6:55-61

Cooper KM, Reid MR (2008) Uranium-series crystal ages. Rev Mineral Geochem 69:479-554

Costa F (2000) The petrology and geochemistry of diverse crustal xenoliths, Tatara-San Pedro Volcanic Complex, Chilean Andes. Terre & Environnement 19, 120 pp. Ph.D. thesis University of Geneva

Costa F (2008) Residence times of silicic magmas associated with calderas. In: Caldera Volcanism: Analysis, Modelling and Response. Gottsmann J, Marti J (eds) Developments in Volcanology 10:1-55

Costa F, Chakraborty S (2004) Decadal time gaps between mafic intrusion and silicic eruption obtained from chemical zoning patterns in olivine. Earth Planet Sci Lett 227:517-530

Costa F, Chakraborty S (2008) The effect of water on Si and O diffusion rates in olivine and their relation to transport properties and processes in the upper mantle. Phys Earth Planet Int 166:11-29

Costa F, Chakraborty S, Dohmen R (2003) Diffusion coupling between trace and major elements and a model for calculation of magma residence times using plagioclase. Geochim Cosmochim Acta 67:2189-2200

Costa F, Chakraborty S, Fedortchouk Y, Canil D (2005) Crystals in Kimberlites: Where are the Phenocrysts? Eos Trans AGU, 86(52), Fall Meet. Suppl., Abstract V13B-0545

Costa F, Coogan L, Chakraborty S (2008) The time scales of magma mixing and mingling at mid-ocean ridges. Geophys Res Abst 10:6229

Costa F, Dungan M (2005) Short time scales of magmatic assimilation from diffusion modelling of multiple elements in olivine. Geology 33:837-840

Costa F, Dungan M, Singer B (2002) Hornblende and phlogopite-bearing gabbroic crustal xenoliths from Volcán San Pedro (36° S), Chilean Andes: evidence for melt and fluid migration and reactions in subduction-related plutons. J Petrol 43: 219-241

Costa F, Streck M (2004) Periodicity and timescales for basaltic magma replenishment at Volcán Arenal (Costa Rica). IAVCEI General Assembly 2004 abstracts

Crank J (1975) The Mathematics of Diffusion. 2nd edition. Oxford Science Publication, Oxford

Davidson JP, Font L, Charlier BLA, Tepley FJ III (2008) Mineral-scale Sr isotope variation in plutonic rocks-a tool for unravelling the evolution of magma systems. Transactions of the Royal Society of Edinburgh: Earth Sciences 97:357-367

Davidson JP, Morgan DJ, Charlier LA (2007a) Isotopic microsampling of magmatic rocks. Elements 3:261-266

Davidson JP, Morgan DJ, Charlier LA, Harlou R, Hora JM (2007b) Microsampling and isotopic analysis of igneous rocks: implications for the study of magmatic systems. Annu Rev Earth Planet Sci 35:273-311

de Capitani C, Brown TH (1987) The computation of chemical equilibrium in complex systems containing non-ideal solutions. Geochim Cosmochim Acta 51:2639-2652

Demouchy S, Jacobsen SD, Gaillard F, Stern CR (2006) Rapid magma ascent recorded by water diffusion profiles in mantle olivine. Geology 34:429-432

Demouchy S, Mackwell S (2006) Mechanisms of hydrogen incorporation and diffusion in iron-bearing olivine. Phys Chem Minerals 33:347-355

Denison C, Carlson WD (1997) Three-dimensional quantitative textural analysis of metamorphic rocks using high-resolution computed X-ray tomography: Part II. Application to natural samples. J Metamorph Geol 15:45-57

Devine JD, Rutherford MJ, Norton GE, Young SR (2003) Magma storage region processes inferred from geochemistry of Fe-Ti oxides in andesite magma, Soufrière Hills volcano, Montserrat, W.I. J Petrol 44:1375-1400

Dimanov A, Sautter V (2000) 'Average' interdiffusion of (Fe, Mn)-Mg in natural diopside. Eur J Mineral 12:749-760

Dodson MH (1973) Closure temperature in cooling geochronological and petrological systems. Contrib Mineral Petrol 40:259-274

Dodson MH (1976) Kinetic processes and thermal cooling of slowly cooling solids. Nature 259:551-553

Dodson MH (1986) Closure profiles in cooling systems. Mater Sci Forum 7:145-154

Dohmen R, Becker HW, Chakraborty S (2007) Fe-Mg diffusion coefficients in olivine, Part I: Experimental determination between 700 and 1200 °C as a function of composition, crystal orientation and oxygen fugacity. Phys Chem Mineral 34:389-407

Dohmen R, Becker HW, Meissner E, Etzel T, Chakraborty S (2002a) Production of silicate thin films using pulsed laser deposition (PLD) and applications to studies in mineral kinetics. Eur J Mineral 14:1155-1168

Dohmen R, Chakraborty S (2003) Mechanism and kinetics of element and isotopic exchange mediated by a fluid phase. Am Mineral 88:1251-1270

Dohmen R, Chakraborty S (2007a) Fe-Mg diffusion in olivine II: point defect chemistry, change of diffusion mechanisms and a model for calculation of diffusion coefficients in natural olivine. Phys Chem Mineral34:409-430

Dohmen R, Chakraborty S (2007b) Fe-Mg diffusion in olivine II: point defect chemistry, change of diffusion mechanisms and a model for calculation of diffusion coefficients in natural olivine (vol 34, pg 409, 2007). Phys Chem Mineral34:597-598

Dohmen R, Chakraborty S, Becker HW (2002b) Si and O diffusion in olivine and implications for characterizing plastic flow in the mantle. Geophys Res Lett 29: doi:10.1029/2002GL015480

Ducea MN, Ganguly J, Rosenberg EJ, Patchett PJ, Cheng W, Isachsen C (2003) Sm-Nd dating of spatially controlled domains of garnet single crystals: a new method of high temperature thermochronology. Earth Planet Sci Lett 213:31-42

Eichelberger JC (1978) Andesitic volcanism and crustal evolution. Nature 275:21-27

Elphick SC, Ganguly J, Loomis TP (1985) Experimental determination of cation diffusivities in aluminosilicate garnets: I. Experimental methods and interdiffusion data. Contrib Mineral Petrol 90:36-44

Eugster HP, Wones DR (1962) Stability relations of the ferruginous biotite, annite. J Petrol 3: 82-125

Fick A (1855) On liquid diffusion. Phil Mag J Sci 10:31-39

Flynn CP (1972) Point Defects and Diffusion. Clarendon Press, Oxford

Freer R, Hauptman Z (1978) An experimental study of magnetite-titanomagnetite interdiffusion. Phys Earth Planet Int 16:223-231

Gagnevin D, Daly JS, Poli G, Morgan D (2005) Microchemical and Sr isotopic investigation of zoned K-feldspar megacrysts: Insights into the petrogenesis of a granitic system and disequilibrium crystal growth. J Petrol 46:1689-1724

Ganguly J (2002) Diffusion kinetics in minerals: principles and applications to tectono-metamorphic processes. Eur Mineral Union 4:271-309

Ganguly J, Bhattacharya RN, Chakraborty S (1988) Convolution effect in the determination of compositional zoning by microprobe step scans. Am Mineral 73:901-909

Ganguly J, Dasgupta S, Cheng W, Neogi S (2000) Exhumation history of a section of the sikkim Himalayas, India: records in the metamorphic mineral equilibria and compositional zoning of garnet. Earth Planet Sci Lett 183:471-486

Ganguly J, Tirone M (1999) Diffusion closure temperature and age of a mineral with arbitrary extent of diffusion: theoretical formulation and applications. Earth Planet Sci Lett 170:131-140

Ganguly J, Tirone M (2001) Relationship between cooling rate and cooling age of a mineral: theory and applications to meteorites. Meteorit Planet Sci 36:167-175

Gérard O, Jaoul O (1989) Oxygen diffusion in San-Carlos olivine. J Geophys Res 94:4119-4128

Ghiorso MS, Sack RO (1995) Chemical mass transfer in magmatic processes IV. A revised and internally consistent thermodynamic model for the interpolation and extrapolation of liquid-solid equilibria in magmatic systems at elevated temperatures and pressures. Contrib Mineral Petrol 119:197-212

Giletti BJ, Casserly JED (1994) Strontium diffusion kinetics in plagioclase feldspars. Geochim Cosmochim Acta 58:3785-3797

Giletti BJ, Shanahan TM (1997) Alkali diffusion in plagioclase feldspar. Chem Geol 139:3-20

Ginibre C, Wörner G, Kronz A (2007) Crystal zoning as an archive for magmatic evolution. Elements 3:261-266

Glazner AF, Bartley JM, Coleman DS, Gray W, Taylor RZ (2004) Are plutons assembled over millions of years by amalgamation from small magma chambers. Geology Today 14:4-11

Glicksman ME (2000) Diffusion in Solids. John Wiley & Sons, Inc. New York

Goldstein JI, Short JM (1967) Cooling rates of 27 iron and stony-iron meteorites. Geochim Cosmochim Acta 31:1001-1023

Grove TL, Baker MB, Kinzler RJ (1984) Coupled CaAl-NaSi diffusion in plagioclase feldspar: experiments and applications to cooling rate speedometry. Geochim Cosmochim Acta 48:2113-2121

Hauzenberger CA, Robl J, Stüwe K (2005) Garnet zoning in high pressure granulite facies metapelites, Mozambique belt, SE-Kenya: constraints on the cooling history. Eur J Mineral 17:43-55

Hawkesworth C, George R, Turner S, Zellmer G (2004) Time scales of magmatic processes. Earth Planet Sci Lett 218:1-16

Hier-Majumder S, Anderson IM, Kohlstedt DL (2005) Influence of protons on Fe-Mg interdiffusion in olivine. J Geophys Res 110, doi: 10.1029/2004JB003292

Hildreth W, Wilson CJN (2007) Compositional zoning of the Bishop Tuff. J Petrol 48:951-999

Holzapfel C, Chakraborty S, Rubie DC, Frost DJ (2007) Effect of pressure on Fe-Mg, Ni and Mn diffusion in $(Fe_xMg_{1-x})_2SiO_4$ olivine. Phys Earth Planet Int 162:186-98

Huebner JS, Sato M (1970) The oxygen fugacity temperature relationships of manganese and nickel oxides buffers. Am Mineral 55:934-952

Ingrin Y, Blanchard M (2006) Diffusion of hydrogen in minerals. Rev Mineral Geochem 62:291-320

Ito M, Ganguly J (2006) Diffusion kinetics of Cr in olivine and 53Mn-53Cr thermochronology of early solar system objects. Geochim Cosmochim Acta 70:799-809

Kirkpatrick RJ, Robinson GR, Hays FJ (1976) Kinetics of crystal growth from silicate melts: anorthite and diopside. J Geophys Res 81:5715-5720

Klügel A (2001) Prolonged reactions between harzburgite xenoliths and silica-undersaturated melt: Implications for dissolution and Fe-Mg interdiffusion rates of orthopyroxene. Contrib Mineral Petrol 141:1-14

Kohlstedt DL, Mackwell SJ (1998) Diffusion of hydrogen and intrinsic point defects in olivine. Z Phys Chem 207:147-162

Kohn SC, Henderson, CMB, Mason (1989) Element zoning trends in olivine phenocrysts from a supposed primary high magnesian andesite: an electron and ion-microprobe study. Contrib Mineral Petrol 103:242-252

Kröger FA, Vink HJ (1965) Relation between the concentrations of imperfections in crystalline solids. Solid State Phys 3:307-435

Larsen ES, Irving J, Gonyer FA, Larsen ES III (1938) Petrologic results of a study of the minerals from the Tertiary volcanic rocks of the San Juan region, Colorado. Am Mineral 23:227-257

Lasaga AC (1979) Multicomponent exchange and diffusion in silicates. Geochim Cosmochim Acta 43:455-469

Lasaga AC (1983) Geospeedometry: an extension of geothermometry. In: Kinetics and Equilibrium in mineral reactions. Advances in Physical Geochemistry, Vol 3. Saxena SK (ed) New-York, Springer-Verlag, p 81-114

Lasaga AC (1998) Kinetic theory in the earth sciences, Princeton University Press, Princeton

Lasaga AC, Jiang J (1995) Thermal history of rocks; P-T-t paths for geospeedometry, petrologic data, and inverse theory techniques. Am J Sci 295:697-741

Lasaga AC, Richardson SM, Holland HD (1977) The mathematics of cation diffusion and exchange between silicate minerals during retrograde metamorphism. In: Energetics of Geological Processes, Saxena SK, Bhattachanji S (eds) p 353-388. Springer-Verlag, New York

LaTourrette T, Wasserburg GJ (1998) Mg diffusion in anorthite: implications for the formation of early solar system planetesimals. Earth Planet Sci Lett 158:91-108

Liang Y, Richter FM, Chaberlin L (1997) Diffusion in silicate melts: III. Empirical models for multicomponent diffusion. Geochim Cosmochim Acta 61:5295-5312

Lofgren G (1972) Temperature induced zoning in synthetic plagioclase feldspar. In: The Feldspars. MacKenzie WS (ed) Univ. of Manchester Press, Manchester, p 367-377

Loomis TP (1982) Numerical simulations of crystallization processes of plagioclase in complex melts: the origin of major and oscillatory zoning in plagioclase. Contrib Mineral Petrol 81:219-229

Loomis TP (1986) Metamorphism of metapelites: calculations of equilibrium assemblages and numerical simulations of the crystallization of garnet. J Metam Geol 4:201-229

Loomis TP, Ganguly J, Elphick SC (1985) Experimental determination of cation diffusivities in aluminosilicate garnets: II. Multicomponent simulation and tracer diffusion coefficients. Contrib Mineral Petrol 90: 45-51

Manning JR (1968) Diffusion Kinetics for Atoms in Crystals. Princeton University Press, Princeton

Milman-Barris MS, Beckett JR, Michael MB, Hofmann AE, Morgan Z, Crowley MR, Vielzeuf D, Stolper E (2008) Zoning of phosphorus in igneous olivine. Contrib Mineral Petrol 155:739-765

Moore JG, Evans BW (1967) The role of olivine in the crystallization of the prehistoric Makaopuhi tholeiitic lava lake, Hawaii. Contrib Mineral Petrol 15:202-223

Morgan DJ, Blake S (2006) Magmatic residence times of zoned phenocrysts: introduction and application of the binary element diffusion modelling (BEDM) technique. Contrib Mineral Petrol 151:58-70

Morgan DJ, Blake S, Rogers NW, DeVivo B, Rolandi G, Davidson J (2006) Magma recharge at Vesuvius in the century prior to the eruption AD 79. Geology 34:845-848

Morgan DJ, Blake S, Rogers NW, DeVivo B, Rolandi G, Macdonald R, Hawkesworth J (2004) Timescales of crystal residence and magma chamber volume from modeling of diffusion profiles in phenocrysts: Vesuvius 1944. Earth Planet Sci Lett 222:933-946

Nakamura A, Schmalzried H (1983) On the nonstoichiometry and point defects of olivine. Phys Chem Minerals 10:27-37

Nakamura M (1995a) Continuous mixing of crystal mush and replenished magma in the on going Unzen eruption. Geology 23:807-810

Nakamura M (1995b) Residence time and crystallization history of nickeliferous olivine phenocrysts from the northern Yatsugatake volcanoes, Central Japan: application of a growth and diffusion model in the system Mg-Fe-Ni. J Volcanol Geotherm Res 66:81-100

Onsager L (1945) Theories and problems of liquid diffusion. Ann N Y Acad Sci 46: 241-265

Ozawa K, Nagahara H (2000) Kinetics of diffusion-controlled evaporation of Fe-Mg olivine: experimental study and implication for stability of Fe-rich olivine in the solar nebula. Geochim Cosmochim Acta 64:939-955

Ozawa K, Nagahara H (2001) Chemical and isotopic fractionations by evaporation and their cosmochemical applications. Geochim Cosmochim Acta 65:2171-2199

Pan Y, Batiza R (2002) Mid-ocean ridge magma chamber processes: Constraints from olivine zonation in lavas from the East Pacific Rise at 9°30′N and 10°30′N. J Geophys Res 107, DOI 10.1029/2001JB000435

Pearce TH (1984) The analysis of zoning in magmatic crystals with emphasis on olivine. Contrib Mineral Petrol 86:149-154

Pearce TH, Kolisnik AM (1990) Observations of plagioclase zoning using interference imaging. Earth Sci Rev 29:9-26

Peate DW, Hawkesworth CJ (2005) U series disequilibria: insights into mantle melting and the timescales of magma differentiation. Rev Geophys 43, 2004RG000154

Perugini D, Ventura G, Petrelli M, Poli G (2004) Kinematic significance of morphological structures generated by mixing of magmas: a case study from Salina Island (southern Italy). Earth Planet Sci Lett 222:1051-1066

Peslier AH, Luhr JF (2006) Hydrogen loss from olivines in mantle xenoliths from Simcoe (USA) and Mexico: mafic alkalic magma ascent rates and water budget of the sub-continental lithosphere. Earth Planet Sci Lett 242:302-319

Petrelli M, Perugini D, Poli G (2006) Time-scales of hybridization of magmatic enclaves in regular and chaotic flow fields: petrologic and volcanological implications. Bull Volcanol 68:285-293

Petry C, Chakraborty S, Palme H (2004) Experimental determination of Ni diffusion coefficients in olivine and their dependence on temperature, composition, oxygen fugacity, and crystallographic orientation. Geochim Cosmochim Acta 68:4179-4188

Philibert J (1991) Atom Movements: Diffusion and Mass Transport in Solids. Les Éditions de Physique. Les Ulis

Pitzer KS, Sterner SM (1994) Equations of state valid continuously from zero to extreme pressures for H_2O and CO_2. J Chem Phys 101:3111-3116

Press WH, Teukolsky SA, Vetterling WT, Flannery BP (2007) Numerical Recipes 3rd Edition: The Art of Scientific Computing, Cambridge University Press, Cambrigde

Ramos FC, Tepley FJ III (2008) Inter- and intracrystalline isotopic disequilibria: techniques and applications. Rev Mineral Geochem 69:403-443

Reid M (2003) Timescales of magma transfer and storage in the crust, In: Treatise on Geochemistry, Volume 3: The Crust. Holland HD, Turekian KK (eds) Elsevier, p 167-193

Reid MR, Coath CD (2000) In situ U-Pb ages of zircons from the Bishop Tuff: no evidence for long crystal residence times. Geology 28:443-446

Ryerson FJ, Durham WB, Cherniak DJ, Lanford WA (1989) Oxygen diffusion in olivine - effect of oxygen fugacity and implications for creep. J Geophys Res 94:4105-4118

Sato (1986) High temperature a.c. electrical properties of olivine single crystal with varying oxygen partial pressure: implications for the point defect chemistry. Phys Earth Planet Inter 41:269-282

Schmalzried H (1981) Solid State Reactions. Verlag Chemie, Weinheim

Shaw CSJ (2004) The temporal evolution of three magmatic systems in the West Eifel volcanic field, Germany. J Volcanol Geotherm Res 131:213-240

Shaw CSJ, Heidelbach F, Dingwell DB (2006) The origin of reaction textures in mantle peridotite xenoliths from Sal Island, Cape Verde: the case for metasomatism by the host lava. Contrib Mineral Petrol 151:681-697

Short M, Goldstein JI (1967) Rapid methods of determining cooling rates of iron and stony iron meteorites. Science 165: 59-61

Sigmarsson O (1996) Short magma chamber residence time at an Icelandic volcano inferred from U-series disequilibria. Nature 391:440-442

Simon JI, Reid MR (2005) The pace of rhyolite differentiation and storage in an 'archetypical' silicic magma system, Long Valley, California. Earth Planet Sci Lett 235:123-140

Singer BS, Dungan MA, Layne GD (1995) Textures and Sr, Ba, Mg, Fe, K, and Ti compositional profiles in volcanic plagioclase: clues to the dynamics of calc-alkaline magma chambers. Am Mineral 80:776-798

Sparks RSJ, Sigurdsson H, Wilson L (1977) Magma mixing: a mechanism for triggering acid explosive eruptions. Nature 267:315-318

Spear FS (2004) Fast cooling and exhumation of the Valhalla metamorphic core complex, southeastern British Columbia. Int Geol Rev 46:193-209

Spear FS, Kohn MJ, Florence F, Menard T (1990) A model for garnet and plagioclase growth in pelitic schist: implications for thermobarometry and P-T path determinations. J Metamorph Geol 8:683-696

Streck MJ (2008) Mineral textures and zoning as evidence for open system processes. Rev Mineral Geochem 69:595-622

Takeda H, Miyamoto M, Ishii T, Lofgren GE (1975) Relative cooling rates of mare basalts at the Apollo 12 and 15 sites as estimated from pyroxene exsolution data. Proc Lunar Sci Conf 6th, 987-996

Taylor LA, McCallister RH, Sardi O (1973) Cooling histories of lunar rocks based on opaque mineral geothermometers. Proc Lunar Sci Conf 4th, 819-828

Tepley FJ III, Davidson JP (2003) Mineral scales Sr-istope constraints on magma evolution and chamber dynamics in the Rum layered intrusion, Scotland. Contrib Mineral Petrol 145:628-641

Tomkeieff SI (1939) Zoned olivines and their petrogenetic significance. Mineral Mag 25:229-251

Trepmann CA, Stöckhert B, Chakraborty S (2004) Oligocene trondhjemitic dikes in the Austroalpine basement of the Pfunderer Berge, Südtirol- level of emplacement and metamorphic overprint. Eur J Mineral 16:641-659

Tsai TL, Dieckmann R (2002) Variation of the oxygen content and point defects in olivines, $(Fe_xMg_{1-x})_2SiO_4$, $0.2 = x = 1.0$. Phys Chem Mineral 29:680-694

Turner S, Costa F (2007) Measuring timescales of magmatic evolution. Elements 3:267-272

Umino S, Horio A (1998) Multistage magma mixing revealed in phenocryst zoning of the Yunokuchi pumice, Akagi volcano, Japan. J Petrol 39:101-124

Van Orman JA, Grove TL, Shimizu N, Graham DL (2001) Rare earth element diffusion in diopside: influence of temperature, pressure and ionic radius, and an elastic model for diffusion in silicates. Contrib Mineral Petrol 141:687-703

Venezky DY, Rutherford MJ (1999) Petrology and Fe-Ti oxide reequilibration of the 1991 Mount Unzen mixed magma. J Volcanol Geotherm Res 89:213-230

Walker D, Longhi J, Lasaga AC, Stolper EM, Grove TL, Hays TL (1977): Slowly cooled microgabbros15555 and 15065. Proc Lunar Sci Conf 8th, 1521-1547

Wallace GS, Bergantz GW (2002) Wavelet-based correlation (WBC) of zoned crystal populations and magma mixing. Earth Planet Sci Lett 202:133-145

Wark DA, Hildreth W, Spear FS, Cherniak DJ, Watson EB (2007) Pre-eruption recharge of the Bishop magma system. Geology 35:235-238

Wark DA, Watson EB (2006) TitaniQ: a titanium-in-quartz geothermometer. Contrib Mineral Petrol 152:743-754
Watson EB (1994) Diffusion in volatile-bearing magmas. Rev Mineral 30:371-411
Watson EB (1996) Dissolution, growth and survival of zircons during crustal fusion: kinetic principles, geological models and implications for isotopic inheritance. Trans R Soc Edinburgh Earth Sci 87:43-56
Watson EB, Baxter EF (2007), Diffusion in solid-Earth systems. Earth Planet Sci Lett 253:307-327
Wolff JA, Balsley SD, Gregory RT (2002) Oxygen isotope disequilibrium between quartz and sanidine from the Bandelier Tuff, New Mexico, consistent with a short residence time of phenocrysts in rhyolitic magma. J Volcanol Geotherm Res 116:119-135
Wood JA (1964) The cooling rates and parent bodies of several iron meteorites. Icarus 3:429-459
Young DA (2003) Mind Over Magma. Princeton University Press, Princeton
Zellmer GF, Blake S, Vance D, Hawkesworth C, Turner S (1999) Plagioclase residence times at two island arc volcanoes (Kameni Islands, Santorini, and Soufriere, St. Vincent) determined by Sr diffusion systematics. Contrib Mineral Petrol 136:345-357
Zellmer GF, Clavero JE (2006) Using trace element correlation patterns to decipher a sanidine crystal growth chronology: an example from Taapaca volcano, Central Andes. J Volcanol Geotherm Res 156:291-301
Zellmer GF, Sparks RSJ, Hawkesworth C, Wiedenbeck M (2003) Magma emplacement and remobilization timescales beneath Montserrat: insights from Sr and Ba zonation in plagioclase phenocrysts. J Petrol 44:1413-1431
Zhao Y-H, Ginsberg SB, Kohlstedt DL (2004) Solubility of hydrogen in olivine: dependence on temperature and iron content. Contrib Mineral Petrol 147:155-161

APPENDIX I

SHORT GUIDE TO THE USE OF THE FINITE DIFFERENCE METHOD FOR NUMERICAL CALCULATIONS WITH THE DIFFUSION EQUATION

The first step is to discretize the space and time coordinates into finite but small increments to make a grid. This is shown in Figure A1. We have a step size Δx on the space coordinate, and Δt on the time coordinate. In the simplest implementations these are taken to be constants, but are not required to be so. These define the smallest variations that can be "recognized" by the computer at that particular point of space and time. For example, if we let $\Delta x = 5$ μm and $\Delta t = 3600$ s, then distances smaller than 5 μm and time scales smaller than 3600 s cannot be resolved by the numerical calculation. Any numerical calculation is a trade off between speed (larger the individual steps, fewer of these are required to get to any particular value) and accuracy (smaller the step, closer it is to being truly infinitesimal, as required by rigorous mathematics). We are concerned with describing concentrations as a function of space and time—the variable $C(x,t)$ used in the text. In the numerical implementation, we can associate each value of concentration with two indices written as subscripts—one for space (i) and the other for time (j), e.g., $C_{i,j}$. If we write, for example, $C_{2,2}$ we are referring to the concentration at grid point number 2 on the spatial scale at time step number 2. For our chosen value of $\Delta x = 5$ μm and $\Delta t = 3600$ s, $C_{2,2}$ is the concentration at a distance of 10 μm from our arbitrarily chosen zero position after 2 hours (Fig. A1).

The next step is to convert the differential expressions into discrete forms using the Taylor Series approximation (see Crank 1975, for a detailed exposition geared to the solution of diffusion problems). There are several levels of approximation that may be used (first order, second order etc.) and different options about the choice of neighborhood in which to carry out the Taylor expansion (forward difference, backward difference, central difference etc.). The simplest of these, in a forward difference scheme, can be written as

$$\frac{\partial C}{\partial x} = \frac{C_{i+1,j} - C_{i,j}}{\Delta x}$$

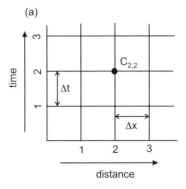

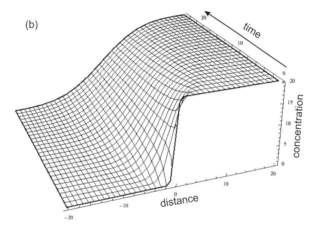

Figure A1. Visualization of the discretization of time and space used in finite difference algorithms. (a) There is a value of concentration associated with each point in space and time. For example, $C_{2,2}$ stands for the value of concentration at a distance of $2 \cdot \Delta x$ at the point of time $2 \cdot \Delta t$. (b) Three-dimensional visualization of how the concentration changes due to diffusion using the discretization of space and time. See also text for more explanation.

The most important point to note here is that this expression converts a differential expression into a simple algebraic expression that can be manipulated by straightforward rules of addition and multiplication. Higher derivatives may be written by repeating the same approach. For example,

$$\frac{\partial^2 C}{\partial x} = \frac{\partial}{\partial x}\left(\frac{\partial C}{\partial x}\right) = \frac{1}{\Delta x}\left(\left(\frac{C_{i+1,j} - C_{i,j}}{\Delta x}\right) - \left(\frac{C_{i,j} - C_{i-1,j}}{\Delta x}\right)\right) = \frac{C_{i+1,j} - 2C_{i,j} + C_{i-1,j}}{(\Delta x)^2}$$

Similarly, for the time derivative we have

$$\frac{\partial C}{\partial t} = \frac{C_{i,j+1} - C_{i,j}}{\Delta t}$$

If we put it all together and move terms around using algebra, the continuity Equation (11a) in the text translates to

$$C_{i,j+1} = C_{i,j} + D\Delta t \left(\frac{C_{i+1,j} - 2C_{i,j} + C_{i-1,j}}{\Delta x^2}\right) \tag{A1}$$

for calculating the concentration at a grid point i at the next time step, $j+1$. This formula can be iterated over all grid points, shifting the index i by 1 each time, to obtain a complete concentration profile at the time step $j+1$. The entire procedure may be then repeated to calculate the concentration profile at the next time step, $j+2$, using the values determined at step $j+1$ as the "old" values. If this is done until the index j reaches the value 2 (with the time step, $\Delta t = 3600$ s, as above), we would have calculated a concentration profile that would be measured after 2 hours of diffusion. This numerical method is called the explicit finite difference method.

These simple formulae and iterations can be implemented on a standard spreadsheet program such as Excel®. On the MSA WWW site (*http://www.minsocam.org/MSA/RIM*), we provide a spreadsheet (olivine_diffu_excel.xls) and a Mathematica® notebook (olivine_diffu_mathema.nb) with an example for Fe-Mg diffusion in olivine, but it is an instructive exercise to consider the various constitutive equations described in the text, derive the corresponding continuity equations, and then digitize them to the discrete numerical form as above. For example, if the diffusion coefficient depends on concentration we need to solve another form of the diffusion equation (Eqn. 11b), which in finite difference form is:

$$C_{i,j+1} = C_{i,j} + \Delta t \left(\frac{D_{i+1,j} - D_{i,j}}{\Delta x}\right)\left(\frac{C_{i+1,j} - C_{i,j}}{\Delta x}\right) + D_{i,j}\Delta t \left(\frac{C_{i+1,j} - 2C_{i,j} + C_{i-1,j}}{\Delta x^2}\right) \tag{A2}$$

For actual numerical implementation it is necessary to pay attention to a few other details. For any explicit numerical scheme to be stable, it is necessary that the so-called Courant condition be fulfilled. For Equation (A1), D, Δx and Δt can be combined together into one parameter, $r = D\Delta t/(\Delta x)^2$. The Courant condition is that $r < 0.5$ (e.g, see Crank 1975). If the scheme has been implemented on a spreadsheet, it is easy to see what happens when r exceeds a value of about 0.5. If D, Δx and Δt are not constant during the course of a calculation (e.g., Eqn. A2 where D change with position or concentration, or for a non-isothermal case, where D changes with temperature, and hence, time), then it is necessary to ensure that the Courant condition is fulfilled at each grid point. With these equations and the boundary and initial conditions for the case at hand we can solve the diffusion equation and fit compositional profiles. For example, for the case of no flux at the boundary, we write $C_{1,j(all)} = C_{2,j(all)}$, and likewise $C_{n,j(all)} = C_{n-1,j(all)}$ (where n is the total number of space grid points and $j(all)$ means for all times; this is the condition used in Fig. 6a). For the case that there is exchange, but a constant composition at the boundary (C_b) we have $C_{1,j(all)} = C_{n,j(all)} = C_b$ (this is the condition used in Fig. 6b, with $C_b = 40$).

Inspection of the finite difference formula provides several insights into the advantages and limitations of the numerical method. Some of these are:

(1) If an explicit method is to be used, we cannot choose D, Δx and Δt arbitrarily. As D and Δx are usually specified by the problem at hand, the choice of Δt is limited by the Courant condition. If it turns out that Δt needs to be 3600 s or less, for example, it means that to calculate a concentration profile after 100 years of diffusion one would need to carry out the steps noted above 876000 times at each spatial grid point. For calculations to millions of years (e.g., cooling of a pluton), the difficulty escalates. This is why it is necessary to use a programming tool or language to carry out realistic calculations and "iteration by hand" on a spreadsheet can become a little too cumbersome. Packages such as Mathematica®, Mathcad®, Maple®, or Matlab® allow such implementation of automated iteration without full-fledged programming using a high level language (e.g., C, Fortran). The latter are in general faster.

(2) Inspection of the formula above shows why it is necessary to have the two boundary

conditions and one initial condition. To calculate the concentration at a grid point i, it is necessary to know the concentrations at grid points $i-1$ and $i+1$. These formulae cannot be used to calculate concentrations at the first (no $i-1$ grid point available) and the last (no $i+1$ grid point available) positions. The two boundary conditions provide the additional information to calculate concentrations at these points. To obtain the concentrations at time step $j+1$, we need information about concentrations at time step j. But these formulae cannot be used for the first time step and therefore we need the initial conditions to start the iterations. It is an instructive exercise to try to implement various forms of boundary conditions (concentration held constant, no flux etc.) in the discrete form.

(3) As these calculations proceed by explicitly determining the concentrations at each position at each time step, it is easy to add variations in the model. For example, to start off a calculation we need to define the concentrations at each grid point (the initial condition). Because each concentration is specified individually, it is irrelevant whether the concentration profile is homogeneous, has a regular geometric form, or is of an arbitrary shape. Similarly, if we wish to change the boundary conditions subject to which diffusion occurs after, say, the 20th time step, we can directly do so. This is where the flexibility of the numerical method noted in the text comes from. This is also how we can take outputs from thermodynamic programs such as MELTS and track diffusion along complex P-T-f_{O_2}-f_{H_2O} paths, with boundary conditions and diffusion coefficients that vary as the coexisting phase assemblage and equilibrium conditions change. More about discretization and numerical solution can be found in for example Press et al. (2007).

APPENDIX II

POINT DEFECT THERMODYNAMICS AND HOW INTENSIVE VARIABLES AFFECT DIFFUSION COEFFICIENTS

For diffusion to occur within a solid it is necessary to have point defects (e.g., see Flynn 1972 for a discussion), the simplest of which are vacancies (e.g., an atom missing in a crystallographic site) and interstitials (an additional atom in a non-regular crystallographic site). The diffusivities and concentration of such point defects is what mainly controls and explains the diffusivities of the cations. Point defects need to exist in solids because thermodynamic constraints show that a lower free energy (through an increase in entropy) is obtained by their presence at temperatures above 0 K (e.g., Flynn 1972). This means that the presence of point defects in single crystals is an equilibrium feature. Point defects can be treated as quasi-chemical species having a chemical potential like any other solid solution component and hence the concentrations of the individual point defects depend, in addition to P and T, on the activities of the thermodynamic components (e.g., Kröger and Vink 1965; Schmalzried 1981).

We consider that diffusion of an atom i occurs by a vacancy mechanism; a treatment for diffusion via interstitials would be analogous. A specific atom i can make a jump only if one of the adjacent sites is vacant. In contrast, from the perspective of a vacancy the probability of a jump to an adjacent occupied site is always high. The vacancy does not have to "wait" for another vacancy to arrive at the adjacent site. We define the vacancy diffusion coefficient as: $D_V = w_V \cdot l^2$. We define the tracer diffusion coefficient of the element by recognizing that the probability of a jump of i to an adjacent site is equal to the jump frequency (w_v) times the probability of i to be next to a vacancy. This latter probability is simply X_V – the fraction of vacancies on the crystallographic sub-lattice of the element i. The tracer diffusion coefficient of i is:

$$D_i^* = f_i \cdot X_V \cdot D_V = f_i \cdot X_V \cdot w_V \cdot l^2 \tag{A3}$$

where f_i is the correlation factor for the diffusion of element i (see main text for the significance of f_i). The correlation factor f_i and the jump distance can be treated as constants for a given diffusion mechanism, but the concentration of the vacancies, X_V, and the jump frequency of the vacancy, w_V, are strongly sensitive to a number of parameters. Temperature is the dominant one, since both variables are thermally activated, and the dependence of the jump frequency can be summarized by an Arrhenian equation (for exceptions see Flynn 1972):

$$w_v = w_v^\circ \cdot \exp\left(-\frac{\Delta H_m + P \cdot \Delta V_m}{R \cdot T}\right) \tag{A4}$$

where ΔH_m and ΔV_m are called **migration enthalpy** and **migration volume**, respectively. The next task is to describe the concentration of vacancies. If n is the number of the thermodynamic components of a crystal, the molar fraction of the vacancy may depend on P and T plus the activities of $n-1$ chemical components:

$$X_V = f(P, T, a_1, a_2, ..., a_{n-1}) \tag{A5}$$

In most cases the influence of several component activities can be effectively ignored and only one component controls the point defect chemistry.

Example of point defect chemistry of olivine and the relation to Fe-Mg diffusion. To define the stoichiometry of $(Fe,Mg)_2SiO_4$-olivine (structural formula $M_2[SiO_4]$, with $M = Fe^{2+}$, Mg^{2+}; it is ignored that the M1 and M2 sites can be distinguished), allowing for the possibility of existence of defects, four chemical components are required and thus three components have to be fixed. These may be, for example, a_{SiO_2}, f_{O_2} and $a_{Fe_2SiO_4}$. The common convention to describe the structural units of a crystal is the Kröger-Vink notation (Kröger and Vink 1965), which for the present case leads to:

V_M'' – vacancy on a M site; $Fe_M^\bullet$–Fe^{3+} on a M site; Fe_M^x– Fe^{2+} on a M site

Here the main symbol stands for the element (exception: V for vacancy), the subscripted symbol stands for the crystallographic site (e.g., M for an octahedral metal site or i for interstitial) and the superscripted symbol stands for the effective charge relative to the ideal crystal, where • stands for one positive effective charge, I stands for one negative effective charge and × is used for a neutral structural unit, e.g., a regular structural unit like Fe_M^x.

To understand how point defects affect the diffusion of ions we write quasi-chemical reactions for the formation of such defects. The terms and dependencies in these reactions explicitly incorporate the relations with the diffusion coefficients. The equations need to conserve mass, charge, and site balance. One kind of positive point defect and one kind of negative point defect are commonly much more abundant than any other point defect and effectively control the point defect chemistry of the crystal. In olivine V_M'' and $Fe_M^\bullet$ are probably the majority point defects (e.g., Nakamura and Schmalzried 1983; Sato 1986). The effective charge balance condition is:

$$2 \cdot X_V = X_{Fe^{3+}} \tag{A6}$$

with X_V and $X_{Fe^{3+}}$ as the molar fractions of vacancies and Fe^{3+} on the metal site, respectively. For the formation of the metal vacancies and Fe^{3+} in olivine the following reaction has been considered (e.g., Nakamura and Schmalzried 1983):

$$6 \cdot Fe_M^x + SiO_2 + O_2\,(g) = 2 \cdot V_M'' + 4 \cdot Fe_M^\bullet + Fe_2SiO_4 \tag{A7}$$

This is a net transfer reaction, where metal vacancies are formed by oxidation of Fe^{2+} on octahedral sites to Fe^{3+} and a new formula unit of fayalite is formed on the surface by transfer of solid SiO_2 (e.g., from orthopyroxene or a melt) and gaseous O_2 from an external source. We introduce activity coefficients of the structural units using

$$\mu_i = \mu_i^\circ + RT \ln(\gamma_i X_i) \tag{A8}$$

where μ_i° is the chemical potential of structural unit i in the reference state—the perfect single crystal. In the reference state the crystallographic sites are fully occupied with the regular structural units without any point defect. For a solid solution like olivine the selection of a reference state for the point defects is not as straightforward within this formalism, an olivine with a fixed forsterite content has to be chosen (see discussion in Dohmen and Chakraborty 2007a for more details). In thermodynamic equilibrium at constant P and T the mass action law of the above point defect reaction (Eqn. A7) can be derived:

$$K = \exp\left(-\frac{\Delta G^\circ(P,T)}{RT}\right) = \frac{a_{V_M''}^2 \cdot a_{Fe_M^\bullet}^4 \cdot a_{Fe_2SiO_4}}{a_{Fe_M^\times}^6 \cdot a_{SiO_2} \cdot a_{O_2}} \tag{A9}$$

Because the point defects are highly diluted we can assume that these activity coefficients are constant for a fixed forsterite content of the olivine. If we assume in addition that Fe and Mg in olivine mixes ideally on the metal site it follows from Equation (A9):

$$K = \exp\left(-\frac{\Delta G^\circ}{RT}\right) = K_\gamma(X_{Fe}) \frac{X_V^2 \cdot X_{Fe^{3+}}^4}{X_{Fe}^4 \cdot a_{SiO_2} \cdot a_{O_2}} \tag{A10}$$

Here a K_γ has been defined, which incorporates the effects of the activity constants and their dependence on the forsterite content (equivalent to the molar fraction of Fe, see Dohmen and Chakraborty 2007a, for an explicit expression for K_γ). In combination with the charge neutrality condition (Eqn. A6) an expression can be derived from Equation (A10), where the molar fraction of vacancies is directly given as a function of P, T, f_{O_2}, a_{SiO_2}, and X_{Fe}:

$$X_V = \exp\left(-\frac{\Delta G^\circ}{6RT}\right) \cdot \frac{X_{Fe}^{2/3}}{16^{1/6} K_\gamma(X_{Fe})^{1/6}} \cdot a_{SiO_2}^{1/6} \cdot a_{O_2}^{1/6} \tag{A11}$$

$$\Rightarrow X_V = \exp\left(-\frac{\Delta H^\circ - T \cdot \Delta S^\circ + P \cdot \Delta V^\circ}{6RT}\right) \cdot \frac{X_{Fe}^{2/3}}{16^{1/6} K_\gamma(X_{Fe})^{1/6}} \cdot a_{SiO_2}^{1/6} \cdot a_{O_2}^{1/6} \tag{A12}$$

$$\Rightarrow X_V = \exp\left(\frac{\Delta S_f}{R}\right) \cdot \exp\left(-\frac{\Delta H_f + P \cdot \Delta V_f}{RT}\right) \cdot \frac{X_{Fe}^{2/3}}{16^{1/6} K_\gamma(X_{Fe})^{1/6}} \cdot a_{SiO_2}^{1/6} \cdot a_{O_2}^{1/6} \tag{A13}$$

with $\Delta H_f = \Delta H^\circ/6$, $\Delta V_f = \Delta V^\circ/6$ and $\Delta S_f = \Delta S^\circ/6$. These quantities are called the **formation enthalpy**, **formation volume** and **formation entropy** of vacancies on the metal site, respectively (note that the specific values of these formation energies depends on the choice of majority point defects). Using Equations (A3, A4, and A13), a final equation for the tracer diffusion coefficient of Mg (and analogously for Fe) in olivine for a given forsterite content is obtained:

$$\Rightarrow D_{Mg}^* = D^\circ(X_{Fe}) \cdot \exp\left(-\frac{(\Delta H_f + \Delta H_m) + P \cdot (\Delta V_f + \Delta V_m)}{RT}\right) \cdot a_{SiO_2}^{1/6} \cdot f_{O_2}^{1/6} \tag{A14}$$

The above equation is strictly applicable if V_M'' and $Fe_M^\bullet$ are the majority point defects, a_{SiO_2}, f_{O_2} are held constant at a given value (i.e., buffered), and X_{Fe} is kept constant. D° depends on the major element composition (X_{Fe}), the migration terms may depend on this as well. Comparing this equation with Equation (20) we can know clearly identify the parameters that are buried within the pre-exponential and exponential factors of the Arrhenius equation.

The above example can be generalized for any crystal with a charge neutrality condition that considers only two point defects, and a fundamental equation for the tracer diffusion coefficient can be derived which summarizes the dependencies on the thermodynamic parameters:

$$D_i^* = D° \cdot \exp\left(-\frac{\Delta H_f + \Delta H_m + P \cdot (\Delta V_f + \Delta V_m)}{RT}\right) \cdot a_1^{m_1} \cdot a_2^{m_2} \cdot ... \cdot a_{n-1}^{m_{n-1}} \qquad (A15)$$

Here the exponents m_i are rational numbers determined by the charge neutrality condition and the stoichiometry of the vacancy forming reaction. In deriving such expressions it is implicitly assumed that the vacancies equilibrate instantaneously compared to the element diffusion time scales. The justification for this can be found in Equation (A3) which shows that diffusion of vacancies is orders of magnitude faster than diffusion of the elements (the factor is simply $1/X_V$ the fraction of vacancies on the respective site). However, the kinetics of point defect equilibration is not always rate limited by their diffusion. The rate limiting factor could, in principle, be the kinetics of the vacancy forming reactions such as Equation (A7). These require a net transfer of elements from or to the crystal. While macroscopic diffusion coefficient measurements as a function of temperature yield combined values of energies and volumes of migration and formation (e.g., $Q = \Delta H_f + \Delta H_m$), computer simulations may provide information on these parameters separately and provide insights on diffusion mechanisms.

Mineral Textures and Zoning as Evidence for Open System Processes

Martin J. Streck

Department of Geology
Portland State University
Portland, Oregon, 97207, U.S.A.

streckm@pdx.edu

INTRODUCTION

Investigations of mineral textures and zoning as evidence for open system processes during magmatic evolution have always been a centerpiece of petrological studies and have provided some of the best evidence for magma mixing and crustal contamination for many decades (e.g., Milch 1905; Kuno 1936; Eichelberger 1975; Sato 1975; Anderson 1976). In fact, evidence from mineral studies was instrumental in the acceptance of magma mixing as important petrological process ever since it was initially proposed by Bunsen (1851). In recent years, mineral studies are invigorated by the development of high-precision, high-resolution analytical instruments and techniques through which textural information and compositional data are combined (e.g., Jerram and Davidson 2007). As minerals respond texturally and compositionally to changing magmatic environments, they preserve in their crystal growth stratigraphy a wealth of information regarding their past history of magmatic processes and compositions (cf. Ginibre et al. 2007). On the other hand, magmatic liquids (melts) are snapshots of current magmatic states, and provide less direct evidence of the processes responsible for their evolution. In addition, liquids may crystallize thereby destroying the direct evidence they provide. With an appropriate set of observations and measurements, we can correlate textures with mineral compositions and thus produce a richer composite picture of magmatic evolution than compositional data of minerals alone. This contribution reviews the zoning and textures of minerals frequently encountered in volcanic rocks (many of which apply equally to plutonic rocks) and the interpretations ascribed to these features in terms of open system magmatic processes.

First, I will address the analytical tools that are used to acquire textural and compositional data, followed by a brief review of open system processes and a summary of the relevant textures. From that point on, I will address some mineral specific aspects followed by a discussion of practices that yielded evidence for open system behavior in magmatic systems based on naturally occurring mineral assemblages and populations.

METHODS TO IMAGE TEXTURES AND QUANTIFY COMPOSITION

Standard transmitted light microscopy remains at the heart of all mineralogical investigations and provides us with important information on mineral abundances, distributions, sizes, and textures. Although the optical microscope reveals crystal zoning, it seldom easily differentiates between different chemical compositions. Standard light microscopes can be equipped with reflected light Nomarski Differential Interference Contrast (NDIC) capabilities. Nomarski interference contrast is a reflected light, beam-splitting technique useful for imaging surface relief. Anderson (1983) was the first to systematically apply this method to the study of textures of plagioclase. In addition to a standard analyzer, the technique requires reflected polarized light and a Nomarksi prism, a double-crystal prism, which can fit either in the "accessory plate" slot or can be placed elsewhere above the objectives. To apply this method, samples need to be etched to develop a micro-relief (Anderson 1983 and references therein). The micro-relief develops because different compositional zones within a crystal etch differentially. The developed micro relief is typically less than 1 μm, and a relief of ~500 Å is sufficient for NDIC microscopy. Etching is performed with concentrated fluoboric (HBF_4) acid in the case of plagioclase (Anderson 1983) and with HF or HCl for pyroxene and olivine (Clark et al. 1986). Typical etch times are 20-40 seconds for plagioclase and several minutes for olivine and pyroxene (Clark et al. 1986). Etching procedures for pyroxene and olivine will strongly damage the rest of the thin section, while etching for plagioclase is short enough in duration to have little impact on other minerals or glasses. This is probably the main reason why use of NDIC microscopy is most commonly applied to plagioclase and very limited with respect to other phases. Etching is performed on polished surfaces (thin section or grain mount), and the better the polish the better the quality of NDIC images. Samples etched for plagioclase for +/− 30 seconds can be used directly for microprobe analysis. I have not found any correlation between a surface showing strong textural variation and low analysis totals. Advantages of NDIC microscopy are: 1) low cost, 2) speed (allows for a survey of large number of crystals), and 3) superb textural resolution, particularly for recognition of dissolution surfaces (see below).

Back-scattered electron (BSE) images obtained with a scanning electron microscope (SEM) or electron microprobe (EMP) provide compositional data since observed intensity of the grayscale images directly reflect differences in mean atomic weight of the material observed (Blundy and Cashman 2008). BSE images are widely used to show zoning patterns of a variety of minerals where compositional changes lead to changes in the mean atomic weight. In addition, high-resolution BSE images of plagioclase can be calibrated to provide high-resolution profiles of Anorthite (An) content profiles (see Ginibre et al. 2002 for description of technique). Catholuminescence (CL) images can in some instances (e.g., alkali feldspar, Ginibre et al. 2007) provide more details on zoning than BSE images or provide zoning information not revealed by other methods (e.g., quartz, Müller et al. 2003; see review of Blundy and Cashman 2008). Elemental X-ray maps obtained most commonly with the EMP can provide superb details concerning compositional zoning and information on the distribution of several elements simultaneously, however, elemental X-ray maps require considerable more time to acquire.

Analytical instruments used to obtain quantitative analyses at spatial resolutions sufficiently small to be compatible with the above techniques are: electron microprobe (for major and minor elements and occasionally trace elements), several laser ablation mass spectrometery techniques (mostly for trace elements but also for major elements and isotopic ratios), secondary ion mass spectrometry (SIMS) (isotopic ratios, trace element and volatile species abundances) and the proton microprobe (PIXE) (see reviews of Ginibre et al. 2007; Jerram and Davidson 2007; Blundy and Cashman 2008; Kent 2008; Ramos and Tepley 2008).

BRIEF REVIEW OF OPEN SYSTEM PROCESSES

Open system magmatic behavior involves a range of processes that lead to mass and energy exchanges between a magma body and its external environment. This includes degassing and loss of volatiles from an ascending magma and migration of volatiles to one magma from another degassing at depth. Here, I focus on evaluating the mineralogical record for open system processes of magmatic recharge (mixing) and the addition of country rock (contamination). Both processes often leave a marked record in the texture and zoning of single minerals and in the observed mineral populations of the resident magmas as crystals and melts are added from recharging magma or from country rock.

Mixing

Mixing of magmas is now recognized to play a major role in the petrogenesis of magmas and to operate at a multitude of spatial scales. In fact, most phyric magmas seem to indicate some form of mixing recorded by their minerals (see Bindeman 2008; Ramos and Tepley 2008). An important question in evaluating the significance of magma mixing processes is to determine, based upon the mineralogical record, whether a recharge event occurred and, if so, to what extent. For example, if a magma batch "self-mixes" (e.g., Couch et al. 2001) due to temperature differences then this may occur without new magma input (i.e., recharge), yet this process is likely to leave traces in the mineral record as an increase in temperature may cause mineral resorption followed by renewed crystallization (e.g., Singer et al. 1995).

Magmas involved in mixing processes often exhibit distinctive ranges in their crystal-liquid proportions and physical properties that influence the mixing process resulting in important consequences for the mineral record. The following are two end-member type scenarios capturing variable conditions of mixing. During the events leading up to the 1991 Pinatubo eruption, crystal-poor basaltic magma intruded into cool, crystal-rich dacitic magma yielding hybrid andesite and causing quenching of the mafic magma as evidenced by quenched magmatic inclusions (Pallister et al. 1996). On the other hand, mixing of high-silica rhyolite and basaltic andesitic magmas to yield hybrid dacite in the Rattlesnake Tuff involved magmas that were both crystal poor and seemingly equally fluid, without immediate quenching of the basaltic andesitic magma (Streck and Grunder 1999). Despite the differences in both of these examples, the ranges of magma compositions involved were quite similar. In general, the greater the dissimilarities in composition, temperature, volatile and crystal content of mixing magmas, the greater the difference in viscosity, which strongly influences the rheological and fluid dynamical aspects of the mixing process (see summary in Russell 1990 and references therein). The physical properties of the mixing magmas and the mixing proportions will also determine the extent of liquid-state blending versus solid-state disintegration of chilled blobs of magma. One extreme case was shown by Feeley and Dungan (1996) in which magma mixing occurred, in part, by disintegration and dispersion of crystals derived from quenched basaltic andesitic inclusions of a mafic magma that intruded into a dacitic magma (also Clynne 1999; Tepley et al. 1999).

Mixing may also involve multiple end-members and multiple mixing events. In some cases, the rock record may allow petrologists to disentangle a simple mixing history into a sequence of events and discrete compositions of end-members, however in many cases this is likely to be just an approximation of what really occurred. For example, Streck et al. (2005b) proposed that the compositionally monotonous yet mineralogically complex basaltic andesites erupting from Arenal volcano during the ongoing 1968-2008 eruption are a product of multi-stage mixing during ascent.

Contamination (assimilation)

The process of contamination or assimilation can be described as the incorporation of country rock into a magma. Most commonly, this is viewed as the incorporation of solids that subsequently melt after incorporation (also known as bulk assimilation, cf. Edwards and Russell 1996a). However, a partially molten country rock mush or extraction of partial melts obtained from the country rock (known as selective assimilation, cf. Edwards and Russell 1996a) may fall into the same category. During bulk assimilation, crystals from the incorporated country rock may melt or dissolve and thus may blend with the resident melt. The country rock may also simply disintegrate and disperse its crystals. Most often the actual process involved will likely be a combination of both (addition of crystals and melts) and will depend on composition and mineralogy of the country rock and the temperature and composition of the formerly uncontaminated magma.

MINERAL RECORD FOR MIXING OF MAGMAS AND COUNTRY ROCK – PHENOCYRST, ANTECRYST, XENOCRYST

Mineralogical responses induced by open system processes can be recognized in contrasting mineral assemblages, or in textural or composition variations including major and trace elements or isotopic ratios within individual crystals. Larger crystals embedded in finer crystalline or glassy matrix are traditionally called phenocrysts. Upon closer inspection, these are defined as phenocrysts, antecrysts, and xenocrysts. Phenocrysts are crystals that grow *in-situ* from a magma. Antecrysts are cognate crystals that originate from a magma genetically related to the one in which they are found, but are older and did not grow from the liquid in which they are found (Hildreth pers. comm.). Finally, xenocrysts are crystals derived from country rock that have no genetic relationship to the magma in question.

Exotic mineral compositions, textures, or mineral types are usually used for determining the incorporation of xenocrystic material, but in practice this can be difficult to apply. For example, positive identification of xenoliths and xenocrysts, based on compositional criteria alone, is complicated or impossible when the xenocrysts are similar in composition to crystals growing from the uncontaminated magma. In this case, older crystal ages may reveal xenocrystic material (Cooper and Reid 2008).

Individual crystals may have multiple contributions. Xenocrystic minerals can be overgrown by new crystal material disguising their origin further (this applies equally to antecrysts) (Fig. 1). Further, to distinguish crystals as antecrysts requires knowing the exact conditions of crystallization for both phenocrysts and antecrysts. Where do we draw the line between phenocrysts, which record crystallization conditions of the resident magma batch, but under variable conditions, and antecrysts? Evidence provided by zoning and textures within single crystals and relationships among crystals of distinct crystal populations should guide our assessment of whether crystals or portions of crystals are phenocyrsts, antecrysts, or xenocrysts.

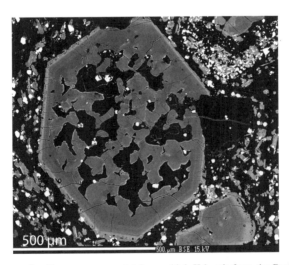

Figure 1. BSE image of a clinopyroxene crystal from a high-K basalt from the Boring volcanic field, Portland, Oregon showing a normally zoned, euhedral 30-80 µm wide overgrowth on resorbed, cellular-textured core. Gray shading inside from the outermost 10 µm wide, light (Fe-richer) rim, corresponds to Mg# 88 for darkest shade and to Mg# 84 for lighter gray. Normally zoned overgrowth is phenocrystic part while resorbed core is likely an older, antecrystic portion as suggested by compositional similarlity between overgrowth and core and highlights the difficulty in classifying single crystals "wholesale" as phenocryst, antecrsyst, or xenocryst.

SUMMARY OF ZONING AND TEXTURES

In this section, I will summarize common types of zoning and textures relevant to the discussion of open system processes. In general, zoning is the expression of compositional variability within a single crystal. Intra-grain mineral textures refer to the internal arrangement of morphological or other features that lead to spatial patterns within single crystals. In some cases, both terms are nearly inseparable and thus are often used synonymously. The summary below draws from a variety of sources with some additions by the author and attempts to contrast terms that are opposites. Notable sources are the works of Pearce and coworkers (Pearce and Kolisnik 1990; Pearce 1994) who have extensively studied plagioclase textures, and Hibbard (1995) and Vernon (2004) who both provide a general treatment of zoning and textures. Other notable contributions to zoning and textures are Vance (1963, 1965) and Anderson (1984) for oscillatory and patchy zoning, Downes (1974) for distinguishing fine (oscillatory) from coarse banding in pyroxene, and Shore and Fowler (1996) for discussing a large array of minerals with oscillatory zoning.

Crystal zoning

Normal zoning vs. reverse zoning. Normal zoning includes compositional changes in the mineral from core to rim (i.e., along the crystal growth stratigraphy) as would be induced by compositional changes in the melt as it follows a liquid line of descent during cooling. The changing mineral composition during crystallization reflects the progressively evolving composition of the melt during the solidification process in a closed system. Reverse zoning is thus any compositional inversion within the growth stratigraphy of single crystals from what would be expected to result from crystallization in a closed system.

Step zoning vs. progressive zoning. Normal or reverse zoning can manifest itself in the form of discrete compositional steps along analytical core-rim traverses. This is called stepzoning. Progressive zoning is the other, more typical case that produces smooth profiles where compositions progressively change with position along an analysis profile. Step zoning is always associated with textural transitions (imaged by one of the techniques mentioned above), but textural transitions can have internally concordant (euhedral) or discordant (anhedral) relationships (Fig. 2). On the other hand, progressive zoning occurs without textural transitions unless superimposed by oscillatory zoning. Another exception may exist if diffusional relaxation modifies an original step zoning into progressive zoning (Costa et al. 2008).

Oscillatory zoning vs. monotonous zoning. Oscillatory zoning is defined as "compositionally varying growth-shells [or layers] that are generally parallel to crystallographic planes of low Miller indices and have thicknesses ranging from tens of nanometers to several tens of micrometers" (Shore and Fowler 1996), and is probably the zoning style that received the most attention. Oscillatory zoning is common to many minerals (Shore and Fowler 1996). Furthermore, oscillatory zoning is divided into "fine banding" and "coarse banding" based on the width of the zoning (Downes 1974). I believe separating "fine" from "coarse" banding is a useful distinction because fine banding on the scale of single microns may in fact have a generally different origin than "coarse banding," as recently suggested by Ginibre et al. (2002) for plagioclase. This distinction is also supported by findings of multiple authors on coarse banding observed in pyroxene (see below). Fine banding may be largely kinetically controlled while coarse banding reflects dynamic magmatic processes. Therefore, for this contribution, oscillatory zoning is taken to be the "fine banding" (i.e., on the scale of 1-10 µm) that is associated with small-scale compositional oscillations of various forms (sinuous or irregular). Coarse banding is also known as "growth zones," "growth bands," and "mantles" (e.g., Duda and Schmincke 1985); the term "mantle," however, should be reserved for a reaction zone or overgrowth of a different mineral as indicated by Hibbard (1981; 1995 p. 125). In plagioclase, fine-banding is found either as straight (euhedral) or convolute conformable layers (Pearce

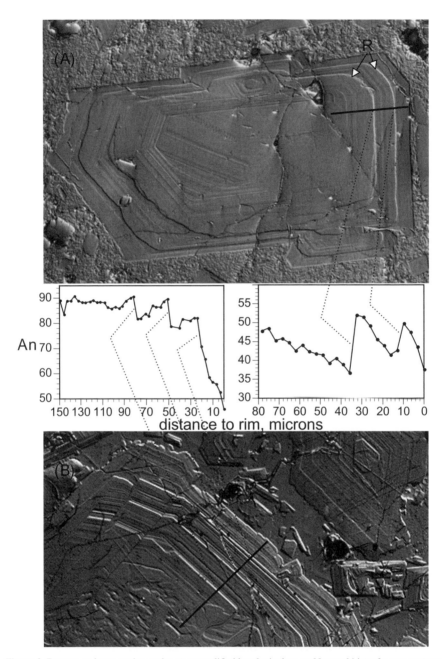

Figure 2. Reverse and step zoning styles as exemplified by plagioclase on Nomarski interference contrast images. (A) Mount St. Helens plagioclase of the 2004-05 dacite with dissolution surfaces (R) and sawtoothed zoning pattern; note reverse zoning is associated with crossing the dissolution surface. (B) Plagioclase from a basaltic andesite of Arenal volcano, Costa Rica showing progressive reverse zoning in growth bands (segments) that are separated by sharp drops in An content without notable dissolution within growth bands and between adjacent growth bands.

1994). The opposite of oscillatory zoning is monotonous (or textureless) zoning. The development of oscillatory vs. monotonous zoning may be directly related to the growth conditions. In the case of plagioclase, it has been suggested, based on the observation that oscillatory zoning is virtually absent in experimentally produced plagioclase, that formation of oscillatory zoning (fine banding) requires growth rates smaller than $\sim 10^{-10}$ m/s and is favored by slow growth rates of 10^{-11} to 10^{-13} m/s (Shore and Fowler 1996). This in turn implies that monotonous (textureless) zoning may result from faster growth rates. This is in accordance with the example of plagioclase from Arenal volcano where monotonically zoned growth bands are located at the outer margin of many crystals that likely grew rapidly during final ascent and solidification (Streck et al. 2005b). A similar correlation of compositions and textural relations has been observed in a number of other locations (Shore and Fowler 1996).

Patchy zoning vs. concentric zoning. Irregularly distributed zoning patterns are commonly referred to as "patchy" zoning (Vance 1962), in contrast to concentric zoning. Terms like "normal" or "reverse" zoning are typically not applied to areas of patchy zoning. Although normal and reverse zoning relate compositional information to clear relative positions within the crystal growth stratigraphy of a crystal, similar relations can be difficult to establish for patchy zoned areas. Patchy zoning can result when crystallization into open spaces of skeletal crystals produced compositions that differ from the ones of the skeletal crystal host. Patchy zoning can also be generated when a crystal is in the process of re-equilibration through diffusion. Determination of which process is responsible for this zoning pattern is a matter of careful analysis of the mineral in question. A possible discriminator between crystal growth and diffusional re-equilibration as cause for patchy zoning is the sharpness of the compositional transition (step vs. progressive) between "patches." If the compositional change associated with going from one patch to the neighboring patch is sharp (step-like), crystal growth is likely (Stewart and Pearce 2004) while a progressive transition suggests diffusional re-equilibration (Tomiya and Takahashi 2005; Streck et al. 2007) (Fig. 3).

Crystal textures

Dissolution surface vs. pervasive resorption. Resorption processes lead to removal of previously crystallized material and they occur in different forms. In one case, resorption leads to dissolution of crystal material progressively from the exterior of euhedral crystals leading to various degrees of rounded, anhedral appearances. The dynamics of dissolution in a variety of studies have been discussed by Edwards and Russell (1996b), and the interested reader is referred to their review. Peripheral dissolution may range from subtle rounding of edges to forming strongly rounded or embayed crystals. Pervasive resorption may affect the entire crystal leading to an open, cellular mineral structure, which can have a wide range of appearances and textures. Also known as "sieve" textures, cellular textures range in size from those that are clearly discernible (on a petrographic microscope at low to medium magnification; e.g., Stewart and Pearce 2004) (Fig. 4) to a crystal containing a nearly sub-microscopic "haze" of variable opacity that represent miniscule glass droplets (vitric or devitrified); for such cases, the term "dusty" texture is also used (e.g., Tsuchiyama 1985) (Fig. 4b). Cellular textures can have a boxy or a spongy appearance (also synonymous with vermicular and wormy). Boxy-cellular textures are typically related to rapid growth and represent a form of quenched textures of which hopper-shaped or dendritic crystals are well known examples (cf. Lofgren 1980). In plagioclase, boxy-cellular texture can also be due to dissolution and recrystallization in a reaction zone (Nakamura and Shimakita 1998). On the other hand, spongy cellular textures are commonly attributed to pervasive dissolution.

Breakdown mantle (reaction rim) vs. growth mantle. Break-down textures result when an existing mineral is out of equilibrium and begins to re-crystallize into a new set of minerals instead of dissolving (e.g., Rutherford and Hill 1993). Breakdown textures may be restricted to the rim areas of crystals producing reaction rims or may affect the entire crystal generating

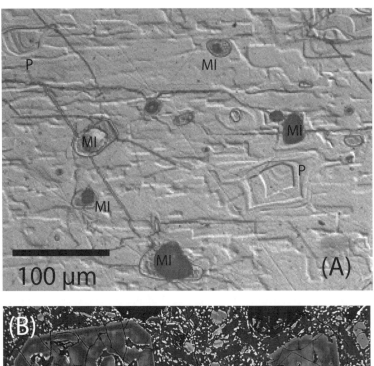
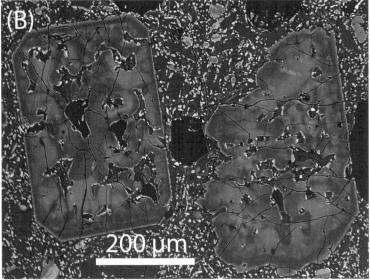

Figure 3. Two examples of patchy zoning. (A) Nomarksi image of the core of a plagioclase crystal from Arenal volcano, Costa Rica shows skeletal higher An plagioclase (high relief; ~An_{89}) infilled by lower An plagioclase (denoted by "P" and lower relief; ~An_{82}). Note sharp transition between infilled and host plagioclase. MI denotes some of the melt inclusions present. (B) BSE image with two orthopyroxene crystals from the Whaleback volcano near Mount Shasta, California show diffuse patchy zoning; lighter shades reflect more iron-rich portions (lightest shade ~Mg# 79) and darker higher Mg# (darkest shade ~Mg#86). Note younger thin, euhedral rim of renewed high Mg# pyroxene, best seen in left crystal.

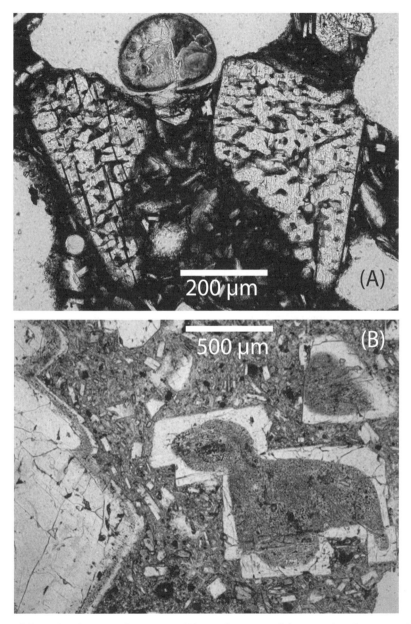

Figure 4. Examples of spongy celluar textures. (A) coarsely spongy cellular texture in orthopyroxene from the Whaleback volcano near Mount Shasta, California showing resorption channels. (B) finely spongy cellular texture (also called "sieved" or "dusty") in center and near the rim of plagioclase crystals from Narcondam volcano, Andaman Sea.

pseudomorphs. On the other hand, growth mantles are overgrowths on existing minerals. Growth occurs either because nucleation on preexisting grains is more energetically favorable than nucleating new crystals or because the dissolution of preexisting grains produces compositional boundary layers that facilitate growth (e.g., Coombs and Gardner 2004).

MINERAL SPECIFIC TEXTURES AND ZONING

There is a vast amount of data on mineral textures and zoning that pertain to the topic discussed here. Hence, it is only possible to showcase some aspects of mineral specific textures and zoning features and how they are commonly interpreted with regard to open system processes.

Plagioclase

Plagioclase displays a wide range of textures and zoning patterns, and these have been explored as tools for investigating magma dynamics for more than a century (e.g., Milch 1905). The diversity and details of preservation of intra- and inter-grain variability of observed zoning and textures in plagioclase, particularly of calc-alkaline andesites to dacites, can be very large that make the study of plagioclase challenging. When plagioclase is present, it is typically very abundant with volumetric phenocryst percentages often between 10-30%, corresponding to a few to many hundreds of crystals per thin section. A proper documentation of textures and zoning of the plagioclase crystals present in a single sample can be a considerable endeavor on its own, requiring considerable investment of time and money. Single plagioclase crystals often exhibit a multi-stage growth and resorption history (e.g., leading to a complex array of textures and compositions) (Fig. 5) that can be difficult to sort into phenocrystic, antecrystic or xenocrystic portions. Kinetic controls on plagioclase growth can also lead to texturally pronounced oscillatory zoning (e.g., Low amplitude oscillations (LAO) of Ginibre et al. 2002) superimposed on zoning and textures induced by other factors (e.g., compositional changes during recharge or temperature increase due to self-mixing) further complicating the assessment of features of open system processes. Here, I focus on textures and zoning that have been ascribed to open system processes based on evidence within single crystals.

Resorption in plagioclase is evident as individual dissolution surfaces (Fig. 2A) or as a pervasive feature resulting in boxy- or spongy-cellular textures with a diverse array of textural forms (Fig. 4B). The reader is referred to Pearce and Kolisnik (1990) and Pearce (1994) for detailed descriptions of these textures. Although open system processes are commonly invoked to explain such features, verifying this through documentation and observation may be nontrivial. When confronted with plagioclase showing cellular textures, the first question should be to what extent the texture might be due to rapid growth. Figure 6 is one case where a resorbed core is overgrown by skeletal plagioclase and thus provides an example where skeletal growth can be distinguished from a resorption feature. When shapes of remaining plagioclase in cellular zones are rounded rather than squared, skeletal growth is unlikely, as rapid growth would favor crystallographic directions leading to straight edges (Lofgren 1980). On the other hand, boxy shapes can still be due to resorption if dissolution was followed by (re)crystallization, leading to square-shaped melt inclusions and a boxy-cellular texture. Nakamura and Shimakita (1989) produced these textures by placing a seed crystal of An_{59} into a melt that was in equilibrium with higher An plagioclase, thus simulating magma mixing or contamination scenarios. Based on their work, this process would result in Ca-richer plagioclase lining the melt inclusion wall and overprinting the original Na-rich plagioclase. An alternative mechanism to induce pervasive resorption is decompression, as postulated by Vance (1965) and experimentally produced by Nelson and Montana (1992) in a dry system. Decompression of water-saturated magmas would crystallize plagioclase as volatiles are exsolved, while resorption is possible during ascent prior to water saturation (Blundy and

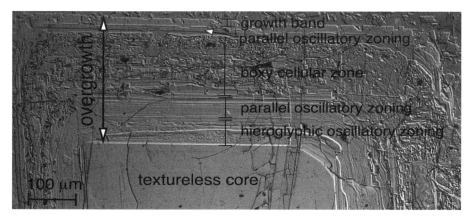

Figure 5. A single plagioclase crystal of the current (1968-2008) eruption of Arenal volcano showing a multi stage crystallization history as expressed by texturally different zones (after Lunney 2001). Compositions associated with zones are as follows: core: An 95-92, hieroglyphic oscillatory z.: An 90-72, parallel oscillatory z: An 72-60-70, boxy cellular zone: An 70-80, rim growth band: An ~60-55.

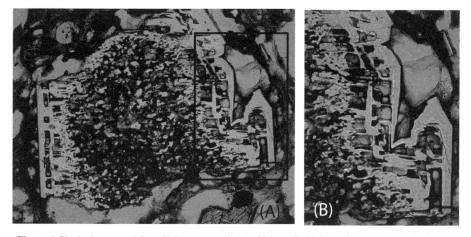

Figure 6. Plagioclase crystal from Kalama age andesite of Mount St. Helens with spongy cellular texture resulting from resorption (center) and with boxy cellular textured overgrowth on left and right side originating from skeletal growth. B highlights marked area by rectangle in (A). Light areas surrounding crystals are vesicles and dark areas are glass. Image A is about 400 microns in horizontal dimension.

Cashman 2001). Finely sieved or coarsely cellular textured resorption zones can be located anywhere between core and rim of crystals or may affect the entire crystal (Figs. 4B, 5)

Dissolution surfaces are found in most plagioclase crystals and interpretation of their relevance regarding open system processes strongly depends on the nature of compositional gradients on either side of a dissolution surface (see below).

Patchy zoning of plagioclase appears to be closely related to cellular textures in as much as spongy-cellular texture may become patchy zoning during continued crystallization. Stewart and Pearce (2004) studied one case where plagioclase of higher An content "infilled" spaces in pre-existing cellular plagioclase. Other studies have found infilling plagioclase to be of lower An content (e.g., Streck et al. 2005b; Humphreys et al. 2006).

Reverse zoning, expressed in terms of increased An content towards the rim of a crystal, can indicate that plagioclase grew from a more calcic melt. In turn, this can be a sign of open system behavior in the form of influx of higher Ca/Na melt, often considered to be synonymous with addition of a more mafic melt. However, reverse zoning can also be induced by increase in temperature or by increase in pressure under water-saturated conditions. On the other hand, increase in pressure under water-undersaturated conditions would decrease the An content (Blundy and Cashman 2001). For example, water-saturated experiments using Mount St. Helens dacites as starting material have shown that raising the temperature by ~50 °C at isobaric conditions increases the An content by about 20 mol% (mainly because of higher Ca/Na of the melt at lower crystallinity), while an increase in pressure of 100 MPa (corresponding to ~3.5 km) would raise the An content by only ~10 mol% (Rutherford et al. 1985; Rutherford and Devine 2008). From this, we see that an increase in temperature is more effective in inducing changes in An content. Recently, Blundy et al. (2006) have also shown that increase in An content related to temperature increase can occur during ascent of water-saturated magmas, where decompression drives crystallization. This in turn causes the temperature to increase due to release of latent heat of crystallization. Other increases in temperature can be induced by self-mixing (Couch et al. 2001) or due to mixing with hotter, recharging magma. To differentiate between options, it is important to correlate changes in An content with minor or trace elements (e.g., Ruprecht and Wörner 2007) or isotopic composition (Tepley et al. 1999; Hart et al. 2007). More specifically, Ruprecht and Wörner (2007) suggest that when increased An content correlates with increased Fe concentrations, it is a sign for magma mixing with more calcic magmas. If Fe concentrations stay constant, then an increase in temperature is the likely cause. On the other hand, increases in Fe concentrations may also be a sign of increased D_{Fe} due to higher oxygen fugacity (Wilke and Behrens 1999) or could be achieved by dissolution of ferromagnesian phases. The evidence for magma mixing is stronger where increases in An content correlate with changes in $^{87}Sr/^{86}Sr$, as this will not be affected by changes in temperature. This behavior has been observed in dacites of Chaos Crags, Lassen volcano (Tepley et al. 1999), and in a number of other cases (see review of Ramos and Tepley 2008). In general, the larger the increase in An content, the less likely that temperature alone could have induced the changes.

Compositional reverse zoning observed in plagioclase comes in two different textural associations. First, the increase in An content occurs as progressive zoning (see above) within one growth band that may be oscillatory textured (expressed in Nomarksi images as parallel lines) (Fig. 2B). There are no major dissolution surfaces found within the growth band, indicating no growth hiatus during which resorption occurred. Several such progressively zoned growth bands may be found adjacent to each other. Transition between growth bands can be conformable or may be bound by a dissolution surface; in either case, An content steps back down before it increases again in the next growth band. Alternatively, such growth bands can be bound by texturally different zones, such as bands of spongy cellular (sieved) plagioclase. In the second form of reverse zoning, the An content increase is abrupt and occurs when stepping across a dissolution surface (Fig. 2A). Anorthite contents outbound from the dissolution surface towards the rim are commonly highest immediately after the dissolution surface and decrease thereafter (e.g., Tepley et al. 2000; Streck et al. 2008). Association of several such zones leads to the well-known saw-tooth zoning pattern of plagioclase (Fig. 2A) (e.g., Pearce 1994; Ginibre et al. 2002; and references therein). Recently, Ginibre (2002) investigated oscillatory zoning of plagioclase and found that saw-toothed shaped An content fluctuations of larger amplitude (~ > 5 mol% An) (STR pattern of Ginibre et al. 2002) are not true oscillations because An content increase happens as a step function associated with an obvious growth hiatus during which an unknown amount of plagioclase was resorbed followed by a progressive decrease before crossing the next dissolution surface. This was also noted by Pearce (1994) for his Type 2 zones. On the other hand, low-amplitude oscillations (e.g., LAO pattern of Ginibre et al. 2002) with An changes of ≤ 2 mol% may possibly be true oscillations, and reflect kinetic effects as opposed to saw tooth zoning that is induced by processes as mentioned above. Furthermore, if indeed reverse zoning associated with crossing a dissolution surface marks growth from a more calcic melt, then the more calcic melt may in fact not entirely be due to influx of new melt. The dissolution surface is evidence that an unknown amount of plagioclase was resorbed. This dissolution of plagioclase could have led to a rise in the Ca/Na ratio of the surrounding melt thus causing higher An plagioclase as crystallization resumes to overgrow dissolution surface (see discussion in Streck et al. 2008).

Pyroxene

Reverse zoning and resorption textures associated with compositional step zoning or progressive zoning are quite common in Ca- and Na-clinopyroxenes and in low-Ca pyroxene. In many instances, these can be interpreted as signs for open system processes, such as magma recharge and contamination, when used in the context of the full range of pyroxene compositions and in the context of compositional data of melts and bulk rock (i.e., data serving as test for equilibrium). In addition, step-zoning in normally zoned segments may also be indicative of more complex growth histories involving open systems processes (e.g., Tomiya and Takahashi 2005). Normal zoning of pyroxene from basaltic to basaltic andesitic magmas typically means core-to-rim changes indicating less magnesian compositions that can be coupled with lower Cr and higher Ti and Mn contents and reflect compositional changes observed in the liquid line of descent of these magmas. The opposite is the case for reverse zoning. On the other hand, normal vs. reverse zoning in pyroxene from intermediate to silicic (felsic) melts is not recorded by the distribution of major and minor elements consistently in the same way, and pyroxenes may not show significant zoning at all. For example, crystallization may lead to increases in Mg contents of pyroxene with differentiation instead of decreases, as exemplified by clinopyroxenes from a zoned high-silica rhyolitic tuff (Streck and Grunder 1997). Addition of the trace-element record may also be needed to unambiguously establish the existence and style of zoning.

Reverse zoning is evident using the optical microscope when compositional contrast is strong and leads to different color intensities with the greater color saturation in the iron-richer pyroxene (Fig. 7). The best examples are found among clinopyroxenes from alkaline magmas, known also as "green core" pyroxenes (e.g., Brooks and Printzlau 1978; Barton and van Bergen 1981; Duda and Schmincke 1985; Dobosi 1989; Barton et al. 1992; see Dobosi for earlier reports on "green core" pyroxenes). The same workers attributed the observed reverse zoning in "green-core" pyroxene to open system processes (magma mixing or contamination). Similar examples of reverse zoning in pyroxenes as expressed by colorless rims (rim growth bands) on colored cores are also found among sub-alkaline varieties (Fig. 7). Anderson (1974) and Pe-Piper (1984) interpreted reverse zoning of clinopyroxene from sub-alkaline magmas as evidence for magma mixing. In less optically evident zoning examples, BSE images provide the evidence for the existence of zoning in major components – as changes in Fe/Mg of the pyroxene often lead to strong contrast differences (X-ray element maps may provide improved resolution of zoning characteristics although these are more time consuming). Contrast differences in BSE images may not be solely due to changes in the Fe/Mg ratio; variations in other elements such as Ca in Ca-pyroxene may contribute to contrast differences, the extent of which can easily be verified by elemental quantification with the electron microprobe.

In single crystals, reverse zoning can take the form of growth bands located anywhere between the core and rim of crystals (e.g., Dobosi 1989; Nakagawa et al. 2002; Streck et al. 2002, 2005b), 2007; Tomiya and Takahashi 2005; Ohba et al. 2007) and have been interpreted to indicate growth from more mafic magmas after initially growing from more differentiated, more silicic (or generally felsic) magmas. Typically, these compositional changes associated with such growth bands occur in a step-zoning fashion, unless subsequently modified by diffusion (cf. Costa et al. 2008). Such growth bands can have euhedral (conformable) or anhedral (unconformable) interfaces with neighboring pyroxene and can repeat themselves in a single

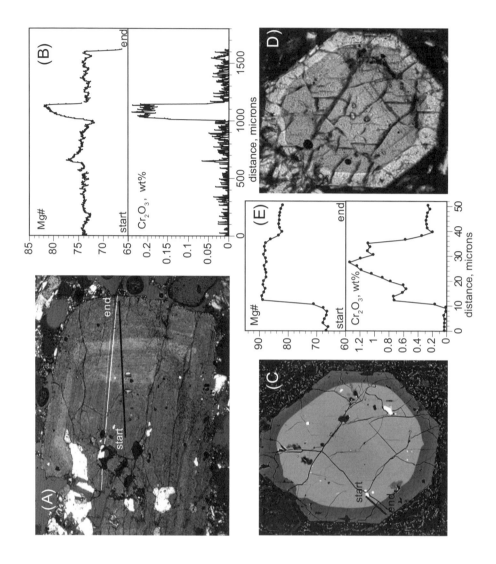

Figure 7. Reverse and step zoning styles in pyroxene observable by various methods. (A) high Mg#, 200 μm wide growth band in ~middle of crystal is seen under cross-polarized light; high Mg# growth band (40 μm wide) at rim of crystal viewed in BSE image in (B) and under plain light in (D). Composition profiles along black line are shown in (B) of crystal in (A) (clinopyroxene (cpx) from Arenal volcano) and in (E) for crystal seen in (C, D) (cpx from Whaleback volcano, California).

crystal, indicating multiple growth events from more mafic magmas, or occur consistently only as rim growth bands (e.g., Duda and Schmincke 1985; Streck et al. 2007). The interpretation that a more magnesian composition corresponds to a more mafic melt composition, and is not primarily due to other factors (e.g., increase in f_{O_2} reducing the amount of Fe^{2+} in the melt), is supported when changes in Mg/Fe correlate with changes of other elements. For example, Mg/Fe can positively correlate with Cr content of the pyroxene providing additional evidence for growth from more mafic (higher Mg, Cr) magmas.

Another zonation type that can occur within single crystals and that may be used as evidence for open system processes is patchy zoning instead of clear growth bands. As with the interpretation of growth bands, the extent of compositional contrast and the correlation of changing Mg/Fe with other chemical parameters needs to be evaluated, whether the zoning records open system processes or not. In the case of patchy zoning, however, it is often more difficult to find evidence as to whether the open system process is magma mixing or contamination with solid crystal debris. Patchy zoning can result when step-zoned crystals with originally continuous growth bands or homogenous core areas are in the process of being slowly eliminated by intra-crystal diffusional processes over time (Fig. 8) (Tomiya and Takahashi 2005). Patchy zoning may also result from melt-crystal re-equilibration processes of originally unzoned crystals. Such re-equilibration processes can take place when single crystals are suspended in liquids with which they are not in equilibrium or where crystals interact with liquids via cracks and small channels along the margins of magmatic reservoirs.

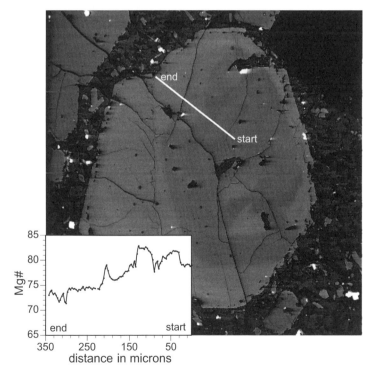

Figure 8. Patchy zoning as revealed on BSE image of a clinopyroxene crystal from Arenal volcano. Lower Mg# cpx seemingly overprints higher Mg# core. Such irregular shaped patchy zoning of higher Mg# core is interpreted to result from diffusional re-equilibration rather than by infilling and overgrowth on spongy cellular crystal.

Tomiya and Takahashi (2005) and Streck et al. (2007) interpreted patchy zoning observed in strongly resorbed orthopyroxene crystals to result from diffusional re-equilibration processes, in the latter example caused by magmas infiltrating country rocks. I'm not aware of a case where patchy zoning resulted from infilling of newly crystallized pyroxene into a framework pyroxene. To decide which is the original pyroxene composition and which is the overprinting one in patchy zoning is an important aspect for temporal relationships. This aspect should be clarified by observations based on multiple crystals through which the overall range of zoning patterns and textures can be established. Textural relationships of individual crystals may be equivocal or apparent crystal stratigraphic aspects (i.e., core = older, rim = younger) may even suggest opposite relationships. For example, crystal stratigraphic aspects of the orthopyroxene crystal of Figure 9 suggest that patchy zoning is due to higher Mg# composition (darker) orthopyroxene overprinting lower Mg# pyroxene because lower Mg# pyroxene also constitutes the core of the crystal. Rather, closer inspection of the textural details of the patchy zoned area suggests the opposite. This is also in accordance with compositional relationships of other patchy zoned crystals of the same sample. The observed "inverted" stratigraphy of the pyroxene of Figure 9 may result in xenocrysts with complex growth histories when they are broken up from country rock and incorporated into a liquid dominated magmatic system.

Resorption can have different textural expressions in pyroxene. Resorption can lead to zones of spongy texture (Fig. 7A), or pyroxene grains can also simply be rounded (e.g., Dobosi

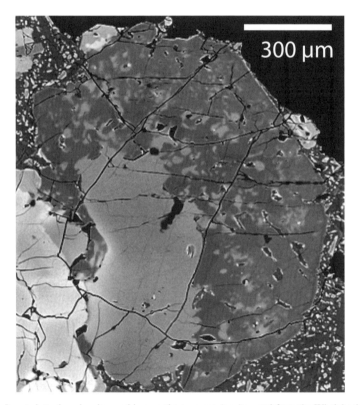

Figure 9. Inverted stratigraphy observed in an orthopyroxene (opx) crystal from the Whaleback volcano, California. Lighter shaded core (~Mg# 80) of crystal grew after darker shaded opx (~Mg# 86) based on superposition of lighter spotted area across high Mg# opx and based on zoning of opx crystals across thin section. Very thin euhedral rim grew last.

1989). Rounding of corners can be readily induced if pyroxene is exposed to a different melt composition (P. Ulmer person. communication). Spongy textured pyroxene can be associated with patchy zoning of the remaining pyroxene (Tomiya and Takahashi 2005; Streck et al. 2007) and this association is probably more common than recognized to date. Resorption in pyroxene in the form of a spongy cellular texture is a clear indication of open system behavior, while dissolution surfaces may only be significant in the context of compositional zoning steps.

Olivine

Single olivine crystals can respond rapidly to non-equilibrium conditions by dissolution (e.g., Donaldson 1990), diffusional re-equilibration (Costa et al. 2008) or by forming reaction rims (commonly orthopyroxene) (Coombs and Gardner 2004). All of the above processes can indicate open system processes. However, dissolution can also be induced by higher temperature. Likewise, profiles of lower Fo content toward the rim do not necessarily need to be diffusion profiles, but could also represent normal zoning induced by fractional crystallization. Skeletal olivine can also form in response to rapid growth and needs to be distinguished from embayed or cellular structures that result from dissolution. The presence of olivine crystals in intermediate to silicic magmas is often taken by itself as evidence for mafic recharge and can be supported by evaluating olivine-liquid equilibria conditions via Roeder and Emslie (1970). Careful analysis of compositional and textural relationships among olivines of entirely basaltic systems can yield evidence for magma mixing and inheritance. For example, Helz (1987) shows two types of inheritance. She attributes strained olivines, similar to the one shown in Figure 10B, to be derived from the mantle while fragmented olivine originated from brittle failure of a prior olivine cumulate within the volcanic edifice. In either case, these are distinct from phenocrystic olivines. Thornber (2001) also in a study on olivine from Kilauea lavas could identify magma mixing between rift-stored and summit-derived basaltic magmas based on the existence of reversely zoned olivines. Compositionally reverse zoning with higher Fo content toward the rim of crystals also occurs in olivine from Stromboli and serves as evidence for open system processes (e.g., Landi et al. 2007), but, as is the case with other ferromagnesian minerals, the correlation with other elements (e.g., Ni, in case of olivine) is important to exclude other factors such as changes in f_{O_2}.

Resorption of olivine most commonly leads to embayed olivine crystals. Although cases of resorption where the olivine is pervasively affected by breakdown reactions occur (Fig. 10A), they are rare with the exception of iddingsite formation, which is a common, mostly low temperature alteration of olivine. There are studies that have found evidence that iddingsite forms also at high temperature (Haggerty and Baker 1967; Clément et al. 2007) but because of the prevalence of iddingsite resulting from surficial processes, it can be challenging to unequivocally prove a high-T origin for these samples.

Amphibole

Amphiboles are compositionally complicated, and the composition reflects dependencies on parameters such as temperature, pressure, composition, and volatile budget. This often makes zoning patterns of amphibole difficult to interpret and difficult to relate to open system processes. For example, higher Mg# in hornblende can readily be induced by higher f_{O_2} in the melt (Scaillet and Evans 1999) instead of growth from higher Mg# melts. Reverse zoning has also been related to temperature (e.g., Bachmann and Dungan 2002; Rutherford and Devine 2003) and pressure fluctuations within the magmatic reservoir (e.g., Rutherford and Devine 2008). Thus, reverse zoning in amphibole may best express magmatic recharge if zoning reflects a temperature increase, as postulated for the Fish Canyon magma (Bachmann and Dungan 2002) and if this increase exceeds values estimated by other processes such as by magma chamber turnover. On the other hand, to infer mafic recharge from increased Mg# requires to exclude the other factors mentioned above. To emphasize this complexity, Rutherford and Devine (2003) found contradictory evidence in experimental amphibole of

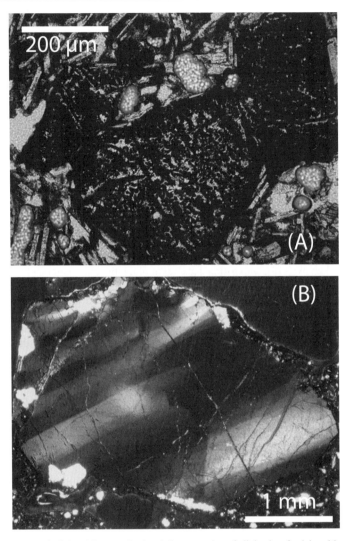

Figure 10. Textures of olivine: (A) pervasive breakdown reaction of olivine in a fresh basaltic sample from the Boring Volcanic Field, Portland without any sign of iddingsite alteration. (B) undulatory extension of olivine xenocryst derived from disintegrated mantle xenoliths in an alkali basalt from Borée, Massif Central, France.

Soufriere Hills magma, where lower temperature amphiboles had higher Mg# and the higher temperature amphiboles lower Mg#.

Resorption or recrystallization expressed as breakdown features are very commonly found in amphibole. Rounding of amphibole or embayment features consistent with dissolution appears less common although it can occur (e.g., Fig. 11b in Humphreys et al. 2006). Breakdown features in amphibole leading to reaction rims of variable thickness or pervasive alteration are very common but are difficult to resolve from those caused by destabilization of amphibole during decompression and ascent from subvolcanic reservoirs (e.g., Rutherford and Hill 1993; Rutherford 2008). Most reaction rims of amphibole found in volcanic rocks are interpreted to result from such ascent-related breakdown reactions and not from open system processes. However, magma recharge can similarly cause disequilibrium conditions by changing temperature or compositional requirements (major element, volatiles) for continued amphibole crystallization. Distinguishing between these two scenarios can be a difficult task. For example, a population of amphiboles in Soufriere Hills magma has unusually thick reaction rims interpreted to result from a heating event of the magmatic reservoir prior to the events leading to the ongoing eruption, since most other amphiboles have thin (mostly <50 μm) or no reaction rims and lower Al contents (Rutherford et al. 1998). Thus, it seems that decompression induced breakdown reactions tend to be near rim features, while amphibole breakdown, due to open system processes, often leads to large scale breakdown that may affect the entire crystal or juxtaposition of nonaltered and altered amphiboles (cf. Rutherford 2008). Another example is made by some andesitic lavas from Narcondam volcano (Streck et al. 2005a), which only contain pseudomorphs after amphibole, whereas other andesitic samples, which are slightly more silicic, contain populations of both fresh and reacted amphiboles and pseudomorphs after amphibole next to each other (Fig. 11). In addition, there is independent mineralogical evidence that magma mixing was important. These samples also record an instance where breakdown reactions are in large part due to magma mixing, and not related to ascent from the eruption-feeding reservoir to the surface.

Overgrowth of clinopyroxene as an optically continuous single crystal on amphibole (Fig. 12), and not as a reaction zone (e.g., Rutherford and Devine 2003), seems to be a clear sign of open system processes. It is unlikely that such a feature forms when amphibole becomes unstable during final ascent, but overgrowth of clinopyroxene on a pseudomorph after amphibole is consistent with crystallization during an up-temperature gradient (Rutherford et al. 1998), possibly combined with changes in volatile saturation level as can be induced by magmatic recharge. In many cases, however, such overgrown amphiboles are often completely decomposed, in which case it is difficult to identify the interior of the clinopyroxene as originally having been an amphibole (Fig. 12). In fact, samples lacking amphibole entirely but containing clinopyroxenes with cores consisting of a microcrystalline patchwork of pyroxene, oxide and feldspar likely indicate that this patchwork was amphibole before the amphibole became unstable and was subsequently overgrown by clinopyroxene. The amphiboles either originated from country rock material or from an earlier magma that crystallized amphibole.

Other major minerals

Other major minerals where zoning has been explored for inferring open system processes are alkali-feldspar and quartz. Repetitious concentric zoning in Barium of alkali feldspar has been imaged on BSE pictures and CL maps (Lipman et al. 1997; Ginibre et al. 2004, 2007). This zoning may be indicative of interaction of variably evolved felsic magmas. Barium (Ba) is rapidly depleted in the felsic melts once alkali feldspar (sanidine or anorthoclase) starts crystallizing, thus a return to high Ba concentrations outward from a more depleted Ba core indicates replenishment of less evolved, Ba-rich melt (cf. Ginibre et al. 2004).

Zoning in quartz is revealed by CL images where brighter zones are correlated with Ti-rich compositions that in turn are a function of higher temperature conditions (Müller et al. 2003; Wark et al. 2007 and references therein). For example, in CL-images of quartz, darker, lower temperature cores overgrown by brighter, higher temperature quartz are interpreted to indicate a magma heating event due to recharge of a hotter rhyolitic magma than the one that crystallized the inner portion of the quartz (e.g., Wark et al. 2007). The recent development of being able to tie zoning patterns in quartz to temperature estimates of crystallization sheds new light on the interpretation of the evolution and interaction of rhyolitic magmas.

Accessory minerals

Accessory phases are also subject to resorption and zoning. Their textural images, combined with compositional information are explored for open system behavior as well. The most

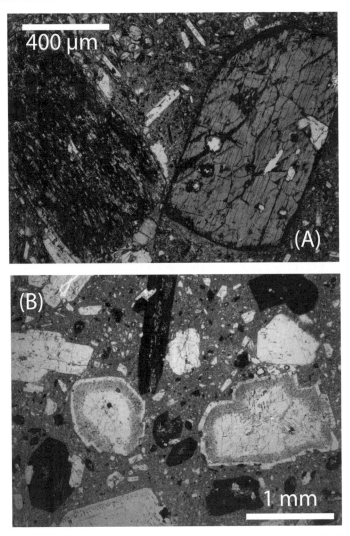

Figure 11. Amphibole in two andesites from Narcondam volcano, Andaman Sea requiring disequlibrium conditions prior final ascent; (A) juxtaposition of fresh amphibole crystal without any reaction rim (right crystal) and amphibole pseudomorph (left crystal) implies magma mixing; (B) sample in which all amphibole crystals are pervasively broken down into a very fine opaque mass while preserving characteristic shape; the extent of reaction is suggestive of prolonged disequilibrium conditions either as magma carrying amphiboles was heated or ascended to a storage depth outside amphibole stability.

widespread accessory mineral used in this regard is clearly zircon, and much has been written about its zonation patterns, which are correlated with crystal age information (e.g., Charlier et al. 2005) or other compositional information (e.g., Claiborne et al. 2006; Bindeman et al. 2007). The interested reader is referred to RiMG volume 53, particularly to the contribution by Cofu et al. (2003) who compile many zoning features of zircons. Other accessory minerals showing zonation that can be explored for open system processes are apatite (e.g., Streck and Dilles 1998; Tepper and Kuehner 1999; Broderick et al. 2007), allanite (Vazquez and Reid 2004; Beard et al. 2006), sphene (titanite) (Robinson and Miller 1999) and others.

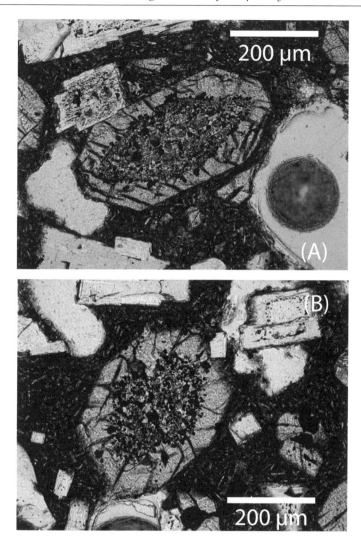

Figure 12. Pseudomorphs after amphibole found in cores of clinopyroxene crystals from one Kalama age andesitic sample of Mount St. Helens; (A) preservation of euhedral shape of former amphibole allows identification of the core of the clinopyroxene as pseudomorph; core of crystal seen in (B), although identical in origin as core in (A), is difficult to identify as former amphibole and instead the crystal patchwork could have been interpreted as disequilibrium feature of a former pyroxene composition (see text).

MINERAL POPULATIONS AS EVIDENCE FOR OPEN SYSTEM BEHAVIOR

While textures and compositional zoning of individual crystals hold a great deal of information on whether, and to what extent, open system processes have occurred, important insight into open system behavior can also be obtained from investigating mineral populations (Fig. 13). Investigations of mineral populations are typically performed in concert with investigations of textures and zoning of individual crystals that are the basis for recognizing different crystal "types." Investigating mineral populations can be pursued for single phases or for phase assemblages where different mineral phases are compositionally (e.g., Mg# of ferromagnesian silicates) or spatially (e.g., in form of mutual mineral inclusions, e.g., Freundt

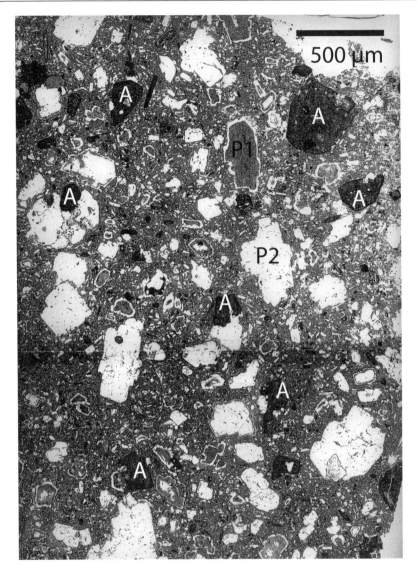

Figure 13. Striking example for two plagioclase populations in an andesite from Narcondam volcano, Andaman Sea. P1 type plagioclase crystals show finely sieved (spongy cellular) textured interiors rimmed by clear plagioclase; P2 type plagioclase crystals lack any sieved portion (with one or two exceptions). "A" denotes some of the amphibole phenocrysts.

and Schmincke 1992) cross-correlated. Streck et al. (2005b) investigated mineral populations in compositionally monotonous basaltic andesites from the current eruption at Arenal volcano by correlating the composition of mineral inclusions with host mineral compositions in the light of extensive data on compositional variability of plagioclase, clinopyroxene, orthopyroxene, spinel, and olivine. Their data suggest the existence of four different mineral assemblages that resulted from variations in liquid compositions and physical conditions of crystallization. Several of these mineral assemblages can be found at different growth stratigraphic positions in single crystals, and thus require open system processes as an explanation. Although Arenal is remarkable for its limited bulk compositional (basaltic andesitic) compositions, similar examples are found at many other arc related intermediate volcanic centers where complex textural and compositional mineral relationships indicate disequilibrium assemblages (e.g., Nixon 1988; Clynne 1999).

The study by Helz (1987) is an example where mostly textural (or morphological) data from a single phase (in this case, olivine), and only to a lesser extent the distinction by chemistry, allowed the author to establish five types of olivine that could be interpreted in terms of open system processes. The different olivines have a relatively narrow compositional range (~Fo_{83-88}) and so textures were used as the basis for interpreting the origins of crystals that include mantle inheritance (xenocrysts), inheritance from cumulitic olivine of older crystallization episodes (~antecrysts), and phenocrystic olivine.

On the other hand, crystal populations can be purely established on compositional grounds. Suppose clinopyroxene crystals in one dacite sample are nearly unzoned but show distinct compositions, e.g., one at higher and one at lower Mg# (or any other compositional parameter, e.g., trace element, isotopic ratio, etc.). In this fictive case, it is the juxtaposition of crystals from both populations, which could not have co-precipitated, that provides the evidence for mixing or for contamination (if one crystal type is derived from country rock). Mixing of two or more unzoned crystal populations occurs in explosive eruptions where mixing of liquids shortly predates or is concurrent with eruption (e.g., Boden 1989; Streck and Grunder 1999; Cathey and Nash 2004). The short time frame during which crystals experience disequilibrium can prevent any re-equilibration or overgrowth of minerals that otherwise occur in systems with compositional hybridization before eruption.

Another discriminator that can be used to establish open system processes based on mineral populations is size of crystals through the technique of crystal size distributions (CSD) (Marsh 1988; Armienti 2008). Mixed crystals populations can be interpreted from kinks, changes in slope, or disturbances (i.e., scatter in data) in the crystal size distribution (e.g., Cashman 1990). A recent example where this was demonstrated is work by Armienti et al. (2007) on clinopyroxenes form Mt. Etna volcano. They show distribution patterns that are disturbed at sizes around 800 and 3000 μm. The authors interpreted this as input of larger crystals as explanation for the disturbance at 3000 μm while the lower size disturbance is less constrained and could signify a burst in the nucleation rate or input of crystals.

When crystals are complexly zoned, it can be difficult to find criteria to be used for evaluation of crystal populations. For these reasons, attempts have been made to find mathematical solutions to express the extent of similarities (e.g., in their zoning profile) that crystals share and to establish a common magmatic heritage (Wallace and Bergantz 2005). Crystal populations have also been investigated using cluster analysis applied to the observed range of mineral compositions (Cortes et al. 2007). Such an approach may be most fruitful where compositional mineral data are investigated without the context of textural information. On the other hand, select textural or compositional criteria can be used to distinguish crystal populations. For example, de Silva et al. (2008) contrasted reversely zoned plagioclase crystals from complexly oscillatory zoned and unzoned crystals.

In general, most, if not all investigations dealing with mineral populations are faced with the difficult task of deciding which individual crystals are being taken as "representative" of a population, which then are analyzed in greater detail. Typically, visual inspection across the spectrum of crystals observed is the basis to select a few "nice" examples for further analysis. The more crystal rich the sample and the more diverse styles of zoning that exist within one mineral type, the more daunting it is to pick representative examples.

To circumvent the limitation of being qualitative regarding the distribution of crystal types in a sample, crystal populations can be quantitatively surveyed using consistent criteria distinguishing crystal types – similar to measuring sizes of crystals for CSD studies but using textural and/or compositional criteria. For example, Streck et al. (2008) distinguished plagioclase types in Mount St. Helens samples of the 2004-05 eruption based on whether crystals contained dissolution surfaces or not, and whether crystals contained acicular orthopyroxene inclusions. Using these criteria, they surveyed pre-selected areas across thin sections and "mapped" each individual plagioclase phenocryst to establish the distribution and proportionalities among crystal types. Their study shows that ~10% of plagioclase phenocysts of the 2004-05 eruption are void of near-rim dissolution surfaces and that those were likely supplied to the erupting magma reservoir during the last magmatic recharge event. In a similar study, Streck and Leeman (2008) mapped all pyroxene crystals (> ~50 µm) of one entire thin section with regards to textural and compositional features. In this study, they were able to show that 50% of pyroxenes (>200 µm) initially grew from silicic melt followed by growth from high-Mg magma while the second most abundant (~37%) population had xenocrystic characteristics; both features combined attest to the important role of open system processes in a magma that was previously identified as being little affected by mixing and contamination (cf. Streck et al. 2007). Depending on the scope of the study, quantitative evaluation of crystal populations, such as the ones just mentioned above, can be used to demonstrate the volumetric importance of a population (e.g., crystals inherited from country rock) or to establish differences in the populations among samples (e.g., magmas). How significant the differences are, in turn, depends on the criteria used to define populations.

OUTLOOK AND CONCLUSIONS

When minerals are present, it is likely that textures and compositions reveal some features that require mixing processes in the form of turnover, magma mixing or mixing with country rock contaminants. Use of the mineralogical features discussed above, as mileposts to interpret the evolutionary paths of magmas that erupt to form volcanic rocks has revealed, in many cases, an incredible degree of complexity. This complexity seems bottomless, because as new analytical tools are developed, additional scales of heterogeneity, which also require explanation, are discovered. There are several areas of research that I believe need to be pursued in tandem to continue progress towards our ultimate goal to understand how magmas form and evolve. One area is to pursue more experimental studies geared toward understanding compositional zoning patterns, e.g., reverse zoning, in the context of growth and resorption textures and toward how resorption textures develop, similar to studies by Tsuchiyama (1985), Nakamura and Shimakita (1998), and Nelson and Montana (1992) to name a few. With the lessons learned from such experiments, we will be in a better position to determine conditions that yield observed features of mixing and contamination. The other area that deserves attention is better quantification of petrographic observations of crystal bearing natural samples, not only regarding size or composition, but to combine textural records as expressed by zoning and resorption features observed in single crystals and crystal populations. No doubt, the minimum requirement for all this is time, if we want to reach beyond a reconnaissance study regarding information preserved in single crystals, within single minerals, and in crystal assemblages. Surely, not every sample can be scrutinized at the same level of detail but those that will be investigated in detail will advance our understanding of how we can end up with what appears sometimes as a nearly unexplainable array of crystals across a thin section of only ~4 cm in length.

ACKNOWLEDGMENTS

I would like to thank the editors Keith Putirka and Frank Tepley for the invitation to contribute to this RiMG issue. Reviews by Shan de Silva, Catherine Ginibre Adam Kent and by the editors are greatly acknowledged and considerably helped to strengthen this review. Editorial suggestions by the reviewers, the editors as well as by Karen Carroll and Sue Wacaster are also much appreciated and improved the English.

REFERENCES

Anderson AT (1974) Evidence for a picritic, volatile-rich magma beneath Mt. Shasta, California. J Petrol 15:243-267
Anderson AT (1976) Magma mixing: petrological process and volcanological tool. J Volcanol Geotherm Res 1:3-33
Anderson AT (1983) Oscillatory zoning of plagioclase: Nomarksi interference contrast microscopy of etched sections. Am Mineral 68:125-129
Anderson AT (1984) Probable relations between plagioclase zoning and magma dynamics, Fuego volcano, Guatemala. Am Mineral 69:660-676
Armienti P (2008) Decryption of igneous rock textures: crystal size distribution tools. Rev Mineral Geochem 69:623-649
Armienti P, Tonarini S, Innocenti F, D'Orzio M (2007) Mount Etna pyroxene as tracer of petrogenetic processes and dynamics of the feeding system. Geol Soc Am Spec Paper 418:265-276
Bachmann O, Dungan MA (2002) Temperature-induced Al-zoning in hornblendes of the Fish Canyon magma, Colorado. Am Mineral 87:1062-1076
Barton M, van Bergen MJ (1981) Green clinopyroxenes and associated phases in a potassium-rich lava from the Leucite Hills, Wyoming. Contrib Mineral Petrol 77:101-114
Barton M, Varekamp JC, van Bergen MJ (1982) Complex zoning of clinopyroxene in the lavas of Vulsini, Latium, Italy: evidence for magma mixing. J Volcanol Geotherm Res 14:361-388
Beard JS, Sorensen SS, Gieré R (2006) *REE* zoning in allanite related to changing partition coefficients during crystallization: implications for *REE* behaviour in an epidote-bearing tonalite. Mineral Mag 70:419-435
Bindeman IN, Watts KE, Schmitt AK, Morgan LA, Shanks PWC (2007) Voluminous low delta^{18}O in the late Miocene Heise volcanic field, Idaho: Implications for the fated of Yellowstone hotspot calderas. Geology 35:1019-1022
Bindeman I (2008) Oxygen isotopes in mantle and crustal magmas as revealed by single crystal analysis. Rev Mineral Geochem 69:445-478
Blundy J, Cashman K (2001) Ascent driven crystallization of dacite magmas at Mount St. Helens, 1980-1986. Contrib Mineral Petrol 140:631-650
Blundy J, Cashman K (2008) Petrologic reconstruction of magmatic system variables and processes. Rev Mineral Geochem 69:179-239
Blundy J, Cashman K, Humhreys M (2006) Magma heating by decompression-driven crystallization beneath andesite volcanoes. Nature 443:76-80
Boden DR (1989) Evidence for step-function zoning of magma and eruptive dynamics, Toquima caldera complex, Nevada. J Volcanol Geotherm Res 37:39-57
Broderick CA, Streck MJ, Halter WE (2007) Sulfur-rich apatites in silicic, calc-alkaline magmas: inherited or not? EOS Trans AGU 88(52), Fall Meet. Suppl., Abstract V11B-0592
Brooks CK, Printzlau I (1978) Magma mixing in mafic alkaline volcanic rocks: the evidence from relict phenocryst phases and other inclusions. J Volcanol Geotherm Res 4:315-331
Bunsen R (1851) Ueber die Prozesse der vulkanischen Gesteinsbildungen von Islands. Annalen Phys Chem 83:197-272
Cashman KV (1990) Textural constraints on the kinetics of crystallization of igneous rocks. Rev Mineral Geochem 24:260-314
Cathey HE, Nash BP (2004) The Cougar Point Tuff: Implications for thermochemical zonation and longevity of high-temperature, large-volume silicic magmas of the Miocene Yellowstone Hotspot. J Petrol 45:27-58
Charlier BLA, Wilson CJN, Lowenstern JB, Blake S, van Calsteren PW, Davidson JP (2005) Magma generation at a large hyperactive silicic volcano (Taupo, New Zealand) revealed by U-Th and U-Pb systematics in zircons. J Petrol 46:3-32
Claiborne LL, Miller CF, Walker BA, Wooden JL, Mazdab FK, Bea F (2006) Tracking magmatic processes through Zr/Hf ratios in rocks and Hf and Ti zoning in zircons: An examples from the Spirit Mountain batholith, Nevada. Mineral Mag 70:517-543
Clark AH, Pearce TH, Roeder PL, Wolfson RI (1986) Oscillatory zoning and other microstructures in magmatic olivine and augite: Nomarski interference contrast observations on etched polished surfaces. Am Mineral 71:734-741

Clément JP, Caroff M, Dudoignon P, Launeau P, Bohn M, Cotten J, Blais S, Guille G (2007) A possible link between gabbros bearing High Temperature Iddingsite alteration and huge pegmatoid intrusions: The Society Islands, French Polynesia. Lithos 96:524-542

Clynne MA (1999) A complex magma mixing origin for rocks erupted in 1915, Lassen Peak, California. J Petrol 40:105-132

Coombs ML, Gardner JE (2004) Reaction rim growth on olivine in silicic melts: Implications for magma mixing. Am Mineral 89:748-758

Cooper KM, Reid MR (2008) Uranium-series crystal ages. Rev Mineral Geochem 69:479-554

Corfu F, Hanchar JM, Hoskin PWO, Kinney P (2003) Atlas of zircon textures. Rev Mineral Geochem 53:469-495

Cortes JA, Palma JL, Wilson M 2007 Deciphering magma mixing: The application of cluster analysis to the mineral chemistry of crystal populations. J Volcanol Geotherm Res 165:163-188

Costa F, Dohmen R, Chakraborty S (2008) Time scales of magmatic processes from modeling the zoning patterns of crystals. Rev Mineral Geochem 69:545-594

Couch S, Sparks RSJ, Carroll MR (2001) Mineral disequilibrium in lavas explained by convective self-mixing in open magma chambers. Nature 411:1037-1039

de Silva S, Salas G, Schubring S (2008) Triggering explosive eruptions – The case for silicic magma recharge at Huaynaputina, southern Peru. Geology 36:387-39

Dobosi G (1989) Clinopyroxene zoning patterns in the young alkali basalts of Hungary and their petrogenetic significance. Contrib Mineral Petrol 101:112-121

Donaldson CH (1990) Forsterite dissolution in superheated basaltic, andesitic and rhyolitic melts. Mineral Mag 54:67-74

Downes MJ (1974) Sector and oscillatory zoning in calcic augites from M. Etna, Sicily. Contrib Mineral Petrol 47:187-196

Duda A, Schmincke H-U (1985) Polybaric differentiation of alkali basaltic magmas: evidence from green core clinopyroxenes (Eifel, FRG). Contrib Mineral Petrol 91:340-353

Edwards BR, Russell JK (1996a) Influence of magmatic assimilation on mineral growth and zoning. Can Mineral 34:1149-1162

Edwards BR, Russell JK (1996b) A review and analysis of silicate mineral dissolution experiments in natural silicate melts. Chem Geol 130:233-245

Eichelberger JC (1975) Origin of andesite and dacite: evidence of mixing at Glass Mountain in California and at other circum-Pacific volcanoes. Geol Soc Am Bull 86:1381-1391

Feeley TC, Dungan MA (1996) Compositional and dynamic controls on mafic-silicic magma interactions at continental arc volcanoes: evidence from Gordón El Guadal, Tatara-San Pedro complex, Chile. J Petrol 37:1547-1577

Freundt A, Schmincke H-U (1992) Mixing of rhyolite, trachyte and basalt magma erupted from a vertically and laterally zoned reservoir, composite flow P1, Gran Canaria. Contrib Mineral Petrol 112:1-19

Ginibre C, Kronz A, Wörner G (2002) High-resolution quantitative imaging of plagioclase composition using accumulated backscattered electron images: new constraints on oscillatory zoning. Contrib Mineral Petrol 142:436-448

Ginibre C, Wörner G, Kronz A (2004) Structure and dynamics of the Laacher See magma chamber (Eifel, Germany) from major and trace element zoning in sanidine: a cathodoluminescene and electron microprobe study. J Petrol 45:2197-2223

Ginibre C, Wörner G, Kronz A (2007) Crystal zoning as an archive for magma evolution. Elements 3:261-266

Haggerty SE, Baker I (1967) The alteration of olivine in basaltic and associated lavas. Contrib Mineral Petrol 16:233-257

Hart GL, Streck MJ, Wolff JA (2007) Sr isotope and compositional zoning of plagioclase in the active Mount St. Helens lava dome. EOS Trans AGU, 88(52), Fall Meet. Suppl., abstract V32C-03

Helz RT (1987) Diverse olivine types in lavas of the 1959 eruption of Kilauea volcano and their bearing on eruption dynamics. *In:* Volcanism in Hawaii. Decker RW, Wright TL, Stauffer PH (eds) US Prof Paper 1350:691-722

Hersum TG, Marsh BD (2007) Igneous textures: on the kinetics behind the words. Elements 3:247-252

Hibbard MJ (1981) The magma mixing origin of mantled feldspars. Contrib Mineral Petrol 76:158-170

Hibbard MJ (1995) Petrography to Petrogenesis. Prentice Hall, New Jersey

Humphreys MCS, Blundy JD, Sparks RSJ (2006) Magma evolution and open-system processes at Shiveluch volcano: insights from phenocryst zoning. J Petrol 47:2303-2334

Jerram DA, Davidson JP (2007) Frontiers in textural and microgeochemical analysis. Elements 3:235-238.

Kent AJR (2008) Melt inclusions in basaltic and related volcanic rocks. Rev Mineral Geochem 69:273-331

Kuno H (1936) Petrological notes on some pyroxene-andesites from Hakone volcano, with special reference to some types of pigeonite phenocrysts. Jap J Geol Geograph 13:107-140

Landi P, Francalanci L, Pompilio M, Rosi M, Corsaro RA, Petrone CM, Nardini I, Miraglia L (2007) The December 2002–July 2003 effusive event at Stromboli volcano, Italy: Insights into the shallow plumbing system by petrochemical studies. J Volcanol Geotherm Res 155:263-284

Lipman PW, Dungan MA, Bachman O (1997) Comagmatic granophyrice granite in the Fish Canyon Tuff, Colorado: implications for magma-chamber processes: Geology 25:915-918

Lofgren G (1980) Experimental studies on the dynamic crystallization of silicate melts. *In:* Physics of Magmatic Processes. Hargraves RB (ed) Princeton University Press, p 487-551

Lunney M (2001) Andesite magma evolution based on textural and compositional analysis of plagioclase phenocrysts of Arenal volcano, Costa Rica. MS thesis, Portland State University, Portland, Oregon

Marsh BD (1988) Crystal size distribution (CSD) in rocks and the kinetics and dynamics of crystallization I. Theory. Contrib Mineral Petrol 99:277-291

Milch L (1905) Über magmatische Resorption und porphyrische Struktur. Neues Jb Mineral Geol Paläont 22:1-32

Müller A, Wiedenbeck M, Van Den Kerkhof AM, Kronz A, Simon K (2003) Trace element in quartz – a combined electron microprobe, secondary ion mass spectrometry, laser-ablation ICP-MS, and cathodoluminescenc study. Eur J Mineral 15:747-763

Nakagawa M, Wada K, Wood CP (2002) Mixed magmas, mush chambers and eruption triggers: evidence from zoned clinopyroxene phenocrysts in andesitic scoria from the 1995 eruptions of Ruapehu volcano, New Zealand. J Petrol 43:2279-2303

Nakamura M, Shimakita S (1998) Dissoultion origin and syn-entrapment compositional change of melt inclusion in plagioclase. Earth Planet Sci Lett 161:1998

Nelson ST, Montana A (1992) Sieve-textured plagioclase in volcanic rocks produced by rapid decompression. Am Mineral 77:1242-1249

Nixon GT (1988) Petrology of the younger andesites and dacites of Iztaccíhuatl volcano Mexico: I. Disequilibrium phenocyrst assemblages as indicators of magma chamber processes. J Petrol 29:213-264

Ohba T, Kimura Y, Fujimaki H (2007) High-magnesian andesite produced by two-stage magma mixing: a case study from Hachimantai, Northern Honshu, Japan. J Petrol 48:627-645

Pallister JS, Hoblitt RP, Meeker GP, Knight RY, Siems DF (1996) Magma mixing at Pinatubo: Petrographic and chemical evidence for the 1991 deposits. *In:* Fire and mud: eruptions and lahars of Mount Pinatubo, Philippine. Newhall CG, Punongbayan RS (eds) PHIVOLCS, Quezon City, Philippines and University of Washington Press, p 687-731

Pearce TH (1994) Recent work on oscillatory zoning in plagioclase. *In:* Feldspars and their Reactions. Parsons I (ed) Kluwer Academic Publisher, p 313-349

Pearce TH, Clark AH (1989) Nomarski interference contrast observations of textural details in volcanic rocks. Geology 17:757-759

Pearce TH, Kolisnik AM (1990) Observations of plagioclase zoning using interference imaging. Earth Sci Rev 29:9-26

Pe-Piper G (1984) Zoned pyroxenes from shoshonite lavas of Lesbos, Greece: Inferences concerning shoshonite petrogenesis. J Petrol 25:453-472

Ramos FC, Tepley FJ III (2008) Inter- and intracrystalline isotopic disequilibria: techniques and applications. Rev Mineral Geochem 69:403-443

Robinson DM, Miller CF (1999) Record of magma chamber processes preserved in accessory mineral assemblages Aztec Wash pluton, Nevada. Am Mineral 84:1346-1353

Roeder PL, Emslie, RF (1970) Olivine-liquid equilibrium. Contrib Mineral Petrol 29:275-289

Ruprecht P, Wörner G (2007) Variable regimes in magma systems documented in plagioclase zoning patterns: El Misti stratovolcano and Andahua monogenetic cones. J Volcanol Geotherm Res 165:142-162

Russell JK (1990) Magma mixing processes: insights and constraints from thermodynamic calculations. Rev Mineral Petrol 24:153-190

Rutherford MJ, Devine D, Barclay J (1998) Changing magma conditions and ascent rates during the Soufriere Hills eruption on Montserrat. GSA Today 8:1-7

Rutherford MJ, Devine JD (2003) Magmatic conditions and magma ascent as indicated by hornblende phase equilibria and reactions in the 1995-2002 Soufriére Hills magma. J Petrol 44:1433-1454

Rutherford MJ, Devine JD III (2008) Magmatic conditions and processes in the storage zone of the 2004–2006 Mount St. Helens dacite. *In:* A Volcano Rekindled: The Renewed Eruption of Mount St. Helens, 2004–2006. Sherrod DR, Scott WE, Stauffer PH (eds) U.S. Geol Survey Prof Paper 1750 (in press)

Rutherford MJ, Hill PM (1993) Magma ascent rates from amphibole breakdown: An experimental study applied to the 1980-1986 Mount St Helens eruptions. J Geophys Res 98(B11):19667-19685

Rutherford MJ, Sigurdson H, Carey S, Davis A (1985) The May 18, 1980 eruption of Mount St. Helens. 1. Melt composition and experimental phase equilibria. J Geophys Res 90:2929-2947

Rutherford MJ (2008) Magma ascent rates. Rev Mineral Geochem 69:241-271

Sato H (1975) Diffusion coronas around quartz xenocrysts in andesite and basalt from Tertiary volcanic region in north eastern Shikoku, Japan. Contrib Mineral Petrol 50:49-64

Scaillet B, Evans BW (1999) The 15 June 1991 eruption of Mount Pinatubo. I. Phase equilibria and pre-eruption P-T-fO2-fH2O conditions of the dacite magma. J Petrol 40:381-411

Shore M, Fowler AD (1996) Oscillatory zoning in minerals; a common phenomenon. Can Mineral 34:1111-1126

Singer BS, Dungan MA, Layne GD (1995) Textures and Sr, Ba, Mg, Fe, K, and Ti compositional profiles in volcanic plagioclase; clues to the dynamics of calc-alkaline magma chambers. Am Mineral 80:776–798

Smith JV, Brown WL (1988) Feldspar Minerals. Springer-Verlag, Berlin

Stewart ML, Pearce TH (2004) Sieve-textured plagioclase in dacitic magma: Interference imaging results. Am Mineral 89:348-351

Streck MJ, Broderick CA, Thornber CR, Clynne MA, Pallister JS (2008) Plagioclase populations and zoning in dacite of the 2004-2005 Mount St. Helens eruption: constraints for magma origin and dynamics. *In:* A Volcano Rekindled: The Renewed Eruption of Mount St. Helens. Sherrod DR, Scott WE, Stauffler PH (eds), USGS Prof Paper 1750 (in press)

Streck MJ, Browning-Craig H, Haldar D, Ramos FC, Duncan R (2005a) Hornblende andesites/dacites in an oceanic arc setting at Narcondam volcano, Andaman Sea, S.E. Asia. Geochim Cosmochim Acta Suppl. 69, issue 10 Goldschmidt Conference A647

Streck MJ, Dilles JH (1998) Sulfur evolution of oxidized arc magmas as recorded in apatite from a porphyry copper batholith. Geology 26:523-526

Streck MJ, Dungan MA, Bussy F, Malavassi E (2005b) Mineral inventory of continuously erupting basaltic andesites at Arenal volcano, Costa Rica: Implications for interpreting monotonous, crystal-rich, mafic arc stratigraphies. J Volcanol Geotherm Res 140:133-155

Streck MJ, Dungan MA, Malavassi E, Reagan MK, Bussy F (2002) The role of basalt replenishment in the generation of basaltic andesites of the ongoing activity at Arenal volcano, Costa Rica: evidence from clinopyroxene and spinel. Bull Volcanol 64:316-327

Streck MJ, Grunder AL (1997) Compositional gradients and gaps in high-silica rhyolites of the Rattlesnake Tuff, Oregon. J Petrol 38:133-163

Streck MJ, Grunder AL (1999) Enrichment of basalt and mixing of dacite in the rootzone of a large rhyolite chamber: inclusions and pumices from the Rattlesnake Tuff, Oregon. Contrib Mineral Petrol 136:193-212

Streck MJ, Leeman WP (2008) Complex crustal assembly of "Mt. Shasta" high-Mg andesite (HMA): Evidence from mineral componentry. Geochim Cosmochim Acta 72:A906-A906 Suppl 1

Streck MJ, Leeman WP, Chesley J (2007) High-magnesian andesite from Mount Shasta: A product of magma mixing and contamination, not a primitive mantle melt. Geology 35:351-354

Tepley FJ III, Davidson JP, Tilling RI, Arth JG (2000) Magma mixing, recharge and eruption histories recorded in plagioclase phenocrysts from El Chichón volcano, Mexico. J Petrol 41:1397-1411, doi:10.1093/petrology/41.9.1397

Tepley FJ III, Davidson JP, Clynne MA (1999) Magmatic interactions as recorded in plagioclase phenocrysts of Chaos Crags, Lassen volcanic center, California. J Petrol 40:787-806

Tepper JH, Kuehner SM (1999) Complex zoning in apatite from the Idaho Batholith; a record of magma mixing and intracrystalline trace element diffusion. Am Mineral 84:581-595

Thornber CR (2001) Olivine-liquid relations of lava erupted by Kilauea volcano from 1994 to 1998: Implications for shallow magmatic processes associated with the ongoing East-Rift-zone eruption. Can Mineral 39:239-266

Tomiya A, Takahashi E (2005) Evolution of the magma chamber beneath Usu volcano since 1663: a natural laboratory for observing changing phenocryst compositions and textures. J Petrol 46:2395-2426.

Tsuchiyama A (1985) Dissolution kinetics of plagioclase in the melt of the system diopside-albite-anorthite, and origin of dusty plagioclase in andesite. Contrib Mineral Petrol 89:1-16

Vance JA (1962) Zoning in igneous plagioclase: normal and oscillatory zoning. Am J Sci 260:746-760

Vance JA (1965) Zoning in igneous plagioclase: patchy zoning. J Geol 73:636-651

Vazquez JA, Reid MR (2004) Probing the accumulation history of the voluminous Toba magma. Science 305:991-994

Vernon RH (2004) A practical guide to rock microstructure. Cambridge Univ. Press, Cambridge

Wallace GS, Bergantz GW (2005) Reconciling heterogeneity in crystal zoning data: An application of shared characteristic diagrams at Chaos Crags, Lassen volcanic center, California. Contrib Mineral Petrol 149:98-112

Wark DA, Hildreth W, Spear FS, Cherniak DJ, Watson EB (2007) Pre-eruption recharge of the Bishop magma system. Geology 35:235-238

Wilke M, Behrens H (1999) The dependence of the partitioning of iron and europium between plagioclase and hydrous tonalitic melt on oxygen fugacity. Contrib Mineral Petrol 137:102-114

Decryption of Igneous Rock Textures: Crystal Size Distribution Tools

Pietro Armienti

Dipartimento di Scienze della Terra
via S. Maria 53
Pisa, Italy
armienti@dst.unipi.it

In this overview it is planned to show how quantitative determination of mineral modes data, by means of Crystal Size Distribution (CSD), may help in deciphering the kinetics of magmatic crystallization both in natural and experimental conditions. Some stereological considerations will be discussed to provide criteria for the choice of numbers and lengths of size intervals for CSD computations. Equations for the balance of number and size of crystals, in steady state crystallization and in dynamic regime, will be used to define the time frame of magmatic processes, in the context of the case studies of Stromboli and Etna volcanoes.

The functional dependence of crystal size (L) and crystal number from time is explicitly derived by interpreting experiments of olivine crystallization that report CSD data, thus obtaining evaluation of thermodynamic parameters from CSD measurements.

CRYSTAL SIZE DISTRIBUTIONS

Modal data obtained from rock sections can be represented as a Particle Size Distribution: $N(L)$ that is the number of objects (crystals, bubbles) of given size L contained in the unit volume. $N(L)$ is usually displayed in semi-logarithmic diagrams vs. a characteristic length of the grains (Fig. 1; Marsh 1988). In the following discussion the size L is the diameter D for spherical particles or the diameter of the sphere with the equivalent volume of the particle. In igneous petrology, Crystal Size Distributions (CSD) and Bubble Size Distributions (BSD) can be related to the cooling and degassing history of a magmatic body since:

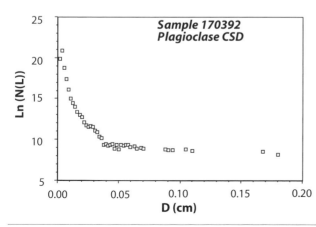

Figure 1. Particle Size Distribution: $N(L)$ represents the number of objects (crystals, bubbles) of given size L contained in the unit volume. $N(L)$ is usually displayed in semi-logarithmic diagrams vs. a characteristic length of the grains. Sample 170392 is taken from the flowing lava flow during the 1991-93 eruption of Mt. Etna (Armienti et al. 1994).

- crystal and bubble populations are generated by nucleation and growth processes that are strongly dependent on the degree of undercooling (or supersaturation) and, ultimately, by the temporal evolution of pressure and temperature paths of magma bodies.
- the number of crystals and bubbles which nucleated and grew in the magma can be balanced to account for the loss or gain of grains due to some mechanism of transport.

This allows the extraction of *temporal information* from lava modes if we know how crystals and bubbles nucleate and grow during magma transport and cooling processes.

Since the number of crystals depends on the process of rock formation, CSDs may be profitably used to describe and model also metamorphic and sedimentary rocks, accounting for the laws that allow grain nucleation and growth or transport and deposition.

DETERMINATION OF THE CURVES OF POPULATION DENSITY

Population density $N(L)$ is defined as the number of particles in each size interval divided by the size of the interval, like in Figure 2—shortening the interval, the histograms tend to a continuous function. Note that $N(L)$ is to be determined for each class of size for the unit volume. So, $N(L)$ will depend on ΔL and will tend to zero for infinitesimal intervals. Often measurements of $N(L)$ for very small classes are difficult, thus it is preferred to determine the cumulative crystal number

$$n(L) = \int_0^L N(L)dL \qquad (1)$$

Obtaining $N(L)$ through derivation of the cumulative curve

$$N(L) = \frac{dn(L)}{dL} \qquad (2)$$

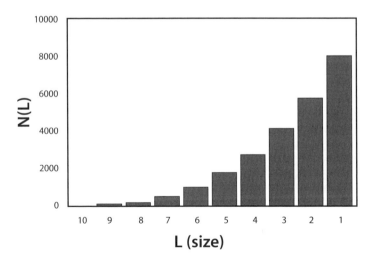

Figure 2. Population density $N(L)$ represents the number of particles in each size interval divided by the size of the interval.

A graphic example of the derivation of the cumulative curve $N(L)$ is reported in Figure 3. The most common methods to obtain cumulative curves of particle sizes are by sieving of loose materials or by *stereologic* correction of a distribution of areas measured in rock sections to get a distribution of sizes in the space.

Planar distributions are measured by collecting data related to features of the particles such as diameters or radii (for spherical particles), cord lengths (for generic shapes), and areas (for generic shapes). All these features may be obtained through handling an image of a rock, in which we distinguish the phase of interest from the groundmass and the other phases. Image analysis software provides methods of *segmentation* of the images which puts in evidence only the phase of interest. Commercial packages like Visilog ® or open source codes like Image-J are particularly suitable to handle images of rocks that may be acquired by cameras mounted on a microscope, or from normal scanners. These may be easily modified to hold polished rock slabs, while film scanners may be equipped with polaroids to acquire images from rock thin sections (Armienti and Tarquini 2002).

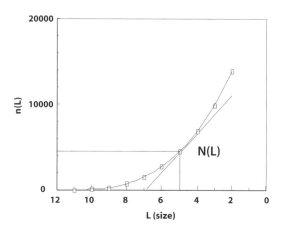

Figure 3. Derivation of the cumulative curve $N(L)$.

Analysis of distributions of areas

Segmentation puts in evidence the pixels belonging to the phase of interest. Segmentation is easier if there is a strong contrast between the phase and the background. In Figure 4 is a BW image of Hawaiitic lava from Mt. Etna where magnetite is black. Selection of a suitable threshold promptly allows acquisition of a binary map of its distribution where pixels belonging to magnetite are set to equal 1 (black) and the remaining are set to equal 0 (white). In the example of Figure 5, plagioclase, olivine, clinopyroxene and vesicles appear to be light colored and the lower and high grey level thresholds were chosen to separate them (white) from the glass and magnetite. More complex situations (e.g., distinction among light colored phases) require some manipulation of images and different techniques of acquisition.

The final goal of measuring size distribution requires a further step, involving assumptions of the grain shapes and a suitable stereologic treatment of planar data must be made to access information in the space.

Stereologic reconstruction of CSD: monodispersed distributions

The reconstruction of the spatial distribution of crystals in rocks is based on the collection of planar data collected on the surface of rock sections. This requires that some assumptions on the grain shapes must be made *a priori* since spatial distributions of particles with different sizes and shapes can generate the same planar size distribution (Fig. 6). We will assume in the following treatment that minerals and bubbles can be approximated as spherical particles, and the analysis which follows is applied to the radii of the circles of equivalent area. This approximation involves relevant deviation from reality only for strongly elongated minerals but is sufficient for most of the minerals occurring in lavas (Higgins 2000, 2002; Morgan and Jerram 2006).

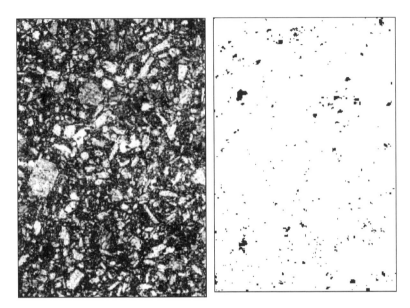

Figure 4. An example of segmentation through thresholding. (*left*) A BW image of Hawaiitic lava from Mt. Etna where magnetite is black. (*right*) Acquisition of a binary map of its distribution where pixels belonging to magnetite are set to equal 1 (black) and the remaining are set to equal 0 (white).

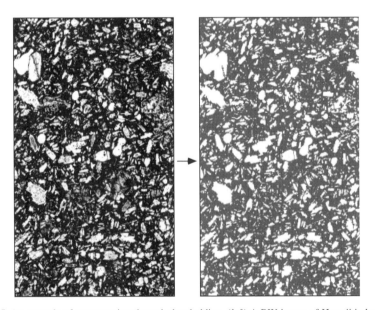

Figure 5. An example of segmentation through thresholding. (*left*) A BW image of Hawaiitic lava from Mt. Etna where magnetite is black. Plagioclase, olivine, clinopyroxene and vesicles appear to be light colored. (*right*) Lower and high grey level thresholds were chosen to separate them (white) from the glass and magnetite.

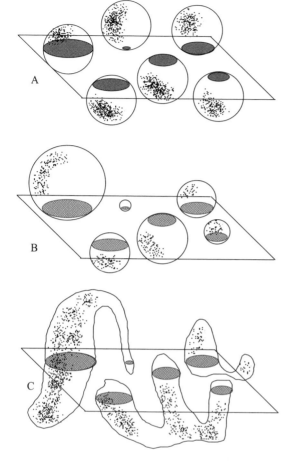

Figure 6. *Case A* is that of a mono-dispersed distribution of spheres (all have the same diameter); this situation may occur in well sorted sedimentary rocks, in thermometamorphic rocks in correspondence of a single nucleation event, as well as in artificial aggregates of materials, like concrete. *Case B* illustrates how the same planar distribution can be generated by a poly-dispersed distribution of diameters. *Case C* illustrates the distribution of areas is the same as in A and B, but only one section cannot give information on the spatial distribution.

For the simple case of a mono-dispersed distribution of spheres of diameter D_j the following relation holds:

$$N_V = N_A / D_j \qquad (3)$$

where N_V is the number of particles per unit volume, N_A is the number of particles per unit area, and D_j is the constant diameter of spheres. In Figure 7 only the spheres whose center is at a distance less than $D_j/2$ are cut by the section (black line) of unit area. The number of sections per unit area is N_A and consequently the number per unit volume results to be N_A/D_j.

Stereologic reconstruction of CSD: poly-dispersed distributions

When dealing with crystals and vesicles nucleated at different times, we can assume that planar distributions are generated by particles of different sizes cut at different distances from their center (case B in Fig. 6). There are several methods to determine, in the space, a CSD of a dispersed phase from 2D information (Russ 1986). A simple conversion from apparent 2D to volumetric size population may be performed considering the planar distribution as if it was collected in the space, by rising N_A to the power 3/2 (Kirkpatrick 1977a)

$$N_V = N_A^{3/2}$$

In this way, the contribution of larger spheres to smaller areas is ignored. In fact, small circles on a section can be originated by small spheres but also by larger ones, cut at distances from their centers almost equal to their radii (Fig. 8). Trivially smaller spheres cannot contribute to larger size classes.

A method of "unfolding" of the areal size distribution was independently by Schwartz (1939) and Saltykov (1949) and is now identified as the Schwartz-Saltykov method (DeHoff and Rhines 1972). It is particularly suitable for digitized images where particle areas can assume only discrete values, as some multiple of the pixel resolution. The algorithm is based on the consideration that the diameter of a particle in a plane depends on where the section cuts it, thus a frequency histogram for circle sizes obtained by random sectioning of a sphere can be computed.

If we measure in the plane a section of size d_i, let:

- $p(i,j)$ be the probability that a sphere of diameter $D_j > d_i$ is cut in such a way to produce a section of diameter d_i.
- $N_A(d_i, D_j)$ be the number of sections of diameter d_i (measured) per unit area obtained from spheres of diameter D_j.

The true number of spheres of diameter D_j per unit area is then given by

$$N_A(D_j) = \frac{N_A(d_j, D_j)}{p(j,j)} \qquad (4)$$

and the problem to solve is to determine the values of p.

Assuming a "suitable" class interval Δ we can generate j classes by cutting a sphere of diameter D_j at a distance Δ from each other. The probability of cutting a section with a

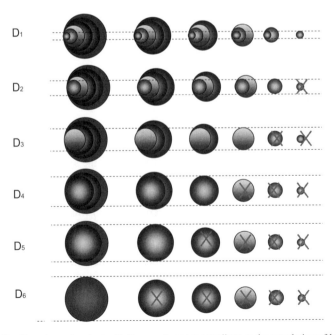

Figure 8. Smaller spheres (marked with the cross) cannot contribute to the population of larger ones.

diameter between $d_i = i \times \Delta$ and $d_{(i-1)} = (i \times 1) \times \Delta$ (with $0 < i \le j$), is directly proportional to the thickness h of the sections and inversely proportional to the radius of the sphere:

$$p(i,j) = \frac{h}{D_j/2} = \frac{h_{(i-1)} - h_i}{D_j/2} \qquad (5)$$

where $h_{(i-1)}$ and h_i are the distances from the center of the sphere to which diameters $d_{(i-1)}$ and d_i respectively correspond (Fig. 9). On the basis of this probability, the Schwartz-Saltykov method to "unfolds" the contribution of larger spheres to the population of smaller ones by using a technique of iterative subtraction.

In order to express Equation (5) as a function of i, j and D_i from Figure (5) above we observe that

$$h_i = \sqrt{\left(\frac{D_j}{2}\right)^2 - \left(\frac{d_i}{2}\right)^2} \qquad (6a)$$

from which

$$h_i = \sqrt{\left(\Delta \cdot \frac{j}{2}\right)^2 - \left(\Delta \cdot \frac{i}{2}\right)^2} \qquad (6b)$$

and after substitution in Equation (5) and a little manipulation

$$p(i,j) = \sqrt{1 - \left(\frac{i-1}{j}\right)^2} - \sqrt{1 - \left(\frac{i}{j}\right)^2} \qquad (7a)$$

$$p(i,j) = \frac{\sqrt{\left(D_j/2\right)^2 - \left(h_{i-1}/2\right)^2} - \sqrt{\left(D_j/2\right)^2 - \left(h_i/2\right)^2}}{D_j/2} \qquad (7b)$$

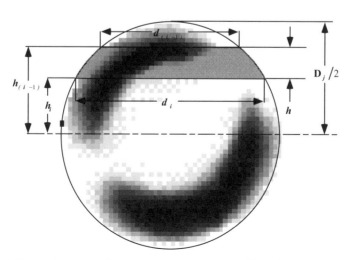

Figure 9. Relations between surfaces cut from a sphere and probability of cutting definite values.

In Figure 10, the values of $p(i,j)$ are computed for twenty distinct classes ($j = 20$). It is evident that the largest contribution of class j is to the class j itself.

By substitution of p, Equation (4) then becomes:

$$N_A(D_j) = \frac{N_A(d_j, D_j)}{p(j,j)} = \frac{N_A(d_j, D_j)}{\sqrt{1-\left(\frac{j-1}{j}\right)^2} - \sqrt{1-\left(\frac{j}{j}\right)^2}} \tag{8}$$

which provides the number of spheres of diameter $d_j = D_j$ per unit volume, while the number of spheres with apparent diameter $d_i < D_j$ provided by spheres of diameter D_j is

$$N_A(d_i) = \frac{N_A(d_i, D_j)}{p(i,j)} = \frac{N_A(d_i, D_j)}{\sqrt{1-\left(\frac{i-1}{j}\right)^2} - \sqrt{1-\left(\frac{i}{j}\right)^2}} \tag{9}$$

That guarantees the possibility of calculating the values on $N_V(D_j)$ for each class given that at least one $N_V(D_j)$ is known. Consider Figure 7 with six classes of spheres $D_1, D_2, \ldots, D_6$: we do know the value of $N_V(D_6)$ of the largest class (the one that contributes only to itself)

$$N_V(D_6) = \frac{N_A(d_6, D_6)}{p(6,6)} \frac{1}{D_6} \tag{10}$$

Equation (8) then allows one to evaluate the contribution of the class D_j to the class D_{j-1} and so on, until the smallest class D_1. If this operation is accomplished for each size interval the unfolding will yield the spatial distribution of each class of spheres. For class 5

$$N_A(d_5) = \frac{N_A(d_5, D_5)}{\sqrt{1-\left(\frac{5-1}{5}\right)^2} - \sqrt{1-\left(\frac{5}{5}\right)^2}} - \frac{N_A(d_5, D_6)}{\sqrt{1-\left(\frac{5-1}{6}\right)^2} - \sqrt{1-\left(\frac{5}{6}\right)^2}} \tag{11}$$

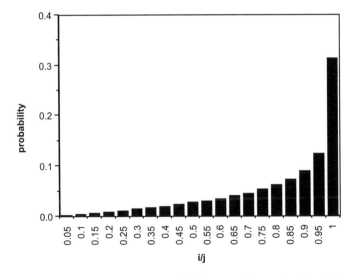

Figure 10. Distribution of probability of a class to contribute to the class itself.

For class 1

$$N_A(d_1) = N_A(d_1,D_1) + N_A(d_1,D_2) + N_A(d_1,D_3) + N_A(d_1,D_4) + N_A(d_1,D_5) + N_A(d_1,D_6) \quad (12)$$

and in general

$$N_A(d_i) = \sum N_A(d_i,D_j) \quad (13)$$

which simply states that the contribution to class i in the section is due the number of sections of class 1 plus the sum of contributions due to larger spheres. Since $N_A(d_6,D_6)$ is measured we can obtain $N_A(d_6,D_6)$ from Equation (10) and Equation (7) provides N_V of the particles of group 6 (Eqn. 10), which in turn allows one to calculate $N_A(d_5,D_5)$. Then

$$N_A(d_5,D_5) = N_A(d_5) - N_A(d_5,D_6)$$

is computed until unfolding is completed for the contribution to the smallest class of all the other particles.

The "best" CSD for a set of unfolded planar data

It is evident that the shape assumed by the CSDs of the same set of planar sections is strongly controlled by the number of intervals and by the size of the first class. Figure 11 and Table 1 report the different CSDs obtained for the same set of 3497 planar sections of crystals (total area of 3693 mm²) measured on a slab of 6587 mm². The Delesse theorem (1847) ensures that the volume fraction of the measured particles corresponds to the area fraction (vol% = 100 × 3693/6587 = 56.10%). On the basis of which elements can one choose the most suitable partitions of a data set?

We can start considering that two fundamental measured quantities must be conserved in the CSD, namely the volume fraction and the total number of particles. However when arranging the binning of our data set, these two values must be conserved. We can obtain these two quantities from CSDs: in each class of size L_i, the number of sections of that size is given by the product of their volumetric number density [$N_V(L_i)$] and the sampled volume for particles of that size [$Area \times L_i$]. Thus the total number of particles of a given CSD is:

$$N_{tot} = Area \sum_i L_i N_V(L_i) \quad (14)$$

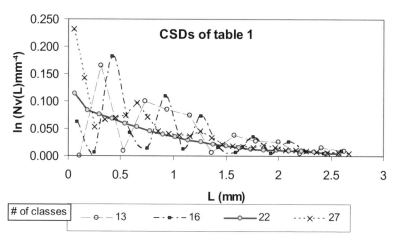

Figure 11. Choice of the best possible distribution.

Table 1. Different CSDs computed from the same areal distributions.

	$\Delta L = 0.210$			$\Delta L = 0.168$			$\Delta L = 0.120$			$\Delta L = 0.101$		
	L	$N_a(L)$	$N_v(L)$	L	$N_a(L)$	$N_v(L)$	L	$N_a(L)$	$N_v(L)$	L	$N_a(L)$	$N_v(L)$
class	(mm)	(mm^{-3})	(mm^{-4})	(mm)	(mm^{-3})	(mm^{-4})	(mm)	(mm^{-3})	(mm^{-4})	(mm)	(mm^{-3})	(mm^{-4})
1	0.105	91	0.000	0.084	91	0.062	0.060	91	0.114	0.050	91	0.233
2	0.315	438	0.165	0.252	190	0.006	0.180	190	0.083	0.151	0	0.142
3	0.525	275	0.010	0.420	523	0.182	0.300	248	0.076	0.252	190	0.052
4	0.735	562	0.100	0.588	283	0.043	0.420	275	0.068	0.352	248	0.067
5	0.945	517	0.085	0.756	279	0.014	0.540	283	0.060	0.453	275	0.071
6	1.154	438	0.075	0.923	517	0.109	0.660	279	0.052	0.554	283	0.074
7	1.364	186	0.006	1.091	230	0.012	0.779	267	0.046	0.655	279	0.097
8	1.574	308	0.037	1.259	394	0.073	0.899	250	0.040	0.755	0	0.071
9	1.784	236	0.027	1.427	164	0.015	1.019	230	0.034	0.856	267	0.046
10	1.994	192	0.025	1.595	144	0.006	1.139	208	0.029	0.957	250	0.042
11	2.204	80	0.005	1.763	236	0.035	1.259	186	0.025	1.057	230	0.038
12	2.414	117	0.015	1.931	101	0.004	1.379	164	0.021	1.158	208	0.037
13	2.624	57	0.009	2.099	171	0.025	1.499	144	0.017	1.259	186	0.045
14				2.267	66	0.007	1.619	126	0.015	1.359	0	0.033
15				2.435	51	0.005	1.739	110	0.012	1.460	164	0.022
16				2.602	57	0.010	1.859	101	0.011	1.561	144	0.019
17							1.978	91	0.010	1.662	126	0.016
18							2.098	80	0.010	1.762	110	0.015
19							2.218	66	0.008	1.863	101	0.020
20							2.338	51	0.007	1.964	0	0.015
21							2.458	33	0.005	2.064	91	0.011
22							2.578	24	0.005	2.165	80	0.010
23										2.266	66	0.009
24										2.366	51	0.008
25										2.467	33	0.007
26										2.568	0	0.006
27										2.669	24	0.005

Computed values

N_{tot}		3497			3497			3497			2937	
% vol		60.70			58.63			53.69			31.8	

Measured values

Area = 6587 mm^2 N_{tot} = 3497   Area Particles = 3693 mm^2 % area = % vol = 56.10%

Notes: $N_{tot} = \Sigma[N_V(L_i) \times L_i] \times Area$; % vol = $100 \times \Sigma[N_V(L_i) \times L_i^3] \times \pi/6$

Similarly, the volume fraction of particles of size L_i is the product of their volume $[4/3\pi(L_i/2)^3]$ and their number in the volume sampled for that class size $[Area \, L_i N_V(L_i)]$ divided the sampled volume $[Area \, L_i]$. Thus the volume fraction of all the particles of a given CSD is:

$$V_f = \sum N_V(L_i) \frac{4}{3}\pi \left(\frac{L_i}{2}\right)^3 = \frac{\pi}{6}\sum L_i^3 N_V(L_i) \tag{15}$$

or

$$V\% = \frac{100\pi}{6}\sum L_i^3 N_V(L_i) \tag{16}$$

Equations (14) and (16) allow an easy choice of the "best" CSD that results so that the difference between the number of computed sections and the number particles derived from

CSD (Eqn. 14) and the difference between the volume fraction derived from the section areas and that derived from CSD (Eqn. 16) are minima.

Truncation effects

CSD data may show truncation effects arising from resolution (left-hand truncation - LHT) of the adopted measurement device and from the size of the specimen (right-hand truncation - RHT). Evaluation of left-hand and right-hand truncation effects on the CSD is essential before attempting any interpretation (Pickering et al. 1995). Left-hand truncation arises from the resolution of the image analysis procedure as a whole and is revealed by a decrease in the number density of the counted grains. This truncation is reduced when data are acquired with a greater resolution (Armienti et al. 1994). It is important to observe that resolution is controlled both by the resolution of the starting image and by errors introduced by image processing. Thus, due to the heavy image manipulations needed for image segmentation, in spite of the high resolution of starting images, often CSDs lack LHT effects only for crystal sizes much larger than resolution (Fig. 12). Any interpretation of a "true" reduction of the number density of the smallest grains as due to some processes require the acquisition of images with a higher resolution.

Right-hand truncation produces the flattening of size distributions on nearly horizontal trends at large size ranges in which not all the classes of size occur in the sections (Fig. 12). This results in a constant value of the number density for classes that contain at most one particle.

A logarithmic binning of size classes (i.e., logarithmic variation of the width of the class size) allows to partly avoid this effect by counting a larger number of grains in larger (and wider) classes, but the effect still appears at the limit of the distribution with a decrease of the slope at the right hand of the distribution. Right-hand truncation effects can be limited by enlarging the size of the measured sample. This is obtained by measuring CSDs of several images of the rock in different positions. In general the statistic reliability of data has always to be checked before attempting any interpretation. In the case of particle size distributions we must be sure to analyze an area that is sufficient to guarantee that the largest particle detected has the "true" number density and is not affected by RHT effects, otherwise the CSD plots will be "sparsely" populated by points at larger diameters. In this case the $N(L)$ does not represent measurement of the number density but is related to "isolated" casual findings of particles of that size in the explored section.

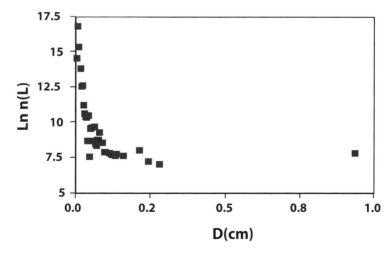

Figure 12. Right-hand truncation effect.

The minimum area to be analyzed depends on the number of particles and the width of the classes L according to the formula (Armienti et al. 1994)

$$Area = \frac{1}{2N(L)\,\Delta L\,\sqrt{L\,\Delta L}} \qquad (17)$$

for the data in Figure 13a, in which larger bubbles have $L = 0.15$ cm, $\Delta L = 0.0036$ cm, and $N(L) \approx e^8$ cm^{-4}. The minimum area must not be less than 2 cm^2 to have a statistical significance. In that case the analyzed area was 3 cm^2. This implies that the continuous distribution up to 0.2 cm is statistically reliable, that the isolated point at about 1 cm must be discarded and that a larger area is needed to obtain the "true" $N(L)$ of bubbles larger than 0.2 cm. The problem can be solved by using different samples for different grain sizes—thin sections, polished bricks and photos of the outcrops—then merging the size distributions obtained at different scales. This was done

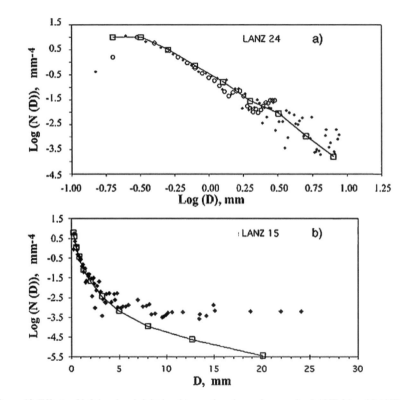

Figure 13. Effects of left-hand and right-hand truncation shown for samples LANZ 24 and LANZ 15. (a) The typical effect of left-hand truncation is to reduce the number of counted individual crystals near the resolution limit of the acquisition procedure. This can be observed both in the CSDs computed from one thin section (circles) and from 10 thin sections (dots). Right-hand truncation arises from the limited size of the explored thin section that allows the detection of only a few (at the limit only one) particles of large size in each size interval: this results in the constant value of $N(D)$ revealed by the flattening of the CSD trend (empty circles). For the same choice of size interval, the effects may be avoided by analyzing a larger population of crystal (dots). A different choice of size intervals would also reduce this effect as is evident in the CSDs obtained from computed logarithmic binning of the size interval (squares). (b) Right-hand truncation may not disappear, even when handling a large population of crystals (small diamonds) if the number of size intervals is too large: this is particularly evident in sample LANZ 15, where CSD is plotted against D. In this case, right-hand truncation is revealed by the constant values of $\log(N(D))$ and by the shift of linearly binned CSD with respect to the logarithmic one. From Armienti and Tarquini (2002).

for the data in Figure 14, which reports the Bubble Size Distributions in a lava flow from Mt Etna. Note the different sizes at which RHT effects appears when larger areas are acquired. The adoption of a logarithmic binning for the calculation of CSD on a single section may also ensure good results since it increases the value of the denominator of Equation (17), but it is important to check that CSDs satisfy the conditions imposed by Equations (14) and (16).

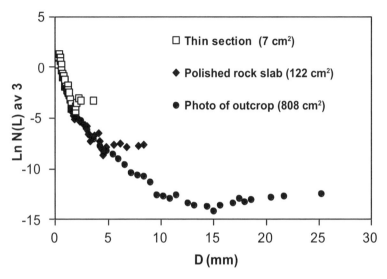

Figure 14. Bubble Size Distributions in a Mt. Etna lava flow collected with different tools (thin section, a polished slab, a photo of the outcrop) that explore different total area, providing different RHT. Note overlap of dominions not affected by under-sampling.

CSDs AND THE PETROGENETIC CONTEXT

Crystal Size Distributions (CSD) are suitable to be interpreted in a petrologic context since i) crystal populations are generated by nucleation and growth processes that are strongly dependent on the degree of undercooling (and thus on the thermal history); ii) the number of crystals which nucleated and grew in the magma can be balanced to account for the loss or gain of grains due to some mechanism of transport (fractionation, mixing).

The first step in the interpretation of CSD diagrams is to write down equations that balance the numbers of crystals

$$\begin{array}{c}\text{Number of crystals}\\\text{in unit volume of size}\\\text{between } L \text{ and } L + \Delta L\end{array} = \begin{array}{c}\text{Growth Input} - \text{Growth Output}\\+\\\text{Flux In} - \text{Flux Out}\end{array}$$

or

$$\left(V_2 N_2 - V_1 N_1\right)\Delta L = \left(G_1 V_1 N_1 - G_2 V_2 N_2\right)\Delta t + \left(Q_i N_i - Q_o N_o\right)\Delta L \Delta t$$

that has the differential form

$$\frac{\partial(VN)}{\partial t} + \frac{\partial(GVN)}{\partial L} = Q_i N_i - Q_o N_o \tag{18}$$

where $N(L)$ = crystal density ($N(L)dL$ = number of crystals per unit volume with size between

L and $L+ dL$), G = crystal growth rate; V = volume; L = crystal size (e.g. diameter); t = time; $Q_i n_i - Q_o n_o$ = Flux in − Flux out.

Equation (18) can be solved under different assumptions which result in different analytical solutions. I will illustrate two different approaches to the solution of Equation (18) which can be related to different petrogenetic mechanisms.

Continuously taped and refilled magma chamber: the Stromboli example

This situation may occur in stationary systems like a fractionating magma chamber, continuously refilled and taped by eruptions. The typical case is represented by the activity of Stromboli (Armienti et al. 2007).

In this case, Equation (18) may be solved with the following assumptions:

a) Crystallization takes place in a magma chamber of volume V
b) which is refilled at the rate Q
c) by a crystal free magma
d) and is continuously taped by eruptions at rate Q.

In this case,

$$In\ flux\ of\ crystals = QN_i = 0 \quad (crystal\ free\ magma \Rightarrow N_i = 0)$$
$$Out\ flux\ of\ crystals = QN_o = VN_o / \tau \quad (\tau = time\ of\ recharge)$$

which allows us to simplify Equation (18) into

$$\frac{\partial(GVN)}{\partial L} = \frac{N}{\tau} \qquad (19)$$

For constant G (Marsh 1988)

$$N = N_o \exp(-L/G\tau) \quad \text{or} \quad \ln(N) = \ln(N_o) - L/G\tau$$

where $\tau = V/Q_{out}$ is defined as the time whereupon the system of volume V is emptied or totally exchanged at a flux rate of Q_{out}, growth rate G does not depend on L, and the integration constant N_o represents the density of crystal nuclei (i.e. the value of N as $L > 0$). In this case, in a plot of $N(L)$ respect to L, the slope of a linear distribution of $N(L)$ is linked to the product $G\tau$ by the relation slope= $-1/(G\tau)$ and, if G or τ are known, the other variable may be estimated by measuring the slope of the distribution. The linear distribution of CSDs provides information on the characteristic time of recharge of the system. In the example of Figure 14 (Stromboli) the slope is -6.6242 mm^{-1} that for a growth rate on the order of 10^{-9} cm sec^{-1} would give a residence time of $\sim 1.5 \times 10^7$ sec, which corresponds to about 174 days.

The linear relationship of Figure 15 is typical of Stromboli and and additional data reveal almost constant slopes for at least 1000 years (Armienti et al. 2007). However, even in such an apparently simple system, the stationary condition of equilibrium is achieved through continuous shifts from equilibrium and determination of the magma residence time through the evaluation of the CSD slope may not be so direct as it appears.

The residence time of the Stromboli shallow feeding system has been estimated to be 19 years on the basis of Sr isotopic data (Francalanci et al. 1999). This estimate of residence time would imply a plagioclase growth rate on the order of 2×10^{-11} cm s^{-1}. This seems too low considering the results obtained on basaltic systems which average 10^{-9}-10^{-10} cm sec^{-1} (Kirkpatrick et al. 1976; Lofgren 1980; Armienti et al. 1994). However, the obtained low growth rate of the plagioclase must be the result of an average behavior which accounts for the peculiar textural and chemical zoning of the plagioclase. In fact, plagioclase crystals in the crystal-rich scoriae show alternating layers of labradoritic and bytownitic composition, that allow one to recognize distinct recurrent phases of growth and dissolution (Landi et al. 2004).

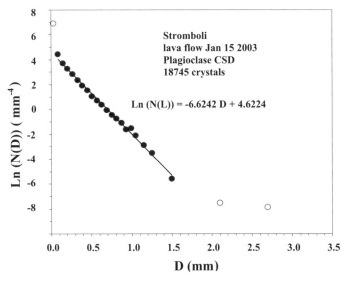

Figure 15. CSD of a Stromboli lava flow. Data were collected on several sections. Note the RHT effect on larger crystals in spite of the arithmetic increase of the binning of size classes.

The process suggested by Figure 16 leads one to assume that the growth rate G derived from CSD data is a mean value resulting from growth and dissolution episodes:

$$\overline{G} = G^+ f^+ + G^- f^- \quad (20)$$

where G^+ and G^- are mean growth and dissolution rates, respectively, and f^+ and f^- are the volume fractions of the system in which crystals grow or dissolve respectively. For a system that behaves like a crystallizing system, we can assume f^+ to be proportional to the volume fraction of magma supplied in a characteristic time:

$$f^+ = (V - Q \times \Delta t)/V \quad (21)$$

where Q is the magma supply rate, which in a steady system is equal to the mass eruption rate, and Δt is a characteristic time during which resorption events may occur due to mixing of undegassed magma with the resident magma.

From the obvious relations $f^+ + f^- = 1$ and $V/Q = \tau$, it follows:

$$f^+ = 1 - \Delta t/\tau \quad \text{and} \quad f^- = \Delta t/\tau \quad (22)$$

from which

$$\overline{G} = G^+\left(1 - \frac{\Delta t}{\tau}\right) + G^-\left(\frac{\Delta t}{\tau}\right) \quad (23)$$

which states that the net growth rate is a mean of positive and negative (i.e., resorption) growth rates weighted for the times in which the two conditions hold. The internal consistency of Crystal Size Distribution and residence times based on isotopic disequilibria may be assessed by means of the above equation. Assigning different values for G^+, the value of G^- may be computed from Equation (23) for different choices of Δt. In correspondence with these characteristic times of dissolution, we can compute the amount of dissolution allowed for crystals ($G^- \times \Delta t$). Thus a plot $\Delta t/\tau$ vs. ($-G^- \times \Delta t$) for different G^+ (Fig. 17) will provide an estimate of the characteristic size of the crystal that will be completely resorbed. It is worth

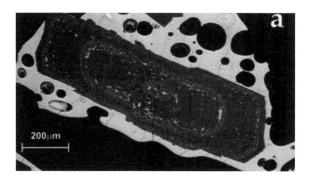
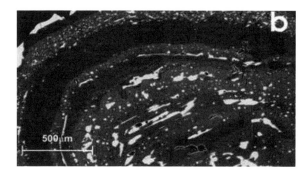

Figure 16. Back-scattered electron images of plagioclase phenocrysts. a) Characteristic zoning of the plagioclase, consisting of alternating layers of labradorite (dark grey) and sieve-textured, bytownite (clear grey). b) Enlargement showing small-scale (1-5 μm) oscillatory zoning in labradoritic layers and sieved textures with abundant glass inclusions in bitownitic layers. A dissolution surface which indents for several tens of μm a labradoritic layer is evident in the external part. From Armienti et al. (2007).

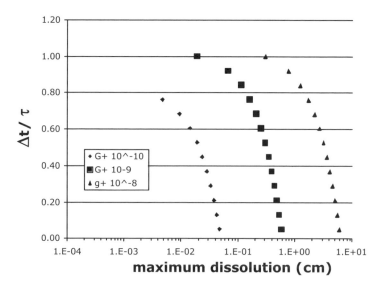

Figure 17. Plot of $\Delta t/\tau$ vs. $-G^- \times \Delta t$ from Equation (23). The diagram provides the maximum size of crystals that can be completely resorbed for given values of $\Delta t/\tau$. $\tau = 19$ years and different values for plagioclase growth rate. See text for major details. From Armienti et al. (2007).

noting that crystals, with sizes smaller than the amount of dissolution allowed, will not show resorption features in so far as they are completely consumed during this interval.

Armienti et al. (2007) report that for $\bar{G} = 10^{-8}$ cm/s crystals of 1 cm size would be completely resorbed. Thus it would be impossible to observe any resorption features in phenocrysts less than 1 cm in diameter. Or, if 1 cm crystls are not completely resorbed, then $\bar{G}$ must be 10^{-9} cm/s for most $\Delta t / \tau$ values. To resorb only millimeter-sized crystals the volume fraction in which the process should take place would be too large. Since only crystals in the size range of a few hundreds of μm do not show resorption, an order of magnitude of $G^+ = 10^{-10}$ cm/s must be assumed for plagioclase growth rates in the Stromboli feeder conduit. This implies also that the plagioclase nucleation rates result from a mean behavior of the system and that the observed intercepts of CSDs are the balance of resorption and nucleation bursts.

Cooling of a magma batch: the case of Mount Etna

CSDs evolve in a completely different way if the crystallizing system is not in a stationary state but evolves due changes of temperature and/or undercooling. Such a system can be described with the following conditions:

a) Magma is a batch of volume V
b) With no migration of crystals (no fractionation).
c) The only input of crystals to a given class is due to nucleation and growth.
d) Growth rate does not depend upon size L of crystals.

Under these limiting assumptions Equation (18) reduces to:

$$\frac{\partial(VN)}{\partial t} + G\frac{\partial(VN)}{\partial L} = JVd(L) \qquad (24)$$

where:

G = growth rate (here not dependent on crystal size L)
J = nucleation rate
$d(L)$ = Dirac function (= 0 for $L > 0$ and =1 for $L = 0$)
(thus accounting for crystal input only due to nucleation effects).

By solving the system

$$dt = \frac{d(L)}{G} = \frac{d(VN)}{J(t)V(t)\,dL} \qquad (25)$$

we find

$$N(t,L) = \frac{J(t_L)V(t_L)}{G(t_L)V(t)} \qquad (26)$$

where t_L is a time such that in the interval t-t_L crystals belonging to the class between L and $L+\Delta L$ at time t, grew from 0 to L. $V(t)$ and $V(t_L)$ are the volumes of the liquid fractions in which crystallization took place at t and t_L respectively.

The ratio $V(t_L)/V(t)$ is a function of the total crystallinity and in many cases can be neglected. For example, for Etnean lavas the Porphiritic Index is always between 20 and 30% (Armienti et al. 1994), thus $1.25 < V(t_L)/V(t) < 1.43$ and in the worst case $\ln(1.43) = 0.36$ that implies a correction of 0.36 for J/G. Thus the effect of the factor $V(t_L)/V(t)$ in most lavas can be safely neglected and Equation (26) is simplified to

$$N(t,L) = \frac{J(t_L)}{G(t_L)} \qquad (27)$$

All the crystals simultaneously undergo the same variations of growth rate and this implies that crystals born in different times cannot have the same size. This trivial observation has many important consequences and allows one to use Equation (27) for petrologic interpretations. In fact, the condition that crystal size depends on time spent to grow and that $N(L)$ reflects the ratio J/G at the time in which crystals of size L appeared, allow one to conclude that: different trends of $N(L)$ vs. L reflect changes in the ratio J/G during crystallization and, due to dependence of J and G on undercooling, a plot of $N(L)$ vs. L corrresponds to a plot of undercooling vs. time.

Dependence of growth and nucleation rates from the undercooling ΔT may be explicitly defined (Cashman 1990; Toramaru 1991):

$$J \propto \exp\left(\frac{-E^*}{R_g T_m}\right) \exp\left(\frac{-16\pi\gamma V_c^2 T_o^2}{3k T_m \Delta h^2 (T_m - T_o)^2}\right) \qquad (28)$$

$$G \propto \exp\left(\frac{-E}{R_g T_m}\right)\left[1 - \exp\left(\frac{\Delta h(T_m - T_o)}{k T_o T_m}\right)\right] \qquad (29)$$

where V_c is the volume of one molecule of the crystallizing phase, R_g is the gas constant, γ is the interfacial tension between crystal and melt, T_m is magma temperature, T_o is the liquidus temperature, $\Delta T = T_m - T_o$, Δh is the enthalpy of fusion per molecule (= $k\Delta H / R_g$, where ΔH is the molar enthalpy of fusion), k is the Boltzman constant, E^* and E are the activation energies per mole. The shape of the two functions is reported in Figure 18. The two expressions rapidly rise reaching maxima which do not necessarily occur at the same value of ΔT, then rapidly drop when the undercooling is very high and the system acquires a large viscosity.

At small undercooling (the stippled area of Fig. 18) Maaløe et al. (1989) suggest, for growth and nucleation rates respectively, empirical laws to substitute for the above equations

$$G = A [\exp(B\Delta T - 1)]$$

and

$$J = C [\exp(D\Delta T - 1)]$$

from which

$$N(L) = \frac{J}{G} = \frac{C[\exp(D\Delta T) - 1]}{A[\exp(B\Delta T) - 1]} \qquad (30)$$

where A, B, C and D are constants. If D > B, then as undercooling increases, so does $N(L)$. As undercooling further increases, the two equations are no longer valid and the shapes of J and G are better described by Eqns. (28) and (29) (Fig. 18). If the maximum of nucleation occurs at the undercooling ΔT_J and that of growth rate at ΔT_G, with $\Delta T_J > \Delta T_G$, then $N(L)$ will display a marked increased after ΔT_J. For undercooling greater than ΔT_G the J/G ratio also will decrease, with increasing undercooling. We can interpret the trends of measured CSDs in Etnean lavas observing that at constant and small undercooling Equation (30) guarantees that $N(L)$ is constant; thus we expect, in principle, a horizontal trend for CSD in systems at constant temperatures. In fact

$$\Delta T = \text{constant, implies } J = \text{constant, } G = \text{constant and } N(L) = \text{constant}$$

The interpretation of porphyritic structures of volcanic rocks identify a population of intratelluric crystals reaching larger size in deep seated conditions, then joined by a population of microphenocrysts and microlites whose occurrence is attributed to the rapid cooling of lavas as a consequence of rise and eruption. In the hypothesis that no crystal is subtracted from a magmatic body by crystal migration mechanisms or added by processes different from

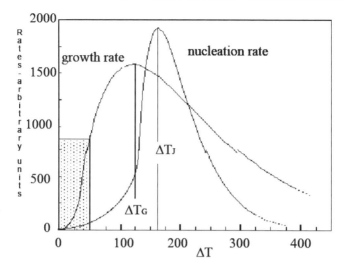

Figure 18. Shapes of the functions representing the dependence of growth (G) and nucleation (J) rates from undercooling. The values of J and G suddenly rise reaching maxima which not necessarily occur at the same value of ΔT, then rapidly drop when the undercooling is very high and the system acquires a large viscosity. At small undercooling (stippled area), Maaløe et al. (1989) suggest that growth and nucleation rates follow exponential laws.

nucleation (e.g., fractionation, aggregation, xenolith addition), Equation (25) directly provides $N(L)$ as a function of the ratio between nucleation and growth rates at the time when crystals of size L were formed. In this way CSDs can be seen as a continuous record of the cooling history of the magma.

Figure 19 displays the CSD of plagioclase resulting from the simultaneous treatment of binary images of feldspar collected on a data set of about 25 thin sections (~150 cm²). This computation allows one to explore the continuity of populations belonging to the largest classes and puts in evidence the main features of CSDs of all the minerals of Mt. Etna lavas. Trend A is referred to a period of deep seated growth at low undercooling. Trend B and C are related to the ascent of magma at increasing undercooling (the kink between B and C is probably due to an increasing rate of undercooling). Trend D isperhaps generated after the emission of lavas.

The effects of different cooling regimes on trend D are more evident in Figure 20, which reports plagioclase CSDs determined on images collected at a higher magnification from samples cooled at various rates. Samples were directly taken from the lava flow while still yellow hot. The rate of cooling cannot be quantified, but increases in cooling rates for samples 140392 and 170392 can be observed. Sample 140392 is a stubby chunk about 20 cm in diameter, cooled in air; sample 170392 is 5-cm in size and was drop quenched in water. Glassy groundmass characterizes the sample quenched in water, while sample 140392 has a microcrystalline groundmass with scarce interstertial brown glsss. As shown in Figure 20, the number of quench crystals is heavily influenced by the mode of quenching (rate of cooling) and by the degree of undercooling reached before the glass transition. In some samples quenched in water, particularly when CSDs are measured at a high resolution (or the glassy groundmass allows detection of even the smallest crystals), it is possible to observe the occurrence of a maximum of CSDs in the size interval D. Pointing out the significance of $N(L)$ as the ratio J/G, this feature can be interpreted as being due to a condition of undercooling during which both the nucleation and growth rates decrease after a stage when J increases and G decreases.

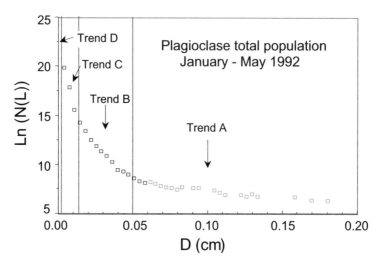

Figure 19. Interpretation of porphyritic structures of volcanic rocks.

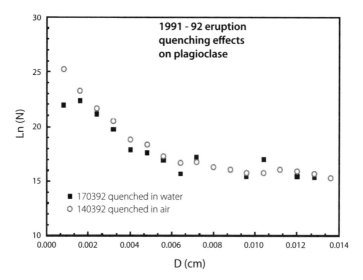

Figure 20. Effect of the quenching procedures on microlite crystallization.

Comparison between the total populations of different crystals in Mt. Etna lavas from the 1991-93 eruption (Fig. 21) reveals that the shapes of the distributions are the same, thus indicating the same cooling history. CSDs displayed in Figure 21 are related to samples collected at the vent and 6 km downhill at the lava front. There is a good overlap for regions A and B and a slight increase for sizes less than 0.003 cm in the sample gathered at the front, except for olivine.

The largest increment in the number of crystals is observed for Ti-magnetite, whose crystallization may be influenced by surface oxidation of lava. Since only the quenching crystals (trend D) show slight changes in their number density, CSDs confirm that the flux in lava tubes takes place at constant T (there is no change in undercooling). This implies that

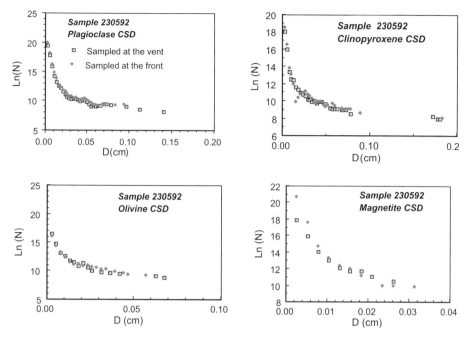

Figure 21. Comparison between the total populations of different crystals in Mt. Etna lavas of the 1991-93 eruption.

major variations of undercooling are met by magma *before* eruption and are such to pervade the whole batch of magma contemporaneously. Otherwise, we should observe strong variations in crystallinity in different portions of the magma batch that achieve different undercooling in response to the thermal gradient. Pressure, through the control exerted on the water solubility (Simakin et al. 1999) may provide such a control on undercooling.

Estimate of mineral growth rates for Mount Etna

An interesting application of Equation (30) is to the estimate of growth rates of minerals and of the time elapsed since the nucleation of the largest crystals.

At small (<10 °C) and constant undercooling J/G is constant and the largest crystals belonging to trend A can be used to assess growth times. This implies that the sizes of the largest crystals may be interpreted to reflect the time they spent in a deeper seated reservoir. The long-lasting 1991-1993 eruption of Mt. Etna allowed this kind of determination since it was sampled, in the first year, every day at the eruptive vents and at the lava fronts 6 km downstream (Armienti et al. 1994). Figure 22 reports a set of measurements made on the largest plagioclase of each sample in a time span of five months. Two sections of each specimen were examined, corresponding to about 15 cm^2 and measurements of the largest linear dimension crystals were performed using digitized images. More sections representing several days were examined to find the largest crystals, since insufficient sections are available for a given day. In fact for $N(L) \approx e^5$ cm^{-4}, $L \approx 0.1$ cm and $\Delta L = 0.002$ cm, the area to investigate must be greater than 5 cm^2 and becomes of the order of 10^2 cm^2 for a density $N(L) \approx e^5$ cm^{-4}. The maximum crystal sizes increased continuously during the five month interval examined in this work. A maximum growth rate at depth can thus be assessed from the envelope of the maximum sizes shown in the figure, under the simplified assumption of a constant growth rate.

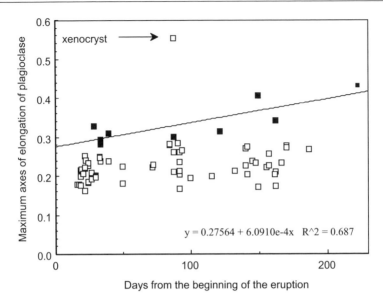

Figure 22. Estimate of plagioclase growth rate in intratelluric conditions. Black squares are the largest crystals detected in the collection of thin sections from the first 6 months of the eruption.

Trends of the envelope give values of 7.1×10^{-9} cm/s for plagioclase. This estimate of growth rate of plagioclase at low undercooling is one order of magnitude greater than those reported by Kirkpatrick (1977b) for much more crystalline basalts, but compares well to those found by Kirkpatrick (1978) for pure anorthite and those estimated by Maaløe et al. (1989) for basalts at undercooling of less than one degree. The obtained value for deep plagioclase growth rate also matches those interpreted by Cashman (1990) for the Mauna Loa 1984 eruption. The obtained growth rate allows an appraisal of the minimum times of the onset of crystallization for plagioclase. To derive this datum we have to purge measured diameters for the growth during the rise, as estimated by the kink of range A. Net deep growth is to be corrected by 0.05 cm. The resulting intercept indicates that plagioclase growth started 470±30% days before the onset of the eruption.

TEMPORAL EVOLUTION OF CSDs: SIZE AS FUNCTION OF TIME

The observation that in the case of a crystallizing batch of magma the equation

$$N(t,L) = \frac{J(t_L)}{G(t_L)}$$

guarantees that a plot of $N(L)$ vs. L is equivalent to a plot of the *temporal* evolution of undercooling, opens the possibility of extracting temporal information from the dependence of crystal size from time. In fact, the length attained by crystals in the time interval $t-t_0$ can be provided by the equation

$$L_t = \int_{t_o}^{t} G(\Delta T(t))\, dt \qquad (31)$$

if we know the functional dependence of G from undercooling ($G(\Delta T)$) and of undercooling from time ($\Delta T(t)$).

The experimental determinations of CSDs provided by Zieg and Lofgren (2006) report a set of data that may illustrate a method of general applicability to obtain this kind of dependence using CSD data. These authors performed experiments of nucleation and growth of olivine in a basaltic melt, starting from liquidus temperature and cooling their experimental charges at a rate of 92 °C/h to induce different final undercooling by simply adopting different time intervals. Their results are summarized as CSDs in Figure 23 where different symbols refer to experiments of different duration and different final size of crystals. The computed CSD and numerical solutions found for time t from Equation (31) are also reported as continuous and dashed lines respectively. The procedure followed to obtain the explicit dependence of L and $N(L)$ from time is illustrated below.

The Zieg and Lofgren experimental setting seems to imply a linear dependence of ΔT from time

$$\Delta T(t) = \alpha + \beta t \tag{32}$$

where α (°C) and β (°C s^{-1}) are constants to be computed. The features of the CSDs may be constrained with Equation (26)

$$N(t,L) = \frac{J(t_L) V(t_L)}{G(t_L) V(t)}$$

and adopting relations derived from Equations (28) and (29) to describe variations of J and G with undercooling (Kirkpatrick 1977a):

$$J = A \exp(-B/T_m) \left(\exp\left(-B'T_0^2 / (T_m \cdot \Delta T^2)\right)\right) \tag{33}$$

$$G = C \exp(-D/T_m) \left(1 - \exp\left(D'\Delta T / (T_0 T_m)\right)\right) \tag{34}$$

the unknown constants A, B, B', C, D, and D' describe the kinetics of crystallization. We can

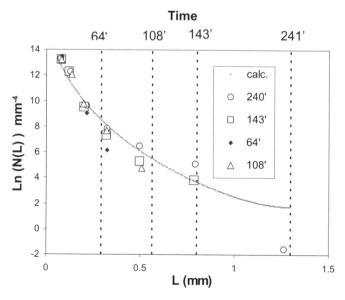

Figure 23. Zieg and Lofgren (2006) CSD measurements for different experimental duration. The fitting of data excludes the largest crystals obtained for experiments of 240 min that could be affected by right hand truncation. The correspondence between size and time is provided in the upper horizontal axis.

write a set of boundary conditions derived from the CSDs of Zieg and Lofgren (2006), namely

$$\int_{t_o}^{t} G\,dt = L_t \quad \text{(maxiumum length attained by crystals from time } t\text{-}t_o) \tag{35}$$

$$\int_{t_o}^{t} J\,dt = N_{tot} \quad \text{(total number of crystals)} \tag{36}$$

$$\int_{t_o}^{t} \frac{J}{G}\frac{4}{3}\pi L^3\,dt = \text{volume fraction} \quad \text{(Marsh 1998)} \tag{37}$$

$$N(L_0) = \frac{J_0}{G_0} \quad \text{at } t = t_o \tag{38}$$

$$N(L\max) = \frac{J_{t\max}}{G_{t\max}} \tag{39}$$

Equations (32)-(39) were numerically solved for the four Zieg and Lofgren (2006) CSD data sets reported in Figure 23, providing solutions and fittings of parameters for CSDs characterization (Table 2 and Figs. 24, 25). The close agreement between the model and the experimental data sets suggests that it is possible to use the obtained constants to estimate relevant thermodynamic and kinetic parameters. For the studied case we get Table 3. These

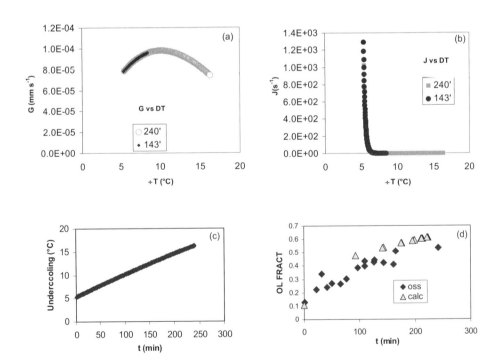

Figure 24. Reconstruction of the Zieg and Lofgren (2006) experiments: a) and b) report the dependence of growth and nucleation rates for the experiments at 240 and 143 minutes. As stated in the paper initial undercooling is in the order of 5° and nuclei practically do not form during the experiments as the early drop of J to 0 clearly shows. c) reports the dependence of undercooling from time. d) compares the observed and computed evolution of crystallinity accounting for the initial crystal fraction of 0.13 reported by the authors.

Table 2. Fitting parameters and comparison between observed and computed experimental data mathematicl procedure of minimization.

Fitting parameters			calculated	observed (240′)
A =	0.361	ΔT_{final} (°C)	16	
B =	1.06×10^{-4}	G_{fin} (mm/sec)	7.43×10^{-5}	
B′ =	0.221	L_{max} (mm)	1.30	1.3
C =	1.61×10^4	t_{tot} (min)	241	240
D =	1.15×10^4	$\ln(N(L_{min}))$	16.38	13.5
D′ =	−1.67	% volume	0.50	0.6
α =	5.35	$L(J/G = 6)$	0.41	0.48
β =	8.21×10^{-4}	$\ln(J_o/G_o)$	1.72	1.72

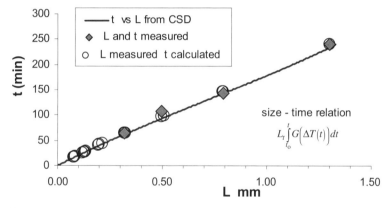

Figure 25. Relation between L and t. Diamonds are the maximum lengths measured at the end of experiments. Circles are lengths obtained from CSD measurements. The curve is obtained by the above discussed mathematical procedure of minimization.

Table 3. Computed thermodynamic parameters

Parameter	From Eqns. (28) and (29)	Value	Units
E^* (nucl.)	= B R_g	= -8.813×10^4	J mol^{-1}
Y_{ol}	= B′ 3 $k \Delta h^2/(16 \pi V_c^2)$	= 4.10×10^{-3}	J m^{-2}
E (growth)	= B′ × R_g	= -9.554×10^4	J mol^{-1}
Δh	= k D′	= -2.310×10^{-23}	J/ molecule

Adopted constants:
   V_c (cell volume for fo94) = 291×10^{-30} m^3
   k (Boltzman's constant) = 1.382×10^{-23} J K^{-1}
   R_g (gas constant) = 8.310 J K mol^{-1}

values (Table 3) also closely compare with literature data for olivine in basaltic melts (Dowty 1980; Cooper and Kohlstedt 1982) confirming, independently, the validity of this new model.

CONCLUDING REMARKS

CSD is an important tool in deciphering the crystallization history of magma, but it derives from heavy data manipulation. Such manipulations require great care when choosing boundary conditions for the system. Boundary conditions can only be derived from a profound petrologic comprehension of the system under examination. This means that we have somehow have *a priori* knowledge of which kind of CSD we expect and this controls the kind of model to apply. CSDs are not able to tell us *which* kind of petrologic processes are under examination, so much as they offer a key to understanding *how* known processes develop.

REFERENCES

Allard B, Sotin C (1988) Determination of mineral phase percentages in granular rocks by image analysis on a microcomputer. Comput Geosci 14:261-269
Armienti A, Francalanci L, Landi P (2007) Textural effects of steady state behaviour of the Stromboli feeding system. J Volcanol Geotherm Res 160:86-98
Armienti P, Innocenti F, Pareschi MT, Pompilio M (1994) Effects of magma storage and ascent on the kinetics of crystal growth. The case of the 1991-92 Mt. Etna eruption. Contrib Mineral Petrol 115:402-414
Armienti P, Innocenti F, Pareschi MT, Pompilio M, Rocchi S (1991) Crystal population density in not stationary volcanic systems: estimate of olivine growth rate in basalts of Lanzarote (Canary Islands). Mineral Petrol 44:181-196
Armienti P, Pareschi MT (1992) Cooling hystory of basalts and its effects on the kinetics of olivine crystallization. Acta Volcanol 2:7-15
Armienti P, Tarquini S (2002) Power-law olivine crystal size distributions in lithospheric mantle xenoliths. Lithos 63(3-4):273-285
Brandeis G, Jaupart C (1987) The kinetics of nucleation and crystal growth and scaling laws for magmatic crystallization. Contrib Mineral Petrol 96:24-34
Cashman KV (1988) Crystallization of Mount St. Helens 1980-1986 dacite: A quantitative textural approach. Bull Volcanol 50:194-209
Cashman KV (1990) Textural constraints on the kinetics of crystallization of igneous rocks. Rev Mineral 24:259-314
Cashman KV (1992) Groundmass crystallization of Mount St. Helens dacite, 1980-1986: a tool for interpreting shallow magmatic processes. Contrib Mineral Petrol 109:431-449
Cashman KV, Marsh BM (1988) Crystal size distribution (CSD) in rocks and the kinetics and dynamics of crystallization. II: Makaopuhi lava lake. Contib Mineral Petrol 99:292-305
Cashman KV, Marsh BM (1988) Crystal size distribution (CSD) in rocks and the kinetics and dynamics of crystallization. II: Makaopuhi lava lake. Contrib Mineral Petrol 99:292-305
Chayes F (1950) On the bias of grain-size measurements made in thin section. J Geol 58:156-160
Chayes F (1956) Petrographic Modal Analysis - An Elementary Statistical Appraisal. John Wiley & Sons, Inc., New York
Cooper RF, Kohlstedt DL (1982) Interfacial energies in the olivine basalt system. In: High-pressure Researches in Geophysics. Akimoto S, Manghnami MH (eds) Center Academic Publication. Tokyo. Adv Earth Planetary Sci 12:217-228
DeHoff RT, Rhines FN (1972) Quantitative Microscopy. Masson et Cie, Paris
Delesse MA (1847) Procédé mécanique pour determiner la composition des roches. Comptes Rendus Hebdomadaires des Sciences de l'Academie de Sciences 25:544-545
Donaldson CH (1976) An experimental investigation of olivine morphology. Contrib Mineral Petrol 57:187-213
Dowty E (1980) Crystal growth and nucleation theory and numerical simulation of igneous crystallization. In: Physics of Magmatic Processes. Hargraves RB (ed) Princeton University Press, p 385-417
Francalanci L, Tommasini S, Conticelli S, Davies GR (1999) Sr isotope evidence for short magma residence time for the 20th century activity at Stromboli volcano, Italy. Earth Planet Sci Lett 167:61-69
Gray N (1970) Crystal growth and nucleation in two large diabase dikes. Can Jour Earth Sci 7:366-375
Gray N (1978) Nucleation and growth of plagioclase, Makaopui and Alae lava lakes, Kilauea Volcano, Hawaii: Discussion. Geol Soc Am Bull 89:797-798

Higgins MD (2000) Measurement of crystal size distributions. Am Mineral 85:1-12
Higgins MD (2002) Closure in crystal size distribution (CSD), verification of CSD calculations and the significance of CSD fans. Am Mineral 87:160-164
Jambon A, Lussiez P, Clocchiatti R, Weisz J, Hernandez J (1992) Olivine growth rate in a tholeiitic basalt: an experimental study of melt inclusions in plgioclase. Chem Geol 96:277-287
Kilburn C (1989) Surface of Aa flow-fields on Mount Etna, Sicily: morphology, rheology, crystallization and scaling phenomena. In: Lava Flows and Domes. Fink J (ed) Springer-Verlag, Berlin, p 129-156
Kirkpatrick RJ (1977a) Crystal growth from melt: a review. Am Mineral 60:798-814
Kirkpatrick RJ (1977b) Nucleation and growth of plagioclase, Makaopuhi and Alae lava lakes, Kilauea Volcano, Hawaii. Geol Soc Am Bull 88:78-84
Kirkpatrick RJ (1978) Nucleation and growth of plagioclase, Makaopui and Alae lava lakes, Kilauea Volcano, Hawaii. Reply. Geol Soc Am Bull 89:799-800
Kirkpatrick RJ, Gilpin RR, Hays JF (1976) Kinetics of crystal growth from silicate melts: anorthite and diopside. J Geophys Res 81:5715-5720
Landi P, Métrich N, Bertagnini A, Rosi M (2004) Dynamics of magma mixing and degassing recorded in plagioclase at Stromboli (Aeolian Archipelago, Italy). Contrib Mineral Petrol 147:213-227
Lippman PW, Banks NG, Rhodes JM (1985) Degassing induced effects on lava rheology. Nature 317:604-607
Lofgren GE (1974) An experimental study of plagioclase crystal morphology: isothermal crystallization. Am J Sci 274:243-273
Lofgren GE (1980) Experimental studies on the dynamic crystallization of silicate melts. In: The Physics of Magmatic Processes. Hargraves RB (ed) Princeton Univ Press, Princeton, p 487-551
Maaløe S, Tumyr O, James D (1989) Population density and zoning of olivine phenocrysts in tholeiites from Kauai, Hawaii. Contrib Mineral Petrol 101:176-186
Mangan MT (1990) Crystal size distribution systematics and the determination of magma storage times: the 1959 eruption of Kilauea volcano, Hawaii. J Volcanol Geotherm Res 44:295-302
Marsh DB (1988) Crystal size distribution (CSD) in rocks and the kinetics and dynamics of crystallisation. I. Theory. Contrib Mineral Petrol 99:277-291
Morgan DJ, Jerram DA (2006) On estimating crystal shape for crystal size distribution analysis. J Volcanol Geotherm Res.154:1-7
Muncill GE, Lasaga AC (1987) Crystal-growth kinetics of plagioclase in igneous systems: one atmosphere experiments and approximation of a simplified growth model. Am Mineral 72:299-311
Pareschi MT, Pompilio M, Innocenti F (1990) Automated evaluation of volumetric grain-size distribution density from thin-section images. Comput Geosci 16:1067-1084
Pickering G, Bull JM, Sanderson DJ (1995) Sampling power law distribution. Tectonophys 248:1-20
Russ JC (1986) Practical Stereology. Plenum Press, New York
Saltykov SA (1949) Calculation of the distribution curves for the size of dispersed grains. Plant Laboratory 15:1317-1319
Schwartz HA (1939) Metallographic determination of the size distribution of tempered carbon nodules. Metals Alloys 5(6):139-140
Simakin AG, Armienti P, Epel'baum MB (1999) Coupled degassing and crystallization: experimental study in continuos pressure drop, with application to volcanic bombs. Bull Volcanol 61:275-287
Toramaru A (1991) Model of nucleation and growth of crystals in cooling magmas. Contrib Mineral Petrol 108:106-117
Zieg M, Lofgren G (2006) An experimental investigation of texture evolution during continuous cooling. J Volcanol Geotherm Res 154:74-88

17

Deciphering Magma Chamber Dynamics from Styles of Compositional Zoning in Large Silicic Ash Flow Sheets

Olivier Bachmann and George W. Bergantz

Department of Earth and Space Sciences
University of Washington
Mailstop 351310
Seattle, Washington, 98195-1310, U.S.A.

bachmano@u.washington.edu, bergantz@u.washington.edu

INTRODUCTION

The understanding of the dynamic processes in magmatic systems has grown and changed markedly in the last decade. Old models for magmatic systems as vats of near-liquidus material have been revised by observations from seismology (Sinton and Detrick 1992), crystal chemistry and zoning (Davidson et al. 2007) and the geochronology and geochemistry of both plutonic and volcanic systems (Hildreth 2004; Charlier et al. 2007; Miller et al. 2007; Peressini et al. 2007; Walker et al. 2007). New views emphasise magmatic systems as temporally dominated by crystal mushes (magma bodies with a high fraction of solid particles; see definition in Miller and Wark 2008), that wax and wane in temperature and crystallinity, and are subject to significant open system processes (Charlier et al. 2007; Hildreth and Wilson 2007; Walker et al. 2007; Bachmann and Bergantz 2008). These processes, such as magma reintrusion, mixing, gas sparging, and subsequent thermal rejuvenation, may be significantly more important in producing the characteristics of a magmatic system than the previous closed-system, near-liquidus behaviour would predict. There are a number of recent reviews that summarize these observations (for example see Eichelberger et al. 2006; Bachmann et al. 2007b; Lipman 2007). Our aim here is to illustrate how this new perspective to magma dynamics is motivated by observations of heterogeneities (or lack thereof) in erupted rocks (and to a lesser amount in plutons).

Owing to rapid withdrawal and quenching of magma during explosive volcanic eruptions (hours to a few days), large-volume (>1 km^3) pyroclastic deposits (also referred to as *ignimbrites* or *ash-flow tuffs*) provide an instant image of the state of the magma chamber evacuated during eruption. A first-order observation that characterizes these pyroclastic deposits of intermediate to silicic composition is that many do not tap into chemically homogeneous reservoirs but show compositional and thermal zoning, with early-erupted material differing significantly from late-erupted material (e.g., Lipman et al. 1966; Lipman 1967; Smith 1979; Hildreth 1981). In detail, nearly every eruptive product is different (see Hildreth 1981 for a comprehensive classification), but three general groups can be defined as follows (Table 1 and Fig. 1): 1) deposits showing quasi monotonic (or linear) gradients in composition and/or temperature, 2) deposits showing abrupt gradients in composition, with gaps that can reach several wt% in major elements over a very narrow stratigraphic interval, and 3) deposits showing no significant compositional or thermal gradients. Our objective is to provide a synopsis of zoning patterns preserved in ignimbrites as a means for understanding magma dynamics in silicic systems. We will argue that retaining and/or producing homogeneity by convective stirring is actually more challenging than preserving gradients, particularly in viscous, *crystal-rich*, silicic magma chambers, and

Table 1. Characteristics of some well-studied ignimbrites.

	Abruptly zoned units	Linearly zoned units	Homogeneous units
Examples	*Normally zoned:* Purico ignimbrite, Crater Lake ignimbrite, Katmai 1912, Los Humeros, Timber Mountain/Oasis Valley caldera complex *Reversely zoned:* Whakamaru ignimbrite, Bonanza Tuff	Bishop Tuff, Bandelier Tuff, Huckleberry Ridge, Tuff La Primavera Tuff, Taylor Creek Rhyolite, Loma Seca Tuff, Laacher See ignimbrite, Campanian Ignimbrite, Aso Ignimbrite	Fish Canyon Tuff, Lund Tuff, Cerro Galan ignimbrite, Atana ignimbrite, Oranui ignimbrite
Gradients in:			
Major element	Abrupt gaps (up to 20 wt% SiO_2)	Generally limited (few wt% at most)	Not measurable
Trace element variations	Significant	Significant	Not measurable
Isotopic ratios	Limited or absent in some cases, and important in others	Significant, and often correlated with Differentiation Index	Some heterogeneities at the whole-rock scale
Crystallinity	Abrupt change (few percent to >30 %). More mafic endmember generally more crystal-rich	Significant increase from early-erupted to late-erupted (~1 to >20 vol%)	Not measurable
Temperature	Up to 100 °C in cases but absent in others	Reported (up to 100 °C)	Not measurable
Gas content	Top richer in volatiles	Top richer in volatiles	Not measurable

Linearly zoned units: Hildreth 1981; Christiansen 1984; Grunder and Mahood 1988; Mahood and Halliday 1988; Johnson 1989; Duffield et al. 1995; Christiensen and Halliday 1996; Davies and Halliday 1998; Knesel et al. 1999, Lipman 1967; Wolff and Storey 1984; Worner and Schmincke 1984b; Worner and Schmincke 1984a; Wolff 1985; Wolff et al. 1990; Civetta et al. 1997

Abruptly zoned units: Whitney et al. 1988; Chesner 1998; Lindsay et al. 2001; Schmitt et al. 2001 Bacon 1983; Bacon and Druitt 1988; Hildreth and Fierstein 2000; Lipman et al. 1966; Varga and Smith 1984; Brown et al. 1998; Bindeman and Valley 2003

Homogeneous units: Francis et al. 1989; Lindsay et al. 2001; Bachmann et al. 2002; Maughan et al. 2002; Dunbar et al. 1989; Wilson et al. 2006.

require two conditions: (1) density filtering of material allowed to enter the growing silicic reservoir, and (2) thermal buffering close to the haplogranite eutectic.

Types of gradients in ignimbrites

Precisely reconstructing the zonation in a magma chamber from erupted deposits is difficult. Low sampling density, complexities in the depositional patterns induced by shifting eruptive vent sites and fluctuations in eruption rates (e.g., Wilson and Hildreth 1997), partial mixing during magma ascent in conduits (Blake and Ivey 1986a,b; Trial and Spera 1992) and the time-dependence of the dispersal of fragmented materials (e.g., Neri et al. 2003) will scramble many of the finer details of any spatial patterns of composition within the chamber prior to eruption. Hence only first-order observations, such as how early-erupted material differs from late-erupted material, can be considered as robust expressions of pre-eruptive

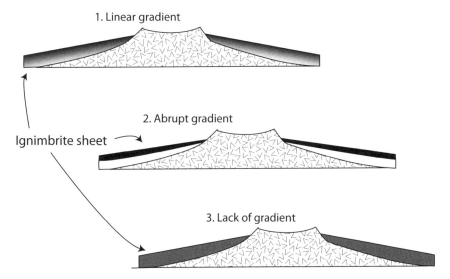

Figure 1. Schematic illustration of the three types of zoning patterns that commonly occur in ignimbrite (or ash-flow sheets) (modified from Bachmann and Bergantz 2008).

conditions for inter-deposit comparison. However, by sampling *unfragmented material* (pumice or lava blocks) from multiple localities in a given ignimbrite, an approximation of the pattern of gradients that existed in the magma chamber (whether abrupt, nearly continuous or absent, Fig. 1) can be obtained.

It is important to appreciate that by using the term "gradient," we are referring to a *horizontally-averaged quantity*. In detail the variations in composition, temperature, and crystallinity associated with convective stirring can locally vary significantly over short distances (envisage mixing two different flavors of honey with different colors), but these intensive variables will on-average produce a vertical gradient in a gravity field if the different phases have variable densities. Below we provide a short synopsis of the different types of gradients, then assess the role of convection in magma chambers and combine the two to explain how distinct types of zoning, or the lack thereof, emerge.

Abrupt gradients

Step-like changes in the whole-rock chemical composition of pumices up to several wt% for certain major elements has been documented in many deposits of explosive volcanic eruptions (e.g., Brophy 1991, Fig. 2). Striking color changes make these abrupt variations visibly manifest (see Bacon and Druitt 1988; Druitt and Bacon 1989; Eichelberger et al. 2000 and Fig. 2 for pictures), and they are commonly associated with large changes in crystallinity (the more mafic deposits being more crystal-rich, Bacon and Druitt 1988; Hildreth and Fierstein 2000).

Linear gradients

When plotted on geochemical variation diagrams, many silicic ignimbrites present a nearly continuous array in pumice composition (e.g., Smith and Bailey 1966; Hildreth 1979; Halliday et al. 1984; Grunder and Mahood 1988; Streck and Grunder 1997; Wolff et al. 1999; Milner et al. 2003, Fig. 3) from most evolved, early-erupted to least-evolved late-erupted deposits, although, in some volcanic units, small compositional gaps have been reported (Fridrich and Mahood 1987; Streck and Grunder 1997). The observed chemical variations are particularly striking for trace elements, but also involve major elements, isotopic rations, crystallinity,

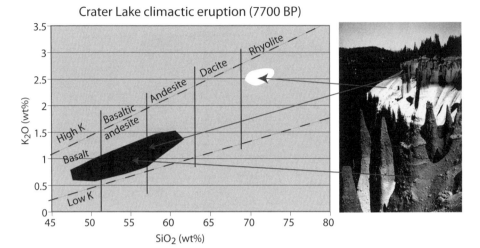

Figure 2. Compositional gap in the Crater Lake ignimbrite, obvious both on chemical plots and in the field. (modified from Bacon and Lanphere 2006).

Figure 3. Compositional spread in pumices of the Bishop Tuff, showing progressive depletion in Ba and FeO* as a function of SiO_2 (modified from Hildreth and Wilson 2007).

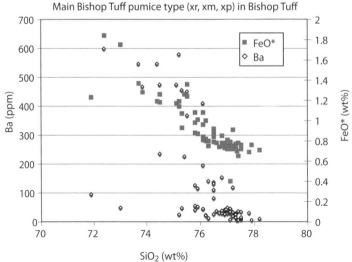

temperature, and gas content (Table 1). These variations in most magmatic tracers imply that fairly continuous compositional and thermal gradients are present in the magma reservoir tapped by the eruption (no information on their spatial distribution though).

As an example of a typical linearly-zoned system, the Bishop Tuff shows: (1) an increase in crystal content and decrease in volatile content from early- to late-erupted material (Hildreth 1979; Wallace et al. 1999), (2) lack of significant major element zoning, but more than twofold variation in incompatible trace elements and extreme depletion in compatible trace element (e.g., Sr, Ba) in early-erupted material (Hildreth 1979; Michael 1983; Hervig and Dunbar 1992; Knesel and Davidson 1997; Anderson et al. 2000), (3) variations in ε_{Nd}, $^{206}Pb/^{204}Pb_i$, and $^{87}Sr/^{86}Sr_i$, but homogeneity in oxygen isotopes (Bindeman and Valley 2002), and (4) a ΔT of ~100 °C from early to late-erupted material, documented with Fe-Ti oxides (Hildreth and Wilson 2007), oxygen isotope thermometry (Bindeman and Valley 2002), and Ti-in-Quartz thermometry (Wark et al. 2007).

Lack of gradients

Among a number of well-described zoned ignimbrites, several examples stand out as having very homogeneous whole-rock characteristics (including major and traces elements, isotopic ratios, crystallinity, and temperature; Fig. 4). Most of these homogeneous units belong to the *Monotonous Intermediates*, a group of large (>1,000 km^3), crystal-rich (up to 45 vol% crystals), dacitic units erupted in mature continental arcs (Hildreth 1981, Francis et al. 1989; de Silva 1991; Chesner 1998; Lindsay et al. 2001; Bachmann et al. 2002; Maughan et al. 2002). However, some homogeneous examples also occur in rhyolitic, crystal-poor units (Dunbar et al. 1989; Wilson et al. 2006).

Figure 4. A section of intracaldera Fish Canyon Tuff (in Canyon Diablo, CO), showing a lack of any kind of gradient from top to bottom. Up to 1 km thick section is exposed on the northern flank of the resurgent dome (La Garita mountains), with no base and top exposed, implying a thickness significantly greater than 1 km (estimated at an average of 2 km for the entire 80×30 km La Garita caldera collapse area; Lipman 2000).

MECHANISMS FOR GENERATING GRADIENTS IN SILICIC MAGMA CHAMBERS

Following the recognition that many large-volume ignimbrites originate in compositionally complex magma reservoirs (Lipman et al. 1966; Lipman 1967), numerous models have been proposed to explain the origin of these gradients. Below we summarize some key elements of these models, and discuss them in light of recent advances in our understanding of the dynamics of viscous, multi-phase mixtures (Bergantz and Ni 1999; Jellinek and Kerr 1999; Jellinek et al. 1999; Bergantz 2000; Bergantz and Breidenthal 2001; Phillips and Woods 2002; Burgisser et al. 2005; Dufek and Bergantz 2005). We will first summarize some aspects of the kinematics of convection in magmatic systems.

The role of convection

As noted by Grout almost a century ago (Grout 1918) and many workers since (Morse 1986; Marsh 1988; Rudman 1992; Simakin et al. 1997; Simakin and Botcharnikov 2001), convection in magma reservoirs is unavoidable as long as a dominantly-liquid (*but not necessarily crystal-poor*) batch of magma is present. The most significant source of density changes that provides the potential energy for convection are associated with phase change; examples include the formation of crystals and/or bubbles. Magmatic systems are also open to new injections (with both similar and different chemical compositions) and can assimilate their wall-rocks (DePaolo 1981; Davidson and Tepley 1997; Bohrson and Spera 2001; Thompson et al. 2002; Beard et al. 2005), leading to additional temperature and crystallinity (and hence buoyancy) fluctuations within the reservoir (Keller 1969; Bergantz 2000; Murphy et al. 2000; Bachmann et al. 2002). Thus, convective dynamics within magma reservoir are the result of both (1) *in situ* density changes arising from cooling and/or volatile exsolution and (2) magma reintrusion and/or volatile flux.

Convective stirring in magmatic systems is likely to be highly intermittent and arise from many or all of the above mentioned buoyancy sources during the lifetime of a reservoir (which can last several hundreds of thousands of years; e.g., Reid et al. 1997; Brown and Fletcher 1999; Vazquez and Reid 2004; Bacon and Lowenstern 2005; Simon and Reid 2005; Bachmann et al. 2007a). Geological data require that any particular model must accommodate the following observations: a) magma bodies cool on conductive timescales and rarely assimilate large amounts of their margins by rapid heat transfer (e.g., Carrigan 1988; Barboza and Bergantz 2000; Dufek and Bergantz 2005; Petcovic and Dufek 2005), b) heterogeneous crystal populations (particularly in age population of zircons) at the margins and core of plutonic suites generally preclude simple, monotonic sidewall and roof crystallization (Eichelberger et al. 2006a; Miller et al. 2007; Walker et al. 2007), c) open system processes and secular cooling produces intermittent convection with Rayleigh numbers high enough ($>10^5$) where conditions for chaotic convection are (at least temporarily) obtained.

In these complex, open-system magma reservoirs, convection has a dual nature: it reduces the scales and intensity of heterogeneities (e.g., Oldenburg et al. 1989; Coltice and Schmalzl 2006), but can *also* produce gradients in intensive variables and composition if the source of buoyancy in itself generates heterogeneities (Bergantz and Ni 1999; Jellinek and Kerr 1999; Jellinek et al. 1999, Figs. 5 to 8). The following paragraphs attempt to clarify this dual nature.

It is widely understood that mixing requires advection of material, which acts as a source of localized high strain, reducing length scales of heterogeneities and allowing a significant volume of material to be serviced by the mixing process. Subsequently, diffusion can operate effectively once the average distance between heterogeneities have been reduced below the diffusive length scale (e.g., Coltice and Schmalzl 2006). The most efficient mixing systems will be those that can produce the greatest amount of strain at the widest range of scales, typically by the rotational motion of eddy-like structures that both entrain fluid and wind them.

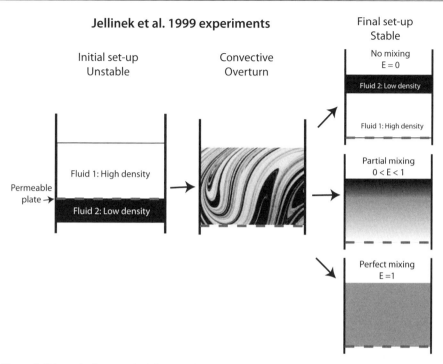

Figure 5. Schematic diagram of mixing experiments described by Jellinek et al. (1999), starting from an initial, gravitationally unstable set-up, inducing an overturn, and leading to a final mixture, whose mixing efficiency will depend on the density and viscosity ratios of the two fluids.

The Reynolds and Rayleigh numbers are the most important diagnostics of the dynamic and kinematic conditions within in a magma body. A very general form of the Rayleigh number is (Jellinek et al. 1999):

$$\text{Ra} = \frac{BH^4}{\nu\kappa^2} \qquad (1)$$

where B is the buoyancy flux, H the height of the system, ν the kinematic viscosity of the ambient magma and κ the diffusivity. The Reynolds number is traditionally defined as:

$$\text{Re} = \frac{uL}{\nu} \qquad (2)$$

where u is a scaled velocity, L the matching length scale and ν the kinematic viscosity. It can also be written as (Jellinek et al. 1999):

$$\text{Re} = \frac{B^{1/3}H^{4/3}}{\nu} \qquad (3)$$

(assuming that the velocity is scaled on a balance of inertia and buoyancy, a problematic assumption for small Reynolds numbers but that does not impact our overall conclusions). The buoyancy flux appearing in both the Rayleigh and Reynolds numbers will be a complex function of the conduction-controlled rate of cooling, the geometry (as it impacts cooling rate), the resulting density changes associated with phase change, and thermal buffering with latent heat (see, for example, Jellinek and Kerr 2001).

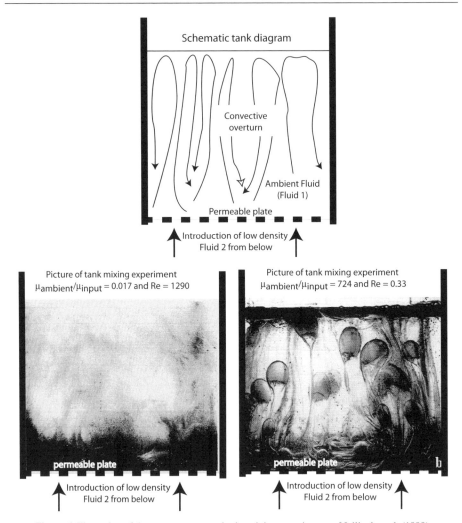

Figure 6. Illustration of the overturn process in the mixing experiments of Jellinek et al. (1999).

There are generally three regimes where the Reynolds number produces a distinct flow template: (1) laminar, (2) transitional and (3) fully turbulent. In the laminar regime (Re <1), buoyancy forces are balanced by viscous forces (viscosity controls the dynamics of the flow), while in the turbulent regime (Re ~>1000), the buoyancy force is balanced by inertial forces (onset of true turbulence). Intermediate values between these two (1 < Re < 1000) indicate that both viscosity and inertia play a role, and this is the most difficult range to generalize because the flows are not self-similar. This means that the structure of the flow is not ordered at the largest scales: a small change in Reynolds number can produce an intermittent flow regime that is not similar to the one that preceded it.

At high values of the Reynolds number (~1000), the mixing transition is reached, and there is a full spectrum of eddy sizes from largest scales set in part by the geometry of the system (initial and boundary conditions), to the smallest scales of eddies set by the viscosity. For systems with Reynolds numbers at or above the mixing transition, mixing is fast and

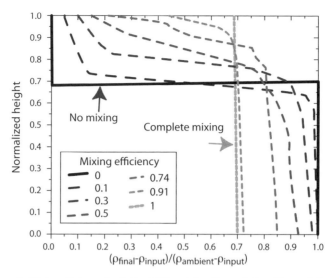

Figure 7. Mixing efficiency for fluids at different densities and viscosities as defined by experiments of Jellinek et al. (1999), illustrated in a 2D cross section of a tank.

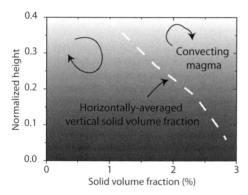

Figure 8. Horizontally-averaged vertical gradient in crystallinity within a reservoir (modified from Bergantz and Ni 1999). These results were obtained by numerically simulating cooling of an originally solid-free liquid, triggering precipitation of dense, solid phases at the roof. Once the dense roof layer reaches a critical crystallinity, solid-rich plumes collapse, leading to the growth of a stable gradient in solid fraction within the convecting, liquid body. See Bergantz and Ni (1999) for details.

thorough, as mixed material is quickly passed among eddies of many sizes (Broadwell and Mungal 1991; Dimotakis 2005), thus efficiently breaking-down existing gradients. Jets and plumes produced in explosive eruptions reach this Reynolds number.

For Re <1000 (low to transitional Reynolds numbers), there is not a statistically complete set of eddy sizes in the flow. Only a few eddy sizes dominate and the high-strain stirring is of limited efficiency. If the Rayleigh number is high enough, such situations are prone to "chaotic" convection (e.g., Ottino 1989; Ferrachat and Ricard 1998; Petrelli et al. 2006). The onset of chaotic conditions can be determined by the appropriate Rayleigh numbers, although the values will depend on the specific system under consideration. Generally, however, as long as the Rayleigh number is above about 10^5, chaotic conditions can be obtained, even with a low Reynolds number (Coltice 2005).

Chaotic mixing leads, on the short term, to islands of well-stirred material in a domain of otherwise poorly stirred material (see overturn illustration in Fig. 5). If kept active long enough, this kind of flow can lead to efficient mixing (unless there are strong contrasts in viscosity and density; Ottino 1989). However, if the source of buoyancy and convective stirring itself induces heterogeneities, then complete homogeneity cannot be reached.

Consider *discrete, mechanically passive*, chemical heterogeneities such as a local patch within the magma system that has a distinct character, such as a small difference in crystallinity, temperature or composition (or all three). If there is chamber-wide overturn driven by cooling, heating or reintrusion from the boundaries, this patch will be homogenized with the rest of the magma reservoir in a few overturns in a chaotically convecting system (e.g., Coltice 2005) even at low Reynolds number; it does not require sustained vigorous fluid circulation. If the viscosity contrast between the patch and the magma reservoir is large, the homogenization may take longer but even crystal multiphase aggregates can be eroded and dissipated by shear (Powell and Mason 1982; Petrelli et al. 2006).

On the other hand, if new heterogeneities are the sources of the convective process itself (such as dense crystal plumes originating from a cold ceiling, or reintrusion of low density magma from below), the system cannot reach perfect homogeneity (except at extremely high mixing efficiencies in high Re number situations). As heat exchange with the surroundings never completely stops, new heterogeneities in temperature, inducing variations in crystal content are constantly created, impeding complete homogeneity in crystallinity (and therefore composition); new gradients are re-established as old ones are consumed. More generally, any process taking place at the chamber boundaries that leads to variations in trace elements, crystal content, or temperature can export that variation into the magma body by convection (e.g., Marsh 1989) and produce a gradient in these quantities, while also homogenizing pre-existing gradients. Therefore, even in a chemically closed (but cooling) system, perfect homogeneity cannot be achieved for waning systems.

As illustrated in numerical and laboratory experiments (Figs. 5 to 8), convective processes can readily generate gradients as they transport material and heat across a magma system. Any density instability that originates at a boundary (e.g., crystal-rich plume falling from the top, a gas-rich region that rises from the bottom, or a reintrusion of less dense material from below) will all produce a gradient as it passes through the system. As material at the boundary rises or falls, this process will entrain surrounding material (see Bergantz and Breidenthal 2001), become diluted, and deliver this now mixed material to the other boundary; it also leaves a "wake" of material that is a mix of the initial falling or rising plume and the surrounding magma. If the process is repeated, it will produce and sustain a (horizontally averaged) gradient, as it has been exemplified for a number of applications to magmas (Bergantz and Ni 1999; Jellinek and Kerr 1999; Jellinek et al. 1999; Ruprecht et al. 2008).

The ability of rising or falling plumes to create gradients in the surrounding magma has been called the "mixing efficiency" (Linden and Redondo 1991; Jellinek and Kerr 1999; Jellinek et al. 1999). It is defined on the basis of an initially unstable stratification, such as a less dense layer entering the bottom, allowed to mix with the over-lying fluid as it penetrates upward. One endmember is that the plumes rise or fall through the ambient magma with no mixing, simply changing position, and the least amount of initial gravitational potential energy is retained in the final configuration. The second endmember is perfect mixing, leaving no gradient but a uniform mixture over the entire vertical span of the system. Between these two extremes, the overturn process produces a stably stratified system and the vertical variation in the density, and hence concentration, from the mixing of the penetrating and ambient fluids produces a gradient (Fig. 7). To quantify this, the mixing efficiency, E, is defined as:

$$E = \frac{P_f - P_{\min}}{P_{\max} - P_{\min}} \quad (4)$$

where $P_{\max}$ is the *final*, maximum value of the potential energy associated with perfect mixing (which yields no gradient), $P_{\min}$ is the *final* value associated with no mixing as the two layers simply exchange places and P_f is the actual final potential energy of the system following overturn:

$$P_f = \int_0^H g(\rho_a - \rho(z))z\,dz \qquad (5)$$

g the scalar acceleration of gravity, ρ_a the density of the ambient, or resident fluid and $\rho(z)$ the horizontally averaged vertical density profile after overturn. It is important to note that this definition of the mixing efficiency is a measure of the (horizontally averaged) vertical gradient. The mixing efficiency has a value for zero if the two fluids slide past each other with no mixing, and unity if the two fluids mix perfectly creating a system with no horizontally-averaged vertical gradient. So a value of the mixing efficiency of say 0.5, does not imply that 50% of fluid A and 50% of fluid B are everywhere well mixed. It means that there is a gradient in density (and hence concentration) from top to bottom with a slope of 0.5 and with equal amounts of the resident and intruding material at only one vertical position.

For silicic magmas, there will be a wide range of buoyancy flux B depending on the source of buoyancy (thermal, compositional, phase change), but reasonable values will vary between 10^{-7} and 10^{-11}. Thus, for a given height, kinematic viscosity and diffusivity, one can estimate the approximate Rayleigh and Reynolds numbers across a range of compositions (using Eqns. 1 and 2). As shown in Figure 9, the Reynolds number generally remains low (<10) even for the largest possible examples (up to 5 km thick) and most dynamic scenarios. Therefore, the mixing efficiency will always be low in magmatic systems, and for heterogeneities related to magma cooling, perfect homogeneity cannot be reached.

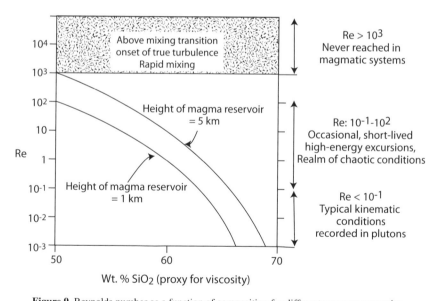

Figure 9. Reynolds number as a function of composition for different magmas reservoirs.

Origin of heterogeneity in silicic magmas: convective stirring vs. phase separation

On the basis of the arguments expressed above, the presence of zoning or heterogeneities in a volcanic unit can represent two endmembers: (1) incomplete, or partial convective mixing of two magma batches or (2) chemical differentiation or distillation by crystal-liquid separation of an initially more homogeneous system. These two scenarios are similar to the concepts of "arrested homogenization" and "progressive unmixing" of Eichelberger et al. (2000), although we stress that convective stirring of different magma batches does not necessarily lead to homogenization (and therefore does not need to be "*arrested*").

Convective interactions of compositionally variable magma batches present in the same reservoir are clearly common in the geological record. Magmatic plumbing systems favour it, and a number of ignimbrites record the mechanical encounters between magmas of from different sources (Smith 1979; Hervig and Dunbar 1992; Mills et al. 1997; Bindeman and Valley 2003; Eichelberger et al. 2006b; Davidson et al. 2007; Knesel and Duffield 2007). However, many volcanic deposits display chemical heterogeneities that have been interpreted as being *dominantly* induced by crystal fractionation within a given reservoir (*in situ* differentiation, e.g., Thompson 1972; Hildreth 1981; Michael 1983; Brophy 1991; Francalanci et al. 1995; Thompson et al. 2001; Hildreth 2004). A magma residing in a cooler reservoir will undergo crystallization and volatile exsolution. As the new phases produced have different densities, the system will have the tendency to differentiate, or unmix, into two or more batches with distinct characteristics. Therefore, zoning patterns present in deposits with an *in situ* differentiation signature represent *progressive unmixing* of an initially more homogeneous magma body (e.g., McBirney 1980; Eichelberger et al. 2000).

Both scenarios of heterogeneization can occur at different time scales depending on the physico-chemical conditions in the magma reservoirs. They are explored below (when applicable) in the development of the different types of gradients in magma chambers.

Mechanisms to generate abrupt gradients

An abrupt juxtaposition of distinct magma compositions (and/or crystal content) can easily be explained by the interaction of two distinct magma batches with different characteristics (e.g., Eichelberger et al. 2006b). However, abrupt compositional gaps are also found in units showing geochemical evidence for *in situ* differentiation (Bacon and Druitt 1988; Druitt and Bacon 1989; Brophy 1991; Hildreth and Fierstein 2000). Due to the small size of crystals in magmatic systems (typically around 0.1 to 5 mm) and the high viscosity of SiO_2-rich melts ($10^4 - 10^6$ Pa·s), separating crystals from its melt is a slow process (Sparks et al. 1984; Reid et al. 1997; Anderson et al. 2000; Eichelberger et al. 2006a), particularly if the magma is undergoing some convective stirring (Martin and Nokes 1989; Burgisser et al. 2005). Therefore, crystal-melt separation is enhanced when convection stops, which usually occurs when a crystal-bearing magma transforms into a locked crystalline mush (at ~50 vol% crystals for low strain rates; Vigneresse et al. 1996; Petford 2003; Rosenberg and Handy 2005; Caricchi et al. 2007; Champallier et al. 2008). Once the crystallinity is high enough (> ~70%), permeability becomes so low that crystal-melt separation by compaction cannot occur on geologically reasonable timescales (McKenzie 1985; Bachmann and Bergantz 2004). Hence, the most favourable crystallinity window appears to be around the rheological threshold (~50-60 vol% crystals), when convection has stopped but the system remains permeable enough for the interstitial melt to be expelled from the compacting crystalline framework (Thompson 1972; Brophy 1991; Thompson et al. 2001). The low-density interstitial melt can then accumulate above the mush, and generate a nearly crystal-free cap (Bacon and Druitt 1988; Bachmann and Bergantz 2004; Hildreth and Wilson 2007; Walker et al. 2007, Fig. 10). In the event of the formation of a crystal-poor cap by interstitial liquid extraction from a crystalline mush, eruption of both the crystal-poor cap and its underlying mush during the same eruptive episode will lead to the observed abrupt gradient present in many systems (Fig. 11).

THE CASE OF THE ABRUPTLY ZONED CRATER LAKE ERUPTION

Despite the wide compositional gap between the nearly-aphyric Crater Lake rhyodacite and the co-erupted crystal-rich andesite and clear evidence for a chemically open-system (e.g.,

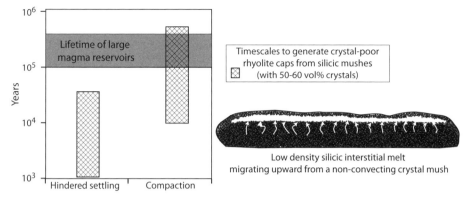

Figure 10. Range of timescales of interstitial liquid extraction (by hindered settling and compaction) for typical silicic mushes with crystallinities of 50-60 vol% (for more details, see Bachmann and Bergantz 2004).

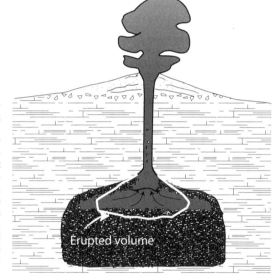

Figure 11. Development of a crystal-poor horizon above a crystalline mush (and below a crystallizing roof layer). When such a systems in tapped by eruption, the least-viscous, crystal-poor material is expected to erupt first. When exhausted, the viscous-crystal-rich portion can be drawn-out, rapidly leading to chocking of the eruption. In natural deposits, such as the 7700 BP Crater Lake ignimbrite, one expects the commonly observed sequence of a crystal-poor base covered by a relatively thin, crystal-rich, co-magmatic layer.

low and high Sr magma types), the two compositional layers have strong affinities, which suggest a genetic link by crystal fractionation (Bacon and Druitt 1988; Druitt and Bacon 1989): (1) the rhyodacite has trace element contents that require extensive fractional crystallization. (2) The interstitial glass trapped within the andesitic layer is very similar in composition to the rhyodacite. (3) Temperature and oxygen fugacity are nearly identical in both compositional layers. These chemical and thermal affinities, in addition to the physical proximity, indicate that the eruption tapped a magma reservoir similar to the setting illustrated in Figure 11. The zoning pattern was induced by crystal-melt separation.

Mechanism to generate linear gradients

As for the abrupt gradient case, the two models for generating linear gradients in magma chambers are: *in situ* differentiation processes (Hildreth 1981; McBirney et al. 1985; Trial and

Spera 1990; Marsh 2002) and magma mixing by new addition from below (Smith 1979; Hervig and Dunbar 1992; Eichelberger et al. 2000; Knesel and Duffield 2007). The dynamic template for the latter hypothesis has been investigated by a number of authors (Sparks and Marshall 1986; Frost and Mahood 1987; Oldenburg et al. 1989); incomplete mixing leads to imperfect chemical blending (often referred to as mingling), and stratification in the chamber (Jellinek et al. 1999). However, the mechanisms leading to stratification by the *in situ* evolution process is more controversial. The most commonly accepted mechanism is the "sidewall crystallization" hypothesis, involving the extraction of interstitial liquid from a crystallizing boundary layer at the cold margin of a magma chamber (Chen and Turner 1980; McBirney 1980; Rice 1981; Huppert and Sparks 1984; Spera et al. 1984; McBirney et al. 1985; Wolff et al. 1990; de Silva and Wolff 1995; Spera et al. 1995). Although this mechanism may play a role in the generation of chemical and physical heterogeneities in some cases, several considerations preclude sidewall crystallization being the dominant differentiation template in crustal magma chambers.

a. Bodies of eruptible magmas in large silicic chambers are sill-like, having a low wall to roof (or floor) ratio, rendering sidewall crystallization inefficient (de Silva and Wolff 1995)

b. Little evidence for sidewall crystallization is preserved in extensively mapped and studied plutonic bodies (e.g., McNulty et al. 2000; Barnes et al. 2001; Zak and Paterson 2005; Eichelberger et al. 2006a; Miller et al. 2007; Walker et al. 2007).

c. The fact that most crystals found in silicic magmas (except for some rhyolites and aplites) are complexly zoned requires that crystals undergo complex transport paths and circulation. This is inconsistent with the monotonic sidewall crystallization hypothesis, where crystals would not be available to circulate and respond to changes in the magmatic environment.

d. The vertical stacking of several convecting magma "layers" is inherently unstable as drag and entrainment should occur at the interfaces between the different magma batches, leading to re-blending at a rate similar to estimated differentiation rate (see Davaille 1999; Gonnermann et al. 2002; Bachmann and Bergantz 2004 for more details in this process).

Thus, the generation of a linear gradient by a "box filling" mechanism of double-diffusive convection in a largely fluid reservoir appears unlikely in magmatic situations. We argue that mixing of heterogeneities by sluggish convection is the most common process of generating linear gradients in magma chambers. Linear gradients will certainly develop when two different magmas interact, but such gradients can also develop *in situ* as crystallization and differentiation continue even during periods of closed-system evolution, generating heterogeneities in crystallinity and composition (e.g., Bergantz and Ni 1999; Bergantz 2000; Couch et al. 2001) that will be progressively smeared over the entire chamber.

THE CASE OF THE LINEARLY ZONED BISHOP TUFF

Numerous models have been proposed to explain the distinctive characteristics of the Bishop Tuff. The most recent and comprehensive treatment of this unit is Hildreth and Wilson (2007). They propose a model for the zonation of the Bishop Tuff in which different pockets of rhyolitic melt are segregated from a large, subjacent crystal mush. As the conditions within the mush vary over time, extracting interstitial liquid at different stages would have lead to rhyolitic pockets with slightly different chemical characteristics. This model appears plausible, but does not take into account the effect of convection, which must have occurred in a system showing a thermal gradient of ~100 °C (Hildreth 1979; Hildreth and Wilson 2007; Ghiorso and Evans 2008).

On the basis of the observations mentioned above and new constraints provided by chemical and thermal heterogeneities in quartz crystals, which require timescales of less than 10^2-10^4 years (Bindeman and Valley 2002; Wark et al. 2007), the characteristics of the Bishop Tuff seem best explained by a model very similar to the one proposed by Hildreth and Wilson (2007), but with the requirement of some late convection stirring in the rhyolite cap induced by a hotter reintrusion (leading to the development of a thermal gradient and the growth of bright rims on quartz from the late-erupted material; Wark et al. 2007). So both convective stirring just prior to eruption, and progressive unmixing to produce the rhyolite, seems to have played a role in the evolution of the Bishop Tuff magma (see Fig. 12).

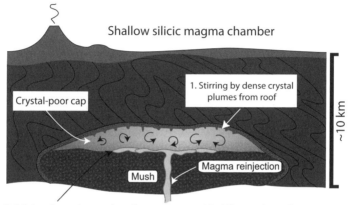

Figure 12. Schematic illustration of the two main possible sources of heterogeneities in shallow, crystal-poor, silicic magma chambers, such as the Bishop magma reservoir.

The generation of homogeneity

In light of the discussion above, homogenizing at the whole-rock scale a crystal-rich silicic magma, the most viscous silicate liquid on Earth (Scaillet et al. 1998), appears unlikely. We consider below possible mechanisms to reach homogeneity in viscous silicic magmas. These ideas are largely based on studies of homogeneous volcanic systems, such as the Fish Canyon magma body (see following section), but stress that more work is needed to better constrain some of these issues.

The largest silicic magmas bodies (10,000+ km³) are certainly constructed incrementally (e.g., Deniel et al. 1987; Petford et al. 2000; Lipman 2007). To obtain these large homogeneous masses of magma in such a clearly open-system situation, new magmatic additions need to be either (1) similar in composition to the growing reservoir, or (2) efficiently blended by mechanical stirring. Furthermore, crystal-melt separation (which will recreate heterogeneities) should be hampered (Fig. 13).

All three requirements can be met if the following conditions are obtained. As a new magma batch intrudes in the upper crust, it will quickly start crystallizing in the cold, low-pressure environment. Once crystallinity increases, further changes in temperature will be slowed by the latent heat release by crystallization, and increasingly ineffective conductive cooling (Koyaguchi and Kaneko 1999). Therefore, magma batches are expected to remain as crystal mushes (with crystallinities > 50 vol%) for most of their time above the solidus.

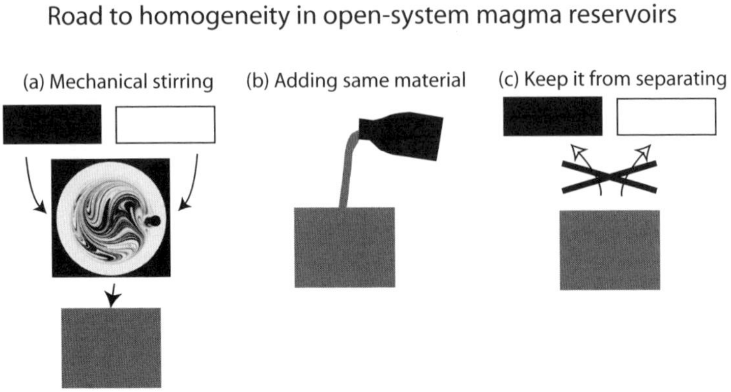

Road to homogeneity in open-system magma reservoirs

(a) Mechanical stirring   (b) Adding same material   (c) Keep it from separating

Figure 13. Illustrations of possible ways to render and/or keep open-system magma bodies homogeneous.

Magmatic additions to this growing *silicic* magma reservoir are restricted to either (1) similar composition (Hervig and Dunbar 1992; Eichelberger et al. 2000), released by a filtering lower crustal MASH zone (Hildreth and Moorbath 1988) or (2) more mafic composition (hotter, less viscous, and generally denser), which will mostly pond beneath the low-density, silicic mush (Miller and Miller 2002; Wiebe et al. 2002; Harper et al. 2004). This latter situation is prone to rejuvenation and self-mixing as the mafic magma acts as a hot plate (Couch et al. 2001; Bachmann and Bergantz 2006). As the magma body never remains for long in a crystal-poor situation, crystal-melt separation can only occur by interstitial melt extraction.

THE CASE OF THE FISH CANYON MAGMA BODY

The Fish Canyon magma erupted during the largest known silicic volcanic eruption, the crystal-rich Fish Canyon Tuff (45 vol% crystals) and its satellite units erupted from the same magma chamber (the Pagosa Peak Dacite and Nutras Creek Dacite, Lipman et al. 1997). This system is well known for its whole-rock homogeneity (at the scale of the hand sample and larger), in major and trace element composition, mineralogy, modal abundance, isotopic ratios, temperature, and water content (Whitney and Stormer 1985; Johnson and Rutherford 1989; Bachmann et al. 2002; Charlier et al. 2007). Similarly homogeneous units have been described in detail in the Andes (Francis et al. 1989; Lindsay et al. 2001), and in the Great Basin, USA (Maughan et al. 2002). However, at the mm scale, The Fish Canyon magma shows an enormous range in composition (both major, trace and isotopic), implying complex open system behaviour and excursions in intensive variables (Bachmann and Dungan 2002; Bachmann et al. 2005; Charlier et al. 2007). Of particular note: (1) different patches of glass analyzed by microdrilling with a mm spacing are heterogeneous in $^{87}Sr/^{86}Sr_i$; also, (2) biotite crystals have $^{87}Sr/^{86}Sr_i$ higher than any other component in the magma (including the glass around them) and thus indicate a provenance from the Precambrian wall rocks surrounding the magma chamber (Charlier et al. 2007). Therefore, as these biotites were derived from the disaggregation of blocks of material from the walls and are now found in the Fish Canyon magma, *some* crystal dispersal by convection is required. On the basis of Sr diffusive reequilibration of these Precambrian biotites, this dispersal had to occur in less than 10,000 years prior to eruption.

One possibility to dissipate gradients in crystal-rich situations, as anticipated by the "defrosting" hypothesis (Mahood 1990), is reheating of locked silicic mushes by interaction with hot, intrusive magmas. As it has been described in multiple natural systems (Murphy et al.

2000; Couch et al. 2001; Bachmann et al. 2002; Bachmann and Bergantz 2006; Hildreth and Wilson 2007), cooling silicic magmas reach a crystal mush state (crystal content >50-60%) that becomes rigid at low strain rates (convecting currents stop). If sufficient heat is added by a more mafic re-intrusion *from below* (a late reheating event of about 50 °C is recorded by the mineral phases in the Fish Canyon magma (Bachmann and Dungan 2002; Bachmann et al. 2002; Bachmann et al. 2005), the crystalline framework will progressively re-melt from the bottom until becoming liquid again (able to sustain convection). The stirring agent in such case is a buoyancy force generated not only by heating, but more importantly by the melting of the high crystallinity barrier at the bottom of the chamber (thermal expansion accounts for only 1-10% of density variation due to crystallinity changes) AND by the injection of volatiles from the commonly gas-rich new magma (Bachmann and Bergantz 2003, 2006). This re-heating can create an inverted density stratification that could evolve to full-scale self-mixing if the reheating event can supply enough energy.

An important observation made in rejuvenated systems is that they appear not to have significantly mixed with the more mafic magma that acted as a heat source; convective mixing is limited to the reactivated mush (self-mixing of Couch et al. 2001). In the Fish Canyon system, only a few mafic enclaves have been preserved in the late erupted intracaldera facies of the Fish Canyon Tuff (Bachmann et al. 2002) although a large basaltic andesite unit (the Huerto Andesite) erupted immediately after the Fish Canyon Tuff eruption (Parat et al. 2005, 2008). This absence of chemical mixing can be understood if buoyancy terms are compared. As density variations due to compositional changes are of the order of a few percent (~10 times larger than buoyancy forces driven by thermal expansion during a reheating of around 100 °C), reheating alone will not be sufficient to invert the compositional density gradients and enable large scale convection involving both mafic and silicic magmas.

THE CASE OF THE HOMOGENEOUS GRANITOIDS

Most large silicic plutons (such as those found in the Sierra Nevada Batholith) are very similar to the Monotonous Intermediates. We consider the largest of them (e.g., the Half Dome Granodiorite in the Tuolumne Intrusive Suite; Bateman and Chappell 1979) as unerupted equivalents of units such as the Fish Canyon magma. We argue that they follow the same incremental growth scenario, undergoing periodic intrusions of (a) evolved magma of similar composition blending in the growing reservoirs and (b) more mafic intrusions ponding at the base of the mush (Wiebe and Collins 1998; Robinson and Miller 1999; Waight et al. 2001; Miller and Miller 2002; Wiebe et al. 2002), which induce convective stirring due to addition of heat and gas from below (Wiebe et al. 2007). The only difference between plutonic and volcanic rocks is that the former cooled slowly to full solidification, allowing time to undergo some local crystal-liquid separation. Such late crystal-melt separation, leading to evolved granitic cupolas on top of large granodioritic bodies, is observed in several cases of well-exposed plutonic sections (Johnson et al. 1990; Barnes et al. 2001; Walker et al. 2007; Wiebe et al. 2007).

CONCLUSIONS

Many caldera-forming (large-volume) ignimbrite sheets display chemical and thermal heterogeneities, reflecting evacuation from a compositionally zoned shallow reservoir. These chemically and thermally complex reservoirs are an expected consequence of open-system processes that are common in magmatic systems. Mixing of magma batches with different physical properties (density, viscosity), assimilation of wall rocks, and internal phase changes (crystallization, gas exsolution) related to cooling and decompression all lead to chemical and thermal gradients within the reservoirs.

In contrast to common belief, convection is not an efficient homogenizing agent at all scales in viscous magmatic systems and can produce long-lasting gradients. Recent experiments and numerical studies have shown that heterogeneities (in composition, crystal content, temperature) will arise (or be preserved) as sluggish convection stirs the system. If left active long enough in a chaotic mode, convection can produce homogeneity, except in the case of thermally-induced heterogeneities, which are continuously re-established in cooling magma bodies. Therefore, in most magmatic situations, gradients are inescapable. This inference is on par with volcanic deposits from explosive eruptions (ignimbrites), which, in many cases, show heterogeneities in their whole-rock composition, crystallinity, temperature and volatile contents. Looking at the ignimbrite record, we conclude that most crystal-poor magma chambers will preserve heterogeneities as the system undergoes sluggish convection.

Paradoxically, the most viscous of these magmas (the crystal-rich Monotonous Intermediates) and many of the exposed fossil magma chambers (upper crustal silicic plutons) are remarkably homogeneous at the hand sample scale. If complete homogenization is the result of stirring by low Re convection, one would expect the most viscous magmas to be the least homogeneous. This apparent contradiction can be resolved by recognizing that homogeneous units are evolved, crystal-rich mush zones. Many silicic magmas persist as near-solidus crystal mushes, due to a combination of slow conductive cooling and latent heat buffering temperature close to the eutectic. We argue that large silicic crystal mushes mostly grow by addition of compositionally similar magmas, released by lower to mid-crustal MASH zones. As a result, the base-line homogeneity is dictated by processes occurring in source regions, and is modified by open-system events in the upper crust. These mushes block denser mafic reintrusions at their bases, and keep them from thorough mixing. This hot underplating triggers periodic stirring through large-scale overturn of the reservoirs by self-mixing and gas sparging. Such periodic chamber-wide stirring is well illustrated in erupted crystal-rich units (the Monotonous Intermediates) that show thorough convective stirring in the last 1,000-10,000 years prior to eruption.

Plutons cannot record a high-energy state, as they must cool slowly to their solidii. They, nonetheless, commonly appear fairly homogeneous at the hand sample scale, and we propose that (1) they are also stirred periodically by reintrusion events, and (2) are kept fairly homogeneous by the sluggishness of crystal-liquid separation in silicic systems. However, plutons do partially unmix by crystal-liquid separation late in their histories, as evidenced by highly evolved granitic caps on top of several large plutonic bodies (Johnson et al. 1990; Barnes et al. 2001; Miller and Miller 2002; Walker et al. 2007).

ACKNOWLEDGMENTS

Swiss NSF grant #200021-111709 provided support to O.B. and NSF grants EAR-0440391 and EAR-0711551 to G.W.B. during the completion of this paper. We thank Christian Huber, Josef Dufek and Philipp Ruprecht for numerous lively discussions on topics covered by this paper. We are grateful to Calvin Miller, Peter Lipman, Guil Gualda and Fidel Costa for comments on earlier drafts of this manuscript, and to Keith Putirka for the time and effort he invested in this contribution and volume.

REFERENCES

Anderson AT, Davis AM, Fangqiong L (2000) Evolution of the Bishop Tuff rhyolitic magma based on melt and magnetite inclusions and zoned phenocrysts. J Petrol 41:449-473

Bachmann O, Berganz GW (2003) Rejuvenation of the Fish Canyon magma body: a window into the evolution of large-volume silicic magma systems. Geology 31:789-792

Bachmann O, Bergantz GW (2004) On the origin of crystal-poor rhyolites: extracted from batholithic crystal mushes. J Petrol 45:1565-1582
Bachmann O, Bergantz GW (2006) Gas percolation in upper-crustal silicic crystal mushes as a mechanism for upward heat advection and rejuvenation of near-solidus magma bodies. J Volcanol Geotherm Res 149:85-102
Bachmann O, Bergantz GW (2008) The magma reservoirs that feed supereruptions. Elements 4:17-21
Bachmann O, Charlier BLA, Lowenstern JB (2007a) Zircon crystallization and recycling in the magma chamber of the rhyolitic Kos Plateau Tuff (Aegean Arc). Geology 35:73-76
Bachmann O, Dungan MA (2002) Temperature-induced Al-zoning in hornblendes of the Fish Canyon magma, Colorado. Am Mineral 87:1062-1076
Bachmann O, Dungan MA, Bussy F (2005) Insights into shallow magmatic processes in large silicic magma bodies: the trace element record in the Fish Canyon magma body, Colorado. Contrib Mineral Petrol 149:338-349
Bachmann O, Dungan MA, Lipman PW (2002) The Fish Canyon magma body, San Juan volcanic field, Colorado: rejuvenation and eruption of an upper crustal batholith. J Petrol 43:1469-1503
Bachmann O, Miller CF, de Silva S (2007b) The volcanic-plutonic connection as a stage for understanding crustal magmatism. J Volcanol Geotherm Res 167:1-23
Bacon CR (1983) Eruptive history of Mount Mazama and Crater Lake caldera, Cascade Range, U.S.A. J Volcanol Geotherm Res 18:57-115
Bacon CR, Druitt TH (1988) Compositional evolution of the zoned calcalkaline magma chamber of Mount Mazama, Crater Lake, Oregon. Contrib Mineral Petrol 98:224-256
Bacon CR, Lanphere MA (2006) Eruptive history and geochronology of Mount Mazama and the Crater Lake region, Oregon. Geol Soc Am Bull 118:1331–1359
Bacon CR, Lowenstern JB (2005) Late Pleistocene granodiorite source for recycled zircon and phenocrysts in rhyodacite lava at Crater Lake, Oregon. Earth Planet Sci Lett 233:277-293
Barboza SA, Bergantz GW (2000) Metamorphism and anatexis in the mafic complex contact aureole, Ivrea Zone, Northern Italy. J Petrol 41:1307-1327
Barnes CG, Burton BR, Burling TC, Wright JE, Karlsson HR (2001) Petrology and Geochemistry of the Late Eocene Harrison Pass Pluton, Ruby Mountains Core Complex, Northeastern Nevada. J Petrol 42:901-929
Bateman PC, Chappell BW (1979) Crystallization, fractionation, and solidification of the Tuolumne Intrusive Series, Yosemite National Park, California. Geol Soc Am Bull 90:465-482
Beard JS, Ragland PC, Crawford ML (2005) Reactive bulk assimilation: A model for crust-mantle mixing in silicic magmas. Geology 33:681-684
Bergantz GW (2000) On the dynamics of magma mixing by reintrusion: implications for pluton assembly processes. J Struct Geol 22:1297-1309
Bergantz GW, Breidenthal RE (2001) Non-stationary entrainment and tunneling eruptions: A dynamic link between eruption processes and magma mixing. Geophys Res Lett 28:3075-3078
Bergantz GW, Ni J (1999) A numerical study of sedimentation by dripping instabilities in viscous fluids. Int J Multiphase Flow 25:307-320
Bindeman IN, Valley JW (2002) Oxygen isotope study of the Long Valley magma system, California: isotope thermometry and convection in large silicic magma bodies. Contrib Mineral Petrol 144:185-205
Bindeman IN, Valley JW (2003) Rapid generation of both high- and low- $\delta^{18}O$, large volume silicic magmas at the Timber Mountain/Oasis Valley caldera complex, Nevada. Geol Soc Am Bull 115:581-595
Blake S, Ivey GN (1986a) Density and viscosity gradients in zoned magma chambers, and their influence withdrawal dynamics. J Volcanol Geotherm Res 30:201-230
Blake S, Ivey GN (1986b) Magma-mixing and the dynamics of withdrawal from stratified reservoirs. J Volcanol Geotherm Res 27:153-178
Bohrson WA, Spera FJ (2001) Energy-constrained open-system magmatic processes. II: Application of energy-constrained assimilation-fractional crystallization (EC-AFC) model to magmatic systems. J Petrol 42:1019-1041
Broadwell JE, Mungal MG (1991) Large-scale structures and molecular mixing. Phys Fluid A 3:1193-1206
Brophy JG (1991) Composition gaps, critical crystallinity, and fractional crystallization in orogenic (calc-alkaline) magmatic systems. Contrib Mineral Petrol 109:173-182
Brown SJA, Fletcher IR (1999) SHRIMP U-Pb dating of the preeruption growth history of zircons from the 340 ka Whakamaru Ignimbrite, New Zealand: Evidence for >250 k.y. magma residence times. Geology 27:1035-1038
Brown SJA, Wilson CJN, Cole JW, Wooden J (1998) The Whakamaru group ignimbrites, Taupo Volcanic Zone, New Zealand: evidence for reverse tapping of a zoned silicic magmatic system. J Volcanol Geotherm Res 84:1-37
Burgisser A, Bergantz GW, Breidenthal RE (2005) Addressing complexity in laboratory experiments: the scaling of dilute multiphase flows in magmatic systems. J Volcanol Geotherm Res 141:245-265

Caricchi L, Burlini L, Ulmer P, Gerya T, Vassali M, Papale P (2007) Non-Newtonian rheology of crystal-bearing magmas and implications for magma ascent dynamics. Earth Planet Sci Lett 264:402-419

Carrigan CR (1988) Biot number and thermos bottle effect: Implications for magma-chamber convection. Geology 16:771-774

Champallier R, Bystricky M, Arbaret L (2008) Experimental investigation of magma rheology at 300 MPa: From pure hydrous melt to 76 vol.% of crystals. Earth Planet Sci Lett 267:571-583

Charlier BLA, Bachmann O, Davidson JP, Dungan MA, Morgan D (2007) The upper crustal evolution of a large silicic magma body: evidence from crystal-scale Rb/Sr isotopic heterogeneities in the Fish Canyon magmatic system, Colorado. J Petrol 48:1875-1894

Chen CF, Turner JS (1980) Crystallization in double-diffusive system. J Geophys Res 85:2573-2593

Chesner CA (1998) Petrogenesis of the Toba Tuffs, Sumatra, Indonesia. J Petrol 39:397-438

Christiansen RL (1984) Yellowstone magmatic evolution: Its bearing on understanding large-volume explosive volcanism. In: Explosive Volcanism: Its Inception, Evolution, and Hazards. Nat Res Council Studies in Geophysics. Nat Acad Press, Washington D.C.:84-95

Christiensen JN, Halliday AN (1996) Rb-Sr and Nd isotopic compositions of melt inclusions from the Bishop Tuff and the generation of silicic magma. Earth Planet Sci Lett 144:547-561

Civetta L, Orsi G, Pappalardo L, Fisher RV, Heiken G, Ort M (1997) Geochemical zoning, mingling, eruptive dynamics and depositional processes-the Campanian Ignimbrite, Campi Flegrei, Italy. J Volcanol Geotherm Res 75:183-219

Coltice N (2005) The role of convective mixing in degassing the Earth's mantle. Earth Planet Sci Lett 234:15-25

Coltice N, Schmalzl J (2006) Mixing times in the mantle of the early Earth derived from 2-D and 3-D numerical simulations of convection. Geophys Res Lett 33, L23304, doi:10.1029/2006GL027707

Couch S, Sparks RSJ, Carroll MR (2001) Mineral disequilibrium in lavas explained by convective self-mixing in open magma chambers. Nature 411:1037-1039

Davaille A (1999) Two-layer thermal convection in miscible fluids. J Fluid Mech 379:223-253

Davidson JP, Morgan DJ, Charlier BLA, Harlou R, Hora JM (2007) Microsampling and isotopic analysis of igneous rocks: implications for the study of magmatic systems. Annu Rev Earth Planet Sci 35:273-311

Davidson JP, Tepley FJ III (1997) Recharge in volcanic systems: evidence from isotope profiles of phenocrysts. Science 275:826-829

Davies GR, Halliday AN (1998) Development of the Long Valley rhyolitic magma system: Strontium and neodymium isotope evidence from glass and individual phenocrysts. Geochim Cosmochim Acta 62:3561-3574

de Silva SL (1991) Styles of zoning in the central Andean ignimbrites: Insights into magma chamber processes. In: Andean Magmatism and its Tectonic Setting. Harmon RS, Rapela CW (eds) Geol Soc Am Spec Paper 265:233-243

de Silva SL, Wolff JA (1995) Zoned magma chambers; the influence of magma chamber geometry on sidewall convective fractionation. J Volcanol Geotherm Res 65:111-118

Deniel C, Vidal P, Fernandez A, Fort P, Peucat J-J (1987) Isotopic study of the Manaslu granite (Himalaya, Nepal): inferences on the age and source of Himalayan leucogranites. Contrib Mineral Petrol 96:78-92

DePaolo DJ (1981) Trace element and isotopic effects of combined wallrock assimilation and fractional crystallization. Earth Planet Sci Lett 53:189-202

Dimotakis PE (2005) Turbulent mixing. Annu Rev Fluid Mech 37:329-356

Druitt TH, Bacon CR (1989) Petrology of the zoned calcalkaline magma chamber of Mount Mazama, Crater Lake, Oregon. Contrib Mineral Petrol 101:245-259

Dufek J, Bergantz GW (2005) Lower crustal magma genesis and preservation: a stochastic framework for the evaluation of basalt–crust interaction. J Petrol 46:2167-2195

Duffield WA, Ruiz J, Webster JD (1995) Roof-rock contamination of magma along the top of the reservoir for the Bishop Tuff. J Volcanol Geotherm Res 69:187-195

Dunbar NW, Kyle PR, Wilson CJN (1989) Evidence for limited zonation in silicic magma systems, Taupo Volcanic Zone, New Zeland. Geology 17:234-236

Eichelberger JC, Chertkoff DG, Dreher ST, Nye CJ (2000) Magmas in collision; rethinking chemical zonation in silicic magmas. Geology 28:603-606

Eichelberger JC, Izbekov PE, Browne BL (2006) Bulk chemical trends at arc volcanoes are not liquid lines of descent. Lithos 87(1-2):135-154

Ferrachat S, Ricard Y (1998) Regular vs. chaotic mantle mixing. Earth Planet Sci Lett 155(1-2):75-86

Francalanci L, Varekamp JC, Vougioukalakis G, Defant MJ, Innocenti F, Manetti P (1995) Crystal retention, fractionation and crustal assimilation in a convecting magma chamber, Nisyros Volcano, Greece. Bull Volcanol 56:601-620

Francis PW, Sparks RSJ, Hawkesworth CJ, Thorpe RS, Pyle DM, Tait SR, Mantovani MS, McDermott F (1989) Petrology and geochemistry of the Cerro Galan caldera, northwest Argentina. Geol Mag 126:515-547

Fridrich CJ, Mahood GA (1987) Compositional layers in the zoned magma chamber of the Grizzly Peak Tuff. Geology 15:299-303
Frost TP, Mahood G (1987) Field, chemical, and physical constraints on mafic-felsic magma interaction in the Lamarck Granodiorite, Sierra Nevada, California. Geol Soc Am Bull 99:272-291
Ghiorso MS, Evans BW (2008) Thermodynamics of rhombohedral oxide solid solutions and a revision of the Fe-Ti oxide geothermometer and oxygen-barometer. Am J Sci (in press)
Gonnermann HM, Manga M, Jellinek AM (2002) Dynamics and longevity of an initially stratified mantle. Geophys Res Lett 29; doi:10.1029/2002GL014851
Grout FF (1918) Two-phase convection in igneous magmas. J Geol 26:481-499
Grunder AL, Mahood GA (1988) Physical and chemical models of zoned silicic magmas: the Loma Seca Tuff and Calabozos caldera, southern Andes. J Petrol 29:831-867
Halliday AN, Fallick AE, Hutchinson J, Hildreth W (1984) A Nd, Sr, O isotopic investigation into the causes of chemical and isotopic zonation in the Bishop Tuff, California. Earth Planet Sci Lett 68:378-391
Harper B, Miller C, Koteas C, Cates N, Wiebe R, Lazzareschi D, Cribb W (2004) Granites, dynamic magma chamber processes and pluton construction: the Aztec Wash pluton, Eldorado Mountains, Nevada, USA. Trans R Soc Edinburgh: Earth Sci 95:277-296
Hervig RL, Dunbar NW (1992) Cause of chemical zoning in the Bishop (California) and Bandelier (New Mexico) magma chambers. Earth Planet Sci Lett 111:97-108
Hildreth W (1979) The Bishop Tuff: evidence for the origin of the compositional zonation in silicic magma chambers. Geol Soc Am Spec Pap 180:43-76
Hildreth W (1981) Gradients in silicic magma chambers: Implications for lithospheric magmatism. J Geophys Res 86:10153-10192
Hildreth W (2004) Volcanological perspectives on Long Valley, Mammoth Mountain, and Mono Craters: several contiguous but discrete systems. J Volcanol Geotherm Res 136:169-198
Hildreth W, Fierstein J (2000) Katmai volcanic cluster and the great eruption of 1912. Geol Soc Am Bull 112:1594-1620
Hildreth WS, Moorbath S (1988) Crustal contributions to arc magmatism in the Andes of Central Chile. Contrib Mineral Petrol 98:455-499
Hildreth WS, Wilson CJN (2007) Compositional zoning in the Bishop Tuff. J Petrol 48:951-999
Huppert HE, Sparks RSJ (1984) Double-diffusive convection due to crystallization in magmas. Annu Rev Earth Planet Sci 12:11-37
Jellinek AM, Kerr RC (1999) Mixing and compositional stratification produced by natural convection: 2. Applications to the differentiation of basaltic and silicic magma chambers and komatiite lava flows. J Geophys Res 104:7203-7218
Jellinek AM, Kerr RC, Griffiths RW (1999) Mixing and compositional stratification produced by natural convection: 1. Experiments and their applications to Earth's core and mantle. J Geophys Res 104:7183-7201
Jellinek AM, Kerr RC (2001) Magma dynamics, crystallization, and chemical differentiation of the 1959 Kilauea Iki lava lake, Hawaii, revisited. J Volcanol Geotherm Res 110:235-263
Johnson CM (1989) Isotopic zonations in silicic magma chambers. Geology 17:1136-1139
Johnson CM, Czamanske GK, Lipman PW (1990) H, O, Sr, Nd, and Pb isotope geochemistry of the Latir volcanic field and cogenetic intrusions, New Mexico, and relations between evolution of a continental magmatic center and modifications of the lithosphere. Contrib Mineral Petrol 104:99-124
Johnson M, Rutherford M (1989) Experimentally determined conditions in the Fish Canyon Tuff, Colorado, magma chamber. J Petrol 30:711-737
Keller J (1969) Origin of rhyolites by anatectic melting of granitic crustal rocks; the example of rhyolitic pumice from the island of Kos (Aegean sea). Bull Volcanol 33:942-959
Knesel KM, Davidson JP (1997) The origin and evolution of large-volume silicic magma systems: Long Valley caldera. Int Geol Rev 39:1033-1052
Knesel KM, Davidson JP, Duffield WA (1999) Evolution of silicic magma through assimilation and subsequent recharge: Evidence from Sr isotopes in sanidine phenocrysts, Taylor Creek Rhyolite, NM. J Petrol 40:773-786
Knesel KM, Duffield WA (2007) Gradients in silicic eruptions caused by rapid inputs from above and below rather than protracted chamber differentiation. J Volcanol Geotherm Res 167:181-197
Koyaguchi T, Kaneko K (1999) A two-stage thermal evolution model of magmas in continental crust. J Petrol 40:241-254
Linden PF, Redondo JM (1991) Molecular mixing in Rayleigh-Taylor instability. Part 1: global mixing. Phys Fluid A 3:1269-1277
Lindsay JM, Schmitt AK, Trumbull RB, De Silva SL, Siebel W, Emmermann R (2001) Magmatic evolution of the La Pacana caldera system, Central Andes, Chile: Compositional variation of two cogenetic, large-volume felsic ignimbrites. J Petrol 42:459-486

Lipman PW (1967) Mineral and chemical variations within an ash-flow sheet from Aso caldera, South Western Japan. Contrib Mineral Petrol 16:300-327
Lipman PW (2000) The central San Juan caldera cluster: Regional volcanic framework. *In:* Ancient Lake Creede: Its Volcano-Tectonic Setting, History of Sedimentation, and Relation of Mineralization in the Creede Mining District. Bethke PM, Hay RL (eds) Geol Soc Am Spec Paper 346:9-69
Lipman PW (2007) Incremental assembly and prolonged consolidation of Cordilleran magma chambers: Evidence from the Southern Rocky Mountain volcanic field. Geosphere 3:1-29
Lipman PW, Christiansen RL, O'Connor JT (1966) A compositionally zoned ash-flow sheet in southern Nevada. USGS Prof Paper 524-F:1-47
Lipman PW, Dungan MA, Bachmann O (1997) Comagmatic granophyric granite in the Fish Canyon Tuff, Colorado: Implications for magma-chamber processes during a large ash-flow eruption. Geology 25:915-918
Mahood GA (1990) Second reply to comment of R.S.J. Sparks, H.E. Huppert, and C.J.N. Wilson on "Evidence for long residence times of rhyolitic magma in the Long Valley magmatic system: the isotopic record in precaldera lavas of Glass Mountain". Earth Planet Sci Lett 99:395-399
Mahood GA, Halliday AN (1988) Generation of high-silica rhyolite: a Nd, Sr, and O isotopic study of Sierra La Primavera, Mexican neovolcanic belt. Contrib Mineral Petrol 100:183-191
Marsh BD (1988) Crystal capture, sorting, and retention in convecting magma. Geol Soc Am Bull 100:1720-1737
Marsh BD (1989) Magma chambers. Annu Rev Earth Planet Sci 17:439-474
Marsh BD (2002) On bimodal differentiation by solidification front instability in basaltic magmas, part 1: basic mechanics. Geochim Cosmochim Acta 66:2211-2229
Martin D, Nokes R (1989) A fluid-dynamical study of crystal settling in convecting magmas. J Petrol 30:1471-1500
Maughan LL, Christiansen EH, Best MG, Gromme CS, Deino AL, Tingey DG (2002) The Oligocene Lund Tuff, Great Basin, USA: a very large volume monotonous intermediate. J Volcanol Geotherm Res 113:129-157
McBirney AR (1980) Mixing and unmixing of magmas. J Volcanol Geotherm Res 7:357-371
McBirney AR, Baker BH, Nilson RH (1985) Liquid fractionation. Part 1: Basic principles and experimental simulations. J Volcanol Geotherm Res 24:1-24
McKenzie DP (1985) The extraction of magma from the crust and mantle. Earth Planet Sci Lett 74:81-91
McNulty BA, Tobish OT, Cruden AR, Gilder S (2000) Multi-stage emplacement of the Mount Givens pluton, central Sierra Nevada batholith, California. Geol Soc Am Bull 112:119-135
Michael PJ (1983) Chemical differentiation of the Bishop Tuff and other high-silica magmas through crystallization processes. Geology 11:31-34
Miller CF, Miller JS (2002) Contrasting stratified plutons exposed in tilt blocks, Eldorado Mountains, Colorado River Rift, NV, USA. Lithos 61:209-224
Miller CF, Wark DA (2008) Supervolcanoes and their explosive supereruptions. Elements 4:11-16
Miller JS, Matzel JEP, Miller CF, Burgess SD, Miller RB (2007) Zircon growth and recycling during the assembly of large, composite arc plutons. J Volcanol Geotherm Res 167:282-299
Mills J, James G., Saltoun BW, Vogel TA (1997) Magma batches in the Timber Mountain magmatic system, Southwestern Nevada Volcanic Field, Nevada, USA. J Volcanol Geotherm Res 78:185-208
Milner DM, Cole JW, Wood CP (2003) Mamaku Ignimbrite: a caldera-forming ignimbrite erupted from a compositionally zoned magma chamber in Taupo Volcanic Zone, New Zealand. J Volcanol Geotherm Res 122:243-264
Morse SA (1986) Thermal structure of crystallizing magma with two-phase convection. Geol Mag 123:205-214
Murphy MD, Sparks RSJ, Barclay J, Carroll MR, Brewer TS (2000) Remobilization of andesitic magma by intrusion of mafic magma at the Soufrière Hills Volcano, Montserrat, West Indies. J Petrol 41:21-42
Neri A, Esposti Ongaro T, Macedonio G, Gidaspow D (2003) Multiparticle simulation of collapsing volcanic columns and pyroclastic flow. J Geophys Res 108(B4):2202, doi:2210.1029/2001JB000508
Oldenburg CM, Spera FJ, Yuen DA, Sewell G (1989) Dynamic mixing in magma bodies: Theory, simulations, and implications. J Geophys Res 94:9215-9236
Ottino JM (1989) The Kinematics of Mixing: Stretching, Chaos, and Transport. Cambridge University Press, Cambridge
Parat F, Dungan MA, Lipman PW (2005) Contemporaneous trachyandesitic and calc-alkaline volcanism of the Huerto Andesite, San Juan Volcanic Field, Colorado, USA. J Petrol 46(5):859-891
Parat F, Holtz F, Feig S (2008) Pre-eruptive Conditions of the Huerto Andesite (Fish Canyon System, San Juan Volcanic Field, Colorado): Influence of volatiles (C-O-H-S) on phase equilibria and mineral composition. J Petrol 49:911-935
Peressini G, Quick JE, Sinigoi S, Hofmann AW, Fanning M (2007) Duration of a large mafic intrusion and heat transfer in the lower crust: a SHRIMP U–Pb zircon Study in the Ivrea–Verbano Zone (Western Alps, Italy). J Petrol 48:1185-1218

Petcovic HL, Dufek J (2005) Modeling of magma flow and cooling dikes: implications for emplacement of Columbia River Flood Basalts. J Geophys Res 110:1-15

Petford N (2003) Rheology of granitic magmas during ascent and emplacement. Annu Rev Earth Planet Sci 31:399-427

Petford N, Cruden AR, McCaffrey KJW, Vigneresse J-L (2000) Granite magma formation, transport and emplacement in the Earth's crust. Nature 408:669-673

Petrelli M, Perugini D, Poli G (2006) Time-scales of hybridisation of magmatic enclaves in regular and chaotic flow fields: petrologic and volcanologic implications. Bull Volcanol 68:285-293

Phillips JC, Woods AW (2002) Suppression of large-scale magma mixing by melt-volatile separation. Earth Planet Sci Lett 204:47-60

Powell RL, Mason SG (1982) Dispersion by laminar flow. AiChE J 28:286-293

Reid MR, Coath CD, Harrison TM, McKeegan KD (1997) Prolonged residence times for the youngest rhyolites associated with Long Valley Caldera: ^{230}Th-^{238}U microprobe dating of young zircons. Earth Planet Sci Lett 150:27-39

Rice A (1981) Convective fractionation: a mechanism to provide cryptic zoning (macrosegregation), layering crescumulates, banded tuffs and explosive volcanism in igneous processes. J Geophys Res 86:405-417

Robinson DM, Miller CF (1999) Record of magma chamber processes preserved in accessory mineral assemblages, Aztec Wash pluton, Nevada. Am Mineral 84:1346-1353

Rosenberg CL, Handy MR (2005) Experimental deformation of partially melted granite revisited: implications for the continental crust. J Metamorph Geol 23:19-28

Rudman M (1992) Two-phase natural convection: implications for crystal settling in magma chambers. Phys Earth Planet Int 72:153-172

Ruprecht P, Bergantz GW, Dufek J (2008) Modeling of Gas-Driven Magmatic Overturn: Tracking of Phenocryst Dispersal and Gathering During Magma Mixing. Geochem. Geophys. Geosyst 9:Q07017, doi:10.1029/2008GC002022

Scaillet B, Holtz F, Pichavant M (1998) Phase equilibrium constraints on the viscosity of silicic magmas 1. Volcanic-plutonic comparison. J Geophys Res 103:27257-27266

Schmitt AK, de Silva SL, Trumbull RB, Emmermann R (2001) Magma evolution in the Purico ignimbrite complex, nothern Chile: evidence for zoning of a dacite magma by injection of rhyolite melts following mafic recharge. Contrib Mineral Petrol 140:680-700

Simakin A, Botcharnikov R (2001) Degassing of stratified magma by compositional convection. J Volcanol Geotherm Res 105:207-224

Simakin A, Schmeling H, Trubissyn V (1997) Convection in melts due to sedimentary crystal flux from above. Phys Earth Planet Int 102:185-200

Simon JI, Reid MR (2005) The pace of rhyolite differentiation and storage in an 'archetypical' silicic magma system, Long Valley, California. Earth Planet Sci Lett 235:123-140

Sinton JM, Detrick RS (1992) Mid-Ocean Ridge Magma Chambers. J Geophys Res 97:197-216

Smith RL (1979) Ash-flow magmatism. Geol Soc Am Spec Paper 180:5-25

Smith RL, Bailey RA (1966) The Bandelier Tuff—A study of ash-flow eruption cycles from zoned magma chambers. Bull Volcanol 29:83-104

Sparks RSJ, Huppert HE, Turner JS (1984) The fluid dynamics of evolving magma chambers. Phil Trans R Soc London 310:511-534

Sparks RSJ, Marshall LA (1986) Thermal and mechanical constraints on mixing between mafic and silicic magmas. J Volcanol Geotherm Res 29:99-124

Spera FJ, Oldenburg CM, Christiensen C, Todesco M (1995) Simulations of convection with crystallization in the system $KAlSi_2O_6$-$CaMgSi_2O_6$: Implications for compositionally zoned magma bodies. Am Mineral 40:1188-1207

Spera FJ, Yuen DA, Kemp DV (1984) Mass transfer rates along vertical walls in magma chambers and marginal upwelling. Nature 310:764-767

Streck MJ, Grunder AL (1997) Compositional gradients and gaps in high-silica rhyolites of the Rattlesnake Tuff, Oregon. J Petrol 38:133-163

Thompson, Smith, Malpas (2001) Origin of oceanic phonolites by crystal fractionation and the problem of the Daly gap: an example from Rarotonga. Contrib Mineral Petrol 142:336-346

Thompson AB, Matile L, Ulmer P (2002) Some thermal constraints on crustal assimilation during fractionation of hydrous, mantle-derived magmas with examples from Central Alpine Batholiths J Petrol 43:403-422

Thompson RN (1972) Evidence for a chemical discontinuity near the basalt-andesite transition in many anorogenic volcanic suites. Nature 236:106-110

Trial AF, Spera FJ (1990) Mechanisms for the generation of compositional heterogeneities in magma chambers. Geol Soc Am Bull 102:353-367

Trial AF, Spera FJ (1992) Simulations of magma withdrawal from compositionally zoned bodies. J Geophys Res 97:6713 6733

Varga RJ, Smith BM (1984) Evolution of the early Oligocene bonanza caldera, Northeast San Juan volcanic field, Colorado. J Geophys Res 89:8679-8694

Vazquez JA, Reid MR (2004) Probing the Accumulation History of the Voluminous Toba Magma. Science 305:991-994

Vigneresse J-L, Barbey P, Cuney M (1996) Rheological transitions during partial melting and crystallization with application to felsic magma segregation and transfer. J Petrol 37:1579-1600

Waight TE, Wiebe RA, Krogstad EJ, Walker RJ (2001) Isotopic responses to basaltic injection into silicic magma chambers: a whole-rock and microsampling study of macrorhythmic units in the Pleasant Bay layered gabbro-diorite complex, Maine, USA. Contrib Mineral Petrol 142:323-335

Walker BJ, Miller CF, Lowery LE, Wooden JL, Miller JS (2007) Geology and geochronology of the Spirit Mountain batholith, southern Nevada: implications for timescales and physical processes of batholith construction. J Volcanol Geotherm Res 167:239-262

Wallace PJ, Anderson AT, Davis AM (1999) Gradients in H_2O, CO_2, and exsolved gas in a large-volume silicic magma chamber: interpreting the record preserved in the melt inclusions from the Bishop Tuff. J Geophys Res 104:20097-20122

Wark DA, Hildreth WS, Spear FS, Cherniak DJ, Watson EB (2007) Pre-eruption recharge in the Bishop Tuff magma chamber. Geology 35:235-238

Whitney JA, Dorais MJ, Stormer JC, Kline SW, Matty DJ (1988) Magmatic conditions and development of chemical zonation in the Carpenter Ridge Tuff, Central San Juan volcanic field, Colorado. Am J Sci 288:16-44

Whitney JA, Stormer JC Jr. (1985) Mineralogy, petrology, and magmatic conditions from the Fish Canyon Tuff, central San Juan volcanic field, Colorado. J Petrol 26:726-762

Wiebe R, Wark D, Hawkins D (2007) Insights from quartz cathodoluminescence zoning into crystallization of the Vinalhaven granite, coastal Maine. Contrib Mineral Petrol 154:439-453

Wiebe RA, Blair KD, Hawkins DP, Sabine CP (2002) Mafic injections, in situ hybridization, and crystal accumulation in the Pyramid Peak granite, California. Geol Soc Am Bull 114:909-920

Wiebe RA, Collins WJ (1998) Depositional features and stratigraphic sections in granitic plutons: implications for the emplacement and crystallization of granitic magma. J Struct Geol 20:1273-1289

Wilson CJN, Blake S, Charlier BLA, Sutton AN (2006) The 26.5 ka Oruanui Eruption, Taupo Volcano, New Zealand: Development, characteristics and evacuation of a large rhyolitic magma body. J Petrol 47:35-69

Wilson CJN, Hildreth W (1997) The Bishop Tuff: new insights from eruptive stratigraphy. J Geol 105:407-439

Wolff JA (1985) Zonation, mixing and eruption of a silica-undersaturated alkaline magma: a case study from Tenerife, Canary Islands. Geol Mag 122:623-640

Wolff JA, Ramos FC, Davidson JP (1999) Sr isotope disequilibrium during differentiation of the Bandelier Tuff: Constraints on the crystallization of a large rhyolitic magma chamber. Geology 27:495-498

Wolff JA, Storey M (1984) Zoning in highly alkaline magma bodies. Geol Mag 121:563-575

Wolff JA, Worner G, Blake S (1990) Gradients in physical parameters in zoned felsic magma bodies: implications for evolution and eruptive withdrawal. J Volcanol Geotherm Res 43:37-55

Worner G, Schmincke H-U (1984a) Mineralogical and chemical zonation of the Laacher See tephra. J Petrol 25:805-835

Worner G, Schmincke H-U (1984b) Petrogenesis of the Laacher See tephra. J Petrol 25: 836-851

Zak J, Paterson SR (2005) Characteristics of internal contacts in the Tuolumne Batholith, central Sierra Nevada, California (USA): Implications for episodic emplacement and physical processes in a continental arc magma chamber. Geol Soc Am Bull 117:1242-1255